Texts in Applied Mathematics

Volume 44

The mathematization of all sciences, the fading of traditional scientific boundaries, the impact of computer technology, the growing importance of computer modelling and the necessity of scientific planning all create the need both in education and research for books that are introductory to and abreast of these developments. The aim of this series is to provide such textbooks in applied mathematics for the student scientist. Books should be well illustrated and have clear exposition and sound pedagogy. Large number of examples and exercises at varying levels are recommended. TAM publishes textbooks suitable for advanced undergraduate and beginning graduate courses, and complements the Applied Mathematical Sciences (AMS) series, which focuses on advanced textbooks and research-level monographs.

More information about this series at https://link.springer.com/bookseries/1214

Peter Knabner · Lutz Angermann

Numerical Methods for Elliptic and Parabolic Partial Differential Equations

With contributions by Andreas Rupp

Second Extended Edition

Peter Knabner
Department of Mathematics
University of Erlangen-Nuremberg
Erlangen, Bayern, Germany

Lutz Angermann
Department of Mathematics
Clausthal University of Technology
Clausthal-Zellerfeld, Germany

ISSN 0939-2475 ISSN 2196-9949 (electronic)
Texts in Applied Mathematics
ISBN 978-3-030-79387-6 ISBN 978-3-030-79385-2 (eBook)
https://doi.org/10.1007/978-3-030-79385-2

Mathematics Subject Classification: 65N30, 65M60, 65N06, 65N08, 65M06, 65M08, 65F10, 65F05, 76Sxx, 35Jxx, 35Kxx

This Springer imprint is published by the registered company Springer Nature Switzerland AG
The registered company address is: Gewerbestrasse 11, 6330 Cham, Switzerland

Preface to the Second English Edition

If a textbook originally dates back 20 years, as the German edition of this book, it is obvious that there are many new developments that could not yet be covered. Therefore we decided to take up the enterprise to describe some of the essential achievements of the last decades. On the one hand, we did not alter the inductive, but rigorous character of the existing parts of the book, on the other hand with more than 300 new pages the amendments are so extensive that this edition actually is a new book. Taking into account that the new sections and chapters belong to the more advanced part of the subject, our style became slightly less detailed, but hopefully still clear. In the centre of our exposition there is still the scalar linear elliptic and parabolic differential equation, but in addition to the diffusive term it always includes a convective and a reactive part and is always supplemented by general linear boundary conditions that allow flow, Dirichlet and mixed conditions to be treated. Its dual (mixed) formulation, the Darcy equation, extends the scope to non-coercive saddle point formulations. Thus we found it natural also to include some aspects of the (Navier–)Stokes equations, without being exhaustive in this respect. Also in a different way than in other, excellent textbooks, we do not restrict ourselves to the stationary elliptic case, but also have treated the time-dependent case for most of spatial discretizations considered. So in summary, many new developments to the best of our knowledge are treated for the first time in a textbook. Concerning the extensions described above of the "Poisson equation with Dirichlet conditions" prototype, again to the best of our knowledge, several of the added results are original and published for the first time.

In more detail, the changes and amendments are following: In the more elementary classical core of Chapters 0 to 3, hardly any changes have been made, apart from some outlooks to the new chapters to come and slight additions in Sections 2.5 and 3.4. The first section of Chapter 4 has been essentially rewritten, and a third section about convergence of adaptive methods has been added. In Chapter 5 the more advanced Sections 5.4 and 5.5 have been enlarged and a new section about space decomposition methods added.

For the remaining chapters a reordering was necessary as seen from the following table:

first edition	second edition
6	8
7	9
8	11
9	10

The new Chapter 6 provides the general framework to deal with nonconforming (beyond the results of Section 3.6) and nonconsistent methods, and with non-coercive saddle point formulations. In the new Chapter 7 mixed discretizations both for the Darcy and the Stokes equations are investigated, as well as various discontinuous Galerkin methods as nonconforming methods, prepared by the treatment of Crouzeix–Raviart elements. In Sections 7.5 and 7.6 we deal with hybridization as a tool to improve the properties of the discrete problem and its approximation quality, especially to ensure local mass conservation by proper postprocessing. In Finite Volume Chapter 8, Sections 8.1 and 8.2 have been revised and new Sections 8.3 to 8.6 added, including cell-oriented methods and the (Navier–)Stokes equation. In the time-dependence Chapter 9, Section 9.2 needed a considerable enlargement to include most of the new spatial discretization methods. The new Sections 9.5 and 9.6 discuss higher order methods. In Chapter 10, Sections 10.5 and 10.6 about stabilization by limiting have been added, and finally in Chapter 11, Sections 11.5 to 11.7 are new, extending the still non-comprehensive treatment of nonlinear problems beyond semilinear versions. Correspondingly also the Appendices have been modified, in particular, Appendix A.1 on the notation has been significantly expanded. Also a large amount of new exercise problems have been added, which are now more uniformly distributed across the text. Moreover, compared to the previous editions, we included programming tasks, too. We have intentionally formulated them neutrally with respect to the programming languages and integrated development environments (IDEs) that can be used, since a corresponding specification of these tasks depends on many aspects such as the auditorium, the curriculum (existing previous knowledge) but also the software and hardware equipment of the facility, etc. In our courses, most of the tasks were actually worked on by students with various programming languages/IDEs (e.g., C/C++, GNU Octave/MATLAB, Python) over the course of time. The choice of the programming language/IDE can be left to more experienced students, and for less experienced students it can be helpful to specify them, possibly even to point out helpful built-in functionalities, or to give links to suitable software packages, including grid generation, visualization tools, etc.

Several of these amendments benefited crucially from the collaboration with Dr. Andreas Rupp. In particular he has contributed the first versions of Sections 7.4, 10.5, and 10.6 and participated intensively in the improvement of the whole text. We are grateful to Tobias Elbinger for providing basic versions of most of the programming projects (and also to many other (former) members of our working groups who contributed), and also the exposition of the general order of convergence estimate in Section 9.2 and its application to mixed problems are coined by results in his thesis [132]. We thank Sebastian Czop and Xingyu Xu for their support in transferring handwritten notes into LaTeX, and express our gratitude to Donna Chernyk from Springer Science Business Media for her efficient handling of the publishing process. As a final acknowledgement, we would like to express our appreciation to the anonymous experts who kindly reviewed the manuscript and provided insightful suggestions and constructive criticisms.

Remarks for the Reader and the Use in Lectures

The size of the text corresponds roughly to four hours of lectures per week over two to three terms. If the course lasts only one term, then a selection is necessary, which should be oriented to the audience. We recommend the following "cuts":

For an introductory course:

Chapter 0 may be skipped if the partial differential equations treated therein are familiar. Section 0.5 should be consulted because of the notation collected there. The same is true for Chapter 1; possibly Section 1.4 may be integrated into Chapter 3 if one wants to deal with Section 3.9 or with Section 9.7.

Chapters 2 and 3 are the core of the book. The inductive presentation that we preferred for some theoretical aspects may be shortened for students of mathematics. To the lecturer's taste and depending on the knowledge of the audience in numerical mathematics Section 2.5 may be skipped. This might impede the treatment of the ILU preconditioning in Section 5.3. Observe that in Sections 2.1–2.3 the treatment of the model problem is merged with basic abstract statements. Skipping the treatment of the model problem, in turn, requires an integration of these statements into Chapter 3. In doing so Section 2.4 may be easily combined with Section 3.5. In Chapter 3 the theoretical kernel consists of Sections 3.1, 3.2.1, 3.3–3.4.

Chapter 4 presents an overview of its subject, not a detailed development, and is an extension of the classical subjects. The classical parts of Chapter 9 are Sections 9.1, 9.3 to 9.4, and the corresponding parts of Sections 9.2 and 9.8. If time permits, Sections 10.1, 10.2 (and 10.4) are recommended. In the extensive Chapter 5 one might focus on special subjects or just consider Sections 5.2, 5.3 (and 5.4) in order to present at least one practically relevant and modern iterative method.

Section 11.2 and the first part of Section 11.3 contain basic knowledge of numerical mathematics and, depending on the audience, may be omitted.

For an advanced course:

After Chapter 6 as basis, Chapters 7 and 8 provide various aspects of mixed or nonconforming finite element methods and of finite volume methods, with various possibilities to select. The time-dependent versions require the parts of Sections 9.2 and 9.8, and possibly Sections 9.5, 9.6, and 9.7, further amended by Sections 10.3 and 10.4 and possibly 10.5 and 10.6.

The appendices are meant only for consultation and may complete the basic lectures, such as in analysis, linear algebra, and advanced mathematics for engineers.

Concerning related textbooks for supplementary use, to the best of our knowledge there is none covering approximately the same topics. Quite a few deal with finite element methods, and the closest one in spirit probably is [40], but also [9] or [10] or [24] has a certain overlap, and also offers additional material not covered here. From the books specialized in finite difference methods, we mention [58] as an example. The (node-oriented) finite volume method is popular in engineering, in particular, in fluid dynamics, but to the best of our knowledge there is no presentation similar to ours in a mathematical textbook. References to textbooks specialized in the topics of Chapters 4, 5, and 11 are given there.

Remarks on the Notation

Printing in *italics* emphasizes definitions of notation, even if this is not carried out as a numbered definition.

Vectors appear in different forms: Besides the "short" space vectors $x \in \mathbb{R}^d$ there are "long" representation vectors $\mathbf{u} \in \mathbb{R}^m$, which describe, in general, the degrees of freedom of a finite element (or volume) approximation or represent the values on grid points of a finite difference method. Here we choose **bold type**, also in order to have a distinctive feature from the generated functions, which frequently have the same notation, or from the grid functions.

We deviate from this rule in Chapters 0, 6, 7, and 8 and the related parts of Chapters 9 and 10, where formulations and discretization methods are treated, where in addition to scalar quantities also vectorial ones (in $\mathbb{R}^d$) appear. For clearer differentiation, also these are printed in bold type, and then consequently also $\boldsymbol{x} \in \mathbb{R}^d$ or $\boldsymbol{n} \in \mathbb{R}^d$ (for the normal vector). There are also deviations in Chapters 5, where the unknowns of linear and nonlinear systems of equations, which are treated in a general manner there, are denoted by $x \in \mathbb{R}^m$.

Components of vectors will be designated by a subindex, creating a double index for indexed quantities. Sequences of vectors will be supplied with a superindex (in parentheses); only in an abstract setting do we use subindices.

Erlangen, Germany
Clausthal-Zellerfeld, Germany
December 2020

Peter Knabner
Lutz Angermann

From the Preface to the First English Edition

Shortly after the appearance of the German edition we were asked by Springer to create an English version of our book, and we gratefully accepted. We took this opportunity not only to correct some misprints and mistakes that have come to our knowledge but also to extend the text at various places. This mainly concerns the role of the finite difference and the finite volume methods, which have gained more attention by a slight extension of Chapters 1 and 8 and by a considerable extension of Chapter 9. Time-dependent problems are now treated with all three approaches (finite differences, finite elements, and finite volumes), doing this in a uniform way as far as possible. This also made a reordering of Chapters 8–11 necessary. Also, the index has been enlarged. To improve the direct usability in courses, exercises now follow each section and should provide enough material for homework.

This new version of the book would not have come into existence without our already mentioned team of helpers, who also carried out first versions of translations of parts of the book. Beyond those already mentioned, the team was enforced by Cecilia David, Basca Jadamba, Dr. Serge Kräutle, Dr. Wilhelm Merz, and Peter Mirsch. Alexander Prechtel now took charge of the difficult modification process. Prof. Paul DuChateau suggested improvements. We want to extend our gratitude to all of them. Finally, we thank Senior Editor Achi Dosanjh, from Springer-Verlag New York, Inc., for her constant encouragement.

⋮

Erlangen, Germany
Clausthal-Zellerfeld, Germany
January 2002

Peter Knabner
Lutz Angermann

Preface to the German Edition

This book resulted from lectures given at the University of Erlangen–Nuremberg and at the University of Magdeburg. On these occasions we often had to deal with the problem of a heterogeneous audience composed of students of mathematics and of different natural or engineering sciences. Thus the expectations of the students concerning the mathematical accuracy and the applicability of the results were widely spread. On the other hand, neither relevant models of partial differential equations nor some knowledge of the (modern) theory of partial differential equations could be assumed among the whole audience. Consequently, in order to overcome the given situation, we have chosen a selection of models and methods relevant for applications (which might be extended) and attempted to illuminate the whole spectrum, extending from the theory to the implementation, without assuming advanced mathematical background. Most of the theoretical obstacles, difficult for nonmathematicians, will be treated in an "inductive" manner. In general, we use an explanatory style without (hopefully) compromising the mathematical accuracy.

We hope to supply especially students of mathematics with the information necessary for the comprehension and implementation of finite element/finite volume methods. For students of the various natural or engineering sciences the text offers, beyond the possibly already existing knowledge concerning the application of the methods in special fields, an introduction into the mathematical foundations, which should facilitate the transformation of specific knowledge to other fields of applications.

We want to express our gratitude, for the valuable help that we received during the writing of this book, to Dr. Markus Bause, Sandro Bitterlich, Dr. Christof Eck, Alexander Prechtel, Joachim Rang, and Dr. Eckhard Schneid who did the proofreading and suggested important improvements. From the anonymous referees we received useful comments. Very special thanks go to Mrs. Magdalena Ihle and Dr. Gerhard Summ. Mrs. Ihle transposed the text quickly and precisely into TEX. Dr. Summ not only worked on the original script and on the TEX-form, but he also organized the complex and distributed rewriting and extension procedure.

The elimination of many inconsistencies is due to him. Additionally he influenced parts of Sections 3.4 and 3.8 by his outstanding diploma thesis. We also want to thank Dr. Chistoph Tapp for the preparation of the graphic of the title and for providing other graphics from his doctoral thesis [221].

Of course, hints concerning (typing) mistakes and general improvements are always welcome.

We thank Springer-Verlag for their constructive collaboration.

Last, but not least, we want to express our gratitude to our families for their understanding and forbearance, which were necessary for us especially during the last months of writing.

Erlangen, Germany
Magdeburg, Germany
February 2000

Peter Knabner
Lutz Angermann

Contents

Chapter 0
For Example: Modelling Processes in Porous Media with Differential Equations

This chapter illustrates the scientific context in which differential equation models may occur, in general, and also in a specific example. Section 0.1 reviews the fundamental equations, for some of them discretization techniques will be developed and investigated in this book. In Sections 0.2–0.4 we focus on reaction and transport processes in porous media. These sections are independent of the remaining parts and may be skipped by the reader. Section 0.5, however, should be consulted because it fixes some notation to be used later on.

0.1 The Basic Partial Differential Equation Models

Partial differential equations, often abbreviated as *PDE(s)*, are equations involving some partial derivatives of an unknown function u in several independent variables. Partial differential equations which arise from the modelling of spatial (and temporal) processes in nature or technology are of particular interest. Therefore, we assume that the variables of u are $x = (x_1, \ldots, x_d)^T \in \mathbb{R}^d$ for $d \geq 1$, representing a spatial point, and possibly $t \in \mathbb{R}$, representing time. Thus the minimal set of variables is (x_1, x_2) or (x_1, t), otherwise we have ordinary differential equations (*ODE(s)* for short). We will assume that $x \in \Omega$, where Ω is a bounded domain, e.g., a metal workpiece, or a groundwater aquifer, and $t \in (0, T]$ for some (time horizon) $T > 0$. Nevertheless also processes acting in the whole $\mathbb{R}^d \times \mathbb{R}$, or in unbounded subsets of it, are of interest. One may consult the Appendix A.1 for notations from analysis etc. used here. Often the function u represents, or is related to, the volume density of an extensive quantity like mass, energy, or momentum, which is conserved. In their original form all quantities have dimensions that we denote in accordance with the International System of Units (SI) and write in square brackets [ ]. Let a be a symbol for the unit of the extensive quantity, then its volume density is assumed to have the form $S = S(u)$, i.e., the unit of $S(u)$ is $\mathrm{a/m^3}$. For example, for mass conservation a = kg, and $S(u)$ is a concentration. For describing the conservation we consider an arbitrary "not too bad" subset $\tilde{\Omega} \subset \Omega$, the *control volume*. The time variation of the

P. Knabner and L. Angermann, *Numerical Methods for Elliptic and Parabolic Partial Differential Equations*, Texts in Applied Mathematics 44,
https://doi.org/10.1007/978-3-030-79385-2_0

total extensive quantity in $\tilde{\Omega}$ is then

$$\partial_t \int_{\tilde{\Omega}} S(u(x,t))dx\,. \tag{0.1}$$

If this function does not vanish, only two reasons are possible due to conservation:

- There is an internally distributed source density $Q = Q(x,t,u)$ [a/m^3/s], being positive if $S(u)$ is produced, and negative if it is destroyed, i.e., one term to balance (0.1) is $\int_{\tilde{\Omega}} Q(x,t,u(x,t))dx$.
- There is a net flux of the extensive quantity over the boundary $\partial\tilde{\Omega}$ of $\tilde{\Omega}$. Let $\boldsymbol{J} = \boldsymbol{J}(x,t)$ [a/m^2/s] denote the flux density, i.e., $\boldsymbol{J}_i$ is the amount, that passes a unit square perpendicular to the ith axis in one second in the direction of the ith axis (if positive), and in the opposite direction otherwise. Then another term to balance (0.1) is given by

  $$-\int_{\partial\Omega} \boldsymbol{J}(x,t)\cdot \boldsymbol{n}(x)d\sigma\,,$$

  where $\boldsymbol{n}$ denotes the outer unit normal on $\partial\Omega$.

Summarizing the conservation reads

$$\partial_t \int_{\tilde{\Omega}} S(u(x,t))dx = -\int_{\partial\tilde{\Omega}} \boldsymbol{J}(x,t)\cdot \boldsymbol{n}(x)d\sigma + \int_{\tilde{\Omega}} Q(x,t,u(x,t))dx\,. \tag{0.2}$$

The integral theorem of Gauss (see (2.3)) and an exchange of time derivative and integral leads to

$$\int_{\tilde{\Omega}} [\partial_t S(u(x,t)) + \nabla\cdot \boldsymbol{J}(x,t) - Q(x,t,u(x,t))]dx = 0\,,$$

and, as $\tilde{\Omega}$ is arbitrary, also to

$$\partial_t S(u(x,t)) + \nabla\cdot \boldsymbol{J}(x,t) = Q(x,t,u(x,t)) \quad \text{for } x\in\Omega,\ t\in(0,T]. \tag{0.3}$$

All manipulations here are formal assuming that the functions involved have the necessary properties. The partial differential equation (0.3) is the basic pointwise conservation equation, (0.2) its corresponding integral form. Equation (0.3) is one requirement for the two unknowns u and $\boldsymbol{J}$, thus it has to be closed by a (phenomenological) *constitutive law*, postulating a relation between $\boldsymbol{J}$ and u.

Assume Ω is a container filled with a fluid in which a substance is dissolved. If u is the concentration of this substance, then $S(u) = u$ and a = kg. The description of $\boldsymbol{J}$ depends on the processes involved. If the fluid is at rest, then flux is only possible due to *molecular diffusion*, i.e., a flux from high to low concentrations due to random motion of the dissolved particles. Experimental evidence leads to

$$\boldsymbol{J}^{(1)} = -K\nabla u \tag{0.4}$$

with a parameter $K > 0$ [m^2/s], the *molecular diffusivity*. Equation (0.4) is called *Fick's law*.

In other situations, like heat conduction in a solid, a similar model occurs. Here, u represents the temperature, and the underlying principle is energy conservation. The constitutive law is *Fourier's law*, which also has the form (0.4), but as K is a material parameter, it may vary with space or, for anisotropic materials, be a matrix-valued function $\boldsymbol{K} : \Omega \to \mathbb{R}^{d,d}$ instead of a scalar function $K : \Omega \to \mathbb{R}$.

Thus we obtain the *diffusion equation*

$$\partial_t u - \nabla \cdot (\boldsymbol{K}\nabla u) = Q\,. \tag{0.5}$$

If the coefficient is scalar and constant, it can be chosen—possibly after a scaling—as $\boldsymbol{K} := \boldsymbol{I}$, where $\boldsymbol{I} \in \mathbb{R}^{d,d}$ is the identity matrix. If, in addition, $f := Q$ is independent of u, the equation simplifies further to

$$\partial_t u - \Delta u = f\,,$$

where $\Delta u := \nabla \cdot (\nabla u)$. We mentioned already that this equation also occurs in the modelling of heat conduction, therefore this equation or (0.5) is also called the *heat equation*.

If the fluid is in motion with a (given) velocity $\boldsymbol{c}$ then *(forced) convection* of the particles takes place, being described by

$$\boldsymbol{J}^{(2)} = u\boldsymbol{c}\,, \tag{0.6}$$

i.e., taking both processes into account, the model takes the form of the *convection–diffusion equation*

$$\partial_t u - \nabla \cdot (\boldsymbol{K}\nabla u - \boldsymbol{c}u) = Q\,. \tag{0.7}$$

The relative strength of the two processes is measured by the Péclet number (defined in Section 0.4). If convection is dominating one may ignore diffusion and only consider the *transport equation*

$$\partial_t u + \nabla \cdot (\boldsymbol{c}u) = Q\,. \tag{0.8}$$

The different nature of the two processes has to be reflected in the models, therefore, adapted discretization techniques will be necessary. In this book we will consider models like (0.7), usually with a significant contribution of diffusion, and the case of dominating convection is studied in Chapter 10. The pure convective case like (0.8) will not be treated.

In more general versions of (0.7) $\partial_t u$ is replaced by $\partial_t S(u)$, where S depends linearly or nonlinearly on u. In the case of heat conduction S is the internal energy density, which is related to the temperature u via the factors mass density and specific heat. For some materials the specific heat depends on the temperature, then S is a nonlinear function of u.

Further aspects come into play by the source term Q if it depends linearly or nonlinearly on u, in particular due to (chemical) reactions. Examples for these cases will be developed in the following sections. Since equation (0.3) and its examples describe conservation in general, it still has to be adapted to a concrete situation to ensure a unique solution u. This is done by the specification of an initial condition

$$S(u(x,0)) = S_0(x) \quad \text{for } x \in \Omega,$$

and by boundary conditions. In the example of the water filled container no mass flux will occur across its walls, therefore, the following boundary condition

$$\boldsymbol{J} \cdot \boldsymbol{n}(x,t) = 0 \quad \text{for } x \in \partial\Omega,\ t \in (0,T) \tag{0.9}$$

is appropriate, which–depending on the definition of $\boldsymbol{J}$–prescribes the normal derivative of u, or a linear combination of it and u. In Section 0.5 additional situations are depicted.

If a process is stationary, i.e., time-independent, then equation (0.3) reduces to

$$\nabla \cdot \boldsymbol{J}(x) = Q(x, u(x)) \quad \text{for } x \in \Omega,$$

which in the case of diffusion and convection is specified to

$$-\nabla \cdot (\boldsymbol{K}\nabla u - \boldsymbol{c}u) = Q\,.$$

For a scalar and constant diffusion coefficient—let $\boldsymbol{K} := \boldsymbol{I}$ by scaling—, $\boldsymbol{c} := \boldsymbol{0}$, and $f := Q$, being independent of u, this equation reduces to

$$-\Delta u = f \quad \text{in } \Omega,$$

the *Poisson equation.*

Instead of the boundary condition (0.9), one can prescribe the values of the function u at the boundary:

$$u(x) = g(x) \quad \text{for } x \in \partial\Omega\,.$$

For models, where u is a concentration or temperature, the physical realization of such a boundary condition may raise questions, but in mechanical models, where u is to interpreted as a displacement, such a boundary condition seems reasonable. The last boundary value problem will be the first model, whose discretization will be discussed in Chapters 1 and 2.

Finally it should be noted that it is advisable to non-dimensionalize the final model before numerical methods are applied. This means that both the independent variables x_i (and t), and the dependent one u, are replaced by $x_i/x_{i,\text{ref}}$, t/t_{ref}, and u/u_{ref}, where $x_{i,\text{ref}}$, t_{ref}, and u_{ref} are fixed reference values of the same dimension as x_i, t, and u, respectively. These reference values are considered to be of typical size for the problems under investigation. This procedure has two advantages: On the one hand, the typical size is now 1, such that there is an absolute scale for (an error in) a quantity to be small or large. On the other hand, if the reference values are chosen appropriately a reduction in the number of equation parameters like $\boldsymbol{K}$ and $\boldsymbol{c}$ in (0.7) might be possible, having only fewer algebraic expressions of the original material parameters in the equation. This facilitates numerical parameter studies.

0.2 Reactions and Transport in Porous Media

A *porous medium* is a heterogeneous material consisting of a *solid matrix* and a *pore space* contained therein. We consider the pore space (of the porous medium) as connected; otherwise, the transport of *fluids* in the pore space would not be possible. Porous media occur in nature and manufactured materials. Soils and aquifers are examples in geosciences; porous catalysts, chromatographic columns, and ceramic foams play important roles in chemical engineering. Even the human skin can be considered a porous medium. In the following we focus on applications in the geosciences. Thus we use a terminology referring to the natural soil as a porous medium. On the *micro* or *pore scale* of a single grain or pore, i.e., in a range of μm to mm, the fluids constitute different phases in the thermodynamic sense. Thus we name this system in the case of k fluids including the solid matrix as $(k + 1)$*-phase system* or we speak of k*-phase flow*.

We distinguish three classes of fluids with different affinities to the solid matrix. These are an aqueous phase, marked with the index "w" for water, a nonaqueous phase liquid (like oil or gasoline as natural resources or contaminants), marked with the index "o," and a gaseous phase, marked with the index "g" (e.g., soil air). Locally, at least one of these phases has always to be present; during a transient process phases can locally disappear or be generated. These fluid phases are in turn *mixtures* of several *components*. In applications of the earth sciences, for example, we do not deal with pure water but encounter different species in true or colloidal solution in the *solvent* water. The wide range of chemical components includes plant nutrients, mineral nutrients from salt domes, organic decomposition products, and various organic and inorganic chemicals. These substances are normally not inert but are subject to reactions and transformation processes. Along with diffusion, *forced convection* induced by the motion of the fluid is the essential driving mechanism for the transport of solutes. But we also encounter *natural convection* by the coupling of the dynamics of the substance to the fluid flow. The description level at the microscale that we have used so far is not suitable for processes at the laboratory or technical scale, which take place in ranges of cm to m, or even for processes in a catchment area with units of km. For those *macroscales* new models have to be developed, which emerge from averaging procedures of the models on the microscale. There may also exist principal differences among the various macroscales that let us expect different models, which arise from each other by *upscaling*. But this aspect will not be investigated here further. For the transition of micro to macro scales the engineering sciences provide the heuristic method of *volume averaging*, and mathematics the rigorous (but of only limited use) approach of *homogenization* (see [65] or [37]). None of the two possibilities can be depicted here completely. Where necessary we will refer to volume averaging for (heuristic) motivation.

Let $\Omega \subset \mathbb{R}^d$ be the domain of interest. All subsequent considerations are formal in the sense that the admissibility of the analytic manipulations is supposed. This can be achieved by the assumption of sufficient smoothness for the corresponding functions and domains.

Let $V \subset \Omega$ be an admissible *representative elementary volume* in the sense of volume averaging around a point $x \in \Omega$. Typically the shape and the size of a representative elementary volume are selected in such a manner that the averaged values of all geometric characteristics of the microstructure of the pore space are independent of the size of V but depend on the location of the point x. Then we obtain for a given variable ω_α in the phase α (after continuation of ω_α with 0 outside of α) the corresponding macroscopic quantities, assigned to the location x, as the *extrinsic phase average*

$$\langle \omega_\alpha \rangle := \frac{1}{|V|} \int_V \omega_\alpha$$

or as the *intrinsic phase average*

$$\langle \omega_\alpha \rangle^\alpha := \frac{1}{|V_\alpha|} \int_{V_\alpha} \omega_\alpha \, .$$

Here V_α denotes the subset of V corresponding to α. Let $t \in (0, T)$ be the time at which the process is observed. The notation $x \in \Omega$ means the vector in Cartesian coordinates, whose coordinates are referred to by x, y, and $z \in \mathbb{R}$. Despite this ambiguity the meaning can always be clearly derived from the context.

Let the index "s" (for solid) stand for the solid phase; then

$$\phi(x) := |V \setminus V_{\mathrm{s}}| \; / \; |V| > 0$$

denotes the *porosity*, and for every liquid phase α,

$$S_\alpha(x, t) := |V_\alpha| \; / \; |V \setminus V_{\mathrm{s}}| \geq 0$$

is the *saturation* of the phase α. Here we suppose that the solid phase is stable and immobile. Thus

$$\langle \omega_\alpha \rangle = \phi S_\alpha \langle \omega_\alpha \rangle^\alpha$$

for a fluid phase α and

$$\sum_{\alpha:\mathrm{fluid}} S_\alpha = 1 \, . \tag{0.10}$$

So if the fluid phases are *immiscible* on the micro scale, they may be miscible on the macro scale, and the immiscibility on the macro scale is an additional assumption for the model.

As in other disciplines the differential equation models are derived here from conservation laws for the *extensive quantities* mass, impulse, and energy, supplemented by *constitutive relationships*, where we want to focus on the mass.

0.3 Fluid Flow in Porous Media

Consider a liquid phase α on the micro scale. In this chapter, for clarity, we write "short" vectors in $\mathbb{R}^d$ also in bold with the exception of the coordinate vector x. Let $\tilde{\varrho}_\alpha$ $[\mathtt{kg/m^3}]$ be the (microscopic) *density*, $\tilde{\boldsymbol{q}}_\alpha := \left(\sum_\eta \tilde{\varrho}_\eta \tilde{\boldsymbol{v}}_\eta\right)/\tilde{\varrho}_\alpha$ $[\mathtt{m/s}]$ the *mass average mixture velocity* based on the *particle velocity* $\tilde{\boldsymbol{v}}_\eta$ of a component η and its concentration in solution $\tilde{\varrho}_\eta$ $[\mathtt{kg/m^3}]$. The transport theorem of Reynolds (see, for example, [15, Sect. 1.1]) leads to the mass conservation law

$$\partial_t \tilde{\varrho}_\alpha + \nabla \cdot \left(\tilde{\varrho}_\alpha \tilde{\boldsymbol{q}}_\alpha\right) = \tilde{f}_\alpha \tag{0.11}$$

with a distributed *mass source density* $\tilde{f}_\alpha$. By averaging we obtain from here the mass conservation law

$$\partial_t(\phi S_\alpha \varrho_\alpha) + \nabla \cdot \left(\varrho_\alpha \boldsymbol{q}_\alpha\right) = f_\alpha \tag{0.12}$$

with ϱ_α, the density of phase α, as the intrinsic phase average of $\tilde{\varrho}_\alpha$ and $\boldsymbol{q}_\alpha$, the *volumetric fluid velocity* or *Darcy velocity* of the phase α, as the extrinsic phase average of $\tilde{\boldsymbol{q}}_\alpha$. Correspondingly, f_α is an average mass source density.

Before we proceed in the general discussion, we want to consider some specific situations: The area between the groundwater table and the impermeable body of an *aquifer* is characterized by the fact that the whole pore space is occupied by a fluid phase, the soil water. The corresponding saturation thus equals 1 everywhere, and with omission of the index equation (0.12) takes the form

$$\partial_t(\phi\varrho) + \nabla \cdot (\varrho \boldsymbol{q}) = f\,. \tag{0.13}$$

If the density of water is assumed to be constant, due to neglecting the mass of solutes and compressibility of water, equation (0.13) simplifies further to the stationary equation

$$\nabla \cdot \boldsymbol{q} = f\,, \tag{0.14}$$

where f has been replaced by the volume source density f/ϱ, keeping the same notation. This equation will be completed by a relationship that can be interpreted as the macroscopic analogue of the conservation of momentum, but should be accounted here only as an experimentally derived constitutive relationship. This relationship is called *Darcy's law*, which reads as

$$\boldsymbol{q} = -\boldsymbol{K}\left(\nabla p + \varrho g \boldsymbol{e}_z\right) \tag{0.15}$$

and can be applied in the range of laminar flow. Here p $[\mathtt{N/m^2}]$ is the intrinsic average of the *water pressure*, g $[\mathtt{m/s^2}]$ the gravitational acceleration, $\boldsymbol{e}_z$ the unit vector in the z-direction oriented against the gravitation,

$$\boldsymbol{K} = \boldsymbol{k}/\mu\,, \tag{0.16}$$

a quantity, which is given by the *permeability* $\boldsymbol{k}$ determined by the solid phase, and the *viscosity* μ determined by the fluid phase. For an *anisotropic* solid, the matrix $\boldsymbol{k} = \boldsymbol{k}(x)$ is a symmetric positive definite matrix.

Inserting (0.15) in (0.14) and replacing $\boldsymbol{K}$ by $\boldsymbol{K}\varrho g$, known as *hydraulic conductivity* in the literature, and keeping the same notation gives the following linear equation for

$$h(x,t) := \frac{1}{\varrho g}\, p(x,t) + z\,,$$

the *piezometric head* h [m]:

$$-\nabla \cdot (\boldsymbol{K}\nabla h) = f\,. \tag{0.17}$$

The resulting equation is stationary and linear. We call a differential equation model *stationary* if it depends only on the location x and not on the time t, and *instationary* otherwise. A differential equation and corresponding boundary conditions (cf. Section 0.5) are called *linear* if the sum or a scalar multiple of a solution again forms a solution for the sum, respectively, the scalar multiple, of the sources.

If we deal with an *isotropic* solid matrix, we have $\boldsymbol{K} = K\boldsymbol{I}$ with a scalar function $K : \Omega \to \mathbb{R}$. Equation (0.17) in this case reads

$$-\nabla \cdot (K\nabla h) = f\,. \tag{0.18}$$

Finally if the solid matrix is homogeneous, i.e., K is constant, we get from division by K and maintaining the notation f the *Poisson equation*

$$-\Delta h = f\,, \tag{0.19}$$

which is termed the *Laplace equation* for $f = 0$. This model and its more general formulations occur in various contexts. If, contrary to the above assumption, the solid matrix is compressible under the pressure of the water, and if we suppose (0.13) to be valid, then we can establish a relationship

$$\phi = \phi(x,t) = \phi_0(x)\phi_f(p)$$

with $\phi_0(x) > 0$ and a monotone increasing ϕ_f such that with $S(p) := \phi_f'(p)$ we get the equation

$$\phi_0\, S(p)\, \partial_t p + \nabla \cdot \boldsymbol{q} = f$$

and the instationary equations corresponding to (0.17)–(0.19), respectively. For constant $S(p) > 0$ this yields the following linear equation:

$$\phi_0\, S\, \partial_t h - \nabla \cdot (\boldsymbol{K}\nabla h) = f\,, \tag{0.20}$$

which also represents a common model in many contexts and is known from corresponding fields of application as the *heat conduction equation*.

We consider single phase flow further, but now we will consider gas as fluid phase. Because of the compressibility, the density is a function of the pressure, which is invertible due to its strict monotonicity to

$$p = P(\varrho)\,.$$

Together with (0.13) and (0.15) we get a nonlinear variant of the heat conduction equation in the unknown ϱ:

$$\partial_t(\phi\varrho) - \nabla\cdot\left(\boldsymbol{K}(\varrho\nabla P(\varrho) + \varrho^2 g\boldsymbol{e}_z)\right) = f\,, \tag{0.21}$$

which also contains derivatives of first order in space. If $P(\varrho) = \ln(\alpha\varrho)$ holds for a constant $\alpha > 0$, then $\varrho\nabla P(\varrho)$ simplifies to $\alpha\nabla\varrho$. Thus for horizontal flow we again encounter the heat conduction equation. For the relationship $P(\varrho) = \alpha\varrho$ suggested by the universal gas law, $\alpha\varrho\nabla\varrho = \frac{1}{2}\alpha\nabla\varrho^2$ remains nonlinear. The choice of the variable $u := \varrho^2$ would result in $u^{1/2}$ in the time derivative as the only nonlinearity. Thus in the formulation in ϱ the coefficient of $\nabla\varrho$ disappears in the divergence of $\varrho = 0$. Correspondingly, the coefficient $S(u) = \frac{1}{2}\phi u^{-1/2}$ of $\partial_t u$ in the formulation in u becomes unbounded for $u = 0$. In both versions the equations are *degenerate*, whose treatment is beyond the scope of this book. A variant of this equation has gained much attention as the *porous medium equation* (with convection) in the field of analysis (see, for example, [85], [62]).

Returning to the general framework, the following generalization of Darcy's law can be justified experimentally for several liquid phases:

$$\boldsymbol{q}_\alpha = -\frac{k_{r\alpha}}{\mu_\alpha}\boldsymbol{k}\left(\nabla p_\alpha + \varrho_\alpha g\boldsymbol{e}_z\right)\,.$$

Here the *relative permeability* $k_{r\alpha}$ of the phase α depends upon the saturations of the present phases and takes values in $[0, 1]$.

At the interface of two liquid phases α_1 and α_2 we observe a difference of the pressures, the so-called *capillary pressure*, that turns out experimentally to be a function of the saturations:

$$p_{c\alpha_1\alpha_2} := p_{\alpha_1} - p_{\alpha_2} = F_{\alpha_1\alpha_2}(S_w, S_o, S_g)\,. \tag{0.22}$$

A general model for multiphase flow, formulated for the moment in terms of the variables p_α, S_α, is thus given by the equations

$$\partial_t(\phi S_\alpha\varrho_\alpha) - \nabla\cdot(\varrho_\alpha\lambda_\alpha\boldsymbol{k}(\nabla p_\alpha + \varrho_\alpha g\boldsymbol{e}_z)) = f_\alpha \tag{0.23}$$

with the *mobilities* $\lambda_\alpha := k_{r\alpha}/\mu_\alpha$, and the equations (0.22) and (0.10), where one of the S_α's can be eliminated. For two liquid phases w and g, e.g., water and air, equations (0.22) and (0.10) for $\alpha = \text{w}, \text{g}$ read $p_c = p_g - p_w = F(S_w)$ and $S_g = 1 - S_w$. Apparently, this is a time-dependent, nonlinear model in the variables p_w, p_g, S_w, where one of the variables can be eliminated. Assuming constant densities ϱ_α, further formulations based on

$$\nabla\cdot\left(\boldsymbol{q}_w + \boldsymbol{q}_g\right) = f_w/\varrho_w + f_g/\varrho_g \tag{0.24}$$

can be given as consequences of (0.10). These equations consist of a stationary equation for a new quantity, the *global pressure*, based on (0.24), and a time-dependent equation for one of the saturations (see Problem 0.2). In many situations it is justified to assume a gaseous phase with constant pressure in the whole domain and to scale this pressure to $p_g = 0$. Thus for $\psi := p_w = -p_c$ we have

$$\phi\partial_t S(\psi) - \nabla \cdot (\lambda(\psi)\boldsymbol{k}(\nabla\psi + \varrho g \boldsymbol{e}_z)) = f_w/\varrho_w \tag{0.25}$$

with constant pressure $\varrho := \varrho_w$, and $S(\psi) := F^{-1}(-\psi)$ as a strictly monotone increasing nonlinearity as well as λ.

With the convention to set the value of the air pressure to 0, the pressure in the aqueous phase is in the *unsaturated state*, where the gaseous phase is also present, and represented by negative values. The water pressure $\psi = 0$ marks the transition from the unsaturated to the *saturated* zone. Thus in the unsaturated zone, equation (0.25) represents a nonlinear variant of the heat conduction equation for $\psi < 0$, the *Richards equation*. As most functional relationships have the property $S'(0) = 0$, the equation degenerates in the absence of a gaseous phase, namely, to a stationary equation in a way that is different from above.

Equation (0.25) with $S(\psi) := 1$ and $\lambda(\psi) := \lambda(0)$ can be continued in a consistent way with (0.14) and (0.15) also for $\psi \geq 0$, i.e., for the case of a sole aqueous phase. The resulting equation is also called Richards equation or a model of *saturated–unsaturated flow*.

0.4 Reactive Solute Transport in Porous Media

In this chapter we will discuss the transport of a single component in a liquid phase and some selected reactions. We will always refer to water as liquid phase explicitly. Although we treat *inhomogeneous reactions* in terms of surface reactions with the solid phase, we want to ignore exchange processes between the fluid phases. On the microscopic scale the mass conservation law for a single component η is, in the notation of (0.11) by omitting the phase index w,

$$\partial_t \tilde{\varrho}_\eta + \nabla \cdot (\tilde{\varrho}_\eta \tilde{\boldsymbol{q}}) + \nabla \cdot \boldsymbol{J}_\eta = \tilde{Q}_\eta,$$

where

$$\boldsymbol{J}_\eta := \tilde{\varrho}_\eta \left(\tilde{\boldsymbol{v}}_\eta - \tilde{\boldsymbol{q}}\right) [\mathtt{kg/m^2/s}] \tag{0.26}$$

represents the *diffusive mass flux* of the component η and $\tilde{Q}_\eta$ $[\mathtt{kg/m^3/s}]$ is its *volumetric production rate*. For a description of reactions via the *mass action law* it is appropriate to choose the mole as the unit of mass. The diffusive mass flux requires a phenomenological description. The assumption that solely binary molecular diffusion, described by *Fick's law*, acts between the component η and the solvent, means that

$$\boldsymbol{J}_\eta = -\tilde{\varrho} D_\eta \nabla \left(\tilde{\varrho}_\eta/\tilde{\varrho}\right) \tag{0.27}$$

with a *molecular diffusivity* $D_\eta > 0$ [m^2/s]. The averaging procedure applied on (0.26), (0.27) leads to

$$\partial_t(\Theta c_\eta) + \nabla \cdot (\boldsymbol{q} c_\eta) + \nabla \cdot \boldsymbol{J}^{(1)} + \nabla \cdot \boldsymbol{J}^{(2)} = Q_\eta^{(1)} + Q_\eta^{(2)}$$

for the *solute concentration* of the component η, c_η [kg/m^3], as intrinsic phase average of $\tilde{\varrho}_\eta$. Here, we have $\boldsymbol{J}^{(1)}$ as the average of $\boldsymbol{J}_\eta$ and $\boldsymbol{J}^{(2)}$, the mass flux due to *mechanical dispersion*, a newly emerging term at the macroscopic scale. Analogously, $Q_\eta^{(1)}$ is the intrinsic phase average of $\tilde{Q}_\eta$, and $Q_\eta^{(2)}$ is a newly emerging term describing the exchange between the liquid and solid phases.

The *volumetric water content* is given by $\Theta := \phi S_{\mathrm{w}}$ with the water saturation S_{w}. Experimentally, the following phenomenological descriptions are suggested:

$$\boldsymbol{J}^{(1)} = -\Theta \tau D_\eta \nabla c_\eta$$

with a *tortuosity factor* $\tau \in (0, 1]$,

$$\boldsymbol{J}^{(2)} = -\Theta \boldsymbol{D}_{\mathrm{mech}} \nabla c_\eta \,, \tag{0.28}$$

and a symmetric positive definite *matrix of mechanical dispersion* $\boldsymbol{D}_{\mathrm{mech}}$, which depends on q/Θ. Consequently, the resulting differential equation reads

$$\partial_t(\Theta c_\eta) + \nabla \cdot \left(\boldsymbol{q} c_\eta - \Theta \boldsymbol{D} \nabla c_\eta\right) = Q_\eta \tag{0.29}$$

with $\boldsymbol{D} := \tau \boldsymbol{D}_\eta + \boldsymbol{D}_{\mathrm{mech}}$, $Q_\eta := Q_\eta^{(1)} + Q_\eta^{(2)}$.

Because the mass flux consists of $\boldsymbol{q} c_\eta$, a part due to *forced convection*, and of $\boldsymbol{J}^{(1)} + \boldsymbol{J}^{(2)}$, a part that corresponds to a generalized Fick's law, an equation like (0.29) is called a *convection–diffusion equation*. Accordingly, for the part with first spatial derivatives like $\nabla \cdot (\boldsymbol{q} c_\eta)$ the term *convective part* is used, and for the part with second spatial derivatives like $-\nabla \cdot (\Theta \boldsymbol{D} \nabla c_\eta)$ the term *diffusive part* is used. If the first term determines the character of the solution, the equation is called *convection-dominated*. The occurrence of such a situation is measured by the quantity Pe, the *global Péclet number*, that has the form $\mathrm{Pe} = \|\boldsymbol{q}\| L / \|\Theta \boldsymbol{D}\|$ [-]. Here L is a characteristic length of the domain Ω. The extreme case of purely convective transport results in a conservation equation of first order. Since the common models for the dispersion matrix lead to a bound for Pe, the reduction to the purely convective transport is not reasonable. However, we have to take convection-dominated problems into consideration.

Likewise, we speak of diffusive parts in (0.17) and (0.20) and of (nonlinear) diffusive and convective parts in (0.21) and (0.25). Also, the multiphase transport equation can be formulated as a nonlinear convection–diffusion equation by use of (0.24) (see Problem 0.2), where convection often dominates. If the production rate Q_η is independent of c_η, equation (0.29) is linear.

In general, in case of a surface reaction of the component η, the kinetics of the reaction have to be described. If this component is not in competition with the other components, one speaks of *adsorption*. The kinetic equation thus takes the general form

$$\partial_t s_\eta(x,t) = k_\eta f_\eta(x, c_\eta(x,t), s_\eta(x,t)) \tag{0.30}$$

with a rate parameter k_η for the *sorbed concentration* s_η [kg/kg], which is given in reference to the mass of the solid matrix. Here, the components in sorbed form are considered spatially immobile. The conservation of the total mass of the component undergoing sorption gives

$$Q_\eta^{(2)} = -\varrho_b \partial_t s_\eta \tag{0.31}$$

with the *bulk density* $\varrho_b = \varrho_s(1-\phi)$, where ϱ_s denotes the density of the solid phase. With (0.30), (0.31) we have a system consisting of an instationary partial and an ordinary differential equation (with $x \in \Omega$ as parameter). A widespread model by *Langmuir* reads

$$f_\eta = k_a c_\eta (\overline{s}_\eta - s_\eta) - k_d s_\eta$$

with constants k_a, k_d that depend upon the temperature (among other factors), and a *saturation concentration* $\overline{s}_\eta$ (cf. for example [44]). If we assume $f_\eta = f_\eta(x, c_\eta)$ for simplicity, we get a scalar nonlinear equation in c_η,

$$\partial_t(\Theta c_\eta) + \nabla \cdot \left(\boldsymbol{q} c_\eta - \Theta \boldsymbol{D} \nabla c_\eta\right) + \varrho_b k_\eta f_\eta(\cdot, c_\eta) = Q_\eta^{(1)}, \tag{0.32}$$

and s_η is decoupled and extracted from (0.30). If the time scales of transport and reaction differ greatly, and the limit case $k_\eta \to \infty$ is reasonable, then (0.30) is replaced by

$$f_\eta(x, c_\eta(x,t), s_\eta(x,t)) = 0\,.$$

If this equation is solvable for s_η, i.e.,

$$s_\eta(x,t) = \varphi_\eta(x, c_\eta(x,t)),$$

the following scalar equation for c_η with a nonlinearity in the time derivative emerges:

$$\partial_t(\Theta c_\eta + \varrho_b \varphi_\eta(\cdot, c_\eta)) + \nabla \cdot \left(\boldsymbol{q} c_\eta - \Theta \boldsymbol{D} \nabla c_\eta\right) = Q_\eta^{(1)}\,.$$

If the component η is in competition with other components in the surface reaction, as, e.g., in ion exchange, then f_η has to be replaced by a nonlinearity that depends on the concentrations of all involved components $c_1, \dots, c_N, s_1, \dots, s_N$. Thus we obtain a coupled system in these variables. Finally, if we encounter *homogeneous reactions* that take place solely in the fluid phase, an analogous statement is true for the source term $Q_\eta^{(1)}$.

Exercises

Problem 0.1 Give a geometric interpretation for the matrix condition of $\boldsymbol{k}$ in (0.16) and $\boldsymbol{D}_{\text{mech}}$ in (0.28).

Problem 0.2 Consider the two-phase flow (with constant ϱ_α, $\alpha \in \{\mathrm{w}, \mathrm{g}\}$)

$$\begin{aligned} \partial_t(\phi S_\alpha) + \nabla \cdot \boldsymbol{q}_\alpha &= f_\alpha\,, \\ \boldsymbol{q}_\alpha &= -\lambda_\alpha \boldsymbol{k}\,(\nabla p_\alpha + \varrho_\alpha g \boldsymbol{e}_z)\,, \\ S_\mathrm{w} + S_\mathrm{g} &= 1\,, \\ p_\mathrm{g} - p_\mathrm{w} &= p_\mathrm{c} \end{aligned}$$

with coefficient functions

$$p_\mathrm{c} = p_\mathrm{c}(S_\mathrm{w}), \quad \lambda_\alpha = \lambda_\alpha(S_\mathrm{w}), \quad \alpha \in \{\mathrm{w}, \mathrm{g}\}.$$

Starting from equation (0.23), perform a transformation to the new variables

$$\boldsymbol{q} = \boldsymbol{q}_\mathrm{w} + \boldsymbol{q}_\mathrm{g} \qquad \text{"total flow"},$$

$$p = \frac{1}{2}(p_\mathrm{w} + p_\mathrm{g}) + \frac{1}{2}\int_{S_\mathrm{c}}^{S} \frac{\lambda_\mathrm{g} - \lambda_\mathrm{w}}{\lambda_\mathrm{g} + \lambda_\mathrm{w}} \frac{dp_\mathrm{c}}{d\xi}\, d\xi \qquad \text{"global pressure"},$$

and the water saturation S_w. Derive a representation of the phase flows in the new variables.

Problem 0.3 A frequently employed model for mechanical dispersion is

$$D_\mathrm{mech} = \lambda_\mathrm{L} |\boldsymbol{v}|_2 P_{\boldsymbol{v}} + \lambda_\mathrm{T} |\boldsymbol{v}|_2 (I - P_{\boldsymbol{v}})$$

with parameters $\lambda_\mathrm{L} > \lambda_\mathrm{T}$, where $\boldsymbol{v} = \boldsymbol{q}/\Theta$ and $P_{\boldsymbol{v}} = \boldsymbol{v}\boldsymbol{v}^T/|\boldsymbol{v}|_2^2$. Here λ_L and λ_T are the *longitudinal* and *transversal dispersion lengths*. Give a geometrical interpretation.

0.5 Boundary and Initial Value Problems

The differential equations that we derived in Sections 0.3 and 0.4 have the common form

$$\partial_t S(u) + \nabla \cdot (\boldsymbol{C}(u) - \boldsymbol{K}(\nabla u)) = Q(u) \tag{0.33}$$

with a *storage term* S, a convective part $\boldsymbol{C}$, a diffusive part $\boldsymbol{K}$, i.e., a total flux $\boldsymbol{C} - \boldsymbol{K}$ and a source term Q, which depend linearly or nonlinearly on the unknown u. For simplification, we assume u to be a scalar. The nonlinearities $S, \boldsymbol{C}, \boldsymbol{K}$, and Q may also depend on x and t, which shall be suppressed in the notation in the following. Such an equation is said to be in *divergence form* or in *conservative form*; a more general formulation is obtained by differentiating $\nabla \cdot \boldsymbol{C}(u) = \frac{\partial}{\partial u}\boldsymbol{C}(u) \cdot \nabla u + (\nabla \cdot \boldsymbol{C})(u)$ or by introducing a generalized "source term" $Q = Q(u, \nabla u)$. Up to now we have considered differential equations pointwise in $x \in \Omega$ (and $t \in (0, T)$) under the assumption that all occurring functions are well defined. Due to the applicability of the integral theorem of Gauss on $\tilde{\Omega} \subset \Omega$ (cf. (3.11)), the *integral form* of the conservation equation follows straightforwardly from the above:

$$\int_{\tilde{\Omega}} \partial_t S(u)\,dx + \int_{\partial\tilde{\Omega}} (\boldsymbol{C}(u) - \boldsymbol{K}(\nabla u)) \cdot \mathfrak{n}\,d\sigma = \int_{\tilde{\Omega}} Q(u, \nabla u)\,dx \tag{0.34}$$

with the outer unit normal $\mathfrak{n}$ (see Theorem 3.9) for a fixed time t or also in t integrated over $(0, T)$. Indeed, this equation (on the microscopic scale) is the primary description of the conservation of an extensive quantity: Changes in time through storage and sources in $\tilde{\Omega}$ are compensated by the normal flux over $\partial\tilde{\Omega}$. Moreover, for $\partial_t S$, $\nabla \cdot (\boldsymbol{C} - \boldsymbol{K})$, and Q continuous on the closure of $\tilde{\Omega}$, (0.33) follows from (0.34). If, on the other hand, F is a hyperplane in $\tilde{\Omega}$ where the material properties may rapidly change, the *jump condition*

$$[\![(\boldsymbol{C}(u) - \boldsymbol{K}(\nabla u)) \cdot \mathfrak{n}]\!] = 0 \tag{0.35}$$

for a fixed unit normal $\mathfrak{n}$ on F follows from (0.34), where $[\![\cdot]\!]$ denotes the difference of the one-sided limits (see Problem 0.4).

Since the differential equation describes conservation only in general, it has to be supplemented by initial and boundary conditions in order to specify a particular situation where a unique solution is expected. Boundary conditions are specifications on $\partial\Omega$, where $\mathfrak{n}$ denotes the outer unit normal

- of the normal component of the flux (inwards):

$$-(\boldsymbol{C}(u) - \boldsymbol{K}(\nabla u)) \cdot \mathfrak{n} = g_1 \quad \text{on } \Gamma_1 \tag{0.36}$$

  (flux boundary condition),
- of a linear combination of the normal flux and the unknown itself:

$$-(\boldsymbol{C}(u) - \boldsymbol{K}(\nabla u)) \cdot \mathfrak{n} + \alpha u = g_2 \quad \text{on } \Gamma_2 \tag{0.37}$$

  (mixed boundary condition),
- of the unknown itself:

$$u = g_3 \quad \text{on } \Gamma_3 \tag{0.38}$$

  (Dirichlet boundary condition).

Here $\Gamma_1, \Gamma_2, \Gamma_3$ form a disjoint decomposition of $\partial\Omega$:

$$\Gamma := \partial\Omega = \Gamma_1 \cup \Gamma_2 \cup \Gamma_3, \tag{0.39}$$

where Γ_3 is supposed to be a relatively closed subset of $\partial\Omega$ (i.e., closed in the induced topology of $\partial\Omega$). The *inhomogeneities* g_i and the factor α in general depend on $x \in \Omega$, and for nonstationary problems (where $S(u) \neq 0$ holds) on $t \in (0, T)$. The boundary conditions are linear if the g_i do not depend (nonlinearly) on u (see below). If the g_i are zero, we speak of *homogeneous*, otherwise of *inhomogeneous, boundary conditions*.

Thus the pointwise formulation of a nonstationary equation (where S does not vanish) requires the validity of the equation in the *space-time cylinder*

$$Q_T := \Omega \times (0, T)$$

and the boundary conditions on the *lateral surface* of the space-time cylinder

$$S_T := \partial\Omega \times (0, T) .$$

Different types of boundary conditions are possible with decompositions of the type (0.39). Additionally, an *initial condition* on the *bottom* of the space-time cylinder is necessary:

$$S(u(x, 0)) = S_0(x) \quad \text{for } x \in \Omega . \tag{0.40}$$

These are so-called *initial boundary value problems*; for stationary problems we speak of *boundary value problems*. As shown in (0.34) and (0.35) flux boundary conditions have a natural relationship with the differential equation (0.33). For a linear diffusive part $\boldsymbol{K}(\nabla u) = \boldsymbol{K}\nabla u$ alternatively we may require

$$\partial_{\mathfrak{n}_K} u := \boldsymbol{K}\nabla u \cdot \mathfrak{n} = g_1 \quad \text{on } \Gamma_1 , \tag{0.41}$$

and an analogous mixed boundary condition. This boundary condition is the so-called *Neumann boundary condition*. Since $\boldsymbol{K}$ is symmetric, $\partial_{\mathfrak{n}_K} u = \nabla u \cdot \boldsymbol{K}\mathfrak{n}$ holds; i.e., $\partial_{\mathfrak{n}_K} u$ is the derivative in direction of the *conormal* $\boldsymbol{K}\mathfrak{n}$. For the special case $\boldsymbol{K} = \boldsymbol{I}$ the normal derivative is given.

In contrast to ordinary differential equations, there is hardly any general theory of partial differential equations. In fact, we have to distinguish different types of differential equations according to the various described physical phenomena. These determine, as discussed, different (initial) boundary value specifications to render the problem *well posed*. *Well-posedness* means that the problem possesses a unique solution (with certain properties yet to be defined) that depends continuously (in appropriate norms) on the data of the problem, in particular on the (initial and) boundary values. There exist also *ill-posed* boundary value problems for partial differential equations, which correspond to physical and technical applications. They require special techniques and shall not be treated here.

The classification into different types is simple if the problem is linear and the differential equation is of second order as in (0.33). By *order* we mean the highest order of the derivative with respect to the variables $(x_1, \ldots, x_d, t)$ that appears, where the time derivative is considered to be like a spatial derivative. Almost all differential equations treated in this book will be of second order, although important models in elasticity theory are of fourth order or certain transport phenomena are modelled by systems of first order.

The differential equation (0.33) is generally *nonlinear* due to the nonlinear relationships $S, \boldsymbol{C}, \boldsymbol{K}$, and Q. Such an equation is called *quasilinear* if all derivatives of the highest order are linear, i.e., we have

$$\boldsymbol{K}(\nabla u) = \boldsymbol{K}\nabla u \tag{0.42}$$

with a matrix $\boldsymbol{K}$, which may also depend (nonlinearly) on x, t, and u. Furthermore, (0.33) is called *semilinear* if nonlinearities are present only in u, but not in the derivatives, i.e., if in addition to (0.42) with $\boldsymbol{K}$ being independent of u, we have

$$S(u) = Su, \quad \boldsymbol{C}(u) = u\boldsymbol{c} \tag{0.43}$$

with scalar and vectorial functions S and $\boldsymbol{c}$, respectively, which may depend on x and t. Such variable factors standing before u or differential terms are called *coefficients* in general.

Finally, the differential equation is *linear* if we have, in addition to the above requirements,

$$Q(u) = -ru + f$$

with functions r and f of x and t.

In the case $f = 0$ the linear differential equation is termed *homogeneous*, otherwise *inhomogeneous*. A linear differential equation obeys the *superposition principle*: Suppose u_1 and u_2 are solutions of (0.33) with the source terms f_1 and f_2 and otherwise identical coefficient functions. Then $u_1 + \gamma u_2$ is a solution of the same differential equation with the source term $f_1 + \gamma f_2$ for arbitrary $\gamma \in \mathbb{R}$. The same holds for linear boundary conditions. The term *solution* of an (initial) boundary value problem is used here in a classical sense, yet to be specified, where all the quantities occurring should satisfy pointwise certain regularity conditions (see Definition 1.1 for the Poisson equation). However, for variational solutions (see Definition 2.2), which are appropriate in the framework of finite element methods, the above statements are also valid.

Linear differential equations of second order in two variables (x, y) (including possibly the time variable) can be classified in different *types* as follows:

To the homogeneous differential equation

$$\begin{aligned} Lu = a(x, y)\frac{\partial^2}{\partial x^2}u + b(x, y)\frac{\partial^2}{\partial x \partial y}u + c(x, y)\frac{\partial^2}{\partial y^2}u \\ + d(x, y)\frac{\partial}{\partial x}u + e(x, y)\frac{\partial}{\partial y}u + f(x, y)u = 0 \end{aligned} \tag{0.44}$$

the following quadratic form is assigned:

$$(\xi, \eta) \mapsto a(x, y)\xi^2 + b(x, y)\xi\eta + c(x, y)\eta^2. \tag{0.45}$$

According to its eigenvalues, i.e., the eigenvalues of the matrix

$$\begin{pmatrix} a(x, y) & \frac{1}{2}b(x, y) \\ \frac{1}{2}b(x, y) & c(x, y) \end{pmatrix}, \tag{0.46}$$

we classify the types. In analogy with the classification of conic sections, which are described by (0.45) (for fixed (x, y)), the differential equation (0.44) is called *at the point* (x, y)

- *elliptic* if the eigenvalues of (0.46) are not 0 and have the same sign,
- *hyperbolic* if one eigenvalue is positive and the other is negative,
- *parabolic* if exactly one eigenvalue is equal to 0.

For the corresponding generalization of the terms for $d + 1$ variables and arbitrary order, the stationary boundary value problems we treat in this book will be elliptic, of second order, and—except in Chapter 11—also linear; the nonstationary initial boundary value problems will be parabolic.

Systems of hyperbolic differential equations of first order require particular approaches, which are beyond the scope of this book. Nevertheless, we dedicate Chapter 10 to convection-dominated problems, i.e., elliptic or parabolic problems close to the hyperbolic limit case.

The different discretization strategies are based on various formulations of the (initial) boundary value problems: The *finite difference method*, which is presented in Section 1, and further outlined for nonstationary problems in Chapter 9, has the pointwise formulation of (0.33), (0.36)–(0.38) (and (0.40)) as a starting point. The *finite element method*, which lies in the focus of our book (Chapters 2, 3, and 9), is based on an integral formulation of (0.33) (which we still have to depict) that incorporates (0.36) and (0.37). The conditions (0.38) and (0.40) have to be enforced additionally. Finally, the *finite volume method* (Chapters 8 and 9) will be derived from the integral formulation (0.34), where also initial and boundary conditions come along as in the finite element approach.

Exercises

Problem 0.4 Derive (formally) (0.35) from (0.34).

Problem 0.5 Derive the orders of the given differential operators and differential equations, and decide in every case whether the operator is linear or nonlinear, and whether the linear equation is homogeneous or inhomogeneous:

a) $Lu := u_{xx} + xu_y$,
b) $Lu := u_x + uu_y$,
c) $Lu := \sqrt{1 + x^2}(\cos y)u_x + u_{yxy} - \left(\arctan \frac{x}{y}\right) u = \ln(x^2 + y^2)$,
d) $Lu := u_t + u_{xxxx} + \sqrt{1 + u} = 0$,
e) $u_{tt} - u_{xx} + x^2 = 0$.

Problem 0.6 a) Determine the type of the given differential operator:

1) $Lu := u_{xx} - u_{xy} + 2u_y + u_{yy} - 3u_{yx} + 4u$,
2) $Lu = 9u_{xx} + 6u_{xy} + u_{yy} + u_x$.

b) Determine the parts of the plane where the differential operator $Lu := yu_{xx} - 2u_{xy} + xu_{yy}$ is elliptic, hyperbolic, or parabolic.

c) 1) Determine the type of $Lu := 3u_y + u_{xy}$.
 2) Compute the general solution of $Lu = 0$.

Problem 0.7 Consider the equation $Lu = f$ with a linear differential operator of second order, defined for functions in d variables ($d \in \mathbb{N}$) in $x \in \Omega \subset \mathbb{R}^d$. The transformation $\Phi : \Omega \to \Omega' \subset \mathbb{R}^d$ has a continuously differentiable, nonsingular Jacobi matrix $D\Phi := \frac{\partial \Phi}{\partial x}$.

Show that the partial differential equation does not change its type if it is written in the new coordinates $\xi = \Phi(x)$.

Chapter 1
For the Beginning: The Finite Difference Method for the Poisson Equation

1.1 The Dirichlet Problem for the Poisson Equation

In this section we want to introduce the finite difference method, frequently abbreviated as *FDM*, using the Poisson equation on a rectangle as an example. By means of this example and generalizations of the problem, advantages and limitations of the approach will be elucidated. Also, in the following section the Poisson equation will be the main topic, but then on an arbitrary domain. For the spatial basic set of the differential equation $\Omega \subset \mathbb{R}^d$ we assume as minimal requirement that Ω is a domain, where a *domain* is a nonempty, open, and connected set. The boundary of this domain will be denoted by $\partial\Omega$, the closure $\Omega \cup \partial\Omega$ by $\overline{\Omega}$ (see Appendix A.2). The *Dirichlet problem for the Poisson equation* is then defined as follows: Given functions $g : \partial\Omega \to \mathbb{R}$ and $f : \Omega \to \mathbb{R}$, we are looking for a function $u : \overline{\Omega} \to \mathbb{R}$ such that

$$-\sum_{i=1}^{d} \frac{\partial^2}{\partial x_i^2} u = f \quad \text{in } \Omega, \tag{1.1}$$

$$u = g \quad \text{on } \partial\Omega. \tag{1.2}$$

This differential equation model has already appeared in (0.19) and (0.38) and beyond this application has an importance in a wide spectrum of disciplines. The unknown function u can be interpreted as an electromagnetic potential, a displacement of an elastic membrane, or a temperature. Similar to the multi-index notation to be introduced in (2.17) (but with indices at the top) from now on for partial derivatives we use the following notation:

For $u : \Omega \subset \mathbb{R}^d \to \mathbb{R}$ we set

P. Knabner and L. Angermann, *Numerical Methods for Elliptic and Parabolic Partial Differential Equations*, Texts in Applied Mathematics 44,
https://doi.org/10.1007/978-3-030-79385-2_1

$$\begin{aligned}
\partial_i u &:= \frac{\partial}{\partial x_i} u \qquad \text{for } i = 1, \dots, d, \\
\partial_{ij} u &:= \frac{\partial^2}{\partial x_i \, \partial x_j} u \quad \text{for } i, j = 1, \cdots, d, \\
\Delta u &:= (\partial_{11} + \dots + \partial_{dd}) \, u \, .
\end{aligned}$$

The expression Δu is called the *Laplace operator*. By means of this, (1.1) can be written in abbreviated form as

$$-\Delta u = f \quad \text{in } \Omega \, . \tag{1.3}$$

We could also define the Laplace operator by

$$\Delta u = \nabla \cdot (\nabla u) \, ,$$

where $\nabla u = (\partial_1 u, \dots, \partial_d u)^T$ denotes the *gradient* of a function u, and $\nabla \cdot \boldsymbol{v} = \partial_1 v_1 + \cdots + \partial_d v_d$ the *divergence* of a vector field $\boldsymbol{v}$. Therefore, an alternative notation exists, which will not be used in the following: $\Delta u = \nabla^2 u$. The incorporation of the minus sign in the left-hand side of (1.3), which looks strange at first glance, is related to the monotonicity and definiteness properties of $-\Delta$ (see Sections 1.4 and 2.1, respectively).

The notion of a solution for (1.1), (1.2) still has to be specified more precisely. Considering the equations in a pointwise sense, which will be pursued in this chapter, the functions in (1.1), (1.2) have to exist, and the equations have to be satisfied pointwise. Since (1.1) is an equation on an open set Ω, there are no implications for the behaviour of u up to the boundary $\partial\Omega$. To have a real requirement due to the boundary condition, u has to be at least continuous up to the boundary, that is, on $\overline{\Omega}$. These requirements can be formulated in a compact way by means of corresponding function spaces. The function spaces are introduced more precisely in Appendix A.5. Some examples are

$$\begin{aligned}
C(\Omega) &:= \left\{ u : \Omega \to \mathbb{R} \;\middle|\; u \text{ continuous in } \Omega \right\}, \\
C^1(\Omega) &:= \Big\{ u : \Omega \to \mathbb{R} \;\Big|\; u \in C(\Omega), \; \partial_i u \text{ exists in } \Omega, \\
&\qquad\qquad \partial_i u \in C(\Omega) \text{ for all } i = 1, \dots, d \Big\}.
\end{aligned}$$

The spaces $C^k(\Omega)$ for $k \in \mathbb{N}$, $C(\overline{\Omega})$, and $C^k(\overline{\Omega})$, as well as $C(\partial\Omega)$, are defined analogously. In general, the requirements related to the (continuous) existence of derivatives are called, a little bit vaguely, *smoothness requirements*.

In the following, in view of the finite difference method, f and g will also be assumed continuous in Ω and $\partial\Omega$, respectively.

Definition 1.1 Assume $f \in C(\Omega)$ and $g \in C(\partial\Omega)$. A function u is called a (*classical*) *solution* of (1.1), (1.2) if $u \in C^2(\Omega) \cap C(\overline{\Omega})$, (1.1) holds for all $x \in \Omega$, and (1.2) holds for all $x \in \partial\Omega$.

1.2 The Finite Difference Method

The finite difference method is based on the following approach: We are looking for an approximation to the solution of a boundary value problem at a finite number of points in $\overline{\Omega}$ (the *grid points*). For this reason we substitute the derivatives in (1.1) by difference quotients, which involve only function values at grid points in Ω and require (1.2) only at grid points. By this we obtain algebraic equations for the approximating values at grid points. In general, such a procedure is called the *discretization* of the boundary value problem. Since the boundary value problem is linear, the system of equations for the approximate values is also linear. In general, for other (differential equation) problems and other discretization approaches we also speak of the *discrete problem* as an *approximation* of the *continuous problem*. The aim of further investigations will be to estimate the resulting error and thus to judge the quality of the approximative solution.

Generation of Grid Points

In the following, for the beginning, we will restrict our attention to problems in two space dimensions ($d = 2$). For simplification we consider the case of a constant *step size* (or *grid width*) $h > 0$ in both space directions. The quantity h here is the *discretization parameter*, which in particular determines the dimension of the discrete problem.

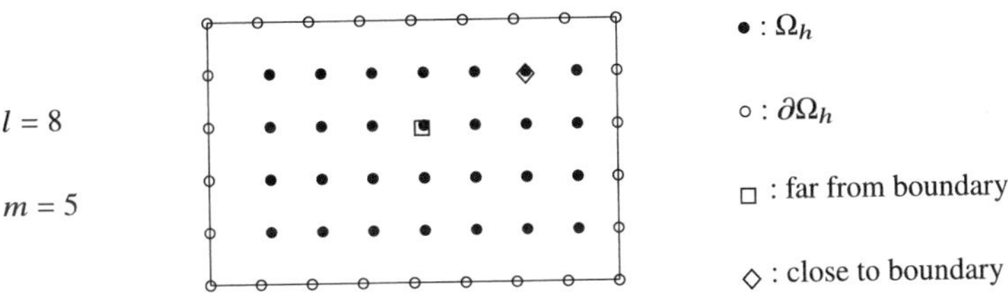

Fig. 1.1: Grid points in a square domain.

For the time being, let Ω be a rectangle, which represents the simplest case for the finite difference method (see Figure 1.1). By translation of the coordinate system the situation can be reduced to $\Omega := (0, a) \times (0, b)$ with $a, b > 0$. We assume that the lengths a, b, and h are such that

$$a = lh, \quad b = mh \quad \text{for certain } l, m \in \mathbb{N}. \tag{1.4}$$

We define

$$\begin{aligned} \Omega_h &:= \left\{(ih, jh) \;\middle|\; i = 1, \ldots, l-1, \ j = 1, \ldots, m-1\right\} \\ &= \left\{(x, y) \in \Omega \;\middle|\; x = ih, \ y = jh \text{ with } i, j \in \mathbb{Z}\right\} \end{aligned} \tag{1.5}$$

as a set of *grid points in* Ω in which an approximation of the differential equation has to be satisfied. In the same way,

$$\begin{aligned}\partial\Omega_h &:= \{(ih, jh) \mid i \in \{0, l\},\ j \in \{0, \dots, m\} \text{ or } i \in \{0, \dots, l\},\ j \in \{0, m\}\} \\ &= \{(x, y) \in \partial\Omega \mid x = ih,\ y = jh \text{ with } i, j \in \mathbb{Z}\}\end{aligned}$$

defines the *grid points on* $\partial\Omega$ in which an approximation of the boundary condition has to be satisfied. The union of grid points will be denoted by

$$\overline{\Omega_h} := \Omega_h \cup \partial\Omega_h .$$

Setup of the System of Equations

Lemma 1.2 *Let* $\Omega := (x - h, x + h)$ *for* $x \in \mathbb{R}$, $h > 0$. *Then there exists a quantity* R, *depending on* u *and* h, *the absolute value of which can be bounded independently of* h *and such that*

1) for $u \in C^2(\overline{\Omega})$:

$$u'(x) = \frac{u(x+h) - u(x)}{h} + hR \quad \textit{and} \quad |R| \le \frac{1}{2}\|u''\|_\infty ,$$

2) for $u \in C^2(\overline{\Omega})$:

$$u'(x) = \frac{u(x) - u(x-h)}{h} + hR \quad \textit{and} \quad |R| \le \frac{1}{2}\|u''\|_\infty ,$$

3) for $u \in C^3(\overline{\Omega})$:

$$u'(x) = \frac{u(x+h) - u(x-h)}{2h} + h^2R \quad \textit{and} \quad |R| \le \frac{1}{6}\|u'''\|_\infty ,$$

4) for $u \in C^4(\overline{\Omega})$:

$$u''(x) = \frac{u(x+h) - 2u(x) + u(x-h)}{h^2} + h^2R \quad \textit{and} \quad |R| \le \frac{1}{12}\|u^{(4)}\|_\infty .$$

Here the maximum norm $\|\cdot\|_\infty$ *(see Appendix A.5) has to be taken over the interval of the involved points* $(x, x + h)$, $(x - h, x)$, *or* $(x - h, x + h)$.

Proof The proof follows immediately by Taylor expansion. As an example we consider statement 3): From

$$u(x \pm h) = u(x) \pm hu'(x) + \frac{h^2}{2}u''(x) \pm \frac{h^3}{6}u'''(x \pm \xi_\pm) \quad \text{for certain } \xi_\pm \in (0, h)$$

the assertion follows by linear combination. □

Notation

The quotient in statement 1) is called the *forward difference quotient*, and it is denoted by $\partial^+ u(x)$. The quotient in statement 2) is called the *backward difference quotient* ($\partial^- u(x)$), and the one in statement 3) the *symmetric difference quotient* ($\partial^0 u(x)$). The quotient appearing in statement 4) can be written as $\partial^- \partial^+ u(x)$ by means of the above notation.

In order to use statement 4) in every space direction for the approximation of $\partial_{11} u$ and $\partial_{22} u$ in a grid point (ih, jh), in addition to the conditions of Definition 1.1, the further smoothness properties $\partial^{(3,0)} u, \partial^{(4,0)} u \in C(\overline{\Omega})$ and analogously for the second coordinate are necessary. Here we use, e.g., the notation $\partial^{(3,0)} u := \partial^3 u / \partial x_1^3$ (see (2.17)).

Using these approximations for the boundary value problem (1.1), (1.2), at each grid point $(ih, jh) \in \Omega_h$ we get

$$-\left(\frac{u((i+1)h, jh) - 2u(ih, jh) + u((i-1)h, jh)}{h^2} \right. \\ \left. + \frac{u(ih, (j+1)h) - 2u(ih, jh) + u(ih, (j-1)h)}{h^2} \right) = f(ih, jh) + R(ih, jh)h^2 . \tag{1.6}$$

Here R is as described in statement 4) of Lemma 1.2, a function depending on the solution u and on the step size h, but the absolute value of which can be bounded independently of h. In cases where we have less smoothness of the solution u, we can nevertheless formulate the approximation (1.6) for $-\Delta u$, but the size of the error in the equation is unclear at the moment.

For the grid points $(ih, jh) \in \partial\Omega_h$ no approximation of the boundary condition is necessary:

$$u(ih, jh) = g(ih, jh) .$$

If we neglect the term Rh^2 in (1.6), we get a system of linear equations for the approximating values u_{ij} for $u(x, y)$ at points $(x, y) = (ih, jh) \in \overline{\Omega}_h$. They have the form

$$\frac{1}{h^2}\big(- u_{i,j-1} - u_{i-1,j} + 4u_{ij} - u_{i+1,j} - u_{i,j+1} \big) = f_{ij} \tag{1.7}$$

$$\text{for } i = 1, \ldots, l-1, \; j = 1, \ldots, m-1,$$

$$u_{ij} = g_{ij} \quad \text{if } i \in \{0, l\},\; j = 0, \ldots, m \text{ or } j \in \{0, m\},\; i = 0, \ldots, l. \tag{1.8}$$

Here we used the abbreviations

$$f_{ij} := f(ih, jh), \quad g_{ij} := g(ih, jh) . \tag{1.9}$$

Therefore, for each unknown grid value u_{ij} we get an equation. The grid points (ih, jh) and the approximating values u_{ij} located at these have a natural two-dimensional indexing.

In equation (1.7) for a grid point (i, j) only the *neighbours* at the four cardinal points of the compass appear, as it is displayed in Figure 1.2. This interconnection is also called the *five-point stencil* of the difference method and the method the *five-point stencil discretization*.

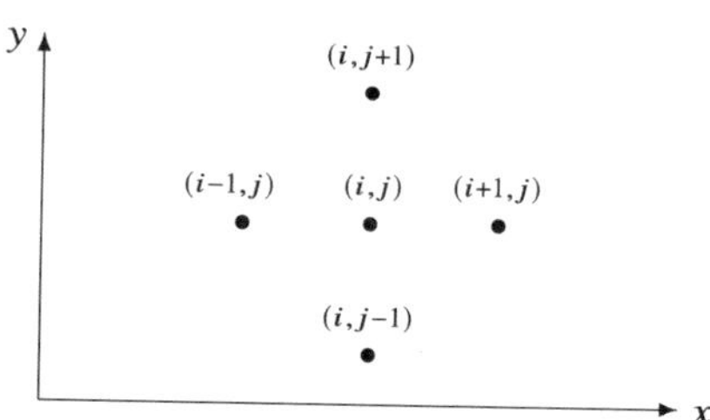

Fig. 1.2: Five-point stencil.

At the interior grid points $(x, y) = (ih, jh) \in \Omega_h$, two cases can be distinguished:

1. (i, j) has a position such that its all neighbouring grid points lie in Ω_h (*far from the boundary*).
2. (i, j) has a position such that at least one neighbouring grid point (r, s) lies on $\partial\Omega_h$ (*close to the boundary*). Then in equation (1.7) the value u_{rs} is known due to (1.8) ($u_{rs} = g_{rs}$), and (1.7) can be modified in the following way:

   Remove the values u_{rs} with $(rh, sh) \in \partial\Omega_h$ in the equations for (i, j) close to the boundary and add the value g_{rs}/h^2 to the right-hand side of (1.7). The set of equations that arises by this elimination of boundary unknowns by means of Dirichlet boundary conditions we call (1.7)*, it is equivalent to (1.7), (1.8).

Instead of considering the values u_{ij}, $i = 1, \ldots, l-1$, $j = 1, \ldots, m-1$, one also speaks of the *grid function* $u_h : \Omega_h \to \mathbb{R}$, where $u_h(ih, jh) = u_{ij}$ for $i = 1, \ldots, l-1$, $j = 1, \ldots, m-1$. Grid functions on $\partial\Omega_h$ or on $\overline{\Omega}_h$ are defined analogously. Thus we can formulate the *finite difference method (FDM)* in the following way:

> Find a grid function u_h on $\overline{\Omega}_h$ such that equations (1.7), (1.8) hold, or, equivalently find a grid function u_h on Ω_h such that equations (1.7)* hold.

Structure of the System of Equations

After choosing an ordering of the u_{ij} for $i = 0, \ldots, l$, $j = 0, \ldots, m$, the system of equations (1.7)* takes the following form:

$$A_h \boldsymbol{u}_h = \boldsymbol{q}_h \tag{1.10}$$

with $A_h \in \mathbb{R}^{M_1,M_1}$ and $\boldsymbol{u}_h, \boldsymbol{q}_h \in \mathbb{R}^{M_1}$, where $M_1 = (l-1)(m-1)$.

This means that nearly identical notations for the grid function and its representing vector are chosen for a fixed numbering of the grid points. The only difference is that the representing vector is printed in bold. The ordering of the grid points may be arbitrary, with the restriction that the points in Ω_h are enumerated by the first M_1 indices, and the points in $\partial\Omega_h$ are labelled with the subsequent $M_2 = 2(l+m)$ indices. The structure of A_h is not influenced by this restriction.

Because of the described elimination process, the right-hand side $\boldsymbol{q}_h$ has the following form:

$$\boldsymbol{q}_h = -\hat{A}_h \boldsymbol{g} + \boldsymbol{f}, \tag{1.11}$$

where $\boldsymbol{g} \in \mathbb{R}^{M_2}$ and $\boldsymbol{f} \in \mathbb{R}^{M_1}$ are the vectors representing the grid functions

$$f_h : \Omega_h \to \mathbb{R} \quad \text{and} \quad g_h : \partial\Omega_h \to \mathbb{R}$$

according to the chosen numbering with the values defined in (1.9). The matrix $\hat{A}_h \in \mathbb{R}^{M_1, M_2}$ has the following form:

$$(\hat{A}_h)_{ij} = \begin{cases} -\dfrac{1}{h^2} & \text{if the node } i \text{ is close to the boundary and} \\ & j \text{ is a neighbour in the five-point stencil,} \\ 0 & \text{otherwise}. \end{cases} \tag{1.12}$$

For any ordering, only the diagonal element and at most four further entries in a row of A_h, defined by (1.7), are different from 0; that is, the matrix is *sparse* in a strict sense, as is assumed in Chapter 5.

An obvious ordering is the *rowwise* numbering of Ω_h according to the following scheme:

$$\begin{array}{cccc}
\underset{(l-1)(m-2)+1}{(h,b-h)} & \underset{(l-1)(m-2)+2}{(2h,b-h)} & \cdots\cdots & \underset{(l-1)(m-1)}{(a-h,b-h)} \\
\underset{(l-1)(m-3)+1}{(h,b-2h)} & \underset{(l-1)(m-3)+2}{(2h,b-2h)} & \cdots\cdots & \underset{(l-1)(m-2)}{(a-h,b-2h)} \\
\vdots & \vdots & \ddots\ \ddots & \vdots \\
\underset{l}{(h,2h)} & \underset{l+1}{(2h,2h)} & \cdots\cdots & \underset{2l-2}{(a-h,2h)} \\
\underset{1}{(h,h)} & \underset{2}{(2h,h)} & \cdots\cdots & \underset{l-1}{(a-h,h)}
\end{array} \quad . \tag{1.13}$$

Another name of the above scheme is *lexicographic* ordering. (However, this name is better suited to the *columnwise* numbering.)

In this case the matrix A_h has the following form of an $(m-1)\times(m-1)$ block tridiagonal matrix:

$$A_h = h^{-2} \begin{pmatrix} T & -I & & & & \\ -I & T & -I & & \mathbf{0} & \\ & \ddots & \ddots & \ddots & & \\ & & \ddots & \ddots & \ddots & \\ & \mathbf{0} & & -I & T & -I \\ & & & & -I & T \end{pmatrix} \tag{1.14}$$

with the identity matrix $I \in \mathbb{R}^{l-1,l-1}$ and

$$T = \begin{pmatrix} 4 & -1 & & & & \\ -1 & 4 & -1 & & 0 & \\ & \ddots & \ddots & \ddots & & \\ & & \ddots & \ddots & \ddots & \\ & 0 & & -1 & 4 & -1 \\ & & & & -1 & 4 \end{pmatrix} \in \mathbb{R}^{l-1,l-1} .$$

We return to the consideration of an arbitrary numbering. In the following we collect several properties of the matrix $A_h \in \mathbb{R}^{M_1,M_1}$ and the extended matrix

$$\tilde{A}_h := \left(A_h \mid \hat{A}_h \right) \in \mathbb{R}^{M_1,M} ,$$

where $M := M_1 + M_2$. The matrix $\tilde{A}_h$ takes into account all the grid points in $\overline{\Omega}_h$. It has no relevance with the resolution of (1.10), but with the stability of the discretization, which will be investigated in Section 1.4.

- $(A_h)_{rr} > 0$ for all $r = 1, \ldots, M_1$,
- $(\tilde{A}_h)_{rs} \le 0$ for all $r = 1, \ldots, M_1,\ s = 1, \ldots, M$ such that $r \ne s$,
- $$\sum_{s=1}^{M_1} (A_h)_{rs} \begin{cases} \ge 0 & \text{for all } r = 1, \ldots, M_1, \\ > 0 & \text{if } r \text{ belongs to a grid point close to the boundary,} \end{cases} \tag{1.15}$$
- $\sum_{s=1}^{M} (\tilde{A}_h)_{rs} = 0$ for all $r = 1, \ldots, M_1$,
- A_h is irreducible,
- A_h is regular.

Therefore, the matrix A_h is weakly row diagonally dominant (see Appendix A.3 for definitions from linear algebra). The irreducibility follows from the fact that two arbitrary grid points may be connected by a path consisting of corresponding neighbours in the five-point stencil. The regularity follows from the irreducible diagonal dominance. From this we can conclude that (1.10) can be solved by Gaussian elimination without pivot search. In particular, if the matrix has a band structure, this will be preserved. This fact will be explained in more detail in Section 2.5.

The matrix A_h has the following further properties:

- A_h is symmetric,
- A_h is positive definite.

It is sufficient to verify these properties for a fixed ordering, for example, the rowwise one, since by a change of the ordering matrix, A_h is transformed to PA_hP^T with some regular matrix P, by which neither symmetry nor positive definiteness is destroyed. Nevertheless, the second assertion is not obvious. One way to verify it is to compute eigenvalues and eigenvectors explicitly, but we refer to Chapter 2, where the assertion follows naturally from Lemma 2.13 and (2.40). The eigenvalues and eigenvectors are specified in (5.24) for the special case $l = m = n$ and also in (9.71). Therefore, (1.10) can be resolved by Cholesky's method, taking into account the band structure.

Quality of the Approximation by the Finite Difference Method

We now address the following question: To what accuracy does the grid function u_h corresponding to the solution $\boldsymbol{u}_h$ of (1.10) approximate the solution u of (1.1), (1.2)?

To this end we consider the grid function $U : \Omega_h \to \mathbb{R}$, which is defined by

$$U(ih, jh) := u(ih, jh). \tag{1.16}$$

To measure the size of $U - u_h$, we need a norm (see Appendix A.4 and also A.5 for the subsequently used definitions). Examples are the *maximum norm*

$$\|u_h - U\|_\infty := \max_{\substack{i=1,\dots,l-1 \\ j=1,\dots,m-1}} |(u_h - U)(ih, jh)| \tag{1.17}$$

and the *discrete L^2-norm*

$$\|u_h - U\|_{0,h} := h\left(\sum_{i=1}^{l-1}\sum_{j=1}^{m-1} ((u_h - U)(ih, jh))^2\right)^{1/2}. \tag{1.18}$$

Both norms can be conceived as the application of the continuous norms $\|\cdot\|_\infty$ of the function space $L^\infty(\Omega)$ or $\|\cdot\|_0$ of the function space $L^2(\Omega)$ to piecewise constant prolongations of the grid functions (with a special treatment of the area close to the boundary). Obviously, we have

$$\|v_h\|_{0,h} \le \sqrt{ab}\,\|v_h\|_\infty$$

for a grid function v_h, but the reverse estimate does not hold uniformly in h, so that $\|\cdot\|_\infty$ is a stronger norm. In general, we are looking for a norm $\|\cdot\|_h$ in the space of grid functions in which the method *converges* in the sense

$$\|u_h - U\|_h \to 0 \quad \text{for } h \to 0$$

or even has an *order of convergence* $p > 0$, by which we mean the existence of a constant $C > 0$ independent of h such that

$$\|u_h - U\|_h \leq C\,h^p\,.$$

Due to the construction of the method, for a solution $u \in C^4(\overline{\Omega})$ we have

$$A_h\boldsymbol{U} = \boldsymbol{q}_h + h^2\boldsymbol{R}\,,$$

where $\boldsymbol{U}$ and $\boldsymbol{R} \in \mathbb{R}^{M_1}$ are the representations of the grid functions U and R according to (1.6) in the selected ordering. Therefore, we have

$$A_h(\boldsymbol{u}_h - \boldsymbol{U}) = -h^2\boldsymbol{R}$$

and thus

$$|A_h(\boldsymbol{u}_h - \boldsymbol{U})|_\infty = h^2|\boldsymbol{R}|_\infty = Ch^2$$

with a constant $C(= |\boldsymbol{R}|_\infty) > 0$ independent of h.

From Lemma 1.2, 4) we conclude that

$$C = \frac{1}{12}\left(\|\partial^{(4,0)}u\|_\infty + \|\partial^{(0,4)}u\|_\infty\right)\,.$$

That is, for a solution $u \in C^4(\overline{\Omega})$ the method is asymptotically consistent with the boundary value problem with an *order of consistency* 2. More generally, the notion takes the following form:

Definition 1.3 Let (1.10) be the system of equations that corresponds to a (finite difference) approximation on the grid points Ω_h with a discretization parameter h. Let $\boldsymbol{U}$ be the representation of the grid function that corresponds to the solution u of the boundary value problem according to (1.16). Furthermore, let $\|\cdot\|_h$ be a norm in the space of grid functions on Ω_h, and let $|\cdot|_h$ be the corresponding vector norm in the space $\mathbb{R}^{M_{1h}}$, where M_{1h} is the number of grid points in Ω_h. The approximation is called *asymptotically consistent* with respect to $\|\cdot\|_h$ if

$$|A_h\boldsymbol{U} - \boldsymbol{q}_h|_h \to 0 \quad \text{for} \quad h \to 0\,.$$

The approximation has the *order of consistency* $p > 0$ if

$$|A_h\boldsymbol{U} - \boldsymbol{q}_h|_h \leq Ch^p$$

with a constant $C > 0$ independent of h.

Thus the *consistency* or *truncation error* $A_h\boldsymbol{U} - \boldsymbol{q}_h$ measures the quality of how the exact solution satisfies the approximating equations. As we have seen, in general it can be determined easily by Taylor expansion, but at the expense of unnaturally high smoothness assumptions. But one has to be careful in expecting the error $|\boldsymbol{u}_h - \boldsymbol{U}|_h$ to behave like the consistency error. We have

$$\left|\boldsymbol{u}_h - \boldsymbol{U}\right|_h = \left|A_h^{-1} A_h(\boldsymbol{u}_h - \boldsymbol{U})\right|_h \leq \left\|A_h^{-1}\right\|_h \left|A_h(\boldsymbol{u}_h - \boldsymbol{U})\right|_h, \tag{1.19}$$

where the matrix norm $\|\cdot\|_h$ has to be chosen to be compatible with the vector norm $|\cdot|_h$. The error behaves like the consistency error asymptotically in h if $\left\|A_h^{-1}\right\|_h$ can be bounded independently of h; that is, if the method is *stable* in the following sense:

Definition 1.4 In the situation of Definition 1.3, the approximation is called *stable* with respect to $\|\cdot\|_h$ if there exists a constant $C > 0$ independent of h such that

$$\left\|A_h^{-1}\right\|_h \leq C .$$

From the above definition we can obviously conclude, with (1.19), the following result:

Theorem 1.5 *An asymptotically consistent and stable method is convergent, and the order of convergence is at least equal to the order of consistency.*

Therefore, specifically for the five-point stencil discretization of (1.1), (1.2) on a rectangle, stability with respect to $\|\cdot\|_\infty$ is desirable. In fact, it follows from the structure of A_h: Namely, we have

$$\left\|A_h^{-1}\right\|_\infty \leq \frac{1}{16}(a^2 + b^2) . \tag{1.20}$$

This follows from more general considerations in Section 1.4 (Theorem 1.15). Putting the results together we have the following theorem:

Theorem 1.6 *Let the solution u of* (1.1), (1.2) *on a rectangle Ω be in $C^4(\overline{\Omega})$. Then the five-point stencil discretization has an order of convergence* 2 *with respect to $\|\cdot\|_\infty$, more precisely,*

$$|\boldsymbol{u}_h - \boldsymbol{U}|_\infty \leq \frac{1}{192}(a^2 + b^2)\left(\|\partial^{(4,0)}u\|_\infty + \|\partial^{(0,4)}u\|_\infty\right) h^2 .$$

Exercises

Problem 1.1 Complete the proof of Lemma 1.2 and also investigate the error of the respective difference quotients, assuming only $u \in C^2[x - h, x + h]$.

Problem 1.2 Generalize the discussion concerning the five-point stencil discretization (including the order of convergence) of (1.1), (1.2) on a rectangle for $h_1 > 0$ in the x_1 direction and $h_2 > 0$ in the x_2 direction.

Problem 1.3 Show that an irreducible weakly row diagonally dominant matrix cannot have vanishing diagonal elements.

Programming project 1.1 Consider the one-dimensional Poisson equation with inhomogeneous Dirichlet boundary conditions

$$\begin{aligned} -u'' &= f \quad \text{in } \Omega := (0,1), \\ u(0) &= g_l, \quad u(1) = g_r, \end{aligned}$$

on an equidistant grid with grid size $h := 1/m$, $m \in \mathbb{N}$, and

a) $g_l := g_r := 0$ and $f := \pi^2 \sin(\pi x)$ with the exact solution $u(x) = \sin(\pi x)$,
b) $g_l := 0$, $g_r := 1$ and $f := -2$ with the exact solution $u(x) = x^2$.

Write a function that solves the problem numerically for a given m using the finite difference method. The input data are the right-hand side f, the boundary data g_l, g_r, the exact solution u, and the number of intervals m. The output data are $\boldsymbol{u}_h \in \mathbb{R}^{m-1}$ and the maximum norm of the error, i.e., $|\boldsymbol{U} - \boldsymbol{u}_h|_\infty$.

Implement a method of your choice to solve the resulting system of linear equations.

Perform a series of experiments for different values of m (e.g., $m = 10, 20, 50, 100$) and analyse the asymptotic behaviour of the computed error w.r.t. h.

Programming project 1.2 Consider the boundary value problem

$$\begin{aligned} -\Delta u &= 4 \qquad \text{in } \Omega := (0,1)^2, \\ u &= x^2 + y^2 \quad \text{on } \partial\Omega. \end{aligned}$$

Let $h := 1/32$, and $\Omega_h := \{(x,y) = (ih, jh) \mid 1 \le i, j \le 31\}$. Use the five-point-stencil to discretize $-\Delta$ and

a) a lexicographical ordering of the unknowns,
b) a *red-black ordering* of the unknowns:

$$\Omega_h^r := \{(x,y) \in \Omega_h \mid (x+y)/h \text{ is odd}\},$$

$$\Omega_h^b := \{(x,y) \in \Omega_h \mid (x+y)/h \text{ is even}\}.$$

First number the "red" nodes $(x,y) \in \Omega_h^b$ lexicographically, then the "black" nodes $(x,y) \in \Omega_h^w$.

Derive the matrix A_h of the red-black ordering.

1.3 Generalizations and Limitations of the Finite Difference Method

We continue to consider the boundary value problem (1.1), (1.2) on a rectangle Ω. The five-point stencil discretization developed may be interpreted as a mapping $-\Delta_h$ from functions on $\overline{\Omega}_h$ into grid functions on Ω_h, which is defined by

$$-\Delta_h v_h(x_1, x_2) := \sum_{i,j=-1}^{1} c_{ij} v_h(x_1 + ih, x_2 + jh), \tag{1.21}$$

where $c_{0,0} = 4/h^2$, $c_{0,1} = c_{1,0} = c_{0,-1} = c_{-1,0} = -1/h^2$, and $c_{ij} = 0$ for all other (i, j). For the description of such a difference stencil as defined in (1.21) the points of the compass (in two space dimensions) may also be involved. In the five-point stencil only the main points of the compass appear.

The question of whether the *weights* c_{ij} can be chosen differently such that we gain an approximation of $-\Delta u$ with higher order in h has to be answered negatively (see Problem 1.7). In this respect the five-point stencil is optimal. This does not exclude that other difference stencils with more entries, but of the same order of convergence, might be worthwhile to consider. An example, which will be derived in Problem 3.11 by means of the finite element method, has the following form:

$$c_{0,0} = \frac{8}{3h^2}, \quad c_{ij} = -\frac{1}{3h^2} \quad \text{for all other } i, j \in \{-1, 0, 1\}. \tag{1.22}$$

This nine-point stencil can be interpreted as a linear combination of the five-point stencil and a five-point stencil for a coordinate system rotated by $\frac{\pi}{4}$ (with step size $\sqrt{2}\,h$), using the weights $\frac{1}{3}$ and $\frac{2}{3}$ in this linear combination. Using a general nine-point stencil a method with order of consistency greater than 2 can be constructed only if the right-hand side f at the point (x_1, x_2) is approximated not by the evaluation $f(x_1, x_2)$, but by applying a more general stencil. The *mehrstellen method* ("Mehrstellenverfahren") defined by Collatz is such an example (see, for example, [32, Sect. 4.6]).

Methods of higher order can be achieved by larger stencils, meaning that the summation indices in (1.21) have to be replaced by k and $-k$, respectively, for $k \in \mathbb{N}$. But already for $k = 2$ such difference stencils cannot be used for grid points close to the boundary, so that there one has to return to approximations of lower order.

If we consider the five-point stencil to be a suitable discretization for the Poisson equation, the high smoothness assumption for the solution in Theorem 1.6 should be noted. This requirement cannot be ignored, since in general it does not hold true. On the one hand, for a smoothly bounded domain (see Appendix A.5 for a definition of a domain with C^l-boundary) the smoothness of the solution is determined only by the smoothness of the data f and g (see, for example, [28, Theorem 6.19]), but on the other hand, corners in the domain reduce this smoothness the more, the more reentrant the corners are. Let us consider the following examples:

For the boundary value problem (1.1), (1.2) on a rectangle $(0, a)\times(0, b)$ we choose $f = 1$ and $g = 0$; this means arbitrarily smooth functions. Nevertheless, for the solution u, the statement $u \in C^2(\overline{\Omega})$ cannot hold, because otherwise, $-\Delta u(0, 0) = 1$ would be true, but on the other hand, we have $\partial_{1,1} u(x, 0) = 0$ because of the boundary condition and hence also $\partial_{1,1} u(0, 0) = 0$ and $\partial_{2,2} u(0, y) = 0$ analogously. Therefore, $\partial_{2,2} u(0, 0) = 0$. Consequently, $-\Delta u(0, 0) = 0$, which contradicts the assumption above. Therefore, Theorem 1.6 is not applicable here.

In the second example we consider the domain with reentrant corner (see Figure 1.3)

$$\Omega = \left\{(x, y) \in \mathbb{R}^2 \;\middle|\; x^2 + y^2 < 1,\ x < 0 \text{ or } y > 0\right\} .$$

In general, if we identify $\mathbb{R}^2$ and $\mathbb{C}$, this means $(x, y) \in \mathbb{R}^2$ and $z = x + iy \in \mathbb{C}$, we have that if $w : \mathbb{C} \to \mathbb{C}$ is analytic (holomorphic), then both the real and the imaginary parts $\operatorname{Re} w, \operatorname{Im} w : \mathbb{C} \to \mathbb{R}$ are *harmonic*, which means that they solve $-\Delta u = 0$.

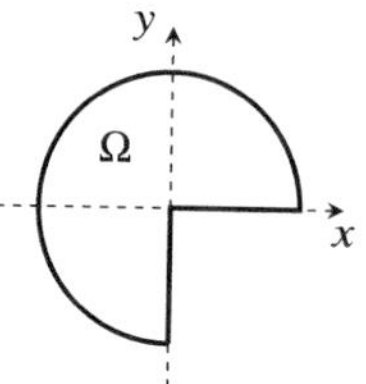

Fig. 1.3: Domain Ω with reentrant corner.

We choose $w(z) := z^{2/3}$. Then the function $u(x, y) := \operatorname{Im}\left((x + iy)^{2/3}\right)$ solves the equation

$$-\Delta u = 0 \quad \text{in} \quad \Omega .$$

In polar coordinates, $x = r \cos\varphi$, $y = r \sin\varphi$, the function u takes the form

$$u(x, y) = \operatorname{Im}\left(\left(re^{i\varphi}\right)^{2/3}\right) = r^{2/3} \sin\left(\frac{2}{3}\varphi\right) .$$

Therefore, u satisfies the boundary conditions

$$\begin{aligned} u\left(e^{i\varphi}\right) &= \sin\left(\frac{2}{3}\varphi\right) \quad \text{for } 0 \le \varphi \le \frac{3\pi}{2} , \\ u(x, y) &= 0 \quad \text{otherwise on } \partial\Omega . \end{aligned} \tag{1.23}$$

But note that $w'(z) = \frac{2}{3} z^{-1/3}$ is unbounded for $z \to 0$, so that $\partial_1 u, \partial_2 u$ are unbounded for $(x, y) \to 0$. Therefore, in this case we do not even have $u \in C^1(\overline{\Omega})$.

The examples do not show that the five-point stencil discretization is not suitable for the boundary value problems considered, but they show the necessity of a theory of convergence, which requires only as much smoothness as was to be expected.

In the following we discuss some generalizations of the boundary value problems considered so far.

General Domains Ω

We continue to consider (1.1), (1.2) but on a general domain in $\mathbb{R}^2$, for which the parts of the boundary are not necessarily aligned to the coordinate axes. Therefore we can keep the second equation in (1.5) as the definition of Ω_h but have to redefine the set of boundary grid points $\partial\Omega_h$.

For example, if for some point $(x, y) \in \Omega_h$ we have

$$(x - h, y) \notin \Omega,$$

then there exists a number $s \in (0, 1]$ such that

$$(x - \vartheta h, y) \in \Omega \quad \text{for all } \vartheta \in [0, s) \quad \text{and} \quad (x - sh, y) \notin \Omega.$$

Then $(x - sh, y) \in \partial\Omega$, and therefore we define

$$(x - sh, y) \in \partial\Omega_h.$$

The other main points of the compass are treated analogously. In this way the grid spacing in the vicinity of the boundary becomes variable; in particular, it can be smaller than h.

For the quality of the approximation we have the following result:

Lemma 1.7 *Let $\Omega = (x - h_1, x + h_2)$ for $x \in \mathbb{R}$, $h_1, h_2 > 0$.*

1) Then for $u \in C^3(\overline{\Omega})$,

$$u''(x) = \frac{2}{h_1 + h_2}\left(\frac{u(x + h_2) - u(x)}{h_2} - \frac{u(x) - u(x - h_1)}{h_1}\right) + \max\{h_1, h_2\} R,$$

where R is bounded independently of h_1, h_2.

2) There are no $\alpha, \beta, \gamma \in \mathbb{R}$ such that

$$u''(x) = \alpha\, u(x - h_1) + \beta\, u(x) + \gamma\, u(x + h_2) + R_1 h_1^2 + R_2 h_2^2$$

for all polynomials u of degree 3 if $h_1 \neq h_2$.

Proof Problems 1.4 and 1.5. □

This leads to a discretization that is difficult to set up and for which the order of consistency and order of convergence are not easily determined.

Other Boundary Conditions

We want to consider the following example. Let $\partial\Omega = \Gamma_1 \cup \Gamma_3$ be divided into two disjoint subsets. We are looking for a function u such that

$$\begin{aligned} -\Delta u &= f && \text{in } \Omega, \\ \partial_{\mathbf{n}} u := \nabla u \cdot \mathbf{n} &= g && \text{on } \Gamma_1, \\ u &= 0 && \text{on } \Gamma_3, \end{aligned} \tag{1.24}$$

where $\mathfrak{n} : \partial\Omega \to \mathbb{R}^d$ is the outer unit normal, and thus $\partial_{\mathfrak{n}} u$ is the normal derivative of u.

For a part of the boundary oriented in a coordinate direction, $\partial_{\mathfrak{n}} u$ is just a positive or negative partial derivative. But if only grid points in $\overline{\Omega}_h$ are to be used, only $\pm\partial^+ u$ and $\pm\partial^- u$ respectively (in the coordinates orthogonal to the direction of the boundary) are available directly from the above approximations with a corresponding reduction of the order of consistency. For a boundary point without these restrictions the question of how to approximate $\partial_{\mathfrak{n}} u$ appropriately is open.

As an example we consider (1.24) for a rectangle $\Omega := (0, a) \times (0, b)$, where

$$\Gamma_1 := \{(a, y) \mid y \in (0, b)\}, \; \Gamma_3 := \Gamma \setminus \Gamma_1 . \tag{1.25}$$

At the boundary grid points (a, jh), $j = 1, \ldots, m-1$, $\partial_2 u = \nabla u \cdot \mathfrak{n}$ is prescribed, which can be approximated directly only by $\partial^- u$. Due to Lemma 1.2, 2 this leads to a reduction in the consistency order (see Problem 1.8). The resulting system of equations may include the Neumann boundary grid points in the set of unknowns, for which an equation with the entries $1/h$ in the diagonal and $-1/h$ in an off-diagonal corresponding to the eastern neighbour $(a - h, jh)$ has to be added. Alternatively, those boundary points can be eliminated, leading for the eastern neighbour to a modified difference stencil (multiplied by h^2)

$$\begin{array}{cc} & -1 \\ -1 & 3 \\ & -1 \end{array} \tag{1.26}$$

for the right-hand side $h^2 f(a-h, jh) + hg(a, jh)$. In both cases the matrix properties of the system of equations as collected in (1.15) still hold, with the exception of $\sum_{s=1}^{M_1} (A_h)_{rs} = 0$, both for the Neumann boundary points and their neighbours, if no Dirichlet boundary point is involved in their stencil. Thus the term "close to the boundary" has to be interpreted as "close to the Dirichlet boundary."

If one wants to take advantage of the symmetric difference quotient $\partial^0 u$, then "artificial" values at new external grid points $(a + h, jh)$ appear.

To keep the balance of unknowns and equations, it can be assumed that the differential equation also holds at (a, jh), and thus it is discretized with the five-point stencil there. If one attributes the discrete boundary condition to the external grid point, then again the properties (1.15) hold with the above-mentioned interpretation. Alternatively, the external grid points can be eliminated, leading to a modified difference stencil (multiplied by h^2) at (a, jh):

$$\begin{array}{cc} & -1 \\ -2 & 4 \\ & -1 \end{array} \tag{1.27}$$

for the right-hand side $h^2 f(a, jh) + 2hg(a, jh)$, with the same interpretation of properties (1.15).

More General Differential Equations

As an example we consider the differential equation

$$-\nabla \cdot (k\,\nabla u) = f \quad \text{on } \Omega \tag{1.28}$$

with a continuous coefficient function $k : \Omega \to \mathbb{R}$, which is bounded from below by a positive constant on Ω. This equation states the conservation of an extensive quantity u whose flux density is $-k\nabla u$ (see Section 0.5). This should be respected by the discretization, and therefore the form of (1.28) obtained by working out the derivatives is not recommended as a basis for the discretization. The differential expression in (1.28) can be discretized by a successive application of central difference quotients, but then again the order of consistency has to be investigated.

In addition, one has to take into account the fact that the smoothness of u depends on the smoothness of k. If processes in heterogeneous materials have to be described, then k is often discontinuous. In the simplest example k is assumed to take two different values: Let $\Omega = \Omega_1 \cup \Omega_2$ and

$$k|_{\Omega_1} = k_1 > 0\,, \quad k|_{\Omega_2} = k_2 > 0$$

with constants $k_1 \neq k_2$.

As worked out in Section 0.5, on the interior boundary $S := \overline{\Omega}_1 \cap \overline{\Omega}_2$ a *transmission condition* has to be imposed:

- u is continuous,
- $(k\nabla u) \cdot \mathbf{n}$ is continuous, where $\mathbf{n}$ is the outer normal on $\partial\Omega_1$, for example.

This leads to the following conditions on u_i, being the restrictions of u on $\overline{\Omega}_i$ for $i = 1, 2$:

$$-k_1 \Delta u_1 = f \quad \text{in } \Omega_1\,, \tag{1.29}$$
$$-k_2 \Delta u_2 = f \quad \text{in } \Omega_2\,,$$
$$u_1 = u_2 \quad \text{on } S\,, \tag{1.30}$$
$$k_1 \partial_{\mathbf{n}} u_1 = k_2 \partial_{\mathbf{n}} u_2 \quad \text{on } S\,.$$

In this case the question of an appropriate discretization is also open.

Summarizing, we have the following catalogue of requirements: We are looking for a notion of solution for (general) boundary value problems with nonsmooth coefficients and right-hand sides such that, for example, the transmission condition is fulfilled automatically.

We are looking for a discretization on general domains such that, for example, the (order of) convergence can also be assured for less smooth solutions and also Neumann boundary conditions as in (1.24) can be treated easily.

The finite element method (FEM) in the subsequent chapters will fulfil these requirements to a large extent. To prepare the transition to FEM let us reconsider the (order of) convergence results obtained so far. Theorem 1.6 indicates that an asymptotically consistent FDM, which is stable, is also convergent. Such a statement, also called the *Lax equivalence theorem* (in the form: Convergence and stability are equivalent for an asymptotically consistent scheme), or the *Lax–Richtmyer theorem*, is typical for FDM, as here only grid functions are compared. A FDM does not provide an approximating function on the domain Ω, but a suitable postprocessing is possible.

We consider for simplicity as in Section 1.2 the case $\Omega := (0, a) \times (0, b)$ and a grid with stepsize $h > 0$. A grid function $u_h : \Omega_h \to \mathbb{R}$ can be attributed to a function $I_h(u_h) := U_h : \Omega \to \mathbb{R}$, which is piecewise bilinear interpolated, i.e., for $(x, y) \in K_{ij} := [ih, (i+1)h] \times [jh, (j+1)h]$ we set

$$U_h(x, y) := \frac{1}{h^2}\Big[u_{ij}((i+1)h - x)(j+1)h - y) + u_{i+1,j}(x - ih)(j+1)h - y) \\ + u_{i+1,j+1}(x - ih)(y - jh) + u_{i,j+1}((i+1)h - x)(y - jh)\Big].$$

The function U_h is affine-linear at each straight piece of K_{ij} and thus affine-linear across such inter-element boundaries, i.e., $U_h \in C(\overline{\Omega})$ (see Lemma 2.10). Thus I_h is a linear mapping from the set of grid functions to $C(\overline{\Omega})$ and—understood as the composition with the canonical restriction (1.16)—also from $C(\overline{\Omega})$ to $C(\overline{\Omega})$. Because of

$$|U_h(x, y)| \le \max_{l,k \in \{0,1\}} \{|u_{i+l}|, |u_{j+k}|\}$$

we have

$$\|I_h(u_h)\|_\infty = \|u_h\|_\infty$$

and thus

$$\|u_h - U\|_\infty = \|I_h(u_h) - I_h(u)\|_\infty .$$

To get an order of convergence estimate, in addition the *approximation error*

$$\|u - I_h(u)\|_\infty$$

has to be estimated. To this end, it is sufficient to estimate the contribution of each K_{ij}, for which an estimate on $\tilde{K} := K_{11} = [0, h] \times [0, h]$ with $I_{\tilde{K}}$ denoting the bilinear interpolant is representative. This is possible thanks to the following error representation.

Lemma 1.8 *For $v \in C^2(\tilde{K})$ it holds*

$$v(x, y) = \big(I_{\tilde{K}}(u)\big)(x, y) + \left(1 - \frac{y}{h}\right)\int_0^y \partial^{(0,1)}v(0, t)dt - \frac{y}{h}\int_0^y \partial^{(0,1)}v(0, t)dt \\ + \frac{x}{h}\left(1 - \frac{y}{h}\right)\int_0^y \int_0^h \partial^{(1,1)}v(s, t)dsdt - \frac{xy}{h^2}\int_y^h \int_0^h \partial^{(1,1)}v(s, t)dsdt \\ - \left(1 - \frac{x}{h}\right)\int_0^x s\partial^{(2,0)}v(s, y)ds - \frac{x}{h}\int_x^h (h - s)\partial^{(2,0)}v(s, y)ds .$$

Proof For $f \in C^2[0,h]$, integration by parts yields

$$f(t) = f(0) + \int_0^t f'(s)ds = f(0) + tf'(t) - \int_0^t sf''(s)ds\,,$$
$$f(t) = f(h) - \int_t^h f'(s)ds = f(h) - (h-t)f'(t) - \int_t^h (h-s)f''(s)ds\,.$$

Setting $\big(I_{[0,h]}(f)\big)(t) := f(0)\frac{h-t}{h} + f(h)\frac{t}{h}$ for the linear interpolant, the convex combination of the above equations gives

$$f(t) = \big(I_{[0,h]}(f)\big)(t) - \frac{h-t}{h}\int_0^t sf''(s)ds - \frac{t}{h}\int_t^h (h-s)f''(s)ds \tag{1.31}$$
$$= \big(I_{[0,h]}(f)\big)(t) + \frac{h-t}{h}\int_0^t f'(s)ds - \frac{t}{h}\int_t^h f'(s)ds\,. \tag{1.32}$$

Applying (1.31) to $u(\cdot, y)$ for fixed y, we obtain

$$\begin{aligned} u(x,y) &= \big(I_{[0,h]}(u(\cdot,y))\big)(x) \\ &\quad - \frac{h-x}{h}\int_0^x s\partial^{(2,0)}v(s,y)ds - \frac{x}{h}\int_x^h (h-s)\partial^{(2,0)}v(s,y)ds\,, \end{aligned} \tag{1.33}$$

and again applying (1.32) to $\big(I_{[0,h]}(u(\cdot,y))\big)(x)$ with respect to y for fixed x gives

$$\begin{aligned} \big(I_{[0,h]}(u(\cdot,y))\big)(x) &= u(0,y)\frac{h-x}{h} + u(h,y)\frac{x}{h} \\ &= \frac{h-x}{h}\Bigg[u(0,0)\frac{h-y}{h} + u(0,h)\frac{y}{h} \\ &\qquad + \frac{h-y}{h}\int_0^y \partial^{(0,1)}v(0,t)dt - \frac{y}{h}\int_y^h \partial^{(0,1)}v(0,t)dt\Bigg] \\ &\quad + \frac{x}{h}\Bigg[u(h,0)\frac{h-y}{h} + u(h,h)\frac{y}{h} \\ &\qquad + \frac{h-y}{h}\int_0^y \partial^{(0,1)}v(h,t)dt - \frac{y}{h}\int_y^h \partial^{(0,1)}v(h,t)dt\Bigg] \\ &= \big(I_{\tilde K}(u)\big)(x,y) \\ &\qquad + \frac{h-y}{h}\int_0^y \partial^{(0,1)}v(0,t)dt - \frac{y}{h}\int_y^h \partial^{(0,1)}v(0,t)dt \\ &\qquad + \frac{x(h-y)}{h^2}\int_0^y\int_0^h \partial^{(1,1)}v(s,t)dsdt \\ &\qquad - \frac{xy}{h^2}\int_y^h\int_0^h \partial^{(1,1)}v(s,t)dsdt\,. \end{aligned}$$

Inserting this relationship into (1.33) concludes the proof. □

This representation shows (the single integrals in the representation are the critical ones) that even for smooth solutions u we can only expect

$$\|u - I_h(u)\|_\infty \le Ch,$$

which gives only first order of convergence for

$$\|u - U_h\|_\infty .$$

If we consider the discrete L^2-norm (1.18) instead, we have to investigate

$$\|u - I_h(u)\|_{0,h} ,$$

thus gaining the missing power of h, and we obtain for smooth solutions u that

$$\|u - U_h\|_{0,h} \le Ch^2 .$$

The same holds true for $\|u - U_h\|_0$, as the quadrature error is of the same order.

To return to the discussion of asymptotic consistency and stability, stability may be viewed as a property which allows to relate the error of the numerical solution to an approximation theoretic error and a consistency error. The latter always reflects the formulation of the discrete approximation. As the FDM is based on a pointwise classical formulation, we may speak of *PDE consistency*. It is also conceivable that the consistency error vanishes (exact consistency, see (2.43) for the FEM). Here this is the case for *collocation methods*, where the exact fulfilment of the PDE (including the boundary conditions) at certain *collocation points* is required. This needs sufficient regularity at the collocation points. We will not discuss such approaches in the following.

To summarize, for an approximation scheme three ingredients have to be considered:

- stability,
- consistency error,
- (space) approximation error.

In the (classical) FEM, which will be dealt with in the next two chapters, the focus is concentrated on the last aspect.

Exercises

Problem 1.4 Prove Lemma 1.7, 1).

Problem 1.5 Under the assumption that $u : \Omega \subset \mathbb{R} \to \mathbb{R}$ is a sufficiently smooth function, determine in the ansatz

$$\alpha u(x - h_1) + \beta u(x) + \gamma u(x + h_2), \quad h_1, h_2 > 0,$$

the coefficients $\alpha = \alpha(h_1, h_2)$, $\beta = \beta(h_1, h_2)$, $\gamma = \gamma(h_1, h_2)$, such that

a) for $x \in \Omega$, $u'(x)$ will be approximated with the order as high as possible,
b) for $x \in \Omega$, $u''(x)$ will be approximated with the order as high as possible,

and in particular, prove 1.7, 2).
Hint: Determine the coefficients such that the formula is exact for polynomials with the degree as high as possible.

Problem 1.6 Let $\Omega \subset \mathbb{R}^2$ be a bounded domain. For a sufficiently smooth function $u : \Omega \to \mathbb{R}$ determine the difference formula with an order as high as possible to approximate $\partial_{11}u(x_1, x_2)$, using the 9 values $u(x_1 + \gamma_1 h, x_2 + \gamma_2 h)$, where $\gamma_1, \gamma_2 \in \{-1, 0, 1\}$.

Problem 1.7 Let $\Omega \subset \mathbb{R}^2$ be a bounded domain. Show that in (1.21) there exists no choice of c_{ij} such that for an arbitrary smooth function $u : \Omega \to \mathbb{R}$,

$$|\Delta u(x) - \Delta_h u(x)| \leq Ch^3$$

is valid with a constant C independent of h.

Problem 1.8 For the example (1.24), (1.25), investigate the order of consistency both for the discretization (1.26) and (1.27) in the maximum norm. Are there improvements possible considering the discrete L^2-norm (see (1.18))?

Problem 1.9 Consider example (1.24) with

$$\begin{aligned} \Gamma_1 &:= \{(a, y) \mid y \in (0, b)\} \cup \{(x, b) \mid x \in (0, a]\}, \\ \Gamma_3 &:= \Gamma \setminus \Gamma_1, \end{aligned}$$

and discuss the applicability of the one-sided and the symmetric difference quotients for the approximation of the Neumann boundary condition, in particular with respect to properties (1.15). In which way does the boundary condition at (a, b), where no unique normal exists, have to be interpreted?

Problem 1.10 Generalize the discussion concerning the five-point stencil discretization (including the order of convergence) to the boundary value problem

$$\begin{aligned} -\Delta u + ru &= f \quad \text{in } \Omega, \\ u &= g \quad \text{on } \partial\Omega, \end{aligned}$$

for $r > 0$ and $\Omega := (0, a) \times (0, b)$. To approximate the reactive term ru, the following schemes in the notation of (1.21) are to be used:

a) $c_{0,0} = 1$, $\quad c_{ij} = 0$ otherwise,
b) $c_{0,0} > 0$, $\quad c_{0,1}, c_{1,0}, c_{0,-1}, c_{-1,0} \geq 0$, $\quad c_{ij} = 0$ otherwise, $\quad$ and $\sum_{i,j=-1}^{1} c_{ij} = 1$.

Programming project 1.3 Consider the two-dimensional Poisson equation with Dirichlet and Neumann boundary conditions

$$\begin{aligned} -\Delta u &= f && \text{in } \Omega := (0,1)^2, \\ \partial_{\mathbf{n}} u &= g_1 && \text{on } \Gamma_1 := (0,1) \times \{0,1\}, \\ u &= g_3 && \text{on } \Gamma_3 := \partial\Omega \setminus \Gamma_1 . \end{aligned}$$

Write a function that solves this problem numerically by means of a second-order finite difference scheme on a quadratic grid with grid size $h := 1/m$, $m \in \mathbb{N}$. The input data are the right-hand side f, the boundary data g_1, g_3, the exact solution u, and the number of intervals m in one coordinate direction. The output data are $\boldsymbol{u}_h \in \mathbb{R}^{m-1,m+1}$ and the maximum norm of the error, i.e., $|\boldsymbol{U} - \boldsymbol{u}_h|_\infty$.

Determine the functions f, g_1, g_3 from the exact solutions

a) $u = \sin(\pi x)\cos(\pi y)$,
b) $u = \cos(7x)\cos(7y)$,

and perform a series of experiments for $m = 10, 20, 50, 100$.
Hints:

- Start by implementing Dirichlet boundary conditions on the whole boundary. Once this works, include the Neumann boundary conditions.
- The resulting system matrix is sparse and you should not store the complete matrix.

Programming project 1.4 Write a function that solves the problem

$$\begin{aligned} -\Delta u &= f && \text{in } \Omega := (0,1)^2, \\ \partial_{\mathbf{n}} u &= g_1 && \text{on } \Gamma_1 := \{0\} \times (0,1), \\ u &= g_3 && \text{on } \Gamma_3 := \partial\Omega \setminus \Gamma_1 , \end{aligned}$$

with a finite difference schema on a quadratic grid with grid size $h := 1/m$, $m \in \mathbb{N}$. The input data are the right-hand side f, the boundary data g_1, g_3, the exact solution u, and the number of intervals m in one coordinate direction. The output data are $\boldsymbol{u}_h \in \mathbb{R}^{m,m-1}$ and the maximum norm of the error, i.e., $|\boldsymbol{U} - \boldsymbol{u}_h|_\infty$.

Use the following methods to implement the Neumann boundary condition:

a) Use a one-sided difference quotient.
b) Use the central difference formula as described in the discussion of the model problem (1.24).

Test your implementation for the data f, g_1, g_3 such that $u(x, y) := \cos(\pi x)\cos(\pi y) + \sin(\pi x)\sin(\pi y)$ is the exact solution.
Hint: If you wish to visualize the error, use a double logarithmic plot and $h = 2^{-i}$, $i = 4, \ldots, 7$.

1.4 Maximum Principles and Stability

In this section the proof of the stability estimate (1.20), which is still missing, will be given. For this reason we develop a more general framework, in which we will then also discuss the finite element method (see Section 3.9) and the time-dependent problems (see Section 9.7). The boundary value problem (1.1), (1.2) satisfies a *(weak) maximum principle* in the following sense: If f is continuous and $f(x) \le 0$ for all $x \in \Omega$ (for short $f \le 0$), then

$$\max_{x \in \overline{\Omega}} u(x) \le \max_{x \in \partial\Omega} u(x) .$$

This *maximum principle* is also *strong* in the following sense: the maximum of u on $\overline{\Omega}$ can be attained in Ω only if u is constant (see, for example, [28, Thm. 3.5], also for the following assertions). By exchanging u, f, g by $-u, -f, -g$, respectively, we see that there is an analogous *(strong) minimum principle*. The same holds for more general linear differential equations as in (1.28), which may also contain convective parts (this means first-order derivatives). But if the equation contains a reactive part (this means without derivatives), as in the example

$$-\Delta u + ru = f \quad \text{in } \Omega$$

with a continuous function $r : \Omega \to \mathbb{R}$ such that $r(x) \ge 0$ for $x \in \Omega$, there is a weak maximum principle only in the following form: If $f \le 0$, then

$$\max_{x \in \overline{\Omega}} u(x) \le \max \left\{ \max_{x \in \partial\Omega} u(x), 0 \right\} .$$

The weak maximum principle directly implies assertions about the dependence of the solution u of the boundary value problem on the data f and g; this means *stability properties*. One can also follow this method in investigating the discretization. For the basic example we have

Theorem 1.9 *Let u_h be a grid function on $\overline{\Omega}_h$ defined by* (1.7), (1.8) *and suppose $f_{ij} \le 0$ for all $i = 1, \ldots, l-1$, $j = 1, \ldots, m-1$. Then if u_h attains its maximum on $\Omega_h \cup \partial\Omega_h^*$ at a point $(i_0 h, j_0 h) \in \Omega_h$, then the following holds:*

$$u_h \text{ is constant on } \Omega_h \cup \partial\Omega_h^* .$$

Here

$$\partial\Omega_h^* := \partial\Omega_h \setminus \{(0,0), (a,0), (0,b), (a,b)\} .$$

In particular, we have

$$\max_{(x,y) \in \Omega_h} u_h(x,y) \le \max_{(x,y) \in \partial\Omega_h^*} u_h(x,y) .$$

Proof Let $\overline{u} := u_h(i_0 h, j_0 h)$. Then because of (1.7) and $f_{ij} \le 0$ we have

$$4\overline{u} \le \sum_{(k,l)\in N_{(i_0,j_0)}} u_h(kh, lh) \le 4\overline{u}\,,$$

since in particular $u_h(kh, lh) \le \overline{u}$ for $(k, l) \in N_{(i_0,j_0)}$. Here we used the notation

$$N_{(i_0,j_0)} = \{((i_0 - 1), j_0), ((i_0 + 1), j_0), (i_0, (j_0 + 1)), (i_0, (j_0 - 1))\}$$

for the set of indices of neighbours of $(i_0 h, j_0 h)$ in the five-point stencil. From these inequalities we conclude that

$$u_h(kh, lh) = \overline{u} \quad \text{for} \quad (k, l) \in N_{(i_0,j_0)}\,.$$

If we apply this argument to the neighbours in $\overline{\Omega}_h$ of the grid points (kh, lh) for $(k, l) \in N_{(i_0,j_0)}$ and then continue in the same way to the sets of neighbours in $\overline{\Omega}_h$ arising in every such step, then finally, for each grid point $(ih, jh) \in \Omega_h \cup \partial\Omega_h^*$ the claimed identity $u_h(ih, jh) = \overline{u}$ is achieved. □

The exceptional set of vertices $\partial\Omega_h \setminus \partial\Omega_h^*$ does not participate in any difference stencil, so that the values there are of no relevance for u_h.

We want to generalize this result and therefore consider a system of equations as in (1.10), (1.11):

$$A_h \boldsymbol{u}_h = \boldsymbol{q}_h = -\hat{A}_h \hat{\boldsymbol{u}}_h + \boldsymbol{f}\,, \tag{1.34}$$

where $A_h \in \mathbb{R}^{M_1,M_1}$ as in (1.10), $\hat{A}_h \in \mathbb{R}^{M_1,M_2}$ as in (1.11), $\boldsymbol{u}_h, \boldsymbol{f} \in \mathbb{R}^{M_1}$, and $\hat{\boldsymbol{u}}_h \in \mathbb{R}^{M_2}$. This may be interpreted as the discretization of a boundary value problem obtained by the finite difference method or any other approach and without restrictions on the dimensionality of the domain. At least on one part of the boundary Dirichlet boundary conditions are required. Then the entries of the vector $\boldsymbol{u}_h$ can be interpreted as the unknown values at the grid points in $\Omega_h \cup \partial\Omega_h^{(1)}$, where $\partial\Omega_h^{(1)}$ correspond to a part of $\partial\Omega$ (with flux or mixed boundary condition). Analogously, the vector $\hat{\boldsymbol{u}}_h$ (indexed from $M_1 + 1$ to $M_1 + M_2$) corresponds to the values fixed by the Dirichlet boundary conditions on $\partial\Omega_h^{(2)}$. Again let $M = M_1 + M_2$ and

$$\tilde{A}_h := \left(A_h \,\middle|\, \hat{A}_h \right) \in \mathbb{R}^{M_1,M}\,.$$

This means in particular that the dimensions M_1 and M_2 are not fixed but are in general unbounded for $h \to 0$.

Oriented on (1.15) we require the following general assumptions for the rest of the section:

$$
\begin{aligned}
&(1)\quad (A_h)_{rr} > 0 \quad \text{for all } r = 1, \dots, M_1\,,\\
&(2)\quad (A_h)_{rs} \le 0 \quad \text{for all } r, s = 1, \dots, M_1 \text{ such that } r \ne s\,,\\
&(3)\quad \text{(i)}\quad \sum_{s=1}^{M_1} (A_h)_{rs} \ge 0 \quad \text{for all } r = 1, \dots, M_1\,,\\
&\qquad\quad \text{(ii)}\quad \text{for at least one index } r \text{ the strict inequality holds},\\
&(4)\quad A_h \text{ is irreducible},\\
&(5)\quad (\hat{A}_h)_{rs} \le 0 \quad \text{for all } r = 1, \dots, M_1,\ s = M_1 + 1, \dots, M\,,\\
&(6)\quad \sum_{s=1}^{M} (\tilde{A}_h)_{rs} \ge 0 \quad \text{for all } r = 1, \dots, M_1\,,\\
&(7)\quad \text{for every } s = M_1 + 1, \dots, M \text{ there exists } r \in \{1, \dots, M_1\}\\
&\qquad \text{such that } (\hat{A}_h)_{rs} \ne 0.
\end{aligned}
\tag{1.35}
$$

Generalizing the notation above for $r \in \{1, \dots, M_1\}$, the indices $s \in \{1, \dots, M\} \setminus \{r\}$ are called *neighbours*, for which $(\tilde{A}_h)_{rs} \ne 0$, and they are assembled to form the set N_r. Therefore, the irreducibility of A_h means that arbitrary $r, s \in \{1, \dots, M_1\}$ can be connected by neighbourhood relationships.

The condition (7) is not a restriction: It only avoids the inclusion of known values $(\hat{\boldsymbol{u}}_h)_s$ that do not influence the solution of (1.34) at all. For the five-point stencil on the rectangle, these are the values at the corner points. Because of the condition (7), every index $r \in \{M_1 + 1, \dots, M\}$ is connected to every index $s \in \{1, \dots, M_1\}$ by means of neighbourhood relationships.

The conditions (2) and (3) imply the weak diagonal dominance of A_h. Note that the conditions are formulated redundantly: The condition (3) also follows from (5) through (7).

To simplify the notation we will use the following conventions, where $\boldsymbol{u}$, $\boldsymbol{v}$ and A, B are vectors and matrices, respectively, of suitable dimensions:

$$
\begin{aligned}
\boldsymbol{u} \ge \boldsymbol{0} &\quad \text{if and only if} \quad (\boldsymbol{u})_i \ge 0 \quad \text{for all indices } i\,,\\
\boldsymbol{u} \ge \boldsymbol{v} &\quad \text{if and only if} \quad \boldsymbol{u} - \boldsymbol{v} \ge 0\,,\\
A \ge 0 &\quad \text{if and only if} \quad (A)_{ij} \ge 0 \quad \text{for all indices } (i, j)\,,\\
A \ge B &\quad \text{if and only if} \quad A - B \ge 0\,.
\end{aligned}
\tag{1.36}
$$

Theorem 1.10 *We consider* (1.34) *under the assumptions* (1.35). *Furthermore, let* $f \le \boldsymbol{0}$. *Then a* strong maximum principle *holds If the components of* $\tilde{\boldsymbol{u}}_h = \binom{\boldsymbol{u}_h}{\hat{\boldsymbol{u}}_h}$ *attain a nonnegative maximum for some index* $r \in \{1, \dots, M_1\}$, *then all the components are equal. In particular, a* weak maximum principle *is fulfilled:*

$$
\max_{r \in \{1, \dots, M\}} (\tilde{\boldsymbol{u}}_h)_r \le \max\left\{0, \max_{r \in \{M_1+1, \dots, M\}} (\hat{\boldsymbol{u}}_h)_r\right\}. \tag{1.37}
$$

Proof Let $\overline{u} = \max_{s\in\{1,\dots,M\}}(\tilde{\boldsymbol{u}}_h)_s$, and $\overline{u} = (\boldsymbol{u}_h)_r$ where $r \in \{1, \dots, M_1\}$. Because of (1.35) (2), (5), (6) the rth row of (1.34) implies

$$\begin{aligned}(A_h)_{rr}\overline{u} &\le - \sum_{s\in N_r} \left(\tilde{A}_h\right)_{rs}(\tilde{\boldsymbol{u}}_h)_s = \sum_{s\in N_r} \left|\left(\tilde{A}_h\right)_{rs}\right| (\tilde{\boldsymbol{u}}_h)_s \\ &\le \sum_{s\in N_r} \left|\left(\tilde{A}_h\right)_{rs}\right| \overline{u} \le (A_h)_{rr}\overline{u}\,,\end{aligned} \tag{1.38}$$

where the assumption $\overline{u} \ge 0$ is used in the last estimate. Therefore, everywhere equality has to hold. Since the second inequality is valid also for every single term and $(\tilde{A}_h)_{rs} \neq 0$ by the definition of N_r, we finally conclude that

$$(\tilde{\boldsymbol{u}}_h)_s = \overline{u} \quad \text{for all } s \in N_r\,.$$

This allows us to apply this argument to all $s \in N_r \cap \{1, \dots, M_1\}$, then to the corresponding sets of neighbours, and so on, until the assertion is proven. □

The requirement of irreducibility can be weakened if instead of (1.35) (6) we have

(6)* $\displaystyle\sum_{s=1}^{M} \left(\tilde{A}_h\right)_{rs} = 0 \quad \text{for all } r = 1, \dots, M_1\,.$

Then condition (1.35) (4) can be replaced by the requirement

(4)* For every $r_1 \in \{1, \dots, M_1\}$ such that

$$\sum_{s=1}^{M_1} (A_h)_{r_1 s} = 0 \tag{1.39}$$

there are indices $r_2, \dots, r_{l+1}$ such that

$$(A_h)_{r_i r_{i+1}} \neq 0 \quad \text{for } i = 1, \dots, l$$

and

$$\sum_{s=1}^{M_1} (A_h)_{r_{l+1} s} > 0\,. \tag{1.40}$$

This collection of modified conditions (1.35) (1)–(3), (4)*, (1.35) (5), (6)* will be denoted by (1.35)*

Motivated by the example above we call a point $r \in \{1, \dots, M_1\}$ *far from the boundary* if (1.39) holds, and *close to the boundary* if (1.40) holds, and the points $r \in \{M_1 + 1, \dots, M\}$ are called *boundary points*.

Theorem 1.11 *We consider* (1.34) *under the assumption* (1.35)*.

If $\boldsymbol{f} \le \boldsymbol{0}$, *then*

$$\max_{r\in\{1,\dots,M\}} (\tilde{\boldsymbol{u}}_h)_r \le \max_{r\in\{M_1+1,\dots,M\}} (\hat{\boldsymbol{u}}_h)_r\,. \tag{1.41}$$

Proof We use the same notation and the same arguments as in the proof of Theorem 1.10. In (1.38) in the last estimate equality holds, so that no sign conditions for $\overline{u}$ are necessary. Because of (4)* the maximum will also be attained at a point close to the boundary and therefore also at its neighbours. Because of (6)* a boundary point also belongs to these neighbours, which proves the assertion. □

From the maximum principles we immediately conclude a *comparison principle*:

Lemma 1.12 *We assume* (1.35) *or* (1.35)*.
Let $\boldsymbol{u}_{h1}, \boldsymbol{u}_{h2} \in \mathbb{R}^{M_1}$ *be solutions of*

$$A_h \boldsymbol{u}_{hi} = -\hat{A}_h \hat{\boldsymbol{u}}_{hi} + \boldsymbol{f}_i \quad \textit{for } i = 1, 2$$

for given $\boldsymbol{f}_1, \boldsymbol{f}_2 \in \mathbb{R}^{M_1}$, $\hat{\boldsymbol{u}}_{h1}, \hat{\boldsymbol{u}}_{h2} \in \mathbb{R}^{M_2}$, *which satisfy* $\boldsymbol{f}_1 \leq \boldsymbol{f}_2$, $\hat{\boldsymbol{u}}_{h1} \leq \hat{\boldsymbol{u}}_{h2}$. *Then*

$$\boldsymbol{u}_{h1} \leq \boldsymbol{u}_{h2} \, .$$

Proof From $A_h(\boldsymbol{u}_{h1} - \boldsymbol{u}_{h2}) = -\hat{A}_h(\hat{\boldsymbol{u}}_{h1} - \hat{\boldsymbol{u}}_{h2}) + \boldsymbol{f}_1 - \boldsymbol{f}_2$ we can conclude with Theorem 1.10 or 1.11 that

$$\max_{r \in \{1,\ldots,M_1\}} (\boldsymbol{u}_{h1} - \boldsymbol{u}_{h2})_r \leq 0 \, .$$

□

This implies in particular the uniqueness of a solution of (1.34) for arbitrary $\hat{\boldsymbol{u}}_h$ and $\boldsymbol{f}$ and also the regularity of A_h.

In the following we denote by $\boldsymbol{0}$ and 0 the zero vector and the zero matrix, respectively, where all components are equal to 0. An immediate consequence of Lemma 1.12 is the following

Theorem 1.13 *Let* $A_h \in \mathbb{R}^{M_1,M_1}$ *be a matrix with the properties* (1.35) (1)–(3)(i), (4)*, *and* $\boldsymbol{u}_h \in \mathbb{R}^{M_1}$. *Then*

$$A_h \boldsymbol{u}_h \geq \boldsymbol{0} \quad \textit{implies} \quad \boldsymbol{u}_h \geq \boldsymbol{0} \, . \tag{1.42}$$

Proof To be able to apply Lemma 1.12, one has to construct a matrix $\hat{A}_h \in \mathbb{R}^{M_1,M_2}$ such that (1.35)* holds. Obviously, this is possible. Then one can choose

$$\begin{aligned} \boldsymbol{u}_{h2} &:= \boldsymbol{u}_h \, , \; \boldsymbol{f}_2 := A_h \boldsymbol{u}_{h2} \, , \; \hat{\boldsymbol{u}}_{h2} := \boldsymbol{0} \, , \\ \boldsymbol{u}_{h1} &:= \boldsymbol{0} , \quad \boldsymbol{f}_1 := \boldsymbol{0} , \qquad \hat{\boldsymbol{u}}_{h1} := \boldsymbol{0} \end{aligned}$$

to conclude the assertion. Because of $\hat{\boldsymbol{u}}_{hi} := \boldsymbol{0}$ for $i = 1, 2$ the specific definition of $\hat{A}_h$ plays no role. □

A matrix with the property (1.42) is called *inverse monotone*. An equivalent requirement is

$$\boldsymbol{v}_h \geq \boldsymbol{0} \quad \Rightarrow \quad A_h^{-1} \boldsymbol{v}_h \geq \boldsymbol{0} \, ,$$

and therefore by choosing the unit vectors as $\boldsymbol{v}_h$,

$$A_h^{-1} \geq 0\,.$$

Inverse monotone matrices that also satisfy (1.35) (1), (2) are called *M-matrices*.

Finally, we can weaken the assumptions for the validity of the comparison principle.

Corollary 1.14 *Suppose that $A_h \in \mathbb{R}^{M_1,M_1}$ is inverse monotone and* (1.35) (5) *holds. Let $\boldsymbol{u}_{h1}, \boldsymbol{u}_{h2} \in \mathbb{R}^{M_1}$ be solutions of*

$$A_h \boldsymbol{u}_{hi} = -\hat{A}_h \hat{\boldsymbol{u}}_{hi} + \boldsymbol{f}_i \quad \textit{for } i = 1, 2$$

for given $\boldsymbol{f}_1, \boldsymbol{f}_2 \in \mathbb{R}^{M_1}$, $\hat{\boldsymbol{u}}_{h1}, \hat{\boldsymbol{u}}_{h2} \in \mathbb{R}^{M_2}$ that satisfy $\boldsymbol{f}_1 \leq \boldsymbol{f}_2$, $\hat{\boldsymbol{u}}_{h1} \leq \hat{\boldsymbol{u}}_{h2}$. Then

$$\boldsymbol{u}_{h1} \leq \boldsymbol{u}_{h2}\,.$$

Proof Multiplying the equation

$$A_h(\boldsymbol{u}_{h1} - \boldsymbol{u}_{h2}) = -\hat{A}_h(\hat{\boldsymbol{u}}_{h1} - \hat{\boldsymbol{u}}_{h2}) + \boldsymbol{f}_1 - \boldsymbol{f}_2$$

from the left by the matrix A_h^{-1}, we get

$$\boldsymbol{u}_{h1} - \boldsymbol{u}_{h2} = -\underbrace{A_h^{-1}}_{\geq 0}\underbrace{\hat{A}_h}_{\leq 0}\underbrace{(\hat{\boldsymbol{u}}_{h1} - \hat{\boldsymbol{u}}_{h2})}_{\leq 0} + \underbrace{A_h^{-1}}_{\geq 0}\underbrace{(\boldsymbol{f}_1 - \boldsymbol{f}_2)}_{\leq 0} \leq 0\,.$$

□

The importance of Corollary 1.14 lies in the fact that there exist discretization methods, for which the matrix $\tilde{A}_h$ does not satisfy, e.g., condition (1.35) (6), or (6)* but $A_h^{-1} \geq 0$. A typical example of such a method is the finite volume method described in Chapter 8.

In the following we denote by $\mathbf{1}$ a vector (of suitable dimension) whose components are *all* equal to 1.

Theorem 1.15 *We assume* (1.35) (1)–(3), (4)*, (5). *Furthermore, let $\boldsymbol{w}_h^{(1)}, \boldsymbol{w}_h^{(2)} \in \mathbb{R}^{M_1}$ be given such that*

$$A_h \boldsymbol{w}_h^{(1)} \geq \mathbf{1}, \quad A_h \boldsymbol{w}_h^{(2)} \geq -\hat{A}_h \mathbf{1}\,. \tag{1.43}$$

Then a solution of $A_h \boldsymbol{u}_h = -\hat{A}_h \hat{\boldsymbol{u}}_h + \boldsymbol{f}$ satisfies

1) $-\left(|\boldsymbol{f}|_\infty \boldsymbol{w}_h^{(1)} + |\hat{\boldsymbol{u}}_h|_\infty \boldsymbol{w}_h^{(2)}\right) \leq \boldsymbol{u}_h \leq |\boldsymbol{f}|_\infty \boldsymbol{w}_h^{(1)} + |\hat{\boldsymbol{u}}_h|_\infty \boldsymbol{w}_h^{(2)}$,
2) $|\boldsymbol{u}_h|_\infty \leq \left|\boldsymbol{w}_h^{(1)}\right|_\infty |\boldsymbol{f}|_\infty + \left|\boldsymbol{w}_h^{(2)}\right|_\infty |\hat{\boldsymbol{u}}_h|_\infty$.

Under the assumptions (1.35) (1)–(3), (4)*, *and* (1.43) *the matrix norm $\|\cdot\|_\infty$ induced by $|\cdot|_\infty$ satisfies*

$$\left\|A_h^{-1}\right\|_\infty \leq \left|\boldsymbol{w}_h^{(1)}\right|_\infty.$$

Proof Since $-|\boldsymbol{f}|_\infty \mathbf{1} \le \boldsymbol{f} \le |\boldsymbol{f}|_\infty \mathbf{1}$ and the analogous statement for $\hat{\boldsymbol{u}}_h$ is valid, the vector $\boldsymbol{v}_h := |\boldsymbol{f}|_\infty \boldsymbol{w}_h^{(1)} + |\hat{\boldsymbol{u}}_h|_\infty \boldsymbol{w}_h^{(2)} - \boldsymbol{u}_h$ satisfies

$$A_h \boldsymbol{v}_h \ge |\boldsymbol{f}|_\infty \mathbf{1} - \boldsymbol{f} - \hat{A}_h \left(|\hat{\boldsymbol{u}}_h|_\infty \mathbf{1} - \hat{\boldsymbol{u}}_h\right) \ge \mathbf{0},$$

where we have also used $-\hat{A}_h \ge 0$ in the last estimate. Therefore, the right inequality of 1) implies from Theorem 1.13 that the left inequality can be proven analogously. The further assertions follow immediately from 1). □

Because of the inverse monotonicity and from (1.35) (5) the vectors postulated in Theorem 1.15 have to satisfy $\boldsymbol{w}_h^{(i)} \ge \mathbf{0}$ necessarily for $i = 1, 2$. Thus stability with respect to $\|\cdot\|_\infty$ of the method defined by (1.34) assuming (1.35) (1)–(3), (4)* is guaranteed if a vector $\mathbf{0} \le \boldsymbol{w}_h \in \mathbb{R}^{M_1}$ and a constant $C > 0$ independent of h can be found such that

$$A_h \boldsymbol{w}_h \ge \mathbf{1} \quad \text{and} \quad |\boldsymbol{w}_h|_\infty \le C\,. \tag{1.44}$$

Finally, this will be proven for the five-point stencil discretization (1.1), (1.2) on the rectangle $\Omega = (0, a) \times (0, b)$ for $C = \frac{1}{16}(a^2 + b^2)$.

For this reason we define polynomials of second degree w_1, w_2 by

$$w_1(x) := \frac{1}{4}\, x(a - x) \quad \text{and} \quad w_2(y) := \frac{1}{4}\, y(b - y)\,. \tag{1.45}$$

It is clear that $w_1(x) \ge 0$ for all $x \in [0, a]$ and $w_2(y) \ge 0$ for all $y \in [0, b]$. Furthermore, we have $w_1(0) = 0 = w_1(a)$ and $w_2(0) = 0 = w_2(b)$, and

$$w_1''(x) = -\frac{1}{2} \quad \text{and} \quad w_2''(y) = -\frac{1}{2}\,.$$

Therefore w_1 and w_2 are strictly concave and attain their maximum in $\frac{a}{2}$ and $\frac{b}{2}$, respectively. Thus the function $w(x, y) := w_1(x) + w_2(x)$ satisfies

$$\begin{aligned} -\Delta w &= 1 \quad \text{in } \Omega\,, \\ w &\ge 0 \quad \text{on } \partial\Omega\,. \end{aligned} \tag{1.46}$$

Now let $\boldsymbol{w}_h \in \mathbb{R}^{M_1}$ be, for a fixed ordering, the representation of the grid function w_h defined by

$$(w_h)(ih, jh) := w(ih, jh) \quad \text{for } i = 1, \ldots, l-1\,,\ j = 1, \ldots, m-1\,.$$

Analogously, let $\hat{\boldsymbol{w}}_h \in \mathbb{R}^{M_2}$ be the representation of the function $\hat{w}_h$ defined on $\partial\Omega_h^*$. As can be seen from the error representation in Lemma 1.2, 4), the difference quotient $\partial^-\partial^+ u(x)$ is exact for polynomials of second degree. Therefore, we conclude from (1.46) that

$$A_h \boldsymbol{w}_h = -\hat{A}_h \hat{\boldsymbol{w}}_h + \mathbf{1} \ge \mathbf{1},$$

which finally implies

$$|\boldsymbol{w}_h|_\infty = \|w_h\|_\infty \le \|w\|_\infty = w_1\left(\frac{a}{2}\right) + w_2\left(\frac{b}{2}\right) = \frac{1}{16}(a^2 + b^2)\,.$$

This example motivates the following general procedure to construct $\boldsymbol{w}_h \in \mathbb{R}^{M_1}$ and a constant C such that (1.44) is fulfilled.

Assume that the boundary value problem under consideration reads in an abstract form

$$\begin{aligned}(Lu)(x) &= f(x) \text{ for } x \in \Omega, \\ (Ru)(x) &= g(x) \text{ for } x \in \partial\Omega.\end{aligned} \tag{1.47}$$

Similar to (1.46) we can consider — in case of existence — a solution w of (1.47) for some f, g, such that $f(x) \geq 1$ for all $x \in \Omega$, $g(x) \geq 0$ for all $x \in \Omega$. If w is bounded on Ω, then

$$(\boldsymbol{w}_h)_i := w(x_i), \quad i = 1, \ldots, M_1,$$

for the (non-Dirichlet) grid points x_i, is a candidate for $\boldsymbol{w}_h$. Obviously,

$$|\boldsymbol{w}_h|_\infty \leq \|w\|_\infty .$$

Correspondingly, we set

$$(\hat{\boldsymbol{w}}_h)_i = w(x_i) \geq 0, \quad i = M_1 + 1, \ldots, M_2,$$

for the Dirichlet-boundary grid points.

The exact fulfilment of the discrete equations by $\boldsymbol{w}_h$ cannot be expected anymore, but in case of consistency the residual can be made arbitrarily small for small h. This leads to

Theorem 1.16 *Assume that a solution $w \in C(\overline{\Omega})$ of (1.47) exists for data $f \geq 1$ and $g \geq 0$. If the discretization of the form (1.34) is asymptotically consistent with (1.47) (for these data), and there exists $\tilde{H} > 0$ so that for some $\tilde{\alpha} > 0$:*

$$-\hat{A}_h \hat{\boldsymbol{w}}_h + \boldsymbol{f} \geq \tilde{\alpha}\mathbf{1} \quad \textit{for} \quad h \leq \tilde{H}, \tag{1.48}$$

then for every $0 < \alpha < \tilde{\alpha}$ there exists $H > 0$, so that

$$A_h \boldsymbol{w}_h \geq \alpha\mathbf{1} \quad \textit{for} \quad h \leq H .$$

Proof Set

$$\boldsymbol{\tau}_h := A_h \boldsymbol{w}_h + \hat{A}_h \hat{\boldsymbol{w}}_h - \boldsymbol{f}$$

for the consistency error, then

$$|\boldsymbol{\tau}_h|_\infty \to 0 \quad \text{for} \quad h \to 0 .$$

Thus

$$\begin{aligned} A_h \boldsymbol{w}_h &= \boldsymbol{\tau}_h - \hat{A}_h \hat{\boldsymbol{w}}_h + \boldsymbol{f} \\ &\geq -|\boldsymbol{\tau}_h|_\infty \mathbf{1} + \tilde{\alpha}\mathbf{1} \quad \text{for} \quad h \leq \tilde{H} \\ &\geq \alpha\mathbf{1} \quad \text{for} \quad h \leq H \end{aligned}$$

and some appropriate $H > 0$. □

Thus a proper choice in (1.44) is

$$\frac{1}{\alpha}\boldsymbol{w}_h \quad \text{and} \quad C := \frac{1}{\alpha}\|w\|_\infty\,. \tag{1.49}$$

The condition (1.48) is not critical: In case of Dirichlet boundary conditions and (1.35) (5) (for corresponding rows i of $\hat{A}_h$) then, due to $(\boldsymbol{f})_i \geq 1$, we can even choose $\tilde{\alpha} = 1$. The discussion of Neumann boundary conditions following (1.24) shows that the same can be expected.

Theorem 1.16 shows that for a discretization with an inverse monotone system matrix consistency already implies stability.

To conclude this section let us discuss the various ingredients of (1.35) or (1.35)* that are sufficient for a range of properties from the inverse monotonicity up to a strong maximum principle: For the five-point stencil on a rectangle all the properties are valid for Dirichlet boundary conditions. If partly Neumann boundary conditions appear, the situation is the same, but now *close* and *far* from the boundary refers to its Dirichlet part. In the interpretation of the implications one has to take into account that the heterogeneities of the Neumann boundary condition are now part of the right-hand side $\boldsymbol{f}$, as seen, e.g., in (1.26). If mixed boundary conditions are applied, as

$$\partial_{\mathbf{n}} u + \alpha u = g \quad \text{on } \Gamma_2 \tag{1.50}$$

for some $\Gamma_2 \subset \Gamma$ and $\alpha = \alpha(x) > 0$, then the situation is the same again if αu is approximated just by evaluation, at the cost that (4)* no longer holds. The situation is similar if reaction terms appear in the differential equation (see Problem 1.10).

Exercises

Problem 1.11 Give an example of a matrix $\hat{A}_h \in \mathbb{R}^{M_1,M_2}$ that can be used in the proof of Theorem 1.13.

Problem 1.12 Show that the transposition of an M-matrix is again an M-matrix.

Problem 1.13 In the assumptions of Theorem 1.10 substitute (1.35) (4) by (4)* and amend (6) to

(6)# Condition (1.35) (6) is valid and

$$\sum_{s=1}^{M_1} (A_h)_{rs} > 0 \Rightarrow \text{there exists } s \in \{M_1, \ldots, M\} \text{ such that } (\hat{A}_h)_{rs} < 0.$$

Under these conditions prove a weak maximum principle as in Theorem 1.10.

Problem 1.14 Assuming the existence of $\boldsymbol{w}_h \in \mathbb{R}^{M_1}$ such that $A_h \boldsymbol{w}_h \geq \mathbf{1}$ and $|\boldsymbol{w}_h|_\infty \leq C$ for some constant C independent of h, show directly (without Theorem 1.15) a refined order of convergence estimate on the basis of an order of consistency estimate in which also the shape of $\boldsymbol{w}_h$ appears.

Chapter 2
The Finite Element Method for the Poisson Equation

The finite element method, frequently abbreviated by FEM, was developed in the fifties in the aircraft industry, after the concept had been independently outlined by mathematicians at an earlier time. Even today, the terminology used reflects that one origin of the development lies in structural mechanics. Shortly after this beginning, the finite element method was applied to problems of heat conduction and fluid mechanics, which form the application background of this book.

An intensive mathematical analysis and further development was started in the later sixties. The basics of this mathematical description and analysis are to be developed in this and the following chapter. The homogeneous Dirichlet boundary value problem for the Poisson equation forms the paradigm of this chapter, but more generally valid considerations will be emphasized. In this way the abstract foundation for the treatment of more general problems in Chapter 3 is provided. In spite of the importance of the finite element method for structural mechanics, the treatment of the linear elasticity equations will be omitted. But we note that only a small expense is necessary for the application of the considerations to these equations. We refer to [19], where this is realized with a very similar notation.

2.1 Variational Formulation for the Model Problem

We will develop a new solution concept for the boundary value problem (1.1), (1.2) as a theoretical foundation for the finite element method. For such a solution, the validity of the differential equation (1.1) is no longer required pointwise but in the sense of some integral average with "arbitrary" weighting functions φ. In the same way, the boundary condition (1.2) will be weakened by the renunciation of its pointwise validity.

For the present, we want to confine the considerations to the case of homogeneous boundary conditions (i.e., $g \equiv 0$), and so we consider the following homogeneous Dirichlet problem for the Poisson equation:

Given a function $f : \Omega \to \mathbb{R}$, find a function $u : \overline{\Omega} \to \mathbb{R}$ such that

P. Knabner and L. Angermann, *Numerical Methods for Elliptic and Parabolic Partial Differential Equations*, Texts in Applied Mathematics 44,
https://doi.org/10.1007/978-3-030-79385-2_2

$$-\Delta u = f \quad \text{in } \Omega, \tag{2.1}$$
$$u = 0 \quad \text{on } \partial\Omega. \tag{2.2}$$

In the following let Ω be a domain such that the integral theorem of Gauss is valid, i.e., for any vector field $\boldsymbol{q} : \Omega \to \mathbb{R}^d$ with components in $C(\overline{\Omega}) \cap C^1(\Omega)$ it holds

$$\int_\Omega \nabla \cdot \boldsymbol{q}(x)\, dx = \int_{\partial\Omega} \boldsymbol{n}(x) \cdot \boldsymbol{q}(x)\, d\sigma . \tag{2.3}$$

Let the function $u : \overline{\Omega} \to \mathbb{R}$ be a classical solution of (2.1), (2.2) in the sense of Definition 1.1, which additionally satisfies $u \in C^1(\overline{\Omega})$ to facilitate the reasoning. Next we consider arbitrary $v \in C_0^\infty(\Omega)$ as so-called *test functions*. The smoothness of these functions allows all operations of differentiation, and furthermore, all derivatives of a function $v \in C_0^\infty(\Omega)$ vanish on the boundary $\partial\Omega$. We multiply equation (2.1) by v, integrate the result over Ω, and obtain

$$\begin{aligned} \langle f, v\rangle_0 := \int_\Omega f(x)v(x)\, dx &= -\int_\Omega \nabla \cdot (\nabla u)(x)\, v(x)\, dx \\ &= \int_\Omega \nabla u(x) \cdot \nabla v(x)\, dx - \int_{\partial\Omega} \nabla u(x) \cdot \boldsymbol{n}(x)\, v(x)\, d\sigma \\ &= \int_\Omega \nabla u(x) \cdot \nabla v(x)\, dx . \end{aligned} \tag{2.4}$$

The equality sign at the beginning of the second line of (2.4) is obtained by integration by parts using the integral theorem of Gauss with $\boldsymbol{q} = v\nabla u$. The boundary integral vanishes because $v = 0$ holds on $\partial\Omega$.

If we define, for $u \in C^1(\overline{\Omega})$, $v \in C_0^\infty(\Omega)$, a real-valued mapping a by

$$a(u, v) := \int_\Omega \nabla u(x) \cdot \nabla v(x)\, dx ,$$

then the classical solution of the boundary value problem satisfies the identity

$$a(u, v) = \langle f, v\rangle_0 \quad \text{for all } v \in C_0^\infty(\Omega) . \tag{2.5}$$

The mapping a defines a scalar product on $C_0^\infty(\Omega)$ that induces the norm

$$\|v\|_a := \sqrt{a(v, v)} = \left(\int_\Omega |\nabla v|^2\, dx\right)^{1/2} \tag{2.6}$$

(see Appendix A.4 for these notions). Most of the properties of a scalar product are obvious. Only the definiteness (A4.7) requires further considerations. Namely, we have to show that

$$a(v, v) = \int_\Omega (\nabla v \cdot \nabla v)(x)\, dx = 0 \iff v = 0.$$

To prove this assertion, first we show that $a(v, v) = 0$ implies $\nabla v(x) = 0$ for all $x \in \Omega$. To do this, we suppose that there exists some point $\overline{x} \in \Omega$ such that $\nabla v(\overline{x}) \neq 0$. Then $(\nabla v \cdot \nabla v)(\overline{x}) = |\nabla v|^2(\overline{x}) > 0$. Because of the continuity of ∇v, a small neighbourhood G of $\overline{x}$ exists with a positive measure $|G|$ and $|\nabla v|(x) \geq \alpha > 0$ for all $x \in G$. Since $|\nabla v|^2(x) \geq 0$ for all $x \in \Omega$, it follows that

$$\int_\Omega |\nabla v|^2 (x)\, dx \geq \alpha^2 |G| > 0,$$

which is in contradiction to $a(v, v) = 0$. Consequently, $\nabla v(x) = 0$ holds for all $x \in \Omega$; i.e., v is constant in Ω. Since $v(x) = 0$ for all $x \in \partial\Omega$, the assertion follows.

Unfortunately, the space $C_0^\infty(\Omega)$ is too small to play the part of the basic space because the solution u does not belong to $C_0^\infty(\Omega)$ in general. The identity (2.4) is to be satisfied for a larger class of functions, which include, as an example for v, the solution u and the finite element approximation to u to be defined later.

Preliminary, insufficient definition of the basic space

For the present we define as the basic space V,

$$V := \big\{ v : \Omega \to \mathbb{R} \,\big|\, v \in C(\overline{\Omega}),\ \partial_i v \text{ exists and is piecewise continuous for all } i = 1, \ldots, d,\ v = 0 \text{ on } \partial\Omega \big\}. \tag{2.7}$$

To say that $\partial_i v$ is *piecewise continuous* means that the domain Ω can be decomposed as follows:

$$\overline{\Omega} = \bigcup_j \overline{\Omega}_j ,$$

with a finite number of open sets Ω_j, with $\Omega_j \cap \Omega_k = \emptyset$ for $j \neq k$, and $\partial_i v$ is continuous on Ω_j and it can continuously be extended on $\overline{\Omega}_j$.

Then the following properties hold:

- a is a scalar product also on V,
- $C_0^\infty(\Omega) \subset V$,
- $C_0^\infty(\Omega)$ is *dense* in V with respect to $\|\cdot\|_a$; i.e., for any $v \in V$ a sequence $(v_n)_{n\in\mathbb{N}}$ in $C_0^\infty(\Omega)$ exists such that $\|v_n - v\|_a \to 0$ for $n \to \infty$, (2.8)
- $C_0^\infty(\Omega)$ is dense in V with respect to $\|\cdot\|_0$. (2.9)

The first and second statements are obvious. The two others require a certain technical effort. A more general statement will be formulated in Theorem 3.8.

With that, we obtain from (2.5) the following result:

Lemma 2.1 *Let u be a classical solution of (2.1), (2.2) and let $u \in C^1(\overline{\Omega})$. Then*

$$a(u, v) = \langle f, v\rangle_0 \quad \textit{for all } v \in V\,. \tag{2.10}$$

Equation (2.10) is also called a variational equation.

Proof Let $v \in V$. Then $v_n \in C_0^\infty(\Omega)$ exist with $v_n \to v$ with respect to $\|\cdot\|_0$ and also to $\|\cdot\|_a$. Therefore, it follows from the continuity of the bilinear form with respect to $\|\cdot\|_a$ (see (A4.24)) and the continuity of the functional defined by the right-hand side $v \mapsto \langle f, v\rangle_0$ with respect to $\|\cdot\|_0$ (because of the Cauchy–Schwarz inequality in $L^2(\Omega)$) that

$$\langle f, v_n\rangle_0 \to \langle f, v\rangle_0 \qquad \text{and} \qquad a(u, v_n) \to a(u, v) \quad \text{for } n \to \infty\,.$$

Since $a(u, v_n) = \langle f, v_n\rangle_0$, we get $a(u, v) = \langle f, v\rangle_0$. □

The space V in the identity (2.10) can be further enlarged as long as (2.8) and (2.9) will remain valid. This fact will be used later to give a correct definition.

Definition 2.2 A function $u \in V$ is called a *weak* (or *variational*) *solution* of (2.1), (2.2) if the following variational equation holds:

$$a(u, v) = \langle f, v\rangle_0 \quad \text{for all } v \in V\,.$$

If u models, e.g., the displacement of a membrane, this relation is called the *principle of virtual work.*

Lemma 2.1 guarantees that a classical solution u is a weak solution.

The weak formulation has the following properties:

- It requires less smoothness: $\partial_i u$ has to be only piecewise continuous.
- The validity of the boundary condition is guaranteed by the definition of the function space V.

We now show that the variational equation (2.10) has exactly the same solution(s) as a minimization problem:

Lemma 2.3 *The variational equation (2.10) has the same solutions $u \in V$ as the minimization problem*

$$F(v) \to \min \quad \textit{for all } v \in V\,, \tag{2.11}$$

where

$$F(v) := \frac{1}{2}a(v, v) - \langle f, v\rangle_0 \quad \left(= \frac{1}{2}\|v\|_a^2 - \langle f, v\rangle_0\right)\,.$$

Proof (2.10) $\Longrightarrow$ (2.11) :

Let u be a solution of (2.10) and let $v \in V$ be chosen arbitrarily. We define $w := v - u \in V$ (because V is a vector space), i.e., $v = u + w$. Then, using the bilinearity and symmetry, we have

$$\begin{aligned} F(v) &= \frac{1}{2}a(u + w, u + w) - \langle f, u + w\rangle_0 \\ &= \frac{1}{2}a(u, u) + a(u, w) + \frac{1}{2}a(w, w) - \langle f, u\rangle_0 - \langle f, w\rangle_0 \\ &= F(u) + \frac{1}{2}a(w, w) \geq F(u), \end{aligned} \tag{2.12}$$

where the last inequality follows from the positivity of a; i.e., (2.11) holds.

(2.10) $\Longleftarrow$ (2.11) :

Let u be a solution of (2.11) and let $v \in V$, $\varepsilon \in \mathbb{R}$ be chosen arbitrarily. We define $g(\varepsilon) := F(u + \varepsilon v)$ for $\varepsilon \in \mathbb{R}$. Then

$$g(\varepsilon) = F(u + \varepsilon v) \geq F(u) = g(0) \quad \text{for all } \varepsilon \in \mathbb{R},$$

because $u + \varepsilon v \in V$; i.e., g has a global minimum at $\varepsilon = 0$.

It follows analogously to (2.12):

$$g(\varepsilon) = \frac{1}{2}a(u, u) - \langle f, u\rangle_0 + \varepsilon \left(a(u, v) - \langle f, v\rangle_0\right) + \frac{\varepsilon^2}{2}a(v, v)\,.$$

Hence the function g is a quadratic polynomial in ε, and in particular, $g \in C^1(\mathbb{R})$ is valid. Therefore we obtain the necessary condition

$$0 = g'(\varepsilon) = a(u, v) - \langle f, v\rangle_0$$

for the existence of a minimum at $\varepsilon = 0$. Thus u solves (2.10), because $v \in V$ has been chosen arbitrarily. □

For applications, e.g., in structural mechanics as above, the minimization problem is called the *principle of minimal potential energy*.

Remark 2.4 Lemma 2.3 holds for general vector spaces V if a is a symmetric, positive bilinear form and the right-hand side $\langle f, v\rangle_0$ is replaced by $\ell(v)$, where $\ell : V \to \mathbb{R}$ is a linear mapping, a *linear functional*. Then the variational equation reads as

$$\text{Find } u \in V \quad \text{with} \quad a(u, v) = \ell(v) \quad \text{for all } v \in V\,, \tag{2.13}$$

and the minimization problem as

$$\text{Find } u \in V \quad \text{with} \quad F(u) = \min_{v \in V} F(v), \tag{2.14}$$

where $\quad F(v) := \dfrac{1}{2}a(v, v) - \ell(v)\,.$

To incorporate inhomogeneous (Dirichlet) boundary conditions, the variational equation has to be changed to

$$\text{Find } u \in \overline{V} \quad \text{with} \quad a(u, v) = b(v) \quad \text{for all } v \in V, \tag{2.15}$$

where $\overline{V} := V + \{u_0\}$ is an affine space to V. Then (2.15) is equivalent to

$$\text{Find } \tilde{u} \in V \quad \text{with} \quad a(\tilde{u}, v) = \tilde{b}(v) := b(v) - a(u_0, v) \quad \text{for all } v \in V$$

and thus (2.15) is equivalent to a minimization problem of the form (2.14), but over space $\overline{V}$.

Lemma 2.5 *The weak solution according to* (2.10) *(or* (2.11)*) is unique.*

Proof Let u_1, u_2 be two weak solutions, i.e.,

$$\left.\begin{aligned} a(u_1, v) &= \langle f, v\rangle_0 \\ a(u_2, v) &= \langle f, v\rangle_0 \end{aligned}\right\} \quad \text{for all } v \in V.$$

By subtraction, it follows that

$$a(u_1 - u_2, v) = 0 \quad \text{for all } v \in V.$$

Choosing $v := u_1 - u_2$ implies $a(u_1 - u_2, u_1 - u_2) = 0$ and consequently $u_1 = u_2$, because a is definite. □

Remark 2.6 Lemma 2.5 is generally valid if a is a definite bilinear form and ℓ is a linear form.

So far, we have defined two different norms on V, namely $\|\cdot\|_a$ and $\|\cdot\|_0$. The difference between these norms is essential because they are not equivalent on the vector space V defined by (2.7), and consequently, they generate different convergence concepts, as will be shown by the following example.

Example 2.7 *Let* $\Omega = (0, 1)$, *i.e.,*

$$a(u, v) := \int_0^1 u'v'\,dx,$$

and let $v_n : \Omega \to \mathbb{R}$ *for* $n \geq 2$ *be defined by (cf. Figure 2.1)*

$$v_n(x) = \begin{cases} nx, & \textit{for} \quad 0 \leq x \leq \frac{1}{n}, \\ 1, & \textit{for} \quad \frac{1}{n} \leq x \leq 1 - \frac{1}{n}, \\ n - nx, & \textit{for} \quad 1 - \frac{1}{n} \leq x \leq 1. \end{cases}$$

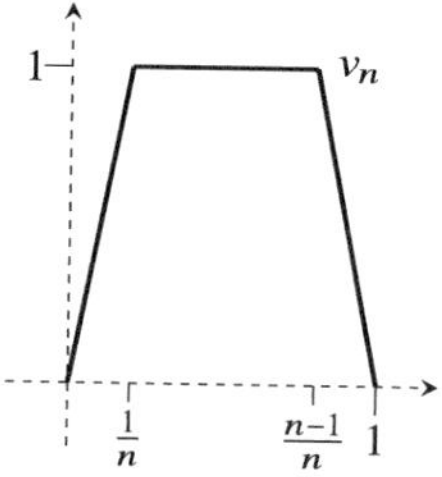

Fig. 2.1: The function v_n.

Then

$$\|v_n\|_0 \le \left(\int_0^1 1\, dx \right)^{1/2} = 1,$$

$$\|v_n\|_a = \left(\int_0^{\frac{1}{n}} n^2\, dx + \int_{1-\frac{1}{n}}^1 n^2\, dx \right)^{1/2} = \sqrt{2n} \to \infty \textit{ for } n \to \infty .$$

Therefore, there exists no constant $C > 0$ such that $\|v\|_a \le C\|v\|_0$ for all $v \in V$.

However, as we will show in Theorem 2.18, there exists a constant $C > 0$ such that the estimate

$$\|v\|_0 \le C\|v\|_a \quad \text{for all } v \in V$$

holds; i.e., $\|\cdot\|_a$ is the stronger norm.

It is possible to enlarge the basic space V without violating the previous statements. The enlargement is also necessary because, for instance, the proof of the existence of a solution of the variational equation (2.13) or the minimization problem (2.14) requires in general the completeness of V. However, the actual definition of V does not imply the completeness, as the following example shows.

Example 2.8 *Let $\Omega = (0, 1)$ again and therefore*

$$a(u, v) := \int_0^1 u'v'\, dx .$$

For $u(x) := x^\alpha(1-x)^\alpha$ with $\alpha \in \left(\frac{1}{2}, 1\right)$ we consider the sequence of functions

$$u_n(x) := \begin{cases} u(x) & \textit{for} \quad x \in \left[\frac{1}{n}, 1 - \frac{1}{n}\right], \\ n\, u(\frac{1}{n})\, x & \textit{for} \quad x \in \left[0, \frac{1}{n}\right], \\ n\, u(1-\frac{1}{n})(1-x) & \textit{for} \quad x \in \left[1 - \frac{1}{n}, 1\right]. \end{cases}$$

Then

$$\|u_n - u_m\|_a \to 0 \quad \textit{for } n, m \to \infty,$$
$$\|u_n - u\|_a \to 0 \quad \textit{for } n \to \infty,$$

but $u \notin V$, where V is defined analogously to (2.7) with $d = 1$.

In Section 3.1 we will see that a vector space $\tilde{V}$ normed with $\|\cdot\|_a$ exists such that $u \in \tilde{V}$ and $V \subset \tilde{V}$. Therefore, V is not complete with respect to $\|\cdot\|_a$; otherwise, $u \in V$ must be valid. In fact, there exists a (unique) completion of V with respect to $\|\cdot\|_a$ (see Appendix A.4, especially (A4.28)), but we have to describe the new "functions" added by this process. Besides, integration by parts must be valid such that a classical solution continues to be also a weak solution (compare with Lemma 2.1). Therefore, the following idea is unsuitable.

Attempt of a correct definition of V:

Let V be the set of all v with the property that $\partial_i v$ exists for all $x \in \Omega$ without any requirements on $\partial_i v$ in the sense of a function.

For instance, there exists *Cantor's function* with the following properties: $f : [0, 1] \to \mathbb{R}$, $f \in C([0, 1])$, $f \neq 0$, f is not constant, $f'(x)$ exists with $f'(x) = 0$ for all $x \in [0, 1]$.

Here the fundamental theorem of calculus, $f(x) = \int_0^x f'(s)\, ds + f(0)$, and thus the principle of integration by parts, are no longer valid.

Consequently, additional conditions for $\partial_i v$ are necessary.

To prepare an adequate definition of the space V, we extend the definition of derivatives by means of their action on averaging procedures. In order to do this, we introduce the *multi-index* notation.

A d-tuple $\alpha = (\alpha_1, \ldots, \alpha_d)$ of nonnegative integers $\alpha_i \in \mathbb{N}_0 := \{0, 1, 2, \ldots\}$ is called a *multi-index*. The number $|\alpha| := \sum_{i=1}^{d} \alpha_i$ denotes the *order* (or *length*) of α. For $x \in \mathbb{R}^d$ let

$$x^\alpha := x_1^{\alpha_1} \cdots x_d^{\alpha_d} . \tag{2.16}$$

A shorthand notation for the differential operations can be adopted by this: For an appropriately differentiable function v let

$$\partial^\alpha v := \partial_1^{\alpha_1} \cdots \partial_d^{\alpha_d} v . \tag{2.17}$$

We can obtain this definition from (2.16) by replacing x by the symbolic vector

$$\nabla := (\partial_1, \ldots, \partial_d)^T$$

of the first partial derivatives.

For example, if $d = 2$ and $\alpha = (1, 2)$, then $|\alpha| = 3$ and

$$\partial^\alpha v = \partial_1 \partial_2^2 v = \frac{\partial^3 v}{\partial x_1 \partial x_2^2} .$$

Now let $\alpha \in \mathbb{N}_0^d$ be a multi-index of length k and let $u \in C^k(\Omega)$. We then obtain for arbitrary test functions $\varphi \in C_0^\infty(\Omega)$ by integration by parts

$$\int_\Omega \partial^\alpha u\, \varphi\, dx = (-1)^k \int_\Omega u\, \partial^\alpha \varphi\, dx .$$

The boundary integrals vanish because $\partial^\beta \varphi = 0$ on $\partial\Omega$ for all multi-indices β. This motivates the following definition.

Definition 2.9 $v \in L^2(\Omega)$ is called the *weak* (or *generalized*) derivative $\partial^\alpha u$ of $u \in L^2(\Omega)$ for the multi-index $\alpha \in \mathbb{N}_0^d$ if

$$\int_\Omega v\, \varphi\, dx = (-1)^{|\alpha|} \int_\Omega u\, \partial^\alpha \varphi\, dx \quad \text{for all } \varphi \in C_0^\infty(\Omega).$$

The weak derivative is well defined because it is unique: Let $v_1, v_2 \in L^2(\Omega)$ be two weak derivatives of u. It follows that

$$\int_\Omega (v_1 - v_2)\,\varphi\,dx = 0 \quad \text{for all } \varphi \in C_0^\infty(\Omega).$$

Since $C_0^\infty(\Omega)$ is dense in $L^2(\Omega)$, we can furthermore conclude that

$$\int_\Omega (v_1 - v_2)\,\varphi\,dx = 0 \quad \text{for all } \varphi \in L^2(\Omega).$$

If we now choose specifically $\varphi := v_1 - v_2$, we obtain

$$\|v_1 - v_2\|_0^2 = \int_\Omega (v_1 - v_2)\,(v_1 - v_2)\;dx = 0\,,$$

and $v_1 = v_2$ (a.e.) follows immediately. In particular, $u \in C^k(\overline{\Omega})$ has weak derivatives $\partial^\alpha u$ for α with $|\alpha| \le k$, and the weak derivatives are identical to the classical (pointwise) derivatives.

Also the differential operators of vector calculus can be given a weak definition analogous to Definition 2.9. For example, for a vector field $\boldsymbol{q}$ with components in $L^2(\Omega)$, $v \in L^2(\Omega)$ is the *weak divergence* $v = \nabla \cdot \boldsymbol{q}$ if for all $\varphi \in C_0^\infty(\Omega)$

$$\int_\Omega v\varphi\,dx = -\int_\Omega \boldsymbol{q} \cdot \nabla\varphi\,dx\,. \tag{2.18}$$

The **correct choice of the space** V is the space $H_0^1(\Omega)$, which will be defined below. First we define

$$\begin{aligned} H^1(\Omega) := \big\{v :\ \Omega \to \mathbb{R}\ \big|\, v \in L^2(\Omega),\ &v \text{ has weak derivatives} \\ &\partial_i v \in L^2(\Omega) \text{ for all } i = 1, \ldots, d\big\}. \end{aligned} \tag{2.19}$$

A scalar product on $H^1(\Omega)$ is defined by

$$\langle u, v\rangle_1 := \int_\Omega u(x)v(x)\,dx + \int_\Omega \nabla u(x) \cdot \nabla v(x)\,dx \tag{2.20}$$

with the norm

$$\|v\|_1 := \sqrt{\langle v, v\rangle_1} = \left(\int_\Omega |v(x)|^2\,dx + \int_\Omega |\nabla v(x)|^2\,dx\right)^{1/2} \tag{2.21}$$

induced by this scalar product.

The above "temporary" definition (2.7) of V takes care of the boundary condition $v = 0$ on $\partial\Omega$ by conditions for the functions. For instance, we want to choose the basic space V analogously as

$$H_0^1(\Omega) := \big\{v \in H^1(\Omega)\,\big|\, v = 0 \quad \text{on } \partial\Omega\big\}\ . \tag{2.22}$$

Here $H^1(\Omega)$ and $H_0^1(\Omega)$ are special cases of so-called *Sobolev spaces*.

For $\Omega \subset \mathbb{R}^d$, $d \geq 2$, $H^1(\Omega)$ may contain unbounded functions. In particular, we have to examine carefully the meaning of $v|_{\partial\Omega}$ ($\partial\Omega$ has the d-dimensional measure 0) and, in particular, $v = 0$ on $\partial\Omega$. This will be described in Section 3.1.

Exercises

Problem 2.1

a) Consider the interval $(-1, 1)$; prove that the function $u(x) := |x|$ has the generalized derivative $u'(x) = \text{sign}(x)$.
b) Does $\text{sign}(x)$ have a generalized derivative?

Problem 2.2 Let $\Omega := (a, b) \subset \mathbb{R}$, $a < b$, be an open and bounded interval.

a) Let $v \in L^2(\Omega)$ have the weak derivative $v'_w \in L^2(\Omega)$ and the classical derivative v'_c in Ω. Prove that $v'_w = v'_c$ almost everywhere.
b) Assuming that $f \in L^2(\Omega)$ with a weak derivative $f' \in L^1(\Omega)$, show that for almost every $x_1, x_2 \in \Omega$ the fundamental theorem of calculus holds, i.e.,

$$f(x_2) - f(x_1) = \int_{x_1}^{x_2} f'(x)dx. \tag{2.23}$$

c) Conclude from (2.23) that there is a function $g \in C(\Omega)$ such that $f = g$ almost everywhere, and that $f \in L^\infty(\Omega)$.
d) If $f \in L^2(\Omega)$, with a weak derivative $f' \in L^p(\Omega)$, $1 < p \leq \infty$, show that g from c) is α-Hölder continuous with $\alpha = 1 - 1/p$.
e) Let $f \in H_0^1(\Omega)$. Use (2.23) to show that

$$\|f\|_0 \leq |b - a| \, \|f'\|_0.$$

Problem 2.3 Let $\overline{\Omega} = \bigcup_{l=1}^N \overline{\Omega}_l$, $N \in \mathbb{N}$, where the bounded subdomains $\Omega_l \subset \mathbb{R}^2$ are pairwise disjoint and possess piecewise smooth boundaries. Show that a function $u \in C(\overline{\Omega})$ with $u|_{\Omega_l} \in C^1(\overline{\Omega}_l)$, $1 \leq l \leq N$, has a weak derivative $\partial_i u \in L^2(\Omega)$, $i = 1, 2$, that coincides in $\bigcup_{l=1}^N \Omega_l$ with the classical one.

Problem 2.4 Let V be the set of functions that are continuous and piecewise continuously differentiable on $[0, 1]$ and that satisfy the additional conditions $v(0) = v(1) = 0$. Show that there exist infinitely many elements in V that minimizes the functional

$$F(v) := \int_0^1 \left(1 - [v'(x)]^2\right)^2 dx.$$

2.2 The Finite Element Method with Linear Elements

The weak formulation of the boundary value problem (2.1), (2.2) leads to particular cases of the following general, here equivalent, problems:

Let V be a real vector space, let $a: V \times V \to \mathbb{R}$ be a bilinear form, and let $\ell: V \to \mathbb{R}$ be a linear form.

Variational equation:

$$\text{Find } u \in V \quad \text{such that} \quad a(u, v) = \ell(v) \quad \text{for all } v \in V. \tag{2.24}$$

Minimization problem:

$$\begin{aligned} \text{Find } u \in V \quad \text{such that} \quad & F(u) = \min_{v \in V} F(v), \\ \text{where} \quad & F(v) := \frac{1}{2} a(v, v) - \ell(v). \end{aligned} \tag{2.25}$$

The *discretization approach* consists in the following procedure: Replace the space V by a finite-dimensional subspace V_h; i.e., solve instead of (2.24) the finite-dimensional variational equation:

$$\text{Find } u_h \in V_h \quad \text{such that} \quad a(u_h, v) = \ell(v) \quad \text{for all } v \in V_h. \tag{2.26}$$

This approach is called the *Galerkin method.* Or solve instead of (2.25) the finite-dimensional minimization problem:

$$\text{Find } u_h \in V_h \quad \text{such that} \quad F(u_h) = \min_{v \in V_h} F(v). \tag{2.27}$$

This approach is called the *Ritz method.*

It is clear from Lemma 2.3 and Remark 2.4 that the Galerkin method and the Ritz method are equivalent for a positive and symmetric bilinear form. The finite-dimensional subspace V_h is called an *ansatz space*.

The finite element method can be interpreted as a Galerkin method (and in our example as a Ritz method, too) for an ansatz space with special properties. In the following, these properties will be extracted by means of the simplest example.

Let V be defined by (2.7) or let $V = H_0^1(\Omega)$.

The weak formulation of the boundary value problem (2.1), (2.2) corresponds to the choice

$$a(u, v) := \int_\Omega \nabla u \cdot \nabla v \, dx, \quad \ell(v) := \int_\Omega f v \, dx.$$

Let $\Omega \subset \mathbb{R}^2$ be a domain with a polygonal boundary; i.e., the boundary Γ of Ω consists of a finite number of straight-line segments as shown in Figure 2.2.

Let $\mathcal{T}_h$ be a partition of Ω into closed triangles K (i.e., including the boundary ∂K) with the following properties:

1) $\overline{\Omega} = \bigcup_{K \in \mathcal{T}_h} K$;

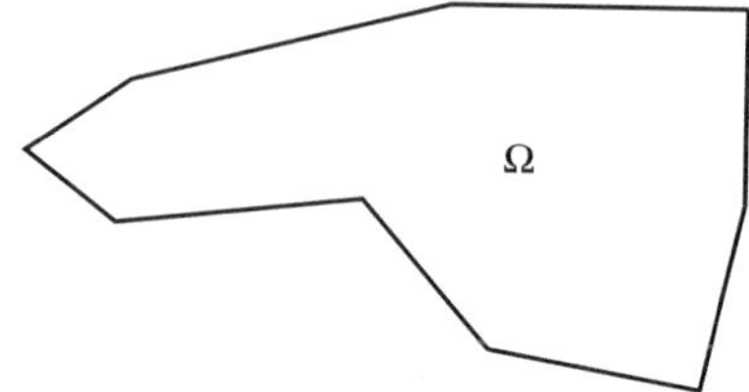

Fig. 2.2: Domain with a polygonal boundary.

2) For $K, K' \in \mathcal{T}_h$, $K \neq K'$,

$$\operatorname{int}(K) \cap \operatorname{int}(K') = \emptyset, \tag{2.28}$$

where $\operatorname{int}(K)$ denotes the open triangle (without the boundary ∂K).

3) If $K \neq K'$ but $K \cap K' \neq \emptyset$, then $K \cap K'$ is either a common point or a common edge of K and K' (cf. Figure 2.3).

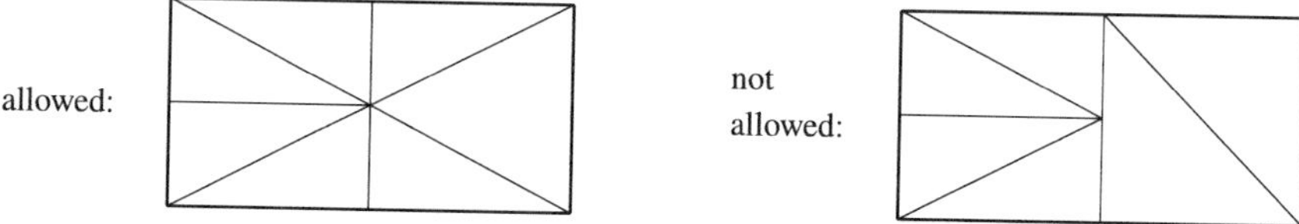

Fig. 2.3: Partitions.

A partition of Ω into triangles with the properties 1), 2) is called a *simplicial partition* or a *triangulation* of Ω. If, in addition, a triangulation of Ω satisfies property 3), it is called *consistent* (or *conforming*) (cf. Figure 2.4).

The triangles of a triangulation will be numbered $K_1, \ldots, K_N$. The subscript h indicates the fineness of the triangulation, e.g.,

$$h := \max_{K \in \mathcal{T}_h} \operatorname{diam}(K),$$

where $\operatorname{diam}(K) := \sup\left\{|x - y| \;\middle|\; x, y \in K\right\}$ denotes the diameter of K. Thus here h is the maximum length of the edges of all the triangles. Sometimes, $K \in \mathcal{T}_h$ is also called a (geometric) *element* of the partition.

The vertices of the triangles are called the *nodes*, and they will be numbered

$$a_1, a_2, \ldots, a_M,$$

i.e., $a_i = (x_i, y_i)$, $i = 1, \ldots, M$, where $M = M_1 + M_2$ and

$$\begin{aligned} a_1, \ldots, a_{M_1} &\in \Omega, \\ a_{M_1+1}, \ldots, a_M &\in \partial\Omega. \end{aligned} \tag{2.29}$$

This kind of arrangement of the nodes is chosen only for the sake of simplicity of the notation and is not essential for the following considerations.

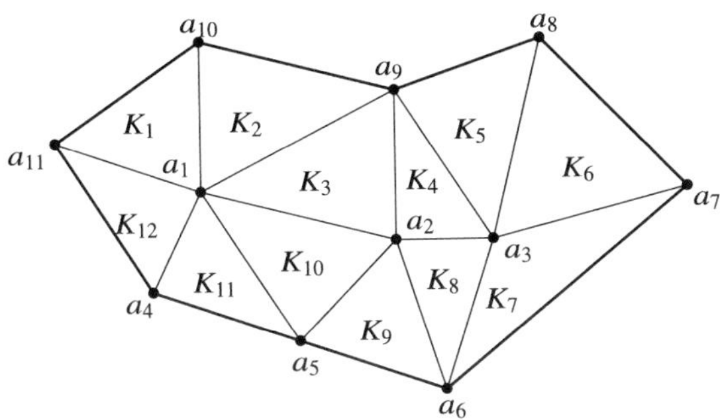

Fig. 2.4: A consistent triangulation with $N = 12$, $M = 11$, $M_1 = 3$, $M_2 = 8$.

An approximation of the boundary value problem (2.1), (2.2) with *linear finite elements* on a given triangulation $\mathcal{T}_h$ of Ω is obtained if the ansatz space V_h is defined as follows:

$$V_h := \left\{ v \in C(\overline{\Omega}) \mid v|_K \in \mathcal{P}_1(K) \text{ for all } K \in \mathcal{T}_h,\ v = 0 \text{ on } \partial\Omega \right\} . \tag{2.30}$$

Here $\mathcal{P}_1(K)$ denotes the set of polynomials of first degree (in 2 variables) on K; i.e., $p \in \mathcal{P}_1(K) \Leftrightarrow p(x, y) = \alpha + \beta x + \gamma y$ for all $(x, y) \in K$ and for fixed $\alpha, \beta, \gamma \in \mathbb{R}$.

Since $p \in \mathcal{P}_1(K)$ is also defined on the space $\mathbb{R}\times\mathbb{R}$, we use the short but inaccurate notation $\mathcal{P}_1 = \mathcal{P}_1(K)$; according to the context, the domain of definition will be given as $\mathbb{R} \times \mathbb{R}$ or as a subset of it.

We have

$$V_h \subset V .$$

This is clear for the case of definition of V by (2.7) because $\partial_x v|_K = \text{const}$, $\partial_y v|_K = \text{const}$ for $K \in \mathcal{T}_h$ for all $v \in V_h$. If $V = H_0^1(\Omega)$, then this inclusion is not so obvious. A proof will be given in Theorem 3.21 below.

An element $v \in V_h$ is determined uniquely by the values $v(a_i)$, $i = 1, \ldots, M_1$ (the *nodal values*).

In particular, the given nodal values already enforce the continuity of the piecewise linear composed functions. Correspondingly, the homogeneous Dirichlet boundary condition is satisfied if the nodal values at the boundary nodes are set to zero.

In the following, we will demonstrate these properties by an unnecessarily involved proof. The reason is that this proof will introduce all of the considerations that will lead to analogous statements for the more general problems of Section 3.4.

Let X_h be the larger ansatz space consisting of continuous, piecewise linear functions but regardless of any boundary conditions, i.e.,

$$X_h := \left\{ v \in C(\overline{\Omega}) \mid v|_K \in \mathcal{P}_1(K) \quad \text{for all } K \in \mathcal{T}_h \right\} . \tag{2.31}$$

Lemma 2.10 *For given values at the nodes $a_1, \ldots, a_M$, the interpolation problem in X_h is uniquely solvable. That is, if the values $v_1, \ldots, v_M$ are given, then there exists a uniquely determined element*

$$v \in X_h \quad \textit{such that} \quad v(a_i) = v_i , \quad i = 1, \ldots, M .$$

If $v_j = 0$ for $j = M_1 + 1, \ldots, M$, then it is even true that

$$v \in V_h .$$

Proof Step 1: For any arbitrary $K \in \mathcal{T}_h$ we consider the *local interpolation problem*:

$$\text{Find } p = p_K \in \mathcal{P}_1 \quad \text{such that} \quad p(a_i) = v_i , \ i = 1, 2, 3, \tag{2.32}$$

where $a_i, i = 1, 2, 3$, denote the vertices of K, and the values $v_i, i = 1, 2, 3$, are given. First we show that problem (2.32) is uniquely solvable for a particular triangle.

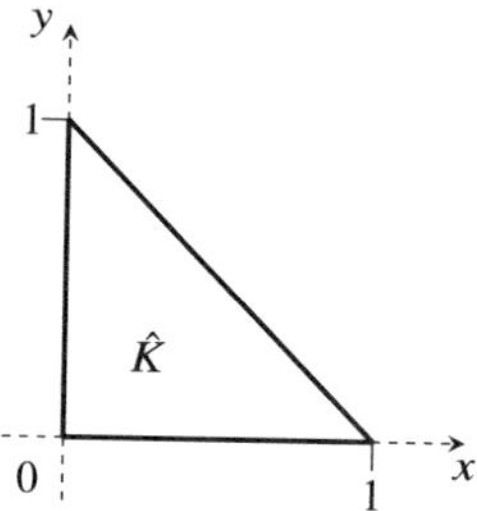

Fig. 2.5: Reference element $\hat{K}$.

A solution of (2.32) for the so-called *reference element* $\hat{K}$ (cf. Figure 2.5) with the vertices $\hat{a}_1 = (0, 0)$, $\hat{a}_2 = (1, 0)$, $\hat{a}_3 = (0, 1)$ is given by

$$p(x, y) = v_1 N_1(x, y) + v_2 N_2(x, y) + v_3 N_3(x, y)$$

with the *shape functions*

$$\begin{aligned} N_1(x, y) &:= 1 - x - y , \\ N_2(x, y) &:= x , \\ N_3(x, y) &:= y . \end{aligned} \tag{2.33}$$

Evidently, $N_i \in \mathcal{P}_1$, and furthermore,

$$N_i\left(\hat{a}_j\right) = \delta_{ij} := \begin{cases} 1 & \text{for } i = j \\ 0 & \text{for } i \neq j \end{cases} \qquad \text{for} \quad i, j = 1, 2, 3,$$

and thus

$$p(\hat{a}_j) = \sum_{i=1}^{3} v_i N_i(\hat{a}_j) = v_j \quad \text{for all} \quad j = 1, 2, 3 .$$

The uniqueness of the solution can be seen in the following way: If p_1, p_2 satisfy the interpolation problem (2.32) for the reference element, then for $p := p_1 - p_2 \in \mathcal{P}_1$ we have

$$p(\hat{a}_i) = 0, \quad i = 1, 2, 3 .$$

Here p is given in the form $p(x, y) = \alpha + \beta x + \gamma y$. If we fix the second variable $y = 0$, we obtain a polynomial function of one variable

$$p(x, 0) = \alpha + \beta x =: q(x) \in \mathcal{P}_1(\mathbb{R}) .$$

The polynomial q satisfies $q(0) = 0 = q(1)$, and $q = 0$ follows by the uniqueness of the polynomial interpolation in one variable; i.e., $\alpha = \beta = 0$. Analogously, we consider

$$q(y) := p(0, y) = \alpha + \gamma y = \gamma y ,$$

and we obtain from $q(1) = 0$ that $\gamma = 0$ and consequently $p = 0$.

In fact, this additional proof of uniqueness is not necessary, because the uniqueness already follows from the solvability of the interpolation problem because of $\dim \mathcal{P}_1 = 3$ (compare with Section 3.3).

Now we turn to the case of a general triangle K. A general triangle K is mapped onto $\hat{K}$ by an affine transformation (cf. Figure 2.6)

$$F : \hat{K} \to K , \quad F(\hat{x}) = B\hat{x} + d , \tag{2.34}$$

where $B \in \mathbb{R}^{2,2}$, $d \in \mathbb{R}^2$ are such that $F(\hat{a}_i) = a_i$.

$B = (b_1, b_2)$ and d are determined by the vertices a_i of K as follows:

$$\begin{aligned} a_1 &= F(\hat{a}_1) = F(0) = d , \\ a_2 &= F(\hat{a}_2) = b_1 + d = b_1 + a_1 , \\ a_3 &= F(\hat{a}_3) = b_2 + d = b_2 + a_1 ; \end{aligned}$$

i.e., $b_1 = a_2 - a_1$ and $b_2 = a_3 - a_1$. The matrix B is regular because $a_2 - a_1$ and $a_3 - a_1$ are linearly independent, ensuring $F(\hat{a}_i) = a_i$.

Since

$$K = \operatorname{conv}\{a_1, a_2, a_3\} := \left\{ \sum_{i=1}^{3} \lambda_i a_i \;\middle|\; 0 \le \lambda_i \le 1 , \sum_{i=1}^{3} \lambda_i = 1 \right\}$$

and especially $\hat{K} = \operatorname{conv}\{\hat{a}_1, \hat{a}_2, \hat{a}_3\}$, $F[\hat{K}] = K$ follows from the fact that the affine-linear mapping F satisfies

$$F\left(\sum_{i=1}^{3}\lambda_i\hat{a}_i\right)=\sum_{i=1}^{3}\lambda_i F(\hat{a}_i)=\sum_{i=1}^{3}\lambda_i a_i$$

for $0\le\lambda_i\le 1$, $\sum_{i=1}^{3}\lambda_i=1$.

In particular, the edges (where one λ_i is equal to 0) of $\hat{K}$ are mapped onto the edges of K.

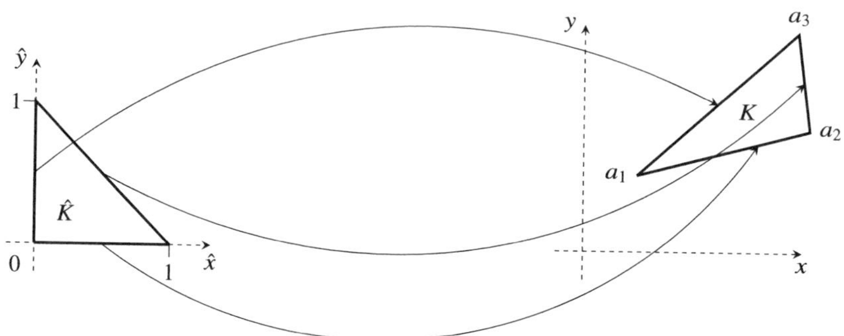

Fig. 2.6: Affine-linear transformation.

Analogously, the considerations can be applied to the space $\mathbb{R}^d$ word for word by replacing the set of indices $\{1,2,3\}$ by $\{1,\ldots,d+1\}$. This will be done in Section 3.3.

The polynomial space $\mathcal{P}_1$ does not change under the affine transformation F.

Step 2: We now prove that the local functions $v|_K$ can be composed continuously:

For every $K\in\mathcal{T}_h$, let $p_K\in\mathcal{P}_1$ be the unique solution of (2.32), where the values v_1,v_2,v_3 are the values v_{i_1},v_{i_2},v_{i_3} ($i_1,i_2,i_3\in\{1,\ldots,M\}$) that have to be interpolated at these nodes.

Let $K,K'\in\mathcal{T}_h$ be two different elements that have a common edge E. Then $p_K=p_{K'}$ on E is to be shown. This is valid because E can be mapped onto $[0,1]\times\{0\}$ by an affine transformation (cf. Figure 2.7). Then $q_1(x)=p_K(x,0)$ and $q_2(x):=p_{K'}(x,0)$ are elements of $\mathcal{P}_1(\mathbb{R})$, and they solve the same interpolation problem at the points $x=0$ and $x=1$; thus $q_1\equiv q_2$.

Therefore, the definition of v by means of

$$v(x)=p_K(x)\quad\text{for } x\in K\in\mathcal{T}_h \tag{2.35}$$

is unique, and this function satisfies $v\in C(\overline{\Omega})$ and $v\in X_h$.

Step 3: Finally, we will show that $v=0$ on $\partial\Omega$ for v defined by (2.35) if $v_i=0$, $i=M_1+1,\ldots,M$, for the boundary nodes.

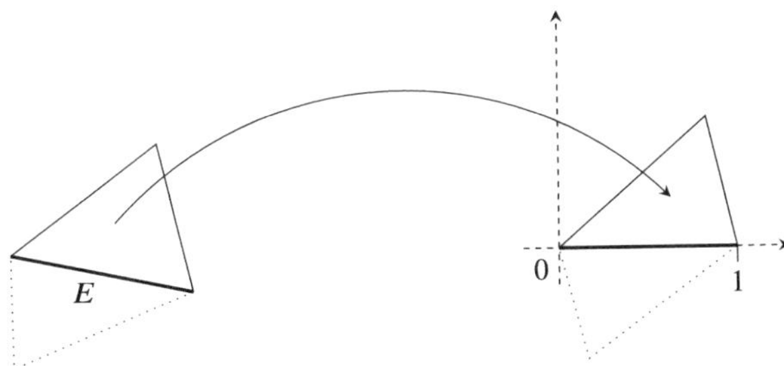

Fig. 2.7: Affine-linear transformation of E on the reference element $[0, 1]$.

The boundary $\partial\Omega$ consists of edges of elements $K \in \mathcal{T}_h$. Let E be such an edge; i.e., E has the vertices a_{i_1}, a_{i_2} with $i_j \in \{M_1 + 1, \ldots, M\}$. The given boundary values yield $v(a_{i_j}) = 0$ for $j = 1, 2$. By means of an affine transformation analogously to the above one we obtain that $v|_E$ is a polynomial of first degree in one variable and that $v|_E$ vanishes at two points. So $v|_E = 0$, and the assertion follows. □

The following statement is an important consequence of the unique solvability of the interpolation problem in X_h irrespective of its particular definition: The interpolation conditions

$$\varphi_i(a_j) = \delta_{ij}, \quad j = 1, \ldots, M, \tag{2.36}$$

uniquely determine functions $\varphi_i \in X_h$ for $i = 1, \ldots, M$. For any $v \in X_h$, we have

$$v(x) = \sum_{i=1}^{M} v(a_i)\varphi_i(x) \quad \text{for} \quad x \in \Omega, \tag{2.37}$$

because both the left-hand side and the right-hand side functions belong to X_h and are equal to $v(a_i)$ at $x = a_i$.

The representation $v = \sum_{i=1}^{M} \alpha_i \varphi_i$ is unique, too, for otherwise, a function $w \in X_h$, $w \neq 0$, such that $w(a_i) = 0$ for all $i = 1, \ldots, M$ would exist. Thus $\{\varphi_1, \ldots, \varphi_M\}$ is a basis of X_h, especially $\dim X_h = M$. This basis is called a *nodal basis* because of (2.37). For the particular case of a piecewise linear ansatz space on triangles, the basis functions are called *pyramidal functions* because of their shape. If the set of indices is restricted to $\{1, \ldots, M_1\}$; i.e., we omit the basis functions corresponding to the boundary nodes, then a basis of V_h will be obtained and $\dim V_h = M_1$.

Summary: the function values $v(a_i)$ at the nodes $a_1, \ldots, a_M$ are the *degrees of freedom* of $v \in X_h$, and the values at the interior points $a_1, \ldots, a_{M_1}$ are the *degrees of freedom* of $v \in V_h$.

The following consideration is valid for an arbitrary ansatz space V_h with a basis $\{\varphi_1, \ldots, \varphi_M\}$. The **Galerkin method** (2.26) reads as follows:

$$\text{Find } u_h = \sum_{i=1}^{M} \xi_i \varphi_i \in V_h \text{ such that } a(u_h, v) = \ell(v) \text{ for all } v \in V_h.$$

Since $v = \sum_{i=1}^{M} \eta_i \varphi_i$ for $\eta_i \in \mathbb{R}$, this is equivalent to

$$\begin{aligned} a(u_h, \varphi_i) &= \ell(\varphi_i) \quad \text{for all } i = 1, \ldots, M \iff \\ a\left(\sum_{j=1}^{M} \xi_j \varphi_j, \varphi_i\right) &= \ell(\varphi_i) \quad \text{for all } i = 1, \ldots, M \iff \\ \sum_{j=1}^{M} a\left(\varphi_j, \varphi_i\right) \xi_j &= \ell(\varphi_i) \quad \text{for all } i = 1, \ldots, M \iff \\ A_h \boldsymbol{\xi} &= \boldsymbol{q}_h \end{aligned} \tag{2.38}$$

with $A_h = \left(a(\varphi_j, \varphi_i)\right)_{ij} \in \mathbb{R}^{M,M}$, $\boldsymbol{\xi} = (\xi_1, \ldots, \xi_M)^T$ and $\boldsymbol{q}_h = (\ell(\varphi_i))_i$. Therefore, the Galerkin method is equivalent to the system of equations (2.38).

The considerations for deriving (2.38) show that, in the case of equivalence of the Galerkin method with the Ritz method, the system of equations (2.38) is equivalent to the minimization problem

$$F_h(\boldsymbol{\xi}) = \min_{\boldsymbol{\eta} \in \mathbb{R}^M} F_h(\boldsymbol{\eta}), \tag{2.39}$$

where

$$F_h(\boldsymbol{\eta}) = \frac{1}{2} \boldsymbol{\eta}^T A_h \boldsymbol{\eta} - \boldsymbol{q}_h^T \boldsymbol{\eta} \, .$$

Because of the symmetry and positive definiteness, the equivalence of (2.38) and (2.39) can be easily proven, and it forms the basis for the CG methods that will be discussed in Section 5.2.

Usually, A_h is called *stiffness matrix*, and $\boldsymbol{q}_h$ is called the *load vector*. These names originated from mechanics. For our model problem, we have

$$\begin{aligned} (A_h)_{ij} &= a(\varphi_j, \varphi_i) = \int_\Omega \nabla \varphi_j \cdot \nabla \varphi_i \, dx \,, \\ \left(\boldsymbol{q}_h\right)_i &= \ell(\varphi_i) = \int_\Omega f \varphi_i \, dx \, . \end{aligned}$$

By applying the finite element method, we thus have to perform the following steps:

1. Determination of A_h, $\boldsymbol{q}_h$. This step is called *assembling*.
2. Solution of $A_h \boldsymbol{\xi} = \boldsymbol{q}_h$.

If the basis functions φ_i have the property $\varphi_i(a_j) = \delta_{ij}$, then the solution of system (2.38) satisfies the relation $\xi_i = u_h(a_i)$, i.e., we obtain the vector of the nodal values of the finite element approximation.

Using only the properties of the bilinear form a, we obtain the following properties of A_h:

- A_h is symmetric for an arbitrary basis $\{\varphi_i\}$ because a is symmetric.
- A_h is positive definite for an arbitrary basis $\{\varphi_i\}$ because for $u = \sum_{i=1}^{M} \xi_i \varphi_i$,

$$\begin{aligned} \boldsymbol{\xi}^T A_h \boldsymbol{\xi} &= \sum_{i,j=1}^M \xi_j a(\varphi_j, \varphi_i)\xi_i = \sum_{j=1}^M \xi_j a\left(\varphi_j, \sum_{i=1}^M \xi_i\varphi_i\right) \\ &= a\left(\sum_{j=1}^M \xi_j\varphi_j, \sum_{i=1}^M \xi_i\varphi_i\right) = a(u,u) > 0 \end{aligned} \tag{2.40}$$

for $\boldsymbol{\xi} \neq 0$ and therefore $u \not\equiv 0$.
Here we have used only the positive definiteness of a.

Thus we have proven the following lemma.

Lemma 2.11 *The Galerkin method (2.26) has a unique solution if a is a symmetric, positive definite bilinear form and if ℓ is a linear form.*

In fact, as we will see in Theorem 3.1, the symmetry of a is not necessary.

- For a special basis (i.e., for a specific finite element method), A_h is a sparse matrix, i.e., only a few entries $(A_h)_{ij}$ do not vanish. Evidently,

$$(A_h)_{ij} \neq 0 \quad \Leftrightarrow \quad \int_\Omega \nabla\varphi_j \cdot \nabla\varphi_i \, dx \neq 0 \,.$$

This can happen only if $\operatorname{supp}\varphi_i \cap \operatorname{supp}\varphi_j \neq \emptyset$, as this property is again necessary for $\operatorname{supp}\nabla\varphi_i \cap \operatorname{supp}\nabla\varphi_j \neq \emptyset$ because of

$$\left(\operatorname{supp}\nabla\varphi_i \cap \operatorname{supp}\nabla\varphi_j\right) \subset \left(\operatorname{supp}\varphi_i \cap \operatorname{supp}\varphi_j\right) \,.$$

The basis function φ_i vanishes on an element that does not contain the node a_i because of the uniqueness of the solution of the local interpolation problem. Therefore,

$$\operatorname{supp}\varphi_i = \bigcup_{\substack{K\in\mathcal{T}_h \\ a_i\in K}} K \,,$$

cf. Figure 2.8, and thus

$$(A_h)_{ij} \neq 0 \quad \Rightarrow \quad a_i, a_j \in K \text{ for some } K \in \mathcal{T}_h \,; \tag{2.41}$$

i.e., a_i, a_j are *neighbouring* nodes.

If we use the piecewise linear ansatz space on triangles and if a_i is an interior node in which L elements meet, then there exist at most L nondiagonal entries in the ith row of A_h. This number is determined only by the type of the triangulation, and it is independent of the fineness h, i.e., of the number of unknowns of the system of equations.

Example 2.12 *We consider again the boundary value problem (2.1), (2.2) on $\Omega = (0, a) \times (0, b)$ again, i.e.,*

$$\begin{aligned} -\Delta u &= f \quad \text{in } \Omega \,, \\ u &= 0 \quad \text{on } \partial\Omega \,, \end{aligned}$$

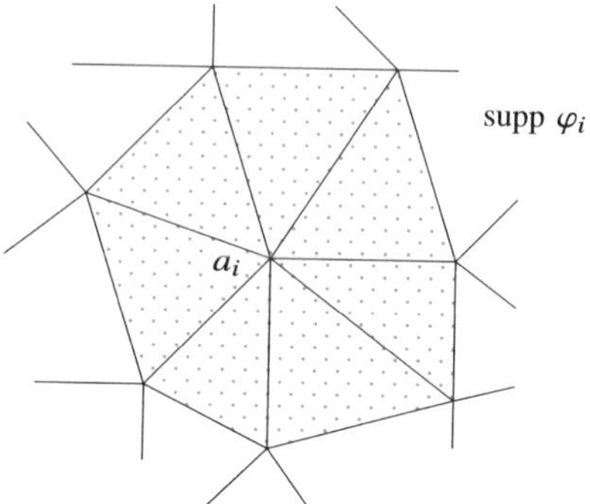

Fig. 2.8: Support of the nodal basis function.

*under the condition (1.4). The triangulation on which the method is based is created by a partition of Ω into squares with edges of length h and by a subsequent uniform division of each square into two triangles according to a fixed rule (*Friedrichs–Keller triangulation*). In order to do this, two possibilities a) and b) (see Figures 2.9 and 2.10) exist.*

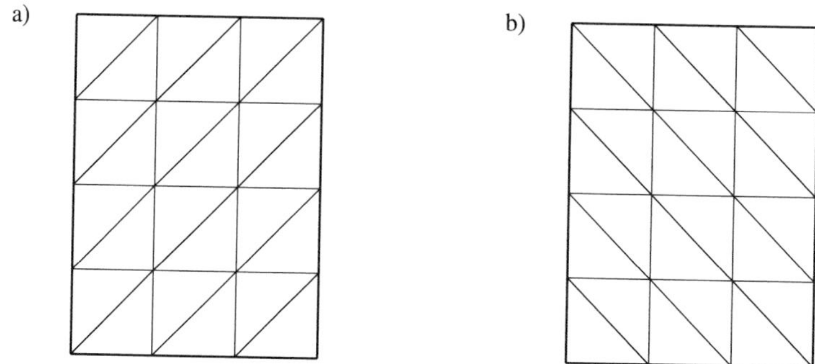

Fig. 2.9: Possibilities of Friedrichs–Keller triangulation.

In both cases, a node a_Z belongs to six elements, and consequently, it has at most six neighbours:

Case a) becomes case b) by the transformation $x \mapsto a - x, y \mapsto y$. This transformation leaves the differential equation or the weak formulation, respectively, unchanged. Thus the Galerkin method with the ansatz space V_h according to (2.30) does not change, because $\mathcal{P}_1$ is invariant with respect to the above transformation. Therefore, the discretization matrices A_h according to (2.38) are seen to be identical by taking into account the renumbering of the nodes by the transformation.

Thus it is sufficient to consider only one case, say b). A node which is far away from the boundary has 6 neighbouring nodes in $\{a_1, \ldots, a_{M_1}\}$, a node close to the boundary has less. The entries of the matrix in the row corresponding to a_Z

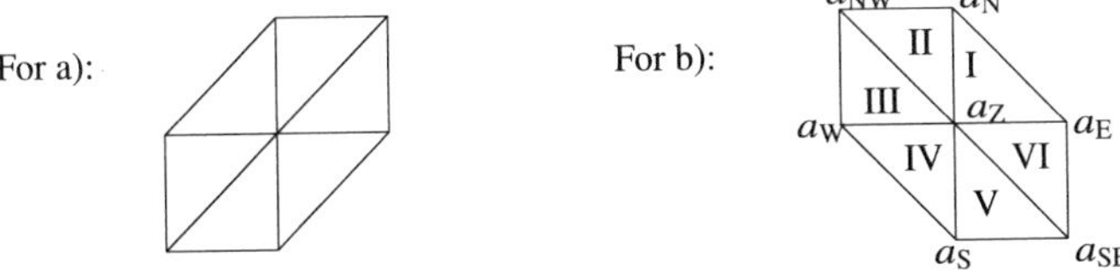

Fig. 2.10: Support of the basis function.

depend on the derivatives of the basis function φ_Z as well as on the derivatives of the basis functions corresponding to the neighbouring nodes. The values of the partial derivatives of φ_Z in elements having the common vertex a_Z are listed in Table 2.1, where these elements are numbered according to Figure 2.10.

	I	II	III	IV	V	VI
$\partial_1\varphi_Z$	$-\frac{1}{h}$	0	$\frac{1}{h}$	$\frac{1}{h}$	0	$-\frac{1}{h}$
$\partial_2\varphi_Z$	$-\frac{1}{h}$	$-\frac{1}{h}$	0	$\frac{1}{h}$	$\frac{1}{h}$	0

Table 2.1: Derivatives of the basis functions.

Thus for the entries of the matrix in the row corresponding to a_Z we have

$$(A_h)_{Z,Z} = a(\varphi_Z, \varphi_Z) = \int_{I\cup\ldots\cup VI} |\nabla\varphi_Z|^2\, dx = 2\int_{I\cup II\cup III} \left[(\partial_1\varphi_Z)^2 + (\partial_2\varphi_Z)^2 \right] dx,$$

because the integrands are equal on I and IV, on II and V, and on III and VI. Therefore

$$(A_h)_{Z,Z} = 2\int_{I\cup III} (\partial_1\varphi_Z)^2\, dx + 2\int_{I\cup II} (\partial_2\varphi_Z)^2\, dx = 2h^{-2}h^2 + 2h^{-2}h^2 = 4,$$

$$(A_h)_{Z,N} = a(\varphi_N, \varphi_Z) = \int_{I\cup II} \nabla\varphi_N \cdot \nabla\varphi_Z\, dx$$

$$= \int_{I\cup II} \partial_2\varphi_N \partial_2\varphi_Z\, dx = \int_{I\cup II} \left(-h^{-1}\right) h^{-1} dx = -1,$$

because $\partial_1\varphi_Z = 0$ on II and $\partial_1\varphi_N = 0$ on I. The element I for φ_N corresponds to the element V for φ_Z; i.e., $\partial_1\varphi_N = 0$ on I, analogously, it follows that $\partial_2\varphi_N = h^{-1}$ on I $\cup$ II. In the same way we get

$$(A_h)_{Z,E} = (A_h)_{Z,W} = (A_h)_{Z,S} = -1$$

as well as

$$(A_h)_{Z,NW} = a(\varphi_{NW}, \varphi_Z) = \int_{II \cup III} \partial_1 \varphi_{NW}\, \partial_1 \varphi_Z + \partial_2 \varphi_{NW}\, \partial_2 \varphi_Z \, dx = 0 .$$

The last identity is due to $\partial_1 \varphi_{NW} = 0$ *on* III *and* $\partial_2 \varphi_{NW} = 0$ *on* III*, because the elements V and VI for* φ_Z *agree with the elements III and II for* φ_{NW}*, respectively.*

Analogously, we obtain for the remaining value

$$(A_h)_{Z,SE} = 0 ,$$

such that only 5 (instead of the maximum 7) nonzero entries per row exist.

The way of assembling the stiffness matrix described above is called node-based *assembling. However, most of the computer programs implementing the finite element method use an* element-based *assembling, which will be considered in Section 2.4.*

If the nodes are numbered rowwise analogously to (1.13) and if the equations are divided by h^2*, then* $h^{-2} A_h$ *coincides with the discretization matrix (1.14), which is known from the finite difference method. But here the right-hand side is given by*

$$h^{-2} \left(\boldsymbol{q}_h\right)_i = h^{-2} \int_\Omega f \varphi_i \, dx = h^{-2} \int_{I \cup \ldots \cup VI} f \varphi_i \, dx$$

for $a_Z = a_i$ *and thus it is not identical to* $f(a_i)$*, the right-hand side of the finite difference method.*

However, if the trapezoidal rule, *which is exact for* $g \in \mathcal{P}_1$*, is applied to approximate the right-hand side according to*

$$\int_K g(x)\, dx \approx \frac{1}{3} \text{vol}\,(K) \sum_{i=1}^{3} g(a_i) \tag{2.42}$$

for a triangle K with the vertices a_i*,* $i = 1, 2, 3$ *and with the area* $\text{vol}\,(K)$*, then*

$$\int_I f \varphi_i \, dx \approx \frac{1}{3} \frac{1}{2} h^2 \left(f(a_Z) \cdot 1 + f(a_O) \cdot 0 + f(a_N) \cdot 0 \right) = \frac{1}{6} h^2 f(a_Z).$$

Analogous results are obtained for the other triangles, and thus

$$h^{-2} \int_{I \cup \ldots \cup VI} f \varphi_i \, dx \approx f(a_Z) .$$

In summary, we have the following result.

Lemma 2.13 *The finite element method with linear finite elements on a triangulation according to Figure 2.9 and with the trapezoidal rule to approximate the right-hand side yields the same discretization as the finite difference method from (1.7), (1.8).*

We now return to the general formulation (2.24)–(2.27). The approach of the Ritz method (2.27), instead of the Galerkin method (2.26), yields an identical approximation because of the following result.

Lemma 2.14 *If a is a symmetric and positive bilinear form and ℓ is a linear form, then the Galerkin method (2.26) and the Ritz method (2.27) have identical solutions.*

Proof Apply Lemma 2.3 with V_h instead of V. □

Hence the finite element method is the Galerkin method (and in our problem the Ritz method, too) for an *ansatz space* V_h with the following properties:

- The coefficients have a local interpretation (here as nodal values).

  The basis functions have a small support such that:

- the discretization matrix is sparse,
- the entries of the matrix can be assembled locally.

Finally, for the boundary value problem (2.1), (2.2) with the corresponding weak formulation, we consider other ansatz spaces, which to some extent do not have these properties:

1) In Section 3.2.1, (3.29), we will show that mixed boundary conditions need not be included in the ansatz space. Then we can choose the finite dimensional polynomial space $V_h = \text{span}\left\{1, x, y, xy, x^2, y^2, \ldots\right\}$ for it. But in this case, A_h is a dense matrix and ill-conditioned. Such ansatz spaces yield the *classical* Ritz–Galerkin methods.
2) Let $V_h = \text{span}\{\varphi_1, \ldots, \varphi_N\}$ and let $\varphi_i \not\equiv 0$ satisfy, for some λ_i,

$$a(\varphi_i, v) = \lambda_i \langle \varphi_i, v \rangle_0 \quad \text{for all } v \in V,$$

   i.e., the weak formulation of the eigenvalue problem

$$\begin{aligned} -\Delta u &= \lambda u \quad && \text{in } \Omega, \\ u &= 0 && \text{on } \partial\Omega, \end{aligned}$$

   for which eigenvalues $0 < \lambda_1 \le \lambda_2 \le \ldots$ and corresponding eigenfunctions φ_i exist such that $\left\langle \varphi_i, \varphi_j \right\rangle_0 = \delta_{ij}$ (e.g., see [25, Sect. 6.5, Thm. 1]). For special domains Ω, (λ_i, φ_i) can be determined explicitly, and

$$(A_h)_{ij} = a(\varphi_j, \varphi_i) = \lambda_j \left\langle \varphi_j, \varphi_i \right\rangle_0 = \lambda_j \delta_{ij}$$

   is obtained. Thus A_h is a diagonal matrix, and the system of equations $A_h \xi = \boldsymbol{q}_h$ can be solved without too great expense. But this kind of assembling is possible with acceptable costs for special cases only.
3) Specific collocation methods (on simple domains) can be interpreted as Galerkin methods on finite element spaces, but with modified (bi)linear forms. Such non-conforming FEM will be considered from Section 3.6 on.

The above examples describe Galerkin methods without having the typical properties of a finite element method.

Exercises

Problem 2.5 Consider the one-dimensional Poisson problem $-u'' = f$ on the interval $\Omega := (a, b)$, $a < b$, with the following boundary conditions:

a) $u(a) = u(b) = 0$.
b) $u'(a) = 0,\ u(b) = 0$.
c) $-u'(a) + \alpha u(a) = 0,\ u(b) = 0$, where $\alpha > 0$.

Let an equidistant partition of Ω be given as $a = x_0 < x_1 < \ldots < x_m = b$, and $x_i - x_{i-1} := h := (b-a)/m$ for $i = 1, \ldots, m$.

Calculate the stiffness matrix $A = \big(a(\varphi_j, \varphi_i)\big)_{ij}$ that results from a conforming finite element discretization with piecewise linear ansatz functions, i.e.,

$$V_h := \left\{ v \in C([a,b]) \;\middle|\; v|_{[x_{i-i}, x_i]} \in \mathcal{P}_1,\ i = 1, \ldots, m,\ v(a) = v(b) = 0 \right\}$$

defined on a nodal basis. Compare it to the corresponding matrix resulting from a classical finite difference approximation.

Problem 2.6 Consider a square domain $\Omega \subset \mathbb{R}^2$ subdivided into square elements with edge length h. A *finite difference* approximation of

$$-\Delta u = f \quad \text{in } \Omega$$

is given in the interior by the 9-point stencil

$$\frac{1}{3h^2}\begin{pmatrix} -1 & -1 & -1 \\ -1 & +8 & -1 \\ -1 & -1 & -1 \end{pmatrix}.$$

A *finite element method* on the same grid can be constructed using bilinear basis functions. On the reference element $\hat{K} := (0,1) \times (0,1)$ they are given by

$$N_1(\hat{x}, \hat{y}) := (1-\hat{x})(1-\hat{y}),\ N_2(\hat{x}, \hat{y}) := \hat{x}(1-\hat{y}),\ N_3(\hat{x}, \hat{y}) := \hat{x}\hat{y},\ N_4(\hat{x}, \hat{y}) := (1-\hat{x})\hat{y}.$$

a) Assemble the local stiffness matrix for one element.
b) Consider a patch of four elements sharing a node and assemble the row of the global system matrix corresponding to this node.
c) Approximate the right-hand side by the two-dimensional versions of

- the trapezoidal rule $\displaystyle\int_a^b f(x)dx \approx (b-a)\frac{f(a)+f(b)}{2}$,
- the midpoint rule $\displaystyle\int_a^b f(x)dx \approx (b-a) f\Big(\frac{a+b}{2}\Big)$,

which of the two resulting finite element discretizations is equivalent to above finite difference method?

Problem 2.7 Show that the following version of the trapezoidal rule on a triangle K with vertices a_i, $i = 1, 2, 3$, is exact for $g \in \mathcal{P}_1$:

$$\int_K g(x)dx \approx \frac{1}{3}\text{vol}\,(K)\sum_{i=1}^{3} g(a_i)\,.$$

For integrands $g = fq$ with $f \in C^2(K)$, $q \in \mathcal{P}_1(K)$, derive an order of convergence estimate.

Problem 2.8 a) Compute the local stiffness matrix of a finite element discretization of the Poisson equation for the triangle with vertices $a_1 := (0, 0)$, $a_2 := (3, -1)$, $a_3 := (2, 1)$, and the triangle with vertices $a_1 := (0, 0)$, $a_2 := (3, 1)$, $a_3 := (2, 1)$.

b) Estimate the asymptotic behaviour of $\|B\|_2\|B^{-1}\|_2$ for $h \to 0$ for both triangle shapes (you may use some software to perform the computations), with B according to (2.34). For the estimate of $h \to 0$ you may assume that all coordinates are multiplied by h, scaling the triangles.

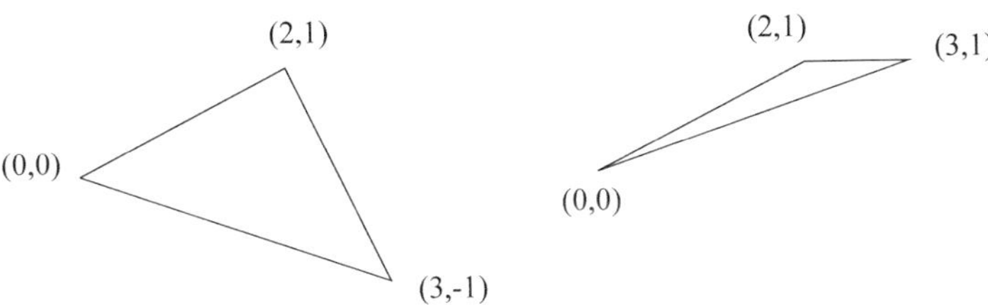

2.3 Stability and Convergence of the Finite Element Method

We consider the general case of a variational equation of the form (2.24) and the Galerkin method (2.26). Here let a be a bilinear form, which is not necessarily symmetric, and let ℓ be a linear form.

Then, if

$$e := u - u_h \ (\in V)$$

denotes the error, the important *error equation*

$$a(e, v) = 0 \quad \text{for all } v \in V_h \tag{2.43}$$

is satisfied. The reason is the *exact consistency* of the FEM in the sense

$$a(u, v) = \ell(v) \quad \text{for all } v \in V_h,$$

so to obtain (2.43) it is sufficient to consider equation (2.24) only for $v \in V_h \subset V$ and then to subtract from the result the Galerkin equation (2.26).

If, in addition, a is symmetric and positive definite, i.e.,

$$a(u,v) = a(v,u), \quad a(v,v) \geq 0, \quad a(v,v) = 0 \Leftrightarrow v = 0 \quad \text{for all } u, v \in V$$

(i.e., a is a scalar product), then the error is orthogonal to the space V_h with respect to the scalar product a.

Therefore, the relation (2.43) is often called the *orthogonality of the error (to the ansatz space)*. In general, the element $u_h \in V_h$ with minimal distance to $u \in V$ with respect to the induced norm $\|\cdot\|_a$ is characterized by (2.43):

Lemma 2.15 *Let $V_h \subset V$ be a subspace, let a be a scalar product on V, and let $\|u\|_a := a(u,u)^{1/2}$ be the norm induced by a. Then for $u_h \in V_h$, it follows that*

$$a(u - u_h, v) = 0 \quad \text{for all } v \in V_h \quad \Longleftrightarrow \tag{2.44}$$

$$\|u - u_h\|_a = \min_{v \in V_h} \|u - v\|_a \,. \tag{2.45}$$

Proof For arbitrary but fixed $u \in V$, let $\ell(v) := a(u,v)$ for $v \in V_h$. Then ℓ is a linear form on V_h, so (2.44) is a variational formulation on V_h. According to Lemma 2.14 or Lemma 2.3, this variational formulation has the same solutions as

$$F(u_h) = \min_{v \in V_h} F(v)$$

$$\text{with} \quad F(v) := \frac{1}{2}a(v,v) - \ell(v) = \frac{1}{2}a(v,v) - a(u,v)\,.$$

Furthermore, F has the same minima as the functional

$$\begin{aligned}\Big(2F(v) + a(u,u)\Big)^{1/2} &= \Big(a(v,v) - 2a(u,v) + a(u,u)\Big)^{1/2} \\ &= \Big(a(u-v, u-v)\Big)^{1/2} = \|u - v\|_a\,,\end{aligned}$$

because the additional term $a(u,u)$ is a constant. Therefore, F has the same minima as (2.45).

□

If an approximation u_h of u is to be sought exclusively in V_h, then the element u_h, determined by the Galerkin method, is the optimal choice with respect to $\|\cdot\|_a$.

A general, not necessarily symmetric, bilinear form a is assumed to satisfy the following conditions, where $\|\cdot\| := \|\cdot\|_V$ denotes a norm on V:

- a is *continuous* with respect to $\|\cdot\|$; i.e., there exists $M > 0$ such that

$$|a(u,v)| \leq M\|u\|\|v\| \quad \text{for all } u, v \in V; \tag{2.46}$$

- a is V-*coercive*; i.e., there exists $\alpha > 0$ such that

$$a(v,v) \geq \alpha\|v\|^2 \quad \text{for all } v \in V. \tag{2.47}$$

If a is a scalar product, then (2.46) with $M = 1$ and (2.47) (as equality) with $\alpha = 1$ are valid for the induced norm $\|\cdot\| := \|\cdot\|_a$ due to the Cauchy–Schwarz inequality.

The V-coercivity is an essential condition for the unique existence of a solution of the variational equation (2.24) and of the boundary value problem described by it, which will be presented in more detail in Sections 3.1 and 3.2. It also implies—without further conditions—the stability of the Galerkin approximation.

Lemma 2.16 *The Galerkin solution u_h according to (2.26) is* stable *in the following sense:*

$$\|u_h\| \leq \frac{1}{\alpha}\|\ell\| \quad \textit{independently of } h\,, \tag{2.48}$$

where

$$\|\ell\| := \sup_{v \in V\setminus\{0\}} \frac{|\ell(v)|}{\|v\|}\,.$$

Proof In the case $u_h = 0$, there is nothing to prove. Otherwise, from $a(u_h, v) = \ell(v)$ for all $v \in V_h$, it follows that

$$\alpha\|u_h\|^2 \leq a(u_h, u_h) = \ell(u_h) \leq \frac{|\ell(u_h)|}{\|u_h\|}\|u_h\| \leq \|\ell\|\,\|u_h\|.$$

Dividing this relation by $\alpha\|u_h\|$, we get the assertion. □

Moreover, the approximation property (2.45) holds up to a constant:

Theorem 2.17 (Céa's lemma) *Assume (2.46), (2.47). Then the following error estimate for the Galerkin solution holds:*

$$\|u - u_h\| \leq \frac{M}{\alpha}\min_{v \in V_h}\|u - v\|\,. \tag{2.49}$$

Proof If $\|u - u_h\| = 0$, then there is nothing to prove. Otherwise, let $v \in V_h$ be arbitrary. Because of the error equation (2.43) and $u_h - v \in V_h$,

$$a(u - u_h, u_h - v) = 0\,.$$

Therefore, using (2.47) we have

$$\begin{aligned}\alpha\|u - u_h\|^2 &\leq a(u - u_h, u - u_h) = a(u - u_h, u - v) - a(u - u_h, u_h - v)\\ &= a(u - u_h, u - v)\,.\end{aligned}$$

Furthermore, by means of (2.46) we obtain

$$\alpha\|u - u_h\|^2 \leq a(u - u_h, u - v) \leq M\|u - u_h\|\,\|u - v\| \text{ for arbitrary } v \in V_h\,.$$

Thus the assertion follows by division by $\alpha\|u - u_h\|$. □

Therefore also in general, in order to get an asymptotic error estimate in h, it is sufficient to estimate the *best approximation error* of V_h, i.e.,

$$\min_{v \in V_h} \|u - v\| .$$

An approximation u_h fulfilling

$$\|u - u_h\| \leq C \min_{v \in V_h} \|u - v\|$$

with a constant $C > 0$ independent of h, is called *quasi-optimal*. If

$$Q_h : V \to V_h, \quad u \mapsto u_h$$

then Q_h is also called quasi-optimal.

However, this consideration is meaningful only in those cases where M/α is not too large. Section 3.2 shows that this condition is no longer satisfied for convection-dominated problems. Therefore, the Galerkin approach has to be modified, which will be described in Chapter 10.

We want to apply the theory developed up to now to the weak formulation of the boundary value problem (2.1), (2.2) with V according to (2.7) or (2.22) and V_h according to (2.30). According to (2.4) the bilinear form a and the linear form ℓ read as

$$a(u, v) := \int_\Omega \nabla u \cdot \nabla v \, dx , \quad \ell(v) := \int_\Omega f v \, dx .$$

In order to guarantee that the linear form ℓ is well defined on V, it is sufficient to assume that the right-hand side f of the boundary value problem belongs to $L^2(\Omega)$.

Since a is a scalar product on V,

$$\|v\| = \|v\|_a = \left(\int_\Omega |\nabla v|^2 \, dx \right)^{1/2}$$

is an appropriate norm. Alternatively, the norm introduced in (2.21) for $V = H_0^1(\Omega)$ can be taken as

$$\|v\|_1 = \left(\int_\Omega |v(x)|^2 \, dx + \int_\Omega |\nabla v(x)|^2 \, dx \right)^{1/2} .$$

In the latter case, the question arises whether the conditions (2.46) and (2.47) are still satisfied. Indeed,

$$|a(u, v)| \leq \|u\|_a \|v\|_a \leq \|u\|_1 \|v\|_1 \quad \text{for all } u, v \in V .$$

The first inequality follows from the Cauchy–Schwarz inequality for the scalar product a, and the second inequality follows from the trivial estimate

$$\|v\|_a = \left(\int_\Omega |\nabla v(x)|^2 \, dx \right)^{1/2} \leq \|v\|_1 \quad \text{for all } v \in V .$$

Thus a is continuous with respect to $\| \cdot \|_1$ with $M = 1$.

The V-coercivity of a, i.e., the property

$$a(v,v) = \|v\|_a^2 \geq \alpha \|v\|_1^2 \quad \text{for some } \alpha > 0 \text{ and all } v \in V\,,$$

is not valid in general for $V = H^1(\Omega)$. However, in the present situation of $V = H_0^1(\Omega)$ it is valid because of the incorporation of the boundary condition into the definition of V :

Theorem 2.18 (Poincaré) *Let $\Omega \subset \mathbb{R}^n$ be open and bounded. Then a constant $C > 0$ exists (depending on Ω) such that*

$$\|v\|_0 \leq C \left(\int_\Omega |\nabla v(x)|^2 \, dx \right)^{1/2} \quad \textit{for all } v \in H_0^1(\Omega)\,.$$

Proof Cf. [28, Sect. 7.8]. For a special case, see Problem 2.10. □

Thus (2.47) is satisfied, for instance, with

$$\alpha := \frac{1}{1 + C^2}\,,$$

(see also (3.27) below) and thus in particular

$$\alpha \|v\|_1^2 \leq a(v,v) = \|v\|_a^2 \leq \|v\|_1^2 \quad \text{for all } v \in V\,, \tag{2.50}$$

i.e., the norms $\|\cdot\|_1$ and $\|\cdot\|_a$ are equivalent on $V = H_0^1(\Omega)$ and therefore they generate the same convergence concept:

$$v_h \longrightarrow v \text{ with respect to } \|\cdot\|_1 \Leftrightarrow \|v_h - v\|_1 \longrightarrow 0$$
$$\Leftrightarrow \|v_h - v\|_a \longrightarrow 0 \Leftrightarrow v_h \longrightarrow v \text{ with respect to } \|\cdot\|_a\,.$$

In summary the estimate (2.49) holds for $\|\cdot\| = \|\cdot\|_1$ with the constant $1/\alpha$.

Because of the Cauchy–Schwarz inequality for the scalar product on $L^2(\Omega)$ and

$$\ell(v) = \int_\Omega f(x) v(x) \, dx\,,$$

i.e., $|\ell(v)| \leq \|f\|_0 \, \|v\|_0 \leq \|f\|_0 \, \|v\|_1$, and thus $\|\ell\| \leq \|f\|_0$, the stability estimate (2.48) for a right-hand side $f \in L^2(\Omega)$ takes the particular form

$$\|u_h\|_1 \leq \frac{1}{\alpha} \|f\|_0\,.$$

Up to now, our considerations have been independent of the special form of V_h. Now we make use of the choice of V_h according to (2.30). In order to obtain an estimate of the approximation error of V_h, it is sufficient to estimate the term $\|u - \overline{v}\|$ for some special element $\overline{v} \in V_h$. For this element $\overline{v} \in V_h$, we choose the interpolant $I_h(u)$,

where

$$
\begin{aligned}
I_h : \Big\{u \in C(\overline{\Omega}) \;\Big|\; u = 0 \text{ on } \partial\Omega\Big\} &\to V_h\,, \\
u \mapsto I_h(u) \text{ with } I_h(u)(a_i) &:= u(a_i)\,.
\end{aligned}
\tag{2.51}
$$

This interpolant exists and is unique (Lemma 2.10). Obviously,

$$\min_{v \in V_h} \|u - v\|_1 \le \|u - I_h(u)\|_1 \quad \text{for } u \in C(\overline{\Omega}) \text{ and } u = 0 \text{ on } \partial\Omega\,.$$

If the weak solution u possesses weak derivatives of second order, then for certain sufficiently fine triangulations $\mathcal{T}_h$, i.e., $0 < h \le \overline{h}$ for some $\overline{h} > 0$, an estimate of the type

$$\|u - I_h(u)\|_1 \le Ch \tag{2.52}$$

holds, where C depends on u but is independent of h (cf. (3.101)). The proof of this estimate will be explained in Section 3.4, where also sufficient conditions on the family of triangulations $(\mathcal{T}_h)_h$ will be specified.

Exercises

Problem 2.9 Let $a(u, v) := \int_0^1 x^2 u' v' dx$ for arbitrary $u, v \in H_0^1(0, 1)$.

a) Show that there is no constant $C_1 > 0$ such that the inequality

$$a(v, v) \ge C_1 \int_0^1 (v')^2\, dx \quad \text{for all } v \in H_0^1(0, 1)$$

is valid.

b) Now let $\mathcal{T}_h := \{(x_{i-1}, x_i)\}_{i=1}^N$, $N \in \mathbb{N}$, be an equidistant partition of $(0, 1)$ with the parameter $h = 1/N$ and $V_h := \text{span}\,\{\varphi_i\}_{i=1}^{N-1}$, where

$$\varphi_i(x) := \begin{cases} (x - x_{i-1})/h_{i-1} & \text{in } (x_{i-1}, x_i]\,, \\ (x_{i+1} - x)/h_i & \text{in } (x_i, x_{i+1})\,, \\ 0 & \text{otherwise}\,. \end{cases}$$

Does there exist a constant $C_2 > 0$ with

$$a(v_h, v_h) \ge C_2 \int_0^1 \left(v_h'\right)^2 dx \quad \text{for all } v_h \in V_h\,?$$

Problem 2.10

a) For $\Omega := (\alpha, \beta) \times (\gamma, \delta)$ and V according to (2.7), prove the *inequality of Poincaré:* There exists a positive constant C with

$$\|v\|_0 \leq C\|v\|_a \quad \text{for all } v \in V\,.$$

Hint: Start with the relation $v(x, y) = \int_\alpha^x \partial_x v(s, y)\, ds$.

b) For $\Omega := (\alpha, \beta)$ and $v \in C([\alpha, \beta])$ with a piecewise continuous derivative v' and $v(\gamma) = 0$ for some $\gamma \in [\alpha, \beta]$, show that

$$\|v\|_0 \leq (\beta - \alpha)\|v'\|_0\,.$$

Problem 2.11 Let $\Omega := (0, 1)\times(0, 1)$. Given $f \in C(\overline{\Omega})$, discretize the boundary value problem $-\Delta u = f$ in Ω, $u = 0$ on $\partial\Omega$, by means of the usual five-point difference stencil as well as by means of the finite element method with linear elements. A quadratic grid as well as the corresponding Friedrichs–Keller triangulation will be used.

Prove the following stability estimates for the matrix of the linear system of equations:

$$\text{a) } \|A_h^{-1}\|_\infty \leq \frac{1}{8}, \quad \text{b) } \|A_h^{-1}\|_2 \leq \frac{1}{16}, \quad \text{c) } \|A_h^{-1}\|_0 \leq 1,$$

where $\|\cdot\|_\infty$, $\|\cdot\|_2$ denote the maximum row sum norm and the spectral norm of a matrix, respectively, and

$$\|A_h^{-1}\|_0 := \sup_{v_h \in V_h\setminus\{0\}} \frac{\|v_h\|_0^2}{\|v_h\|_a^2} \quad \text{with} \quad \|v_h\|_a^2 := \int_\Omega |\nabla v_h|^2\, dx\,.$$

Comment: The constant in c) is not optimal.

Problem 2.12 Let Ω be a domain with polygonal boundary and let $\mathcal{T}_h$ be a consistent triangulation of Ω. The nodes a_i of the triangulation are enumerated from 1 to M.

Let the triangulation satisfy the following assumption: There exist constants $C_1, C_2 > 0$ such that for all triangles $K \in \mathcal{T}_h$ the relation

$$C_1 h^2 \leq \operatorname{vol}(K) \leq C_2 h^2$$

is satisfied. h denotes the maximum of the diameters of all elements of $\mathcal{T}_h$.

a) Show the equivalence of the following norms for $u_h \in V_h$ in the space V_h of continuous, piecewise linear functions over Ω:

$$\|u_h\|_0 := \left(\int_\Omega |u_h|^2\, dx\right)^{1/2}, \quad \|u_h\|_{0,h} := h\left(\sum_{i=1}^M u_h^2(a_i)\right)^{1/2}.$$

b) Consider the special case $\Omega := (0, 1)\times(0, 1)$ with the Friedrichs–Keller triangulation as well as the subspace $V_h \cap H_0^1(\Omega)$ and find "as good as possible" constants in the corresponding equivalence estimate.

2.4 The Implementation of the Finite Element Method: Part 1

In this section we will consider some aspects of the implementation of the finite element method using linear ansatz functions on triangles for the model boundary value problem (1.1), (1.2) on a polygonally bounded domain $\Omega \subset \mathbb{R}^2$. The case of inhomogeneous Dirichlet boundary conditions will be treated also to a certain extent as far as it is possible up to now.

2.4.1 Preprocessor

The main task of the preprocessor is to create a partition of the domain Ω.

An input file might have the following format:

Let the number of variables (including also the boundary nodes for Dirichlet boundary conditions) be M. We generate the following list (*nodal coordinate matrix*):

$$\begin{array}{cc} x\text{-coordinate of node } 1 & y\text{-coordinate of node } 1 \\ \vdots & \vdots \\ x\text{-coordinate of node } M & y\text{-coordinate of node } M \end{array}$$

Let the number of (triangular) elements be N. These elements will be listed in the *element-node table* or *element connectivity matrix*. Here, every element is characterized by the indices of the nodes corresponding to this element in a well-defined order (e.g., counterclockwise); cf. Figure 2.11.

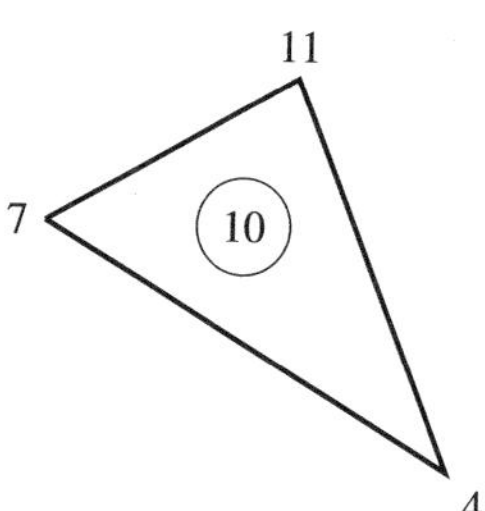

Fig. 2.11: Element no. 10 with nodes nos. 4, 11, 7.

For example, the 10th row of the element-node table contains the entry

$$4 \qquad 11 \qquad 7\,.$$

Usually, a partition is generated by a grid generation algorithm that performs the following steps: 1. Geometrical description of the domain; often this is a CAD output. 2. Boundary grid generation. Some grid generation algorithms require, as an initial step, a partition of the boundary $\partial\Omega$ of the domain. 3. Domain grid generation, i.e., the generation of a partition $\mathcal{T}_h$ of Ω. A short overview on methods for the grid generation will be given in Section 4.1. One of the simplest versions of a grid generation algorithm has the following structure (cf. Figure 2.12):

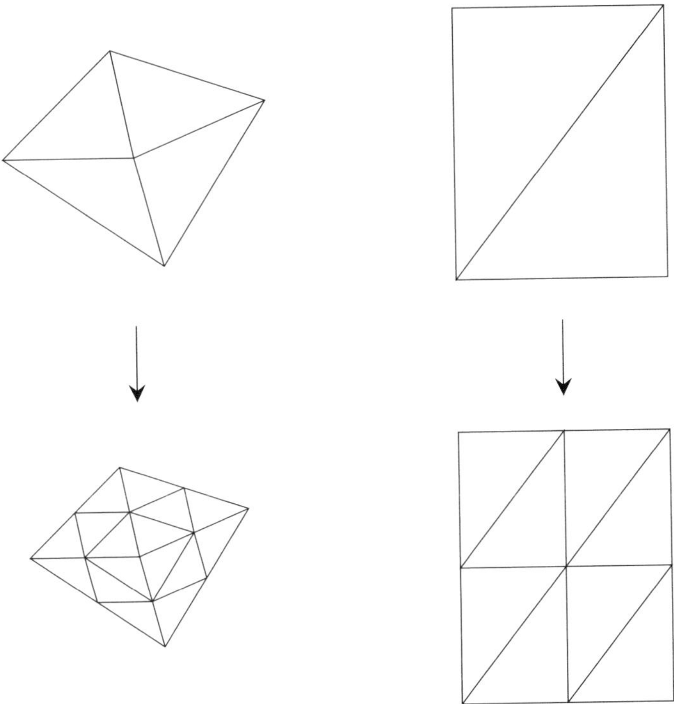

Fig. 2.12: Refinement by quartering.

Prescribe a coarse triangulation (according to the above format) and refine this triangulation (repeatedly) by subdividing a triangle into 4 congruent triangles by connecting the midpoints of the edges with straight lines.

If this uniform refinement (quadrisection) is done globally, i.e., for all triangles of the coarse grid, then triangles are created that have the same interior angles as the elements of the coarse triangulation. Thus the quality of the triangulation, indicated, for example, by the ratios of the diameters of an element and of its inscribed circle (see Definition 3.31), does not change. However, if the subdivision is performed only locally, the resulting triangulation is no longer consistent, in general. Such an inconsistent triangulation can be corrected by bisection of the corresponding neighbouring (unrefined) triangles. But this implies that some of the interior angles are bisected and consequently, the quality of the triangulation becomes poorer if the bisection step is performed too frequently. The following algorithm, which was implemented in the PLTMG package already at the end of the 1970s [91], circumvents the described problem (although now (after version 8.0) it has been replaced by the so-called longest edge bisection (see Subsection 4.1.5 later) in conjunction with a modified data structure [5]).

A Possible Refinement Algorithm

Let a (uniform) triangulation $\mathcal{T}$ be given (e.g., by repeated uniform refinement of a coarse triangulation). The edges of this triangulation are called *red edges*.

1. Subdivide the edges according to a certain local refinement criterion (introduction of new nodes) by successive bisection (cf. Figure 2.13).
2. If a triangle $K \in \mathcal{T}$ has on its edges in addition to the vertices two or more nodes, then subdivide K into four congruent triangles.
   Iterate over step 2 (cf. Figure 2.14).
3. Subdivide the triangles with nodes at the midpoints of the edges into 2 triangles by bisection. This step introduces the so-called *green edges*.
4. If the refinement is to be continued, first remove the green edges.

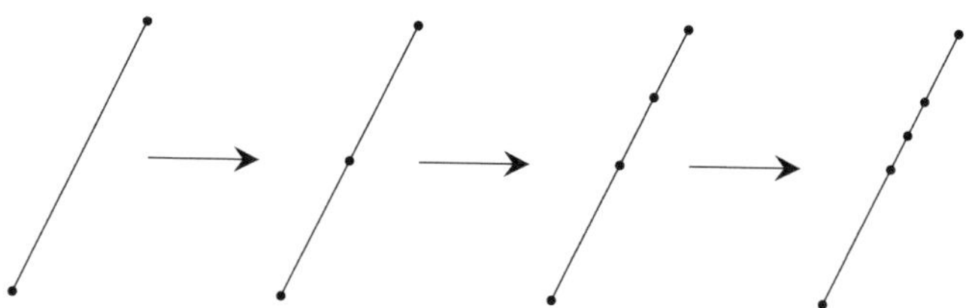

Fig. 2.13: New nodes on edges.

2.4.2 Assembling

Denote by $\varphi_1, \dots, \varphi_M$ the global basis functions. Then the stiffness matrix A_h has the following entries:

$$(A_h)_{ij} = \int_\Omega \nabla\varphi_j \cdot \nabla\varphi_i \, dx = \sum_{m=1}^{N} A_{ij}^{(m)}$$

with

$$A_{ij}^{(m)} = \int_{K_m} \nabla\varphi_j \cdot \nabla\varphi_i \, dx \, .$$

Let $a_1, \dots, a_M$ denote the nodes of the triangulation. Because of the implication

$$A_{ij}^{(m)} \neq 0 \ \Rightarrow \ a_i, a_j \in K_m$$

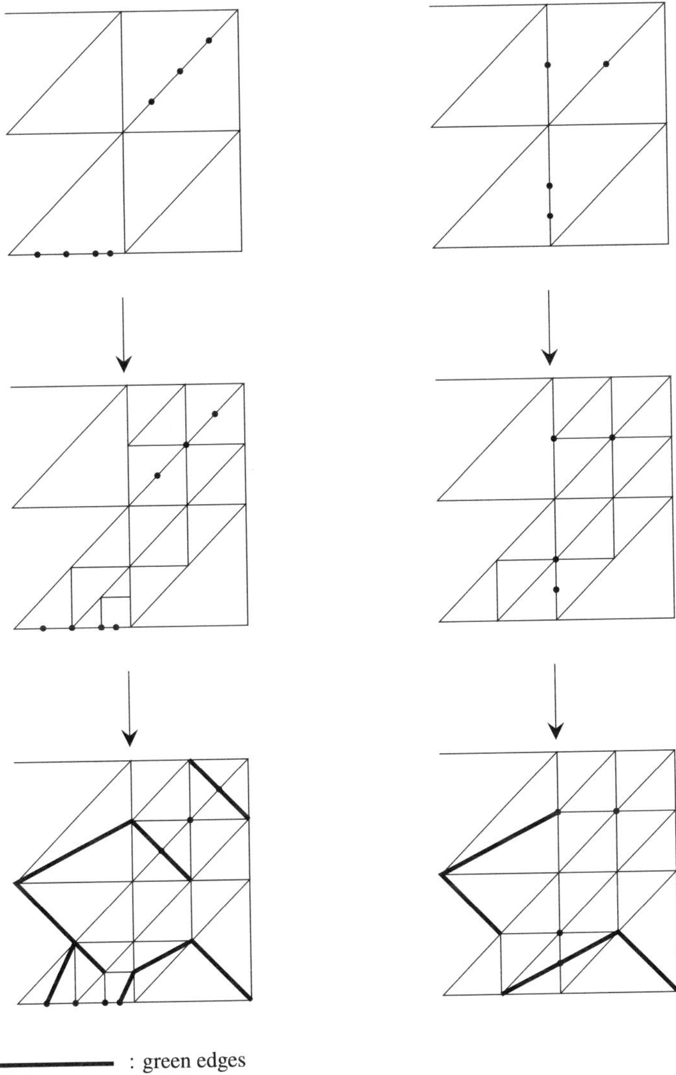

Fig. 2.14: Two refinement sequences.

(cf. (2.41)), the element K_m yields nonzero contributions for $A_{ij}^{(m)}$ only if $a_i, a_j \in K_m$ at best. Such nonzero contributions are called *element entries* of A_h. They add up to the *entries* of A_h.

In Example 2.12 we explained a node-based assembling of the stiffness matrix. In contrast to this and on the basis of the above observations, in the following we will perform an *element-based assembling* of the stiffness matrix.

To assemble the entries of $A^{(m)}$, we will start from a local numbering (cf. Figure 2.15) of the nodes by assigning the local numbers 1, 2, 3 to the global node numbers r_1, r_2, r_3 (numbered counterclockwise). In contrast to the usual notation adopted in this book, here indices of vectors according to the local numbering are included in parentheses and written as superscripts.

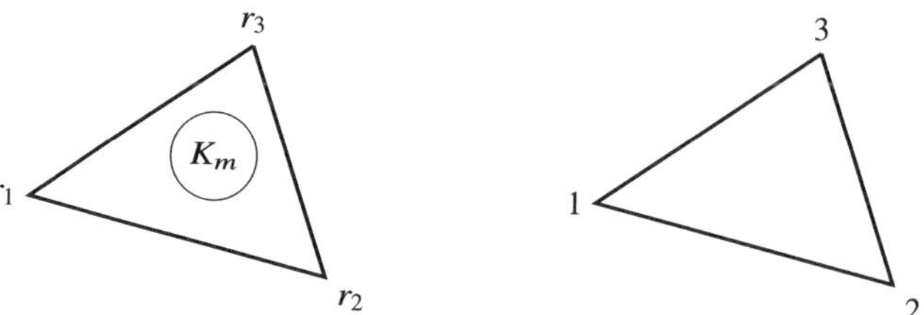

Fig. 2.15: Global (left) and local numbering.

Thus in fact, we generate

$$\left(A^{(m)}_{r_i r_j}\right)_{i,j=1,2,3} \quad \text{as} \quad \left(\tilde{A}^{(m)}_{ij}\right)_{i,j=1,2,3} .$$

To do this, we first perform a transformation of K_m onto some reference element and then we evaluate the integral on this element exactly.

Hence the entry of the *element stiffness matrix* reads as

$$\tilde{A}^{(m)}_{ij} = \int_{K_m} \nabla \varphi_{r_j} \cdot \nabla \varphi_{r_i} \, dx .$$

The reference element $\hat{K}$ is transformed onto the global element K_m by means of the relation $F(\hat{x}) = B\hat{x} + d$, therefore

$$D_{\hat{x}} u(F(\hat{x})) = D_x u(F(\hat{x}))\, D_{\hat{x}} F(\hat{x}) = D_x u(F(\hat{x}))\, B ,$$

where $D_x u$ denotes the row vector $(\partial_1 u, \partial_2 u)$, i.e., the corresponding differential operator. Using the more standard notation in terms of gradients and taking into consideration the relation $B^{-T} := (B^{-1})^T$, we obtain

$$\nabla_x u\,(F(\hat{x})) = B^{-T} \nabla_{\hat{x}}\,(u\,(F(\hat{x}))) \tag{2.53}$$

and thus

$$\begin{aligned} \tilde{A}^{(m)}_{ij} &= \int_{\hat{K}} \nabla_x \varphi_{r_j}\,(F(\hat{x})) \cdot \nabla_x \varphi_{r_i}\,(F(\hat{x}))\,|\det(DF(\hat{x}))|\; d\hat{x} \\ &= \int_{\hat{K}} B^{-T} \nabla_{\hat{x}} \left(\varphi_{r_j}\,(F(\hat{x}))\right) \cdot B^{-T} \nabla_{\hat{x}} \left(\varphi_{r_i}\,(F(\hat{x}))\right) |\det(B)|\; d\hat{x} \\ &= \int_{\hat{K}} B^{-T} \nabla_{\hat{x}} \hat{\varphi}_{r_j}(\hat{x}) \cdot B^{-T} \nabla_{\hat{x}} \hat{\varphi}_{r_i}(\hat{x})\,|\det(B)|\; d\hat{x} \\ &= \int_{\hat{K}} B^{-T} \nabla_{\hat{x}} N_j(\hat{x}) \cdot B^{-T} \nabla_{\hat{x}} N_i(\hat{x})\,|\det(B)|\; d\hat{x} , \end{aligned} \tag{2.54}$$

where the transformed basis functions $\hat{\varphi}_{r_i}$, $\hat{\varphi}(\hat{x}) := \varphi(F(\hat{x}))$ coincide with the local basis functions on $\hat{K}$, i.e., with the shape functions N_i:

$$\hat{\varphi}_{r_i}(\hat{x}) = N_i(\hat{x}) \quad \text{for } \hat{x} \in \hat{K} \,.$$

The shape functions N_i have been defined in (2.33) (where (x, y) there must be replaced by $(\hat{x}_1, \hat{x}_2)$ here) for the standard reference element defined there.

Introducing the matrix $C := \left(B^{-1}\right)\left(B^{-1}\right)^T = \left(B^T B\right)^{-1}$, we can write

$$\tilde{A}_{ij}^{(m)} = \int_{\hat{K}} C\, \nabla_{\hat{x}} N_j(\hat{x}) \cdot \nabla_{\hat{x}} N_i(\hat{x})\, |\det(B)|\, d\hat{x} \,. \tag{2.55}$$

Denoting the matrix B by $B = \left(b^{(1)}, b^{(2)}\right)$, then it follows that

$$C = \begin{pmatrix} b^{(1)} \cdot b^{(1)} & b^{(1)} \cdot b^{(2)} \\ b^{(1)} \cdot b^{(2)} & b^{(2)} \cdot b^{(2)} \end{pmatrix}^{-1} = \frac{1}{\det(B)^2} \begin{pmatrix} b^{(2)} \cdot b^{(2)} & -b^{(1)} \cdot b^{(2)} \\ -b^{(1)} \cdot b^{(2)} & b^{(1)} \cdot b^{(1)} \end{pmatrix}$$

because $\det(B^T B) = \det(B)^2$. The previous considerations can be easily extended to the computation of the stiffness matrices of more general differential operators like

$$\int_{\Omega} K(x) \nabla \varphi_j(x) \cdot \nabla \varphi_i(x)\, dx$$

(cf. Section 3.5). For the standard reference element, which we use from now on, we have $b^{(1)} = a^{(2)} - a^{(1)}$, $b^{(2)} = a^{(3)} - a^{(1)}$. Here $a^{(i)}$, $i = 1, 2, 3$, are the locally numbered nodes of K interpreted as vectors of $\mathbb{R}^2$.

From now on we make also use of the special form of the stiffness matrix and obtain

$$\begin{aligned} \tilde{A}_{ij}^{(m)} = \gamma_1 & \int_{\hat{K}} \partial_{\hat{x}_1} N_j\, \partial_{\hat{x}_1} N_i\, d\hat{x} \\ & + \gamma_2 \int_{\hat{K}} \partial_{\hat{x}_1} N_j\, \partial_{\hat{x}_2} N_i + \partial_{\hat{x}_2} N_j\, \partial_{\hat{x}_1} N_i\, d\hat{x} \\ & + \gamma_3 \int_{\hat{K}} \partial_{\hat{x}_2} N_j\, \partial_{\hat{x}_2} N_i\, d\hat{x} \end{aligned} \tag{2.56}$$

with

$$\begin{aligned} \gamma_1 &:= c_{11} |\det(B)| = \frac{1}{|\det(B)|} \left(a^{(3)} - a^{(1)}\right) \cdot \left(a^{(3)} - a^{(1)}\right), \\ \gamma_2 &:= c_{12} |\det(B)| = -\frac{1}{|\det(B)|} \left(a^{(2)} - a^{(1)}\right) \cdot \left(a^{(3)} - a^{(1)}\right), \\ \gamma_3 &:= c_{22} |\det(B)| = \frac{1}{|\det(B)|} \left(a^{(2)} - a^{(1)}\right) \cdot \left(a^{(2)} - a^{(1)}\right). \end{aligned}$$

In the implementation it is advisable to compute the values γ_i just once from the local geometrical information given in the form of the vertices $a^{(i)} = a_{r_i}$ and to store them permanently.

Thus we obtain for the local stiffness matrix

$$\tilde{A}^{(m)} = \gamma_1 S_1 + \gamma_2 S_2 + \gamma_3 S_3 \tag{2.57}$$

with

$$S_1 := \left(\int_{\hat{K}} \partial_{\hat{x}_1} N_j \partial_{\hat{x}_1} N_i \, d\hat{x} \right)_{ij} ,$$

$$S_2 := \left(\int_{\hat{K}} \partial_{\hat{x}_1} N_j \partial_{\hat{x}_2} N_i + \partial_{\hat{x}_2} N_j \partial_{\hat{x}_1} N_i \, d\hat{x} \right)_{ij} ,$$

$$S_3 := \left(\int_{\hat{K}} \partial_{\hat{x}_2} N_j \partial_{\hat{x}_2} N_i \, d\hat{x} \right)_{ij} .$$

An explicit computation of the matrices S_i is possible because the integrands are constant, and also these matrices can be stored permanently:

$$S_1 = \frac{1}{2} \begin{pmatrix} 1 & -1 & 0 \\ -1 & 1 & 0 \\ 0 & 0 & 0 \end{pmatrix}, \; S_2 = \frac{1}{2} \begin{pmatrix} 2 & -1 & -1 \\ -1 & 0 & 1 \\ -1 & 1 & 0 \end{pmatrix}, \; S_3 = \frac{1}{2} \begin{pmatrix} 1 & 0 & -1 \\ 0 & 0 & 0 \\ -1 & 0 & 1 \end{pmatrix}.$$

The right-hand side $(\boldsymbol{q}_h)_i = \int_\Omega f(x)\varphi_i(x)\,dx$ can be treated in a similar manner:

$$(\boldsymbol{q}_h)_i = \sum_{m=1}^{N} \left(\boldsymbol{q}^{(m)}\right)_i$$

with

$$\left(\boldsymbol{q}^{(m)}\right)_i = \int_{K_m} f(x)\varphi_i(x)\,dx \quad (\neq 0 \implies a_i \in K_m) .$$

Again, we transform the global numbering $\left(q_{r_i}^{(m)}\right)_{i=1,2,3}$ for the triangle $K_m = \operatorname{conv}\left\{a_{r_1}, a_{r_2}, a_{r_3}\right\}$ into the local numbering $\left(\tilde{q}_i^{(m)}\right)_{i=1,2,3}$. Analogously to the determination of the entries of the stiffness matrix, we have

$$\begin{aligned} \tilde{q}_i^{(m)} &= \int_{\hat{K}} f\left(F(\hat{x})\right) \varphi_{r_i}\left(F(\hat{x})\right) |\det(B)| \, d\hat{x} \\ &= \int_{\hat{K}} \hat{f}(\hat{x})\, N_i(\hat{x})\, |\det(B)| \, d\hat{x} , \end{aligned}$$

where $\hat{f}(\hat{x}) := f(F(\hat{x}))$ for $\hat{x} \in \hat{K}$.

In general, this integral cannot be evaluated exactly. Therefore, it has to be approximated by a quadrature rule.

A *quadrature rule* for $\int_{\hat{K}} g(\hat{x})\,d\hat{x}$ is of the type

$$\sum_{k=1}^{R} \omega_k \, g\big(\hat{\ell}^{(k)}\big)$$

with certain *weights* ω_k and *quadrature points* $\hat{\ell}^{(k)}$. As an example, we take the *trapezoidal rule* (cf. (2.42)), where

$$\hat{\ell}^{(1)} = \hat{a}_1 = (0,0)\,, \quad \hat{\ell}^{(2)} = \hat{a}_2 = (1,0)\,, \quad \hat{\ell}^{(3)} = \hat{a}_3 = (0,1)\,,$$

$$\omega_k = \frac{1}{6}\,, \quad k = 1,2,3\,.$$

Thus for arbitrary but fixed quadrature rules, we have

$$\tilde{q}_i^{(m)} \approx \sum_{k=1}^{R} \omega_k \, \hat{f}\big(\hat{\ell}^{(k)}\big) \, N_i\big(\hat{\ell}^{(k)}\big) \, |\det(B)|\,. \tag{2.58}$$

Of course, the application of different quadrature rules on different elements is possible, too. The values $N_i\big(\hat{\ell}^{(k)}\big)$, $i = 1,2,3$, $k = 1,\dots,R$, should be evaluated just once and should be stored. The discussion on the use of quadrature rules will be continued in Sections 3.5.2 and 3.6.

In summary, the following algorithm provides the assembling of the stiffness matrix and the right-hand side:

Loop over all elements $m = 1,\dots,N$:

1. Allocating a local numbering to the nodes based on the element-node table: $1 \mapsto r_1$, $2 \mapsto r_2$, $3 \mapsto r_3$.
2. Assembling of the element stiffness matrix $\tilde{A}^{(m)}$ according to (2.55) or (2.57). Assembling of the right-hand side according to (2.58).
3. Loop over $i, j = 1,2,3$:

$$\begin{aligned} (A_h)_{r_i r_j} &:= (A_h)_{r_i r_j} + \tilde{A}_{ij}^{(m)}\,, \\ (q_h)_{r_i} &:= (q_h)_{r_i} + \tilde{q}_i^{(m)}\,. \end{aligned}$$

For the sake of efficiency of this algorithm, it is necessary to adjust the memory structure to the particular situation; we will see how this can be done in Section 2.5.

2.4.3 Realization of Dirichlet Boundary Conditions: Part 1

Nodes where a Dirichlet boundary condition is prescribed must be labelled specially, here, for instance, by the convention $M = M_1 + M_2$, where the nodes numbered from $M_1 + 1$ to M correspond to the Dirichlet boundary nodes. In more general cases, other realizations are to be preferred.

In the first step of assembling of stiffness matrix and the load vector, the Dirichlet nodes are treated like all the other ones. After this, the Dirichlet nodes are considered separately. If such a node has the number j, the boundary condition is included by the following procedure:

Replace the jth row of A_h by the jth unit vector and $(q_h)_j$ by $g(a_j)$, if $u(x) = g(x)$ is prescribed for $x \in \partial\Omega$. To conserve the symmetry of the matrix A_h, modify the right-hand side $(q_h)_i$ for $i \neq j$ to $(q_h)_i - (A_h)_{ij} g(a_j)$, and after that replace the jth column of A_h by the jth unit vector. In other words, the contributions caused by the Dirichlet boundary condition are included into the right-hand side. This is exactly the elimination that led to the form (1.10), (1.11) in Chapter 1.

2.4.4 Notes on Software

Based on the above information, it is possible to develop a (very basic) finite element code on one's own. This is the way we suggest learning finite element programming (and later finite volume programming) and continuing this through various programming projects that will follow. As indicated in the introduction, this is in principle independent of the IDE chosen. It may be tedious or restricting to develop appropriate tools for geometric partitioning (in particular, in three space dimensions) or for the efficient solution of the systems of linear algebraic equations (for specific algorithms see Section 2.5 and Chapter 5). Here the chosen IDE may already provide tools (e.g., the backslash operator in GNU Octave/MATLAB or NumPy [147], a core package of SciPy [224], using Python).

If the solver of the systems of linear algebraic equations is the bottleneck, the usage of one of the many linear algebra packages my be advisable (if not included in the used IDE). For an overview see [227]. If the focus is more on the scientific/technological application, and this is complex, it may be advisable to use a toolbox dedicated to the (finite element) approximation of the respective class of PDE problems. Partitioning tools are typically also provided here, and various element types are available.

Most open-source simulation tools for approximating the solutions of PDEs have been implemented in C++ with the most prominent examples currently being deal.ii [82], DUNE [96], FEniCS [74], MFEM [75]. However, there are significant differences between the several packages.

While deal.II describes itself as a state-of-the-art finite element library focused on generality, dimension-independent programming, parallelism, and extensibility, its overall strength lies in a thorough documentation (including written and video tutorials) and a diverse worldwide community contributing to the library. The latter aspects helping users to get along with the library quite easily. As opposed to this, deal.ii is a very large software package and, therefore, not a lightweight package. The sheer amount of aspects and components of the software easily overwhelms users that are not used to work with large software packages.

DUNE takes a different approach: it maintains a clear structure providing all the tools to solve numerical problems while using heavy template magic to both, gain optimal performance as well as separating data structures from algorithms. By this approach, DUNE is split up into different modules (such as common, geometry, grid, fem, pdelab, dumux, ...), which provide application specific functionalities. This overall structure makes DUNE a very general and highly performant simulation software, while making it tough for beginners to get along with the software (due to the use of very advanced C++ techniques).

The FeniCS Project consists of a collection of interoperable software components and provides high-level Python and C++ interfaces to facilitate the use its components to beginners. Thus, it vastly tries to automate the solution of PDEs (see also the title of the FeniCS book [180]). However, the online tutorials and documentation are not very detailed as compared to other FE software packages.

MFEM tries to be a lightweight, flexible, and scalable C++ library. Thus, it also has a very modular structure and heavily relies on the library HYPRE [151] to provide the interfaces for a heavily parallel linear algebra (including several algebraic multigrid approaches). Its tutorials are very colourful, but they do not really enable a starter to use the software.

If only a specific field of applications is of relevance, there is a variety of closed-source (commercial) and open-source tools, in which the numerical methods are not necessarily clearly visible anymore ("black boxes"). An open-source tool for fluid flow problems (computational fluid dynamics, CFD) is openFOAM [199], which also includes finite volume discretization schemes. As a commercial tool, which tries to cover features of various application fields and a PDE software, we mention COMSOL Multiphysics [128]. Since COMSOL Multiphysics emerged from FEMLAB and the latter in turn from a MATLAB toolbox, it is not surprising that it offers a well-functioning bidirectional interface to MATLAB and Simulink.

2.4.5 Testing Numerical Methods and Software

To test numerical software, problems with a known, easily computable to sufficient accuracy solution have to be available. For PDE problems as treated here this is only the case for very specific boundary value problems (or, in Chapter 9, initial boundary value problems). Typically such solutions are smooth, i.e., the order of convergence can be tested but not the necessary regularity of the solution. In principle any appropriately regular function can be made a solution of the PDE problem under consideration by inserting it into the defining equations (pointwise or in the variational form) and the essential (Dirichlet) boundary conditions to form the corresponding right-hand sides (source terms). This method is sometimes called the *method of manufactured solutions*. The disadvantage of such a procedure lies in the fact that even "atypical" solutions can be chosen leading, for instance, to strong and sign changing source terms to compensate for this, so that the problems generated are often not representative of real-world problems.

Finally we want to briefly comment the practical determination of the asymptotic convergence rate. If the exact solution is known, we can perform some computational experiments to obtain a sequence of pairs $(h_i, e(h_i))_{i=0}^m$, where $h_0 > \ldots > h_i > h_{i+1} > \ldots > h_m$ are (grid) parameters and $e(h_i)$ is the computed error corresponding to the parameter h_i, e.g., $e(h_i) := \|u_{h_i} - u\|_{0,h_i}$ as in Programming project 2.4. Based upon the theoretical investigations of the numerical method, an error model of the form $e(h) \approx Ch^p$ with two unknown parameters $C, p > 0$ is assumed.

Now the following sequence $(p^{(i)})$ can be computed:

$$p^{(i)} := \frac{\ln(e(h_{i+1})/e(h_i))}{\ln(h_{i+1}/h_i)}, \qquad i = 0, \dots, m-1.$$

If the error model fits, the sequence $(p^{(i)})$ converges to p, and $\tilde{p} := p^{(m-1)}$ is a sufficiently accurate approximation to p.

In case of an unknown exact solution, we can estimate the convergence rate at least pointwise as follows. We perform a sequence of computational experiments using the parameters $h_i := \varrho^i h_0$, $i = 0, \dots, m$, $m \geq 2$, for some $h_0 > 0$, $\varrho \in (0, 1)$. Assuming the error model $u_h(x) - u(x) \approx Ch^p$ at some fixed point of continuity $x \in \overline{\Omega}$, the relationship $u_{h_{i+1}}(x) - u_{h_i}(x) \approx Ch_{i+1}^p - Ch_i^p = C(\varrho^p - 1)(\varrho^i h_0)^p$ motivates to compute the sequence $(p^{(i)})$ as follows:

$$p^{(i)} := \frac{1}{\ln \varrho} \ln \left(\frac{u_{h_{i+2}}(x) - u_{h_{i+1}}(x)}{u_{h_{i+1}}(x) - u_{h_i}(x)} \right), \qquad i = 0, \dots, m-2,$$

and $\tilde{p} := p^{(m-2)}$ is taken as an approximation to p.

2.5 Solving Sparse Systems of Linear Equations by Direct Methods

Let A be an $M \times M$ matrix. Given a vector $\mathbf{q} \in \mathbb{R}^M$, we consider the system of linear equations

$$A\xi = \mathbf{q}\,.$$

The matrices arising from the finite element discretization are *sparse*; i.e., they have a bounded number of nonzero entries per row independent of the dimension of the system of equations. For the simple example of Section 2.2, this bound is determined by the number of neighbouring nodes (see (2.41)). Methods for solving systems of equations should take advantage of the sparse structure. For iterative methods, which will be examined in Chapter 5, this is easier to reach than for direct methods. Therefore, the importance of direct methods has decreased. Nevertheless, in adapted form and for small or medium size problems, they are still the method of choice.

Elimination without Pivoting using Band Structure

In the general case, where the matrix A is assumed only to be nonsingular, there exist matrices $P, L, U \in \mathbb{R}^{M,M}$ such that

$$PA = LU\,.$$

Here P is a permutation matrix, L is a scaled lower triangular matrix, and U is an upper triangular matrix; i.e., they have the form

$$L = \begin{pmatrix} 1 & & 0 \\ & \ddots & \\ l_{ij} & & 1 \end{pmatrix}, \qquad U = \begin{pmatrix} u_{11} & & u_{ij} \\ & \ddots & \\ 0 & & u_{MM} \end{pmatrix}.$$

This factorization corresponds to the Gaussian elimination method with pivoting. The method is very easy and has favourable properties with respect to the sparse structure, if pivoting is not necessary (i.e., $P = I$, $A = LU$). Then the matrix A is called *LU factorizable*.

Denote by A_k the leading principal submatrix of A of dimension $k \times k$, i.e.,

$$A_k := \begin{pmatrix} a_{11} & \cdots & a_{1k} \\ \vdots & \ddots & \vdots \\ a_{k1} & \cdots & a_{kk} \end{pmatrix},$$

and suppose that it already has been factorized as $A_k = L_k U_k$. This is obviously possible for $k = 1$: $A_1 = (a_{11}) = (1)(a_{11})$. The matrix A_{k+1} can be represented in the form of a block matrix

$$A_{k+1} = \left(\begin{array}{c|c} A_k & b \\ \hline c^T & d \end{array}\right)$$

with $b, c \in \mathbb{R}^k$, $d \in \mathbb{R}$.

Using the ansatz

$$L_{k+1} = \left(\begin{array}{c|c} L_k & 0 \\ \hline l^T & 1 \end{array}\right), \qquad U_{k+1} = \left(\begin{array}{c|c} U_k & u \\ \hline 0 & s \end{array}\right)$$

with unknown vectors $u, l \in \mathbb{R}^k$ and $s \in \mathbb{R}$, it follows that

$$A_{k+1} = L_{k+1} U_{k+1} \iff L_k u = b\,,\ U_k^T l = c\,,\ l^T u + s = d\,. \tag{2.59}$$

From this, we have the following result:

$$\begin{aligned}&\text{Let } A \text{ be nonsingular. Then lower and upper triangular matrices}\\ &L, U \text{ exist with } A = LU \text{ if and only if } A_k \text{ is nonsingular for all}\\ &1 \le k \le M. \text{ For this case, } L \text{ and } U \text{ are determined uniquely.}\end{aligned} \tag{2.60}$$

Furthermore, from (2.59) we have the following important consequences: If the first l components of the vector b are equal to zero, then this is valid for the vector u, too:

$$\text{If } b = \begin{pmatrix} 0 \\ - \\ \beta \end{pmatrix}, \text{ then } u \text{ also has the structure } u = \begin{pmatrix} 0 \\ - \\ \varrho \end{pmatrix}.$$

Similarly,

$$c = \begin{pmatrix} 0 \\ - \\ \gamma \end{pmatrix} \text{ implies the structure } l = \begin{pmatrix} 0 \\ - \\ \lambda \end{pmatrix}.$$

For example, if the matrix A has a structure as shown in Figure 2.16, then the zeros outside of the surrounded entries are preserved after the LU factorization. Before we introduce appropriate definitions to generalize these results, we want to consider the special case of symmetric matrices.

$$A = \begin{pmatrix} * & 0 & * & 0 & 0 \\ 0 & * & * & 0 & * \\ * & * & * & * & * \\ 0 & 0 & * & * & 0 \\ 0 & * & * & 0 & * \end{pmatrix}$$

Fig. 2.16: Profile of a matrix.

If A is as before nonsingular and LU factorizable, then $U = DL^T$ with a diagonal matrix $D = \operatorname{diag}(d_i)$, and therefore

$$A = LDL^T .$$

This is true because A has the form $A = LD\tilde{U}$, where the upper triangular matrix $\tilde{U}$ satisfies the scaling condition $\tilde{u}_{ii} = 1$ for all $i = 1, \ldots, M$. Such a factorization is unique, and thus

$$A = A^T \text{ implies } L^T = \tilde{U}, \text{ therefore } A = LDL^T .$$

If in particular A is symmetric and positive definite, then also $d_i > 0$ is valid. Thus exactly one matrix $\tilde{L}$ of the form

$$\tilde{L} = \begin{pmatrix} l_{11} & & 0 \\ & \ddots & \\ l_{ij} & & l_{MM} \end{pmatrix} \quad \text{with } l_{ii} > 0 \quad \text{for all } i$$

exists such that

$$A = \tilde{L}\tilde{L}^T, \quad \text{the so-called } \mathit{Cholesky\ factorization}.$$

We have

$$\tilde{L}_{\text{Chol}} = L_{\text{Gauss}} \sqrt{D}, \quad \text{where} \quad \sqrt{D} := \operatorname{diag}\left(\sqrt{d_i}\right).$$

This shows that the Cholesky method for the determination of the Cholesky factor $\tilde{L}$ also preserves certain zeros of A in the same way as the Gaussian elimination without pivoting.

In what follows, we want to specify the set of zeros that is preserved by Gaussian elimination without pivoting. We will not consider a symmetric matrix; but for the sake of simplicity we will consider a matrix with a symmetric distribution of its entries.

Definition 2.19 Let $A \in \mathbb{R}^{M,M}$ be a matrix such that $a_{ii} \neq 0$ for $i = 1, \ldots, M$ and

$$a_{ij} \neq 0 \quad \text{if and only if} \quad a_{ji} \neq 0 \quad \text{for all } i, j = 1, \ldots, M. \tag{2.61}$$

We define, for $i = 1, \ldots, M$,

$$f_i(A) := \min \left\{ j \mid a_{ij} \neq 0,\ 1 \leq j \leq i \right\}.$$

Then

$$m_i(A) := i - f_i(A)$$

is called the *ith (left-hand side) row bandwidth* of A.

The *bandwidth* of a matrix A that satisfies (2.61) is the number

$$m(A) := \max_{1 \leq i \leq M} m_i(A) = \max \left\{ i - j \mid a_{ij} \neq 0,\ 1 \leq j \leq i \leq M \right\}.$$

The *band* of the matrix A is

$$B(A) := \left\{ (i,j), (j,i) \mid i - m(A) \leq j \leq i,\ 1 \leq i \leq M \right\}.$$

The set

$$\mathrm{Env}\,(A) := \left\{ (i,j), (j,i) \mid f_i(A) \leq j \leq i,\ 1 \leq i \leq M \right\}$$

is called the *hull* or *envelope* of A. The number

$$p(A) := M + 2 \sum_{i=1}^{M} m_i(A)$$

is called the *profile* of A.

The profile is the number of elements of $\mathrm{Env}(A)$.

For the matrix A in Figure 2.16 we have $(m_1(A), \ldots, m_5(A)) = (0, 0, 2, 1, 3)$, $m(A) = 3$, and $p(A) = 17$.

Summarizing the above considerations, we have proved the following theorem:

Theorem 2.20 *Let A be a matrix with the symmetric structure (2.61). Then the Cholesky method or the Gaussian elimination without pivoting preserves the hull and in particular the bandwidth.*

The hull may contain zeros that will be replaced by (nonzero) entries during the factorization process. Therefore, in order to keep this *fill-in* small, the profile should be as small as possible.

Furthermore, in order to exploit the matrix structure for an efficient assembling and storage, this structure (or some estimate of it) should be known in advance, before the computation of the matrix entries is started.

For example, if A is a stiffness matrix with the entries

$$a_{ij} = a(\varphi_j, \varphi_i) = \int_\Omega \nabla\varphi_j \cdot \nabla\varphi_i \, dx \,,$$

then the property

$$a_{ij} \neq 0 \quad \Rightarrow \quad a_i, a_j \text{ are neighbouring nodes}$$

can be used for the definition of an (eventually too large) symmetric matrix structure. This is also valid for the case of a nonsymmetric bilinear form and thus a nonsymmetric stiffness matrix. Also in this case, the definition of $f_i(A)$ can be replaced by

$$f_i(A) := \min\left\{ j \mid 1 \leq j \leq i\,,\ j \text{ is a neighbouring node of } i \right\}.$$

Since the characterization (2.60) of the possibility of the Gaussian elimination without pivoting cannot be checked directly, we have to specify sufficient conditions. Examples for such conditions are the following (see [61]):

- A is symmetric and positive definite,
- A is an M-matrix.
  Sufficient conditions for this property were given in (1.35) and (1.35)*. In Section 3.9, geometrical conditions for the family of triangulations $(\mathcal{T}_h)_h$ will be derived that guarantee that the finite element discretization considered here creates an M-matrix.

Data Structures

For sparse matrices, it is appropriate to store only the components within the band or the hull. A symmetric matrix $A \in \mathbb{R}^{M,M}$ with bandwidth m can be stored in $M(m+1)$ memory positions. By means of the index conversion $a_{ik} \rightsquigarrow b_{i,k-i+m+1}$ for $k \leq i$, the matrix

$$A = \begin{pmatrix} a_{11} & a_{12} & \cdots & a_{1,m+1} & & & \\ a_{21} & a_{22} & \cdots & \vdots & \ddots & 0 & \\ \vdots & \vdots & \ddots & \vdots & \ddots & \ddots & \\ a_{m+1,1} & a_{m+1,2} & \cdots & a_{m+1,m+1} & \ddots & \ddots & \ddots \\ & \ddots & \ddots & \ddots & \ddots & \ddots & \ddots \\ & 0 & \ddots & \ddots & \ddots & \ddots & \ddots \\ & & & a_{M,M-m} & \cdots & a_{M,M-1} & a_{M,M} \end{pmatrix} \in \mathbb{R}^{M,M}$$

is mapped to the matrix

$$B = \begin{pmatrix} 0 & \cdots & \cdots & 0 & a_{11} \\ 0 & \cdots & 0 & a_{21} & a_{22} \\ \vdots & & & \vdots & \vdots \\ 0 & a_{m,1} & \cdots & \cdots & a_{m,m} \\ a_{m+1,1} & \cdots & \cdots & a_{m+1,m} & a_{m+1,m+1} \\ \vdots & \vdots & \vdots & \vdots & \vdots \\ \vdots & \vdots & \vdots & \vdots & \vdots \\ a_{M,M-m} & \cdots & \cdots & a_{M,M-1} & a_{M,M} \end{pmatrix} \in \mathbb{R}^{M,m+1}.$$

The unused elements of B, i.e., $(B)_{ij}$ for $i = 1, \ldots, m$, $j = 1, \ldots, m+1-i$, are here filled with zeros.

For a general band matrix, the matrix $B \in \mathbb{R}^{M,2m+1}$ obtained by the above conversion has the following form:

$$B = \begin{pmatrix} 0 & \cdots & 0 & a_{11} & a_{12} & \cdots & a_{1,m+1} \\ 0 & \cdots & a_{21} & a_{22} & \cdots & \cdots & a_{2,m+2} \\ \vdots & & \vdots & \vdots & \vdots & \vdots & \vdots \\ 0 & a_{m,1} & \cdots & \cdots & \cdots & \cdots & a_{m,2m} \\ a_{m+1,1} & \cdots & \cdots & \cdots & \cdots & \cdots & a_{m+1,2m+1} \\ \vdots & \vdots & \vdots & \vdots & \vdots & \vdots & \vdots \\ a_{M-m,M-2m} & \cdots & \cdots & \cdots & \cdots & \cdots & a_{M-m,M} \\ a_{M-m+1,M-2m+1} & \cdots & \cdots & \cdots & \cdots & a_{M-m+1,M} & 0 \\ \vdots & \vdots & \vdots & \vdots & \vdots & & \vdots \\ a_{M,M-m} & \cdots & \cdots & a_{M,M} & 0 & \cdots & 0 \end{pmatrix}.$$

Here, in the right lower part of the matrix, a further sector of unused elements arose, which is also filled with zeros.

If the storage is based on the hull, additionally a pointer field is needed, which points to the diagonal elements, for example. If the matrix is symmetric, again the storage of the lower triangular matrix is sufficient. For the matrix A from Figure 2.16

under the assumption that A is symmetric, the pointer field could act as shown in Figure 2.17.

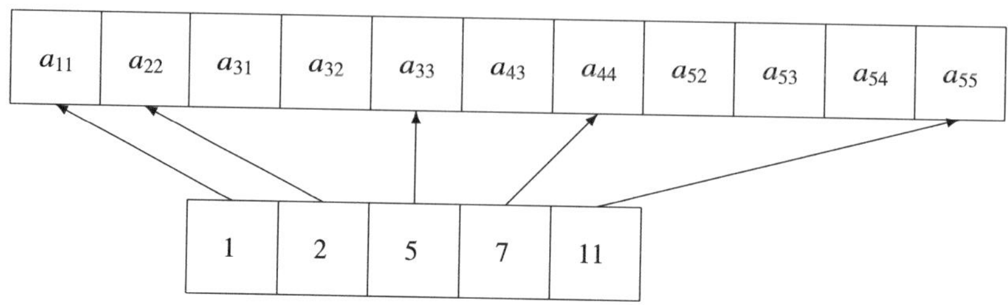

Fig. 2.17: Linear storage of the hull.

Coupled Assembling and Factorization

A formerly popular method, the so-called *frontal method*, performs simultaneously assembling and the Cholesky factorization.

We consider this method for the example of the stiffness matrix $A_h = \left(a_{ij}\right) \in \mathbb{R}^{M,M}$ with bandwidth m (with the original numbering).

The method is based on the kth step of the Gaussian or Cholesky method (cf. Figure 2.18).

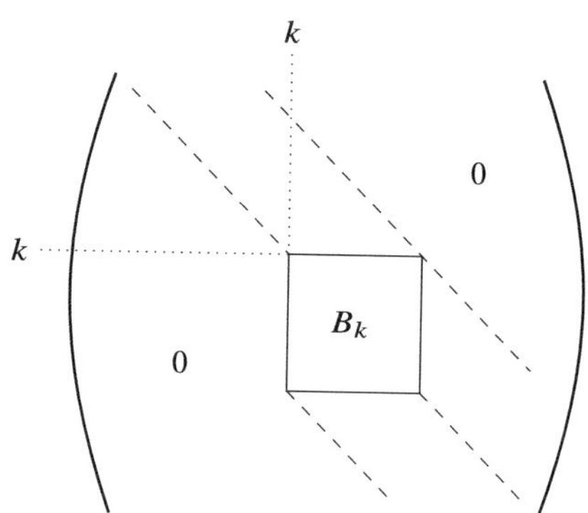

Fig. 2.18: kth step of the Cholesky method.

Only the entries of B_k are to be changed, i.e., only those elements a_{ij} with $k \leq i, j \leq k + m$. The corresponding formula is

$$a_{ij}^{(k+1)} = a_{ij}^{(k)} - \frac{a_{ik}^{(k)}}{a_{kk}^{(k)}} a_{kj}^{(k)}, \quad i, j = k+1, \ldots, k+m. \tag{2.62}$$

Here, the upper indices indicate the steps of the elimination method, which we store in a_{ij}. The entries a_{ij} are generated by summation of entries of the element stiffness matrix of those elements K that contain nodes with the indices i, j.

Furthermore, to perform the elimination step (2.62), only $a_{ik}^{(k)}, a_{kj}^{(k)}$ for $i, j = k, \ldots, k+m$ must be completely assembled; $a_{ij}^{(k)}, i, j = k+1, \ldots, k+m$, can be replaced by $\tilde{a}_{ij}^{(k)}$ if $a_{ij}^{(k+1)}$ is later defined by $a_{ij}^{(k+1)} := \tilde{a}_{ij}^{(k+1)} + a_{ij}^{(k)} - \tilde{a}_{ij}^{(k)}$. That is, for the present, a_{ij} needs to consist of only a few contributions of elements K with nodes i, j in K.

From these observations, the following algorithm is obtained. The kth step for $k = 1, \ldots, M$ reads as follows:

1. Assemble all of the missing contributions of elements K that contain the node with index k.
2. Compute $A^{(k+1)}$ by modification of the entries of B_k according to (2.62).
3. Store the kth row of $A^{(k+1)}$, also out of the main memory.
4. Define B_{k+1} (by a south-east shift).

Here the assembling is node-based and not element-based.

The advantage of this method is that A_h need not be completely assembled and stored in the main memory, but only a matrix $B_k \in \mathbb{R}^{m+1,m+1}$. Of course, if M is not too large, there may be no advantage.

Bandwidth Reduction

The *complexity*, i.e., the number of operations, is crucial for the application of a particular method:

The Cholesky method, applied to a symmetric matrix $A \in \mathbb{R}^{M,M}$ with bandwidth m, requires $O(m^2 M)$ operations in order to compute L.

However, the bandwidth m of the stiffness matrix depends on the numbering of the nodes. Therefore, a numbering is to be found where the number m is as small as possible.

We want to consider this again for the example of the Poisson equation on the rectangle with the discretization according to Figure 2.9. Let the interior nodes have the coordinates (ih, jh) with $i = 1, \ldots k-1$, $j = 1, \ldots, l-1$. The discretization corresponds to the finite difference method introduced beginning with (1.10); i.e., the bandwidth is equal to $k-1$ for a rowwise numbering or $l-1$ for a columnwise numbering.

For $k \ll l$ or $k \gg l$, this fact results in a large difference of the bandwidth m or of the profile (of the left triangle), which is of size $(k-1)(l-1)(m+1)$ except for a term of m^2. Therefore, the columnwise numbering is preferred for $k \gg l$; the rowwise numbering is preferred for $k \ll l$.

For a general domain Ω, a numbering algorithm based on a given partition $\mathcal{T}_h$ and on a basis $\{\varphi_i\}$ of V_h is necessary with the following properties:

The structure of A resulting from the numbering must be such that the band or the profile of A is as small as possible. Furthermore, the numbering algorithm should yield the numbers $m(A)$ or $f_i(A)$, $m_i(A)$ such that the matrix A can also be assembled using the element matrices $A^{(k)}$.

Given a partition $\mathcal{T}_h$ and a corresponding basis $\{\varphi_i \mid 1 \le i \le M\}$ of V_h, we start with the assignment of some graph G to this partition as follows:

The nodes of G coincide with the nodes $\{a_1, \dots, a_M\}$ of the partition. The definition of its edges is

$$\begin{gathered}\overline{a_i a_j} \text{ is an edge of } G \\ \Longleftrightarrow \\ \text{there exists a } K \in \mathcal{T}_h \text{ such that } \varphi_i|_K \not\equiv 0,\ \varphi_j|_K \not\equiv 0.\end{gathered}$$

In Figure 2.19 some examples are given, where the example 2) will be introduced in Section 3.3.

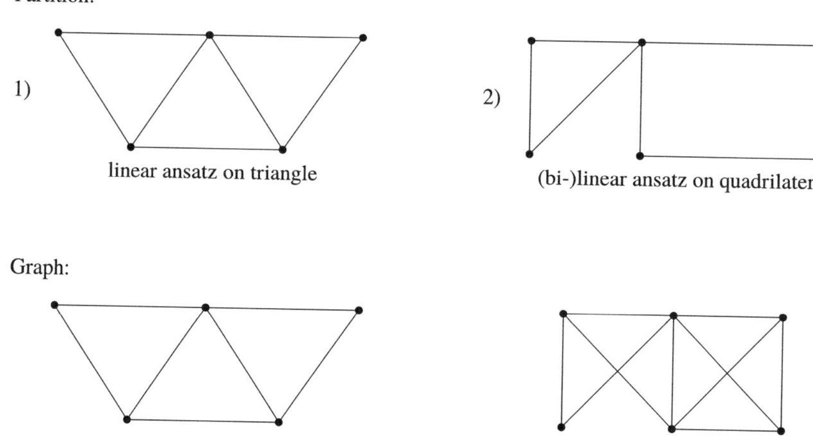

Fig. 2.19: Partitions and assigned graphs.

If several degrees of freedom are assigned to some node of the partition $\mathcal{T}_h$, then also in G several nodes are assigned to it. This is the case, for example, if so-called Hermite elements are considered, which will be introduced in Section 3.3. The costs of administration are small if the same number of degrees of freedom is assigned to all nodes of the partition.

An often used numbering algorithm is the *Cuthill–McKee method*. This algorithm operates on the graph G just defined. Two nodes a_i, a_j of G are called *neighbouring* if $\overline{a_i a_j}$ is an edge of G.

The *degree* of a node a_i of G is defined as the number of neighbours of a_i.

The kth step of the algorithm for $k = 1, \dots, M$ has the following form:

1. $k = 1$: Choose a starting node, which gets the number 1. This starting node forms the level 1.
2. $k > 1$: If all nodes are already numbered, the algorithm is terminated. Otherwise, the level k is formed by taking all the nodes that are not numbered yet and that are neighbours of a node of level $k - 1$. The nodes of level k will be consecutively numbered.

Within a level, we can sort, for example, by the degree, where the node with the smallest degree is numbered first.

The *reverse Cuthill–McKee method* consists of the above method and the inversion of the numbering at the end; i.e.,

$$\text{new node number} = M + 1 - \text{old node number}.$$

This corresponds to a reflection of the matrix at the counter diagonal. The bandwidth does not change by the inversion, but the profile may diminish drastically for many examples (cf. Figure 2.20).

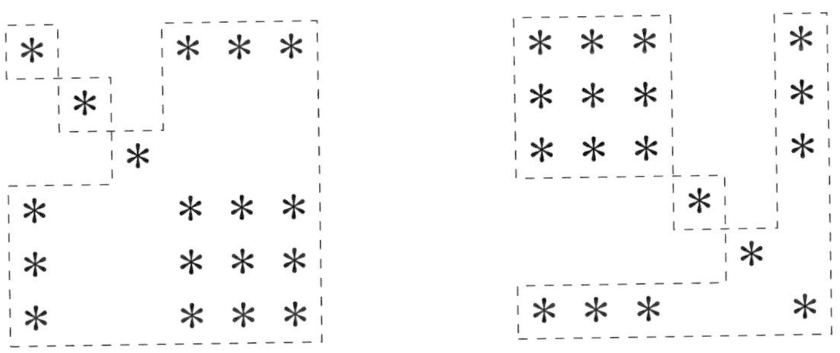

Fig. 2.20: Change of the hull by reflection at the counterdiagonal.

The following estimate holds for the bandwidth m of the numbering created by the Cuthill–McKee algorithm:

$$\frac{D+i}{2} \leq m \leq \max_{2 \leq k \leq l} \left(N_{k-1} + N_k - 1\right).$$

Here D is the maximum degree of a node of G, l is the number of levels, and N_k is the number of nodes of level k. The number i is equal to 0 if D is even, and i is equal to 1 if D is odd. The left-hand side of the above inequality is easy to understand by means of the following argument: To reach a minimal bandwidth, all nodes that are neighbours of a_i in the graph G should also be neighbours of a_i in the numbering. Then the best situation is given if the neighbouring nodes would appear uniformly immediately before and after a_i. If D is odd, then one side has one node more than the other.

To verify the right-hand side, consider a node a_i that belongs to level $k-1$ as well as a node a_j that is a neighbour of a_i in the graph G and that is not yet numbered in level $k-1$. Therefore, a_j will get a number in the kth step. The largest bandwidth is obtained if a_i is the first node of the numbering of level $k-1$ and if a_j is the last node of level k. Hence exactly $(N_{k-1}-1)+(N_k-1)$ nodes lie between both of these; i.e., their distance in the numbering is $N_{k-1}+N_k-1$.

It is favourable if the number l of levels is as large as possible and if all the numbers N_k are of the same size, if possible. Therefore, the starting node should be chosen "at one end" of the graph G if possible; if all the starting nodes are to be checked, the expense will be $O(M\tilde{M})$, where $\tilde{M}$ is the number of edges of G. One possibility consists in choosing a node with minimum degree for the starting node. Another possibility is to let the algorithm run once and then to choose the last-numbered node as the starting node.

If a numbering is created by the (reverse) Cuthill–McKee algorithm, we can try to improve it "locally", i.e., by exchanging particular nodes.

Direct Methods for General Sparse Matrices

Iterative methods based on matrix-vector multiplications avoid the fill-in problem and have been considered to be preferable for some decades, as they also do not require to store the full matrix. Therefore we give these methods a broad consideration in Chapter 5. Nevertheless developments both in computer hardware and in algorithms have led to a renaissance of direct methods. We will only sketch a few aspects of this development. Although computer memory meanwhile is hardly a restriction, cache locality of algorithms has become an important issue for efficiency, besides parallelizability. Dealing with (sparse) matrices (in the formats described above), a row- or column-oriented storage should be reflected in a corresponding data access of the algorithm. For the same reason the concentration of the computation load to small, densely populated matrices is advisable. Basic operations are preferably vector operations (on contiguously stored data) in the form of an AXPY (scalar times vector plus vector) operation $\alpha x + y \in \mathbb{R}^M$, scalar product $x^T y \in \mathbb{R}$, or an outer product $xy^T \in \mathbb{R}^{M,M}$. Considering the LU factorization with its three loops k, i, j, by reordering of the loops different equivalent algorithms with different data access and vector operations are possible. In the following we will concentrate for simplicity on a symmetric, positive definite A and on the Cholesky method. The Cholesky method in the kji-form is depicted in the Table 2.2, assuming that the lower half of A is stored in L initially and overwritten with the correct L.

The update is here by adding a lower triangular $(M-k)\times(M-k)$ outer product and corresponds in the first step to

$$A =: \begin{pmatrix} a_{11} & v^T \\ v & B \end{pmatrix} = L_1 \begin{pmatrix} 1 & 0 \\ 0 & B - \frac{1}{a_{11}} v v^T \end{pmatrix} L_1^T \tag{2.63}$$

$k = 1, \dots, M:$
$\quad l_{kk} := l_{kk}^{1/2}$
$\quad s = k+1, \dots, M:$
$\quad\quad l_{sk} := l_{sk}/l_{kk}$
$\quad\quad j = k+1, \dots, M:$
$\quad\quad\quad i = j, \dots, M:$
$\quad\quad\quad l_{ij} := l_{ij} - l_{ik}l_{jk}$

Table 2.2: Cholesky method kji-formulation.

with $B \in \mathbb{R}^{M-1,M-1}$, $L_1 := \begin{pmatrix} \beta & 0 \\ \frac{1}{\beta}v & I_{M-1} \end{pmatrix}$, $\beta := a_{11}^{1/2}$. Repeated application to the right lower submatrix leads to $L = L_1 \dots L_{M-1}$, where the ith column (from the diagonal on) is just the first column of L_i. For the Cholesky method the fill-in can be determined exactly by means of a *symbolic Cholesky factorization*. In the following algorithm in Table 2.3 we also determine the *elimination tree* $T(A)$. This is an acyclic undirected graph with the nodes $\{1, \dots, M\}$, where node M is the root and a parent function π (and by this the full graph) is defined by

$$\pi(j) := \min\{i > j \mid l_{ij} \neq 0\}.$$

To assure that this graph is connected, i.e., a *tree*, we assume from now on that A is irreducible.

Let $\mathcal{A}_i, \mathcal{L}_i \subset \{i, \dots, M\}$ be the set of indices of the column $a^{(i)}, l^{(i)}$, respectively, of nonzero entries, π the *parent function*.

$i = 1, \dots, M : \pi(i) := 0$
$i = 1, \dots, M:$
$\quad \mathcal{L}_i := \mathcal{A}_i,$
$\quad$ for $j \in \{1, \dots, M\}$ such that $\pi(j) = i$:
$\quad\quad \mathcal{L}_i := (\mathcal{L}_i \cup \mathcal{L}_j) \setminus \{j\}$
$\quad \pi(i) := \min \mathcal{L}_i \setminus \{i\}$

Table 2.3: Nonzero pattern of L and elimination tree.

Based on π, we will also speak of *children* i of j ($\pi(i) = j$) and of *descendants* i of j ($\pi^k(i) = j$ for some k). Every subgraph consisting of a node j and its descendants will also be a tree and denoted by $T[j]$, i.e., $T(A) = T[M]$. Based on the elimination tree (or similar graphs), *multifrontal algorithms* can be developed, which reduce updates to the manipulation of small dense matrices (and also detect possible parallelization). We will develop only the basic idea.

The Multifrontal Cholesky Method

We need the following result.

Lemma 2.21 *1) If node k is a descendant of j in $T(A)$, then $\mathcal{L}_k \subset \mathcal{L}_j$.*
2) If $l_{jk} \neq 0$ and $k < j$, then k is a descendant of j in $T(A)$.

Proof See [177]. □

Consider the jth column $l^{(j)}$ of L and $\mathcal{L}_j := \{i_0, \ldots, i_r\}$ with $i_0 := j$. For a vector $v \in \mathbb{R}^{M-j+1}$, indexed from j to M, denote by $v^* \in \mathbb{R}^{r+1}$

$$v^* := \begin{pmatrix} v_j \\ v_{i1} \\ \vdots \\ v_{ir} \end{pmatrix}.$$

We define the *subtree update matrix* of column j by

$$U_j := - \sum_{k \in T[j] \setminus \{j\}} l^{(k)*} l^{(k)*^T} . \tag{2.64}$$

According to Lemma 2.21, U_j contains all outer-product updates of all descendant columns of j.

The *jth frontal matrix* F_j for A is defined by

$$F_j := \begin{pmatrix} a_{jj} & a_{ji_1} & \cdots & a_{ji_r} \\ a_{i_1 j} & & & \\ \vdots & & 0 & \\ a_{i_r j} & & & \end{pmatrix} + U_j . \tag{2.65}$$

As the first row/column of F_j result from the full update process of $a^{(j)}$, we have (compare (2.63))

$$F_j = \begin{pmatrix} l^{(j)*} & 0 \\ & I_r \end{pmatrix} \begin{pmatrix} 1 & 0 \\ 0 & U_j \end{pmatrix} \begin{pmatrix} l^{(j)*^T} & \\ 0 & I_r \end{pmatrix} \tag{2.66}$$

with the *update matrix* U_j from column j. Comparing this equation with (2.65) leads to

Lemma 2.22

$$U_j = \sum_{k \in T[j]} \begin{pmatrix} l_{i_1 k} \\ \vdots \\ l_{i_r k} \end{pmatrix} \begin{pmatrix} l_{i_1 k} & \cdots & l_{i_r k} \end{pmatrix} .$$

Proof Problem 2.17. □

We define a *sparse addition* of (dense) matrices $A \in \mathbb{R}^{r,r}$, $B \in \mathbb{R}^{s,s}$ with indices $i_1, \ldots, i_r$ and $j_1, \ldots, j_s$ as a matrix $C \in \mathbb{R}^{t,t}$, where the indices are $\{k_1, \ldots, k_t\} =$

$\{i_1, \ldots, i_r\} \cup \{j_1, \ldots, j_s\}$, i.e., $t \leq r + s$, by extending A, B with zero entries to matrices $\widetilde{A}, \widetilde{B} \in \mathbb{R}^{t,t}$, and

$$C := A +_{\text{ext}} B := \widetilde{A} + \widetilde{B}.$$

Lemma 2.23 *Let $c_1, \ldots, c_s$ be the children of node j in the elimination tree, then*

$$F_j = \begin{pmatrix} a_{jj} & a_{i_1 j} & \cdots & a_{i_r j} \\ a_{i_1 j} & & & \\ \vdots & & 0 & \\ a_{i_r j} & & & \end{pmatrix} +_{\text{ext}} U_{c_1} +_{\text{ext}} U_{c_s}. \tag{2.67}$$

Proof Problem 2.18. □

Thus the Cholesky method reads as in Table 2.4.

$j = 1, \ldots, n$:

Let $\mathcal{L}_j = \{j, i_1, \ldots, i_r\}$.

Let $c_1, \ldots, c_s$ be the children of j in the elimination tree.

Define F_j as in (2.67).

Factor F_j as in (2.66).

Table 2.4: Cholesky method via frontal and update matrices.

Remark 2.24 1) In this presentation we followed closely [177].
2) Update matrices may not be used immediately after their generation, no temporary storage is necessary.
3) A reordering of the nodes may achieve that only few U_k have to be stored only for a few steps. Such a reordering should preserve the elimination tree (up to renaming of the nodes). Strategies are discussed in [177].

Exercises

Problem 2.13 Investigate the fill-in for an LU factorization without pivoting of the following "arrow" matrices

$$\begin{pmatrix} * & \cdots & \cdots & * \\ \vdots & \ddots & & \\ \vdots & & \ddots & \\ * & & & * \end{pmatrix}, \quad \begin{pmatrix} * & & & * \\ & \ddots & & \vdots \\ & & \ddots & \vdots \\ * & \cdots & \cdots & * \end{pmatrix}.$$

Problem 2.14 Show that the number of arithmetic operations for the Cholesky method for an $M \times M$ matrix with bandwidth m has order $Mm^2/2$; additionally, M square roots have to be calculated.

Problem 2.15 Find other forms of the Cholesky method (e.g., a jki and a ijk form) and discuss data access and vector operations.

Problem 2.16 Verify the algorithm in Table 2.3.

Problem 2.17 Show (2.66) and Lemma 2.22.

Problem 2.18 Show Lemma 2.23.

Programming project 2.1 Consider the one-dimensional Poisson equation with inhomogeneous Dirichlet boundary conditions

$$\begin{aligned} -u'' &= f \quad \text{in } \Omega := (0,1), \\ u(0) &= g_l, \quad u(1) = g_r\,. \end{aligned}$$

Write a function that solves the problem numerically for a given N, $N \in \mathbb{N}$, using the finite element method. The input data are the right-hand side f, the boundary data g_l, g_r, the exact solution u, and the number of intervals N. The output data are $\xi \in \mathbb{R}^{N-1}$ and the maximum norm of the error, i.e., $\|u - u_h\|_\infty$.

Proceed step by step according to the following scheme:

1. Derive the variational formulation for the problem.
2. Choose a partition $\mathcal{T}_h := \{(x_{i-1}, x_i)\}_{i=1}^N$ of the interval $(0,1)$ into N intervals, where $0 =: x_0 < x_1 < \ldots < x_N := 1$ and

$$h_i := x_{i+1} - x_i := \begin{cases} \frac{1}{N} + \frac{1}{N^2} & \text{if } i \text{ is even}, \\ \frac{1}{N} - \frac{1}{N^2} & \text{if } i \text{ is odd}. \end{cases}$$

   Let $X_h := \left\{v \in C[0,1] \;\middle|\; v|_{(x_{i-1},x_i)} \in \mathcal{P}_1(x_{i-1},x_i),\ i = 1,\ldots,N\right\}$ and $V_h := \left\{v \in X_h \;\middle|\; v(x_0) = v(x_N) = 0\right\}$.

   Replace the variational formulation by a sequence of finite-dimensional problems of the form

   Find $u_h \in X_h$ such that $u_h(x_0) = g_l$, $u_h(x_N) = g_r$ and

$$a(u_h, v_h) = \langle f, v_h \rangle_0 \quad \text{for all } v_h \in V_h \tag{2.68}$$

3. Choose "hat functions" as a basis for X_h:

$$\varphi_i(x) := \begin{cases} (x - x_{i-1})/h_{i-1} & \text{in } (x_{i-1}, x_i], \\ (x_{i+1} - x)/h_i & \text{in } (x_i, x_{i+1}), \\ 0 & \text{otherwise}, \end{cases}$$

   and assemble problem (2.68) into a linear system of equations (using an *elementwise* approach)

$$\hat{A}_h \hat{\boldsymbol{\xi}} = \boldsymbol{q}_h,$$

   with unknowns $\hat{\boldsymbol{\xi}} = (u_0, \ldots, u_N)^T$ such that $u_h(x) = \sum_{i=0}^{N} u_i \varphi_i(x)$.

   Use the trapezoidal rule

$$\int_{x_i}^{x_{i+1}} g(x)dx \approx \frac{x_{i+1} - x_i}{2} (g(x_{i+1}) + g(x_i))$$

   to evaluate $\langle f, v_h \rangle_0$ and use sparse data structures where applicable.
4. Apply the Dirichlet boundary conditions (either by elimination or modification of the appropriate equations).
5. Solve the resulting linear system with respect to $\boldsymbol{\xi} := (u_1, \ldots, u_{N-1})^T$ using a method of your choice.

Test the convergence order of your method for

$$f(x) := \pi^2 \sin(\pi x), \quad g_l := g_r := 0$$

(i.e., $u(x) = \sin(\pi x)$) and $N = 10, 20, 40, 80, 160$.

Programming project 2.2 Implement a finite element method for the two-dimensional Poisson problem

$$\begin{aligned} -\Delta u &= f \quad \text{in } \Omega := (0, 1)^2, \\ u &= 0 \quad \text{on } \partial\Omega. \end{aligned}$$

Use the same steps as given in Project 2.1, where the partition of Ω consists of $m \times m$ rectangles $K_{ij} := [x_{i,j}, x_{i+1,j}] \times [x_{i,j}, x_{i,j+1}]$, $i, j = 0, \ldots, m-1$, $m \in \mathbb{N}$, where

$$x_{i+1,j} - x_{i,j} = \begin{cases} \frac{1}{m} + \frac{1}{m^2} & \text{if } i \text{ is even}, \\ \frac{1}{m} - \frac{1}{m^2} & \text{if } i \text{ is odd}, \end{cases}$$

and

$$x_{i,j+1} - x_{i,j} = \begin{cases} \frac{1}{m} + \frac{1}{m^2} & \text{if } j \text{ is even}, \\ \frac{1}{m} - \frac{1}{m^2} & \text{if } j \text{ is odd}. \end{cases}$$

Form the basis of V_h as a tensor product of the univariate hat functions given in Project 2.1.

Test your implementation for $f(x) := 2\pi^2 \sin(\pi x_1)\sin(\pi x_2)$ and compare it to the exact solution u.

Programming project 2.3 Consider the Dirchlet boundary value problem for the two-dimensional Poisson problem

$$\begin{aligned} -\Delta u &= f \quad \text{in } \Omega, \\ u &= g \quad \text{on } \partial\Omega, \end{aligned}$$

where $\Omega \subset \mathbb{R}^2$ is a polygonally bounded domain given by $l \in \mathbb{N}$ vertices in mathematically positive order (that is, counterclockwise).

a) Write a function that refines a given (coarse) triangulation of Ω by subdividing each triangle into four congruent triangles. For the grids use the data structure described in Section 2.4.1, i.e., a nodal coordinate matrix, the number of rows is equal to the number of nodes and whose two columns contain the x- and y-coordinates of the nodes, and an element connectivity matrix, the number of rows is equal to the number of triangles and whose three colums contain the (ordered) indices of the nodes corresponding to the triangles.
The input data are the nodal coordinate and element connectivity tables of a coarse initial triangulation of Ω and a grid parameter $h > 0$. The output data are the nodal coordinate and element connectivity tables of the refined triangulation such that the diameters of all triangles are less than h.
b) Based on the grid refinement function from a) and the test and ansatz spaces according to (2.30) and (2.31), respectively, follow the scheme given in Project 2.1 to create a function that solves the boundary value problem by means of the finite element method. Apply the method described in Section 2.4.3 for realizing the Dirichlet boundary conditions. The input data are the right-hand side f, the boundary function g, the exact solution u, a coarse initial triangulation of Ω and a grid parameter $h > 0$. The output data are $\xi \in \mathbb{R}^M$ and the maximum norm of the error, i.e., $\|u - u_h\|_\infty$.
c) Consider the case $\Omega := (0, 1)^2$ and a simple square Friedrichs–Keller triangulation as an initial triangulation, i.e., divide Ω into two triangles. You may select among the following exact solutions to test certain aspects of your implementation:

1) $u(x, y) := x + y$ (zero right-hand side),
2) $u(x, y) := \sin(\pi x)\sin(\pi y)$ (zero boundary values),
3) $u(x, y) := \cos(7x)\cos(7y)$ (nonzero right-hand side and boundary values).

Programming project 2.4 Use the finite element solver of Project 2.3 to solve the equation

$$-\Delta u = 0 \quad \text{in } \Omega$$

with Dirichlet boundary conditions (given in polar coordinates, $x = r\cos\varphi$, $y = r\sin\varphi$, $\varphi \in [0, 2\pi)$)

$$u(r\cos\varphi, r\sin\varphi) = r^{2/3}\sin\left(\frac{2}{3}\varphi\right)$$

on the following domain with a reentrant corner (the thin lines indicate the initial triangulation):

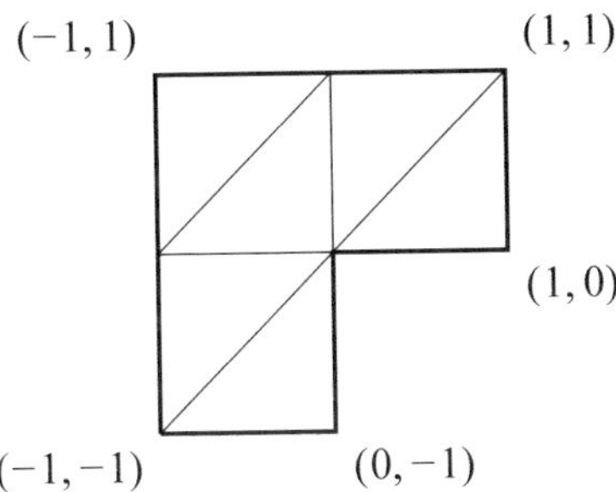

The exact solution is $u(r\cos\varphi, r\sin\varphi) = r^{2/3}\sin\left(\frac{2}{3}\varphi\right)$.

a) Solve this problem on 2...5 times uniformly refined grids and compute the l^2-errors

$$\|u - u_h\|_{0,h} := \left(\frac{1}{M}\sum_{i=1}^{M}\left(u_h(a_i) - u(a_i)\right)^2\right)^{1/2}.$$

b) Compute the numerical convergence rate, i.e., the slope of the straight line that fits the logarithms of the l^2-errors to the logarithms of the corresponding step sizes in a least squares sense. Comment on which order of convergence you are getting and why order two cannot be expected.

Programming project 2.5 Extend your implementation from Project 2.3 to a scenario with homogeneous Neumann boundary conditions:

$$\begin{aligned} -\Delta u &= f \quad \text{in } \Omega, \\ \partial_{\mathfrak{n}} u &= 0 \quad \text{on } \Gamma_1, \\ u &= g_3 \quad \text{on } \Gamma_3 := \partial\Omega \setminus \Gamma_1 . \end{aligned}$$

a) Extend the grid refinement function of Project 2.3 to a function that can additionally handle boundary conditions. Add a data structure which represents the boundary conditions, for instance, a matrix which has three columns. Every row should contain a label (or a *flag*) for the type of boundary condition (say 1 for Neumann and 3 for Dirichlet boundary conditions) and the indices of the end-points of one edge. In every refinement step the edges on the boundary have to be refined.
b) Based on the grid refinement function from a) modify the assembling procedure from Project 2.3 b) in order to take the homogeneous Neumann boundary conditions into account.
c) Test your implementation for the case $\Omega := (0, 1)^2$, $\Gamma_1 := (0, 1) \times \{0\}$ and the exact solution $u(x, y) := \sin(\pi x)\cos(\pi y)$.

Chapter 3
The Finite Element Method for Linear Elliptic Boundary Value Problems of Second Order

3.1 Variational Equations and Sobolev Spaces

We now continue the definition and analysis of the "correct" function spaces that we began in (2.19)–(2.22). An essential assumption ensuring the existence of a solution of the variational equation (2.13) is the completeness of the basic space $(V, \|\cdot\|)$. In the concrete case of the Poisson equation the "preliminary" function space V according to (2.7) can be equipped with the norm $\|\cdot\|_1$, defined in (2.21), which has been shown to be equivalent to the norm $\|\cdot\|_a$, given in (2.6) (see (2.50)). If we consider the minimization problem (2.14), which is equivalent to the variational equation, the functional F is bounded from below such that the infimum assumes a finite value and there exists a *minimal sequence* $(v_n)_n$ in V, that is, a sequence with the property

$$\lim_{n\to\infty} F(v_n) = \inf_{v\in V} F(v) .$$

The form of F also implies that $(v_n)_n$ is a Cauchy sequence. If this sequence converges to an element $v \in V$, then, due to the continuity of F with respect to $\|\cdot\|$, it follows that v is a solution of the minimization problem. This completeness of V with respect to $\|\cdot\|_a$, and hence with respect to $\|\cdot\|_1$, is not satisfied in the definition (2.7), as Example 2.8 has shown. Therefore, an extension of the basic space V, as formulated in (2.22), is necessary. This space will turn out to be "correct", since it is complete with respect to $\|\cdot\|_1$.

In what follows we use the following **general assumption**:

$$\begin{aligned}
&V \text{ is a real vector space with scalar product } \langle\cdot,\cdot\rangle \text{ and the norm } \|\cdot\| \\
&\quad\text{induced by } \langle\cdot,\cdot\rangle \text{ (for this, } \|v\| := \langle v, v\rangle^{1/2} \text{ for all } v \in V \text{ is satisfied)}\,; \\
&V \text{ is complete with respect to } \|\cdot\|, \text{ i.e., a Hilbert space}\,; \\
&a : V \times V \to \mathbb{R} \text{ is a (not necessarily symmetric) bilinear form}\,; \\
&\ell : V \to \mathbb{R} \text{ is a linear form}\,.
\end{aligned} \tag{3.1}$$

P. Knabner and L. Angermann, *Numerical Methods for Elliptic and Parabolic Partial Differential Equations*, Texts in Applied Mathematics 44,
https://doi.org/10.1007/978-3-030-79385-2_3

The following theorem generalizes the above consideration to nonsymmetric bilinear forms:

Theorem 3.1 (Lax–Milgram) *Suppose the following conditions are satisfied:*

1) The bilinear form a is continuous (cf. (2.46)), that is, there exists some constant $M > 0$ such that

$$|a(u, v)| \leq M\|u\| \, \|v\| \quad \text{for all } u, v \in V; \tag{3.2}$$

*2) The bilinear form a is V-*coercive *(cf. (2.47)), that is, there exists some constant $\alpha > 0$ such that*

$$a(v, v) \geq \alpha\|v\|^2 \quad \text{for all } v \in V; \tag{3.3}$$

3) The linear form ℓ is continuous, that is, there exists some constant $C > 0$ such that

$$|\ell(v)| \leq C\|v\| \quad \text{for all } v \in V. \tag{3.4}$$

Then the variational equation (2.24), namely,

$$\text{find } \overline{u} \in V \text{ such that} \quad a(\overline{u}, v) = \ell(v) \quad \text{for all } v \in V, \tag{3.5}$$

has one and only one solution.

The solution $\overline{u}$ is stable in the sense that

$$\|\overline{u}\| \leq \frac{C}{\alpha}$$

Here, one cannot avoid assumptions (3.1) and (3.2)–(3.4) in general.

Proof See, for example, [51]; for an alternative proof see Problem 3.1. □

Now returning to the example above, assumptions (3.2) and (3.3) are obviously satisfied for $\|\cdot\| = \|\cdot\|_a$. However, the "preliminary" definition of the function space V of (2.7) with norm $\|\cdot\|_a$ defined in (2.21) is insufficient, since $(V, \|\cdot\|_a)$ is not complete. Therefore, the space V must be extended. Indeed, it is not the norm on V that has been chosen incorrectly, since V is also not complete with respect to another norm $\|\cdot\|$ that satisfies (3.2) and (3.3). In this case the norms $\|\cdot\|$ and $\|\cdot\|_a$ would be equivalent (cf. (2.50)), and consequently

$$(V, \|\cdot\|_a) \text{ complete} \quad \Longleftrightarrow \quad (V, \|\cdot\|) \text{ complete.}$$

Now we extend the space V and thereby generalize definition (2.19).

Definition 3.2 Suppose $\Omega \subset \mathbb{R}^d$ is a (bounded) domain.
The *Sobolev space* $H^k(\Omega)$ is defined by

$$H^k(\Omega) := \big\{ v : \Omega \to \mathbb{R} \,\big|\, v \in L^2(\Omega) \text{ and the weak derivatives } \partial^\alpha v \text{ exist} \\ \text{in } L^2(\Omega) \text{ for all multi-indices } \alpha \in \mathbb{N}_0^d \text{ with } |\alpha| \leq k \big\} .$$

A scalar product $\langle\cdot,\cdot\rangle_k$ and the resulting norm $\|\cdot\|_k$ in $H^k(\Omega)$ are defined as follows:

$$\langle v, w\rangle_k := \int_\Omega \sum_{\alpha\in\mathbb{N}_0^d:\ |\alpha|\le k} \partial^\alpha v\, \partial^\alpha w\, dx\,, \tag{3.6}$$

$$\|v\|_k := \langle v, v\rangle_k^{1/2} = \left(\int_\Omega \sum_{\alpha\in\mathbb{N}_0^d:\ |\alpha|\le k} |\partial^\alpha v|^2\, dx\right)^{1/2} \tag{3.7}$$

$$= \left(\sum_{\alpha\in\mathbb{N}_0^d:\ |\alpha|\le k} \int_\Omega |\partial^\alpha v|^2\, dx\right)^{1/2} = \left(\sum_{\alpha\in\mathbb{N}_0^d:\ |\alpha|\le k} \|\partial^\alpha v\|_0^2\right)^{1/2}. \tag{3.8}$$

Greater flexibility with respect to the smoothness properties of the functions that are contained in the definition is obtained by requiring that v and its weak derivatives should belong not to $L^2(\Omega)$ but to $L^p(\Omega)$. In the norm denoted by $\|\cdot\|_{k,p}$ the $L^2(\Omega)$ and ℓ_2 norms (for the vector of the derivative norms) have to be replaced by the $L^p(\Omega)$ and ℓ_p norms, respectively (see Appendices A.3 and A.5). However, the resulting space, denoted by $W_p^k(\Omega)$, can no longer be equipped with a scalar product for $p \neq 2$. Although these spaces offer greater flexibility, we will not use them except in Sections 3.6, 8.2, and 10.3.

Besides the norms $\|\cdot\|_k$, there are seminorms $|\cdot|_l$ for $0 \le l \le k$ in $H^k(\Omega)$, defined by

$$|v|_l := \left(\sum_{\alpha\in\mathbb{N}_0^d:\ |\alpha|\le l} \|\partial^\alpha v\|_0^2\right)^{1/2},$$

such that

$$\|v\|_k = \left(\sum_{l=0}^k |v|_l^2\right)^{1/2}.$$

In particular, these definitions are compatible with those in (2.20),

$$\langle v, w\rangle_1 := \int_\Omega vw + \nabla v\cdot\nabla w\, dx\,,$$

and with the notation $\|\cdot\|_0$ for the $L^2(\Omega)$ norm, giving a meaning to this one.

The above definition contains some assertions that are formulated in the following theorem.

Theorem 3.3 *The bilinear form $\langle\cdot,\cdot\rangle_k$ is a scalar product on $H^k(\Omega)$, that is, $\|\cdot\|_k$ is a norm on $H^k(\Omega)$.*

$H^k(\Omega)$ is complete with respect to $\|\cdot\|_k$, and is thus a Hilbert space.

Proof See, for example, [66]. □

Obviously,

$$H^k(\Omega) \subset H^l(\Omega) \quad \text{for } k \ge l\,,$$

and the embedding is continuous, since

$$\|v\|_l \leq \|v\|_k \quad \text{for all } v \in H^k(\Omega)\,. \tag{3.9}$$

In the one-dimensional case ($d = 1$) $v \in H^1(\Omega)$ is necessarily continuous:

Lemma 3.4

$$H^1(a, b) \subset C[a, b]\,,$$

and the embedding is continuous, where $C[a, b]$ is equipped with the norm $\|\cdot\|_\infty$, that is, there exists some constant $C > 0$ such that

$$\|v\|_\infty \leq C\|v\|_1 \quad \textit{for all } v \in H^1(a, b)\,. \tag{3.10}$$

Proof See Problem 3.2. □

Since the elements of $H^k(\Omega)$ are first of all only square integrable functions, they are determined only up to points of a set of (d-dimensional) measure zero. Therefore, a result as in Lemma 3.4 means that the function is allowed to have removable discontinuities at points of such a set of measure zero that vanish by modifying the function values.

However, in general, $H^1(\Omega) \not\subset C(\overline{\Omega})$.

As an example for this, we consider a circular domain in dimension $d = 2$:

$$\Omega = B_R(0) = \left\{x \in \mathbb{R}^2 \;\middle|\; |x| < R\right\}, \quad R < 1\,.$$

Then the function

$$v(x) := |\log |x|\,|^\gamma \quad \text{for some } \gamma < \frac{1}{2}$$

is in $H^1(\Omega)$, but not in $C(\overline{\Omega})$ (see Problem 3.3).

The following problem now arises: In general, one cannot speak of a value $v(x)$ for some $x \in \Omega$ because a set of one point $\{x\}$ has (Lebesgue) measure zero. How do we then have to interpret the Dirichlet boundary conditions? A way out is to consider the boundary (pieces of the boundary, respectively) not as arbitrary points but as $(d-1)$-dimensional "spaces" (manifolds).

The above question can therefore be reformulated as follows: Is it possible to interpret v on $\partial\Omega$ as a function of $L^2(\partial\Omega)$ ($\partial\Omega$ "$\subset$" $\mathbb{R}^{d-1}$)?

It is indeed possible if we have some minimal regularity of $\partial\Omega$ in the following sense: It has to be possible to choose locally, for some boundary point $x \in \partial\Omega$, a coordinate system in such a way that the boundary is locally a hyperplane in this coordinate system and the domain lies on one side. Depending on the smoothness of the parametrization of the hypersurface we then speak of *Lipschitz*, *C^k- (for $k \in \mathbb{N}$)*, and *C^∞- domains* (for an exact definition see Appendix A.5).

Example 3.5 *1) A circle $\Omega := \left\{x \in \mathbb{R}^d \;\middle|\; |x - x_0| < R\right\}$ is a C^k-domain for all $k \in \mathbb{N}$, and hence a C^∞-domain.*

2) A rectangle $\Omega := \{x \in \mathbb{R}^d \mid 0 < x_i < a_i,\ i = 1, \ldots, d\}$ *is a Lipschitz domain, but not a* C^1*-domain.*

3) A circle with a cut $\Omega := \{x \in \mathbb{R}^d \mid |x - x_0| < R,\ x \neq x_0 + \lambda e_1 \text{ for } 0 \leq \lambda < R\}$ *is not a Lipschitz domain, since* Ω *does not lie on one side of* $\partial\Omega$ *(see Figure 3.1).*

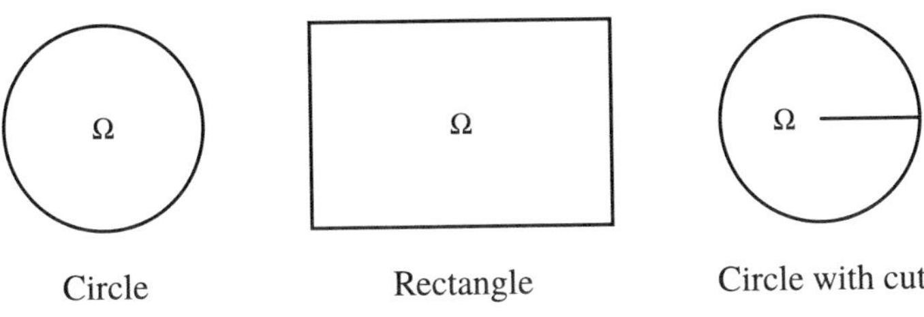

Fig. 3.1: Domains of different smoothness.

Hence, suppose Ω is a Lipschitz domain. Since only a finite number of overlapping coordinate systems are sufficient for the description of $\partial\Omega$, using these, it is possible to introduce a $(d-1)$-dimensional measure on $\partial\Omega$ and define the space $L^2(\partial\Omega)$ of square integrable functions with respect to this measure (see Appendix A.5 or [66] for an extensive description). In the following, let $\partial\Omega$ be equipped with this $(d-1)$-dimensional measure $d\sigma$, and integrals over the boundary are to be interpreted accordingly. This also holds for Lipschitz subdomains of Ω, as they are given by the finite element partition, for example.

Theorem 3.6 (Trace Theorem) *Suppose* Ω *is a bounded Lipschitz domain. We define*

$$C^\infty(\mathbb{R}^d)|_\Omega := \left\{v : \Omega \to \mathbb{R} \,\middle|\, v \text{ can be extended to } \tilde{v} : \mathbb{R}^d \to \mathbb{R} \text{ and } \tilde{v} \in C^\infty(\mathbb{R}^d)\right\}.$$

Then, $C^\infty(\mathbb{R}^d)|_\Omega$ *is* dense *in* $H^1(\Omega)$*, that is, with respect to* $\|\cdot\|_1$ *an arbitrary* $w \in H^1(\Omega)$ *can be approximated arbitrarily well by some* $v \in C^\infty(\mathbb{R}^d)|_\Omega$*.*

The mapping that restricts v *to* $\partial\Omega$*,*

$$\gamma_0 : \left(C^\infty(\mathbb{R}^d)|_\Omega, \|\cdot\|_1\right) \to \left(L^2(\partial\Omega), \|\cdot\|_0\right), \quad v \mapsto v|_{\partial\Omega},$$

is continuous.

Thus there exists a unique, linear, and continuous extension

$$\gamma_0 : \left(H^1(\Omega), \|\cdot\|_1\right) \to \left(L^2(\partial\Omega), \|\cdot\|_0\right).$$

Proof See, for example, [66]. □

Therefore, in short form, $\gamma_0(v) \in L^2(\partial\Omega)$, and there exists some constant $C > 0$ such that

$$\|\gamma_0(v)\|_0 \le C\|v\|_1 \quad \text{for all } v \in H^1(\Omega).$$

Here $\gamma_0(v) \in L^2(\partial\Omega)$ is called the *trace* of $v \in H^1(\Omega)$.

The mapping γ_0 is not surjective, that is, $\{\gamma_0(v) \mid v \in H^1(\Omega)\}$ is a real subset of $L^2(\partial\Omega)$. For all $v \in C^\infty(\mathbb{R}^d)|_\Omega$ we have

$$\gamma_0(v) = v|_{\partial\Omega}.$$

In the following we will use again $v|_{\partial\Omega}$ or "v on $\partial\Omega$" for $\gamma_0(v)$, but in the sense of Theorem 3.6. According to this theorem, definition (2.22) is well defined with the interpretation of u on $\partial\Omega$ as the trace:

Definition 3.7 $H_0^1(\Omega) := \{v \in H^1(\Omega) \mid \gamma_0(v) = 0 \text{ (as a function on } \partial\Omega)\}$.

Theorem 3.8 *Suppose $\Omega \subset \mathbb{R}^d$ is a bounded Lipschitz domain. Then $C_0^\infty(\Omega)$ is dense in $H_0^1(\Omega)$.*

Proof See [66]. □

The assertion of Theorem 3.6, that $C^\infty(\mathbb{R}^d)|_\Omega$ is dense in $H^1(\Omega)$, has severe consequences for the treatment of functions in $H^1(\Omega)$ which are in general not very smooth. It is possible to consider them as smooth functions if at the end only relations involving continuous expressions in $\|\cdot\|_1$ (and not requiring something like $\|\partial_i v\|_\infty$) arise. Then, by some "density argument" the result can be transferred to $H^1(\Omega)$ or, as for the trace term, new terms can be defined for functions in $H^1(\Omega)$. Thus, for the proof of Lemma 3.4 it is necessary simply to verify estimate (3.10), for example, for $v \in C^1[a, b]$. By virtue of Theorem 3.8, analogous results hold for $H_0^1(\Omega)$.
Hence, for $v \in H^1(\Omega)$ integration by parts is possible.

Theorem 3.9 *Suppose $\Omega \subset \mathbb{R}^d$ is a bounded Lipschitz domain. The outer unit normal vector $\mathfrak{n} = (\mathfrak{n}_i)_{i=1,\dots,d} : \partial\Omega \to \mathbb{R}^d$ is defined almost everywhere and $\mathfrak{n}_i \in L^\infty(\partial\Omega)$.*
For $v, w \in H^1(\Omega)$ and $i = 1, \dots, d$,

$$\int_\Omega \partial_i v \, w \, dx = -\int_\Omega v \, \partial_i w \, dx + \int_{\partial\Omega} v \, w \, \mathfrak{n}_i \, d\sigma.$$

Proof See, for example, [29] or [66]. □

If $v \in H^2(\Omega)$, then due to the above theorem, $v|_{\partial\Omega} := \gamma_0(v) \in L^2(\partial\Omega)$ and $\partial_i v|_{\partial\Omega} := \gamma_0(\partial_i v) \in L^2(\partial\Omega)$, since also $\partial_i v \in H^1(\Omega)$. Hence, the *normal derivative*

$$\partial_{\mathfrak{n}} v|_{\partial\Omega} := \sum_{i=1}^{d} \partial_i v|_{\partial\Omega} \, \mathfrak{n}_i$$

is well defined and belongs to $L^2(\partial\Omega)$.

Thus, the trace mapping

$$\gamma : H^2(\Omega) \to L^2(\partial\Omega) \times L^2(\partial\Omega), \; v \mapsto (v|_{\partial\Omega}, \partial_{\mathfrak{n}} v|_{\partial\Omega})$$

is well defined and continuous. The continuity of this mapping follows from the fact that it is a composition of continuous mappings:

$$\begin{aligned} v \in H^2(\Omega) \overset{\text{continuous}}{\mapsto} \partial_i v \in H^1(\Omega) &\overset{\text{continuous}}{\mapsto} \partial_i v|_{\partial\Omega} \in L^2(\partial\Omega) \\ &\overset{\text{continuous}}{\mapsto} \partial_i v|_{\partial\Omega}\, \mathfrak{n}_i \in L^2(\partial\Omega)\,. \end{aligned}$$

Corollary 3.10 *Suppose $\Omega \subset \mathbb{R}^d$ is a bounded Lipschitz domain.*

1) Let $w \in H^1(\Omega)$, $q_i \in H^1(\Omega)$, $i = 1, \ldots, d$. Then

$$\int_\Omega q \cdot \nabla w \, dx = -\int_\Omega \nabla \cdot q \, w \, dx + \int_{\partial\Omega} q \cdot \mathfrak{n} \, w \, d\sigma\,. \tag{3.11}$$

2) Let $v \in H^2(\Omega)$, $w \in H^1(\Omega)$. Then

$$\int_\Omega \nabla v \cdot \nabla w \, dx = -\int_\Omega \Delta v \, w \, dx + \int_{\partial\Omega} \partial_{\mathfrak{n}} v \, w \, d\sigma\,.$$

The integration by parts formulas also hold more generally if only it is ensured that the function whose trace has to be formed belongs to $H^1(\Omega)$. For example, if $\boldsymbol{K} = \left(k_{ij}\right)_{ij}$, where $k_{ij} \in W^1_\infty(\Omega)$ and $v \in H^2(\Omega)$, $w \in H^1(\Omega)$, it follows that

$$\int_\Omega \boldsymbol{K}\nabla v \cdot \nabla w \, dx = -\int_\Omega \nabla \cdot (\boldsymbol{K}\nabla v) \, w \, dx + \int_{\partial\Omega} \boldsymbol{K}\nabla v \cdot \mathfrak{n} \, w \, d\sigma \tag{3.12}$$

with *conormal derivative* (see (0.41))

$$\partial_{\mathfrak{n}_{\boldsymbol{K}}} v := \boldsymbol{K}\nabla v \cdot \mathfrak{n} = \nabla v \cdot \boldsymbol{K}^T \mathfrak{n} = \sum_{i,j=1}^{d} k_{ij} \partial_j v \, \mathfrak{n}_i\,.$$

Here it is important that the components of $\boldsymbol{K}\nabla v$ belong to $H^1(\Omega)$, using the fact that for $v \in L^2(\Omega)$, $k \in L^\infty(\Omega)$,

$$kv \in L^2(\Omega) \quad \text{and} \quad \|kv\|_0 \le \|k\|_\infty \|v\|_0\,.$$

Theorem 3.11 *Suppose $\Omega \subset \mathbb{R}^d$ is a bounded Lipschitz domain.*

If $k > d/2$, then

$$H^k(\Omega) \subset C(\overline{\Omega}),$$

and the embedding is continuous.

Proof See, for example, [66]. □

For dimension $d = 2$ this requires $k > 1$, and for dimension $d = 3$ we need $k > \frac{3}{2}$. Therefore, in both cases $k = 2$ satisfies the assumption of the above theorem.

Exercises

Problem 3.1 Prove the Lax–Milgram theorem in the following way:

a) Show, by using the Riesz representation theorem, the equivalence of (3.5) with the operator equation
$$A\overline{u} = f$$
for $A \in L[V, V]$ and $f \in V$.

b) Show, for $T_\varepsilon : V \to V$, $T_\varepsilon v := v - \varepsilon(Av - f)$ and $\varepsilon > 0$, that for some $\varepsilon > 0$, the operator T_ε is a contraction on V. Then conclude the assertion by Banach's fixed-point theorem (in the Banach space setting, cf. Remark 11.5).

Problem 3.2 Prove estimate (3.10) by showing that even for $v \in H^1(a, b)$,
$$|v(x) - v(y)| \le |v|_1 |x - y|^{1/2} \quad \text{for all } x, y \in (a, b)\,.$$

Problem 3.3 Suppose $\Omega \subset \mathbb{R}^2$ is the open disc with radius $\frac{1}{2}$ and centre 0. Prove that for the function $u(x) := \big| \ln |x| \big|^\alpha$, $x \in \Omega \setminus \{0\}$, $\alpha \in (0, \frac{1}{2})$ we have $u \in H^1(\Omega)$, but u cannot be extended continuously to $x = 0$.

Problem 3.4 Suppose $\Omega \subset \mathbb{R}^2$ is the open unit disc. Prove that each $u \in H^1(\Omega)$ has a trace $u|_{\partial\Omega} \in L_2(\partial\Omega)$ satisfying $\|u\|_{0,\partial\Omega} \le \sqrt[4]{8}\, \|u\|_{1,\Omega}$.

3.2 Elliptic Boundary Value Problems of Second Order

In this section we integrate boundary value problems for the linear, stationary case of the differential equation (0.33) into the general theory of Section 3.1.

Concerning the domain we will assume that Ω is a bounded Lipschitz domain. We consider the equation
$$(Lu)(x) := -\nabla \cdot (\boldsymbol{K}(x)\nabla u(x)) + \boldsymbol{c}(x) \cdot \nabla u(x) + r(x)u(x) = f(x) \text{ for } x \in \Omega \quad (3.13)$$
with the data
$$\boldsymbol{K} : \Omega \to \mathbb{R}^{d,d}, \quad \boldsymbol{c} : \Omega \to \mathbb{R}^d, \quad r, f : \Omega \to \mathbb{R}.$$

Assumptions about the Coefficients and the Right-Hand Side

For an interpretation of (3.13) in the classical sense, we need

$$\partial_i k_{ij}, c_i, r, f \in C(\overline{\Omega}), \quad i, j \in \{1, \ldots, d\}, \tag{3.14}$$

and for an interpretation in the sense of $L^2(\Omega)$ with weak derivatives, and hence for a solution in $H^2(\Omega)$,

$$\partial_i k_{ij}, c_i, r \in L^\infty(\Omega), \; f \in L^2(\Omega), \quad i, j \in \{1, \ldots, d\}. \tag{3.15}$$

Once we have obtained the variational formulation, weaker assumptions about the smoothness of the coefficients will be sufficient for the verification of the properties (3.2)–(3.4), which are required by the Lax–Milgram theorem, namely,

$$\begin{aligned} k_{ij}, c_i, \nabla \cdot \boldsymbol{c}, r \in L^\infty(\Omega), \; f \in L^2(\Omega), \quad i, j \in \{1, \ldots, d\}, \\ \text{and if } |\Gamma_1 \cup \Gamma_2|_{d-1} > 0, \quad \boldsymbol{n} \cdot \boldsymbol{c} \in L^\infty(\Gamma_1 \cup \Gamma_2). \end{aligned} \tag{3.16}$$

Here we refer to a definition of the boundary conditions as in (0.36)–(0.39) (see also below). Furthermore, the *uniform coercivity* of L is assumed: There exists some constant $k_0 > 0$ such that for (almost) every $x \in \Omega$,

$$\sum_{i,j=1}^{d} k_{ij}(x)\xi_i\xi_j \geq k_0|\xi|^2 \quad \text{for all } \xi \in \mathbb{R}^d \tag{3.17}$$

(that is, the coefficient matrix $\boldsymbol{K}$ is positive definite uniformly in x). Moreover, $\boldsymbol{K}$ should be symmetric.

If $\boldsymbol{K}$ is a diagonal matrix, that is, $k_{ij}(x) = k_i(x)\delta_{ij}$ (this is, in particular, the case if $k_i(x) = k(x)$ with $k : \Omega \to \mathbb{R}$, $i \in \{1, \ldots, d\}$, where $\boldsymbol{K}\nabla u$ becomes $k\nabla u$), this means that

$$(3.17) \quad \Leftrightarrow \quad k_i(x) \geq k_0 \text{ for (almost) every } x \in \Omega, \quad i \in \{1, \ldots, d\}.$$

Finally, there exists a constant $r_0 \geq 0$ such that

$$r(x) - \frac{1}{2}\nabla \cdot \boldsymbol{c}(x) \geq r_0 \quad \text{for (almost) every } x \in \Omega. \tag{3.18}$$

Boundary Conditions

As in Section 0.5, suppose $\Gamma_1, \Gamma_2, \Gamma_3$ is a disjoint decomposition of the boundary $\partial\Omega$ (cf. (0.39)):

$$\partial\Omega = \Gamma_1 \cup \Gamma_2 \cup \Gamma_3,$$

where Γ_3 is a relatively closed subset of the boundary (i.e., closed in the induced topology of $\partial\Omega$). For given functions $g_j : \Gamma_j \to \mathbb{R}$, $j = 1, 2, 3$, and $\alpha : \Gamma_2 \to \mathbb{R}$ we assume on $\partial\Omega$

- Neumann boundary condition (cf. (0.41) or (0.36))

$$\boldsymbol{K}\nabla u \cdot \mathfrak{n} = \partial_{\mathfrak{n}_{\boldsymbol{K}}} u = g_1 \quad \text{on } \Gamma_1 , \tag{3.19}$$

- mixed boundary condition (cf. (0.37))

$$\boldsymbol{K}\nabla u \cdot \mathfrak{n} + \alpha u = \partial_{\mathfrak{n}_{\boldsymbol{K}}} u + \alpha u = g_2 \quad \text{on } \Gamma_2 , \tag{3.20}$$

- Dirichlet boundary condition (cf. (0.38))

$$u = g_3 \quad \text{on } \Gamma_3 . \tag{3.21}$$

Concerning the boundary data the following is assumed: For the classical approach we need

$$g_j \in C(\overline{\Gamma}_j), \quad j = 1, 2, 3, \quad \alpha \in C(\overline{\Gamma}_2), \tag{3.22}$$

whereas for the variational interpretation,

$$g_j \in L^2(\Gamma_j), \quad j = 1, 2, 3, \quad \alpha \in L^\infty(\Gamma_2) \tag{3.23}$$

is sufficient.

3.2.1 Variational Formulation of Special Cases

The basic strategy for the derivation of the variational formulation of boundary value problems (3.13) has already been demonstrated in Section 2.1. Assuming the existence of a classical solution of (3.13) the following steps are performed in general:

Step 1: Multiplication of the differential equation by test functions that are chosen compatible with the type of boundary condition and subsequent integration over the domain Ω.

Step 2: Integration by parts under incorporation of the boundary conditions in order to derive a suitable bilinear form.

Step 3: Verification of the required properties like coercivity and continuity.

In the following the above steps will be described for some important special cases.

(I) Homogeneous Dirichlet Boundary Conditions: $\partial\Omega := \Gamma_3$, $g_3 := 0$, $V := H_0^1(\Omega)$

Suppose u is a solution of (3.13), (3.21), that is, in the sense of classical solutions let $u \in C^2(\Omega) \cap C(\overline{\Omega})$ and the differential equation (3.13) be satisfied pointwise in Ω under assumptions (3.14) as well as $u = 0$ pointwise on $\partial\Omega$. However, the weaker case in which $u \in H^2(\Omega) \cap V$ and the differential equation is satisfied in the sense of $L^2(\Omega)$, now under the assumptions (3.15), can also be considered.

Multiplying (3.13) by $v \in C_0^\infty(\Omega)$ (in the classical case) or by $v \in V$, respectively, then integrating by parts according to (3.12) and taking into account that $v = 0$ on $\partial\Omega$ by virtue of the definition of $C_0^\infty(\Omega)$ and $H_0^1(\Omega)$, respectively, we obtain

$$\begin{aligned} a(u, v) &:= \int_\Omega \{\boldsymbol{K}\nabla u \cdot \nabla v + \boldsymbol{c} \cdot \nabla u\, v + r\, uv\}\, dx \\ &= \ell(v) := \int_\Omega f v\, dx \quad \text{for all } v \in C_0^\infty(\Omega) \text{ or } v \in V . \end{aligned} \tag{3.24}$$

The bilinear form a is symmetric if $\boldsymbol{c}$ vanishes (almost everywhere).

For $f \in L^2(\Omega)$,

$$\ell \text{ is continuous on } (V, \|\cdot\|_1) . \tag{3.25}$$

This follows directly from the Cauchy–Schwarz inequality, since

$$|\ell(v)| \leq \int_\Omega |f|\, |v|\, dx \leq \|f\|_0\, \|v\|_0 \leq \|f\|_0\, \|v\|_1 \quad \text{for } v \in V .$$

Further, by (3.16),

$$a \text{ is continuous on } (V, \|\cdot\|_1) . \tag{3.26}$$

Proof First, we obtain

$$|a(u, v)| \leq \int_\Omega \{|\boldsymbol{K}\nabla u|\, |\nabla v| + |\boldsymbol{c}|\, |\nabla u||v| + |r|\, |u|\, |v|\}\, dx .$$

Here $|\cdot|$ denotes the absolute value of a real number or the Euclidean norm of a vector. Using also $\|\cdot\|_2$ for the (associated) spectral norm, and $\|\cdot\|_\infty$ for the $L^\infty(\Omega)$ norm of a function, we further introduce the following notation:

$$C_1 := \max\left\{\left\|\|\boldsymbol{K}\|_2\right\|_\infty, \|r\|_\infty\right\} < \infty, \quad C_2 := \left\||\boldsymbol{c}|\right\|_\infty < \infty .$$

By virtue of

$$|\boldsymbol{K}(x)\nabla u(x)| \leq \|\boldsymbol{K}(x)\|_2\, |\nabla u(x)|,$$

we continue to estimate as follows:

$$|a(u,v)| \leq \underbrace{C_1 \int_\Omega \{|\nabla u|\,|\nabla v| + |u|\,|v|\}\,dx}_{=:A_1} + \underbrace{C_2 \int_\Omega |\nabla u|\,|v|\,dx}_{=:A_2} .$$

The integrand of the first addend is estimated by the Cauchy–Schwarz inequality for $\mathbb{R}^2$, and then the Cauchy–Schwarz inequality for $L^2(\Omega)$ is applied:

$$\begin{aligned} A_1 &\leq C_1 \int_\Omega \left(|\nabla u|^2 + |u|^2\right)^{1/2} \left(|\nabla v|^2 + |v|^2\right)^{1/2} dx \\ &\leq C_1 \left(\int_\Omega |u|^2 + |\nabla u|^2\,dx\right)^{1/2} \left(\int_\Omega |v|^2 + |\nabla v|^2\,dx\right)^{1/2} = C_1 \|u\|_1\, \|v\|_1 . \end{aligned}$$

Dealing with A_2, we can employ the Cauchy–Schwarz inequality for $L^2(\Omega)$ directly:

$$\begin{aligned} A_2 &\leq C_2 \left(\int_\Omega |\nabla u|^2\,dx\right)^{1/2} \left(\int_\Omega |v|^2\,dx\right)^{1/2} \\ &\leq C_2 \|u\|_1\, \|v\|_0 \leq C_2 \|u\|_1\, \|v\|_1 \quad \text{for all } u, v \in V . \end{aligned}$$

Thus, the assertion follows. □

Remark 3.12 In the proof of propositions (3.25) and (3.26) it has not been used that the functions u, v satisfy homogeneous Dirichlet boundary conditions. Therefore, under the assumptions (3.16) these properties hold for every subspace $V \subset H^1(\Omega)$.

Conditions for the V-Coercivity of a

(A) a is symmetric, that is, $\boldsymbol{c} = \boldsymbol{0}$ (a.e.): Condition (3.18) then has the simple form $r(x) \geq r_0$ for almost all $x \in \Omega$.

(A1) $\boldsymbol{c} = \boldsymbol{0}$, $r_0 > 0$:

Because of (3.17) we directly get

$$a(v,v) \geq \int_\Omega \{k_0|\nabla v|^2 + r_0|v|^2\}\,dx \geq C_3\|v\|_1^2 \quad \text{for all } v \in V,$$

where $C_3 := \min\{k_0, r_0\}$. This also holds for every subspace $V \subset H^1(\Omega)$.

(A2) $\boldsymbol{c} = \boldsymbol{0}$, $r_0 \geq 0$:

According to the Poincaré inequality (Theorem 2.18), there exists some constant $C_P > 0$, independent of v, such that

$$\|v\|_0 \leq C_P \left(\int_\Omega |\nabla v|^2\,dx\right)^{1/2} \quad \text{for all } v \in H_0^1(\Omega).$$

Taking into account (3.17) and using the simple decomposition $k_0 = \frac{k_0}{1+C_P^2} + \frac{C_P^2}{1+C_P^2}\,k_0$ we can further conclude that

$$
\begin{aligned}
a(v,v) &\geq \int_\Omega k_0 |\nabla v|^2\,dx \\
&\geq \frac{k_0}{1+C_P^2}\int_\Omega |\nabla v|^2\,dx + \frac{C_P^2}{1+C_P^2}\,k_0 \frac{1}{C_P^2}\int_\Omega |v|^2\,dx = C_4\|v\|_1^2\,,
\end{aligned}
\tag{3.27}
$$

where $C_4 := \dfrac{k_0}{1+C_P^2} > 0$.

For this estimate it is essential that v satisfies the homogeneous Dirichlet boundary condition.

(B) $\big\||\boldsymbol{c}|\big\|_\infty > 0$:
First of all, we consider a smooth function $v \in C_0^\infty(\Omega)$. From $v\nabla v = \frac{1}{2}\nabla v^2$ we get by integrating by parts

$$
\int_\Omega \boldsymbol{c}\cdot\nabla v\, v\,dx = \frac{1}{2}\int_\Omega \boldsymbol{c}\cdot\nabla v^2\,dx = -\frac{1}{2}\int_\Omega \nabla\cdot\boldsymbol{c}\, v^2\,dx\,.
$$

Since according to Theorem 3.8 the space $C_0^\infty(\Omega)$ is dense in V, the above relation also holds for $v \in V$. Consequently, by virtue of (3.17) and (3.18) we obtain

$$
\begin{aligned}
a(v,v) &= \int_\Omega \left\{\boldsymbol{K}\nabla v\cdot\nabla v + \left(r - \frac{1}{2}\nabla\cdot\boldsymbol{c}\right)v^2\right\}dx \\
&\geq \int_\Omega \{k_0|\nabla v|^2 + r_0|v|^2\}\,dx \quad \text{for all } v \in V\,.
\end{aligned}
\tag{3.28}
$$

Hence, a distinction concerning r_0 as in **(A)** with the same results (constants) is possible.

Summarizing, we have therefore proven the following application of the Lax–Milgram theorem (Theorem 3.1).

Theorem 3.13 *Suppose $\Omega \subset \mathbb{R}^d$ is a bounded Lipschitz domain. Under the assumptions (3.16)–(3.18) the homogeneous Dirichlet problem has one and only one weak solution $u \in H_0^1(\Omega)$.*

(II) Mixed Boundary Conditions: $\partial\Omega := \Gamma_2\,,\ V := H^1(\Omega)$

Suppose u is a solution of (3.13), (3.20), that is, in the classical sense let $u \in C^2(\Omega)\cap C^1(\overline{\Omega})$ and the differential equation (3.13) be satisfied pointwise in Ω and (3.20) pointwise on $\partial\Omega$ under assumptions (3.14), (3.22). However, the weaker case can again be considered, now under the assumptions (3.15), (3.23), that $u \in H^2(\Omega)$

and the differential equation is satisfied in the sense of $L^2(\Omega)$ as well as the boundary condition (3.20) in the sense of $L^2(\partial\Omega)$.

As in **(I)**, according to (3.12),

$$\begin{aligned} a(u, v) &:= \int_\Omega \{\boldsymbol{K}\nabla u \cdot \nabla v + \boldsymbol{c} \cdot \nabla u\, v + r\, uv\}\, dx + \int_{\partial\Omega} \alpha\, uv\, d\sigma \\ &= \ell(v) := \int_\Omega f v\, dx + \int_{\partial\Omega} g_2 v\, d\sigma \quad \text{for all } v \in V\,. \end{aligned} \tag{3.29}$$

Under the assumptions (3.16), (3.23) the continuity of ℓ and a, respectively, ((3.25) and (3.26)) can easily be shown. The additional new terms can be estimated, for instance, under assumptions (3.16), (3.23), by the Cauchy–Schwarz inequality and the Trace Theorem 3.6 as follows:

$$\left|\int_{\partial\Omega} g_2 v\, d\sigma\right| \le \|g_2\|_{0,\partial\Omega} \|v|_{\partial\Omega}\|_{0,\partial\Omega} \le C\|g_2\|_{0,\partial\Omega}\|v\|_1 \quad \text{for all } v \in V$$

and

$$\begin{aligned} \left|\int_{\partial\Omega} \alpha uv\, d\sigma\right| &\le \|\alpha\|_{\infty,\partial\Omega} \|u|_{\partial\Omega}\|_{0,\partial\Omega} \|v|_{\partial\Omega}\|_{0,\partial\Omega} \\ &\le C^2 \|\alpha\|_{\infty,\partial\Omega}\|u\|_1\|v\|_1 \quad \text{for all } u, v \in V, \end{aligned}$$

where $C > 0$ denotes the constant appearing in the Trace theorem.

Conditions for the V-Coercivity of a

For the proof of the V-coercivity we proceed similarly to **(I)(B)**, but now taking into account the mixed boundary conditions. For the convective term we have

$$\int_\Omega \boldsymbol{c} \cdot \nabla v\, v\, dx = \frac{1}{2}\int_\Omega \boldsymbol{c} \cdot \nabla v^2\, dx = -\frac{1}{2}\int_\Omega \nabla \cdot \boldsymbol{c}\, v^2\, dx + \frac{1}{2}\int_{\partial\Omega} \boldsymbol{\nu} \cdot \boldsymbol{c}\, v^2\, d\sigma\,,$$

and thus

$$a(v, v) = \int_\Omega \left\{\boldsymbol{K}\nabla v \cdot \nabla v + \left(r - \frac{1}{2}\nabla \cdot \boldsymbol{c}\right) v^2\right\} dx + \int_{\partial\Omega} \left(\alpha + \frac{1}{2}\boldsymbol{\nu} \cdot \boldsymbol{c}\right) v^2\, d\sigma\,.$$

This shows that $\alpha + \frac{1}{2}\boldsymbol{\nu} \cdot \boldsymbol{c} \ge 0$ on $\partial\Omega$ should additionally be assumed. If $r_0 > 0$ in (3.18), then the V-coercivity of a follows directly. However, if only $r_0 \ge 0$ is valid, then the so-called *Friedrichs' inequality*, a refined version of the Poincaré inequality, helps (see [46, Theorem 1.9]).

Theorem 3.14 *Suppose $\Omega \subset \mathbb{R}^d$ is a bounded Lipschitz domain and let the set $\widetilde{\Gamma} \subset \partial\Omega$ have a positive $(d-1)$-dimensional measure. Then there exists some constant $C_F > 0$ such that*

$$\|v\|_1 \leq C_F \left(\int_{\widetilde{\Gamma}} v^2 \, d\sigma + \int_\Omega |\nabla v|^2 \, dx \right)^{1/2} \quad \textit{for all } v \in H^1(\Omega). \tag{3.30}$$

If $\alpha + \frac{1}{2}\mathbf{n} \cdot \boldsymbol{c} \geq \alpha_0 > 0$ for $x \in \widetilde{\Gamma} \subset \Gamma_2$ and $\widetilde{\Gamma}$ has a positive $(d-1)$-dimensional measure, then $r_0 \geq 0$ is already sufficient for the V-coercivity. Indeed, using Theorem 3.14, we have

$$a(v, v) \geq k_0 |v|_1^2 + \alpha_0 \int_{\widetilde{\Gamma}} v^2 \, d\sigma \geq \min\{k_0, \alpha_0\} \left(|v|_1^2 + \int_{\widetilde{\Gamma}} v^2 \, d\sigma \right) \geq C_5 \|v\|_1^2$$

with $C_5 := C_F^{-2} \min\{k_0, \alpha_0\}$. Therefore, we obtain the existence and uniqueness of a solution analogously to Theorem 3.13.

(III) The General Case

First, we consider the case of a **homogeneous Dirichlet boundary condition** on Γ_3 with $|\Gamma_3|_{d-1} > 0$. For this, we define

$$V := \left\{ v \in H^1(\Omega) \mid \gamma_0(v) = 0 \text{ on } \Gamma_3 \right\} . \tag{3.31}$$

Here V is a closed subspace of $H^1(\Omega)$, since the trace mapping $\gamma_0 : H^1(\Omega) \to L^2(\partial\Omega)$ and the restriction of a function from $L^2(\partial\Omega)$ to $L^2(\Gamma_3)$ are continuous.

Suppose u is a solution of (3.13), (3.19)–(3.21), that is, in the sense of classical solutions let $u \in C^2(\Omega) \cap C^1(\overline{\Omega})$ and the differential equation (3.13) be satisfied pointwise in Ω and the boundary conditions (3.19)–(3.21) pointwise on their respective parts of $\partial\Omega$ under assumptions (3.14), (3.22). However, the weaker case that $u \in H^2(\Omega)$ and the differential equation is satisfied in the sense of $L^2(\Omega)$ and the boundary conditions (3.19)–(3.21) are satisfied in the sense of $L^2(\Gamma_j)$, $j = 1, 2, 3$, under assumptions (3.15), (3.23) can also be considered here.

As in **(I)**, according to (3.12),

$$\begin{aligned} a(u, v) &:= \int_\Omega \{ \boldsymbol{K} \nabla u \cdot \nabla v + \boldsymbol{c} \cdot \nabla u \, v + r \, uv \} \, dx + \int_{\Gamma_2} \alpha \, uv \, d\sigma \\ &= \ell(v) := \int_\Omega f v \, dx + \int_{\Gamma_1} g_1 v \, d\sigma + \int_{\Gamma_2} g_2 v \, d\sigma \quad \text{for all } v \in V. \end{aligned} \tag{3.32}$$

Under assumptions (3.16), (3.23) the continuity of a and ℓ, (3.26)) and ((3.25) can be proven analogously to **(II)**.

Conditions for V-Coercivity of a

For the verification of the V-coercivity we again proceed similarly to **(II)**, but now the boundary conditions are more complicated. Here we have for the convective term

$$\int_\Omega \boldsymbol{c} \cdot \nabla v \, v \, dx = -\frac{1}{2} \int_\Omega \nabla \cdot \boldsymbol{c} \, v^2 \, dx + \frac{1}{2} \int_{\Gamma_1 \cup \Gamma_2} \mathfrak{n} \cdot \boldsymbol{c} v^2 \, d\sigma , \tag{3.33}$$

and therefore

$$\begin{aligned} a(v,v) = & \int_\Omega \left\{ \boldsymbol{K} \nabla v \cdot \nabla v + \left(r - \frac{1}{2} \nabla \cdot \boldsymbol{c} \right) v^2 \right\} dx \\ & + \frac{1}{2} \int_{\Gamma_1} \mathfrak{n} \cdot \boldsymbol{c} \, v^2 \, d\sigma + \int_{\Gamma_2} \left(\alpha + \frac{1}{2} \mathfrak{n} \cdot \boldsymbol{c} \right) v^2 \, d\sigma . \end{aligned}$$

In order to ensure the V-coercivity of a we need, besides the obvious conditions

$$\mathfrak{n} \cdot \boldsymbol{c} \geq 0 \quad \text{on } \Gamma_1 \quad \text{and} \quad \alpha + \frac{1}{2} \mathfrak{n} \cdot \boldsymbol{c} \geq 0 \quad \text{on } \Gamma_2 , \tag{3.34}$$

the following corollary from Theorem 3.14.

Corollary 3.15 *Suppose $\Omega \subset \mathbb{R}^d$ is a bounded Lipschitz domain and $\widetilde{\Gamma} \subset \partial\Omega$ has a positive $(d-1)$-dimensional measure. Then there exists some constant $C_F > 0$ such that*

$$\|v\|_0 \leq C_F \left(\int_\Omega |\nabla v|^2 \, dx \right)^{1/2} = C_F |v|_1 \quad \textit{for all } v \in H^1(\Omega) \textit{ with } v|_{\widetilde{\Gamma}} = 0 .$$

This corollary yields the same results as in the case of homogeneous Dirichlet boundary conditions on the whole of $\partial\Omega$.

If $|\Gamma_3|_{d-1} = 0$, then by tightening conditions (3.34) for c and α, the application of Theorem 3.14 as done in **(II)** may be successful.

Summary

We will now present a summary of our considerations for the case of homogeneous Dirichlet boundary conditions.

Theorem 3.16 *Suppose $\Omega \subset \mathbb{R}^d$ is a bounded Lipschitz domain. Under assumptions (3.16), (3.17), (3.23) with $g_3 = 0$, the boundary value problem (3.13), (3.19)–(3.21) has one and only one* weak solution *$u \in V$, if*

1) $r - \frac{1}{2} \nabla \cdot \boldsymbol{c} \geq 0$ in Ω.
2) $\mathfrak{n} \cdot \boldsymbol{c} \geq 0$ on Γ_1.
3) $\alpha + \frac{1}{2} \mathfrak{n} \cdot \boldsymbol{c} \geq 0$ on Γ_2.
4) Additionally, one of the following conditions is satisfied:

a) $|\Gamma_3|_{d-1} > 0$.
b) There exists some $\tilde{\Omega} \subset \Omega$ with $|\tilde{\Omega}|_d > 0$ and $r_0 > 0$ such that $r - \frac{1}{2} \nabla \cdot \boldsymbol{c} \geq r_0$ on $\tilde{\Omega}$.
c) There exists some $\widetilde{\Gamma}_1 \subset \Gamma_1$ with $|\widetilde{\Gamma}_1|_{d-1} > 0$ and $c_0 > 0$ such that $\mathfrak{n} \cdot \boldsymbol{c} \geq c_0$ on $\widetilde{\Gamma}_1$.

d) There exists some $\widetilde{\Gamma}_2 \subset \Gamma_2$ *with* $|\widetilde{\Gamma}_2|_{d-1} > 0$ *and* $\alpha_0 > 0$ *such that* $\alpha + \frac{1}{2}\mathfrak{n}\cdot\boldsymbol{c} \geq \alpha_0$ *on* $\widetilde{\Gamma}_2$.

Remark 3.17 We point out that by using different techniques in the proof, it is possible to weaken conditions 4)b)–d) in such a way that only the following has to be assumed:

a) $\left|\left\{x \in \Omega : r - \frac{1}{2}\nabla\cdot\boldsymbol{c} > 0\right\}\right|_d > 0,$
b) $\left|\{x \in \Gamma_1 : \mathfrak{n}\cdot\boldsymbol{c} > 0\}\right|_{d-1} > 0,$
c) $\left|\left\{x \in \Gamma_2 : \alpha + \frac{1}{2}\mathfrak{n}\cdot\boldsymbol{c} > 0\right\}\right|_{d-1} > 0\,.$

However, we stress that the conditions of Theorem 3.16 are only sufficient, since concerning the V-coercivity, it might also be possible to balance an indefinite addend by some "particular definite" addend. But this would require conditions in which the constants C_P and C_F are involved.

As one of the many examples we mention: Assume there is a number $\tilde{\alpha} > 0$ such that

$$\tilde{\alpha}\|v\|_1^2 \leq a(v,v) - \int_\Omega c\cdot\nabla v\, v\,dx \quad \text{for all } v \in V.$$

Then V-coercivity holds if $\alpha := \tilde{\alpha} - \frac{1}{2}\|c\|_\infty > 0$ or, based on (3.33), similar smallness requirements for $\nabla\cdot c$ can be postulated.

Note that the pure Neumann problem for the Poisson equation

$$\begin{aligned} -\Delta u &= f \quad \text{in } \Omega, \\ \partial_{\mathfrak{n}} u &= g \quad \text{on } \partial\Omega \end{aligned} \tag{3.35}$$

is excluded by the conditions of Theorem 3.16. This is consistent with the fact that not always a solution of (3.35) exists, and if a solution exists, it obviously is not unique (see Problem 3.8).

Before we investigate inhomogeneous Dirichlet boundary conditions, we introduce an alternative formulation that comes closer to a natural situation described in Chapter 0.

The linear stationary case of the differential equation (0.33) is in the form

$$-\nabla\cdot(\boldsymbol{K}\nabla u - \boldsymbol{c}\,u) + \tilde{r}\,u = f. \tag{3.36}$$

Such a formulation is called *of divergence form*. Considering the total flux density $\boldsymbol{p} := -\boldsymbol{K}\nabla u + \boldsymbol{c}\,u$ as the relevant physical quantity, we assume boundary conditions of the form

$$(\boldsymbol{K}\nabla u - \boldsymbol{c}\,u)\cdot\mathfrak{n} = \tilde{g}_1 \quad \text{on } \widetilde{\Gamma}_1, \tag{3.37}$$

$$(\boldsymbol{K}\nabla u - \boldsymbol{c}\,u)\cdot\mathfrak{n} + \tilde{\alpha}u = \tilde{g}_2 \quad \text{on } \widetilde{\Gamma}_2, \tag{3.38}$$

$$u = \tilde{g}_3 \quad \text{on } \widetilde{\Gamma}_3\,. \tag{3.39}$$

Here $\widetilde{\Gamma}_1, \widetilde{\Gamma}_2, \widetilde{\Gamma}_3$ is a subdivision of $\partial\Omega$ with the same properties as those given by $\Gamma_1, \Gamma_2, \Gamma_3$. The variational formulations take the form

$$\hat{a}(u, v) = \hat{\ell}(v) \quad \text{for all } u, v \in V$$

with V as in (3.31) (Γ_3 substituted by $\widetilde{\Gamma}_3$), $\hat{\ell}$ as ℓ in (3.32) (with Γ_i, g_i substituted by $\widetilde{\Gamma}_i$, $\tilde{g}_i$, $i = 1, 2$). The bilinear form reads as

$$\hat{a}(u, v) := \int_\Omega \big[\boldsymbol{K}\nabla u \nabla v + u\, \tilde{\boldsymbol{c}} \cdot \nabla v + ruv\big] dx + \int_{\widetilde{\Gamma}_2} \tilde{\alpha} uv d\sigma \,, \tag{3.40}$$

where $\tilde{\boldsymbol{c}} := -\boldsymbol{c}$. The relationship between both formulations can be investigated for classical solutions, the same relationship holds in the variational context. We obtain, by differentiating and rearranging the convective term,

$$-\nabla \cdot (\boldsymbol{K}\nabla u) + \boldsymbol{c} \cdot \nabla u + (\nabla \cdot \boldsymbol{c} + \tilde{r})\, u = f \,,$$

which gives the form (3.13) with $r := \nabla \cdot \boldsymbol{c} + \tilde{r}$. The boundary conditions on $\widetilde{\Gamma}_1$ correspond to the mixed boundary conditions (3.20), i.e., $\widetilde{\Gamma}_1 \subset \Gamma_2$, $\alpha := -\mathfrak{n} \cdot \boldsymbol{c}$, those on $\widetilde{\Gamma}_2$ also have the form of (3.20) with $\alpha := -\mathfrak{n} \cdot \boldsymbol{c} + \tilde{\alpha}$, i.e., $\widetilde{\Gamma}_2$ has to be subdivided into $\widetilde{\Gamma}_{2,1} := \{x \in \widetilde{\Gamma}_2 \mid \alpha(x) = 0\}$, $\widetilde{\Gamma}_{2,2} := \{x \in \widetilde{\Gamma}_2 \mid \alpha(x) \neq 0\}$ and then

$$\Gamma_2 = \widetilde{\Gamma}_1 \cup \widetilde{\Gamma}_{2,2}, \quad \Gamma_1 = \widetilde{\Gamma}_{2,1} . \tag{3.41}$$

The same holds true in the variational formulations as follows:

$$\begin{aligned} \int_\Omega u\, \tilde{\boldsymbol{c}} \cdot \nabla v \, dx &= \int_\Omega u \nabla \cdot (\tilde{\boldsymbol{c}} v) dx - \int_\Omega \nabla \cdot \tilde{\boldsymbol{c}}\, uv dx \\ &= -\int_{\widetilde{\Gamma}_1 \cup \widetilde{\Gamma}_2} \mathfrak{n} \cdot \boldsymbol{c}\, uv d\sigma + \int_\Omega \boldsymbol{c} \cdot \nabla u\, v dx + \int_\Omega \nabla \cdot \boldsymbol{c}\, uv dx \,. \end{aligned} \tag{3.42}$$

In summary, conditions 1)–3) of Theorem 3.16 for the divergence form (3.36), (3.37)–(3.39) read as follows:

$$\begin{aligned} &1)\ \tilde{r} + \frac{1}{2}\nabla \cdot \boldsymbol{c} \geq 0 \quad \text{in } \Omega, \\ &2)\ \mathfrak{n} \cdot \boldsymbol{c} \geq 0 \quad \text{on } \widetilde{\Gamma}_{2,1} := \{x \in \widetilde{\Gamma}_2 \mid \tilde{\alpha}(x) = \mathfrak{n} \cdot \boldsymbol{c}\}, \\ &3)\ \tilde{\alpha} - \frac{1}{2}\mathfrak{n} \cdot \boldsymbol{c} \geq 0 \quad \text{on } \widetilde{\Gamma}_2 \setminus \widetilde{\Gamma}_{2,1}, \quad \mathfrak{n} \cdot \boldsymbol{c} \leq 0 \quad \text{on } \widetilde{\Gamma}_1 . \end{aligned} \tag{3.43}$$

Let us consider the following typical example. The boundary $\partial\Omega$ consists only of two parts $\widetilde{\Gamma}_1$ and $\widetilde{\Gamma}_2$. Therein, $\widetilde{\Gamma}_2$ is an *outflow boundary* and $\widetilde{\Gamma}_1$ an *inflow boundary,* that is, the conditions

$$\mathfrak{n} \cdot \boldsymbol{c} < 0 \quad \text{on } \widetilde{\Gamma}_1 \quad \text{and} \quad \mathfrak{n} \cdot \boldsymbol{c} \geq 0 \quad \text{on } \widetilde{\Gamma}_2$$

hold. Frequently prescribed boundary conditions are

$$\begin{aligned} -(\boldsymbol{c}\, u - \boldsymbol{K}\nabla u) \cdot \mathfrak{n} &= g_2 && \text{on } \widetilde{\Gamma}_1 , \\ -(\boldsymbol{c}\, u - \boldsymbol{K}\nabla u) \cdot \mathfrak{n} &= -\mathfrak{n} \cdot \boldsymbol{c}\, u && \text{on } \widetilde{\Gamma}_2 . \end{aligned}$$

That is, on $\widetilde{\Gamma}_2$ we have $\tilde{\alpha} = \mathfrak{n} \cdot \boldsymbol{c}$. They are based on the following assumptions: On the inflow boundary $\widetilde{\Gamma}_1$ the normal component of the total (mass) flux density is prescribed, but on the outflow boundary $\widetilde{\Gamma}_2$, on which in the extreme case $\boldsymbol{K} = 0$ the boundary conditions would drop out, only the following is required:

- the normal component of the total (mass) flux density is continuous over $\widetilde{\Gamma}_2$,
- the ambient mass flux density that is outside Ω consists only of a convective part,
- the extensive variable (for example, the concentration) is continuous over $\widetilde{\Gamma}_2$, that is, the ambient concentration in x is also equal to $u(x)$.

Therefore, as indicated, we get $\Gamma_1 = \widetilde{\Gamma}_2, \Gamma_2 = \widetilde{\Gamma}_1$, and the Neumann boundary condition (3.19), and the mixed boundary condition (3.20),

$$\begin{aligned} \boldsymbol{K}\nabla u \cdot \mathfrak{n} &= 0 \qquad \text{on } \Gamma_1 , \\ \boldsymbol{K}\nabla u \cdot \mathfrak{n} + \alpha\, u &= g_2 \qquad \text{on } \Gamma_2 , \end{aligned}$$

where $\alpha = -\mathfrak{n} \cdot \boldsymbol{c}$.

Now the conditions of Theorem 3.16 can be checked:

We have $r - \frac{1}{2}\nabla \cdot \boldsymbol{c} = \tilde{r} + \frac{1}{2}\nabla \cdot \boldsymbol{c}$; therefore, for the latter term the inequalities in 1) and 4)b) have to be satisfied. Further, the condition $\mathfrak{n} \cdot \boldsymbol{c} \geq 0$ on Γ_1 holds due to the characterization of the outflow boundary. Because of $\alpha + \frac{1}{2}\mathfrak{n} \cdot \boldsymbol{c} = -\frac{1}{2}\mathfrak{n} \cdot \boldsymbol{c}$, the condition 3) is satisfied due to the definition of the inflow boundary.

Now we address the case of **inhomogeneous Dirichlet boundary conditions** ($|\Gamma_3|_{d-1} > 0$).

This situation can be reduced to the case of homogeneous Dirichlet boundary conditions, if we are able to choose some (fixed) element $w \in H^1(\Omega)$ in such a way that (in the sense of traces) we have

$$\gamma_0(w) = g_3 \qquad \text{on } \Gamma_3 . \tag{3.44}$$

The existence of such an element w is a necessary assumption for the existence of a solution $\tilde{u} \in H^1(\Omega)$. On the other hand, such an element w can exist only if g_3 belongs to the range of the mapping

$$H^1(\Omega) \ni v \;\mapsto\; \gamma_0(v)|_{\Gamma_3} \in L^2(\Gamma_3).$$

However, this is not valid for all $g_3 \in L^2(\Gamma_3)$, since the range of the trace operator of $H^1(\Omega)$ is a proper subset of $L^2(\partial\Omega)$ (for more details see Subsection 6.2.1).

Therefore, we assume the existence of such an element $w \in H^1(\Omega)$. To facilitate the relationship between the two formulations above and the adjoint problem later (Definition 3.39), we also assume that

$$\gamma_0(w) = 0 \quad \text{on } \Gamma_1 \cup \Gamma_2. \tag{3.45}$$

Since only the homogeneity of the Dirichlet boundary conditions of the test functions plays a role in derivation (3.32) of the bilinear form a and the linear form ℓ, we first obtain with the space V, defined in (3.31), and

$$\widetilde{V} := \{v \in H^1(\Omega) \mid \gamma_0(v) = g_3 \text{ on } \Gamma_3\} = \{v \in H^1(\Omega) \mid v - w \in V\}$$

the following variational formulation:

Find $\tilde{u} \in \widetilde{V}$ such that

$$a(\tilde{u}, v) = \ell(v) \quad \text{for all } v \in V.$$

However, this formulation does not fit into the theoretical concept of Section 3.1 since the set $\widetilde{V}$ is not a linear space.

However, if we put $\tilde{u} := u + w$, then this is equivalent to the following problem:

Find $u \in V$ such that

$$a(u, v) = \ell(v) - a(w, v) =: \tilde{\ell}(v) \quad \text{for all } v \in V. \tag{3.46}$$

Now we have a variational formulation for the case of inhomogeneous Dirichlet boundary conditions that has the form required in the theory. The relationship between the formulations (3.13), (3.19)–(3.21) and (3.35), (3.37)–(3.39) remains as above, as the term $-\hat{a}(w, v)$ transforms according to (3.42) to $-a(w, v)$ (for $r := \tilde{r} + \nabla \cdot \boldsymbol{c}$), as $\int_{\partial\Omega} \tilde{\boldsymbol{c}} \cdot \boldsymbol{n}\, wv d\sigma$ vanishes due to the boundary condition of v on $\widetilde{\Gamma}_3$ and of w on $\Gamma_1 \cup \Gamma_2$ from (3.45).

Remark 3.18 In the existence result of Theorem 3.1, the only assumption is that ℓ has to be a continuous linear form on V.

For $d := 1$ and $\Omega := (a, b)$ this is also satisfied, for instance, for the special linear form

$$\delta_\gamma(v) := v(\gamma) \quad \text{for } v \in H^1(a, b),$$

where $\gamma \in (a, b)$ is arbitrary but fixed, since by Lemma 3.4 the space $H^1(a, b)$ is continuously embedded in the space $C[a, b]$. Thus, for $d = 1$ point sources ($\ell = \delta_\gamma$) are also allowed. However, for $d \geq 2$ this does not hold since $H^1(\Omega) \not\subset C(\overline{\Omega})$.

Finally, we will once again state the **general assumptions** under which the variational formulation of the boundary value problem (3.13), (3.19)–(3.21) in the space (3.31),

$$V = \{v \in H^1(\Omega) \mid \gamma_0(v) = 0 \text{ on } \Gamma_3\}$$

has properties that satisfy the conditions of the Lax–Milgram theorem (Theorem 3.1):

- $\Omega \subset \mathbb{R}^d$ is a bounded Lipschitz domain.
- $k_{ij}, c_i, \nabla \cdot \boldsymbol{c}, r \in L^\infty(\Omega)$, $f \in L^2(\Omega)$, $i, j \in \{1, \ldots, d\}$, and, if $|\Gamma_1 \cup \Gamma_2|_{d-1} > 0$, $\boldsymbol{n} \cdot \boldsymbol{c} \in L^\infty(\Gamma_1 \cup \Gamma_2)$ (i.e., (3.16)).
- There exists some constant $k_0 > 0$ such that a.e. in Ω we have $\xi \cdot \boldsymbol{K}(x)\xi \geq k_0|\xi|^2$ for all $\xi \in \mathbb{R}^d$ (i.e., (3.17)),
- $g_j \in L^2(\Gamma_j)$, $j = 1, 2, 3$, $\alpha \in L^\infty(\Gamma_2)$ (i.e., (3.23)).
- The following hold:

1) $r - \frac{1}{2}\nabla \cdot \boldsymbol{c} \geq 0$ in Ω.
2) $\mathfrak{n} \cdot \boldsymbol{c} \geq 0$ on Γ_1.
3) $\alpha + \frac{1}{2}\mathfrak{n} \cdot \boldsymbol{c} \geq 0$ on Γ_2.
4) Additionally, one of the following conditions is satisfied:
   a) $|\Gamma_3|_{d-1} > 0$.
   b) There exists some $\tilde{\Omega} \subset \Omega$ with $|\tilde{\Omega}|_d > 0$ and $r_0 > 0$ such that $r - \frac{1}{2}\nabla \cdot \boldsymbol{c} \geq r_0$ on $\tilde{\Omega}$.
   c) There exists some $\widetilde{\Gamma}_1 \subset \Gamma_1$ with $|\widetilde{\Gamma}_1|_{d-1} > 0$ and $c_0 > 0$ such that $\mathfrak{n} \cdot \boldsymbol{c} \geq c_0$ on $\widetilde{\Gamma}_1$.
   d) There exists some $\widetilde{\Gamma}_2 \subset \Gamma_2$ with $|\widetilde{\Gamma}_2|_{d-1} > 0$ and $\alpha_0 > 0$ such that $\alpha + \frac{1}{2}\mathfrak{n} \cdot \boldsymbol{c} \geq \alpha_0$ on $\widetilde{\Gamma}_2$.

- If $|\Gamma_3|_{d-1} > 0$, then there exists some $w \in H^1(\Omega)$ with $\gamma_0(w) = g_3$ on Γ_3 (i.e., (3.44)).

3.2.2 An Example of a Boundary Value Problem of Fourth Order

The Dirichlet problem for the *biharmonic equation* reads as follows:

Find $u \in C^4(\Omega) \cap C^1(\bar{\Omega})$ such that

$$\begin{aligned} \Delta^2 u &= f \qquad \text{in } \Omega, \\ \partial_{\mathfrak{n}} u = u &= 0 \text{ on } \partial\Omega, \end{aligned} \tag{3.47}$$

where

$$\Delta^2 u := \Delta(\Delta u) = \sum_{i,j=1}^{d} \partial_i^2 \left(\partial_j^2 u\right).$$

In the case $d = 1$ this collapses to $\Delta^2 u = u^{(4)}$.

For $u, v \in H^2(\Omega)$ it follows from Corollary 3.10 that

$$\int_\Omega (u\,\Delta v - \Delta u\, v)\, dx = \int_{\partial\Omega} \{u\,\partial_{\mathfrak{n}} v - \partial_{\mathfrak{n}} u\, v\} d\sigma$$

and hence for $u \in H^4(\Omega)$, $v \in H^2(\Omega)$ (by replacing u with Δu in the above equation),

$$\int_\Omega \Delta u\,\Delta v\, dx = \int_\Omega \Delta^2 u\, v\, dx - \int_{\partial\Omega} \partial_{\mathfrak{n}} \Delta u\, v\, d\sigma + \int_{\partial\Omega} \Delta u\, \partial_{\mathfrak{n}} v\, d\sigma.$$

For a Lipschitz domain Ω we define

$$H_0^2(\Omega) := \left\{ v \in H^2(\Omega) \;\middle|\; v = \partial_{\mathfrak{n}} v = 0 \text{ on } \partial\Omega \right\}$$

and obtain the variational formulation of (3.47) in the space $V := H_0^2(\Omega)$:

Find $u \in V$, such that

$$a(u, v) := \int_\Omega \Delta u \, \Delta v \, dx = \ell(v) := \int_\Omega f v \, dx \quad \text{for all } v \in V.$$

More general, for a boundary value problem of order $2m$ in conservative form, we obtain a variational formulation in $H^m(\Omega)$ or $H_0^m(\Omega)$.

3.2.3 Regularity of Boundary Value Problems

In Section 3.2.1 we stated conditions under which the linear elliptic boundary value problem admits a unique solution u ($\tilde{u}$, respectively) in some subspace V of $H^1(\Omega)$. In many cases, for instance, for the interpolation of the solution or in the context of error estimates (also in norms other than the $\|\cdot\|_V$ norm) it is not sufficient that u ($\tilde{u}$, respectively) have only first weak derivatives in $L^2(\Omega)$.

Therefore, within the framework of the so-called regularity theory, the question of the assumptions under which the weak solution belongs to $H^2(\Omega)$, for instance, has to be answered. These additional conditions contain conditions about

- the smoothness of the boundary of the domain,
- the shape of the domain,
- the smoothness of the coefficients and the right-hand side of the differential equation and the boundary conditions,
- the kind of the transition of boundary conditions in those points, where the type is changing,

which can be quite restrictive as a whole. Therefore, in what follows we often assume only the required smoothness. Here we cite as an example one regularity result ([28, Theorem 8.12]).

Theorem 3.19 *Suppose Ω is a bounded C^2-domain and $\Gamma_3 = \partial\Omega$. Further, assume that $k_{ij} \in C^1(\bar{\Omega})$, $c_i, r \in L^\infty(\Omega)$, $f \in L^2(\Omega)$, $i, j \in \{1, \ldots, d\}$, as well as (3.17). Suppose there exists some function $w \in H^2(\Omega)$ with $\gamma_0(w) = g_3$ on Γ_3. Let $\tilde{u} = u + w$ and let u be a solution of (3.46). Then $\tilde{u} \in H^2(\Omega)$ and*

$$\|\tilde{u}\|_2 \leq C\{\|u\|_0 + \|f\|_0 + \|w\|_2\}$$

with a constant $C > 0$ independent of u, f, and w.

One drawback of the above result is that it excludes polyhedral domains. If the convexity of Ω is additionally assumed, then it can be transferred to this case. Simple examples of boundary value problems in domains with reentrant corners show that one cannot avoid such additional assumptions (see Problem 3.5).

Exercises

Problem 3.5 Consider the boundary value problem (1.1), (1.2) for $f = 0$ in the sector $\Omega := \{(x, y) \in \mathbb{R}^2 \mid x = r\cos\varphi,\ y = r\sin\varphi \text{ with } 0 < r < 1,\ 0 < \varphi < \alpha\}$ for some $0 < \alpha < 2\pi$, thus with the interior angle α. Derive as in (1.23), by using the ansatz $w(z) := z^{1/\alpha}$, a solution $u(x, y) = \operatorname{Im} w(x + iy)$ for an appropriate boundary function g. Then check the regularity of u, that is, $u \in H^k(\Omega)$, in dependence of α.

Problem 3.6 Consider the problem (1.29) with the transmission condition (1.30) and, for example, Dirichlet boundary conditions and derive a variational formulation for this.

Problem 3.7 Consider the variational formulation:

Find $u \in H^1(\Omega)$ such that

$$\int_\Omega \nabla u \cdot \nabla v \, dx = \int_\Omega f v \, dx + \int_{\partial\Omega} g v \, d\sigma \quad \text{for all } v \in H^1(\Omega), \tag{3.48}$$

where Ω is a bounded Lipschitz domain, $f \in L^2(\Omega)$ and $g \in L^2(\partial\Omega)$.

a) Let $u \in H^1(\Omega)$ be a solution of this problem. Show that $-\Delta u$ exists in the weak sense in $L^2(\Omega)$ and

$$-\Delta u = f\,.$$

b) If additionally $u \in H^2(\Omega)$, then $\partial_{\mathfrak{n}} u|_{\partial\Omega}$ exists in the sense of trace in $L^2(\partial\Omega)$ and

$$\partial_{\mathfrak{n}} u = g$$

where this equality is to be understood as

$$\int_{\partial\Omega} (\partial_{\mathfrak{n}} u - g) v \, d\sigma = 0 \quad \text{for all } v \in H^1(\Omega)\,.$$

Problem 3.8 Consider the variational equation (3.48) for the Neumann problem for the Poisson equation as in Problem 3.7.

a) If a solution $u \in H^1(\Omega)$ exists, then the compatibility condition

$$\int_\Omega f \, dx + \int_{\partial\Omega} g \, d\sigma = 0 \tag{3.49}$$

has to be fulfilled.

b) Consider the following bilinear form on $H^1(\Omega)$:

$$\tilde{a}(u, v) := \int_\Omega \nabla u \cdot \nabla v \, dx + \left(\int_\Omega u \, dx\right)\left(\int_\Omega v \, dx\right).$$

Show that $\tilde{a}$ is V-coercive on $H^1(\Omega)$.
Hint: Do it by contradiction using the fact that a bounded sequence in $H^1(\Omega)$ possesses a subsequence converging in $L^2(\Omega)$ (see, e.g., [66]).

c) Consider the unique solution $\tilde{u} \in H^1(\Omega)$ of

$$\tilde{a}(u, v) = \int_\Omega f v \, dx + \int_{\partial\Omega} g v \, d\sigma \quad \text{for all } v \in H^1(\Omega) .$$

Then:

$$|\Omega| \int_\Omega \tilde{u} \, dx = \int_\Omega f \, dx + \int_{\partial\Omega} g \, d\sigma .$$

Furthermore, if (3.49) is valid, then $\tilde{u}$ is a solution of (3.48) (with $\int_\Omega \tilde{u} \, dx = 0$).

Problem 3.9 Show analogously to Problem 3.7: A weak solution $u \in V \subset H^1(\Omega)$ of (3.32), where V is defined in (3.31), with data satisfying (3.15) and (3.23), fulfills a differential equation in $L^2(\Omega)$. The boundary conditions are fulfilled in the following sense:

$$\int_{\Gamma_1} \partial_{\mathfrak{n}_K} u \, v \, d\sigma + \int_{\Gamma_2} (\partial_{\mathfrak{n}_K} u + \alpha \, u) v \, d\sigma = \int_{\Gamma_1} g_1 v \, d\sigma + \int_{\Gamma_2} g_2 v \, d\sigma \quad \text{for all } v \in V .$$

3.3 Element Types and Affine Equivalent Partitions

In order to be able to exploit the theory developed in Sections 3.1 and 3.2 we make the assumption that Ω is a Lipschitz domain.

The finite element discretization of the boundary value problem (3.13) with the boundary conditions (3.19)–(3.21) corresponds to performing a Galerkin approximation (cf. (2.26)) of the variational equation (3.46) with the bilinear form a and the linear form ℓ, supposed to be defined as in (3.32), and some $w \in H^1(\Omega)$ with the property $w = g_3$ on Γ_3. The solution of the weak formulation of the boundary value problem is then given by $\tilde{u} := u + w$, if u denotes the solution of the variational equation (3.46).

Since the bilinear form a is, in general, not symmetric, (2.24) and (2.26), respectively (the variational equation), are no longer equivalent to (2.25) and (2.27), respectively (the minimization problem), so that in the following we pursue only the first, more general, ansatz.

The **Galerkin approximation** of the variational equation (3.46) reads as follows:

Find some $u \in V_h$ such that

$$a(u_h, v) = \ell(v) - a(w, v) =: \tilde{\ell}(v) \quad \text{for all } v \in V_h . \tag{3.50}$$

The space V_h that is to be defined has to satisfy $V_h \subset V$. Therefore, we speak of a *conforming* finite element discretization, whereas for a *nonconforming* discretization this property, for instance, can be violated. The ansatz space is defined piecewise with respect to a partition $\mathcal{T}_h$ of Ω with the goal of getting small supports for the basis functions. A partition in two space dimensions consisting of triangles (i.e., a triangulation) has already been defined in definition (2.28). The generalization to d space dimensions reads as follows:

Definition 3.20 A *partition* $\mathcal{T}_h$ of a set $\Omega \subset \mathbb{R}^d$ consists of a finite number of subsets K of Ω with the following properties:

(T1) Every $K \in \mathcal{T}_h$ is closed.
(T2) For every $K \in \mathcal{T}_h$ its nonempty interior $\mathrm{int}\,(K)$ is a Lipschitz domain.
(T3) $\overline{\Omega} = \cup_{K \in \mathcal{T}_h} K$.
(T4) For different K_1 and K_2 of $\mathcal{T}_h$ the intersection of $\mathrm{int}\,(K_1)$ and $\mathrm{int}\,(K_2)$ is empty.

The sets $K \in \mathcal{T}_h$, which are called somewhat inaccurately *elements* in the following, form a nonoverlapping decomposition of $\overline{\Omega}$. Here the formulation is chosen in such a general way, since in Section 3.8 elements with curved boundaries will also be considered. In Definition 3.20 some condition, which corresponds to the property 3) of definition (2.28), is still missing. In the following this will be formulated specifically for each element type. The parameter h is a measure for the size of all elements and mostly chosen as

$$h := \max_{K \in \mathcal{T}_h} \mathrm{diam}\,(K),$$

that is, for instance, for triangles h is the length of the triangle's longest edge.

For a given vector space V_h let

$$P_K := \{v|_K \mid v \in V_h\} \quad \text{for } K \in \mathcal{T}_h, \tag{3.51}$$

that is,

$$V_h \subset \left\{v : \Omega \to \mathbb{R} \;\middle|\; v|_K \in P_K \text{ for all } K \in \mathcal{T}_h\right\}.$$

In the example of "linear triangles" in (2.30) we have $P_K = \mathcal{P}_1$, the set of polynomials of first order. In the following definitions the space P_K will always consist of polynomials or of smooth "polynomial-like" functions, such that we can assume $P_K \subset H^1(K) \cap C(K)$. Here, $H^1(K)$ is an abbreviation for $H^1(\mathrm{int}\,(K))$. The same holds for similar notation.

As the following theorem shows, elements $v \in V_h$ of a conforming ansatz space $V_h \subset V$ have therefore to be continuous:

Theorem 3.21 *Suppose $P_K \subset H^1(K) \cap C(K)$ for all $K \in \mathcal{T}_h$. Then*

$$V_h \subset C(\overline{\Omega}) \iff V_h \subset H^1(\Omega)$$

and, respectively, for $V_{0h} := \left\{v \in V_h \;\middle|\; v = 0 \text{ on } \partial\Omega\right\}$,

$$V_{0h} \subset C(\overline{\Omega}) \iff V_{0h} \subset H_0^1(\Omega).$$

Proof See, for example, [16, Theorem 5.1 (p. 62)] or also Problem 3.10. □

If $V_h \subset C(\overline{\Omega})$, then we also speak of *C^0-elements*. Hence with this notion we do not mean only the $K \in \mathcal{T}_h$, but these provided with the local ansatz space P_K (and the degrees of freedom still to be introduced). For a boundary value problem of fourth order, $V_h \subset H^2(\Omega)$ and hence the requirement $V_h \subset C^1(\overline{\Omega})$ are necessary for a conforming finite element ansatz. Therefore, this requires, analogously to Theorem 3.21, so-called *C^1-elements*. By *degrees of freedom* we denote a finite number of values that are obtained for some $v \in P_K$ from evaluating linear functionals on P_K. The set of these functionals is denoted by Σ_K. In the following, these will basically be the function values in fixed points of the element K, as in the example of (2.30). We refer to these points as *nodes*. (Sometimes, this term is used only for the vertices of the elements, which at least in our examples are always nodes.) If the degrees of freedom are only function values, then we speak of *Lagrange elements* and specify Σ by the corresponding nodes of the element. Other possible degrees of freedom are values of derivatives in fixed nodes or also integrals. Values of derivatives are necessary if we want to obtain C^1-elements.

As in the example of (2.30) (cf. Lemma 2.10), V_h is defined by specifying P_K and the degrees of freedom on K for $K \in \mathcal{T}_h$. These have to be chosen such that, on the one hand, they enforce the continuity of $v \in V_h$ and, on the other hand, the satisfaction of the homogeneous Dirichlet boundary conditions at the nodes. For this purpose, compatibility between the Dirichlet boundary condition and the partition is necessary, as it will be required in (T6).

As can be seen from the proof of Lemma 2.10, it is essential

(F1) that the interpolation problem, locally defined on $K \in \mathcal{T}_h$ by the degrees of freedom, is uniquely solvable in P_K, (3.52)

(F2) that this also holds on the $(d-1)$-dimensional boundary surfaces F of $K \in \mathcal{T}_h$ for the degrees of freedom from F and the functions $v|_F$ where $v \in P_K$; this then ensures the continuity of $v \in V_h$, if P_K and $P_{K'}$ match in the sense of $P_K|_F = P_{K'}|_F$ for $K, K' \in \mathcal{T}_h$ intersecting in F (see Figure 3.2). (3.53)

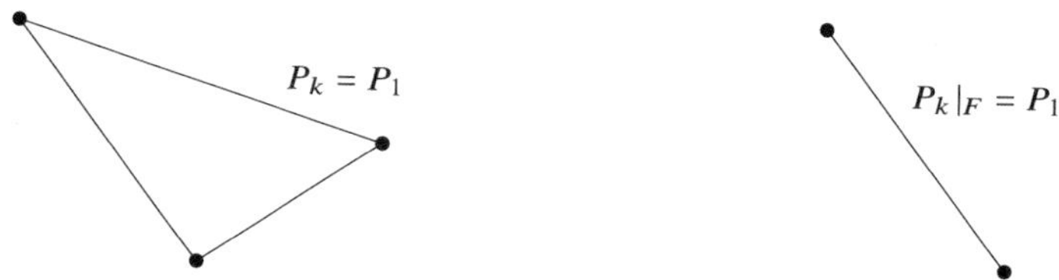

Fig. 3.2: Compatibility of the ansatz space on the boundary surface and the degrees of freedom there.

The following *finite elements* defined by their basic domain $K(\in \mathcal{T}_h)$, the local ansatz space P_K, and the degrees of freedom Σ_K satisfy these properties.

For this, let $\mathcal{P}_k(K)$ be the set of mappings $p : K \to \mathbb{R}$ of the following form:

$$p(x) = p(x_1, \ldots, x_d) = \sum_{|\alpha| \le k} \gamma_{\alpha_1 \ldots \alpha_d} x_1^{\alpha_1} \cdots x_d^{\alpha_d} = \sum_{|\alpha| \le k} \gamma_\alpha x^\alpha , \tag{3.54}$$

hence the polynomials of order k in d variables. The set $\mathcal{P}_k(K)$ forms a vector space, and since $p \in \mathcal{P}_k(K)$ is differentiable arbitrarily often, $\mathcal{P}_k(K)$ is a subset of all function spaces introduced so far (provided that the boundary conditions do not belong to their definition).

For both, $K \in \mathcal{T}_h$ and $K = \mathbb{R}^d$ we have

$$\dim \mathcal{P}_k(K) = \dim \mathcal{P}_k(\mathbb{R}^d) = \binom{d+k}{k}, \tag{3.55}$$

as even $\mathcal{P}_k(\mathbb{R}^d)|_K = \mathcal{P}_k(K)$ (see Problem 3.12). Therefore, for short we will use the notation $\mathcal{P}_1 = \mathcal{P}_1(K)$ if the dimension of the basic space is fixed.

We start with *simplicial finite elements*, that is, elements whose basic domain is a regular d-simplex of $\mathbb{R}^d$. By this we mean the following:

Definition 3.22 A set $K \subset \mathbb{R}^d$ is called a *regular d-simplex* if there exist $d + 1$ distinct points $a_1, \ldots, a_{d+1} \in \mathbb{R}^d$, the vertices of K, such that

$$a_2 - a_1, \ldots, a_{d+1} - a_1 \qquad \text{are linearly independent} \tag{3.56}$$

(that is, $a_1, \ldots, a_{d+1}$ do not lie in a hyperplane) and

$$\begin{aligned} K &= \operatorname{conv}\{a_1, \ldots, a_{d+1}\} \\ &:= \left\{ x = \sum_{i=1}^{d+1} \lambda_i a_i \;\middle|\; 0 \le \lambda_i (\le 1),\ \sum_{i=1}^{d+1} \lambda_i = 1 \right\} \\ &= \left\{ x = a_1 + \sum_{i=2}^{d+1} \lambda_i (a_i - a_1) \;\middle|\; \lambda_i \ge 0,\ \sum_{i=2}^{d+1} \lambda_i \le 1 \right\}. \end{aligned} \tag{3.57}$$

A *face* of K is a $(d-1)$-simplex defined by d points of $\{a_1, \ldots, a_{d+1}\}$. The particular d-simplex

$$\hat{K} := \operatorname{conv}\{\hat{a}_1, \ldots, \hat{a}_{d+1}\} \text{ with } \hat{a}_1 = 0,\ \hat{a}_{i+1} = e_i,\ i = 1, \ldots, d \tag{3.58}$$

is called the *standard simplicial reference element.*

From (3.136) we conclude in particular

$$\operatorname{vol}(\hat{K}) = 1/d!\,.$$

In the case $d = 2$ we get a triangle with $\dim \mathcal{P}_1 = 3$ (cf. Lemma 2.10). The faces are the three edges of the triangle. In the case $d = 3$ we get a tetrahedron with

$\dim \mathcal{P}_1 = 4$, the faces are the four triangle surfaces, and finally, in the case $d = 1$, it is a line segment with $\dim \mathcal{P}_1 = 2$ and the two boundary points as faces.

More precisely, a face is not interpreted as a subset of $\mathbb{R}^d$, but of a $(d-1)$-dimensional space that, for instance, is spanned by the vectors $a_2 - a_1, \ldots, a_d - a_1$ in the case of the defining points $a_1, \ldots, a_d$.

Sometimes, we also consider *degenerate d-simplices*, where assumption (3.56) of linear independence is dropped. We consider, for instance, a line segment in the two-dimensional space as it arises as an edge of a triangular element. In the one-dimensional parametrization it is a regular 1-simplex, but in $\mathbb{R}^2$ a degenerate 2-simplex.

The unique coefficients $\lambda_i = \lambda_i(x)$, $i = 1, \ldots, d+1$, in (3.57), are called *barycentric coordinates* of x. This defines mappings $\lambda_i : K \to \mathbb{R}$, $i = 1, \ldots, d+1$.

We consider a_j as a column of a matrix, that is, for $j = 1, \ldots, d+1$, $a_j = (a_{ij})_{i=1,\ldots,d}$. The defining conditions for $\lambda_i = \lambda_i(x)$ can be written as a $(d+1)\times(d+1)$ system of equations:

$$\left.\begin{aligned} \sum_{j=1}^{d+1} a_{ij}\lambda_j &= x_i \\ \sum_{j=1}^{d+1} \lambda_j &= 1 \end{aligned}\right\} \iff B\lambda = \begin{pmatrix} x \\ 1 \end{pmatrix} \tag{3.59}$$

for

$$B = \begin{pmatrix} a_{11} & \cdots & a_{1,d+1} \\ \vdots & \ddots & \vdots \\ a_{d1} & \cdots & a_{d,d+1} \\ 1 & \cdots & 1 \end{pmatrix}. \tag{3.60}$$

The matrix B is nonsingular due to assumption (3.56), that is, $\lambda(x) = B^{-1}\binom{x}{1}$, and hence

$$\lambda_i(x) = \sum_{j=1}^{d} c_{ij}x_j + c_{i,d+1} \quad \text{for all} \quad i = 1, \ldots, d+1,$$

where $C = (c_{ij})_{ij} := B^{-1}$.

Consequently, the λ_i are affine-linear, and hence $\lambda_i \in \mathcal{P}_1$. The level surfaces $\{x \in K \mid \lambda_i(x) = \mu\}$ correspond to intersections of hyperplanes with the simplex K (see Figure 3.3). The level surfaces for distinct μ_1 and μ_2 are parallel to each other, that is, in particular, to the level surface for $\mu = 0$, which corresponds to the triangle face spanned by all the vertices apart of a_i.

By (3.59), the barycentric coordinates can be defined for arbitrary $x \in \mathbb{R}^d$ (with respect to some fixed d-simplex K). Then

$$x \in K \iff 0 \le \lambda_i(x) \le 1 \quad \text{for all } i = 1, \ldots, d+1.$$

Applying Cramer's rule to the system $B\lambda = \binom{x}{1}$, we get for the ith barycentric coordinate

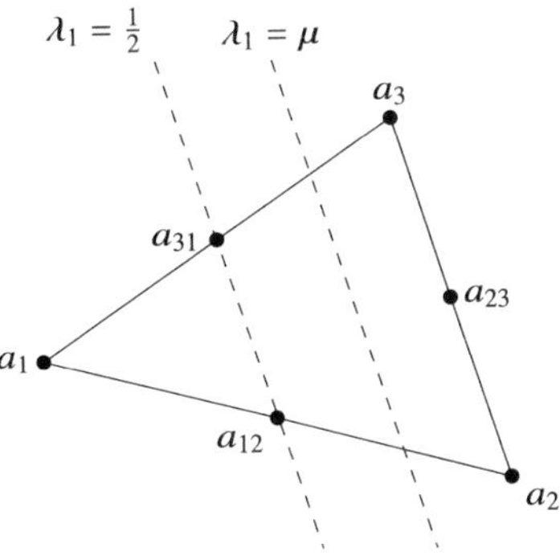

Fig. 3.3: Barycentric coordinates and hyperplanes.

$$\lambda_i(x) = \frac{1}{\det(B)} \det \begin{pmatrix} a_{11} & \cdots & x_1 & \cdots & a_{1,d+1} \\ \vdots & & \vdots & & \vdots \\ a_{d1} & \cdots & x_d & \cdots & a_{d,d+1} \\ 1 & \cdots & 1 & \cdots & 1 \end{pmatrix}.$$

Here, in the ith column a_i has been replaced with x. Since, in general,

$$\mathrm{vol}\,(K) = \mathrm{vol}\,(\hat{K})\,|\det(B)| \tag{3.61}$$

for the reference simplex $\hat{K}$ defined by (3.58) (cf. (2.54)), we have for the volume of the d-simplex $K := \mathrm{conv}\,\{a_1, \ldots, a_{d+1}\}$,

$$\mathrm{vol}\,(K) = \frac{1}{d!} \left| \det \begin{pmatrix} a_{11} & \cdots & a_{1,d+1} \\ \vdots & \ddots & \vdots \\ a_{d1} & \cdots & a_{d,d+1} \\ 1 & \cdots & 1 \end{pmatrix} \right|,$$

and from this,

$$\lambda_i(x) = \pm \frac{\mathrm{vol}\ (\mathrm{conv}\ \{a_1, \ldots, x, \ldots, a_{d+1}\})}{\mathrm{vol}\ (\mathrm{conv}\ \{a_1, \ldots, a_i, \ldots, a_{d+1}\})}\,. \tag{3.62}$$

The sign is determined by the arrangement of the coordinates.

In the case $d = 2$, for example, we have

$$\mathrm{vol}\,(K) = \det(B)/2$$
$$\Longleftrightarrow\ a_1, a_2, a_3 \text{ are ordered positively (that is, counterclockwise).}$$

Here, $\mathrm{conv}\ \{a_1, \ldots, x, \ldots, a_{d+1}\}$ is the d-simplex that is generated by replacing a_i with x and is possibly degenerate if x lies on a face of K (then $\lambda_i(x) = 0$). Hence, in the case $d = 2$ we have for $x \in K$ that the barycentric coordinates $\lambda_i(x)$ are the relative areas of the triangles that are spanned by x and the vertices other than a_i. Therefore, we also speak of *area coordinates* (see Figure 3.4). Analogous interpretations hold for $d = 3$. Using the barycentric coordinates, we can now easily specify points that admit a geometric characterization. The midpoint $a_{ij} := \frac{1}{2}\left(a_i + a_j\right)$ of a line segment that is given by a_i and a_j satisfies, for instance,

$$\lambda_i(x) = \lambda_j(x) = \frac{1}{2}\,.$$

By the *barycentre* of a d-simplex we mean

$$a_S := \frac{1}{d+1}\sum_{i=1}^{d+1} a_i\,; \quad \text{thus} \quad \lambda_i(a_S) = \frac{1}{d+1} \quad \text{for all } i = 1,\ldots,d+1. \tag{3.63}$$

A geometric interpretation follows directly from the above considerations.

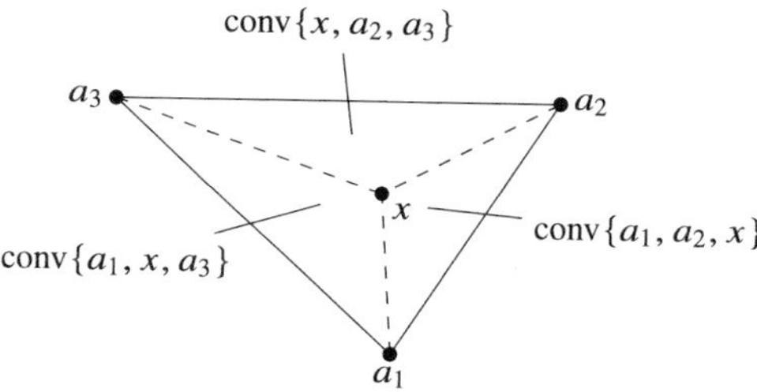

Fig. 3.4: Barycentric coordinates as area coordinates.

In the following suppose conv $\{a_1,\ldots,a_{d+1}\}$ to be a regular d-simplex. We make the following definition:

Finite Element: Linear Ansatz on the Simplex

$$\begin{aligned} K &= \text{conv}\ \{a_1,\ldots,a_{d+1}\}\,,\\ P &= \mathcal{P}_1(K),\\ \Sigma &= \{p(a_i),\ i = 1,\ldots,d+1\}\,. \end{aligned} \tag{3.64}$$

The *local interpolation problem in P, given by the degrees of freedom* Σ, namely,

$$\begin{aligned} &\text{find some } p \in P \text{ for } u_1,\ldots,u_{d+1} \in \mathbb{R} \text{ such that}\\ &\qquad p(a_i) = u_i \quad \text{for all } i = 1,\ldots,d+1, \end{aligned}$$

can be interpreted as the question of finding the inverse image of a linear mapping from P to $\mathbb{R}^{|\Sigma|}$. By virtue of (3.55),

$$|\Sigma| = d + 1 = \dim P\,.$$

Since both vector spaces have the same dimension, the solvability of the interpolation problem is equivalent to the uniqueness of the solution. This consideration holds independently of the type of the degrees of freedom (as far as they are linear functionals on P). Therefore, we need only to ensure the solvability of the interpolation problem. This is obtained by specifying

$$N_1, \ldots, N_{d+1} \in P \quad \text{with } N_i\left(a_j\right) = \delta_{ij} \quad \text{for all } i, j = 1, \ldots, d+1\,,$$

the so-called *shape functions* (see (2.33) for $d = 2$). Then the solution of the interpolation problem is given by

$$p(x) = \sum_{i=1}^{d+1} u_i N_i(x) \tag{3.65}$$

and analogously in the following, that is, the shape functions form a basis of P and the coefficients in the representation of the interpolating function are exactly the degrees of freedom $u_1, \ldots, u_{d+1}$.

Due to the above considerations, the specification of the shape functions can easily be done by choosing

$$N_i = \lambda_i\,.$$

Finite Element: Quadratic Ansatz on the Simplex

Here, we have

$$\begin{aligned} K &= \operatorname{conv}\left\{a_1, \ldots, a_{d+1}\right\}, \\ P &= \mathcal{P}_2(K), \\ \Sigma &= \left\{p\left(a_i\right), p\left(a_{ij}\right), \quad i = 1, \ldots, d+1,\ i < j \le d+1\right\}, \end{aligned} \tag{3.66}$$

where the a_{ij} denote the midpoints of the edges (see Figure 3.5).

Since here we have

$$|\Sigma| = \frac{(d+1)(d+2)}{2} = \dim P\,,$$

it also suffices to specify the shape functions. They are given by

$$\begin{aligned} \lambda_i\left(2\lambda_i - 1\right), \quad & i = 1, \ldots, d+1\,, \\ 4\lambda_i\lambda_j, \quad & i, j = 1, \ldots, d+1\,,\ i < j\,. \end{aligned}$$

Fig. 3.5: Quadratic simplicial elements.

If we want to have polynomials of higher degree as local ansatz functions, but still Lagrange elements, then degrees of freedom also arise in the interior of K:

Finite Element: Cubic Ansatz on the Simplex

$$\begin{aligned} K &= \operatorname{conv}\{a_1, \ldots, a_{d+1}\}, \\ P &= \mathcal{P}_3(K), \\ \Sigma &= \{p(a_i), p(a_{i,i,j}), p(a_{i,j,k})\}, \end{aligned} \tag{3.67}$$

where

$$a_{i,i,j} := \frac{2}{3}a_i + \frac{1}{3}a_j \qquad \text{for} \quad i, j = 1, \ldots, d+1, \ i \neq j,$$

$$a_{i,j,k} := \frac{1}{3}(a_i + a_j + a_k) \ \text{ for } \quad i, j, k = 1, \ldots, d+1, \ i < j < k.$$

Since here $|\Sigma| = \dim P$ also holds, it is sufficient to specify the shape functions, which is possible by

$$\begin{aligned} &\frac{1}{2}\lambda_i(3\lambda_i - 1)(3\lambda_i - 2), && i = 1, \ldots, d+1, \\ &\frac{9}{2}\lambda_i\lambda_j(3\lambda_i - 1), && i, j = 1, \ldots, d+1, \ i \neq j, \\ &27\lambda_i\lambda_j\lambda_k, && i, j, k = 1, \ldots, d+1, \ i < j < k. \end{aligned}$$

Thus for $d = 2$ the value at the barycentre arises as a degree of freedom. This, and in general the $a_{i,j,k}, i < j < k$, can be dropped if the ansatz space P is reduced (see [16, p. 70]).

All finite elements discussed so far have degrees of freedom that are defined in convex combinations of the vertices. On the other hand, two regular d-simplices can be mapped bijectively onto each other by a unique affine-linear F, that is, $F \in \mathcal{P}_1$ such that as defining condition, the vertices of the simplices should be mapped onto each other. If we choose, besides the general simplex K, the standard reference element $\hat{K}$ defined by (3.58), then $F = F_K : \hat{K} \to K$ is defined by

$$F(\hat{x}) = B\hat{x} + a_1 , \tag{3.68}$$

where $B = (a_2 - a_1, \ldots, a_{d+1} - a_1)$.

Since for F we have

$$F\left(\sum_{i=1}^{d+1} \lambda_i \hat{a}_i\right) = \sum_{i=1}^{d+1} \lambda_i F(\hat{a}_i) \quad \text{for} \quad \lambda_i \geq 0, \; \sum_{i=1}^{d+1} \lambda_i = 1 ,$$

F is indeed a bijection that maps the degrees of freedom onto each other as well as the faces of the simplices. Since the ansatz spaces P and $\hat{P}$ remain invariant under the transformation F_K, the finite elements introduced so far are (in their respective classes) *affine equivalent* to each other and to the *reference element*.

Definition 3.23 Two Lagrange elements $(K, P, \Sigma), (\hat{K}, \hat{P}, \hat{\Sigma})$ are called *equivalent* if there exists a bijective $F : \hat{K} \to K$ such that

$$\begin{aligned} &\left\{F(\hat{a}) \,\middle|\, \hat{a} \in \hat{K} \text{ generates a degree of freedom on } \hat{K}\right\} \\ &\quad = \left\{a \,\middle|\, a \in K \text{ generates a degree of freedom on } K\right\} \\ \text{and} \\ &P = \left\{p : K \to \mathbb{R} \,\middle|\, p \circ F \in \hat{P}\right\} . \end{aligned} \tag{3.69}$$

They are called *affine equivalent* if F is affine-linear.

Here we have formulated the definition in a more general way, since in Section 3.8 elements with more general F will be introduced: For *isoparametric* elements the same functions F as in the ansatz space are admissible for the transformation. From the elements discussed so far only the simplex with linear ansatz is thus isoparametric. Hence, in the (affine) equivalent case a transformation not only of the points is defined by

$$\hat{x} = F^{-1}(x),$$

but also of the mappings, defined on K and $\hat{K}$, (not only of P and $\hat{P}$) is given by

$$\hat{v} : \hat{K} \to \mathbb{R}, \quad \hat{v}(\hat{x}) := v(F(\hat{x}))$$

for $v : K \to \mathbb{R}$ and vice versa.

We can also use the techniques developed so far in such a way that only the reference element is defined, and then a general element is obtained from this by an affine-linear transformation. As an example of this, we consider elements on a cube.

Suppose $\hat{K} := [0, 1]^d = \left\{x \in \mathbb{R}^d \,\middle|\, 0 \leq x_i \leq 1,\, i = 1, \ldots, d\right\}$ is the unit cube. The *faces* of $\hat{K}$ are defined by setting a coordinate to 0 or 1; thus, for instance,

$$\prod_{i=1}^{j-1}[0,1] \times \{0\} \times \prod_{j+1}^{d}[0,1] .$$

Let $Q_k(K)$ denote the set of polynomials on K that are of the form

$$p(x) = \sum_{\substack{0\le\alpha_i\le k\\ i=1,\dots,d}} \gamma_{\alpha_1,\dots,\alpha_d} x_1^{\alpha_1} \cdots x_d^{\alpha_d} .$$

Hence, we have $\mathcal{P}_k \subset Q_k \subset \mathcal{P}_{dk}$.

Therefore, we define a reference element generally for $k \in \mathbb{N}$ as follows:

Finite Element: *d*-polynomial Ansatz on the Cuboid

$$\begin{aligned}
\hat{K} &= [0,1]^d , \\
\hat{P} &= Q_k(\hat{K}), \\
\hat{\Sigma} &= \left\{ p(\hat{x}) \;\middle|\; \hat{x} = \left(\frac{i_1}{k},\dots,\frac{i_d}{k}\right),\ i_j \in \{0,\dots,k\},\ j = 1,\dots,d \right\},
\end{aligned} \tag{3.70}$$

which is depicted in Figure 3.6. Again, we have $|\hat{\Sigma}| = \dim \hat{P}$, such that for the unique solvability of the local interpolation problem we have only to specify the shape functions. They are obtained on $\hat{K}$ as the product of the corresponding shape functions for the case $d = 1$, thus of the *Lagrange basis polynomials*

$$p_{i_1,\dots,i_d}(\hat{x}) := \prod_{j=1}^{d} \left(\prod_{\substack{i'_j=0\\ i'_j\neq i_j}}^{k} \frac{k\hat{x}_j - i'_j}{i_j - i'_j} \right). \tag{3.71}$$

Interior degrees of freedom arise from $k = 2$ onward. Hence the ansatz space on the general element K is, according to the definition above,

$$P = \left\{ \hat{p} \circ F_K^{-1} \;\middle|\; \hat{p} \in Q_k(\hat{K}) \right\} .$$

In the case of a general rectangular cuboid, that is, if B in (3.68) is a diagonal matrix, then $P = Q_k(K)$ holds, analogously to the simplices. However, for a general B additional polynomial terms arise that do not belong to Q_k (see Problem 3.14).

An affine-linear transformation does not generate general cuboids but only d-epipeds, thus for $d = 3$ parallelepipeds and for $d = 2$ only parallelograms. To map the unit square to an arbitrary general convex quadrilateral, we need some transformation of Q_1, that is, isoparametric elements (see (3.161)).

Let $\mathcal{T}_h$ be a partition of d-simplices or of affinely transformed d-unit cubes. In particular, $\Omega = \operatorname{int}(\cup_{K\in\mathcal{T}_h} K)$ is polygonally bounded. Condition (F1) in (3.52) is

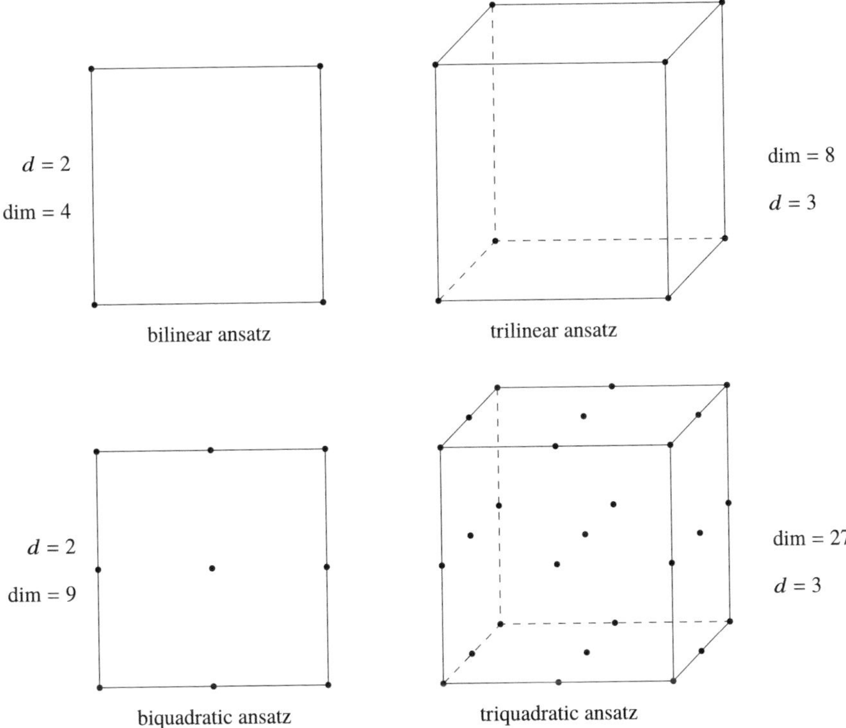

Fig. 3.6: Quadratic and cubic elements on the cube.

always satisfied. In order to be able to satisfy the condition (F2) in (3.53) as well, a further assumption in addition to (T1)–(T4) has to be made about the partition:

(T5) Every face of some $K \in \mathcal{T}_h$ is either a subset of the boundary Γ of Ω or identical to a face of another $\tilde{K} \in \mathcal{T}_h$.

In order to ensure the validity of the homogeneous Dirichlet boundary condition on Γ_3 for the $v_h \in V_h$ that have to be defined, we additionally assume the following:

(T6) The boundary sets $\overline{\Gamma}_1, \overline{\Gamma}_2, \Gamma_3$ decompose into faces of elements $K \in \mathcal{T}_h$.

A face F of $K \in \mathcal{T}_h$ that is lying on $\partial\Omega$ is therefore only allowed to contain a point from the intersection $\overline{\Gamma}_i \cap \overline{\Gamma}_j$ for $i \neq j$, if and only if the point is a boundary point of F. We recall that the set Γ_3 has been defined as being closed in $\partial\Omega$.

In the following, we suppose that these conditions are always satisfied. A partition that also satisfies (T5) and (T6) is called *consistent* (or *conforming*).

Then, for all of the above finite elements,

- If $K, K' \in \mathcal{T}_h$ have a common face F, then the degrees of freedom of K and K' coincide on F. (3.72)
- F itself becomes a finite element (that is, the local interpolation problem is uniquely solvable) with the ansatz space $P_K|_F$ and the degrees of freedom on F. (3.73)

We now choose V_h as follows:

$$\begin{aligned} V_h := \big\{ v : \Omega \to \mathbb{R} \,\big|\, v|_K \in P_K \text{ for all } K \in \mathcal{T}_h \text{ and} \\ v \text{ is uniquely determined by the degrees of freedom} \big\} . \end{aligned} \tag{3.74}$$

Analogously to the proof of Lemma 2.10, we can see that $v \in V_h$ is continuous over the face of an element; thus $V_h \subset C(\overline{\Omega})$, that is, $V_h \subset H^1(\Omega)$ according to Theorem 3.21.

Further, $u|_F = 0$ if F is a face of $K \in \mathcal{T}_h$ with $F \subset \partial\Omega$ and the specifications in the degrees of freedom of F are zero (Dirichlet boundary conditions only in the nodes), that is, the homogeneous Dirichlet boundary conditions are satisfied by enforcing them in the degrees of freedom. Due to assumption (T6), the boundary set Γ_3 is fully taken into account in this way.

Consequently, we have the following theorem:

Theorem 3.24 *Suppose $\mathcal{T}_h$ is a consistent partition of d-simplices or d-epipeds of a domain $\Omega \subset \mathbb{R}^d$. The elements are defined as in one of Examples (3.64), (3.66), (3.67), (3.70).*

Let the degrees of freedom be given in the nodes $a_1, \dots, a_M$. Suppose they are numbered in such a way that $a_1, \dots, a_{M_1} \in \Omega \cup \Gamma_1 \cup \Gamma_2$ and $a_{M_1+1}, \dots, a_M \in \Gamma_3$. If the ansatz space V_h is defined by (3.74), then an element $v \in V_h$ is determined uniquely by specifying $v(a_i)$, $i = 1, \dots, M$, and

$$v \in H^1(\Omega) .$$

If $v(a_i) = 0$ for $i = M_1 + 1, \dots, M$, then we also have

$$v = 0 \quad \text{on } \Gamma_3 .$$

Exactly as in Section 2.2 (see (2.36)), functions $\varphi_i \in V_h$ are uniquely determined by the interpolation condition

$$\varphi_i(a_j) = \delta_{ij}, \quad i, j = 1, \dots, M .$$

By the same consideration as there and as for the shape functions (see (3.65)) we observe that the φ_i form a basis of V_h, the *nodal basis*, since each $v \in V_h$ has a unique representation

$$v(x) = \sum_{i=1}^{M} v(a_i)\varphi_i(x) . \tag{3.75}$$

If for Dirichlet boundary conditions, the values in the boundary nodes $a_i, i = M_1 + 1, \ldots, M$, are given as zero, then the index has to run only up to M_1.

The support $\operatorname{supp} \varphi_i$ of the basis functions thus consists of all elements that contain the node a_i, since in all other elements φ_i assumes the value 0 in the degrees of freedom and hence vanishes identically. In particular, for an interior degree of freedom, that is, for some a_i with $a_i \in \operatorname{int}(K)$ for an element $K \in \mathcal{T}_h$, we have $\operatorname{supp} \varphi_i = K$.

Different element types can also be combined (see Figure 3.7) if only (3.72) is satisfied, thus, for instance, for $d = 2$ (3.70), $k = 1$, can be combined with (3.64) or (3.70), $k = 2$, with (3.66).

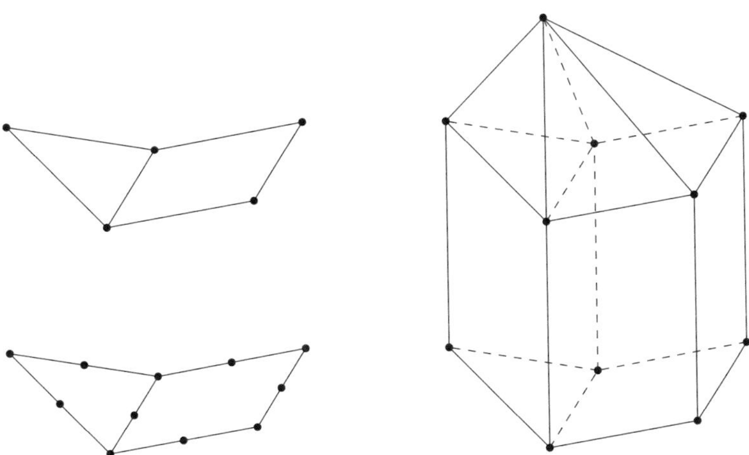

Fig. 3.7: Conforming combination of different element types.

For $d = 3$ a combination of simplices and parallelepipeds is not possible, since they have different types of faces. Tetrahedra can be combined with prisms at their two triangular surfaces, whereas their three quadrilateral surfaces (see Problem 3.17) allow for a combination of prisms with parallelepipeds. Possibly also pyramids are necessary as transition elements (see [162]).

So far, the degrees of freedom have always been function values (*Lagrange elements*). If, additionally, derivative values are specified, then we speak of *Hermite elements*. As an example, we present the following:

Finite Element: Cubic Hermite Ansatz on the Simplex

$$\begin{aligned}
K &= \operatorname{conv}\{a_1, \ldots, a_{d+1}\}, \\
P &= \mathcal{P}_3(K), \\
\Sigma &= \big\{p(a_i),\ i = 1, \ldots, d+1,\ p(a_{i,j,k}),\ i, j, k = 1, \ldots, d+1,\ i < j < k, \\
&\qquad \nabla p(a_i) \cdot (a_j - a_i),\ i, j = 1, \ldots, d+1,\ i \neq j\big\}.
\end{aligned} \tag{3.76}$$

Instead of the directional derivatives we could also have chosen the partial derivatives as degrees of freedom, but would not have generated affine equivalent elements in that way. In order to ensure that directional derivatives in the directions ξ and $\hat{\xi}$ are mapped onto each other by the transformation, the directions have to satisfy

$$\xi = B\hat{\xi},$$

where B is the linear part of the transformation F according to (3.68). This is satisfied for (3.76), but would be violated for the partial derivatives, that is, $\xi = \hat{\xi} = e_i$ (Fig. 3.8). This has also to be taken into account for the question of which degrees of freedom have to be chosen for Dirichlet boundary conditions (see Problem 3.19). Thus, the desired property that the degrees of freedom be defined "globally" is lost here. Nevertheless, we do not have a C^1-element: The ansatz (3.76) ensures only the continuity of the tangential, not of the normal derivative over a face.

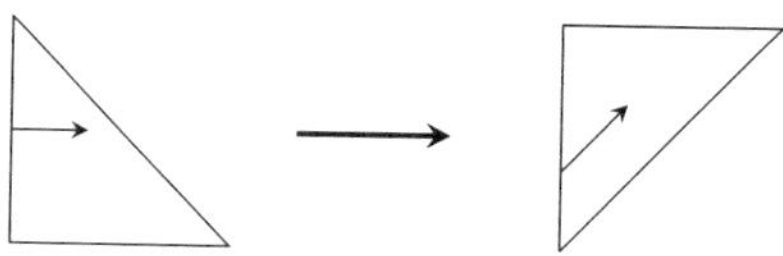

Fig. 3.8: Affine transformation of the reference triangle and an inner normal according to $\hat{\xi}_1 \mapsto \xi_1$, $\hat{\xi}_1 + \hat{\xi}_2 \mapsto \xi_2$.

Finite Element: Bogner–Fox–Schmit Rectangle

The simplest C^1-*element* is for $d = 2$:

$$\begin{aligned} \hat{K} &= [0, 1]^2, \\ \hat{P} &= Q_3(\hat{K}), \\ \hat{\Sigma} &= \{p(a),\ \partial_1 p(a),\ \partial_2 p(a),\ \partial_{12} p(a) \quad \text{for all vertices } a\}, \end{aligned} \tag{3.77}$$

that is, the element has 16 degrees of freedom.

In the case of Hermite elements, the above propositions concerning the nodal basis hold analogously with an appropriate extension of the identity (3.75).

Further, all considerations of Section 2.2 concerning the determination of the Galerkin approximation as a solution of a system of equations (2.38) also hold, since there only the (bi)linearity of the forms is supposed. Therefore using the nodal basis, the quantity $a(\varphi_j, \varphi_i)$ has to be computed as the (i, j)th matrix entry of the system of equations that has to be set up for the bilinear form a. The form of the bilinear form (3.32) shows that the consideration of Section 2.2, concerning that there is at most a nonzero entry at position (i, j) if,

$$\operatorname{supp}\varphi_i \cap \operatorname{supp}\varphi_j \neq \emptyset\,, \tag{3.78}$$

still holds.

Since in the examples discussed, $\operatorname{supp}\varphi_i$ consists of at most of those elements containing the node a_i (see Figure 3.11), the nodes have to be *adjacent*, for the validity of (3.78), that is, they should belong to some common element. In particular, an interior degree of freedom of some element is connected only with the nodes of the same element: This can be used to eliminate such nodes from the beginning (*static condensation*).

The following consideration can be helpful for the choice of the element type: An increase in the size of polynomial ansatz spaces increases the (computational) cost by an increase in the number of nodes and an increase in the population of the matrix.

As an example for $d = 2$ we consider triangles with linear a) and quadratic b) ansatz (see Figure 3.9).

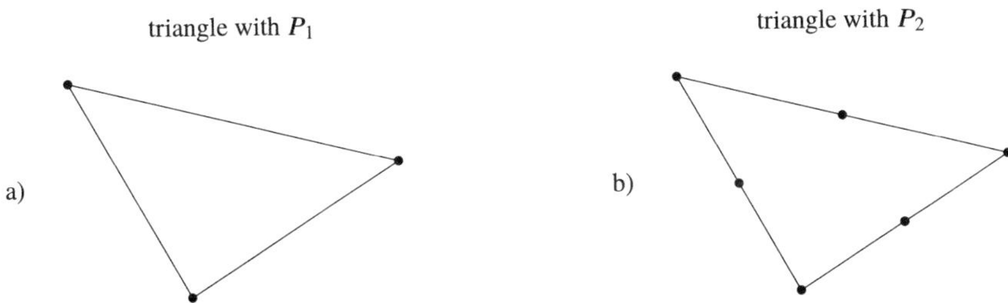

Fig. 3.9: Comparison between linear and quadratic triangles.

In order to have the same number of nodes we compare b) with the discretization parameter h with a) with the discretization parameter $h/2$ (one step of "red refinement") (see Figure 3.10).

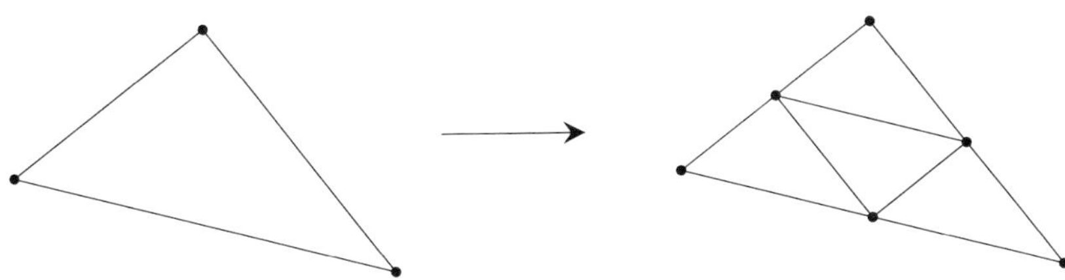

Fig. 3.10: Generation of the same number of nodes.

However, this shows that we have a denser population in b) than in a).

To have still an advantage by using the higher polynomial order, the ansatz b) has to have a higher convergence rate. In Theorem 3.32 we will prove the following estimate for a shape-regular family of partitions $\mathcal{T}_h$ (see Definition 3.31):

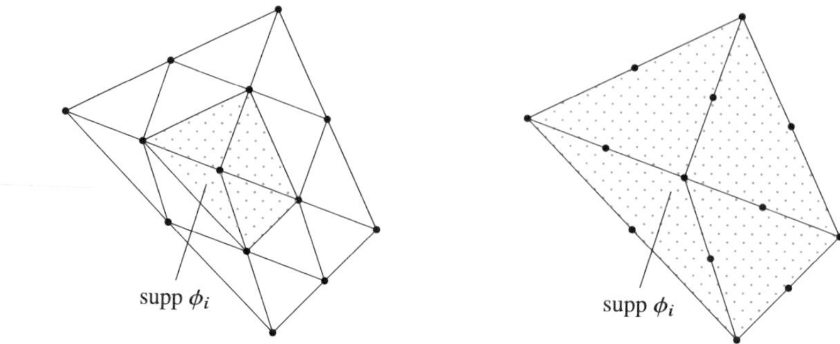

Fig. 3.11: Supports of the basis functions.

- If $u \in H^2(\Omega)$, then for a) and b) we have the estimate

$$\|u - u_h\|_1 \leq C_1 h\,. \tag{3.79}$$

- If $u \in H^3(\Omega)$, then for b) but not for a) we have the estimate

$$\|u - u_h\|_1 \leq C_2 h^2\,. \tag{3.80}$$

For the constants we may, in general, expect $C_2 > C_1$.

In order to be able to make a comparison between the variants a) and b), we consider in the following the case of a rectangle $\Omega = (0, a) \times (0, b)$. The number of the nodes is then proportional to $1/h^2$ if the elements are all "essentially" of the same size.

However, if we consider the number of nodes M as given, then h is proportional to $1/\sqrt{M}$.

Using this in the estimate (3.79), we get for a solution $u \in H^2(\Omega)$,

$$\begin{aligned}
&\text{in the case a) for } h/2: \quad && \|u - u_{h/2}\|_1 \leq C_1 \frac{1}{2\sqrt{M}}\,,\\
&\text{in the case b) for } h: && \|u - u_h\|_1 \leq \overline{C}_1 \frac{1}{\sqrt{M}}\,.
\end{aligned}$$

If both constants are the same, this means an advantage for the variant a).

On the other hand, if the solution is smoother and satisfies $u \in H^3(\Omega)$, then the estimate (3.80), which can be applied only to the variant b), yields

$$\begin{aligned}
&\text{in the case a) for } h/2: \quad && \|u - u_{h/2}\|_1 \leq C_1 \frac{1}{2\sqrt{M}}\,,\\
&\text{in the case b) for } h: && \|u - u_h\|_1 \leq C_2 \frac{1}{M}\,.
\end{aligned}$$

By an elementary reformulation, we get

$$C_2 \frac{1}{M} < (<) C_1 \frac{1}{2\sqrt{M}} \iff M > (>) 4 \frac{C_2^2}{C_1^2},$$

which gives an advantage for b) if the number of variables M is chosen, depending on C_2/C_1, sufficiently large. However, the denser population of the matrix in b) has to be confronted with this.

Hence, a higher order polynomial ansatz has an advantage only if the smoothness of the solution leads to a higher convergence rate. Especially for nonlinear problems with less-smooth solutions, a possible advantage of the higher order ansatz has to be examined critically.

For further usage we can give an abstract notion of a finite element.

Definition 3.25 A *finite element* is a triple (K, P, Σ), such that

1) $K \subset \mathbb{R}^d$ is compact, connected, and with a Lipschitz boundary, with nonempty interior.
2) P is a linear space of functions $p : K \to \mathbb{R}^M$.
3) $\Sigma = \{\varphi_1, \cdots, \varphi_M\} \subset P'$ such that the mapping from P to $\mathbb{R}^M$ defined by

$$p \mapsto (\varphi_1(p), \cdots, \varphi_M(p))^T$$

is bijective.

Remark 3.26 1) For a finite element, if $\dim P = M$ is known, either injectivity ($\varphi_i(p) = 0$ for $i = 1, \cdots, M \Rightarrow p = 0$) or surjectivity has to be checked. This can be done by specifying *local shape function* $p_i \in P$ with the property $\varphi_i(p_j) = \delta_{ij}, i, j = 1, \cdots, M$.
2) For a finite element, local shape functions exist uniquely and form a basis of P.
3) The linear functionals $\varphi_1, \cdots, \varphi_M$ are linear independent for a finite element, i.e., a basis of span (Σ), the dual basis to the basis $p_1, \cdots, p_M$.

Later we will call such a "H^1-based" finite element approach also the *primal finite element method*, as opposed to approaches based on a dual formulation, in particular, the dual mixed approaches of Sections 7.2, 7.3, or a *continuous finite element method*, as opposed to discontinuous Galerkin approaches discussed in Section 7.4.

Exercises

Problem 3.10 Prove the implication "$\Rightarrow$" in Theorem 3.21.
Hint: For $v \in V_h$ define a function w_i by $w_i|_{\text{int}(K)} := \partial_i v$, $i = 1, \ldots, d$, and show that w_i is the ith partial derivative of v.

Problem 3.11 Construct the element stiffness matrix for the Poisson equation on a rectangle with quadratic bilinear rectangular elements. Verify that this finite element discretization of the Laplace operator can be interpreted as a finite difference method with the difference stencil according to (1.22).

Problem 3.12 Prove that:

a) $\dim \mathcal{P}_k(\mathbb{R}^d) = \binom{d+k}{k}$.
b) $\mathcal{P}_k(\mathbb{R}^d)|_K = \mathcal{P}_k(K)$ if $\operatorname{int}(K) \neq \emptyset$.

Problem 3.13 Prove for given vectors $a_1, \ldots, a_{d+1} \in \mathbb{R}^d$ that $a_2 - a_1, \ldots, a_{d+1} - a_1$ are linear independent if and only if $a_1 - a_i, \ldots, a_{i-1} - a_i, a_{i+1} - a_i, \ldots, a_{d+1} - a_i$ are linearly independent for some $i \in \{2, \ldots, d\}$.

Problem 3.14 Determine for the polynomial ansatz on the cuboid as reference element (3.70) the ansatz space P that is obtained by an affine-linear transformation to a d-epiped.

Problem 3.15 Suppose K is a rectangle with the (counterclockwise numbered) vertices $a_1, \ldots, a_4$ and the corresponding edge midpoints $a_{12}, a_{23}, a_{34}, a_{41}$. Show that the elements f of $Q_1(K)$ are not determined uniquely by the degrees of freedom $f(a_{12}), f(a_{23}), f(a_{34}), f(a_{41})$.

Problem 3.16 Check the given shape functions for (3.66) and (3.67).

Problem 3.17 Define a reference element in $\mathbb{R}^3$ by

$$\begin{aligned}
\hat{K} &:= \operatorname{conv}\{\hat{a}_1, \hat{a}_2, \hat{a}_3\} \times [0,1] \text{ with } \hat{a}_1 := \begin{pmatrix}0\\0\end{pmatrix},\ \hat{a}_2 := \begin{pmatrix}1\\0\end{pmatrix},\ \hat{a}_3 := \begin{pmatrix}0\\1\end{pmatrix},\\
\hat{P} &:= \left\{p_1(x_1, x_2)\, p_2(x_3) \;\middle|\; p_1 \in \mathcal{P}_1(\mathbb{R}^2),\ p_2 \in \mathcal{P}_1(\mathbb{R})\right\},\\
\hat{\Sigma} &:= \left\{p(\hat{x}) \;\middle|\; \hat{x} = (\hat{a}_i, j),\ i = 0, 1, 2,\ j = 0, 1\right\}.
\end{aligned}$$

Show the unique solvability of the local interpolation problem and describe the elements obtained by affine-linear transformation.

Problem 3.18 Suppose $d+1$ points a_j, $j = 1, \ldots, d+1$, in $\mathbb{R}^d$ are given with the property as in Problem 3.13. Additionally, we define as in (3.59), (3.60) the barycentric coordinates $\lambda_j = \lambda_j(x; S)$ of x with respect to the d-simplex S generated by the points a_j. Show that for each bijective affine-linear mapping $\mathcal{L} : \mathbb{R}^d \to \mathbb{R}^d$, $\lambda_j(x; S) = \lambda_j(\mathcal{L}(x); \mathcal{L}(S))$, which means that the barycentric coordinates are invariant under such transformations.

Problem 3.19 Discuss for the cubic Hermite ansatz (3.76) and Dirichlet boundary conditions the choice of the degrees of freedom with regard to the angle between two edges of boundary elements that is either $\alpha \neq 2\pi$ or $\alpha = 2\pi$.

Problem 3.20 Construct a nodal basis for the Bogner–Fox–Schmit element in (3.77).

3.4 Convergence Rate Estimates

In this section we consider further a finite element approximation in the framework described in the previous section: The bounded basic domain $\Omega \subset \mathbb{R}^d$ of the boundary value problem is decomposed into consistent partitions $\mathcal{T}_h$, which may also consist of different types of elements. Here, by an element we mean not only the set $K \in \mathcal{T}_h$, but this equipped with some ansatz space P_K and degrees of freedom Σ_K. However, the elements are supposed to decompose into a fixed number of subsets, independent of h, each consisting of elements that are affine equivalent to each other. Different elements have to be compatible with each other such that the ansatz space V_h, introduced in (3.74), is well defined. The smoothness of the functions arising in this way has to be consistent with the boundary value problem, in so far as $V_h \subset V$ is guaranteed. In the following we consider only one element type; the generalization to the more general situation will be obvious. The goal is to prove *a priori estimates* of the form

$$\|u - u_h\| \leq C|u|h^\alpha \tag{3.81}$$

with constants $C > 0$, $\alpha > 0$, and norms and seminorms $\|\cdot\|$ and $|\cdot|$, respectively.

We do not attempt to give the constant C explicitly, although, in principle, this is possible (with other techniques of proof). In particular, in the following C has to be understood generically, that is, by C we denote at different places different values, which, however, are independent of h. Therefore, estimate (3.81) does not serve only to estimate numerically the error for a fixed partition $\mathcal{T}_h$. It is rather useful for estimating what gain in accuracy can be expected by increasing the effort, which then corresponds to the reduction of h by some refinement (see the discussion around (3.79)). Independently of the *convergence rate* α, (3.81) provides the certainty that an arbitrary accuracy in the desired norm $\|\cdot\|$ can be obtained at all. In the following, we will impose some geometric conditions on the family $(\mathcal{T}_h)_h$, which have always to be understood uniformly in h. For a fixed partition these conditions are always trivially satisfied, since here we have a finite number of elements. For a family $(\mathcal{T}_h)_h$ with $h \to 0$, thus for increasing refinement, this number becomes unbounded. In the following estimates we have therefore to distinguish between "variable" values like the number of nodes $M = M(h)$ of $\mathcal{T}_h$, and "fixed" values like the dimension d or the dimension of P_K or equivalence constants in the renorming of P_K, which can all be included in the generic constant C.

3.4.1 Energy Norm Estimates

If we want to derive estimates in the norm of the Hilbert space V underlying the variational equation for the boundary value problem, concretely, in the norm of Sobolev spaces, then Céa's lemma (Theorem 2.17) shows that for this purpose it is necessary only to specify a comparison element $v_h \in V_h$ for which the inequality

$$\|u - v_h\| \le C|u|h^\alpha \tag{3.82}$$

holds. For $\|\cdot\| = \|\cdot\|_1$, these estimates are called *energy norm estimates* due to the equivalence of $\|\cdot\|_1$ and $\|\cdot\|_a$ (cf. (2.50)) in the symmetric case. Therefore, the comparison element v_h has to approximate u as well as possible, and, in general, it is specified as the image of a linear operator I_h:

$$v_h = I_h(u)\,.$$

The classical approach consists in choosing for I_h the *interpolation operator* with respect to the degrees of freedom. To simplify the notation, we restrict ourselves in the following to Lagrange elements, the generalization to Hermite elements is also easily possible.

We suppose that the partition $\mathcal{T}_h$ has its degrees of freedom in the nodes $a_1, \dots, a_M$ with the corresponding nodal basis $\varphi_1, \dots, \varphi_M$. Then let

$$I_h(u) := \sum_{i=1}^{M} u(a_i)\varphi_i \;\in V_h\,. \tag{3.83}$$

For the sake of $I_h(u)$ being well defined, $u \in C(\overline{\Omega})$ has to be assumed in order to ensure that u can be evaluated in the nodes. This requires a certain smoothness assumption about the solution u, which we formulate as

$$u \in H^{k+1}(\Omega)\,.$$

Thus, if we assume again $d \le 3$ for the sake of simplicity, the embedding theorem (Theorem 3.11) ensures that I_h is well defined on $H^{k+1}(\Omega)$ for $k \ge 1$. For the considered C^0-elements, we have $I_h(u) \in H^1(\Omega)$ by virtue of Theorem 3.21. Therefore, we can substantiate the desired estimate (3.82) to

$$\|u - I_h(u)\|_1 \le Ch^\alpha |u|_{k+1}\,. \tag{3.84}$$

Sobolev (semi) norms can be decomposed into expressions over subsets of Ω, thus, for instance, the elements of $\mathcal{T}_h$,

$$|u|_l^2 = \int_\Omega \sum_{|\alpha|=l} |\partial^\alpha u|^2\,dx = \sum_{K\in\mathcal{T}_h} \int_K \sum_{|\alpha|=l} |\partial^\alpha u|^2\,dx = \sum_{K\in\mathcal{T}_h} |u|_{l,K}^2\,,$$

and, correspondingly,

$$\|u\|_l^2 = \sum_{K\in\mathcal{T}_h} \|u\|_{l,K}^2\,,$$

where, if Ω is not basic domain, this will be included in the indices of the norm. Since the elements K are considered as being closed, K should more precisely be replaced by $\operatorname{int}(K)$. By virtue of this decomposition, it is sufficient to prove the estimate (3.84) for the elements K. This has some analogy to the (elementwise) assembling described in Section 2.4.2, which is also to be seen in the following. On K, the operator I_h reduces to the analogously defined local interpolation operator. Suppose the nodes of the degrees of freedom on K are $a_{i_1}, \ldots, a_{i_L}$, where $L \in \mathbb{N}$ is the same for all $K \in \mathcal{T}_h$ due to the equivalence of elements. Then

$$I_h(u)|_K = I_K(u|_K) \quad \text{for } u \in C(\overline{\Omega}),$$

$$I_K(u) := \sum_{j=1}^{L} u(a_{i_j})\varphi_{i_j} \quad \text{for } u \in C(K), \tag{3.85}$$

since both functions of P_K solve the same interpolation problem on K (cf. Lemma 2.10). Since we have an (affine) equivalent partition, the proof of the local estimate

$$\|u - I_K(u)\|_{m,K} \le Ch^\alpha |u|_{k+1,K} \tag{3.86}$$

is generally done in three steps:

- Transformation to some reference element $\hat{K}$,
- Proof of (3.86) on $\hat{K}$,
- Back-transformation to the element K.

To be precise, estimate (3.86) will even be proved with h_K instead of h, where

$$h_K := \operatorname{diam}(K) \quad \text{for } K \in \mathcal{T}_h,$$

and in the second step, the fixed value $h_{\hat{K}}$ is incorporated in the constant. The powers of h_K are due to the transformation steps.

Therefore, let some reference element $\hat{K}$ with the nodes $\hat{a}_1, \ldots, \hat{a}_L$ be chosen as fixed. By assumption, there exists some bijective, affine-linear mapping

$$\begin{aligned} F = F_K \,:\, \hat{K} &\to K, \\ F(\hat{x}) &= B\hat{x} + d, \end{aligned} \tag{3.87}$$

(cf. (2.34) and (3.68)). By this transformation, functions $v : K \to \mathbb{R}$ are mapped to functions $\hat{v} : \hat{K} \to \mathbb{R}$ by

$$\hat{v}(\hat{x}) := v(F(\hat{x}))\,. \tag{3.88}$$

This transformation is also *compatible* with the local interpolation operator in the following sense:

$$\widehat{I_K(v)} = I_{\hat{K}}(\hat{v}) \quad \text{for } v \in C(K)\,. \tag{3.89}$$

This follows from the fact that the nodes of the elements as well as the shape functions are mapped onto each other by F.

For a classically differentiable function the chain rule (see (2.53)) implies

$$\nabla_x v(F(\hat{x})) = B^{-T}\nabla_{\hat{x}}\hat{v}(\hat{x}), \tag{3.90}$$

and corresponding formulas for higher order derivatives, for instance,

$$D_x^2 v(F(\hat{x})) = B^{-T} D_{\hat{x}}^2 \hat{v}(\hat{x}) B^{-1},$$

where $D_x^2 v(x)$ denotes the matrix of the second-order derivatives. These chain rules hold also for corresponding $v \in H^l(K)$ (Problem 3.21).

The situation becomes particularly simple in one space dimension ($d = 1$). The considered elements reduce to a polynomial ansatz on intervals. Thus

$$F : \hat{K} = [0, 1] \to K := [a_{i_1}, a_{i_2}], \; \hat{x} \mapsto h_K \hat{x} + a_{i_1},$$

where $h_K := a_{i_2} - a_{i_1}$ denotes the length of the element. Hence, for $l \in \mathbb{N}$,

$$\partial_x^l v(F(\hat{x})) = h_K^{-l} \partial_{\hat{x}}^l \hat{v}(\hat{x}).$$

By the substitution rule for integrals (cf. (2.54)) an additional factor $|\det(B)| = h_K$ arises such that, for $v \in H^l(K)$, we have

$$|v|_{l,K}^2 = \left(\frac{1}{h_K}\right)^{2l-1} |\hat{v}|_{l,\hat{K}}^2.$$

Hence, for $0 \le m \le k+1$ it follows by (3.89) that

$$|v - I_K(v)|_{m,K}^2 = \left(\frac{1}{h_K}\right)^{2m-1} |\hat{v} - I_{\hat{K}}(\hat{v})|_{m,\hat{K}}^2.$$

Thus, what is missing is an estimate of the type

$$|\hat{v} - I_{\hat{K}}(\hat{v})|_{m,\hat{K}} \le C |\hat{v}|_{k+1,\hat{K}} \tag{3.91}$$

for $\hat{v} \in H^{k+1}(\hat{K})$. In specific cases this can partly be proven directly but in the following a general proof, which is also independent of $d = 1$, will be sketched. For this, the mapping

$$G : H^{k+1}(\hat{K}) \to H^m(\hat{K}), \; \hat{v} \mapsto \hat{v} - I_{\hat{K}}(\hat{v}) \tag{3.92}$$

is considered. The mapping is linear but also continuous, since

$$\begin{aligned} \|I_{\hat{K}}(\hat{v})\|_{m,\hat{K}} &\le \left\| \sum_{i=1}^{L} \hat{v}(\hat{a}_i)\hat{\varphi}_i \right\|_{k+1,\hat{K}} \\ &\le \sum_{i=1}^{L} \|\hat{\varphi}_i\|_{k+1,\hat{K}} \|\hat{v}\|_{\infty,\hat{K}} \le C \|\hat{v}\|_{k+1,\hat{K}}, \end{aligned} \tag{3.93}$$

where the continuity of the embedding of $H^{k+1}(\hat{K})$ in $H^m(\hat{K})$ (see (3.9)) and of $H^{k+1}(\hat{K})$ in $C(\hat{K})$ (Theorem 3.11) is used, and the norm contribution from the fixed basis functions $\hat{\varphi}_i$ is included in the constant.

If the ansatz space $\hat{P}$ is chosen in such a way that $\mathcal{P}_k \subset \hat{P}$, then G has the additional property

$$G(p) = 0 \quad \text{for } p \in \mathcal{P}_k ,$$

since these polynomials are interpolated then exactly. Such mappings satisfy the Bramble–Hilbert lemma, which will directly be formulated, for further use, in a more general way.

Theorem 3.27 (Bramble–Hilbert lemma)
Suppose $K \subset \mathbb{R}^d$ is open, $k \in \mathbb{N}_0$, $1 \le p \le \infty$, and $G : W_p^{k+1}(K) \to \mathbb{R}$ is a continuous linear functional that satisfies

$$G(q) = 0 \quad \textit{for all } q \in \mathcal{P}_k . \tag{3.94}$$

Then there exists some constant $C > 0$ independent of G such that for all $v \in W_p^{k+1}(K)$

$$|G(v)| \le C \, \|G\| \, |v|_{k+1,p,K} .$$

Proof See [16, Theorem 28.1]. □

Here $\|G\|$ denotes the operator norm of G (see (A4.27)). The estimate with the full norm $\|\cdot\|_{k+1,p,K}$ on the right-hand side (and $C = 1$) would hence only be the operator norm's definition. Condition (3.94) allows the reduction to the highest seminorm.

For the application of the Bramble–Hilbert lemma (Theorem 3.27), which was formulated only for functionals, to the operator G according to (3.92) an additional argument is required (alternatively, Theorem 3.27 could be generalized):
Generally, for $\hat{w} \in H^m(\hat{K})$ (as in every normed space) we have

$$\|\hat{w}\|_{m,\hat{K}} = \sup_{\substack{\varphi \in (H^m(\hat{K}))' \\ \|\varphi\| \le 1}} \varphi(\hat{w}) , \tag{3.95}$$

where the norm applying to φ is the operator norm defined in (A4.27).

For any fixed $\varphi \in (H^m(\hat{K}))'$ the linear functional on $H^{k+1}(\hat{K})$ is defined by

$$\tilde{G}(\hat{v}) := \varphi(G(\hat{v})) \quad \text{for} \quad \hat{v} \in H^{k+1}(\hat{K}) . \tag{3.96}$$

According to (3.93), $\tilde{G}$ is continuous and it follows that

$$\|\tilde{G}\| \le \|\varphi\| \, \|G\| .$$

Theorem 3.27 is applicable to $\tilde{G}$ and yields

$$|\tilde{G}(\hat{v})| \le C \, \|\varphi\| \, \|G\| \, |\hat{v}|_{k+1,\hat{K}} .$$

By means of (3.95) it follows that

$$\|G(\hat{v})\|_{m,\hat{K}} \le C \, \|G\| \, |\hat{v}|_{k+1,\hat{K}} .$$

The same proof can also be used in the proof of Theorem 3.34 (3.112).

Applied to G defined in (3.92), the estimate (3.93) shows that the operator norm $\|\mathrm{Id} - I_{\hat{K}}\|$ can be estimated independently from m (but dependent on k and the $\hat{\varphi}_i$) and can be incorporated in the constant that gives (3.91), in general, independent of the one-dimensional case.

Therefore, in the one-dimensional case, we can continue with the estimation and get

$$|v - I_K(v)|_{m,K}^2 \leq \left(\frac{1}{h_K}\right)^{2m-1} C|\hat{v}|_{k+1,\hat{K}}^2 \leq C(h_K)^{1-2m+2(k+1)-1}|v|_{k+1,K}^2 .$$

Since due to $I_h(v) \in H^1(\Omega)$ we have for $m = 0, 1$

$$\sum_{K\in\mathcal{T}_h} |v - I_K(v)|_{m,K}^2 = |v - I_h(v)|_m^2 ,$$

we have proven the following Theorem:

Theorem 3.28 *Consider in one space dimension $\Omega = (a, b)$ the polynomial Lagrange ansatz on elements with maximum length h and suppose that for the respective local ansatz spaces P, the inclusion $\mathcal{P}_k \subset P$ is satisfied for some $k \in \mathbb{N}_0$. Then there exists some constant $C > 0$ such that for all $v \in H^{k+1}(\Omega)$ and $0 \leq m \leq k + 1$,*

$$\left(\sum_{K\in\mathcal{T}_h} |v - I_K(v)|_{m,K}^2\right)^{1/2} \leq Ch^{k+1-m}|v|_{k+1} .$$

If the solution u of the boundary value problem (3.13), (3.19)–(3.21) belongs to $H^{k+1}(\Omega)$, then we have for the finite element approximation u_h according to (3.50),

$$\|u - u_h\|_1 \leq Ch^k|u|_{k+1} .$$

Note that for $d = 1$ a direct proof is also possible (see Problem 3.22).

Now we address to the general d-dimensional situation: The seminorm $|\cdot|_1$ is transformed, for instance, as follows (cf. (2.53)):

$$|v|_{1,K}^2 = \int_K |\nabla_x v|^2\,dx = \int_{\hat{K}} B^{-T}\nabla_{\hat{x}}\hat{v} \cdot B^{-T}\nabla_{\hat{x}}\hat{v}\,|\det(B)|\,d\hat{x} . \tag{3.97}$$

From this, it follows for $\hat{v} \in H^1(\hat{K})$ that

$$|v|_{1,K} \leq C\,\|B^{-1}\|\,|\det(B)|^{1/2}\,|\hat{v}|_{1,\hat{K}} .$$

Since d is one of the mentioned "fixed" quantities and all norms on $\mathbb{R}^{d,d}$ are equivalent, the matrix norm $\|\cdot\|$ can be chosen arbitrarily, and it is also possible to change between such norms. In the above considerations K and $\hat{K}$ had equal rights; thus similarly for $v \in H^1(K)$, we have

$$|\hat{v}|_{1,\hat{K}} \le C\,\|B\|\,|\det(B)|^{-1/2}\,|v|_{1,K}\,.$$

In general, we have the following theorem:

Theorem 3.29 *Suppose K and $\hat{K}$ are bounded domains in $\mathbb{R}^d$ that are mapped onto each other by an affine bijective linear mapping F, defined in (3.87). If $v \in W_p^l(K)$ for $l \in \mathbb{N}$ and $p \in [1,\infty]$, then we have for $\hat{v}$ (defined in (3.88)), $\hat{v} \in W_p^l(\hat{K})$, and for some constant $C > 0$ independent of v,*

$$|\hat{v}|_{l,p,\hat{K}} \le C\,\|B\|^l\,|\det(B)|^{-1/p}\,|v|_{l,p,K}\,, \tag{3.98}$$

$$|v|_{l,p,K} \le C\,\|B^{-1}\|^l\,|\det(B)|^{1/p}\,|\hat{v}|_{l,p,\hat{K}}\,. \tag{3.99}$$

Proof See [16, Theorem 15.1]. □

For further use, also this theorem has been formulated in a more general way than would be necessary here. Here, only the case $p = 2$ is relevant.

Hence, if we use the estimate of Theorem 3.27, then the value $\|B\|$ (for some matrix norm) has to be related to the geometry of K. For this, let for $K \in \mathcal{T}_h$,

$$\varrho_K := \sup\left\{\operatorname{diam}(S) \;\middle|\; S \text{ is a ball in } \mathbb{R}^d \text{ and } S \subset K\right\}.$$

Hence, in the case of a triangle, h_K denotes the longest edge and ϱ_K the diameter of the inscribed circle. Similarly, the reference element has its (fixed) parameters $\hat{h}$ and $\hat{\varrho}$. For example, for the reference triangle with the vertices $\hat{a}_1 = (0,0)$, $\hat{a}_2 = (1,0)$, $\hat{a}_3 = (0,1)$ we have that $\hat{h} = 2^{1/2}$ and $\hat{\varrho} = 2 - 2^{1/2}$.

Theorem 3.30 *For $F = F_K$ according to (3.87), in the spectral norm $\|\cdot\|_2$, we have*

$$\|B\|_2 \le \frac{h_K}{\hat{\varrho}} \quad \text{and} \quad \|B^{-1}\|_2 \le \frac{\hat{h}}{\varrho_K}\,.$$

Proof Since K and $\hat{K}$ have equal rights in the assertion, it suffices to prove one of the statements: We have (cf. (A4.27))

$$\|B\|_2 = \sup_{|\xi|_2=\hat{\varrho}} \left|B\left(\frac{1}{\hat{\varrho}}\xi\right)\right|_2 = \frac{1}{\hat{\varrho}} \sup_{|\xi|_2=\hat{\varrho}} |B\xi|_2\,.$$

For every $\xi \in \mathbb{R}^d$ with $|\xi|_2 = \hat{\varrho}$ there exist some points $\hat{y}, \hat{z} \in \hat{K}$ such that $\hat{y} - \hat{z} = \xi$. Since $B\xi = F(\hat{y}) - F(\hat{z})$ and $F(\hat{y}), F(\hat{z}) \in K$, we have $|B\xi|_2 \le h_K$. Consequently, by the above identity we get the first inequality. □

If we combine the local estimates of (3.91), Theorem 3.29, and Theorem 3.30, we obtain for $v \in H^{k+1}(K)$ and $0 \le m \le k+1$,

$$|v - I_K(v)|_{m,K} \le C\left(\frac{h_K}{\varrho_K}\right)^m h_K^{k+1-m}\,|v|_{k+1,K}\,, \tag{3.100}$$

where $\hat{\varrho}$ and $\hat{h}$ are included in the constant C. In order to obtain some convergence rate result, we have to control the term h_K/ϱ_K. If this term is bounded (uniformly for all partitions), we get the same estimate as in the one-dimensional case (where even $h_K/\varrho_K = 1$). Conditions of the form

$$\varrho_K \geq \sigma h_K^{1+\alpha}$$

for some $\sigma > 0$ and $0 \leq \alpha < \frac{k+1}{m} - 1$ for $m \geq 1$ would also lead to convergence rate results. Here we pursue only the case $\alpha = 0$.

Definition 3.31 A family of partitions $(\mathcal{T}_h)_h$ is called *shape-regular* if there exists some $\sigma > 0$ such that for all $h > 0$ and all $K \in \mathcal{T}_h$,

$$\varrho_K \geq \sigma h_K \, .$$

From estimate (3.100) we conclude directly the following theorem:

Theorem 3.32 *Consider a family of Lagrange finite element discretizations in* $\mathbb{R}^d$ *for* $d \leq 3$ *on a shape-regular family of partitions* $(\mathcal{T}_h)_h$ *in the generality described at the very beginning. For the respective local ansatz spaces* P *suppose* $\mathcal{P}_k \subset P$ *for some* $k \in \mathbb{N}_0$.

Then there exists some constant $C > 0$ *such that for all* $v \in H^{k+1}(\Omega)$ *and* $0 \leq m \leq k+1$,

$$|v - I_h(v)|_{m,\mathcal{T}_h} = \left(\sum_{K \in \mathcal{T}_h} |v - I_K(v)|_{m,K}^2 \right)^{1/2} \leq C h^{k+1-m} |v|_{k+1} \, . \qquad (3.101)$$

If the solution u *of the boundary value problem (3.13), (3.19)–(3.21) belongs to* $H^{k+1}(\Omega)$, *then for the finite element approximation* u_h *defined in (3.50), it follows that*

$$\|u - u_h\|_1 \leq C h^k |u|_{k+1} \, . \qquad (3.102)$$

Here we used the *broken Sobolev seminorm*

$$|v|_{l,p,\mathcal{T}_h} := \left(\sum_{K \in \mathcal{T}_h} |v|_{l,p,K}^p \right)^{1/p}$$

for $l \in \mathbb{N}$ and $p \in [1, \infty)$ or $p = \infty$ (correspondingly modified), and analogously $\|v\|_{l,p,\mathcal{T}_h}$. For $p = 2$, this index is omitted.

Remark 3.33 1) Indeed, here and also in Theorem 3.28 a sharper estimate has been shown, which, for instance, for (3.102), has the following form:

$$\|u - u_h\|_1 \leq C \left(\sum_{K \in \mathcal{T}_h} h_K^{2k} |u|_{k+1,K}^2 \right)^{1/2} \leq h^k |u|_{k+1,\mathcal{T}_h} \, , \qquad (3.103)$$

and only $u \in H^2(\Omega) \cap H^{k+1}(\mathcal{T}_h)$ is required.

The discussion of solution and data regularity necessary will be put more in the focus of Chapters 7 and 10.

2) Analysing the above proofs, we see that actually the following general result can be shown.
Let $\hat{K} \subset \mathbb{R}^d$ be a bounded domain, such that the embedding

$$W_\alpha^{p+1}(\hat{K}) \hookrightarrow W_\beta^m(\hat{K})$$

is continuous, for $\alpha, \beta \in [1, \infty]$, $p, m \geq 0$.
Let $\hat{\Pi}$ be a continuous linear mapping from $W_\alpha^{p+1}(\hat{K})$ to $W_\beta^m(\hat{K})$ such that

$$\hat{\Pi}\hat{v} = \hat{v} \quad \text{for } \hat{v} \in \mathcal{P}_k(\hat{K}). \tag{3.104}$$

Set $K = F(\hat{K})$, with F from (3.87),

$$\Pi_K v(x) := \hat{\Pi}\hat{v}(F^{-1}(x)) \quad \text{for } x \in K$$

for $v \in W_\alpha^{p+1}(K)$: Then there exists a constant $C = C(\hat{\Pi}, \hat{K})$ such that for $m \leq p+1$

$$\begin{aligned} |\hat{\Pi}\hat{v} - \hat{v}|_{m,\beta,\hat{K}} &\leq C|\hat{v}|_{p+1,\alpha,\hat{K}}, \\ |v - \Pi_K v|_{m,\beta,K} &\leq C|K|^{1/\beta-1/\alpha} h_K^{p+1} \varrho_k^{-m} |v|_{p+1,\alpha,K}. \end{aligned} \tag{3.105}$$

In particular, $v \in H^{k+1}(\Omega)$ can be substituted by $v \in H^{k+1}(\mathcal{T}_h)$.
From this local estimate (3.101) can be regained for $\hat{\Pi} = I_{\hat{K}}$, assuming shape regularity, $\alpha = \beta = 2, 0 \leq m \leq l \leq k+1$. The continuity is shown in (3.93), condition (3.104) is obvious.

3) Another choice of $\hat{\Pi}$ is the L^2-orthogonal projection

$$\int_{\hat{K}} (\hat{\Pi}\hat{v} - \hat{v})\hat{w}\, dx = 0 \quad \text{for } \hat{w} \in \mathcal{P}_k(\hat{K}) \tag{3.106}$$

such that (3.104) is clear and the continuity follows, as for each orthogonal projection

$$\|\hat{\Pi}\hat{v}\|_{0,\hat{K}} \leq \|\hat{v}\|_{0,\hat{K}} \leq \|\hat{v}\|_{p+1,\alpha,\hat{K}}$$

and on $\hat{\Pi}[L^2(\hat{K})]$, being finite-dimensional, all norms are equivalent. Π_K is the L^2-orthogonal projection on $\mathcal{P}_k(K)$ and the global mapping Π (corresponding the transition from I_K to I_h) is then the L^2-orthogonal projection on

$$\mathcal{P}_k(\mathcal{T}_h) := \{w_h \in L^2(\Omega) \mid w_h|_K \in \mathcal{P}_k(K) \quad \text{for all } K \in \mathcal{T}_h\} \tag{3.107}$$

to be used in Section 7.4 for the *Discontinuous Galerkin Methods*.
The analogue of Theorem 3.28 (now including $k = 0$) reads

$$\|v - \Pi v\|_{m,\mathcal{T}_h} \leq Ch^{k+1-m}|v|_{k+1,\mathcal{T}_h}. \tag{3.108}$$

For $k = 0$ we have

$$\Pi_K v = \frac{1}{|K|} \int_K v dx.$$

4) The type and position of the degrees of freedom induce certain regularity conditions to the global ansatz space V_h according to (3.74), and thus decide whether $V_h \subset V$ holds or not. The latter, nonconforming case, will be considered from Section 6.1 on.
5) On a regular d-simplex K or on a d-parallelepiped K, the following is also a finite element:

$$P := \mathcal{P}_0,\ \Sigma := \{p(\overline{a})\} \quad \text{or} \quad \Sigma := \left\{ \int_K p(x)dx \right\},$$

where $\overline{a} \in K$ is arbitrary but fixed. However, the global space has only the property

$$V_h \subset L^\infty(\Omega).$$

In the following we will discuss what the regularity assumption means in the two simplest cases.

For a rectangle and the cuboid K, whose edge lengths can be assumed, without any loss of generality, to be of order $h_1 \leq h_2 [\leq h_3]$, we have

$$\frac{h_K}{\varrho_K} = \left(1 + \left(\frac{h_2}{h_1} \right)^2 \left[+ \left(\frac{h_3}{h_1} \right)^2 \right] \right)^{1/2} .$$

This term is uniformly bounded if and only if there exists some constant $\alpha (\geq 1)$ such that

$$\begin{aligned} h_1 &\leq h_2 \leq \alpha h_1 , \\ h_1 &\leq h_3 \leq \alpha h_1 . \end{aligned} \tag{3.109}$$

In order to satisfy this condition, a refinement in one space direction has to imply a corresponding one in the other directions, although in certain *anisotropic* situations only the refinement in one space direction is recommendable. If, for instance, the boundary value problem (3.13), (3.19)–(3.21) with $c = r = 0$, but space-dependent conductivity K, is interpreted as the simplest ground water model (see (0.18)), then it is typical that K varies discontinuously due to some *layering* or more complex geological structures (see Figure 3.12).

If thin layers arise in such a case, on the one hand, they have to be *resolved*, that is, the partition has to be compatible with the layering and there have to be sufficiently many elements in this layer. On the other hand, the solution often changes less strongly in the direction of the layering than over the boundaries of the layer, which suggests an *anisotropic* partition, that is, a strongly varying dimensioning of the elements. The restriction (3.109) is not compatible with this, but in the case of rectangles this is due only to the techniques of proof. In this simple situation, the local interpolation error estimate can be performed directly, at least for $P = Q_1(K)$, without any transformation such that the estimate (3.102) (for $k = 1$) is obtained without any restrictions like (3.109).

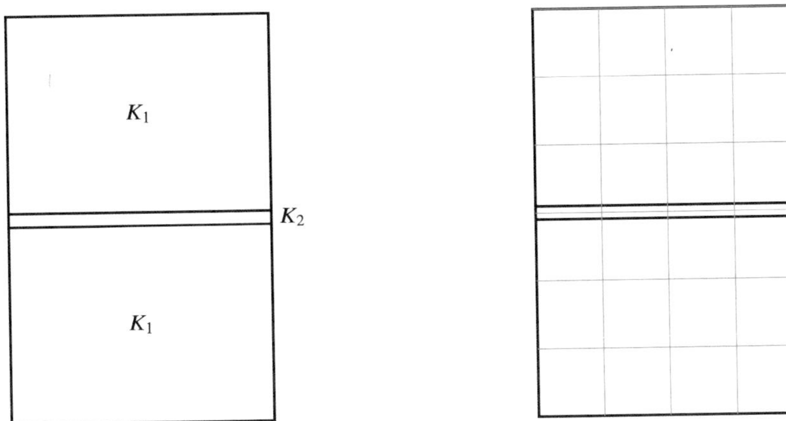

Fig. 3.12: Layering and anisotropic partition.

The next simple example is a triangle K: The smallest angle $\alpha_{\min} = \alpha_{\min}(K)$ includes the longest edge h_K, and without loss of generality, the situation is as illustrated in Figure 3.13.

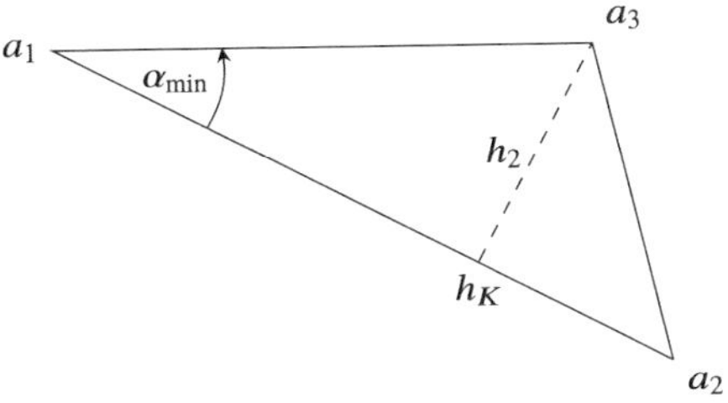

Fig. 3.13: Triangle with the longest edge and the height as parameters.

For the 2×2 matrix $B = (a_2 - a_1\,,\ a_3 - a_1)$, in the Frobenius norm $\|\cdot\|_F$ (see (A3.5)) we have

$$\|B^{-1}\|_F = \frac{1}{|\det(B)|}\|B\|_F\,,$$

and further, with the height h_2 over h_K,

$$\det(B) = h_K h_2\,, \tag{3.110}$$

since $\det(B)/2$ is the area of the triangle, as well as

$$\|B\|_F^2 = |a_2 - a_1|_2^2 + |a_3 - a_1|_2^2 \geq h_K^2\,,$$

such that

$$\|B\|_F\|B^{-1}\|_F \geq h_K/h_2\,,$$

and thus by virtue of $\cot \alpha_{\min} < h_K/h_2$,

$$\|B\|_F \|B^{-1}\|_F > \cot \alpha_{\min} .$$

Since we get by analogous estimates

$$\|B\|_F \|B^{-1}\|_F \leq 4 \cot \alpha_{\min} ,$$

it follows that $\cot \alpha_{\min}$ describes the asymptotic behaviour of $\|B\|\|B^{-1}\|$ for a fixed chosen arbitrary matrix norm. Therefore, from Theorem 3.30 we get the existence of some constant $C > 0$ independent of h such that for all $K \in \mathcal{T}_h$,

$$\frac{h_K}{\varrho_K} \geq C \cot \alpha_{\min}(K) . \tag{3.111}$$

Consequently, a family of partitions $(\mathcal{T}_h)_h$ of triangles can only be shape-regular if all angles of the triangles are uniformly bounded from below by some positive constant. This condition sometimes is called the *minimum angle condition*. In the situation of Figure 3.12 it would thus not be allowed to decompose the flat rectangles in the thin layer by means of a Friedrichs–Keller partition. Obviously, using directly the estimates of Theorem 3.29 we see that the minimum angle condition is sufficient for the estimates of Theorem 3.32. This still leaves the possibility open that less severe conditions are also sufficient.

3.4.2 The Maximum Angle Condition on Triangles

In what follows we show that condition (3.111) is due only to the techniques of proof, and at least in the case of the linear ansatz, it has indeed only to be ensured that the largest angle is uniformly bounded away from π. Therefore, this allows the application of the described approach in the layer example of Figure 3.12.

Estimate (3.100) shows that for $m = 0$ the crucial part does not arise; hence only for $m = k = 1$ do the estimates have to be investigated. It turns out to be useful to prove the following sharper form of the estimate (3.91):

Theorem 3.34 *For the reference triangle $\hat{K}$ with linear ansatz functions there exists some constant $C > 0$ such that for all $\hat{v} \in H^2(\hat{K})$ and $j = 1, 2$,*

$$\left\| \frac{\partial}{\partial \hat{x}_j} \left(\hat{v} - I_{\hat{K}}(\hat{v}) \right) \right\|_{0,\hat{K}} \leq C \left| \frac{\partial}{\partial \hat{x}_j} \hat{v} \right|_{1,\hat{K}} .$$

Proof In order to simplify the notation, we drop the hat ^ in the notation of the reference situation in the proof. Hence, we have $K = \operatorname{conv}\{a_1, a_2, a_3\}$ with $a_1 = (0,0)^T$, $a_2 = (1,0)^T$, and $a_3 = (0,1)^T$. We consider the following linear mappings: $F_1 : H^1(K) \to L^2(K)$ is defined by

$$F_1(w) := \int_0^1 w(s,0)\,ds\,,$$

and, analogously, F_2 as the integral over the boundary part conv $\{a_1, a_3\}$. The image is taken as constant function on K. By virtue of the Trace Theorem 3.6, and the continuous embedding of $L^2(0,1)$ in $L^1(0,1)$, the F_i are well defined and continuous. Since we have for $w \in \mathcal{P}_0(K)$,

$$F_i(w) = w\,,$$

the Bramble–Hilbert lemma (Theorem 3.27) implies the existence of some constant $C > 0$ such that for $w \in H^1(K)$,

$$\|F_i(w) - w\|_{0,K} \le C|w|_{1,K}\,. \tag{3.112}$$

This can be seen in the following way: Let $v \in H^1(K)$ be arbitrary but fixed, and for this, consider on $H^1(K)$ the functional

$$G(w) := \langle F_i(w) - w, F_i(v) - v\rangle \quad \text{for } w \in H^1(K)\,.$$

We have $G(w) = 0$ for $w \in \mathcal{P}_0(K)$ and

$$|G(w)| \le \|F_i(w) - w\|_{0,K}\|F_i(v) - v\|_{0,K} \le C\|F_i(v) - v\|_{0,K}\|w\|_{1,K}$$

by the above consideration. Thus by Theorem 3.27,

$$|G(w)| \le C\,\|F_i(v) - v\|_{0,K}\,|w|_{1,K}\,.$$

For $v = w$ this implies (3.112). On the other hand, for $w := \partial_1 v$ it follows that

$$\begin{aligned} F_1(\partial_1 v) &= v(1,0) - v(0,0) = (I_K(v))(1,0) - (I_K(v))(0,0) \\ &= \partial_1(I_K(v))(x_1,x_2) \end{aligned}$$

for $(x_1, x_2) \in K$ and, analogously, $F_2(\partial_2 v) = \partial_2(I_K(v))(x_1, x_2)$. This, substituted into (3.112), gives the assertion. □

Compared with estimate (3.91), for example, in the case $j = 1$ the term $\frac{\partial^2}{\partial \hat{x}_2^2}\hat{v}$ does not arise on the right-hand side: The derivatives and thus the space directions are therefore treated "more separately".

Next, the effect of the transformation will be estimated more precisely. For this, let $\alpha_{\max} = \alpha_{\max}(K)$ be the largest angle arising in $K \in \mathcal{T}_h$, supposed to include the vertex a_1, and let $h_1 = h_{1K} := |a_2 - a_1|_2$, $h_2 = h_{2K} := |a_3 - a_1|$ (see Figure 3.14).

As a variant of (3.99) (for $l = 1$) we have the following:

Theorem 3.35 *Suppose K is a general triangle. With the above notation for $v \in H^1(K)$ and the transformed $\hat{v} \in H^1(\hat{K})$,*

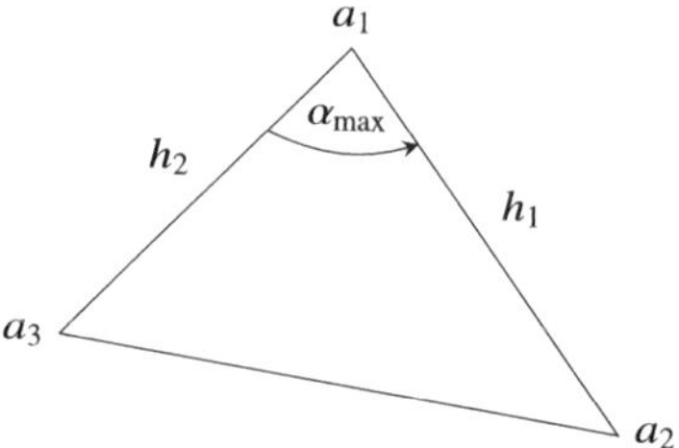

Fig. 3.14: A general triangle.

$$|v|_{1,K} \le \sqrt{2}\,|\det(B)|^{-1/2} \left(h_2^2 \left\| \frac{\partial}{\partial \hat{x}_1} \hat{v} \right\|_{0,\hat{K}}^2 + h_1^2 \left\| \frac{\partial}{\partial \hat{x}_2} \hat{v} \right\|_{0,\hat{K}}^2 \right)^{1/2} .$$

Proof We have

$$B = (a_2 - a_1, a_3 - a_1) =: \begin{pmatrix} b_{11} & b_{12} \\ b_{21} & b_{22} \end{pmatrix}$$

and hence

$$\left| \begin{pmatrix} b_{11} \\ b_{21} \end{pmatrix} \right| = h_1 , \qquad \left| \begin{pmatrix} b_{12} \\ b_{22} \end{pmatrix} \right| = h_2 . \tag{3.113}$$

From

$$B^{-T} = \frac{1}{\det(B)} \begin{pmatrix} b_{22} & -b_{21} \\ -b_{12} & b_{11} \end{pmatrix}$$

and (3.97) it thus follows that

$$|v|_{1,K}^2 = \frac{1}{|\det(B)|} \int_{\hat{K}} \left| \begin{pmatrix} b_{22} \\ -b_{12} \end{pmatrix} \frac{\partial}{\partial \hat{x}_1} \hat{v} + \begin{pmatrix} -b_{21} \\ b_{11} \end{pmatrix} \frac{\partial}{\partial \hat{x}_2} \hat{v} \right|^2 d\hat{x}$$

and from this the assertion. □

In modification of the estimate (3.98) (for $l = 2$) we prove the following result:

Theorem 3.36 *Suppose K is a general triangle with diameter $h_K = \mathrm{diam}\,(K)$. With the above notation for $\hat{v} \in H^2(\hat{K})$ and the transformed $v \in H^2(K)$,*

$$\left| \frac{\partial}{\partial \hat{x}_i} \hat{v} \right|_{1,\hat{K}} \le 4 |\det(B)|^{-1/2} h_i h_K |v|_{2,K} \quad \textit{for } i = 1, 2 .$$

Proof According to (3.97) we get by exchanging K and $\hat{K}$,

$$|\hat{w}|_{1,\hat{K}}^2 = \int_K B^T \nabla_x w \cdot B^T \nabla_x w \, dx \, |\det(B)|^{-1}$$

and, consequently, for $\hat{w} = \frac{\partial}{\partial \hat{x}_i} \hat{v}$, thus by (3.90) for $w = (B^T \nabla_x v)_i$,

$$\left|\frac{\partial}{\partial \hat{x}_i}\hat{v}\right|^2_{1,\hat{K}} = \int_K \left|B^T \nabla_x \left(\left(B^T \nabla_x v\right)_i\right)\right|^2 dx \, |\det(B)|^{-1} .$$

According to (3.113), the norm of the ith row vector of B^T is equal to h_i, which implies the assertion. □

Instead of the shape regularity of the family of partitions and hence the uniform bound for $\cot \alpha_{\min}(K)$ (see (3.111)) we require the following definition.

Definition 3.37 A family of plane triangulations $(\mathcal{T}_h)_h$ satisfies the *maximum angle condition* if there exists some constant $\overline{\alpha} < \pi$ such that for all $h > 0$ and $K \in \mathcal{T}_h$ the maximum angle $\alpha_{\max}(K)$ of K satisfies

$$\alpha_{\max}(K) \leq \overline{\alpha} .$$

Since $\alpha_{\max}(K) \geq \pi/3$ is always satisfied, the maximum angle condition is equivalent to the existence of some constant $\tilde{s} > 0$, such that

$$\sin(\alpha_{\max}(K)) \geq \tilde{s} \quad \text{for all } K \in \mathcal{T}_h \text{ and } h > 0 . \tag{3.114}$$

The relation of this condition to the above estimates is given by (cf. (3.110))

$$\det(B) = h_1 h_2 \sin \alpha_{\max} . \tag{3.115}$$

Inserting the estimates of Theorem 3.35 (for $v - I_K(v)$), Theorem 3.34, and Theorem 3.36 into each other and recalling (3.114), (3.115), the following theorem follows from Céa's lemma (Theorem 2.17).

Theorem 3.38 *Consider the linear ansatz (3.64) on a family of plane triangulations $(\mathcal{T}_h)_h$ that satisfies the maximum angle condition. Then there exists some constant $C > 0$ such that for $v \in H^2(\Omega)$,*

$$\|v - I_h(v)\|_1 \leq C\, h\, |v|_2 .$$

If the solution u of the boundary value problem (3.13), (3.19)–(3.21) belongs to $H^2(\Omega)$, then for the finite element approximation u_h defined in (3.50) we have the estimate

$$\|u - u_h\|_1 \leq Ch|u|_2 . \tag{3.116}$$

Problem 3.27 shows the necessity of the maximum angle condition. Again, a remark analogous to Remark 3.33 holds. For an analogous investigation of tetrahedra we refer to [167].

With a modification of the above considerations and an additional condition *anisotropic error estimates* of the form

$$|v - I_h(v)|_1 \leq C \sum_{i=1}^{d} h_i \, |\partial_i v|_1$$

can be proven for $v \in H^2(\Omega)$, where the h_i denote length parameter depending on the element type. In the case of triangles, these are the longest edge ($h_1 = h_K$) and the height on it as shown in Figure 3.13 (see [81]).

3.4.3 L^2 Error Estimates

The error estimate (3.102) also contains a result about the approximation of the gradient (and hence of the flux density), but it is linear only for $k = 1$, in contrast to the error estimate of Chapter 1 (Theorem 1.6). The question is whether an improvement of the convergence rate is possible if we strive only for an estimate of the function values. The *duality argument* of Aubin and Nitsche shows that this is correct, if the adjoint boundary value problem is regular, where we have the following definition.

Definition 3.39 The *adjoint boundary value problem* for (3.13), (3.19)–(3.21) is defined by the bilinear form

$$(u, v) \mapsto a(v, u) \quad \text{for } u, v \in V$$

with V from (3.31). It is called *regular* if for every $f \in L^2(\Omega)$ there exists a unique solution $u = u_f \in V$ of the adjoint boundary value problem

$$a(v, u) = \langle f, v\rangle_0 \quad \text{for all } v \in V$$

and even $u_f \in H^2(\Omega)$ is satisfied, and for some constant $C > 0$ a stability estimate of the form

$$|u_f|_2 \le C\|f\|_0 \quad \text{for given } f \in L^2(\Omega)$$

is satisfied.

The V-coercivity and the continuity of the bilinear form (3.2), (3.3) directly carry over from (3.32) to the adjoint boundary value problem, so that in this case the unique existence of $u_f \in V$ is ensured. More precisely, the adjoint boundary value problem is obtained by an exchange of the arguments in the bilinear form, which does not effect any change in its symmetric parts. The nonsymmetric part of (3.32) is $\int_\Omega \boldsymbol{c} \cdot \nabla u\, v\, dx$, which becomes $\int_\Omega \boldsymbol{c} \cdot \nabla v\, u\, dx$.

Therefore the adjoint problem to a problem that is not in divergence form ((3.13), (3.19)–(3.21)) is a problem in divergence form ((3.36)–(3.39)), for $\tilde{\boldsymbol{c}} := -\boldsymbol{c}$ and $\widetilde{\Gamma}_i := \Gamma_i$, $i = 1, 2, 3$ (but note the different meaning of the boundary conditions) (and vice versa).

In general an adjoint problem has the same solvability conditions as the original one, with the same stability as due to the Closed Range Theorem (see Appendix A.4) we have

$$A^{-T} \text{ exists} \Leftrightarrow A^{-1} \text{ exists and } \|A^{-T}\| = \|A^{-1}\|.$$

This has two consequences:

- Comparing a problem and its adjoint one, different regularity of solutions is only due to different regularity of the right-hand sides.
- Solvability and stability of the formulation in divergence form and not in divergence form transform to each other, as one is the adjoint of the other.

For a regular adjoint problem we get an improvement of the convergence rate in $\|\cdot\|_0$:

Theorem 3.40 (Aubin and Nitsche) *Consider the situation of Theorem 3.32 or Theorem 3.38 and suppose the adjoint boundary value problem is regular. Then there exists some constant $C > 0$ such that for the solution u of the boundary value problem (3.13), (3.19)–(3.21) and its finite element approximation u_h defined by (3.50),*

1) $\|u - u_h\|_0 \leq Ch\|u - u_h\|_1$,
2) $\|u - u_h\|_0 \leq Ch\|u\|_1$,
3) $\|u - u_h\|_0 \leq Ch^{k+1}|u|_{k+1}$, *if* $u \in H^{k+1}(\Omega)$.

Proof Assertions 2) and 3) follow directly from 1). On the one hand, by using $\|u-u_h\|_1 \leq \|u\|_1 + \|u_h\|_1$ and the stability estimate (2.48), on the other hand directly from (3.102) and (3.116), respectively.

For the proof of 1), we consider the solution u_f of the adjoint problem with the right-hand side $f = u - u_h \in V \subset L^2(\Omega)$. Choosing the test function $u - u_h$ and using the error equation (2.43) gives

$$\|u - u_h\|_0^2 = \langle u - u_h, u - u_h\rangle_0 = a(u - u_h, u_f) = a(u - u_h, u_f - v_h)$$

for all $v_h \in V_h$. If we choose specifically $v_h := I_h(u_f)$, then from the continuity of the bilinear form, Theorem 3.32, and Theorem 3.38, and the regularity assumption it follows that

$$\begin{aligned}\|u - u_h\|_0^2 &\leq C\|u - u_h\|_1\|u_f - I_h(u_f)\|_1 \\ &\leq C\|u - u_h\|_1 h|u_f|_2 \leq C\|u - u_h\|_1 h\|u - u_h\|_0 .\end{aligned}$$

Division by $\|u - u_h\|_0$ gives the assertion, which is trivial in the case $\|u - u_h\|_0 = 0$. □

Thus, if a rough right-hand side in (3.13) prevents convergence from being ensured by Theorem 3.32 or Theorem 3.38, then the estimate 2) can still be used to get a convergence estimate (of lower order).

In the light of the considerations from Section 1.2, the result of Theorem 3.40 is surprising, since we have only (pointwise) consistency of first order. On the other hand, Theorem 1.6 also raises the question of convergence rate results in $\|\cdot\|_\infty$ which then would give a result stronger, in many respects, than Theorem 1.6. Although the considerations described here (as in Section 3.9) can be the starting point of such L^∞ estimates, we get the most far-reaching results with the weighted norm technique (see [16, pp. 155]), whose description is not presented here.

The above theorems contain convergence rate results under regularity assumptions that may often, even though only locally, be violated. In fact, there also exist (weaker) results with less regularity assumptions. However, the following observation seems to be meaningful: Estimate (3.103) indicates that on subdomains, where the solution has less regularity, on which the (semi) norms of the solutions thus become large, local refinement is advantageous (without improving the convergence rate by this). Adaptive grid refinement strategies on the basis of a posteriori error estimates described in Chapter 4 provide a systematical approach in this direction.

Exercises

Problem 3.21 Prove the chain rule (3.90) for $v \in H^1(K)$.

Problem 3.22 a) Consider for $d := 1$ the linear finite element ansatz (3.64) and prove that for $K \in \mathcal{T}_h$ the following estimate holds:

$$|v - I_K(v)|_{1,K} \leq h_K |v|_{2,K} \quad \text{for all } v \in H^2(K)\,.$$

Hint: Rolle's theorem and Problem 2.10 b) (Poincaré inequality).

b) Generalize the considerations to an arbitrary (one-dimensional) polynomial ansatz $P = \mathcal{P}_k(K)$ by proving

$$|v - I_K(v)|_{1,K} \leq h_K^k |v|_{k+1,K} \quad \text{for all } v \in H^{k+1}(K)\,.$$

Problem 3.23 For $k \in \mathbb{N}$ show that the following definition provides a stronger norm on $H^{k+1}(\hat{K})$ for the reference simplex $\hat{K} \subset \mathbb{R}^d$, $d \leq 3$,

$$|||v|||_{k+1} := |v|_{k+1,\hat{K}} + \sum_{\varphi \in \Sigma_{\hat{K}}} |\varphi(v)|,$$

where $\Sigma_{\hat{K}}$ are the degrees of freedom of a finite element with $P_{\hat{K}} \supset \mathcal{P}_k(\hat{K})$. Substitute the usage of the Bramble–Hilbert lemma by this result.
Hint: Use the compact embedding $H^{k+1}(\hat{K}) \hookrightarrow H^k(\hat{K})$.

Problem 3.24 Show for linear elements, the reference triangle $\hat{K} \subset \mathbb{R}^2$ and $v \in H^2(\hat{K})$:

$$|v - I_{\hat{K}} v|_{1,\hat{K}} \leq \frac{5}{2} |v|_{2,\hat{K}}\,,$$
$$\|v - I_K v\|_{0,\hat{K}} \leq \sqrt{10} |v|_{2,\hat{K}}\,.$$

Problem 3.25 Show for bilinear elements, the reference square $\hat{K} \subset \mathbb{R}^2$ and $v \in H^2(\hat{K})$:

$$|v - I_K v|_{1,\hat{K}} \leq \frac{1}{\sqrt{3}} |v|_{2,\hat{K}} \, .$$

Hint: Start from the identity of Lemma 1.8.

Problem 3.26 Let a triangle K with the vertices a_1, a_2, a_3 and a function $v \in C^2(K)$ be given. Show that if v is interpolated by a linear polynomial $I_K(v)$ with $(I_K(v))(a_i) = v(a_i)$, $i = 1, 2, 3$, then the error the estimate

$$\|v - I_K(v)\|_{\infty,K} + h\|\nabla(v - I_K(v))\|_{\infty,K} \leq 2M \frac{h^2}{\cos(\alpha/2)}$$

holds, where h denotes the diameter, α the size of the largest interior angle of K, and M an upper bound for the maximum of the norm of the Hessian matrix of v on K.

Problem 3.27 Consider a triangle K with the vertices $a_1 := (-h, 0)$, $a_2 := (h, 0)$, $a_3 := (0, \varepsilon)$, and $h, \varepsilon > 0$. Suppose that the function $v(x) := x_1^2$ is linearly interpolated on K such that $(I_h(v))(a_i) = v(a_i)$ for $i = 1, 2, 3$.
Determine $\|\partial_2(I_h(v) - v)\|_{2,K}$ as well as $\|\partial_2(I_h(v) - v)\|_{\infty,K}$ and discuss the consequences for different orders of magnitude of h and ε.

Problem 3.28 Suppose that no further regularity properties are known for the solution $u \in V$ of the boundary value problem (3.13). Show under the assumptions of Section 3.4 that for the finite element approximation $u_h \in V_h$

$$\|u - u_h\|_1 \to 0 \quad \text{for } h \to 0 \, .$$

Problem 3.29 Derive analogously to Theorem 3.32 a convergence rate result for the Hermite elements (3.76) and (3.77) (Bogner–Fox–Schmit element) and the boundary value problem (3.13) with Dirichlet boundary conditions.

Problem 3.30 Derive analogously to Theorem 3.32 a convergence rate result for the Bogner–Fox–Schmit element (3.77) and the boundary value problem (3.47).

3.5 The Implementation of the Finite Element Method: Part 2

3.5.1 Incorporation of Dirichlet Boundary Conditions: Part 2

In the theoretical analysis of boundary value problems with inhomogeneous Dirichlet boundary conditions $u = g_3$ on Γ_3, the existence of a function $w \in H^1(\Omega)$ with

$w = g_3$ on Γ_3 has been assumed so far. The solution $u \in V$ (with homogeneous Dirichlet boundary conditions) is then defined according to (3.32) such that $\tilde{u} = u+w$ satisfies the variational equation with test functions in V:

$$a(u + w, v) = \ell(v) \quad \text{for all } v \in V\,. \tag{3.117}$$

For the Galerkin approximation u_h, which has been analysed in Section 3.4, this means that the parts $-a(w, \varphi_i)$ with nodal basis functions φ_i, $i = 1, \dots, M_1$, go into the right-hand side of the system of equations (2.38), and then $\tilde{u}_h := u_h + w$ has to be considered as the solution of the inhomogeneous problem

$$a(u_h + w, v) = \ell(v) \quad \text{for all } v \in V_h\,. \tag{3.118}$$

If we complete the basis of V_h by the basis functions $\varphi_{M_1+1}, \dots, \varphi_M$ for the Dirichlet boundary nodes $a_{M_1+1}, \dots, a_M$ and denote the generated space by X_h,

$$X_h := \operatorname{span}\left\{\varphi_1, \dots, \varphi_{M_1}, \varphi_{M_1+1}, \dots, \varphi_M\right\}, \tag{3.119}$$

that is, the ansatz space without taking into account boundary conditions, then, in particular, $\tilde{u}_h \in X_h$ does not hold in general. This approach does not correspond to the practice described in Section 2.4.3. That practice, applied to a general variational equation, reads as follows.

For all degrees of freedom $1, \dots, M_1, M_1 + 1, \dots, M$ the system of equations is built with the components

$$a(\varphi_j, \varphi_i), \quad i, j = 1, \dots, M\,, \tag{3.120}$$

for the stiffness matrix and

$$\ell(\varphi_i), \quad i = 1, \dots, M\,, \tag{3.121}$$

for the load vector. The vector of unknowns is therefore

$$\tilde{\xi} = \begin{pmatrix} \xi \\ \hat{\xi} \end{pmatrix} \quad \text{with} \quad \xi \in \mathbb{R}^{M_1},\ \hat{\xi} \in \mathbb{R}^{M_2}.$$

For Dirichlet boundary conditions the equations $M_1 + 1, \dots, M$ are replaced by

$$\tilde{\xi}_i = g_3(a_i), \quad i = M_1 + 1, \dots, M\,,$$

and the concerned variables are eliminated in equations $1, \dots, M_1$. Of course, it is assumed here that $g_3 \in C(\Gamma_3)$. This procedure can also be interpreted in the following way: If we set

$$A_h := \left(a(\varphi_j, \varphi_i)\right)_{i,j=1,\dots,M_1}, \quad \hat{A}_h := \left(a(\varphi_j, \varphi_i)\right)_{i=1,\dots,M_1,\ j=M_1+1,\dots,M},$$

then the first M_1 equations of the generated system of equations are

$$A_h \boldsymbol{\xi} + \hat{A}_h \hat{\boldsymbol{\xi}} = \boldsymbol{q}_h ,$$

where $\boldsymbol{q}_h \in \mathbb{R}^{M_1}$ consists of the first M_1 components according to (3.121). Hence the elimination leads to

$$A_h \boldsymbol{\xi} = \boldsymbol{q}_h - \hat{A}_h \hat{\boldsymbol{\xi}} \tag{3.122}$$

with $\hat{\boldsymbol{\xi}} = (g_3(a_i))_{i=M_1+1,\dots,M_2}$. Suppose

$$w_h := \sum_{i=M_1+1}^{M} g_3(a_i)\,\varphi_i \in X_h \tag{3.123}$$

is the ansatz function that satisfies the boundary conditions in the Dirichlet nodes and assumes the value 0 in all other nodes. The system of equations (3.122) is then equivalent to

$$a(\check{u}_h + w_h, v) = \ell(v) \quad \text{for all } v \in V_h \tag{3.124}$$

for $\check{u}_h = \sum_{i=1}^{M_1} \xi_i \varphi_i \in V_h$ (that is, the "real" solution), in contrast to the variational equation (3.118) was used in the analysis. This consideration also holds if another h-dependent bilinear form a_h and analogously a linear form ℓ_h instead of the linear form ℓ is used for assembling. In the following we assume that there exists some function $w \in C(\overline{\Omega})$ that satisfies the boundary condition on Γ_3. Instead of (3.124), we consider the finite-dimensional auxiliary problem of finding some $\check{\check{u}}_h \in V_h$, such that

$$a(\check{\check{u}}_h + \bar{I}_h(w), v) = \ell(v) \quad \text{for all } v \in V_h . \tag{3.125}$$

Here $\bar{I}_h : C(\overline{\Omega}) \to X_h$ is the interpolation operator with respect to all degrees of freedom,

$$\bar{I}_h(v) := \sum_{i=1}^{M_1+M_2} v(a_i)\varphi_i ,$$

whereas in Section 3.4 we considered the interpolation operator I_h for functions that vanish on Γ_3. In the following, when analysing the effect of quadrature, we will show that —also for some approximation of a and ℓ—

$$\tilde{u}_h := \check{\check{u}}_h + \bar{I}_h(w) \in X_h \tag{3.126}$$

is an approximation of $u + w$ of the quality established in Theorem 3.32 (see Theorem 3.45). We have $w_h - \bar{I}_h(w) \in V_h$ and hence also $\check{u}_h + w_h - \bar{I}_h(w) \in V_h$. If (3.125) is uniquely solvable, which follows from the general assumption of the V-coercivity of a (3.3), we have

$$\check{u}_h + w_h - \bar{I}_h(w) = \check{\check{u}}_h$$

and hence for $\tilde{u}_h$, according to (3.126),

$$\tilde{u}_h = \check{u}_h + w_h . \tag{3.127}$$

In this way the described implementation practice for Dirichlet boundary conditions is justified.

3.5.2 Numerical Quadrature

We consider again a boundary value problem in the variational formulation (3.32) and a finite element discretization in the general form described in Sections 3.3 and 3.4. If we step through Section 2.4.2 describing the assembling within a finite element code, we notice that the general element-to-element approach with transformation to the reference element is here also possible, with the exception that due to the general coefficient functions K, c, r, and f, the arising integrals cannot be evaluated exactly in general. If K_m is a general element with degrees of freedom in $a_{r_1}, \ldots, a_{r_L}$, then the components of the element stiffness matrix for $i, j = 1, \ldots, L$ are

$$\begin{aligned} A_{ij}^{(m)} &= \int_{K_m} K\nabla\varphi_{r_j} \cdot \nabla\varphi_{r_i} + \boldsymbol{c} \cdot \nabla\varphi_{r_j}\varphi_{r_i} + r\varphi_{r_j}\varphi_{r_i}\, dx + \int_{K_m\cap\Gamma_2} \alpha\varphi_{r_j}\varphi_{r_i}\, d\sigma \\ &=: \int_{K_m} v_{ij}(x)\, dx + \int_{K_m\cap\Gamma_2} w_{ij}(\sigma)\, d\sigma \\ &= \int_{\hat{K}} \hat{v}_{ij}(\hat{x})\, d\hat{x}\, |\det(B)| + \int_{\hat{K}'} \hat{w}_{ij}(\hat{\sigma})\, d\hat{\sigma}\, \sqrt{|\det(\tilde{B}^T\tilde{B})|}\,. \end{aligned} \tag{3.128}$$

Here, K_m is affine equivalent to the reference element $\hat{K}$ by the mapping $F(\hat{x}) = B\hat{x} + d$. By virtue of the conformity of the partition (T6), the boundary part $K_m \cap \bar{\Gamma}_2$ consists of none, one, or more complete faces of K_m. For every face of K_m there is a parametrization in $\mathbb{R}^{d-1}$ given by a mapping $\tilde{F}$ from $\mathbb{R}^{d-1}$ to $\mathbb{R}^d$. For simplicity, we restrict ourselves to the case of one face that is affine equivalent to the reference element $\hat{K}'$ by some mapping $\tilde{F}(\hat{\sigma}) = \tilde{B}\hat{\sigma} + \tilde{d}$ (cf. (3.53)). The generalization to the other cases is obvious. The functions $\hat{v}_{ij}$ and analogously $\hat{w}_{ij}$ are the transformed functions defined in (3.88).

Correspondingly, we get as components for the right-hand side of the system of equations, that is, for the load vector,

$$\begin{aligned} \left(\boldsymbol{q}^{(m)}\right)_i &= \int_{\hat{K}} \hat{f}(\hat{x})N_i(\hat{x})\, d\hat{x}\, |\det(B)| \\ &\quad + \int_{\hat{K}_1'} \hat{g}_1(\hat{\sigma})N_i(\hat{\sigma})\, d\hat{\sigma}\, |\det(\tilde{B}_1)| + \int_{\hat{K}_2'} \hat{g}_2(\hat{\sigma})N_i(\hat{\sigma})\, d\hat{\sigma}\, |\det(\tilde{B}_2)|, \end{aligned} \tag{3.129}$$

$i = 1, \ldots, L$. Here, the N_i, $i = 1, \ldots, L$, are the shape functions, that is, the local nodal basis functions on $\hat{K}$.

If the transformed integrands contain derivatives with respect to x, they can be transformed into derivatives with respect to $\hat{x}$. For instance, for the first addend in $A_{ij}^{(m)}$ we get, as an extension of (2.54),

$$\int_{\hat{K}} K(F(\hat{x}))B^{-T}\nabla_{\hat{x}}N_j(\hat{x}) \cdot B^{-T}\nabla_{\hat{x}}N_i(\hat{x})\, d\hat{x}\, |\det(B)|\,.$$

The shape functions, their derivatives, and their integrals over $\hat{K}$ are known which has been used in (2.56) for the exact integration. Since general coefficient functions arise, this is in general, but also in the remaining special cases no longer possible, for example, for polynomial $K(x)$ it is also not recommendable due to the corresponding effort. Instead, one should approximate these integrals (and, analogously, also the boundary integrals) by using some *quadrature formula*.

A quadrature formula on $\hat{K}$ for the approximation of $\int_{\hat{K}} \hat{v}(\hat{x})\, d\hat{x}$ has the form

$$\sum_{i=1}^{R} \hat{\omega}_i\, \hat{v}(\hat{b}_i) \tag{3.130}$$

with *weights* $\hat{\omega}_i$ and *quadrature* or *integration points* $\hat{b}_i \in \hat{K}$. Hence, applying (3.130) assumes the evaluability of $\hat{v}$ in $\hat{b}_i$, which is in the following ensured by the continuity of $\hat{v}$. This implies the same assumption for the coefficients, since the shape functions N_i and their derivatives are continuous. In order to ensure the numerical stability of a quadrature formula, it is usually required that

$$\hat{\omega}_i > 0 \quad \text{for all} \quad i = 1, \ldots, R, \tag{3.131}$$

which we will also do. Since all the considered finite elements are such that their faces with the enclosed degrees of freedom represent again a finite element (in $\mathbb{R}^{d-1}$) (see (3.53)), the boundary integrals are included in a general discussion. In principle, different quadrature formulas can be applied for each of the above integrals, but here we will disregard this possibility (with the exception of distinguishing between volume and boundary integrals because of their different dimensions).

A quadrature formula on $\hat{K}$ generates a quadrature formula on a general element K, recalling

$$\int_K v(x)\, dx = \int_{\hat{K}} \hat{v}(\hat{x})\, d\hat{x}\, |\det(B)|$$

by

$$\sum_{i=1}^{R} \omega_{i,K}\, v(b_{i,K}),$$

where $\omega_i = \omega_{i,K} = \hat{\omega}_i |\det(B)|$ and $b_i = b_{i,K} := F(\hat{b}_i)$ are dependent on K. The positivity of the weights is preserved. Here, again $F(\hat{x}) = B\hat{x} + d$ denotes the affine-linear transformation from $\hat{K}$ to K. The errors of the quadrature formulas

$$\begin{aligned} \hat{E}(\hat{v}) &:= \int_{\hat{K}} \hat{v}(\hat{x})\, d\hat{x} - \sum_{i=1}^{R} \hat{\omega}_i\, \hat{v}(\hat{b}_i), \\ E_K(v) &:= \int_K v(x)\, dx - \sum_{i=1}^{R} \omega_i\, v(b_i) \end{aligned} \tag{3.132}$$

are related to each other by

$$E_K(v) = |\det(B)|\hat{E}(\hat{v}) . \tag{3.133}$$

The *accuracy* of a quadrature formula will be defined by the requirement that for l as large as possible,

$$\hat{E}(\hat{p}) = 0 \quad \text{for } \hat{p} \in \mathcal{P}_l(\hat{K})$$

is satisfied, which transfers directly to the integration over K. A quadrature formula should further provide the desired accuracy by using quadrature nodes as less as possible, since the evaluation of the coefficient functions is often expensive. In contrast, for the shape functions and their derivatives a single evaluation is sufficient. In the following we discuss some examples of quadrature formulas for the elements that have been introduced in Section 3.3.

The most obvious approach consists in using *nodal quadrature formulas*, which have the nodes $\hat{a}_1, \ldots, \hat{a}_L$ of the reference element $(\hat{K}, \hat{P}, \hat{\Sigma})$ as quadrature nodes. The requirement of exactness in $\hat{P}$ is then equivalent to

$$\hat{\omega}_i = \int_{\hat{K}} N_i(\hat{x})\, d\hat{x} , \tag{3.134}$$

so that the question of the validity of (3.131) remains.

We start with the **unit simplex $\hat{K}$** defined in (3.58). Here, the weights of the quadrature formulas can be given directly on a general simplex K: If the shape functions are expressed by their barycentric coordinates λ_i, the integrals can be computed by

$$\int_K \lambda_1^{\alpha_1} \lambda_2^{\alpha_2} \cdots \lambda_{d+1}^{\alpha_{d+1}}(x)\, dx = \frac{\alpha_1! \alpha_2! \cdots \alpha_{d+1}!}{(\alpha_1 + \alpha_2 + \cdots + \alpha_{d+1} + d)!} \frac{\mathrm{vol}\,(K)}{\mathrm{vol}\,(\hat{K})} \tag{3.135}$$

(see Problem 3.31).

If $\boldsymbol{P = \mathcal{P}_1(K)}$ and thus the quadrature nodes are the vertices, it follows that

$$\omega_i = \int_K \lambda_i(x)\, dx = \frac{1}{d+1} \mathrm{vol}\,(K) \quad \text{for all } i = 1, \ldots, d+1 . \tag{3.136}$$

For $\boldsymbol{P = \mathcal{P}_2(K)}$ and $d = 2$ we get, by the shape functions $\lambda_i(2\lambda_i - 1)$, the weights 0 for the nodes a_i and, by the shape functions $4\lambda_i\lambda_j$, the weights

$$\omega_i = \frac{1}{3} \mathrm{vol}\,(K) \quad \text{for} \quad b_i = a_{ij} ,\ i, j = 1, \ldots, 3,\ i > j ,$$

so that we have obtained here a quadrature formula that is superior to (3.136) (for $d = 2$). However, for $d \geq 3$ this ansatz leads to negative weights and is thus useless. We can also get the exactness in $\boldsymbol{\mathcal{P}_1(K)}$ by a single quadrature node, by the barycentre (see (3.63)):

$$\omega_1 = \mathrm{vol}\,(K) \quad \text{and} \quad b_1 = a_S = \frac{1}{d+1} \sum_{i=1}^{d+1} a_i , \tag{3.137}$$

which is obvious due to (3.135).

As a formula that is exact for $\mathcal{P}_2(K)$ and $d = 3$ (see [145]) we present $R = 4$, $\omega_i = \frac{1}{4}\,\mathrm{vol}\,(K)$, and the b_i are obtained by cyclic exchange of the barycentric coordinates:

$$\left(\frac{5-\sqrt{5}}{20}, \frac{5-\sqrt{5}}{20}, \frac{5-\sqrt{5}}{20}, \frac{5+3\sqrt{5}}{20}\right).$$

On the **unit cuboid $\hat{K}$** we obtain nodal quadrature formulas, which are exact for $\mathcal{Q}_k(\hat{K})$, from the Newton–Côtes formulas in the one-dimensional situation by

$$\hat{\omega}_{i_1\ldots i_d} = \hat{\omega}_{i_1} \cdots \hat{\omega}_{i_d} \quad \text{for} \quad \hat{b}_{i_1\ldots i_d} = \left(\frac{i_1}{k}, \ldots, \frac{i_d}{k}\right) \qquad (3.138)$$
$$\text{for } i_j \in \{0, \ldots, k\} \text{ and } j = 1, \ldots, d\,.$$

Here the $\hat{\omega}_{i_j}$ are the weights of the Newton–Côtes formula for $\int_0^1 f(x)dx$ (see [56, p. 148]). As in (3.136), for $k = 1$ we have here a generalization of the *trapezoidal rule* (cf. (2.42), (11.43)) with the weights 2^{-d} in the 2^d vertices. From $k = 8$ on, negative weights arise. This can be avoided and the accuracy for a given number of points increased if the Newton–Côtes integration is replaced by the *Gauss–(Legendre) integration*: In (3.138), i_j/k has to be replaced by the jth node of the kth Gauss–Legendre formula (see [56, p. 178] there on $[-1, 1]$) and analogously $\hat{\omega}_{i_j}$. In this way, by $(k+1)^d$ quadrature nodes the exactness in $\mathcal{Q}_{2k+1}(\hat{K})$, not only in $\mathcal{Q}_k(\hat{K})$, is obtained.

Now the question as to which quadrature formula should be chosen arises. For this, different criteria can be considered (see also (11.41)). Here, we require that the convergence rate result that was proved in Theorem 3.32 should not be deteriorated. In order to investigate this question we have to clarify which problem is solved by the approximation $\bar{u}_h \in V_h$ based on quadrature. To simplify the notation, from now on we do not consider boundary integrals, that is, only Dirichlet and homogeneous Neumann boundary conditions are allowed. However, the generalization should be clear. Replacing the integrals in (3.128) and (3.129) by quadrature formulas $\sum_{i=1}^{R} \hat{\omega}_i \hat{v}(\hat{b}_i)$ leads to some approximation $\overline{A}_h$ of the stiffness matrix and $\overline{\boldsymbol{q}}_h$ of the load vector in the form

$$\overline{A}_h = \left(a_h(\varphi_j, \varphi_i)\right)_{i,j}, \quad \overline{\boldsymbol{q}}_h = (\ell_h(\varphi_i))_i\,,$$

for $i, j = 1, \ldots, M$. Here the φ_i are the basis functions of X_h (see (3.119)) without taking into account the Dirichlet boundary condition and

$$a_h(v, w) := \sum_{K \in \mathcal{T}_h} \sum_{l=1}^{R} \omega_{l,K} (K\nabla v \cdot \nabla w)(b_{l,K})$$

$$+ \sum_{K \in \mathcal{T}_h} \sum_{l=1}^{R} \omega_{l,K} (\boldsymbol{c} \cdot \nabla v w)(b_{l,K}) + \sum_{K \in \mathcal{T}_h} \sum_{l=1} \omega_{l,K} (r v w)(b_{l,K}), \quad (3.139)$$

$$\ell_h(v) := \sum_{K \in \mathcal{T}_h} \sum_{l=1} \omega_{l,K} (f v)(b_{l,K}) \qquad \text{for all } v, w \in X_h .$$

The above-given mappings a_h and ℓ_h are well defined on $X_h \times X_h$ and X_h, respectively, if the coefficient functions can be evaluated in the quadrature nodes. Here we take into account that for some element K, ∇v for $v \in X_h$ can have jump discontinuities on ∂K. Thus, for the quadrature nodes $b_{l,K} \in \partial K$ in $\nabla v(b_{l,K})$ we have to choose the value "belonging to $b_{l,K}$" that corresponds to the limit of sequences in the interior of K. We recall that, in general, a_h and ℓ_h are not defined for functions of V. Obviously, a_h is bilinear and ℓ_h is linear. If we take into account the analysis of incorporating the Dirichlet boundary conditions in (3.117)–(3.124), we get a system of equations for the degrees of freedom $\overline{\boldsymbol{\xi}} = (\xi_1, \ldots, \xi_{M_1})^T$, which is equivalent to the variational equation on V_h for $\overline{u}_h = \sum_{i=1}^{M_1} \overline{\xi}_i \varphi_i \in V_h$:

$$a_h(\overline{u}_h, v) = \ell_h(v) - a_h(w_h, v) \quad \text{for all } v \in V_h \tag{3.140}$$

with w_h according to (3.123). As has been shown in (3.127), (3.140) is equivalent, in the sense of the total approximation $\overline{u}_h + w_h$ of $u + w$, to the variational equation for $\overline{\overline{u}}_h \in V_h$,

$$a_h(\overline{\overline{u}}_h, v) = \overline{\ell}_h(v) := \ell_h(v) - a_h(\overline{I}_h(w), v) \quad \text{for all } v \in V_h\,, \tag{3.141}$$

if this system of equations is uniquely solvable.

Exercises

Problem 3.31 Prove Equation (3.135) by first proving the equation for $K = \hat{K}$ and then deducing from this the assertion for the general simplex by Problem 3.18.

Problem 3.32 Let K be a nondegenerate triangle with vertices a_1, a_2, a_3. Further, let a_{12}, a_{13}, a_{23} denote the corresponding edge midpoints, a_{123} the barycentre and $|K|$ the area of K. Check that the quadrature formula

$$\frac{|K|}{60}\left[3\sum_{i=1}^{3} v(a_i) + 8\sum_{i<j} v(a_{ij}) + 27 v(a_{123})\right]$$

computes the integral $\int_K v\,dx$ exactly for polynomials of third degree.

Programming project 3.1 Extend the finite element solver of Project 2.5 to the boundary value problem

$$\begin{aligned} -\nabla \cdot (\boldsymbol{K}\nabla u) + \boldsymbol{c} \cdot \nabla u + ru &= f \quad \text{in } \Omega, \\ \boldsymbol{K}\nabla u \cdot \mathrm{n} &= g_1 \quad \text{on } \Gamma_1, \\ u &= g_3 \quad \text{on } \Gamma_3 := \partial\Omega \setminus \Gamma_1 . \end{aligned}$$

a) Start with the case of a homogeneous boundary condition on Γ_1, i.e., $g_1 = 0$. Assemble the stiffness matrix A_h for this more general boundary problem by the help of a quadrature formula approach as described in Section 3.5.2, where the parameters of the one-point quadrature formula on the reference triangle $\hat{K}$ are $\hat{b} := (1/3, 1/3)$, $\hat{\omega} := 1/2$.
Test your implementation for the following particular boundary value problem: $\Omega := (0,2) \times (0,1)$, $\Gamma_1 := (0,2) \times \{0,1\}$,

$$\boldsymbol{K} := k\boldsymbol{I} \quad \text{with} \quad k(x,y) := \begin{cases} 0.1 & \text{for } x \le 1, \\ 1 & \text{otherwise,} \end{cases} \qquad \boldsymbol{c} := (0.1, 1)^T,\ r := 0,\ f := 0,$$

$$\text{and} \quad g_3(x,y) := \begin{cases} 0 & \text{for } x = 0, \\ e^{0.1} - e^{-1} & \text{for } x = 2 . \end{cases}$$

The exact solution is $u(x,y) = \begin{cases} e^{x-1} - e^{-1} & \text{for } x < 1 \\ e^{0.1(x-1)} - e^{-1} & \text{otherwise.} \end{cases}$

b) Extend the code from part a) to inhomogeneous Neumann boundary conditions, i.e., implement an approximation of the boundary integral

$$\int_{\Gamma_1} g_1 v_h \, d\sigma$$

on the right-hand side of the discretization. Do the assembling as described in Section 3.5.2 and use the trapezoidal rule on the reference element $\hat{K}' := (0,1)$, i.e. $\hat{b}'_1 := 0$, $\hat{b}'_2 := 1$, $\hat{\omega}'_1 := \hat{\omega}'_2 := 0.5$.
Test your implementation for the following particular boundary value problem: $\Omega := (0,2) \times (0,1)$, $\Gamma_3 := \{0\} \times [0,1]$,

$$\boldsymbol{K} := k\boldsymbol{I} \quad \text{with} \quad k(x,y) := \begin{cases} 0.1 & \text{for } (0.875 < x < 1.125) \wedge (y > 0.125), \\ 1 & \text{otherwise,} \end{cases}$$

$$\boldsymbol{c} := \boldsymbol{0},\ r := 0,\ f := 0,\ g_3 := 0, \text{ and} \quad g_1(x,y) := \begin{cases} y & \text{for } x = 2, \\ 0 & \text{otherwise on } \Gamma_1 . \end{cases}$$

Visualize the numerical solution of this problem.

3.6 Convergence Rate Results in the Case of Quadrature and Interpolation

The purpose of this section is to analyse the approximation quality of a solution $\bar{\bar{u}}_h + \bar{I}_h(w)$ according to (3.141) and thus of $\bar{u}_h + w_h$ according to (3.140) of the boundary value problem (3.13), (3.19)–(3.21).

Hence, we have left the field of Galerkin methods, and we have to investigate the influence of the errors

$$a - a_h\,, \quad \ell - a(w, \cdot) - \ell_h + a_h(\bar{I}_h(w), \cdot).$$

To this end, we consider, in general, the variational equation in a normed space $(V, \|\cdot\|)$

$$u \in V \text{ satisfies} \qquad a(u, v) = \tilde{\ell}(v) \quad \text{for all } v \in V, \tag{3.142}$$

and the approximation in subspaces $V_h \subset V$ for $h > 0$,

$$u_h \in V_h \text{ satisfies} \qquad a_h(u_h, v) = \tilde{\ell}_h(v) \quad \text{for all } v \in V_h\,. \tag{3.143}$$

Here a and a_h are bilinear forms on $V \times V$ and $V_h \times V_h$, respectively, and $\tilde{\ell}, \tilde{\ell}_h$ are linear forms on V and V_h, respectively. Then we have the following theorem

Theorem 3.41 (First Lemma of Strang)
Suppose there exists some $\alpha > 0$ such that for all $h > 0$ and $v \in V_h$,

$$\alpha \|v\|^2 \le a_h(v, v), \tag{3.144}$$

and let a be continuous on $V \times V$.

Then, there exists a constant $C > 0$ independent of h such that

$$\begin{aligned} \|u - u_h\| \le C \Bigg\{ \inf_{v \in V_h} \Bigg\{ \|u - v\| + \sup_{w \in V_h \setminus \{0\}} \frac{|a(v, w) - a_h(v, w)|}{\|w\|} \Bigg\} \\ + \sup_{w \in V_h \setminus \{0\}} \frac{|\tilde{\ell}(w) - \tilde{\ell}_h(w)|}{\|w\|} \Bigg\}. \end{aligned} \tag{3.145}$$

Proof Let $v \in V_h$ be arbitrary. Then it follows from (3.142)–(3.144) that

$$\begin{aligned} \alpha \|u_h - v\|^2 &\le a_h(u_h - v, u_h - v) \\ &= a(u - v, u_h - v) + \big(a(v, u_h - v) - a_h(v, u_h - v)\big) \\ &\quad + \big(\tilde{\ell}_h(u_h - v) - \tilde{\ell}(u_h - v)\big) \end{aligned}$$

and moreover, by the continuity of a (cf. (3.2)),

$$\alpha \|u_h - v\| \le M \|u - v\| + \sup_{w \in V_h \setminus \{0\}} \frac{|a(v, w) - a_h(v, w)|}{\|w\|}$$

$$+ \sup_{w \in V_h \setminus \{0\}} \frac{|\tilde{\ell}_h(w) - \tilde{\ell}(w)|}{\|w\|} \quad \text{for all } v \in V_h \,.$$

By means of $\|u - u_h\| \leq \|u - v\| + \|u_h - v\|$ and taking the infimum over all $v \in V_h$, the assertion follows. □

For $a_h = a$ and $\tilde{\ell}_h = \tilde{\ell}$ the assertion reduces to Céa's lemma (Theorem 2.17), which was the initial point for the analysis of the convergence rate in Section 3.4. Here we can proceed analogously. For that purpose, the following conditions must be fulfilled additionally:

- The *uniform V_h-coercivity* of a_h according to (3.144) must be ensured.
- For the *consistency errors*

$$A_h(v) := \sup_{w \in V_h \setminus \{0\}} \frac{|a(v, w) - a_h(v, w)|}{\|w\|} \tag{3.146}$$

  for an arbitrarily chosen comparison function $v \in V_h$ and for

$$\sup_{w \in V_h \setminus \{0\}} \frac{|\tilde{\ell}(w) - \tilde{\ell}_h(w)|}{\|w\|}$$

  the behaviour in h has to be analysed.

The first requirement is not crucial if only a itself is V-coercive and A_h tends suitably to 0 for $h \to 0$:

Lemma 3.42 *Suppose the bilinear form a is V-coercive and there exists some non-negative function $C = C(h)$ with $C(h) \to 0$ for $h \to 0$ such that*

$$A_h(v) \leq C(h)\, \|v\| \quad \textit{for all } v \in V_h \,.$$

Then there exists some $\overline{h} > 0$ such that a_h is uniformly V_h-coercive for $h \leq \overline{h}$.

Proof By assumption, there exists some $\alpha > 0$ such that for $v \in V_h$,

$$\alpha \|v\|^2 \leq a_h(v, v) + a(v, v) - a_h(v, v)$$

and

$$|a(v, v) - a_h(v, v)| \leq A_h(v)\|v\| \leq C(h)\|v\|^2 \,.$$

Therefore, for instance, choose $\overline{h}$ such that $C(h) \leq \alpha/2$ for $h \leq \overline{h}$. □

We concretely address the analysis of the influence of numerical quadrature, that is, a_h is defined as in (3.139) and $\tilde{\ell}_h$ corresponds to $\overline{\ell}_h$ in (3.141) with the approximate linear form ℓ_h according to (3.139). Since this is an extension of the convergence results (in $\|\cdot\|_1$) given in Section 3.4, the assumptions about the finite element discretization are as summarized there at the beginning. In particular, the partitions $\mathcal{T}_h$ consist of elements that are affine equivalent to each other. Furthermore, for a simplification of the notation, let again $d \leq 3$ and only Lagrange elements are considered. In particular, let the general assumptions about the boundary value problems which are specified at the end of Section 3.2.1 be satisfied.

According to Theorem 3.41, the uniform V_h-coercivity of a_h must be ensured and the consistency errors (for an appropriate comparison element $v \in V_h$) must have the correct convergence behaviour. If the step size h is small enough, the first proposition is implied by the second proposition by virtue of Lemma 3.42. Now, simple criteria that are independent of this restriction will be presented. The quadrature formulas satisfy the properties (3.130), (3.131) introduced in Section 3.5, in particular, the weights are positive.

Lemma 3.43 *Suppose the coefficient function* $\boldsymbol{K}$ *satisfies (3.17) and let* $\boldsymbol{c} = \boldsymbol{0}$ *in* Ω, *let* $|\Gamma_3|_{d-1} > 0$, *and let* $r \geq 0$ *in* Ω. *If* $P \subset \mathcal{P}_k(K)$ *for the ansatz space and if the quadrature formula is exact for* $\mathcal{P}_{2k-2}(K)$, *then* a_h *is uniformly* V_h*-coercive.*

Proof Let $\alpha > 0$ be the constant of the uniform positive definiteness of $\boldsymbol{K}(x)$. Then we have for $v \in V_h$:

$$a_h(v, v) \geq \alpha \sum_{K \in \mathcal{T}_h} \sum_{l=1}^{R} \omega_{l,K} \, |\nabla v|^2(b_{l,K}) = \alpha \int_\Omega |\nabla v|^2(x)\, dx = \alpha |v|_1^2 \,,$$

since $|\nabla v|^2\big|_K \in \mathcal{P}_{2k-2}(K)$. The assertion follows from Corollary 3.15. □

Further results of this type can be found in [16, pp. 194]. To investigate the consistency error we can proceed similarly to the estimation of the interpolation error in Section 3.4: The error is split into the sum of the errors over the elements $K \in \mathcal{T}_h$ and there transformed by means of (3.133) into the error over the reference element $\hat{K}$. The derivatives (in $\hat{x}$) arising in the error estimation over $\hat{K}$ are back-transformed by using Theorem 3.29 and Theorem 3.30, which leads to the desired h_K-factors. But note that powers of $\|B^{-1}\|$ or similar terms do not arise. If the powers of $\det(B)$ arising in both transformation steps cancel each other (which will happen), in this way no condition about the geometric quality of the family of partitions arises. Of course, these results must be combined with estimates for the approximation error of V_h, for which, in particular, both approaches of Section 3.4 (either shape regularity or maximum angle condition) are admissible.

For the sake of simplicity, we restrict our attention in the following to the case of the polynomial ansatz space $P = \mathcal{P}_k(K)$. More general results of similar type, in particular, for partitions with the cuboid element and $\hat{P} = \mathcal{Q}_k(\hat{K})$ as reference element, are summarized in [16, p. 207].

We recall the notation and the relations introduced in (3.132), (3.133) for the local errors. In the following theorems we make use of the Sobolev spaces W^l_∞ on Ω and on K with the norms $\|\cdot\|_{l,\infty}$ and $\|\cdot\|_{l,\infty,K}$, respectively, and the seminorms $|\cdot|_{l,\infty}$ and $|\cdot|_{l,\infty,K}$, respectively. The essential local assertion is the following:

Theorem 3.44 *Suppose* $k \in \mathbb{N}$ *and* $\hat{P} = \mathcal{P}_k(\hat{K})$ *and the quadrature formula is exact for* $\mathcal{P}_{2k-2}(\hat{K})$:

$$\hat{E}(\hat{v}) = 0 \quad \textit{for all } \hat{v} \in \mathcal{P}_{2k-2}(\hat{K})\,. \tag{3.147}$$

Then there exist some constant $C > 0$ independent of $h > 0$ and $K \in \mathcal{T}_h$ such that for $l \in \{1, k\}$ the following estimates are given:

1) $|E_K(apq)| \leq Ch_K^l \|a\|_{k,\infty,K} \|p\|_{l-1,K} \|q\|_{0,K}$
   for all $a \in W_\infty^k(K)$, $p, q \in \mathcal{P}_{k-1}(K)$,

2) $|E_K(cpq)| \leq Ch_K^l \|c\|_{k,\infty,K} \|p\|_{l-1,K} \|q\|_{1,K}$
   for all $c \in W_\infty^k(K)$, $p \in \mathcal{P}_{k-1}(K)$, $q \in \mathcal{P}_k(K)$,

3) $|E_K(rpq)| \leq Ch_K^l \|r\|_{k,\infty,K} \|p\|_{l,K} \|q\|_{1,K}$
   for all $r \in W_\infty^k(K)$, $p, q \in \mathcal{P}_k(K)$,

4) $|E_K(fq)| \leq Ch_K^k \|f\|_{k,\infty,K} \operatorname{vol}(K)^{1/2} \|q\|_{1,K}$
   for all $f \in W_\infty^k(K)$, $q \in \mathcal{P}_k(K)$.

The (unnecessarily varied) notation of the coefficients already indicates the field of application of the respective estimate. The smoothness assumption concerning the coefficients in 1)–3) can be weakened to some extent. We prove only assertion 1). However, a direct application of this proof to assertions 2)–4) leads to a loss of convergence rate (or higher exactness conditions for the quadrature are needed). Here, quite technical considerations including the insertion of projections are necessary, which can be found to some extent in [16, pp. 201–203]. In the following proof we intensively make use of the fact that all norms are equivalent on the "fixed" finite-dimensional ansatz space $\mathcal{P}_k(\hat{K})$. The assumption (3.147) is equivalent to the same condition on a general element. However, the formulation already indicates an assumption that is also sufficient in more general cases.

Proof (of Theorem 3.44, 1)) We consider a general element $K \in \mathcal{T}_h$ and mappings $a \in W_\infty^k(K)$, $p, q \in \mathcal{P}_{k-1}(K)$ on it and, moreover, mappings $\hat{a} \in W_\infty^k(\hat{K})$, $\hat{p}, \hat{q} \in \mathcal{P}_{k-1}(\hat{K})$ defined according to (3.88). First, the proof is done for $l = k$. On the reference element $\hat{K}$, for $\hat{v} \in W_\infty^k(\hat{K})$ and $\hat{q} \in \mathcal{P}_{k-1}(\hat{K})$), we have

$$\left|\hat{E}(\hat{v}\hat{q})\right| = \left|\int_{\hat{K}} \hat{v}\hat{q}\, d\hat{x} - \sum_{l=1}^{R} \hat{\omega}_l\, (\hat{v}\hat{q})(\hat{\ell}_l)\right| \leq C\, \|\hat{v}\hat{q}\|_{\infty,\hat{K}} \leq C\, \|\hat{v}\|_{\infty,\hat{K}}\, \|\hat{q}\|_{\infty,\hat{K}}\,,$$

where the continuity of the embedding of $W_\infty^k(\hat{K})$ in $C(\hat{K})$ is used (see [11, p. 181]). Therefore, by the equivalence of $\|\cdot\|_{\infty,\hat{K}}$ and $\|\cdot\|_{0,\hat{K}}$ on $\mathcal{P}_{k-1}(\hat{K})$, it follows that

$$\left|\hat{E}(\hat{v}\hat{q})\right| \leq C\, \|\hat{v}\|_{k,\infty,\hat{K}}\, \|\hat{q}\|_{0,\hat{K}}\,.$$

If a fixed $\hat{q} \in \mathcal{P}_{k-1}(\hat{K})$ is chosen, then a linear continuous functional G is defined on $W_\infty^k(\hat{K})$ by $\hat{v} \mapsto \hat{E}(\hat{v}\hat{q})$ that has the following properties:

$$\|G\| \leq C\|\hat{q}\|_{0,\hat{K}} \quad \text{and} \quad G(\hat{v}) = 0 \quad \text{for} \quad \hat{v} \in \mathcal{P}_{k-1}(\hat{K})$$

by virtue of (3.147).

The Bramble–Hilbert lemma (Theorem 3.27) implies

$$\left|\hat{E}(\hat{v}\hat{q})\right| \le C\,|\hat{v}|_{k,\infty,\hat{K}}\,\|\hat{q}\|_{0,\hat{K}}\,.$$

According to the assertion we now choose

$$\hat{v} = \hat{a}\hat{p} \quad \text{for} \quad \hat{a} \in W^{k,\infty}(\hat{K}),\ \hat{p} \in \mathcal{P}_{k-1}(\hat{K}),$$

and we have to estimate $|\hat{a}\hat{p}|_{k,\infty\hat{K}}$ (thanks to the Bramble–Hilbert lemma not $\|\hat{a}\hat{p}\|_{k,\infty,\hat{K}}$). The Leibniz rule for the differentiation of products implies the estimate

$$|\hat{a}\hat{p}|_{k,\infty,\hat{K}} \le C\sum_{j=0}^{k}|\hat{a}|_{k-j,\infty,\hat{K}}\,|\hat{p}|_{j,\infty,\hat{K}}\,. \tag{3.148}$$

Here the constant C depends only on k, but not on the domain $\hat{K}$.

Since $\hat{p} \in \mathcal{P}_{k-1}(\hat{K})$, the last term of the sum in (3.148) can be omitted. Therefore, we have obtained the following estimate holding for $\hat{a} \in W^k_\infty(\hat{K})$, $\hat{p}, \hat{q} \in \mathcal{P}_{k-1}(\hat{K})$:

$$\begin{aligned}\left|\hat{E}(\hat{a}\hat{p}\hat{q})\right| &\le C\left(\sum_{j=0}^{k-1}|\hat{a}|_{k-j,\infty,\hat{K}}\,|\hat{p}|_{j,\infty,\hat{K}}\right)\|\hat{q}\|_{0,\hat{K}}\\ &\le C\left(\sum_{j=0}^{k-1}|\hat{a}|_{k-j,\infty,\hat{K}}\,|\hat{p}|_{j,\hat{K}}\right)\|\hat{q}\|_{0,\hat{K}}\,.\end{aligned} \tag{3.149}$$

The last estimate uses the equivalence of $\|\cdot\|_\infty$ and $\|\cdot\|_0$ on $\mathcal{P}_{k-1}(\hat{K})$.

We suppose that the transformation F of $\hat{K}$ to the general element K has, as usual, the linear part B. The first transformation step yields the factor $|\det(B)|$ according to (3.133), and for the back-transformation it follows from Theorem 3.29 and Theorem 3.30 that

$$\begin{aligned}|\hat{a}|_{k-j,\infty,\hat{K}} &\le C\,h_K^{k-j}\,|a|_{k-j,\infty,K}\,,\\ |\hat{p}|_{j,\hat{K}} &\le C\,h_K^j\,|\det(B)|^{-1/2}\,|p|_{j,K}\,,\\ \|\hat{q}\|_{0,\hat{K}} &\le C\,|\det(B)|^{-1/2}\,\|q\|_{0,K}\end{aligned} \tag{3.150}$$

for $0 \le j \le k-1$. Here a, p, q are the mappings $\hat{a}, \hat{p}, \hat{q}$ (back)transformed according to (3.88). Substituting these estimates into (3.149) therefore yields

$$\left|E_K(apq)\right| \le C\,h_K^k\left(\sum_{j=0}^{k-1}|a|_{k-j,\infty,K}\,|p|_{j,K}\right)\|q\|_{0,K}$$

and from this, assertion 1) follows for $l = k$.

If $l = 1$, we modify the proof as follows. Again, in (3.149) we estimate by using the equivalence of norms:

$$\left|\hat{E}(\hat{a}\hat{p}\hat{q})\right| \leq C\left(\sum_{j=0}^{k-1} |\hat{a}|_{k-j,\infty,\hat{K}} \, \|\hat{p}\|_{j,\infty,\hat{K}}\right) \|\hat{q}\|_{0,\hat{K}}$$

$$\leq C\left(\sum_{j=0}^{k-1} |\hat{a}|_{k-j,\infty,\hat{K}}\right) \|\hat{p}\|_{0,\hat{K}} \, \|\hat{q}\|_{0,\hat{K}} \, .$$

The first and the third estimates of (3.150) remain applicable; the second estimate is replaced with the third such that we have

$$\left|E_K(apq)\right| \leq C\, h_K \left(\sum_{j=0}^{k-1} |a|_{k-j,\infty,K}\right) \|p\|_{0,K} \, \|q\|_{0,K}$$

since the lowest h_K-power arises for $j = k - 1$. This estimate yields the assertion 1) for $l = 1$. □

Finally, we can now verify the assumptions of Theorem 3.41 with the following result.

Theorem 3.45 *Consider a family of affine equivalent Lagrange finite element discretizations in $\mathbb{R}^d$, $d \leq 3$, with $P = \mathcal{P}_k$ for some $k \in \mathbb{N}$ as local ansatz space. Suppose that the family of partitions is shape-regular or satisfies the maximum angle condition in the case of triangles with $k = 1$. Suppose that the applied quadrature formulas are exact for $\mathcal{P}_{2k-2}$. Let the function w satisfying the Dirichlet boundary condition and let the solution u of the boundary value problem (3.13), (3.19)–(3.21) (with $g_3 := 0$) belong to $H^{k+1}(\Omega)$.*

Then there exist some constants $C > 0$, $\overline{h} > 0$ independent of u and w such that for the finite element approximation $\overline{u}_h + w_h$ according to (3.123), (3.140), it follows for $h \leq \overline{h}$ that

$$\left\|u + w - (\overline{u}_h + w_h)\right\|_1 \leq C\, h^k \left\{ |u|_{k+1} + |w|_{k+1} \right.$$

$$\left. + \left(\sum_{i,j=1}^{d} \|k_{ij}\|_{k,\infty} + \sum_{i=1}^{d} \|c_i\|_{k,\infty} + \|r\|_{k,\infty}\right)\left(\|u\|_{k+1} + \|w\|_{k+1}\right) + \|f\|_{k,\infty} \right\} .$$

Proof According to (3.126), we aim at estimating $\left\|u + w - (\overline{\overline{u}}_h + \overline{I}_h(w))\right\|_1$, where $\overline{\overline{u}}_h$ satisfies (3.141).

By virtue of Theorem 3.32 or Theorem 3.38 (set formally $\Gamma_3 := \emptyset$) we have

$$\|w - \overline{I}_h(w)\|_1 \leq C h^k |w|_{k+1} \, . \tag{3.151}$$

For the bilinear form a_h defined in (3.139), it follows from Theorem 3.44 for $v, w \in V_h$ and $l \in \{0, k\}$ that

$$
\begin{aligned}
&\left|a(v,w)-a_h(v,w)\right| \\
&\le \sum_{K\in\mathcal{T}_h}\left(\sum_{i,j=1}^{d}\left|E_K(k_{ij}\partial_j(v|_K)\partial_i(w|_K))\right| + \sum_{i=1}^{d}\left|E_K(c_i\partial_i(v|_K)w)\right| + \left|E_K(rvw)\right|\right) \\
&\le C\sum_{K\in\mathcal{T}_h} h_K^l\left(\sum_{i,j=1}^{d}\|k_{ij}\|_{k,\infty,K} + \sum_{i=1}^{d}\|c_i\|_{k,\infty,K} + \|r\|_{k,\infty,K}\right)\|v\|_{l,K}\|w\|_{1,K} \\
&\le C\,h^l\left(\sum_{i,j=1}^{d}\|k_{ij}\|_{k,\infty} + \sum_{i=1}^{d}\|c_i\|_{k,\infty} + \|r\|_{k,\infty}\right)\left(\sum_{K\in\mathcal{T}_h}\|v\|_{l,K}^2\right)^{1/2}\|w\|_1,
\end{aligned}
\tag{3.152}
$$

by estimating the $\|\cdot\|_{k,\infty,K}$-norms in terms of norms on the domain Ω and then applying the Cauchy–Schwarz inequality with "index" $K\in\mathcal{T}_h$.

From this we obtain for $l=1$ an estimate of the form

$$
|a(v,w)-a_h(v,w)| \le Ch\|v\|_1\|w\|_1
$$

such that the estimate required in Lemma 3.42 holds (with $C(h)=C\cdot h$). Therefore, there exists some $\overline{h}>0$ such that a_h is uniformly V_h-coercive for $h\le\overline{h}$. Hence, estimate (3.145) is applicable, and the first addend, the approximation error, behaves as asserted according to Theorem 3.32 or Theorem 3.38 (again, choose $v := I_h(u)$ for the comparison element).

In order to estimate the consistency error of a_h, a comparison element $v\in V_h$ has to be found for which the corresponding part of the norm in (3.152) is uniformly bounded. This is satisfied for $v := I_h(u)$, since

$$
\begin{aligned}
\left(\sum_{K\in\mathcal{T}_h}\|I_h(u)\|_{k,K}^2\right)^{1/2} &\le \|u\|_k + \left(\sum_{K\in\mathcal{T}_h}\|u-I_h(u)\|_{k,K}^2\right)^{1/2} \\
&\le \|u\|_k + Ch|u|_{k+1} \le \|u\|_{k+1}
\end{aligned}
$$

due to Theorem 3.32 or Theorem 3.38.

Hence, the consistency error in a behaves as asserted according to (3.152), so that only the consistency error of l has to be investigated: We have

$$
\tilde{\ell}-\tilde{\ell}_h = \ell-\ell_h - a(w,\cdot) + a_h(\overline{I}_h(w),\cdot),
$$

where ℓ_h is defined in (3.139).

If $v\in V_h$, then

$$
\left|a(w,v)-a_h(\overline{I}_h(w),v)\right| \le \left|a(w,v)-a(\overline{I}_h(w),v)\right| + \left|a(\overline{I}_h(w),v)-a_h(\overline{I}_h(w),v)\right|.
$$

For the first addend the continuity of a implies

$$
\left|a(w,v)-a(\overline{I}_h(w),v)\right| \le C\left\|w-\overline{I}_h(w)\right\|_1\|v\|_1,
$$

so that the corresponding consistency error part behaves like $\|w - \bar{I}_h(w)\|_1$, which has already been estimated in (3.151). The second addend just corresponds to the estimate used for the consistency error in a (here, the difference between I_h and $\bar{I}_h$ is irrelevant), so that the same contribution to the convergence rate, now with $\|u\|_{k+1}$ replaced by $\|w\|_{k+1}$, arises. Finally, Theorem 3.44, 4) yields for $v \in V_h$,

$$\begin{aligned} |\ell(v) - \ell(v_h)| &\leq \sum_{K \in \mathcal{T}_h} |E_K(fv)| \leq C \sum_{K \in \mathcal{T}_h} h_K^k \operatorname{vol}(K)^{1/2} \|f\|_{k,\infty,K} \|v\|_{1,K} \\ &\leq h^k |\Omega|^{1/2} \|f\|_{k,\infty} \|v\|_1 \end{aligned}$$

by proceeding as in (3.152). This implies the last part of the asserted estimate. □

If the uniform V_h-coercivity of a_h is ensured in a different way (perhaps by Lemma 3.43), one can dispense with the smallness assumption about h. If estimates as given in Theorem 3.44 are also available for other types of elements, then partitions consisting of combinations of various elements can also be considered.

Exercises

Programming project 3.2 Extend the finite element solver of Project 3.1 to more general quadrature rules. The code shall work for Newton–Côtes quadrature rules of accuracy $l = 0, 1, 2$. In the implementation of the right-hand side, the quadrature rules for triangles and the quadrature rules for edges should have equal accuracy. The corresponding parameters are

order	0	1	2
weights (triangle)	0.5	$\frac{1}{6}, \frac{1}{6}, \frac{1}{6}$	$\frac{1}{6}, \frac{1}{6}, \frac{1}{6}$
quadrature points (triangle)	$\left(\frac{1}{3}, \frac{1}{3}\right)$	vertices	edge midpoints
weights (interval)	1	0.5 , 0.5	$\frac{1}{6}, \frac{4}{6}, \frac{1}{6}$
quadrature points (interval)	0.5	0 , 1	0 , 0.5 , 1

Test your implementation for $\Omega := (0, 2) \times (0, 1)$, $\Gamma_1 := \{2\} \times (0, 1)$,

$$\boldsymbol{K} := k\boldsymbol{I} \quad \text{with} \quad k(x, y) := \begin{cases} 0.1 & \text{for } x \leq 1, \\ 1 & \text{otherwise,} \end{cases} \qquad \boldsymbol{c} := (0.1, 1)^T,\ r := 0,\ f := 0,$$

$$\text{and} \quad g_1(x, y) := -0.1 e^{0.1(x-1)},\ g_3(x, y) := \begin{cases} e^{x-1} - e^{-1} & \text{if } x < 1, \\ e^{0.1(x-1)} - e^{-1} & \text{otherwise.} \end{cases}$$

The Dirichlet boundary value coincides with the restriction of the exact solution u to Γ_3.

Programming project 3.3 Extend the finite element solver of Project 3.2 to quadratic finite elements. The main difference to linear elements is that additional degrees of freedoms have to be taken into account, namely, the function values in the edge midpoints of the triangles. To this end the element-node table should be a matrix with six columns where every row contains the indices of the nodes of one triangle. Use the following local numbering for the nodes of one triangle:

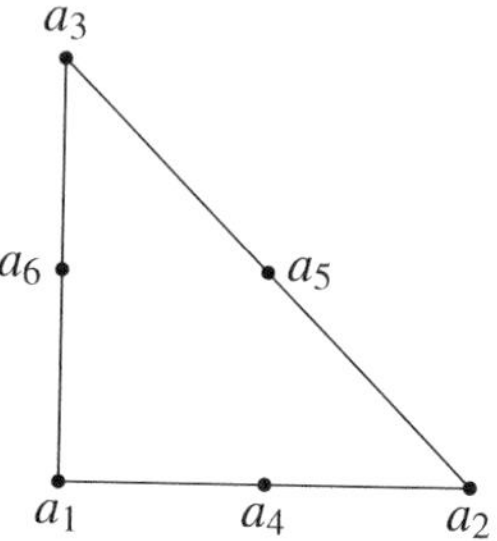

Accordingly, boundary nodes are not only the endpoints of the respective edge.

For the reference triangle, the shape functions of a quadratic triangular element are

$$\begin{aligned} \hat{N}_1(\hat{x},\hat{y}) &= (1-\hat{x}-\hat{y})(1-2\hat{x}-2\hat{y}), & \hat{N}_4(\hat{x},\hat{y}) &= 4\hat{x}(1-\hat{x}-\hat{y}),\\ \hat{N}_2(\hat{x},\hat{y}) &= \hat{x}(2\hat{x}-1), & \hat{N}_5(\hat{x},\hat{y}) &= 4\hat{x}\hat{y},\\ \hat{N}_3(\hat{x},\hat{y}) &= \hat{y}(2\hat{y}-1), & \hat{N}_6(\hat{x},\hat{y}) &= 4\hat{y}(1-\hat{x}-\hat{y}). \end{aligned}$$

Apply the trapezoidal rule for numerical integration.

To test your implementation, consider the Poisson equation on the unit square with $\Gamma_1 := \{0,1\} \times (0,1)$ and the exact solution $u(x,y) = \cos(\pi x)\sin(\pi y)$.

3.7 The Condition Number of Finite Element Matrices

The stability of solution algorithms for linear systems of equations as described in Section 2.5 depends on the condition number of the system matrix (see [54, Sect. 1.13]). The condition number also plays an important role for the convergence behaviour of iterative methods, which will be discussed in Chapter 5. Therefore, in this section we shall estimate the spectral condition number (see Appendix A.3) of the stiffness matrix

$$A = \big(a(\varphi_j,\varphi_i)\big)_{i,j=1,\ldots,M} \tag{3.153}$$

and also of the mass matrix (see (9.51))

$$B = \big(\langle\varphi_j,\varphi_i\rangle_0\big)_{i,j=1,\ldots,M}\,, \tag{3.154}$$

which is of importance for time-dependent problems. Again, we consider a finite element discretization in the general form of Section 3.4 restricted to Lagrange elements. In order to simplify the notation, we assume the affine equivalence of all elements. Further we suppose that

- the family $(\mathcal{T}_h)_h$ of partitions is shape-regular.

We assume that the variational formulation of the boundary value problem leads to a bilinear form a that is V-coercive and continuous on $V \subset H^1(\Omega)$.

As a modification of definition (1.18), let the following norm (which is also induced by a scalar product) be defined on the ansatz space $V_h = \operatorname{span}\{\varphi_1, \ldots, \varphi_M\}$:

$$\|v\|_{0,h} := \left(\sum_{K \in \mathcal{T}_h} h_K^d \sum_{a_i \in K} |v(a_i)|^2 \right)^{1/2} . \tag{3.155}$$

Here, $a_1, \ldots, a_M$ denote the nodes of the degrees of freedom, where in order to simplify the notation, M instead of M_1 is used for the number of degrees of freedom. The norm properties follow directly from the corresponding properties of $|\cdot|_2$ except for the definiteness. But the definiteness follows from the uniqueness of the interpolation problem in V_h with respect to degrees of freedom a_i.

Theorem 3.46 *1) There exist constants $C_1, C_2 > 0$ independent of h such that*

$$C_1 \|v\|_0 \le \|v\|_{0,h} \le C_2 \|v\|_0 \quad \textit{for all } v \in V_h .$$

2) There exists a constant $C > 0$ independent of h such that

$$\|v\|_1 \le C \left(\min_{K \in \mathcal{T}_h} h_K \right)^{-1} \|v\|_0 \quad \textit{for all } v \in V_h .$$

Proof As already known from Sections 3.4 and 3.6, the proof is done locally in $K \in \mathcal{T}_h$ and there transformed to the reference element $\hat{K}$ by means of $F(\hat{x}) = B\hat{x} + d$.

1): All norms are equivalent on the local ansatz space $\hat{P}$, thus also $\|\cdot\|_{0,\hat{K}}$ and the Euclidean norm in the degrees of freedom. Hence, there exist some $\hat{C}_1, \hat{C}_2 > 0$ such that for $\hat{v} \in \hat{P}$,

$$\hat{C}_1 \|\hat{v}\|_{0,\hat{K}} \le \left(\sum_{i=1}^{L} |\hat{v}(\hat{a}_i)|^2 \right)^{1/2} \le \hat{C}_2 \|\hat{v}\|_{0,\hat{K}} .$$

Here, $\hat{a}_1, \ldots, \hat{a}_L$ are the degrees of freedom in $\hat{K}$. Due to (3.61) we have

$$\operatorname{vol}(K) = \operatorname{vol}(\hat{K})\, |\det(B)| ,$$

and according to the definition of h_K and the shape regularity of the family $(\mathcal{T}_h)_h$, there exist constants $\tilde{C}_i > 0$ independent of h such that

$$\tilde{C}_1 h_K^d \le \tilde{C}_3\, \varrho_K^d \le |\det(B)| \le \tilde{C}_2 h_K^d .$$

By the transformation rule we thus obtain for $v \in P_K$, the ansatz space on K, that

$$\begin{aligned}
\hat{C}_1 \|v\|_{0,K} &= \hat{C}_1 \,|\det(B)|^{1/2} \,\|\hat{v}\|_{0,\hat{K}} \le \left(\tilde{C}_2 \, h_K^d\right)^{1/2} \left(\sum_{i=1}^{L} |\hat{v}(\hat{a}_i)|^2\right)^{1/2} \\
&= \tilde{C}_2^{1/2} \left(\sum_{a_i \in K} h_K^d |v(a_i)|^2\right)^{1/2} = \left(\tilde{C}_2 \, h_K^d\right)^{1/2} \left(\sum_{i=1}^{L} |\hat{v}(\hat{a}_i)|^2\right)^{1/2} \\
&\le \left(\tilde{C}_2 \, h_K^d\right)^{1/2} \hat{C}_2 \,\|\hat{v}\|_{0,\hat{K}} = \left(\tilde{C}_2 \, h_K^d\right)^{1/2} \hat{C}_2 \,|\det(B)|^{-1/2} \,\|v\|_{0,K} \\
&\le \tilde{C}_2^{1/2} \, \hat{C}_2 \, \tilde{C}_1^{-1/2} \,\|v\|_{0,K} .
\end{aligned}$$

This implies assertion 1).

2): Arguing as before, now using the equivalence of $\|\cdot\|_{1,\hat{K}}$ and $\|\cdot\|_{0,\hat{K}}$ in $\hat{P}$, it follows by virtue of (3.99) for $v \in P_K$ (with the generic constant C) that

$$\|v\|_{1,K} \le C \,|\det(B)|^{1/2} \,\|B^{-1}\|_2 \,\|\hat{v}\|_{0,\hat{K}} \le C \,\|B^{-1}\|_2 \,\|v\|_{0,K} \le C \, h_K^{-1} \,\|v\|_{0,K}$$

by Theorem 3.30 and the shape regularity of $(\mathcal{T}_h)_h$, and from this, the assertion 2).□

An estimate as in Theorem 3.46, 2), where on a finite-dimensional space a higher order norm is estimated by a lower order norm to the expense of a negative power of h, is called an *inverse estimate*.

In order to make the norm $\|\cdot\|_{0,h}$ comparable with the (weighted) Euclidean norm we assume in the following:

- There exists a constant $C_A > 0$ independent of h such that for every node of $\mathcal{T}_h$, the number of elements to which this node belongs is bounded by C_A. (3.156)

This condition is (partly) redundant: For $d = 2$ and triangular elements, the condition follows from the uniform lower bound (3.111) for the smallest angle as an implication of the shape regularity. Note that the condition need not be satisfied if only the maximum angle condition is required.

In general, if $C \in \mathbb{R}^{M,M}$ is a matrix with real eigenvalues $\lambda_1 \le \cdots \le \lambda_M$ and an orthonormal basis of eigenvectors $\boldsymbol{\xi}_1, \ldots, \boldsymbol{\xi}_M$, for instance, a symmetric matrix, then it follows for $\boldsymbol{\xi} \in \mathbb{R}^M \setminus \{\mathbf{0}\}$ that

$$\lambda_1 \le \frac{\boldsymbol{\xi}^T C \boldsymbol{\xi}}{\boldsymbol{\xi}^T \boldsymbol{\xi}} \le \lambda_M , \tag{3.157}$$

and the bounds are assumed for $\boldsymbol{\xi} = \boldsymbol{\xi}_1$ and $\boldsymbol{\xi} = \boldsymbol{\xi}_M$.

Theorem 3.47 *There exists a constant $C > 0$ independent of h such that we have*

$$\kappa(B) \le C \left(\frac{h}{\min_{K \in \mathcal{T}_h} h_K}\right)^d$$

for the spectral condition number of the mass matrix B according to (3.154).

Proof $\kappa(B) = \lambda_M/\lambda_1$ must be determined. For arbitrary $\boldsymbol{\xi} \in \mathbb{R}^M \setminus \{\mathbf{0}\}$ we have

$$\frac{\boldsymbol{\xi}^T B\boldsymbol{\xi}}{\boldsymbol{\xi}^T\boldsymbol{\xi}} = \frac{\boldsymbol{\xi}^T B\boldsymbol{\xi}}{\|v\|_{0,h}^2}\,\frac{\|v\|_{0,h}^2}{\boldsymbol{\xi}^T\boldsymbol{\xi}},$$

where $v := \sum_{i=1}^M \xi_i\varphi_i \in V_h$. By virtue of $\boldsymbol{\xi}^T B\boldsymbol{\xi} = \langle v, v\rangle_0$, the first factor on the right-hand side is uniformly bounded from above and below according to Theorem 3.46. Further, by (3.156) and $\boldsymbol{\xi} = (v(a_1), \ldots, v(a_M))^T$ it follows that

$$\min_{K\in\mathcal{T}_h} h_K^d\,|\boldsymbol{\xi}|^2 \le \|v\|_{0,h}^2 \le C_A\,h^d\,|\boldsymbol{\xi}|^2,$$

and thus the second factor is estimated from above and below. This leads to estimates of the type

$$\lambda_1 \ge C_1 \min_{K\in\mathcal{T}_h} h_K^d, \quad \lambda_M \le C_2\,h^d,$$

and from this, the assertion follows. □

Therefore, if the family of partitions $(\mathcal{T}_h)_h$ is *quasi-uniform* in the sense that there exists a constant $C > 0$ independent of h such that

$$h \le C\,h_K \quad \text{for all } K \in \mathcal{T}_h, \tag{3.158}$$

then $\kappa(B)$ is uniformly bounded (Fig. 3.15).

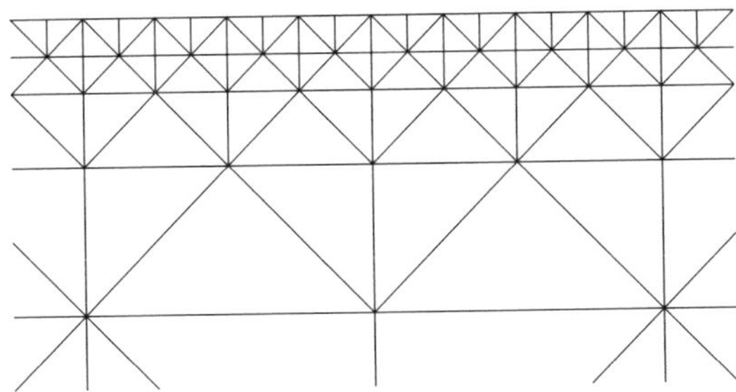

Fig. 3.15: A shape-regular triangulation, which is not quasi-uniform (in the limit).

In order to be able to argue analogously for the stiffness matrix, we assume that we stay close to the symmetric case:

Theorem 3.48 *Suppose the stiffness matrix A (3.153) admits real eigenvalues and a basis of eigenvectors. Then there exists a constant $C > 0$ independent of h such that the following estimates for the spectral condition number κ hold:*

$$\kappa(B^{-1}A) \le C \left(\min_{K \in \mathcal{T}_h} h_K \right)^{-2},$$

$$\kappa(A) \le C \left(\min_{K \in \mathcal{T}_h} h_K \right)^{-2} \kappa(B).$$

Proof With the notation of (3.157), we proceed analogously to the proof of Theorem 3.47. Since

$$\frac{\xi^T A \xi}{\xi^T \xi} = \frac{\xi^T A \xi}{\xi^T B \xi} \frac{\xi^T B \xi}{\xi^T \xi},$$

it suffices to bound the first factor on the right-hand side from above and below. This also yields a result for the eigenvalues of $B^{-1}A$, since we have for the variable $\eta := B^{1/2}\xi$,

$$\frac{\xi^T A \xi}{\xi^T B \xi} = \frac{\eta^T B^{-1/2} A B^{-1/2} \eta}{\eta^T \eta},$$

and the matrix $B^{-1/2}AB^{-1/2}$ possesses the same eigenvalues as $B^{-1}A$ by virtue of $B^{-1/2}(B^{-1/2}AB^{-1/2})B^{1/2} = B^{-1}A$. Here, $B^{1/2}$ is the symmetric positive definite matrix that satisfies $B^{1/2}B^{1/2} = B$, and $B^{-1/2}$ is its inverse.

Since $\xi^T A \xi / \xi^T B \xi = a(v,v)/\langle v, v\rangle_0$ and

$$\begin{aligned} a(v,v) &\ge \alpha \|v\|_1^2 \ge \alpha \|v\|_0^2, \\ a(v,v) &\le C\|v\|_1^2 \le C \left(\min_{K \in \mathcal{T}_h} h_K \right)^{-2} \|v\|_0^2, \end{aligned} \tag{3.159}$$

with a generic constant $C > 0$ (the last estimate is due to Theorem 3.46, 2)), it follows that

$$\alpha \le \frac{a(v,v)}{\langle v, v\rangle_0} = \frac{\xi^T A \xi}{\xi^T B \xi} = \frac{a(v,v)}{\langle v, v\rangle_0} \le C \left(\min_{K \in \mathcal{T}_h} h_K \right)^{-2}, \tag{3.160}$$

and from this the assertion. □

The analysis of the eigenvalues of the model problem in Example 2.12 shows that the above-given estimates are not too pessimistic.

Exercises

Problem 3.33 Let $G \subset \mathbb{R}^d$ be a bounded domain, set $h := \operatorname{diam}(G)$, $\widehat{G} := \{h^{-1}x \mid x \in G\}$ and $\hat{v}(\hat{x}) := v(h\hat{x})$ for $\hat{x} \in \widehat{G}$. Show for $k \in \mathbb{N}_0$, $1 \le r \le \infty$, and $v \in W_r^k(G)$:

$$|\hat{v}|_{k,r,\widehat{G}} = h^{k-d/r} |v|_{k,r,G} \,.$$

Problem 3.34 Under the assumptions of Problem 3.33, let $h \le 1$ and P be a finite-dimensional subspace of $W_p^l(G) \cap W_p^m(G)$, $l, m \in \mathbb{N}_0$, $m \le l$, $1 \le p \le \infty$, $1 \le q \le \infty$.

Prove that there is a constant $C > 0$, independent of h, such that the following *local inverse estimate* holds:

$$\|v\|_{l,p,G} \leq C h^{m-l+d/p-d/q} \|v\|_{m,q,G} \quad \text{for all } v \in P.$$

Problem 3.35 Under the parameter setting of Problem 3.34, let $(\mathcal{T}_h)_h$ be a quasi-uniform partition of a polyhedral domain $\Omega \subset \mathbb{R}^d$, affine equivalent to the reference element $(\widehat{K}, P_{\widehat{K}}, \Sigma_{\widehat{K}})$ such that $P_{\widehat{K}} \subset W_p^l(\widehat{K}) \cap W_q^m(\widehat{K})$. Consider the *broken ansatz space*

$$P(\mathcal{T}_h) := \left\{ w \in L^1(\Omega) \;\middle|\; w|_K \in P_K \text{ for all } K \in \mathcal{T}_h \right\}.$$

Show that there exists a constant $C > 0$, independent of h, such that the following *global inverse estimate* holds:

$$\|v\|_{l,p,\mathcal{T}_h} \leq C h^{m-l+\min\{0,d/p-d/q\}} \|v\|_{m,q,\mathcal{T}_h} \quad \text{for all } v \in P(\mathcal{T}_h).$$

3.8 General Domains and Isoparametric Elements

All elements considered so far are bounded by straight lines or plane surfaces. Therefore, only polyhedral domains can be decomposed exactly by means of a partition. Depending on the application, domains with a curved boundary may appear. With the available elements the obvious way of dealing with such domains is the following (in the two-dimensional case): for elements K that are close to the boundary put only the nodes of one edge on the boundary $\partial\Omega$. This implies an approximation error for the domain, for $\Omega_h := \bigcup_{K \in \mathcal{T}_h} K$, there holds, in general, neither $\Omega \subset \Omega_h$ nor $\Omega_h \subset \Omega$ (see Figure 3.16).

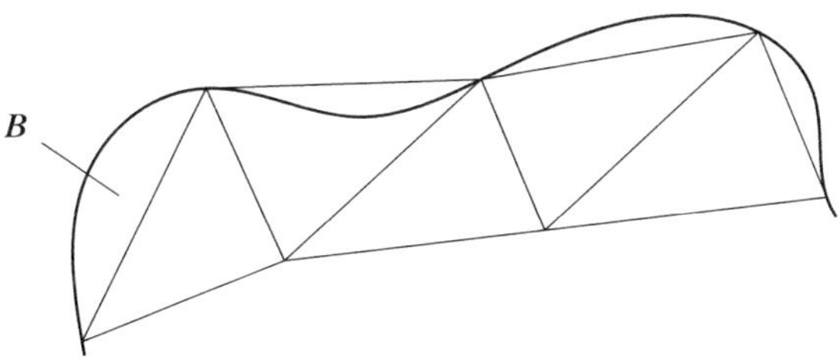

Fig. 3.16: Ω and Ω_h.

As the simplest example, we consider homogeneous Dirichlet boundary conditions, thus $V = H_0^1(\Omega)$, on a convex domain for which therefore $\Omega_h \subset \Omega$ is satisfied. If an ansatz space V_h is introduced as in Section 3.3, then functions defined on Ω_h are generated. Therefore, these functions must be extended to Ω in such a way that they vanish on $\partial\Omega$, and, consequently, for the generated function space $\tilde{V}_h$, $\tilde{V}_h \subset V$. This is supposed to be done by adding the domains B whose boundary consists of a boundary part of some element $K \in \mathcal{T}_h$ close to the boundary and a subset of $\partial\Omega$ to the set of elements with the ansatz space $P(B) = \{0\}$. Céa's lemma (Theorem 2.17) can still be applied, so that for an error estimate in $\|\cdot\|_1$ the question of how to choose a comparison element $v \in \tilde{V}_h$ arises. The ansatz $v = \tilde{I}_h(u)$, where $\tilde{I}_h(u)$ denotes the interpolation on Ω_h extended by 0 on the domains B, is admissible only for the (multi-)linear ansatz: Only in this case are all nodes of an edge "close to the boundary" located on $\partial\Omega$ and therefore have homogeneous degrees of freedom, so that the continuity on these edges is ensured. For the present, let us restrict our attention to this case, so that $\|u - \tilde{I}_h(u)\|_1$ has to be estimated where u is the solution of the boundary value problem.

The techniques of Section 3.4 can be applied to all $K \in \mathcal{T}_h$, and by the conditions assumed there about the partition, this yields

$$\begin{aligned}\|u - u_h\|_1 &\leq C\big(\|u - I_h(u)\|_{1,\Omega_h} + \|u\|_{1,\Omega\setminus\Omega_h}\big)\\ &\leq C\big(h|u|_{2,\Omega_h} + \|u\|_{1,\Omega\setminus\Omega_h}\big).\end{aligned}$$

If $\partial\Omega \in C^2$, then we have the estimate

$$\|u\|_{1,\Omega\setminus\Omega_h} \leq Ch\|u\|_{2,\Omega}$$

for the new error part due to the approximation of the domain, and thus the convergence rate is preserved. Already for a quadratic ansatz this is no longer satisfied, where only

$$\|u - u_h\|_1 \leq Ch^{3/2}\|u\|_3$$

holds instead of the order $O(h^2)$ of Theorem 3.32 (see [57, Sect. 4.4]). One may expect that this decrease of the approximation quality arises only locally close to the boundary; however, one may also try to obtain a better approximation of the domain by using curved elements. Such elements can be defined on the basis of the reference elements $(\hat{K}, \hat{P}, \hat{\Sigma})$ of Lagrange type introduced in Section 3.3 if a general element is obtained from this one by an *isoparametric transformation*, that is, choose an

$$F \in \hat{P}^d \tag{3.161}$$

that is injective and then

$$K := F(\hat{K}), \quad P := \big\{\hat{p} \circ F^{-1} \,\big|\, \hat{p} \in \hat{P}\big\}, \quad \Sigma := \big\{F(\hat{a}) \,\big|\, \hat{a} \in \hat{\Sigma}\big\}.$$

Since the bijectivity of $F : \hat{K} \to K$ is ensured by requirement, a finite element is thus defined in terms of (3.69). By virtue of the unique solvability of the interpolation problem, F can be defined by prescribing $a_1, \ldots, a_L$, $L = |\hat{\Sigma}|$, and requiring

$$F(\hat{a}_i) = a_i\,, \quad i = 1, \ldots, L\,.$$

However, this does not, in general, ensure the injectivity. Since, on the other hand, in the grid generation process elements are created by defining the nodes (see Section 4.1), geometric conditions about their positions that characterize the injectivity of F are desirable. A typical curved element that can be used for the approximation of the boundary can be generated on the basis of the unit simplex with $\hat{P} = \mathcal{P}_2(\hat{K})$ (see Figure 3.17).

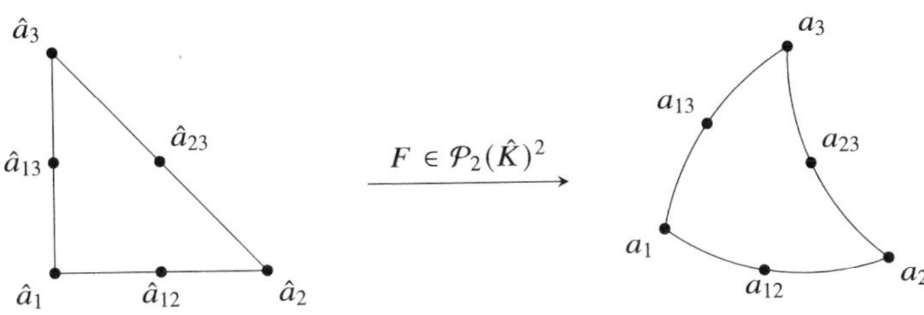

Fig. 3.17: Isoparametric element: quadratic ansatz on triangle.

Elements with, in general, one curved edge and otherwise straight edges thus are suggested for the problem of boundary approximation. They are combined with affine "quadratic triangles" in the interior of the domain. *Subparametric elements* can be generated analogously to the *isoparametric elements* if (the components of) the transformations in (3.161) are restricted to some subspace $\hat{P}_T \subset \hat{P}$. If $\hat{P}_T = \mathcal{P}_1(\hat{K})$, we again obtain the affine equivalent elements.

However, isoparametric elements are also important if, for instance, the unit square or cube is supposed to be the reference element. Only the isoparametric transformation allows for "general" quadrilaterals and hexahedra, respectively, which are preferable in anisotropic cases (for instance, in generalization of Figure 3.12) to simplices due to their adaptability to local coordinates. In what follows, let $\hat{K} := [0,1]^d$, $\hat{P} := Q_1(\hat{K})$.

In general, since also a finite element (in $\mathbb{R}^{d-1}$) is defined for every face $\hat{S}$ of $\hat{K}$ with $\hat{P}|_{\hat{S}}$ and $\hat{\Sigma}|_{\hat{S}}$, the "faces" of K, that is, $F[\hat{S}]$, are already uniquely defined by the related nodes.

Consequently, if $d = 2$, the edges of the general quadrilateral are straight lines (see Figure 3.18), but if $d = 3$, we have to expect curved surfaces (hyperbolic paraboloids) for a general hexahedron.

A geometric characterization of the injectivity of F is still unknown (to our knowledge) for $d = 3$, but it can be easily derived for $d = 2$: Let the nodes a_1, a_2, a_3, a_4 be numbered counterclockwise and suppose that they are not on a straight line, and thus (by rearranging) $T = \operatorname{conv}(a_1, a_2, a_4)$ forms a triangle such that

$$2\,\mathrm{vol}\,(T) = \det(B) > 0\,.$$

Here $F_T(\hat{x}) = B\hat{x} + d$ is the affine-linear mapping that maps the reference triangle $\operatorname{conv}(\hat{a}_1, \hat{a}_2, \hat{a}_4)$ bijectively to T. If $\tilde{a}_3 := F_T^{-1}(a_3)$, then the quadrilateral $\tilde{K}$ with the vertices $\hat{a}_1, \hat{a}_2, \tilde{a}_3, \hat{a}_4$ is mapped bijectively to K by F_T.

The transformation F can be decomposed into

$$F = F_T \circ F_Q\,,$$

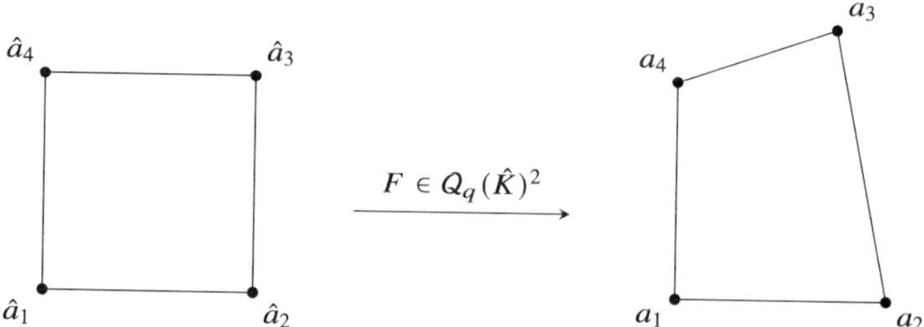

Fig. 3.18: Isoparametric element: bilinear ansatz on rectangle.

where $F_Q \in Q_1(\hat{K})^2$ denotes the mapping defined by

$$F_Q(\hat{a}_i) = \hat{a}_i\,, \quad i = 1, 2, 4, \quad F_Q(\hat{a}_3) = \tilde{a}_3$$

(see Figure 3.19).

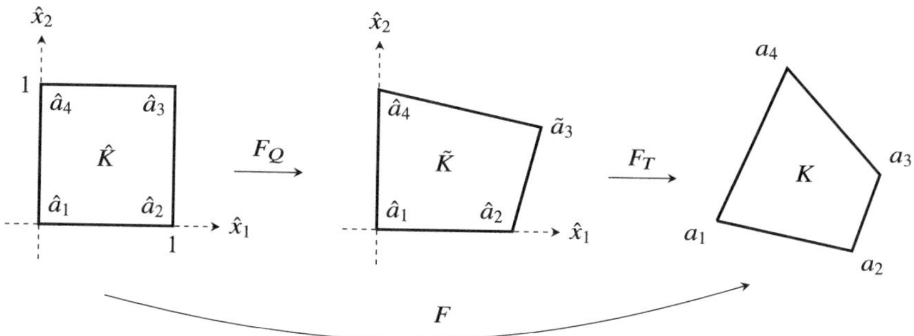

Fig. 3.19: Decomposition of the bilinear isoparametric mapping.

Therefore, the bijectivity of F is equivalent to the bijectivity of F_Q.

We characterize a "uniform" bijectivity which is defined by $\det(DF(\hat{x}_1, \hat{x}_2)) \neq 0$ for the functional matrix $DF(\hat{x}_1, \hat{x}_2)$:

Theorem 3.49 *Suppose Q is a quadrilateral with the vertices $a_1, \ldots, a_4$ (numbered counterclockwise). Then,*

$$\begin{aligned} & \det(DF(\hat{x}_1, \hat{x}_2)) \neq 0 \quad \textit{for all } (\hat{x}_1, \hat{x}_2) \in [0,1]^2 \\ \Longleftrightarrow\ & \det(DF(\hat{x}_1, \hat{x}_2)) > 0 \quad \textit{for all } (\hat{x}_1, \hat{x}_2) \in [0,1]^2 \\ \Longleftrightarrow\ & Q \textit{ is convex and does not degenerate into a triangle or straight line.} \end{aligned}$$

Proof By virtue of

$$\det\left(DF(\hat{x}_1, \hat{x}_2)\right) = \det(B) \det\left(DF_Q(\hat{x}_1, \hat{x}_2)\right)$$

and $\det(B) > 0$, F can be replaced with F_Q in the assertion. Since

$$F_Q(\hat{x}_1, \hat{x}_2) = \begin{pmatrix} \hat{x}_1 \\ \hat{x}_2 \end{pmatrix} + \begin{pmatrix} \tilde{a}_{3,1} - 1 \\ \tilde{a}_{3,2} - 1 \end{pmatrix} \hat{x}_1 \hat{x}_2 ,$$

it follows by some simple calculations that

$$\det\left(DF_Q(\hat{x}_1, \hat{x}_2)\right) = 1 + (\tilde{a}_{3,2} - 1)\hat{x}_1 + (\tilde{a}_{3,1} - 1)\hat{x}_2$$

is an affine-linear mapping because the quadratic parts just cancel each other. This mapping assumes its extrema on $[0, 1]^2$ at the 4 vertices, where we have the following values:

$$(0, 0) : 1, \quad (1, 0) : \tilde{a}_{3,2}, \quad (0, 1) : \tilde{a}_{3,1}, \quad (1, 1) : \tilde{a}_{3,1} + \tilde{a}_{3,2} - 1 .$$

A uniform sign is thus obtained if and only if the function is everywhere positive. This is the case if and only if

$$\tilde{a}_{3,1}, \tilde{a}_{3,2}, \tilde{a}_{3,1} + \tilde{a}_{3,2} - 1 > 0,$$

which just characterizes the convexity and the nondegeneration of $\tilde{K}$. By the transformation F_T this also holds for K. □

According to this theorem it is not allowed that a quadrilateral degenerates into a triangle (now with linear ansatz). But a more careful analysis [156] shows that this does not affect negatively the quality of the approximation.

In general, for isoparametric elements we have the following:

From the point of view of implementation, only slight modifications have to be made: In the integrals (3.128), (3.129) transformed to the reference element or their approximation by quadrature (3.139), $|\det B|$ has to be replaced with $|\det(DF(\hat{x}))|$ (in the integrand).

The analysis of the order of convergence can be done along the same lines as in Section 3.4 (and 3.6); however, the transformation rules for the integrals become more complex (see [16, pp. 237]).

Exercises

Problem 3.36 Determine $F_K \in Q_1^2$ with $F_K(\hat{K}) = K$, $\det(DF_K) > 0$, and $\hat{K}$, K from the following figure, where the origin is mapped to the point $(0, 2h)$. Is there any restriction for $h > 0$?

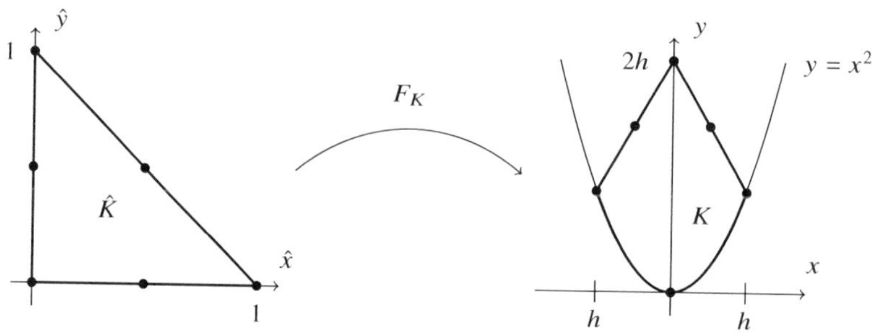

Programming project 3.4 Consider the Dirichlet boundary value problem for the two-dimensional Poisson problem

$$\begin{aligned} -\Delta u &= f \text{ in } \Omega, \\ u &= g \text{ on } \partial\Omega, \end{aligned}$$

where $\Omega \subset \mathbb{R}^2$ is a piecewise parabolically bounded domain:

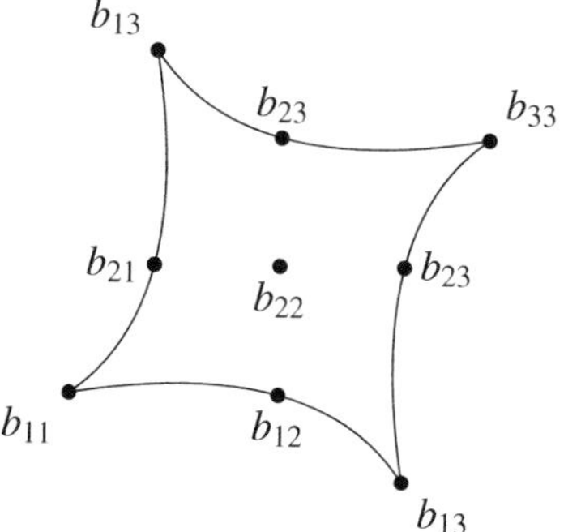

This domain is the biquadratic image of $(-1, 1)^2$, where

	b_{11}	b_{12}	b_{13}	b_{21}	b_{22}	b_{23}	b_{31}	b_{32}	b_{33}
$b_{ij,1}$	-1.25	0	0.75	-0.75	0	0.75	-0.75	0	1.25
$b_{ij,2}$	-0.75	-0.75	-1.25	0	0	0	1.25	0.75	0.75

a) Write a function that generates a logically quadrilateral partition of Ω by subdividing $\hat{\Omega}$ into a quadratic "reference partition" with $m \times m$ squares and mapping it onto Ω. The number $m \in \mathbb{N}$ is determined by the requirement that all elements of the resulting partition of Ω should have a diameter less than $h > 0$. The input data are the coordinates of $b_{11}, \ldots, b_{33}$ and the grid parameter h. The output data are the nodal coordinate and element connectivity tables (defined analogously to the triangular case, see Programming project 2.3 or Section 2.4.1) of the obtained partition.

b) Making use of a), write a function that solves the boundary value problem by means of isoparametric biquadratic elements. The input data are the right-hand side f, the boundary function g, the exact solution u, the coordinates of $b_{11}, \ldots, b_{33}$, and the fineness h. The output data are $\boldsymbol{\xi} \in \mathbb{R}^M$ and the maximum norm of the error, i.e., $\|u - u_h\|_\infty$.
c) Test your implementation for the following cases:

1) $u(x, y) := x^2 + y^2$,
2) $u(x, y) := \cos(7x)\cos(7y)$.

3.9 The Maximum Principle for Finite Element Methods

In this section maximum and comparison principles that have been introduced for the finite difference method are outlined for the finite element method.

In the case of two-dimensional domains Ω the situation has been well investigated for linear elliptic boundary value problems of second-order and linear elements. For higher dimensional problems ($d > 2$) as well as other types of elements, the corresponding assumptions are much more complex, or there does not necessarily exist any maximum principle.

From now on, let $\Omega \subset \mathbb{R}^2$ be a polygonally bounded domain and let X_h denote the finite element space of continuous, piecewise linear functions for a consistent triangulation $\mathcal{T}_h$ of Ω where the function values in the nodes on the Dirichlet boundary Γ_3 are included in the degrees of freedom. First, we consider the discretization developed for the Poisson equation $-\Delta u = f$ with $f \in L^2(\Omega)$. The algebraization of the method is done according to the scheme described in Section 2.4.3. According to this, first all nodes inside Ω and on Γ_1 and Γ_2 are numbered consecutively from 1 to a number M_1. The nodal values $u_h(a_r)$ for $r = 1, \ldots, M_1$ are arranged in the vector $\boldsymbol{u}_h$. Then, the nodes that belong to the Dirichlet boundary are numbered from $M_1 + 1$ to some number $M_1 + M_2$, the corresponding nodal values generate the vector $\hat{\boldsymbol{u}}_h$. The combination of $\boldsymbol{u}_h$ and $\hat{\boldsymbol{u}}_h$ gives the vector of all nodal values $\tilde{\boldsymbol{u}}_h = \binom{\boldsymbol{u}_h}{\hat{\boldsymbol{u}}_h} \in \mathbb{R}^M$, $M = M_1 + M_2$.

This leads to a linear system of equations of the form (1.34) described in Section 1.4:

$$A_h \boldsymbol{u}_h = -\hat{A}_h \hat{\boldsymbol{u}}_h + \boldsymbol{f}$$

with $A_h \in \mathbb{R}^{M_1, M_1}$, $\hat{A}_h \in \mathbb{R}^{M_1, M_2}$, $\boldsymbol{u}_h, \boldsymbol{f} \in \mathbb{R}^{M_1}$ and $\hat{\boldsymbol{u}}_h \in \mathbb{R}^{M_2}$.

Recalling the support properties of the basis functions $\varphi_i, \varphi_j \in X_h$, we obtain for a general element of the (extended) stiffness matrix $\widetilde{A}_h := \left(A_h \mid \hat{A}_h\right) \in \mathbb{R}^{M_1, M}$ the relation

$$(\widetilde{A}_h)_{ij} = \int_\Omega \nabla\varphi_j \cdot \nabla\varphi_i \, dx = \int_{\operatorname{supp}\varphi_i \cap \operatorname{supp}\varphi_j} \nabla\varphi_j \cdot \nabla\varphi_i \, dx\,.$$

Therefore, if $i \neq j$, the actual domain of integration consists of at most two triangles. Hence, for the present it is reasonable to consider only one triangle as the domain of integration.

Lemma 3.50 *Suppose $\mathcal{T}_h$ is a consistent triangulation of Ω. Then for an arbitrary triangle $K \in \mathcal{T}_h$ with the vertices a_i, a_j $(i \neq j)$, the following relation holds:*

$$\int_K \nabla\varphi_j \cdot \nabla\varphi_i \, dx = -\frac{1}{2} \cot \alpha_{ij}^K ,$$

where α_{ij}^K denotes the interior angle of K that is opposite to the edge with the boundary points a_i, a_j.

Proof Suppose the triangle K has the vertices a_i, a_j, a_k (see Figure 3.20). On the edge opposite to the point a_j, we have

$$\varphi_j \equiv 0 .$$

Therefore, $\nabla\varphi_j$ has the direction of a normal vector to this edge and—by considering in which direction φ_j increases—the orientation opposite to the outward normal vector $\mathfrak{n}_{ki}$, that is,

$$\nabla\varphi_j = -\left|\nabla\varphi_j\right| \mathfrak{n}_{ki} \quad \text{with} \quad |\mathfrak{n}_{ki}| = 1 . \tag{3.162}$$

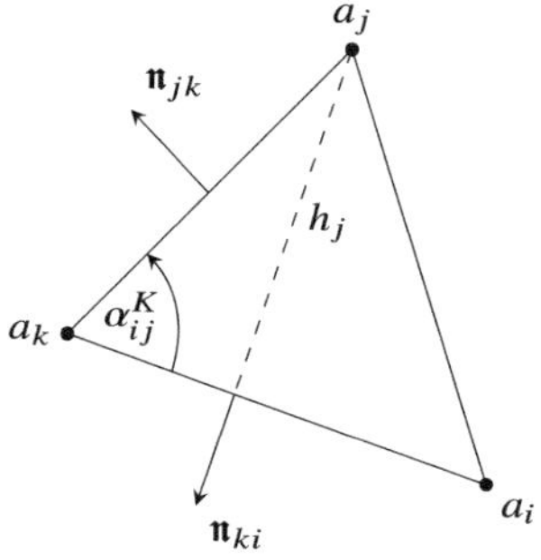

Fig. 3.20: Notation for the proof of Lemma 3.50.

In order to calculate $\left|\nabla\varphi_j\right|$ we use the following: From (3.162) we obtain

$$\left|\nabla\varphi_j\right| = -\nabla\varphi_j \cdot \mathfrak{n}_{ki} ,$$

that is, we have to compute a directional derivative. By virtue of $\varphi_j(a_j) = 1$, we have

$$\nabla\varphi_j \cdot \mathfrak{n}_{ki} = \frac{0-1}{h_j} = -\frac{1}{h_j} ,$$

where h_j denotes the height of K with respect to the edge opposite a_j. Thus we have obtained the relation

$$\nabla \varphi_j = -\frac{1}{h_j} \mathfrak{n}_{ki} .$$

Hence we have

$$\nabla \varphi_j \cdot \nabla \varphi_i = \frac{\mathfrak{n}_{ki} \cdot \mathfrak{n}_{jk}}{h_j\, h_i} = -\frac{\cos \alpha_{ij}^K}{h_j\, h_i} .$$

Since

$$2\,|K| = h_j\,|a_k - a_i| = h_i\,|a_j - a_k| = |a_k - a_i|\,|a_j - a_k|\,\sin \alpha_{ij}^K ,$$

we obtain

$$\nabla \varphi_j \cdot \nabla \varphi_i = -\frac{\cos \alpha_{ij}^K}{4\,|K|^2}\,|a_k - a_i|\,|a_j - a_k| = -\frac{1}{2} \cot \alpha_{ij}^K \,\frac{1}{|K|} ,$$

so that the assertion follows by integration. □

Corollary 3.51 *If K and K' are two triangles of $\mathcal{T}_h$ which have a common edge spanned by the nodes a_i, a_j, then*

$$(\widetilde{A}_h)_{ij} = \int_{K \cup K'} \nabla \varphi_j \cdot \nabla \varphi_i \, dx = -\frac{1}{2} \frac{\sin(\alpha_{ij}^K + \alpha_{ij}^{K'})}{(\sin \alpha_{ij}^K)(\sin \alpha_{ij}^{K'})} .$$

Proof The formula follows from the addition theorem for the cotangent function. □

Lemma 3.50 and Corollary 3.51 are the basis for the proof of the assumption (1.35)* in the case of the extended system matrix $\widetilde{A}_h$. Indeed, additional assumptions about the triangulation $\mathcal{T}_h$ are necessary:

Angle condition: For any two triangles of $\mathcal{T}_h$ with a common edge, the sum of the interior angles opposite to this edge does not exceed the value π. If a triangle has an edge on the boundary part Γ_1 or Γ_2, then the angle opposite this edge must not be obtuse.

Connectivity condition: For every pair of nodes both belonging to $\Omega \cup \Gamma_1 \cup \Gamma_2$ there exists a polygonal line between these two nodes such that the polygonal line consists only of triangle edges whose boundary points also belong to $\Omega \cup \Gamma_1 \cup \Gamma_2$ (see Figure 3.21).

Discussion of assumption (1.35)*: The proof of (1), (2), (5), (6)* is rather elementary. For the "diagonal elements",

$$(A_h)_{rr} = \int_\Omega |\nabla \varphi_r|^2 \, dx = \sum_{K \subset \operatorname{supp} \varphi_r} \int_K |\nabla \varphi_r|^2 \, dx > 0, \quad r = 1, \ldots, M_1 ,$$

which already is (1). Checking the sign conditions (2) and (5) for the "nondiagonal elements" of $\widetilde{A}_h$ requires the analysis of two cases:

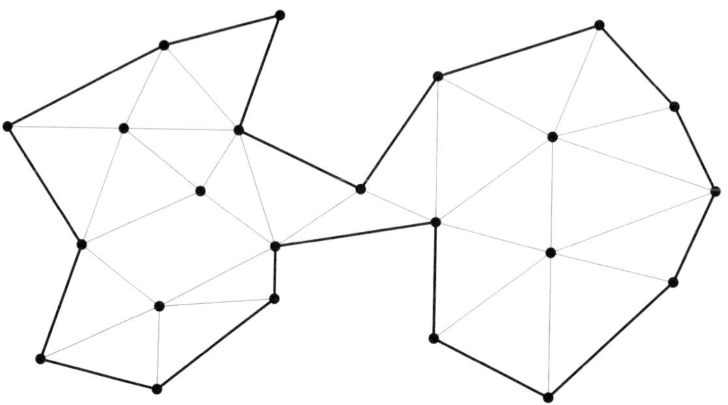

Fig. 3.21: Example of a nonconnected triangulation ($\Gamma_3 = \partial\Omega$).

1. For $r = 1, \ldots, M_1$ and $s = 1, \ldots, M$ with $r \neq s$, there exist two triangles that have the common vertices a_r, a_s.
2. There exists only one triangle that has a_r as well as a_s as vertices.

In case 1, Corollary 3.51 can be applied, since if K, K' just denote the two triangles with a common edge spanned by a_r, a_s, then $0 < \alpha_{rs}^K + \alpha_{rs}^{K'} \leq \pi$ and thus $(\widetilde{A}_h)_{rs} \leq 0$, $r \neq s$. In case 2, Lemma 3.50, due to the part of the angle condition that refers to the boundary triangles, can be applied directly yielding the assertion.

Further, since $\sum_{s=1}^{M} \varphi_s = 1$ in Ω, we obtain

$$\sum_{s=1}^{M} (\widetilde{A}_h)_{rs} = \sum_{s=1}^{M} \int_\Omega \nabla\varphi_s \cdot \nabla\varphi_r \, dx = \int_\Omega \nabla \left(\sum_{s=1}^{M} \varphi_s \right) \cdot \nabla\varphi_r \, dx = 0 .$$

This is (6)*.

The sign condition in (3) now follows from (6)* and (5), since we have

$$\sum_{s=1}^{M_1} (A_h)_{rs} = \underbrace{\sum_{s=1}^{M} (\widetilde{A}_h)_{rs}}_{=0} - \sum_{s=M_1+1}^{M} (\hat{A}_h)_{rs} \geq 0 . \tag{3.163}$$

The difficult part of the proof of (3) consists in showing that at least one of these inequalities (3.163) is satisfied strictly. This is equivalent to the fact that at least one element $(\hat{A}_h)_{rs}$, $r = 1, \ldots, M_1$ and $s = M_1 + 1, \ldots, M$, is negative, which can be shown in terms of an indirect proof by using Lemma 3.50 and Corollary 3.51, but is not done here in order to save space. Simultaneously, this also proves condition (7).

The remaining condition (4)* is proved similarly. First, due to the connectivity condition, the existence of geometric connections between pairs of nodes by polygonal lines consisting of edges is obvious. It is more difficult to prove that under all possible connections there exists one along which the corresponding matrix elements do not vanish. This can be done by the same technique of proof as used in the second part of (3), which, however, is not presented here.

If the angle condition given above is replaced with a stronger angle condition in which stretched and right angles are excluded, then the proof of (3) and (4)* becomes trivial.

Recalling the relations

$$\max_{x\in\overline{\Omega}} u_h(x) = \max_{r\in\{1,\dots,M\}} (\tilde{\boldsymbol{u}}_h)_r$$

and

$$\max_{x\in\Gamma_3} u_h(x) = \max_{r\in\{M_1+1,\dots,M\}} (\hat{\boldsymbol{u}}_h)_r\,,$$

which hold for linear elements, the following result can be derived from Theorem 1.11.

Theorem 3.52 *If the triangulation $\mathcal{T}_h$ satisfies the angle condition and the connectivity condition, then we have the following estimate for the finite element solution u_h of the Poisson equation in the space of linear elements for a nonpositive right-hand side $f \in L^2(\Omega)$:*

$$\max_{x\in\overline{\Omega}} u_h(x) \le \max_{x\in\Gamma_3} u_h(x)\,.$$

Finally, we make two remarks concerning the case of more general differential equations.

If an equation with a variable scalar diffusion coefficient $k:\ \Omega\to\mathbb{R}$ is considered instead of the Poisson equation, then the relation in Corollary 3.51 loses its purely geometric character. Even if the diffusion coefficient is supposed to be elementwise constant, the data-dependent relation

$$(\widetilde{A}_h)_{ij} = -\frac{1}{2}\left\{k_K \cot\alpha_{ij}^K + k_{K'}\cot\alpha_{ij}^{K'}\right\}$$

would arise, where k_K and $k_{K'}$ denote the constant restriction of k to the triangles K and K', respectively. The case of matrix-valued coefficients $\boldsymbol{K}:\ \Omega\to\mathbb{R}^{d,d}$ is even more problematic.

The second remark concerns differential expressions that also contain lower order terms, that is, convective and reactive parts. If the diffusive term $-\nabla\cdot(\boldsymbol{K}\nabla u)$ can be discretized in such a way that a maximum principle holds, then this maximum principle is preserved if the discretization of the other terms leads to matrices whose "diagonal elements" are nonnegative and whose "nondiagonal elements" are nonpositive. These matrix properties are much simpler than conditions (1.35) and (1.35)*. However, satisfying these properties causes difficulties in special cases, e.g., for convection-dominated equations (see Chapter 10), unless additional restrictive assumptions are made or special discretization schemes are used.

Exercise

Problem 3.37 Investigate the validity of a weak maximum principle (in the sense of Theorem 3.52) for quadratic elements and the cases

a) of a Friedrichs–Keller triangulation,
b) of a triangulation consisting only of isosceles triangles.

What is to be expected in general under the assumptions of Theorem 3.52?

Chapter 4
Grid Generation and A Posteriori Error Estimation

4.1 Grid Generation

An essential step in the implementation of the finite element method (and also of the finite volume method as described in Chapter 8) is to create an initial "geometric discretization" of the domain Ω.

This part of a finite element code is usually included in the so-called *preprocessor* (see also Section 2.4.1). In general, a stand-alone finite element code consists further of the intrinsic *kernel* (*assembling* of the finite-dimensional system of algebraic equations, rearrangement of data (if necessary), and solution of the algebraic problem) and the *postprocessor* (editing of the results, extraction of intermediate results, preparation for graphic output, and a posteriori error estimation). Adaptive finite element codes have an additional feedback control between the postprocessor and the preprocessor.

4.1.1 Classification of Grids

Grids can be categorized according to different criteria: One criterion considers the geometric shape of the elements (triangles, quadrilaterals, tetrahedra, hexahedra, prisms/wedges (pentahedra), and pyramids; possibly with curved boundaries). A further criterion distinguishes the logical structure of the grid (structured or unstructured grids). Besides these rough classes, in practice one can find a large number of variants combining grids of different classes (combined grids).

A *structured grid in the strict sense* is characterized by a regular arrangement of the grid points (nodes), that is, the connectivity pattern between neighbouring nodes (the *node-adjacency structure*) is identical everywhere in the interior of the grid. The only exceptions to that pattern may occur near the boundary of the domain Ω.

P. Knabner and L. Angermann, *Numerical Methods for Elliptic and Parabolic Partial Differential Equations*, Texts in Applied Mathematics 44,
https://doi.org/10.1007/978-3-030-79385-2_4

Typical examples of structured grids are rectangular Cartesian two- or three-dimensional grids as they are also used within the framework of the finite difference methods described in Chapter 1 (see, e.g., Figure 1.1).

A *structured grid in the wider sense* is obtained by the application of a piecewise smooth bijective transformation to some "reference grid", which is a structured grid in the strict sense. Grids of this type are also called logically structured, because only the logical structure of the connectivity pattern is fixed in the interior of the grid. However, the edges or faces of the geometric elements of a logically structured grid are not necessarily straight or plain, respectively.

Logically structured grids have the advantage of simple implementation, because the pattern already defines the neighbours of a given node. Furthermore, there exist efficient methods for the solution of the algebraic system resulting from the discretization, including parallelized resolution algorithms.

In contrast to structured grids, unstructured grids do not have a self-repeating node pattern. Moreover, elements of different geometric types can be combined in unstructured grids. Unstructured grids are suitable tools for the modelling of complex geometries of Ω and for the adjustment of the grid to the numerical solution (local grid adaptation).

Although "the mesh generation methodology is becoming increasingly recognized as a subject in its own right" [43, p. xvii], a detailed explanation is out of the scope of this book. Nevertheless we want to give a brief overview on a few methods for creating and modifying unstructured grids in the subsequent sections. More comprehensive descriptions can be found, for instance, in the books [26], [43], or [8].

4.1.2 Generation of Simplicial Grids

A simplicial grid consists of triangles (in two dimensions) or tetrahedra (in three dimensions). To generate simplicial grids, the following three fundamentally distinctive types of methods are widely used, often even in combination:

- tree-based methods,
- Delaunay triangulation methods, and
- advancing front methods.

Tree-based Methods

The so-called *quadtree* (in two dimensions) or *octree methods* (in three dimensions) start from a rectangular or hexahedral set (the *bounding box* or *root cell*) that covers the domain and is partitioned by a structured grid (the *overlay* or *background grid*), typically a comparatively coarse rectangular Cartesian two- or three-dimensional grid.

The substantial part of the algorithm consists of fitting routines for those parts of the initial grid that are located near the boundary, and of simplicial subdivisions of the obtained geometric elements. The fitting procedures perform recursive subdivisions of the near-boundary rectangles or rectangular parallelepipeds in such a way that at the end every geometric element contains at most one geometry defining point (i.e., a vertex of Ω or a point of $\partial\Omega$, where the type of boundary conditions changes). After that, the resulting rectangles or rectangular parallelepipeds are split into simplices according to a set of refinement templates. Finally, a so-called *smoothing step*, which optimizes the grid with respect to certain regularity criteria, can be supplemented; see Section 4.1.4.

The method is named after the tree data structure for managing the recursive subdivisions. Typically, grids generated by tree-based methods are close to structured grids in the interior of the domain. Near the boundary, they lose the structure. Further details can be found in the references [228], [216], and [174].

Delaunay Triangulation Methods

The core algorithm of these methods generates, for a given cloud of isolated points (nodes), a triangulation of their convex hull. Therefore, a grid generator based on this principle has to include a procedure for the generation of this point set (for example, the points resulting from a tree-based method) as well as certain fitting procedures (to cover, for example, nonconvex domains, too). A pretty comprehensive book about this topic is [14].

The Delaunay triangulation of the convex hull of a given point set in $\mathbb{R}^d$ is characterized by the following property (the *empty sphere criterion*, Figure 4.1): Any open d-ball, the boundary of which contains $d + 1$ points from the given set, does not contain any other points from that set. The triangulation can be generated from the so-called *Voronoi tessellation* of $\mathbb{R}^d$ for the given point set. In two dimensions, this procedure is described in Chapter 8, which deals with finite volume methods (Section 8.2.1). However, practical algorithms ([111], [226]) apply the empty sphere criterion more directly.

The interesting theoretical properties of Delaunay triangulations are one of the reasons for the "popularity" of this method. In two dimensions, the so-called *max-min-angle property* is valid: Among all triangulations of the convex hull $\overline{G}$ of a given point set in the plane, the Delaunay triangulation maximizes the minimal interior angle over all triangles. No generalization of this fact is known for higher dimensions, at least in terms of some angular measures of simplices. In contrast to the case $d = 2$, even poorly shaped elements (the so-called *sliver elements*) may occur. Since a Delaunay triangulation possesses many optimality properties in two dimensions, geometers longly suspected that it should optimize something in three dimensions. Rajan [204, 205] discovered the first such result (which is true even for arbitrary dimensions). A further important property of a two-dimensional Delaunay triangulation is that the sum of two angles opposite to an interior edge is not more

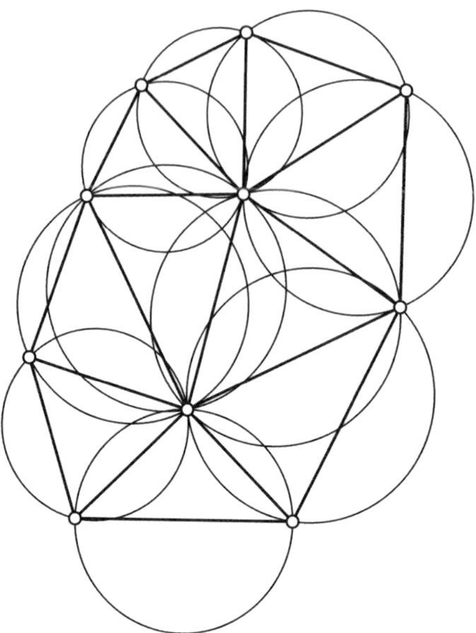

Fig. 4.1: Empty sphere criterion in two dimensions ($d = 2$).

than π. For example, such a requirement is a part of the angle condition formulated in Section 3.9.

Advancing Front Methods

The basic idea of these methods is to build a triangulation incrementally starting from a discretization of the current boundary (see, e.g., [120], [157], [194], and [203]). In particular, the methods require an already existing partition of the boundary of $G_0 := \Omega$. For $d = 2$, this "initial front" is made up of patches of curves, whereas in $d = 3$ it consists of patches of curved surfaces (the so-called "2.5-dimensional partition"). Typically these boundary curves or surfaces are represented by means of B-splines or nonuniform rational B-splines (NURBS) (see, e.g., [26]), in the simplest cases by polygonal lines or triangular facets. However we will not explain the details of 2.5-dimensional partitioning here. Many techniques developed for planar grid generation are also principally suitable for surface grid generation, but additional difficulties on curved surfaces may appear that require modifications of the algorithms.

The advancing front method consists of an iteration of the following general step (Figure 4.2): An element of the current front (i.e., a straight-line segment or a triangle) is taken and then, either generating a new inner point or taking an already existing point, a new simplex K_j that belongs to $\overline{G}_{j-1}$ is defined. After the data of the new simplex are stored, the simplex is deleted from $\overline{G}_{j-1}$. In this way, a smaller domain G_j with a new boundary ∂G_j (a new "current front") results. The general step is repeated until the current front is empty. Often, the grid generation process is supplemented by a smoothing step; see Section 4.1.4.

4.1.3 Generation of Quadrilateral and Hexahedral Grids

Grids consisting of quadrilaterals or hexahedra can also be generated by means of tree-based methods (e.g., [153], [86]), where—roughly speaking in terms of the description given in Section 4.1.2—the final steps of simplicial splitting are omitted, or *advancing front methods* (e.g., [107], [108]). However, especially the automatic generation of high-quality, fully hexahedral grids remains an unsatisfactorily solved and therefore still intense research task.

An interesting application of simplicial *advancing front methods* in the two-dimensional case is given in the paper [231]. The method, which belongs to the class of so-called *indirect methods*, is based on the simple fact that any two triangles sharing a common edge form a quadrilateral. Obviously, a necessary condition for the success of the method is that the underlying triangulation should consist of an even number of triangles. Unfortunately, the generalization of the method to the three-dimensional situation is difficult, because a comparatively large number of adjacent tetrahedra should be merged to form a hexahedron. However, this idea can be successfully used to create grids that are predominantly composed of hexahedra; see the subsequent paragraph about mixed grids.

Multiblock Methods

The basic idea of these methods is to partition the domain into a small number of subdomains ("blocks") of simple shape (quadrilaterals, hexahedra, as well as triangles, tetrahedra, prisms, pyramids, etc.) and then generate structured or logically structured grids in the individual subdomains (e.g., [60], [201]).

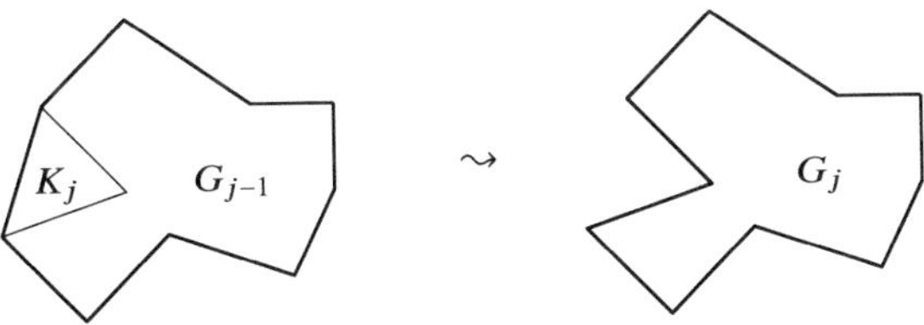

Fig. 4.2: Step j of the advancing front method: The new simplex K_j is deleted from the domain G_{j-1}.

In multiblock grids, special attention has to be devoted to the treatment of common boundaries of adjacent blocks. Unless special discretization methods such as, for example, the so-called *mortar finite element method* (cf. [102]) are used in this situation, there may be a conflict between certain compatibility conditions at the common block interfaces (to ensure, e.g., the continuity of the finite element functions across the interfaces) on the one hand and the output directives of an error estimation procedure that may advise to refine a block-internal grid locally on the other hand.

Mixed Grids

Especially in three-dimensional situations, the generation of "purely" hexahedral (*all-hexahedral*) grids may be very difficult for complicated geometries of the domain. Therefore, the so-called *mixed* (or *combined* or *hexahedral-dominant*) grids that consist of hexahedral grids in geometrically simple subdomains and tetrahedral, prismatic, pyramidal, etc. grids in more critical subregions are used (e.g., [200]). In this context indirect methods that combine tetrahedra to hexahedra are attractive (e.g., [188], [202]).

Chimera Grids

These grids are also called *overset grids* (see, e.g., [121]). In contrast to the multiblock grids described above, here the domain is covered by a comparatively small number of domains of simple shape, and then structured or logically structured grids are generated on the individual domains. That is, a certain overlapping of the blocks and thus of the subgrids is admitted.

4.1.4 Grid Optimization

Many grid generation codes include "smoothing algorithms" that optimize the grid with respect to certain grid quality measures or regularity criteria. In the so-called *r-method* (relocation method) the nodes are slightly moved, keeping the logical structure (the connectivity) of the grid fixed (see, e. g., [83]). Another approach is to improve the node-adjacency structure of the grid itself. In principle, optimization-based smoothing algorithms differ in the type of grid being smoothed, the so-called *distortion metric* selected to construct an appropriate objective function, and the optimization method used. However, grid optimization on its own in terms of grid quality may not be very useful if, for instance, the underlying (initial-)boundary value problem is of anisotropic nature.

A typical example for r-methods is given by the so-called *Laplacian smoothing* (or *barycentric smoothing*), where any inner grid point is moved into the barycentre of its neighbours (e.g., [120]). A local weighting of selected neighbours can also be used (*weighted barycentric smoothing*). From a formal point of view, the application of the Laplacian smoothing corresponds to the solution of a system of linear algebraic equations that is obtained from the equations of the arithmetic (or weighted) average of the nodes. The matrix of this system is large but sparse. The structure of this matrix is very similar to the one that results from a finite volume discretization of the Poisson equation as described in Section 8.2 (see the corresponding special case of (8.11)). In general, there is no need to solve this system exactly. Typically, only one to three steps of a simple iterative solver (as presented in Section 5.1) are performed. When the domain is almost convex, Laplacian smoothing will produce good results. It is also clear that for strongly nonconvex domains or other special situations, the method may produce invalid grids (see, e.g., [152]).

Among the methods to optimize the grid connectivities, the so-called *2:1-rule* and, in the two-dimensional case, the *edge swap* (or *diagonal swap*, [173]) are well known. The 2:1-rule is used within the quadtree or octree method to reduce the difference of the refinement levels between neighbouring quadrilaterals or hexahedra to one by means of additional refinement steps; see Figure 4.3.

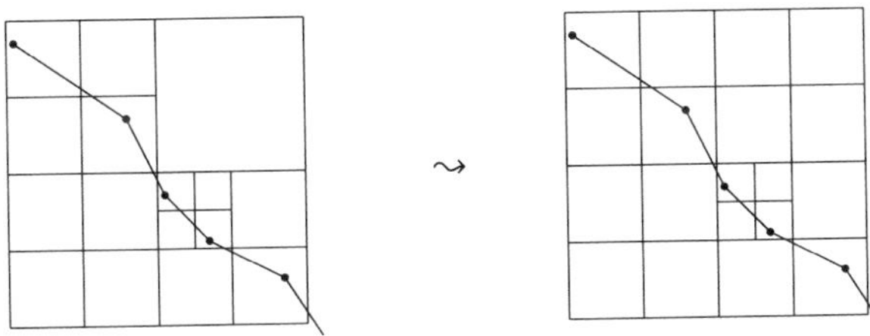

Fig. 4.3: Illustration of the 2:1-rule.

In the edge swap method, a triangular grid is improved. Since any two triangles sharing an edge form a convex quadrilateral, the method decides which of the two diagonals of the quadrilateral optimizes a given criterion. If the optimal diagonal does not coincide with the common edge, the other configuration will be taken; i.e., the edge will be swapped.

Finally, it should be mentioned that there exist grid optimization methods that delete nodes or even complete elements from the grid.

4.1.5 Grid Refinement

A typical grid refinement algorithm for a triangular grid, the so-called *red/green refinement*, has previously been introduced in Section 2.4.1. This refinement method was successfully extended to tetrahedral grids, see, for instance, [229] (red refinements only), [104], [110], and [105], and to simplicial grids of any dimension $d \geq 2$ in [143]. It consists of three components—the red and the green refinement rules, and a global refinement step, which coordinates the application of the two local refinement rules [104].

The red refinement of a tetrahedron K (see Figure 4.4) yields a partition of K into eight subtetrahedra (a so-called *octasection*) with the following properties: All vertices of the subtetrahedra coincide either with vertices or with edge midpoints of K. At all the faces of K, the two-dimensional red refinement scheme can be observed.

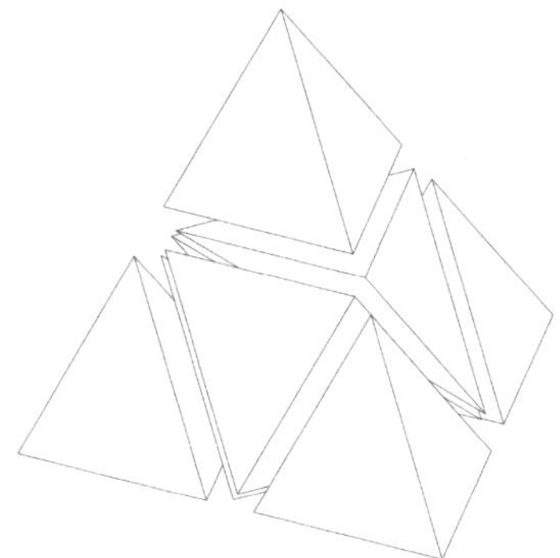

Fig. 4.4: The red refinement of a tetrahedron.

A further class of methods is based on bisection, that is, a simplex is cut by a hyperplane passing through the midpoint of an edge and through the vertices that do not belong to this edge (see Figure 4.5).

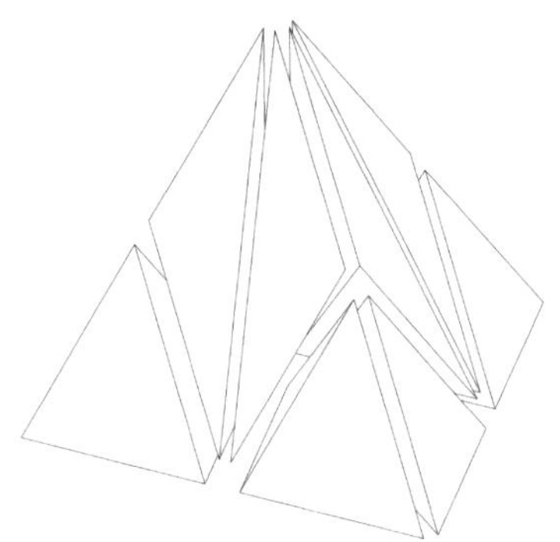

Fig. 4.5: A multiply bisected tetrahedron.

Bisection methods are based on simplicial grids and are characterized by the number of bisections used within one refinement step (the *stage number of the method*) as well as by the criterion of how to select the edge where the new node is to be located. Among the bisection methods, the *longest edge bisection* [209] and the *newest vertex bisection* [190], [192] are well developed. Both bisection methods have been extended to the multi-dimensional case. The generalization of the longest edge bisection method to three dimensions was made by Rivara herself [210]. A first paper on the general case $d \geq 2$ is [90].

The first three-dimensional versions of Mitchell's method can be found in [89], [184], [165], and [178]/[179]. In [185], [223], and [84] generalizations to arbitrary dimensions $d \geq 2$ were given; for an overview we refer to [193].

In the (conventional) longest edge bisection method, the general (recursive) refinement step for some simplex K is of the following form:

1. Find a longest edge E in K.
2. Bisect the simplex K by a hyperplane passing through the midpoint of E and the vertices of K that do not belong to E.

However, this type of refinement may generate nonconsistent triangulations. There exists also a consistent version of longest edge bisection method [192], [163], where the general (recursive) refinement step is of the following form:

1. Find a longest edge E in a given consistent triangulation $\mathcal{T}$ (or in a subset $\mathcal{M} \subset \mathcal{T}$).
2. Bisect each simplex $K \in \mathcal{T}$ sharing this edge by a hyperplane passing through the midpoint of E and the vertices of K that do not belong to E.

In this version, all generated triangulations are consistent.

A step of the newest vertex bisection method always starts from a labelled (or marked) triangulation, that is, for each simplex one vertex is labelled as the so-called *newest vertex*. In the two-dimensional case, this vertex uniquely defines an edge opposite to this vertex, and this edge is called the *refinement edge*. In dimensions greater than 2, there is not a unique edge opposite to the newest vertex. Therefore the algorithm has to select a refinement edge, too, and this is the reason why some authors prefer to call methods of this type in the higher dimensional case $d \geq 3$ rather *recursive bisection*. The existing newest vertex bisection methods differ in the way how this edge selection is achieved. The general refinement step includes the following:

1. A simplex is bisected by creating a new vertex at the midpoint of the refinement edge and cutting the simplex by a hyperplane that passes through the new vertex and through the vertices that do not belong to this edge.
2. The new vertex created at the midpoint of the refinement edge is assigned to be the newest vertex of both of the resulting simplices.

Thanks to the second rule, the subsequent triangulations inherit the label and the bisection process can be continued.

In general, there are two main aspects in designing a good refinement method:

1) Consistency (or conformity; see the conditions (T5) and (T6) in Subsection 3.3),
2) Shape regularity of the generated family of triangulations (see Definition 3.31).

In what follows we briefly comment on some of these aspects.

In special cases, e. g. for the so-called uniform refinement of a complete consistent triangulation $\mathcal{T}$, the resulting refined triangulation $\mathcal{T}'$ is consistent, too. Examples are the global uniform refinement by quadrisection mentioned in Subsection 2.4.1 or [218, Thm. 4.3]. A further example is the consistent longest edge bisection method mentioned above. In general, however, nonconsistent refined partitions arise. Furthermore, in adaptive finite element methods only certain proper subsets $\mathcal{M} \subset \mathcal{T}$ are to be refined, and the sole refinement of the elements of $\mathcal{M}$ leads to nonconsistent partitions. Therefore, in order to restore the consistency of a refined partition $\mathcal{T}'$, so-called *completion* (or *closure*) *rules* are applied. That is, the set $\mathcal{M}$ is enlarged to a set of elements $\mathcal{R} \supset \mathcal{M}$, the refinement of which yields a consistent grid. Since the refinement of an element $K \in \mathcal{T}$ means that K is split into at least two and at most $C_{\text{son}} \geq 2$ smaller elements (children), it holds that

$$|\mathcal{M}| \leq |\mathcal{T}'| - |\mathcal{T}| \leq (C_{\text{son}} - 1)|\mathcal{T}|.$$

In addition to the regularity problem already mentioned above, the completion process causes additional problems, namely its termination after a limited number of completion steps and the nestedness of the refined partition. Under the latter, the following is understood.

Definition 4.1 Given two partitions $\mathcal{T}, \mathcal{T}'$ with corresponding node sets $\overline{\mathcal{N}}, \overline{\mathcal{N}}'$. The partition $\mathcal{T}'$ is called a *grid-nested refinement* of the partition $\mathcal{T}$ if each element $K' \in \mathcal{T}'$ is covered by exactly one element $K \in \mathcal{T}$ and the following strong inclusion holds: $\overline{\mathcal{N}} \subset \overline{\mathcal{N}}', \overline{\mathcal{N}}' \neq \overline{\mathcal{N}}$.

The two bisection methods, along with their completion rules, produce grid-nested refinements, while the red/green refinement does not. An example of the latter can be found in [142, Fig. 3.2].

The consistency of the red/green refinement has already been discussed in Section 2.4.1 for $d = 2$. The rollback step 4 in refinement algorithm given there shows that the minimum angle in a refined triangulation is not less than half the minimum angle of the initial triangulation. In the three-dimensional case, the completion of a refined tetrahedron can be done in a similar way as for $d = 2$ (see Section 2.4.1). If all four edges of the tetrahedron are marked, a red refinement (octasection) can take place. However, in dependence on whether one, two, or three edges are marked, three different types of green completions (see e. g. [110, Figs. 6-8]) are required. The paper [168] ventilates the shape regularity in the three-dimensional case. In the general case $d \geq 2$, the green refinement rule described in [143] yields a consistent triangulation (see [143, Thm. 3.15]), introduces no additional vertices, and is able to handle every possible refinement pattern. It produces only a finite number of similarity classes of simplices. [143, Thm. 4.6] shows that there is no avalanche or domino effect which inevitably spreads out the red refinement over the grid.

For the longest edge bisection method in the case $d = 2$, a consistency and shape regularity result was demonstrated in [210, Thm. 11]. The termination of the completion process has been shown in [208]. The shape regularity of the two-dimensional consistent longest edge bisection method has been investigated in [163]. The properties of consistency and finite termination of a three-dimensional local longest edge bisection refinement have been demonstrated in [179]. The verification of the shape regularity of longest edge bisection method for $d \geq 3$ is a difficult problem, although there are many computer experiments that show that this type of local refinement can possess this property for both triangular and tetrahedral grids; for an overview see [164].

The consistency and shape regularity of the newest vertex bisection method in the case $d = 2$ have already been demonstrated in [191]. Bänsch [89] considered the consistency and shape regularity for the dimensions $d = 2, 3$. They also showed that the completion process will terminate. First consistency and shape regularity results in the general case $d \geq 2$ can be found in [185], [223], and in [218] a result from [106] has been generalized to $d \geq 2$ dimensions: The total number of d-simplices in the partition at termination of the method can be bounded by some absolute multiple of the number of d-simplices that were marked for refinement in all iterations. For more information, we refer to [73].

Regarding the consistent refinement of quadrilateral and hexahedral grids, the algorithms and their theoretical properties are not as well developed and understood as in the simplicial case. We mention only a few works. In [230], an adaptive quadrilateral grid refinement procedure without hanging nodes has been proposed. It can be regarded as an extension of the usual red/green refinement to quadrilateral grids. Especially in the case $d = 3$, *refinement templates* or *transition elements* are often used; see, e. g., [220], [43, Sections 5.8.14/15].

As a final remark and without going into more details we mention that there are also algorithms that deal with the problem of *grid de-refinement*. For instance, in time-dependent problems, those regions of the grid may also vary in which refinement is required. Then, an efficient coarsening of grid patches that no longer have to be fine can reduce the overall computational effort. An example of an efficient implementation of a de-refinement algorithm using the newest vertex bisection method in $\mathbb{R}^2$ can be found in [122].

Exercises

Problem 4.1 For a given triangle K, the circumcentre can be computed by finding the intersection of the perpendicular bisectors associated with two edges of K. This can be achieved by solving a linear system of equations with respect to the coordinates of the circumcentre.

a) Formulate such a system.
b) How can the radius of the circumcircle be obtained from this solution?

Problem 4.2 Given a triangle K, denote by h_i the length of edge i, $i \in \{1, 2, 3\}$. Prove that the following expression equals the radius of the circumcircle (without using the circumcentre!):

$$\frac{h_1 h_2 h_3}{4|K|} .$$

Problem 4.3 Let K_1, K_2 be two triangles sharing an edge.

a) Show the equivalence of the following edge swap criteria:
Angle criterion: Select the diagonal of the so-formed quadrilateral that maximizes the minimum of the six interior angles among the two configurations.
Circle criterion: Choose the diagonal of the quadrilateral for which the open circumcircle disks to the resulting triangles do not contain any of the remaining vertices.

b) If α_1, α_2 denote the two interior angles that are located opposite to the common edge of the triangles K_1 and K_2, respectively, then the circle criterion states that an edge swap is to be performed if

$$\alpha_1 + \alpha_2 > \pi .$$

Prove this assertion.

c) The criterion in b) is numerically expensive. Show that the following test is equivalent:

$$\begin{aligned}
&[(a_{1,1} - a_{3,1})(a_{2,1} - a_{3,1}) + (a_{1,2} - a_{3,2})(a_{2,2} - a_{3,2})] \\
&\quad \times [(a_{2,1} - a_{4,1})(a_{1,2} - a_{4,2}) - (a_{1,1} - a_{4,1})(a_{2,2} - a_{4,2})] \\
&< [(a_{2,1} - a_{4,1})(a_{1,1} - a_{4,1}) + (a_{2,2} - a_{4,2})(a_{1,2} - a_{4,2})] \\
&\quad \times [(a_{2,1} - a_{3,1})(a_{1,2} - a_{3,2}) - (a_{1,1} - a_{3,1})(a_{2,2} - a_{3,2})] .
\end{aligned}$$

Here $a_i = (a_{i,1}, a_{i,2})^T$, $i \in \{1, 2, 3\}$, denote the vertices of a triangle ordered clockwise, and $a_4 = (a_{4,1}, a_{4,2})^T$ is the remaining vertex of the quadrilateral, the position of which is tested in relation to the circumcircle defined by a_1, a_2, a_3.
Hint: Addition theorems for the sin function.

4.2 A Posteriori Error Estimates

In the practical application of discretization methods to partial differential equations, ideally an approximation should have been calculated, the error of which is known or even within prescribed bounds. The error estimates described in the previous chapters cannot be used to achieve these goals because they contain information about the unknown weak solution u and indefinite constants (such as the seminorm $|u|_{k+1}$ and the constant C in (3.102)). Such estimates only yield information on the asymptotic error behaviour and require regularity conditions to the weak solution which may be not satisfied, for instance, in the presence of singularities.

Therefore, methods are needed for estimating and quantifying the occurring error, which allow to extract the required information *a posteriori* from the computed numerical solution and the given data of the problem. To achieve a guaranteed accuracy, an additional adaptive control of the parameters of the numerical process, for instance, a grid control, is required. The latter will be the topic of the next section.

For elliptic or parabolic differential equations of order two, a typical norm to quantify the error $u - u_h$ is the energy norm (respectively an equivalent norm) or the L^2-norm. Some practically important problems involve the approximation of so-called derived quantities which can be mathematically interpreted in terms of values of certain linear functionals $u \mapsto J(u)$ (*goal functionals*) of the solution u. In such a case, an estimate of the corresponding error is also of interest.

Example 4.2 (goal functionals)

$$J(u) = \int_{\Gamma_0} \mathfrak{n}\cdot\nabla u \, d\sigma : \qquad \textit{flux of } u \textit{ through a part of the boundary } \Gamma_0 \subset \partial\Omega,$$

$$J(u) = \int_{\Omega_0} u \, dx : \qquad \textit{integral mean of } u \textit{ on some subdomain } \Omega_0 \subset \Omega.$$

In the following we will consider some estimates for a norm $\|\cdot\|$ of the error $u-u_h$ and explain the corresponding terminology. Similar statements remain true if $\|u - u_h\|$ is replaced by $|J(u) - J(u_h)|$.

The error estimates given in the previous chapters are characterized by the fact that they do not require any information about the computed solution u_h. Estimates of this type are called *a priori* error estimates.

For example, consider a variational equation with a bilinear form that satisfies (for some space V such that $H_0^1(\Omega) \subset V \subset H^1(\Omega)$ and $\|\cdot\| := \|\cdot\|_1$) the assumptions (2.46), (2.47) and use numerically piecewise linear, continuous finite elements. Then Céa's lemma (Theorem 2.17) together with the interpolation error estimate from Theorem 3.32 implies the estimate

$$\|u - u_h\|_1 \leq \frac{M}{\alpha} \|u - I_h(u)\|_1 \leq \frac{M}{\alpha} Ch, \tag{4.1}$$

where the constant C depends on the weak solution u of the variational equation.

Here C has the special form

$$C = \bar{C} \left(\int_\Omega \sum_{|\alpha|=2} |\partial^\alpha u|^2 \, dx \right)^{1/2} \tag{4.2}$$

with $\bar{C} > 0$ independent of u. Unfortunately, as mentioned above, the structure of the bound (4.2) does not allow the immediate numerical application of (4.1).

But even if the constant C could be estimated and (4.1) could be used to determine the discretization parameter h (maximum diameter of the triangles in $\mathcal{T}_h$) for a prescribed tolerance, in general this would lead to a grid that is too fine. This corresponds to an algebraic problem that is too large. The reason is that the described approach determines a global parameter, whereas the true error measure may have different magnitudes in different regions of the domain Ω.

So we should aim at error estimates of type

$$\|u - u_h\| \leq C_{\text{rel}}\, \eta \tag{4.3}$$

or

$$C_{\text{eff}}\, \eta \leq \|u - u_h\| \leq C_{\text{rel}}\, \eta\,, \tag{4.4}$$

where the constants C_{rel}, $C_{\text{eff}} > 0$ do not depend on the discretization parameters and

$$\eta = \left(\sum_{K \in \mathcal{T}_h} \eta_K^2 \right)^{1/2} . \tag{4.5}$$

Here the quantities η_K should be computable using only the data—including possibly $u_h|_K$—which are known on the particular element K.

If the bounds η (or the terms η_K, respectively) in (4.3) (respectively (4.4)) depend on u_h, i.e., they can be evaluated only if u_h is known, then they are called (local) *a posteriori error estimators* in the wider sense.

Often the bounds also depend on the weak solution u of the variational equation, so in fact, they cannot be evaluated immediately. In such a case they should be replaced by computable quantities that do not depend on u in a direct way. So, if the bounds can be evaluated without knowing u but using possibly u_h, then they are called (local) a posteriori error estimators in the strict sense.

Inequalities of the form (4.3) guarantee, for a given tolerance $\varepsilon > 0$, that the inequality $\eta \leq \varepsilon$ implies that the error measure does not exceed ε up to a multiplicative constant. In this sense the error estimator η is called *reliable*. Now, if the computed approximative solution u_h is sufficiently precise in the described sense, then the computation can be finished. If u_h is such that $\eta > \varepsilon$, then the question of how to modify the discretization in order to achieve the tolerance or, if the computer resources are nearly exhausted, how to minimize the overshooting of η arises. That is, the information given by the evaluation of the bounds has to be used to adapt the discretization and then to perform a new run of the solution process. A typical modification is to refine or to de-refine the grid.

Error estimators may overestimate the real error measure significantly; thus a grid adaptation procedure based on such an error estimate generates a grid that is too fine, and consequently, the corresponding algebraic problem is too large.

This effect can be reduced or even avoided if the error estimator satisfies a two-sided inequality like (4.4). Then the ratio $C_{\text{rel}}/C_{\text{eff}}$ is a measure of the *efficiency* of the error estimator.

An error estimator η is called *asymptotically exact* if for an arbitrary convergent sequence of approximations (u_h) with $\|u - u_h\| \to 0$ the following limit is valid:

$$\frac{\eta}{\|u - u_h\|} \to 1\,.$$

Usually, a posteriori error estimators are designed for a well-defined class of boundary or initial boundary value problems. Within a given class of problems, the question regarding the sensitivity of the constants C_{eff}, C_{rel} in (4.3)/(4.4), with respect to the particular data of the problem (e.g., coefficients, inhomogeneities, geometry of the domain, grid geometry, ...), arises. If this dependence of the data is not crucial, then the error estimator is called *robust* within this class.

Design of A Posteriori Error Estimators

In the following, three basic principles of the design of a posteriori error estimators will be described. In order to illustrate the underlying ideas and to avoid unnecessary technical difficulties, a model problem will be treated: Consider a diffusion–reaction equation on a polygonally bounded domain $\Omega \subset \mathbb{R}^2$ with homogeneous Dirichlet boundary conditions

$$\begin{aligned} -\Delta u + ru &= f \quad \text{in } \Omega, \\ u &= 0 \quad \text{on } \partial\Omega, \end{aligned} \tag{4.6}$$

where $f \in L^2(\Omega)$ and $r \in C(\overline{\Omega})$ with $r(x) \geq 0$ for all $x \in \Omega$. The problem is discretized using piecewise linear, continuous finite element functions as described in Section 2.2.

Setting $a(u, v) := \int_\Omega (\nabla u \cdot \nabla v + ruv)\, dx$ for $u, v \in V := H_0^1(\Omega)$, we have the following variational (weak) formulation:

$$\text{Find } u \in V \text{ such that } a(u, v) = \langle f, v \rangle_0 \text{ for all } v \in V. \tag{4.7}$$

The corresponding finite element method reads as follows:

$$\text{Find } u_h \in V_h \text{ such that } a(u_h, v_h) = \langle f, v_h \rangle_0 \text{ for all } v_h \in V_h. \tag{4.8}$$

Residual Error Estimators

Similar to the derivation of the a priori error estimate in the proof of Céa's lemma (Theorem 2.17), the V-coercivity of a (2.47) implies that

$$\alpha \|u - u_h\|_1^2 \leq a(u - u_h, u - u_h)\,.$$

Without loss of generality we may suppose $u - u_h \in V \setminus \{0\}$, hence

$$\|u - u_h\|_1 \leq \frac{1}{\alpha} \frac{a(u - u_h, u - u_h)}{\|u - u_h\|_1} \leq \frac{1}{\alpha} \sup_{v \in V} \frac{a(u - u_h, v)}{\|v\|_1}\,. \tag{4.9}$$

We observe that the term

$$a(u - u_h, v) = a(u, v) - a(u_h, v) = \langle f, v \rangle_0 - a(u_h, v) \tag{4.10}$$

is the *residual* of the variational equation; i.e., the right-hand side of inequality (4.9) can be interpreted as a certain norm of the variational residual.

In a next step, the variational residual will be split into local terms according to the given grid, and these terms are transformed by means of integration by parts. For arbitrary $v \in V$, from (4.10) it follows that

$$\begin{aligned} a(u-u_h, v) &= \sum_{K\in\mathcal{T}_h} \left(\int_K f v\,dx - \int_K (\nabla u_h \cdot \nabla v + r u_h v)\,dx \right) \\ &= \sum_{K\in\mathcal{T}_h} \left(\int_K [f - (-\Delta u_h + r u_h)] v\,dx - \int_{\partial K} \mathfrak{n}_K \cdot \nabla u_h v\,d\sigma \right), \end{aligned}$$

where $\mathfrak{n}_K$ denotes the outer unit normal on ∂K. The first factor in the integrals over the elements K is the classical elementwise residual of the differential equation:

$$r_K(u_h) := [f - (-\Delta u_h + r u_h)]\big|_K .$$

All quantities entering $r_K(u_h)$ are known. In the case considered here we even have $-\Delta u_h = 0$ on K, hence $r_K(u_h) = [f - r u_h]\big|_K$.

The integrals over the boundary of the elements K are further split into a sum over the integrals along the element edges $F \subset \partial K$:

$$\int_{\partial K} \mathfrak{n}_K \cdot \nabla u_h v\,d\sigma = \sum_{F\subset\partial K} \int_F \mathfrak{n}_K \cdot \nabla u_h v\,d\sigma .$$

Since $v = 0$ on $\partial\Omega$, only the integrals along edges lying in Ω contribute to the sum. Denoting by $\mathcal{F}_h$ the set of all interior edges of all elements $K \in \mathcal{T}_h$, we see that in the summation of the split boundary integrals over all $K \in \mathcal{T}_h$ there occur exactly two integrals along one and the same edge $F = K \cap K' \in \mathcal{F}_h$. This observation results in the relation

$$\sum_{K\in\mathcal{T}_h} \int_{\partial K} \mathfrak{n}_K \cdot \nabla u_h v\,d\sigma = \sum_{F\in\mathcal{F}_h} \int_F [\![\nabla u_h]\!]_F\, v\,d\sigma ,$$

where, in general for a piecewise continuous vector field $\boldsymbol{q} : \Omega \to \mathbb{R}^d$, the term

$$[\![\boldsymbol{q}]\!] := [\![\boldsymbol{q}]\!]_F := \boldsymbol{q}|_K \cdot \mathfrak{n}_K + \boldsymbol{q}|_{K'} \cdot \mathfrak{n}_{K'} \tag{4.11}$$

denotes the *jump* of the normal components of $\boldsymbol{q}$ across the edge F common to the elements $K, K' \in \mathcal{T}_h$.

In summary, we have the following relation:

$$a(u-u_h, v) = \sum_{K\in\mathcal{T}_h} \int_K r_K(u_h) v\,dx - \sum_{F\in\mathcal{F}_h} \int_F [\![\nabla u_h]\!]_F\, v\,d\sigma .$$

Using the error equation (2.43), we obtain for an arbitrary element $v_h \in V_h$ the fundamental identity

$$
\begin{aligned}
a(u-u_h,v) &= a(u-u_h, v-v_h) \\
&= \sum_{K\in\mathcal{T}_h}\int_K r_K(u_h)(v-v_h)\,dx \\
&\quad - \sum_{F\in\mathcal{F}_h}\int_F [\![\nabla u_h]\!]_F\,(v-v_h)\,d\sigma\,,
\end{aligned}
$$

which is the starting point for the construction of further estimates.

First we see that the Cauchy–Schwarz inequality immediately implies

$$
\begin{aligned}
a(u-u_h,v-v_h) \le \sum_{K\in\mathcal{T}_h} &\|r_K(u_h)\|_{0,K}\|v-v_h\|_{0,K} \\
+\sum_{F\in\mathcal{F}_h} &\big\|[\![\nabla u_h]\!]\big\|_{0,F}\|v-v_h\|_{0,F}\,.
\end{aligned}
\tag{4.12}
$$

To get this bound as small as possible, the function $v_h \in V_h$ is chosen such that the element $v \in V$ is approximated adequately in both spaces $L^2(K)$ and $L^2(F)$. One suggestion is the use of an interpolating function according to (2.51). However, since $V \not\subset C(\overline{\Omega})$, this interpolant is not defined. Therefore other approximation procedures have to be applied. Roughly speaking, suitable approximation principles, due to Clément [125] or Scott and Zhang [214], are based on taking certain local integral means. However, at this place we cannot go further into these details and refer to the cited literature. In fact, for our purposes it is important only that such approximations exist. Their particular design is of minor interest.

We will formulate the relevant facts as a lemma. To do so, we need some additional notations (see Figure 4.6):

triangular neighbourhood of a triangle K : $\quad \Delta(K) := \bigcup_{K':\,K'\cap K\neq\emptyset} K'$, (4.13)

triangular neighbourhood of an edge F : $\quad \Delta(F) := \bigcup_{K':\,K'\cap F\neq\emptyset} K'$. (4.14)

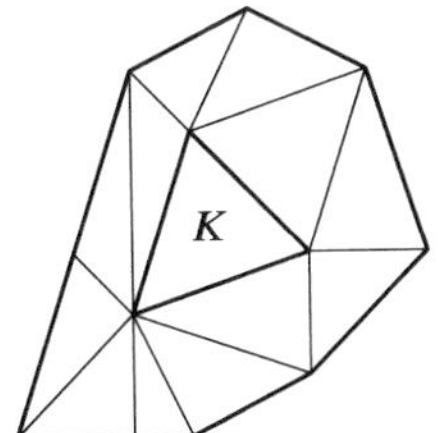

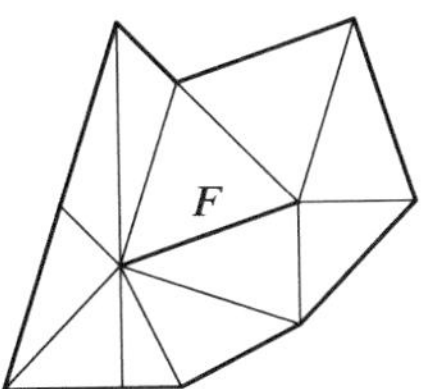

Fig. 4.6: The triangular neighbourhoods $\Delta(K)$ (left) and $\Delta(F)$ (right).

Thus $\Delta(K)$ consists of the union of the supports of those nodal basis functions that are associated with the vertices of K, whereas $\Delta(F)$ is formed by the union of those nodal basis functions that are associated with the boundary points of F. Furthermore, the length of the edge F is denoted by $h_F := |F|$.

Lemma 4.3 *Let a shape-regular family $(\mathcal{T}_h)$ of triangulations of the domain Ω be given. Then for any $v \in V$ there exists an element $Q_h v \in V_h$ such that for all triangles $K \in \mathcal{T}_h$ and all edges $F \in \mathcal{F}_h$ the following estimates are valid:*

$$\begin{aligned} \|v - Q_h v\|_{0,K} &\le C h_K |v|_{1,\Delta(K)}, \\ \|v - Q_h v\|_{0,F} &\le C\sqrt{h_F}\, |v|_{1,\Delta(F)}, \end{aligned}$$

where the constant $C > 0$ depends only on the family of triangulations.

Now, setting $v_h = Q_h v$ in (4.12), the discrete Cauchy–Schwarz inequality yields

$$\begin{aligned} a(u - u_h, v) \le{}& C \sum_{K\in\mathcal{T}_h} h_K \|r_K(u_h)\|_{0,K} |v|_{1,\Delta(K)} \\ &+ C \sum_{F\in\mathcal{F}_h} \sqrt{h_F} \big\| [\![\nabla u_h]\!] \big\|_{0,F} |v|_{1,\Delta(F)} \\ \le{}& C \Big(\sum_{K\in\mathcal{T}_h} h_K^2 \|r_K(u_h)\|_{0,K}^2 \Big)^{1/2} \Big(\sum_{K\in\mathcal{T}_h} |v|_{1,\Delta(K)}^2 \Big)^{1/2} \\ &+ C \Big(\sum_{F\in\mathcal{F}_h} h_F \big\| [\![\nabla u_h]\!] \big\|_{0,F}^2 \Big)^{1/2} \Big(\sum_{F\in\mathcal{F}_h} |v|_{1,\Delta(F)}^2 \Big)^{1/2} . \end{aligned}$$

A detailed investigation of the two second factors shows that we can decompose the integrals over $\Delta(K)$, $\Delta(F)$, according to

$$\int_{\Delta(K)} \cdots = \sum_{K'\subset\Delta(K)} \int_{K'} \cdots, \qquad \int_{\Delta(F)} \cdots = \sum_{K'\subset\Delta(F)} \int_{K'} \cdots .$$

This leads to a repeated summation of the integrals over the single elements K. However, due to the shape regularity of the family of triangulations, the multiplicity of these summations is bounded independently of the particular triangulation (see (3.111)). So we arrive at the estimates

$$\sum_{K\in\mathcal{T}_h} |v|_{1,\Delta(K)}^2 \le C|v|_1^2, \qquad \sum_{F\in\mathcal{F}_h} |v|_{1,\Delta(F)}^2 \le C|v|_1^2 . \tag{4.15}$$

Using the inequality $a + b \le \sqrt{2(a^2 + b^2)}$ for $a, b \in \mathbb{R}$, we get

$$\begin{aligned} &a(u - u_h, v) \\ &\le C \Big(\sum_{K\in\mathcal{T}_h} h_K^2 \|r_K(u_h)\|_{0,K}^2 + \sum_{F\in\mathcal{F}_h} h_F \| [\![\nabla u_h]\!]_F \|_{0,F}^2 \Big)^{1/2} |v|_1 . \end{aligned}$$

Finally, (4.9) yields

$$\|u - u_h\|_1 \leq D\eta \qquad \text{with} \quad \eta^2 := \sum_{K \in \mathcal{T}_h} \eta_K^2 \tag{4.16}$$

and

$$\eta_K^2 := h_K^2 \|f - ru_h\|_{0,K}^2 + \frac{1}{2} \sum_{F \subset \partial K \setminus \partial\Omega} h_F \left\| [\![\nabla u_h]\!] \right\|_{0,F}^2 . \tag{4.17}$$

Here we have taken into account that in the transformation of the edge sum

$$\sum_{F \in \mathcal{F}_h} \cdots \qquad \text{into the double sum} \qquad \sum_{K \in \mathcal{T}_h} \sum_{F \subset \partial K \setminus \partial\Omega} \cdots$$

the latter sums up every interior edge twice.

In summary, we have obtained an a posteriori error estimate of the form (4.3). Unfortunately, the estimate (4.16) may considerably overestimate the quantity $\|u - u_h\|_1$ in many cases, for instance, if the two constants in the estimates from (4.15) significantly differ. Such overestimations were actually observed in numerical tests, e. g., [115]. Regardless, in practice the quantities η_K are often used, but not to calculate a guaranteed upper bound of the global error, but as *error indicators* that give a feeling for the distribution of the local errors in Ω.

By means of refined arguments it is also possible to derive lower bounds for $\|u - u_h\|_1$. For details, we refer to the literature, for example, [63].

Error Estimation by Gradient Recovery

If we are interested in an estimate of the error $u - u_h \in V = H_0^1(\Omega)$ measured in the H^1- or energy norm $\|\cdot\|$, this problem can be simplified by means of the fact that both norms are equivalent on V to the H^1-seminorm

$$|u - u_h|_1 = \left(\int_\Omega |\nabla u - \nabla u_h|^2 \, dx \right)^{1/2} =: \|\nabla u - \nabla u_h\|_0 .$$

This is a simple consequence of the definitions and the Poincaré inequality (see Theorem 2.18). That is, there exist constants $C_1, C_2 > 0$ independent of h such that

$$C_1 |u - u_h|_1 \leq \|u - u_h\| \leq C_2 |u - u_h|_1 \tag{4.18}$$

(cf. Problem 4.7). Consequently, ∇u remains the only unknown quantity in the error bound.

The idea of error estimation by means of gradient recovery is to replace the unknown gradient of the weak solution u by a suitable quantity $R_h u_h$ that is computable from the approximative solution u_h at moderate expense. A popular example of such a technique is the so-called Z^2*estimate*. Here we will describe a simple version of it. Further applications can be found in the original papers by Zienkiewicz and Zhu, e.g., [232].

Similar to the notation introduced in the preceding subsection, for a given node a the set

$$\Delta(a) := \bigcup_{K' :\, a \in \partial K'} K' \tag{4.19}$$

denotes the triangular neighbourhood of a (see Figure 4.7).

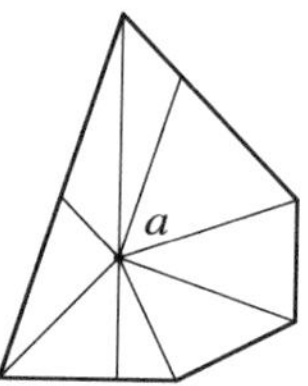

Fig. 4.7: The triangular neighbourhood $\Delta(a)$.

This set coincides with the support of the piecewise linear, continuous basis function associated with that node.

The gradient ∇u_h of a piecewise linear, continuous finite element function u_h is constant on every triangle K. This suggests that at any node a of the triangulation $\mathcal{T}_h$ we define the average $R_h u_h(a)$ of the values of the gradients on those triangles sharing the vertex a:

$$R_h u_h(a) := \frac{1}{|\Delta(a)|} \sum_{K \subset \Delta(a)} \nabla u_h|_K \, |K| \, .$$

Interpolating the two components of these nodal values of $R_h u_h$ separately in V_h, we get a *recovery operator* $R_h : V_h \to V_h \times V_h$.

Now a local error estimator can be defined by the simple restriction of the quantity $\overline{\eta} := \|R_h u_h - \nabla u_h\|_0$ onto a single element K:

$$\overline{\eta}_K := \|R_h u_h - \nabla u_h\|_{0,K} \, .$$

A nice insight into the properties of this local estimator was given by Rodríguez ([211], see also [116]), who compared it with the corresponding residual estimator (4.17). Namely, neglecting in the residual estimator just the residual part, i.e., setting

$$\tilde{\eta}_K^2 := \frac{1}{2} \sum_{F \subset \partial K \setminus \partial\Omega} h_F \big\| [\![\nabla u_h]\!] \big\|_{0,F}^2 \qquad \text{and} \quad \tilde{\eta}^2 := \sum_{K \in \mathcal{T}_h} \tilde{\eta}_K^2 \, ,$$

then the following result is true:

Theorem 4.4 *There exist two constants $c_1, c_2 > 0$ depending only on the family of triangulations such that*

$$c_1\tilde{\eta} \le \overline{\eta} \le c_2\tilde{\eta}\,.$$

The motivation for the method of gradient recovery is to be seen in the fact that $R_h u_h$ possesses special convergence properties. Namely, under certain assumptions the recovered gradient $R_h u_h$ converges asymptotically to ∇u faster than ∇u_h does. In such a case $R_h u_h$ is said to be a *superconvergent approximation* to ∇u.

If superconvergence holds, the simple decomposition

$$\nabla u - \nabla u_h = R_h u_h - \nabla u_h + \nabla u - R_h u_h$$

demonstrates that the first difference on the right-hand side represents the asymptotically dominating, computable part of the gradient error $\nabla u - \nabla u_h$. In other words, if we could define, for the class of problems under consideration, a superconvergent gradient recovery $R_h u_h$ that is computable with moderate expense, then the quantities $\overline{\eta}_K$ and $\overline{\eta}$ defined above may serve as a tool for a posteriori error estimation.

Although such superconvergence properties are valid only under rather restrictive assumptions (especially with respect to the grid and to the regularity of the weak solution), it can theoretically and numerically be justified (see, e.g., [116]) that averaging techniques are far more robust with respect to local grid asymmetry and lack of superconvergence.

The following example, which is due to Repin [206], shows that a recovered gradient does not necessarily reflect the actual behaviour of the error.

Example 4.5 *Consider the following boundary value problem for $d = 1$ and $\Omega = (0, 1)$:*

$$-u'' = f \quad \text{in } \Omega, \quad u(0) = u(1) - 1 = 0\,.$$

If $f \neq 0$ is constant, the exact solution reads $u(x) = x(2 + (1 - x)f)/2$. Suppose we have found the function $v_h(x) = x$ as an approximate solution. For an arbitrary partition of Ω into subintervals, this function is piecewise linear and it satisfies the boundary conditions formulated above. Now let R_h be an arbitrary gradient recovery operator that is able to reproduce at least constants. Since $v_h' = 1$, we have $v_h' - R_h v_h = 0$, whereas the real error is $v_h' - u' = (x - \frac{1}{2})f$.

An interpretation of this effect is that the function v_h from the example does not satisfy the corresponding discrete (Galerkin) equations. But such a property is used for the proof of superconvergence. This property also plays an important role in the derivation of the residual error estimates, because the error equation is used therein.

Dual-Weighted Residual Error Estimators

The aforementioned a posteriori error estimates have two disadvantages: On the one hand, certain global constants, which are not known in general, are part of the bounds. Typical examples are α^{-1} in (4.9) and the constants C_1, C_2 in the equivalence relation (4.18). On the other hand, we obtained scaling factors like h_K and $\sqrt{h_F}$ simply by using a particular approximation operator.

In the following, we will outline a method that attempts to circumvent these drawbacks. It is especially appropriate for the estimation of errors of functionals depending linearly on the solution.

So let $J : V \to \mathbb{R}$ denote a linear, continuous functional. We are interested in an estimate of $|J(u) - J(u_h)|$. Therefore, the following auxiliary *dual* problem is considered:

$$\text{Find } w \in V \text{ such that } a(v, w) = J(v) \text{ for all } v \in V.$$

Taking $v = u - u_h$, we get immediately

$$J(u) - J(u_h) = J(u - u_h) = a(u - u_h, w).$$

If $w_h \in V_h$ is an arbitrary element, the error equation (2.43) yields

$$J(u) - J(u_h) = a(u - u_h, w - w_h).$$

Obviously, the right-hand side is of the same structure as in the derivation of the estimate (4.12). Consequently, by using the same arguments it follows that

$$\begin{aligned} |J(u) - J(u_h)| \le & \sum_{K \in \mathcal{T}_h} \|r_K(u_h)\|_{0,K} \|w - w_h\|_{0,K} \\ & + \sum_{F \in \mathcal{F}_h} \big\| [\![\nabla u_h]\!] \big\|_{0,F} \|w - w_h\|_{0,F} . \end{aligned}$$

In contrast to the previous approaches, here the norms of $w - w_h$ will not be theoretically analysed but numerically approximated. This can be done by an approximation of the dual solution w. There are several (more or less heuristic) ways to do this.

1) Estimation of the approximation error: Here, the norms of $w - w_h$ are estimated as in the case of residual error estimators. Since the result depends on the unknown H^1-seminorm of w, which is equivalent to the L^2-norm of ∇w, the finite element solution $w_h \in V_h$ of the auxiliary problem is used to approximate ∇w. It is a great disadvantage of this approach that again global constants enter in the final estimate through the estimation of the approximation error. Furthermore, the discrete auxiliary problem is of similar complexity to that of the original discrete problem.
2) Higher order discretizations of the auxiliary problem: The auxiliary problem is solved numerically by using a method that is more accurate than the original method to determine a solution in V_h. Then w is replaced by that solution and $w_h \in V_h$ by an interpolant of that solution. Unfortunately, since the discrete auxiliary problem is of comparatively large dimension, this approach is rather expensive.
3) Approximation by means of higher order recovery: This method works similarly to the approach described in the previous subsection; w is replaced by an element that is recovered from the finite element solution $w_h \in V_h$ of the auxiliary problem. The recovered element approximates w with higher order in both norms than w_h

does. This method exhibits two problems: On the one hand, the auxiliary problem has to be solved numerically, and on the other hand, ensuring the corresponding superconvergence properties may be difficult.

At the end of this section we want to mention how the method could be used to estimate certain norms of the error. In the case where the norms are induced by particular scalar products, there is a simple, formal way. For example, for the L^2-norm we have

$$\|u - u_h\|_0 = \frac{\langle u - u_h, u - u_h\rangle_0}{\|u - u_h\|_0}\,.$$

Keeping u and u_h fixed, we get with the definition

$$J(v) := \frac{\langle v, u - u_h\rangle_0}{\|u - u_h\|_0}$$

a linear, continuous functional $J : H^1(\Omega) \to \mathbb{R}$ such that $J(u) - J(u_h) = \|u - u_h\|_0$.

The practical difficulty of this approach consists in the fact that to be able to find the solution w of the auxiliary problem we have to know the values of J, but they depend on the unknown element $u - u_h$. The idea of approximating these values immediately implies two problems: There is additional expense, and the influence of the approximation quality on the accuracy of the obtained bounds has to be analysed.

Exercises

Problem 4.4 Let $\Omega \subset \mathbb{R}^2$ be a bounded domain with a polygonal, Lipschitz continuous boundary and $V := H_0^1(\Omega)$. Now consider a V-coercive, continuous bilinear form a and a continuous linear form b. The problem

$$u \in V : \quad a(u, v) = b(v) \quad \text{for all } v \in V$$

is discretized using piecewise linear, continuous finite elements. If E_i denotes the support of the nodal basis functions of V_h associated with the vertex a_i, show that the abstract local error indicators

$$\eta_i := \sup_{v \in H_0^1(E_i)} \frac{a(e, v)}{\|v\|}$$

can be estimated by means of the solutions $e_i \in H_0^1(E_i)$ of the local boundary value problems

$$e_i \in H_0^1(E_i) : \quad a(e_i, v) = b(v) - a(u_h, v) \quad \text{for all } v \in H_0^1(E_i)$$

as follows (M and α denote the constants appearing in the continuity and coercivity conditions on a):

$$\alpha\|e_i\| \le \eta_i \le M\|e_i\|\,.$$

If necessary, the elements of $H_0^1(E_i)$ are extended by zero to the whole domain Ω.

Problem 4.5 A linear polynomial on some triangle is uniquely defined either by its values at the vertices or by its values at the edge midpoints. For a fixed triangulation of a polygonally bounded, simply connected domain $\Omega \subset \mathbb{R}^2$, there can be defined two finite element spaces by identifying common degrees of freedom of adjacent triangles.

a) Show that the dimension of the space defined by the degrees of freedom located at the vertices is less than the dimension of the other space (provided that the triangulation consists of more than one triangle).
b) How can one explain this "loss of degrees of freedom"?

Problem 4.6 Let a shape-regular family of triangulations $(\mathcal{T}_h)$ of a domain $\Omega \subset \mathbb{R}^2$ be given. Show that there exist constants $C > 0$ that depend only on the family $(\mathcal{T}_h)$ such that

$$\sum_{K \in \mathcal{T}_h} |v|^2_{0,\Delta(K)} \leq C\|v\|_0^2 \qquad \text{for all } v \in L^2(\Omega),$$

$$\sum_{F \in \mathcal{F}_h} |v|^2_{0,\Delta(F)} \leq C\|v\|_0^2 \qquad \text{for all } v \in L^2(\Omega).$$

Problem 4.7 Let $\Omega \subset \mathbb{R}^d$ be a bounded domain. Show that there are constants $C_1, C_2 > 0$ such that for all $v \in H_0^1(\Omega)$,

$$C_1|v|_1 \leq \|v\|_1 \leq C_2|v|_1 .$$

Problem 4.8 Formulate the Galerkin equations for the problem in Example 4.5 and show that the function $v_h(x) = x$ does not satisfy them except that $f = 0$.

4.3 Convergence of Adaptive Methods

In the previous subsection we have described examples of computable error estimators, depending on the discrete solution and the data, that can be used to evaluate the accuracy of the discrete solution. These quantities were used over a quite long period (about 25 years) in rather heuristic algorithms to balance the accuracy and the computational effort.

For instance, let us assume that the local error estimators η_K composing an efficient error estimator η for an approximate solution u_h on some grid $\mathcal{T}_h$ really reflect the error on the element K and that this local error can be improved by a refinement of K (e.g., following the principles of Section 4.1.5). Then the following grid adaptation strategies can be applied until the given tolerance ε is reached or the computer resources are exhausted.

Equidistribution strategy: The objective of the grid adaptation (refinement of elements or de-refinement of element patches) is to get a new grid $\mathcal{T}_h^{\text{new}}$ such that the local error estimators η_K^{new} for this new grid take one and the same value for all elements $K \in \mathcal{T}_h^{\text{new}}$; that is, (cf. (4.5))

$$\eta_K^{\text{new}} \approx \frac{\varepsilon}{\sqrt{|\mathcal{T}_h^{\text{new}}|}} \quad \text{for all} \quad K \in \mathcal{T}_h^{\text{new}} .$$

Since the number of elements of the new grid enters the right-hand side of this criterion, the strategy is an implicit method. In practical use, it is approximated iteratively.

Cut-off strategy: Given a parameter $\kappa \in (0, 1)$, a threshold value $\kappa\eta$ is defined. Then the elements K with $\eta_K > \kappa\eta$ will be refined.

Reduction strategy: Given a parameter $\kappa \in (0, 1)$, an auxiliary tolerance $\varepsilon_\eta := \kappa\eta$ is defined. Then a couple of steps following the equidistribution strategy with the tolerance ε_η are performed.

Among all three strategies, the reduction strategy has demonstrated to be a good compromise in many situations. However, it is not clear right off that the resulting sequence of discrete solutions converges to the weak solution u. Despite their practical success, it has been shown only in the past years that adaptive methods of this kind converge and even possess certain optimality properties. We therefore want to outline these considerations in what follows.

In this subsection, based on the paper [118], the main aspects of adaptive finite element methods (AFEM) will be explained, whereby we will restrict ourselves to the diffusion–reaction problem with Dirichlet boundary conditions (4.6) as in the previous subsection.

We start with the observation that, in adaptive methods, the previously used concept of convergence, in which a family of partitions $(\mathcal{T}_h)_h$ with $h \to 0$ as in Section 3.4 was considered, is not appropriate. Already in the work [87, Rem. 6.1] it was emphasized that the convergence of finite element solutions can be achieved without the assumption that the maximum size h of the partition elements approaches 0. Therefore we consider a shape-regular family (or sequence) of partitions $(\mathcal{T}_k)_{k\in\mathbb{N}_0}$ generated by the adaptive method, where the corresponding grid parameters $h_k := \max\left\{\operatorname{diam}(K) \,\middle|\, K \in \mathcal{T}_k\right\}$ do not necessarily tend to 0. Typically $(\mathcal{T}_k)_{k\in\mathbb{N}_0}$ consists of consecutive partitions such that $\mathcal{T}_{k+1}$ is a nested refinement of $\mathcal{T}_k$ (cf. Def. 4.1).

The discussion of convergence requires some generalizations in the notation. For a general partition $\mathcal{T}$ in the sense of Def. 3.20, we define the finite element space $V_\mathcal{T}$ in analogy to (3.74). For simplicity, we assume that $V_\mathcal{T} \subset V$, and the space $V_\mathcal{T}$ is equipped with the induced norm $\|\cdot\| := \|\cdot\|_V$ of V. By $u_\mathcal{T} \in V_\mathcal{T}$ we denote the solution of the finite-dimensional variational equation (4.8). Finally, in order to clearly express the dependence on the concrete partition $\mathcal{T}$ and the solution $u_\mathcal{T} \in V_\mathcal{T}$, the global and local error estimators η_K, η in formula (4.5) are replaced by $\eta_{\mathcal{T},K}(u_\mathcal{T})$ and $\eta_\mathcal{T}(u_\mathcal{T})$, respectively.

A general adaptive algorithm has loops of the following form:

$$\text{Solve} \longrightarrow \text{Estimate} \longrightarrow \text{Mark} \longrightarrow \text{Refine/De-refine}.$$

Here we will give a formulation based on local grid refinement only (Table 4.1).

Input: Data of the boundary value problem, tolerance ε, adaptivity parameter $\theta \in (0, 1]$, relative cardinality threshold $C_{\min} \geq 1$.

Core:

1. Construct an initial consistent partition $\mathcal{T}_0$ which resolves the geometry and the data of the problem.
2. For $k = 0, 1, \ldots$ do the following loop:

S: Solve the discrete problem (4.8) corresponding to $\mathcal{T}_k$.
E1: For $K \in \mathcal{T}_k$ do the following loop:
Compute an estimate $\eta_{\mathcal{T}_k,K}(u_{\mathcal{T}_k})$ of the error on K.
E2: Compute the global estimate $\eta_{\mathcal{T}_k}(u_{\mathcal{T}_k}) = \left(\sum_{K\in\mathcal{T}_k} \eta_{\mathcal{T}_k,K}(u_{\mathcal{T}_k})^2 \right)^{1/2}$.
E3: If $\eta_{\mathcal{T}_k}(u_{\mathcal{T}_k}) \leq \varepsilon$ then terminate the loop (desired accuracy attained).
M: Based on $(\eta_{\mathcal{T}_k,K}(u_{\mathcal{T}_k}))_{K\in\mathcal{T}_k}$, select among all subsets $\mathcal{K}_k \subset \mathcal{T}_k$ satisfying the condition

$$\theta\, \eta_{\mathcal{T}_k}(u_{\mathcal{T}_k})^2 \leq \sum_{K\in\mathcal{K}_k} \eta_{\mathcal{T}_k,K}(u_{\mathcal{T}_k})^2 \tag{4.20}$$

a set $\mathcal{M}_k$ such that $|\mathcal{M}_k| \leq C_{\min}|\mathcal{K}_k|$.
R: Based on $\mathcal{M}_k$ determine a consistent refinement $\mathcal{T}_{k+1}$ of $\mathcal{T}_k$.

Output: Approximate solution to the boundary value problem with error less than ε.

Table 4.1: Adaptive algorithm.

Remark 4.6 1) Nowadays the condition (4.20) is often called *Dörfler's marking criterion* [130].
2) The parameter $C_{\min} \geq 1$ controls a possible deviation of the cardinality of the set $\mathcal{M}_k$ from the minimal cardinality of all sets satisfying (4.20).
3) The adaptivity parameter θ allows to vary between only a few marked elements to be refined ($0 < \theta \ll 1$) and nearly all marked elements ($\theta = 1$).

Next we describe a set of axioms that are sufficient for the convergence of the sequence of estimators $(\eta_{\mathcal{T}_k}(u_{\mathcal{T}_k}))_{k\in\mathbb{N}_0}$ computed by means of the above algorithm, provided the step E3 is skipped. In this description, we partially omit the subscript k and consider the discrete solution $u_{\mathcal{T}} \in V_{\mathcal{T}}$ for any given grid $\mathcal{T}$. The constants occurring in the subsequently listed axioms satisfy $C_{\text{stab}}, C_{\text{red}}, C_{\text{rel}} \geq 1$ as well as $0 < \rho_{\text{red}} < 1$ and depend solely on

$$\mathbb{T} := \{\mathcal{T} \mid \mathcal{T} \text{ is a consistent refinement of } \mathcal{T}_0\}.$$

For reasons of space we simplify the original setting of [118] by omitting two of the four axioms because they are easy to satisfy for the model problem (4.6). The omitted axioms are generalizations of an orthogonality identity (see (4.23) later in the proof) and the reliability property (4.3) of the estimator.

(A1) **Stability on non-refined element domains:** For all refinements $\mathcal{T}' \in \mathbb{T}$ of a partition $\mathcal{T} \in \mathbb{T}$, for all subsets $\mathcal{K} \subset \mathcal{T} \cap \mathcal{T}'$ of *non-refined* element domains, and for all corresponding discrete functions $v_{\mathcal{T}'} \in V_{\mathcal{T}'}$ and $v_{\mathcal{T}} \in V_{\mathcal{T}}$ (not necessarily the approximations to the weak solution), it holds that

$$\left| \left(\sum_{K \in \mathcal{K}} \eta^2_{\mathcal{T}',K}(v_{\mathcal{T}'}) \right)^{1/2} - \left(\sum_{K \in \mathcal{K}} \eta^2_{\mathcal{T},K}(v_{\mathcal{T}}) \right)^{1/2} \right| \leq C_{\text{stab}} \|v_{\mathcal{T}'} - v_{\mathcal{T}}\|.$$

(A2) **Reduction property on refined element domains:** Any refinement $\mathcal{T}' \in \mathbb{T}$ of a partition $\mathcal{T} \in \mathbb{T}$ satisfies

$$\sum_{K \in \mathcal{T}' \setminus \mathcal{T}} \eta^2_{\mathcal{T}',K}(u_{\mathcal{T}'}) \leq \rho_{\text{red}} \sum_{K \in \mathcal{T} \setminus \mathcal{T}'} \eta^2_{\mathcal{T},K}(u_{\mathcal{T}}) + C_{\text{red}} \|u_{\mathcal{T}'} - u_{\mathcal{T}}\|^2.$$

The main result of this section states the convergence of the adaptive algorithm in the sense that the sequence of error estimators converges, which implies a convergence result for the corresponding norm of the error $u - u_{\mathcal{T}_k}$, where $u \in V$ is the solution of the variational equation (4.7).

Theorem 4.7 *Suppose (A1) and (A2). Then, for all $\theta \in (0, 1]$ there exist constants $\rho_{\text{conv}} \in (0, 1)$ and $C_{\text{conv}} > 0$ such that the adaptive algorithm (with skipped step E3) generates a sequence of estimators $(\eta_{\mathcal{T}_k}(u_{\mathcal{T}_k}))_{k \in \mathbb{N}_0}$ satisfying*

$$\eta^2_{\mathcal{T}_{k+j}}(u_{\mathcal{T}_{k+j}}) \leq C_{\text{conv}} \rho^j_{\text{conv}} \eta^2_{\mathcal{T}_k}(u_{\mathcal{T}_k}) \quad \textit{for all } k, j \in \mathbb{N}_0.$$

In particular,

$$\|u - u_{\mathcal{T}_k}\| \leq C^{1/2}_{\text{conv}} \rho^{k/2}_{\text{conv}} C_{\text{rel}}\, \eta_{\mathcal{T}_0}(u_{\mathcal{T}_0}) \quad \textit{for all } k \in \mathbb{N}_0.$$

The constants $\rho_{\text{conv}}, C_{\text{conv}}$ depend only on $\rho_{\text{red}}, C_{\text{stab}}, C_{\text{red}}, C_{\text{rel}},$ and θ.

Proof Step 1 (Perturbed contraction of the error estimator): We start from the following elementwise decomposition of $\eta^2_{\mathcal{T}_{k+1}}(u_{\mathcal{T}_{k+1}})$:

$$\eta^2_{\mathcal{T}_{k+1}}(u_{\mathcal{T}_{k+1}}) = \sum_{K \in \mathcal{T}_{k+1} \setminus \mathcal{T}_k} \eta^2_{\mathcal{T}_{k+1},K}(u_{\mathcal{T}_{k+1}}) + \sum_{K \in \mathcal{T}_{k+1} \cap \mathcal{T}_k} \eta^2_{\mathcal{T}_{k+1},K}(u_{\mathcal{T}_{k+1}}). \tag{4.21}$$

Using the elementary inequality $ab \leq \varepsilon a^2 + \varepsilon^{-1} b^2$ for arbitrary $a, b \in \mathbb{R}$ and $\varepsilon > 0$, the second sum can be estimated by means of (A1) as follows:

$$\begin{aligned}
&\sum_{K \in \mathcal{T}_{k+1} \cap \mathcal{T}_k} \eta^2_{\mathcal{T}_{k+1},K}(u_{\mathcal{T}_{k+1}}) \\
&\leq \left(\left(\sum_{K \in \mathcal{T}_{k+1} \cap \mathcal{T}_k} \eta^2_{\mathcal{T}_k,K}(u_{\mathcal{T}_k}) \right)^{1/2} + C_{\text{stab}} \|u_{\mathcal{T}_{k+1}} - u_{\mathcal{T}_k}\| \right)^2 \\
&\leq (1 + \varepsilon) \sum_{K \in \mathcal{T}_{k+1} \cap \mathcal{T}_k} \eta^2_{\mathcal{T}_k,K}(u_{\mathcal{T}_k}) + (1 + \varepsilon^{-1}) C^2_{\text{stab}} \|u_{\mathcal{T}_{k+1}} - u_{\mathcal{T}_k}\|^2.
\end{aligned}$$

Applying (A2) to the first term in (4.21), we obtain

$$\eta^2_{\mathcal{T}_{k+1}}(u_{\mathcal{T}_{k+1}}) \\ \leq \rho_{\text{red}} \sum_{K \in \mathcal{T}_k \setminus \mathcal{T}_{k+1}} \eta^2_{\mathcal{T}_k,K}(u_{\mathcal{T}_k}) + (1+\varepsilon) \sum_{K \in \mathcal{T}_{k+1} \cap \mathcal{T}_k} \eta^2_{\mathcal{T}_k,K}(u_{\mathcal{T}_k}) + C_{\text{est}} \|u_{\mathcal{T}_{k+1}} - u_{\mathcal{T}_k}\|^2,$$

where $C_{\text{est}} := C_{\text{red}} + (1 + \varepsilon^{-1})C^2_{\text{stab}}$. Using a similar decomposition as (4.21) for $\eta^2_{\mathcal{T}_k}(u_{\mathcal{T}_k})$, we see that

$$\eta^2_{\mathcal{T}_{k+1}}(u_{\mathcal{T}_{k+1}}) \\ \leq (1+\varepsilon)\eta^2_{\mathcal{T}_k}(u_{\mathcal{T}_k}) + \big(\rho_{\text{red}} - (1+\varepsilon)\big) \sum_{K \in \mathcal{T}_k \setminus \mathcal{T}_{k+1}} \eta^2_{\mathcal{T}_k,K}(u_{\mathcal{T}_k}) + C_{\text{est}} \|u_{\mathcal{T}_{k+1}} - u_{\mathcal{T}_k}\|^2.$$

Since $\rho_{\text{red}} - (1+\varepsilon) \leq (1+\varepsilon)(\rho_{\text{red}} - 1)$, it follows that

$$\begin{aligned} &\eta^2_{\mathcal{T}_{k+1}}(u_{\mathcal{T}_{k+1}}) \\ &\leq (1+\varepsilon)\left(\eta^2_{\mathcal{T}_k}(u_{\mathcal{T}_k}) - (1-\rho_{\text{red}}) \sum_{K \in \mathcal{T}_k \setminus \mathcal{T}_{k+1}} \eta^2_{\mathcal{T}_k,K}(u_{\mathcal{T}_k})\right) + C_{\text{est}} \|u_{\mathcal{T}_{k+1}} - u_{\mathcal{T}_k}\|^2 \\ &\leq (1+\varepsilon)\left(\eta^2_{\mathcal{T}_k}(u_{\mathcal{T}_k}) - (1-\rho_{\text{red}}) \sum_{K \in \mathcal{M}_k} \eta^2_{\mathcal{T}_k,K}(u_{\mathcal{T}_k})\right) + C_{\text{est}} \|u_{\mathcal{T}_{k+1}} - u_{\mathcal{T}_k}\|^2, \end{aligned}$$

where we have used the inclusion $\mathcal{M}_k \subset \mathcal{T}_k \setminus \mathcal{T}_{k+1}$ in the last step. It remains to apply Dörfler's marking criterion (4.20):

$$\eta^2_{\mathcal{T}_{k+1}}(u_{\mathcal{T}_{k+1}}) \leq \rho_{\text{est}}\eta^2_{\mathcal{T}_k}(u_{\mathcal{T}_k}) + C_{\text{est}} \|u_{\mathcal{T}_{k+1}} - u_{\mathcal{T}_k}\|^2, \tag{4.22}$$

where $\rho_{\text{est}} := (1+\varepsilon)\big(1 - (1-\rho_{\text{red}})\theta\big) > 0$. Selecting a sufficiently small $\varepsilon > 0$, we can also get $\rho_{\text{est}} < 1$.

Step 2 (Uniform summability of the error estimator): It is not difficult to demonstrate (Problem 4.23) that under the conditions of the model problem (4.6) the following orthogonality identity (Pythagoras theorem) holds:

$$\|u - u_{\mathcal{T}_{k+1}}\|^2 + \|u_{\mathcal{T}_{k+1}} - u_{\mathcal{T}_k}\|^2 = \|u - u_{\mathcal{T}_k}\|^2 \quad \text{for all } k \in \mathbb{N}_0. \tag{4.23}$$

The application of this relation to the last term on the right-hand side of (4.22) yields

$$\eta^2_{\mathcal{T}_{k+1}}(u_{\mathcal{T}_{k+1}}) \leq \rho_{\text{est}}\eta^2_{\mathcal{T}_k}(u_{\mathcal{T}_k}) + C_{\text{est}} \left(\|u - u_{\mathcal{T}_k}\|^2 - \|u - u_{\mathcal{T}_{k+1}}\|^2\right).$$

Summing up this estimate from $k+1$ to some $N > k$, we have that

$$\begin{aligned}\sum_{l=k+1}^{N} \eta_{\mathcal{T}_l}^2(u_{\mathcal{T}_l}) &\le \sum_{l=k+1}^{N} \left(\rho_{\text{est}}\eta_{\mathcal{T}_{l-1}}^2(u_{\mathcal{T}_{l-1}}) + C_{\text{est}}\left(\|u-u_{\mathcal{T}_{l-1}}\|^2 - \|u-u_{\mathcal{T}_l}\|^2\right)\right)\\ &= \rho_{\text{est}} \sum_{l=k+1}^{N} \eta_{\mathcal{T}_{l-1}}^2(u_{\mathcal{T}_{l-1}}) + C_{\text{est}}\left(\|u-u_{\mathcal{T}_k}\|^2 - \|u-u_{\mathcal{T}_N}\|^2\right)\\ &\le \rho_{\text{est}} \sum_{l=k+1}^{N} \eta_{\mathcal{T}_{l-1}}^2(u_{\mathcal{T}_{l-1}}) + C_{\text{est}}\|u-u_{\mathcal{T}_k}\|^2.\end{aligned}$$

Now we let $N \to \infty$ and see that

$$\sum_{l=k+1}^{\infty} \eta_{\mathcal{T}_l}^2(u_{\mathcal{T}_l}) \le \rho_{\text{est}} \sum_{l=k+1}^{\infty} \eta_{\mathcal{T}_{l-1}}^2(u_{\mathcal{T}_{l-1}}) + C_{\text{est}}\|u-u_{\mathcal{T}_k}\|^2.$$

The reliability of the error estimator (4.3) implies

$$\begin{aligned}\sum_{l=k+1}^{\infty} \eta_{\mathcal{T}_l}^2(u_{\mathcal{T}_l}) &\le \rho_{\text{est}} \sum_{l=k+1}^{\infty} \eta_{\mathcal{T}_{l-1}}^2(u_{\mathcal{T}_{l-1}}) + C_{\text{est}}C_{\text{rel}}^2\, \eta_{\mathcal{T}_k}(u_{\mathcal{T}_k})^2\\ &\le \rho_{\text{est}} \sum_{l=k+1}^{\infty} \eta_{\mathcal{T}_l}^2(u_{\mathcal{T}_l}) + \left(\rho_{\text{est}} + C_{\text{est}}C_{\text{rel}}^2\right)\eta_{\mathcal{T}_k}(u_{\mathcal{T}_k})^2,\end{aligned}$$

and this results in

$$\sum_{l=k+1}^{\infty} \eta_{\mathcal{T}_l}^2(u_{\mathcal{T}_l}) \le C_{\text{us}}\, \eta_{\mathcal{T}_k}(u_{\mathcal{T}_k})^2 \tag{4.24}$$

with $C_{\text{us}} := \left(\rho_{\text{est}} + C_{\text{est}}C_{\text{rel}}^2\right)/\left(1-\rho_{\text{est}}\right)$.

Step 3 (Uniform R-linear convergence on any level $j \in \mathbb{N}_0$): Dividing the above estimate (4.24) by C_{us} and adding to both sides of the result the term $\sum_{l=k+1}^{\infty} \eta_{\mathcal{T}_l}^2(u_{\mathcal{T}_l})$, we obtain the relation

$$\left(1 + C_{\text{us}}^{-1}\right) \sum_{l=k+1}^{\infty} \eta_{\mathcal{T}_l}^2(u_{\mathcal{T}_l}) \le \sum_{l=k}^{\infty} \eta_{\mathcal{T}_l}^2(u_{\mathcal{T}_l}).$$

Then the repeated application of (4.24) leads to

$$\begin{aligned}\eta_{\mathcal{T}_{k+j}}^2(u_{\mathcal{T}_{k+j}}) &\le \sum_{l=k+j}^{\infty} \eta_{\mathcal{T}_l}^2(u_{\mathcal{T}_l}) \le \left(1 + C_{\text{us}}^{-1}\right)^{-1} \sum_{l=k+j-1}^{\infty} \eta_{\mathcal{T}_l}^2(u_{\mathcal{T}_l})\\ &\le \ldots \le \left(1 + C_{\text{us}}^{-1}\right)^{-j} \sum_{l=k}^{\infty} \eta_{\mathcal{T}_l}^2(u_{\mathcal{T}_l}) \le \left(1 + C_{\text{us}}\right)\left(1 + C_{\text{us}}^{-1}\right)^{-j} \eta_{\mathcal{T}_k}(u_{\mathcal{T}_k})^2.\end{aligned}$$

This proves the first statement of the theorem with $C_{\text{conv}} := 1 + C_{\text{us}}$ and $\rho_{\text{conv}} := \left(1 + C_{\text{us}}^{-1}\right)^{-1}$. The second statement follows from the reliability of the error estimator (4.3) in conjunction with the above estimate, where $k := 0$ and j is replaced by k:

$$\|u-u_{\mathcal{T}_k}\|^2 \le C_{\text{rel}}^2\, \eta_{\mathcal{T}_k}(u_{\mathcal{T}_k}) \le C_{\text{conv}}\rho_{\text{conv}}^k C_{\text{rel}}^2\, \eta_{\mathcal{T}_0}^2(u_{\mathcal{T}_0}).$$

□

In order to complete the discussion of the model problem (4.6), we mention that the verification of the axioms (A1)–(A2) for the newest vertex bisection refinement (see Section 4.1.5) can be found in the proof of [119, Corollary 3.4].

We conclude this chapter with some final comments about further aspects. Most of the contributions to the mathematical understanding of adaptive algorithms have so far been mainly in the context of conforming finite element methods. The theoretically founded development of a posteriori error estimates and adaptive algorithms for finite volume methods (see Chapter 8) and discontinuous Galerkin methods (see Section 7.4) has been slower, with significant progress only in recent years. For orientation, only a few works will be mentioned.

Results about a posteriori error estimates for finite volume methods and adaptive algorithms can be found, e. g., in [76], [80] and [133], [134]. In [158], the current theory of a posteriori error estimation for discontinuous Galerkin methods is reviewed, in particular some results from [117] are improved.

A last remark concerns the error measures with respect to which adaptive methods can be investigated. Especially in the case when lower bounds or questions of optimality are to be included in the investigation, an estimate like (4.4) is difficult to obtain. This led to theories where the so-called *quasi-error* (which is a weighted sum of the error plus the error estimator, e. g. [119]) or the *total error* (which is a weighted sum of the error plus an *oscillation term*, e. g. [187]) are estimated. The fact of unavoidable appearance of oscillation terms in a posteriori error estimates for finite element methods has already been observed in [88]. Data approximation terms, nowadays falling under the concept of oscillation terms, already appeared in [76] in the context of finite volume methods.

Exercise

Problem 4.9 Verify the orthogonality identity (4.23) for the model problem (4.6) under consideration.

Chapter 5
Iterative Methods for Systems of Linear Equations

We consider again the system of linear equations

$$Ax = b \tag{5.1}$$

with nonsingular matrix $A \in \mathbb{R}^{m,m}$, right-hand side $b \in \mathbb{R}^m$, and solution $x \in \mathbb{R}^m$. As shown in Chapters 2 and 3, such systems of equations arise from finite element discretizations of elliptic boundary value problems. The matrix A is the stiffness matrix and thus sparse, as can be seen from (2.41). A *sparse* matrix is vaguely a matrix with so many vanishing elements that using this structure in the solution of (5.1) is advantageous. Taking advantage of a band or hull structure was discussed in Section 2.5. More precisely, if (5.1) represents a finite element discretization, then it is not sufficient to know the properties of the solution method for a fixed m. It is on the contrary necessary to study a sequence of problems with increasing dimension m, as it appears by the refinement of a triangulation. In the strict sense, we understand by the notion *sparse matrices* a sequence of matrices in which the number of nonzero elements per row is bounded independently of the dimension. This is the case for the stiffness matrices due to (2.41) if the underlying sequence of triangulations is shape-regular in the sense of Definition 3.31, for example. In finite element discretizations of time-dependent problems (Chapter 9) as well as in finite volume discretizations (Chapter 8) systems of equations of equal properties arise, so that the following considerations can be also applied there.

The described matrix structure is best applied in iterative methods that have the operation matrix $\times$ vector as an essential module, where either the system matrix A or a matrix of similar structure derived from it is concerned. If the matrix is sparse in the strict sense, then $O(m)$ elementary operations are necessary. In particular, list-oriented storage schemes can be of use, as pointed out in Section 2.5.

The effort for the approximative solution of (5.1) by an iterative method is determined by the number of elementary operations per iteration step and the number of iterations k that are necessary in order to reach the desired *relative error level* $\varepsilon > 0$, i.e., to meet the requirement

P. Knabner and L. Angermann, *Numerical Methods for Elliptic and Parabolic Partial Differential Equations*, Texts in Applied Mathematics 44,
https://doi.org/10.1007/978-3-030-79385-2_5

$$\left\|x^{(k)} - x\right\| \leq \varepsilon \left\|x^{(0)} - x\right\| . \tag{5.2}$$

Here $\left(x^{(k)}\right)_k$ is the sequence of iterates for the initial value $x^{(0)}$, $\|\cdot\|$ a fixed norm in $\mathbb{R}^m$, and $x = A^{-1}b$ the exact solution of (5.1).

For all methods to be discussed we will have *linear convergence* of the kind

$$\left\|x^{(k)} - x\right\| \leq \varrho^k \left\|x^{(0)} - x\right\| \tag{5.3}$$

with a *contraction number* $\varrho \in (0, 1)$, which in general depends on the dimension m. To satisfy (5.2), k iterations are thus sufficient, with

$$k \geq \left(\ln \frac{1}{\varepsilon}\right) \Big/ \left(\ln \frac{1}{\varrho}\right) . \tag{5.4}$$

The computational effort of a method obviously depends on the size of ε, although this will be seen as fixed and only the dependence on the dimension m is considered: often ε will be omitted in the corresponding Landau's symbols. The methods differ therefore by their convergence behaviour, described by the contraction number ϱ and especially by its dependence on m (for specific classes of matrices and boundary value problems). A method is *(asymptotically) optimal* if the contraction numbers are bounded independently of m:

$$\varrho(m) \leq \overline{\varrho} < 1 . \tag{5.5}$$

In this case the total effort for a sparse matrix is $O(m)$ elementary operations, as for a matrix $\times$ vector step. Of course, for a more exact comparison, the corresponding constants, which also reflect the effort of an iteration step, have to be exactly estimated.

While direct methods solve the system of equations (5.1) with machine precision, provided it is solvable in a stable manner, one can freely choose the accuracy with iterative methods. If (5.1) is generated by the discretization of a boundary value problem, it is recommended to solve it only with that accuracy with which (5.1) approximates the boundary value problem. Asymptotic statements hereto have, among others, been developed in (3.102), (9.199) and give an estimation of the approximation error by Ch^α, with constants $C, \alpha > 0$, whereby h is the grid size of the corresponding triangulation. Since the constants in these estimates are usually unknown, the error level can be adapted only asymptotically in m, in order to gain an *algorithmic error* of equal asymptotics compared to the error of approximation. Although this contradicts the above-described point of view of a constant error level, it does not alter anything in the comparison of the methods: The respective effort always has to be multiplied by a factor $O(\ln m)$ if in d space dimensions $m \sim h^{-d}$ is valid, and the relations between the methods compared remain the same.

Furthermore, the choice of the error level ε will be influenced by the quality of the initial iterate. Generally, statements about the initial iterate are only possible for special situations: For parabolic initial boundary value problems (Chapter 9) and a one-step time discretization, it is recommended to use the approximation of the old time level as initial iterate. In the case of a hierarchy of space discretizations, a *nested iteration* is possible (Section 5.6), where the initial iterates will naturally result.

5.1 Linear Stationary Iterative Methods

5.1.1 General Theory

We begin with the study of the following class of affine-linear iteration functions,

$$\Phi(x) := Mx + Nb, \tag{5.6}$$

with matrices $M, N \in \mathbb{R}^{m,m}$ to be specified later. By means of Φ an iteration sequence $x^{(0)}, x^{(1)}, x^{(2)}, \ldots$ is defined through a *fixed-point iteration*

$$x^{(k+1)} := \Phi\big(x^{(k)}\big), \quad k = 0, 1, \ldots, \tag{5.7}$$

from an initial approximation $x^{(0)}$. Methods of this kind are called *linear stationary*, because of their form (5.6) with a fixed *iteration matrix* M. The function $\Phi : \mathbb{R}^m \to \mathbb{R}^m$ is continuous, so that in case of convergence of $x^{(k)}$ for $k \to \infty$, for the limit x we have

$$x = \Phi(x) = Mx + Nb .$$

In order to achieve that the fixed-point iteration defined by (5.6) is *consistent* with $Ax = b$, i.e., each solution of (5.1) is also a *fixed point*, we must require

$$A^{-1}b = MA^{-1}b + Nb \quad \text{for arbitrary } b \in \mathbb{R}^m ,$$

i.e., $A^{-1} = MA^{-1} + N$, and thus

$$I = M + NA . \tag{5.8}$$

On the other hand, if N is nonsingular, which will always be the case in the following, then (5.8) also implies that a fixed point of (5.6) solves the system of equations.

Assuming the validity of (5.8), the fixed-point iteration for (5.6) can also be written as

$$x^{(k+1)} = x^{(k)} - N\big(Ax^{(k)} - b\big), \tag{5.9}$$

because

$$Mx^{(k)} + Nb = (I - NA)\, x^{(k)} + Nb .$$

If N is nonsingular, we have additionally an equivalent form given by

$$W\big(x^{(k+1)} - x^{(k)}\big) = -\big(Ax^{(k)} - b\big) \tag{5.10}$$

with $W := N^{-1}$. The *correction* $x^{(k+1)} - x^{(k)}$ for $x^{(k)}$ is given by the *residual*

$$g^{(k)} := Ax^{(k)} - b$$

through (5.9) or (5.10), possibly by solving a system of equations. In order to compete with the direct method, the solution of (5.10) should require one order in m fewer elementary operations. For dense matrices no more operations than $O(m^2)$ should be necessary as are already necessary for the calculation of $g^{(k)}$. The same holds for sparse matrices, for example, band matrices. On the other side the method should converge, and that as quickly as possible.

In the form (5.6) Φ is Lipschitz continuous for a given norm $\|\cdot\|$ on $\mathbb{R}^m$ with Lipschitz constant $\|M\|$, where $\|\cdot\|$ is a norm on $\mathbb{R}^{m,m}$ that is consistent with the vector norm (see (A3.9)).

More precisely, for a consistent iteration the *error*

$$e^{(k)} := x^{(k)} - x\,,$$

with $x = A^{-1}b$ still denoting the exact solution, even satisfies

$$e^{(k+1)} = Me^{(k)}\,,$$

because (5.7) and (5.8) imply

$$e^{(k+1)} = x^{(k+1)} - x = Mx^{(k)} + Nb - Mx - NAx = Me^{(k)}\,. \tag{5.11}$$

The *spectral radius of* M, that is, the maximum of the absolute values of the (complex) eigenvalues of M, will be denoted by $\varrho(M)$.

The following general convergence theorem holds:

Theorem 5.1 *A fixed-point iteration given by* (5.6) *to solve* $Ax = b$ *is globally and linearly convergent if*

$$\varrho(M) < 1\,. \tag{5.12}$$

This is satisfied if for a matrix norm $\|\cdot\|$ *on* $\mathbb{R}^{m,m}$ *induced by a norm* $\|\cdot\|$ *on* $\mathbb{R}^m$ *we have*

$$\|M\| < 1\,. \tag{5.13}$$

If the consistency condition (5.8) *holds and the matrix and vector norms applied are consistent, then the convergence is monotone in the following sense:*

$$\left\|e^{(k+1)}\right\| \le \|M\|\left\|e^{(k)}\right\|. \tag{5.14}$$

Proof Assuming (5.12), then for $\varepsilon = (1 - \varrho(M))/2 > 0$ there is a norm $\|\cdot\|_S$ on $\mathbb{R}^m$ such that the induced norm $\|\cdot\|_S$ on $\mathbb{R}^{m,m}$ satisfies

$$\|M\|_S \le \varrho(M) + \varepsilon < 1$$

(see [33, Lemma B.26]). The function Φ is a contraction with respect to this special norm on $\mathbb{R}^m$. Therefore, Banach's fixed-point theorem (Theorem 11.4) can be applied on $X = (\mathbb{R}^m, \|\cdot\|_S)$, which ensures the global convergence of the sequence $\left(x^{(k)}\right)_k$ to a fixed point $\bar{x}$ of Φ.

If (5.13) holds, Φ is a contraction even with respect to the norm $\|\cdot\|$ on $\mathbb{R}^m$, and $\|M\|$ is the Lipschitz constant. Finally relation (5.14) follows from (5.11). □

In any case, we have convergence in any norm on $\mathbb{R}^m$, since they are all equivalent. Linear convergence for (5.12) holds only in the generally not available norm $\|\cdot\|_S$ with $\|M\|_S$ as contraction number.

As termination criterion for the concrete iterative methods to be introduced, often

$$\left\|g^{(k)}\right\| \leq \delta \left\|g^{(0)}\right\| \tag{5.15}$$

is used with a control parameter $\delta > 0$, abbreviated as $\|g^{(k)}\| = 0$. The connection to the desired reduction of the relative error according to (5.2) is given by

$$\frac{\left\|e^{(k)}\right\|}{\left\|e^{(0)}\right\|} \leq \kappa(A)\,\frac{\left\|g^{(k)}\right\|}{\left\|g^{(0)}\right\|}, \tag{5.16}$$

where the *condition number* $\kappa(A) := \|A\|\|A^{-1}\|$ is to be computed with respect to a matrix norm that is consistent with the chosen vector norm. Relation (5.16) follows from

$$\begin{aligned}
\left\|e^{(k)}\right\| &= \left\|A^{-1}g^{(k)}\right\| \leq \left\|A^{-1}\right\|\left\|g^{(k)}\right\|, \\
\left\|g^{(0)}\right\| &= \left\|Ae^{(0)}\right\| \leq \left\|A\right\|\left\|e^{(0)}\right\|.
\end{aligned}$$

Therefore, for the selection of δ in (5.15), we have to take into account the behaviour of the condition number.

For the iteration matrix M, according to (5.8), we have

$$M = I - NA,$$

or according to (5.10) with nonsingular W,

$$M = I - W^{-1}A.$$

To improve the convergence, i.e., to reduce $\varrho(M)$ (or $\|M\|$), we need

$$N \approx A^{-1} \text{ and } W \approx A,$$

which is in contradiction to the fast solvability of (5.10).

5.1.2 Classical Methods

The fast solvability of (5.10) (in $O(m)$ operations) is ensured by choosing

$$W := D, \tag{5.17}$$

where $A = L + D + R$ is the unique partition of A, with a strictly lower triangular matrix L, a strictly upper triangular matrix R, and the diagonal matrix D:

$$L := \begin{pmatrix} 0 & \cdots & \cdots & 0 \\ a_{2,1} & 0 & \cdots & 0 \\ \vdots & \ddots & \ddots & \vdots \\ a_{m,1} & \cdots & a_{m,m-1} & 0 \end{pmatrix}, \quad R := \begin{pmatrix} 0 & a_{1,2} & \cdots & a_{1,m} \\ \vdots & \ddots & \ddots & \vdots \\ 0 & \cdots & 0 & a_{m-1,m} \\ 0 & \cdots & \cdots & 0 \end{pmatrix}, \tag{5.18}$$

$$D := \begin{pmatrix} a_{11} & & & \\ & a_{22} & & \mathbf{0} \\ \mathbf{0} & & \ddots & \\ & & & a_{mm} \end{pmatrix}.$$

Assume $a_{ii} \neq 0$ for all $i = 1, \ldots, m$, or equivalently that D is nonsingular, which can be achieved by row and column permutation.

The choice of (5.17) is called the *method of simultaneous displacements* or *Jacobi's method*. In the formulation form (5.6), we have

$$\begin{aligned} N &= D^{-1}, \\ M_{\mathrm{J}} &= I - NA = I - D^{-1}A = -D^{-1}(L + R). \end{aligned}$$

Therefore, the iteration can be written as

$$D\big(x^{(k+1)} - x^{(k)}\big) = -\big(Ax^{(k)} - b\big)$$

or

$$x^{(k+1)} = D^{-1}\big(- Lx^{(k)} - Rx^{(k)} + b\big) \tag{5.19}$$

or

$$x_i^{(k+1)} = \frac{1}{a_{ii}}\left(- \sum_{j=1}^{i-1} a_{ij}x_j^{(k)} - \sum_{j=i+1}^{m} a_{ij}x_j^{(k)} + b_i \right) \text{ for all } i = 1, \ldots, m.$$

On the right side in the first sum it is reasonable to use the new iterate $x^{(k+1)}$ where it is already calculated. This leads us to the iteration

$$x^{(k+1)} = D^{-1}\big(- Lx^{(k+1)} - Rx^{(k)} + b\big) \tag{5.20}$$

or

$$(D + L)\, x^{(k+1)} = -Rx^{(k)} + b$$

or

$$(D + L)\, \big(x^{(k+1)} - x^{(k)}\big) = -\big(Ax^{(k)} - b\big), \tag{5.21}$$

the so-called *method of successive displacements* or *Gauss–Seidel method*. According to (5.21), we have here a consistent iteration with

$$W = D + L\,.$$

Since D is nonsingular, W is nonsingular. Written in the form (5.6) the method is defined by

$$N = W^{-1} = (D + L)^{-1}\,,$$
$$M_{\mathrm{GS}} = I - NA = I - (D + L)^{-1} A = -(D + L)^{-1} R\,.$$

In contrast to the Jacobi iteration, the Gauss–Seidel iteration depends on the order of the equations. However, the derivation (5.20) shows that the number of operations per iteration step is equal,

Jacobi becomes Gauss–Seidel,

if $x^{(k+1)}$ is stored on the same vector as $x^{(k)}$.

A sufficient convergence condition is given by the following theorem:

Theorem 5.2 *Jacobi's method and the Gauss–Seidel method converge globally and monotonically with respect to* $\|\cdot\|_\infty$ *if the* strict row sum criterion

$$\sum_{\substack{j=1\\ j\neq i}}^{m} |a_{ij}| < |a_{ii}| \quad \textit{for all } i = 1, \ldots, m \tag{5.22}$$

is satisfied.

Proof The proof here is given only for the Jacobi iteration. For the other method see, for example, [33].

The inequality (5.22) is equivalent to $\|M_{\mathrm{J}}\|_\infty < 1$ because of $M_{\mathrm{J}} = -D^{-1}(L + R)$ if $\|\cdot\|_\infty$ is the matrix norm that is induced by $\|\cdot\|_\infty$, which means the maximum-row-sum norm (see (A3.6)). □

It can be shown that the Gauss–Seidel method converges "better" than Jacobi's method, as expected: Under the assumption of (5.22) for the respective iteration matrices,

$$\|M_{\mathrm{GS}}\|_\infty \leq \|M_{\mathrm{J}}\|_\infty < 1$$

(see, for example, [33]).

Theorem 5.3 *If A is symmetric and positive definite, then the Gauss–Seidel method converges globally. The convergence is monotone in the energy norm $\|\cdot\|_A$, where $\|x\|_A := \left(x^T A x\right)^{1/2}$ for $x \in \mathbb{R}^m$.*

Proof See [33, Thm. 3.39]. □

If the differential operator, and therefore the bilinear form, is symmetric, that is, if (3.13) holds with $c = 0$, then Theorem 5.3 can be applied. Concerning the applicability of Theorem 5.2, even for the Poisson equation with Dirichlet boundary conditions (1.1), (1.2) requirements for the finite element discretization are necessary in order to satisfy at least a weaker version of (5.22). This example then satisfies the *weak* row sum criterion only in the following sense:

$$\begin{aligned} &\sum_{\substack{j=1\\j\neq i}}^{m} |a_{ij}| \leq |a_{ii}| \quad \text{for all } i = 1, \ldots, m\,; \\ &\text{“} < \text{” holds for at least one } i \in \{1, \ldots, m\}\,. \end{aligned} \tag{5.23}$$

In the case of the finite difference method (1.7) for the rectangular domain or the finite element method from Section 2.2, which leads to the same discretization matrix, (5.23) is satisfied. For a general triangulation with linear ansatz functions, conditions for the angles of the elements must be required (see the angle condition in Section 3.9). The condition (5.23) is also sufficient, if A is irreducible (see Appendix A.3).

Theorem 5.4 *If A satisfies the condition* (5.23) *and is irreducible, then Jacobi's method converges globally.*

Proof See [54, Thm. 4.9]. □

The qualitative statement of convergence does not say anything about the usefulness of Jacobi's and the Gauss–Seidel method for finite element discretizations. As an example we consider the Dirichlet problem for the Poisson equation on a rectangular domain as in (1.5), with the five-point stencil discretization introduced in Section 1.2. We restrict ourselves to an equal number of nodes in both space directions for simplicity of the notation. This number is denoted by $n+1$, differently than in Chapter 1. Therefore, $A \in \mathbb{R}^{m,m}$ according to (1.14), with $m = (n-1)^2$ being the number of interior nodes. The factor h^{-2} can be omitted by multiplying the equation by h^2.

In the above example the eigenvalues and therefore the spectral radius can be calculated explicitly. Due to $D = 4I$ we have for Jacobi's method

$$M = -\frac{1}{4}(A - 4I) = I - \frac{1}{4}A\,,$$

and therefore A and M have the same eigenvectors, namely,

$$\left(z^{k,l}\right)_{ij} = \sin\frac{ik\pi}{n}\,\sin\frac{jl\pi}{n}\,, \qquad 1 \leq i, j, k, l \leq n-1\,,$$

with the eigenvalues

$$2\left(2 - \cos\frac{k\pi}{n} - \cos\frac{l\pi}{n}\right) \tag{5.24}$$

for A and

$$\frac{1}{2}\cos\frac{k\pi}{n} + \frac{1}{2}\cos\frac{l\pi}{n} \tag{5.25}$$

for M with $1 \le k,\ l \le n-1$. This can be proven directly with the help of trigonometric identities (see, for example, [32, Sect. 4.4]). Thus we have

$$\varrho(M) = -\cos\frac{(n-1)\pi}{n} = \cos\frac{\pi}{n} = 1 - \frac{\pi^2}{2n^2} + O\left(n^{-4}\right). \tag{5.26}$$

With growing n the rate of convergence becomes worse. The effort to gain an approximative solution, which means to reduce the error level below a given threshold ε, is proportional to the number of iterations $\times$ operations for an iteration, as we discussed at the beginning of this chapter. Due to (5.4) and (5.12), the number of necessary operations is calculated as follows:

$$\frac{\ln(1/\varepsilon)}{-\ln(\varrho(M))} \cdot O(m) = \ln\frac{1}{\varepsilon} \cdot O\left(n^2\right) \cdot O(m) = \ln\frac{1}{\varepsilon}\, O(m^2)\,.$$

Here the well-known expansion $\ln(1+x) = x + O(x^2)$ is employed in the determination of the leading term of $-1/(\ln(\varrho(M))$. An analogous result with better constants holds for the Gauss–Seidel method.

In comparison to this, the elimination or the Cholesky method requires

$$O\left(\text{band-width}^2 \cdot m\right) = O(m^2)$$

operations; i.e., they are of the same complexity. Therefore, both methods are of use for only moderately large m.

An iterative method has a superior complexity to that of the Cholesky method if

$$\varrho(M) = 1 - O(n^{-\alpha}) \tag{5.27}$$

with $\alpha < 2$. In the ideal case (5.5) holds; then the method needs $O(m)$ operations, which is asymptotically optimal.

In the following we will present a sequence of methods with increasingly better convergence properties for systems of equations that arise from finite element discretizations.

The simplest iteration is the *Richardson method*, defined by

$$M = I - A, \quad \text{i.e.,} \quad N = W = I\,. \tag{5.28}$$

For this method we have

$$\varrho(M) = \max\left\{|1 - \lambda_{\max}(A)|, |1 - \lambda_{\min}(A)|\right\},$$

with $\lambda_{\max}(A)$ and $\lambda_{\min}(A)$ being the largest and smallest eigenvalues of A, respectively. Therefore, this method is convergent for special matrices only. In the case of a nonsingular D, the Richardson method for the transformed system of equations

$$D^{-1}Ax = D^{-1}b$$

is equivalent to Jacobi's method.

More generally, the following can be shown: If a consistent method is defined by M, N with $I = M + NA$, and N nonsingular, then it is equivalent to the Richardson method applied to

$$NAx = Nb\,. \tag{5.29}$$

The Richardson method for (5.29) has the form

$$x^{(k+1)} - x^{(k)} = -\tilde{N}\left(NAx^{(k)} - Nb\right)$$

with $\tilde{N} = I$, which means the form (5.9), and vice versa.

Equation (5.29) can also be interpreted as a *preconditioning* of the system of equations (5.1), with the aim to reduce the spectral condition number $\kappa(A)$ of the system matrix, since this is essential for the convergence behaviour. This will be further specified in the following considerations (5.33), (5.74). As already seen in the aforementioned examples, the matrix NA will not be constructed explicitly, since N is in general dense, even if N^{-1} is sparse. The evaluation of $y = NAx$ therefore means solving the auxiliary system of equations

$$N^{-1}y = Ax\,.$$

Obviously, we have the following:

Lemma 5.5 *If the matrix A is symmetric and positive definite, then for the Richardson method all eigenvalues of M are real and smaller than* 1.

5.1.3 Relaxation

We continue to assume that A is symmetric and positive definite. Therefore, divergence of the procedure can be caused only by negative eigenvalues of $I - A$ less than or equal to -1. In general, bad or nonconvergent iterative methods can be improved in their convergence behaviour by *relaxation* if they meet certain conditions.

For an iterative method, given in the form (5.6), (5.7), the corresponding *relaxation method* with relaxation parameter $\omega > 0$ is defined by

$$x^{(k+1)} := \omega\big(Mx^{(k)} + Nb\big) + (1-\omega)x^{(k)}, \tag{5.30}$$

which means

$$M_\omega := \omega M + (1-\omega)I\,, \quad N_\omega := \omega N\,, \tag{5.31}$$

or if the condition of consistency $M = I - NA$ holds,

$$\begin{aligned} x^{(k+1)} &= \omega\left(x^{(k)} - N\left(Ax^{(k)} - b\right)\right) + (1-\omega)x^{(k)} \\ &= x^{(k)} - \omega N\left(Ax^{(k)} - b\right)\,. \end{aligned}$$

Let us assume for the procedure (5.6) that all eigenvalues of M are real. For the smallest one $\lambda_{\min}$ and the largest one $\lambda_{\max}$ we assume

$$\lambda_{\min} \le \lambda_{\max} < 1\,;$$

this is, for example, the case for the Richardson method. Then also the eigenvalues of M_ω are real, and we conclude that

$$\lambda_i(M_\omega) = \omega\lambda_i(M) + 1 - \omega = 1 - \omega\big(1 - \lambda_i(M)\big)$$

if the $\lambda_i(B)$ are the eigenvalues of B in an arbitrary ordering. Hence

$$\varrho(M_\omega) = \max\left\{ |1 - \omega\,(1 - \lambda_{\min}(M))|\,, |1 - \omega\,(1 - \lambda_{\max}(M))| \right\},$$

since $f(\lambda) := 1 - \omega(1 - \lambda)$ is a straight line for a fixed ω (with $f(1) = 1$ and $f(0) = 1 - \omega$).

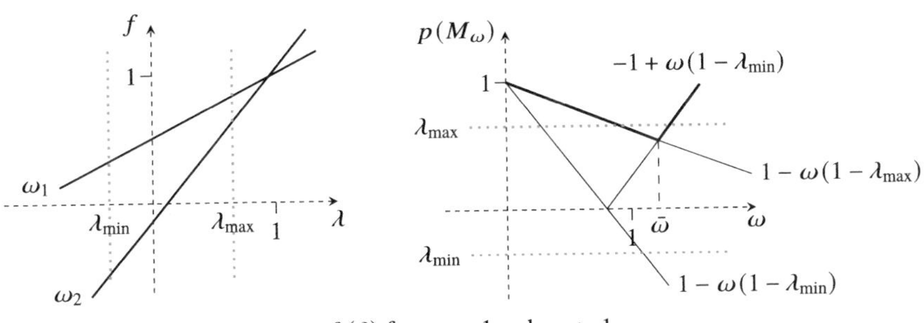

Fig. 5.1: Calculation of $\bar{\omega}$.

For the *optimal* parameter $\bar{\omega}$, i.e., $\bar{\omega}$ with

$$\varrho(M_{\bar{\omega}}) = \min_{\omega > 0} \varrho(M_\omega),$$

we therefore have, as can be seen from Figure 5.1,

$$\begin{aligned} & 1 - \bar{\omega}\,(1 - \lambda_{\max}(M)) = -1 + \bar{\omega}\,(1 - \lambda_{\min}(M)) \\ \Longleftrightarrow\quad & \bar{\omega} = \frac{2}{2 - \lambda_{\max}(M) - \lambda_{\min}(M)}\,. \end{aligned}$$

Hence $\bar{\omega} > 0$ and

$$\varrho(M_{\bar{\omega}}) = 1 - \bar{\omega}(1 - \lambda_{\max}(M)) < 1;$$

consequently, the method converges with optimal ω even in cases where it would not converge for $\omega = 1$. But keep in mind that one needs the eigenvalues of M to determine $\bar{\omega}$.

Moreover, we have

$$\bar{\omega} < 1 \quad \Leftrightarrow \quad \lambda_{\max}(M) + \lambda_{\min}(M) < 0 .$$

If $\lambda_{\min}(M) \neq -\lambda_{\max}(M)$, that is, $\bar{\omega} \neq 1$, we will achieve an improvement by relaxation:

$$\varrho(M_{\bar{\omega}}) < \varrho(M) .$$

The case of $\omega < 1$ is called *underrelaxation*, whereas in the case of $\omega > 1$ we speak of an *overrelaxation*.

In particular, for the Richardson method with the iteration matrix $M = I - A$, due to $\lambda_{\min}(M) = 1 - \lambda_{\max}(A)$ and $\lambda_{\max}(M) = 1 - \lambda_{\min}(A)$, the optimal $\bar{\omega}$ is given by

$$\bar{\omega} = \frac{2}{\lambda_{\min}(A) + \lambda_{\max}(A)} . \tag{5.32}$$

Hence

$$\varrho(M_{\bar{\omega}}) = 1 - \bar{\omega}\lambda_{\min}(A) = \frac{\lambda_{\max}(A) - \lambda_{\min}(A)}{\lambda_{\min}(A) + \lambda_{\max}(A)} = \frac{\kappa(A) - 1}{\kappa(A) + 1} < 1 , \tag{5.33}$$

with the spectral *condition number* of A

$$\kappa(A) := \frac{\lambda_{\max}(A)}{\lambda_{\min}(A)}$$

(see Appendix A.3).

For large $\kappa(A)$ we have

$$\varrho(M_{\bar{\omega}}) = \frac{\kappa(A) - 1}{\kappa(A) + 1} \approx 1 - \frac{2}{\kappa(A)} ,$$

the variable of the proportionality being $\kappa(A)$. For the example of the five-point stencil discretization, due to (5.24),

$$\lambda_{\min}(A) + \lambda_{\max}(A) = 4\left(2 - \cos\frac{n-1}{n}\pi - \cos\frac{\pi}{n}\right) = 8 ,$$

and thus due to (5.32),

$$\bar{\omega} = \frac{1}{4} .$$

Hence the iteration matrix $M_{\bar{\omega}} = I - \frac{1}{4}A$ is identical to the Jacobi iteration: We have rediscovered Jacobi's method.

By means of (5.33) we can estimate the contraction number, since we know from (5.24) that

$$\kappa(A) = \frac{4\left(1 - \cos\frac{n-1}{n}\pi\right)}{4\left(1 - \cos\frac{\pi}{n}\right)} = \frac{1 + \cos\frac{\pi}{n}}{1 - \cos\frac{\pi}{n}} \approx \frac{4n^2}{\pi^2} . \tag{5.34}$$

This shows the stringency of Theorem 3.48, and again we can conclude that

$$\varrho(M_{\bar{\omega}}) = \cos\frac{\pi}{n} \approx 1 - \frac{\pi^2}{2n^2}. \tag{5.35}$$

Due to Theorem 3.48, the convergence behaviour seen for the model problem is also valid in general for quasi-uniform triangulations.

5.1.4 SOR and Block-Iteration Methods

We assume again that A is a general nonsingular matrix. For the relaxation of the Gauss–Seidel method, we use it in the form

$$Dx^{(k+1)} = -Lx^{(k+1)} - Rx^{(k)} + b,$$

instead of the resolved form (5.20).

The relaxed method is then

$$Dx^{(k+1)} = \omega\big(-Lx^{(k+1)} - Rx^{(k)} + b\big) + (1-\omega)Dx^{(k)} \tag{5.36}$$

with a relaxation parameter $\omega > 0$. This is equivalent to

$$(D + \omega L)\,x^{(k+1)} = (-\omega R + (1-\omega)D)\,x^{(k)} + \omega b. \tag{5.37}$$

Hence

$$\begin{aligned} M_\omega &:= (D+\omega L)^{-1}\,(-\omega R + (1-\omega)D)\,, \\ N_\omega &:= (D+\omega L)^{-1}\,\omega\,. \end{aligned}$$

In the application to discretizations of boundary value problems, normally we choose $\omega > 1$, which means overrelaxation. This explains the name of the *SOR method* as an abbreviation of *successive overrelaxation*. The effort to execute an iteration step is hardly higher than for the Gauss–Seidel method. Although we have to add $3m$ operations to the evaluation of the right side of (5.36), the forward substitution to solve the auxiliary system of equations in (5.37) is already part of the form (5.36).

The calculation of the optimal $\bar{\omega}$ here is more difficult, because M_ω depends nonlinearly on ω. Only for special classes of matrices can the optimal $\bar{\omega}$ minimizing $\varrho(M_\omega)$ be calculated explicitly in dependence on $\varrho(M_1)$, the convergence rate of the (nonrelaxed) Gauss–Seidel method. Before we sketch this, we will look at some further variants of this procedure:

The matrix N_ω is nonsymmetric even for symmetric A. One gets a symmetric N_ω if after one SOR step another one is performed in which the indices are run through in reverse order $m, m-1, \ldots, 2, 1$, which means that L and R are exchanged. The two half steps

$$Dx^{(k+\frac{1}{2})} = \omega\big(-Lx^{(k+\frac{1}{2})} - Rx^{(k)} + b\big) + (1-\omega)Dx^{(k)},$$

$$Dx^{(k+1)} = \omega\big(-Lx^{(k+\frac{1}{2})} - Rx^{(k+1)} + b\big) + (1-\omega)Dx^{(k+\frac{1}{2})},$$

form one step of the *symmetric SOR*, the *SSOR method* for short. A special case is the *symmetric Gauss–Seidel method* for $\omega = 1$.

We write down the procedure for symmetric A, i.e., $R = L^T$ in the form (5.6), in which the symmetry of N becomes obvious:

$$\begin{aligned} M &= \left(D + \omega L^T\right)^{-1} \left[(1-\omega)D - \omega L\right] (D+\omega L)^{-1} \left[(1-\omega)D - \omega L^T\right], \\ N &= \omega(2-\omega)\left(D + \omega L^T\right)^{-1} D\,(D+\omega L)^{-1}. \end{aligned} \tag{5.38}$$

The effort for SSOR is only slightly higher than for SOR if the vectors already calculated in the half steps are stored and used again, as, for example, $Lx^{(k+1/2)}$.

Other variants of these procedures are created if the procedures are not applied to the matrix itself but to a block partitioning

$$A = (A_{ij})_{i,j} \quad \text{with } A_{ij} \in \mathbb{R}^{m_i,m_j}, \quad i,j = 1,\dots,p, \tag{5.39}$$

with $\sum_{i=1}^{p} m_i = m$. As an example we get the *block-Jacobi method*, which is analogous to (5.19) and has the form

$$\xi_i^{(k+1)} = A_{ii}^{-1}\left(-\sum_{j=1}^{i-1} A_{ij}\xi_j^{(k)} - \sum_{j=i+1}^{p} A_{ij}\xi_j^{(k)} + \beta_i\right) \text{ for all } i = 1,\dots,p. \tag{5.40}$$

Here $x = (\xi_1,\dots,\xi_p)^T$ and $b = (\beta_1,\dots,\beta_p)^T$, respectively, are corresponding partitions of the vectors. By exchanging $\xi_j^{(k)}$ with $\xi_j^{(k+1)}$ in the first sum one gets the *block-Gauss–Seidel method* and then in the same way the relaxed variants. The iteration (5.40) includes p vector equations. For each of them we have to solve a system of equations with system matrix A_{ii}. To get an advantage compared to the pointwise method a much lower effort should be necessary than for the solution of the total system. This can require—if at all possible—a rearranging of the variables and equations. The necessary permutations will not be noted explicitly here. Such methods are applied in finite difference methods or other methods with structured grids (see Section 4.1) if an ordering of nodes is possible such that the matrices A_{ii} are diagonal or tridiagonal and therefore the systems of equations are solvable with $O(m_i)$ operations.

As an example we again discuss the five-point stencil discretization of the Poisson equation on a square with $n+1$ nodes per space dimension. The matrix A then has the form (1.14) with $l = m = n$. If the nodes are numbered rowwise and we choose one block for each line, which means $p = n-1$ and $m_i = n-1$ for all $i = 1,\dots,p$, then the matrices A_{ii} are tridiagonal. On the other hand, if one chooses a partition of the indices of the nodes in subsets S_i such that a node with index in S_i has neighbours only in other index sets, then for such a selection and arbitrary ordering within the index sets the matrices A_{ii} become diagonal. *Neighbours* here denote the nodes within a difference stencil or more generally, those nodes that contribute to the corresponding row of the discretization matrix. In the example of the five-point stencil, starting with rowwise numbering, one can combine all odd indices to a block S_1 (the "red nodes") and all even indices to a block S_2 (the "black" nodes). Here we have $p = 2$. We call this a *red-black ordering* (see Figure 5.2). If two "colours" are not sufficient, one can choose $p > 2$.

$$\begin{pmatrix} 4 & -1 & 0 & -1 & 0 & 0 & 0 & 0 & 0 \\ -1 & 4 & -1 & 0 & -1 & 0 & 0 & 0 & 0 \\ 0 & -1 & 4 & 0 & 0 & -1 & 0 & 0 & 0 \\ -1 & 0 & 0 & 4 & -1 & 0 & -1 & 0 & 0 \\ 0 & -1 & 0 & -1 & 4 & -1 & 0 & -1 & 0 \\ 0 & 0 & -1 & 0 & -1 & 4 & 0 & 0 & -1 \\ 0 & 0 & 0 & -1 & 0 & 0 & 4 & -1 & 0 \\ 0 & 0 & 0 & 0 & -1 & 0 & -1 & 4 & -1 \\ 0 & 0 & 0 & 0 & 0 & -1 & 0 & -1 & 4 \end{pmatrix}$$

$m = 3 \times 3$: rowwise ordering.

$$\left(\begin{array}{ccccc|cccc} 4 & 0 & 0 & 0 & 0 & -1 & -1 & 0 & 0 \\ 0 & 4 & 0 & 0 & 0 & -1 & 0 & -1 & 0 \\ 0 & 0 & 4 & 0 & 0 & -1 & -1 & -1 & -1 \\ 0 & 0 & 0 & 4 & 0 & 0 & -1 & 0 & -1 \\ 0 & 0 & 0 & 0 & 4 & 0 & 0 & -1 & -1 \\ \hline -1 & -1 & -1 & 0 & 0 & 4 & 0 & 0 & 0 \\ -1 & 0 & -1 & -1 & 0 & 0 & 4 & 0 & 0 \\ 0 & -1 & -1 & 0 & -1 & 0 & 0 & 4 & 0 \\ 0 & 0 & -1 & -1 & -1 & 0 & 0 & 0 & 4 \end{array}\right)$$

red-black ordering:
red: node 1, 3, 5, 7, 9 from rowwise ordering
black: node 2, 4, 6, 8 from rowwise ordering

Fig. 5.2: Comparison of orderings.

We return to the SOR method and its convergence: In the following the iteration matrix will be denoted by $M_{\mathrm{SOR}(\omega)}$ with the relaxation parameter ω. Likewise, M_{J} and M_{GS} are the iteration matrices of Jacobi's and the Gauss–Seidel method, respectively. General propositions are summarized in the following theorem:

Theorem 5.6 (of Kahan; Ostrowski and Reich)

1) $\varrho\left(M_{\mathrm{SOR}(\omega)}\right) \geq |1-\omega|$ *for* $\omega \neq 0$.
2) If A is symmetric and positive definite, then

$$\varrho\left(M_{\mathrm{SOR}(\omega)}\right) < 1 \quad \textit{for } \omega \in (0,2)\,.$$

Proof See [33, Lemma 3.40; Thm. 3.41]. □

Therefore, we use only $\omega \in (0, 2)$. For a useful procedure we need more information about the optimal relaxation parameter ω_{opt}, given by

$$\varrho\left(M_{\mathrm{SOR}(\omega_{\mathrm{opt}})}\right) = \min_{0<\omega<2} \varrho\left(M_{\mathrm{SOR}(\omega)}\right),$$

and about the size of the contraction number. This is possible only if the ordering of equations and unknowns has certain properties:

Definition 5.7 A matrix $A \in \mathbb{R}^{m,m}$ is *consistently ordered* if for the partition (5.18), D is nonsingular and

$$C(\alpha) := \alpha^{-1} D^{-1} L + \alpha D^{-1} R$$

has eigenvalues independent of α for $\alpha \in \mathbb{C}\backslash\{0\}$.

There is a connection to the possibility of a multi-colour ordering, because a matrix in the block form (5.39) is consistently ordered if it is block-tridiagonal (i.e., $A_{ij} = 0$ for $|i - j| > 1$) and the diagonal blocks A_{ii} are nonsingular diagonal matrices (see [54, Sect. 4.2.5]).

In the case of a consistently ordered matrix one can prove a relation between the eigenvalues of M_{J}, M_{GS}, and $M_{\mathrm{SOR}(\omega)}$. From this we can see how much faster the Gauss–Seidel method converges than Jacobi's method:

Theorem 5.8 *If A is consistently ordered, then*

$$\varrho(M_{\mathrm{J}})^2 = \varrho(M_{\mathrm{GS}}).$$

Proof For a special case see [33, Sect. 4.5]. □

Due to (5.4) we can expect a halving of the number of iteration steps, but this does not change the asymptotic statement (5.27).

Finally, in the case that Jacobi's method converges the following theorem holds:

Theorem 5.9 *Let A be consistently ordered with nonsingular diagonal matrix D, the eigenvalues of M_{J} being real and $\beta := \varrho(M_{\mathrm{J}}) < 1$. Then we have for the SOR method:*

1) $\omega_{\mathrm{opt}} = \dfrac{2}{1 + (1 - \beta^2)^{1/2}}$,

2) $$\varrho(M_{\mathrm{SOR}(\omega)}) = \begin{cases} 1 - \omega + \frac{1}{2}\omega^2\beta^2 + \omega\beta\left(1 - \omega + \frac{\omega^2\beta^2}{4}\right)^{1/2} & \text{for } 0 < \omega < \omega_{\mathrm{opt}}, \\ \omega - 1 & \text{for } \omega_{\mathrm{opt}} \le \omega < 2, \end{cases}$$

3) $\varrho\left(M_{\mathrm{SOR}(\omega_{\mathrm{opt}})}\right) = \dfrac{\beta^2}{(1 + (1 - \beta^2)^{1/2})^2}$.

Proof See [35, p. 216]. □

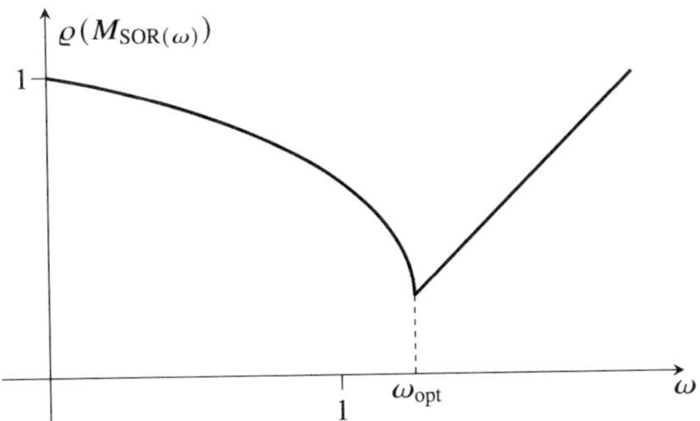

Fig. 5.3: Dependence of $\varrho\left(M_{\mathrm{SOR}(\omega)}\right)$ on ω.

If $\varrho(M_{\mathrm{J}})$ is known for Jacobi's method, then ω_{opt} can be calculated. This is the case in the example of the five-point stencil discretization on a square: From (5.28) and Theorem 5.8, it follows that

$$\varrho(M_{\mathrm{GS}}) = \left(\cos\frac{\pi}{n}\right)^2 = 1 - \frac{\pi^2}{n^2} + O(n^{-4})\,;$$

hence

$$\omega_{\mathrm{opt}} = 2/\left(1 + \sin\tfrac{\pi}{n}\right),$$

$$\varrho\left(M_{\mathrm{SOR}(\omega_{\mathrm{opt}})}\right) = \omega_{\mathrm{opt}} - 1 = 1 - 2\tfrac{\pi}{n} + O(n^{-2})\,.$$

Therefore, the optimal SOR method has a lower complexity than all methods described up to now.

Correspondingly, the number of operations to reach the relative error level $\varepsilon > 0$ is reduced to $\ln\frac{1}{\varepsilon}\,O(m^{3/2})$ operations in comparison to $\ln\frac{1}{\varepsilon}\,O(m^2)$ operations for the previous procedures.

Table 5.1 gives an impression of the convergence for the model problem. It displays the theoretically to be expected values for the numbers of iterations of the Gauss–Seidel method (m_{GS}), as well as for the SOR method with optimal relaxation parameter (m_{SOR}). Here we use the very moderate termination criterion $\varepsilon = 10^{-3}$ measured in the Euclidean norm.

n	m_{GS}	m_{SOR}
8	43	8
16	178	17
32	715	35
64	2865	70
128	11466	140
256	45867	281

Table 5.1: Gauss–Seidel and optimal SOR method for the model problem.

The optimal SOR method is superior, even if we take into account the almost doubled effort per iteration step. But generally, ω_{opt} is not known explicitly. Figure 5.3 shows that it is probably better to overestimate ω_{opt} instead of underestimating. More generally, one can try to improve the relaxation parameter during the iteration:

If $\varrho(M_{\mathrm{J}})$ is a simple eigenvalue, then this also holds true for the spectral radius $\varrho(M_{\mathrm{SOR}(\omega)})$. The spectral radius can thus be approximated by the power method on the basis of the iterates. By Theorem 5.9, 3) one can approximate $\varrho(M_{\mathrm{J}})$, and by Theorem 5.9, 1) then also ω_{opt}.

This basic principle can be extended to an algorithm (see, for example, [35, Section 9.5]), but the upcoming overall procedure is no longer a linear stationary method.

5.1.5 Extrapolation Methods

Another possibility for an extension of the linear stationary methods, related to the adaption of the relaxation parameter, is the following: Starting with a linear stationary basic iteration $\tilde{x}^{k+1} := \Phi\left(\tilde{x}^k\right)$ we define a new iteration by

$$x^{(k+1)} := \omega_k \Phi\big(x^{(k)}\big) + (1-\omega_k)x^{(k)}, \tag{5.41}$$

with *extrapolation factors* ω_k to be chosen. A generalization of this definition is to start with the iterates of the basic iteration $\tilde{x}^{(0)}, \tilde{x}^{(1)}, \ldots$. The iterates of the new method are to be determined by

$$x^{(k)} := \sum_{j=0}^{k} \alpha_{k_j} \tilde{x}^{(j)},$$

with α_{k_j} defined by a polynomial $p_k \in \mathcal{P}_k$, with the property $p_k(t) = \sum_{j=0}^{k} \alpha_{k_j} t^j$ and $p_k(1) = 1$. For an appropriate definition of such *extrapolation* or *semi-iterative methods* we need to know the spectrum of the basic iteration matrix M, since the error $e^{(k)} = x^{(k)} - x$ satisfies

$$e^{(k)} = p_k(M)e^{(0)},$$

where M is the iteration matrix of the basic iteration. This matrix should be normal, for example, such that

$$\|p_k(M)\|_2 = \varrho(p_k(M))$$

holds. Then we have the obvious estimation

$$\left|e^{(k)}\right|_2 \le \left|p_k(M)e^{(0)}\right|_2 \le \left\|p_k(M)\right\|_2 \left|e^{(0)}\right|_2 \le \varrho(p_k(M))\left|e^{(0)}\right|_2. \tag{5.42}$$

If the method is to be defined in such a way that

$$\varrho(p_k(M)) = \max\left\{|p_k(\lambda)| \mid \lambda \in \sigma(M)\right\}$$

is minimized by choosing p_k, then the knowledge of the spectrum $\sigma(M)$ is necessary. Generally, instead of this, we assume that suitable supersets are known: If $\sigma(M)$ is real and

$$\underline{\sigma} \le \lambda \le \overline{\sigma} \quad \text{for all } \lambda \in \sigma(M),$$

then, due to

$$\left|e^{(k)}\right|_2 \le \max_{\lambda \in [\underline{\sigma},\overline{\sigma}]} \left|p_k(\lambda)\right| \left|e^{(0)}\right|_2,$$

it makes sense to determine the polynomials p_k as a solution of the minimization problem on $[\underline{\sigma}, \overline{\sigma}]$,

$$\max_{\lambda \in [\underline{\sigma},\overline{\sigma}]} |p_k(\lambda)| \to \min \quad \text{for all} \quad p \in \mathcal{P}_k \quad \text{with } p(1) = 1\,. \tag{5.43}$$

In the following sections we will introduce methods with an analogous convergence behaviour, without control parameters necessary for their definition.

For further information on semi-iterative methods see, for example, [33, Chapter 8].

Exercises

Problem 5.1 Investigate Jacobi's method and the Gauss–Seidel method for solving the linear system of equations $Ax = b$ with respect to their convergence if we have the following system matrices:

$$\text{a)} \quad A = \begin{pmatrix} 1 & 2 & -2 \\ 1 & 1 & 1 \\ 2 & 2 & 1 \end{pmatrix}, \qquad \text{b)} \quad A = \frac{1}{2}\begin{pmatrix} 2 & -1 & 1 \\ 2 & 2 & 2 \\ -1 & -1 & 2 \end{pmatrix}.$$

Problem 5.2 We consider the system of linear equations $Ax = b$ with

$$A = \begin{pmatrix} 2 & 1 & 0 \\ 0 & 1 & 0.5 \\ 0.5 & 0 & 1 \end{pmatrix}.$$

a) Does the Gauss–Seidel method converge for the above problem? If yes, for which right-hand sides b and which initial values $x^{(0)}$?
b) Calculate a contraction number ϱ

$$\|e^{(n+1)}\| \le \varrho \|e^{(n)}\| \quad \text{for all } n \in \mathbb{N}$$

for the Jacobi's method applied on the above system for a chosen norm.

How many steps of the Jacobi's method are necessary to reduce the initial error by a factor of 10^{-6}?

Problem 5.3 Prove the consistency of the SOR method.

Problem 5.4 Solve $Ax = b$ for

$$A = \begin{pmatrix} 6 & -2 & -2 & 0 \\ -2 & 6 & 0 & -2 \\ -2 & 0 & 6 & -2 \\ 0 & -2 & -2 & 6 \end{pmatrix}, \qquad b = \begin{pmatrix} -0.42 \\ 0.246 \\ 0.912 \\ 1.578 \end{pmatrix}$$

using Jacobi's and the Gauss-Seidel method.

Choose $x^{(0)} = (0, 0, 0, 0)^T$ as initial vector and $\|Ax^{(n)} - b\|_2 \leq 10^{-6}$ as stop criterion. Argue for both methods to converge. Why does relaxation of Jacobi's method not improve convergence?

Using the permutation matrix

$$P = \begin{pmatrix} 1 & 0 & 0 & 0 \\ 0 & 0 & 0 & 1 \\ 0 & 0 & 1 & 0 \\ 0 & 1 & 0 & 0 \end{pmatrix}$$

one receives a consistently ordered matrix $\hat{A} = P^{-1}AP$. We want to solve a system of linear equations $\hat{A}y = d$ which is equivalent to the original system. Calculate the optimal relaxation parameter of the SOR method with respect to $\hat{A}$ and solve the auxiliary system using the SOR method under the same conditions as above.

Problem 5.5 Prove Theorem 5.6, 1).

Problem 5.6 a) Let $A \in \mathbb{R}^{m,m}$ be a nonsingular matrix with

$$\sum_{j=1}^{m} |a_{ij}| = 1 \quad \text{for all } 1 \leq i \leq m\,.$$

Show that for every diagonal matrix $D \in \mathbb{R}^{m,m}$, $D = \operatorname{diag}(d_i)$, $d_i \neq 0$, $i = 1, \ldots, m$, the following holds

$$\kappa(DA) \geq \kappa(A),$$

where κ is the condition with respect to the matrix norm induced by $\|\cdot\|_\infty$.

b) Now let $Au = b$ be the linear system resulting from a finite element discretization. What would be a good diagonal preconditioner to improve the condition number with respect to the matrix norm induced by $\|\cdot\|_\infty$?

Programming project 5.1 Consider the boundary value problem and its discretizations described in Project 1.2.

Solve the resulting linear system of equations for both cases using the Gauss–Seidel iteration scheme. As initial value for the iteration take $u_{ij}^{(0)} := 0$. Compute the iterates $u_{16,16}^{(k)}$ in the point $(0.5, 0.5)$, the errors $e^{(k)}$ and the quantities $e^{(k-1)}/e^{(k)}$ for $k = 1, 2, 10, 100, 200, 300$, where

$$e^{(k)} := \max\{|u_{ij}^{(k)} - (i^2 + j^2)h^2| \mid 1 \leq i, j \leq 31\}$$

(note that the exact solution is $u(x, y) = x^2 + y^2$).

Interpret your results.

5.2 Gradient and Conjugate Gradient Methods

In this section let $A \in \mathbb{R}^{m,m}$ be symmetric and positive definite. Then the system of equations $A\bar{x} = b$ is equivalent to the problem

$$\text{Minimize} \quad f(x) := \frac{1}{2}x^T Ax - b^T x \quad \text{for } x \in \mathbb{R}^m, \tag{5.44}$$

since for such a functional the minima and stationary points coincide, where a *stationary point* is an x satisfying

$$0 = \nabla f(x) = Ax - b\,. \tag{5.45}$$

In contrast to the notation $x \cdot y$ for the "short" space vectors $x, y \in \mathbb{R}^d$ we write here the Euclidean scalar product as matrix product $x^T y$.

For the finite element discretization this corresponds to the equivalence of the Galerkin method (2.26) with the Ritz method (2.27) if A is the stiffness matrix and b the load vector (see (2.38) and (2.39)). More generally, Lemma 2.3 implies the equivalence of (5.44) and (5.45), if as bilinear form the so-called *energy scalar product*

$$\langle x, y\rangle_A := x^T Ay \tag{5.46}$$

is chosen.

A general iterative method to solve (5.44) has the following structure:

Define a *search direction* $d^{(k)}$.

$$\text{Minimize} \quad \alpha \mapsto \tilde{f}(\alpha) := f\big(x^{(k)} + \alpha d^{(k)}\big) \tag{5.47}$$

exactly or approximately, with the solution α_k.

$$\text{Define} \qquad x^{(k+1)} := x^{(k)} + \alpha_k d^{(k)}\,. \tag{5.48}$$

If f is defined as in (5.44), the exact value of α_k can be computed from the condition $\tilde{f}'(\alpha) = 0$ and

$$\tilde{f}'(\alpha) = \nabla f\left(x^{(k)} + \alpha d^{(k)}\right)^T d^{(k)}$$

as

$$\alpha_k = -\frac{g^{(k)^T} d^{(k)}}{d^{(k)^T} A d^{(k)}}, \tag{5.49}$$

where

$$g^{(k)} := Ax^{(k)} - b = \nabla f\left(x^{(k)}\right). \tag{5.50}$$

The error of the kth iterate is denoted by $e^{(k)}$:

$$e^{(k)} := x^{(k)} - \bar{x}.$$

Some relations that are valid in this general framework are the following: Due to the one-dimensional minimization of f, we have

$$g^{(k+1)^T} d^{(k)} = 0, \tag{5.51}$$

and from (5.50) we can conclude immediately that

$$Ae^{(k)} = g^{(k)}, \; e^{(k+1)} = e^{(k)} + \alpha_k d^{(k)}, \tag{5.52}$$

$$g^{(k+1)} = g^{(k)} + \alpha_k A d^{(k)}. \tag{5.53}$$

We consider the *energy norm*

$$\|x\|_A := \left(x^T A x\right)^{1/2} \tag{5.54}$$

induced by the energy scalar product. For a finite element stiffness matrix A with a bilinear form a, we have the correspondence

$$\|x\|_A = a(u, u)^{1/2} = \|u\|_a$$

for $u = \sum_{i=1}^m x_i \varphi_i$ if the φ_i are the underlying basis functions. Comparing the solution $x = A^{-1}b$ with an arbitrary $y \in \mathbb{R}^m$ leads to

$$f(y) = f(x) + \frac{1}{2}\|y - x\|_A^2, \tag{5.55}$$

so that condition (5.44) also minimizes the distance to x in $\|\cdot\|_A$. The energy norm will therefore have a special importance. Measured in the energy norm we have, due to (5.52),

$$\left\|e^{(k)}\right\|_A^2 = e^{(k)^T} g^{(k)} = g^{(k)^T} A^{-1} g^{(k)},$$

and therefore due to (5.52) and (5.51),

$$\left\|e^{(k+1)}\right\|_A^2 = g^{(k+1)^T} e^{(k)}.$$

The vector $-\nabla f\left(x^{(k)}\right)$ in $x^{(k)}$ points in the direction of the locally steepest descent, which motivates the *gradient method*, i.e.,

$$d^{(k)} := -g^{(k)}, \tag{5.56}$$

and thus

$$\alpha_k = \frac{d^{(k)^T} d^{(k)}}{d^{(k)^T} A d^{(k)}} . \tag{5.57}$$

The above identities imply for the gradient method

$$\left\|e^{(k+1)}\right\|_A^2 = \left(g^{(k)} + \alpha_k A d^{(k)}\right)^T e^{(k)} = \|e^{(k)}\|_A^2 \left(1 - \alpha_k \frac{d^{(k)^T} d^{(k)}}{d^{(k)^T} A^{-1} d^{(k)}}\right)$$

and thus by means of the definition of α_k from (5.57)

$$\left\|x^{(k+1)} - x\right\|_A^2 = \left\|x^{(k)} - x\right\|_A^2 \left(1 - \frac{\left(d^{(k)^T} d^{(k)}\right)^2}{d^{(k)^T} A d^{(k)}\, d^{(k)^T} A^{-1} d^{(k)}}\right).$$

With the *inequality of Kantorovich* (see, for example, [54, Lemma 5.8]),

$$\frac{x^T A x\, x^T A^{-1} x}{\left(x^T x\right)^2} \le \left(\frac{1}{2}\kappa^{1/2} + \frac{1}{2}\kappa^{-1/2}\right)^2 ,$$

where $\kappa := \kappa(A)$ is the spectral condition number, and the relation

$$1 - \frac{4}{\left(a^{1/2} + a^{-1/2}\right)^2} = \frac{(a-1)^2}{(a+1)^2} \quad \text{for } a > 0,$$

we obtain the following theorem:

Theorem 5.10 *For the gradient method we have*

$$\left\|x^{(k)} - x\right\|_A \le \left(\frac{\kappa - 1}{\kappa + 1}\right)^k \left\|x^{(0)} - x\right\|_A . \tag{5.58}$$

This is the same estimate as for the optimally relaxed Richardson method (with the sharper estimate $\|M\|_A \le \frac{\kappa-1}{\kappa+1}$ instead of $\varrho(M) \le \frac{\kappa-1}{\kappa+1}$). The essential difference lies in the fact that this is possible without knowledge of the spectrum of A.

Nevertheless, for finite element discretizations we have the same poor convergence rate as for Jacobi's or similar methods. The reason for this deficiency lies in the fact that due to (5.51), we have $g^{(k+1)^T} g^{(k)} = 0$, but in general not $g^{(k+2)^T} g^{(k)} = 0$. On the contrary, these search directions are very often almost parallel, as can be seen from Figure 5.4.

The reason for this problem is the fact that for large κ the search directions $g^{(k)}$ and $g^{(k+1)}$ can be almost parallel with respect to the scalar products $\langle \cdot, \cdot \rangle_A$ (see

$m = 2$:

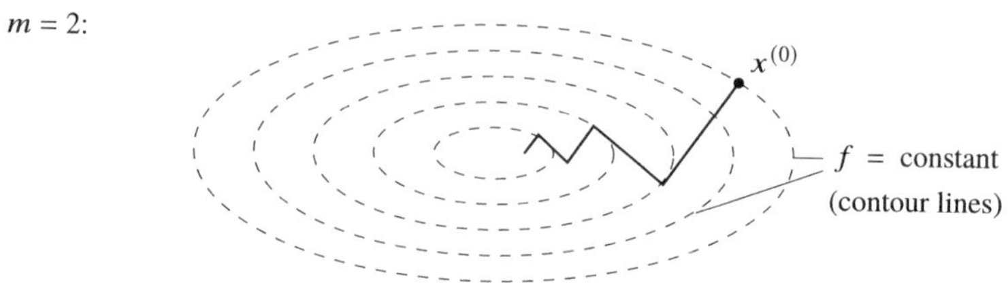

Fig. 5.4: Zigzag behaviour of the gradient method.

Problem 5.9), but with respect to $\|\cdot\|_A$ the distance to the solution will be minimized (see (5.55)).

The search directions $d^{(k)}$ should be orthogonal with respect to $\langle\cdot,\cdot\rangle_A$, which we call *conjugate*.

Definition 5.11 The vectors $d^{(0)},\ldots,d^{(l)} \in \mathbb{R}^m$ are *conjugate* if they satisfy

$$\left\langle d^{(i)}, d^{(j)}\right\rangle_A = 0 \quad \text{for } i,j = 0,\ldots,l,\ i \neq j\,.$$

If the search directions of a method defined according to (5.48), (5.49) are chosen as conjugate, it is called a *method of conjugate directions*.

Let $d^{(0)},\ldots,d^{(m-1)}$ be conjugate directions. Then they are also linearly independent and thus form a basis in which the solution x of (5.1) can be represented, say by the coefficients γ_k:

$$x = \sum_{k=0}^{m-1} \gamma_k d^{(k)}\,.$$

Since the $d^{(k)}$ are conjugate and $Ax = b$ holds, we have

$$\gamma_k = \frac{d^{(k)^T} b}{d^{(k)^T} A d^{(k)}}\,, \tag{5.59}$$

and the γ_k can be calculated without knowledge of x. If the $d^{(k)}$ would by given a priori, for example, by orthogonalization of a basis with respect to $\langle\cdot,\cdot\rangle_A$, then x would be determined by (5.59).

If we apply (5.59) to determine the coefficients for $x - x^{(0)}$ in the form

$$x - x^{(0)} = \sum_{k=0}^{m-1} \gamma_k d^{(k)}\,,$$

which means replacing b with $b - Ax^{(0)}$ in (5.59), then we get

$$\gamma_k = -\frac{g^{(0)^T} d^{(k)}}{d^{(k)^T} A d^{(k)}}\,.$$

For the kth iterate we have, according to (5.48);

$$x^{(k)} = x^{(0)} + \sum_{i=0}^{k-1} \alpha_i d^{(i)}$$

and therefore (see (5.50))

$$g^{(k)} = g^{(0)} + \sum_{i=0}^{k-1} \alpha_i A d^{(i)} .$$

For a method of conjugate directions this implies

$$g^{(k)^T} d^{(k)} = g^{(0)^T} d^{(k)}$$

and therefore

$$\gamma_k = -\frac{g^{(k)^T} d^{(k)}}{d^{(k)^T} A d^{(k)}} = \alpha_k ,$$

which means that $x = x^{(m)}$. A method of conjugate directions therefore is exact after at most m steps. Under certain conditions such a method may terminate before reaching this step number with $g^{(k)} = 0$ and the final iterate $x^{(k)} = x$. If m is very large, this exactness of a method of conjugate directions is less important than the fact that the iterates can be interpreted as the solution of a minimization problem approximating (5.44):

Theorem 5.12 *The iterates $x^{(k)}$ that are determined by a method of conjugate directions minimize the functional f from* (5.44) *as well as the error $\left\|x^{(k)} - x\right\|_A$ on $x^{(0)} + K_k(A; g^{(0)})$, where*

$$K_k(A; g^{(0)}) := \text{span}\left\{d^{(0)}, \ldots, d^{(k-1)}\right\} .$$

This is due to

$$g^{(k)^T} d^{(i)} = 0 \quad \textit{for } i = 0, \ldots, k-1 . \tag{5.60}$$

Proof It is sufficient to prove (5.60). Due to the one-dimensional minimization this holds for $k = 1$ and for $i = k-1$ (see (5.51) applied to $k-1$). To conclude the assertion for k from its knowledge for $k-1$, we note that (5.53) implies, for $0 \le i < k-1$,

$$d^{(i)^T}\left(g^{(k)} - g^{(k-1)}\right) = \alpha_{k-1} d^{(i)^T} A d^{(k-1)} = 0 .$$

□

In the *method of conjugate gradients*, or *CG method*, the $d^{(k)}$ are determined during the iteration by the ansatz

$$d^{(k+1)} := -g^{(k+1)} + \beta_k d^{(k)} . \tag{5.61}$$

Then we have to clarify whether

$$\left\langle d^{(k)}, d^{(i)} \right\rangle_A = 0 \quad \text{for } k > i$$

can be obtained. The necessary requirement $\langle d^{(k+1)}, d^{(k)} \rangle_A = 0$ leads to

$$-\left\langle g^{(k+1)}, d^{(k)} \right\rangle_A + \beta_k \left\langle d^{(k)}, d^{(k)} \right\rangle_A = 0 \iff \beta_k = \frac{g^{(k+1)^T} A d^{(k)}}{d^{(k)^T} A d^{(k)}} . \qquad (5.62)$$

In applying the method it is recommended not to compute $g^{(k+1)}$ directly but to use (5.53) instead, because $Ad^{(k)}$ is already necessary to determine α_k and β_k.

The following equivalences hold:

Theorem 5.13 *In case the CG method does not terminate prematurely with $x^{(k-1)}$ being the solution of* (5.1)*, then we have for* $1 \le k \le m$

$$\begin{aligned} K_k(A; g^{(0)}) &= \text{span}\left\{g^{(0)}, Ag^{(0)}, \dots, A^{k-1} g^{(0)}\right\} \\ &= \text{span}\left\{g^{(0)}, \dots, g^{(k-1)}\right\} . \end{aligned} \qquad (5.63)$$

Furthermore,

$$\begin{aligned} g^{(k)^T} g^{(i)} &= 0 \quad \textit{for } i = 0, \dots, k-1, \quad \textit{and} \\ \dim K_k(A; g^{(0)}) &= k . \end{aligned} \qquad (5.64)$$

The space $K_k(A; g^{(0)}) = \text{span}\{g^{(0)}, Ag^{(0)}, \dots, A^{k-1} g^{(0)}\}$ is called the *Krylov (sub)space* of dimension k of A with respect to $g^{(0)}$.

Proof The identities (5.64) are immediate consequences of (5.63) and Theorem 5.12. The proof of (5.63) is given by induction:
For $k = 1$ the assertion is trivial. Let us assume that for $k \ge 1$ the identity (5.63) holds and therefore also (5.64) does. Due to (5.53) (applied to $k-1$) it follows that

$$g^{(k)} \in A\left[K_k\left(A; g^{(0)}\right)\right] + g^{(k-1)} \subset \text{span}\left\{g^{(0)}, \dots, A^k g^{(0)}\right\}$$

and thus

$$\text{span}\left\{g^{(0)}, \dots, g^{(k)}\right\} = \text{span}\left\{g^{(0)}, \dots, A^k g^{(0)}\right\},$$

because the left space is contained in the right one and the dimension of the left subspace is maximal ($= k+1$) due to (5.64) and $g^{(i)} \neq 0$ for all $i = 0, \dots, k$. The identity

$$\text{span}\left\{d^{(0)}, \dots, d^{(k)}\right\} = \text{span}\left\{g^{(0)}, \dots, g^{(k)}\right\}$$

follows from the induction hypothesis and (5.61). □

The number of operations per iteration can be reduced to one matrix-vector, two scalar products, and three AXPY operations, if the following equivalent terms are used:

$$\alpha_k = \frac{g^{(k)^T} g^{(k)}}{d^{(k)^T} A d^{(k)}} , \qquad \beta_k = \frac{g^{(k+1)^T} g^{(k+1)}}{g^{(k)^T} g^{(k)}} . \qquad (5.65)$$

The identities (5.65) can be seen as follows: Concerning α_k we note that because of (5.51) and (5.61),

$$-g^{(k)^T} d^{(k)} = -g^{(k)^T} \left(- g^{(k)} + \beta_{k-1} d^{(k-1)} \right) = g^{(k)^T} g^{(k)} ,$$

and concerning β_k, because of (5.53), (5.64), (5.62), and the identity (5.49) for α_k, we have

$$\begin{aligned} g^{(k+1)^T} g^{(k+1)} = g^{(k+1)^T} \left(g^{(k)} + \alpha_k A d^{(k)} \right) &= \frac{d^{(k)^T} d^{(k)}}{d^{(k)^T} A d^{(k)}} g^{(k+1)^T} A d^{(k)} \\ &= \frac{g^{(k+1)^T} A d^{(k)}}{d^{(k)^T} A d^{(k)}} g^{(k)^T} g^{(k)} \end{aligned}$$

and hence the assumption. The algorithm is summarized in Table 5.2.

Choose any $x^{(0)} \in \mathbb{R}^m$ and calculate

$$d^{(0)} := -g^{(0)} = b - Ax^{(0)} .$$

For $k = 0, 1, \dots$ put

$$\begin{aligned} \alpha_k &= \frac{g^{(k)^T} g^{(k)}}{d^{(k)^T} A d^{(k)}} , \\ x^{(k+1)} &= x^{(k)} + \alpha_k d^{(k)} , \\ g^{(k+1)} &= g^{(k)} + \alpha_k A d^{(k)} , \\ \beta_k &= \frac{g^{(k+1)^T} g^{(k+1)}}{g^{(k)^T} g^{(k)}} , \\ d^{(k+1)} &= -g^{(k+1)} + \beta_k d^{(k)} , \end{aligned}$$

until the termination criterion ("$|g^{(k+1)}|_2 = 0$") is fulfilled.

Table 5.2: CG method.

Indeed, the algorithm defines conjugate directions:

Theorem 5.14 *If $g^{(k-1)} \neq 0$, then $d^{(k-1)} \neq 0$ and the $d^{(0)}, \dots, d^{(k-1)}$ are conjugate.*

Proof The proof is done by induction:

The case $k = 1$ is clear. Assume that $d^{(0)}, \dots, d^{(k-1)}$ are all nonzero and conjugate. Thus according to Theorem 5.12 and Theorem 5.13, the identities (5.60)–(5.64) hold up to index k. Let us first prove that $d^{(k)} \neq 0$:

Due to $g^{(k)} + d^{(k)} = \beta_{k-1} d^{(k-1)} \in K_k(A; g^{(0)})$ the assertion $d^{(k)} = 0$ would imply directly $g^{(k)} \in K_k(A; g^{(0)})$. But relations (5.63) and (5.64) imply for the index k,

$$g^{(k)^T} x = 0 \quad \text{for all } x \in K_k(A; g^{(0)}),$$

which contradicts $g^{(k)} \neq 0$.

In order to prove $d^{(k)^T} A d^{(i)} = 0$ for $i = 0, \dots, k-1$, according to (5.62) we have to prove only the case $i \leq k-2$. We have

$$d^{(i)^T} Ad^{(k)} = -d^{(i)^T} Ag^{(k)} + \beta_{k-1} d^{(i)^T} Ad^{(k-1)} . \tag{5.66}$$

The first term disappears due to $Ad^{(i)} \in A\left(\mathcal{K}_{k-1}\left(A; g^{(0)}\right)\right) \subset \mathcal{K}_k\left(A; g^{(0)}\right)$, which means that $Ad^{(i)} \in \operatorname{span}\left\{d^{(0)}, \dots, d^{(k-1)}\right\}$, and (5.60). The second term disappears because of the induction hypothesis. □

Methods that aim at minimizing the error or residual on $\mathcal{K}_k\left(A; g^{(0)}\right)$ with respect to a norm $\|\cdot\|$ are called *Krylov subspace methods*. Here the error will be minimized in the energy norm $\|\cdot\| = \|\cdot\|_A$ according to (5.55) and Theorem 5.12.

Due to the representation of the Krylov space in Theorem 5.13 the elements $y \in x^{(0)} + \mathcal{K}_k\left(A; g^{(0)}\right)$ are exactly the vectors of the form $y = x^{(0)} + q(A)g^{(0)}$, for any $q \in \mathcal{P}_{k-1}$ (for the notation $q(A)$ see Appendix A.3). Hence it follows that

$$y - x = x^{(0)} - x + q(A)A\left(x^{(0)} - x\right) = p(A)\left(x^{(0)} - x\right),$$

with $p(z) = 1 + q(z)z$, i.e., $p \in \mathcal{P}_k$ and $p(0) = 1$. Polynomials of this type are also called residual polynomials. On the other hand, any such polynomial can be represented in the given form (define q by $q(z) = (p(z) - 1)/z$). Thus Theorem 5.12 implies

$$\left\|x^{(k)} - x\right\|_A \le \|y - x\|_A = \left\|p(A)\left(x^{(0)} - x\right)\right\|_A \tag{5.67}$$

for any $p \in \mathcal{P}_k$ with $p(0) = 1$.

Let $z_1, \dots, z_m$ be an orthonormal basis of eigenvectors, that is,

$$Az_j = \lambda_j z_j \quad \text{and} \quad z_i^T z_j = \delta_{ij} \quad \text{for } i, j = 1, \dots, m . \tag{5.68}$$

Then we have $x^{(0)} - x = \sum_{j=1}^m c_j z_j$ for certain $c_j \in \mathbb{R}$, and hence

$$p(A)\left(x^{(0)} - x\right) = \sum_{j=1}^m p\left(\lambda_j\right) c_j z_j$$

and therefore

$$\left\|x^{(0)} - x\right\|_A^2 = \left(x^{(0)} - x\right)^T A\left(x^{(0)} - x\right) = \sum_{i,j=1}^m c_i c_j z_i^T A z_j = \sum_{j=1}^m \lambda_j \left|c_j\right|^2$$

and analogously

$$\left\|p(A)\left(x^{(0)} - x\right)\right\|_A^2 = \sum_{j=1}^m \lambda_j \left|c_j p(\lambda_j)\right|^2 \le \left(\max_{i=1,\dots,m} |p(\lambda_i)|\right)^2 \left\|x^{(0)} - x\right\|_A^2 . \tag{5.69}$$

Relations (5.67), (5.69) imply the following theorem:

Theorem 5.15 *For the CG method and any $p \in \mathcal{P}_k$ satisfying $p(0) = 1$, we have*

$$\left\|x^{(k)} - x\right\|_A \le \max_{i=1,\dots,m} |p(\lambda_i)| \left\|x^{(0)} - x\right\|_A ,$$

with the eigenvalues $\lambda_1, \dots, \lambda_m$ of A.

If the eigenvalues of A are not known, but their location is, i.e., if one knows $\underline{\sigma}, \overline{\sigma} \in \mathbb{R}$ such that

$$\underline{\sigma} \leq \lambda_1, \ldots, \lambda_m \leq \overline{\sigma}, \tag{5.70}$$

then only the following weaker estimate can be used:

$$\left\| x^{(k)} - x \right\|_A \leq \max_{\lambda \in [\underline{\sigma}, \overline{\sigma}]} |p(\lambda)| \left\| x^{(0)} - x \right\|_A . \tag{5.71}$$

Therefore, we have to find $p \in \mathcal{P}_m$ with $p(0) = 1$ that minimizes $\max \left\{ |p(\lambda)| \,\middle|\, \lambda \in [\underline{\sigma}, \overline{\sigma}] \right\}$.

This approximation problem in the maximum norm appeared already in (5.43), because there is a bijection between the sets $\left\{ p \in P_k \,\middle|\, p(1) = 1 \right\}$ and $\left\{ p \in P_k \,\middle|\, p(0) = 1 \right\}$ through

$$p \mapsto \tilde{p} \quad , \quad \tilde{p}(t) := p(1 - t) . \tag{5.72}$$

Its solution can represented by using the Chebyshev polynomials of the first kind (see, for example, [68, p. 302]). They are recursively defined by

$$T_0(x) := 1, \quad T_1(x) := x, \quad T_{k+1}(x) := 2xT_k(x) - T_{k-1}(x) \quad \text{for } x \in \mathbb{R}$$

and have the representation

$$T_k(x) = \cos(k \arccos(x))$$

for $|x| \leq 1$. This immediately implies

$$|T_k(x)| \leq 1 \quad \text{for } |x| \leq 1 .$$

A further representation, valid for $x \in \mathbb{R}$, is

$$T_k(x) = \frac{1}{2} \left(\left(x + \left(x^2 - 1 \right)^{1/2} \right)^k + \left(x - \left(x^2 - 1 \right)^{1/2} \right)^k \right) . \tag{5.73}$$

The optimal polynomial in (5.71) is then defined by

$$p(z) := \frac{T_k\left((\overline{\sigma} + \underline{\sigma} - 2z)/(\overline{\sigma} - \underline{\sigma}) \right)}{T_k\left((\overline{\sigma} + \underline{\sigma})/(\overline{\sigma} - \underline{\sigma}) \right)} \quad \text{for } z \in \mathbb{R} .$$

This implies the following result:

Theorem 5.16 *Let κ be the spectral condition number of A and assume $\kappa > 1$. Then*

$$\left\| x^{(k)} - x \right\|_A \leq \frac{1}{T_k\left(\frac{\kappa+1}{\kappa-1} \right)} \left\| x^{(0)} - x \right\|_A \leq 2 \left(\frac{\kappa^{1/2} - 1}{\kappa^{1/2} + 1} \right)^k \left\| x^{(0)} - x \right\|_A . \tag{5.74}$$

Proof Choose $\underline{\sigma}$ as the smallest eigenvalue $\lambda_{\min}$ and $\overline{\sigma}$ as the largest one $\lambda_{\max}$.

The first inequality follows immediately from (5.71) and $\kappa = \overline{\sigma}/\underline{\sigma}$. For the second inequality note that due to $(\kappa+1)/(\kappa-1) = 1 + 2/(\kappa-1) =: 1 + 2\eta \geq 1$, (5.73) implies

$$T_k\left(\frac{\kappa+1}{\kappa-1}\right) \geq \frac{1}{2}\left(1 + 2\eta + \left((1+2\eta)^2 - 1\right)^{1/2}\right)^k$$
$$= \frac{1}{2}\left(1 + 2\eta + 2\left(\eta(\eta+1)\right)^{1/2}\right)^k .$$

Finally,

$$1 + 2\eta + 2\left(\eta(\eta+1)\right)^{1/2} = \left(\eta^{1/2} + (\eta+1)^{1/2}\right)^2 = \frac{(\eta+1)^{1/2} + \eta^{1/2}}{(\eta+1)^{1/2} - \eta^{1/2}}$$
$$= \frac{(1+1/\eta)^{1/2} + 1}{(1+1/\eta)^{1/2} - 1},$$

which concludes the proof because of $1 + 1/\eta = \kappa$. □

For large κ we have again

$$\frac{\kappa^{1/2} - 1}{\kappa^{1/2} + 1} \approx 1 - \frac{2}{\kappa^{1/2}} .$$

Compared with (5.58), κ has been improved to $\kappa^{1/2}$.

From (5.4) and (5.34), the complexity of the five-point stencil discretization of the Poisson equation on the square results in

$$\ln\left(\frac{1}{\varepsilon}\right) O\left(\kappa^{1/2}\right) O(m) = O(n)\, O(m) = O\left(m^{3/2}\right) .$$

This is the same behaviour as that of the SOR method with optimal relaxation parameter. The advantage of the above method lies in the fact that the determination of parameters is not necessary for applying the CG method. For quasi-uniform triangulations, Theorem 3.48 implies an analogous general statement.

A relation to the semi-iterative methods follows from (5.77): The estimate (5.67) can also be expressed as

$$\left\|e^{(k)}\right\|_A \leq \left\|p(I-A)e^{(0)}\right\|_A \tag{5.75}$$

for any $p \in \mathcal{P}_k$ with $p(1) = 1$.

This is the same estimate as (5.42) for the Richardson iteration (5.28) as basis method, with the Euclidean norm $|\cdot|_2$ replaced by the energy norm $\|\cdot\|_A$. While the semi-iterative methods are defined by minimization of upper bounds in (5.42), the CG method is optimal in the sense of (5.75), without knowledge of the spectrum $\sigma(I - A)$. In this manner the CG method can be seen as an (optimal) acceleration method for the Richardson iteration.

Exercises

Problem 5.7 Let $A \in \mathbb{R}^{n,n}$ be regular and $b \in \mathbb{R}^n$. Show that, for any $m \geq 1$, the following properties are equivalent:

(i) $b, Ab, \ldots, A^m b$ are linearly dependent.
(ii) $K_m(A; b) = K_{m+1}(A; b)$.
(iii) $A(K_m(A; b)) \subset K_m(A; b)$.
(iv) There is a linear subspace $\mathcal{M} \subset \mathbb{R}^n$ with $\dim(\mathcal{M}) \leq m$ and $b \in \mathcal{M}$ which is invariant with respect to A, i.e., $A(\mathcal{M}) \subset \mathcal{M}$.
(v) For $x = A^{-1}b$ we have $x \in K_m(A; b)$.

Problem 5.8 Show that the representations of the step sizes of the CG methods

$$\alpha_k = \frac{g^{(k)^T} g^{(k)}}{d^{(k)^T} A d^{(k)}}, \qquad \beta_k = \frac{g^{(k+1)^T} g^{(k+1)}}{g^{(k)^T} g^{(k)}}$$

are equivalent to

$$\alpha_k = -\frac{g^{(k)^T} d^{(k)}}{d^{(k)^T} A d^{(k)}}, \qquad \beta_k = \frac{g^{(k+1)^T} A d^{(k)}}{d^{(k)^T} A d^{(k)}}.$$

Problem 5.9 Let $A \in \mathbb{R}^{m,m}$ be a symmetric positive definite matrix.

a) Show that for x, y with $x^T y = 0$ we have

$$\frac{\langle x, y \rangle_A}{\|x\|_A \|y\|_A} \leq \frac{\kappa - 1}{\kappa + 1},$$

where κ denotes the spectral condition number of A.

Hint: Represent x, y in terms of an orthonormal basis consisting of eigenvectors of A.

b) Show using the example $m = 2$ that this estimate is sharp. To this end, look for a positive definite symmetric matrix $A \in \mathbb{R}^{2,2}$ as well as vectors $x, y \in \mathbb{R}^2$ with $x^T y = 0$ and

$$\frac{\langle x, y \rangle_A}{\|x\|_A \|y\|_A} = \frac{\kappa - 1}{\kappa + 1}.$$

Problem 5.10 Prove that the computation of the conjugate direction in the CG method in the general step $k \geq 2$ is equivalent to the three-term recursion formula

$$d^{(k+1)} = [\alpha_k A + (\beta_k + 1)I]\, d^{(k)} - \beta_{k-1} d^{(k-1)}.$$

Problem 5.11 Let $A \in \mathbb{R}^{m,m}$ be a symmetric positive definite matrix with spectral condition number κ and $x := A^{-1}b$. Suppose that the spectrum $\sigma(A)$ of the matrix A satisfies $\sigma_0 \in \sigma(A)$ as well as $\sigma(A) \setminus \{\sigma_0\} \subset [\underline{\sigma}, \overline{\sigma}]$ with $0 < \sigma_0 < \underline{\sigma} \leq \overline{\sigma}$.

Show that this yields the following convergence estimate for the CG method:

$$\|x^{(k)} - x\|_A \leq 2\, \frac{\overline{\sigma} - \sigma_0}{\sigma_0} \left(\frac{\hat{\kappa}^{1/2} - 1}{\hat{\kappa}^{1/2} + 1} \right)^{k-1} \|x^{(0)} - x\|_A ,$$

where $\hat{\kappa} := \overline{\sigma}/\underline{\sigma}\ (< \kappa)$.

What changes for $\underline{\sigma} \leq \overline{\sigma} < \sigma_0$?

5.3 Preconditioned Conjugate Gradient Method

Due to Theorem 5.16, $\kappa(A)$ should be small or only weakly growing in m, which is not true for a finite element stiffness matrix.

The technique of preconditioning is used—as already discussed in Section 5.1—to transform the system of equations in such a way that the condition number of the system matrix is reduced without increasing the effort in the evaluation of the matrix-vector product too much.

In a *preconditioning from the left*, the system of equations is transformed to

$$C^{-1}Ax = C^{-1}b$$

with a *preconditioner* C; in a *preconditioning from the right* it is transformed to

$$AC^{-1}y = b,$$

such that $x = C^{-1}y$ is the solution of (5.1). Since the matrices are generally sparse, this always has to be interpreted as a solution of the system of equations $Cx = y$.

If A is symmetric and positive definite, then this property is generally violated by the transformed matrix for both variants, even for a symmetric positive definite C. We assume for a moment to have a decomposition of C with a nonsingular matrix W as

$$C = WW^T .$$

Then $Ax = b$ can be transformed to $W^{-1}AW^{-T}W^Tx = W^{-1}b$, i.e., to

$$By = c \quad \text{with} \quad B = W^{-1}AW^{-T}, \; c = W^{-1}b . \tag{5.76}$$

The matrix B is symmetric and positive definite. The solution x is then given by $x = W^{-T}y$. This procedure is called *split preconditioning*.

Due to $W^{-T}BW^T = C^{-1}A$ and $WBW^{-1} = AC^{-1}$, $B, C^{-1}A$ and AC^{-1} have the same eigenvalues, and therefore also the same spectral condition number κ. Therefore, C should be "close" to A in order to reduce the condition number. The CG method, applied to (5.76) and then back transformed, leads to the *preconditioned conjugate gradient (PCG) method*:

The terms of the CG method applied to (5.76) will all be marked by ~, with the exception of α_k and β_k.

Due to the back transformation

$$x = W^{-T}\tilde{x}$$

the algorithm has the search direction

$$d^{(k)} := W^{-T}\tilde{d}^{(k)}$$

for the transformed iterate

$$x^{(k)} := W^{-T}\tilde{x}^{(k)} . \tag{5.77}$$

The gradient $g^{(k)}$ of (5.44) in $x^{(k)}$ is given by

$$g^{(k)} := Ax^{(k)} - b = W\left(B\tilde{x}^{(k)} - c\right) = W\tilde{g}^{(k)},$$

and hence

$$g^{(k+1)} = g^{(k)} + \alpha_k WB\tilde{d}^{(k)} = g^{(k)} + \alpha_k Ad^{(k)},$$

so that this formula remains unchanged compared with the CG method with a new interpretation of the search direction. The search directions are updated by

$$d^{(k+1)} = -W^{-T}W^{-1}g^{(k+1)} + \beta_k d^{(k)} = -C^{-1}g^{(k+1)} + \beta_k d^{(k)},$$

so that in each iteration step additionally the system of equations $Ch^{(k+1)} = g^{(k+1)}$ has to be solved.

Finally, we have

$$\tilde{g}^{(k)^T}\tilde{g}^{(k)} = g^{(k)^T}C^{-1}g^{(k)} = g^{(k)^T}h^{(k)}$$

and

$$\tilde{d}^{(k)^T}B\tilde{d}^{(k)} = d^{(k)^T}Ad^{(k)},$$

so that the algorithm takes the form of Table 5.3.

The solution of the additional systems of equations for sparse matrices should have the complexity $O(m)$, in order not to worsen the complexity for an iteration. It is not necessary to know a decomposition $C = WW^T$.

Alternatively, the PCG method can be established by noting that $C^{-1}A$ is self-adjoint and positive definite with respect to the energy scalar product $\langle\cdot,\cdot\rangle_C$ defined by C:

$$\langle C^{-1}Ax, y\rangle_C = \left(C^{-1}Ax\right)^T Cy = x^T Ay = x^T C(C^{-1}Ay) = \langle x, C^{-1}Ay\rangle_C$$

and hence also $\langle C^{-1}Ax, x\rangle_C > 0$ for $x \neq 0$.

Choose any $x^{(0)} \in \mathbb{R}^m$ and calculate

$$g^{(0)} = Ax^{(0)} - b, \quad d^{(0)} := -h^{(0)} := -C^{-1}g^{(0)}.$$

For $k = 0, 1, \dots$ put

$$\begin{aligned}
\alpha_k &= \frac{g^{(k)^T} h^{(k)}}{d^{(k)^T} A d^{(k)}}, \\
x^{(k+1)} &= x^{(k)} + \alpha_k d^{(k)}, \\
g^{(k+1)} &= g^{(k)} + \alpha_k A d^{(k)}, \\
h^{(k+1)} &= C^{-1} g^{(k+1)}, \\
\beta_k &= \frac{g^{(k+1)^T} h^{(k+1)}}{g^{(k)^T} h^{(k)}}, \\
d^{(k+1)} &= -h^{(k+1)} + \beta_k d^{(k)},
\end{aligned}$$

up to the termination criterion ("$|g^{(k+1)}|_2 = 0$").

Table 5.3: PCG method.

Choosing the CG method for (5.76) with respect to $\langle \cdot, \cdot \rangle_C$, we obtain precisely the above method.

In case the termination criterion "$\left|g^{(k+1)}\right|_2 = 0$" is used for the iteration, the scalar product must be additionally calculated. Alternatively, we may use "$\left|g^{(k+1)^T} h^{(k+1)}\right| = 0$". Then the residual is measured in the norm $\|\cdot\|_{C^{-1}}$.

Following the reasoning at the end of Section 5.2, the PCG method can be interpreted as an acceleration of a linear stationary method with iteration matrix

$$M = I - C^{-1}A.$$

For a consistent method, we have $N = C^{-1}$ or, in the formulation (5.10), $W = C$. This observation can be extended in such a way that the CG method can be used for the acceleration of iterative methods, for example, also for the multigrid method, which will be introduced in Section 5.5 and due to the deduction of the preconditioned CG method and the identity

$$\left\|x^{(k)} - x\right\|_A = \left\|\tilde{x}^{(k)} - \tilde{x}\right\|_B,$$

which results from the transformation (5.77), the approximation properties for the CG method also hold for the PCG method if the spectral condition number $\kappa(A)$ is replaced by $\kappa(B) = \kappa(C^{-1}A)$. Therefore,

$$\left\|x^{(k)} - x\right\|_A \le 2 \left(\frac{\kappa^{1/2} - 1}{\kappa^{1/2} + 1}\right)^k \left\|x^{(0)} - x\right\|_A$$

with $\kappa = \kappa(C^{-1}A)$.

There is a close relation between those preconditioning matrices C, which keep $\kappa(C^{-1}A)$ small, and well-convergent linear stationary iterative methods with $N = C^{-1}$ (and $M = I - C^{-1}A$) if N is symmetric and positive definite. Indeed,

$$\kappa(C^{-1}A) \le (1 + \varrho(M))/(1 - \varrho(M)) \tag{5.78}$$

if the method defined by M and N is convergent and N is symmetric for symmetric A (see Problem 5.12).

From the considered linear stationary methods because of the required symmetry we may take

- Jacobi's method:

This corresponds exactly to the *diagonal scaling*, which means the division of each equation by its diagonal element. Indeed, from the decomposition (5.18) we have $C = N^{-1} = D$, and the PCG method is equivalent to the preconditioning from the left by the matrix C^{-1} in combination with the usage of the energy scalar product $\langle\cdot,\cdot\rangle_C$.

- The SSOR method:

According to (5.38) we have for fixed $w \in (0, 2)$

$$C = \omega^{-1}(2 - \omega)^{-1}(D + \omega L)D^{-1}(D + \omega L^T)\,.$$

Hence C is symmetric and positive definite. The solution of the auxiliary systems of equations needs only forward and backward substitutions with the same structure of the matrix as for the system matrix, so that the requirement of lower complexity is also fulfilled. An exact estimation of $\kappa(C^{-1}A)$ shows (see [4, pp. 328]) that under certain requirements for A, which reflect properties of the boundary value problem and the discretization, we find a considerable improvement of the conditioning by using an estimate of the type

$$\kappa(C^{-1}A) \le \text{const}(\kappa(A)^{1/2} + 1)\,.$$

The choice of the relaxation parameter ω is not critical. Instead of trying to choose an optimal one for the contraction number of the SSOR method, we can minimize an estimation for $\kappa(C^{-1}A)$ (see [4, p. 337]), which recommends a choice of ω in $[1.2, 1.6]$.

For the five-point stencil discretization of the Poisson equation on the square we have, according to (5.34), $\kappa(A) = O(n^2)$, and the above conditions are fulfilled (see [4, pp. 330 f.]). By SSOR preconditioning this is improved to $\kappa(C^{-1}A) = O(n)$, and therefore the complexity of the method is

$$\ln\left(\frac{1}{\varepsilon}\right) O\big(\kappa^{1/2}\big)O(m) = \ln\left(\frac{1}{\varepsilon}\right) O\big(n^{1/2}\big)O(m) = O\big(m^{5/4}\big)\,. \tag{5.79}$$

As discussed in Section 2.5, direct elimination methods are not suitable in conjunction with the discretization of boundary value problems with large node numbers, because in general fill-in occurs. As discussed in Section 2.5, $L = (l_{ij})$ describes a lower triangular matrix with $l_{ii} = 1$ for all $i = 1, \ldots, m$ (the dimension is described there with the number of degrees of freedom M) and $U = (u_{ij})$ an upper triangular matrix. The idea of the *incomplete LU factorization*, or *ILU factorization*, is to allow only certain *patterns* $\mathcal{E} \in \{1, \ldots, m\}^2$ for the entries of L and U, and instead of $A = LU$, in general we can require only

$$A = LU - R.$$

Here the remainder $R = (r_{ij}) \in \mathbb{R}^{m,m}$ has to satisfy

$$r_{ij} = 0 \quad \text{for } (i, j) \in \mathcal{E}\,. \tag{5.80}$$

The requirements

$$a_{ij} = \sum_{k=1}^{m} l_{ik} u_{kj} \quad \text{for } (i, j) \in \mathcal{E} \tag{5.81}$$

mean $|\mathcal{E}|$ equations for the $|\mathcal{E}|$ entries of the matrices L and U. (Notice that $l_{ii} = 1$ for all i.) The existence of such factorizations will be discussed later.

Analogously to the close connection between the LU factorization and an $\mathrm{LDL^T}$ or $\mathrm{LL^T}$ factorization for symmetric or symmetric positive definite matrices, as defined in Section 2.5, we can use the *IC factorization* (*incomplete Cholesky factorization*) for such matrices. The IC factorization needs a representation in the following form:

$$A = LL^T - R\,.$$

Based on an ILU factorization a linear stationary method is defined by $N = (LU)^{-1}$ (and $M = I - NA$), the *ILU iteration*. We thus have an expansion of the old method of *iterative refinement*.

Using $C = N^{-1} = LU$ for the preconditioning, the complexity of the auxiliary systems depends on the choice of the matrix pattern $\mathcal{E}$. In general, the following is required:

$$\mathcal{E}' := \big\{(i, j) \,\big|\, a_{ij} \neq 0,\ i, j = 1, \ldots, m\big\} \subset \mathcal{E}, \quad \big\{(i, i) \,\big|\, i = 1, \ldots, m\big\} \subset \mathcal{E}\,. \tag{5.82}$$

The requirement of equality $\mathcal{E}' = \mathcal{E}$ is most often used. Then, and also in the case of fixed expansions of $\mathcal{E}'$, it is ensured that for a sequence of systems of equations with a matrix A that is sparse in the strict sense, this will also hold for L and U. All in all, only $O(m)$ operations are necessary, including the calculation of L and U, as in the case of the SSOR preconditioning for the auxiliary system of equations. On the other hand, the remainder R should be rather small in order to ensure a good convergence of the ILU iteration and also to ensure a small spectral condition number $\kappa(C^{-1}A)$. Possible matrix patterns $\mathcal{E}$ are shown, for example, in [54, Sect. 10.3.2], where a more specific structure of L and U is discussed if the matrix A is created by a discretization on a structured grid, for example, by a finite difference method.

The question of the existence (and stability) of an ILU factorization remains to be discussed. It is known from (2.60) that also for the existence of an LU factorization certain conditions are necessary, as, for example, the M-matrix property. This is even sufficient for an ILU factorization.

Theorem 5.17 *Let $A \in \mathbb{R}^{m,m}$ be an M-matrix. Then for a given pattern $\mathcal{E}$ that satisfies* (5.82)*, an ILU factorization exists. The hereby defined decomposition of A as $A = LU - R$ is* regular *in the following sense:*

$$\left((LU)^{-1}\right)_{ij} \geq 0, \quad (R)_{ij} \geq 0 \quad \textit{for all } i, j = 1, \dots, m\,.$$

Proof See [33, Thm. 7.35]. □

An ILU (or IC) factorization can be defined by solving the equations (5.80) for l_{ij} and u_{ij} in an appropriate order. Alternatively, the elimination or Cholesky method can be used in its original form on the pattern $\mathcal{E}$.

An improvement of the eigenvalue distribution of $C^{-1}A$ is sometimes possible by using an MIC factorization (***m****odified* ***i****ncomplete* ***C****holesky factorization*) instead of an IC factorization. In contrast to (5.81) the updates in the elimination method for positions outside the pattern are not ignored here but have to be performed for the corresponding diagonal element.

Concerning the reduction of the condition number by the ILU or IC preconditioning for the model problem, we have the same situation as for the SSOR preconditioning. In particular (5.79) holds, too.

The auxiliary system of equations with $C = N^{-1}$, which means that

$$h^{(k+1)} = Ng^{(k+1)}\,,$$

can also be interpreted as an iteration step of the iterative method defined by N with initial value $z^{(0)} = 0$ and right-hand side $g^{(k+1)}$. An expansion of the discussed possibilities for preconditioning is therefore obtained by using a fixed number of iteration steps instead of only one.

Exercises

Problem 5.12 Let $A_1, A_2, \dots, A_k, C_1, C_2, \dots, C_k \in \mathbb{R}^{m,m}$ be symmetric positive semidefinite matrices with the property

$$\underline{\sigma} x^T C_i x \leq x^T A_i x \leq \overline{\sigma} x^T C_i x \quad \text{for all } x \in \mathbb{R}^m,\ i = 1, \dots, k, \text{ and } 0 < \underline{\sigma} \leq \overline{\sigma}\,.$$

Prove: If $A := \sum_{i=1}^k A_i$ and $C := \sum_{i=1}^k C_i$ are positive definite, then the spectral condition number κ of $C^{-1}A$ satisfies

$$\kappa(C^{-1}A) \leq \frac{\overline{\sigma}}{\underline{\sigma}}\,.$$

Problem 5.13 Investigate the convergence of the (P)CG method on the basis of Theorem 3.48 and distinguish between $d = 2$ and $d = 3$.

5.4 Krylov Subspace Methods for Nonsymmetric Systems of Equations

With the different variants of the PCG method we have methods that are quite appropriate—regarding their complexity—for those systems of equations that arise from the discretization of boundary value problems. However, this holds only under the assumption that the system matrix is symmetric and positive definite, reducing the possibilities of application, for example, to finite element discretizations of purely diffusive processes without convective transport mechanism (see (3.24)). Exceptions for time-dependent problems are only the (semi-)explicit time discretization (compare (9.98)) and the Lagrange–Galerkin method (see Section 10.4). In all other cases the occurring system matrices are always nonsymmetric and possibly positive real, whereby a matrix $A \in \mathbb{R}^{m,m}$ is called *positive real* if

$$A + A^T \quad \text{is positive definite.}$$

This in turn is equivalent to

$$(Ax)^T x > 0 \quad \text{for } x \in \mathbb{R}^m, x \neq 0$$

(see Problem 5.14) and thus a consequence of the V-coercivity of the underlying variational formulation for which sufficient conditions are formulated in Theorem 3.16, Remark 3.17.

It is desirable to generalize the (P)CG methods for such matrices. The CG method is characterized by two properties:

- The iterate $x^{(k)}$ minimizes $f(\cdot) = \|\cdot - x\|_A$ on $x^{(0)} + K_k\left(A; g^{(0)}\right)$, where $x = A^{-1}b$.
- The basis vectors $d^{(i)}$, $i = 0, \ldots, k-1$, of $K_k\left(A; g^{(0)}\right)$ do not have to be calculated in advance (and stored in the computer), but will be calculated by a *three-term recursion* (5.61) during the iteration. An analogous relation holds by definition for $x^{(k)}$ (see (5.48)).

The first property can be preserved in the following, whereby the norm of the error or residual minimization varies in each method. The second property is partially lost, because generally all basis vectors $d^{(0)}, \ldots, d^{(k-1)}$ are necessary for the calculation of $x^{(k)}$. This will result in memory space problems for large k. As for the CG methods, preconditioning will be necessary for an acceptable convergence of the methods. The conditions for the preconditioning matrices are the same as for the CG method with the exception of symmetry and positive definiteness. All three methods of preconditioning are in principle possible. From now on we assume that $A \in \mathbb{R}^{m,m}$ to be invertible, possibly with further properties, if noted.

The simplest approach is the application of the CG method to a system of equations with symmetric positive definite matrix equivalent to (5.1). This is the case for the *normal equations*

$$A^T Ax = A^T b\,. \tag{5.83}$$

The approach is called *CGNR* (***C****onjugate* ***G****radient* ***N****ormal* ***R****esidual*), because here the iterate $x^{(k)}$ minimizes the Euclidean norm of the residual on $x^{(0)} + K_k\left(A^T A; g^{(0)}\right)$ with $g^{(0)} = A^T\left(Ax^{(0)} - b\right)$. This follows from the equation

$$\|y - x\|^2_{A^T A} = (Ay - b)^T (Ay - b) = |Ay - b|_2^2 \tag{5.84}$$

for any $y \in \mathbb{R}^m$ and the solution $x = A^{-1}b$.

All advantages of the CG method are preserved, although in (5.53) and (5.65) $Ad^{(k)}$ is to be replaced by $A^T Ad^{(k)}$. Additionally to the doubling of the number of operations this may be a disadvantage if $\kappa_2(A)$ is large, since $\kappa_2(A^T A) = \kappa_2(A)^2$ can lead to problems of stability and convergence. Due to (5.34) this is to be expected for a large number of degrees of freedom.

Furthermore, in the case of list-based storage one of the operations Ay and $A^T y$ is always very expensive due to searching. It is even possible that we do not explicitly know the matrix A but that only the mapping $y \mapsto Ay$ can be evaluated, which then disqualifies this method completely (see Problem 11.10).

The same drawback occurs if

$$AA^T \tilde{x} = b \tag{5.85}$$

with the solution $\tilde{x} = A^{-T}x$ taken instead of (5.83). If $\tilde{x}^{(k)}$ is the kth iterate of the CG method applied to (5.85), then the $x^{(k)} := A^T \tilde{x}^{(k)}$ minimizes the error in the Euclidean norm on $x_0 + A^T\left[K_k\left(AA^T; g^{(0)}\right)\right]$: Note that

$$\|\tilde{y} - \tilde{x}\|^2_{AA^T} = \left(A^T\tilde{y} - x\right)^T \left(A^T\tilde{y} - x\right) = \left|A^T\tilde{y} - x\right|_2^2$$

holds for any $\tilde{y} \in \mathbb{R}^m$ and $g^{(0)} = Ax^{(0)} - b$. This explains the terminology CGNE (with *E* for ***E****rror*).

Whether a method minimizes the error or the residual obviously depends on the norm used. For a symmetric positive definite $B \in \mathbb{R}^{m,m}$, any $y \in \mathbb{R}^m$, and $x = A^{-1}b$, we have

$$\|Ay - b\|_B = \|y - x\|_{A^T BA}\,.$$

For $B = A^{-T}$ and a symmetric positive definite A we get the situation of the CG method:

$$\|Ay - b\|_{A^{-T}} = \|y - x\|_A\,.$$

For $B = I$ we get again (5.84):

$$|Ay - b|_2 = \|y - x\|_{A^T A}\,.$$

The minimization of this functional on $x^{(0)} + K_k(A; g^{(0)})$ (not $K_k(A^T A; g^{(0)})$) leads to the GMRES method (***G**eneralized **M**inimum **RES**idual*).

There are several methods based on the minimization of the Euclidean norm of the residuum, from which GMRES is the modern form of the full Krylov subspace method. So instead of (5.44) we consider

$$f(x) = |Ax - b|_2^2 \tag{5.86}$$

and follow the reasoning of Section 5.2. We can define a method of the form (5.47), (5.48) by substituting (5.49) by

$$\alpha_k = \frac{g^{(k)^T} Ad^{(k)}}{(Ad^{(k)})^T Ad^{(k)}} \tag{5.87}$$

because of

$$\nabla f(x) = 2A^T(Ax - b). \tag{5.88}$$

Although we keep the notation for further comparison, $g(\tilde{x}) = A\tilde{x} - b$ is not the gradient of the functional anymore, but the residuum. The relation (5.51) then has to be substituted by

$$g^{(k+1)^T} Ad^{(k)} = 0, \tag{5.89}$$

and (5.52) , (5.53) remain valid.

For the choice (5.56) we get an analogue of the gradient method, sometimes called *Orthomin(1)*.

As in general,

$$\tilde{f}(\alpha) = \left|A\left(x^{(k)} + \alpha d^{(k)}\right) - b\right|_2^2 = \left|g^{(k)} + \alpha Ad^{(k)}\right|_2^2,$$

$-\alpha_k Ad^{(k)}$ is the orthogonal projection of $g^{(k)}$ onto the corresponding one-dimensional space, and for $d^{(k)} = -g^{(k)}$ it holds that

$$\left|g^{(k+1)}\right|_2^2 = \tilde{f}(\alpha_k) \le \tilde{f}(\alpha) = \left|(I - \alpha A)\, g^{(k)}\right|_2^2 \quad \text{for all } \alpha \in \mathbb{R}, \tag{5.90}$$

therefore

$$\left|g^{(k+1)}\right|_2 \le \|I - \alpha A\|_2 \left|g^{(k)}\right|_2 .$$

Assume that A is normal with eigenvalues $\lambda_i \in \mathbb{C}$, $i = 1, \ldots, m$, then

$$\left|g^{(k+1)}\right|_2 \le \max_{i=1,\ldots,m} |1 - \alpha\lambda_i| \left|g^{(k)}\right|_2 \quad \text{for all } \alpha \in \mathbb{R}.$$

If all eigenvalues are real, $\lambda_{\min} \le \lambda \le \lambda_{\max}$, then we can choose $\alpha := 2(\lambda_{\min} + \lambda_{\max})$ and reach at

$$\left|g^{(k+1)}\right|_2 \le \left(\frac{\kappa-1}{\kappa+1}\right)\left|g^{(k)}\right|_2 \tag{5.91}$$

with the condition number $\kappa = \lambda_{\max}/\lambda_{\min}$. This is the same convergence behaviour (for the residuum, not for the error) as for the gradient method in Theorem 5.10. If A is symmetric and positive definite, then it is a convergence estimate in $\|\cdot\|_{A^{1/2}}$.

If A is positive real, then another (worse) estimate is possible. Namely, there exists a constant $\tau > 0$ such that

$$(Ax)^T x \ge \tau |x|^2 \quad \text{for } x \in \mathbb{R}^m, \tag{5.92}$$

see Problem 5.14. Starting from (5.90) we have

$$\begin{aligned}\left|g^{(k+1)}\right|_2^2 &\le \left|g^{(k)}\right|_2^2 - (2\alpha A^T g^{(k)})^T g^{(k)} + \alpha^2 \left|Ag^{(k)}\right|_2^2 \\ &= \left|g^{(k)}\right|_2^2 - \frac{\left[(Ag^{(k)})^T g^{(k)}\right]^2}{|Ag^{(k)}|_2^2} \quad \text{for the choice } \alpha := \frac{(A^T g^{(k)})^T g^{(k)}}{|Ag^{(k)}|_2^2} \\ &\le \left|g^{(k)}\right|_2^2 - \frac{\tau^2}{\|A\|^2}\left|g^{(k)}\right|_2^2 = \left(1 - \frac{\tau^2}{\|A\|^2}\right)\left|g^{(k)}\right|_2^2,\end{aligned} \tag{5.93}$$

where $\|A\|$ is the matrix norm induced by $|\cdot|_2$, i.e., the spectral norm. Under the symmetry assumption for A—a stronger assumption than for (5.91)—the estimate reads

$$\left|g^{(k+1)}\right|_2 \le \left(1-\kappa^{-2}\right)^{1/2}\left|g^{(k)}\right|_2,$$

i.e., it is worse.

As these convergence rates are not satisfactory, we try to mimic the CG method: The search directions are defined by (5.61) with the requirement

$$\left(Ad^{(k+1)}\right)^T Ad^{(k)} = 0, \tag{5.94}$$

which gives instead of (5.62):

$$\beta_k = \frac{\left(Ag^{(k+1)}\right)^T Ad^{(k)}}{\left(Ag^{(k)}\right)^T Ad^{(k)}}. \tag{5.95}$$

Then not only (5.89) holds true, but also

$$g^{(k+1)^T} Ad^{(k-1)} = g^{(k)^T} Ad^{(k-1)} - \alpha_k \left(Ad^{(k)}\right)^T Ad^{(k-1)} = 0$$

(as can be proved by induction), so that $g^{(k)}$ is not projected onto span $\{Ad^{(k)}\}$, but onto span $\{Ad^{(k)}, Ad^{(k-1)}\}$. Thus we have a method, similar to the CG method (see Table 5.2, with α_k, β_k given by (5.87), (5.95)), called *Orthomin(2)*.

However, in general this method is not advisable, as it may be even ill-defined: If x_0 is such that $g^{(0)^T} Ag^{(0)} = 0$, i.e., $\alpha_0 = 0$ and $x_1 = x_0$, then $d^{(1)} = 0$. The picture changes for symmetric A (not necessarily positive definite).

Theorem 5.18 *Assume that A is symmetric and in Orthomin(2) all step sizes α_i, $i = 0, \ldots, k-1$, are well defined, $\alpha_i \neq 0$ and $g^{(k-1)} \neq 0$. Then*

$$g^{(k)^T} Ad^{(i)} = \left(Ad^{(k)}\right)^T Ad^{(i)} = 0 \quad \textit{for } i = 0, \ldots, k-1,$$

and then $x^{(k)}$ is such that $g^{(k)}$ is norm-minimal on

$$g^{(0)} + A\left[K_k\left(A; g^{(0)}\right)\right].$$

In other words: $x^{(k)}$ minimizes $|Ay - b|_2^2$ on $x^{(0)} + K_k\left(A; g^{(0)}\right)$.

Proof The proof can be accomplished by repeating the proofs of Theorem 5.13 and Theorem 5.14, which do not use the specific expressions for α_i, β_i, but only the symmetry of A in (5.66). □

This concept can be expanded to an Orthomin(j) method, $j \in \mathbb{N}$. In the MINRES approach, the requirement to use only a fixed number of basis vectors of the Krylov subspace and then to have a $(j+1)$-term recursive relation is given up in favour of directly minimizing over the full Krylov subspace. The specific form of the GMRES (and other) methods are founded algorithmically on the recursive construction of orthonormal bases of $K_k\left(A; g^{(0)}\right)$ by *Arnoldi's method*. This method combines the generation of a basis according to (5.61) and Schmidt's orthonormalization (see Table 5.4).

Let $g^{(0)} \in \mathbb{R}^m$, $g^{(0)} \neq 0$ be given, Set

$$v_1 := g^{(0)} / |g^{(0)}|_2 .$$

For $j = 1, \ldots, k$ calculate

$$h_{ij} := v_i^T Av_j \quad \text{for } i = 1, \ldots, j,$$

$$w_j := Av_j - \sum_{i=1}^{j} h_{ij} v_i ,$$

$$h_{j+1,j} := |w_j|_2 .$$

If $h_{j+1,j} = 0$, termination; otherwise, set

$$v_{j+1} := w_j / h_{j+1,j} .$$

Table 5.4: Arnoldi algorithm.

If Arnoldi's method can be performed up to the index k, then

$$
\begin{aligned}
h_{ij} &:= 0 \quad \text{for } j = 1, \ldots, k,\ i = j+2, \ldots, k+1\,, \\
H_k &:= \left(h_{ij}\right)_{ij} \in \mathbb{R}^{k,k}\,, \\
\bar{H}_k &:= \left(h_{ij}\right)_{ij} \in \mathbb{R}^{k+1,k}\,, \\
V_{k+1} &:= (v_1, \ldots, v_{k+1}) \in \mathbb{R}^{m,k+1}\,.
\end{aligned}
$$

The matrix H_k is an upper Hessenberg matrix (see Appendix A.3). The basis for the GMRES method is the following theorem:

Theorem 5.19 *If Arnoldi's method can be performed up to the index k, then*

1) $v_1, \ldots, v_{k+1}$ form an orthonormal basis of $K_{k+1}(A; g^{(0)})$.
2)

$$
AV_k = V_k H_k + w_k \tilde{e}_k^T = V_{k+1} \bar{H}_k\,, \tag{5.96}
$$

with $\tilde{e}_k = (0, \ldots, 0, 1)^T \in \mathbb{R}^k$,

$$
V_k^T A V_k = H_k\,. \tag{5.97}
$$

3) The problem

$$
\text{Minimize} \quad |Ay - b|_2 \quad \text{for} \quad y \in x^{(0)} + K_k(A; g^{(0)})
$$

with minimum $x^{(k)}$ is equivalent to

$$
\text{Minimize} \quad \left|\bar{H}_k \xi - \beta e_1\right|_2 \quad \text{for} \quad \xi \in \mathbb{R}^k \tag{5.98}
$$

with $\beta := -\left|g^{(0)}\right|_2$ and minimum $\xi^{(k)}$, and we have

$$
x^{(k)} = x^{(0)} + V_k \xi^{(k)}\,.
$$

If Arnoldi's method terminates at the index k, then

$$
x^{(k)} = x = A^{-1} b\,.
$$

Proof 1): The vectors $v_1, \ldots, v_{k+1}$ are orthonormal by construction; hence we have only to prove $v_i \in K_{k+1}\left(A; g^{(0)}\right)$ for $i = 1, \ldots, k+1$. This follows from the representation

$$
v_i = q_{i-1}(A) v_1 \quad \text{with polynomials} \quad q_{i-1} \in \mathcal{P}_{i-1}\,.
$$

In this form we can prove the statement by induction with respect to k. For $k = 0$ the assertion is trivial. Let the statement hold for $k - 1$. The validity for k then follows from

$$
h_{k+1,k} v_{k+1} = A v_k - \sum_{i=1}^{k} h_{ik} v_i = \left(A q_{k-1}(A) - \sum_{i=1}^{k} h_{ik} q_{i-1}(A) \right) v_1\,.
$$

2): Relation (5.97) follows from (5.96) by multiplication by V_k^T, since $V_k^T V_k = I$ and $V_k^T w_k = h_{k+1,k} V_k^T v_{k+1} = 0$ due to the orthonormality of the v_i.

The relation in (5.96) is the matrix representation of

$$Av_j = \sum_{i=1}^{j} h_{ij} v_i + w_j = \sum_{i=1}^{j+1} h_{ij} v_i \quad \text{for} \quad j = 1, \ldots, k\,.$$

3): Due to 1), the space $x^{(0)} + K_k\left(A; g^{(0)}\right)$ has the parametrization

$$y = x^{(0)} + V_k \xi \quad \text{with} \quad \xi \in \mathbb{R}^k\,. \tag{5.99}$$

The assertion is a consequence of the identity

$$\begin{aligned} Ay - b &= A\left(x^{(0)} + V_k \xi\right) - b = A V_k \xi + g^{(0)} \\ &= V_{k+1} \bar{H}_k \xi - \beta v_1 = V_{k+1}\left(\bar{H}_k \xi - \beta e_1\right), \end{aligned}$$

which follows from 2), since it implies

$$|Ay - b|_2 = \left|V_{k+1}(\bar{H}_k \xi - \beta e_1)\right|_2 = \left|\bar{H}_k \xi - \beta e_1\right|_2$$

due to the orthogonality of V_{k+1}. The last assertion finally can be seen in this way: If Arnoldi's method breaks down at the index k, then relation 2) becomes

$$A V_k = V_k H_k\,,$$

and

$$A V_k = V_{k+1} \bar{H}_k$$

will further hold with v_{k+1} chosen arbitrarily (due to $h_{k+1,k} = 0$). Since A is nonsingular, this also holds for H_k. Hence the choice

$$\xi := H_k^{-1}(\beta e_1),$$

which satisfies

$$\left|\bar{H}_k \xi - \beta e_1\right|_2 = |H_k \xi - \beta e_1|_2 = 0,$$

is possible. Hence the corresponding $y \in \mathbb{R}^m$ defined by (5.99) fulfils $y = x^{(k)} = x$. □

The solution of a least squares problem like (5.98) can be accomplished by a QR decomposition of $\bar{H}_k$, where $Q \in \mathbb{R}^{k+1,k+1}$ is orthogonal and $R \in \mathbb{R}^{k+1,k}$ is upper triangular, as then

$$\left|\bar{H}_k \xi - \beta e_1\right|_2 = \left|R\xi - \beta Q^T e_1\right|_2,$$

i.e., a backward substitution has to be performed in k variables, with the residuum $\beta q_{1,k+1}$. As $\bar{H}_k$ is an upper Hessenberg matrix, the QR decomposition can be accomplished by only one Givens rotation per column l to eliminate the $(l+1, l)$ entry. As only this column and the right-hand side is involved, this procedure can be intertwined with the setup of the columns, i.e., with the Arnoldi process. The size of the residuum in (5.98), i.e., $|\beta q_{1,k+1}|$, can be used as a stopping criterion in the iteration (Table 5.5).

Let $x^{(0)} \in \mathbb{R}^m$, $g^{(0)} := Ax^{(0)} - b \neq 0$ be given, $\beta := -\left|g^{(0)}\right|_2$, $f := \beta e_1$.

Set $v_1 := g^{(0)} / \left|g^{(0)}\right|_2$.

For $k = 1, \dots$

$$V_k = (v_1, \dots, v_k),$$
$$v_{k+1} := Av_k .$$

For $j = 1, \dots, k$

$$h_{jk} := v_{k+1}^T v_j ,$$
$$v_{k+1} := v_{k+1} - h_{jk} v_j ,$$

$h_{k+1,k} := |v_{k+1}|_2$.

Test for loss of orthogonality and reorthogonalize, if necessary

$$v_{k+1} := v_{k+1} / h_{k+1,k} .$$

If $k > 1$:

$$\nu := \left(h_{kk}^2 + h_{k+1,k}^2\right)^{1/2} ,$$
$$c_k := h_{kk}/\nu, \; s_k := -h_{k+1,k}/\nu.$$

Defining the Givens rotation $G_k(c_k, s_k)$:

$$h_{kk} := c_k h_{kk} - s_k h_{k+1,k}, \; h_{k+1,k} := 0,$$
$$f := G_k(s_k, c_k) f .$$

Set $r_{ij} := h_{ij}, \; 1 \leq i, j \leq k,$

$w_i := f_i, \; 1 \leq i \leq k.$

Do backward substitution to solve

$$Ry^k = w,$$

$x_k := x_0 + V_k y^k$.

Table 5.5: The GMRES method with Givens rotation.

If we apply the above procedure to a symmetric matrix, then due to (5.97) also H_k is symmetric, i.e., tridiagonal such that only the entries $\alpha_i = H_{i,i}$ and $\beta_i = H_{i+1,i}$ have to be computed. The adopted GMRES method is also called *MINRES* (***MIN****imal* ***RES****idual*).

One problem of Arnoldi's method is that the orthogonality of the v_i is easily lost due to rounding errors. If the assignment

$$w_j := Av_j - \sum_{i=1}^{j} h_{ij} v_i$$

in Table 5.4 is replaced by the operations

$$
\begin{aligned}
&w_j := A v_j\,, \\
&\text{For } i = 1, \dots, j \text{ calculate} \\
&\qquad h_{ij} := w_j^T v_i\,, \\
&\qquad w_j := w_j - h_{ij} v_i\,,
\end{aligned}
$$

which define the same vector, we get the *modified Arnoldi's method*. From this relation and from (5.98), the GMRES method is constructed in its basic form. Alternatively, Schmidt's orthonormalization can be replaced by the Householder method (see [54, Sect. 6.5]). In exact arithmetic the GMRES algorithm terminates only after reaching the exact solution (with $h_{k+1,k} = 0$). This is not always the case for alternative methods of the same class. For an increasing iteration index k and large problem dimensions m, there may be lack of enough memory for storing the basis vectors $v_1, \dots, v_k$. A remedy is offered by working with a fixed number n of iterations and then to restart the algorithm with $x^{(0)} := x^{(n)}$ and $g^{(0)} := Ax^{(0)} - b$, until finally the convergence criterion is fulfilled (*GMRES method with restart*).

The Orthomin(j) methods thus can be viewed as truncated versions of the GMRES method and should be implemented accordingly. Concerning convergence estimates, assume that A is diagonalizable and $A = V\Lambda V^{-1}$ an eigendecomposition, where $\Lambda = \operatorname{diag}(\lambda_1, \dots, \lambda_m)$ is composed of the eigenvalues $\lambda_i \in \mathbb{C}$.

Similar to (5.66) we get for $y \in x^{(0)} + K_k\left(A; g^{(0)}\right)$

$$
Ay - b = A(y - x) = g^{(0)} + Aq(A)g^{(0)} \quad \text{for all } q \in \mathcal{P}_{k-1},
$$

i.e.

$$
Ay - b = p(A)g^{(0)} \quad \text{for all } p \in \mathcal{P}_k \text{ such that } p(0) = 1.
$$

Therefore, by definition for all such residual polynomials it holds

$$
\left|g^{(k)}\right|_2 \le \left|p(A)g^{(0)}\right|_2 \le \|p(A)\|_2 \left|g^{(0)}\right|_2,
$$

and with the assumptions about A:

$$
\le \kappa(V)\|p(\Lambda)\|_2 |g^{(0)}|_2 < \kappa(V) \max_{i=1,\dots,m} |p(\lambda_i)| \left|g^{(0)}\right|_2 . \tag{5.100}
$$

Thus we have achieved (for the norm of the gradient) the analogues of Theorem 5.15, but now with the additional factor $\kappa(V)$. To achieve as above an error estimate for all initial iterates, i.e., of worst case type, the two factors in (5.100) have to be estimated, both reflecting the distribution of eigenvalues. Analogously to Theorem 5.16 we get

Theorem 5.20 *For the GMRES method we have for a diagonalizable $A \in \mathbb{R}^{m,m}$, with $A = V\Lambda V^{-1}$ and $\Lambda = \operatorname{diag}(\lambda_1, \dots, \lambda_n)$, $\lambda_i \in \mathbb{C}$:*

$$\left|Ax^{(k)} - b\right|_2 \leq 2\kappa(V) \left(\frac{\kappa^{1/2} - 1}{\kappa^{1/2} + 1}\right)^k \left|Ax^{(0)} - b\right|_2$$

with $\kappa = \kappa(A)$.

Thus normal matrices seem to be an advantage, and scaling of the eigenbasis may help to tighten the above estimate. For A being positive real, also the estimate (5.93) holds true, as $g^{(k)} + \alpha A g^{(k)} \in g^{(0)} + A\left[K_k\left(A; g^{(0)}\right)\right]$.

Nevertheless, for a general matrix and initial iterates, the worst case

$$x_0 = \ldots = x_{m-1}, \quad x_m = x \tag{5.101}$$

is possible regardless of the eigenvalues. Consider a matrix, being derived from a lower triangular matrix by shifting the last column to the first position (such a pattern cannot occur in stiffness matrices). If $g^{(0)} = \gamma_1 e_1$, then $Ag^{(0)} = \gamma_2 e_m$, $A^2 g^{(0)} = \gamma_3 e_{m-1} + \gamma_4 e_m$ etc., such that all these vectors are orthogonal to $g^{(0)}$ until $k = m - 1$, and thus (5.101), i.e., the GMRES method makes no progress until the last step, which has to give the exact solution. As the class of matrices described above contains the companion matrices for a general polynomial, every desired set of eigenvalues is possible. Nothing will be changed for matrices being unitary similar to the ones from above.

The minimization of the error in the energy norm (on a vector space K) as with the CG method makes sense only for symmetric positive definite matrices A. However, the variational equation

$$(Ay - b)^T z = 0 \quad \text{for all } z \in K, \tag{5.102}$$

which characterizes this minimum in general, can be taken as defining condition for y. Further variants of Krylov subspace methods rely on this. Another large class of such methods is founded on the *Lanczos biorthogonalization*, in which apart from a basis $v_1, \ldots, v_k$ of $K_k(A; v_1)$ another basis $w_1, \ldots, w_k$ of $K_k(A^T; w_1)$ is constructed, such that

$$v_j^T w_i = \delta_{ij} \quad \text{for} \quad i, j = 1, \ldots, k\,.$$

This can be accomplished as indicated in Table 5.6:

Theorem 5.21 *If the two-side Lanczos algorithm can be performed up to* $j = k$, *i.e.* $\tilde{v}_j \neq 0$ *and* $v_j^T w_j \neq 0$ *for* $j = 1, \ldots, k$, *then* $v_i^T w_j = 0$ *for* $i \neq j$, $1 \leq i, j \leq k$.

Proof See Problem 5.17. □

The algorithm can breakdown, if $\tilde{v}_{j+1} = 0$ or $\tilde{w}_{j+1} = 0$ (*regular termination*) or if $\tilde{v}_{j+1}^T w_{j+1} = 0$ (*serious termination*). In the latter case one of the desired vectors does not exist, but it may be for further iterations. This is used in the *look-ahead* versions of the algorithm. Although breakdown cannot be excluded in general, it will be not be considered in the following.

The *Biconjugate Gradient Method* (BiCG) uses the basis $\{v_1 \ldots, v_k\}$ of $K(A; g^{(0)})$ and $\{w_1, \ldots, w_k\}$ of $K(A^T; \hat{g}^{(0)})$ to solve the variational equation (5.102) in

Given $g^{(0)} = Ax^{(0)} - b$ and $\hat{g}^{(0)}$ such that $g^{(0)^T} \hat{g}^{(0)} \neq 0$,

set $v_1 := g^{(0)} / \left|g^{(0)}\right|_2$, $w_1 := \hat{g}^{(0)} / v_1^T \hat{g}^{(0)}$.

Set $\beta_0 := \gamma_0 := 0$, $v_0 := w_0 := 0$.

For $j = 1, \ldots, k$:

$$\begin{aligned}
\alpha_j &:= (Av_j)^T w_j, \\
\tilde{v}_{j+1} &:= Av_j - \alpha_j v_j - \beta_{j-1} v_{j-1}, \\
\tilde{w}_{j+1} &:= A^T w_j - \alpha_j w_j - \gamma_{j-1} w_{j-1}, \\
\gamma_j &:= |\tilde{v}_{j+1}|_2, \qquad \text{if } \gamma_j \neq 0: \ v_{j+1} := \tilde{v}_{j+1}/\gamma_j \text{ otherwise STOP}, \\
\beta_j &:= v_{j+1}^T \tilde{w}_{j+1}, \ \text{if } \beta_j \neq 0: \ w_{j+1} := \tilde{w}_{j+1}/\beta_j \text{ otherwise STOP}.
\end{aligned}$$

Table 5.6: Two-sided Lanczos method.

$x^{(0)} + K(A; g^{(0)})$ with the "test space" $K = K(A^T; \tilde{g}^{(0)})$. Similar to the Arnoldi process, the two-sided Lanczos method can be written in matrix form as

$$\begin{aligned}
AV_k &= V_{k+1} \bar{H}_k, = V_k H_k + \gamma_k v_{k+1} e_k^T, \\
A^T W_k &= W_{k+1} \hat{H}_k = W_k H_k^T + \beta_k w_{k+1} e_k^T, \ W_k^T A V_k = H_k,
\end{aligned}$$

where $V_k = (v_1, \ldots, v_k)$, $W_k = (w_1, \ldots, w_k)$. The matrices $\bar{H}_k, \hat{H}_k \in \mathbb{R}^{k+1,k}$ are such that the upper left part is $H_k \in \mathbb{R}^{k,k}$ or H_k^T, respectively, and the $(k+1)$th row only has the entry $h_{k+1,k} = \gamma_k$ or β_k, respectively. Finally

$$H_k = \begin{pmatrix} \alpha_1 & \beta_1 & 0 \\ \gamma_1 & \ddots & \beta_{k-1} \\ 0 & \gamma_{k-1} & \alpha_k \end{pmatrix}.$$

The biorthogonality is expressed by

$$V_k^T W_k = I.$$

Thus $x^{(k)} = x^{(0)} + V_k y_k \in x^{(0)} + K_k(A; g^{(0)})$ solves the equation (5.102) if and only if

$$0 = W_k^T g^{(k)} = W_k^T g^{(0)} + W_k^T A V_k y_k = \beta e_1 + H_k y_k$$

with $\beta := |g^0|_2$. This tridiagonal system may be singular rendering a further type of breakdown. In the nonsingular case an LDU-factorization can be applied to H_k leading to the following form of the algorithm. In fact, in the same way the CG method can be rediscovered from the GMRES method, as in the case of a symmetric, positive definite matrix the upper Hessenberg matrix H_k in Arnoldi's method is symmetric, i.e., tridiagonal. With the method here we are back to the short recursion of the CG method (in fact for symmetric A it is the CG method with all operations twice

Given $x^{(0)}$, $g^{(0)} := Ax^{(0)} - b$, $d^{(0)} := -g^{(0)}$, $\hat{g}^{(0)}$ such that

$$g^{(0)^T}\hat{g}^{(0)} \neq 0,\ \hat{d}^{(0)} := -\hat{g}^{(0)}.$$

For $k = 1, \ldots$:

$$\begin{aligned}
\alpha_{k-1} &:= g^{(k-1)^T}\hat{g}^{(k-1)}/(Ad^{(k-1)})^T\hat{d}^{(k+1)},\\
x^{(k)} &:= x^{(k-1)} + \alpha_{k-1}d^{(k-1)},\\
g^{(k)} &= g^{(k-1)} + \alpha_{k-1}Ad^{(k-1)},\\
\hat{g}^{(k)} &= \hat{g}^{(k-1)} + \alpha_{k-1}A^T d^{(k-1)},\\
\beta_{k-1} &:= g^{(k)^T}\hat{g}^{(k)}/g^{(k-1)^T}\hat{g}^{(k-1)},\\
d^{(k)} &:= -g^{(k)} + \beta_{k-1}d^{(k-1)},\\
\hat{d}^{(k)} &:= -\hat{g}^{(k)} + \beta_{k-1}\hat{d}^{(k-1)}.
\end{aligned}$$

– The breakdown treatment is not explicitly incorporated –

Table 5.7: Biconjugate gradient method (BiCG).

performed), but the occurrence of a breakdown is possible and another matrix-vector operation with A^T is necessary.

This can be avoided in the *Conjugate Gradient Squared (CGS)* method, where instead the matrix-vector operations with A are doubled. The basis is the following observation:

$$g^{(k)} = p_k(A)g^{(0)}, \quad \hat{g}^{(k)} = p_k(A^T)g^{(0)}$$

with $p_k \in \mathcal{P}_k$, and then

$$g^{(k)^T}\hat{g}^{(k)} = \left(p_k(A)g^{(0)}\right)^T p_k(A^T)g^{(0)} = \left(p_k(A)^2 g^{(0)}\right)^T g^{(0)}.$$

Finally a variant in order to damp oscillations in the convergence curve leads to the BiCGstab method. For further discussion of this topic see, for example, [54].

Concerning preconditioning, continuing the considerations of Section 5.3, in the absence of symmetry of A the preconditioning matrix C need not be symmetric, and preconditioning from the left is sufficient. As in Section 5.3, we concentrate on preconditioning by applying a linear stationary iteration, given by the matrices M, N in the form

$$y^{(i+1)} = My^{(i)} + Nb, \tag{5.103}$$

where (5.8) is supposed to hold, i.e., we have $N = (I - M)A^{-1}$. The use of the preconditioner $C^{-1} = N$ means for a solution step:

$$\text{Solve} \quad Ch = g,$$

or equivalently:

$$\text{Perform one step of (5.103) for } b = g \text{ and } y^{(0)} = 0.$$

Let now B play the role of C^{-1} in the preconditioning (from left). An obvious extension consists in doing k iteration steps, i.e., the matrix C^{-1} is given as the representation matrix of a linear operator $B : \mathbb{R}^m \to \mathbb{R}^m$ defined by the following procedure:

$$\begin{aligned} &\text{For } x \in \mathbb{R}^m \text{ set } b := x,\ y^{(0)} := 0.\\ &\text{Perform } k \text{ steps of (5.103).}\\ &\text{Set } Bx := y^{(k)}. \end{aligned} \tag{5.104}$$

This can be expressed as one step of a linear stationary method

$$x = M^k y^{(0)} + \hat{N}x = \hat{N}x$$

with $\hat{N} := \left(\sum_{j=0}^{k-1} M^j\right) N$ and thus

$$B = \hat{N} = (I - M^k)A^{-1}.$$

The solution of a system of equations $Ch = g$ is substituted by $h = Bg$, but B need not be computed, but rather the matrix-vector product is evaluated by (5.104). Similar to (5.78) we get

Theorem 5.22 *Let the preconditioning from left be defined by k iteration steps as in (5.103), assume $\|M\| =: \xi < 1$ in some submultiplicative matrix norm. Then:*

$$\|BA\| \le 1 + \xi^k, \quad \kappa(BA) \le \frac{1 + \xi^k}{1 - \xi^k}, \quad (BAx)^T x \ge \left(1 - \xi^k\right) |x|^2,$$

where the condition number is with respect to $|\cdot|$.

Proof The first estimate follows from

$$\|BA\| = \|I - M^k\| \le \|I\| + \|M^k\| \le 1 + \xi^k.$$

Since $\|M^k\| \le \xi^k < 1$, the invertibility of $I - M^k$ is a consequence of the perturbation lemma (see Appendix A.3). In particular, (A3.11) implies that

$$\|(I - M^k)^{-1}\| \le \frac{1}{1 - \|M^k\|} \le \frac{1}{1 - \xi^k},$$

giving the second assertion. Finally:

$$(BAx)^T x = |x|_2^2 - (M^k x)^T x \ge \left(1 - \xi^k\right) |x|^2.$$

□

Typically, the preconditioned Krylov subspace method has a better rate of convergence than the basic linear stationary method (also for $k = 1$). If the complexity of one step of the Krylov subspace method is considered comparable to the preconditioning step, one also speaks of an *acceleration* of the basic method by the Krylov subspace approach. To illustrate this situation, consider a symmetric positive definite matrix A and a symmetric preconditioner B.

According to Theorem 5.16, the convergence factor of the CG method is

$$\frac{\kappa(AB)^{1/2} - 1}{\kappa(AB)^{1/2} + 1} = \frac{1 - (1 - \xi^{2k})^{1/2}}{\xi^k}$$

by using $\kappa(AB) = (1 + \xi^k)/(1 - \xi^k)$, and thus

$$\frac{\kappa(AB)^{1/2} - 1}{\kappa(AB)^{1/2} + 1} < \xi^k .$$

Similarly, the GMRES method fulfils the pessimistic estimate (5.93), as the assumption (5.92) is fulfilled due to Theorem 5.22. For the preconditioned method this means a contraction factor of the residuum norm of

$$\left(1 - \frac{\tau^2}{\|A\|^2}\right)^{1/2} = \frac{2\sqrt{\xi}^k}{1 + \xi^k},$$

i.e., the method is convergent.

Exercises

Problem 5.14 Let $A \in \mathbb{R}^{m,m}$. Show that

$$\left(\alpha A + (1 - \alpha)A^T\right)^T x = (Ax)^T x \quad \text{for all } x \in \mathbb{R}^m,\ \alpha \in \mathbb{R}.$$

Problem 5.15 Consider the linear system $Ax = b$, where $A = \alpha Q$ for some $\alpha \in \mathbb{R} \setminus \{0\}$ and some orthogonal matrix Q. Show that, for an arbitrary initial iterate $x^{(0)}$, the CGNE method terminates after one step with the exact solution.

Problem 5.16 Provided that Arnoldi's method can be performed up to the index k, show that it is possible to incorporate a convergence test of the GMRES method without computing the approximate solution explicitly, i.e., prove the following formulas:

$$\begin{aligned} g^{(k)} &:= Ax^{(k)} - b = h_{k+1,k}\tilde{e}_k^T \xi^{(k)} v_{k+1}, \\ |g^{(k)}|_2 &= h_{k+1,k}|\tilde{e}_k^T \xi^{(k)}| . \end{aligned}$$

Problem 5.17 Show Theorem 5.21.

Problem 5.18 Verify the BiCG algorithm by doing an LDU-factorization of the matrix H_k.

Programming project 5.2 The purpose of this exercise is to compare different iterative linear solvers and preconditioners. Many iterative linear solvers (e.g., GMRES, BiCGstab) are available in free software environments, but also, for example, in MATLAB, which is why this task does not focus on the implementation of the solvers themselves, but on their actual behaviour when applied to a specific test problem. So the reader is referred to the corresponding documentations and code sources.

As model problem, consider the linear system which is obtained by discretization of the boundary value problem

$$\begin{aligned} -0.01\Delta u + \boldsymbol{c} \cdot \nabla u &= 0 \qquad && \text{in } \Omega := (0,1)^2, \\ u(x,y) &= x^2 + y^2 && \text{on } \partial\Omega \end{aligned}$$

with $\boldsymbol{c} := (\sqrt{2}/2, \sqrt{2}/2)^T$ using conforming linear elements on triangles.

As grid, use a square Friedrichs–Keller triangulation with refinement level 7. Compare the linear solvers GMRES (with restart after say 30 steps) and BiCGstab with and without preconditioner. Use Jacobi, symmetric Gauss-Seidel and ILU as preconditioners. Solve the linear system of the model problem for all available combinations of linear solver and preconditioner. Make two plots (one for each linear solver) with the number of iteration steps on the x-axis and the logarithm of the tolerance on the y-axis. Plot the number of iteration steps needed to compute the solution with no preconditioner and with the used preconditioners for the tolerances 10^{-i}, $i = 4, \ldots, 12$.

5.5 The Multigrid Method

5.5.1 The Idea of the Multigrid Method

We discuss again the model problem of the five-point stencil discretization for the Poisson equation on the square and use the relaxed Jacobi's method. Then due to (5.31) the iteration matrix is

$$M = \omega M_{\mathrm{J}} + (1 - \omega)I = I - \frac{\omega}{4} A,$$

with A being the stiffness matrix according to (1.14). For $\tilde{\omega} = \omega/4$ this coincides with the relaxed Richardson method, which according to (5.35) has the poor convergence behaviour of Jacobi's method, even for optimal choice of the parameter. Nevertheless, for a suitable ω the method has positive properties. Due to (5.25) the eigenvalues of M are

$$\lambda_{k,l} = 1 - \omega + \frac{\omega}{2}\left(\cos\frac{k\pi}{n} + \cos\frac{l\pi}{n}\right), \quad 1 \le k, l \le n-1 .$$

This shows that there is a relation between the size of the eigenvalues and the position of the frequency of the assigned eigenfunction depending on the choice of ω: For $\omega = 1$, which is Jacobi's method, $\varrho(M) = \lambda_{1,1} = -\lambda_{n-1,n-1}$. Thus the eigenvalues are large if k and l are close to 1 or n. Hence there are large eigenvalues for eigenfunctions with low frequency as well as for eigenfunctions with high frequency. For $\omega = \frac{1}{2}$, however, we have $\varrho(M) = \lambda_{1,1}$, and the eigenvalues are large only in the case that k and l are near to 1, which means that the eigenfunctions have low frequency.

In general, if the error of the iterate $e^{(k)}$ had a representation in terms of orthonormal eigenvectors z_ν with small eigenvalues, as, for example, $|\lambda_\nu| \le \frac{1}{2}$,

$$e^{(k)} = \sum_{\nu:\,|\lambda_\nu| \le \frac{1}{2}} c_\nu z_\nu ,$$

then according to (5.11) it would follow for the error measured in the Euclidean vector norm $|\cdot|_2$ that

$$\begin{aligned} \left|e^{(k+1)}\right|_2 &= \left|\sum_{\nu:\,|\lambda_\nu| \le \frac{1}{2}} \lambda_\nu c_\nu z_\nu\right|_2 = \left(\sum_{\nu:\,|\lambda_\nu| \le \frac{1}{2}} \lambda_\nu^2 c_\nu^2\right)^{1/2} \\ &\le \frac{1}{2}\left(\sum_{\nu:\,|\lambda_\nu| \le \frac{1}{2}} c_\nu^2\right)^{1/2} = \frac{1}{2}\left|e^{(k)}\right|_2 \end{aligned}$$

if the eigenvectors are chosen orthonormal with respect to the Euclidean scalar product (compare (5.68)). For such an initial error and in exact arithmetic the method would thus have a "small" contraction number independent of the discretization.

For Jacobi's method damped by $\omega = \frac{1}{2}$ this means that if the initial error consists of functions of high frequency only (in the sense of an eigenvector expansion only of eigenvectors with k or l distant to 1), then the above considerations hold. But already due to rounding errors we will always find functions of low frequency in the error such that the above statement of convergence indeed does not hold, but instead the *smoothing property* for the damped Jacobi's method is valid: A few steps only lead to a low reduction of the error but smooth the error in the sense that the parts of high frequency are reduced considerably.

The very idea of the multigrid method lies in the approximative calculation of this remaining error on a coarse grid. The smooth error can still be represented on the coarser grid and should be approximated there. Generally, the dimension of the problem is greatly reduced in this way. Since the finite element discretizations are a central topic of this book, we develop the idea of multigrid methods for such an example. But it will turn out that the multigrid method can be used as well for both the finite difference and the finite volume methods. Multigrid methods have even been successfully used in areas other than the discretization of differential equations. *Algebraic multigrid methods* are generally applicable to systems of linear equations (5.1) and generate by themselves an abstract analogy of a "grid hierarchy" (see, for example, [212]).

5.5.2 Multigrid Method for Finite Element Discretizations

Let $\mathcal{T}_l = \mathcal{T}_h$ be a triangulation that originates from a coarse triangulation $\mathcal{T}_0$ by l applications of a refinement strategy, for example the strategy of Section 2.4.1. As we will see, it is not necessary that, for example, in two space dimensions going from $\mathcal{T}_k$ to $\mathcal{T}_{k+1}$ each triangle will be partitioned into four triangles. Only the relation

$$V_k \subset V_{k+1}\,, \quad k = 1, \ldots, l-1\,,$$

has to hold for finite-dimensional approximation spaces $V_0, V_1, \ldots, V_l = V_h$ generated by a fixed ansatz; i.e., the approximation spaces have to be *nested*. This holds for all approximation spaces discussed in Section 3.3 if $\mathcal{T}_{k+1}$ is still a consistent triangulation and results from $\mathcal{T}_k$ by partitioning of $K \in \mathcal{T}_k$ into a possibly varying number of elements of equal kind.

The nodes of $\mathcal{T}_k$, which are the degrees of freedom of the discretization (possibly multiple in a Hermite ansatz), are denoted by

$$a_i^k\,, \quad i = 1, \ldots, M_k\,,$$

and the corresponding basis functions of V_k are denoted by

$$\varphi_i^k\,, \quad i = 1, \ldots, M_k\,,$$

with the index $k = 0, \ldots, l$. For a quadratic ansatz on a triangle and Dirichlet boundary conditions, the a_i^k are just the vertices and midpoints of the edges in the interior of the domain. Let the underlying variational equation (2.24) be defined by the bilinear form a and the linear form ℓ on the function space V. The system of equations to be solved is

$$A_l x_l = b_l\,. \tag{5.105}$$

In addition, we have to consider auxiliary problems

$$A_k \overline{x}_k = \overline{b}_k$$

for $k = 0, \ldots, l-1$. For the discretization matrix on each refinement level we have, according to (2.38),

$$(A_k)_{ij} = a\big(\varphi_j^k, \varphi_i^k\big)\,, \quad i, j = 1, \ldots, M_k\,, \quad k = 0, \ldots, l\,,$$

and for the right side of the problem to be solved

$$(b_l)_i = \ell\left(\varphi_i^l\right), \quad i = 1, \ldots, M_l \,.$$

In Section 2.2, x_l is denoted by $\boldsymbol{\xi}$, and b_l is denoted by $\boldsymbol{q}_h$.

First we discuss the finite element discretization of a variational equation with symmetric bilinear form, so that in reference to Lemma 2.14 the Galerkin method to be solved is equivalent to the Ritz method, i.e., to the minimization of

$$F_l(x_l) := \frac{1}{2} x_l^T A_l x_l - b_l^T x_l \,.$$

Note that l indicates the discretization level and is *not* an index of a component or an iteration step.

We distinguish between the function $u_l \in V_l$ and the representation vector $x_l \in \mathbb{R}^{M_l}$, so that

$$u_l = \sum_{i=1}^{M_l} x_{l,i}\, \varphi_i^l \,. \tag{5.106}$$

For a Lagrange ansatz we have

$$x_{l,i} = u_l\left(a_i^l\right), \quad i = 1, \ldots, M_l \,,$$

as illustrated by Figure 5.5.

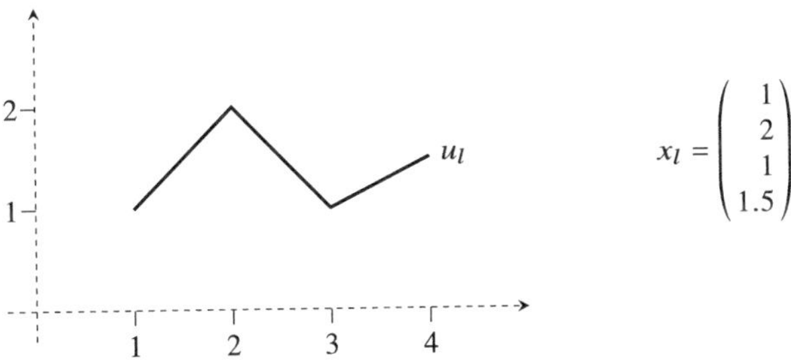

Fig. 5.5: u_l and x_l.

Relation (5.106) defines a linear bijective mapping

$$P_l: \ \mathbb{R}^{M_l} \to V_l \,. \tag{5.107}$$

Thus for $z_l \in \mathbb{R}^{M_l}$ (compare (2.39)),

$$F_l(z_l) = \frac{1}{2} z_l^T A_l z_l - b_l^T z_l = \frac{1}{2} a\left(P_l z_l, P_l z_l\right) - \ell(P_l z_l) = F\left(P_l z_l\right),$$

where

$$F(u) := \frac{1}{2} a(u, u) - \ell(u) \quad \text{for } u \in V$$

is the energy functional for the variational equation.

If $\overline{x}_l$ is an approximation of x_l, then the error $y_l := x_l - \overline{x}_l$ satisfies the *error equation*

$$A_l y_l = b_l - A_l \overline{x}_l \,. \tag{5.108}$$

This equation is equivalent to the minimization problem

$$F_l\left(\overline{x}_l + y_l\right) = \min_{y \in \mathbb{R}^{M_l}} F_l\left(\overline{x}_l + y\right)$$

and therefore to

$$F\left(P_l \overline{x}_l + v_l\right) = \min_{v \in V_l} F\left(P_l \overline{x}_l + v\right), \tag{5.109}$$

with $v_l = P_l y_l$.

If the error y_l is "smooth" in the sense that it can be well approximated also in the lower dimensional space V_{l-1}, one can solve the error equation (5.108) approximately as part of an iteration step by solving the minimization problem (5.109) only on V_{l-1}. The starting condition of a "smooth" error will be ensured by the application of a fixed number of steps of a smoothing iterative method. Let S_l denote the application of such a smoothing operation, for example, the damped Jacobi's method

$$S_l x = x - \omega D_l^{-1}\left(A_l x - b_l\right)$$

with the diagonal matrix D_l corresponding to A_l according to (5.18).

Thus we get the algorithm of the *two-grid iteration*, whose $(k+1)$th step is described in Table 5.8. Problem (5.110) from Table 5.8 is equivalent to (compare with Lemma 2.3)

Let $x_l^{(k)}$ be the kth iterate to the solution of (5.105).

1) **Smoothing step:** For fixed $\nu \in \{1, 2, \ldots\}$ calculate

$$x_l^{(k+1/2)} := S_l^{\nu} x_l^{(k)} .$$

Let the corresponding function be:

$$u_l^{(k+1/2)} := P_l x_l^{(k+1/2)} \in V_l .$$

2) **Coarse grid correction:** Solve (exactly)

$$F\left(u_l^{(k+1/2)} + v\right) \to \min \tag{5.110}$$

varying $v \in V_{l-1}$, with solution $\overline{v}_{l-1}$. Then set

$$x_l^{(k+1)} := P_l^{-1}\left(u_l^{(k+1/2)} + \overline{v}_{l-1}\right) = x_l^{(k+1/2)} + P_l^{-1}\overline{v}_{l-1} .$$

Table 5.8: $(k+1)$th step of the two-grid iteration.

$$a\left(u_l^{(k+1/2)} + v, w\right) = \ell(w) \quad \text{for all } w \in V_{l-1} \tag{5.111}$$

and thus again to the Galerkin discretization of a variational equation with V_{l-1} instead of V, with the same bilinear form and with a linear form defined by

$$w \mapsto \ell(w) - a\left(u_l^{(k+1/2)}, w\right) \quad \text{for } w \in V_{l-1}\,.$$

Hence we can ignore the assumption of symmetry for the bilinear form a and find the approximative solution of the error equation (5.108) on grid level $l-1$ by solving the variational equation (5.111). The equivalent system of equations will be derived in the following. On the one hand, this problem has a lower dimension than the original problem, but it also must be solved for each iteration. This suggests the following recursive procedure: If we have more than two-grid levels, we again approximate this variational equation by μ multigrid iterations; in the same way we treat the hereby created Galerkin discretization on level $l-2$ until level 0 is reached, where we solve exactly. Furthermore, to conclude each iteration step smoothing steps should be performed. This leads to the algorithm of the multigrid iteration. The $(k+1)$th step of the *multigrid iteration on level l for the bilinear form a, linear form ℓ*, and starting iteration $x_l^{(k)}$ is described in Table 5.9.

1) **A priori smoothing:** Perform ν_1 smoothing steps:

$$x_l^{(k+1/3)} = S_l^{\nu_1} x_l^{(k)},$$

where $\nu_1 \in \{1, 2, \ldots\}$ is fixed. Let the corresponding function be

$$u_l^{(k+1/3)} := P_l x_l^{(k+1/3)}.$$

2) **Coarse grid correction:** Solve on V_{l-1} the Galerkin discretization

$$a(\overline{v}_{l-1}, w) = \tilde{\ell}(w) \quad \text{for all } w \in V_{l-1} \tag{5.112}$$

with the bilinear form a and the linear form

$$\tilde{\ell}(w) := \ell(w) - a\left(u_l^{(k+1/3)}, w\right)$$

a. for $l = 1$ exactly,
b. for $l > 1$ by μ steps of a multigrid iteration on level $l-1$ for a and $\tilde{\ell}$ and for the start approximation 0.

Set
$$x_l^{(k+2/3)} = x_l^{(k+1/3)} + P_l^{-1}\overline{v}_{l-1}.$$

3) **A posteriori smoothing:** Perform ν_2 smoothing steps

$$x_l^{(k+1)} = S_l^{\nu_2} x_l^{(k+2/3)},$$

with $\nu_2 \in \{1, 2, \ldots\}$ fixed.

Table 5.9: $(k+1)$th step of the multigrid iteration on level l for bilinear form a and linear form ℓ.

In general, $\nu_1 = \nu_2$ is used. In a convergence analysis, it turns out that only the sum of smoothing steps is important. Despite the recursive definition of a multigrid iteration we have here a finite method, because the level 0 is reached after at most l recursions, where the auxiliary problem will be solved exactly. For μ usually only the values $\mu = 1$ or $\mu = 2$ are used. The terms *V-cycle* for $\mu = 1$ and *W-cycle* for $\mu = 2$ are commonly used, because for an iteration, the sequence of levels on which operations are executed have the shape of these letters (see Figure 5.6).

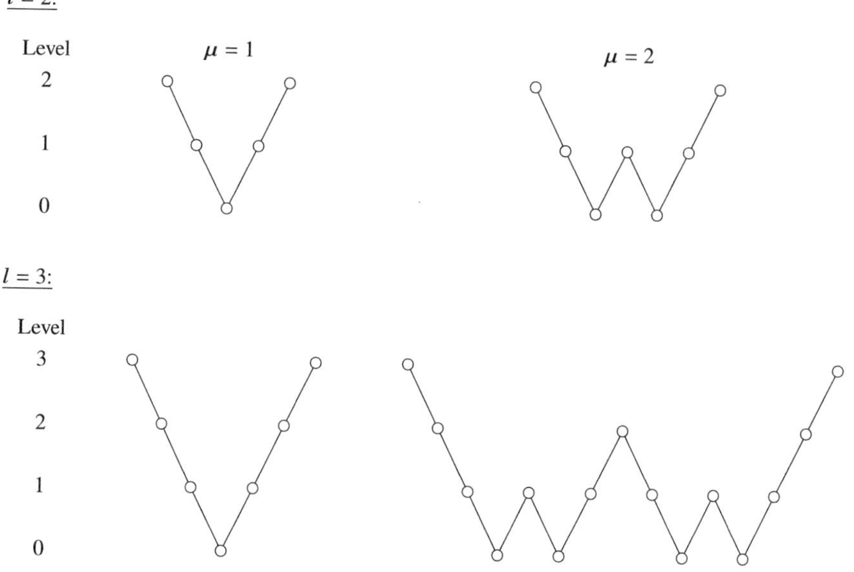

Fig. 5.6: Grid levels for the V-cycle ($\mu = 1$) and the W-cycle ($\mu = 2$).

The problems in (5.111) and (5.112) (see Table 5.9) have the form

$$a\,(u + v, w) = \ell(w) \quad \text{for all } w \in V_{l-1}\,, \tag{5.113}$$

where $v \in V_{l-1}$ is unknown and $u \in V_l$ is known. An equivalent system of equations arises by inserting the basis functions φ_j^{l-1}, $j = 1, \ldots, M_{l-1}$, for w and an appropriate representation for v. If we again take the representation with respect to φ_j^{l-1}, we get as in (2.38)

$$A_{l-1} P_{l-1}^{-1} v = d_{l-1}\,. \tag{5.114}$$

Here the *residual* $d_k \in \mathbb{R}^{M_k}$ of u on the different levels $k = 0, \ldots, l$ is defined by

$$d_{k,i} := \ell\big(\varphi_i^k\big) - a\big(u, \varphi_i^k\big), \quad i = 1, \ldots, M_k\,.$$

We now develop an alternative representation for (5.114) and the coarse grid correction for possible generalizations beyond the Galerkin approximations. Therefore, let $R \in \mathbb{R}^{M_{l-1}, M_l}$ be the matrix that arises through the unique representation of the basis functions φ_j^{l-1} with respect to the basis φ_i^l, which means the elements r_{ji} of R are determined by the equations

$$\varphi_j^{l-1} = \sum_{i=1}^{M_l} r_{ji}\, \varphi_i^l, \quad j = 1, \dots, M_{l-1}.$$

Then (5.113) is equivalent to

$$\begin{aligned} & a(v, w) = \ell(w) - a(u, w) \quad \text{for all } w \in V_{l-1} \\ \Leftrightarrow\ & a\left(\sum_{s=1}^{M_{l-1}} \left(P_{l-1}^{-1} v \right)_s \varphi_s^{l-1}, \varphi_j^{l-1} \right) = \ell(\varphi_j^{l-1}) - a\left(u, \varphi_j^{l-1}\right), \ j = 1, \dots, M_{l-1} \\ \Leftrightarrow\ & \sum_{s=1}^{M_{l-1}} \left(P_{l-1}^{-1} v \right)_s a\left(\sum_{t=1}^{M_l} r_{st} \varphi_t^l, \sum_{i=1}^{M_l} r_{ji} \varphi_i^l \right) = \sum_{i=1}^{M_l} r_{ji} \left(\ell(\varphi_i^l) - a\left(u, \varphi_i^l\right) \right) \\ \Leftrightarrow\ & \sum_{s=1}^{M_{l-1}} \sum_{i,t=1}^{M_l} r_{ji} a\left(\varphi_t^l, \varphi_i^l\right) r_{st} \left(P_{l-1}^{-1} v \right)_s = (R d_l)_j, \quad j = 1, \dots, M_{l-1}. \end{aligned}$$

Hence the system of equations has the form

$$R A_l R^T \left(P_{l-1}^{-1} v \right) = R d_l. \tag{5.115}$$

The matrix R is easy to calculate for a node-based basis φ_i^l satisfying $\varphi_i^l\left(a_j^l\right) = \delta_{ij}$, since in this case we have for $v \in V_l$,

$$v = \sum_{i=1}^{M_l} v\left(a_i^l\right) \varphi_i^l,$$

and therefore in particular,

$$\varphi_j^{l-1} = \sum_{i=1}^{M_l} \varphi_j^{l-1}\left(a_i^l\right) \varphi_i^l$$

and thus

$$r_{ji} = \varphi_j^{l-1}\left(a_i^l\right).$$

For the linear ansatz in one space dimension with Dirichlet boundary conditions (i.e., with $V = H_0^1(a, b)$ as basic space) this means that

$$R = \begin{pmatrix} \frac{1}{2} & 1 & \frac{1}{2} & & & & & \\ & & \frac{1}{2} & 1 & \frac{1}{2} & & & \\ & & & & \ddots & & & \\ & & & & & \frac{1}{2} & 1 & \frac{1}{2} \end{pmatrix} . \tag{5.116}$$

The representation (5.115) can also be interpreted in this way:

Due to $V_{l-1} \subset V_l$ the identity defines a natural *prolongation* from V_{l-1} to V_l, which means that

$$\tilde{p} : \; V_{l-1} \to V_l \, , \; v \mapsto v \, ,$$

as illustrated by Figure 5.7.

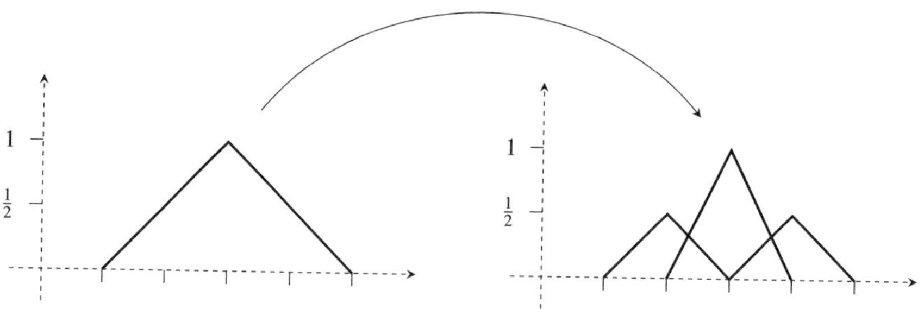

Fig. 5.7: Prolongation.

This prolongation corresponds to a prolongation p from $\mathbb{R}^{M_{l-1}}$ to $\mathbb{R}^{M_l}$, the *canonical prolongation*, through the transition to the representation vectors (5.107). It is given by

$$p := P_l^{-1} P_{l-1} \, , \tag{5.117}$$

since for $x_{l-1} \in \mathbb{R}^{M_{l-1}}$, p can be composed as follows:

$$x_{l-1} \mapsto P_{l-1} x_{l-1} \overset{\tilde{p}}{\mapsto} P_{l-1} x_{l-1} \mapsto P_l^{-1} P_{l-1} x_{l-1} \, .$$

Obviously, p is linear and can be identified with its matrix representation in $\mathbb{R}^{M_l, M_{l-1}}$. Then

$$p = R^T \tag{5.118}$$

holds, because

$$P_{l-1} y = \sum_{j=1}^{M_{l-1}} y_j \varphi_j^{l-1} = \sum_{i=1}^{M_l} \sum_{j=1}^{M_{l-1}} y_j r_{ji} \varphi_i^l \, ,$$

i.e., $R^T y = P_l^{-1} (P_{l-1} y)$ for any $y \in \mathbb{R}^{M_{l-1}}$.

In the following $\mathbb{R}^{M_l}$ will be endowed with a scalar product $\langle \cdot , \cdot \rangle^{(l)}$, which is an Euclidean scalar product scaled by a factor S_l,

$$\langle x_l, y_l \rangle^{(l)} := S_l \sum_{i=1}^{M_l} x_{l,i}\, y_{l,i}\,. \tag{5.119}$$

The scaling factor is to be chosen such that for the induced norm $\|\cdot\|_{(l)}$ and the $L^2(\Omega)$-norm on V_l,

$$C_1 \,\|P_l x_l\|_0 \le \|x_l\|_{(l)} \le C_2 \,\|P_l x_l\|_0 \tag{5.120}$$

for $x \in \mathbb{R}^{M_l}$, $l = 0, 1, \ldots$, with constants $C_1, C_2 > 0$ independent of l: If the triangulations are members of a shape-regular and quasi-uniform family $\mathcal{T}_h$ (see Definition 3.31), then in d space dimensions one can choose $S_l = h_l^d$, with h_l being the maximal diameter of $K \in \mathcal{T}_l$ (see Theorem 3.46).

Let $r : \mathbb{R}^{M_l} \to \mathbb{R}^{M_{l-1}}$ be defined by

$$r = p^*\,, \tag{5.121}$$

with the *adjoint* p^* defined with respect to the scalar products $\langle \cdot, \cdot \rangle^{(l-1)}$ and $\langle \cdot, \cdot \rangle^{(l)}$; that is,

$$\langle r\, x_l\,,\, y_{l-1} \rangle^{(l-1)} = \langle p^*\, x_l\,,\, y_{l-1} \rangle^{(l-1)} = \langle x_l\,,\, p\, y_{l-1} \rangle^{(l)}\,.$$

If p is the canonical prolongation, then r is called the *canonical restriction*. For the representation matrices,

$$\frac{S_{l-1}}{S_l} r = p^T = R\,. \tag{5.122}$$

In example (5.119) for $d = 2$ with $h_l = h_{l-1}/2$ we have $S_{l-1}/S_l = 4$. Due to $P_l p = P_{l-1}$, the canonical restriction of $\mathbb{R}^{M_l}$ on $\mathbb{R}^{M_{l-1}}$ satisfies

$$r R_l = R_{l-1}\,,$$

where $R_l : V_l \to \mathbb{R}^{M_l}$ is defined as the adjoint of P_l,

$$\langle P_l x_l, v_l \rangle_0 = \langle x_l, R_l v_l \rangle^{(l)} \quad \text{for all } x_l \in \mathbb{R}^{M_l}\,,\ v_l \in V_l\,,$$

because for any $y_{l-1} \in \mathbb{R}^{M_{l-1}}$ and for $v_{l-1} \in V_{l-1} \subset V_l$,

$$\begin{aligned} \langle r R_l v_{l-1}\,,\, y_{l-1} \rangle^{(l-1)} &= \langle R_l v_{l-1}\,,\, p y_{l-1} \rangle^{(l)} = \langle v_{l-1}\,,\, P_l p y_{l-1} \rangle_0 \\ &= \langle v_{l-1}, P_{l-1} y_{l-1} \rangle_0 \;= \langle R_{l-1} v_{l-1}, y_{l-1} \rangle^{(l-1)}\,. \end{aligned}$$

Using (5.122) we see the equivalence of equation (5.115) to

$$(r A_l p) y_{l-1} = r d_l\,. \tag{5.123}$$

Setting $v := P_{l-1} \tilde{y}_{l-1}$ for a perhaps only approximative solution $\tilde{y}_{l-1}$ of (5.123), the coarse grid correction will be finished by addition of $P_l^{-1} v$. Due to

$$P_l^{-1} v = P_l^{-1} P_{l-1} \left(P_{l-1}^{-1} v \right) = p \left(P_{l-1}^{-1} v \right),$$

the coarse grid correction is

$$x_l^{(k+2/3)} = x_l^{(k+1/3)} + p(\tilde{y}_{l-1}) .$$

The above-mentioned facts suggest the following structure of a general multigrid method: For discretizations defining a hierarchy of discrete problems,

$$A_l x_l = b_l ,$$

one needs *prolongations*

$$p : \mathbb{R}^{M_{k-1}} \to \mathbb{R}^{M_k}$$

and *restrictions*

$$r : \mathbb{R}^{M_k} \to \mathbb{R}^{M_{k-1}}$$

for $k = 1, \ldots, l$ and the matrices $\tilde{A}_{k-1}$ for the error equations. The coarse grid correction steps (5.110) and (5.112) hence take the following form:

Solve (with μ steps of the multigrid method)

$$\tilde{A}_{l-1} y_{l-1} = r\left(b_l - A_l x_l^{(k+1/3)}\right)$$

and set

$$x_l^{(k+2/3)} = x_l^{(k+1/3)} + p y_{l-1} .$$

The above choice

$$\tilde{A}_{l-1} = r A_l p$$

is called the *Galerkin product*. For Galerkin approximations this coincides with the discretization matrix of the same type on the grid of level $l-1$ due to (5.114). This is also a common choice for other discretizations and then an alternative to the Galerkin product. In view of the choice of p and r we should observe the validity of (5.121). An interpolatory definition of the prolongation on the basis of (finite element) basis functions as, for example (5.118) (see also example (5.116)), is also common in other discretizations. In more difficult problems, as, for example, those with (dominant) convection in addition to diffusive transport processes, nonsymmetric problems arise with a small constant of V-coercivity. Here the use of *matrix-dependent*, that means A_l-dependent, prolongations, and restrictions is recommended.

5.5.3 Effort and Convergence Behaviour

In order to judge the efficiency of a multigrid method the number of operations per iteration and the number of iterations (required to reach an error level ε, see (5.4)) has to be estimated. Due to the recursive structure, the first number is not immediately clear. The aim is to have only the optimal amount of $O(M_l)$ operations for sparse matrices. For this the dimensions of the auxiliary problems have to decrease sufficiently. This is expressed by the following:

There exists a constant $C > 1$ such that

$$M_{l-1} \leq M_l / C \quad \text{for } l \in \mathbb{N}. \tag{5.124}$$

Hence we assume an infinite hierarchy of problems and/or grids, which also corresponds to the asymptotic point of view of a discretization from Section 3.4. Relation (5.124) is thus a condition for a refinement strategy. For the model problem of the Friedrichs–Keller triangulation of a rectangle (see Figure 2.9) in the case of a regular "red" refinement we have $h_l = h_{l-1}/2$. Thus $C = 4$, and for analogous constructions in d space dimensions $C = 2^d$. The matrices that appear should be sparse, so that for level l the following holds:

smoothing step	$\widehat{=}\ C_S M_l$ operations,
error calculation and restrictions	$\widehat{=}\ C_D M_l$ operations,
prolongation and correction	$\widehat{=}\ C_C M_l$ operations.

Then we can prove the following (see [33, Thm. 11.16]):

If the number μ of multigrid steps in the recursion satisfies

$$\mu < C, \tag{5.125}$$

then the number of operations for an iteration step for a problem on level l can be estimated by

$$C(\nu) M_l\,. \tag{5.126}$$

Here ν is the number of a priori and a posteriori smoothing steps and

$$C(\nu) = \frac{\nu C_S + C_D + C_S}{1 - \mu/C} + O\big((\mu/C)^l\big)\,.$$

The requirement (5.125) will be satisfied in general through the restriction to $\mu = 1$, $\mu = 2$. Analogously, the memory requirement is $O(M_l)$, since

$$\sum_{k=0}^{l} M_k \leq \frac{C}{C-1} M_l\,.$$

Whether this larger effort (of equal complexity) in comparison to other methods discussed is justified will be decided by the rate of convergence. The multigrid method is a linear stationary method. The iteration matrix M_l^{TGM} of the two-grid method with ν_1 a priori and ν_2 a posteriori steps and $\tilde{S}_l$ denoting an affine-linear operator and S_l the linear part of this operator results from

$$\begin{aligned} x_l^{(k+1/2)} &= \tilde{S}_l^{\nu_1} x_l^{(k)}, \\ x_l^{(k+1)} &= \tilde{S}_l^{\nu_2} \left(x_l^{(k+1/2)} + p \left(A_{l-1}^{-1} \left(r \left(b_l - A_l x_l^{(k+1/2)} \right) \right) \right) \right), \end{aligned}$$

therefore

$$x_l^{(k+1)} = \tilde{S}_l^{\nu_2}\left(\tilde{S}_l^{\nu_1} x_l^{(k)} + p\left(A_{l-1}^{-1}\left(r\left(b_l - A_l \tilde{S}_l^{\nu_1}\left(x_l^{(k)}\right)\right)\right)\right)\right). \tag{5.127}$$

This shows the representation

$$M_l^{TGM} = S_l^{\nu_2}(I - pA_{l-1}^{-1} r A_l) S_l^{\nu_1} \tag{5.128}$$

for the iteration matrix and also the consistency of the two-grid method, provided the smoother is consistent:

Let $x_l^{(k)} = x_l$, i.e. $\tilde{S}_l^{\nu} x_l^{(k)} = x_l^{(k)}$ and thus

$$\begin{aligned} x_l^{(k+1)} &= \tilde{S}_l^{\nu}\left(x_l^{(k)} + p\left(A_{l-1}^{-1}\left(r\left(b_l - A_l x_l^{(k)}\right)\right)\right)\right) \\ &= \tilde{S}_l^{\nu}\left(x_l^{(k)} + 0\right) = x_l^{(k)}. \end{aligned}$$

Let $\gamma \geq 1$ be fixed and

$$M_j(\nu_1, \nu_2)$$

be the iteration matrix (if it is indeed a linear stationary method) of the multigrid iteration at level j with γ multigrid cycles for the coarse grid iteration. Denote by ν_1, ν_2 the a priori/a posteriori smoothing steps, and by r_j, p_j the restrictions/prolongations at level j.

Lemma 5.23 *The multigrid method is a linear stationary iterative method and the iteration matrix satisfies*

$$M_j(\nu_1, \nu_2) = S_j^{\nu_2}\left(I - p_j\left(I - M_{j-1}(\nu_1, \nu_2)^{\gamma}\right) A_{j-1}^{-1} r_j A_j\right) S_j^{\nu_1}, \quad j = 1, \ldots, l,$$

where $M_0(\nu_1, \nu_2) := 0$,

Proof The proof is given by induction over j. For $j = 1$ we have

$$M_1(\nu_1, \nu_2) = M_1^{TGM}$$

and thus the method is linear stationary, consistent and the identity holds because of (5.128).

Let all assertions hold true for $j - 1$. For the multigrid method all substeps are the same as for the two-grid method apart of the coarse grid correction. Instead of exactly solving the coarse grid error equation, i.e., applying A_{j-1}^{-1}, γ steps of the multigrid method at level $j - 1$ for the initial iterate 0 are performed. Hereby by assumption this is a linear stationary consistent method with iteration matrix M_{j-1}. Therefore the described operation due to (5.8) is given by the operation

$$\left(I - M_{j-1}(\nu_1, \nu_2)^{\gamma}\right) A_{j-1}^{-1}, \tag{5.129}$$

which shows that also at level j the multigrid method is linear stationary. In the same way as for the two-grid method we also get (5.127) (with j instead of l) and

A_{j-1}^{-1} substituted by (5.129). Thus the proof of consistency at level j is the same as above. □

Lemma 5.23 shows the following relationship

$$M_l(\nu_1, \nu_2) = M_l^{TGM}(\nu_1, \nu_2) + S_l^{\nu_2}\, p_l M_{l-1}^{\gamma}(\nu_1, \nu_2)\, A_{l-1}^{-1}\, r_l\, A_l\, S_l^{\nu_1}. \tag{5.130}$$

If the second term turns out to be a perturbation of the first, the convergence of the multigrid method can be deduced from the convergence of the two-grid method. This is indeed the case at least for the W-cycle.

Theorem 5.24 *Let $\|\cdot\|^{(l)}$ be a norm on $\mathbb{R}^{M_l}$ and $\|\cdot\|^{(l_1,l_2)}$ the induced matrix norm. Consider the multigrid method for the parameters ν_1, ν_2, γ and assume there is a constant $C_1 > 0$ independent of l such that*

$$C_1^{-1}\|x\|^{(l-1)} \le \|p_l x\|^{(l)} \le C_1\|x\|^{(l-1)} \quad \text{for all } x \in \mathbb{R}^{M_{l-1}},$$
$$\|S_l^{\nu}\|^{(l,l)} \le C_1.$$

If the two-grid method is convergent and satisfies

$$\|TGM^l(\nu, 0)\|^{(l,l)} \le \eta(\nu) < 1$$

with a function $\eta : \mathbb{R}_+ \to \mathbb{R}_+$ independent of l such that $\eta(\nu) \to 0$ for $\nu \to \infty$, then there exists a number $\tilde{\nu} > 0$ such that for $\nu \ge \tilde{\nu}$ and $\gamma \ge 2$

$$\|M_l(\nu, 0)\|^{(l,l)} \le \frac{\gamma}{\gamma - 1}\,\eta(\nu).$$

Proof Due to (5.130) we have to investigate the second term there. By (5.128), we have

$$A_{l-1}^{-1} r_l A_l S_l^{\nu_1} = p_l^{-1}\left(-M_l^{TGM}(\nu, 0) + S_l^{\nu_1}\right),$$

thus

$$\|A_{l-1}^{-1} r_l A_l S_l^{\nu_1}\|^{(l,l-1)} \le \|p_l^{-1}\|^{(l-1,l)}\left(\|M_l^{TGM}(\nu, 0)\|^{(l,l)} + \|S_l^{\nu_1}\|^{(l,l)}\right) \le C_2$$

by assumption with a constant $C_2 > 0$ independent of l, and therefore by (5.130)

$$\|M_l(\nu, 0)\|^{(l,l)} \le \eta(\nu) + C_1^2\|M_{l-1}(\nu, 0)\|^{\gamma} C_2. \tag{5.131}$$

Set $\xi_j := \|M_j(\nu, 0)\|^{(j,j)}$. The assertion thus reads

$$\xi_j \le \frac{\gamma}{\gamma - 1}\,\eta(\nu).$$

To prove it by induction over j, note that

$$\xi_1 \le \eta(\nu) \le \frac{\gamma}{\gamma - 1}\,\eta(\nu).$$

If the estimate holds true for $j-1$, then by (5.131)

$$\xi_j \le \eta(\nu) + C^*\xi_{j-1}^\gamma \le \eta(\nu) + C^*\left(\frac{\gamma}{\gamma-1}\eta(\nu)\right)^\gamma .$$

Therefore the estimate

$$\xi_j \le \frac{\gamma}{\gamma-1}\eta(\nu)$$

holds true if

$$\eta + C^*\left(\frac{\gamma}{\gamma-1}\eta\right)^\gamma \le \frac{\gamma}{\gamma-1}\eta \iff C^*\tilde{\eta}^\gamma \le \frac{1}{\gamma}\tilde{\eta} \quad \text{for} \quad \tilde{\eta} := \frac{\gamma}{\gamma-1}\eta,$$

and the latter condition can be satisfied for η small enough, i.e., for $\nu \ge \tilde{\nu}$ for some $\tilde{\nu} > 0$.
□

Remark 5.25 1) The conditions in Theorem 5.24 are met if the smoother is a convergent method and for instance in the situation of (5.118) the estimate (5.120) holds, which can be ensured by the weighted norms induced by (5.119).
2) The assertion can be extended to ν_1 a priori and ν_2 a posteriori smoothing steps, and then $\nu = \nu_1 + \nu_2$.

For a large class of a priori and a posteriori smoothing operators as well as of restrictions and prolongations it can be shown that there exists a constant $\overline{\varrho} \in (0,1)$ independent of the discretization parameter h_l such that $\varrho(M^{TGM}) \le \overline{\varrho}$ (see [33, Thm. 11.35]). Combined with Theorem 5.24 this shows that multigrid methods are optimal in their complexity. It also shows their potential superiority compared with all other methods described.

In the following we will only indicate the schematic procedure to prove this assertion. We restrict the consideration to $\nu_1 := \nu$ a priori and $\nu_2 := 0$ a posteriori smoothing steps. It is sufficient to prove the following two properties, where the spectral norm is used as the matrix norm:

1) *Smoothing property:* There exists a function $C_S : \mathbb{R}_+ \to \mathbb{R}_+$ independent of l with $C_S(\nu) \to 0$ for $\nu \to \infty$ such that

$$\left\|A_l S_l^\nu\right\| \le C_S(\nu)\,\|A_l\|.$$

2) *Approximation property:* There exists a constant $C_A > 0$ independent of l such that

$$\left\|A_l^{-1} - pA_{l-1}^{-1}r\right\| \le C_A\|A_l\|^{-1} . \tag{5.132}$$

With the above choice of the smoothing parameters, from (5.128) it follows that

$$M_l^{TGM} = (A_l^{-1} - pA_{l-1}^{-1}r)A_l S_l^\nu .$$

Hence we can conclude that

$$\left\|M_l^{TGM}\right\| \le \left\|A_l^{-1} - pA_{l-1}^{-1}r\right\| \left\|A_l S_l^\nu\right\| \le C_S(\nu)C_A ,$$

which implies, for sufficiently large ν,

$$\|M_l^{TGM}\| \le \overline{\varrho} < 1$$

with $\overline{\varrho}$ independent of l.

Note that the formulation of the smoothing and the approximation property contains some ambiguity. This refers to the choice of the norm and the appearance of the quantities $\|A_l\|$, $\|A_l\|^{-1}$, which could be substituted by arbitrary parameters α and α^{-1}. Furthermore there is still a gap concerning the convergence of the multigrid method with the V-cycle. We will close this gap in the symmetric case.

Before proving this result, we summarize some facts that are used in the proof.

Lemma 5.26 *Let $A, A_1, A_2, B, M, W \in \mathbb{R}^{m,m}$, $A = A^T > 0$, $W = W^T$.*
Then:

1) $\|M\|_A = \|A^{1/2} M A^{-1/2}\|_2$,
2) $\|A\|_2 \le C$ is equivalent with $A \le CI$,
3) $A_1 \le A_2$ implies $BA_1B^T \le BA_2B^T$,
4) If $W \ge A$, then $\lambda \le 1$ for the eigenvalues λ of $W^{-1}A$,
5) $\|A\|_2 \le \|W\|_2^{-1}$ implies $A \le W^{-1}$.

Proof Problem 5.23. □

To simplify notation we will write Euclidean norms instead of weighted ones as in (5.108). Moreover, the notation $A \ge 0$ for some matrix $A \in \mathbb{R}^{m,m}$ means that it is symmetric, positive semidefinite, and $A \ge B$ for two matrices $A, B \in \mathbb{R}^{m,m}$ means $A - B \ge 0$.

Theorem 5.27 *Let $A := A_l$ be symmetric and positive definite and*

$$r_j = p_j^*, \; A_{j-1} = r_j A_j p_j \quad \textit{for } j = 1, \ldots, l.$$

Furthermore, let S_l with $\varrho(S_l) < 1$ be of the form

$$S_l = I - W_l^{-1} A_l \textit{ with } W_l = W_l^* \ge A$$

(smoothing property),

and assume the existence of a constant $C_A > 0$ such that

$$\|A_j^{-1} - p_j A_{j-1}^{-1} r_j\|_2 \le C_A \|W_j\|_2^{-1} \quad \textit{for } j = 1, \ldots, l$$

(approximation property).

Then the multigrid method with $\gamma := 1$ and $\nu_1 := \nu_2 := \nu/2$, $\nu = 2, 4, \ldots$ satisfies

$$M_l(\frac{\nu}{2}, \frac{\nu}{2}) \ge 0 \quad \textit{and} \quad \|M_l\left(\frac{\nu}{2}, \frac{\nu}{2}\right)\|_A \le \frac{C_A}{C_A + \nu},$$

i.e., the method converges.

Proof By Lemma 5.26, 1) we have to estimate the spectral norm of the matrix

$$M_j' = A_j^{1/2} M_j \left(\frac{\nu}{2}, \frac{\nu}{2}\right) A_j^{-1/2}.$$

Thus due to Lemma 5.26, 2) the assertion is equivalent to

$$0 \le M_j' \le \frac{C_A}{C_A + \nu} I, \quad \text{for } j = 1, \ldots, l. \tag{5.133}$$

To show these inequalities by induction over j we verify the implication:

$$\begin{aligned} &\text{If} \quad && M_{j-1}' \le \tilde{\xi} I < I \quad \text{for some } 0 < \tilde{\xi} < 1, \\ &\text{then} \quad && M_j' \le \left[\max_{\xi \in [0,1]} (1-\xi)^\nu \left(\tilde{\xi} + (1-\tilde{\xi}) C_A \xi\right)\right] I. \end{aligned} \tag{5.134}$$

Indeed, if (5.134) holds true, then (5.133) is trivially fulfilled for $j = 0$, and if it is fulfilled for $j - 1$, we conclude from (5.134) by choosing $\tilde{\xi} := C_A/(C_A + \nu)$ that

$$M_j' \le \max_{\xi \in [0,1]} (1-\xi)^\nu \, \tilde{\xi} \, (1 + \nu \xi) = \tilde{\xi}$$

as the maximum is attained at $\xi = 0$. To show the implication (5.134) and $M_j \ge 0$ we write

$$\begin{aligned} M_j' &= A_j^{1/2} S_j^{\nu/2} \left(I - p_j \left(I - M_{j-1}\right) A_{j-1}^{-1} r_j A_j\right) S_j^{\nu/2} A_j^{-1/2} \\ &= \left(A_j^{1/2} S_j^{\nu/2} A_j^{-1/2}\right) \left(I - A_j^{1/2} p_j (I - M_{j-1}) A_{j-1}^{-1} r_j A_j^{1/2}\right) \left(A_j^{1/2} S_j^{\nu/2} A_j^{-1/2}\right). \end{aligned}$$

This is an equivalence transformation (see below) of the middle term defined by

$$A_j^{1/2} S_j^{\nu/2} A_j^{-1/2}.$$

This in turn is a similarity transformation of a power and thus

$$\begin{aligned} A_j^{1/2} S_j^{\nu/2} A_j^{-1/2} &= \left(A_j^{1/2} S_j A_j^{-1/2}\right)^{\nu/2} = (I - X_j)^{\nu/2} \\ &\text{with } X_j := A_j^{1/2} W_j^{-1} A_j^{1/2}. \end{aligned}$$

The middle matrix can be rewritten using

$$\begin{aligned} A_j^{1/2} p_j \left(I - M_{j-1}\right) A_{j-1}^{-1} r_j A_j^{1/2} &= \left(A_j^{1/2} p_j A_{j-1}^{-1/2}\right) \left(I - A_{j-1}^{1/2} M_{j-1} A_{j-1}^{-1/2}\right) \left(A_{j-1}^{-1/2} r_j A_j^{1/2}\right) \\ &= p_j' \left(I - M_{j-1}'\right) r_j' \\ &\text{with } p_j' := A_j^{1/2} p_j A_{j-1}^{1/2} \text{ and } r_j' \text{ correspondingly.} \end{aligned}$$

Because of $p_j^* = r_j$, $A_j^* = A_j$ we have $p_j'^* = r_j'$, i.e., an equivalence transformation, which due to Lemma 5.26, 3) does not influence the ordering. In summary, we have

$$M_j' = (I - X_j)^{\nu/2} \left(I - p' \left(I - M_{j-1}'\right) r'\right) (I - X_j)^{\nu/2}.$$

Therefore (by induction over j) M_j' and thus $M_j(\frac{\nu}{2}, \frac{\nu}{2})$ are symmetric. The matrix X_j and thus $(I - X_j)^{\nu/2}$ are symmetric, i.e., the first transformation is an equivalence transformation. Lemma 5.26, 3) with $B := (I - X_j)^{\nu/2} p'$ and the use of $0 \le M_{j-1}' \le \tilde{\xi} I$ lead to

$$\begin{aligned} 0 \le M_j' &\le (1 - X_j)^{\nu/2} \left(I - \left(1 - \tilde{\xi}\right) p' r'\right) (1 - X_j)^{\nu/2} \\ &= (1 - X_j)^{\nu/2} \left(\left(1 - \tilde{\xi}\right)(I - p' r') + \tilde{\xi} I\right) (1 - X_j)^{\nu/2} \\ &= (I - X_j)^{\nu/2} \left(\left(1 - \tilde{\xi}\right) Q_j + \tilde{\xi} I\right), \end{aligned}$$

where

$$Q_j := A_j^{1/2} \left(A_j^{-1} - p_j A_{j-1}^{-1} r_j\right) A_j^{1/2} \le C_A A_j^{1/2} W_j^{-1} A_j^{1/2} = C_A X_j$$

by the approximation property and Lemma 5.26, 5), and therefore

$$M_j' \le (I - X_j)^{\nu/2} \left(\left(1 - \tilde{\xi}\right) C_A X_j + \tilde{\xi} I\right) (1 - X_j)^{\nu/2}. \tag{5.135}$$

Using the smoothing property we get

$$\sigma\left(W_j^{-1} A_j\right) \subset [0, 1]$$

because of Lemma 5.26, 4), and as the eigenvalues of $W_j^{-1} A_j$ are positive because of $\varrho(S_j) < 1$. Hence also

$$\sigma(X_j) = \sigma\left(A_j^{1/2} W_j^{-1} A_j^{1/2}\right) = \sigma\left(W_j^{-1} A_j\right) \subset [0, 1].$$

This together with (5.135) shows the implication in (5.134). □

We return to the discussion of the smoothing and approximation property as in (5.132).

The smoothing property is of an algebraic nature, but for the proof of the approximation property we will use—at least indirectly—the original variational formulation of the boundary value problem and the corresponding error estimate. Therefore, we discuss first the smoothing property for, as an example, the relaxed Richardson method for a symmetric positive definite matrix A_l, i.e.,

$$S_l = I_l - \omega A_l \quad \text{with} \quad \omega \in \left(0, \frac{1}{\lambda_{\max}(A_l)}\right].$$

Let $\{z_i\}_{i=1}^{M_l}$ be an orthonormal basis of eigenvectors of A_l. For any initial vector $x^{(0)}$ represented in this basis as $x^{(0)} = \sum_{i=1}^{M_l} c_i z_i$ it follows that (compare (5.69))

$$\|A_l S_l^\nu x^{(0)}\|^2 = \sum_{i=1}^{M_l} \lambda_i^2 (1-\lambda_i\omega)^{2\nu} c_i^2 = \omega^{-2} \sum_{i=1}^{M_l} (\lambda_i\omega)^2 (1-\lambda_i\omega)^{2\nu} c_i^2$$
$$\le \omega^{-2} \left[\max_{\xi\in[0,1]} \xi(1-\xi)^\nu \right]^2 \sum_{i=1}^{M_l} c_i^2 .$$

The function $\xi \mapsto \xi(1-\xi)^\nu$ has its maximum at $\xi_{\max} = (\nu+1)^{-1}$; thus

$$\xi_{\max}(1-\xi_{\max})^\nu = \frac{1}{\nu+1}\left(1-\frac{1}{\nu+1}\right)^\nu = \frac{1}{\nu}\left(\frac{\nu}{\nu+1}\right)^{\nu+1} \le \frac{1}{e\nu} .$$

Hence

$$\|A_l S_l^\nu x^{(0)}\| \le \frac{1}{\omega e\nu} \|x^{(0)}\| ,$$

which implies

$$\|A_l S_l^\nu\| \le \frac{1}{\omega e\nu} .$$

Since the inclusion $\omega \in (0, 1/\lambda_{\max}(A_l)]$ can be written in the form $\omega = \sigma/\|A_l\|$ with $\sigma \in (0,1]$, we have $C_S := C_S(\nu) := 1/(\sigma e\nu)$.

The approximation property can be motivated in the following way. The fine grid solution x_l of $A_l x_l = b_l$ is replaced in the coarse grid correction by py_{l-1} from $A_{l-1}y_{l-1} = d_{l-1} := rd_l$. Therefore, $py_{l-1} \approx A_l^{-1} d_l$ should hold. The formulation (5.132) thus is just a quantitative version of this requirement. Since in the symmetric case $\|A_l\|^{-1}$ is simply the reciprocal value of the largest eigenvalue, (3.159) in Theorem 3.48 establishes the relation to the statements of convergence in Section 3.4.

More precisely, we try to give conditions for (5.132) for $\|\cdot\| = \|\cdot\|_2$ in the context of a Galerkin discretization as discussed above: (5.132) is equivalent to

$$\|(A_l^{-1} - pA_{l-1}^{-1} r)b\|_2 \le C_A \|A_l\|_2^{-1} \|b\|_2 \quad \text{for all } b \in \mathbb{R}^{M_l} .$$

Define $z_l \in \mathbb{R}^{M_l}$, $z_{l-1} \in \mathbb{R}^{M_{l-1}}$ by

$$A_l z_l = b, \; A_{l-1} z_{l-1} = rb, \tag{5.136}$$

then the assertion reads

$$\|z_l - pz_{l-1}\|_2 \le C_A \|A_l\|_2^{-1} \|b\|_2 . \tag{5.137}$$

Let $(P_l^{-1})^* : \mathbb{R}^{M_l} \to V_l$ be the adjoint of the inverse representation operator from (5.107), $\mathbb{R}^{M_l}$ bearing the scalar product from (5.119), and consider the Galerkin approximations

$$\begin{aligned} a(u_l, v) &= \langle f_l, v\rangle_0 \quad \text{for all } v \in V_l, \\ a(u_{l-1}, v) &= \langle f_l, v\rangle_0 \quad \text{for all } v \in V_{l-1}, \end{aligned} \tag{5.138}$$

where a is the underlying bilinear form and $f_l := (P_l^{-1})^* b$, meaning

$$\langle f_l, v \rangle_0 = \left\langle b, P_l^{-1} v \right\rangle^{(l)} .$$

Therefore

$$z_l = P_l^{-1} u_l \,, \; p z_{l-1} = P_l^{-1} u_{l-1},$$

noting for the last identity that the right-hand side in (5.138) is

$$\begin{aligned} \left\langle b, P_l^{-1} v \right\rangle^{(l)} &= \left\langle b, P_l^{-1} P_{l-1} c \right\rangle^{(l)} \quad \text{with } c \in \mathbb{R}^{M_{l-1}} \\ &= \langle b, pc \rangle^{(l)} = \langle rb, c \rangle^{(l-1)} , \end{aligned}$$

and thus

$$z_{l-1} = P_l^{-1} u_{l-1} \quad \text{implying} \quad p z_{l-1} = P_l^{-1} P_{l-1} P_{l-1}^{-1} u_{l-1} = P_l^{-1} u_{l-1} .$$

Assuming (5.120), a condition sufficient for (5.137) is

$$\|u_l - u_{l-1}\|_0 \leq C_2^{-2} C_A \|A_l\|_2^{-1} \|f_l\|_0$$

with C_2 from (5.120).

Concerning $\|A_l\|_2$ we refer to Section 3.7. Under the conditions stated, there exist constants $\overline{C}_1, \overline{C}_2 > 0$ independent of l such that

$$\overline{C}_1 \leq \|A_l\|_2 \leq \overline{C}_2 \hat{h}_l^{-2}, \quad \text{where } \hat{h}_l := \min_{K \in \mathcal{T}_h} h_{l,K} .$$

Therefore, furthermore assuming quasi-uniformity of the triangulation (see (3.158)), it is sufficient to have an estimate of the form

$$\|u_l - u_{l-1}\|_0 \leq C h_l^2 \|f_l\|_0 .$$

This finally can be conducted from the a priori estimates (see Theorem 3.32 together with Theorem 3.40), provided the problem is regular in the sense of Definition 3.39 and the triangulation satisfies

$$h_{l-1} \leq C h_l$$

for a constant $C > 0$ independent of l.

For a more exact analysis of convergence and a more detailed description of this topic, we refer to the cited literature and [34].

Exercises

Problem 5.19 Determine the prolongation and restriction according to (5.118) and (5.121) for the case of a linear ansatz on a Friedrichs–Keller triangulation.

Problem 5.20 Consider the problem

$$-(k(x)u')' = f(x) \quad \text{in } (0,1), \quad u(0) = u(1) = 0$$

and P_1-conforming finite elements on equidistant grids with $h_i := 2^{-i}$, $i = 2, \ldots, L$. Consider the two-grid iteration with damped Jacobi smoothing, $\omega := \frac{1}{2}$, with ν a priori steps and Galerkin restriction/prolongation. Show the convergence by Fourier analysis.

Let $A \in \mathbb{R}^{m,m}$ be the discretization matrix, M the iteration matrix.

a) Compute the eigenvectors ψ^k, $k = 1, \ldots, m$, of A. Let now $U^k := \operatorname{span}\{\psi^k, \psi^{m-k}\}$, $k = 1, \ldots, \frac{m-1}{2}$ (where it is assumed that the eigenvalues are ordered such that $\lambda_k < \lambda_{k+1}$).

Show that U^k is invariant under M: Let $M^{(k)}$ be the corresponding representing 2×2 blocks, then

$$M^{(k)} = \begin{pmatrix} s_k^2 & c_k^2 \\ s_k^2 & c_k^2 \end{pmatrix} = \begin{pmatrix} s_k^{2\nu} & 0 \\ 0 & s_k^{2\nu} \end{pmatrix} \qquad \text{for } k = 1, \ldots, \frac{m-1}{2},$$

$$M^{(k)} = 2^{-\nu} \qquad \text{for } k = \frac{m+1}{2},$$

where $s_k{}^2 := \sin^2(\pi k \frac{h_L}{2})$, $c_k{}^2 := \cos^2(\pi k \frac{h_L}{2})$ and ν is the number of smoothing steps.
Interpret this representation!

b) Show for the eigenvalues λ of M:

$$|\lambda| \le \frac{1}{\nu+1}\left(1 - \frac{1}{1+\nu}\right)^{\nu}\left(\frac{1}{2}\right)^{\nu+1}$$

and thus

$$|\lambda| \le \frac{1}{2}, \quad |\lambda| \sim \frac{1}{e}\frac{1}{\nu} \quad \text{for } \nu \to \infty.$$

c) Show that

$$\|M\|_2 \le \frac{1}{2}, \quad \|M\|_2 \sim \frac{\sqrt{2}}{e}\frac{1}{\nu} \quad \text{for } \nu \to \infty.$$

Problem 5.21 Try to extend Theorem 5.24 for the two-grid method with ν_1 a priori and ν_2 a posteriori smoothing steps.

Problem 5.22 Consider the two-grid method for the variational setting with canonical restrictions and prolongations (i.e. (5.117) holds).

a) Show that the coarse grid correction operator

$$P := I - pA_{l-1}^{-1} r A_l$$

is a projection (in $\langle \cdot, \cdot \rangle^{(l)}$, $\langle \cdot, \cdot \rangle^{(l-1)}$).

b) Find weighted norms (and conditions) such that this projection is orthogonal.

Problem 5.23 Prove Lemma 5.26.

5.6 Nested Iterations

As in Section 5.5 we assume that besides the system of equations

$$A_l x_l = b_l$$

with M_l unknowns there are given analogous low-dimensional systems of equations

$$A_k x_k = b_k\,, \quad k = 0, \ldots, l-1, \tag{5.139}$$

with M_k unknowns, where $M_0 < M_1 < \cdots < M_l$. Let all systems of equations be an approximation of the same continuous problem such that an error estimate of the type

$$\|u - P_l x_l\| \le C_A h_l^\alpha$$

holds, with P_l according to (5.107) and $\alpha > 0$. Here $\|\cdot\|$ is a norm on the basic space V, and the constant C_A generally depends on the solution u of the continuous problem. The discretization parameter h_l determines the dimension M_l: In the simplest case of a uniform refinement, $h_l^d \sim 1/M_l$ holds in d space dimensions. One may also expect that for the discrete solution,

$$\|p x_{k-1} - x_k\|_k \le C_1 C_A h_k^\alpha\,, \quad k = 1, \ldots, l,$$

holds with a constant $C_1 > 0$. Here $\|\cdot\|_k$ is a norm on $\mathbb{R}^{M_k}$, and the mapping $p = p_{k-1,k} : \mathbb{R}^{M_{k-1}} \to \mathbb{R}^{M_k}$ is a prolongation, for example the canonical prolongation introduced in Section 5.5. In this case the estimate can be rigorously proven with the definition of the canonical prolongation $p = P_k^{-1} P_{k-1}$:

$$\begin{aligned}
\|p x_{k-1} - x_k\|_k &= \left\|P_k^{-1}\left(P_{k-1} x_{k-1} - P_k x_k\right)\right\|_k \\
&\le \left\|P_k^{-1}\right\|_{L[V_k, \mathbb{R}^{M_k}]} \left\|P_{k-1} x_{k-1} - P_k x_k\right\| \\
&\le \left\|P_k^{-1}\right\|_{L[V_k, \mathbb{R}^{M_k}]} \left(C_A h_k^\alpha + C_A h_{k-1}^\alpha\right) \le C_1 C_A h_k^\alpha
\end{aligned}$$

with

$$C_1 := \max_{j=1,\ldots,l} \left\{ \left\|P_j^{-1}\right\|_{L[V_j, \mathbb{R}^{M_j}]} \left(1 + \left(\frac{h_{j-1}}{h_j}\right)^\alpha\right) \right\}.$$

Let the system of equations be solved with an iterative method given by the fixed-point mapping Φ_k, $k = 0, \ldots, l$, which means that x_k according to (5.139) satisfies $x_k = \Phi_k(x_k, b_k)$. Then it is sufficient to determine an iterate $\tilde{x}_l$ with an accuracy

$$\|\tilde{x}_l - x_l\|_l \le \tilde{C}_A h_l^\alpha \tag{5.140}$$

with $\tilde{C}_A := C_A/\|P_l\|_{L[\mathbb{R}^{M_l},V]}$, because then we also have

$$\|P_l\tilde{x}_l - P_l x_l\| \le C_A h_l^\alpha .$$

If one does not have a good initial iterate from the concrete context, the algorithm of *nested iterations* explained in Table 5.10 can be used. It is indeed a finite process.

Choose m_k, $k = 1, \ldots, l$.
Let $\tilde{x}_0$ be an approximation of x_0,

for example $\tilde{x}_0 := x_0 := A_0^{-1} b_0$.

For $k = 1, \ldots, l$:

$$\tilde{x}_k^{(0)} := p\,\tilde{x}_{k-1} .$$

Perform m_k iterations:

$$\tilde{x}_k^{(i)} := \Phi_k\big(\tilde{x}_k^{(i-1)}, b_k\big),\ i = 1, \ldots, m_k .$$

Set $\tilde{x}_k := \tilde{x}_k^{(m_k)}$.

Table 5.10: Nested Iteration.

The question is how to choose the iteration numbers m_k such that (5.140) finally holds, and whether the arising overall effort is acceptable. An answer to this question is provided by the following theorem:

Theorem 5.28 *Let the iterative method Φ_k have the contraction number ϱ_k with respect to $\|\cdot\|_k$. Assume that there exist constants $C_2, C_3 > 0$ such that*

$$\|p\|_{L[\mathbb{R}^{M_{k-1}},\mathbb{R}^{M_k}]} \le C_2 ,$$
$$h_{k-1} \le C_3 h_k ,$$

for all $k = 1, \ldots, l$. If the iteration numbers m_k for the nested iterations are chosen in such a way that

$$\varrho_k^{m_k} \le 1/(C_2 C_3^\alpha + C_1\|P_l\|), \tag{5.141}$$

then

$$\|\tilde{x}_k - x_k\|_k \le \tilde{C}_A h_k^\alpha \quad \text{for all } k = 1, \ldots, l,$$

provided that this estimate holds for $k = 0$.

Proof The proof is given by induction over k. Assume that the assertion is true for $k - 1$. This induces

$$\begin{aligned}
\|\tilde{x}_k - x_k\|_k &\le \varrho_k^{m_k}\ \|p\tilde{x}_{k-1} - x_k\|_k \\
&\le \varrho_k^{m_k}\ \big(\|p(\tilde{x}_{k-1} - x_{k-1})\|_k + \|p x_{k-1} - x_k\|_k\big) \\
&\le \varrho_k^{m_k}\ \big(C_2\tilde{C}_A h_{k-1}^\alpha + C_1 C_A h_k^\alpha\big) \\
&\le \varrho_k^{m_k}\ \big(C_2 C_3^\alpha + C_1\|P_l\|\big)\,\tilde{C}_A h_k^\alpha .
\end{aligned}$$

□

Theorem 5.28 allows the calculation of the necessary number of iterations for the inner iteration from the norms $\|p\|_{L[\mathbb{R}^{M_{k-1}},\mathbb{R}^{M_k}]}$, $\|P_k^{-1}\|_{L[V_k,\mathbb{R}^{M_k}]}$ and the constants $\frac{h_{k-1}}{h_k}$ for $k = 1, \ldots, l$, as well as the order of convergence α of the discretization.

In order to estimate the necessary effort according to (5.141) more exactly, the dependence of ϱ_k of k must be known. In the following we consider only the situation, known from the multigrid method, of a method of optimal complexity

$$\varrho_k \le \overline{\varrho} < 1 .$$

Here, in contrast to other methods, the number of iterations can be chosen constant ($m_k = m$ for all $k = 1, \ldots, l$). If, furthermore, the estimate (5.124) holds with the constant C, then analogously to the consideration in Section 5.5 the total number of operations for the nested iteration can be estimated by

$$m \frac{C}{C-1} \overline{C} M_l .$$

Here $\overline{C}M_k$ is the number of operations for an iteration with the iterative method Φ_k.

In the model problem of the Friedrichs–Keller triangulation with uniform refinement we have $C/(C-1) = 4/3$ and $C_3 = 2$. For $\|\cdot\| = \|\cdot\|_0$ as basic norm, $\alpha = 2$ is a typical case according to Theorem 3.40. The existence of the constant C_2 will hereby finally be ensured consistently by the condition (5.120), observing (5.117). Assuming also that the constants $C_1, C_2, \|P_l\|$ are "small" and the iterative method has a "small" contraction number ϱ, only a small number of iterations m is necessary, in the ideal case $m = 1$. At least in this situation we can count on only a small increase of the necessary effort through the process of nested iterations, which provides an "appropriate" approximation $\tilde{x}_k$ on all levels k of discretization.

If the basic iteration is the multigrid method (with fixed a priori and a posteriori smoothing steps ν_1, ν_2 and μ multigrid cycles for the coarse grid correction) as noted above, the number of iterations m at each level should be the same. One then speaks of the *full multigrid method (FMG)*. The grid levels involved are depicted in Fig. 5.8. The prolongation step from Table 5.7 is indicated by a double line, and may be different form the one used in the multigrid method.

Finally, it is to be observed that the sequence of the discrete problems has to be defined only during the process of the nested iteration. This offers the possibility to combine it with a posteriori error estimators as discussed in Section 4.2, in order to develop a grid $\mathcal{T}_{k+1}$ on which the discrete problem of level $k+1$ is determined, on the basis of $\tilde{x}_k$ as a refinement of $\mathcal{T}_k$.

$l = 3$ and $\mu = 1$:

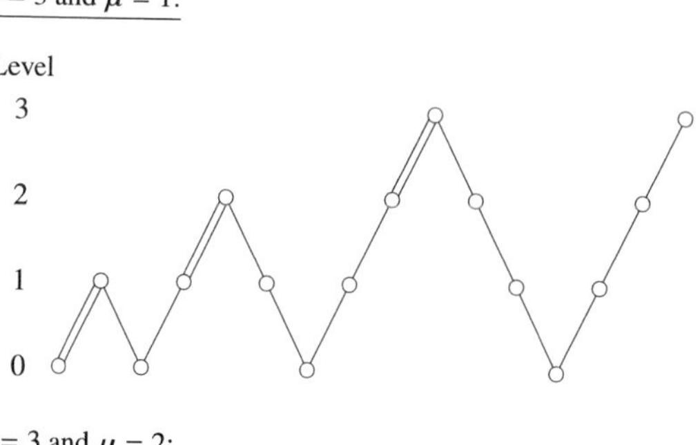

$l = 3$ and $\mu = 2$:

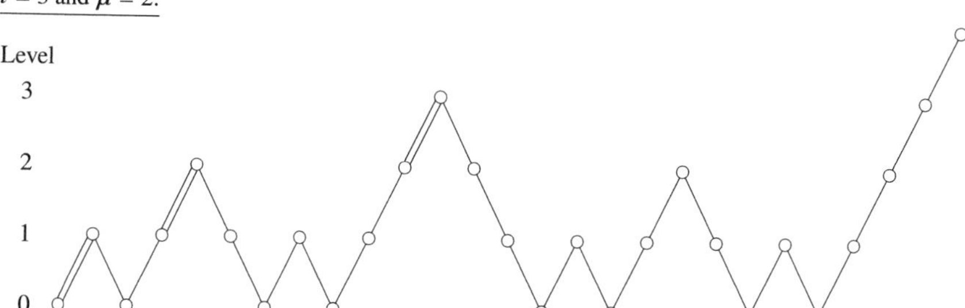

Fig. 5.8: Grid levels in the FMG with $m = 1$.

5.7 Space (Domain) Decomposition Methods

Going back to the most simple linear stationary methods, the Jacobi method

$$
\begin{aligned}
x_i^{(k+1)} &:= x_i^{(k)} - a_{ii}^{-1}\left(Ax^{(k)} - b\right)_i, \quad i = 1, \ldots, m, \\
\text{or} \quad x^{(k+1)} &:= x^{(k)} - \sum_{i=1}^{m} a_{ii}^{-1}\left(Ax^{(k)} - b\right)_i e_i
\end{aligned}
$$

and the Gauss-Seidel method, which may be written (inefficiently) as

$$
\begin{aligned}
x^{(k+1,0)} &:= x^{(k)}, \\
\text{for } i = 1, \ldots, m: \quad x^{(k+1,i)} &:= x^{(k+1,i-1)} - a_{ii}^{-1}\left(Ax^{(k+1,i-1)} - b\right)_i e_i,
\end{aligned}
\tag{5.142}
$$

we may see some *parallelism*, i.e., operations that are independent of each other in the Jacobi method, namely in the assembly of the second term in each component equation, i.e., projection of the residual to a subspace (here span $\{e_i\}$), and the approximate solution of the equation there. Finally these "in parallel" assembled terms are summed up to perform the correction. Therefore one also speaks of an *additive method*. Contrary, in the Gauss–Seidel method the same substeps are performed, but sequentially one after other and dependent on each other. In terms of (iteration) matrices this corresponds to a product, therefore one speaks of a *multiplicative method*. As nowadays not only high performance computers gain their speed by multicore architectures, algorithms which can exploit these (more easily), are desirable. Therefore such additive methods have been extensively developed in recent decades (see, e.g., [233], [30]). The following exposition is largely inspired by [47]. Here we give a short introduction in such methods based on the *space decomposition* approach, also called *iterative substructuring*. The space decomposition can come from a nested structure of finite element ansatz spaces as in Section 5.5 or from a subdivision of the domain of the underlying partial differential equation. The second case leads to *domain decomposition* methods, the goal of the first line of development may be seen in making the multigrid method, which is multiplicative, more additive without deteriorate (too much) the optimal convergence behaviour. As is in the previous sections, the methods to be developed may be viewed as iterative methods, but it is more advisable to use them as preconditioners within Krylov subspace methods.

5.7.1 Preconditioning by Space Decomposition

We restrict the discussion to the self-adjoint case. Let $A \in \mathbb{R}^{m,m}$ be symmetric, positive definite, i.e., $V := \mathbb{R}^m$ is the underlying vector space and $x := A^{-1}b$ for some $b \in \mathbb{R}^m$ the exact solution. Assume there is a (not necessarily direct) decomposition in subspaces $V_i \subset V$, $i = 0, \ldots, l$:

$$V = V_0 + \ldots + V_l, \tag{5.143}$$

i.e., $v \in V$ can be represented as

$$v = \sum_{i=0}^{l} v_i \qquad \text{with } v_i \in V_i\,.$$

Let $Q_i : V \to V_i$ be the orthogonal projection with respect to the Euclidean scalar product $\langle \cdot, \cdot \rangle$, and Q_i^A with respect to the energy scalar product $\langle \cdot, \cdot \rangle_A$ (see (5.46)). We define for $i = 0, \ldots, l$ a *subspace operator* $A_i : V_i \to V_i$ by

$$\langle A_i x_i, v_i \rangle = \langle A x_i, v_i \rangle \quad \text{for all } x_i, v_i \in V_i,$$

i.e., A_i is the orthogonal projection of AV_i to V_i. We have

$$A_i Q_i^A = Q_i A \tag{5.144}$$

because of

$$\langle A_i Q_i^A v, v_i \rangle = \langle A Q_i^A v, v_i \rangle = \langle v, v_i \rangle_A$$
$$= \langle Av, v_i \rangle = \langle Q_i A v, v_i \rangle \quad \text{for all } v \in V,\ v_i \in V_i\,.$$

Therefore $x_i := Q_i^A x$ satisfies the *subspace equation*

$$A_i x_i = b_i := Q_i b. \tag{5.145}$$

Note that

$$\tilde{x} = \sum_{i=0}^{l} x_i = \sum_{i=0}^{l} A_i^{-1} Q_i b$$

is in general not an exact solution of $Ax = b$, as this would require that the subspaces are orthogonal with respect to $\langle \cdot, \cdot \rangle_A$ (see Problem 5.24). However, having an approximation $x^{(k)}$, a better one (?) $x^{(k+1)}$ can be achieved in this way, i.e., by applying the reasoning to the error equation

$$Ae^{(k)} = g^{(k)} := Ax^{(k)} - b.$$

Solving the corresponding subspace equations

$$A_i e_i = Q_i g^{(k)}, \quad i = 0, \dots, m,$$

leads to the approximation

$$x^{(k+1)} := x^{(k)} - \sum_{i=0}^{l} e_i = x^{(k)} - \left(\sum_{i=0}^{l} A_i^{-1} Q_i \right) \left(Ax^{(k)} - b \right).$$

If the subspace equations are not solved exactly, but A_i^{-1} is approximated by an operator

$$R_i : V_i \to V_i\,,$$

then

$$x^{(k+1)} = x^{(k)} - B \left(Ax^{(k)} - b \right) \tag{5.146}$$

with

$$B := \sum_{i=0}^{l} R_i Q_i\,. \tag{5.147}$$

This can be viewed as the definition of a consistent linear stationary iterative method with $N := B$, $M := I - BA$, and therefore B can also be viewed as a preconditioner to be used with a Krylov subspace method.

Lemma 5.29 *If all operators R_i are symmetric, positive definite, then also the operators $R_i Q_i$, $i = 0, \dots, l$, are symmetric and positive semidefinite, and the operators $T_i := R_i Q_i A$ are self-adjoint and positive semidefinite with respect to $\langle \cdot, \cdot \rangle_A$. The* *operator* B *from* (5.147) *is symmetric, positive definite, and* BA *is self-adjoint, positive definite with respect to* $\langle\cdot,\cdot\rangle_A$.

Proof Problem 5.27. □

In (5.146) we have defined an additive method, the corresponding multiplicative method reads, using the given ordering of the subspaces, and generalizing (5.142) to

$$\begin{aligned} x^{(k+1,0)} &:= x^{(k)}, \\ \text{for } i = 0,\dots,l-1: \quad x^{(k+1,i+1)} &:= x^{(k+1,i)} - R_i Q_i \left(A x^{(k+1,i)} - b \right). \end{aligned} \tag{5.148}$$

Comparing the iteration matrices, in (5.146) we have

$$M_l = I - \sum_{i=0}^{l} R_i Q_i A = I - \sum_{i=0}^{l} R_i A_i Q_i^A$$

and in (5.148):

$$M_l = \prod_{i=0}^{l} \left(I - R_i A_i Q_i^A \right).$$

Again the iteration (5.148) can be used to define a preconditioner by performing one or more iteration steps (see (5.104)).

If the operator B from (5.147) is directly used as a preconditioner, then an estimate of the condition number $\kappa(BA)$, ideally independent of m, is required. This can be based on the following concept.

Let H be a Hilbert space and $H_i \subset H$, $i = 1, 2$, be subspaces. Then the inequality

$$\langle u_1, u_2 \rangle_H \le \gamma \|u_1\|_H \|u_2\|_H \quad \text{for all } u_i \in H_i\,,\ i = 1, 2, \tag{5.149}$$

obviously holds for $\gamma = 1$, which is also optimal for $H_1 = H_2$. On the other hand, if the subspaces H_1 and H_2 are orthogonal, also $\gamma = 0$ is valid. If (5.149) holds with $\gamma \in [0, 1)$, we speak of a *strengthened Cauchy–Schwarz inequality*. In the following we will always mean the minimal value, that is

$$\gamma := \sup_{\substack{u_i \in H_i \\ i=1,2}} |\langle u_1, u_2 \rangle|_H \,/\, \|u_1\|_H\, \|u_2\|_H$$

(the principal value of $\arccos \gamma$ is called the *angle between* H_1 *and* H_2).

Considering the situation of (5.143) with $\langle\cdot,\cdot\rangle_A$ and denoting the values from (5.149) for the pairs V_i, V_j by $\gamma_{ij} \in [0, 1]$, then

$$G := (\gamma_{ij})_{ij} \in \mathbb{R}^{l+1,l+1}$$

is a symmetric matrix, i.e., $\|G\|_2 = \varrho(G)$, and $\gamma_{ii} = 1$ for $i = 0,\dots,l$. Consider the quadratic form defined by G, and use $\langle\cdot,\cdot\rangle$ for the Euclidean scalar product on $\mathbb{R}^{l+1}$, then

$$\langle Gz, w\rangle \le \varrho(G)|z|_2|w|_2 \quad \text{for all } z, w \in \mathbb{R}^{l+1}, \tag{5.150}$$

and—more refined for the lower triangular part of G—

$$\sum_{i=1}^{l}\sum_{j=0}^{i-1} \gamma_{ij} z_i w_j = \frac{1}{2}\langle (G-I)\, z, w\rangle \le \frac{1}{2}\varrho(G)|z|_2|w|_2 \,. \tag{5.151}$$

Here we use $\|G - I\|_2 \le \|G\|_2$, which can be shown using the Perron–Frobenius theorem (see, e.g., [6, Cor. 2.1.5]) (Problem 5.28). As an immediate consequence we have for $y_i, v_i \in V_i$, $i = 0, \ldots, l$:

$$\sum_{i,j=0}^{l} \langle y_i, v_j\rangle_A \le \varrho(G)\left(\sum_{i=0}^{l}\|y_i\|_A^2\right)^{1/2}\left(\sum_{i=0}^{l}\|v_i\|_A^2\right)^{1/2}, \tag{5.152}$$

and for $\sum_{i=1}^{l}\sum_{j=0}^{i-1}$ the same estimate but with the factor $\frac{1}{2}$.

By the Gershgorin circle theorem (see, e.g., [32, Criterion 4.3.4], [61, Thm. 1.1.11]) we have: If in each row of G at most l_0 entries are nonzero, then

$$\varrho(G) \le l_0 \tag{5.153}$$

(Problem 5.26).

According to Lemma 5.29, the estimation of $\kappa(BA)$ in the spectral norm is equivalent to looking for upper and lower bounds for the eigenvalues of BA.

Theorem 5.30 *Let the matrices A and R_i be symmetric, positive definite. Assume that for all $v \in V$ there is a decomposition $v = \sum_{i=0}^{l} v_i$, $v_i \in V_i$, such that for some $\alpha > 0$*

$$\alpha \sum_{i=0}^{l} \langle R_i^{-1} v_i, v_i\rangle \le \langle Av, v\rangle \,. \tag{5.154}$$

Let β be the maximum of all eigenvalues of $R_i A_i$, $i = 0, \ldots, l$, then the additive space domain preconditioner from (5.147) *fulfils*

$$\sigma(BA) \subset [\alpha, \beta\varrho(G)].$$

Proof We use the abbreviations

$$T_i := R_i Q_i A = R_i A_i Q_i^A,\ T := BA = \sum_{i=0}^{l} T_i \,.$$

1) $\varrho(T) \le \beta \varrho(B)$

This is equivalent to the upper bound of the assertion, as the eigenvalues of T and B are positive due to Lemma 5.29.

To estimate the spectral radius, it suffices to do so with any induced matrix norm, for which we choose $\|\cdot\|_A$. Because of

$$\|Tv\|_A^2 = \sum_{i,j=0}^{l} \langle T_i v, T_j v\rangle_A$$

we estimate the latter term by (5.151) and obtain

$$\begin{aligned}\|Tv\|_A^2 &\le \varrho(G)\sum_{i=0}^{l}\langle AT_i v, T_i v\rangle = \varrho(G)\sum_{i=0}^{l}\langle A_i T_i v, T_i v\rangle \\ &= \varrho(G)\sum_{i=0}^{l}\left\langle R_i^{-1}R_i A_i T_i v, T_i v\right\rangle \le \beta\varrho(G)\sum_{i=0}^{l}\left\langle R_i^{-1} T_i v, T_i v\right\rangle,\end{aligned} \tag{5.155}$$

making use of the extremal property of the maximal eigenvalue of $R_i A_i$ in the norm $\|\cdot\|_{R_i^{-1}}$ (see [54, Thm. 1.21]). As A, R_i are symmetric, we have

$$\begin{aligned}\left\langle R_i^{-1}T_i v, T_i v\right\rangle &= \left\langle R_i^{-1}R_i Q_i A v, R_i Q_i A v\right\rangle = \langle R_i Q_i A v, Q_i A v\rangle \\ &= \langle R_i Q_i A v, A v\rangle = \langle A R_i Q_i A v, v\rangle = \langle T_i v, v\rangle_A,\end{aligned}$$

and therefore the estimate finally reads

$$\|Tv\|_A^2 \le \beta\varrho(G)\sum_{i=0}^{l}\langle T_i v, v\rangle_A = \beta\varrho(G)\,\langle Tv, v\rangle_A \le \beta\varrho(G)\,\|Tv\|_A\|v\|_A,$$

which shows the assertion.

2) Let $\lambda_{\min}$ be the minimal eigenvalue of BA, then $\lambda_{\min} \ge \alpha$

It is sufficient to prove

$$\langle Tv, v\rangle_A \ge \alpha\|v\|_A^2 \quad \text{for all } v \in V.$$

We have

$$\langle v, v\rangle_A = \sum_{i=0}^{l}\langle v_i, v\rangle = \sum_{i=0}^{l}\langle v_i, Av\rangle = \sum_{i=0}^{l}\langle v_i, Q_i A v\rangle$$

and

$$\sum_{i=0}^{l}\langle v_i, Q_i A v\rangle = \sum_{i=0}^{l}\left\langle R_i^{-1/2}v, R_i^{1/2}Q_i A v\right\rangle \le \sum_{i=0}^{l}\left\langle R_i^{-1}v_i, v_i\right\rangle^{1/2}\langle R_i Q_i A v, Q_i A v\rangle^{1/2}$$

by the Cauchy–Schwarz inequality with respect to $|\cdot| := \langle \cdot, \cdot \rangle^{1/2}$. In the continuation of this estimate follows

$$\begin{aligned}\sum_{i=0}^{l} \langle v_i, Q_i A v \rangle &\leq \sum_{i=0}^{l} \left\langle R_i^{-1} v_i, v_i \right\rangle^{1/2} \langle R_i Q_i A v, v \rangle_A^{1/2} \\ &\leq \left(\sum_{i=0}^{l} \left\langle R_i^{-1} v_i, v_i \right\rangle \right)^{1/2} \left(\sum_{i=0}^{l} \langle T_i v, v \rangle_A \right)^{1/2}\end{aligned}$$

by the Cauchy–Schwarz inequality with respect to $|\cdot|_2$ in $\mathbb{R}^{l+1}$, and thus

$$\|v\|_A^2 \leq \alpha^{-1/2} \|v\|_A \langle T v, v \rangle_A^{1/2},$$

i.e., the assertion. □

In order to use the multiplicative version (5.148) as a linear stationary method directly to define a preconditioner by k steps, $k \in \mathbb{N}$, we need an estimate of a norm of the iteration matrix M_l. This is given by the following result.

Theorem 5.31 *Under the assumption of Theorem 5.30 and additionally with $\beta < 2$ we have*

$$\|M_l\|_A^2 \leq 1 - \frac{\alpha(2-\beta)}{1 + \frac{1}{2}\beta \varrho(G)}.$$

Proof We want to estimate

$$\left\| \prod_{i=0}^{l} (I - T_i) \right\|_A,$$

using the notation of Theorem 5.30. Considering the hierarchy

$$M_k = \prod_{i=0}^{k} (I - T_i), \ k = 0, \ldots, l,$$

we have, setting $M_{-1} := I$:

$$\begin{aligned} M_{j-1} - M_j &= T_j M_{j-1}, \ j = 0, \ldots, l, \\ I - M_i &= \sum_{j=0}^{i} T_j M_{j-1}, \ i = 0, \ldots, l. \end{aligned} \tag{5.156}$$

In what follows we will make frequent use of Lemma 5.29 without mentioning.

In particular, the mapping $(v, w) \mapsto \langle T_i v, w \rangle_A$ satisfies the properties of a scalar product apart from definiteness, i.e., there is a Cauchy–Schwarz inequality with the induced seminorm. Proceeding as in (5.155) but applying now (5.151) to $T_i w_i$ and $T_j v$, we get

$$\sum_{i=1}^{l}\sum_{j=0}^{i-1}\left\langle T_i w_i, T_j v\right\rangle_A \le \frac{1}{2}\beta\varrho(G)\sum_{i=0}^{l}(\langle T_i w_i, w_i\rangle_A)^{1/2}\langle Tv, v\rangle_A^{1/2}. \tag{5.157}$$

By the Cauchy–Schwarz inequality (as mentioned above and in $\mathbb{R}^{l+1}$) we get

$$\sum_{i=0}^{l}\langle T_i w_i, v\rangle_A \le \left(\sum_{i=0}^{l}\langle T_i w_i, w_i\rangle_A\right)^{1/2}\langle Tv, v\rangle_A^{1/2}. \tag{5.158}$$

Using (5.156) in the decomposition $I = M_{i-1} + I - M_{i-1}$, we can rewrite

$$\langle Tv, v\rangle_A = \sum_{i=0}^{l}\langle T_i v, v\rangle_A = \sum_{i=0}^{l}\langle T_i v, M_{i-1}v\rangle_A + \sum_{i=1}^{l}\sum_{j=0}^{i-1}\left\langle T_i v, T_j M_{j-1} v\right\rangle_A,$$

and using the above estimates (with $w_i := M_{i-1}v$) and Lemma 5.29 we get

$$\langle Tv, v\rangle_A \le \left(1 + \frac{1}{2}\beta\varrho(G)\right)\left(\sum_{i=0}^{l}\langle T_i M_{i-1}v, M_{i-1}v\rangle_A\right)^{1/2}\langle Tv, v\rangle_A^{1/2}.$$

From Theorem 5.30 we know that

$$\alpha\|v\|_A^2 \le \langle Tv, v\rangle_A \le \left(1 + \frac{1}{2}\beta\varrho(G)\right)^2\sum_{i=0}^{l}\langle T_i M_{i-1}v, M_{i-1}v\rangle_A. \tag{5.159}$$

The right-hand side can be further transformed by $M_{i-1} - M_i = T_i M_{i-1}$ leading to

$$\begin{aligned}\|M_{i-1}v\|_A^2 - \|M_i v\|_A^2 &= \|T_i M_{i-1}v\|_A^2 + 2\langle T_i M_{i-1}v, M_i v\rangle_A\\ &= \langle T_i M_{i-1}v, T_i M_{i-1}v\rangle_A + 2\langle T_i(I - T_i)M_{i-1}v, M_{i-1}v\rangle_A\\ &= \langle (2I - T_i)T_i M_{i-1}v, M_{i-1}v\rangle_A\\ &\ge (2-\beta)\langle T_i M_{i-1}v, M_{i-1}v\rangle_A.\end{aligned}$$

For the last estimate note that, due to Lemma 5.29, the mapping $u \mapsto \langle T_i u, u\rangle_A$ is a seminorm on V and that $R_i A_i$ on V_i has the same nonzero eigenvalues as $T_i = R_i A_i Q_i^A$ on V, as Q_i^A is a projection, so that the extremal property of the maximal eigenvalue can be applied (see [54, Thm. 1.21]).

Summarizing up these inequalities, we obtain

$$\sum_{i=0}^{l}\langle T_i M_{i-1}v, M_{i-1}v\rangle_A \le (2-\beta)^{-1}\left(\|v\|_A^2 - \|M_l v\|_A^2\right).$$

The estimate (5.159) finally gives

$$\alpha\|v\|_A^2 \le \left(1 + \frac{1}{2}\beta\varrho(G)\right)^2(2-\beta)^{-1}\left(\|v\|_A^2 - \|M_l v\|_l^2\right),$$

and a rearrangement leads to

$$\|M_l v\|_A^2 \le \left(1 - \frac{\alpha(2-\beta)}{1 + \frac{1}{2}\beta\varrho(G)}\right) \|v\|_A^2,$$

i.e. the assertion. □

5.7.2 Grid Decomposition Methods

In finite element discretizations quite often not only a single triangulation is considered together with the corresponding ansatz space, but rather a (nested) hierarchy of ansatz spaces

$$V_0 \subset \ldots \subset V_l = V.$$

This is the case for the multigrid method from Section 5.6 and also in adaptive methods with (local) grid refinement (see Section 4.2). The multigrid method using such a hierarchy has turned out to be optimal, but is a multiplicative method. The aim is to develop additive methods with a performance that comes close to the multigrid method.

Additive multigrid method

We again use the notation of Section 5.6 with the tuple spaces $\mathbb{R}^{M_i}$, $i = 0, \ldots, l$, the isomorphisms $P_i : \mathbb{R}^{M_i} \to V_i$, restrictions $r_i : \mathbb{R}^{M_i} \to \mathbb{R}^{M_{i-1}}$ and prolongations $p_i : \mathbb{R}^{M_{i-1}} \to \mathbb{R}^{M_i}$, $i = 1, \ldots, l$, identifying the latter with their representation matrices. The *additive multigrid method* proceeds similar to Table 5.6 with the following modifications. The smoothing and coarse grid corrections become decoupled by applying the coarse grid correction to the nonsmoothed iterate $x_l^{(k)}$, and the coarse grid correction reads:

> Perform for the old approximation one additive multigrid step for the error equation at level $l-1$ with right-hand side
>
> $$d := r_l(b - A_l x^{(k)}), \quad \text{giving } \tilde{x}_{l-1} \quad \text{and } x^{(k+1)} := S_l^\nu x_l^{(k)} + \Theta p_l \tilde{x}_{l-1}$$
>
> with a weighting parameter Θ.

A scheme similar to Fig. 5.6 is depicted in Fig. 5.9, indicating the independent, i.e., to be performed in parallel, portions of the algorithm.

Let us consider the two-grid situation. Then compared to (5.111) the iteration matrix has the form

$$M_l^A := S_l^\nu + \Theta p_l A_{l-1}^{-1} r_l A_l = I - \left(-\Theta p_l A_{l-1}^{-1} r_l + (I - S_l^\nu) A_l^{-1}\right) A_l\,, \tag{5.160}$$

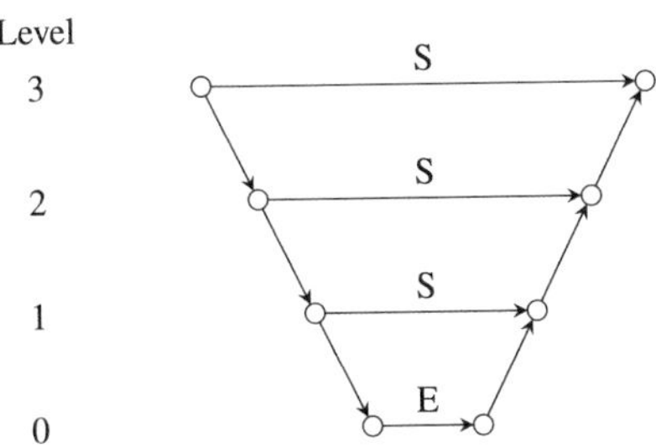

Fig. 5.9: Grid levels for the additive multigrid method: S = smoothing, E = exact solution.

i.e., even if the smoothing method is convergent, this method is not consistent. Therefore we do not consider the additive multigrid method as an independent iterative method, but rather as a preconditioner. The possible preconditioner B_A, i.e., the linear part of one step of the additive multigrid method in the variable b can be written as (Problem 5.29):

$$B_A := \Theta^l \tilde{p}_0 A_0^{-1} \tilde{\gamma}_0 + \sum_{k=1}^{l} \Theta^{l-k} \tilde{p}_k (I - S_k^\nu) A_k^{-1} \tilde{r}_k \,, \tag{5.161}$$

where $\tilde{p}_l := I$, $\tilde{p}_k := p_l \cdots p_{k+1}$, $\tilde{r}_k := \tilde{p}_k^T$, i.e. for $\Theta := 1$, $\nu := 1$, and for $S_k := I - W_k^{-1} A_k$:

$$B_A = \tilde{p}_0 A_0^{-1} \tilde{\gamma}_0 + \sum_{k=1}^{l} \tilde{p}_k W_k^{-1} \tilde{r}_k \,. \tag{5.162}$$

This indeed can be related to a preconditioner of the form (5.147), the so-called *BPX preconditioner*.

The BPX preconditioner

The following space decomposition is named after James Bramble, Josef Pasciak, and Jinchao Xu [233]. It will turn out to be equivalent with the additive multigrid method, as preconditioner in the sense of (5.167). We assume in addition that the Problem $A_l x_l = b_l$ comes from a Galerkin approximation on $V := V_l$ of a variational equation with a bilinear form a with the corresponding equations in V_k, $k < l$. Thus the problem to be solved is given by $\tilde{A} : V \to V'$ and $A_l := \left(\left\langle \tilde{A} \varphi_j^l, \varphi_i^l \right\rangle_0 \right)_{ij}$.

That is, in contrast to (5.143) we consider $V = V_l$ as the basic space and later have to identify the matrix representation. The space decomposition $V_l = \sum_{i=0}^{l} V_i$ is chosen according to the hierarchical ansatz spaces as in Section 5.5.2. On each subspace V_k the approximation of A_k^{-1} is given by $\tilde{R}_k : V_k \to V_k$, $k = 1, \ldots, l$, defined by

$$\tilde{R}_k v_k = \sum_{i=1}^{M_k} \frac{\langle v_k, \varphi_i^k \rangle_0}{a(\varphi_i^k, \varphi_i^k)} \varphi_i^k \tag{5.163}$$

(see Section 5.5.2). On the coarsest grid the equation is solved exactly, i.e., $\tilde{R}_0 := \tilde{A}_0^{-1}$, with

$$\langle \tilde{A}_0 u, v \rangle_0 = a(u, v) \quad \text{for all } u, v \in V_0 .$$

The subspace preconditioner is defined by

$$\tilde{B} := \tilde{R}_0 \tilde{Q}_0 + \sum_{k=1}^{l} \tilde{R}_k \tilde{Q}_k = \tilde{R}_0 \tilde{Q}_0 + \sum_{k=1}^{l} \tilde{R}_k, \tag{5.164}$$

as in $\tilde{R}_k \tilde{Q}_k v$ the term $\langle \tilde{Q}_k v, \varphi_i^k \rangle_0$ is identical with $\langle v, \varphi_i^k \rangle_0$, where $\tilde{Q}_k : V = V_l \to V_k$ is the L^2-orthogonal projection.

To transform the mapping $B : V \to V$ to one from $\mathbb{R}^{M_l}$ to $\mathbb{R}^{M_l}$ and finding its matrix representation, consider for a linear operator $\tilde{C} : V \to V$ the corresponding matrix $C_M \in \mathbb{R}^{M_l, M_l}$ defined by

$$\langle C_M u, v \rangle = \langle \tilde{C} P_l u, P_l v \rangle_0 \quad \text{for all } u, v \in \mathbb{R}^{M_l} . \tag{5.165}$$

Then we can show for $\tilde{B}$ from (5.164) and B_A from (5.162) with $W_k := \operatorname{diag}(A_k)$:

$$(\tilde{A}\tilde{B}\tilde{A})_M = A_l B_A A_l . \tag{5.166}$$

Assume this identity holds, then we conclude for the condition number with respect to the energy norms that

$$\kappa(\tilde{B}\tilde{A}) = \kappa(B_A A_l), \tag{5.167}$$

i.e., both approaches in terms of preconditioning are equivalent. Equation (5.167) can be seen from the following equivalencies for constants $\underline{c}, \overline{c} > 0$:

$$\underline{c}\, a(v_h, v_h) = \underline{c} \langle \tilde{A} v_h, v_h \rangle_0 \le \langle \tilde{A}\tilde{B}\tilde{A} v_h, v_h \rangle_0 \le \overline{c} \langle \tilde{A} v_h, v_h \rangle_0 \quad \text{for all } v_h \in V_l$$

$$\Longleftrightarrow$$

$$\underline{c} \langle A_l v, v \rangle \le \langle A_l B_A A_l v, v \rangle \le \overline{c} \langle A_l v, v \rangle \quad \text{for all } v \in \mathbb{R}_{M_l} .$$

To show (5.166), note that for arbitrary linear operators $\tilde{C}, \tilde{D} : V_l \to V_l$:

$$(\tilde{C}\tilde{D})_M = C_M M_l^{-1} D_M , \tag{5.168}$$

where $M_l := \left(\langle \varphi_j^l, \varphi_i^l \rangle_0 \right)_{ij}$ denotes the mass matrix because of $M_l = P_l^T P_l$ (Problem 5.30). It is sufficient to show (5.166) for each of the terms $k = 0, \ldots, l$ in (5.164) and (5.162), respectively, meaning with (5.168):

$$A_l M_l^{-1} (\tilde{R}_0 \tilde{Q}_0)_M M_l^{-1} A_l = A_l \tilde{p}_0 \tilde{A}_0^{-1} \tilde{r}_0 A_l , \tag{5.169}$$

and for $k = 1, \ldots, l$:

$$A_l M_k^{-1}(\tilde{R}_k)_M M_k^{-1} A_l = A_l \tilde{p}_k W_k^{-1} \tilde{r}_k A_l \,.$$

The last one is a consequence of

$$(\tilde{R}_k)_M = M_l \tilde{p}_k W_k^{-1} \tilde{r}_k M_l \,,$$

and the first of

$$(\tilde{R}_0 \tilde{Q}_0)_M = M_l \tilde{p}_0 A_0^{-1} \tilde{r}_0 M_l \,.$$

Note that $p_l = P_l^{-1} P_{l-1}$ and thus $\tilde{p}_k = P_l^{-1} P_k$ and $\tilde{r}_k = P_k^T P_l^{-T}$, i.e. $\tilde{p}_k, \tilde{r}_k$ shift the representation from the basis $\{\varphi_i^k\}$ to $\{\varphi_i^l\}$ and vice versa. This verifies the above identities.

Theorem 5.32 *Consider the model problem from Sections 2.1, 2.2 (Poisson equation with homogeneous Dirichlet boundary conditions, P_1-elements on triangles in $\mathbb{R}^2$) with a grid hierarchy given by uniform refinement, i.e. $h_k = 2^{l-k} h_l$. Let $u \in H^2(\Omega)$ and the assumptions of Section 3.7 hold true, then the BPX preconditioner is optimal in the sense*

$$\kappa(BA) \leq C$$

with a constant $C \geq 1$ independent of l and h_l.

Proof We want to verify the conditions of Theorem 5.30. To reduce the complexity of notation, we will write

$$\alpha \lesssim \beta \quad \text{instead of} \quad \alpha \leq C\beta,$$

where $C > 0$ is a constant independent of l and h. We first prove (5.154) and by this a lower bound of $\sigma(BA)$. Now let V_k, $k = 0, \ldots, l$, denote the hierarchy of ansatz spaces. By means of the Ritz projections $R_h^{(k)} : V \to V_k$ (see Definition 9.8) we define a decomposition of $v \in V$ as follows. Let $u_k := R_h^{(k)} v$, $k = 0, \ldots, l$, then

$$v_0 := u_0, \quad v_k := u_k - u_{k-1}, \quad k = 1, \ldots, l,$$

and thus

$$v = u_l = \sum_{k=0}^{l} v_k \,.$$

In this way we can replace the decomposition

$$V = V_0 + \ldots + V_l$$

by the direct sum

$$V = W_0 \oplus \ldots \oplus W_l$$

with $v_k \in W_k$, and $W_k \subset V_k$, $k = 0, \ldots, l$. If necessary, we can restrict ourselves to this decomposition. In particular

$$a(u_{k-1}, w) = a(v, w) = a(u_k, w) \quad \text{for all } w \in V_{k-1}, \tag{5.170}$$

and therefore by the Aubin–Nitsche lemma (Theorem 3.40) and (3.158):

$$\|u_k - u_{k-1}\|_0 \lesssim h_{k-1}|\nabla(u_k - u_{k-1})|_1 \lesssim h_k|\nabla(u_k - u_{k-1})|_1 .$$

This can be written equivalently as

$$\|v_k\|_0 \lesssim h_k\, a(v_k, v_k)^{1/2} = h_k \left\langle \tilde{A}v_k, v_k \right\rangle^{1/2} . \tag{5.171}$$

Furthermore from (5.170) we get an orthogonality relation with respect to $a(\cdot,\cdot)$ (*a-orthogonality*)

$$a(v_k, v_m) = 0 \quad \text{for } 0 \le k, m \le l,\ k \ne m,$$

showing

$$\left\langle \tilde{A}v, v \right\rangle_0 = \sum_{k=0}^{l} \left\langle \tilde{A}v_k, v_k \right\rangle_0 \quad \text{for all } v \in V_l .$$

Hence the estimate (5.171) yields

$$\left\langle \tilde{A}v, v \right\rangle_0 \gtrsim \sum_{k=0}^{l} h_k^{-2}\|v_k\|_0^2 \gtrsim \sum_{k=0}^{l} \left\langle \tilde{R}^{-1}v_k, v_k \right\rangle_0 ,$$

and therefore the coefficient α from (5.154) is a positive constant independent of l, h. Here we have used the estimate

$$\left\langle \tilde{R}^{-1}w_k, w_k \right\rangle_0 \lesssim h_k^{-2}\|w_k\|_0^2 \quad \text{for all } w_k \in V_k . \tag{5.172}$$

This can be seen as follows. Consider a matrix representation of $\tilde{R}$ as in (5.165), but with respect to the isomorphism $\tilde{P}_k : V_k \to \mathbb{R}^{M_k},\ v \mapsto \left(\left\langle v, \varphi_i^k \right\rangle_0\right)_{i=1,\ldots,M_k}$. Then

$$(\tilde{R}_k)_M = \operatorname{diag}\left(1/a(\varphi_i^k, \varphi_i^k)\right) .$$

Because of

$$(\tilde{R}_k^{-1})_M = (R_k)_M^{-1} = \operatorname{diag}\left(a(\varphi_i^k, \varphi_i^k)\right) \le C h_k^{-2} \operatorname{diag}\left(\left\langle \varphi_i^k, \varphi_i^k \right\rangle_0\right)$$

for some constant $C > 0$ independent of l, h_l due to Theorem 3.48 and Lemma 5.26, 2), the estimate (5.172) holds true.

To find the spectral radius of $R_k A_k$ or $\tilde{R}_k \tilde{A}_k$, note that we also have the estimate

$$\|A_k\|_2 \lesssim 1 \lesssim \|R_k\|_2^{-1}$$

from Theorem 3.48, since $a\left(\varphi_i^k, \varphi_i^k\right) \gtrsim 1$, $i = 1, \ldots, M_k$, and thus

$$\|A_k R_k\|_2 \lesssim \|A_k\|_2 \|R_k\|_2 \lesssim 1.$$

So only $\varrho(G)$ has to be bounded independently of l, h.

We need the following auxiliary result for $u \in V_k$, $v \in V_m$, $0 \le k \le m \le l$:

$$a(u, v) \lesssim \left(\frac{h_m}{h_k}\right)^{1/2} |u|_1 h_m^{-1} \|v\|_0 . \tag{5.173}$$

For using this estimate with $V = \bigoplus_{k=0}^{l} W_k$, note that for $v = \sum_{i=0}^{l} v_i$, $v_i \in W_i$, by $v_i := R_h^{(i)} v - R_h^{(i-1)} v$ we have $v_i = v_i - R_h^{(i)} v_i$ and therefore by Theorems 3.40 and 3.46:

$$h_i^{-1} \|v_i\|_0 = h_i^{-1} \|v_i - R_h^{(i)} v_i\|_0 \lesssim |v_i| \lesssim h_i^{-1} \|v_i\|_0 .$$

Applying this to (5.173) means that

$$a(v_k, v_m) \le \left(\frac{h_m}{h_k}\right)^{1/2} |\nabla v_k|_1 |\nabla v_m|_1 = \gamma_{mk} a(v_k, v_k)^{1/2} a(v_m, v_m)^{1/2}$$

with

$$\gamma_{mk} \lesssim \min\left\{\left(\frac{h_m}{h_k}\right)^{1/2}, \left(\frac{h_k}{h_m}\right)^{1/2}\right\},$$

and therefore

$$\varrho(G) \le \|G\|_\infty = \max_i \sum_j |\gamma_{ij}| \lesssim 1 \quad \text{for } h_k = 2^{l-k} h_l .$$

For proving (5.173), it is sufficient to do so on a triangle $K \in \mathcal{T}_k$, since the validity of its local version implies

$$\begin{aligned} a(u, v) = \sum_{K \in \mathcal{T}_k} \langle \nabla u, \nabla v \rangle_{0,K} &\le \sum_{K \in \mathcal{T}_k} \left(\frac{h_m}{h_k}\right)^{1/2} |u|_{1,K} h_m^{-1} \|v\|_{0,K} \\ &\le \left(\frac{h_m}{h_k}\right)^{1/2} h_m^{-1} \left(\sum_{K \in \mathcal{T}_k} |u|_{1,K}^2\right)^{1/2} \left(\sum_{K \in \mathcal{T}_k} \|v\|_{0,K}^2\right)^{1/2} . \end{aligned}$$

To show (5.173) on an element $K \in \mathcal{T}_k$, which is subdivided in several elements $K' \in \mathcal{T}_m$ forming $\mathcal{T}_m^K$, we write v on K as $v = v_0 + v_1$, where v_0 is piecewise linear on $\mathcal{T}_m^K$, having the original degrees of freedom on ∂K, and equal to 0 otherwise.

Then $v_1 = 0$ on ∂K and thus

$$\langle \nabla u, \nabla v_1 \rangle_{0,K} = -\langle \Delta u, v_1 \rangle_{0,K} = 0,$$

because the affine linearity of u implies $\Delta u = 0$ on K. Now let S be the union of those elements from $\mathcal{T}_m^K$, where v_0 does not vanish, then $|S| \le Ch_m$ and therefore

$$\begin{aligned}\langle \nabla u, \nabla v_0 \rangle_{0,K} &= \langle \nabla u, \nabla v_0 \rangle_{0,S} \le |u|_{1,S} |v_0|_{1,S} \\ &\le \left(\frac{|S|}{|K|} \right)^{1/2} |u|_{1,K} |v_0|_{1,K} \quad \text{because of the affine linearity of } u \text{ on } K \\ &\le C \left(\frac{h_m}{h_k} \right)^{1/2} |u|_{1,K} |v_0|_{1,K}.\end{aligned}$$

Applying the inverse estimate of Theorem 3.46, 2) to v_0 on $\mathcal{T}_m^K$ gives the required local version of (5.173).

□

Local Grid Refinement and the Hierarchical Basis Preconditioner

In the proof of Theorem 5.32 we needed a uniform global refinement for the definition of the grid hierarchy, similar to having optimality in the multiplicative multigrid method. If the grid hierarchy is defined by local refinement, it seems reasonable to modify the definition in (5.163) to

$$\tilde{R}_k v_k = \sum_{i=1}^{M_k}{}^{*} \frac{\langle v_k, \varphi_i^k \rangle_0}{a(\varphi_i^k, \varphi_i^k)} \varphi_i^k,$$

where we assume for the node numbering that new nodes are added, i.e., being indexed by $M_{k-1} + 1, \ldots, M_k$ from level $k - 1$ to k, and setting

$$\sum_{i=1}^{M_k}{}^{*} := \sum_{\substack{i=1 \\ \varphi_i^k \neq \varphi_i^{k-1}}}^{M_{k-1}} + \sum_{i=M_{k-1}+1}^{M_k},$$

i.e., only the new nodes are considered and those, where the basis function changes. The reduction of the number of operations is significant in particular for local refinements. In this case typically the condition (5.125), ensuring the optimality of the (multiplicative) multigrid method, is violated (Problem 5.25).

A variant is the *hierarchical basis preconditioner*, where the nodal bases on each level, i.e., $\{\varphi_i^{(k)} | i = 1, \ldots, M_k\}$, are substituted by a hierarchical basis

$$\begin{aligned}&\hat{\varphi}_i, \quad i = 1, \ldots, M_0, \\ &\hat{\varphi}_i, \quad i = 1, \ldots, M_0, \; M_0 + 1, \ldots, M_1,\end{aligned}$$

and so on such that the preconditioner further simplifies to

$$\tilde{R}_k v_k = \sum_{i=M_{k-1}+1}^{M_k} \frac{\langle v_k, \hat{\varphi}_i \rangle_0}{a(\hat{\varphi}_i, \hat{\varphi}_i)} \hat{\varphi}_i \tag{5.174}$$

in (5.163), i.e.,

$$\sum_{k=1}^{l} \tilde{R}_k \tilde{Q}_k v = \sum_{k=1}^{l} \tilde{R}_k v = \sum_{i=M_0+1}^{M_l} \frac{\langle v, \hat{\varphi}_i \rangle_0}{a(\hat{\varphi}_i, \hat{\varphi}_i)} \hat{\varphi}_i \, .$$

The hierarchical basis $\{\hat{\varphi}_i \,|\, i = 1, \ldots, M_l\}$ starts with the nodal basis at level 0, and adds for each new level the nodal function for the new nodes. For $\|v\| := |v|_{1,\Omega}$ and P_1-elements this means to add to V_k its orthogonal complement in V_{k+1} (see Fig. 5.10).

Compared to the nonhierarchical basis we observe the loss of the sparse structure of A_l.

Theorem 5.33 *Under the same assumptions as in Theorem 5.32 the estimate*

$$\kappa(BA) \le C\,(1 + |\ln h_l|)$$

holds with a constant $C > 0$ independent of l, h_l.

Proof The proof is similar to the one of Theorem 5.32 (see [47, Sect. 4.4.2] or [10, Sect. 7.2] for more details). We work again on a direct sum

$$V = W_0 \oplus \ldots \oplus W_l \,, \tag{5.175}$$

where

$$W_k := \{v \in V_k \mid v(p) = 0 \quad \text{for all vertices } p \in \mathcal{T}_{k-1}\}, \; k = 1, \ldots, l,$$

and therefore

$$V_k = V_{k-1} \oplus W_k,$$

i.e., the direct sum in (5.175) holds true. The decomposition is given explicitly by

$$v = I_h^{(0)} v + \sum_{\nu=1}^{l} \left(I_h^{(k)} - I_h^{(k-1)} \right) v,$$

where $I_h^{(k)}$ denotes the nodal interpolation operator according to (3.83) on $\mathcal{T}_k$, the level k. Now we estimate the quantities α, $\lambda_{\max}(R_i A_i)$, and $\varrho(G)$ from Theorem 5.30. Concerning $\lambda_{\max}(R_i A_i)$ and $\varrho(G)$, i.e., the upper bound of the eigenvalues of BA, the proof of Theorem 5.32 can be repeated. To estimate α, i.e., to find a lower bound, we want to show

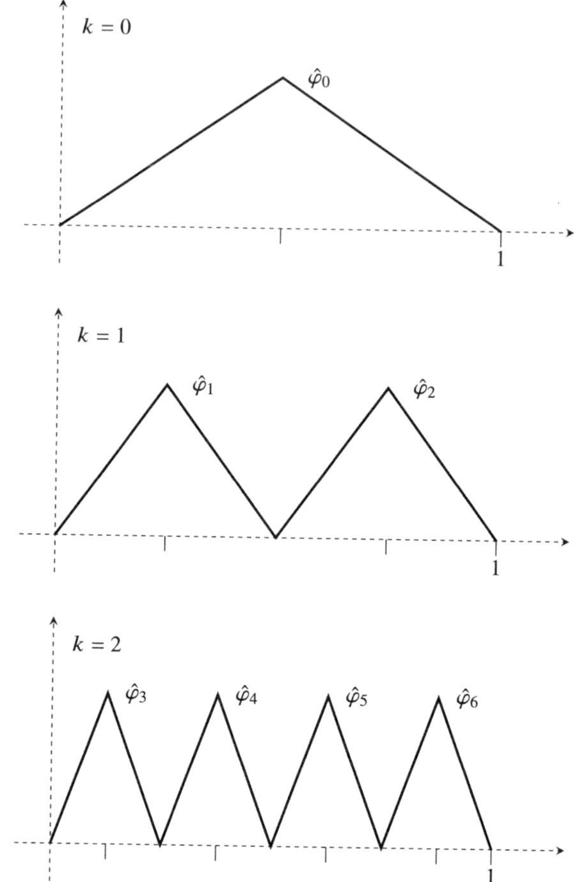

Fig. 5.10: Hierarchical basis in one dimension.

$$\left\langle \sum_{k=0}^{l} \tilde{R}_k^{-1} v_k, v_k \right\rangle \lesssim (1 + |\ln h_l|) \left\langle \tilde{A}u, u \right\rangle \qquad \text{for all } u \in V \text{ with } u = \sum_{k=0}^{l} v_k,\ v_k \in W_k. \tag{5.176}$$

Due to (5.172) it is sufficient to show

$$\sum_{k=0}^{l} h_k^{-2} \|v_k\|_0^2 \lesssim (1 + |\ln h_l|) \left\langle \tilde{A}u, u \right\rangle .$$

According to Theorem 3.32 and (3.158), we have (with $I_h^{(-1)} := 0$):

$$\|v_k\|_0 = \|I_h^{(k)}u - I_h^{(k-1)}u\|_0 = \|I_h^{(k)}u - I_h^{(k-1)}I_h^{(k)}u\|_0$$
$$\lesssim h_{k-1}\|\nabla I_h^{(k)}u\|_0 \lesssim h_k\|\nabla I_h^{(k)}u\|_0 \,.$$

Using the notation $u_K := |K|^{-1}\int_K u\,dx$, we get

$$\|\nabla I_h^{(k)}u\|_0^2 = \sum_{K\in\mathcal{T}_k} \|\nabla\big(I_h^{(k)}u - u_K\big)\|_{0,K}^2 \lesssim \sum_{K\in\mathcal{T}_k} \|I_h^{(k)}u - u_K\|_{L^\infty(K)}^2,$$

where the last estimate easily follows from the fact that $\nabla\big(I_h^{(k)}u - u_K\big)$ is constant on $K \in \mathcal{T}_k$. Now we make use of the following assertion for $u \in V_l$, $K \in \mathcal{T}_k$ (Problem 5.31):

$$\|I_h^{(k)}u - u_K\|_{L^\infty(K)} \lesssim \left(1 + \ln\frac{h_k}{h_l}\right)^{1/2} |u|_{1,K}\,. \tag{5.177}$$

Then we conclude

$$\|\nabla I_h^{(k)}u\|_0^2 \lesssim \sum_{K\in\mathcal{T}_k}\left(1 + \ln\frac{h_k}{h_l}\right)|u|_{1,K}^2 \lesssim \left(1 + \ln\frac{h_k}{h_l}\right)|u|_1^2.$$

Combining the estimates and using the relations of h_l and h_k gives

$$\sum_{k=0}^{l} h_k^{-2}\|v_k\|_0^2 \lesssim \sum_{k=0}^{l}\|\nabla I_h^{(k)}u\|_0^2 \lesssim \left(1 + \ln\frac{h_k}{h_l}\right)|u|_1^2 \lesssim (1 + |\ln h_l|)\big\langle\tilde{A}u, u\big\rangle\,.$$

□

5.7.3 Domain Decomposition Methods

Domain decomposition methods rely on the subdivision of the domain Ω, on which the equation is defined, in ("simpler") subdomains Ω_i. The original boundary value problem is restricted to Ω_i, and the restricted problems are solved exactly or approximatively, with appropriately amended boundary conditions on the inner boundaries $\partial\Omega_i \cap \Omega$. Multiplicative versions can be formulated as iterative methods in infinite-dimensional function spaces, what is in fact the origin of these methods.

In the *alternating Schwarz method* $\Omega = \Omega_1 \cup \Omega_2$ and $\Omega_1 \cap \Omega_2 \neq \emptyset$, i.e., the subdomains are overlapping and an iterative method can be set up for a given iterate $u^{(k)}$ by solving the boundary value problem on Ω_1 with boundary conditions amended by $u = u^{(k)}$ on $\partial\Omega_1 \cap \Omega$, leading to u_1 on Ω_1, and then the boundary value problem on Ω_2 with boundary conditions amended by $u = u_1$ on $\partial\Omega_2 \cap \Omega$, leading to u_2 and

$$u^{(k+1)} = \begin{cases} u_2 & \text{on } \Omega_2\,, \\ u_1 & \text{on } \Omega\setminus\Omega_1\,. \end{cases}$$

Overlapping Domains

This method can be cast into the space decomposition framework (with the infinite-dimensional basis space $V = H_0^1(\Omega)$ in case of the model problem as considered in Section 5.7.2) and

$$\begin{aligned} V_k &:= \{v_k \in V \mid v_k = 0 \quad \text{on } \Omega \setminus \Omega_k\}, \\ \langle \tilde{A}_k v_k, w_k \rangle &= a(v_k, w_k) \quad \text{for all } v_k, w_k \in V_k, \ k = 1, 2, \end{aligned} \tag{5.178}$$

and the multiplicative iteration coincides with the alternating Schwarz method (Problem 5.32). To verify the decomposition

$$V = V_1 + V_2 ,$$

the existence of a *partition of unity* is sufficient, i.e. $\chi_k \in W_\infty^1(\Omega)$, $k = 1, 2$, such that

$$\chi_1 + \chi_2 = 1, \quad \chi_k \geq 0, \quad \chi_k = 0 \quad \text{on } \Gamma_k := \partial\Omega_k \cap \Omega_{k^*},$$

where $1^* = 2$, $2^* = 1$, because then $u = \chi_1 u + \chi_2 u =: u_1 + u_2 \in V_1 + V_2$.

To transform this and similar approaches into a numerical finite-dimensional method, a further discretization of the original problem and of the subdomain problems is necessary. To this end we always assume that the underlying triangulation $\mathcal{T}_h$ is *compatible with the decomposition* $\Omega = \bigcup_{i=1}^l \Omega_i$ in the sense that the internal boundaries are composed by faces of the elements. Thus

$$\mathcal{T}_i := \mathcal{T}_h \cap \Omega_i := \{K \in \mathcal{T}_h \mid K \subset \overline{\Omega}_i\}$$

defines triangulations of Ω_i, on which the subspace equations require an exact or approximate solution of the Galerkin approximation to the boundary value problem with homogeneous Dirichlet boundary conditions on Ω_i.

More precisely we assume that, for each $k = 1, \ldots, l$, there is an index $j \in \{1, \ldots, l\} \setminus \{k\}$ such that $\Omega_k \cap \Omega_j \neq \emptyset$. Furthermore we assume the existence of elements $\chi_k \in W_\infty^1(\Omega)$, $k = 1, \ldots, l$, and a number $\delta > 0$ such that

$$\begin{gathered} \chi_k \geq 0, \quad \chi_k = 0 \quad \text{on } \Omega \setminus \overline{\Omega}_k, \quad \|\nabla \chi_k\|_{\infty,\Omega} \lesssim \frac{1}{\delta}, \quad k = 1, \ldots, l, \\ \sum_{k=1}^{l} \chi_k = 1 \quad \text{in } \Omega. \end{gathered}$$

The finite-dimensional ansatz space $V \subset H_0^1(\Omega)$ is subdivided as follows:

$$V_k := \{v_k \in V \mid v_k = 0 \quad \text{on } \Omega \setminus \Omega_k\}.$$

Then

$$V = V_1 + \dots V_l, \tag{5.179}$$

as for $v \in V$ we have

$$v = I_h v = I_h \left(\sum_{k=1}^{l} \chi_k v \right) = \sum_{k=1}^{l} v_k \tag{5.180}$$

with $v_k := I_h(\chi_k v) \in V_k$, where I_h denotes the nodal interpolation operator into V. We have

$$\langle \tilde{A}_k v_k, w_k \rangle_0 = a(v_k, w_k) \quad \text{for all } v_k, w_k \in V_k,\ k = 1, \dots, l.$$

We can equally well write the method in matrix form, now with

$$\mathbb{R}^M = \hat{V}_1 + \dots + \hat{V}_l,$$

where $M := \dim V$ and $\hat{V}_k := P^{-1}(V_k)$, $P : \mathbb{R}^M \to V$ being the isomorphism defined in (5.107) and A_k being the matrix representation of $\tilde{A}_k$.

The additive or multiplicative methods according to (5.146) or (5.148), respectively, either exactly with A_k^{-1} or a symmetric, positive definite approximation R_k, is called the *additive* or *multiplicative Schwarz preconditioner* (or *iteration*). Concerning R_k we assume *spectral equivalence* in the sense that there are constants $c_R, C_R > 0$ such that

$$c_R R_k^{-1} \le A_k \le C_R R_k^{-1} \quad \text{for } k = 1, \dots, l. \tag{5.181}$$

This means in particular for the eigenvalues λ of $R_k A_k$ that

$$\lambda \le C_R .$$

This can be seen as follows. From Lemma 5.29 and $R_k^T = R_k$ we conclude $R_k A_k R_k \le C_R R_k$, i.e. $\langle R_k A_k v, R_k v \rangle \le C_R \langle R_k v, v \rangle$ for $v \in V_k$. If $v \in V_k$ and $R_k A_k v = \lambda v_k$, then $\lambda \langle v, R_k v \rangle \le C_R \langle v, R_k v \rangle$ and $\langle v, R_k v \rangle \ne 0$.

Theorem 5.34 *Under the assumptions of Theorem 5.30 we have for the additive Schwarz preconditioner:*

$$\kappa(BA) \le C \left(1 + \frac{C_R}{c_R} \left(\frac{l_0}{\delta} \right)^2 \right), \tag{5.182}$$

where l_0 denotes the maximum number of overlapping domains, and c_R, C_R are the constants from (5.181).

Proof Again we have to estimate the quantities from Theorem 5.30. We can use (5.154), i.e., $\varrho(G) \le l_0$, and get from (5.181) that $\lambda_{\max}(R_i A_i) \le C_R$, giving an upper bound $C_R l_0$ for $\lambda_{\max}(BA)$. For a lower bound, i.e., α, we note that (5.181) means

$$\left\langle \tilde{R}_k^{-1} v_k, v_k \right\rangle_0 \le \frac{1}{c_R} \langle \nabla v_k, \nabla v_k \rangle_0 ,$$

hence only an estimate of the type

$$\sum_{k=1}^{l} \langle \nabla v_k, \nabla v_k \rangle_0 \le C \|\nabla v\|_0^2 \tag{5.183}$$

for $v = \sum_{k=1}^{l} v_k$ has to be shown.

We start from

$$\sum_{k=1}^{l} \langle \nabla v_k, \nabla v_k \rangle_0 \le \sum_{k=1}^{l} \|\nabla v_k\|_{0,\Omega_k}^2 = \sum_{k=1}^{l} \sum_{K \in \mathcal{T}_k} \|\nabla v_k\|_{0,K}^2$$

and observe that, for any $K \in \mathcal{T}_k$,

$$\begin{aligned} \|\nabla v_k\|_{0,K}^2 &= \|\nabla I_h (\chi_k v)\|_{0,K}^2 \lesssim \|\nabla (\chi_k v)\|_{0,K}^2 \\ &\lesssim \|\chi_h \nabla v\|_{0,K}^2 + \|v \nabla \chi_k\|_{0,K}^2 \lesssim \left(1 + \delta^{-2}\right) \|v\|_{1,K}^2, \end{aligned}$$

where the first estimate is a consequence of stability of the nodal interpolation operator, which can be seen from the back-transformation of (3.93) (with $m = k + 1$ there) to the original spaces taking into account Theorem 3.29. Summing up gives

$$\begin{aligned} \sum_{k=1}^{l} \langle \nabla v_k, \nabla v_k \rangle_0 &\lesssim \sum_{k=1}^{l} \sum_{K \in \mathcal{T}_k} \left(1 + \delta^{-2}\right) \|v\|_{1,K}^2 \lesssim \left(1 + \delta^{-2}\right) l_0 \sum_{K \in \mathcal{T}_h} \|v\|_{1,K}^2 \\ &= \left(1 + \delta^{-2}\right) l_0 \|v\|_1^2 \lesssim \left(1 + \delta^{-2}\right) l_0 \|\nabla v\|_0^2, \end{aligned}$$

where the last estimate follows from Theorem 2.18. Hence

$$\sum_{k=1}^{l} \left\langle \tilde{R}_k^{-1} v_k, v_k \right\rangle_0 \lesssim \frac{l_0}{c_R} \left(1 + \frac{1}{\delta^2}\right) \|\nabla v\|_0^2 \lesssim \left(1 + \frac{1}{c_R} \frac{l_0}{\delta^2}\right) \|\nabla v\|_0^2 .$$

□

Theorem 5.35 *Under the assumptions of Theorem 5.34 and assuming* $\beta = \lambda_{\max}(R_i A_i) < 2$*, the iteration matrix of the multiplicative Schwarz method satisfies*

$$\|M\|_A^2 \le 1 - \frac{2 - \beta}{\frac{C}{c_R} l_0 (1 + \frac{1}{2} \beta l_0)}$$

with a sufficiently large constant $C \ge 1$.

Proof The estimate follows directly from Theorem 5.31 and the proof of Theorem 5.34. □

Remark 5.36 The appearance of of the quantity l_0/δ is not satisfactory, because it indicates that the overlapping domain must be sufficiently large.

The role of the "communication region" can also be played by a coarse grid discretization (on a compatible triangulation $\mathcal{T}_H$ of Ω), leading to the *two-level additive Schwarz preconditioner*, i.e., we add similar as in (5.164)

$$\begin{aligned}&\text{a space } V_0, \text{ being the approximation space on } \mathcal{T}_H,\\ &\left\langle \tilde{A}_0 v_0, w_0 \right\rangle = a(v_0, w_0) \quad \text{for all } v_0, w_0 \in V_0.\end{aligned}$$

We extend (5.179) to $l = 0$.

Compatible triangulations can be constructed as follows:

Starting from a coarse grid $\mathcal{T}_H$, a non-overlapping decomposition $\tilde{\Omega}_k$, $k = 1, \ldots, l$, is constructed such that $\mathcal{T}_H$ is compatible with it, and then $\mathcal{T}_H$ is refined to $\mathcal{T}_h$. The domains $\tilde{\Omega}_k$ are enlarged to Ω_k by a band of $K \in \mathcal{T}_h$ of width $\delta > 0$, such that $\mathcal{T}_h$ remains compatible to Ω_k.

Theorem 5.37 *Under the assumptions of Theorem 5.34 we have for the two-level additive Schwarz preconditioner:*

$$\kappa(BA) \le C\left(1 + \frac{C_R}{c_R}\left(\frac{l_0 H}{\delta}\right)^2\right)$$

with a constant $C > 0$.

Proof We follow the proof of Theorem 5.34 with the following modifications: The decomposition is

$$v = R_H v + \sum_{k=1}^{l} I_h\left(\chi_k\left(v - R_h v\right)\right) =: \sum_{k=0}^{l} v_k,$$

where $R_H : V \to V_0$ is the Ritz projection. We want to show the estimate (5.183) (where the sum runs from 0 to l now) with the additional term H^2 in the resulting constant.

We have (see Theorem 3.40):

$$|v_0|_1 \lesssim |v|_1, \ \|v - v_0\|_0 \lesssim H|v|_1,$$

thus

$$\left\langle \tilde{A}_0 v_0, v_0 \right\rangle_0 = |v_0|_1^2 \lesssim |v|_1^2 = a(v, v),$$

and for $k = 1, \ldots, l$, setting $w := v - R_h v$:

$$\langle \nabla v_k, \nabla v_k \rangle_0 = \sum_{\substack{K \in \mathcal{T}_h \\ K \subset \Omega_k}} |v_k|_{1,K}^2.$$

Denoting by a_i, $i \in \{1, 2, 3\}$, the vertices of a triangle $K \subset \Omega_k$, we easily see that

$$|v_k|_{1,K}^2 \lesssim |v_k(a_2) - v_k(a_1)|^2 + |v_k(a_3) - v_k(a_1)|^2$$

and

$$v_k(a_2) - v_k(a_1) = (\chi_k(a_2) - \chi_k(a_1))\, w(a_2) + \chi_k(a_1)\,(w(a_2) - w(a_1)),$$

leading to

$$\begin{aligned}|v_k(a_2) - v_k(a_1)|^2 &\lesssim |\chi_k(a_2) - \chi_k(a_1)|^2|w(a_2)|^2 + |\chi_k(a_1)|^2|w(a_2) - w(a_1)|^2\\ &\lesssim \frac{h^2}{\delta^2}|w(a_2)|^2 + |w(a_2) - w(a_1)|^2.\end{aligned}$$

An analogous estimate can be obtained with a_2, a_1 exchanged by a_3, a_1. Hence we get

$$\begin{aligned}|v_k|_{1,K}^2 &\lesssim \frac{h^2}{\delta^2}\left(w^2(a_2) + w^2(a_3)\right) + |w(a_2) - w(a_1)|^2 + |w(a_3) - w(a_1)|^2\\ &\lesssim \frac{1}{\delta^2}\|w\|_{0,K}^2 + |w|_{1,K}^2,\end{aligned}$$

making use of the norm equivalencies from Theorem 3.46. Summing up leads to

$$\begin{aligned}\sum_{k=1}^{l}\langle\nabla v_k, \nabla v_k\rangle_0 &\lesssim \sum_{k=1}^{l}\sum_{\substack{K\in\mathcal{T}_h\\ K\subset\Omega_k}}\left(\frac{1}{\delta^2}\|w\|_{0,K}^2 + |w|_{1,K}^2\right)\\ &\lesssim l_0\frac{1}{\delta^2}\|v - R_H v\|_0^2 + |v - R_H v|_1^2 \lesssim \left(1 + l_0\frac{H^2}{\delta^2}\right)|v|_1^2\end{aligned}$$

because of Theorem 3.40. □

Non-overlapping Domains

We divide the polygonal domain $\Omega \subset \mathbb{R}^2$ into polygonal subdomains Ω_k, $k = 1, \ldots, l$, such that

- $\Omega_i \cap \Omega_j = \emptyset$ for $i \neq j$ and $i, j = 1, \ldots, l$,
- $\overline{\Omega} = \bigcup_{k=1}^{l}\overline{\Omega}_k$,
- $\partial\Omega_i \cap \partial\Omega_j$ is empty, a vertex or an edge for $i \neq j$ and $i, j = 1, \ldots, l$,

and $\mathcal{T}_h$ is a quasi-uniform triangulation of Ω, compatible with this decomposition.

In the overlapping case the continuity of the composed solution is sufficient, here the continuity of the normal derivatives is also required. For ease of exposition we restrict ourselves for a moment to the case $l = 2$ and homogeneous Dirichlet boundary conditions on $\partial\Omega$. To cast this problem into a multi-domain formulation, we introduce the following notation: $V := H_0^1(\Omega)$, $\widetilde{V}_i := \{v_i \in H^1(\Omega_i) \mid v_i|_{\partial\Omega\cap\partial\Omega_i} = 0\}$, $V_i := H_0^1(\Omega_i)$, and $\Gamma := \partial\Omega_1 \cap \partial\Omega_2$. Furthermore let $V_\Gamma := \{\eta = v|_\Gamma \mid v \in V\}$, where the restriction is to be understood in the sense of traces, and $R_i : V_\Gamma \to \widetilde{V}_i$ be linear extension operators, i.e.

$$\left(R_i\eta\right)\Big|_\Gamma = \eta \quad \text{for all } \eta \in V_\Gamma .$$

(For a more detailed explanation about traces see Subsection 6.2.1)

Then it can be shown that the following problems are equivalent (see, e.g., [49, Lemma 1.2.1]):

Find $u \in V$ such that

$$a(u, v) := \int_\Omega \nabla u \cdot \nabla v dx = \langle f, v\rangle_0 \quad \text{for all } v \in V. \tag{5.184}$$

$$\Longleftrightarrow$$

Find $u_i = u|_{\Omega_i} \in \widetilde{V}_i$ such that

$$\begin{aligned} a_i(u_i, v_i) := \int_{\Omega_i} \nabla u_i \cdot \nabla v_i dx = \langle f, v_i\rangle_{0,\Omega_i} \quad & \text{for all } v_i \in V_i,\ i = 1, 2, \\ u_1|_\Gamma = u_2|_\Gamma \qquad & \text{(in the sense of traces)}, \\ a_2(u_2, R_2\mu) - \langle f, R_2\mu\rangle_{0,\Omega_2} = \langle f, R_1\mu\rangle_{0,\Omega_1} - a_1(u_1, R_1\mu) \quad & \text{for all } \mu \in V_\Gamma . \end{aligned} \tag{5.185}$$

The problems (5.184) and (5.185) can be considered from a further point of view: Let the functions $w_i \in V_i$ satisfy (5.185), and $\tilde{u}_i \in \widetilde{V}_i$ satisfy (5.185) with the trivial right-hand side $f = 0$. We also want to assume for the moment that the boundary value problems are more regular, so that w_i and $\tilde{u}_i$ even belong to $H^2(\Omega_i)$. Then, by Corollary 3.10, 2),

$$\begin{aligned} a_i(w_i, R_i\mu) &= \left\langle \partial_{n_i} w_i, R_i\mu\right\rangle_{0,\Gamma} - \langle \Delta w_i, R_i\mu\rangle_{0,\Omega_i} = \left\langle \partial_{n_i} w_i, \mu\right\rangle_{0,\Gamma} + \langle f, R_i\mu\rangle_{0,\Omega_i} , \\ a_i(\tilde{u}_i, R_i\mu) &= \left\langle \partial_{n_i} \tilde{u}_i, R_i\mu\right\rangle_{0,\Gamma} - \langle \Delta \tilde{u}_i, R_i\mu\rangle_{0,\Omega_i} = \left\langle \partial_{n_i} \tilde{u}_i, \mu\right\rangle_{0,\Gamma} , \end{aligned}$$

where the last equality signs in both lines result from the fact that V_i is dense in $L^2(\Omega_i)$. Hence

$$\begin{aligned} & a_1(w_1 + \tilde{u}_1, R_1\mu) + a_2(w_2 + \tilde{u}_2, R_2\mu) \\ &= \left\langle \partial_{n_1}(w_1 + \tilde{u}_1), \mu\right\rangle_{0,\Gamma} + \left\langle \partial_{n_2}(w_2 + \tilde{u}_2), \mu\right\rangle_{0,\Gamma} + \langle f, R_1\mu\rangle_{0,\Omega_1} + \langle f, R_2\mu\rangle_{0,\Omega_2} \\ &= \left\langle \partial_{n_1}(w_1 + \tilde{u}_1) + \partial_{n_2}(w_2 + \tilde{u}_2), \mu\right\rangle_{0,\Gamma} + \langle f, R_1\mu\rangle_{0,\Omega_1} + \langle f, R_2\mu\rangle_{0,\Omega_2} . \end{aligned}$$

Now, if we want $w_i + \tilde{u}_i$ to be a solution of (5.185), then

$$\left\langle \partial_{n_1}(w_1 + \tilde{u}_1) + \partial_{n_2}(w_2 + \tilde{u}_2), \mu\right\rangle_{0,\Gamma} = 0 \quad \text{for all } \mu \in V_\Gamma$$

has to be satisfied. Using the notation

$$\tilde{u} := \begin{cases} \tilde{u}_1 & \text{on } \Omega_1 \\ \tilde{u}_2 & \text{on } \Omega_2 \end{cases}, \qquad w := \begin{cases} w_1 & \text{on } \Omega_1 \\ w_2 & \text{on } \Omega_2 \end{cases} \tag{5.186}$$

and

$$[\![\partial_n(w + \tilde{u})]\!]\,|_\Gamma := \left[\partial_{n_1}(w_1 + \tilde{u}_1) + \partial_{n_2}(w_2 + \tilde{u}_2)\right]|_\Gamma,$$

this condition can be written as

$$\langle [\![\partial_n(w + \tilde{u})]\!], \mu\rangle_{0,\Gamma} = 0 \quad \text{for all } \mu \in V_\Gamma\,. \tag{5.187}$$

By our temporary regularity assumption and Theorems 3.6, 3.9, the element $[\![\partial_n(w + \tilde{u})]\!]\,|_\Gamma$ belongs to $L^2(\Gamma)$, i.e., the left-hand side of (5.187) can be regarded as a linear continuous functional on V_Γ. A more refined analysis (using the trace operators introduced in Subsection 6.2.1, for example) shows that this remains also correct for $w_i \in V_i$ and $\tilde{u}_i \in \widetilde{V}_i$ only, i.e., if the temporary assumption no longer applies.

Keeping the notation, the condition (5.187) can be written as $[\![\partial_n(w + \tilde{u})]\!]\,|_\Gamma = 0$ in V_Γ'. Furthermore, if $\tilde{A}^{-1} : V_\Gamma \to V,\ \lambda \mapsto \tilde{A}^{-1}\lambda := \tilde{u}$ denotes the solution operator of (5.185) with the trivial right-hand side $f = 0$ and the common trace $\lambda := \tilde{u}_1|_\Gamma = \tilde{u}_2|_\Gamma$ $(= u_1|_\Gamma = u_2|_\Gamma)$, then the condition (5.187) admits the form

$$S_\Gamma \lambda = \chi, \tag{5.188}$$

where $S_\Gamma : V_\Gamma \to V_\Gamma',\ \lambda \mapsto S_\Gamma\lambda := [\![\partial_n \tilde{A}^{-1}\lambda]\!]\big|_\Gamma$ is the so-called *Poincaré–Steklov interface operator* and $\chi := -\left(\partial_{n_1} w_1 + \partial_{n_2} w_2\right)|_\Gamma$. It involves the so-called *Dirichlet-to-Neumann map* $\tilde{u}_i|_\Gamma \mapsto \partial_{n_i}\tilde{u}_i|_\Gamma$.

The functions $\tilde{u}_i$ can be understood as the harmonic extensions to Ω_i of λ to correct the functions w_i. Thus we have subdivided the solution u of (5.184) into the addends (5.186) with

$$a(\tilde{u}, v) = 0 \quad \text{for all } v \in V_i \quad (v \text{ extended by 0 to } \Omega).$$

Note that

$$a(\tilde{u}, v) = \int_\Omega \nabla\tilde{u} \cdot \nabla v dx = \int_\Gamma [\![\partial_n \tilde{u}]\!]\, v d\sigma = 0$$

and $V_1 + V_2 = \{v \in H_0^1(\Omega) \,|\, v|_\Gamma = 0\}$.

In generalizing this concept, we can define a subspace decomposition of $V := H_0^1(\Omega)$ in the following way. Set $\Gamma := \bigcup_{i=1}^l \partial\Omega_i \setminus \partial\Omega$ and

$$V^0 := \{v \in V \,|\, v|_\Gamma = 0\},$$

then the space V^0 can be decomposed directly into $V^0 = \bigoplus_{i=1}^{l} V_i$, where

$$V_i := \{v_i \in V \,|\, v_i|_{\Omega_i} \in H_0^1(\Omega_i) \text{ and } v_i|_{\Omega\setminus\Omega_i} = 0\}, \quad i = 1, \ldots, l.$$

The spaces V_i are orthogonal with respect to the scalar product generated by the bilinear form a, i.e., they are *a-orthogonal*. Defining

$$V_0 := (V^0)^\perp \quad \text{w.r.t. the bilinear form } a,$$

we arrive at the a-orthogonal decomposition $v = \sum_{i=0}^{l} v_i$ and thus the variational equation

$$a(u, v) = \ell(v) \quad \text{for all } v \in V \tag{5.189}$$

with $\ell \in V'$ is equivalent to

$$a(u_i, v_i) = \ell(v_i) \quad \text{for all } v_i \in V_i, \quad i = 0, \dots, l, \tag{5.190}$$

where the problems for $i = 1, \dots, l$ are the same boundary value problems on Ω_i as on Ω which can be solved independently (in parallel). If the *Steklov operator* $S_0 : V_0 \to V_0',\ z \mapsto S_0 z$ is defined by

$$\langle S_0 z, v\rangle = a(z, v) \quad \text{for all } z, v \in V_0\,,$$

then the equation

$$S_0 u_0 = \ell_0 \quad \text{with} \quad \ell_0 \in V_0' \quad \text{defined by } \ell_0 = \ell|_{V_0}$$

is equivalent to the first equation in (5.190).

This operator is related to the Poincaré–Steklov interface operator introduced above, as the following considerations for the case $\ell(v) := \langle f, v\rangle_0$ with $f \in L^2(\Omega)$ show. We apply an analogous argumentation as in the above case of two domains, i.e., we obtain (formally, omitting the intermediate regularity consideration w.r.t. u) from the variational equation (5.189)

$$\begin{aligned}
\langle f, v\rangle_0 &= \sum_{i=1}^{l} \langle \nabla u, \nabla v\rangle_{0,\Omega_i} = \sum_{i=1}^{l} \left[-\langle \Delta u, v\rangle_{0,\Omega_i} + \left\langle \partial_{n_i} u, v\right\rangle_{0,\partial\Omega_i} \right] \\
&= \sum_{i=1}^{l} \left[\langle f, v\rangle_{0,\Omega_i} + \left\langle \partial_{n_i} u, v\right\rangle_{0,\partial\Omega_i} \right] \\
&= \langle f, v\rangle_0 + \langle [\![\partial_n u]\!]\,|_\Gamma, v\rangle_{0,\Gamma} \quad \text{for all } v \in V.
\end{aligned}$$

Hence

$$0 = \langle [\![\partial_n u]\!]\,|_\Gamma, v\rangle_{0,\Gamma} = \langle [\![\partial_n u]\!]\,|_\Gamma, v_0\rangle_{0,\Gamma} = \langle [\![\partial_n (u_0 + w)]\!]\,|_\Gamma, v_0\rangle_{0,\Gamma}\,,$$

where $u = u_0 + w$ with $w := \sum_{i=1}^{l} u_i$, which is in accordance with (5.187).

Now we turn to a conforming finite element discretization of (5.189). Let $V_h \subset V$ be the finite-dimensional space on $\mathcal{T}_h$ consisting of P_1-elements. Thanks to the compatibility assumption for $\mathcal{T}_h$, the a-orthogonal decomposition of V^0 carries over to V_h by restriction, i.e. $V_h^0 = V^0 \cap V_h$ and $V_{hi} = V_i \cap V_h$, $i = 1, \dots, l$. The space $V_{0h} := (V_h^0)^\perp$ is the a-orthogonal complement of V_h^0 in V_h, i.e. we have

$$V_h = \bigoplus_{i=0}^{l} V_{hi} \, .$$

An element $v_{0h} \in V_{0h}$ is uniquely determined by its values on $\mathcal{N}_\Gamma := \{p \,|\, p \in \Gamma \text{ is a node of } \mathcal{T}_h\}$, as can be seen from the following result.

Lemma 5.38 *An element $v_{0h} \in V_{0h}$ is uniquely determined by its values on $\mathcal{N}_\Gamma$, using the relationships* (5.192), (5.193) *given below.*

Proof Since an element of V_h is uniquely determined by its values at the interior nodes of $\mathcal{T}_h$, we define, for $v_{0h} \in V_{0h}$, an element $\lambda \in V_h$ by

$$\lambda(p) := \begin{cases} v_{0h}(p) & \text{for } p \in \mathcal{N}_\Gamma \, , \\ 0 & \text{otherwise.} \end{cases} \tag{5.191}$$

As $\hat{v} := v_{0h} - \lambda \in V_h^0$, we get by the a-orthogonality that

$$a(\hat{v}, w) = -a(\lambda, w) \quad \text{for all } w \in V_h^0, \tag{5.192}$$

which uniquely determines $\hat{v}$ and thus

$$v_{0h} = \hat{v} + \lambda. \tag{5.193}$$

□

The representation (5.193) indicates that the nodal basis functions of V_{0h} do not have compact support.

Restricting the first equation in (5.190) to V_{0h}, we obtain the *discrete Steklov operator*: $S_{0h} : V_{0h} \to V_{0h}'$, $z \mapsto S_{0h} z$, defined by

$$\langle S_{0h} z, v \rangle = a(z, v) \quad \text{for all } z, v \in V_{0h} \, .$$

So the aim is to compute the matrix representation of S_{0h} exactly and to solve the (moderately sized) problem or to find a preconditioner for S_{0h}.

For simplicity we return to the case $l = 2$ and consider the resulting matrix structure of the problem by ordering the nodes in the sequence $\Omega_1, \Gamma, \Omega_2$. Then for the unknowns partioned into the vectors u_1, u_Γ, u_2, we get

$$Au = \begin{pmatrix} A_{11} & A_{1\Gamma} & \mathbf{0} \\ A_{\Gamma 1} & A_\Gamma & A_{\Gamma 2} \\ \mathbf{0} & A_{2\Gamma} & A_{22} \end{pmatrix} \begin{pmatrix} u_1 \\ u_\Gamma \\ u_2 \end{pmatrix} = \begin{pmatrix} f_1 \\ f_\Gamma \\ f_2 \end{pmatrix} .$$

All block matrices have sparse structure with exception of the submatrix A_Γ which is densely filled, as the basis functions have no compact support. The analogue of the discrete Steklov operator is the static condensation by a block-Gauss elimination to reduce only to the unknown u_Γ; the procedure is also called the *Schur-complement method*. This leads to the algebraic equation

$$Su_\Gamma = g \tag{5.194}$$

with

$$\begin{aligned} S &:= A_\Gamma - A_{\Gamma 1} A_{11}^{-1} A_{1\Gamma} - A_{\Gamma 2} A_{22}^{-1} A_{2\Gamma}\,, \\ g &:= f_\Gamma - A_{\Gamma 1} A_{11}^{-1} f_1 - A_{\Gamma 2} A_{22}^{-1} f_2\,, \end{aligned}$$

showing the necessity to solve subdomain problems exactly or approximately.

Rewriting this gives the middle row of A^{-1}, and resolving the first and last block for u_1 and u_2, respectively, completes the representation from Fig. 5.11: Preconditioners are obtained by replacing $A_{11}^{-1}, A_{22}^{-1}, S^{-1}$ by corresponding preconditioners.

$$A^{-1} = \begin{pmatrix} A_{11}^{-1} + A_{11}^{-1} A_{1\Gamma} S^{-1} A_{\Gamma 1} A_{11}^{-1} & -A_{11}^{-1} A_{1\Gamma} S^{-1} & \mathbf{0} \\ -S^{-1} A_{\Gamma 1} A_{11}^{-1} & S^{-1} & -S^{-1} A_{\Gamma 2} A_{22}^{-1} \\ \mathbf{0} & -A_{22}^{-1} A_{\Gamma 2} S^{-1} & A_{22}^{-1} + A_{22}^{-1} A_{2\Gamma} S^{-1} A_{\Gamma 2} A_{22}^{-1} \end{pmatrix}$$

Fig. 5.11: Exact preconditioner by Schur-complement method.

Based on the preconditioners of S we have a look at the so-called *Neumann–Neumann preconditioner*. Again we consider only the case $l = 2$. Here A_Γ from (5.194) is subdivided into

$$A_\Gamma = A_\Gamma^{(1)} + A_\Gamma^{(2)},$$

where $a_{ij}^{(k)} := \left\langle \nabla\varphi_j, \nabla\varphi_i \right\rangle_{0,\Omega_k}, \varphi_i, \varphi_j \in V_{0h}$ being the nodal basis functions (note that they do not have compact support), which gives the decomposition

$$S = S_1 + S_2, \quad S_k := A_\Gamma^{(k)} - A_{\Gamma k} A_{kk}^{-1} A_{k\Gamma}, \quad k = 1, 2.$$

The advantage of S_k compared to S lies in the fact that the computation is restricted to Ω_k: $\tilde{u}_p = S_1^{-1} v_\Gamma$ is the second component of the solution of

$$\begin{pmatrix} A_{11} & A_{1\Gamma} \\ A_{\Gamma 1} & A_\Gamma^{(1)} \end{pmatrix} \begin{pmatrix} \tilde{u}_1 \\ \tilde{u}_\Gamma \end{pmatrix} = \begin{pmatrix} \mathbf{0} \\ v_\Gamma \end{pmatrix},$$

again referring to the block-Gauss elimination. This system can be interpreted as a discretization of

$$-\Delta u = 0 \quad \text{in } \Omega_1,\ u|_{\partial\Omega_1\setminus\Gamma} = 0,\ \partial_n u|_\Gamma = v_\Gamma,$$

i.e., with Neumann boundary conditions. This approach can be extended to l domains: $A_{k\Gamma}$ is reduced to its nonzero columns giving $\hat{A}_{k\Gamma}$, i.e., only to the nodes in Γ belonging to $\partial\Omega_k$ are considered, and analogously $\hat{A}_{\Gamma k}$ and $\hat{A}_\Gamma^{(k)}$ are defined. Then S decomposes into S_k, where

$$S_k := \hat{A}_\Gamma^{(k)} - \hat{A}_{\Gamma k} A_{kk}^{-1} \hat{A}_{k\Gamma},$$

and the preconditioner of S is given by

$$B_{NN} := \sum_{k=1}^{l} W_k S_k^{-1} N_k,$$

where $W_k := \operatorname{diag}(1/m(p))_p$ and $m(p)$ subdomains share the node p.

For $l = 2$ the spectral equivalence of S and S_k can be shown (see [47, Sect. 4.4.4]), hence the preconditioning is optimal due to $\alpha = 1$ and $\varrho(G) = 1$, as the V_k are a-orthogonal. In the general case similar to Theorem 5.37 a coarse grid approximation with step size $H > 0$ has to be incorporated leading to an estimate of the type

$$\kappa(\tilde{B}_{NN} A) \lesssim \left(1 + \ln \frac{H}{h}\right)^2$$

(see [10, Sect. 7.7]).

Exercises

Problem 5.24 Let $A \in \mathbb{R}^{m,m}$ be a symmetric positive definite matrix and V_i, $i = 0, \ldots, \ell$, subspaces of $\mathbb{R}^m$. Show: If the subspaces V_i are pairwise orthogonal with respect to $\langle \cdot, \cdot \rangle_A$, then

$$v = \sum_{i=0}^{\ell} Q_i^A v \quad \text{for all } v \in \mathbb{R}^m$$

and $Q_i^A : V \to V_i$ is the orthogonal projection with respect to $\langle \cdot, \cdot \rangle_A$. Give an example that this is not the case for a general direct decomposition.

Problem 5.25 Consider a local adaptive refinement in the sense that in each refinement step half of the previously refined areas is refined. Consider a simple scenario (rectangle, red refinement) and estimate M_l and C from (5.124) and discuss the validity of (5.125).

Problem 5.26 Show (5.134).

Problem 5.27 Show Lemma 5.29.

Problem 5.28 Let $G := (\gamma_{ij})_{ij} \in \mathbb{R}^{m,m}$ with $\gamma_{ij} \in [0, 1]$, $\gamma_{ii} = 1$ for $i, j = 1, \ldots, m$. Show that $\|G - I\|_2 \le \|G\|_2$.
Hint: Use the Perron–Frobenius theorem.

Problem 5.29 Show (5.161).

Problem 5.30 Show (5.168).

Problem 5.31 Show (5.177).

Problem 5.32 Verify in detail the statement that, in the space decomposition framework (with the infinite-dimensional basis space $V = H_0^1(\Omega)$ in case of the model problem as considered in Section 5.7.2) and (5.178), the multiplicative iteration coincides with the alternating Schwarz method.

Chapter 6
Beyond Coercivity, Consistency, and Conformity

6.1 General Variational Equations

Until now, the general framework has been as follows: Let V be a Hilbert space with the scalar product $\langle \cdot, \cdot \rangle_V$ and induced norm $\| \cdot \|_V$, $a : V \times V \to \mathbb{R}$ a bilinear form, and $\ell \in V'$. The continuous problem is the corresponding variational equation, i.e., find $u \in V$ such that

$$a(u, v) = \ell(v) \quad \text{for all } v \in V. \tag{6.1}$$

Besides the standard assumption of continuity (3.2) the decisive condition to ensure unique and stable solvability of (6.1) was V-coercivity (3.3) ensuring the Lax–Milgram Theorem (Theorem 3.1). More precisely, unique solvability for each right-hand side ℓ is meant, with *stability* meaning Lipschitz continuous dependence of the solution u on the data ℓ, i.e., there is a constant $C > 0$, independent of ℓ, such that

$$\|u\|_V \le C\|\ell\|_{V'} .$$

In fact, the assumptions of the Lax–Milgram Theorem imply that (for α being the coercivity constant)

$$\alpha\|u\|_V^2 \le a(u, u) = \ell(u) \le \|\ell\|_{V'}\|u\|_V \quad \Longrightarrow \quad \|u\|_V \le \alpha^{-1}\|\ell\|_{V'}. \tag{6.2}$$

In the equivalent operator notation (6.1) can be formulated as

$$Au = \ell, \tag{6.3}$$

where $A : V \to V'$ is defined pointwise by $(Au)(v) := a(u, v)$ for all $u, v \in V$. Thus, A is linear and continuous because of (3.2), i.e., $A \in L[V, V']$. Then, (6.2) reads

$$\|A^{-1}\ell\|_V \le \alpha^{-1}\|\ell\|_{V'},$$

i.e., A is invertible and A^{-1} is continuous. Thus, $A^{-1} \in L[V', V]$ and

P. Knabner and L. Angermann, *Numerical Methods for Elliptic and Parabolic Partial Differential Equations*, Texts in Applied Mathematics 44,
https://doi.org/10.1007/978-3-030-79385-2_6

$$\|A^{-1}\|_{V',V} \le \alpha^{-1}.$$

Here, $\|\cdot\|_{V,W}$ denotes the operator norm on $L[V,W]$. We extend by

$$\langle \varphi, v \rangle := \varphi(v) \quad \text{for all } \varphi \in V',\ v \in V$$

the scalar product $\langle \cdot, \cdot \rangle_V$ on $V \times V$ to the *duality pairing* and thus have

$$(Au)v = \langle Au, v \rangle .$$

Due to the Riesz representation theorem [67, Sect. III.6] the operator equation (6.3) can be rewritten by means of the *Riesz-isomorphism* $\iota_V : V' \to V$ as

$$\widetilde{A}u := (\iota_V \circ A)u = \tilde{\ell} := \iota_V(\ell), \tag{6.4}$$

i.e., $\widetilde{A} \in L[V,V]$, $\tilde{\ell} \in V$.

In particular

$$(Au)(v) = \langle Au, v \rangle = \left\langle \widetilde{A}u, v \right\rangle_V .$$

As ι_V is also an isometry implying $\|\iota_V\|_{V',V} = 1$, we also have

$$\|A\|_{V,V'} = \|\widetilde{A}\|_{V,V}, \quad \|A^{-1}\|_{V',V} = \|\widetilde{A}^{-1}\|_{V,V}.$$

Therefore, it is legitimate to suppress the symbol ι_V in the notation, as done in the following, which allows us to switch freely between the variational equation (6.1), the operator equation (6.3), and the operator equation (6.4), written as

$$Au = \ell, \quad \text{with} \quad A \in L[V,V], \quad \ell \in V.$$

This identification means that

$$(Au)(v) = \langle Au, v \rangle_V = a(u,v) \quad \text{for all } u, v \in V.$$

The same applies in the more general situation of distinct *solution* and *test spaces* U and V, respectively. That is, from now on we consider two Hilbert spaces U, V as well as a bilinear form $a : U \times V \to \mathbb{R}$ and investigate the following problem:

$$\text{Find } u \in U \text{ such that} \quad a(u,v) = \ell(v) \quad \text{for all } v \in V. \tag{6.1'}$$

Without notice we assume a to be continuous, i.e., there exists a constant $M > 0$ such that

$$|a(u,v)| \le M\|u\|_U\|v\|_V \quad \text{for all } u \in U, v \in V. \tag{6.5}$$

The operator equations (6.3), (6.4) are formulated analogously using the equivalent formulations

$$Au = \ell \ \text{ with } \ A \in L[U,V'],\ \ell \in V', \tag{6.3'}$$

$$\text{or} \quad Au = \ell \ \text{ with } \ A \in L[U, V], \ \ell \in V\,. \tag{6.4'}$$

Most of the subsequent theory is also valid for the more general situation, where U and V are only Banach spaces, but then the distinction of a space and its dual has to be maintained.

In the case of problems with $U \neq V$, the difficulty arises that the coercivity condition previously used no longer makes sense, and therefore we need more general conditions to ensure the invertibility of A and the continuity of A^{-1}. In finite-dimensional spaces, the last requirement follows automatically. Fortunately, the same holds true for Banach spaces due to the Open Mapping Theorem (see Appendix A.4).

Thus, we can relax from Hilbert to Banach spaces, but the completeness of the (function) spaces is indispensable. Therefore, we only have to provide conditions for injectivity and surjectivity of A. To make the stability more quantitative, we additionally need an estimate of

$$\|A^{-1}\|_{V',U} \qquad \text{or} \qquad \|A^{-1}\|_{V,U}$$

in terms of the problem's parameters for (6.3') or (6.4'), respectively. To find characterizing conditions for injectivity and surjectivity, we start with $A : U \to V$ for finite-dimensional spaces. Here we have the following equivalencies:

(i) A is *injective*
(ii) A^{-1}: $\mathrm{im}(A) \to U$ exists and is linear and bounded, i.e., continuous.
(iii) There exists a constant $\alpha > 0$ such that

$$\|Au\|_V \geq \alpha \|u\|_U \quad \text{for all } u \in U\,. \tag{6.6}$$

In the general case of Banach spaces U, V, the first statement has to be substituted by

$$A \text{ is injective} \quad \text{and} \quad \mathrm{im}(A) \text{ is closed,}$$

and then the two equivalences hold true, which can be seen by concluding (i) $\Rightarrow$ (ii) $\Rightarrow$ (iii) $\Rightarrow$ (i) for the true statements. A closed subspace of a Banach space is a Banach space (equipped with the induced norm), and the operator $A \in L[U, \mathrm{im}(A)]$ is surjective. Since by assumption $A \in L[U, \mathrm{im}(A)]$ is injective, too, the inverse operator exists and the Open Mapping Theorem (see Appendix A.4) tells that the inverse operator is continuous (cf. (A4.19)). This is the first implication. The second implication follows if (A4.22) is applied to the inverse operator. Finally, if (6.6) is satisfied, then the norm of every element of $\ker(A)$ must vanish, hence $\ker(A) = \{0\}$ which is the injectivity of A. The closedness of $\mathrm{im}(A)$ follows directly by definition considering a convergent sequence $(Au^{(n)})_n$. This closes the equivalence chain.

In the case (6.1')/(6.3')/(6.4') the norm $\|Au\|$ can be expressed in various ways. Considering A as mapping from U to V', then

$$\|Au\|_{V'} = \sup_{v \in V\setminus\{0\}} \frac{|(Au)(v)|}{\|v\|_V} = \sup_{v \in V\setminus\{0\}} \frac{|a(u,v)|}{\|v\|_V} = \sup_{v \in V\setminus\{0\}} \frac{a(u,v)}{\|v\|_V}, \tag{6.7}$$

and (6.6) reads

$$\sup_{v \in V\setminus\{0\}} \frac{a(u,v)}{\|u\|_U \|v\|_V} \geq \alpha \text{ for all } u \in U \setminus \{0\} \iff \inf_{u \in U\setminus\{0\}} \sup_{v \in V\setminus\{0\}} \frac{a(u,v)}{\|u\|_U \|v\|_V} \geq \alpha. \tag{6.8}$$

To discuss the surjectivity of A we introduce at first the notion of the *adjoint operator*.

Definition 6.1 (Adjoint operator) Given two Hilbert spaces U, V, for $A : U \to V'$, the *adjoint operator* $A^T : V \to U'$ is defined by the identity

$$\langle A^T v, u \rangle = (A^T v)(u) = a(u, v) \quad \text{for all } u \in U,\ v \in V.$$

As above for the injectivity, for the surjectivity of A we again start with considering finite-dimensional spaces:

A is *surjective*

$$\iff \ker(A^T) = \operatorname{im}(A)^{\perp} = \{0\} \iff A^T \text{is injective.} \tag{6.9}$$

Thus, the list of equivalences for the surjectivity of A can be continued with

$$\iff (A^T v = 0 \implies v = 0 \quad \text{for } v \in V)$$

$$\iff (\|A^T v\|_{U'} = 0 \implies v = 0 \quad \text{for } v \in V)$$

and furthermore

$$0 = \|A^T v\|_{U'} = \sup_{u \in U\setminus\{0\}} \frac{|(A^T v)(u)|}{\|u\|_U} = \sup_{u \in U\setminus\{0\}} \frac{|a(u,v)|}{\|u\|_U} \iff a(u,v) = 0 \quad \text{for all } u \in U.$$

Therefore, the equivalence chain can be completed by

$$\iff (a(u,v) = 0 \text{ for all } u \in U \implies v = 0). \tag{6.10}$$

Since the identity in (6.9) also holds true in Banach spaces after adding "im(A) is closed" to the equivalences and the following statements (where the Closed Range Theorem (see Appendix A.4) in connection with (A4.31) needs to be utilized additionally), the reasoning holds true for Hilbert spaces with this modification.

Thus, we have proven:

Theorem 6.2 *Let U, V be Hilbert spaces and consider the variational/operator equation* (6.1')/(6.3')/(6.4'). *If A is bounded, the unique solvability for every right-hand side is equivalent to* (6.10) *and* (6.8), *i.e.,*

$$a(u,v) = 0 \quad \textit{for all } u \in U \implies v = 0 \quad \textit{for } v \in V, \tag{NB1}$$

$$\alpha := \inf_{u \in U \setminus \{0\}} \sup_{v \in V \setminus \{0\}} \frac{a(u, v)}{\|u\|_U \|v\|_V} > 0. \qquad \text{(NB2)}$$

If (NB1), (NB2) *are fulfilled, the problem is stable in the sense that*

$$\|u\|_U \le \alpha^{-1} \|\ell\|_{V'}.$$

Proof According to the above considerations we have that

(A bijective and A^{-1} bounded) $\Longleftrightarrow$ (A surjective and A injective and im(A) closed) $\Longleftrightarrow$ ((NB1) and (NB2)).
Finally we have for the solution u

$$\alpha \|u\|_U \le \sup_{v \in V} \frac{a(u, v)}{\|v\|_V} = \sup_{v \in V} \frac{\ell(v)}{\|v\|_V} = \|\ell\|_{V'}.$$

□

Remark 6.3 1) In particular $\|A^{-1}\| = \alpha^{-1}$. Indeed, consider A as mapping from U to V, then with (6.7):

$$\|A^{-1}\| = \sup_{v \in V \setminus \{0\}} \frac{\|A^{-1} v\|_U}{\|v\|_V} = \sup_{u \in U \setminus \{0\}} \frac{\|u\|_U}{\|Au\|_V}$$

$$= \left(\inf_{u \in U \setminus \{0\}} \frac{\|Au\|_V}{\|u\|_U} \right)^{-1} = \left(\inf_{u \in U \setminus \{0\}} \sup_{v \in V \setminus \{0\}} \frac{a(u, v)}{\|u\|_U \|v\|_V} \right)^{-1}.$$

2) Theorem 6.2 is attributed to Ivo Babuška and Jindřich Nečas. As seen, it relies heavily on the Open Mapping and the Closed Range Theorems of Stefan Banach. Therefore, it is sometimes named after these people. (NB2) is called *inf-sup condition* or *Ladyžhenskaya–Babuška–Brezzi (LBB) condition*.
3) In the case $V = U$, the V-coercivity (3.3) implies (NB1) and (NB2). Thus, the Lax–Milgram Theorem (Theorem 3.1) is a consequence:

$$\text{(NB1)} : a(u, v) = 0 \text{ for all } u \in V \Longrightarrow a(v, v) = 0 \Longrightarrow v = 0,$$

$$\text{(NB2)} : \sup_{v \in V \setminus \{0\}} \frac{a(u, v)}{\|u\|_U \|v\|_U} \ge \frac{a(u, u)}{\|u\|_U^2} \ge \alpha \text{ for all } u \in U$$

with α from (3.3).
4) Condition (NB1) can be substituted by a seemingly stronger condition more in line with (NB2). By (6.9) and the Closed Range Theorem, respectively, we have A is surjective $\Longleftrightarrow$ A^T is injective and im(A^T) is closed. Therefore, the consideration for "A is injective" applied to A^T leads to the alternative characterization

$$\alpha' = \inf_{v \in V \setminus \{0\}} \sup_{u \in U \setminus \{0\}} \frac{a(u, v)}{\|u\|_U \|v\|_V} > 0. \qquad \text{(NB1')}$$

5) In fact, if (NB2) holds for a, also $\alpha' = \alpha$ can be chosen in (NB1') and vice versa. This will be important considering a family of problems (in the discretization parameter h). Then the uniformity in h for one constant is transferred to the other.
"$\alpha' = \alpha$" can be seen as follows. As seen, (NB2) is equivalent to

$$\|Au\|_V \geq \alpha \|u\|_U \quad \text{for all } u \in U, \tag{6.11}$$

and therefore (NB1') is equivalent to

$$\|A^T v\|_U \geq \alpha' \|v\|_V \quad \text{for all } v \in V.$$

Let A be surjective and (NB2) be satisfied. Let $v \in V$ be arbitrary and $u = u_v \in V$ such that $Au = v$, then

$$\begin{aligned}\|v\|_V^2 &= \langle v, v\rangle_V = \langle v, Au\rangle_V = \langle A^T v, u\rangle_U \\ &\leq \|A^T v\|_U \|u\|_U \leq \|A^T v\|_U \alpha^{-1} \|Au\|_V = \alpha^{-1} \|v\|_V \|A^T v\|_U, \quad \text{i.e.,}\end{aligned}$$
$$\|A^T v\|_U \geq \alpha \|v\|_V.$$

Hence (NB1') holds with $\alpha' = \alpha$. Because of $(A^T)^T = A$ also the following holds: Let A be injective and im(A) closed, i.e., A^T be surjective, and let (NB1') hold, then also (NB2) holds with $\alpha = \alpha'$.

We now turn to Galerkin approximations of (6.1')/(6.3')/(6.4'). In Chapter 3, for $U = V$, we have studied the choice of the finite-dimensional space $U_h = V_h$ as solution and test space such that $U_h \subset U$, i.e., a *conforming* method and have used the same forms to define the approximation problem, i.e., a *consistent* method in the form:

$$\text{Find } u_h \in U_h \text{ such that} \quad a(u_h, v_h) = \ell(v_h) \quad \text{for all } v_h \in V_h. \tag{6.12}$$

In this ideal situation we have the *Galerkin orthogonality* (see (2.43))

$$a(u - u_h, v_h) = 0 \quad \text{for all } v_h \in V_h$$

on which the proof of quasi-optimality is based thanks to Céa's lemma (Thm. 2.17).

In Section 3.6 we discussed that in reality we can only deal with approximate forms, i.e.,

$$\text{Find } u_h \in U_h \text{ such that} \quad a_h(u_h, v_h) = \ell_h(v_h) \quad \text{for all } v_h \in U_h. \tag{6.13}$$

Thus, an additional consistency error has to be handled, as done by the First Lemma of Strang (Theorem 3.41). Therefore from now on we consider a more general situation as follows.

Let U_h, V_h be finite-dimensional Hilbert spaces (which is not a restriction, since all finite-dimensional real normed spaces of same dimension, say m, are isometrically isomorphic each to the other, in particular to the Hilbert space $\mathbb{R}^m$ equipped with the Euclidean scalar product) and $a_h : U_h \times V_h \to \mathbb{R}$, $\ell_h : V_h \to \mathbb{R}$ be bilinear and linear forms, respectively. Then the discrete problem is

$$\text{Find } u_h \in U_h \text{ such that} \quad a_h(u_h, v_h) = \ell_h(v_h) \quad \text{for all } v_h \in V_h. \tag{6.14}$$

The corresponding operator equation reads as

$$A_h u_h = \ell_h \quad \text{for} \quad A_h : U_h \to V_h' \quad \text{or} \quad A_h : U_h \to V_h. \tag{6.15}$$

For $U_h = V_h$, we speak of *(standard) Galerkin methods* (sometimes also referred as *Bubnov–Galerkin methods*), otherwise of *Petrov–Galerkin methods*, which can also make sense if $U = V$ (see Section 10.2).

Remark 6.4 1) Since (6.14), (6.15) (for a fixed choice of U_h, V_h, a_h, ℓ_h) are of the form (6.1'), (6.3'), (6.4'), the characterization of unique solvability and stability can also be applied here leading to the conditions (since continuity of A_h is obvious)

$$a_h(u_h, v_h) = 0 \text{ for all } u_h \in U_h \Longrightarrow v_h = 0 \text{ for all } v_h \in V_h, \tag{NB1h}$$

$$\alpha_h := \inf_{u_h \in U_h \setminus \{0\}} \sup_{v_h \in V_h \setminus \{0\}} \frac{a_h(u_h, v_h)}{\|u_h\|_{U_h} \|v_h\|_{V_h}} > 0. \tag{NB2h}$$

As $\|A_h^{-1}\|_{V_h, U_h} = \alpha_h^{-1}$, we expect the necessity of a uniform lower bound for α_h if considering a family of approximations.

2) If arbitrary bases of U_h and V_h are chosen, say $U_h := \text{span}\{\varphi_1, \ldots, \varphi_n\}$, $V_h := \text{span}\{\psi_1, \ldots, \psi_m\}$ with $n = n(h) := \dim U_h$, $m := m(h) = \dim V_h$, then similar to Chapter 2 and Chapter 3, the problems (6.14) and (6.15) are equivalent to the system of linear equations

$$\mathcal{A}_h \boldsymbol{\xi} = \boldsymbol{q}_h,$$

where $\mathcal{A}_h = (a_{ij})_{i=1,\ldots,m, j=1,\ldots,n}$, $a_{ij} = a_h(\varphi_j, \psi_i)$, $\boldsymbol{q}_{hi} = \ell_h(\psi_i)$.

3) We have

$$\begin{aligned} &(\text{NB1h}) \Longleftrightarrow \text{rank}\,\mathcal{A}_h = \dim V_h = m, \\ &(\text{NB2h}) \Longleftrightarrow \ker \mathcal{A}_h = \{0\}, \end{aligned}$$

If $\dim U_h = \dim V_h$, then

$$A_h \text{ is injective} \Longleftrightarrow A_h \text{ is surjective}$$

and therefore

$$(\text{NB1h}) \Longleftrightarrow (\text{NB2h}),$$

as the condition "im(A_h) is closed" is always satisfied.

4) Note that contrary to the coercivity property neither (NB1h) nor (NB2h) are in general induced by (NB1) or (NB2), respectively.

5) (NB2h) can be written as

$$\alpha_h = \inf_{u_h \in U_h \setminus \{0\}} \sup_{v_h \in V_h \setminus \{0\}} \frac{\langle \mathcal{A}_h \boldsymbol{\xi}, \boldsymbol{\eta} \rangle}{\|u_h\|_{U_h} \|v_h\|_{V_h}} > 0,$$

$$\text{where} \quad u_h = \sum_{i=1}^{n} \xi_i \varphi_i, \; v_h = \sum_{j=1}^{m} \eta_j \psi_j$$

and $\langle \cdot, \cdot \rangle$ denotes the Euclidean scalar product (here in $\mathbb{R}^m$). By the isomorphisms $u_h \leftrightarrow \boldsymbol{\xi}$, $v_h \leftrightarrow \boldsymbol{\eta}$, norms $\| \cdot \|_U^{(n)}$, $\| \cdot \|_V^{(m)}$ on $\mathbb{R}^n$ and $\mathbb{R}^m$ are defined and then

$$\alpha_h = \inf_{\boldsymbol{\xi} \in \mathbb{R}^n \setminus \{\mathbf{0}\}} \sup_{\boldsymbol{\eta} \in \mathbb{R}^n \setminus \{\mathbf{0}\}} \frac{\langle \mathcal{A}_h \boldsymbol{\xi}, \boldsymbol{\eta} \rangle}{\|\boldsymbol{\xi}\|_U^{(n)} \|\boldsymbol{\eta}\|_V^{(m)}} = \inf_{\boldsymbol{\xi} \in \mathbb{R}^n \setminus \{\mathbf{0}\}} \sup_{\boldsymbol{\eta} \in \mathbb{R}^n \setminus \{\mathbf{0}\}} \frac{\langle \mathcal{A}_h \boldsymbol{\xi}, \boldsymbol{\eta} \rangle}{\|C\boldsymbol{\xi}\|_{\mathbb{R}^n} \|D\boldsymbol{\eta}\|_{\mathbb{R}^m}}$$

with symmetric, positive definite matrices $C \in \mathbb{R}^{n,n}$, $D \in \mathbb{R}^{m,m}$ (see Problem 6.4). If $\| \cdot \|_{\mathbb{R}^k}$ denotes the Euclidean norm in $\mathbb{R}^k$, then

$$\alpha_h = \inf_{\boldsymbol{\xi} \in \mathbb{R}^n \setminus \{\mathbf{0}\}} \sup_{\boldsymbol{\eta} \in \mathbb{R}^n \setminus \{\mathbf{0}\}} \frac{\left\langle D^{-1} \mathcal{A}_h C^{-1} C \boldsymbol{\xi}, D\boldsymbol{\eta} \right\rangle}{\|C\boldsymbol{\xi}\|_{\mathbb{R}^n} \|D\boldsymbol{\eta}\|_{\mathbb{R}^m}} = \inf_{\widetilde{\boldsymbol{\xi}} \in \mathbb{R}^n \setminus \{\mathbf{0}\}} \frac{\|D^{-1} \mathcal{A}_h C^{-1} \widetilde{\boldsymbol{\xi}}\|_{\mathbb{R}^m}}{\|\widetilde{\boldsymbol{\xi}}\|_{\mathbb{R}^n}}$$

by the Riesz representation theorem [67, Sect. III.6]. Therefore we finally obtain:

$$\alpha_h \text{ is the root of the smallest eigenvalue of } C^{-1} \mathcal{A}_h^T D^{-2} \mathcal{A}_h C^{-1}$$

(see Problem 6.5).

In correspondence to our previous conventions (see Sect. 3.3) we call the discretization (6.14) or (6.15) *conforming*, if $U_h \subset U$ and $V_h \subset V$, and *nonconforming* otherwise. Unless differently specified, we set $\| \cdot \|_{U_h} := \| \cdot \|_U$, $\| \cdot \|_{V_h} := \| \cdot \|_V$ in the conforming case. Also in the conforming case neither (NB1h) nor (NB2h) is implied by (NB1) or (NB2), respectively, opposite to the coercivity condition (see (3.3)). Therefore, these conditions are independent requirements for the choice of U_h and V_h.

To ensure that elements of U or elements of V can also be compared with elements of U_h or V_h, respectively, in the nonconforming case, we introduce the *augmented spaces*

$$U(h) := U + U_h, \; V(h) := V + V_h.$$

In this framework we assume the existence of norms $\| \cdot \|_{U(h)}$ in $U(h)$ and $\| \cdot \|_{V(h)}$ in $V(h)$. Furthermore, in some cases it can be useful to reduce the original solution space U to a proper subspace, for example, by incorporating additional knowledge about the regularity of the solution.

Definition 6.5 The approximation (6.14) or (6.15) is called (exactly) *consistent* on the solution space W, W being a subspace of U, if the bilinear form a_h can be extended to $(W + U_h) \times V_h$ such that the exact solution $u \in W$ of (6.1')/(6.3')/(6.4') satisfies the approximate problem, i.e.,

$$a_h(u, v_h) = \ell_h(v_h) \quad \text{for all } v_h \in V_h,$$

and *nonconsistent* otherwise.

In the nonconsistent case, additionally a *consistency error* occurs. In the consistent case, the *Galerkin orthogonality* holds true, i.e.,

$$a_h(u - u_h, v_h) = 0 \quad \text{for all } v_h \in V_h. \tag{6.16}$$

The convergence analysis done so far can now be extended.

The Consistent, Conforming Case

Theorem 6.6 (Céa's Lemma revisited) *Assume that*

1) (NB2h) *holds with the constant* α_h *and* $\dim U_h = \dim V_h$,
2) the bilinear form a_h *is bounded on* $(U, \|\cdot\|_U) \times (V_h, \|\cdot\|_V)$*, i.e.,*

$$|a_h(w, v)| \le M_h \|w\|_U \|v\|_V \quad \textit{for all } w \in U, v \in V_h.$$

Then

$$\|u - u_h\|_U \le \left(1 + \frac{M_h}{\alpha_h}\right) \inf_{w_h \in U_h} \|u - w_h\|_U$$

for the solution u *of* (6.1')/(6.3')/(6.4') *and the consistent solution* u_h *of* (6.14)/(6.15).

Proof We have $\|u - u_h\|_U \le \|u - w_h\|_U + \|w_h - u_h\|_U$ for $w_h \in U_h$ and by (NB2h), (6.16):

$$\alpha_h \|u_h - w_h\|_U \le \sup_{v_h \in V_h \setminus \{0\}} \frac{a_h(u_h - w_h, v_h)}{\|v_h\|_V} = \sup_{v_h \in V_h \setminus \{0\}} \frac{a_h(u - w_h, v_h)}{\|v_h\|_V} \le M_h \|u - w_h\|_U.$$

Hence

$$\|u - u_h\|_U \le \left(1 + \frac{M_h}{\alpha_h}\right) \|u - w_h\|_U.$$

□

The Nonconsistent, Conforming Case

Theorem 6.7 (First Lemma of Strang revisited) *Under the first condition of Theorem 6.6 it holds:*

$$\begin{aligned} \|u - u_h\|_U \le{}& \frac{C_h}{\alpha_h} \sup_{v_h \in V_h \setminus \{0\}} \frac{|\ell(v_h) - \ell_h(v_h)|}{\|v_h\|_{V_h}} \\ &+ \inf_{w_h \in U_h} \left[\left(1 + \frac{C_h^2 M}{\alpha_h}\right) \|u - w_h\|_U + \frac{C_h}{\alpha_h} \sup_{v_h \in V_h \setminus \{0\}} \frac{|a(w_h, v_h) - a_h(w_h, v_h)|}{\|v_h\|_{V_h}} \right], \end{aligned}$$

where $C_h > 0$ *is such that* $\|w_h\|_U \le C_h \|w_h\|_{U_h}$ *for all* $w_h \in U_h$ *and* $\|v_h\|_V \le C_h \|v_h\|_{V_h}$ *for all* $v_h \in V_h$.

Proof Since

$$a_h(u_h - w_h, v_h) = a(u - w_h, v_h) + a(w_h, v_h) - a_h(w_h, v_h) + \ell_h(v_h) - \ell(v_h),$$

it follows from (NB2h) that

$$\begin{aligned}\alpha_h \|u_h - w_h\|_{U_h} \le & \sup_{v_h \in V_h \setminus \{0\}} \frac{a_h(u_h - w_h, v_h)}{\|v_h\|_{V_h}} \\ \le & C_h M \|u - w_h\|_U + \sup_{v_h \in V_h \setminus \{0\}} \frac{|a(w_h, v_h) - a_h(w_h, v_h)|}{\|v_h\|_{V_h}} + \sup_{v_h \in V_h \setminus \{0\}} \frac{|\ell_h(v_h) - \ell(v)|}{\|v_h\|_{V_h}}.\end{aligned}$$

The proof can be finished as for Theorem 6.6. □

The Consistent, Nonconforming Case

Theorem 6.8 *Assume that the solution u of* (6.1')/(6.3')/(6.4') *belongs to some subspace $W \subset U$, the norm $\|\cdot\|_{U_h}$ extends to a norm on $W + U_h$, and the first condition of Theorem 6.6 (but without $U_h \subset U$, $V_h \subset V$) is satisfied. Then, if a_h can be continuously extended onto $(W + U_h, \|\cdot\|_{U(h)}) \times (V_h, \|\cdot\|_{V_h})$, i.e., there exists a constant $\widetilde{M}_h \ge 0$ such that*

$$|a_h(w, v_h)| \le \widetilde{M}_h \|w\|_{U(h)} \|v_h\|_{V_h} \quad \textit{for all } w \in W + U_h, v \in V_h,$$

it holds

$$\|u - u_h\|_{U_h} \le \inf_{w_h \in U_h} \left(\frac{\widetilde{M}_h}{\alpha_h} \|u - w_h\|_{U(h)} + \|u - w_h\|_{U_h} \right)$$

for the consistent solution u_h of (6.14)/(6.15).

Proof Due to (6.16) we have

$$\begin{aligned}\alpha_h \|u_h - w_h\|_{U_h} \le & \sup_{v_h \in V_h \setminus \{0\}} \frac{a_h(u_h - w_h, v_h)}{\|v_h\|_{V_h}} \\ = & \sup_{v_h \in V_h \setminus \{0\}} \frac{a_h(u - w_h, v_h)}{\|v_h\|_{V_h}} \le \widetilde{M}_h \|u - w_h\|_{U(h)}\end{aligned}$$

and the proof can be finished as above. □

Remark 6.9 1) Under the assumptions of Thm. 6.8, if there exists a constant $\widetilde{C}_h > 0$ such that

$$\|w_h\|_{U(h)} \le \widetilde{C}_h \|w_h\|_{U_h} \quad \text{for all } w_h \in U_h,$$

we can conclude analogously

$$\|u - u_h\|_{U(h)} \leq \left(1 + \frac{\widetilde{C}_h \widetilde{M}_h}{\alpha_h}\right) \inf_{w_h \in U_h} \|u - w_h\|_{U(h)}.$$

2) Under the assumptions of Thm. 6.8, if there exists a constant $\widetilde{C}_h > 0$ such that

$$\|w_h\|_{U_h} \leq \widetilde{C}_h \|w_h\|_{U(h)} \quad \text{for all } w_h \in U_h,$$

then we have

$$\|u - u_h\|_{U_h} \leq \left(\widetilde{C}_h + \frac{\widetilde{M}_h}{\alpha_h}\right) \inf_{w_h \in U_h} \|u - w_h\|_{U(h)}.$$

The Nonconsistent, Nonconforming Case

Theorem 6.10 (Second Lemma of Strang) *Under the conditions of Theorem 6.8 it holds:*

$$\begin{aligned} \|u - u_h\|_{U_h} \leq & \inf_{w_h \in U_h} \left(\frac{\widetilde{M}_h}{\alpha_h} \|u - w_h\|_{U(h)} + \|u - w_h\|_{U_h} \right) \\ & + \frac{1}{\alpha_h} \sup_{v_h \in V_h \setminus \{0\}} \frac{|a_h(u, v_h) - \ell_h(v_h)|}{\|v_h\|_{V_h}}. \end{aligned}$$

Proof From

$$\begin{aligned} a_h(u_h - w_h, v_h) &= a_h(u_h - u, v_h) + a_h(u - w_h, v_h) \\ &= \ell_h(v_h) - a_h(u, v_h) + a_h(u - w_h, v_h) \end{aligned}$$

and (NB2h) it follows that

$$\alpha_h \|u_h - w_h\|_{U_h} \leq \sup_{v_h \in V_h \setminus \{0\}} \frac{|a_h(u, v_h) - \ell_h(v_h)|}{\|v_h\|_{V_h}} + \widetilde{M}_h \|u - w_h\|_{U(h)},$$

and the proof can be finished as above. □

The variants as in Remark 6.9 are possible here, too.

Error Estimates in Weaker Norms

In Section 3.4.3 we have already seen with the help of a duality argument that, by considering the adjoint variational problem

Find $v \in V$ such that

$$a(w, v) = \tilde{\ell}(w) \quad \text{for all } w \in U \tag{6.17}$$

for specific right-hand sides $\tilde{\ell} \in U'$, it is possible to improve the order of convergence in norms that are weaker than $\|\cdot\|_V$, provided the solution of the adjoint problem is sufficiently regular. This requirement can be expressed by the assumption that there exists a suitably chosen subspace $Y \subset V$ in which the solution v of (6.17) lies.

In the subsequent considerations, the following general assumptions will be used: Let Z be a Hilbert space such that $U \subset Z'$, $U_h \subset Z'$. By $\langle \cdot, \cdot \rangle_{Z''}$ we denote the duality pairing on Z' and $Z'' \cong Z$.

Now let $g \in Z$ be given and consider the adjoint problem (6.17) with the right-hand side

$$\tilde{\ell}(w) := \tilde{\ell}_g(w) := \langle w, g \rangle_{Z''} \quad \text{for all } w \in U. \tag{6.18}$$

In addition, it is assumed that the Galerkin approximation $v_{gh} \in V_h$ to the solution $v_g \in Y$ of (6.17), (6.18) exists uniquely, i.e., v_{gh} satisfies

$$a_h(w_h, v_{gh}) = \langle w_h, g \rangle_{Z''} \quad \text{for all } w_h \in U_h. \tag{6.19}$$

Lemma 6.11 *Consider a general Galerkin approximation* (6.14)/(6.15) *of the variational form* (6.1')/(6.3')/(6.4') *with solutions $u \in W$ and $u_h \in U_h$, respectively, where W is a suitably chosen subspace of U. Assume that the bilinear form a_h can be extended to $(W + U_h) \times (Y + V_h)$ such that it holds, for a constant $\widetilde{M}_h \geq 0$,*

$$|a_h(w, z)| \leq \widetilde{M}_h \|w\|_{U(h)} \|z\|_{V(h)} \quad \textit{for all } w \in W + U_h, z \in Y + V_h,$$

and that ℓ_h can be extended to $Y + V_h$. Then:

$$\begin{aligned} \|u - u_h\|_{Z'} \leq \sup_{g \in Z \setminus \{0\}} \frac{1}{\|g\|_Z} \Big\{ & \widetilde{M}_h \|u - u_h\|_{U(h)} \|v_g - v_{gh}\|_{V(h)} \\ & - \big[a_h(u - u_h, v_g) - \langle u - u_h, g \rangle_{Z''}\big] \\ & - \big[a_h(u, v_g - v_{gh}) - \ell_h(v_g - v_{gh})\big] \\ & + \big[(a_h - a)(u, v_g) - (\ell_h - \ell)(v_g)\big] \Big\}. \end{aligned} \tag{6.20}$$

Proof Since

$$\|u - u_h\|_{Z'} = \sup_{g \in Z \setminus \{0\}} \frac{1}{\|g\|_Z} \langle u - u_h, g \rangle_{Z''}$$

holds it is sufficient to consider

$$\begin{aligned} \langle u - u_h, g \rangle_{Z''} &= a(u, v_g) - a_h(u_h, v_{gh}) \\ &= a_h(u, v_g) - a_h(u_h, v_{gh}) + (a - a_h)(u, v_g) \\ &=: I_1 + (a - a_h)(u, v_g). \end{aligned}$$

The first term can be decomposed as

$$I_1 = a_h(u - u_h, v_g - v_{gh}) + a_h(u_h, v_g - v_{gh}) + a_h(u - u_h, v_{gh}) =: I_2 + I_3 + I_4,$$

where I_2 can be estimated immediately by means of the assumed continuity property of a_h:

$$|I_2| \le \widetilde{M}_h \|u - u_h\|_{U(h)} \|v_g - v_{gh}\|_{V(h)}.$$

The terms I_3, I_4 can be rewritten as follows:

$$\begin{aligned}
I_3 &= -a_h(u - u_h, v_g) + a_h(u, v_g) - a_h(u_h, v_{gh}) \\
&= -a_h(u - u_h, v_g) + a(u, v_g) - a_h(u_h, v_{gh}) + (a_h - a)(u, v_g) \\
&= -[a_h(u - u_h, v_g) - \langle u, g\rangle_{Z''} + \langle u_h, g\rangle_{Z''}] + (a_h - a)(u, v_g), \\
&= -[a_h(u - u_h, v_g) - \langle u - u_h, g\rangle_{Z''}] + (a_h - a)(u, v_g), \\
I_4 &= -a_h(u, v_g - v_{gh}) + a_h(u, v_g) - a_h(u_h, v_{gh}) \\
&= -[a_h(u, v_g - v_{gh}) - a(u, v_g) + \ell_h(v_{gh})] + (a_h - a)(u, v_g) \\
&= -[a_h(u, v_g - v_{gh}) - \ell_h(v_g - v_{gh})] + (a_h - a)(u, v_g) - (\ell_h - \ell)(v_g).
\end{aligned}$$

In summary, the statement follows. □

In the following we want to discuss the four contributions in the estimate (6.20). All terms must allow an estimate with the factor $\|g\|_Z$ to achieve an improved order of convergence result. The order of the first term is the product of the convergence order of the original problem according to Theorems 6.6, 6.7, 6.8, 6.10 and the corresponding order of convergence for the adjoint problem.

To be more specific assume that $Y \subset V$ is a Hilbert space equipped with a (semi)norm $\|\cdot\|_Y$ such that the following regularity and convergence properties hold:

- The solution v_g of (6.17), (6.18) even belongs to the space Y and there is a constant $C_s \ge 0$ independent of v_g such that

$$\|v_g\|_Y \le C_s \|g\|_Z. \tag{6.21}$$

- There are constants $C_a > 0, q > 0$ such that the Galerkin approximation $v_{gh} \in V_h$ from (6.19) to the solution $v_g \in Y$ of (6.17), (6.18) satisfies

$$\|v_g - v_{gh}\|_{V(h)} \le C_a h^q \|v_g\|_Y. \tag{6.22}$$

The estimate (6.22) is again a consequence of the above theorems reflecting the order of the approximation of v_g by elements in V_h and of the consistency errors. Under the assumptions (6.21), (6.22) the first term of (6.20) obviously can be estimated by

$$\widetilde{M}_h C_a C_s \|u - u_h\|_{U(h)} h^q.$$

The second term is a consistency error for the adjoint problem in the test function $u - u_h$. We aim at a consistency error estimate for $v_g \in Y$ of the type

$$|a_h(w, v_g) - \langle w, g\rangle_{Z''}| \le C_{ca} \|w\|_{U(h)} \|v_g\|_Y h^\alpha \le C_{ca} C_s \|w\|_{U(h)} \|g\|_Z h^\alpha \tag{6.23}$$

with constants $C_{ca} \ge 0, \alpha > 0$, giving the following estimate for the second term:

$$C_{ca}C_s\|u - u_h\|_{U(h)}h^\alpha.$$

Analogously the third therm is a consistency error for the original problem in the test function $v_g - v_{gh}$. Assuming analogously to (6.23) an estimate of the type

$$|a_h(u, z) - \ell_h(z)| \le C_{cp}\|z\|_{V(h)}\|u\|_Y h^\beta \tag{6.24}$$

with constants $C_{cp} \ge 0$, $\beta > 0$, assuming without loss of generality the same regularity for the original and adjoint problems, in particular $u \in Y$, then the third term together with (6.22) allows the following estimate:

$$C_{cp}C_\alpha C_s h^{q+\beta}.$$

Finally the last term describes approximation errors. Here we assume an estimate of the type

$$|(a_h - a)(u, v_g) - (\ell_h - \ell)(v_g)| \le C_q\|v_g\|_Z h^\gamma \le C_q C_s\|g\|_Z h^\gamma \tag{6.25}$$

with constants $C_q \ge 0$, $\gamma > 0$. If finally the order of convergence of the original problem is $p > 0$, then the achieved order of the bound of $\|\cdot\|_{Z'}$ is

$$\min\{p + q,\ p + \alpha,\ q + \beta,\ \gamma\}.$$

In the conforming case (both for the original and the adjoint problem, see the following table) the last three terms in (6.20) can be rearranged to

$$a(u - u_h, v_g - v_{gh}) - (a - a_h)(u_h, v_{gh}) - (\ell_h - \ell)(v_{gh}). \tag{6.26}$$

In the consistent case (both for the original and the adjoint problem) the last three terms in (6.20) can be rearranged to

$$(a - a_h)(u, v_g). \tag{6.27}$$

Concerning the choice of Z, we concentrate here on formulations with $U \subset H^1(\Omega)$, and $U = V, U_h = V_h$ being $\mathcal{P}_k$-elements according to (3.74) as discussed in Chapter 3 and also in Section 7.2. Then an obvious choice is

$$Z := L^2(\Omega),$$

and (6.21), (6.22) can be fulfilled with $Y := H^2(\Omega)$, $q = 1$, for sufficient smooth data as seen in Section 3.4.3. However, $Z := H^{k-1}(\Omega)$ can also be selected, which may enable the choice $Y := H^{k+1}(\Omega)$ and also $q = k$, and thus leads to the optimal case of order doubling (provided the consistency errors behave correspondingly) in the norm $\|\cdot\|_{-k+1}$.

Such negative norm estimates then also allow for error estimates of functionals $\varphi \in U(h)'$: Assuming that φ is given as a smooth function in the sense that even

$$\varphi \in Z\ (\cong Z'')$$

holds, then

$$\varphi(u - u_h) \le \|\varphi\|_Z \|u - u_h\|_{Z'} .$$

In the following table we list the properties of the variational formulations used, whereby the properties of the second and third columns apply in addition to the first column. Here as above $u, v_g \in V$ are the weak solutions, $u_h, v_{gh} \in V_h$ are their numerical approximations, and $v \in V, v_h \in V_h, w \in U, w_h \in U_h$ are the test functions.

General case	Conforming case	Consistent case
$a(u, v) = \ell(v)$		
	$a(u, v_h) = \ell(v_h)$	
$a_h(u_h, v_h) = \ell_h(v_h)$		
		$a_h(u, v_h) = \ell_h(v_h)$
$a(w, v_g) = \langle w, g\rangle_{Z''}$		
	$a(w_h, v_g) = \langle w_h, g\rangle_{Z''}$	
$a_h(w_h, v_{gh}) = \langle w_h, g\rangle_{Z''}$		
		$a_h(w_h, v_g) = \langle w_h, g\rangle_{Z''}$

Remark 6.12 In the conforming case with $U = V$ an alternative estimate is possible. Under the assumption of Lemma 6.11 we have (using the notation of the proof):

$$\begin{aligned}\langle u - u_h, g\rangle_{Z''} &= a(u - u_h, v_g) \\ &= a(u - u_h, v_g - \Pi_h v_g) + \ell(\Pi_h v_g) - a(u_h, \Pi_h v_g)\end{aligned}$$

with $\Pi_h v_g \in V_h$, leading to the estimate

$$\begin{aligned}\|u - u_h\|_{Z'} \le \sup_{g \in Z \setminus \{0\}} \frac{1}{\|g\|_Z} \Big\{ &\widetilde{M} \|u - u_h\|_V \|v_g - \Pi_h v_g\|_V \\ &- [a(u_h, \Pi_h v_g) - \ell(\Pi_h v_g)] \Big\}.\end{aligned}$$

For $Z := V$ and Π_h being the orthogonal projection of V_h to V, the second term, a "unvariational" consistency error term, can be estimated by

$$| \big\langle a(u_h, \cdot) - a_h(u_h, \cdot), \Pi_h v_g \big\rangle_V | + \|v_g\|_V \|\ell_h - \Pi_h \ell\|_{V'},$$

as by decomposing the term by introducing $\ell_h := a_h(u_h, \cdot)$, the second summand reads by the symmetry of Π_h

$$\big\langle a_h(u_h, \cdot) - \ell, \Pi_h v_g \big\rangle_V = \big\langle \Pi_h(a_h(u_h, \cdot) - \ell), v_g \big\rangle_V = \big\langle \ell_h - \Pi_h \ell, v_g \big\rangle .$$

Exercises

Problem 6.1 Let V, W be Hilbert spaces and $a : V \times V \to \mathbb{R}$, $c : W \times W \to \mathbb{R}$ be symmetric, bounded, and coercive bilinear forms. Furthermore, let $b : V \times W \to \mathbb{R}$

be a bounded bilinear form. Prove that for $f \in V'$, $g \in W'$ there is a unique solution $(u, p) \in V \times W$ such that

$$\begin{aligned} a(u, v) + b(v, p) &= \langle f, v \rangle \qquad \text{for all } v \in V, \\ b(u, q) - c(p, q) &= \langle g, q \rangle \qquad \text{for all } q \in W. \end{aligned}$$

Prove that the following stability estimate holds with α and γ representing coercivity constants:

$$\alpha \|u\|_V^2 + \gamma \|p\|_W^2 \le \alpha^{-1} \|f\|_{V'}^2 + \gamma^{-1} \|g\|_{W'}^2 .$$

Problem 6.2 Let V be a Banach space and $G \subset V$, $Q \subset V'$ subsets. Show:

a) The annihilators $G^\perp$ and $Q_\perp$ (see (A4.31)) are closed subspaces.
b) If G is a subspace of V, then $\overline{G} = (G^\perp)_\perp$.

Problem 6.3 Let U, V be Banach spaces.

a) Let $A \in L[U, V]$. Prove that the statement "im(A) is closed" is equivalent to the existence of a number $\alpha > 0$ such that

$$\forall v \in \operatorname{im}(A)\ \exists u_v \in U : \ Au_v = v \quad \text{and} \quad \alpha \|u_v\|_U \le \|v\|_V .$$

b) Extend the proof for the Banach–Nečas–Babuška Theorem 6.2 to general Banach spaces. You will additionally need that one of the two spaces is reflexive. Which one and why?

Problem 6.4 Let $(\cdot, \cdot)$ be a scalar product on $\mathbb{R}^n$, then there exists a unique symmetric, positive definite matrix $C \in \mathbb{R}^{n,n}$ such that

$$(\boldsymbol{x}, \boldsymbol{y}) = \langle C\boldsymbol{x}, C\boldsymbol{y} \rangle \quad \text{for all } \boldsymbol{x}, \boldsymbol{y} \in \mathbb{R}^n,$$

where $\langle \cdot, \cdot \rangle$ denotes the Euclidean scalar product.

Problem 6.5 Let $C \in \mathbb{R}^{n,n}$ be symmetric, then

$$\inf_{\boldsymbol{x} \in \mathbb{R}^n \setminus \{0\}} \frac{\langle C\boldsymbol{x}, \boldsymbol{x} \rangle}{|\boldsymbol{x}|^2}$$

is the smallest eigenvalue of C.

6.2 Saddle Point Problems

The boundary value problems considered so far have a scalar function u as the unknown representing temperature, concentration, pressure, etc. Remembering the derivation of such models in Chapter 0, they stem from a *conservation law* (like conservation of mass (0.2)) for a vector field $\boldsymbol{q} : \Omega \to \mathbb{R}^n$ (being the *flux density* of a conserved quantity u) and a *constitutive law* (like Fick's law (0.4)) connecting u and $\boldsymbol{q}$. In many applications, $\boldsymbol{q}$ is of the same or even higher interest than u: it can be reconstructed by the constitutive law from u—also only having an approximation u_h for u, but this might deteriorate either the accuracy of approximation or desired qualitative properties (see Section 7.6). Therefore, it is desirable to have methods which (also) approximate $\boldsymbol{q}$ directly.

From now on we write $\boldsymbol{p}, \boldsymbol{q}, \ldots$ for vector fields and use a bold face print $\boldsymbol{p}, \boldsymbol{q}$ to make this character more evident. This consequently also applies to the (given) flow field $\boldsymbol{c}$ and the normal $\boldsymbol{\mathfrak{n}}$.

Remembering the formulation (3.13), (3.19), (3.20), (3.21) for a general elliptic boundary value problem, it can be rewritten as

$$\begin{aligned} \nabla \cdot \boldsymbol{p} + \boldsymbol{c} \cdot \nabla u + ru &= f \quad \text{in } \Omega, && \text{(3.13')} \\ -\boldsymbol{p} \cdot \boldsymbol{\mathfrak{n}} &= g_1 \quad \text{on } \Gamma_1, && \text{(3.19')} \\ -\boldsymbol{p} \cdot \boldsymbol{\mathfrak{n}} + \alpha u &= g_2 \quad \text{on } \Gamma_2, && \text{(3.20')} \\ u &= g_3 \quad \text{on } \Gamma_3, && \text{(3.21')} \end{aligned}$$

with the definition

$$\boldsymbol{p} = -\boldsymbol{K} \nabla u. \tag{6.28}$$

The latter quantity can be interpreted as the *diffusive* part of the total flux density. If it is of primary interest, this formulation could be the starting point. We prefer to write the equation in divergence form (see (3.36), (3.37)–(3.39)), meaning

$$-\nabla \cdot (\boldsymbol{K} \nabla u - \boldsymbol{c} u) + ru = f, \tag{6.29}$$

i.e., the equations now read (if we are interested in the total flux density)

$$\nabla \cdot \boldsymbol{p} + ru = f \quad \text{in } \Omega, \tag{6.30}$$

$$\text{with} \quad \boldsymbol{p} = -\boldsymbol{K} \nabla u + \boldsymbol{c} u. \tag{6.31}$$

The above boundary conditions read:

$$\begin{aligned} -\boldsymbol{p} \cdot \boldsymbol{\mathfrak{n}} &= \widetilde{g_1} \quad \text{on } \widetilde{\Gamma}_1, && (6.32) \\ -\boldsymbol{p} \cdot \boldsymbol{\mathfrak{n}} + \alpha u &= \widetilde{g_2} \quad \text{on } \widetilde{\Gamma}_2, && (6.33) \\ u &= \widetilde{g_3} \quad \text{on } \widetilde{\Gamma}_3. && (6.34) \end{aligned}$$

Again $\widetilde{\Gamma}_1, \widetilde{\Gamma}_2 \widetilde{\Gamma}_3$ is a subdivision of the boundary similar to $\Gamma_1, \Gamma_2, \Gamma_3$. Note that there is a correspondence between $\widetilde{\Gamma}_1$ and Γ_2, and $\widetilde{\Gamma}_2$ and Γ_1, respectively (see (3.36) ff.).

6.2.1 Traces on Subsets of the Boundary

At this point, it is necessary to deepen the discussion of the term *trace* of elements of $H^1(\Omega)$ that started in Section 3.1. Up to now it was sufficient to know that the image of the trace operator

$$\gamma_0 : \left(H^1(\Omega), \|\cdot\|_1\right) \to \left(L^2(\partial\Omega), \|\cdot\|_0\right)$$

(cf. Thm. 3.6) is a certain subspace of $L^2(\partial\Omega)$ but from now on we need the surjectivity of the trace operator. Therefore we define the *trace space* of this operator as

$$H^{1/2}(\partial\Omega) := \mathrm{im}(\gamma_0) = \gamma_0\left(H^1(\Omega)\right)$$

and equip it with the norm induced by $H^1(\Omega)$, that is

$$\|v\|_{H^{1/2}(\partial\Omega)} := \inf_{w\in H^1(\Omega):\ \gamma_0(w)=v} \|w\|_1,$$

so that the trace operator considered as

$$\gamma_0 : \left(H^1(\Omega), \|\cdot\|_1\right) \to \left(H^{1/2}(\partial\Omega), \|\cdot\|_{H^{1/2}(\partial\Omega)}\right) \tag{6.35}$$

is still continuous. An equivalent norm is

$$\|v\|_{H^{1/2}(\partial\Omega)} := \left(\|v\|^2_{L^2(\partial\Omega)} + \int_{\partial\Omega}\int_{\partial\Omega} \frac{|v(x)-v(y)|^2}{|x-y|^d}\, d\sigma_x d\sigma_y\right)^{1/2}$$

(see, e.g., [45, Ch. 3] for more details). The space $H^{1/2}(\partial\Omega)$ is a representative of the so-called *fractional Sobolev spaces* or *Sobolev–Slobodeckij spaces*, which play an important role in the theory of traces.

In this way we have obtained a refined trace theorem compared to Thm. 3.6 in the sense that the trace operator (6.35) is continuous and surjective. Moreover, the so-called *inverse trace theorem* states that this operator has a continuous right inverse operator $\gamma_0^- : \left(H^{1/2}(\partial\Omega), \|\cdot\|_{H^{1/2}(\partial\Omega)}\right) \to (H^1(\Omega), \|\cdot\|_1)$ satisfying $(\gamma_0 \circ \gamma_0^-)(v) = v$ for all $v \in H^{1/2}(\partial\Omega)$. This result goes back to [138].

The dual of $H^{1/2}(\partial\Omega)$ is denoted by $H^{-1/2}(\partial\Omega) := \left(H^{1/2}(\partial\Omega)\right)'$.

When investigating boundary value problems with boundary conditions of mixed type, we need so-called *localized* trace spaces. Let $\Gamma \subset \partial\Omega$ be a proper relatively open subset of positive $(d-1)$-measure and assume that the boundary $\partial\Gamma$ of Γ (considered as a submanifold of $\partial\Omega$ of dimension $d-2$) is sufficiently smooth (actually Lipschitzian). The idea behind is to interpret Γ as a subset of $\mathbb{R}^{d-1}$ after flattening (namely, as the image of a Lipschitz domain in $\mathbb{R}^{d-1}$ under a Lipschitz-continuous mapping into $\mathbb{R}^d$) such that the theory of (though fractional) Sobolev spaces in the lower dimension can be applied.

The space $H^{1/2}(\Gamma)$ is defined as follows:

$$H^{1/2}(\Gamma) := \{v \in L^2(\Gamma) \mid \|v\|_{H^{1/2}(\Gamma)} < \infty, \},$$

where

$$\|v\|_{H^{1/2}(\Gamma)} = \left(\|v\|_{L^2(\Gamma)}^2 + \int_\Gamma \int_\Gamma \frac{|v(x) - v(y)|^2}{|x - y|^d}\, d\sigma_x d\sigma_y\right)^{1/2}.$$

The dual of $H^{1/2}(\Gamma)$ is denoted by $\tilde{H}^{-1/2}(\Gamma) := \left(H^{1/2}(\Gamma)\right)'$.

Given an element $v \in L^2(\Gamma)$, we define $\tilde{v} \in L^2(\partial\Omega)$ as the *extension by zero* of v:

$$\tilde{v}(x) := \begin{cases} v(x) & \text{on } \Gamma, \\ 0 & \text{on } \partial\Omega \setminus \Gamma. \end{cases}$$

The space $H_{00}^{1/2}(\Gamma)$ is defined by

$$H_{00}^{1/2}(\Gamma) := \{v \in L^2(\Gamma) \mid \tilde{v} \in H^{1/2}(\partial\Omega)\}$$

and equipped with the norm

$$\|v\|_{H_{00}^{1/2}(\Gamma)} := \left(\|v\|_{H^{1/2}(\Gamma)}^2 + \int_\Gamma \frac{v^2(x)}{\operatorname{dist}(x, \partial\Gamma)}\, d\sigma\right)^{1/2}.$$

The dual of $H_{00}^{1/2}(\Gamma)$ is denoted by $H^{-1/2}(\Gamma) := \left(H_{00}^{1/2}(\Gamma)\right)'$.

Fortunately, the refined trace theorem analogously holds true for the localized trace operator

$$\gamma_{0,\Gamma} : \left(H_{\Gamma^c}^1(\Omega), \|\cdot\|_1\right) \to \left(H_{00}^{1/2}(\Gamma), \|\cdot\|_{H_{00}^{1/2}(\Gamma)}\right), \tag{6.36}$$

where $\Gamma^c := \partial\Omega \setminus \Gamma$ and

$$H_{\Gamma^c}^1(\Omega) := \{v \in H^1(\Omega) \mid \gamma_0(v) = 0 \text{ on } \Gamma^c\} \tag{6.37}$$

(cf. (3.31)). That is, the trace operator (6.36) is continuous and surjective and it has a continuous right inverse operator $\gamma_{0,\Gamma}^- : \left(H_{00}^{1/2}(\Gamma), \|\cdot\|_{H_{00}^{1/2}(\Gamma)}\right) \to \left(H_{\Gamma^c}^1(\Omega), \|\cdot\|_1\right)$ satisfying $(\gamma_0 \circ \gamma_{0,\Gamma}^-)(v) = v$ for all $v \in H_{00}^{1/2}(\Gamma)$. For details we refer to [31].

In the next subsection we will operate with a complete subspace of $L^2(\Omega)^d$ in which $g := \nabla \cdot \boldsymbol{q}$ exists weakly in $L^2(\Omega)$, i.e.,

$$-\int_\Omega \boldsymbol{q} \cdot \nabla\varphi\, dx = \int_\Omega g\varphi\, dx \quad \text{for all } \varphi \in C_0^\infty(\Omega).$$

It is defined by

$$H(\operatorname{div};\Omega) := \left\{ \boldsymbol{q} \in L^2(\Omega)^d \mid \nabla \cdot \boldsymbol{q} \text{ exists weakly in } L^2(\Omega) \right\},$$
$$\|\boldsymbol{q}\|_{H(\operatorname{div})} := \left(\sum_{i=1}^d \|q_i\|_0^2 + \|\nabla \cdot q\|_0^2 \right)^{1/2}. \tag{6.38}$$

On this space a *normal component trace operator*

$$\gamma_{\mathfrak{n},\Gamma} : \left(H(\operatorname{div};\Omega), \|\cdot\|_{H(\operatorname{div})}\right) \to \left(H^{-1/2}(\Gamma), \|\cdot\|_{H^{-1/2}(\Gamma)}\right) \qquad \boldsymbol{q} \mapsto \boldsymbol{q}\cdot\mathfrak{n}|_\Gamma \tag{6.39}$$

can be introduced such that for $\boldsymbol{q} \in H(\operatorname{div};\Omega) \cap C(\overline{\Omega})^d$ its value coincides with the normal component of $\boldsymbol{q}$ on Ω:

$$\gamma_{\mathfrak{n},\Gamma}(\boldsymbol{q}) = \boldsymbol{q}\cdot\mathfrak{n} \quad \text{on } \Gamma.$$

To see that this operator is bounded, we consider for a moment only smooth arguments $\boldsymbol{q} \in H(\operatorname{div};\Omega) \cap C^1(\overline{\Omega})^d$. Then, using the notation $\langle \mu, v\rangle_X$ for the duality pairing in $X' \times X$ for a Banach space X, i.e., $\langle \mu, v\rangle_X := \mu(v)$, we have

$$\|\gamma_{\mathfrak{n},\Gamma}(\boldsymbol{q})\|_{H^{-1/2}(\Gamma)} = \sup_{v\in H_{00}^{1/2}(\Gamma)} \frac{\langle \boldsymbol{q}\cdot\mathfrak{n}, v\rangle_{H_{00}^{1/2}(\Gamma)}}{\|v\|_{H_{00}^{1/2}(\Gamma)}} \le C \sup_{v\in H_{\Gamma^c}^1(\Omega)} \frac{\left\langle \boldsymbol{q}\cdot\mathfrak{n}, \gamma_{0,\Gamma}(v)\right\rangle_{H_{00}^{1/2}(\Gamma)}}{\|v\|_1},$$

where we have used that the operator $\gamma_{0,\Gamma}^-$ is continuous. Note that due to the continuous embedding

$$H_{00}^{1/2}(\Gamma) \hookrightarrow L^2(\Gamma) \hookrightarrow H^{-1/2}(\Gamma)$$

we have due to the assumed regularity

$$\left\langle \boldsymbol{q}\cdot\mathfrak{n}, \gamma_{0,\Gamma}(v)\right\rangle_{H_{00}^{1/2}(\Gamma)} = \int_\Gamma \boldsymbol{q}\cdot\mathfrak{n}\gamma_{0,\Gamma}(v)\,d\sigma$$

Therefore by the divergence theorem, it follows that

$$\begin{aligned}
\|\gamma_{\mathfrak{n},\Gamma}(\boldsymbol{q})\|_{H^{-1/2}(\Gamma)} &\le C \sup_{v\in H_{\Gamma^c}^1(\Omega)} \frac{\langle \gamma_0(v)\boldsymbol{q}, \mathfrak{n}\rangle_0}{\|v\|_1} = C \sup_{v\in H_{\Gamma^c}^1(\Omega)} \frac{\langle \nabla\cdot(v\boldsymbol{q}), 1\rangle_0}{\|v\|_1} \\
&= C \sup_{v\in H_{\Gamma^c}^1(\Omega)} \frac{\langle \nabla v\cdot \boldsymbol{q}, 1\rangle_0 + \langle v\nabla\cdot\boldsymbol{q}, 1\rangle_0}{\|v\|_1} \\
&\le C \sup_{v\in H_{\Gamma^c}^1(\Omega)} \frac{\|\nabla v\|_0\|\boldsymbol{q}\|_0 + \|v\|_0\|\nabla\cdot\boldsymbol{q}\|_0}{\|v\|_1} \le C\|\boldsymbol{q}\|_{H(\operatorname{div})}.
\end{aligned}$$

Since $H(\operatorname{div};\Omega)\cap C(\overline{\Omega})^d$ is dense in $H(\operatorname{div};\Omega)$ [18, Ch. 9, §1, Thm. 1], the continuity of the normal component trace operator is proved.

The above considerations allow to formulate a refined version of Theorem 3.9, which will be useful in the following section.

Theorem 6.13 *Let $\Omega \subset \mathbb{R}^d$ be a domain, $\Gamma \subset \partial\Omega$ as described above. Then:*

1) For $\boldsymbol{q} \in H(\mathrm{div};\Omega)$, $v \in H^1(\Omega)$:

$$\int_\Omega v\nabla\cdot\boldsymbol{q}\,dx = -\int_\Omega \boldsymbol{q}\cdot\nabla v\,dx + \langle \boldsymbol{q}\cdot\mathfrak{n}, \gamma_0(v)\rangle_{H^{1/2}(\partial\Omega)}\,.$$

2) For $\boldsymbol{q} \in H(\mathrm{div};\Omega)$, $v \in H^1_{\Gamma^c}(\Omega)$:

$$\int_\Omega v\nabla\cdot\boldsymbol{q}\,dx = -\int_\Omega \boldsymbol{q}\cdot\nabla v\,dx + \big\langle \boldsymbol{q}\cdot\mathfrak{n}, \gamma_{0,\Gamma}(v)\big\rangle_{H^{1/2}_{00}(\Gamma)}\,.$$

Proof See Problem 6.7. □

6.2.2 Mixed Variational Formulations

Now we return to the boundary value problem (6.30)–(6.34). Note that expositions of this subject in other textbooks often use the variables $\boldsymbol{p} = \boldsymbol{K}\nabla u$ or $\boldsymbol{p} = \boldsymbol{K}\nabla u - \boldsymbol{c}u$ leading to sign changes in the equations to be developed. Here we prefer the real physical quantities as unknowns.

We keep the general assumption of $\boldsymbol{K} = (k_{ij})$ in (3.16) and in particular (3.17), i.e., $\boldsymbol{K}(x)$ is uniformly bounded and (symmetric), positive definite in $x \in \Omega$. Therefore, $\boldsymbol{L}$ defined by

$$\boldsymbol{L}(x) := \boldsymbol{K}^{-1}(x) \quad \text{for all } x \in \Omega \tag{6.40}$$

has the same properties. The weak formulations treated till now are based on (6.29) with the Dirichlet boundary condition (6.34) as side condition. Now, we start with (6.31) and consider (6.30) (and possible flux boundary conditions) as side conditions, i.e., (assuming for the beginning that $r = 0$) we search for $\boldsymbol{p}$ in

$$W_g^f := \{\boldsymbol{q} \in \mathcal{H}(\mathrm{div};\Omega) \,|\, \nabla\cdot\boldsymbol{q} = f \text{ in } \Omega, \quad -\boldsymbol{q}\cdot\mathfrak{n} = \widetilde{g_1} \text{ on } \widetilde{\Gamma}_1\} \tag{6.41}$$

with the space $\mathcal{H}(\mathrm{div};\Omega)$ to be chosen appropriately (see later). We know that for

$$\boldsymbol{q} \in H^1(\Omega)^d := \{\boldsymbol{q} = (q_i)_i \,|\, q_i \in H^1(\Omega)\},$$

$\nabla\cdot\boldsymbol{q} \in L^2(\Omega)$ and $\boldsymbol{q}\cdot\mathfrak{n} \in H^{1/2}(\partial\Omega) \subset L^2(\partial\Omega)$ (see [66, Theorem 8.7]). We only need a larger complete subspace of $L^2(\Omega)^d$ such that $\nabla\cdot\boldsymbol{q}$ exists weakly in $L^2(\Omega)$. This motivates the use of the space $\big(H(\mathrm{div};\Omega), \|\cdot\|_{H(\mathrm{div})}\big)$ introduced in (6.38). The trace mapping (6.39) is sufficient to treat both Dirichlet- and flux boundary conditions. As we will see later, in the case of mixed boundary conditions ($|\widetilde{\Gamma}_2| \neq 0$) more regularity for the variational formulation is needed, namely, the subspace

$$\mathcal{H}(\mathrm{div};\Omega) := \{\boldsymbol{q} \in H(\mathrm{div};\Omega) \;\big|\; \boldsymbol{q}\cdot\mathfrak{n} \in L^2(\widetilde{\Gamma}_2)\}$$

with the norm

$$\|\boldsymbol{q}\|_{\mathcal{H}(\mathrm{div})} := (\|\boldsymbol{q}\|^2_{H(\mathrm{div})} + \|\boldsymbol{q}\cdot\mathfrak{n}\|^2_{L^2(\widetilde{\Gamma}_2)})^{1/2},$$

which is complete with respect to its norm.

Thus under the assumption $f \in L^2(\Omega)$, $\widetilde{g}_1 \in H^{-1/2}(\widetilde{\Gamma}_1)$, the set W^f_g is well defined and an affine subspace of $\mathcal{H}(\mathrm{div};\Omega)$. All the newly introduced (Sobolev) spaces are Hilbert spaces with obvious scalar products, inducing the given norms. Therefore, we can identify the spaces with their duals as discussed in Section 6.1.

In (6.41), we anticipated already that now the flux boundary condition becomes *essential* (has to be prescribed in the ansatz space) and the Dirichlet boundary condition becomes *natural* (can be incorporated in the variational formulation). Indeed, the following calculations show that this is the case. Namely, it holds

$$\boldsymbol{p} = -\boldsymbol{K}\nabla u + \boldsymbol{c}u \qquad \Longleftrightarrow \qquad \boldsymbol{L}\boldsymbol{p} = -\nabla u + \tilde{\boldsymbol{c}}u, \quad \text{where } \tilde{\boldsymbol{c}} := \boldsymbol{L}c.$$

We test the right relationship with

$$\boldsymbol{q} \in W^0_0 = \{\boldsymbol{q} \in \mathcal{H}(\mathrm{div};\Omega) \mid \nabla\cdot\boldsymbol{q} = 0 \text{ in } \Omega, \quad \boldsymbol{q}\cdot\mathfrak{n} = 0 \text{ on } \widetilde{\Gamma}_1\}.$$

The meaning of the first condition is clear because of $\nabla\cdot\boldsymbol{q} \in L^2(\Omega)$; the second one is to be understood in the sense of the trace operator (6.39), i.e., $\gamma_{\mathfrak{n},\widetilde{\Gamma}_1}(\boldsymbol{q}) = 0$.

Thus, we receive

$$\int_\Omega \boldsymbol{L}\boldsymbol{p}\cdot\boldsymbol{q}dx = -\int_\Omega \nabla u\cdot\boldsymbol{q}dx + \int_\Omega \tilde{\boldsymbol{c}}u\cdot\boldsymbol{q}dx =: I_1 + I_2,$$

and integration by parts (see Thm. 6.13) yields

$$I_1 = \int_\Omega u\nabla\cdot\boldsymbol{q}dx - \langle\boldsymbol{q}\cdot\mathfrak{n}, \gamma_0(u)\rangle_{H^{1/2}(\partial\Omega)}. \tag{6.42}$$

Assuming $u|_{\widetilde{\Gamma}_i} \in H^{1/2}_{00}(\widetilde{\Gamma}_i)$, $i = 1,2,3$, we have

$$\begin{aligned}\langle\boldsymbol{q}\cdot\mathfrak{n}, \gamma_0(u)\rangle_{H^{1/2}(\partial\Omega)} &= \sum_{i=1}^{3}\left\langle\boldsymbol{q}\cdot\mathfrak{n}, \gamma_{0,\widetilde{\Gamma}_i}(u)\right\rangle_{H^{1/2}_{00}(\widetilde{\Gamma}_i)}\\ &= \sum_{i=2}^{3}\left\langle\boldsymbol{q}\cdot\mathfrak{n}, \gamma_{0,\widetilde{\Gamma}_i}(u)\right\rangle_{H^{1/2}_{00}(\widetilde{\Gamma}_i)}.\end{aligned}$$

Returning to the more suggestive integral notation and taking the equation for $\boldsymbol{q}$ and (6.33), (6.34) into account leads to

$$I_1 = -\int_{\widetilde{\Gamma}_2}(\widetilde{g}_2 + \boldsymbol{p}\cdot\mathfrak{n})\alpha^{-1}\boldsymbol{q}\cdot\mathfrak{n}\,d\sigma - \int_{\widetilde{\Gamma}_3}\widetilde{g}_3\boldsymbol{q}\cdot\mathfrak{n}\,d\sigma$$

provided that $\alpha(x) \neq 0$ for all $x \in \widetilde{\Gamma}_2$ to distinguish the *mixed boundary condition* (6.33) from the *flux boundary condition* (6.32). Thus, using the notation

$$a((\boldsymbol{p},u),\boldsymbol{q}) := \int_\Omega \boldsymbol{L}\boldsymbol{p}\cdot\boldsymbol{q}\,dx - \int_\Omega \boldsymbol{L}\boldsymbol{c}u\cdot\boldsymbol{q}\,dx + \int_{\widetilde{\Gamma}_2} \alpha^{-1}\boldsymbol{p}\cdot\boldsymbol{n}\boldsymbol{q}\cdot\boldsymbol{n}\,d\sigma,$$

$$\ell(\boldsymbol{q}) := -\int_{\widetilde{\Gamma}_2} \alpha^{-1}\widetilde{g}_2\boldsymbol{q}\cdot\boldsymbol{n}\,d\sigma - \int_{\widetilde{\Gamma}_3} \widetilde{g}_3\boldsymbol{q}\cdot\boldsymbol{n}\,d\sigma,$$

the new variational problem reads

$$\text{Find } (\boldsymbol{p},u) \in W_g^f \times L^2(\Omega) \text{ such that } \quad a((\boldsymbol{p},u),\boldsymbol{q}) = \ell(\boldsymbol{q}) \quad \text{for all } \boldsymbol{q} \in W_0^0. \tag{6.43}$$

If we additionally assume that

$$\alpha^{-1} \in L^\infty(\widetilde{\Gamma}_2), \qquad \widetilde{g}_2 \in L^2(\widetilde{\Gamma}_2), \qquad \widetilde{g}_3 \in H^{1/2}(\widetilde{\Gamma}_3), \qquad \boldsymbol{c} \in L^\infty(\Omega)^d,$$

all terms are well defined. Additionally we assume $\alpha > 0$.

In the pure diffusion case $\boldsymbol{c} = \boldsymbol{0}$, the unknown function u does not appear. In this case the bilinear form $a(\boldsymbol{p},\boldsymbol{q})$ is symmetric and positive definite on W_0^0, and the variational equation (6.43) is equivalent to the following minimization problem:

$$\text{Find } \boldsymbol{p} \in W_g^f \text{ such that } \quad I(\boldsymbol{p}) = \min\{I(\boldsymbol{q}) | \boldsymbol{q} \in W_g^f\}, \tag{6.44}$$

where

$$\begin{aligned} I(\boldsymbol{q}) : &= \frac{1}{2}a(\boldsymbol{q},\boldsymbol{q}) - \ell(\boldsymbol{q}) \\ &= \frac{1}{2}\int_\Omega \boldsymbol{q}\cdot\boldsymbol{L}\boldsymbol{q}\,dx + \int_{\widetilde{\Gamma}_2} \alpha^{-1}(\boldsymbol{q}\cdot\boldsymbol{n})^2\,d\sigma + \int_{\widetilde{\Gamma}_2} \alpha^{-1}\widetilde{g}_2\boldsymbol{q}\cdot\boldsymbol{n}\,db + \int_{\widetilde{\Gamma}_3} \widetilde{g}_3\boldsymbol{q}\cdot\boldsymbol{n}\,d\sigma \end{aligned} \tag{6.45}$$

(note Lemma 2.3, Remark 2.4).

The formulation of Chapter 3 for the variable u is called *primal* and the above one (for the variable $\boldsymbol{p}$) *dual*. In fact, the minimization problems are primal and dual in the sense of convex analysis. In mechanical applications the primal approach is called *displacement approach*, the functional (from (2.14)) *potential energy*. Here one speaks of the *equilibrium approach* and the functional (6.45) is called *complementary energy* (if $\widetilde{\Gamma}_3 = \partial\Omega$).

A necessary solvability condition is the existence of $\bar{\boldsymbol{q}} \in \mathcal{H}(\mathrm{div};\Omega)$ such that

$$\nabla\cdot\bar{\boldsymbol{q}} = f \quad \text{in } \Omega, \qquad -\bar{\boldsymbol{q}}\cdot\boldsymbol{n} = \widetilde{g}_1 \quad \text{on } \widetilde{\Gamma}_1. \tag{6.46}$$

The following lemma gives a sufficient condition.

Lemma 6.14 *For $f \in L^2(\Omega)$, $\widetilde{g}_1 \in H^{-1/2}(\widetilde{\Gamma}_1)$ and assuming the compatibility conditions*

$$\int_\Omega f\,dx + \int_{\partial\Omega} \widetilde{g}_1\,d\sigma = 0 \tag{6.47}$$

for the case $\widetilde{\Gamma}_2 = \widetilde{\Gamma}_3 = \emptyset$, there exist a solution of problem (6.46) *and a constant $C > 0$ such that*

$$\|\boldsymbol{q}\|_{H(\mathrm{div})} \le C(\|f\|_0 + \|\widetilde{g}_1\|_{H^{-1/2}(\widetilde{\Gamma}_1)}).$$

Proof First let $\boldsymbol{q}_1 \in H(\text{div};\Omega)$ be such that

$$\|\boldsymbol{q}_1\|_{H(\text{div})} \le C_1 \|\widetilde{g}_1\|_{H^{-1/2}(\widetilde{\Gamma}_1)},$$

see Problem 6.6.

Next let $w \in V := \{v \in H^1(\Omega) \mid v = 0 \text{ on } \widetilde{\Gamma}_2 \cup \widetilde{\Gamma}_3\}$ be a solution of

$$\alpha(w,v) := \int_\Omega \nabla w \nabla v dx = \ell(v) := \int_\Omega f v dx - \int_\Omega \nabla \cdot \boldsymbol{q}_1 v dx \quad \text{for all } v \in V.$$

The existence is assured by Theorem 3.16 and Problem 3.8. Furthermore it holds that

$$|w|_1 \le C_2 \left(\|f\|_0 + \|\nabla \cdot \boldsymbol{q}_1\|_0\right)$$

for some constant $C_2 > 0$. Therefore

$$\boldsymbol{q}_2 := \nabla w \in H(\text{div};\Omega) \quad \text{with} \quad \nabla \cdot \boldsymbol{q}_2 = f - \nabla \cdot \boldsymbol{q}_1, \quad \boldsymbol{q}_2 \cdot \mathfrak{n} = 0 \quad \text{on } \widetilde{\Gamma}_1.$$

Thus

$$\boldsymbol{q} := \boldsymbol{q}_1 + \boldsymbol{q}_2$$

satisfies (6.46) and

$$\|\boldsymbol{q}\|^2_{H(\text{div})} \le C_2^2 \left(\|\boldsymbol{q}\|_0 + \|\nabla \cdot \boldsymbol{q}_1\|_0\right)^2 + \|f - \nabla \cdot \boldsymbol{q}_1\|_0^2 \le C^2 \left(\|f\|_0 + \|\widetilde{g}_1\|_{H^{-1/2}(\widetilde{\Gamma}_1)}\right)^2$$

for an appropriate constant $C > 0$. □

Therefore for $\boldsymbol{c} = \boldsymbol{0}$ there exists a unique solution $\boldsymbol{q} \in W_g^f$, assured by the Lax–Milgram Theorem (Theorem 3.1). A definition of u is possible by (6.42), but now for test functions

$$\boldsymbol{q} \in \widetilde{W} := \{\boldsymbol{q} \in \mathcal{H}(\text{div};\Omega) \mid \boldsymbol{q} \cdot \mathfrak{n} = 0 \text{ on } \widetilde{\Gamma}_1\}$$

leading to

$$\begin{aligned} &\int_\Omega u \nabla \cdot \boldsymbol{q} dx + \int_\Omega u \tilde{\boldsymbol{c}} \cdot \boldsymbol{q} dx \\ &= \int_\Omega \boldsymbol{L}\boldsymbol{p} \cdot \boldsymbol{q} dx + \int_{\Gamma_2} \alpha^{-1} \boldsymbol{p} \cdot \mathfrak{n} \cdot \boldsymbol{q} \cdot \mathfrak{n} d\sigma - \ell(\boldsymbol{q}) =: \tilde{\ell}(\boldsymbol{q}) \quad \text{for all } \boldsymbol{q} \in \widetilde{W}. \end{aligned} \tag{6.48}$$

We will investigate the solvability with respect to u only in the pure diffusive case.

Lemma 6.15 *Assume* $\boldsymbol{c} = \boldsymbol{0}$. *Under the assumptions of Lemma 6.14, there is a solution* $u \in L^2(\Omega)$ *of* (6.48).

Proof We have $\tilde{\ell} \in \widetilde{W}'$ for a fixed (solution) $\boldsymbol{p} \in W_g^f$. According to Lemma 6.14, the operator

$$\text{div} : \widetilde{W} \to L^2(\Omega)$$

is continuous and surjective. From the Closed Range Theorem (see Appendix A.4) we conclude for the adjoint operator

$$\operatorname{div}^T : L^2(\Omega) \to \widetilde{W}' : \quad \operatorname{im}(\operatorname{div}^T) = \ker(\operatorname{div})^\perp,$$

where $\cdot^\perp$ denotes the annihilator (see (A4.31)). By definition $\tilde{\ell}(\boldsymbol{q}) = 0$ for $\boldsymbol{q} \in W_0^0$, i.e., $\tilde{\ell} \in \ker(\operatorname{div})^\perp$ and thus (6.46) is solvable. □

If one also wants to include the reaction term, the affine subspace W_g^f has to be modified to

$$W_g^f := \{(\boldsymbol{q}, v) \in \mathcal{H}(\operatorname{div};\Omega) \times L^2(\Omega) \mid \nabla \cdot \boldsymbol{q} + rv = f \quad \text{in } \Omega, \quad -\boldsymbol{q} \cdot \boldsymbol{n} = \widetilde{g}_1 \quad \text{on } \widetilde{\Gamma}_1\}$$

and the bilinear form a according to (6.42) to

$$a((\boldsymbol{p}, u), (\boldsymbol{q}, v)) := a((\boldsymbol{p}, u), \boldsymbol{q}) + \int_\Omega ruv dx. \tag{6.49}$$

In the case $\boldsymbol{c} = \boldsymbol{0}$, the unique existence of a solution $\boldsymbol{p}$ can be justified by the Lax–Milgram Theorem (Theorem 3.1), since a is coercive on W_0^0 assuming, for example, $r(x) \geq r_0 > 0$ and $\alpha \in L^\infty(\widetilde{\Gamma}_2)$ to assure for some constant $C > 0$:

$$\begin{aligned} a((\boldsymbol{p}, u), (\boldsymbol{p}, u)) &= \|\boldsymbol{L}^{1/2}\boldsymbol{p}\|_0^2 + \|\alpha^{-1/2}\boldsymbol{p} \cdot \boldsymbol{n}\|^2_{L^2(\widetilde{\Gamma}_2)} + \|\sqrt{r}u\|_0^2 \\ &\geq C\left(\|\boldsymbol{p}\|_0^2 + \|\nabla \cdot \boldsymbol{p}\|_0^2 + \|\boldsymbol{p} \cdot \boldsymbol{n}\|^2_{L^2(\widetilde{\Gamma}_2)} + \|u\|_0^2\right), \end{aligned} \tag{6.50}$$

where the expression in brackets is the squared norm of the space W_0^0, and the inequality is obtained due to

$$\nabla \cdot \boldsymbol{p} = -ru \quad \Longrightarrow \quad (\nabla \cdot \boldsymbol{p})^2 = -ru(\nabla \cdot \boldsymbol{p}) = r^2u^2.$$

Similarly there are unique existence results for further cases including $\boldsymbol{c} \neq \boldsymbol{0}$ (with structural or size condition) similar to Theorem 3.16 ff. (Problem 6.10). In general, existence also can be concluded from the existence of a solution $u \in H^1(\Omega)$ of the primal formulation, for simplicity only with Dirichlet boundary conditions (see Chapter 3):

Setting $\boldsymbol{p} := -\boldsymbol{K}\nabla u + \boldsymbol{c}u$ implies $\nabla \cdot \boldsymbol{p} \in L^2(\Omega)$ and both $\nabla \cdot \boldsymbol{p} + ru = f$ in $L^2(\Omega)$ and the variational equation holds, i.e., $(\boldsymbol{p}, u)$ is also a solution of the dual problem, with $u \in H^1(\Omega)$ (see Problem 6.11).

Consider a conforming Galerkin discretization, for example, for pure Dirichlet boundary conditions and $r = 0$. The dual approach has the advantage that the Dirichlet boundary condition becomes natural, but the severe disadvantage that $W_h \subset W_0^0$ means that the approximation functions must be exactly divergence-free, which limits their construction.

A possibility to circumvent this disadvantage consists in incorporating the requirement

$$\int_\Omega (\nabla \cdot \boldsymbol{p} + ru)v\, dx = \int_\Omega fv dx \quad \text{for all } v \in L^2(\Omega)$$

into the variational equation, which only keeps the flux boundary condition as "hard" equality constraint in the space. This means that instead of W_g^f the affine space

$$W_g := \{(\boldsymbol{q}, v) \in \mathcal{H}(\mathrm{div}; \Omega) \times L^2(\Omega) \mid \boldsymbol{q} \cdot \boldsymbol{n} = \widetilde{g}_1 \text{ on } \widetilde{\Gamma}_1\}$$

is used. The variational equations (6.43) (or (6.49)) have to be modified—since now $\nabla \cdot \boldsymbol{q}$ cannot be substituted by the imposed equation, i.e., they change to

$$\begin{aligned} a_1((\boldsymbol{p}, u), \boldsymbol{q}) &:= \int_\Omega (\boldsymbol{L}\boldsymbol{p} \cdot \boldsymbol{q} - u\nabla \cdot \boldsymbol{q} - \boldsymbol{L}\boldsymbol{c}u \cdot \boldsymbol{q})\, dx + \int_{\widetilde{\Gamma}_2} \alpha^{-1}(\boldsymbol{p} \cdot \boldsymbol{n})(\boldsymbol{q} \cdot \boldsymbol{n}) d\sigma \\ &= \ell_1(\boldsymbol{q}), \qquad \text{with } \ell_1 := \ell \text{ from (6.43)} \end{aligned} \tag{6.51}$$

and additionally

$$a_2((\boldsymbol{p}, u), v) := -\int_\Omega (\nabla \cdot \boldsymbol{p} v + ruv) dx = -\int_\Omega fv dx =: \ell_2(v). \tag{6.52}$$

This new variational formulation is called *mixed*, so here *mixed dual*. It has no coercive structure anymore. In fact, setting (and changing the notations)

$$X := L^2(\Omega), \qquad M_g := \{\boldsymbol{q} \in \mathcal{H}(\mathrm{div}; \Omega) \mid \boldsymbol{q} \cdot \boldsymbol{n} = \widetilde{g}_1 \text{ on } \widetilde{\Gamma}_1\}, \qquad M := M_0 \tag{6.53}$$

(i.e., we have $W_g = M_g \times X$) and the bilinear forms (where vector fields belong to M_g or M and scalar functions to X)

$$\begin{aligned} a(\boldsymbol{p}, \boldsymbol{q}) &:= \int_\Omega \boldsymbol{L}\boldsymbol{p} \cdot \boldsymbol{q} dx + \int_{\widetilde{\Gamma}_2} \alpha^{-1}(\boldsymbol{p} \cdot \boldsymbol{n})(\boldsymbol{q} \cdot \boldsymbol{n}) d\sigma, \qquad & b(\boldsymbol{q}, u) &:= -\int_\Omega u\nabla \cdot \boldsymbol{q} dx, \\ c(u, \boldsymbol{q}) &:= -\int_\Omega \boldsymbol{L}\boldsymbol{c}u \cdot \boldsymbol{q} dx, & d(u, v) &:= \int_\Omega ruv dx, \end{aligned} \tag{6.54}$$

the variational equation on $M_g \times X$ has the form

Find $(\boldsymbol{p}, u) \in M_g \times X$ such that

$$\begin{aligned} a(\boldsymbol{p}, \boldsymbol{q}) + b(\boldsymbol{q}, u) + c(u, \boldsymbol{q}) &= \ell_1(\boldsymbol{q}) \quad \text{for all } \boldsymbol{q} \in M, \\ b(\boldsymbol{p}, v) - d(u, v) &= \ell_2(v) \quad \text{for all } v \in X. \end{aligned} \tag{6.55}$$

We call a variational problem of this structure to be of *saddle point form*. In general, such a problem is not coercive and requires the general conditions of Section 6.1.

More specifically, we call the problem (6.55) with (6.54) and ℓ_1, ℓ_2 according to (6.43), (6.52) respectively, i.e., the dual mixed formulation of (6.30)–(6.34), or also more general (6.30)–(6.34) itself as *Darcy equation*. As seen, this formulation not only includes the model from (0.14)–(0.15) for single phase flow in porous media, with the fluid velocity as intrinsic variable, but any model of diffusion/dispersion, convection and (single component) reaction, with the flux density as intrinsic variable.

Writing (6.55) in the equivalent operator form (see (6.14) and (6.15)), then

$$\begin{aligned} A\boldsymbol{p} + B^T u + Cu &= \ell_1, \\ B\boldsymbol{p} - Du &= \ell_2, \end{aligned} \tag{6.56}$$

where $B^T : X$ (or X'') $\to M$ (or M') is the adjoint operator to $B : M \to X$ (or X').

The notion *saddle point problem* stems from the fact that in the pure diffusive case ($\boldsymbol{c} = \boldsymbol{0}$, $r = 0$) not only the primal and dual minimization problems hold, but—as in general in quadratic or convex optimization—the corresponding *Lagrange multiplier formulation* for the dual problem:

Find $\boldsymbol{p} \in M_g$ and (a *Lagrange multiplier*) u such that the *Lagrange functional*

$$L(\boldsymbol{q}, v) := I(\boldsymbol{q}) + \int_\Omega (\nabla \cdot \boldsymbol{q} - f) v dx$$

with I from (6.45) has a *saddle point* in $(\boldsymbol{p}, u)$, i.e.,

$$L(\boldsymbol{p}, v) \le L(\boldsymbol{p}, u) \le L(\boldsymbol{q}, u) \quad \text{for all } \boldsymbol{q} \in M_g, v \in X.$$

In a saddle point necessarily the constraint $\nabla \cdot \boldsymbol{q} = f$ holds as otherwise for the sequence $v_t := t(\nabla \cdot \boldsymbol{q} - f)$ we would have $L(\boldsymbol{p}, v_t) \to +\infty$ for $t \to +\infty$.

In a mechanics context this formulation is called the *Hellinger–Reissner principle*. The property of a saddle point to be a critical point, i.e., where the (Gâteaux) derivatives according to $\boldsymbol{q}$ (for fixed u) and v (for fixed $\boldsymbol{p}$) vanish (*Euler–Lagrange equations* in the calculus of variations or *Karush–Kuhn–Tucker conditions* in optimization), leads then to (6.55).

In the primal formulation, only the Dirichlet boundary condition acts as a constraint. The corresponding Lagrange formulation (in the pure diffusive case) gives rise to the *primal mixed* version of the form:

Find $u \in V := H^1(\Omega)$, $\mu \in H^{-1/2}(\Gamma_3)$ (μ somewhat represents the flux density on the boundary) such that

$$\begin{aligned} a(u, v) - \langle \mu, v \rangle_{H^{1/2}(\Gamma_3)} &= \ell(v) && \text{for all } v \in V, \\ \langle \eta, u - g_3 \rangle_{H^{1/2}(\Gamma_3)} &= 0 && \text{for all } \eta \in H^{-1/2}(\Gamma_3), \end{aligned} \tag{6.57}$$

for a and ℓ as developed in Section 3.1. This way of treating Dirichlet boundary conditions is sometimes attributed to Joachim Nitsche.

In Chapter 7, we will study Galerkin approaches based on dual (mixed) formulations, i.e., one of the basic spaces is a subspace of $H(\text{div}; \Omega)$ and approximating finite-dimensional subspaces $V_h \subset V$ are required. According to the approach, which is well known from conforming finite elements, V_h will be defined locally based on a partition $\mathcal{T}_h$ such that an analogue of Theorem 3.21 is necessary. This is given by using the notation of Sections 3.3, 4.2 (with the corresponding modifications for the case of vector fields).

Theorem 6.16 *Let $\mathcal{T}_h$ be a partition satisfying (T1)–(T6). Suppose $P_K \subset H(\mathrm{div}; K) \cap C(K)^d$ for all $K \in \mathcal{T}_h$. Then $\boldsymbol{q} \in V_h$ can be characterized as follows:*

$$\boldsymbol{q} \in H(\mathrm{div}; \Omega) \iff [\![\boldsymbol{q}]\!]_F = 0 \quad \text{for all } F = K \cap K' \in \mathcal{F}_h, \tag{6.58}$$

where $\mathcal{F}_h$ is the set of all interior faces of all elements $K \in \mathcal{T}_h$ and $[\![\boldsymbol{q}]\!]_F = \boldsymbol{q}|_K \cdot \boldsymbol{n}_K + \boldsymbol{q}|_{K'} \cdot \boldsymbol{n}_{K'}$ denotes the jump of the normal components of $\boldsymbol{q}$ across the edge F common to the elements $K, K' \in \mathcal{T}_h$ (see (4.11)).

Proof We have $\nabla \cdot \boldsymbol{q}|_K \in L^2(K)$ for each $K \in \mathcal{T}_h$. Set

$$g(x) := \nabla \cdot \boldsymbol{q}(x) \quad \text{for } x \in \mathrm{int}(K),\ K \in \mathcal{T}_h,$$

then $g \in L^2(\Omega)$ (as only a set of measure 0 is missing, where the function can be defined arbitrarily). Integrating by parts in $H(\mathrm{div}; K)$, we get for $\varphi \in C_0^\infty(\Omega)$:

$$\begin{aligned}\int_\Omega g\varphi dx &= \sum_{K \in \mathcal{T}_h} \int_K g\varphi dx = \sum_{K \in \mathcal{T}_h} \int_K \nabla \cdot \boldsymbol{q}\varphi dx \\ &= -\sum_{K \in \mathcal{T}_h} \int_K \boldsymbol{q} \cdot \nabla\varphi\, dx + \sum_{K \in \mathcal{T}_h} \int_{\partial K} \boldsymbol{q} \cdot \boldsymbol{n}\varphi d\sigma \\ &= -\int_\Omega \boldsymbol{q} \cdot \nabla\varphi dx + \sum_{K \in \mathcal{T}_h} \int_{\partial K} \boldsymbol{q} \cdot \boldsymbol{n}\varphi d\sigma.\end{aligned}$$

If the jump condition in (6.58) holds, then $\nabla \cdot \boldsymbol{q} = g$ because the second term vanishes. Indeed, let $K \in \mathcal{T}_h$ and $F \subset \partial K$ be a face. If $F \subset \partial\Omega$, then $\varphi = 0$. Otherwise, due to (T5), there is exactly one neighbouring element K' of K such that $F := K \cap K'$ is a common face. Therefore, the integral contributions cancel.

If otherwise $\boldsymbol{q} \in H(\mathrm{div}; \Omega)$, then this sum has to vanish for each $\varphi \in C_0^\infty(\Omega)$ (cf. (2.18)). Choosing functions φ such that

$$\mathrm{supp}\, \varphi \cap \bigcup_{K \in \mathcal{T}_h} \partial K \subset F, \tag{6.59}$$

it reduces to

$$\int_F [\![\boldsymbol{q}]\!]_F\, \varphi d\sigma = 0.$$

Now we suppose that there is a point $x \in F$ such that $[\![\boldsymbol{q}]\!]_F(x) \neq 0$, say positive. Since $[\![\boldsymbol{q}]\!]_F$ is continuous on F by assumption we can find a d-ball $B_\delta(x)$, $\delta > 0$, such that

$$B_\delta(x) \cap \bigcup_{K \in \mathcal{T}_h} \partial K \subset F \quad \text{and} \quad [\![\boldsymbol{q}]\!]_F \text{ is positive on } B_\delta(x) \cap F.$$

It remains to choose a function φ among the functions satisfying (6.59) such that $\varphi(x) = 1$ and $\operatorname{supp}\varphi = B_\delta(x)$, leading to the contradictory property

$$\int_F [\![\boldsymbol{q}]\!]_F\, \varphi d\sigma > 0.$$

□

On the one hand, Theorem 6.16 shows that a Galerkin method based on the (mixed) dual formulation is locally mass conservative with respect to the underlying partition (see Section 7.6) due to the interface continuity of the normal flux density. On the other hand, as seen in Chapter 3, such a continuity requirement influences the size of the support of the basis function and thus the sparsity pattern of the discrete problem to be solved. Therefore (and for other reasons to be seen in Section 7.5), it might be desirable to reformulate the continuity requirements of the basic spaces. This leads to so-called *hybrid formulations*:

Let $\mathcal{T}$ be a partition of Ω with the aforementioned properties, $\mathcal{F}_K$ denotes the set of faces of $K \in \mathcal{T}$ and

$$\mathcal{F} := \bigcup_{K\in\mathcal{T}} \mathcal{F}_K \setminus \partial\Omega$$

the set of all interior faces of all elements $K \in \mathcal{T}$. The *dual mixed hybrid* formulation with respect to $\mathcal{T}$ enlarges the space M_g to

$$\widetilde{M}_g := \{\boldsymbol{q} \in L^2(\Omega)^d \mid \boldsymbol{q}|_K \in \mathcal{H}(\mathrm{div}; K) \text{ for all } K \in \mathcal{T}, \quad -\boldsymbol{q}\cdot\mathfrak{n} = \widetilde{g}_1 \text{ on } \widetilde{\Gamma}_1\}.$$

Here $\mathcal{H}(\mathrm{div}; K)$ is to be understood to require $\boldsymbol{q}\cdot\mathfrak{n}|_{\partial K} \in L^2(\partial K)$. Now the continuity of the normal flux densities at the inter-element boundaries does not hold anymore, and the bilinear form b should be modified in the first equation of (6.55) to

$$\widetilde{b}(\boldsymbol{q}, u) := -\int_\Omega u\nabla\cdot\boldsymbol{q}dx + \sum_{F\in\mathcal{F}} \int_F [\![\boldsymbol{q}\cdot\mathfrak{n}]\!] u d\sigma,$$

as integration by parts can only be performed locally on an element. Keeping the form (6.55) of the variational equation with $\widetilde{b}$ instead of b reintroduces the inter-element continuity of the normal flux densities, i.e., $\boldsymbol{p} \in \mathcal{H}(\mathrm{div}; \Omega)$ can be seen by localizing a test function v to the vicinity of an element face F.

The bilinear form $\widetilde{b}$ and the above reasoning is only well defined if X is replaced by a smaller space ensuring $u|_F \in H^{1/2}(F)$ for $F \in \mathcal{F}$, e.g.,

$$\widetilde{X} := H^1(\Omega).$$

Alternatively, the non-existing $u|_F$, $F \in \mathcal{F}$, can be incorporated as a new unknown λ into the problem

$$\lambda \in L^2(\widetilde{\mathcal{F}}) =: Y, \quad \widetilde{\mathcal{F}} := \bigcup_{F\in\mathcal{F}} F,$$

where instead of the above bilinear form $\widetilde{b}$:

$$\widetilde{b}(\boldsymbol{q}, u, \lambda) := -\int_\Omega u\nabla \cdot \boldsymbol{q}dx + \sum_{F\in\mathcal{F}} \int_F [\![\boldsymbol{q}\cdot\mathfrak{n}]\!]\lambda d\sigma.$$

The second equation of (6.55) is kept as it is, and to reintroduce the inter-element continuity of $\boldsymbol{p}\cdot\mathfrak{n}$ the following equations are added:

$$e(\boldsymbol{p}, \mu) := \sum_{F\in\mathcal{F}} \int_F [\![\boldsymbol{p}\cdot\mathfrak{n}]\!]\mu d\sigma = 0 \quad \text{for } \mu \in Y.$$

Then the variational problem now has the form:

Find $(\boldsymbol{p}, u, \lambda) \in \widetilde{M}_g \times X \times Y$ such that

$$\begin{aligned} a(\boldsymbol{p}, \boldsymbol{q}) + b(\boldsymbol{q}, u) + c(u, \boldsymbol{q}) + e(\boldsymbol{q}, \lambda) &= \ell_1(\boldsymbol{q}) && \text{for all } \boldsymbol{q} \in \widetilde{M}_0, \\ b(\boldsymbol{p}, v) - d(u, v) &= \ell_2(v) && \text{for all } v \in X, \\ e(\boldsymbol{p}, \mu) &= 0 && \text{for all } \mu \in Y. \end{aligned} \tag{6.60}$$

Also a *dual hybrid* formulation is possible. The space W_g^f in (6.49) has to be modified accordingly, and the two ways from above can be followed, e.g., substituting X by $\widetilde{X} := H^1(\Omega)$ and considering the variational equation (with a from (6.49)):

$$\begin{aligned} a((\boldsymbol{p}, u), (\boldsymbol{q}, v)) + \sum_{F\in\mathcal{F}} \int_F [\![\boldsymbol{q}\cdot\mathfrak{n}]\!]u d\sigma &= \ell(\boldsymbol{q}) && \text{for all } \boldsymbol{q} \in \widetilde{W}_0^0, \\ \sum_{F\in\mathcal{F}} \int_F [\![\boldsymbol{p}\cdot\mathfrak{n}]\!]v d\sigma &= 0 && \text{for all } v \in \widetilde{X}. \end{aligned}$$

Finally in a *primal (mixed) hybrid* formulation the affine space $V_g := \{v \in H^1(\Omega) \mid v|_{\Gamma_3} = g_3\}$ is relaxed to

$$\widetilde{V_g} := \{v \in L^2(\Omega) \mid v|_K \in H^1(K) \text{ for all } K \in \mathcal{T}, v|_{\Gamma_3} = g_3\},$$

and the jumps of $v \in V$ across $F \in \mathcal{F}$ are incorporated into the variational formulation (Problem 6.12).

We now want to discuss well-posedness results for the general formulation (6.55)/(6.56). We focus on the classical case of pure diffusion, i.e., $\boldsymbol{c} = \boldsymbol{0}$, $r = 0$, where the theory goes back to Franco Brezzi.

Theorem 6.17 *Let M, X be Hilbert spaces, $a: M \times M \to \mathbb{R}$ and $b: M \times X \to \mathbb{R}$ continuous bilinear forms, A and B denote the corresponding continuous operators with the continuity constants $C_a, C_b > 0$. Let $N := \ker B$. The requirements*

1) a satisfies the conditions (NB1), (NB2) *(from Theorem 6.2) on $N \times N$ with a constant $\alpha > 0$*

2) $\widetilde{b}$ satisfies the condition (NB2) *on $X \times N^\perp$ with a constant $\beta > 0$, where $\widetilde{b}(v, \boldsymbol{q}) := b(\boldsymbol{q}, v)$*

are equivalent to the unique existence of a solution $(\boldsymbol{p}, u) \in M \times X$ *of the variational equations*

$$
\begin{aligned}
a(\boldsymbol{p}, \boldsymbol{q}) + b(\boldsymbol{q}, u) &= \ell_1(\boldsymbol{q}) \qquad \text{for all } \boldsymbol{q} \in M,\\
b(\boldsymbol{p}, v) &= \ell_2(v) \qquad \text{for all } v \in X
\end{aligned}
\tag{6.61}
$$

for arbitrary $\ell_1 \in M'$, $\ell_2 \in X'$, *and then there is a constant* $C = C(C_a, C_b, \alpha^{-1}, \beta^{-1}) > 0$ *such that the following stability estimate holds:*

$$
\|\boldsymbol{p}\|_M + \|u\|_X \le C(\|\ell_1\|_{M'} + \|\ell_2\|_{X'}).
$$

Proof We rewrite (6.61) equivalently as

$$
A\boldsymbol{p} + B^T u = \ell_1, \qquad B\boldsymbol{p} = \ell_2\,. \tag{6.62}
$$

Since N is closed, we have $M = N \oplus N^\perp$, and $\boldsymbol{q} = \boldsymbol{q}^N + \boldsymbol{q}^\circ$ is the corresponding decomposition for $\boldsymbol{q} \in M$, q^N being the orthogonal projection onto N. Then (6.62) is equivalent to

$$
\begin{aligned}
A\boldsymbol{p}^N + A\boldsymbol{p}^\circ + B^T u &= \ell_1,\\
B\boldsymbol{p}^N + B\boldsymbol{p}^\circ &= \ell_2,
\end{aligned}
$$

and again to

$$
\begin{aligned}
\left\langle B^T u, \boldsymbol{q}^\circ \right\rangle + \langle A\boldsymbol{p}, \boldsymbol{q}^\circ \rangle &= \ell_1(\boldsymbol{q}^\circ), && \text{(TMS1)}\\
\left\langle A\boldsymbol{p}^N, \boldsymbol{q}^N \right\rangle + \left\langle A\boldsymbol{p}^\circ, \boldsymbol{q}^N \right\rangle &= \ell_1(\boldsymbol{q}^N), && \text{(TMS2)}\\
B\boldsymbol{p}^\circ &= \ell_2, && \text{(TMS3)}
\end{aligned}
$$

where the second equation (TMS2) holds because of $\left\langle B^T u, \boldsymbol{q}^N \right\rangle = \left\langle u, B\boldsymbol{q}^N \right\rangle = 0$. Here the space indices are omitted for simplicity.

This is a triangular system in the equations (TMS3), (TMS2), (TMS1) and the variables $\boldsymbol{p}^\circ, \boldsymbol{p}^N, u$, correspondingly, which can be solved uniquely if and only if each of the subsystems can be solved uniquely with respect to the "diagonal variable".

According to Theorem 6.2 condition 2) is equivalent to

$$
B^T : X \to N^\perp \ \text{ is an isomorphism,} \tag{6.63}
$$

as (NB1) reading $B\boldsymbol{q} = 0 \Longrightarrow \boldsymbol{q} = \boldsymbol{0}$ for $\boldsymbol{q} \in N^\perp$ is trivially fulfilled. This again is equivalent to

$$
B : N^\perp \to X \text{ is an isomorphism,}
$$

meaning the unique solvability of (TMS3). As

$$
\tilde{\ell}_1 := (\ell_1 - A\boldsymbol{p}^\circ)|_N \in N',
$$

the subsystem (TMS2) is uniquely solvable for $\boldsymbol{p}^N$, if and only if A is an isomorphism on N, which is equivalent to 1) and analogously subsystem (TMS1) is unique solvable, if and only if $B^T : X \to N^\perp$ is an isomorphism, i.e., if and only if 2) holds true.

Taking into account that $\|A\| \le C_a$, $\|B\|, \|B^T\| \le C_b$, $\|(A|_N)^{-1}\| = \alpha$, $\|(B^T)^{-1}\| = \beta$ (being the operators from (6.63)) and also $\|(B|_{N^\perp})^{-1}\| = \beta$ (see Remark 6.3, 4) the above reasoning also gives the stability estimate.

Remark 6.18 1) Note that contrary to the specific examples here the solution space is not an affine space due to the side conditions incorporated, but a linear space identical with the test space. This means that the side conditions are either homogeneous or by means of functions fulfilling them transferred to the linear forms (see Chapter 2 for the treatment of Dirichlet boundary conditions in the primal formulation, and also Lemma 6.14).

2) As noted in Remark 6.3, 3), the N-coercivity of $a|_{N\times N}$ is sufficient for the conditions 1).

3) The condition 2) of Theorem 6.17 can be expressed in various ways. The aim is to characterize the surjectivity of $B : M \to X$. There are two equivalent formulations using isomorphisms:

a) $$B|_{N^\perp} : N^\perp \to X \quad \text{is an isomorphism,}$$
as $N^\perp$ is closed, and therefore complete (see Appendix A.4), and thus

$$\beta' := \inf_{\boldsymbol{q}\in N^\perp\setminus\{\boldsymbol{0}\}} \sup_{v\in X\setminus\{0\}} \frac{b(\boldsymbol{q}, v)}{\|\boldsymbol{q}\|_M \|v\|_X} > 0 \tag{6.64}$$

by Remark 6.3, 4). Furthermore we have

$$\beta := \inf_{v\in X\setminus\{0\}} \sup_{\boldsymbol{q}\in N^\perp\setminus\{\boldsymbol{0}\}} \frac{b(\boldsymbol{q}, v)}{\|\boldsymbol{q}\|_M \|v\|_X} > 0 \tag{6.65}$$

as expression of the injectivity of $B|_{N^\perp}$.
Then, if the conditions are valid, it holds that $\beta = \beta'$.
Let

$$\beta'' := \inf_{v\in X\setminus\{0\}} \sup_{\boldsymbol{q}\in M\setminus\{\boldsymbol{0}\}} \frac{b(\boldsymbol{q}, v)}{\|\boldsymbol{q}\|_M \|v\|_X} > 0, \tag{6.66}$$

then if one of the assertions (6.65) or (6.66) is valid, then $\beta'' = \beta$.
Indeed, for $\boldsymbol{q} \in M$ we have

$$\boldsymbol{q} = \boldsymbol{q}^N + \boldsymbol{q}^\circ, \quad \boldsymbol{q}^N \in N,\ \boldsymbol{q}^\circ \in N^\perp, \quad \text{and}$$
$$\|\boldsymbol{q}\|_M^2 = \|\boldsymbol{q}^N\|_M^2 + \|\boldsymbol{q}^\circ\|_M^2,$$

thus

$$\frac{b(\boldsymbol{q}, v)}{\|\boldsymbol{q}\|} \le \frac{b(\boldsymbol{q}^\circ, v)}{\|\boldsymbol{q}^\circ\|}$$

and thus $\beta'' \leq \beta$. The reverse inequality $\beta'' \geq \beta$ holds just because of $M \supset N^\perp$.

b) $$\widetilde{B}: X/N \to M, [x] \mapsto Bx \quad \text{is an isomorphism,}$$
where $[x] := \{y \in X \mid x - y \in N\}$ is the *equivalence class* of the element $x \in X$ and $X/N := \{[x] \mid x \in X\}$ is the *quotient space* equipped with the norm $\|[x]\|_{X/N} := \inf\{\|y\|_X \mid y \in [x]\}$ (see Appendix A.4). Therefore the above-mentioned formulations with β and β' are equivalent formulations with $N^\perp$ substituted by X/N and $\|\cdot\|$ by $\|\cdot\|_{X/N}$.

The more general variational equations (6.55)/(6.56) require more complex conditions, as—similar to the primal formulation in Section 3.1—the interplay between the convection term c and the reaction term d (which may result from the discretization of a time-dependent problem (see Section 9.3)) has to be taken into account to achieve sharp existence conditions, excluding the cases of non-unique existence (e.g., the pure Neumann case requiring (6.47)) on the other hand. We give an example due to P. Ciarlet Jr. et al. [124], slightly modified.

Theorem 6.19 *Let M, X be Hilbert spaces and consider the variational equations* (6.55) *for bilinear forms $a: M \times M \to \mathbb{R}$, $b: M \times X \to \mathbb{R}$, $c: X \times M \to \mathbb{R}$, $d: X \times X \to \mathbb{R}$, all continuous. Assume that the following conditions are satisfied:*

1) a is coercive on M with a constant $\alpha > 0$,
2) $d(v, v) \geq -\gamma\|v\|_X^2$ for some $\gamma \in \mathbb{R}$ such that $\gamma < \alpha\|A\|^{-2}\beta^2$,
3) $\widetilde{b}$ satisfies (NB2) *on $X \times M$ with a constant $\beta > 0$, where $\widetilde{b}(v, \boldsymbol{q}) := b(\boldsymbol{q}, v)$.*

If

$$\delta := \frac{\alpha^{-1}\|B\|\|C\|}{\alpha\|A\|^{-2}\beta^2 - \gamma} < 1, \tag{6.67}$$

then there exists a unique solution $(\boldsymbol{p}, u) \in M \times X$ and there is a stability estimate with the constant depending on the parameters mentioned, in particular on $(1-\delta)^{-1}$.

Proof The proof is based on the following result for $c = 0$ (see [124, Thm. 2.2]): There exists a unique solution $(\boldsymbol{p}, u) \in M \times X$, and

$$\|u\|_X \leq \frac{\alpha^{-1}\|B\|\|\ell_1\|_{M'} + \|\ell_2\|_{X'}}{\alpha\|A\|^{-2}\beta^2 - \gamma}, \tag{6.68}$$

$$\|\boldsymbol{p}\|_M \leq \alpha^{-1}(\|\ell_1\|_{M'} + \|B\|\|u\|_X). \tag{6.69}$$

We write the variational equations (6.55) as a fixed-point operator equation and apply a fixed-point iteration, i.e., with $u^{(0)} := 0$, $\boldsymbol{p}^{(0)} := \boldsymbol{0}$:

$$\begin{aligned} A\boldsymbol{p}^{(n+1)} + B^T u^{(n+1)} &= \ell_1 - Cu^{(n)}, \\ B\boldsymbol{p}^{(n+1)} - Du^{(n+1)} &= \ell_2. \end{aligned} \tag{6.70}$$

Then the estimate (6.68) implies

$$\|u^{(n+1)} - u^{(n)}\|_X \le \frac{\alpha^{-1}\|B\|\|C(u^{(n)} - u^{(n-1)})\|_{M'}}{\alpha\|A\|^{-2}\beta^2 - \gamma} \le \delta\|u^{(n)} - u^{(n-1)}\|_X \qquad (6.71)$$

by the definition (6.67) of δ. Hence we get, for $n \in \mathbb{N}$,

$$\|u^{(n+1)} - u^{(n)}\|_X \le \delta^n \|u^{(1)}\|_X,$$

and for $m, n \in \mathbb{N}$, $m > n$:

$$\|u^{(m)} - u^{(n)}\|_X \le \sum_{i=n}^{m-1} \|u^{(i+1)} - u^{(i)}\|_X \le \frac{\delta^n}{1-\delta}\|u^{(1)}\|_X. \qquad (6.72)$$

This shows that $(u^{(n)})_n$ is a Cauchy sequence, and therefore an element $u \in X$ exists such that $u^{(n)} \to u$ in X for $n \to \infty$.

The estimate (6.69) together with (6.71) yields

$$\begin{aligned}\|\boldsymbol{p}^{(n+1)} - \boldsymbol{p}^{(n)}\|_M &\le \alpha^{-1}(\|C(u^{(n)} - u^{(n-1)})\|_{M'} + \|B\|\|u^{(n+1)} - u^{(n)}\|_X) \\ &\le \alpha^{-1}(\|C\| + \delta\|B\|)\|u^{(n)} - u^{(n-1)}\|_X \\ &\le \alpha^{-1}(\delta^{-1}\|C\| + \|B\|)\delta^n\|u^{(1)}\|_X .\end{aligned} \qquad (6.73)$$

With an analogous argument as for $(u^{(n)})_n$ we conclude from (6.73) that also $(\boldsymbol{p}^{(n)})_n$ is a Cauchy sequence, with the limit $\boldsymbol{p} \in M$, and $(\boldsymbol{p}, u)$ is a solution of (6.55). This solution is unique, as for two solutions $(\boldsymbol{p}_1, u_1)$, $(\boldsymbol{p}_2, u_2)$ we have for $\boldsymbol{p} := \boldsymbol{p}_1 - \boldsymbol{p}_2$, $u := u_1 - u_2$:

$$\begin{aligned} A\boldsymbol{p} + B^T u &= -Cu, \\ B\boldsymbol{p} - Du &= \boldsymbol{0}, \end{aligned}$$

and therefore, applying the estimate (6.68) again,

$$\|u\|_X \le \delta\|u\|_X .$$

This implies $u = 0$ and consequently $\boldsymbol{p} = \boldsymbol{0}$ because of (6.69).

Now, setting $n := 1$ and letting m tend to infinity in (6.72) gives us

$$\|u - u^{(1)}\|_X \le \frac{\delta}{1-\delta}\|u^{(1)}\|_X,$$

so that

$$\|u\|_X \le \|u - u^{(1)}\|_X + \|u^{(1)}\|_X \le \frac{1}{1-\delta}\|u^{(1)}\|_X .$$

The norm of $u^{(1)}$ can be estimated by applying (6.68) to the first step of (6.70) (note $u^{(0)} = 0$). In this way we obtain a stability estimate of the claimed type. An estimate of $\|\boldsymbol{p}\|_M$ follows analogously. □

The existence of a stable solution may thus be impeded by an intense convection $\boldsymbol{c}$, the influence of which can in turn be compensated by a strong reaction or a time discretization with a small time step $(-\gamma \gg 0)$.

Remark 6.20 Assumption 1) restricts the usage of Theorem 6.19 to extensions of the Stokes equation, but excludes the Darcy equation. In [7, Sect. II.1.2] also the general case with $c = 0$ is treated. A special case of the general result [7, Thm. 1.2] assures unique existence, if a is positive (see (A4.6) in Appendix A.4), d is continuous and X-coercive, with similar stability estimates as (6.68)/(6.69). This allows for a repetition of the proof of Theorem 6.19, leading to a general existence result with a not so explicit smallness condition as in (6.67).

The purpose of the new formulation is to pose (conforming) Galerkin methods for the approximation of its solution. For the formulation (6.55) this means:

Choose finite-dimensional subspaces $M_h \subset M$, $X_h \subset X$ and find $(\boldsymbol{p}_h, u_h) \in M_h \times X_h$ such that

$$\begin{aligned} a_h(\boldsymbol{p}_h, \boldsymbol{q}_h) + b_h(\boldsymbol{q}_h, u_h) + c_h(u_h, \boldsymbol{q}_h) &= \ell_{1h}(\boldsymbol{q}_h) \qquad \text{for all } \boldsymbol{q} \in M_h, \\ b_h(\boldsymbol{p}_h, v_h) - d_h(u_h, v_h) &= \ell_{2h}(v_h) \qquad \text{for all } v_h \in X_h \end{aligned}$$

with corresponding approximating bilinear forms a_h, b_h, c_h, d_h and linear forms ℓ_{1h}, ℓ_{2h}.

It is clear that the results for unique existence of a solution can in particular be applied to the finite-dimensional case. For the standard formulation (6.62) the discrete problem reads:

Find $(\boldsymbol{p}_h, \boldsymbol{u}_h) \in M_h \times X_h$ such that

$$\begin{aligned} a_h(\boldsymbol{p}_h, \boldsymbol{q}_h) + b_h(\boldsymbol{q}_h, u_h) &= \ell_{1h}(\boldsymbol{q}_h) \qquad \text{for all } \boldsymbol{q}_h \in M_h, \\ b_h(\boldsymbol{p}_h, v_h) &= \ell_{2h}(v_h) \qquad \text{for all } v_h \in X_h. \end{aligned} \tag{6.74}$$

Theorem 6.21 *Let M_h, X_h be finite-dimensional vector spaces, $a_h : M_h \times M_h \to \mathbb{R}$ and $b_h : M_h \times X_h \to \mathbb{R}$ be bilinear forms with corresponding operators A_h, B_h, the norms of which are bounded by constants $C_{ah}, C_{bh} > 0$, and $\ell_{1h} \in M_h'$, $\ell_{2h} \in X_h'$. Let $N_h :=$ ker B_h. The requirements*

1) a_h satisfies the condition (NB2) *on $N_h \times N_h$ with a constant $\alpha_h > 0$*
2) $\widetilde{b}_h$ satisfies the condition (NB2) *on $X_h \times N_h^\perp$ with a constant $\beta_h > 0$, where* $\widetilde{b}_h(v_h, \boldsymbol{q}_h) := b_h(\boldsymbol{q}, v)$

are equivalent to the unique existence of a solution $(\boldsymbol{p}_h, u_h)$ of (6.74) *which satisfies a stability estimate with the constant depending on $C_{ah}, C_{bh}, \alpha_h^{-1}, \beta_h^{-1}$.*

Proof See Theorem 6.17, as $A_h : N_h \to N_h'$ is injective if and only if it is surjective, and therefore (NB1) $\Longleftrightarrow$ (NB2). □

Remark 6.22 1) If arbitrary bases of M_h and X_h are chosen, similar to Remark 6.4, 2), the discrete variational equation (6.74) is equivalent to the system of linear equations

$$\begin{pmatrix} \mathcal{A}_h & \mathcal{B}_h^T \\ \mathcal{B}_h & 0 \end{pmatrix} \begin{pmatrix} \boldsymbol{\xi} \\ \boldsymbol{\eta} \end{pmatrix} = \boldsymbol{q}_h$$

with $\mathcal{A}_h \in \mathbb{R}^{n,n}$, $n := \dim M_h$, $\mathcal{B}_h \in \mathbb{R}^{m,n}$, $m := \dim X_h$. The condition (NB2) for $\widetilde{b}_h$ indicates the injectivity of $\mathcal{B}_h^T$, i.e., full column rank of $\mathcal{B}_h^T$ or full row rank of $\mathcal{B}_h$. Therefore, necessarily $\dim M_h \geq \dim X_h$ has to hold. This is called the *rank condition*.

2) The conditions (NB2) for a_h or b_h are called discrete *inf-sup (or LBB) conditions*. To have stability, bounds for $C_{ah}, C_{bh}, \alpha_h^{-1}, \beta_h^{-1}$ uniform in h are necessary.

3) Contrary to the coercive situation, even in the conforming case $a_h = a, b_h = b$ the conditions (NB1), (NB2) for a or b on the pair (M, X) do not imply the conditions for a or b on (M_h, X_h). Not even $N_h \subset N$ holds in general. Therefore, also the coercivity of a on $N \times N$ is not sufficient. Hence, the interplay of the discrete spaces is decisive for existence and stability.

4) Similar considerations are possible for the more general case on the basis of Theorem 6.19, leading to a system of linear equations of the type

$$\begin{pmatrix} \mathcal{A}_h & \mathcal{B}_h^T + C_h \\ \mathcal{B}_h & -\mathcal{D}_h \end{pmatrix} \begin{pmatrix} \boldsymbol{\xi} \\ \boldsymbol{\eta} \end{pmatrix} = \boldsymbol{q}_h$$

with $\mathcal{A}_h \in \mathbb{R}^{n,n}, C_h \in \mathbb{R}^{n,m}, \mathcal{B}_h \in \mathbb{R}^{m,n}, \mathcal{D}_h \in \mathbb{R}^{m,m}$. In this situation, the existence conditions of Theorem 6.21 read: coercivity of $\mathcal{A}_h$, *relaxed* coercivity of $\mathcal{D}_h$ (see there), full row rank of $\mathcal{B}_h$, and the smallness condition analogous to (6.67).

5) The application of Theorems 6.6, 6.7, 6.8 leads to corresponding general approximation results.

6) Typically, the inf-sup condition 1) from Theorem 6.17 is fulfilled because of coercivity, sometimes even on the full space M (see Section 7.3), so the (uniform) validity of the inf-sup condition 2) is the critical part.

Beyond Remark 6.22, 5) let us have a closer look at the approximation properties for the formulation (6.62) and its approximation (6.74) for the conforming case $a_h = a, b_h = b$. Obviously, we have the following necessary conditions for the existence of a solution:

$$\begin{aligned} W_l &:= \{\boldsymbol{q} \in M \mid B\boldsymbol{q} = \ell_2\} \neq \emptyset, \\ W_{hl} &:= \{\boldsymbol{q}_h \in M_h \mid B_h\boldsymbol{q} = \ell_{2h}\} \neq \emptyset. \end{aligned}$$

Ignoring a consistency error in ℓ_2, i.e.,

$$\ell_{2h}(v_h) := \ell_2(v_h) \quad \text{for all } v_h \in X_h,$$

the question is whether the conclusion $\ell_2 \in \operatorname{im}(B) \Longrightarrow W_{hl} \neq \emptyset$ holds. This property can be characterized by the following result.

Lemma 6.23 *Under the above general assumptions the following statements are equivalent:*

(i) For each $\ell_2 \in \operatorname{im}(B) : W_{hl} \neq \emptyset$.
(ii) For each $\boldsymbol{q} \in M$ *there exists an element* $\boldsymbol{q}_h =: \Pi_h \boldsymbol{q} \in M_h$ *such that* $b(\boldsymbol{q} - \Pi_h \boldsymbol{q}, v_h) = 0$ *for all* $v_h \in X_h$.

They imply $\ker B_h^T = \ker B^T \cap X_h$, *i.e.,* $\ker B_h^T \subset \ker B^T$.

Proof See Problem 6.14. □

On the other hand, to conclude the validity of an inf-sup condition for a_h from the validity for a, the inclusion

$$\ker B_h \subset \ker B$$

is desirable. Explicitly, this means for $\boldsymbol{q}_h \in M_h$:

$$b(\boldsymbol{q}_h, v_h) = 0 \quad \text{for all } v_h \in X_h \implies b(\boldsymbol{q}_h, v) = 0 \quad \text{for all } v \in X.$$

Due to Lemma 6.23 this can be characterized as follows.

Lemma 6.24 *Under the above general assumptions it holds:*
If for each $v \in X$ *there exists an element* $v_h =: \Pi_h v \in X_h$ *such that*

$$b(\boldsymbol{q}_h, v - \Pi_h v) = 0 \quad \text{for all } \boldsymbol{q}_h \in M_h,$$

then $\ker B_h = \ker B \cap M_h$, *i.e.,* $\ker B_h \subset \ker B$.

Theorem 6.25 *Consider the variational equations* (6.61) *and* (6.74), $a_h = a$, $b_h = b$, $\ell_{1h} = \ell_1$, $\ell_{2h} = \ell_2$. *Under the conditions 1), 2) of Theorem 6.21 and Theorem 6.17 there exist constants* $C_1 = C(C_a, \alpha_h^{-1}, C_b, \beta_h^{-1})$, $C_2 = C_2(C_b, \alpha_h^{-1}) > 0$ *such that*

$$\|\boldsymbol{p} - \boldsymbol{p}_h\|_M \leq C_1 \inf_{\boldsymbol{q}_h \in M_h} \|\boldsymbol{p} - \boldsymbol{q}_h\|_M + C_2 \inf_{v_h \in X_h} \|u - v_h\|_X.$$

If additionally $\ker B_h \subset \ker B$ *is valid, then* $C_2 = 0$ *is possible. Furthermore:*

$$\|u - u_h\|_X \leq \left(1 + \frac{C_b}{\beta_h}\right) \inf_{v_h \in X \setminus \{0\}} \|u - v_h\|_X + \frac{C_a}{\beta_h} \|\boldsymbol{p} - \boldsymbol{p}_h\|_M.$$

Proof Due to Theorems 6.17 and 6.21 solutions $(\boldsymbol{p}, u)$ and $(\boldsymbol{p}_h, u_h)$, respectively, exists uniquely; in particular $W_{hl} \neq \emptyset$. We use the notation of these theorems. Due to (NB2) for a on $N_h \times N_h$ we have for $\boldsymbol{r}_h \in W_{hl}$, i.e., $\boldsymbol{r}_h - \boldsymbol{p}_h \in \ker B_h$:

$$\alpha_h \|\boldsymbol{r}_h - \boldsymbol{p}_h\|_M \leq \sup_{\boldsymbol{q}_h \in N_h \setminus \{\boldsymbol{0}\}} \frac{a(\boldsymbol{r}_h - \boldsymbol{p}, \boldsymbol{q}_h) + a(\boldsymbol{p} - \boldsymbol{p}_h, \boldsymbol{q}_h)}{\|\boldsymbol{q}_h\|_M},$$

and consistency implies

$$a(\boldsymbol{p} - \boldsymbol{p}_h, \boldsymbol{q}_h) + b(\boldsymbol{q}_h, u - u_h) = 0.$$

Therefore,

$$\alpha_h \|\boldsymbol{r}_h - \boldsymbol{p}_h\|_M \le \sup_{\boldsymbol{q}_h \in N_h \setminus \{\boldsymbol{0}\}} \frac{a(\boldsymbol{r}_h - \boldsymbol{p}, \boldsymbol{q}_h) - b(\boldsymbol{q}_h, u - u_h)}{\|\boldsymbol{q}_h\|_M}.$$

If $\ker B_h \subset \ker B$ the last term vanishes and

$$\alpha_h \|\boldsymbol{r}_h - \boldsymbol{p}_h\|_M \le C_a \|\boldsymbol{r}_h - \boldsymbol{p}\|_M . \tag{6.75}$$

Thus

$$\|\boldsymbol{p} - \boldsymbol{p}_h\|_M \le \inf_{\boldsymbol{r}_h \in W_{hl}} \left(\|\boldsymbol{p} - \boldsymbol{r}_h\|_M + \|\boldsymbol{r}_h - \boldsymbol{p}_h\|_M\right) \le \left(1 + \frac{C_a}{\alpha_h}\right) \inf_{\boldsymbol{r}_h \in W_{hl}} \|\boldsymbol{p} - \boldsymbol{r}_h\|_M .$$

In the general case the additional term can be estimated by

$$|b(\boldsymbol{q}_h, u - u_h)| = |b(\boldsymbol{q}_h, u - v_h)| \quad \text{for all } v_h \in X_h$$

as $\boldsymbol{q}_h \in N_h = \ker B_h$ and therefore $C_b \|u - v_h\|_X$ appears as additional term in (6.75) as asserted.

The approximation error over W_{hl} still has to be converted into an approximation error over M_h.

Let $\boldsymbol{q}_h \in M_h$ be arbitrary, then by consistency

$$b(\boldsymbol{p} - \boldsymbol{q}_h, v_h) = b(\boldsymbol{p}_h - \boldsymbol{q}_h, v_h) \quad \text{for all } v_h \in X_h,$$

therefore $\boldsymbol{w}_h \in \boldsymbol{p}_h - \boldsymbol{q}_h + N_h$ satisfies the relation

$$b(\boldsymbol{w}_h, v_h) = b(\boldsymbol{p} - \boldsymbol{q}_h, v_h),$$

and an element $\boldsymbol{w}_h \in N_h^\perp$ can be selected. The condition (NB2) for $\widetilde{b}$ on $X_h \times N_h^\perp$ induces the same condition for b on $N_h^\perp \times X_h$ with the same constant β_h (see Remark 6.18, 3)). This implies

$$\beta_h \|\boldsymbol{w}_h\|_M \le \sup_{v_h \in X_h \setminus \{0\}} \frac{b(\boldsymbol{p} - \boldsymbol{q}_h, v_h)}{\|v_h\|_X} \le C_b \|\boldsymbol{p} - \boldsymbol{q}_h\|_M,$$

and we have $\boldsymbol{w}_h + \boldsymbol{q}_h \in W_{hl}$, thus

$$\inf_{\boldsymbol{r}_h \in W_{hl}} \|\boldsymbol{p} - \boldsymbol{r}_h\|_M \le \|\boldsymbol{p} - \boldsymbol{q}_h - \boldsymbol{w}_h\|_M \le (1 + \frac{C_b}{\beta_h}) \|\boldsymbol{p} - \boldsymbol{q}_h\|_M,$$

and finally for the first part

$$\inf_{\boldsymbol{r}_h \in W_{hl}} \|\boldsymbol{p} - \boldsymbol{r}_h\|_M \le (1 + \frac{C_b}{\beta_h}) \inf_{\boldsymbol{q}_h \in M_h} \|\boldsymbol{p} - \boldsymbol{q}_h\|_M .$$

For the error estimate concerning u we have

$$a(\boldsymbol{p} - \boldsymbol{p}_h, \boldsymbol{q}_h) + b(\boldsymbol{q}_h, u - u_h) = 0$$

and therefore

$$b(\boldsymbol{q}_h, v_h - u_h) = -a(\boldsymbol{p} - \boldsymbol{p}_h, \boldsymbol{q}_h) + b(\boldsymbol{q}_h, v_h - u).$$

Then

$$\begin{aligned}\beta_h \|v_h - u_h\|_X &\le \sup_{\boldsymbol{q}_h \in M_h \setminus \{\boldsymbol{0}\}} \frac{b(\boldsymbol{q}_h, v_h - u_h)}{\|\boldsymbol{q}_h\|_M} \\ &\le \sup_{\boldsymbol{q}_h \in M_h \setminus \{\boldsymbol{0}\}} \frac{a(\boldsymbol{p}_h - \boldsymbol{p}, \boldsymbol{q}_h)}{\|\boldsymbol{q}_h\|_M} + \sup_{\boldsymbol{q}_h \in M_h \setminus \{\boldsymbol{0}\}} \frac{b(\boldsymbol{q}_h, v_h - u)}{\|\boldsymbol{q}_h\|_M}.\end{aligned}$$

The continuity estimates of a and b and the triangle inequality conclude the proof. □

Remark 6.26 In the error estimate for $\boldsymbol{p}$ the term $1/\beta$ enters linearly, in the one for u quadratically.

In the case of $\ker B_h^T \subset \ker B^T$, the inf-sup condition for $\widetilde{b}$ on $X \times (\ker B^T)^\perp$ induces an inf-sup condition for $\widetilde{b}$ on $X \times (\ker B_h^T)^\perp$. The inclusion is implied by the existence of a projection operator Π_h due to Lemma 6.23. Therefore, we have the following generalization.

Theorem 6.27 *Let $\widetilde{M}$ be a continuously embedded subspace of M, and S a vector space with seminorm $|\cdot|$ such that $S \cap X_h \subset X$. If there is a uniformly continuous operator $\Pi_h : \widetilde{M} \to M_h \subset M$ such that*

$$\begin{aligned} b(\Pi_h \boldsymbol{q} - \boldsymbol{q}, v_h) &= 0 \qquad \text{for all } v_h \in S \cap X_h \quad \text{and} \\ \|\Pi_h \boldsymbol{q}\|_M &\le C\|\boldsymbol{q}\|_M \text{ for all } \boldsymbol{q} \in \widetilde{M} \end{aligned}$$

for some constant $C > 0$, then an inf-sup condition of the form

$$\sup_{\boldsymbol{q} \in \widetilde{M} \setminus \{\boldsymbol{0}\}} \frac{b(\boldsymbol{q}, v_h)}{\|\boldsymbol{q}\|_M} \ge \beta |v_h| \quad \text{for all } v_h \in S \cap X_h$$

implies the same one for M_h with constant $\widetilde{\beta} := \beta/C$.

Proof

$$\sup_{\boldsymbol{q}_h \in M_h \setminus \{\boldsymbol{0}\}} \frac{b(\boldsymbol{q}_h, v_h)}{\|\boldsymbol{q}_h\|_M} \ge \sup_{\boldsymbol{q} \in \widetilde{M} \setminus \{\boldsymbol{0}\}} \frac{b(\Pi_h \boldsymbol{q}, v_h)}{\|\Pi_h \boldsymbol{q}\|_M} \ge \frac{1}{C} \sup_{\boldsymbol{q} \in \widetilde{M} \setminus \{\boldsymbol{0}\}} \frac{b(\boldsymbol{q}, v_h)}{\|\boldsymbol{q}\|_M}.$$

Note that we have $\boldsymbol{q} \neq \boldsymbol{0} \implies \Pi_h \boldsymbol{q} \neq \boldsymbol{0}$, as $\Pi_h \boldsymbol{q} = \boldsymbol{0} \implies b(\boldsymbol{q}, v_h) = 0$ for all $v_h \in S \cap X_h$, and then the inf-sup condition gives a contradiction for $\boldsymbol{q} \neq \boldsymbol{0}$. □

This result is also called *Fortin's criterion* after Michel Fortin.

Remark 6.28 The requirements of Theorem 6.27 can also be reviewed as the commutativity of the diagram in Fig. 6.1 with ι being the corresponding embedding.

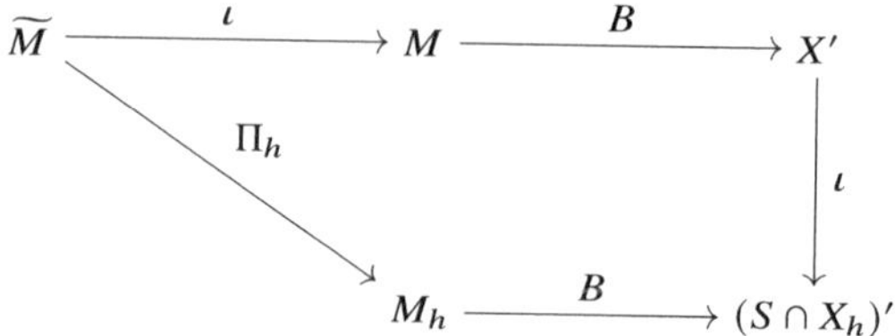

Fig. 6.1: Commutativity in Fortin's criterion.

Exercises

Problem 6.6 Show that the normal component trace operator $\gamma_{\mathfrak{n},\Gamma}$ (6.39) is surjective and has a continuous inverse.

Hint: Given $g \in H^{-1/2}(\Gamma)$, prove along the lines of Section 3.2 that the inhomogeneous Neumann boundary value problem

$$\begin{aligned} -\Delta w + w &= 0 \quad \text{in } \Omega, \\ \mathfrak{n} \cdot \nabla w &= g \quad \text{on } \Gamma, \\ w &= 0 \quad \text{on } \Gamma^c, \end{aligned}$$

has a unique solution $w \in H^1_{\Gamma^c}(\Omega)$ and set $\boldsymbol{q} := \nabla w$.

Problem 6.7 Prove Theorem 6.13.

Problem 6.8 Given two Hilbert spaces M and X, a continuous, coercive, and symmetric bilinear form a on M, a continuous bilinear form $b : M \times X \to \mathbb{R}$, and functionals $f \in M'$, $\chi \in X'$, define $J : M \to \mathbb{R}$ by $J(\boldsymbol{p}) := \frac{1}{2}a(\boldsymbol{p}, \boldsymbol{p}) - \langle f, \boldsymbol{p}\rangle$ and $L : M \times X \to \mathbb{R}$ by

$$L(\boldsymbol{p}, \lambda) := J(\boldsymbol{p}) + b(\boldsymbol{p}, \lambda) - \langle \chi, \lambda\rangle.$$

The bilinear form b defines a linear operator $B : M \to X'$ by

$$\langle B\boldsymbol{q}, \mu\rangle = b(\boldsymbol{q}, \mu) \quad \text{for all } \mu \in X\,.$$

Show that $\boldsymbol{p} \in M$ solves the minimization problem

$$J(\boldsymbol{p}) = \inf_{\substack{\boldsymbol{q} \in M \\ B\boldsymbol{q} = \chi}} J(\boldsymbol{q})$$

if and only if there is a Lagrange multiplier $\lambda \in X$ such that $(\boldsymbol{p}, \lambda) \in M \times X$ is a saddle point of L, i.e.,

$$L(\boldsymbol{p}, \mu) \le L(\boldsymbol{p}, \lambda) \le L(\boldsymbol{q}, \lambda) \quad \text{for all } (\boldsymbol{q}, \mu) \in M \times X.$$

Can the coercivity condition be weakened?

Find the relation to (6.61) and conclude conditions for unique existence of $\boldsymbol{p}$ and λ.

Is the surjectivity of B sufficient?

Problem 6.9 We consider the biharmonic equation

$$\Delta^2 u = f \text{ in } \Omega, \quad u = \Delta u = 0 \text{ on } \partial\Omega.$$

a) Rewrite the above equation such that it becomes a saddle point problem in appropriate spaces.
b) Show unique existence of a solution.

Problem 6.10 Find conditions to $\boldsymbol{c}$ and α to ensure the unique existence of a solution of (6.43).

Problem 6.11 Let $u \in H^1(\Omega)$ be a solution of (3.36), (3.37)–(3.39). Find conditions such that $(\boldsymbol{p}, u)$ with $\boldsymbol{p} := -\boldsymbol{K}\nabla u + \boldsymbol{c}u$ is a solution of problem (6.43) but with the bilinear form (6.49) or of (6.51), (6.52).

Problem 6.12 Work out the exact formulation of a primal hybrid and a primal mixed hybrid formulation of (3.13), (3.19)–(3.21).

Problem 6.13 Consider the discrete saddle point problem

$$\begin{pmatrix} \mathcal{A}_h & \mathcal{B}_h^T \\ \mathcal{B}_h & 0 \end{pmatrix} \begin{pmatrix} \boldsymbol{\xi} \\ \boldsymbol{\eta} \end{pmatrix} = \boldsymbol{q}_h \tag{6.76}$$

(cf. Remark 6.22, 1)), where $\mathcal{A}_h \in \mathbb{R}^{n,n}$, $\mathcal{B}_h \in \mathbb{R}^{m,n}$, $n \geq m$, are given matrices with full row rank, and $\boldsymbol{q}_h \in \mathbb{R}^{n+m}$ is a given vector. Assume that $\mathcal{A}_h$ is symmetric and positive definite.

a) Eliminate the vector $\boldsymbol{\xi}$ to obtain an equation for $\boldsymbol{\eta}$ of the form $\mathcal{S}_h\boldsymbol{\eta} = \hat{\boldsymbol{q}}_\eta$. Give explicit expressions for the matrix $\mathcal{S}_h$ and the right-hand side $\hat{\boldsymbol{q}}_\eta$.
b) Prove the following statements:

    1) The matrix $\mathcal{S}_h$ is positive definite.
    2) The problem (6.76) has a unique solution.

c) The application of the Richardson iteration with a sufficiently small relaxation parameter $\omega > 0$ to the Schur complement system $\mathcal{S}_h\boldsymbol{\eta} = \hat{\boldsymbol{q}}_\eta$ reads as $\boldsymbol{\eta}^{(k+1)} := \boldsymbol{\eta}^{(k)} - \omega\left(\mathcal{S}_h\boldsymbol{\eta}^{(k)} - \hat{\boldsymbol{q}}_\eta\right)$, cf. (5.30). Formulate a variant of this iterative procedure that avoids the explicit use of $\mathcal{S}_h$. (The resulting procedure is a so-called *Uzawa-type algorithm*.)
(d) How the iterates $\boldsymbol{\xi}^{(k)}$ can be obtained?
(e) Prove the convergence of the Uzawa algorithm provided that $0 < \omega < 2\|\mathcal{B}_h^T\mathcal{A}_h\mathcal{B}_h^T\|^{-1}$.

Problem 6.14 Prove Lemma 6.23.

6.3 Fluid Mechanics: Laminar Flows

In Chapter 0 models for flow and reactive transport in porous media have been discussed. These models are on the (practical) *macroscale*, where the complex geometric microstructure only enters via *effective parameters*. These macroscopic models can be derived by means of averaging processes applied to models at the *microscale* derived by continuum mechanics. The derivation of these models is carried out here for the example of laminar flows. We follow the exposition in [23] and refer to this book for further details.

We start with the description of flows in a domain $\Omega \subset \mathbb{R}^d$. As in Chapter 0 we consider a control volume $\widetilde{\Omega} \subset \Omega$, but allow for its time dependence, i.e., we take the point of view of *Eulerian coordinates*, where the position of the observer is fixed, opposite to *Lagrangian coordinates*, where the observer follows the motion (say of a particle), i.e., if $X \in \Omega$ is the Lagrangian and $x = x(t, X)$, t being time, the Eulerian coordinate, then we have the control volume

$$\widetilde{\Omega}(t) := \{x(t, X) \mid X \in \widetilde{\Omega}\}$$

with the velocity

$$\boldsymbol{v}(t, x) := \frac{\partial x}{\partial t}(t, X(t, x)).$$

If $\rho = \rho(t, x)$ denotes the density of the considered body, then *mass conservation* reads

$$\frac{d}{dt}\int_{\widetilde{\Omega}(t)} \rho(t, x) dx = 0$$

and by means of the Reynolds transport theorem (see [15, Sect. 1.1]) the pointwise version (compare (0.3)) is

$$\partial_t \rho + \nabla \cdot (\rho \boldsymbol{v}) = 0\,. \tag{6.77}$$

This equation is also called *continuity equation*, and it can be rewritten as

$$D_t \rho + \rho \nabla \cdot \boldsymbol{v} = 0$$

using the abbreviation

$$D_t \varphi := \partial_t \varphi + \boldsymbol{v} \cdot \nabla \varphi$$

for the *material derivative* for any scalar field $\varphi : \Omega \to \mathbb{R}$. By applying the operator componentwise, also for every vector field $\Psi : \Omega \to \mathbb{R}^d$ we set

$$D_t \Psi := \partial_t \Psi + \boldsymbol{v} \cdot \nabla \Psi, := (\partial_t \Psi_i)_i + (\boldsymbol{v} \cdot \nabla \Psi_i)_i\,.$$

In Section 0.3 *balance of linear momentum* took the form of Darcy's law (0.15). Here in general we have

$$\frac{d}{dt}\int_{\widetilde{\Omega}(t)} \rho \boldsymbol{v}(t, x) dx = \int_{\widetilde{\Omega}(t)} \rho \boldsymbol{f}(t, x) dx + \int_{\partial\widetilde{\Omega}(t)} b(x; \boldsymbol{v}) d\sigma_x \tag{6.78}$$

assuming a distributed force density f (per unit mass) and a boundary force density b (per unit surface), i.e., making use of the *Cauchy axiom*. As a consequence of (6.78) for "any smooth control volume" and smooth function, b is linear in $\boldsymbol{v}$:

$$b(x;\boldsymbol{v}) = \sigma(x)\boldsymbol{v}$$

with the *stress tensor* $\sigma : \Omega \to \mathbb{R}^{d,d}$. If furthermore conservation of *angular momentum* holds, then $\sigma(x)$ necessarily is symmetric. Again using the Reynolds transport theorem and (6.77), we reach the general equation for conservation of linear momentum:

$$\rho D_t \boldsymbol{v} - \nabla \cdot \sigma = \rho \boldsymbol{f}.$$

To close the model, a *constitutive law* for σ is needed expressing the nature of the fluid.

For an *inviscid* (or non-viscous) fluid we have

$$\sigma = -pI,$$

where $I \in \mathbb{R}^{d,d}$ is the identity matrix and $p : \Omega \to \mathbb{R}$ represents the *pressure*. This form is a consequence of isotropy and invariance under a change of observer if the tensor only depends on pressure (and temperature T). This gives the (isothermal) *Euler equations*

$$\partial_t \rho + \nabla \cdot (\rho \boldsymbol{v}) = 0, \qquad \text{(ENS1)}$$

$$\rho D_t \boldsymbol{v} + \nabla p = \rho \boldsymbol{f}. \qquad \text{(ENS2a)}$$

To close this system, a constitutive law of the form

$$p = \hat{p}(\rho)$$

is necessary. The special case of an *incompressible non-viscous flow* requires instead of (ENS1) $\nabla \cdot \boldsymbol{v} = 0$, which follows for constant density, i.e.,

$$\nabla \cdot \boldsymbol{v} = 0, \qquad \text{(ENS1a)}$$

$$D_t \boldsymbol{v} + \frac{1}{\rho}\nabla p = \boldsymbol{f}. \qquad \text{(ENS2a)}$$

These systems, containing only first-order derivatives, are in general of hyperbolic type and out of scope of the methods considered here.

For a *viscous fluid* a relationship

$$\sigma = \hat{\sigma}(p, \varepsilon(\boldsymbol{v}))$$

is assumed, where

$$\varepsilon(\boldsymbol{v}) := \frac{1}{2}(D\boldsymbol{v} + (D\boldsymbol{v})^T)$$

is the *symmetric strain tensor* and $D\boldsymbol{v}$ is the Jacobian of $\boldsymbol{v}$. If the model is *thermodynamically consistent*, i.e., satisfies the second law of thermodynamics, we have

$$\sigma = -pI + \hat{S}(\rho, T, \nabla T, D\boldsymbol{v})$$

with $p = \hat{p}(\rho, T)$,

$$\hat{S} = 2\mu\varepsilon(\boldsymbol{v}) + \lambda\nabla \cdot \boldsymbol{v}I$$

with parameters $\mu > 0$, the *shear viscosity*, and $\lambda > 0$, where $\mu + \frac{2}{3}\lambda$ is the *volume viscosity*. Such a fluid is called *Newtonian*.

In the case of an *(isothermal) compressible viscous fluid*, conservation of mass and linear momentum give the *compressible Navier–Stokes equations*: (ENS1) and

$$\rho D_t\boldsymbol{v} - \mu\Delta\boldsymbol{v} - (\lambda + \mu)\nabla\nabla \cdot \boldsymbol{v} + \nabla p = \rho\boldsymbol{f}. \tag{ENS2}$$

Here $\Delta\boldsymbol{v} = (\Delta v_i)_i$ is the matrix Laplacian and finally in the incompressible case, the *incompressible Navier–Stokes equation* mean (ENS1a) and

$$\partial_t\boldsymbol{v} + (\boldsymbol{v} \cdot \nabla)\boldsymbol{v} - \eta\Delta\boldsymbol{v} + \frac{1}{\rho}\nabla p = \boldsymbol{f}. \tag{ENS2b}$$

Here, $\eta := \mu/\rho$ is the *kinematic viscosity*. An important characteristic number is the *Reynolds number R*

$$R := \frac{Vl}{\eta},$$

where V is a characteristic velocity size and l a characteristic length. For small R the convective part can be neglected and we arrive at the *(time-dependent) Stokes equation*

$$\nabla \cdot \boldsymbol{v} = 0, \tag{ENS1a}$$

$$\partial_t\boldsymbol{v} - \eta\Delta\boldsymbol{v} + \frac{1}{\rho}\nabla p = \boldsymbol{f}. \tag{ENS2c}$$

All the models (at least with viscous forces) have their time-independent counterpart, i.e., for stationary solutions, which may appear as large time limits of solutions of the models above. We only note explicitly for the last case the *stationary Stokes equations*

$$\begin{aligned} \nabla \cdot \boldsymbol{v} &= 0, \\ -\eta\Delta\boldsymbol{v} + \frac{1}{\rho}\nabla p &= \boldsymbol{f}. \end{aligned} \tag{6.79}$$

Up to now temperature did not play a role. If the dynamics of temperature matters or is the primary concern, (also) conservation of energy has to be considered:

$$\begin{aligned} \frac{d}{dt}\int_{\tilde{\Omega}(t)} \rho(\frac{1}{2}|\boldsymbol{v}|^2 + u)dx = &\int_{\tilde{\Omega}(t)} \rho\boldsymbol{f} \cdot \boldsymbol{v}dx + \int_{\partial\tilde{\Omega}(t)} \sigma\mathfrak{n} \cdot \boldsymbol{v}d\sigma \\ &- \int_{\partial\tilde{\Omega}(t)} \boldsymbol{q} \cdot \mathfrak{n}d\sigma + \int_{\tilde{\Omega}(t)} \rho g dx. \end{aligned}$$

Here, the left-hand side describes the change of energy composed of the *kinetic energy* and the *internal energy* with the *internal energy density* u. The terms at the right-hand side describe the work caused by volume $\boldsymbol{f}$ and surface forces σ, the gain and loss due to the *heat flux density* $\boldsymbol{q}$ and finally the energy rate density of distributed sources g. Rewriting the equation as above and using the other conservation equations leads to

$$\rho D_t u - \sigma : D\boldsymbol{v} + \nabla \cdot \boldsymbol{q} = \rho g. \tag{ENS3}$$

An often used constitutive law for $\boldsymbol{q}$ is *Fouriers's law*, suggested by isotropy and invariance under a change of observer, with sign specification due to the second law of thermodynamics

$$\boldsymbol{q} = \hat{\boldsymbol{q}}(\rho, T, \nabla T) = -\hat{k}(\rho, T, |\nabla T|)\nabla T, \tag{ENS4}$$

where T denotes the temperature, with thermal conductivity $\hat{k} > 0$.

This gives the following specific form of (ENS3) for inviscid fluids:

$$\rho D_t u + p\nabla \cdot \boldsymbol{v} - \nabla \cdot (\hat{k}\nabla T) = \rho g$$

to be added to the Euler equation (ENS1), (ENS2a) with $p = \hat{p}(\rho, T)$, $u = \hat{u}(\rho, T)$ and similarly for the (in-)compressible (Navier–)Stokes equation. For incompressible inviscid flow we recover the heat equation (0.7) with a convective term.

Since it is one of the simplest models, we have a look at the analysis of (6.79) to ensure unique existence of a solution. We assume ρ to be a constant and scaled to 1 by a change of variables. As boundary conditions we consider *no slip boundary conditions*, i.e.,

$$\boldsymbol{v} = \boldsymbol{0} \quad \text{on } \partial\Omega. \tag{6.82}$$

The problem (6.79) admits two weak formulations: In the first the incompressibility condition is incorporated into the space similar to (6.41), i.e.,

$$W := \{\boldsymbol{w} \in H_0^1(\Omega)^d \mid \nabla \cdot \boldsymbol{w} = 0 \text{ in } \Omega\},$$

and by integration by parts we conclude from (6.79) for $\boldsymbol{w} \in W$:

$$\begin{aligned} \ell(\boldsymbol{w}) := \int_\Omega \boldsymbol{f} \cdot \boldsymbol{w}\, dx &= -\eta \int_\Omega \Delta\boldsymbol{v} \cdot \boldsymbol{w}\, dx + \int_\Omega \nabla p \cdot \boldsymbol{w}\, dx \\ &= \eta \int_\Omega D\boldsymbol{v} : D\boldsymbol{w}\, dx - \int_\Omega p\nabla \cdot \boldsymbol{w}\, dx \\ &= \eta \int_\Omega D\boldsymbol{v} : D\boldsymbol{w}\, dx =: a(\boldsymbol{v}, \boldsymbol{w}). \end{aligned} \tag{6.83}$$

Therefore we have

Formulation I:

Find $\boldsymbol{v} \in W$ such that

$$a(\boldsymbol{v}, \boldsymbol{w}) = \ell(\boldsymbol{w}) \quad \text{for all } \boldsymbol{w} \in W.$$

As a is W-coercive, the Lax–Milgram Theorem (Theorem 3.1) implies the unique existence of $\boldsymbol{v}$ and by Lemma 2.3 there is the equivalent formulation as a restricted quadratic minimization problem.

Unfortunately, this notion does not provide a pressure p. Note that since only ∇p appears in the equation, p can only be unique up to a constant. Therefore an ansatz space for p is

$$X := L_0^2(\Omega) := \{q \in L^2(\Omega) \mid \int_\Omega q dx = 0\}.$$

Another formulation corresponding to the mixed formulation of (6.79) arises from setting

$$M := H_0^1(\Omega)^d,$$

and (6.82) shows that we have for $\boldsymbol{v} \in M$ and $p \in X$,

$$\begin{aligned} a(\boldsymbol{v}, \boldsymbol{w}) - \int_\Omega p \nabla \cdot \boldsymbol{w} dx &= \ell(\boldsymbol{w}) \quad \text{for all } \boldsymbol{w} \in M, \\ -\int_\Omega q \nabla \cdot \boldsymbol{v} dx &= 0 \qquad\quad \text{for all } q \in X. \end{aligned} \tag{6.84}$$

Therefore, setting

$$b(\boldsymbol{v}, q) := -\int_\Omega q \nabla \cdot \boldsymbol{v} dx$$

we have

Formulation II:

Find $(\boldsymbol{v}, p) \in M \times X$ such that

$$\begin{aligned} a(\boldsymbol{v}, \boldsymbol{w}) + b(\boldsymbol{w}, p) &= \ell(\boldsymbol{w}) \qquad \text{for all } \boldsymbol{w} \in M, \\ b(\boldsymbol{v}, q) &= 0 \qquad\quad \text{for all } q \in X. \end{aligned}$$

Thus we arrive at the same structure as in Section 6.2, even with the same form b (but different spaces). (Note the switch in notation: $(\boldsymbol{p}, u)$ there corresponding to $(\boldsymbol{v}, p)$ here.)

A solution of Formulation II is also a solution of Formulation I. To demonstrate the contrary—and by this also unique existence for Formulation II—for Formulation I a pressure p has to be constructed.

Alternatively, we can apply Theorem 6.17. The form a is coercive on $M \times M$, a fortiori on $N \times N$, $N := \ker B$. The inf-sup condition for $\widetilde{b}$ on $X \times N^\perp$ with

$\widetilde{b}(q, \boldsymbol{w}) := b(\boldsymbol{w}, q)$ can be verified similar to (7.35), but due to the smaller space with considerably more technical effort. This shows unique and stable existence of a solution.

Theorem 6.29 *Let $\Omega \subset \mathbb{R}^d$ be a bounded Lipschitz domain, $\boldsymbol{f} \in (H_0^1(\Omega)^d)'$. Then the stationary Stokes equation* (6.79) *with no slip boundary condition* (6.82) *has a unique solution $(\boldsymbol{v}, p) \in H_0^1(\Omega)^d \times L_0^2(\Omega)$ and there is a constant $C > 0$, independent of the solution, such that*

$$\|\boldsymbol{v}\|_1 + \|p\|_0 \leq C\|\boldsymbol{f}\|_{-1}.$$

Proof To verify the aforementioned inf-sup condition for $\widetilde{b}$, let $q \in X$ and $\boldsymbol{w}_q \in M$ be such that

$$\nabla \cdot \boldsymbol{w}_q = -q$$

with $\|\boldsymbol{w}_q\|_1 \leq C\|q\|_0$ for some constant $C > 0$ independent of q according to the subsequent Theorem 6.30. Then

$$\sup_{\boldsymbol{w} \in M \setminus \{\boldsymbol{0}\}} \frac{b(\boldsymbol{w}_q, q)}{\|\boldsymbol{w}_q\|_M} \geq C^{-1} \int_\Omega (\nabla \cdot \boldsymbol{w}_q)^2 dx / \|q\|_0 = C^{-1} \|q\|_X. \tag{6.85}$$

Theorem 6.17 implies the assertion. □

Theorem 6.30 *Let $\Omega \subset \mathbb{R}^d$ be a bounded Lipschitz domain. There exists a constant $C > 0$ and for each $q \in L_0^2(\Omega)$ a function $\boldsymbol{w} \in H_0^1(\Omega)^d$ such that*

$$\nabla \cdot \boldsymbol{w} = q \quad \text{and} \quad \|\boldsymbol{w}\|_1 \leq C\|q\|_0.$$

Proof (Sketch) Consider the mapping

$$\operatorname{div} : \; M \to X, \quad \boldsymbol{w} \mapsto -\nabla \cdot \boldsymbol{w},$$

which is well defined, linear and continuous (note that $-\int\limits_\Omega \nabla \cdot \boldsymbol{w} dx = \int\limits_{\partial\Omega} \boldsymbol{w} \cdot \boldsymbol{n} d\sigma = 0$).

Set

$$W := \ker(\operatorname{div}).$$

It is sufficient to show that div is surjective. Then

$$\widetilde{\operatorname{div}} : \; M/W \to X$$

is bijective. Then also

$$\widetilde{\operatorname{div}} : \; M|_{W^\perp} \to X$$

is a bijective linear mapping between Banach spaces. Therefore by the Open Mapping Theorem (see Appendix A.4) also its inverse is continuous leading to the assertion. Here the orthogonality has to be understood with respect to the scalar product of $H_0^1(\Omega)^d$, i.e., due to the Poincaré inequality (Theorem 2.18) an equivalent scalar product is

$$\langle \boldsymbol{v}, \boldsymbol{w} \rangle := \int_\Omega D\boldsymbol{v} : D\boldsymbol{w} dx .$$

To show the surjectivity, consider its dual mapping, namely,

$$\text{grad}: X \to M', \quad q \mapsto \nabla q$$

(using the identification $X' = X$).

This duality can be seen from

$$\langle \text{grad}\, q, \boldsymbol{w} \rangle = -\int_\Omega q \nabla \cdot \boldsymbol{w}\, dx = \langle q, \text{div}\, \boldsymbol{w} \rangle$$

(in the sense of weak derivatives), and as a consequence of

$$\text{im}(\text{grad}) \subset W^\perp \quad \left(= \{\psi \in M' \mid \psi(\boldsymbol{w}) = 0 \text{ for all } \boldsymbol{w} \in W\}\right)$$

and $b(\boldsymbol{v}, q + q_0) = b(\boldsymbol{v}, q)$ for all $q_0 \in \mathbb{R}$. By means of the Closed Range Theorem (see Appendix A.4) the surjectivity of div could be concluded from

$$\text{im}(\text{div}) = \ker(\text{grad})^\perp = \{0\}^\perp = X$$

because $\nabla q = \boldsymbol{0}, \int_\Omega q dx = 0$ implies $q = 0$.

The crucial assumption for the validity of the first identity is that im(div) is closed. Again by the Closed Range Theorem this is equivalent to

$$\text{im}(\text{grad}) \quad \text{is closed.}$$

This can be proven making use of the Lions-Poincaré Lemma (see, e.g., [17, Thm. 6.14-1]). □

The boundary condition (6.82) can be generalized to an inhomogeneous Dirichlet condition

$$\boldsymbol{v} = \boldsymbol{g} \quad \text{on } \partial\Omega \tag{6.86}$$

provided $\boldsymbol{g} \in H^{1/2}(\partial\Omega)^d$ and satisfies the necessary compatibility condition

$$\int_{\partial\Omega} \boldsymbol{g} \cdot \boldsymbol{n} d\sigma = \int_{\partial\Omega} \boldsymbol{v} \cdot \boldsymbol{n} d\sigma = \int_\Omega \nabla \cdot \boldsymbol{v} dx = 0.$$

Also the boundary condition can be generalized to (with a subdivision $\partial\Omega = \Gamma_1 \cup \Gamma_3$)

$$\begin{aligned} -(\eta D\boldsymbol{v} - pI) \cdot \boldsymbol{n} &= g_1 \quad \text{on } \Gamma_1, \\ \boldsymbol{v} &= \boldsymbol{g}_3 \quad \text{on } \Gamma_3, \end{aligned} \tag{6.87}$$

having the same variational formulation substituting $H_0^1(\Omega)^d$ by $\{\boldsymbol{v} \in H^1(\Omega)^d \mid \boldsymbol{v} = 0 \text{ on } \Gamma_2\}$ but conditions for the well-posedness have to be investigated anew.

We now consider Galerkin approximations. The space $X = L_0^2(\Omega)$ seems to incorporate a nonlocal condition in the ansatz space. But note that

$$L_0^2(\Omega) \cong L^2(\Omega)/\mathbb{R} \tag{6.88}$$

and the isomorphism is also isometric (Problem 6.19). Therefore also $X_h \subset L^2(\Omega)/\mathbb{R}$ can be considered in the form that one (nodal) value is fixed and then the corresponding equivalence class is part of the solution if the representative with vanishing integral is chosen. The more severe difficulties arise in the choice of $M_h \subset M$, meaning that the ansatz functions have to be (weakly) divergence-free. We will only sketch a possible approach. For $d = 2$ we can use that the *(vector) rotation*

$$\operatorname{curl} f := (\partial_2 f, -\partial_1 f)^T \tag{6.89}$$

defines a divergence-free field. Thus, for an "appropriate" finite-dimensional vector space Y_h we can choose

$$W_h := \operatorname{curl} Y_h .$$

This can be related to rewriting the (Navier–)Stokes equation to an equation for the *stream function* and considering a Galerkin approximation of this equation. To construct an approximate pressure p_h and for approximate velocity $\boldsymbol{v}_h$ we can solve

$$b(\boldsymbol{w}_h, p_h) = \ell(\boldsymbol{w}_h) - a(\boldsymbol{v}_h, \boldsymbol{w}_h) \quad \text{for all } \boldsymbol{w}_h \in M_h . \tag{6.90}$$

Here we have assumed that as in the continuous case

$$W_h := \{\boldsymbol{w}_h \in M_h \mid \nabla \cdot \boldsymbol{w}_h = 0\},$$

giving an implicit definition of W_h.

Consider additionally

$$\widetilde{W}_h := \left\{\boldsymbol{w}_h \in M_h \mid \langle \nabla \cdot \boldsymbol{w}_h, q_h \rangle_0 = 0 \quad \text{for all } q_h \in X_h \right\}, \tag{6.91}$$

i.e., the space of *discretely divergence-free* elements of M_h. We have the embedding

$$W_h \subset \widetilde{W}_h,$$

but $\widetilde{W}_h = W_h$ only for special pairings of M_h and X_h, as in (7.69) for the Raviart–Thomas element (Darcy equation) or the Taylor–Hood element (⑤ of Section 7.3).

Lemma 6.31 *Let $\boldsymbol{v}_h \in \widetilde{W}_h$ be a solution of the variational equation*

$$a_h(\boldsymbol{v}_h, \boldsymbol{w}_h) = \ell(\boldsymbol{w}_h) \quad \textit{for all } \boldsymbol{w}_h \in \widetilde{W}_h . \tag{6.92}$$

Then, for any choice of X_h and M_h, the pressure equation (6.90) *is solvable.*

Proof Selecting a basis of M_h and X_h,

$$M_h = \operatorname{span}\left\{\boldsymbol{w}_1, \ldots, \boldsymbol{w}_{N_M}\right\}, \quad X_h = \operatorname{span}\left\{\varphi_1, \ldots, \varphi_{N_X}\right\},$$

and setting

$$B := \left(b(\boldsymbol{w}_i, \varphi_j)\right)_{i,j}, \quad A := \left(a(\boldsymbol{w}_j, \boldsymbol{w}_i)\right)_{i,j}, \quad \boldsymbol{b} := \left(\ell(\boldsymbol{w}_i)\right)_i,$$

we see that the problem (6.90) is algebraically equivalent to

$$B\boldsymbol{\eta} = -A\boldsymbol{\xi} + \boldsymbol{b},$$

where $\boldsymbol{\xi} \in \mathbb{R}^{N_M}$, $\boldsymbol{\eta} \in \mathbb{R}^{N_X}$ denote the representation vectors of $\boldsymbol{v}_h$ and p_h with respect to the above bases.

The existence of a discrete pressure is ensured if

$$-A\boldsymbol{\xi} + \boldsymbol{b} \in \operatorname{im}(B) = \left(\ker B^T\right)^{\perp}.$$

To characterize an element $\widetilde{\boldsymbol{\xi}} \in \ker B^T \subset \mathbb{R}^{N_M}$ we observe that

$$B^T\widetilde{\boldsymbol{\xi}} = 0 \iff 0 = \langle B^T\widetilde{\boldsymbol{\xi}}, \widetilde{\boldsymbol{\eta}}\rangle = b(\boldsymbol{w}_h, q_h) \quad \text{for all } \widetilde{\boldsymbol{\eta}} \in \mathbb{R}^{N_X}, \tag{6.93}$$

where $\boldsymbol{w}_h := \sum_i \widetilde{\xi}_i \boldsymbol{w}_i$ and $q_h := \sum_j \widetilde{\eta}_j \varphi_j$, i.e.,

$$\widetilde{\boldsymbol{\xi}} \in \ker B^T \iff \boldsymbol{w}_h \in \widetilde{W}_h.$$

Hence the solvability condition turns into

$$\langle A\boldsymbol{\xi} - \boldsymbol{b}, \widetilde{\boldsymbol{\xi}}\rangle = 0 \quad \text{for all } \widetilde{\boldsymbol{\xi}} \in \ker B^T$$

which is equivalent to

$$a(\boldsymbol{v}_h, \boldsymbol{w}_h) - \ell(\boldsymbol{w}_h) = 0 \quad \text{for all } \boldsymbol{w}_h \in \widetilde{W}_h.$$

The velocity $\boldsymbol{v}_h$ satisfies the latter condition by assumption (6.92). □

In general the equation (6.92) is a nonconforming approximation of (6.84), conforming only for $W_h = \widetilde{W}_h$.

Nevertheless, to achieve stability and thus (order of) convergence estimates, X_h has to be chosen such that the discrete inf-sup condition holds uniformly (see condition 2) of Theorem 6.17). Then, by Theorem 6.25, an error estimate for the pressure follows in terms of the approximation error of the discrete pressure space and the convergence error of the velocity approximation.

Exercises

Problem 6.15 A solution $\boldsymbol{v} \in C^2(\Omega) \cap C(\overline{\Omega}), p \in C^1(\Omega)$ of Formulation II is a classical solution of (6.79), (6.82).

Problem 6.16 Let $\Omega \subset \mathbb{R}^3$ be a domain and $(\boldsymbol{v}, p)$ a smooth solution of the incompressible Navier–Stokes equations (ENS2b), (ENS1a). For constant ρ and η show

$$\frac{d}{dt}\Big(\int_\Omega \rho|\boldsymbol{v}|^2 dx\Big) + \int_\Omega \rho\eta|D\boldsymbol{v}|^2 dx = 0,$$

where $|D\boldsymbol{v}|^2 := \sum_{i=1}^3 |\nabla v_i|^2$.

Problem 6.17 a) Let ρ be a constant density, $\boldsymbol{v}$ be a stationary, incompressible flow field in the variables x_1, x_2, being curl-free, i.e., curl $\boldsymbol{v} = 0$ (see (6.89)). Show that $\boldsymbol{v}$ is a solution of the Euler equation (ENS2a) with $p = -\rho|\boldsymbol{v}|/2$.

b) In $\mathbb{R}^2 \setminus \{(0,0)\}$, the velocity field

$$\boldsymbol{v}(x_1, x_2) = \frac{1}{x_1^2 + x_2^2} \begin{pmatrix} -x_2 \\ x_1 \end{pmatrix}$$

fulfils the assumptions of a).

c) The velocity $\boldsymbol{v}$ of b) does not describe a potential flow, i.e., there is no function φ such that $\boldsymbol{v} = \nabla\varphi$.

Problem 6.18 Conclude from Theorem 6.30 *de Rham's Theorem*: Linear functionals on $H_0^1(\Omega)^d$ that are zero on $\ker(\nabla\cdot)$ are gradients of functions in $L_0^2(\Omega)$.

Problem 6.19 Show the relation (6.88) and that an isometric isomorphism exists.

Chapter 7
Mixed and Nonconforming Discretization Methods

The FEM studied in Chapter 3 and the node-oriented FVM in Chapter 8 are all conforming with a variational formulation based on $V \subset H^1(\Omega)$, and thus exhibit inter-element continuity (see Theorem 3.21). In the following we will call such methods *continuous Galerkin methods* (*CGM*). On the positive side, this reduces the degree of freedom, but also has several negative consequences.

- A consistent partition is required (see (T5), (T6)) making refinement more complex and less local.
- The reconstruction of a flux density to render the method locally conservative is not obvious.

To address the last point (with a conforming method), the dual mixed formulation has been developed in Section 6.2, giving up the coercivity of the variational problem, for which FEM will be developed in Section 7.2. As an alternative, the primal, i.e., coercive formulation of Chapter 3, can be maintained, but the conformity is given up by relaxing or abandoning the inter-element continuity. To its full extent, this is done in the *discontinuous Galerkin method* (*DG method* or *DGM*), to be discussed in Section 7.4. Here we will discuss the *Crouzeix–Raviart* element, not so much because of its practical importance, but because it elucidates the transition from CGM to DGM and also has relations to the *mixed FEM* (*MFEM*).

7.1 Nonconforming Finite Element Methods I: The Crouzeix–Raviart Element

Based on the framework of Chapter 3, we want to construct a finite element on a regular d-simplex $K = \operatorname{conv}\{a_1, \dots, a_{d+1}\}$. We start with the local approximation space

$$P = P_K := \mathcal{P}_1(K)$$

as in (3.64) but make a different choice of degrees of freedom $\psi_i \in \Sigma, i = 1, \dots, d+1$:

P. Knabner and L. Angermann, *Numerical Methods for Elliptic and Parabolic Partial Differential Equations*, Texts in Applied Mathematics 44,
https://doi.org/10.1007/978-3-030-79385-2_7

If $F_i := \text{conv}\,\{a_j | j \neq i\} \in \mathcal{F}_K$ denote the faces of K, then

$$\psi_i(p) := \frac{1}{|F_i|_{d-1}} \int_{F_i} p \, d\sigma \quad \text{for all } p \in P. \tag{7.1}$$

It remains to show that the corresponding local interpolation problem is uniquely solvable. As $\dim P = d + 1 = |\Sigma|$, it is sufficient to specify the shape functions. Using barycentric coordinates λ_i (see (3.57)), we set

$$N_i(x) := d\Big(\frac{1}{d} - \lambda_i(x)\Big), \quad i = 0, \dots, d. \tag{7.2}$$

Then

$$N_i \in \mathcal{P}_1(K) = P$$

and

$$N_i|_{F_i} = 1, \; N_i(a_i) = 1 - d,$$

and thus

$$\psi_i(N_i) = \frac{1}{|F_i|_{d-1}} \int_{F_i} 1 \, d\sigma = 1,$$
$$\psi_i(N_j) = \frac{1}{|F_i|_{d-1}} \Big(|F_i|_{d-1} - d \int_{F_i} \lambda_j \, d\sigma\Big), \quad j \neq i.$$

According to (3.135) we have (note the space dimension $d - 1$, and that λ_j, $j \neq i$, are the barycentric coordinates of the $(d - 1)$-simplex F_i)

$$\int_{F_i} \lambda_j \, d\sigma = \frac{1}{d!} \frac{|F_i|_{d-1}}{|\widehat{F}|_{d-1}} = \frac{1}{d} |F_i|_{d-1}$$

and therefore $\psi_i(N_j) = 0$. The finite element is called *Crouzeix–Raviart element*.

Let $a_{S,i}$ be the barycentre of F_i, i.e.,

$$\lambda_i(a_{S,i}) = 0, \quad \lambda_j(a_{S,i}) = \frac{1}{d} \quad \text{for } j \neq i.$$

As we have (see (3.136) for K replaced by the subsimplex F_i)

$$\int_{F_i} p \, d\sigma = |F|_{d-1} p(a_{S,i}) \quad \text{for all } p \in P, \tag{7.3}$$

an equivalent formulation to (7.1) is

$$\psi_i(p) = p(a_{S,i}). \tag{7.4}$$

Note, however, that (7.1) for nonsmooth functions v (instead of polynomials p) only requires $v \in L^1(F_i)$, whereas (7.4) requires $v \in C(F_i)$ to be well defined. Thus the local interpolation operator is well defined, e.g., on $H^1(K)$, given by

$$I_K^{CR}(v) := \sum_{i=0}^{d} \psi_i(v) N_i \,. \tag{7.5}$$

Let $\mathcal{T}_h$ be a consistent triangulation of a polygonally bounded domain Ω and consider the finite elements (K, P_K, Σ_K) from above. In generalizing Definition 3.23 to arbitrary degrees of freedom we have an affine equivalent triangulation, as

$$\frac{1}{|\widehat{F}|_{d-1}} \int_{\widehat{F}} \hat{p}\, d\hat{s} = \frac{1}{|F|_{d-1}} \int_F p\, d\sigma$$

for the faces $F, \widehat{F}$ of $K, \widehat{K}$, respectively, mapped each to the other according to an affine-linear bijection (see (3.68)).

Note that condition (F1) (see (3.52)) is satisfied, but not (F2) (see (3.53)), as $P|_F \in \mathcal{P}_1$ for a face F of the simplex K is not determined by just one degree of freedom. Therefore the approximation space, the global *Crouzeix–Raviart space* V_h (of degree 1), defined by (3.67), has the form

$$\begin{aligned} CR_1(\Omega) := V_h := \{ v \in L^1(\Omega) \;\big|\; & v|_K \in \mathcal{P}_1, \\ & \int_F [\![v]\!]\, d\sigma = 0 \text{ for each interior face } F \subset K,\ K \in \mathcal{T}_h \}. \end{aligned} \tag{7.6}$$

Here we use the notation $[\![\cdot]\!]$ for the jump across $F \in \mathcal{F}$, where $\mathcal{F}$ is the set of interior faces, i.e., for two simplices $K, K' \in \mathcal{T}_h$ with a common face $F = K \cap K'$ we set

$$[\![v]\!] := [\![v]\!]_F := v|_K\, \mathbf{n}_K + v|_{K'}\, \mathbf{n}_{K'} \,. \tag{7.7}$$

Note that, in general, $v \in V_h$ is bivalued on $\mathcal{F}$. For a boundary face $F \in \partial\mathcal{F}$, $F \subset K \in \mathcal{T}_h$, we define

$$[\![v]\!]_F := v|_K\, \mathbf{n}_K \,. \tag{7.8}$$

To include homogeneous Dirichlet boundary conditions on $\widetilde{\Gamma}_3 \subset \partial\Omega$, we assume that each of the boundary subsets $\widetilde{\Gamma}_i$, $i = 1, 2, 3$, is related to a set of faces $\mathcal{F}_i$ as follows:

$$\widetilde{\Gamma}_i = \bigcup_{F \in \mathcal{F}_i} F, \quad i = 1, 2, \quad \widetilde{\Gamma}_3 = \operatorname{int}\Big(\bigcup_{F \in \mathcal{F}_3} F \Big), \tag{7.9}$$

and then

$$V_{h,0} := \{ v \in V_h \mid \int_F v\, d\sigma = 0 \quad \text{for each } F \in \mathcal{F}_3 \} \,. \tag{7.10}$$

Let M be the total number of degrees of freedom, i.e., $M := |\overline{\mathcal{F}}| = |\mathcal{F} \cup \partial\mathcal{F}| = |\mathcal{F} \cup \mathcal{F}_1 \cup \mathcal{F}_2 \cup \mathcal{F}_3|$. Then we can choose an enumeration of all faces in such a way that the elements of $\mathcal{F} \cup \mathcal{F}_1 \cup \mathcal{F}_2$ are enumerated from 1 to $M_1 := |\mathcal{F} \cup \mathcal{F}_1 \cup \mathcal{F}_2|$ and the elements of $\mathcal{F}_3$ from $M_1 + 1$ to M. To obtain a global interpolation operator into V_h (compare Theorem 3.24), it is sufficient to find a nodal basis $\varphi_1, \ldots, \varphi_M$ of V_h such that

$$\psi_i(\varphi_j) = \delta_{ij}, \quad i, j = 1, \ldots, M, \tag{7.11}$$

for the global degrees of freedom

$$\psi_i : H^1(\mathcal{T}_h) \to \mathbb{R},\ v \mapsto \frac{1}{|F_i|_{d-1}} \int_{F_i} v\, d\sigma, \quad i = 1, \ldots, M, \tag{7.12}$$

where $H^k(\mathcal{T}_h)$, $k \in \mathbb{N}$, is the *broken Sobolev space*

$$H^k(\mathcal{T}_h) := \{v \in L^2(\Omega) \mid v|_K \in H^k(K) \quad \text{for all } K \in \mathcal{T}_h\}$$

with the norm

$$\|v\|_{k,\mathcal{T}_h} := \Big(\sum_{K \in \mathcal{T}_h} \|v\|_{k,K}^2 \Big)^{1/2},$$

which already has been used in Section 3.4 (see Theorem 3.32). The functions φ_i are defined as follows. Let $F_i \in \mathcal{F}$, i.e., $F_i = K \cap K'$, then φ_i is defined on K, K' according to N_i from (7.2) applied to K and K', respectively, and $\varphi_i = 0$ for all others elements of $\mathcal{T}_h$. If $F_i \in \mathcal{F}_1 \cup \mathcal{F}_2 \subset \partial\mathcal{F}$, then φ_i is defined as above but only nonzero on the element K with $F_i \subset K$. Hence we observe a reduction in the size of the support of a nodal basis function (see Figure 7.1 (right)) compared to the conforming situation (see Figure 7.1 (left) and also Figure 3.11) to at most two elements.

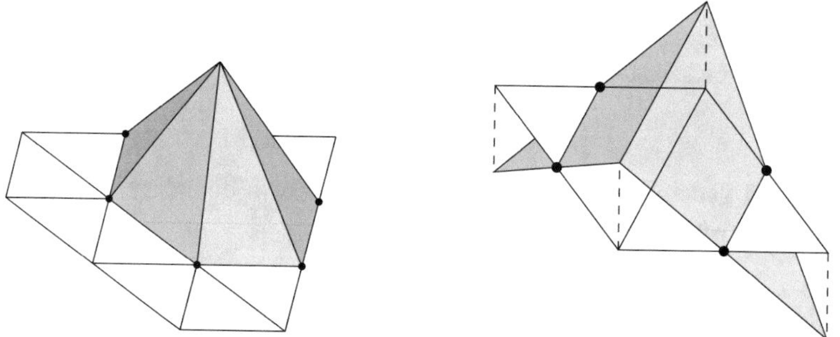

Fig. 7.1: Global shape function for the conforming P_1-approximation space (left) and a global shape function for the discontinuous P_1-approximation space (right).

Lemma 7.1 *The functions φ_i, $i = 1, \ldots, M$, defined above by* (7.11)*, form a nodal basis of V_h. The functions φ_i, $i = 1, \ldots, M_1$, related to $F_i \in \mathcal{F} \cup \mathcal{F}_1 \cup \mathcal{F}_2$, are nodal basis functions of $V_{h,0}$.*

Proof For $F_i \in \mathcal{F}$, $F_i = K \cap K'$, we have by definition (7.11), (7.12)

$$\int_{F_i} \varphi_{i|_K}\, d\sigma = \int_{F_i} \varphi_{i|_{K'}} d\sigma.$$

If φ_i is integrated over $F_j \in \mathcal{F}$, $j \neq i$, both integrals vanish, either by definition of the shape function or by the zero extension. On $F_i \in \mathcal{F}_3$, by the same reasons the degree of freedom vanishes, i.e., $\varphi_i \in V_{h,0}$. By definition, we have

$$\psi_i(\varphi_j) = \delta_{ij}, \quad i, j = 1, \dots, M, \tag{7.13}$$

and therefore for each $v \in V_h$

$$v = \sum_{i=1}^{M} \psi_i(v)\varphi_i \,. \tag{7.14}$$

This can be seen by restricting (7.14) to $K \in \mathcal{T}_h$. Then $v|_K$ is determined by the local degrees of freedom. Applying one of these, say with the global index j, to the right-hand side, gives due to (7.13) also $\psi_j(v)$, i.e., the functions coincide on K and therefore also on Ω. The identity (7.14) shows that φ_i, $i = 1, \dots, M$, is a basis of V_h, in fact the dual basis to the basis $\psi_1, \dots, \psi_M$ of span $\{\psi_1, \dots, \psi_M\} \subset V_h'$. □

Therefore also the global interpolation operator I_h^{CK} is well defined on $H^1(\mathcal{T}_h)$.

Remark 7.2 1) The jump condition $\int_F [\![v]\!] \, d\sigma = \mathbf{0}$ for all $v \in \mathcal{P}_1(K)$ may be seen as a relaxed inter-element continuity condition, as it is equivalent to the orthogonality relation

$$\int_F [\![v]\!] \cdot \boldsymbol{q} \, d\sigma = 0 \quad \text{for all } \boldsymbol{q} \in \mathcal{P}_0(F)^d .$$

2) By 1) the following generalization to degree $k \in \mathbb{N}$ is suggested:

$$X_h := \Big\{ v \in L^1(\Omega) \;\Big|\; v|_K \in \mathcal{P}_k, \; \int_F [\![v]\!] \cdot \mathbf{q} \, d\sigma = 0 \quad \text{for all } \boldsymbol{q} \in \mathcal{P}_{k-1}(F)^d \text{ and each interior face } F \subset K,\; K \in \mathcal{T}_h \Big\}. \tag{7.15}$$

This requirement of the continuity of interface moments up to order $k - 1$ is also called the *patch test*.

3) For $d = 2$, i.e., one-dimensional faces, the integrals in the degrees of freedom, i.e., for integrands from $\mathcal{P}_{2k-1}$, are exactly evaluable by Gauss quadrature in k nodes. Let $x_1^F, \dots, x_k^F$ be the Gauss quadrature points on the face F, thus

$$[\![v(x_i^F)]\!] = 0 \quad \text{for } i = 1, \dots, k \text{ and each interior face } F \in \mathcal{F}$$

is sufficient for an element $v \in L^1(\Omega)$, $v|_K \in \mathcal{P}_k$ for $K \in \mathcal{T}_h$, to be in V_h. But the point functionals $p \mapsto p(x_i^F)$ can only be used as degrees of freedom (with further internal points) for k being odd. For even k there is no linear independence of these functionals as the following example of a nontrivial function $b_K \in \mathcal{P}_2$, which vanishes at all Gauss points, shows:

$$b_K(x) := 2 - 3(\lambda_{1,K}^2 + \lambda_{2,K}^2 + \lambda_{2,K}^3).$$

4) For $d = 2$ and $k \in \mathbb{N}$ odd we define the *Crouzeix–Raviart space* $CR_k(\Omega)$ *of degree* k as in (7.15). The local degrees of freedom related to $K \in \mathcal{T}_h$ are given by the functionals

$$\psi_{K,F,i}(v) := \int_F v w_i^{(k)} dx$$

where $w_i^{(k)}$ span $\mathcal{P}_{k-1}(\mathbb{R})$, and

$$\psi_{K,K,i}(v) := \int_K v \tilde{w}_i^{(k)} dx,$$

where $\tilde{w}_i^{(k)}$ span $\mathcal{P}_{k-3}(\mathbb{R}^2)$ (for $k \geq 3$) (Problem 7.1).
In this way we define local and global interpolation operators

$$\begin{aligned} I_K : H^1(K) &\to \mathcal{P}_k(K), \\ I_h : H^1(\mathcal{T}_h) &\to CR_k(\Omega) \end{aligned}$$

(the domains of definition can be enlarged). Analogously to the proof of Theorem 3.32 we can show the estimate

$$\|I_h v - v\|_0 \leq C h^{k+1} \|v\|_{k+1,\mathcal{T}_h} \tag{7.16}$$

(Problem 7.3), and by scaling we get, for $K \in \mathcal{T}_h$,

$$\|I_K v\|_{0,K} \leq C\big(\|v\|_{0,K} + h_K^{1/2} \|v\|_{0,\partial K}\big) \tag{7.17}$$

(Problem 7.4).

We now consider the elliptic boundary value problem in divergence form (3.36), (3.37)–(3.39) and start with the symmetric case $\boldsymbol{c} = \boldsymbol{0}$. We assume that it has a weak solution $u \in H^1(\Omega)$.

Regularity of the solution

To investigate the consistency error, additional regularity of u is necessary. We need

(1)
$$\boldsymbol{p}(u) = -\boldsymbol{K}\nabla u + \boldsymbol{c}u \in H(\mathrm{div}; \mathcal{T}_h), \tag{7.18}$$

where

$$H(\mathrm{div}; \mathcal{T}_h) := \{\boldsymbol{q} \in L^2(\Omega)^d \mid \boldsymbol{q}|_K \in H(\mathrm{div}; K) \text{ for all } K \in \mathcal{T}_h\},$$

such that integration by parts is possible:

$$\int_K \nabla \cdot \boldsymbol{p}(u) v dx = -\int_K \boldsymbol{p}(u) \cdot \nabla v \, dx + \int_{\partial K} \boldsymbol{p}(u) \cdot \mathfrak{n} \, v \, d\sigma$$
$$\text{for all } v \in H^1(K),\ K \in \mathcal{T}_h,$$

(2)

$$\boldsymbol{p}(u)|_K \cdot \mathfrak{n}|_K + \boldsymbol{p}(u)|_{K'} \cdot \mathfrak{n}|_{K'} = 0, \quad F = K \cap K',\ K, K' \in \mathcal{T}_h, \tag{7.19}$$

i.e., there is the inter-element continuity of the normal flux densities or local mass conservation,

(3)

$$\begin{aligned} \nabla \cdot \boldsymbol{p}(u) + ru - f &\in L^2(\Omega), \\ -\boldsymbol{p}(u) \cdot \mathfrak{n} - \widetilde{g}_1 &\in L^2(\widetilde{\Gamma}_1), \\ -\boldsymbol{p}(u) \cdot \mathfrak{n} + \tilde{\alpha} u - \widetilde{g}_2 &\in L^2(\widetilde{\Gamma}_2), \\ u - \widetilde{g}_3 &\in L^2(\widetilde{\Gamma}_3). \end{aligned} \tag{7.20}$$

For sufficiently smooth coefficients and source terms, $u \in H^2(\Omega)$ is sufficient, but actually only the smoothness of the coefficients and source terms on $K \in \mathcal{T}_h$ or $F \in \overline{\mathcal{F}}$, respectively, is required, allowing, e.g., for conductivities with jumps.

As X_h is not a subspace of $H^1(\Omega)$, but only of $H^1(\mathcal{T}_h)$, to set up a nonconsistent finite element method we define

$$\begin{aligned} a_h(w, v) &:= \sum_{K \in \mathcal{T}_h} \int_K \boldsymbol{K} \nabla w \nabla v \, dx + \int_\Omega r w v \, dx + \int_{\widetilde{\Gamma}_2} \tilde{\alpha} w v \, d\sigma, \\ \ell_h(v) &:= \ell(v) \qquad \text{for all } w, v \in H^1(\mathcal{T}_h) \end{aligned} \tag{7.21}$$

(cf. (3.32)), and the discretization method (for homogeneous Dirichlet data, with $V_{h,0}$ defined in (7.10)) reads as follows:

Find $u_h \in V_{h,0}$ such that

$$a_h(u_h, v) = \ell_h(v) \quad \text{for all } v \in V_{h,0}\,. \tag{7.22}$$

To derive an order of convergence estimate, we use Theorem 6.10 and have to estimate the approximation order of $V_{h,0}$ and the consistency error. Because of $V_h^c \subset V_h$, the corresponding conforming approximation based on $\mathcal{P}_1$ (see (3.64)), we can conclude from Theorem 3.32 and Remark 3.33, 1) assuming that $(\mathcal{T}_h)_h$ is shape-regular

$$\inf_{v_h \in V_h} \|u - v_h\|_{1,\mathcal{T}_h} \le Ch|u|_{2,\mathcal{T}_h}\,, \tag{7.23}$$

but here we need $d \le 3$. Otherwise we can use the Crouzeix–Raviart interpolation operator from Remark 7.2, 4). We have $U(h) := U_h := H^1(\mathcal{T}_h)$. The coercivity of a_h follows from the reasoning for Theorem 3.16, applied to each element $K \in \mathcal{T}_h$, and gives a uniform constant α under the conditions stated there, and correspondingly for the continuity of a_h, ℓ_h.

We remind and extend the definition of jumps and averages:

Let $F = K \cap K' \in \mathcal{F}$ be an interior face of the triangulation $\mathcal{T}_h$, $K, K' \in \mathcal{T}_h$. For scalar functions v such that $v|_K \in H^1(K)$, $v|_{K'} \in H^1(K')$ we set

$$\begin{aligned} [\![v]\!] := [\![v]\!]_F &:= v|_K\, \mathfrak{n}_K + v|_{K'}\, \mathfrak{n}_{K'}, \\ \{\!\!\{v\}\!\!\} := \{\!\!\{v\}\!\!\}_F &:= \frac{1}{2}(v|_K + v|_{K'}); \end{aligned} \tag{7.24}$$

for vector fields $\boldsymbol{p}$ such that $\boldsymbol{p}|_K \in H(\mathrm{div}; K)$, $\boldsymbol{p}|_{K'} \in H(\mathrm{div}; K')$ we define

$$\begin{aligned} [\![\boldsymbol{p}]\!] := [\![\boldsymbol{p}]\!]_F &:= \boldsymbol{p}|_K \cdot \mathfrak{n}_K + \boldsymbol{p}|_{K'} \cdot \mathfrak{n}_{K'}, \\ \{\!\!\{\boldsymbol{p}\}\!\!\} := \{\!\!\{\boldsymbol{p}\}\!\!\}_F &:= \frac{1}{2}(\boldsymbol{p}|_K + \boldsymbol{p}|_{K'}), \end{aligned} \tag{7.25}$$

with corresponding modifications for $F \in \partial\mathcal{F}$, $F \subset K$ as in (7.8):

$$\{\!\!\{v\}\!\!\}_F := v|_K, \quad [\![\boldsymbol{p}]\!]_F := \boldsymbol{p}_K \cdot \mathfrak{n}_K, \quad \{\!\!\{\boldsymbol{p}\}\!\!\}_F := \boldsymbol{p}|_K.$$

Note that jumps of scalar functions are vector fields and vice versa, but averages keep the type.

Then by integration by parts we get for $v \in H^2(\mathcal{T}_h)$, $w \in H^1(\mathcal{T}_h)$

$$\begin{aligned} \sum_{K \in \mathcal{T}_h} \int_K \boldsymbol{K}\nabla v \cdot \nabla w \, dx &= \sum_{F \in \mathcal{F} \cup \mathcal{F}_3} \int_F \{\!\!\{\boldsymbol{K}\nabla v\}\!\!\}_F \cdot [\![w]\!]_F \, d\sigma \\ &+ \sum_{F \in \mathcal{F} \cup \mathcal{F}_1 \cup \mathcal{F}_2} \int_F [\![\boldsymbol{K}\nabla v]\!]_F \{\!\!\{w\}\!\!\}_F \, d\sigma - \sum_{K \in \mathcal{T}_h} \int_K \nabla \cdot (\boldsymbol{K}\nabla v) w \, dx \, . \end{aligned} \tag{7.26}$$

This formula results from elementwise integration by parts of the left-hand side and

$$\begin{aligned} \sum_{K \in \mathcal{T}_h} \int_{\partial K} \boldsymbol{K}\nabla v \cdot \mathfrak{n}_K w \, d\sigma = &\sum_{\substack{F \in \mathcal{F} \\ F = K \cap K'}} \int_F \left[\boldsymbol{K}\nabla v|_K \cdot \mathfrak{n}_K w|_K + \boldsymbol{K}\nabla v|_{K'} \cdot \mathfrak{n}_{K'} w|_{K'} \right] d\sigma \\ &+ \sum_{\substack{F \in \partial\mathcal{F} \\ F \subset K}} \int_F \boldsymbol{K}\nabla v|_K \cdot \mathfrak{n}_K w|_K \, d\sigma, \end{aligned}$$

where the first term on the right-hand side can be split into the first and second sum for $F \in \mathcal{F}$, the second term gives the contribution of $\mathcal{F}_3$ to the first and of $\mathcal{F}_1 \cup \mathcal{F}_2$ to the second sum in (7.26), taking $\mathfrak{n}_{K'} = -\mathfrak{n}_K$ into account (Problem 7.5).

Thus the consistency error can be rewritten as follows, using (7.19), i.e., $[\![\boldsymbol{K}\nabla u]\!]_F = 0$, for an interior face $F \in \mathcal{F}$:

$$
\begin{aligned}
a_h(u, v_h) - \ell_h(v_h) = &\int_\Omega (-\nabla \cdot (\boldsymbol{K}\nabla u) + ru - f)v_h \, dx \\
&+ \sum_{F \in \mathcal{F} \cup \mathcal{F}_3} \int_F \{\!\!\{\boldsymbol{K}\nabla u\}\!\!\} \cdot [\![v_h]\!]_F \, d\sigma + \sum_{F \in \mathcal{F}_1 \cup \mathcal{F}_2} \int_F [\![\boldsymbol{K}\nabla u]\!]_F \{\!\!\{v_h\}\!\!\}_F \, d\sigma \\
&+ \int_{\mathcal{F}_2} \tilde{\alpha} u v_h \, d\sigma - \int_{\mathcal{F}_1} \tilde{g}_1 v_h \, d\sigma - \int_{\mathcal{F}_2} \tilde{g}_2 v_h \, d\sigma \quad \text{for all } v_h \in V_{h,0} \, .
\end{aligned} \tag{7.27}
$$

Because the differential equation and the boundary conditions are fulfilled in $L^2(\Omega)$ and $L^2(\mathcal{F}_1 \cup \mathcal{F}_2)$, respectively, the first term and the last four terms vanish, hence

$$
\begin{aligned}
a_h(u, v_h) - \ell_h(v_h) &= \sum_{F \in \mathcal{F} \cup \mathcal{F}_3} \int_F \boldsymbol{K}\nabla u \cdot [\![v_h]\!]_F \, d\sigma \\
&= \sum_{F \in \mathcal{F} \cup \mathcal{F}_3} \int_F (\boldsymbol{K}\nabla u - \Pi_K(\boldsymbol{K}\nabla u)) \cdot [\![v_h]\!]_F \, d\sigma \, .
\end{aligned}
$$

Here $\Pi_K(\boldsymbol{K}\nabla u) := \int_K \boldsymbol{K}\nabla u \, d\sigma / |K|$, $K \supset F$, is a constant vector and thus the added contribution vanishes due to the weak continuity condition in V_h (see Remark 7.2, 1)) and the Dirichlet boundary conditions. Therefore, by the Cauchy–Schwarz inequality,

$$
\begin{aligned}
|a_h(u, v_h) - \ell_h(v_h)| \le &\Big(\sum_{F \in \mathcal{F} \cup \mathcal{F}_3} h_F \big\| \boldsymbol{K}\nabla u - \Pi_K(\boldsymbol{K}\nabla u) \big\|_{0,F}^2 \Big)^{1/2} \\
&\times \Big(\sum_{F \in \mathcal{F} \cup \mathcal{F}_3} h_F^{-1} \big\| [\![v_h]\!] \big\|_{0,F}^2 \Big)^{1/2} .
\end{aligned} \tag{7.28}
$$

To complete the proof, we develop—also for further use—several relationships between functions on simplices and their traces on faces. The general assumption is a shape-regular family $(\mathcal{T}_h)_h$ of consistent triangulations, where the constant in Definition 3.31 is denoted by $\sigma > 0$. As always h_K, h_F denote the diameters of the simplices K, F in $\mathbb{R}^d$ or $\mathbb{R}^{d-1}$, respectively.

Lemma 7.3 *1) There exist constants $C_1, C_2 > 0$ independent of h_K such that*

$$
C_1 h_K \le h_F \le C_2 h_K \quad \textit{for all } K \in \mathcal{T}_h, \; F \in \overline{\mathcal{F}}, \; F \subset K.
$$

2) If $B \in \mathbb{R}^{d,d}$ denotes the matrix of the affine-linear mapping (3.87) *of $\hat{K}$ to K in $\mathbb{R}^d$, then there exist constants $C_1, C_2 > 0$ independent of h_K such that*

$$
C_1 h_K^d \le |\det(B)| \le C_2 h_K^d .
$$

Proof For 1), see [22, Lemma 2.5 (i)]. The second statement follows from (3.61) and the shape regularity of $(\mathcal{T}_h)_h$. □

Lemma 7.4 (Multiplicative trace inequality) *There exists a constant $C > 0$ such that*

$$\|v\|_{0,\partial K}^2 \le C\left(\|v\|_{0,K}|v|_{1,K} + h_K^{-1}\|v\|_{0,K}^2\right) \quad \textit{for all } K \in \mathcal{T}_h,\ v \in H^1(K).$$

Proof See [22, Lemma 2.19]. □

Finally we need a scaled embedding estimate.

Lemma 7.5 *Let $v \in C(K) \cap H^1(K)$ be such that $v(x) = 0$ for some $x \in F$, $F \subset \partial K$. Then there exists a constant $C > 0$ such that*

$$\|v\|_{0,F} \le Ch_K^{1/2}|v|_{1,K} \quad \textit{for all } v \in H^1(K).$$

Proof For $K = \hat{K}$ the statement follows from the Trace Theorem 3.6 and the Poincaré inequality (cf. Theorem 2.18, having in mind that for $v \in C(K)$ the assumption $v \in H_0^1(K)$ can be replaced by $v(x) = 0$ for some $x \in K$).

For general K it follows from the transformation estimates of Theorem 3.29 applied to K in $\mathbb{R}^d$ and F in $\mathbb{R}^{d-1}$, taking Lemma 7.3, 2) into account. □

Theorem 7.6 *Let $d \le 3$, $(\mathcal{T}_h)_h$ be shape-regular, the weak solution $u \in H^1(\Omega) \cap H^2(\mathcal{T}_h)$ be such that (7.18)–(7.20) are satisfied, and $\boldsymbol{K}$ so smooth that $\boldsymbol{K}\nabla u \in H^1(\mathcal{T}_h)^d$. Then the Crouzeix–Raviart discretization of order 1, i.e., V_h given by (7.10) and the discrete problem by (7.21), satisfies the error estimate*

$$\|u - u_h\|_{1,\mathcal{T}_h} \le Ch(|u|_{2,\mathcal{T}_h} + |\boldsymbol{K}\nabla u|_{1,\mathcal{T}_h}),$$

where $C > 0$ is a constant independent of h.

Proof We continue with the estimate (7.28). According to Lemma 7.4, Remark 3.33, 3) for $m = k = 0$, and Lemma 7.3, 1), the first factor allows the estimate

$$Ch|\boldsymbol{K}\nabla u|_{1,\mathcal{T}_h}.$$

The summand of the second factor can be written as

$$h_F^{-1}\left\|[\![v_h - v(a_{S,F})]\!]\right\|_{0,F}^2 =: h_F^{-1}\left\|[\![\tilde{v}_h]\!]\right\|_{0,F}^2,$$

so that $\tilde{v}_{h|K}$ and $\tilde{v}_{h|K'}$ vanish at the same point in F, $F = K \cap K'$. Thus Lemma 7.5 is applicable, leading together with Lemma 7.3, 1) to the following estimate of the second factor:

$$C\Bigg(\sum_{\substack{F\in\mathcal{F}\\ F=K\cap K'}}\left[h_K^{-1}h_K|v_h|_{1,K}^2 + h_{K'}^{-1}h_{K'}|v_h|_{1,K}^2\right] + \sum_{\substack{F\in\mathcal{F}_3\\ F\subset K}} h_K^{-1}h_K|v_h|_{1,K}^2\Bigg)^{1/2} \le C|v_h|_{1,\mathcal{T}_h}.$$

□

We now include the convective term, i.e., we consider the equation

$$-\nabla\cdot(\boldsymbol{K}\nabla u - \boldsymbol{c}u) + ru = f \quad \text{in } \Omega.$$

The bilinear form has to be chosen such that both the uniform V_h-coercivity (or inf-sup condition) holds and the consistency error can be estimated appropriately. We define

$$\begin{aligned} a_h(u,v) := &\sum_{K\in\mathcal{T}_h}\int_K \left[\boldsymbol{K}\nabla u\cdot\nabla v - u\,\boldsymbol{c}\cdot\nabla v\right]dx + \int_\Omega ruv\,dx \\ &+ \sum_{F\in\mathcal{F}\cup\mathcal{F}_3}\int_F \boldsymbol{c}_{\mathrm{upw}}(u)\cdot[\![v]\!]\,d\sigma + \int_{\widetilde{\Gamma}_2}\tilde{\alpha}uv\,d\sigma. \end{aligned} \tag{7.29}$$

Here the *upwind evaluation* of $\boldsymbol{c}w$ at $F\in\mathcal{F}$, $F=K\cap K'$, for $w\in H^1(K\cup K')$ is defined as follows:

$$\boldsymbol{c}_{\mathrm{upw}}(w) := \begin{cases} \boldsymbol{c}w|_K & \text{for } \boldsymbol{c}\cdot\boldsymbol{n}_K > 0,\\ \boldsymbol{c}w|_{K'} & \text{for } \boldsymbol{c}\cdot\boldsymbol{n}_K \le 0, \end{cases} \tag{7.30}$$

and for a face $F\in\partial\mathcal{F}$:

$$\boldsymbol{c}_{\mathrm{upw}}(w) := \begin{cases} \boldsymbol{c}w|_K & \text{for } \boldsymbol{c}\cdot\boldsymbol{n}_K > 0,\\ 0 & \text{for } \boldsymbol{c}\cdot\boldsymbol{n}_K \le 0. \end{cases} \tag{7.31}$$

Concerning the V_h-coercivity we have, for $u_h\in V_h$,

$$\begin{aligned} &-\sum_{K\in\mathcal{T}_h}\int_K u_h\,\boldsymbol{c}\cdot\nabla u_h\,dx + \sum_{F\in\mathcal{F}\cup\mathcal{F}_3}\int_F \boldsymbol{c}_{\mathrm{upw}}(u_h)\cdot[\![u_h]\!]\,d\sigma \\ =&\frac{1}{2}\sum_{K\in\mathcal{T}_h}\int_{K\in\mathcal{T}_h}\nabla\cdot\boldsymbol{c}\,u_h^2\,dx - \frac{1}{2}\sum_{\substack{F\in\partial\mathcal{F}\\ F\subset K}}\int_F \boldsymbol{c}\cdot\boldsymbol{n}_K u_h^2\,d\sigma \\ &+\sum_{F\in\mathcal{F}}\int_F\left[\boldsymbol{c}_{\mathrm{upw}}(u_h)\cdot[\![u_h]\!] - \frac{1}{2}\boldsymbol{c}\cdot[\![u_h^2]\!]\right]d\sigma + \sum_{F\in\mathcal{F}_3}\int_F \boldsymbol{c}_{\mathrm{upw}}(u_h)\cdot[\![u_h]\!]\,d\sigma. \end{aligned} \tag{7.32}$$

The constituent parts of the second term belonging to $\mathcal{F}_1$ and $\mathcal{F}_2$, together with the corresponding boundary integral from (7.29), are nonnegative due to assumption (3.43), 2), 3). By means of some algebraic manipulation the last but one term can be rewritten as

$$\frac{1}{2}\sum_{F\in\mathcal{F}}\int_F |\boldsymbol{c}\cdot\boldsymbol{n}|[\![u_h^2]\!]\,d\sigma$$

so that under the assumption (3.43), 1) and the additional assumption to handle the contributions from $\mathcal{F}_3$

$$\boldsymbol{c}\cdot\boldsymbol{n}\le 0 \quad \text{on } \widetilde{\Gamma}_3 \tag{7.33}$$

all new terms contribute to nonnegative terms such that the uniform V_h-coercivity is maintained. Condition (7.33) can be avoided for $u_h = 0$ on $\widetilde{\Gamma}_3$. For a weak solution $u \in H^2(\Omega)$, taking into account that $\boldsymbol{c}_{\text{upw}}(u) = \boldsymbol{c}u$ due to Theorem 3.11, the possible amendment to the consistency error is

$$\begin{aligned} &-\sum_{K\in\mathcal{T}_h}\int_K u\,\boldsymbol{c}\cdot\nabla v_h\,dx + \sum_{F\in\mathcal{F}\cup\mathcal{F}_3}\int_F u\,\boldsymbol{c}\cdot[\![v_h]\!]\,d\sigma \\ &= \int_\Omega \nabla\cdot(\boldsymbol{c}u)v_h\,dx - \sum_{F\in\mathcal{F}_1\cup\mathcal{F}_2}\int_F u\,\boldsymbol{c}\cdot\boldsymbol{n}v_h\,d\sigma. \end{aligned} \tag{7.34}$$

So the first term enters the vanishing equation error together with $-\nabla\cdot(\boldsymbol{K}\nabla u) + ru - f$, and the last two terms enter the vanishing boundary residua together with the inhomogeneous boundary terms; in sum there is no new contribution to the consistency error.

Theorem 7.7 *Under the assumptions* (3.43) *and* (7.33), *the estimate of Theorem 7.6 also holds true for the approximation of the problem in divergence form (3.36), (3.37)–(3.39) for $V_{h,0}$ according to* (7.10) *and a_h according to* (7.29).

We see that the consistency error can be reduced by including inter-element boundary terms into the variational formulation. In Section 7.5 this will be developed to the ultimate, rendering a consistent method on $\mathcal{P}_k(\mathcal{T}_h)$ (see (7.99)), namely a *discontinuous Galerkin method.*

Order of Convergence Estimates in Weaker Norms

To apply Lemma 6.11, the adjoint problem and its discretization have to be studied. In Section 3.4.3 we have seen that the adjoint problem to (3.13)–(3.16) is (3.36)–(3.39) and vice versa, but with $\tilde{\boldsymbol{c}} := -\boldsymbol{c}$. Therefore we define the discrete adjoint problem as the adjoint of the discrete problem with the bilinear form (7.29). This preserves the V_h-coercivity with a uniform constant. The consistency error again reduces to the one of the symmetric part, as can be seen directly from

$$\sum_{K\in\mathcal{T}_h}\int_K \tilde{\boldsymbol{c}}\cdot\nabla u v_h\,dx + \sum_{F\in\mathcal{F}\cup\mathcal{F}_3}\int_F \tilde{\boldsymbol{c}}\cdot[\![u]\!]v_h\,d\sigma.$$

The second sum vanishes, and the first one contributes directly to the strong form of the differential equation, i.e., no further consistency error appears. Thus the error estimate of Theorem 7.6 also holds true for the adjoint problem. We want to apply Lemma 6.11 for $Z := L^2(\Omega)$. Therefore, provided the adjoint problem is regular in the sense of Definition 3.39, the first term in the estimate of Lemma 6.11 can be estimated by $Ch^2|u|_{2,\Omega}$, and the fourth term vanishes. In fact, this regularity requirement could be weakened so that it is compatible to (7.18)–(7.20), which we do not want to explain here in detail. The third term being a consistency error for the original problem can be estimated similar to (7.28) and the following considerations, but with v_h substituted by $v_g - v_{gh}$, the approximation error of the adjoint problems (6.17), (6.19).

Thus the final estimate reads as follows (see the proof of Theorem 7.6):

$$Ch|u|_{2,\Omega}\,|v_g - v_{gh}|_{1,\mathcal{T}_h} \le Ch^2|u|_{2,\Omega}\,|v_g|_{2,\Omega} \le Ch^2|u|_{2,\Omega}\|g\|_{0,\Omega},$$

giving the desired estimate.

Finally the second term, being a consistency error of the adjoint problem, can be handled similarly with exchanged roles $u \leftrightarrow v_g$, $v_g - v_{gh} \leftrightarrow u - u_h$, leading to the same estimate. In summary,

Theorem 7.8 *Let $d \le 3$, $(\mathcal{T}_h)_h$ be shape-regular, the weak solution $u \in H^1(\Omega) \cap H^2(\mathcal{T}_h)$ be such that* (7.18)–(7.20) *are satisfied,* **K** *so smooth that $\boldsymbol{K}\nabla u \in H^1(\mathcal{T}_h)^d$, and the solution of the adjoint problem* (6.17) *satisfies*

$$\|v_g\|_{2,\Omega} \le \tilde{C}\|g\|_{0,\Omega} \quad \textit{for some constant } \tilde{C} > 0.$$

Then the Crouzeix–Raviart discretization of order 1 *satisfies the error estimate*

$$\|u - u_h\|_{0,\Omega} \le Ch^2(|u|_{2,\mathcal{T}_h} + |\boldsymbol{K}\nabla u|_{1,\mathcal{T}_h}),$$

where $C > 0$ is a constant independent of h.

Exercises

Problem 7.1 Show that the functionals $\psi_{K,F,i}$, $\psi_{K,K,i}$ introduced in Remark 7.2, 4) are local degrees of freedom for the Crouzeix–Raviart space $CR_k(\Omega)$.

Problem 7.2 a) Let $\Omega := (0,1)^2$ and K_{ij} be squares defined by

$$K_{ij} := [ih, (i+1)h] \times [jh, (j+1)h], \quad i, j = 0, 1, \ldots, n-1,$$

with $h := 1/n$ such that $\overline{\Omega} = \bigcup_{i,j=0}^{n-1} K_{ij}$. Moreover, let

$$X_h := \left\{ v_h \in C(\overline{\Omega}) \;\middle|\; v_h|_K \in Q_1(K) \text{ for all } K \in \mathcal{T}_h,\ v_h|_{\partial\Omega} = 0 \right\}.$$

Calculate the stiffness matrix of the Laplacian as finite difference stencil.

b) Can analogous stencils be deduced for the Crouzeix–Raviart element?

Problem 7.3 Prove the estimate (7.16) in Remark 7.2, 4) following the scheme of the proof of Theorem 3.32.

Problem 7.4 Prove the estimate (7.17) in Remark 7.2, 4) using the usual reference transformation technique as in Section 3.4.1.

Problem 7.5 Given a partition $\mathcal{T}_h$ of the domain $\Omega \subset \mathbb{R}^d$, functions $\boldsymbol{K} : \Omega \to \mathbb{R}^{d,d}$ and $v, w : \Omega \to \mathbb{R}$ such that the subsequent terms are well defined.

a) Let $F \in \mathcal{F}$, $F = K \cap K'$ for $K, K' \in \mathcal{T}_h$ and $\mathfrak{n}_K = -\mathfrak{n}_{K'}$. Show that

$$\begin{aligned}&\int_F \left[\boldsymbol{K}\nabla v|_K \cdot \mathfrak{n}_K w|_K + \boldsymbol{K}\nabla v|_{K'} \cdot \mathfrak{n}_K w|_{K'}\right] d\sigma \\ &= \int_F \{\!\!\{\boldsymbol{K}\nabla v\}\!\!\}_F \cdot [\![w]\!]_F d\sigma + \int_F [\![\boldsymbol{K}\nabla v]\!]_F \cdot \{\!\!\{w\}\!\!\}_F d\sigma.\end{aligned}$$

Hint: Consider for $\boldsymbol{a}_K, \boldsymbol{a}_{K'}, \boldsymbol{c}_K, \boldsymbol{c}_{K'} \in \mathbb{R}^d$, $b_K, b_{K'} \in \mathbb{R}$ the expression

$$(\boldsymbol{a}_K + \boldsymbol{a}_{K'}) \cdot (b_K \boldsymbol{c}_K + b_{K'} \boldsymbol{c}_{K'}) + (b_K + b_{K'})(\boldsymbol{a}_K \cdot \boldsymbol{c}_K + \boldsymbol{a}_{K'} \cdot \boldsymbol{c}_{K'}).$$

b) Let $F \in \partial F$, $F \subset K$. Show that

$$\int_F \boldsymbol{K}\nabla v \cdot \mathfrak{n}_K w|_K d\sigma = \int_F \{\!\!\{\boldsymbol{K}\nabla v\}\!\!\}_F \cdot [\![w]\!]_F d\sigma = \int_F [\![\boldsymbol{K}\nabla v]\!]_F \{\!\!\{w\}\!\!\}_F d\sigma.$$

Programming project 7.1 Discretize the boundary value problem from Project 2.3 by means of the nonconforming Crouzeix–Raviart elements.

a) Based on parts of the code from Project 2.3 (e.g., the grid refinement function) write a function that solves the boundary value problem by means of the Crouzeix–Raviart elements $CR_1(\Omega)$ (see (7.6)) with the corresponding constraint for the Dirichlet boundary condition.
b) To test the implementation, consider a square domain Ω with a sector-shaped incision for various complementary opening angles $\alpha = 2\pi/3, 4\pi/3, 3\pi/2, 2\pi$:

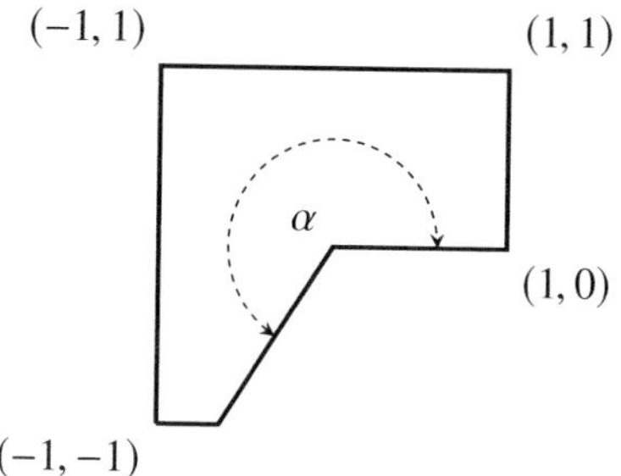

The boundary values are obtained from the exact solution (given in polar coordinates)

$$u(r\cos\varphi, r\sin\varphi) = r^{\pi/\alpha} \sin\left(\frac{\pi}{\alpha}\varphi\right).$$

Compute the l^2-norm of the error for grid sizes $h_i := 1/2^i$, $i = 2, \ldots, 7$, and plot the experimental (minimal) orders of convergence

$$\frac{\ln\left(\|u_{h_{i+1}} - u\|_{0,h_{i+1}} / \|u_{h_i} - u\|_{0,h_i}\right)}{\ln\left(h_{i+1}/h_i\right)}$$

for each specific domain.

c) In case you have completed Project 2.4, compare those results with the case $\alpha = 3\pi/2$.

Hint: Special attention has to be paid to the case of complementary opening angle $\alpha = 2\pi$, where a square with a slit results. The edges of the triangulation must not coincide on top and bottom of the slit.

7.2 Mixed Methods for the Darcy Equation

7.2.1 Dual Formulations in $H(\text{div};\Omega)$

We focus on the mixed dual formulation (6.55) and try to verify the conditions of Theorem 6.17 and Theorem 6.21, both ensuring unique and stable existence. First, let us consider the case of the pure diffusion, i.e., $\boldsymbol{c} = \boldsymbol{0}$, $r = 0$. To cast the problem into the form of (6.55) but with the same solution and test spaces or (6.61), suppose there exists an element $\overline{\boldsymbol{p}} \in H(\text{div};\Omega)$ such that $\overline{\boldsymbol{p}} \cdot \boldsymbol{n} = \widetilde{g}_1$ on $\widetilde{\Gamma}_1$; cf. Remark 6.18, 1). As the image of the normal component trace operator $\gamma_{\boldsymbol{n},\widetilde{\Gamma}_1}$ is $H^{-1/2}(\widetilde{\Gamma}_1)$ according to (6.39), the assumption $\widetilde{g}_1 \in H^{-1/2}(\widetilde{\Gamma}_1)$ is necessary for the existence of a solution and also sufficient to transform the problem into the form (6.61).

To ensure condition (2) of Theorem 6.17 for $v \in X := L^2(\Omega)$ we consider an element $\boldsymbol{q}_v \in M$ according to Lemma 6.14 such that, with the constant $C > 0$ from Lemma 6.14,

$$\nabla \cdot \boldsymbol{q}_v = -v \in L^2(\Omega), \quad \boldsymbol{q}_v \cdot \boldsymbol{n} = 0 \quad \text{on } \widetilde{\Gamma}_1, \quad \|\boldsymbol{q}_v\|_{H(\text{div})} \le C\|v\|_0.$$

Therefore

$$\sup_{\boldsymbol{q} \in M\setminus\{\boldsymbol{0}\}} \frac{b(\boldsymbol{q}, v)}{\|\boldsymbol{q}\|_M} \ge \frac{b(\boldsymbol{q}_v, v)}{\|\boldsymbol{q}_v\|_M} = \frac{\|v\|_X^2}{\|\boldsymbol{q}_v\|_M} \ge C^{-1}\|v\|_X, \tag{7.35}$$

i.e., condition (2) is fulfilled.

As $\boldsymbol{q} \in M$ is allowed, and not only $\boldsymbol{q} \in N^{\perp}$ with $N := \ker B$ (but see Remark 6.18, 3)), this also ensures (2) of Theorem 6.21.

Concerning condition (1) of Theorem 6.17, $a(\boldsymbol{q}, \boldsymbol{q}) = \|\boldsymbol{q}\|_0^2 = \|\boldsymbol{q}\|_M^2$ for $\boldsymbol{q} \in N$, but for the condition (1) of Theorem 6.21, as in Lemma 6.24, a condition like

$$N_h = \ker B_h \subset \ker B = N$$

is necessary.

For the general Darcy equations we refer to Theorem 6.19.

7.2.2 Simplicial Finite Elements in $H(\mathrm{div};\Omega)$

We consider a triangulation $\mathcal{T}_h$ of $\Omega \subset \mathbb{R}^d$ by regular d-simplices (see Definition 3.22) which satisfies (T1)–(T6). Remember that a finite element subspace M_h of M is constructed of finite elements, being a triple of $K \in \mathcal{T}_h$, the finite-dimensional (local) ansatz space $P = P_K$, and the degrees of freedom $\Sigma \subset (P_K)'$ (see Definition 3.23). The space M_h is defined such that

$$P_K = \{\boldsymbol{q}|_K \mid \boldsymbol{q} \in M_h\}. \tag{7.36}$$

Since furthermore the condition $M_h \subset H(\mathrm{div};\Omega)$ has to be satisfied, according to Theorem 6.16, the normal component traces of $\boldsymbol{q}$ across inter-element boundaries have to coincide. Therefore, the normal component traces of these flux densities are expected to be related to the degrees of freedom, namely in the following way.

Suppose $\Sigma_K = \{\psi_1, \ldots, \psi_L\}$, $L = L(K)$, then the *local interpolation problem*

For $d_i \in \mathbb{R}$, $i = 1, \ldots, L$ given, find an element $\boldsymbol{r} \in P_K$ such that it holds

$$\psi_i(\boldsymbol{r}) = d_i \quad \text{for all } i = 1, \ldots, L, \tag{7.37}$$

should be uniquely solvable. If so, also a *local interpolation operator* $I_K : Y_K \to P_K$ is defined, where the local function space Y_K has to be chosen such that the functionals ψ_i are well defined on Y_K, and $I_K(\boldsymbol{q}) \in P_K$ satisfies

$$\psi_i(I_K(\boldsymbol{q})) = \psi_i(\boldsymbol{q}) \quad \text{for all } \boldsymbol{q} \in Y_K. \tag{7.38}$$

Unfortunately, the space Y_K usually has to be chosen slightly smaller than

$$M_K := \{\boldsymbol{q}|_K \mid \boldsymbol{q} \in M\}.$$

In the primal, conforming case of Chapter 3 we took $Y_K = H^2(K)$ and $d \le 3$ instead of $Y_K = H^1(K)$ to ensure $Y_K \subset C(K)$ for the nodal degrees of freedom to be well defined.

Finally, on $Y := \{\boldsymbol{q} \mid \boldsymbol{q}|_K \in Y|_K \text{ for all } K \in \mathcal{T}_h\}$ the *global interpolation operator*

$$I_h : Y \to M_h,$$

defined by

$$I_h(\boldsymbol{q})|_K = I_h(\boldsymbol{q}_K) \quad \text{for all } K \in \mathcal{T}_h,$$

has to be well defined, which means that matching values of the degrees of freedom guarantee the property $I_h(\boldsymbol{q}) \in M_h$ by ensuring the continuity of the normal fluxes.

All the degrees of freedom discussed in Chapter 3 are pointwise evaluations of functions (or their derivatives), but this is not necessary. In Chapter 3 we have chosen $P_K = \mathcal{P}_k(K)$ finally to achieve the order k (in $H^1(\Omega)$). Here, for the same aim, we make the following ansatzes.

Raviart–Thomas (RT) finite elements

For $k \in \mathbb{N}$ set

$$P_K := \mathcal{D}_k(K) := \mathcal{P}_{k-1}(K)^d + \boldsymbol{x}\mathcal{P}_{k-1}(K). \tag{7.39}$$

This space—also called the *(local) Raviart–Thomas space of order* $k-1$—is also denoted by $RT_{k-1}(K)$ (note the shift in notation). Obviously

$$\mathcal{P}_{k-1}(K)^d \subset \mathcal{D}_k(K) \subset \mathcal{P}_k(K)^d,$$

and $\mathcal{D}_k(K)$ is chosen such that

$$\nabla \cdot \boldsymbol{q} \in \mathcal{P}_{k-1}(K) \quad \text{for } \boldsymbol{q} \in \mathcal{D}_k(K). \tag{7.40}$$

This will be demonstrated later. The representation for $\boldsymbol{q} \in \mathcal{D}_k(K)$

$$q_i(x) = q_i^*(x) + x_i q_0^*(x) \quad \text{with } q_i^*, q_0^* \in \mathcal{P}_{k-1}(K),\ \boldsymbol{q}^* = (q_j), \quad i = 1, \ldots, d,$$

is not unique, as, e.g., for $i = 1$ and $k \geq 2$, polynomials $q_0^* \in \mathcal{P}_{k-2}$, multiplied by x_1, reappear in q_1^*. Thus $\mathcal{P}_{k-1}$ for q_0^* has to be substituted by

$$\mathcal{P}_{k-1}(K) \setminus \mathcal{P}_{k-2}(K) =: \mathcal{P}_{k-1}^*(K). \tag{7.41}$$

The set $\mathcal{P}_{k-1}^*(K)$ contains those polynomials where all terms are of exact order $k-1$, i.e., $p \in \mathcal{P}_{k-1}^*(K)$ is *homogeneous of order* $k-1$, that is,

$$p(\lambda x) = \lambda^{k-1} p(x) \quad \text{for } p \in \mathcal{P}_{k-1}^*(K),\ \lambda \in \mathbb{R}. \tag{7.42}$$

Thus, we have a unique representation of the elements of $\mathcal{D}_k(K)$ by

$$q_i(x) = q_i^*(x) + x_i q_0^*(x), \quad q_i^* \in \mathcal{P}_{k-1}(K), \quad q_0^* \in \mathcal{P}_{k-1}^*(K), \tag{7.43}$$

and therefore

$$\dim \mathcal{D}_k(K) = (d+1)\dim \mathcal{P}_{k-1}(K) - \begin{cases} \dim \mathcal{P}_{k-2}(K) & \text{for } k \geq 2, \\ 0 & \text{for } k = 1. \end{cases}$$

Using (3.55) we get

$$\dim \mathcal{D}_k(K) = (d+k)\frac{(d+k-2)!}{(d-1)!(k-1)!}. \tag{7.44}$$

In particular,

$$\dim \mathcal{D}_k(K) = \begin{cases} k(k+2) & \text{for } d = 2, \\ k(k+1)(k+3)/2 & \text{for } d = 3. \end{cases}$$

Furthermore, in general we have for each face $F \subset \partial K$ and its normal $\boldsymbol{n}$ that

$$\boldsymbol{q} \cdot \mathfrak{n} \in \mathcal{P}_{k-1}(F). \tag{7.45}$$

Indeed, since F is a subset of a hyperplane $\boldsymbol{x} \cdot \mathfrak{n} = s$ for some $s \in \mathbb{R}$, for $\boldsymbol{x} \in F$ the relation

$$\boldsymbol{q}(\boldsymbol{x}) \cdot \mathfrak{n} = \boldsymbol{q}^*(\boldsymbol{x}) \cdot \mathfrak{n} + \boldsymbol{x} \cdot \mathfrak{n} q_0^*(\boldsymbol{x}) = \boldsymbol{q}^*(\boldsymbol{x}) \cdot \mathfrak{n} + s q_0^*(\boldsymbol{x})$$

holds, hence $\boldsymbol{q} \cdot \mathfrak{n} \in \mathcal{P}_{k-1}(F)$.

The most often used case is $k = 1$, i.e., the space $RT_0(K)$ with

$$RT_0(K) = \mathcal{D}_1(K) = \mathcal{P}_0(K)^d \oplus \boldsymbol{x}\mathcal{P}_0(K),$$

i.e., its elements have the form

$$\boldsymbol{q}(\boldsymbol{x}) = \boldsymbol{a}_1 + a_0\boldsymbol{x}, \quad \boldsymbol{x} \in K, \quad \text{with } \boldsymbol{a}_1 \in \mathbb{R}^d, \ a_0 \in \mathbb{R}. \tag{7.46}$$

By (7.45), the normal component of elements from RT_0 is constant on each face. Furthermore we can verify the inclusion (7.40), since in the notation of (7.43)

$$\nabla \cdot \boldsymbol{q} = \nabla \cdot \boldsymbol{q}^* + \boldsymbol{x} \cdot \nabla q_0^* + d q_0^* = \nabla \cdot \boldsymbol{q}^* + (k - 1 + d) q_0^*. \tag{7.47}$$

In the last step we used that

$$\boldsymbol{x} \cdot \nabla q_0^* = (k-1) q_0^*,$$

which can be seen directly from $q_0^* \in \mathcal{P}_{k-1}^*(K)$, or by applying Euler's Theorem for homogeneous functions.

Furthermore, the mapping

$$\nabla \cdot : \ \mathcal{D}_k(K) \to \mathcal{P}_{k-1}(K) \tag{7.48}$$

is surjective (see Problem 7.10), explaining the definition (7.39) of $\mathcal{D}_k(K)$.

To achieve (7.37) the same number of degrees of freedom as the dimension of $\mathcal{D}_k(K)$ is required. The degrees of freedom Σ_K are defined as the $(k-1)$th moments of $\boldsymbol{q} \cdot \mathfrak{n}$ at the faces of K, i.e., $\boldsymbol{q} \cdot \mathfrak{n}$ tested with polynomials of degree at most $k-1$, and similarly for $\boldsymbol{q}$ on the element K.

In detail, let $a_k := \dim \mathcal{P}_{k-1}(\mathbb{R}^{d-1}) = \dfrac{(d+k-2)!}{(d-1)!(k-1)!}$ and $\mathcal{P}_{k-1}(\mathbb{R}^{d-1}) = \operatorname{span}\{w_1^{(k)}, \ldots, w_{a_k}^{(k)}\}$ with an arbitrary basis, then with $\mathfrak{n} = \mathfrak{n}_F$ being the normal on F,

$$\psi(\boldsymbol{q}) := \psi_{K,F,i}(\boldsymbol{q}) := \int_F \boldsymbol{q} \cdot \mathfrak{n}\, w_i^{(k)} d\sigma, \quad i = 1, \ldots, a_k\,. \tag{7.49}$$

For $k \geq 2$ let

$$b_k := \dim \mathcal{P}_{k-2}(\mathbb{R}^d)^d = \frac{(d+k-2)!}{(d-1)!(k-2)!}$$

and

$$\mathcal{P}_{k-2}(\mathbb{R}^d)^d = \operatorname{span}\{\boldsymbol{w}_1^{(k)}, \ldots, \boldsymbol{w}_{b_k}^{(k)}\}$$

with an arbitrary basis, then

$$\psi(\boldsymbol{q}) := \psi_{K,K,i}(\boldsymbol{q}) := \int_K \boldsymbol{q} \cdot \boldsymbol{w}_i^{(k)} \, dx, \quad i = 1, \ldots, b_k \,. \tag{7.50}$$

Therefore, we have $(d+1)a_k + b_k = \dim \mathcal{D}_k(K)$ degrees of freedom. Thus, for the unique solvability of the local interpolation problem (7.37) only existence or uniqueness has to be checked (compare Section 3.3). The answer will be affirmative. Similar to (3.83) we need a little more regularity by using

$$Y_K := \mathcal{H}(\text{div}; K) := \{\boldsymbol{q} \in H(\text{div}; K) \mid \boldsymbol{q} \cdot \boldsymbol{n} \in L^2(\partial K)\}. \tag{7.51}$$

Theorem 7.9 *Let K be a regular d-simplex and $k \in \mathbb{N}$.*

1) For $L := \dim P_K$ given real numbers, there is a unique function in P_K for which the degrees of freedom (7.49), (7.50) assume these values.
2) For each $\boldsymbol{q} \in \mathcal{H}(\text{div}; K)$, there is a unique function in P_K, denoted by $I_K(\boldsymbol{q})$, such that the values of the degrees of freedom of $\boldsymbol{q}$ and $I_K(\boldsymbol{q})$ coincide.
3) $\nabla \cdot (I_K(\boldsymbol{q})) = \Pi_K^{k-1}(\nabla \cdot \boldsymbol{q})$ for $\boldsymbol{q} \in \mathcal{H}(\text{div}; K)$, where $\Pi_K^{k-1} : L^2(K) \to \mathcal{P}_{k-1}(K)$ denotes the orthogonal projection with respect to the L^2-scalar product, i.e., the following diagram is commutative:

$$\begin{array}{ccc} \mathcal{H}(\text{div}; K) & \xrightarrow{\nabla\cdot} & L^2(K) \\ \downarrow I_K & & \downarrow P_K^{k-1} \\ \mathcal{D}_k(K) & \xrightarrow{\nabla\cdot} & \mathcal{P}_{k-1}(K) \end{array}$$

Proof 2) follows directly from 1) since the degrees of freedom can be evaluated for $\boldsymbol{q} \in \mathcal{H}(\text{div}; K)$, and for 1) only uniqueness has to be proven. So let $\boldsymbol{q} \in P_K$ such that $\psi(\boldsymbol{q}) = 0$ for all ψ from (7.49), (7.50). Due to (7.49) and (7.45) we have for each face $F \subset \partial K$ that

$$\boldsymbol{q} \cdot \boldsymbol{n} = 0 \quad \text{on } F. \tag{7.52}$$

We also have $\nabla \cdot \boldsymbol{q} \in \mathcal{P}_{k-1}(K)$ (see (7.40)) and $\nabla(\nabla \cdot \boldsymbol{q}) \in \mathcal{P}_{k-2}(K)^d$ (or $\nabla(\nabla \cdot \boldsymbol{q}) = \boldsymbol{0}$ for $k = 1$). Therefore, $\nabla(\nabla \cdot \mathbf{q})$ is a suitable test function in (7.50), and by integration by parts and (7.49), (7.50)

$$\int_K (\nabla \cdot \boldsymbol{q})^2 dx = -\int_K \boldsymbol{q} \cdot \nabla(\nabla \cdot \boldsymbol{q}) dx + \int_{\partial K} \boldsymbol{q} \cdot \boldsymbol{n} \nabla \cdot \boldsymbol{q} \, d\sigma = 0.$$

Hence

$$\nabla \cdot \boldsymbol{q} = 0 \quad \text{in } K\,. \tag{7.53}$$

From (7.47) we conclude that also $\nabla \cdot \boldsymbol{q}^* = 0$, $q_0^* = 0$, as the first term is in $\mathcal{P}_{k-2}(K)$ (or vanishes for $k = 1$), the second in $\mathcal{P}_{k-1}^*(K)$, therefore

$$\boldsymbol{q} \in \mathcal{P}_{k-1}(K)^d. \tag{7.54}$$

For $k = 1$, $\boldsymbol{q}$ is constant on K due to (7.54). Since d linearly independent vectors exist among the $d + 1$ normal vectors of the faces of a regular simplex, the property (7.52) implies $\boldsymbol{q} = \mathbf{0}$. This proves the statement 1) for $k = 1$.

In the general case $k \geq 2$, we argue as follows.

For each face $F \subset \partial K$ with corresponding outward unit normal $\boldsymbol{\mathfrak{n}}$, (7.54) and (7.52) show that

$$\boldsymbol{q} \cdot \boldsymbol{\mathfrak{n}} \in \mathcal{P}_{k-1}(K) \quad \text{and} \quad \boldsymbol{q} \cdot \boldsymbol{\mathfrak{n}} = 0 \quad \text{on } F.$$

Note that this does not imply that $\boldsymbol{q} = \mathbf{0}$ on F in general. Expressing rather $\boldsymbol{q} \cdot \boldsymbol{\mathfrak{n}}$ in barycentric coordinates $\lambda_i = \lambda_i(x)$, there is an index $j \in \{1, \ldots, d + 1\}$ such that $\lambda_j = 0$ on F and therefore, for some $\tilde{q} \in \mathcal{P}_{k-2}$,

$$\boldsymbol{q} \cdot \boldsymbol{\mathfrak{n}} = \lambda_j \tilde{q}(\boldsymbol{\lambda}).$$

Then the vanishing degrees of freedom from (7.50) imply

$$0 = \int_K \boldsymbol{q} \cdot \boldsymbol{\mathfrak{n}}\, p \, dx \quad \text{for } p \in \mathcal{P}_{k-2}(K),$$

and choosing $p := \tilde{q}(\boldsymbol{\lambda})$

$$0 = \int_K \lambda_j \tilde{q}(\boldsymbol{\lambda})^2 dx.$$

As $\lambda_j = \lambda_j(x)$ does not change its sign on K, we have $\tilde{q}(\boldsymbol{\lambda}) = 0$ on K and thus, for all $d + 1$ faces with normal $\boldsymbol{\mathfrak{n}}_i$, $i = 1, \ldots, d + 1$,

$$\boldsymbol{q} \cdot \boldsymbol{\mathfrak{n}}_i = 0 \quad \text{on } K \text{ for } i = 1, \ldots, d + 1.$$

Again, since d normals form a basis of $\mathbb{R}^d$, $\boldsymbol{q} = \mathbf{0}$ on K follows.

To verify the identity in 3) we have to show that

$$\int_K \nabla \cdot (\boldsymbol{q} - (I_K(\boldsymbol{q})))\, w dx = 0 \quad \text{for all } w \in \mathcal{P}_{k-1}(K).$$

For $w \in \mathcal{P}_{k-1}(K)$ it holds that $\nabla w \in \mathcal{P}_{k-2}(K)^d$ (or $\nabla w = \mathbf{0}$ for $k = 1$). By integration by parts and (7.49), (7.50) we obtain

$$\begin{aligned}\int_K \nabla \cdot I_K(\boldsymbol{q}) w dx &= -\int_K I_K(\boldsymbol{q}) \cdot \nabla w dx + \int_{\partial K} I_K(\boldsymbol{q}) \cdot \boldsymbol{\mathfrak{n}} w d\sigma \\ &= -\int_K \boldsymbol{q} \cdot \nabla w dx + \int_{\partial K} \boldsymbol{q} \cdot \boldsymbol{\mathfrak{n}} w d\sigma = \int_K \nabla \cdot \boldsymbol{q} w dx\,.\end{aligned}$$

□

Remark 7.10 Due to Theorem 7.9, 3) for $\boldsymbol{q} \in \mathcal{H}(\text{div}; K)$ holds

$$\nabla \cdot \boldsymbol{q} = 0 \text{ in } K \quad \text{implies that} \quad \nabla \cdot I_K(\boldsymbol{q}) = 0 \text{ in } K.$$

Now we set

$$\begin{aligned} \mathcal{H}(\text{div}; \mathcal{T}_h) := \{\boldsymbol{q} \in L^2(\Omega)^d \;|\; &\boldsymbol{q}|_K \in H(\text{div}; K), \\ &\boldsymbol{q}|_K \cdot \boldsymbol{n}_K \in L^2(\partial K) \text{ for all } K \in \mathcal{T}_h\}. \end{aligned} \tag{7.55}$$

For the global approximation space, the *global Raviart–Thomas space of degree* $k - 1$, we have

$$\begin{aligned} M_h &:= \{\boldsymbol{q} \in H(\text{div}; \Omega) \;|\; \boldsymbol{q}|_K \in \mathcal{D}_k(K) \text{ for all } K \in \mathcal{T}_h\} \\ &= \{\boldsymbol{q} \in \mathcal{H}(\text{div}; \mathcal{T}_h) \;|\; \boldsymbol{q}|_K \in \mathcal{D}_k(K) \text{ for all } K \in \mathcal{T}_h, \; [\![\boldsymbol{q} \cdot \boldsymbol{n}]\!] = 0 \text{ on} \\ &\qquad \partial K_1 \cap \partial K_2 \text{ for all } K_1, K_2 \in \mathcal{T}_h\} \\ &= \{\boldsymbol{q} \in \mathcal{H}(\text{div}; \mathcal{T}_h) \;|\; \boldsymbol{q}_K \in \mathcal{D}_k(K) \text{ for all } K \in \mathcal{T}_h \text{ and the degrees of free-} \\ &\qquad \text{dom (7.49) at the faces } F \text{ of all elements are} \\ &\qquad \text{uniquely defined}\}. \end{aligned} \tag{7.56}$$

This space is also denoted by $RT_{k-1}(\Omega)$. We will also use the *broken Raviart–Thomas space*

$$RT_{k-1}(\mathcal{T}_h) := \{\boldsymbol{q} \in L^2(\Omega)^d \;|\; \boldsymbol{q}_{|K} \in RT_{k-1}(K) \text{ for } K \in \mathcal{T}_h\}.$$

Note that the second equation in (7.56) is due to Theorem 6.16 and the last requirement means for $K_1, K_2 \in \mathcal{T}_h$, $F = K_1 \cap K_2$

$$\int_F \left(\boldsymbol{q} \cdot \boldsymbol{n}_{K_1} + \boldsymbol{q} \cdot \boldsymbol{n}_{K_2}\right) \gamma d\sigma = 0 \quad \text{for all } \gamma \in \mathcal{P}_{k-1}(\mathbb{R}^{d-1}).$$

Analogously to the derivation of (7.52), this is equivalent to

$$[\![\boldsymbol{q} \cdot \boldsymbol{n}]\!] = 0 \quad \text{on } K_1 \cap K_2.$$

Therefore the global interpolation operator

$$I_h : H(\text{div}; \Omega) \cap \mathcal{H}(\text{div}; \mathcal{T}_h) \to M_h, \quad I_h(\boldsymbol{q})|_K = I_K(\boldsymbol{q}|_K) \tag{7.57}$$

is well defined.

Now the norm of the *approximation error* of M_h, needed to apply Theorem 6.6 or Theorem 6.25 for the derivation of (order of) convergence results, can be estimated via the norm of the *interpolation error* $\|\boldsymbol{q} - I_h(\boldsymbol{q})\|$.

Here, we want to repeat the strategy of the proof of Theorem 3.28: The local error on $K \in \mathcal{T}_h$ is handled by transformation to a reference element $\hat{K}$, dealing with the interpolation error there using the Bramble–Hilbert lemma (Theorem 3.28) and transforming back. Considering simplicial elements we have $\hat{K}$ given by conv $\{\mathbf{0}, \boldsymbol{e}_1 \ldots, \boldsymbol{e}_d\}$, and the affine-linear transformation from (3.87), i.e.,

$$F = F_K : \hat{K} \to K, \quad F(\hat{x}) = B\hat{x} + \boldsymbol{d}, \tag{7.58}$$

again using the notation from Section 3.4.1 with $\hat{\cdot}$ referring to the reference element. We would like to have the compatibility condition

$$\widehat{I_K(\boldsymbol{q}|_K)} = I_{\hat{K}}(\hat{\boldsymbol{q}}), \tag{7.59}$$

where the left-hand side refers to a *pullback* of a function on K to a function on $\hat{K}$. For a scalar function the definition (3.88) was and still is sufficient for our purposes; in the vectorial case we use the so-called (covariant) *Piola transformation*, since the standard transformation (3.88) does not preserve normal component traces of vector fields:

$$\hat{\boldsymbol{q}}(\hat{x}) := |\det DF(\hat{x})| DF(\hat{x})^{-1}\boldsymbol{q}(x) = JB^{-1}\boldsymbol{q}(x) \tag{7.60}$$

with $J := |\det B|$.

Lemma 7.11 *Let K, $\hat{K} \subset \mathbb{R}^d$ be regular d-simplices and the affine-linear transformation $F : \hat{K} \to K$ given by* (7.58). *If $\varphi \in H^1(K)$ is transformed according to* (3.88) *and $\boldsymbol{q} \in H(\mathrm{div}; K)$ is transformed according to* (7.60), *we have for $\boldsymbol{q} \in H(\mathrm{div}; K)$, $\varphi \in H^1(K)$*

1) $\displaystyle\int_K \nabla_x \cdot \boldsymbol{q}\, \varphi \, dx = \int_{\hat{K}} \nabla_{\hat{x}} \cdot \hat{\boldsymbol{q}}\, \hat{\varphi}\, d\hat{x},$

2) $\displaystyle\int_K \boldsymbol{q} \cdot \nabla_x \varphi \, dx = \int_{\hat{K}} \hat{\boldsymbol{q}} \cdot \nabla_{\hat{x}} \hat{\varphi}\, d\hat{x},$

3) $\displaystyle\int_{\partial K} \boldsymbol{q} \cdot \mathfrak{n}_x\, \varphi \, d\sigma = \int_{\partial\hat{K}} \hat{\boldsymbol{q}} \cdot \mathfrak{n}_{\hat{x}}\, \hat{\varphi}\, d\hat{\sigma}.$

Proof 2) By (3.90), $\nabla_x \varphi(F(\hat{x})) = B^{-T} \nabla_{\hat{x}} \hat{\varphi}(\hat{x})$ and therefore

$$\int_{\hat{K}} \hat{\boldsymbol{q}} \cdot \nabla_{\hat{x}} \hat{\varphi}\, d\hat{x} = \int_{\hat{K}} B^{-1}\boldsymbol{q}(F(\hat{x})) \cdot B^T \nabla_x \varphi(F(\hat{x})) |\det B|\, d\hat{x} = \int_K \boldsymbol{q} \cdot \nabla_x \varphi \, dx\,.$$

1) We start with a sufficiently smooth function, e.g., $\boldsymbol{q} \in H^1(\Omega)^d$, and finally apply an approximation argument. We have

$$D_x \boldsymbol{q}(x) = \frac{1}{J} B D_x \hat{\boldsymbol{q}}(F^{-1}(x)) = \frac{1}{J} B D_{\hat{x}} \hat{\boldsymbol{q}}(F^{-1}(x)) DF^{-1}(x) = \frac{1}{J} B D_{\hat{x}} \hat{\boldsymbol{q}}(\hat{x}) B^{-1}$$

and thus

$$\nabla_x \cdot \boldsymbol{q}(x) = \sum_{j=1}^{d} (D_x \boldsymbol{q})_{jj} = \frac{1}{J} \sum_{j=1}^{d} (D_{\hat{x}} \hat{\boldsymbol{q}})_{jj} = \frac{1}{J} \nabla_{\hat{x}} \cdot \hat{\boldsymbol{q}}(\hat{x}). \tag{7.61}$$

Hence 1) follows by the variable transformation. The remaining part of the proof of this statement is left to reader (Problem 7.7).

Finally, 3) follows from 1) and 2) by integration by parts:

$$\int_{\partial K} \boldsymbol{q} \cdot \mathfrak{n}_x \, \varphi \, d\sigma = \int_K [\boldsymbol{q} \cdot \nabla_x \varphi + \nabla_x \cdot \boldsymbol{q} \varphi] dx =$$

$$\int_{\hat{K}} [\hat{\boldsymbol{q}} \cdot \nabla_{\hat{x}} \hat{\varphi} + \nabla_{\hat{x}} \cdot \hat{\boldsymbol{q}} \hat{\varphi}] d\hat{x} = \int_{\partial K} \hat{\boldsymbol{q}} \cdot \mathfrak{n}_{\hat{x}} \, \hat{\varphi} \, d\hat{\sigma} \, .$$

□

Note that (7.61) implies the required property of the Piola transformation. We have

$$\boldsymbol{q} \in \mathcal{D}_k(K) \iff \hat{\boldsymbol{q}} \in \mathcal{D}_k(\hat{K}), \tag{7.62}$$

since the affine-linear transformations F and F^{-1} leave $\mathcal{P}_l(\mathbb{R}^d)$ invariant, and in the representation (7.43), by (7.60),

$$\widehat{\boldsymbol{x} q_0(x)} = J B^{-1} \boldsymbol{x} q_0(x) = J \hat{\boldsymbol{x}} q_0(F(\hat{x})) + J B^{-1} \boldsymbol{d} \, q_0(F(\hat{x})) \in \mathcal{D}_k(\hat{K}).$$

Now we are ready to prove (7.59).

Lemma 7.12 *Let $K, \hat{K}$ be as in Lemma 7.11, then*

$$\widehat{I_K(\boldsymbol{q})} = I_{\hat{K}} \hat{\boldsymbol{q}} \quad \textit{for all } \boldsymbol{q} \in \mathcal{H}(\mathrm{div}; K). \tag{7.63}$$

Proof Because of the uniqueness of the interpolating function and (7.62), it only has to be verified that $\widehat{I_K(\boldsymbol{q}|_K)}$ takes the correct values of the degrees of freedom on $\hat{K}$.

Concerning the face moments (7.49), let $\hat{F}$ be a face of $\hat{K}$, $\tilde{F} := F(\hat{F})$ the corresponding face of K, and $p \in \mathcal{P}_{k-1}(\mathbb{R}^{d-1})$.

First of all we have to clarify how to transform surface normal vectors. Let x_0 be a point of the hyperplane containing $\tilde{F}$. Then, for all points x of this hyperplane, it holds that

$$0 = (x - x_0) \cdot \mathfrak{n} = B(\hat{x} - \hat{x}_0) \cdot \mathfrak{n} = (\hat{x} - \hat{x}_0) \cdot (B^T \mathfrak{n}),$$

where $\hat{x}, \hat{x}_0$ lie in the hyperplane containing $\hat{F}$. This shows that the transformed normal vector is of the form

$$\hat{\mathfrak{n}} = \alpha B^T \mathfrak{n} \quad \text{with } \alpha \in \mathbb{R} \text{ such that } |\alpha| = |B^T \mathfrak{n}|^{-1}.$$

Therefore, noting that the elementary surface differentials are related according to $d\sigma = J_{\tilde{F}} d\hat{\sigma}$, where $J_{\tilde{F}}$ is a positive constant, we get

$$\int_{\hat{F}} \widehat{I_K(\boldsymbol{q})} \cdot \hat{\mathfrak{n}}\hat{p}\, d\hat{\sigma} \overset{(7.60)}{=} \int_{\hat{F}} I_K(\boldsymbol{q} \circ F) \cdot B^{-T} B^T \mathfrak{n}(p \circ F) \alpha J\, d\hat{\sigma}$$

$$= \alpha J (J_{\tilde{F}})^{-1} \int_{\tilde{F}} I_K(\boldsymbol{q}) \cdot \mathfrak{n} p\, d\sigma = \alpha J (J_{\tilde{F}})^{-1} \int_{\tilde{F}} \boldsymbol{q} \cdot \mathfrak{n} p\, d\sigma \overset{(7.60)}{=} \int_{\hat{F}} \hat{\boldsymbol{q}} \cdot \hat{\mathfrak{n}} \hat{p}\, d\hat{\sigma}\,.$$

Concerning the moments (7.50), let $\boldsymbol{p} \in \mathcal{P}_{k-2}(\mathbb{R}^d)^d$ for $k \geq 2$. Then

$$\int_{\hat{K}} \widehat{I_K(\boldsymbol{q})} \cdot \hat{\boldsymbol{p}}\, d\hat{x} = \int_{\hat{K}} I_K(\boldsymbol{q} \circ F) \cdot J B^{-T} B^{-1} (\boldsymbol{p} \circ F) J\, d\hat{x} = \int_{K} I_K(\boldsymbol{q}) \cdot \tilde{\boldsymbol{p}}\, dx\,,$$

where $\tilde{\boldsymbol{p}} := J(B^T B)^{-1} \boldsymbol{p}(x) \in \mathcal{P}_{k-2}(\mathbb{R}^d)^d$, and therefore

$$\int_{\hat{K}} \widehat{I_K(\boldsymbol{q})} \cdot \hat{\boldsymbol{p}}\, d\hat{x} = \int_{K} \boldsymbol{q} \cdot \tilde{\boldsymbol{p}}\, dx = \int_{\hat{K}} \hat{\boldsymbol{q}} \cdot \hat{\boldsymbol{p}}\, d\hat{x}\,.$$

□

Theorem 7.13 *Let $(\mathcal{T}_h)_h$ be a shape-regular family of triangulations of the domain $\Omega \subset \mathbb{R}^d$, which satisfies (T1)–(T6), and $k \in \mathbb{N}$. Then there exists a constant $C > 0$ independent of h such that for $\boldsymbol{q} \in H^l(\Omega)^d$, $1 \leq l \leq k$,*

$$\|\boldsymbol{q} - I_h(\boldsymbol{q})\|_0 \leq C h^l |\boldsymbol{q}|_l\,.$$

In addition, if $\nabla \cdot \boldsymbol{q} \in H^l(\Omega)$, then

$$\|\nabla \cdot (\boldsymbol{q} - I_h(\boldsymbol{q}))\|_0 \leq C h^l |\nabla \cdot \boldsymbol{q}|_l\,,$$

where $I_h \boldsymbol{q}$ is the interpolation operator from (7.57) *for Raviart–Thomas elements of order $k-1$.*

Proof We follow the reasoning of Section 3.4.1 (see in particular Remark 3.33, 2)). It is sufficient to show that, for $K \in \mathcal{T}_h$,

$$\|\boldsymbol{q} - I_K(\boldsymbol{q})\|_{0,K} \leq C h_K^l |\boldsymbol{q}|_{l,K}\,. \tag{7.64}$$

This can be done by demonstrating this estimate for the reference element $\hat{K}$ and applying the transformation (7.58) to $\hat{K}$ for general $K \in \mathcal{T}_h$. The reference case again relies on the Bramble–Hilbert lemma (Theorem 3.27). To do so, let

$$G:\ H^l(\hat{K})^d \to L^2(\hat{K})^d, \quad \hat{\boldsymbol{q}} \mapsto \hat{\boldsymbol{q}} - I_{\hat{K}}(\hat{\boldsymbol{q}}).$$

Using that $P_{\hat{K}}$ is finite-dimensional and therefore all norms on $P_{\hat{K}}$ are equivalent, it can be shown that this linear operator is bounded by proving the estimate

$$\|I_{\hat{K}}(\hat{\boldsymbol{q}})\|_{0,\hat{K}} \leq C \|\hat{\boldsymbol{q}}\|_{l,\hat{K}}\,.$$

According to Theorem 7.9, $I_K(\hat{\boldsymbol{q}}) = \hat{\boldsymbol{q}}$ for $\hat{\boldsymbol{q}} \in P_{\hat{K}}$, i.e., $G(\hat{\boldsymbol{q}}) = \boldsymbol{0}$. In particular,

$$G(\boldsymbol{q}) = \boldsymbol{0} \quad \text{for } \boldsymbol{q} \in \mathcal{P}_{k-1}(\mathbb{R}^d)^d,$$

and therefore, by Theorem 3.27,

$$\|\hat{\boldsymbol{q}} - I_{\hat{K}}(\hat{\boldsymbol{q}})\|_{0,\hat{K}} \leq \tilde{C}|\hat{\boldsymbol{q}}|_{l,\hat{K}} \,.$$

For a general element see Theorem 3.29 and the reasoning for Theorem 3.32.

Concerning the second statement, Theorem 7.9, 3) implies that locally

$$\|\nabla\cdot(\boldsymbol{q} - I_h(\boldsymbol{q}))\|_{0,K} = \|\nabla\cdot\boldsymbol{q} - \Pi_K^{k-1}(\nabla\cdot\boldsymbol{q})\|_{0,K} \leq \|\nabla\cdot\boldsymbol{q} - \tilde{I}_K(\nabla\cdot\boldsymbol{q})\|_{0,K} \text{ for } K \in \mathcal{T}_h$$

holds, where $\tilde{I}_K$ is the local interpolation operator (3.85) from Section 3.4, for which the estimate is known from Theorem 3.32 or Remark 3.33, 2) for $k = 1$. □

An approximation space for the scalar variable is given by

$$X_h := \{u \in L^2(\Omega) \mid u|_K \in \mathcal{P}_{k-1}(K) \text{ for all } K \in \mathcal{T}_h\}. \tag{7.65}$$

For $k = 1$ this means u being piecewise constant, where the degree of freedom can be chosen as the function value at an arbitrary point in K or as $\frac{1}{|K|}\int_K u dx$. For $k \geq 2$ these are the polynomial spaces from Section 3.3, whereby the continuity of $u \in X_h$ can be ensured by choosing the degrees of freedom as the point values indicated there.

Since Theorem 3.32 is based only on the local interpolation operator I_K we are free in the choice of the degrees of freedom, only requiring the unique solvability of the local interpolation problem.

In order to obtain a stable and convergent discretization, the discrete inf-sup condition (NB2h) has to be verified in a uniform manner. We have

$$\nabla\cdot\boldsymbol{q}_h \in X_h \quad \text{for all } \boldsymbol{q}_h \in M_h,$$

in particular

$$\int_\Omega \nabla\cdot\boldsymbol{q}_h v_h \, dx = 0 \quad \text{for all } v_h \in X_h \quad \text{implies} \quad \nabla\cdot\boldsymbol{q}_h = 0.$$

That is, in the notation of Lemma 6.24,

$$\ker B_h \subset \ker B. \tag{7.66}$$

To apply Theorem 6.27 we take

$$\Pi_h := I_h \quad \text{(the interpolation operator)},$$
$$S := L^2(\Omega) \quad \text{(not restricting)},$$

$$\tilde{M} := \{\boldsymbol{q} \in H^1(\Omega)^d \mid \boldsymbol{q} \cdot \boldsymbol{n} = 0 \text{ on } \widetilde{\Gamma}_1\}.$$

Then Theorem 7.9, 3) shows

$$b(\Pi_h \boldsymbol{q} - \boldsymbol{q}, v) = 0 \quad \text{for all } v \in X_h,$$

but from Theorem 7.13 for $\boldsymbol{q} \in \tilde{M}$ only the estimate

$$\|I_h(\boldsymbol{q})\|_{H(\mathrm{div})} \le C_1 \left(h|\boldsymbol{q}|_1 + \|\boldsymbol{q}\|_{H(\mathrm{div})}\right)$$

can be concluded. Nevertheless, a uniform inf-sup condition can be shown by modifying the proof of Theorem 6.27. Namely, due to Lemma 6.14 the inf-sup condition holds true for M and X, hence there exists an element $\overline{\boldsymbol{q}} \in M$ (with $\|\overline{\boldsymbol{q}}\|_M = 1$) such that

$$b(\overline{\boldsymbol{q}}, v_h) \ge \frac{\beta}{2}\|v_h\|_0 \quad \text{for } v_h \in X = L^2(\Omega).$$

Since the space $C^1(\overline{\Omega})^d$ is dense in $H(\mathrm{div};\Omega)$ [18, Ch. 9, §1, Thm. 1], we find functions $\overline{\boldsymbol{q}}_\varepsilon \in C^1(\overline{\Omega})^d$ such that

$$\|\overline{\boldsymbol{q}} - \overline{\boldsymbol{q}}_\varepsilon\|_{H(\mathrm{div})} =: a_\varepsilon \to 0, \quad |\overline{\boldsymbol{q}}_\varepsilon|_1 \le C_2\varepsilon^{-1} \quad \text{for } \varepsilon \to 0.$$

Then we have, for $v_h \in X_h$,

$$\sup_{\boldsymbol{q}_h \in M_h\setminus\{\boldsymbol{0}\}} \frac{b(\boldsymbol{q}_h, v_h)}{\|\boldsymbol{q}_h\|_M} \ge \frac{b(I_h(\overline{\boldsymbol{q}}_\varepsilon), v_h)}{\|I_h(\overline{\boldsymbol{q}}_\varepsilon)\|_M} = \frac{b(\overline{\boldsymbol{q}}_\varepsilon, v_h)}{\|I_h(\overline{\boldsymbol{q}}_\varepsilon)\|_M}.$$

The numerator and denominator can be estimated separately as follows:

$$\begin{aligned}
\|I_h\overline{\boldsymbol{q}}_\varepsilon\|_M &\le C_1(h|\overline{\boldsymbol{q}}_\varepsilon|_1 + \|\overline{\boldsymbol{q}}_e\|_{H(\mathrm{div})}) \le C_3(C_2 h\varepsilon^{-1} + 1),\\
b(\overline{\boldsymbol{q}}_\varepsilon, v_h) &= b(\overline{\boldsymbol{q}}, v_h) + b(\overline{\boldsymbol{q}}_\varepsilon - \overline{\boldsymbol{q}}, v_h) \ge b(\overline{\boldsymbol{q}}, v_h) - C_4 a_\varepsilon\|v_h\|_0.
\end{aligned}$$

Therefore we obtain

$$\sup_{\boldsymbol{q}_h \in M_h\setminus\{\boldsymbol{0}\}} \frac{b(\boldsymbol{q}_h, v_h)}{\|\boldsymbol{q}_h\|_M} \ge \frac{1}{C_3(C_2 h\varepsilon^{-1} + 1)}\left(\frac{\beta}{2} - C_4 a_\varepsilon\right)\|v_h\|_0.$$

Choosing $\varepsilon = h$ and h sufficiently small such that $a_\varepsilon = a_h \le \dfrac{\beta}{4C_4}$, we finally arrive at the uniform estimate

$$\sup_{\boldsymbol{q}_h \in M_h\setminus\{\boldsymbol{0}\}} \frac{b(\boldsymbol{q}_h, v_h)}{\|\boldsymbol{q}_h\|_M} \ge \frac{\beta}{4C_3(C_2 + 1)}\|v_h\|_0.$$

Therefore we have the following result.

Theorem 7.14 *The spaces M_h, X_h according to* (7.56), (7.65) *satisfy a uniform discrete inf-sup condition for* (6.54) *provided the inf-sup condition holds there. If the solution $(\boldsymbol{p}, u)$ of* (6.55) *possesses the properties $\boldsymbol{p} \in H^k(\Omega)^d$ and $\nabla \cdot \boldsymbol{p} \in H^k(\Omega)$,* *then the following error estimate holds with a constant $C > 0$ independent of h:*

$$\|\boldsymbol{p} - \boldsymbol{p}_h\|_{H(\mathrm{div})} \leq Ch^k(|\boldsymbol{p}|_k + |\nabla \cdot \boldsymbol{p}|_k).$$

If additionally $u \in H^k(\Omega)$, then

$$\|u - u_h\|_0 \leq Ch^k(|\boldsymbol{p}|_k + |\nabla \cdot \boldsymbol{p}|_k + |u|_k).$$

Proof The first estimate follows from Theorem 6.25 using (7.66) and Theorem 7.13, as well as the second estimate additionally with Theorem 3.32. □

A higher order of convergence is not possible for u_h due to the polynomial degree, but in Section 7.5 we will see that a postprocessing is possible that leads to superconvergence. This is based on the following result.

Theorem 7.15 *Let the assumptions of Theorem 7.14 hold true. Assume that the dual problem (7.67), specified at the begin of the proof, has a weak solution ϕ with the properties*

$$z := -\boldsymbol{K}\nabla\phi \in H^1(\Omega)^d, \quad \phi,\ r\phi,\ \tilde{\boldsymbol{c}} \cdot z \in H^1(\Omega)$$

such that

$$|z|_1 \leq C\|\overline{u}_h - u_h\|_0, \quad |\phi|_1 \leq C\|\overline{u}_h - u_h\|_0,$$

where $\overline{u}_h =: \Pi u \in X_h$ is the orthogonal L^2-projection of u and $C > 0$ is a constant independent of h, $\overline{u}_h$, and u_h. Then there exists a constant $\tilde{C} > 0$ independent of h such that

$$\|\overline{u}_h - u_h\|_0 \leq \tilde{C}h^{k+1}(|\boldsymbol{p}|_k + |\nabla \cdot \boldsymbol{p}|_k + |u|_k).$$

Proof The proof is based on duality arguments (see Lemma 6.11). Let ϕ be weak the solution of

$$\begin{aligned} -\nabla \cdot (\boldsymbol{K}\nabla\phi) - \boldsymbol{c} \cdot \nabla\phi + r\phi &= \overline{u}_h - u_h && \text{in } \Omega, \\ \phi &= 0 && \text{on } \partial\Omega, \end{aligned} \tag{7.67}$$

then

$$\begin{aligned} \|\overline{u}_h - u_h\|_0^2 &= \langle \Pi(\nabla \cdot z) + \tilde{\boldsymbol{c}} \cdot z + r\phi, \overline{u}_h - u_h \rangle_0 \\ &= \langle \nabla \cdot I_h(z) + \tilde{\boldsymbol{c}} \cdot z + r\phi, \overline{u}_h - u_h \rangle_0 \end{aligned} \tag{7.68}$$

using the commutativity property 3) from Theorem 7.9. Since $\mathrm{div}(M_h) \subset X_h$, we have

$$\langle \nabla \cdot I_h(z), \overline{u}_h - u_h \rangle_0 = \langle \nabla \cdot I_h(z), u - u_h \rangle_0,$$

and thus the error equation of $\boldsymbol{p}, u$ implies

$$\langle \nabla \cdot I_h(z), \overline{u}_h - u_h \rangle_0 = a(\boldsymbol{p} - \boldsymbol{p}_h, I_h(z)) + c(u - u_h, I_h(z)) =: I_1 + I_2\,.$$

Furthermore we decompose I_1 as

$$I_1 = a(\boldsymbol{p} - \boldsymbol{p}_h, I_h(z) - z) + I_3,$$

where

$$\begin{aligned} I_3 &:= a(\boldsymbol{p} - \boldsymbol{p}_h, -K\nabla\phi) = -\int_\Omega (\boldsymbol{p} - \boldsymbol{p}_h) \cdot \nabla\phi dx \\ &= \int_\Omega \nabla \cdot (\boldsymbol{p} - \boldsymbol{p}_h)\phi dx = \int_\Omega \nabla \cdot (\boldsymbol{p} - \boldsymbol{p}_h)(\phi - \overline{\phi}_h)dx + I_4 \end{aligned}$$

with $\overline{\phi}_h := \Pi\phi$ and

$$I_4 := -b(\boldsymbol{p} - \boldsymbol{p}_h, \overline{\phi}_h) = -d(u - u_h, \overline{\phi}_h).$$

Combining I_2 with the second term of (7.68) and using the symmetry property of an orthogonal projection, we obtain

$$\langle \Pi(\tilde{\boldsymbol{c}} \cdot \boldsymbol{z}), u - u_h \rangle_0 - \langle \tilde{\boldsymbol{c}} \cdot I_h(\boldsymbol{z}), u - u_h \rangle_0 = I_5 + I_6,$$

where

$$\begin{aligned} I_5 &:= \langle \tilde{\boldsymbol{c}} \cdot (\boldsymbol{z} - I_h(\boldsymbol{z})), u - u_h \rangle_0, \\ I_6 &:= \langle \Pi(\tilde{\boldsymbol{c}} \cdot \boldsymbol{z}) - \tilde{\boldsymbol{c}} \cdot \boldsymbol{z}, u - u_h \rangle_0. \end{aligned}$$

Similarly the combination of I_4 with the third term yields

$$\langle \Pi(r\phi), u - u_h \rangle_0 - \langle r\overline{\phi}_h, u - u_h \rangle_0 = I_7 + I_8$$

with

$$\begin{aligned} I_7 &:= \langle r(\phi - \overline{\phi}_h), u - u_h \rangle_0, \\ I_8 &:= \langle \Pi(r\phi) - r\phi, u - u_h \rangle_0. \end{aligned}$$

This concludes the proof, as, for instance, the first summand of I_1 allows the estimate

$$\begin{aligned} C\|\boldsymbol{p} - \boldsymbol{p}_h\|_0 \|I_h(\boldsymbol{z}) - \boldsymbol{z}\|_0 &\le Ch^k(|\boldsymbol{p}|_k + |\nabla \cdot \boldsymbol{p}|_k h(|\boldsymbol{z}|_1 + |\nabla \cdot \boldsymbol{z}|_1)) \\ &\le Ch^{k+1}(|\boldsymbol{p}|_k + |\nabla \cdot \boldsymbol{p}|_k)\|\overline{u}_h - u_h\|_0, \end{aligned}$$

and similarly the same bound results for $I_3 - I_5, I_7$ and also I_6, I_8. □

Brezzi–Douglas–Marini (BDM) finite elements

For $k \in \mathbb{N}$, $k \ge 2$, set $P_K := BDM_{k-1}(K) := \mathcal{P}_{k-1}(K)^d$, i.e., compared to RT-elements the space is reduced to the dimension $\dfrac{(d+k-1)!}{(d-1)!(k-1)!}$. As degrees of freedom the functionals from (7.49) are used, but instead of (7.50) the corresponding functionals are not defined by the full space $\mathcal{P}_{k-2}(\mathbb{R}^d)^d$, but only those functionals are used, which are defined by

- ∇w for $w \in \mathcal{P}_{k-2}(\mathbb{R}^d)$,
- $\boldsymbol{w} \in \mathcal{P}_{k-2}(\mathbb{R}^d)^d$ such that $\boldsymbol{w} \cdot \boldsymbol{n} = 0$ on ∂K, $\nabla \cdot \boldsymbol{w} = 0$.

The theory developed for *RT*-elements can be presented in parallel for *BDM*-elements. In Theorem 7.9, 3) the space P_K^{k-1} has to be substituted by P_K^{k-2}. Concerning the order of convergence estimate for the interpolation operator (Theorem 7.13), the estimate for the error in the $\|\cdot\|_0$-norm remains the same, but the second estimate has to be substituted by

$$\|\nabla \cdot (\boldsymbol{q} - I_h \boldsymbol{q})\|_0 \le C h^l |\nabla \cdot \boldsymbol{q}|_l \quad \text{for } 0 \le l \le k-1.$$

This gives the possibility of a flux density approximation of the same order for less degrees of freedom, but a worse estimate for $\nabla \cdot \boldsymbol{q}$ (and the scalar variable u).

The corresponding sketches in Figure 7.2 visualize selected elements. An arrow indicates a degree of freedom for the normal component at the boundary, and interior degrees of freedom are indicated by their numbers.

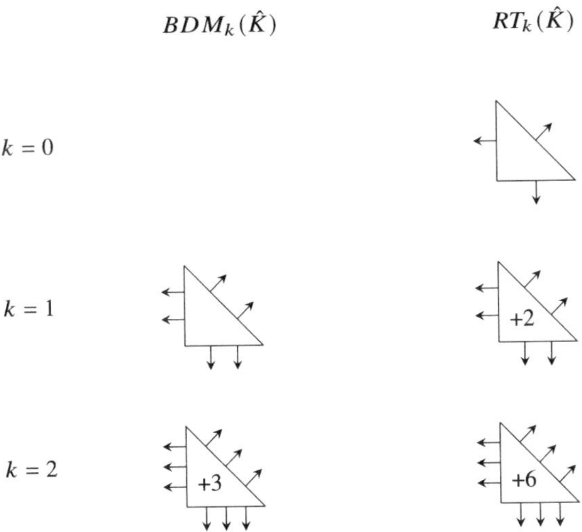

Fig. 7.2: $H(\text{div})$-elements on a triangle.

7.2.3 Finite Elements in $H(\text{div}; \Omega)$ on Quadrangles and Hexahedra

In the following, we first consider the reference element $\hat{K} := [0, 1]^d$ (and restrict ourselves to $d = 2, 3$). We generalize the notation $Q_k(\hat{K})$ (see (3.70)) to the anisotropic version

$$\mathcal{P}_{k_1,\ldots,k_d}(\hat{K}) := \left\{ \hat{p} \mid \hat{p}(\hat{x}) = \sum_{0 \le \alpha_i \le k_i} \gamma_\alpha \hat{x}_1^{\alpha_1} \ldots \hat{x}_d^{\alpha_d}, \quad \gamma_\alpha \in \mathbb{R}, \alpha \in \mathbb{N}_0^d \right\}, \quad (7.69)$$

i.e., we have

$$\mathcal{Q}_k(\hat{K}) = \mathcal{P}_{k\ldots k}(\hat{K}) \quad \text{and} \quad \mathcal{P}_{k_1,\ldots,k_d} \subset \mathcal{P}_{\tilde{k}}, \ \tilde{k} := \sum_{i=1}^{d} k_i.$$

Raviart–Thomas Elements on Quadrangles and Hexahedra

Let $k \in \mathbb{N}$ and define

$$P_{\hat{K}} := \begin{cases} \mathcal{P}_{k,k-1} \times \mathcal{P}_{k-1,k} & \text{for } d = 2, \\ \mathcal{P}_{k,k-1,k-1} \times \mathcal{P}_{k-1,k,k-1} \times \mathcal{P}_{k-1,k-1,k} & \text{for } d = 3, \end{cases}$$

always on $\hat{K}$, then

$$\dim P_{\hat{K}} = \begin{cases} 2k(k+1) & \text{for } d = 2, \\ 3k^2(k+1) & \text{for } d = 3. \end{cases} \quad (7.70)$$

Another notation is $RT_{[k-1]}(\hat{K})$. Furthermore, for $\hat{\boldsymbol{q}} \in P_{\hat{K}}$,

$$\nabla_{\hat{x}} \cdot \hat{\boldsymbol{q}} \in \mathcal{Q}_{k-1}(\hat{K}), \quad (7.71)$$

and for the faces $\hat{F} \subset \partial\hat{K}$,

$$\hat{\boldsymbol{q}} \cdot \mathfrak{n}|_{\hat{F}} \in \begin{cases} \mathcal{P}_{k-1}(\hat{F}) & \text{for } d = 2, \\ \mathcal{Q}_{k-1}(\hat{F}) & \text{for } d = 3. \end{cases} \quad (7.72)$$

Let for $k \ge 2$

$$\mathcal{R}_k(\hat{K}) := \begin{cases} \mathcal{P}_{k-2,k-1} \times \mathcal{P}_{k-1,k-2} & \text{for } d = 2, \\ \mathcal{P}_{k-2,k-1,k-1} \times \mathcal{P}_{k-1,k-2,k-1} \times \mathcal{P}_{k-1,k-1,k-2} & \text{for } d = 3, \end{cases}$$

all polynomial spaces restricted to $\hat{K}$.

Then, the degrees of freedom $\Sigma_{\hat{K}}$ are defined by the following functionals:

- For the degrees of freedom on the faces we take the same functionals as in (7.49), now

$$\text{for a basis of} \quad \mathcal{P}_{k-1}(\hat{F}) \quad \text{for } d = 2 \quad \text{or} \quad \mathcal{Q}_{k-1}(\hat{F}) \quad \text{for } d = 3. \quad (7.73)$$

- For the degrees of freedom on the element we take the same functionals as in (7.50), now

$$\text{for a basis of} \quad \mathcal{R}_k(\hat{K})\,. \quad (7.74)$$

Since the number of degrees of freedom is equal to $\dim P_{\hat{K}}$ (see Problem 7.11), only the linear independence of the functionals as above has to be shown to ensure the unique solvability of the local interpolation problem.

Lemma 7.16 *Let* $\hat{\boldsymbol{q}} \in P_{\hat{K}}$ *be such* $\hat{\psi}(\hat{\boldsymbol{q}}) = 0$ *for all* $\hat{\psi}$ *as defined above by* (7.73) *and* (7.74), *then* $\hat{\boldsymbol{q}} = \boldsymbol{0}$.

Proof Problem 7.12. □

Similar to the conforming (primal) elements of Section 3.3 we consider a partition with elements equivalent to $\hat{K}$. If the elements are equivalent in terms of affine-linear transformations, i.e., we restrict ourselves to parallelepipeds as general elements, the theory for simplices can be repeated leading to corresponding order of convergence estimates. For more general elements (convex quadrangles for $d = 2$ and a certain class of hexahedra for $d = 3$) a transformation $F \in \mathcal{Q}_{k-1}(\hat{K})^d$ is necessary (see Section 3.8) leading to a non-constant Jacobian. This makes the repetition of the theory more complex.

BDM-Elements on Quadrangles and Hexahedra

For $k \in \mathbb{N}$ and $d = 2$ we define

$$P_{\hat{K}} := BDM_{[k]}(\hat{K}) := \{\hat{\boldsymbol{q}} \mid \hat{\boldsymbol{q}}(\hat{x}_1, \hat{x}_2) = \hat{\boldsymbol{p}}(\hat{x}_1, \hat{x}_2) + \alpha \operatorname{curl}_{\hat{x}}(\hat{x}_1^{k+1}\hat{x}_2) + \beta \operatorname{curl}_{\hat{x}}(\hat{x}_1\hat{x}_2^{k+1})$$
$$\text{for } \hat{\boldsymbol{p}} \in \mathcal{P}_k(\hat{K})^2,\ \alpha,\ \beta \in \mathbb{R}\},$$

where curl is the operator defined in (6.89) (but here with respect to the reference variable $\hat{x}$). Similar to (7.71), (7.72) we have

$$\nabla_{\hat{x}} \cdot \hat{\boldsymbol{q}} \in \mathcal{P}_{k-1}(\hat{K}), \quad \hat{\boldsymbol{q}} \cdot \boldsymbol{n}|_{\hat{F}} \in \mathcal{P}_k(\hat{F}) \quad \text{for all faces } \hat{F} \subset \partial\hat{K}. \tag{7.75}$$

The rotation terms are supposed to add enough degrees of freedom on the boundary without disturbing $\nabla_{\hat{x}} \cdot \hat{\boldsymbol{q}}$ (see (6.90) ff.). A possible corresponding construction for $d = 3$ reads as follows:

$$P_{\hat{K}} := BDM_{[k]}(\hat{K}) := \Big\{\hat{\boldsymbol{q}} \mid \hat{\boldsymbol{q}}(\hat{x}_1, \hat{x}_2, \hat{x}_3) = \hat{\boldsymbol{p}}(\hat{x}_1, \hat{x}_2, \hat{x}_3)$$
$$+ \sum_{i=0}^{k} \big[\alpha_i \operatorname{curl}_{\hat{x}}(0, 0, \hat{x}_1\hat{x}_2^{i+1}\hat{x}_3^{k-i}) + \beta_i \operatorname{curl}_{\hat{x}}(\hat{x}_2\hat{x}_3^{i+1}\hat{x}_1^{k-i}, 0, 0)$$
$$+ \gamma_i \operatorname{curl}_{\hat{x}}(0, \hat{x}_3\hat{x}_1^{i+1}\hat{x}_2^{k-i}, 0)\big] \quad \text{for } \hat{\boldsymbol{p}} \in \mathcal{P}_k(\hat{K})^3,\ \alpha_i, \beta_i, \gamma_i \in \mathbb{R},\ i = 0, \ldots, k\Big\}.$$

Consult the definition of curl before (8.80). Then

$$\dim P_k = \begin{cases} k^2 + 3k + 4 & \text{for } d = 2, \\ \frac{1}{2}((k+1)(k+2)(k+3) + 6(k+1)) & \text{for } d = 3, \end{cases}$$

and the same number of degrees of freedom $\Sigma_{\hat{K}}$ is defined by the functionals as above, but this time for a basis of $\mathcal{P}_k(\hat{F})$ or $\mathcal{P}_{k-2}(\hat{K})^d$ (for $k \geq 2$), respectively. The corresponding sketches in Figure 7.3 visualize the elements.

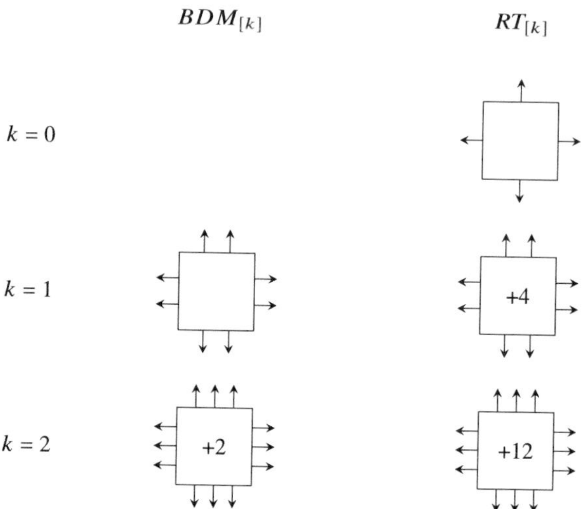

Fig. 7.3: $H(\text{div})$-elements on a square.

Exercises

Problem 7.6 Let K be a triangle with vertices $a_1, a_2, a_3 \in \mathbb{R}^2$. Denote by F_i the edge opposite to a_i and by $\mathfrak{n}_i$ its outer normal. Define functions $\boldsymbol{\tau}_i \in RT_0(K)$ by

$$\boldsymbol{\tau}_i(x) := \frac{|F_i|}{2|K|}(x - a_i), \quad i \in \{1, 2, 3\}.$$

Let $h := \max\{|F_1|, |F_2|, |F_3|\}$ be the maximal edge length and $c > 0$ be chosen such that $h^2 \leq c|K|$.

a) Prove that $\boldsymbol{\tau}_i(x) \cdot \mathfrak{n}(x) = \chi_{F_i}(x)$ for $x \in \partial K$ a.e., where χ_{F_i} denotes the indicator function of F_i, and conclude that $\{\boldsymbol{\tau}_1, \boldsymbol{\tau}_2, \boldsymbol{\tau}_3\}$ is a basis of $RT_0(K)$.
b) Prove that for any $u \in H^1(K)$ and $i \in \{1, 2, 3\}$ there holds

$$\|u\|_{0,F_i}^2 \leq \|\nabla \cdot \boldsymbol{\tau}_i\|_{L^\infty(K)} \|u\|_{0,K}^2 + 2\|\boldsymbol{\tau}_i\|_{L^\infty(K)} \|u\|_{0,K} \|\nabla u\|_{0,K}.$$

Hint: Use a) and apply the divergence theorem: $\int_{\partial\Omega} \boldsymbol{v} \cdot \mathfrak{n} d\sigma = \int_\Omega \nabla \cdot \boldsymbol{v} dx$.
c) Prove that for any $i \in \{1, 2, 3\}$ there holds

$$\|\boldsymbol{\tau}_i\|_{L^\infty(K)} \leq \frac{c}{2} \quad \text{and} \quad \|\nabla \cdot \boldsymbol{\tau}_i\|_{L^\infty(K)} \leq \frac{c}{h}.$$

d) Finally, prove the multiplicative trace inequality of Lemma 7.4.

Problem 7.7 Complete the proof of Lemma 7.11, 1).

Problem 7.8 Make the statement (7.63) of Lemma 7.12 more precise.

Problem 7.9 Work out the details of the proof of Theorem 7.13.

Problem 7.10 Show the surjectivity of the divergence operator from $\mathcal{D}_k(K)$ to $\mathcal{P}_{k-1}(K)$ for $k \in \mathbb{N}$ and a regular d-simplex K.

Problem 7.11 Show for the Raviart–Thomas elements on $\hat{K} = [0,1]^d$, $d = 2, 3$, the number of degrees of freedom defined by (7.73), (7.74) is equal to $\dim P_{\hat{K}}$.

Problem 7.12 Prove Lemma 7.16.

7.3 Mixed Methods for the Stokes Equation

Here we discuss the Galerkin discretization of the stationary Stokes equation (6.79), but—contrary to Section 6.3—not the constrained form (Formulation I, p. 386), but the mixed form (Formulation II, p. 386). Comparing with the general theory of Section 6.2, here we have the saddle point form (6.61), the bilinear form b is the same as for the Darcy equation (6.54), but the spaces are modified to

$$M := H_0^1(\Omega)^d, \quad X := L_0^2(\Omega),$$

and the bilinear form a is even M-coercive on the whole space. Therefore, from the conditions for unique and stable existence of Theorem 6.19 and Theorem 6.25, the conditions (NB1), (NB2) regarding the bilinear form a are also met for a conforming discretization, and only the condition (NB2) with respect to the bilinear form $\tilde{b}$, reflecting the interplay between M_h and X_h, has to be investigated. Obviously, this inf-sup condition is the easier to satisfy the larger ("richer") M_h and the smaller ("poorer") X_h is chosen. The art of "balancing" consists in choosing stable pairs of spaces without "wasting" degrees of freedom in M_h concerning the achievable order of convergence. No new finite element spaces are necessary here. For the components of the velocity finite elements in $H^1(\Omega)$ are required, which have been investigated in Chapter 3. Therefore we can use—note the exchange of notation compared to Section 7.2, namely $(\boldsymbol{u}, p)$ for $(\boldsymbol{p}, u)$ to be in accordance with the standard notation of fluid mechanics—the space

$$M_h := \{\boldsymbol{u}_h \in M \mid \boldsymbol{u}_h|_K \in \mathcal{P}_k(K)^d \quad \text{for all } K \in \mathcal{T}_h\} \tag{7.76}$$

for some $k \in \mathbb{N}$ and an underlying partition $\mathcal{T}_h$, meaning C^0-elements for the velocity components. For X_h ($\subset L_0^2(\Omega)$) no inter-element continuity is required. For instance, the following elements can be used to discretize the pressure p.

Discontinuous elements

For $l \in \mathbb{N}_0$, set

$$X_h := \{p_h \in X \mid p_h|_K \in \mathcal{P}_l(K) \text{ for all } K \in \mathcal{T}_h\}, \tag{7.77}$$

see Figure 7.4. This requires no new analysis of the approximation error for appropriate degrees of freedom.

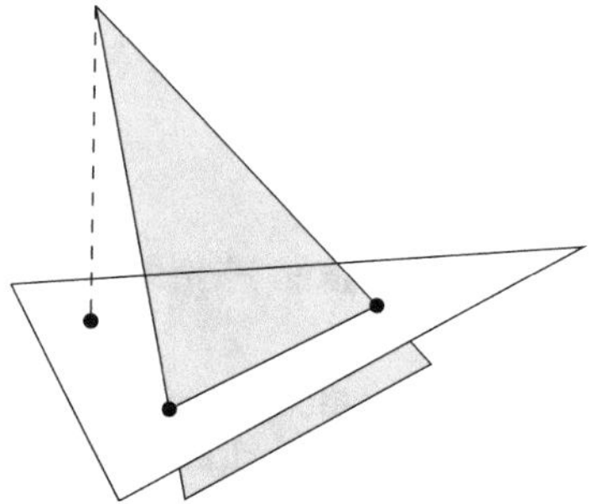

Fig. 7.4: Local shape function for the discontinuous P_1-approximation space ($d = 2$, $l = 1$).

Continuous elements

For $l \in \mathbb{N}$, set

$$X_h = \{p_h \in H^1(\Omega) \cap X \mid p_h|_K \in \mathcal{P}_l(K) \text{ for all } K \in \mathcal{T}_h\} \tag{7.78}$$

with known approximation properties.

For a partition generated by affine-linear or isoparametric transformations of $[0, 1]^d$, corresponding spaces based on $Q_l(\hat{K})$ instead of $\mathcal{P}_l(\hat{K})$ can be considered.

Discontinuous elements, more precisely spaces X_h such that

$$\tilde{X}_h \subset X_h,$$

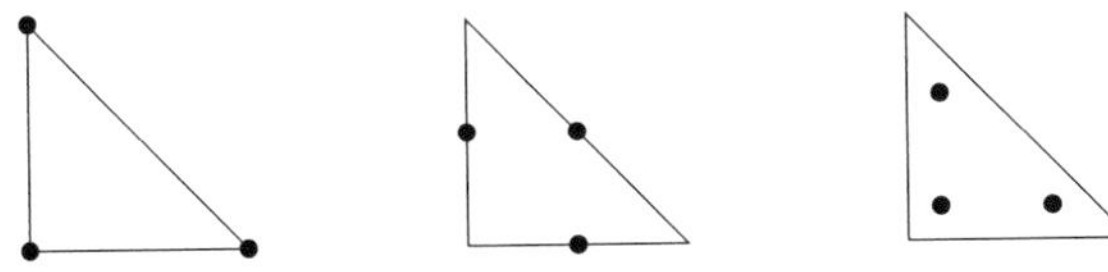

Fig. 7.5: The P_1-elements ($d = 2$).

where $\tilde{X}_h$ correspondents to (7.77) for $l = 0$, the piecewise constant elements (Fig. 7.5), have the advantageous property of local mass conservation: From

$$0 = b(\boldsymbol{v}_h, q_h) = -\int_\Omega q_h \nabla \cdot \boldsymbol{v}_h \, dx \quad \text{for all } q_h \in X_h$$

it follows that, for each $K \in \mathcal{T}_h$,

$$\int_{\partial K} \boldsymbol{u}_h \cdot \mathfrak{n} \, d\sigma = \int_K \nabla \cdot \boldsymbol{u}_h \, dx = 0. \tag{7.79}$$

As even $\boldsymbol{u}_h \in H^1(\Omega)^d$, $[\![u_h]\!]_F = \mathbf{0}$ for all $F \in \mathcal{F}$ for the underlying partition $\mathcal{T}_h$ such that local mass conservation is fulfilled exactly (compare Definition 7.58). In the following we will consider the stability properties of various pairs of spaces, i.e., the validity of a (uniform) inf-sup condition for $\tilde{b}$. While $\ker B_h$ has no influence on the validity of the inf-sup condition, this is the case for $\ker B_h^T$. Namely, if

$$\ker B_h^T \neq \emptyset, \tag{7.80}$$

an inf-sup condition cannot hold, as for $\tilde{q}_h \in \ker B_h^T$

$$\sup_{\boldsymbol{v}_h \in M_h \setminus \mathbf{0}} \frac{b(\boldsymbol{v}_h, \tilde{q}_h)}{\|\boldsymbol{v}_h\|_M} = 0.$$

The elements of $\ker B_h^T$ are called *spurious modes*. If $(\boldsymbol{u}_h, p_h)$ is a solution, then also $(\boldsymbol{u}_h, p_h + q_h)$ is a solution for every $q_h \in \ker B_h^T$.

① $\underline{d = 2,\ \text{squares},\ k = 1, l = 0 :\ (Q_1)^2/\mathcal{P}_0}$

This is (one of) the most simple approache(s) and locally mass conservative, but suffers from spurious modes, as the following example shows.

Consider $\Omega := (0,1)^2$ and $\mathcal{T}_h$ as in the finite difference method (see (1.5)) with step size $h := 1/N$, $N \in \mathbb{N}$. Because of the homogeneous Dirichlet conditions the degrees of freedom of M_h are the nodal values $\boldsymbol{u}_{ij} = (v_{ij}, w_{ij})$ at the points (ih, jh), $i, j = 1, \ldots, N-1$, and the degrees of freedom of X_h are the values

$$p_{i+\frac{1}{2}, j+\frac{1}{2}} \text{ at the elements } K_{ij} := [ih, (i+1)h] \times [jh, (i+1)h],\ i, j = 0, \ldots, N-1,$$

see Figure 7.6.

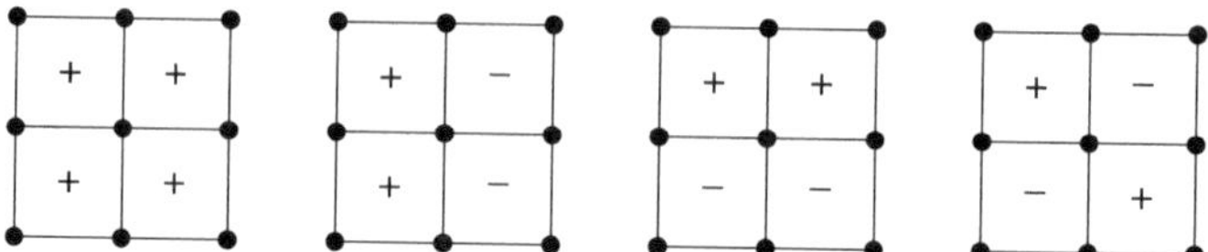

Fig. 7.6: Basis functions of X_h in a macroelelement.

We show that there exist nontrivial elements $p_h \in \ker B_h^T$, i.e.,

$$0 = \int_\Omega p_h \nabla \cdot \boldsymbol{u}_h \, dx \quad \text{for all } \boldsymbol{u}_h \in M_h. \tag{7.81}$$

From

$$\int_{K_{ij}} p_h \nabla \cdot \boldsymbol{u}_h \, dx = p_{i+\frac{1}{2},j+\frac{1}{2}} \int_{\partial K_{ij}} \boldsymbol{u}_h \cdot \mathfrak{n} \, d\sigma$$

$$= p_{i+\frac{1}{2},j+\frac{1}{2}} \, h \frac{1}{2}(v_{i+1,j} + v_{i+1,j+1} + w_{i,j+1} + w_{i+1,j+1} - v_{ij} - v_{i,j+1} - w_{ij} - w_{i+1,j}), \text{ i.e.,}$$

$$\int_{K_{ij}} p_h \nabla \cdot \boldsymbol{u}_h \, dx = -h^2 \sum_{1 \le i,j \le N-1} v_{ij}(\partial_1 p)_{ij} + w_{ij}(\partial_2 p)_{ij}$$

with the “discrete derivatives”

$$\begin{aligned}
(\partial_1 p)_{ij} &:= \frac{1}{2h}(p_{i+\frac{1}{2},j+\frac{1}{2}} + p_{i+\frac{1}{2},j-\frac{1}{2}} - p_{i-\frac{1}{2},j+\frac{1}{2}} - p_{i-\frac{1}{2},j-\frac{1}{2}}), \\
(\partial_2 p)_{ij} &:= \frac{1}{2h}(p_{i+\frac{1}{2},j+\frac{1}{2}} + p_{i-\frac{1}{2},j+\frac{1}{2}} - p_{i+\frac{1}{2},j-\frac{1}{2}} - p_{i-\frac{1}{2},j-\frac{1}{2}})
\end{aligned}$$

we see that the requirement (7.81) is fulfilled if

$$(\partial_1 p)_{ij} = (\partial_2 p)_{ij} = 0 \quad \text{for } i, j = 1, \ldots, N-1,$$

and this in turn is equivalent to the set of homogeneous linear equations

$$p_{i+\frac{1}{2},j+\frac{1}{2}} = p_{i-\frac{1}{2},j-\frac{1}{2}}, \quad p_{i-\frac{1}{2},j+\frac{1}{2}} = p_{i+\frac{1}{2},j-\frac{1}{2}}, \quad i, j = 1, \ldots, N-1.$$

Now it is not difficult to see that the solution space of this system is two-dimensional. Namely, the first condition requires that all elements on “diagonal lines” have the same value, and the second is the same for all “counterdiagonal lines”, which together gives so-called *checkerboard type* solutions with two real parameters $p^{(1)}, p^{(2)}$. The further requirement

$$\int_\Omega p_h(x) dx = 0 \tag{7.82}$$

relates these two values, for even N it is equivalent to

$$p^{(1)} = -p^{(2)}.$$

In summary, we get for even N a one-dimensional space $\ker B_h^T$ with a basis function given by $p^{(1)} = 1, p^{(2)} = -1$. Therefore one also speaks of a *checkerboard instability*. Restricting X_h to $(\ker B_h^T)^\perp$ restores the inf-sup condition, but not uniformly, as then

$$C_1 h \leq \beta_h \leq C_2 h,$$

see [109].

② $d = 2$, triangles, $k = 1, l = 0 : (\mathcal{P}_1)^2/\mathcal{P}_0$

Consider a polygonally bounded domain $\Omega \subset \mathbb{R}^2$ with a consistent triangulation. Let $|\mathcal{T}|, |\mathcal{N}|, |\partial\mathcal{F}|$ denote the numbers of elements, interior nodes, and boundary faces, respectively. Then, by Euler's relations (see [27, Sect. 1.5]),

$$|\mathcal{T}| = 2|\mathcal{N}| + |\partial\mathcal{F}| - 2. \tag{7.83}$$

Since

$$\begin{aligned} \dim(X_h) &= |\mathcal{T}| - 1 \quad \text{because of (7.82)}, \\ \dim(M_h) &= 2|\mathcal{N}|, \end{aligned}$$

we have

$$\begin{aligned} \dim(\ker B_h^T) &= \dim(X_h) - \dim(\operatorname{im}(B_h^T)) \\ &\geq \dim(X_h) - \dim(M_h) = |\mathcal{T}| - 1 - 2|\mathcal{N}| = |\partial\mathcal{F}| - 3 \end{aligned}$$

because of (7.83), and thus spurious modes.

③ $d = 2$, triangles, $k = 1 = l : (\mathcal{P}_1)^2/\mathcal{P}_1$ acc. to (7.78)

Consider $\Omega := (0, 1)^2$ with a Friedrichs–Keller triangulation (see Figure 2.9), $h := 1/N, N \in \mathbb{N}$. Then, with a notation similar to example ①,

$$\int_\Omega p_h \nabla \boldsymbol{u}_h \, dx = \sum_{K \in \mathcal{T}_h} (\nabla \cdot \boldsymbol{u}_h)|_K \frac{|K|}{3} \sum_{(ih,jh)\in K} p_{ij}$$

because of the exactness of the trapezoidal rule for $\int_K p_h \, dx$ (see (2.42)). Therefore, as $(\nabla \cdot \boldsymbol{u}_h)|_K$ is arbitrary in $\mathbb{R}$ for arbitrary $\boldsymbol{u}_h \in M_h$, it follows that

$$p_h \in \ker B_h^T \iff \sum_{(ih,jh)\in K} p_{i,j} = 0 \quad \text{for all } K \in \mathcal{T}_h,$$

which again can be accomplished by a checkerboard construction (see Figure 7.7) also satisfying (7.82).

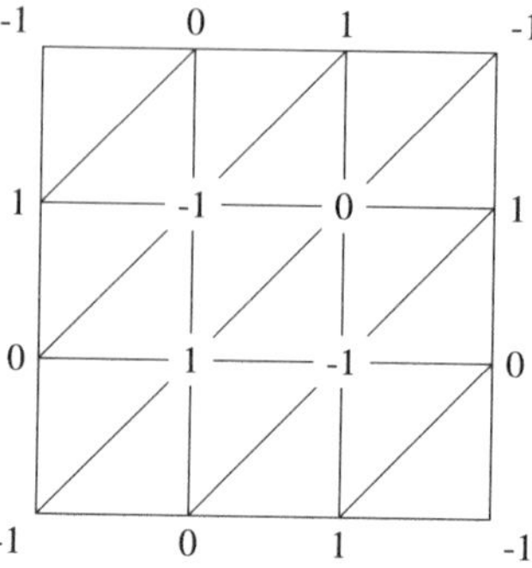

Fig. 7.7: Checkerboard patterns for the $(\mathcal{P}_1)^2/\mathcal{P}_1$-pair.

Therefore an enrichment of M_h is necessary, leading to the most simple stable element, the so-called *mini-element*.

④ $d \in \mathbb{N}$, simplices, $k = 1$, enriched with a bubble function, $l = 1$: $(\mathcal{P}_1 + \text{bubble})^d/\mathcal{P}_1$ acc. to (7.77)

We add an additional degree of freedom to each element $K \in \mathcal{T}_h$ as the nodal value at the barycentre $a_{S,K}$ of K by adding a basis function $b = b_K \in H_0^1(K)$ such that

$$0 \le b \le 1, \quad b(a_{S,K}) = 1, \tag{7.84}$$

which is called a *bubble function*. That is,

$$M_h|_K := P_K := (\mathcal{P}_1(K) \oplus \operatorname{span}(b_K))^d,$$

see Figure 7.8 (left). There are various choices of bubble functions. A possible definition is

$$b_K(x) := (d+1)^{d+1} \prod_{i=1}^{d+1} \lambda_i(x), \tag{7.85}$$

where λ_i are the barycentric coordinates with respect to K (see (3.59)) because of

$$x = a_{S,K} \Longleftrightarrow \lambda_i(x) = \frac{1}{d+1}, \quad i = 1, \dots, d+1.$$

Theorem 7.17 *Let M_h, X_h be defined as above for a shape-regular family $(\mathcal{T}_h)_h$ of triangulations satisfying (T1)–(T6). Then the inf-sup condition holds for $\tilde{b}_h$, $\tilde{b}_h(q_h, \boldsymbol{v}_h) := b_h(\boldsymbol{v}_h, q_h)$, with an uniform positive lower bound for the constant.*

Proof As the inf-sup condition is verified for the continuous problem (Formulation II, p. 386) (see (6.85)), we try to apply Fortin's criterion (Theorem 6.27) by defining a mapping $\Pi_h : H_0^1(\Omega)^d \to M_h$ such that

$$\int_\Omega q_h \nabla \cdot (\Pi_h \boldsymbol{v}) dx = \int_\Omega q_h \nabla \cdot \boldsymbol{v}\, dx \quad \text{for all } q_h \in X_h \tag{7.86}$$

and

$$\|\Pi_h \boldsymbol{v}\|_1 \le C \|\boldsymbol{v}\|_1 \quad \text{for all } \boldsymbol{v} \in H_0^1(\Omega)^d \tag{7.87}$$

for some constant $C > 0$ independent of h. We make the ansatz

$$\Pi_h \boldsymbol{v} := Q_h \boldsymbol{v} + \widetilde{\Pi}_h \boldsymbol{v}, \tag{7.88}$$

where $Q_h : H_0^1(\Omega)^d \to M_h$ is a so-called *quasi-interpolation operator* still to be specified and

$$\widetilde{\Pi}_h \boldsymbol{v} := \sum_{K \in \mathcal{T}_h} b_K \boldsymbol{\alpha}_K(\boldsymbol{v})$$

with local vector fields $\boldsymbol{\alpha}_K(\boldsymbol{v})$, the choice of which will be discussed first.

The equation (7.86) is equivalent to

$$\sum_{K \in \mathcal{T}_h} \int_K (\Pi_h \boldsymbol{v} - \boldsymbol{v}) \nabla q_h \, dx = 0 \quad \text{for all } q_h \in X_h$$

due to $\Pi_h \boldsymbol{v}, \boldsymbol{v} \in H_0^1(\Omega)^d$ and Theorem 3.21, and this again is equivalent to

$$\int_K \Pi_h \boldsymbol{v}\, dx = \int_K \boldsymbol{v}\, dx \quad \text{for all } K \in \mathcal{T}_h$$

as $\nabla q_h|_K$ runs through $\mathbb{R}$.

The latter requirement can be enforced by choosing

$$\boldsymbol{\alpha}_K(\boldsymbol{v}) := \left(\int_K b_K dx \right)^{-1} \int_K (\boldsymbol{v} - Q_h \boldsymbol{v})\, dx \,. \tag{7.89}$$

Unfortunately, for Q_h the nodal interpolation operator from (3.83) cannot be chosen, therefore we apply the so-called *Clément interpolation operator* (see the end of the section, in particular (7.95)), which is well defined here. Since this operator is uniformly stable (see Theorem 7.23, 1)), it remains to prove (7.87) for the second addend in (7.88).

The first integral in the definition (7.89) of $\boldsymbol{\alpha}_K$ can be expressed as follows:

$$\int_K b_K(x) dx = \frac{1}{|\hat{K}|} \int_{\hat{K}} b_{\hat{K}}(\hat{x})\, d\hat{x} |K| =: C_1 |K|,$$

where we have used that $|\det(B)| = |K|/|\hat{K}|$ by (3.61). Thus, by the Cauchy–Schwarz inequality and Theorem 7.23, 2),

$$|\alpha_K(\boldsymbol{v})|^2 \le \frac{1}{(C_1|K|)^2}\|\boldsymbol{v}-Q_h\boldsymbol{v}\|_{0,K}^2|K| \le C_2\frac{h_K^2}{|K|}\|\boldsymbol{v}\|_{1,\Delta(K)}^2,$$

where the *simplicial neighbourhood*

$$\Delta(K) := \bigcup_{K':\, K'\cap K\neq\emptyset} K' \tag{7.90}$$

of K is already known from the two-dimensional case mentioned in Section 4.2; see Figure 4.6.

Furthermore, by Theorems 3.29, 3.30 we have

$$\|b_K\|_{1,K}^2 \le C_3\|B^{-1}\|_2^2\,|\det B|\,\|b_{\hat{K}}\|_{1,\hat{K}}^2 \le C_3\frac{h_{\hat{K}}^2}{\varrho_K^2}\frac{|K|}{|\hat{K}|}\|b_{\hat{K}}\|_{1,\hat{K}}^2 \le \frac{C_4}{h_K^2}|K|$$

using the shape regularity (Definition 3.31) of $(\mathcal{T}_h)_h$.

Putting the estimates of $|\alpha_K(\boldsymbol{v})|$ and $\|b_K\|_{1,K}$ together, we arrive at

$$\begin{aligned}\|\widetilde{\Pi}_h\boldsymbol{v}\|_{1,\Omega}^2 &= \sum_{K\in\mathcal{T}_h}\|\widetilde{\Pi}_h\boldsymbol{v}\|_{1,K}^2 = \sum_{K\in\mathcal{T}_h}|\alpha_K(\boldsymbol{v})|^2\|b_K\|_{1,K}^2\\ &\le C_2C_4\sum_{K\in\mathcal{T}_h}\|\boldsymbol{v}\|_{1,\Delta(K)}^2 \le C_5\|\boldsymbol{v}\|_{1,\Omega}^2,\end{aligned}$$

where the last step follows from the fact that the number of simplices in $\Delta(K)$ is bounded uniformly in h (see Problem 7.14). □

Corollary 7.18 (Error estimate for the mini-element) *Let* $(\boldsymbol{u},p)$ *be a solution of the stationary Stokes equation* (6.78), (6.79) *such that* $\boldsymbol{u}\in(H^2(\Omega)\cap H_0^1(\Omega))^d$, $p\in H^1(\Omega)\cap L_0^2(\Omega)$. *Then* $(u_h,p_h)\in M_h\times X_h$, *the consistent finite element approximation according to* ④, *exists uniquely and satisfies, for some constant* $C>0$ *independent of* h, *the estimate:*

$$\|\boldsymbol{u}-\boldsymbol{u}_h\|_{1,\Omega}+\|p-p_h\|_{0,\Omega}\le Ch(\|\boldsymbol{u}\|_{2,\Omega}+\|p\|_{1,\Omega}).$$

Proof Based on Theorem 7.17, we can conclude the unique existence of the numerical solution from Theorem 6.19 and the error estimate from Theorem 6.25, taking into account the approximation error estimate from Theorem 3.32.

Concerning the error in the pressure, note that $X = L_0^2(\Omega)\cong L^2(\Omega)/\mathbb{R}$ and correspondingly $X_h\cong\{u\in L^2(\Omega)\mid u|_K\in\mathcal{P}_1(K) \text{ for all } K\in\mathcal{T}_h\}/\mathbb{R}$. Denoting by $[\cdot]$ the equivalence classes, we have for $q_h\in X_h$ the relation $\|p-q_h\|_{0,\Omega} = \|[\tilde{p}-\tilde{q}_h]\|_X = \inf_{c\in\mathbb{R}}\|\tilde{p}-(\tilde{q}_h+c)\|_{0,\Omega}$, to which Theorem 3.32 can be applied. □

To increase the order of convergence, the following element, called the *Taylor–Hood element*, is popular.

⑤ $d \in \mathbb{N}$, simplices, $k = 2, l = 1$: $(\mathcal{P}_2)^d/\mathcal{P}_1$ acc. to (7.78)

The setting is as in ④ but we restrict our analysis to $d = 2, 3$; see Figure 7.8 (right) for the case $d = 2$.

Fig. 7.8: Mini-element (left) and Taylor–Hood element (right), •: degrees of freedom for $\boldsymbol{u}$, ×: degrees of freedom for p.

A first step towards a uniform inf-sup estimate is the following result.

Lemma 7.19 *Let M_h, X_h be defined as above for a shape-regular family $(\mathcal{T}_h)_h$ of triangulations satisfying (T1)–(T6) and having the property that the number n_K of interior edges of each simplex K is not less than d, i.e., $n_K \geq d$. Then the estimate*

$$\sup_{\boldsymbol{v}_h \in M_h} \frac{\int_\Omega q_h \nabla \cdot \boldsymbol{v}_h \, dx}{\|\boldsymbol{v}_h\|_1} \geq C\Big(\sum_{K \in \mathcal{T}_h} h_K^2 |q_h|_{1,K}^2 \Big)^{1/2} \quad \text{for all } q_h \in X_h$$

holds, where $C > 0$ is a constant independent of h.

Proof We restrict ourselves to the case $d = 3$. Obviously it is sufficient to show the assertion for a particular element $\boldsymbol{v}_h \in M_h$ adopted to q_h. In general, we have

$$\int_\Omega q_h \nabla \cdot \boldsymbol{v}_h \, dx = - \sum_{K \in \mathcal{T}_h} \int_K \boldsymbol{v}_h \cdot \nabla q_h \, dx + \sum_{K \in \mathcal{T}_h} \int_{\partial K} q_h \boldsymbol{v}_h \cdot \boldsymbol{n} \, d\sigma \,. \tag{7.91}$$

The last term vanishes as $q_h \in C(\overline{\Omega})$ (see Theorem 3.21) and $[\![\boldsymbol{v}_h \cdot \boldsymbol{n}]\!]_F = 0$ for every face F because of Theorem 6.16 and the boundary condition. On $K := \operatorname{conv}\{a_1^K, a_2^K, a_3^K, a_4^K\}$, each component of $\boldsymbol{v}_h$ is uniquely determined by the values at the nodes a_i^K, $i = 1, \ldots, 4$, and the edge midpoints b_i^K, $i = 1, \ldots, 6$ (see (3.66), note the change in the notation for the midpoints), whereas q_h is determined by the values at the nodes a_i^K (see (3.64)).

The integrand can be exactly evaluated by (see Problem 7.15)

$$\int_K f(x) dx = |K| \left[\sum_{i=1}^{6} \frac{1}{5} f(b_i^K) - \sum_{i=1}^{4} \frac{1}{20} f(a_i^K) \right] \quad \text{for all } f \in \mathcal{P}_2(K).$$

If $\boldsymbol{v}_h$ is chosen such that $\boldsymbol{v}_h(a_i^K) = 0$ for every element, then (7.91) reduces to

$$\int_\Omega q_h \nabla \cdot \boldsymbol{v}_h \, dx = -\frac{1}{5} \sum_{K \in \mathcal{T}_h} \sum_{i=1}^{6} \boldsymbol{v}_h(b_i^K) \cdot \nabla q_h(b_i^K) |K| . \tag{7.92}$$

To continue the construction of $\boldsymbol{v}_h$, let E_i be the edge of K with midpoint b_i^K, $|E_i|$ its length, and $\boldsymbol{\tau}_i$ a (fixed) unit tangential vector of E_i. Define the nodal values of $\boldsymbol{v}_h$ on K by

$$\begin{aligned} \boldsymbol{v}_h(a_i^K) &:= 0, \quad i = 1, \ldots, 4, \\ \boldsymbol{v}_h(b_i^K) &:= \begin{cases} -|E_i|^2 \operatorname{sign}(\partial_{\boldsymbol{\tau}_i} q_h)|\partial_{\boldsymbol{\tau}_i} q_h|\boldsymbol{\tau}_i, & \text{if } b_i^K \in \Omega, \\ 0, & \text{otherwise,} \end{cases} \quad i = 1, \ldots, 6. \end{aligned}$$

This determines an element in M_h, as due to $q_h|_{E_i} \in \mathcal{P}_1(E_i)$ also the tangential derivative $\partial_{\boldsymbol{\tau}_i} q_h$ is uniquely determined by the nodal values at the vertices and independent of which of the neighbouring elements is considered. If n_K denotes the number of interior edges of the element K, equation (7.92) can be continued as

$$\begin{aligned} \int_\Omega q_h \nabla \cdot \boldsymbol{v}_h \, dx &= \frac{1}{5} \sum_{K \in \mathcal{T}_h} \sum_{i=1}^{n_K} |E_i|^2 \operatorname{sign}(\partial_{\boldsymbol{\tau}_i} q_h) \, |\partial_{\boldsymbol{\tau}_i} q_h| \, \partial_{\boldsymbol{\tau}_i} q_h \, |K| \\ &\geq C_1 \sum_{K \in \mathcal{T}_h} h_K^2 \sum_{i=1}^{n_K} |\partial_{\boldsymbol{\tau}_i} q_h|^2 \, |K| \, . \end{aligned}$$

In the last step we have used the estimate $|E_i| \geq c h_K$, $c > 0$ independent of K, which is a consequence of the shape regularity.

If, as assumed, $n_K \geq 3$ for all $K \in \mathcal{T}_h$, the constant vector $\nabla q_h|_K$ can be expressed by means of tangential derivatives along interior edges. In more detail, let us consider only the situation where three interior edges E_i, $i = 1, 2, 3$, have a common vertex, say a_1^K, and the further vertices of the edges are labelled by a_2^K, a_3^K, a_4^K. Then the reference transformation F according to (3.87) is such that

$$\partial_{\hat{x}} \hat{v} = \partial_{\boldsymbol{\tau}_i} \hat{v}.$$

From (3.90) and Theorem 3.30 we conclude the estimate

$$|\nabla_x q_h| \leq \|B^{-1}\| |(\partial_{\boldsymbol{\tau}_i} q_h)_i| \leq \frac{\hat{h}_K}{\varrho_K} |(\partial_{\boldsymbol{\tau}_i} q_h)_i| \quad \text{on } K,$$

and the shape regularity implies

$$\sum_{i=1}^{n_K} |\partial_{\boldsymbol{\tau}_i} q_h|^2 \, |K| \geq C_2 h_K^2 |q_h|_{1,K}^2 .$$

The proof is completed by the estimate

$$\|\boldsymbol{v}_h\|_{1,K}^2 \leq C_3 h_K^2 |q_h|_{1,K}^2 \tag{7.93}$$

(see Problem 7.16). A sketch is as follows: It should be noted that, if only $|\boldsymbol{v}_h|_{1,K}^2$ is considered, this integral can be evaluated by the above quadrature rule, and the term $|\partial_{\boldsymbol{\tau}_i} \boldsymbol{v}_h|^2$ at the midpoint nodes is of the order

$$|E_i|^4 |\partial_{\tau_i} q_h|^2 h_K^{-2}.$$

Then, with the above estimate and its reversed version $|(\partial_{\tau_i} q_h)_i| \leq h_K \varrho_{\hat{K}}^{-1} |\nabla_x q_h|$ it follows that

$$\|v_h\|_{1,K}^2 \leq C \max \left\{ |E_i|^4 h_K^{-2} h_K^{-2} h_K^2 |K| |\nabla q_h|_K|^2 \right\}.$$

□

Theorem 7.20 *Under the assumptions of Lemma 7.19, the inf-sup condition for $\tilde{b}$ on $X_h \times M_h$ holds uniformly in h.*

Proof We would like to argue as in the proof of Theorem 7.17, but here the Clément interpolation cannot be amended to fulfil (7.86). Therefore we modify the arguments of the proof of Theorem 6.27 as follows:

$$\begin{aligned}
\sup_{v_h \in M_h} \frac{\int_\Omega q_h \nabla \cdot v_h \, dx}{\|v_h\|_1} &\geq \frac{1}{\|Q_h v\|_1} \int_\Omega q_h \nabla \cdot Q_h v \, dx \geq \frac{C_1}{\|v\|_1} \int_\Omega q_h \nabla \cdot Q_h v \, dx \\
&= \frac{C_1}{\|v\|_1} \int_\Omega q_h \nabla \cdot v \, dx + \frac{C_1}{\|v\|_1} \int_\Omega q_h \nabla \cdot (Q_h v - v) \, dx.
\end{aligned}$$

Here we have used the uniform continuity of the Clément operator (Theorem 7.23, 1)). Because of the inf-sup condition for the Stokes equation (6.85), the first term can be estimated, for some $\beta > 0$ independent of h, from below by

$$\beta \|q_h\|_0. \tag{7.94}$$

The absolute value of the integral in the second term allows the estimate (see (7.91))

$$\left| \int_\Omega q_h \nabla \cdot (Q_h v - v) \, dx \right| \leq \sum_{K \in \mathcal{T}_h} |q_h|_{1,K} \|Q_h v - v\|_{0,K} \leq \sum_{K \in \mathcal{T}_h} |q_h|_{1,K} \, C_2 h_K \|v\|_{1,\Delta(K)}$$

according to Theorem 7.23, 2), and thus

$$\left| \int_\Omega q_h \nabla \cdot (Q_h v - v) \, dx \right| \leq C_3 \sum_{K \in \mathcal{T}_h} |q_h|_{1,K} h_K \|v\|_{1,K}$$

by the same reasoning as in the proof of Theorem 7.17 using the shape regularity. From the Cauchy–Schwarz inequality we see that

$$\begin{aligned}
\sum_{K \in \mathcal{T}_h} |q_h|_{1,K} h_K \|v\|_{1,K} &\leq \Big(\sum_{K \in \mathcal{T}_h} h_K^2 \|q_h\|_{1,K}^2 \Big)^{1/2} \Big(\sum_{K \in \mathcal{T}_h} \|v\|_{1,K}^2 \Big)^{1/2} \\
&= \Big(\sum_{K \in \mathcal{T}_h} h_K^2 \|q_h\|_{1,K}^2 \Big)^{1/2} \|v\|_1 ,
\end{aligned}$$

hence

$$\frac{1}{\|\boldsymbol{v}\|_1}\left|\int_\Omega q_h \nabla\cdot(Q_h\boldsymbol{v}-\boldsymbol{v})\,dx\right| \le C_3\Big(\sum_{K\in\mathcal{T}_h} h_K^2\|q_h\|_{1,K}^2\Big)^{1/2}$$

and by Lemma 7.19

$$\frac{1}{\|\boldsymbol{v}\|_1}\left|\int_\Omega q_h \nabla\cdot(Q_h\boldsymbol{v}-\boldsymbol{v})\,dx\right| \le C_4 \sup_{\boldsymbol{v}_h\in M_h}\frac{\int_\Omega q_h\nabla\cdot\boldsymbol{v}_h\,dx}{\|\boldsymbol{v}_h\|_1},$$

which together with (7.94) concludes the proof (with $\tilde{\beta}=\beta/(1+C_1C_4)$). □

The above reasoning is attributed to Rüdiger Verführth (*Verführth's trick*).

Corollary 7.21 (Error estimate for the Taylor–Hood element) *Let* $(\boldsymbol{u},p)$ *be a solution of the stationary Stokes equation* (6.78), (6.79) *such that* $\boldsymbol{u}\in(H^3(\Omega)\cap H_0^1(\Omega))^d$ *and* $p\in H^2(\Omega)\cap L_0^2(\Omega)$. *Then* $(\boldsymbol{u}_h,p_h)$, *the consistent finite element approximation in* $M_h\times X_h$ *according to* ⑤, *exists uniquely and satisfies, for some constant* $C>0$ *independent of h, the estimate*

$$\|\boldsymbol{u}-\boldsymbol{u}_h\|_{1,\Omega}+\|p-p_h\|_{0,\Omega}\le Ch^2(|\boldsymbol{u}|_{3,\Omega}+|p|_{2,\Omega}).$$

Proof As for Corollary 7.18. □

Remark 7.22 1) The Taylor–Hood ansatz can be generalized to $(\mathcal{P}_k)^d/\mathcal{P}_{k-1}$ or $(Q_k)^d/Q_{k-1}$ with analogous results (order of convergence h^{k+1} for $\boldsymbol{u}\in H^{k+1}(\Omega)^d$, $p\in H^k(\Omega)$).

2) A further example of a stable $M_h\times X_h$ pair is the $(\mathcal{P}_1\text{-iso-}\mathcal{P}_2)^d/\mathcal{P}_1$ element. Here for a considered triangulation, $(\mathcal{P}_2)^d$ for $\boldsymbol{u}_h$ is substituted by $(\mathcal{P}_1)^d$, but on a refined triangulation, achieved by a red refinement step (see Figure 3.10), and first-order convergence in $\boldsymbol{u}$ and p can be achieved.

3) The Taylor–Hood element exhibits a high order of convergence, but is not locally mass conservative.

Appendix: The Clément Interpolation Operator

In Chapter 3 the interpolation operator I_h (see (3.83)) was well defined at least on the space $H^k(\Omega)$, $k\ge 2$ and $d\le 3$, and it furnished the approximation error estimate Theorem 3.32, which in particular renders the operator stable uniformly in h. In Section 4.2 it was also mentioned that there still exist other "interpolation operators" for discontinuous functions with analogue properties like those of Philippe Clément.

Here we want to discuss the basic principles of the Clément operator for scalar functions. Let $(\mathcal{T}_h)_h$ be a shape-regular family of triangulations satisfying (T1)–(T6). Let V_h be a finite element space based on Lagrangian elements with the global degrees of freedoms at the nodes $a_1,\ldots,a_M$ and the corresponding nodal basis functions $\varphi_1,\ldots,\varphi_M$ (cf., e.g., (3.119)), with no requirement of $V_h\subset V$.

Given an element $v \in L^1(\Omega)$, the Clément operator $Q_h : L_1(\Omega) \to V_h$ is basically defined by means of local orthogonal projections with respect to the L^2-scalar product as follows. For each simplicial neighbourhood

$$\Delta(a_i) := \bigcup_{K:\, a_i \in \partial K} K, \quad i = 1, \ldots, M,$$

of the nodes a_i (see Figure 4.7 for the case $d = 2$), a polynomial $p_i \in \mathcal{P}_1(\Delta(a_i))$ can be uniquely determined by

$$\langle v - p_i, q \rangle_{\Delta(a_i)} = 0 \qquad \text{for all } q \in \mathcal{P}_1(\Delta(a_i)).$$

Then we set

$$Q_h v := \sum_{i=1}^{M} p_i(a_i)\varphi_i \,. \tag{7.95}$$

A disadvantage of this definition consists in the fact that it does not conserve Dirichlet boundary conditions. That is, even if the value $v(a_i)$ is defined at some boundary node $a_i \in \partial\Omega$, then $(Q_h v)(a_i) \neq v(a_i)$, in general. In case of homogeneous Dirichlet boundary conditions this problem can be circumvented by simply omitting in (7.95) those basis functions that correspond to the Dirichlet-type boundary nodes.

The properties of this operator are summarized as follows. Note that the subsequent used simplicial neighbourhood $\Delta(K)$ of an element $K \in \mathcal{T}_h$ was already introduced in the proof of Theorem 7.17; see (7.90).

Theorem 7.23 *Under the above assumptions, let the ansatz space V_h be build up by polynomials of degree k (i.e., $P_K \supset \mathcal{P}_k(K)$). Then there are constants $C > 0$ independent of h such that*

1) $\|Q_h v\|_{m,\Omega} \le C\|v\|_{m,\Omega}$ for $m = 0, 1$.
2) For $K \in \mathcal{T}_h$, $v \in H^l(\Delta(K))$ it holds

$$\|v - Q_h v\|_{m,K} \le C h_K^{l-m} \|v\|_{l,\Delta(K)} \quad \text{for } 0 \le m \le l \le k+1,$$

for $K \in \mathcal{T}_h$, $F \subset K$ being a face, $v \in H^l(\Delta(K))$,

$$\|v - Q_h v\|_{m,F} \le C h^{l-m-\frac{1}{2}} \|v\|_{l,\Delta(K)} \quad \text{for } m < l \le k+1.$$

3) For $v \in H^l(\Omega)$, $1 \le l \le k+1$, it holds

$$\inf_{v_h \in V_h} \|v - v_h\|_{m,\Omega} \le C h^{l-m} \|v\|_{l,\Omega} \quad \text{for } 0 \le m \le l.$$

Proof For 1), 2) see [101, 103, 125], and for 3) we take $v_h := Q_h v$ and argue similar to the proof to Theorem 7.17 thanks to the shape regularity. □

Exercises

Problem 7.13 Let $\Omega \subset \mathbb{R}^d$ for $d = 2, 3$ and $(\mathcal{T}_h)_h$ a shape-regular family of triangulations of Ω. We consider the nonconforming discretization of Stokes equation with homogeneous Dirichlet conditions using Crouzeix–Raviart elements for the velocity and piecewise constant, discontinuous elements with vanishing average for the pressure:

$$V_h := \left\{ \boldsymbol{v}_h \in CR_k(\Omega)^d \;\middle|\; \int_F \boldsymbol{v}_h = 0 \quad \text{for } F \in \partial\mathcal{F} \right\}, \quad W_h := \mathcal{P}_0(\mathcal{T}_h) \cap L_0^2(\Omega).$$

Since $V_h \not\subset V := H_0^1(\Omega)^d$ this ansatz yields a nonconforming finite element discretization, while $W_h \subset W := L_0^2(\Omega)$. We consider the broken bilinear forms

$$a_h(\boldsymbol{v}_h, \boldsymbol{w}_h) := \sum_{K \in \mathcal{T}_h} \int_K \nabla \boldsymbol{v}_h : \nabla \boldsymbol{w}_h dx, \quad b_h(\boldsymbol{v}_h, q_h) := -\sum_{K \in \mathcal{T}_h} \int_T q_h \nabla \cdot \boldsymbol{v}_h dx,$$

where $A : B := \sum_{ij} a_{ij} b_{ij}$ for two matrices $A = (a_{ij})$ and $B = (b_{ij})$. For $\boldsymbol{f} \in L^2(\Omega)^d$ and $g \in L^2(\Omega)$ we look for $\boldsymbol{u}_h \in V_h$ and $p_h \in W_h$ such that

$$\begin{aligned} a_h(\boldsymbol{u}_h, \boldsymbol{v}_h) + b_h(\boldsymbol{v}_h, p_h) &= (\boldsymbol{f}, \boldsymbol{v}_h) && \text{for all } \boldsymbol{v}_h \in V_h, \\ b_h(\boldsymbol{u}_h, q_h) &= -(g, q_h) && \text{for all } q_h \in W_h. \end{aligned}$$

a) Show an inf-sup condition uniform in h.
Let $\boldsymbol{u} \in V$ and $p \in W$ denote the solution of the continuous equations. Then we want to show that

$$\|\boldsymbol{u} - \boldsymbol{u}_h\|_{1,\mathcal{T}_h} + \|p - p_h\|_0 \le Ch \left(|\boldsymbol{u}|_{2,\mathcal{T}_h} + |p|_{1,\mathcal{T}_h} \right)$$

for $\boldsymbol{u}$ and p sufficiently regular, where $\|\cdot\|_{1,\mathcal{T}_h}$ denotes the broken norm on V_h induced by a_h. Proceed as follows:

b) Prove the following conformity error estimate (see (7.28), proof of Theorem 7.6):

$$\mathcal{R} := \sup_{\boldsymbol{v}_h \in V_h \setminus \{\boldsymbol{0}\}} \frac{|(\boldsymbol{f}, \boldsymbol{v}_h) - a_h(\boldsymbol{u}, \boldsymbol{v}_h) - b_h(\boldsymbol{v}_h, p)|}{\|\boldsymbol{v}_h\|_{1,\mathcal{T}_h}} \le Ch \left(|\boldsymbol{u}|_{2,\mathcal{T}_h} + |p|_{1,\mathcal{T}_h} \right).$$

c) Construct a projection $\Pi_h : W \to W_h$ such that for $q \in W \cap H^1(\Omega)$ it holds

$$|q - \Pi_h q| \le Ch|q|_{1,\mathcal{T}_h}.$$

d) Prove that

$$\sup_{\boldsymbol{v}_h \in V_h \setminus \{\boldsymbol{0}\}} \frac{|a_h(\boldsymbol{u}, \boldsymbol{v}_h) - a_h(\boldsymbol{u}_h, \boldsymbol{v}_h)|}{\|\boldsymbol{v}_h\|_{1,\mathcal{T}_h}} \le \mathcal{R} + C \inf_{q_h \in W_h} \|p - q_h\|_0.$$

e) Prove the following velocity error estimate (see Lemma 6.24):

$$\|\boldsymbol{u} - \boldsymbol{u}_h\|_{1,\mathcal{T}_h} \leq C\left(\mathcal{R} + \inf_{\boldsymbol{v}_h \in V_h} \|\boldsymbol{u} - \boldsymbol{v}_h\|_{1,\mathcal{T}_h} + \inf_{q_h \in W_h} \|p - q_h\|_0\right).$$

f) Conclude that

$$\|\boldsymbol{u} - \boldsymbol{u}_h\|_{1,\mathcal{T}_h} \leq Ch\left(|\boldsymbol{u}|_{2,\mathcal{T}_h} + |p|_{1,\mathcal{T}_h}\right).$$

g) Use the discrete inf-sup condition to prove

$$\|p - p_h\|_0 \leq Ch\left(|\boldsymbol{u}|_{2,\mathcal{T}_h} + |p|_{1,\mathcal{T}_h}\right).$$

Remark: Using that $W_h \subset W$ one can verify that the results for the discrete problem from Problem 6.8 are also valid in the nonconforming setting with appropriate modifications.

Problem 7.14 Given a shape-regular family $(\mathcal{T}_h)_h$ of triangulations of a domain Ω, show that there exists a constant $C > 0$ depending only on the parameter σ from Definition 3.31 such that

$$\left|\Delta(K)\right| := \left|\left\{K' \mid K' \in \Delta(K)\right\}\right| \leq C \quad \text{for all } K \in \mathcal{T}_h \text{ and all } \mathcal{T}_h \in (\mathcal{T}_h)_h,$$

that is, the number of elements in all the simplicial neighbourhoods $\Delta(K)$ (see (7.90)) is bounded uniformly in h.

Problem 7.15 Let K be a nondegenerate tetrahedron with vertices a_1, a_2, a_3, a_4. Further, let $b_1, \ldots, b_6$ denote its edge midpoints. Check that the quadrature formula

$$\frac{|K|}{20}\left[4\sum_{i=1}^{6} u(b_i) - \sum_{i=1}^{4} u(a_i)\right]$$

computes the integral $\int_K u\,dx$ exactly for polynomials of second degree.

Problem 7.16 Complete the proof of Lemma 7.19 by demonstrating the estimate (7.93).
Hint: In a first step, prove the estimate for the seminorm $|\boldsymbol{v}_h|_{1,K}$.

Programming project 7.2 Based on the code from Project 3.3 create a function that solves Stokes equations (see (6.79)/(6.87))

$$\begin{aligned}
-\Delta \boldsymbol{v} + \nabla p &= \boldsymbol{f} && \text{in } \Omega,\\
\nabla \cdot \boldsymbol{v} &= 0 && \text{in } \Omega,\\
-(D\boldsymbol{v} - pI) \cdot \boldsymbol{n} &= g_1 && \text{on } \Gamma_1,\\
\boldsymbol{v} &= \boldsymbol{g}_3 && \text{on } \Gamma_3 := \partial\Omega \setminus \Gamma_1
\end{aligned}$$

numerically by means of the (stable) $M_h \times X_h$-pair $(\mathcal{P}_2)^2/\mathcal{P}_0$. To solve the discrete problem, implement the Uzawa algorithm explained in Problem 6.13.

a) Test the implementation for the example of a simple flow channel described by $\Omega := (0,3) \times (0,1)$ and $\Gamma_1 := \{3\} \times (0,1)$ with the following boundary data:

$$g_1 := 0 \quad (\textit{do nothing condition}), \qquad \boldsymbol{g}_3 := \begin{cases} (4y(1-y), 0)^T & \text{on } \{0\} \times (0,1), \\ 0 & \text{on } (0,3) \times \{0,1\}. \end{cases}$$

The exact velocity field is $\boldsymbol{v} = \big(4y(1-y), 0\big)^T$.

b) If you wish to see the failure of an unstable pair, implement the $M_h \times X_h$-pair $(\mathcal{P}_1)^2/\mathcal{P}_1$ and run some tests.

7.4 Nonconforming Finite Element Methods II: Discontinuous Galerkin Methods

In Sections 7.2 and 7.3, we have studied dual mixed formulations to be intrinsically, i.e., as consistent and conforming finite element methods, locally mass conservative to the expense of losing coercivity. The FVMs of Chapter 8 also exhibit this property, to be studied furthermore in Section 7.6, by definition and have turned out in the node-oriented case to be nonconsistent continuous FEM. Taking into account that local mass conservation requires a certain locality in the ansatz, we will try to combine the advantages of the classical, i.e., continuous, finite element method (FEM; cf. Chapter 3) with the advantages of the finite volume method (FVM; cf. Chapter 8) to form the *discontinuous Galerkin method* (*DGM*) within a coercive framework. In Section 7.1, we have seen that a relaxation of the inter-element continuity requirements are possible and that the consistency error of the emerging nonconforming method can be influenced by using boundary integral terms in the variational form. That is, we want to create a numerical scheme to discretize the diffusion–advection–reaction equation in divergence form, that

- is based on the formulation of a variational equation (6.1),
- can be analysed using techniques as presented in Chapter 3 and Chapter 6,
- allows for higher order trial and test functions and therefore also for higher order rates of convergence,

which clearly are properties of the FEM. Moreover, this scheme should have some of the favourable properties of FVMs, namely

- a very local character of the degrees of freedom yielding
  - good parallelization properties,
  - the possibility to easily apply multi-grid methods and the efficient family of *additive Schwarz preconditioners* to the linear system of equations resulting from the discretization,
- the use of numerical flux densities (which approximate the flux density of the continuous model),

- good stability properties for convection-dominated problems,
- local conservation of mass, and
- the ability to deal with nonconsistent partitions and hanging nodes.

Beyond this, the DGM also allows for hp-adaptivity, i.e., we can locally adapt the grid size h and the polynomial degree of the test and trial spaces p. These advantages are opposed by a (significantly) increased number of degrees of freedom (cf. Figure 7.9 for a graphical illustration), more technicalities formulating the DGM, and a slightly more complex analysis. For an extensive discussion of the DGM, the reader may also consult the books of Di Pietro and Ern [21], Riviere [52], and Dolejši and Feistauer [22].

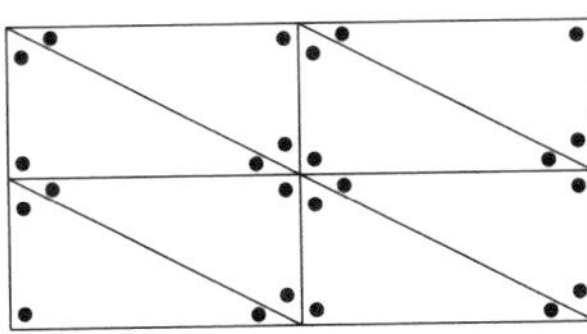
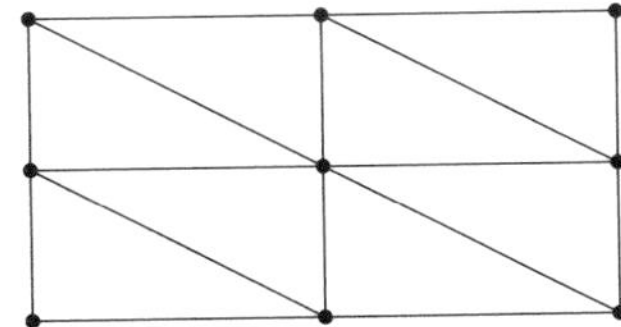

Fig. 7.9: Illustration of the degrees of freedom (bullets) for $\mathcal{P}_1$-based DGM and continuous FEM.

The equation we investigate in the rest of this section is the diffusion–advection–reaction equation in divergence form with the boundary conditions (3.37)–(3.39) on a polygonally bounded, Lipschitz domain Ω:

$$\begin{aligned} -\nabla \cdot (\boldsymbol{K}\nabla u - \boldsymbol{c}\, u) + \tilde{r}\, u &= f && \text{in } \Omega, \\ (\boldsymbol{K}\nabla u - \boldsymbol{c}\, u) \cdot \boldsymbol{\mathfrak{n}} &= \tilde{g}_1 && \text{on } \widetilde{\Gamma}_1, \\ (\boldsymbol{K}\nabla u - \boldsymbol{c}\, u) \cdot \boldsymbol{\mathfrak{n}} + \tilde{\alpha} u &= \tilde{g}_2 && \text{on } \widetilde{\Gamma}_2, \\ u &= \tilde{g}_3 && \text{on } \widetilde{\Gamma}_3 . \end{aligned} \tag{7.96}$$

We assume that the conditions (3.43) are satisfied which contribute to the unique existence of a weak solution $u \in V$ according to Theorem 3.16. Furthermore we refer to the discussion of the necessary regularity in Section 7.1, and assume the conditions (7.18), (7.19), and (7.20).

7.4.1 Interior Penalty Discontinuous Galerkin Methods

We start with formulating a DGM for (3.36)–(3.39) that does not introduce an auxiliary flux variable. The family of such methods for second-order differential equations is generally referred to as *interior penalty DGM* (*IPG*). Doing so, we begin with restricting ourselves to a partition element $K \in \mathcal{T}_h$ which is not adjacent to the boundary, i.e., the faces of K do not intersect the boundary $\partial\Omega$ of Ω. Thus, we locally reconstruct the *primal mixed* version of (7.96) (where the boundary conditions do not act as constraints, cf. (6.57)) by multiplying the differential equation by a test function $v \in H^1(K)$ (extended by zero to the whole domain Ω), integrating over K, and integrating by parts. This gives

$$\int_K [\boldsymbol{K}\nabla u \cdot \nabla v - u\,\boldsymbol{c} \cdot \nabla v + ruv]\,dx - \int_{\partial K} [\boldsymbol{K}\nabla u v - \boldsymbol{c}uv] \cdot \boldsymbol{n}\,d\sigma = \int_K fv dx.$$

Note that by assumption the trace of $\boldsymbol{K}\nabla u$ is well defined. Solutions of lower regularity, i.e., $u \in V$ only, will be discussed later.

To come from this continuous formulation to a discrete version of the method, we assume that the trial and test spaces are polynomial spaces on K. Thus, $u_h|_K, v_h|_K \in \mathbb{P}_k(K) \in \{\mathcal{P}_k(K), \mathcal{Q}_k(K)\}$, and a direct application of this thought would lead to

$$\int_K [\boldsymbol{K}\nabla u_h \cdot \nabla v_h - u_h\,\boldsymbol{c} \cdot \nabla v_h + ru_h v_h]\,dx \\ - \int_{\partial K} [\boldsymbol{K}\nabla u_h v_h - \boldsymbol{c}u_h v_h] \cdot \boldsymbol{n}\,d\sigma = \int_K f v_h dx.$$

The aim is not to include inter-element continuity requirements into the definitions of V_h (and U_h, should the situation arise). Thus, the elements of V_h are discontinuous across the faces $F \in \mathcal{F}$, i.e., they are largely decoupled across the element boundaries, and the formulation lacks local mass conservation.

The solution to this problem is to choose suitable numerical flux densities as done in Section 7.1 for the inter-element boundary integral. Thus, following the lines of Section 7.1, we naively choose $\boldsymbol{c}u_h$ to be the upwind value $\boldsymbol{c}_{\text{upw}}(u_h)$ for the advective part, and the componentwise arithmetic mean $\{\!\{\boldsymbol{K}\nabla u_h\}\!\}$ for the diffusive part of the equation. This results in a reasonable, but not yet coercive formulation:

$$\int_K [\boldsymbol{K}\nabla u_h \cdot \nabla v_h - u_h\,\boldsymbol{c} \cdot \nabla v_h + ru_h v_h]\,dx \\ - \int_{\partial K} \left[\{\!\{\boldsymbol{K}\nabla u_h\}\!\} v_h - \boldsymbol{c}_{\text{upw}}(u_h) v_h\right] \cdot \boldsymbol{n}\,d\sigma = \int_K f v_h dx. \tag{7.97}$$

This scheme has a chance to be locally mass conservative since $\{\!\{\boldsymbol{K}\nabla u_h\}\!\} v_h - \boldsymbol{c}_{\text{upw}}(u_h) v_h$ on $F = K \cap K'$ is symmetric with respect to interchanging the roles of K, K'.

However, considering the purely diffusive case ($\boldsymbol{c} = \boldsymbol{0}$, $r = 0$) with $\boldsymbol{K} = \boldsymbol{I}$, $f = 0$, $\partial\Omega = \widetilde{\Gamma}_3$, $\tilde{g}_3 = 0$, the weak solution is $u = 0$. However, if $\mathcal{T}_h$ is sufficiently fine such that there exists some $K \in \mathcal{T}_h$ with $K \cap \partial\Omega = \emptyset$, then any nontrivial function u_h which is piecewise constant w.r.t. $\mathcal{T}_h$ satisfies the discrete problem (7.97). Thus, the scheme cannot be coercive and the reason for this is the fact that it, loosely speaking, should support overall smooth solutions if diffusion is present. This motivates to additionally penalize the jumps $[\![u_h]\!] = u_h|_K \boldsymbol{n}_K + u_h|_{K'} \boldsymbol{n}_{K'}$ in the solution yielding

$$\int_K [\boldsymbol{K}\nabla u_h \cdot \nabla v_h - u_h\,\boldsymbol{c} \cdot \nabla v_h + ru_h v_h]\,dx \\ + \sum_{F \in \mathcal{F}_K} \int_F \left[- \{\!\{\boldsymbol{K}\nabla u_h\}\!\} v_h + \frac{\eta}{h_F} [\![u_h]\!] v_h + \boldsymbol{c}_{\text{upw}}(u_h) v_h \right] \cdot \boldsymbol{n}\,d\sigma = \int_K f v_h dx,$$

where $\eta > 0$ is called *penalty parameter* and h_F denotes the *face diameter*. This scheme is called *incomplete IPG* (*IIPG*) method. If η is chosen large enough it is coercive. It can be interpreted as a variant of *Nitsche's method* to weakly enforce boundary conditions (cf. (6.57)), but in this case not a boundary value but the continuity across an interface is enforced in a primal mixed formulation. Thus, the method can also be understood as based on the *primal mixed hybrid formulation* (cf. Problem 6.12) which is the most accurate characterization.

For less general boundary value problems it can be shown that the bilinear form is coercive. However, even for a diffusion–reaction equation it is not symmetric with respect to u_h and v_h. Since it can be beneficial to have some kind of symmetry, an additional term can be added to symmetrize the bilinear form, ideally without disturbing its consistency properties. This gives reason to the following (somewhat more general) formulation:

$$\int_K [\boldsymbol{K}\nabla u_h \cdot \nabla v_h - \boldsymbol{c}\, u_h \cdot \nabla v_h + r u_h v_h]\, dx + \sum_{F\in\mathcal{F}_K} \int_F \boldsymbol{c}_{\mathrm{upw}}(u_h)\cdot \mathfrak{n} v_h d\sigma$$
$$+ \sum_{F\in\mathcal{F}_K} \int_F \left[\theta \{\!\!\{\boldsymbol{K}\nabla v_h\}\!\!\} u_h - \{\!\!\{\boldsymbol{K}\nabla u_h\}\!\!\} v_h + \frac{\eta}{h_F} [\![u_h]\!] v_h\right]\cdot \mathfrak{n} d\sigma = \int_K f v_h dx, \tag{7.98}$$

where $\theta \in \{0, \pm 1\}$ is a *symmetrization parameter*. The choice $\theta = 0$ provides the already known IIPG method, while $\theta = 1$ yields the *nonsymmetric IPG* (*NIPG*) and $\theta = -1$ the *symmetric IPG* (*SIPG*) method. The symmetrization term neither destroys the consistency (cf. Lemma 7.25) of the method nor introduces an artificial numerical flux on interior faces.

We still have to enforce boundary conditions as attributed in (7.96). To do so (and to be able to give a formal definition of the scheme in the next paragraph), we assume that $\mathcal{T}_h$ suffices Definition 3.20, and (T5), (T6) from Section 3.3. This actually is not necessary for DGMs, since one of their advantages is that they work on much more general partitions, but will significantly simplify the notation. However, all results can be transferred to much more general partitions using technical arguments as analysed in [21, Ch. 1]. As in Section 7.1 we distinguish between *interior faces* $F \in \mathcal{F}$, separating two partition elements, and the boundary faces $F \in \mathcal{F}_i$, $i = 1, 2, 3$; see (7.9). In addition, the set $\mathcal{F}_2$ is subdivided into the set of *Neumann boundary faces* $F \in \mathcal{F}_{2,1}$ with $F \subset \widetilde{\Gamma}_{2,1}$, i.e., where $\tilde{\alpha} = \boldsymbol{c}\cdot\mathfrak{n}$, and *mixed boundary faces* $F \in \mathcal{F}_{2,2}$ with $F \subset \widetilde{\Gamma}_2 \setminus \widetilde{\Gamma}_{2,1}$. Those terms in (7.98) that are related to inner elements are treated as described. For flux faces $F \in \mathcal{F}_1$ we substitute all boundary integrals by

$$\int_F \boldsymbol{c}_{\mathrm{upw}}(u_h)\cdot\mathfrak{n} v_h d\sigma + \int_F \left[\theta \;\ldots\; v_h\right]\cdot\mathfrak{n} d\sigma \quad \rightsquigarrow \quad -\int_F \tilde{g}_1 v_h d\sigma,$$

for Neumann faces $F \in \mathcal{F}_{2,1}$, $F \subset \widetilde{\Gamma}_{2,1}$, we only replace the boundary integral in the second line of (7.98) by

$$\int_F \left[\theta \{\!\!\{\boldsymbol{K}\nabla v_h\}\!\!\} u_h - \{\!\!\{\boldsymbol{K}\nabla u_h\}\!\!\} v_h + \frac{\eta}{h_F} [\![u_h]\!] v_h\right]\cdot\mathfrak{n} d\sigma \quad \rightsquigarrow \quad -\int_F \tilde{g}_2 v_h d\sigma.$$

For mixed boundary faces $F \in \mathcal{F}_{2,2}$ with $F \subset \widetilde{\Gamma}_2 \setminus \widetilde{\Gamma}_{2,1}$ the substituting term is

$$-\int_F \tilde{g}_2 v_h d\sigma + \int_F \tilde{\alpha} u_h v_h d\sigma.$$

If the face is a Dirichlet face $F \subset \mathcal{F}_3$, we have to be a little more careful: While substituting $\boldsymbol{c}_{\mathrm{upw}}(u_h) \rightsquigarrow \boldsymbol{c}\tilde{g}_3$ and penalizing the difference of $u_h - \tilde{g}_3$ instead of the jump $[\![u_h]\!]$ is suggested, we have to ensure that for $\theta = \pm 1$ the good properties of the method still hold, while we cannot allow an artificial numerical flux across the Dirichlet boundary (or violating consistency of the method) due to symmetrization aspects. Thus, we also have to apply Nitsche's method (cf. (6.57)) for the symmetrization term. This yields

$$\int_F \boldsymbol{c}_{\mathrm{upw}}(u_h) \cdot \mathfrak{n} v_h d\sigma + \int_F \left[\theta \ \dots \ v_h\right] \cdot \mathfrak{n} d\sigma \quad \rightsquigarrow$$

$$\int_F \left[[\boldsymbol{c}\tilde{g}_3 v_h + \theta \boldsymbol{K} \nabla v_h (u_h - \tilde{g}_3) - \boldsymbol{K} \nabla u_h v_h] \cdot \mathfrak{n} + \frac{\eta}{h_F}(u_h - \tilde{g}_3) v_h \right] d\sigma.$$

Different bilinear forms for the IPG methods

Next, we formulate the IPG in the form of (6.1) globally (summed over all elements) and locally (with respect to one element). The global version generally is used for analysis purposes, while the local formulation is primary intended for implementation purposes. With the above considerations in mind, we can formulate the local bilinear and linear forms for an element $K \in \mathcal{T}_h$, $u_h, v_h \in \mathbb{P}_k(K)$ utilizing the previous definitions of upwind, average, and jump with respect to a face $F = K \cap K' \in \mathcal{F}$ of a function w_h with $w_h|_K \in H^1(K)$ and $w_h|_{K'} \in H^1(K')$; see (7.24). Note that all definitions are given with respect to a face and are symmetric with respect to elements, since interchanging the roles of K and K' does not change their values. Thus antisymmetry in the face integrals of (7.98) is given by multiplication with the normal $\mathfrak{n} = \mathfrak{n}_K$ which is assumed to be the outward unit normal with respect to K. This symmetry and antisymmetry of the definitions and of the boundary integrals, respectively, allows for local conservation of mass (with respect to the numerical flux densities as defined in the integrals). Hence, we search for $u_h \in \mathbb{P}_k(K)$ such that on all $K \in \mathcal{T}_h$, we have

$$a_h^{\mathrm{loc}}(u_h, v_h) = \ell_h^{\mathrm{loc}}(v_h) \quad \text{for all } v_h \in \mathbb{P}_k(K)$$

with a linear form ℓ_h and a bilinear form a_h defined as

$$\begin{aligned}
\ell_h^{\mathrm{loc}}(v_h) := & \int_K f v_h dx + \sum_{F \in \mathcal{F}_K \cap \mathcal{F}_1} \int_F \tilde{g}_1 v_h d\sigma + \sum_{F \in \mathcal{F}_K \cap \mathcal{F}_2} \int_F \tilde{g}_2 v_h d\sigma \\
& + \sum_{F \in \mathcal{F}_K \cap \mathcal{F}_3} \int_F \left[\frac{\eta}{h_F} \tilde{g}_3 v_h + \theta \boldsymbol{K} \nabla v_h \cdot \mathfrak{n} \tilde{g}_3 - \boldsymbol{c} \cdot \mathfrak{n} \tilde{g}_3 v_h \right] d\sigma, \\
a_h^{\mathrm{loc}}(u_h, v_h) := & \int_K [\boldsymbol{K} \nabla u_h \cdot \nabla v_h - u_h \boldsymbol{c} \cdot \nabla v_h + r u_h v_h]\, dx \\
& + \sum_{F \in \mathcal{F}_K \cap (\mathcal{F} \cup \mathcal{F}_3)} \int_F \left[\theta \{\!\{\boldsymbol{K} \nabla v_h\}\!\} u_h - \{\!\{\boldsymbol{K} \nabla u_h\}\!\} v_h + \frac{\eta}{h_F} [\![u_h]\!] v_h \right] \cdot \mathfrak{n} d\sigma \\
& + \sum_{F \in \mathcal{F}_K \cap (\mathcal{F} \cup \mathcal{F}_{2,1})} \int_F \boldsymbol{c}_{\mathrm{upw}}(u_h) \cdot \mathfrak{n} v_h d\sigma + \sum_{F \in \mathcal{F}_K \cap \mathcal{F}_{2,2}} \int_F \tilde{\alpha} u_h v_h d\sigma,
\end{aligned}$$

or equivalently

$$
\begin{aligned}
a_h^{\mathrm{loc}}(u_h, v_h) := & \int_K [\boldsymbol{K}\nabla u_h \cdot \nabla v_h - u_h\, \boldsymbol{c} \cdot \nabla v_h + r u_h v_h]\, dx \\
& + \sum_{F \in \mathcal{F}_K \cap (\mathcal{F} \cup \mathcal{F}_3)} \int_F \left[-\{\!\!\{\boldsymbol{K}\nabla u_h\}\!\!\} v_h + \frac{\eta}{h_F} [\![u_h]\!] v_h \right] \cdot \mathfrak{n}\, d\sigma \\
& + \sum_{F \in \mathcal{F}_K \cap (\mathcal{F} \cup \mathcal{F}_3)} \int_F \frac{1}{2} \theta \boldsymbol{K} \nabla v_h \cdot [\![u_h]\!]\, d\sigma \\
& + \sum_{F \in \mathcal{F}_K \cap (\mathcal{F} \cup \mathcal{F}_{2,1})} \int_F \boldsymbol{c}_{\mathrm{upw}}(u_h) \cdot \mathfrak{n} v_h\, d\sigma + \sum_{F \in \mathcal{F}_K \cap \mathcal{F}_{2,2}} \int_F \tilde{\alpha} u_h v_h\, d\sigma,
\end{aligned}
$$

where again $\mathfrak{n} = \mathfrak{n}_K$. Summing up over all elements $K \in \mathcal{T}_h$, the two local bilinear forms yield the same global bilinear form, and therefore can be considered as equivalent. Together with the definition

$$
V_h := \mathbb{P}_k(\mathcal{T}_h) := \{ w_h \in L^2(\Omega) \mid w_h|_K \in \mathbb{P}_k(K) \quad \text{for all } K \in \mathcal{T}_h \} \tag{7.99}
$$

of the finite element space V_h as the *broken polynomial space* of maximum degree k, we obtain the following global formulation:

Find $u_h \in V_h$ such that

$$
a_h(u_h, v_h) = \ell_h(v_h) \quad \text{for all } v_h \in V_h, \tag{7.100}
$$

where

$$
\begin{aligned}
a_h(u_h, v_h) := & \sum_{K \in \mathcal{T}_h} \int_K \left[\boldsymbol{K}\nabla u_h \cdot \nabla v_h - u_h\, \boldsymbol{c} \cdot \nabla v_h + r u_h v_h \right] dx \\
& + \sum_{F \in \mathcal{F} \cup \mathcal{F}_3} \int_F \left[\theta \{\!\!\{\boldsymbol{K}\nabla v_h\}\!\!\} \cdot [\![u_h]\!] - \{\!\!\{\boldsymbol{K}\nabla u_h\}\!\!\} \cdot [\![v_h]\!] \right] d\sigma \\
& + \sum_{F \in \mathcal{F} \cup \mathcal{F}_3} \int_F \frac{\eta}{h_F} [\![u_h]\!] \cdot [\![v_h]\!]\, d\sigma + \sum_{F \in \mathcal{F} \cup \mathcal{F}_{2,1}} \int_F \boldsymbol{c}_{\mathrm{upw}}(u_h) \cdot [\![v_h]\!]\, d\sigma \\
& + \sum_{F \in \mathcal{F}_{2,2}} \int_F \tilde{\alpha} u_h v_h\, d\sigma, \\
\ell_h(v_h) := & \int_\Omega f v_h\, dx + \sum_{F \in \mathcal{F}_1} \int_F \tilde{g}_1 v_h\, d\sigma + \sum_{F \in \mathcal{F}_2} \int_F \tilde{g}_2 v_h\, d\sigma \\
& + \sum_{F \in \mathcal{F}_3} \int_F \left[\frac{\eta}{h_F} \tilde{g}_3 v_h + \theta \boldsymbol{K} \nabla v_h \cdot \mathfrak{n} \tilde{g}_3 - \boldsymbol{c} \cdot \mathfrak{n} \tilde{g}_3 v_h \right] d\sigma.
\end{aligned}
\tag{7.101}
$$

In summary, we use the following terminology.

Definition 7.24 The *Interior Penalty Galerkin method* (*IPG*) to approximate the boundary value problem in divergence form (7.96) is given by the variational problem defined by (7.101) on $\mathbb{P}_k(\mathcal{T}_h)$ for $k \in \mathbb{N}_0, \theta \in \{0, \pm 1\}$. For $\theta = 0$ the method is called *incomplete IPG* (*IIPG*), for $\theta = 1$ the *nonsymmetric IPG* (*NIPG*), and for $\theta = -1$ the *symmetric IPG* (*SIPG*).

Now we want to demonstrate a basic result that characterizes the IPG methods as consistent but clearly nonconforming methods (cf. Section 6.1). Similar to Section 7.1 to this end we assume sufficient regularity for the solution expressed by (7.18), (7.19), and (7.20).

Lemma 7.25 (Consistency of IPG methods) *If $u \in H^1(\Omega)$ is the weak solution of the problem* (7.96) *fulfilling* (7.18), (7.19), *and* (7.20), *the IPG methods are consistent for $k \in \mathbb{N}$ in the sense that*

$$a_h(u, v_h) = \ell_h(v_h) \quad \textit{for all } v_h \in V_h \quad (= \mathbb{P}_k(\mathcal{T}_h)).$$

Proof For $v_h \in V_h$ we have that

$$\begin{aligned}
&a_h(u, v_h) - \ell_h(v_h) \\
&= \sum_{K \in \mathcal{T}_h} \int_K [\boldsymbol{K}\nabla u \cdot \nabla v_h - u\, \boldsymbol{c} \cdot \nabla v_h + r u v_h]\, dx - \sum_{F \in \mathcal{F} \cup \mathcal{F}_3} \int_F \boldsymbol{K}\nabla u \cdot [\![v_h]\!] d\sigma \\
&\quad + \sum_{F \in \mathcal{F} \cup \mathcal{F}_{2,1}} \int_F u\, \boldsymbol{c} \cdot [\![v_h]\!] d\sigma + \sum_{F \in \mathcal{F}_3} \int_F \left[\theta \boldsymbol{K}\nabla v_h \cdot \mathfrak{n} u + \frac{\eta}{h_F} u v_h\right] d\sigma \\
&\quad - \int_\Omega f v_h dx - \sum_{F \in \mathcal{F}_1} \int_F \tilde{g}_1 v_h d\sigma - \sum_{F \in \mathcal{F}_2} \int_F \tilde{g}_2 v_h d\sigma \\
&\quad - \sum_{F \in \mathcal{F}_3} \int_F \left[\frac{\eta}{h_F} \tilde{g}_3 v_h + \theta \boldsymbol{K}\nabla v_h \cdot \mathfrak{n} \tilde{g}_3 - \boldsymbol{c} \cdot \mathfrak{n} \tilde{g}_3 v_h\right] d\sigma + \sum_{F \in \mathcal{F}_{2,2}} \int_F \tilde{\alpha} u v_h d\sigma \\
&= \sum_{K \in \mathcal{T}_h} \int_K [\boldsymbol{K}\nabla u \cdot \nabla v_h - u\, \boldsymbol{c} \cdot \nabla v_h + r u v_h]\, dx + \sum_{F \in \mathcal{F}_{2,2}} \int_F \tilde{\alpha} u v_h d\sigma \\
&\quad - \sum_{F \in \mathcal{F} \cup \mathcal{F}_3} \int_F \boldsymbol{K}\nabla u \cdot [\![v_h]\!] d\sigma + \sum_{F \in \mathcal{F} \cup \mathcal{F}_{2,1}} \int_F u\, \boldsymbol{c} \cdot [\![v_h]\!] d\sigma \\
&\quad - \int_\Omega f v_h dx - \sum_{F \in \mathcal{F}_1} \int_F \tilde{g}_1 v_h d\sigma - \sum_{F \in \mathcal{F}_2} \int_F \tilde{g}_2 v_h d\sigma + \sum_{F \in \mathcal{F}_3} \int_F \boldsymbol{c} \cdot \mathfrak{n} \tilde{g}_3 v_h d\sigma,
\end{aligned}$$

where the first equality uses that the traces of u coincide on $F \in \mathcal{F}$ and the second equality holds due to the fact that $u = \tilde{g}_3$ on $F \in \mathcal{F}_3$. Now, (elementwise) integration by parts, as in (7.27) and (7.34), and the use of the boundary conditions in (7.96) result in

$$\begin{aligned}
a_h(u, v_h) - \ell_h(v_h) &= \int_\Omega [-\nabla \cdot (\boldsymbol{K}\nabla u - \boldsymbol{c} u) + r u - f]\, v_h dx \\
&\quad + \int_{\Gamma_1} [(\boldsymbol{K}\nabla u - \boldsymbol{c} u) \cdot \mathfrak{n} - \tilde{g}_1]\, v_h d\sigma + \int_{\Gamma_2} [(\boldsymbol{K}\nabla u - \boldsymbol{c} u) \cdot \mathfrak{n} + \tilde{\alpha} u - \tilde{g}_2]\, v_h d\sigma.
\end{aligned}$$

The right-hand side vanishes since the differential equation in (7.96) is satisfied in the sense of $L^2(\Omega)$. □

Remark 7.26 (Relation to FEM, FVM, and CR-elements)

1) Using the IPG bilinear and linear forms and the standard (continuous) finite element space $V_h \cap C^0(\Omega)$ results in a classical finite element method with a different treatment of boundary conditions, since all jumps on interfaces vanish due to the continuity of test and trial functions. This cancels all integrals with respect to interfaces and leaves only domain boundary integrals. If we furthermore only allow for test and trial functions whose traces are zero on Dirichlet boundary faces, we have arrived in the setting of Section 3.2 (independent of the choices of θ and η).
2) Using the space $\mathbb{P}_0(\mathcal{T}_h)$ cancels all integrals containing gradients. Thus, the element and upwind integrals are the same as described in Section 8.3. If $\boldsymbol{K} = k\boldsymbol{I}$ with a scalar function k, the face integrals for diffusion are the same if $\eta \rightsquigarrow \mu_{i,j}$ is face dependent and describes the average of k, and $h_F \rightsquigarrow d_{i,j}$ is the distance between the barycentres of two elements. Thus, for a constant scalar k with $\eta = k$ and a partition of equilateral triangles or squares, we receive the FVM.
3) From the above considerations, we immediately deduce that the choice $k = 0$ in general does not make sense for IPG methods trying to approximate equations with diffusion, since we lose all information about $\boldsymbol{K}$ and cannot hope for any convergence to the correct solution.
4) Compared to Section 7.1, the term
$$\sum_{F\in\mathcal{F}\cup\mathcal{F}_3} \int_F \boldsymbol{K}\nabla u [\![v_h]\!] d\sigma$$
has been removed from the consistency error by including
$$-\sum_{F\in\mathcal{F}\cup\mathcal{F}_3} \int_F \{\!\{\boldsymbol{K}\nabla u_h\}\!\} [\![v_h]\!] d\sigma$$
to the bilinear form.

Stability of IPG methods

Next, we have to ensure that the problem (7.100) allows for a unique solution we can solve for in a stable manner. We try to show the uniform coercivity and the boundedness of the bilinear form a_h in terms of an energy norm $\|\cdot\|_{V_h}$. To construct this norm, we consider the NIPG bilinear form with $\theta = 1$, since it contains the minimum amount of summands which all can be rewritten to be obviously coercive (remember that $\eta > 0$). We have the identity

$$
\begin{aligned}
a_h^{\mathrm{NIPG}}(u_h, u_h) = & \sum_{K\in\mathcal{T}_h} \|\boldsymbol{K}^{1/2}\nabla u_h\|_{0,K}^2 + \sum_{F\in\mathcal{F}\cup\mathcal{F}_3} \frac{\eta}{h_F}\|[\![u_h]\!]\|_{0,F}^2 \\
& + \frac{1}{2}\int_\Omega [2r + \nabla\cdot\boldsymbol{c}]\, u_h^2 dx + \frac{1}{2}\sum_{F\in\overline{\mathcal{F}}\setminus\mathcal{F}_{2,2}} \int_F |\boldsymbol{c}\cdot\mathfrak{n}|[\![u_h]\!]^2 d\sigma \\
& + \sum_{F\in\mathcal{F}_{2,2}} \int_F (\tilde{\alpha} - \frac{1}{2}\mathfrak{n}\cdot\boldsymbol{c}) u_h^2 d\sigma \quad =: \|u_h\|_{V_h}^2,
\end{aligned}
\tag{7.102}
$$

which can be seen as follows

$$
\begin{aligned}
& -\sum_{K\in\mathcal{T}_h} \int_K u_h\, \boldsymbol{c}\cdot\nabla u_h dx + \sum_{F\in\mathcal{F}\cup\mathcal{F}_{2,1}} \int_F \boldsymbol{c}_{\mathrm{upw}}(u_h)\cdot[\![u_h]\!] d\sigma \\
= & -\frac{1}{2}\sum_{K\in\mathcal{T}_h} \int_K \boldsymbol{c}\cdot\nabla u_h^2 dx + \sum_{F\in\mathcal{F}\cup\mathcal{F}_{2,1}} \int_F \boldsymbol{c}_{\mathrm{upw}}(u_h)\cdot[\![u_h]\!] d\sigma \\
= & \frac{1}{2}\sum_{K\in\mathcal{T}_h} \int_K \nabla\cdot\boldsymbol{c}\, u_h^2 dx - \frac{1}{2}\sum_{F\in\mathcal{F}_1\cup\mathcal{F}_{2,2}\cup\mathcal{F}_3} \int_F \boldsymbol{c}\cdot\mathfrak{n} u_h^2 d\sigma \\
& + \sum_{F\in\mathcal{F}\cup\mathcal{F}_{2,1}} \int_F \left[\boldsymbol{c}_{\mathrm{upw}}(u_h)\cdot[\![u_h]\!] - \frac{1}{2}\boldsymbol{c}\cdot[\![u_h^2]\!]\right] d\sigma \\
= & \frac{1}{2}\sum_{K\in\mathcal{T}_h} \int_K \nabla\cdot\boldsymbol{c}\, u_h^2 dx + \frac{1}{2}\sum_{F\in\overline{\mathcal{F}}\setminus\mathcal{F}_{2,2}} \int_F |\boldsymbol{c}\cdot\mathfrak{n}|[\![u_h]\!]^2 d\sigma - \frac{1}{2}\sum_{F\in\mathcal{F}_{2,2}} \int_F \boldsymbol{c}\cdot\mathfrak{n} u_h^2 d\sigma.
\end{aligned}
\tag{7.103}
$$

The last step requires the same manipulations as in (7.32) and proper sign conditions for the boundary integrals, namely (7.33) and (3.43), 2),3).

Lemma 7.27 *Under the assumptions of Theorem 3.16, adapted to the divergence form (see* (3.43)*), such that the conditions 4)b) for $\tilde{\Omega} := \Omega$ (positive reaction term) or 4)a) (Dirichlet boundary condition with $|\Gamma_3|_{d-1} > 0$) are satisfied, and if h is sufficiently small, $\|\cdot\|_{V_h}$ defines a norm on $H^1(\mathcal{T}_h)$ such that, for some constant $C > 0$ independent of h,*

$$
\|v\|_{1,\mathcal{T}_h} \le C\|v\|_{V_h} \quad \textit{for all } v\in H^1(\mathcal{T}_h).
$$

Proof By assumptions, all terms in (7.102) are nonnegative, and it holds, for some $\alpha > 0$ independent of h,

$$
\sum_{K\in\mathcal{T}_h} \|\boldsymbol{K}^{1/2}\nabla v\|_{0,K}^2 \ge \alpha |v|_{1,\mathcal{T}_h}^2 .
$$

Thus, $\|\cdot\|_{V_h}$ is a seminorm and $\|v\|_{V_h} = 0$ renders v to be piecewise constant. The various conditions of Theorem 3.16 force such functions to be zero: In case of 4)b) for at least one element, where the zero value is enforced in the neighbouring and thus in all elements by the second term in $\|\cdot\|_{V_h}$. In the case of 4)a), this is similar starting from the Dirichlet boundary condition, and in 4)c) or 4)d) analogously.

Furthermore, 4)b) for $\tilde{\Omega} := \Omega$ (note that the form is now $r + \frac{1}{2}\nabla \cdot \boldsymbol{c} \geq r_0 > 0$) ensures

$$\|v\|_{V_h}^2 \geq \alpha \|v\|_{0,\Omega}^2,$$

and for 4)a) the same is true by the following refined Poincaré inequality. □

Lemma 7.28 *There is a constant $C_{\text{Poin}} > 0$ independent of $h > 0$ such that*

$$\|v_h\|_0 \leq C_{\text{Poin}} \|v_h\|_{V_h} \quad \text{for all } v_h \in V_h.$$

Proof See [21, Thm. 5.3]. □

To show the stability of the IPG methods, we additionally need the following result.

Lemma 7.29 (Discrete Trace Inequality) *For a shape-regular family of partitions $(\mathcal{T}_h)_h$, there is a constant $C_{\text{tr}} = C(d, k, \sigma) > 0$ such that for all $h > 0$, $p \in \mathbb{P}_k(\mathcal{T}_h)$, $K \in \mathcal{T}_h$, and $F \in \mathcal{F}_K$*

1) $h_F^{1/2}\|p\|_{0,F} \leq h_K^{1/2}\|p\|_{0,F} \leq C_{\text{tr}}\|p\|_{0,K}$,
2) $h_F^{1/2}\|\nabla p\|_{0,F} \leq h_K^{1/2}\|\nabla p\|_{0,F} \leq C_{\text{tr}}\|\nabla p\|_{0,K}$.

Proof The second inequality can directly be deduced from the first inequality. Thus, it is sufficient to consider $p \in \mathbb{P}_k(\mathcal{T}_h)$ on a single face F with neighbouring element $K \in \mathcal{T}_h$. Here, we have

$$\frac{|\hat{F}|}{|F|}\|p\|_{0,F}^2 = \|p\|_{0,\hat{F}}^2 \leq C_{\text{eq}}\|p\|_{0,\hat{K}}^2 = C_{\text{eq}}\frac{|\hat{K}|}{|K|}\|p\|_{0,K}^2,$$

where the inequality is due to the fact that seminorms are weaker than norms on the finite-dimensional space $\mathbb{P}_k(\hat{K})$. Here, $\hat{F}$ is the face of the reference element $\hat{K}$ that is the image of F under the isomorphic mapping $\hat{K} \mapsto K$. For a simplicial element, with $\tilde{h}_F$ being the height on F, the result follows from

$$\frac{|K|}{|F|} = \frac{1}{d}\tilde{h}_F \geq \frac{1}{d}\sigma h_K,$$

where the equality is the volume formula for simplices and the inequality is a consequence of the shape regularity. Finally, we note that $h_K \geq h_F$. The lemma also holds for more general partitions [21, Lemma 1.46]. □

Now we are prepared to summarize the essential properties of the linear and bilinear forms.

Lemma 7.30 *The linear form ℓ_h and bilinear form a_h are bounded on V_h with respect to $\|\cdot\|_{V_h}$, possibly depending on h. The bilinear form a_h is uniformly coercive with respect to $\|\cdot\|_{V_h}$ with parameter $\alpha = \alpha_h$ independent of h provided that*

$$\eta > \frac{(1-\theta)^2 C_{\text{tr}}^2 (d+1)\|\boldsymbol{K}\|_\infty^2}{4k_0}. \tag{7.104}$$

The NIPG method is uniformly coercive with $\alpha = 1$ if $\eta > 0$.

Thus, Theorem 3.1 can be applied ensuring that there is a unique solution of (7.100).

Proof To verify the coercivity of a_h consider $u_h \in \mathbb{P}_k(\mathcal{T}_h)$ and observe that

$$a_h(u_h, u_h) = \|u_h\|_{V_h}^2 - (1-\theta) \sum_{F \in \mathcal{F} \cup \mathcal{F}_3} \int_F \{\!\!\{ \boldsymbol{K} \nabla u_h \}\!\!\} \cdot [\![u_h]\!] d\sigma =: \|u_h\|_{V_h}^2 - \Psi_h(u_h).$$

Thus, it is sufficient to bound the absolute value of $\Psi_h(u_h)$ by means of a small multiple of $\|u_h\|_{V_h}^2$. This can be done using that $\boldsymbol{K}$ is bounded and symmetric positive definite a.e. (see (3.16), (3.17)), Lemma 7.29, and the discrete Cauchy–Schwarz inequality:

$$\begin{aligned}
|\Psi_h(u_h)| &\le (1-\theta) \sum_{F \in \mathcal{F} \cup \mathcal{F}_3} \|\boldsymbol{K}\|_\infty \|\{\!\!\{ \nabla u_h \}\!\!\}\|_{0,F} \|[\![u_h]\!]\|_{0,F} \\
&\le (1-\theta) \sum_{F \in \mathcal{F} \cup \mathcal{F}_3} \|\boldsymbol{K}\|_\infty C_{\mathrm{tr}} \|\nabla u_h\|_{0,\mathcal{T}_F} h_F^{-1/2} \|[\![u_h]\!]\|_{0,F} \\
&\le \delta \sum_{F \in \mathcal{F} \cup \mathcal{F}_3} \|\boldsymbol{K}^{1/2} \nabla u_h\|_{0,\mathcal{T}_F}^2 + \sum_{F \in \mathcal{F} \cup \mathcal{F}_3} \frac{(1-\theta)^2 C_{\mathrm{tr}}^2 \|\boldsymbol{K}\|_\infty^2}{4\delta k_0 h_F} \|[\![u_h]\!]\|_{0,F}^2,
\end{aligned}$$

where $\mathcal{T}_F$ denotes the (union of the) element(s) with face F, and k_0 is the constant from (3.17). If δ is such that $0 < \delta < 1/|\mathcal{F}_K| = 1/(d+1)$, the first term can be absorbed by the first term of (7.102), since an element K has $d+1$ faces and can therefore occur at most as often. Condition (7.104) ensures that the second term can be absorbed by the second term of (7.102). The boundedness is obvious since all bilinear and linear forms on finite-dimensional spaces are bounded. □

Remark 7.31 1) Lemma 7.30 also holds true for certain nonsimplicial and nonconsistent partitions, with $d+1$ substituted by the maximum number of faces of an element.

2) More elaborate techniques can be used to show that the NIPG method for $\eta = 0$ is also stable. This method is tributed to Oden, Babuška, and Baumann [198], and is referred to as the *OBB method*.

While the above coercivity constant $\alpha = \alpha_h$ is uniform in h, but depends on η, the constants for boundedness are allowed to be h-dependent. This justifies the use of Remark 6.9 provided we are able to find an appropriate normed space $(V(h), \|\cdot\|_{V(h)})$ where a_h is bounded.

Convergence analysis for the full problem

Assume for the solution $u \in H^2(\mathcal{T}_h)$. Next, we justify the assumptions of Remark 6.9 by first constructing a space $V(h) \supset \mathbb{P}_k(\mathcal{T}_h) + \mathrm{span}\,(u)$ such that $\|\cdot\|_{V_h}$ is a norm on

$V(h)$. A suitable choice clearly is

$$V(h) = H^2(\mathcal{T}_h) := \{v \in L^2(\Omega) \mid v \in H^2(K) \quad \text{for all } K \in \mathcal{T}_h\}.$$

On $H^2(\mathcal{T}_h)$ we cannot apply Lemma 7.29 anymore to control $\{\!\{\boldsymbol{K}\nabla u_h\}\!\}$ on the faces $F \in \mathcal{F} \cup \mathcal{F}_3$, as in the proof of Lemma 7.30. Thus, $\|\cdot\|_{V_h}$ has to be extended to both control this term and to preserve the uniform boundedness of a_h on $V(h) \times V_h$. This motivates to define for $w_h \in V(h)$

$$\|w_h\|^2_{V(h)} := \|w_h\|^2_{V_h} + \sum_{F\in\mathcal{F}\cup\mathcal{F}_3} \frac{h_F}{\eta} \|\{\!\{\boldsymbol{K}^{1/2}\nabla w_h\}\!\}\|^2_{0,F} + \sum_{F\in\overline{\mathcal{F}}} \frac{h_F}{\eta} \|\boldsymbol{c} w_h\|^2_{0,F}, \quad (7.105)$$

where in the last term both traces are considered. That is, this term is evaluated twice for interfaces.

Lemma 7.32 *Under the assumptions of Theorem 3.16 adopted to the divergence form* (3.43), *there is a constant* $\widetilde{M}_h > 0$ *independent of h such that*

$$a_h(w_h, v_h) \le \widetilde{M}_h \|w_h\|_{V(h)} \|v_h\|_{V_h} \quad \textit{for all } w_h \in V(h),\ v_h \in V_h.$$

Proof Consider the eight terms defining $a_h(w_h, v_h)$ according to (7.101). Taking the assumptions into account, the first three terms, the sixth, and the eighth term can be estimated using the Cauchy–Schwarz inequality (using only $\|\cdot\|_{V_h}$). To handle the fourth and the fifth terms, introduce $(\eta/h_F)^{1/2}(h_F/\eta)^{1/2}$ before applying the Cauchy–Schwarz inequality to control the terms to the expense of the new terms in $\|\cdot\|_{V(h)}$. For the seventh term and $w_h, v_h \in V_h$, we proceed similar to the fourth term taking advantage of Lemma 7.29. For $w_h \in V(h)$, $v_h \in V_h$, integration by parts with the second term and adding the seventh term lead to

$$\sum_{K\in\mathcal{T}_h} \int_K \nabla\cdot(\boldsymbol{c} w_h) v_h dx - \sum_{F\in\overline{\mathcal{F}}} \int_F \boldsymbol{c}\cdot [\![w_h v_h]\!] d\sigma + \sum_{F\in\mathcal{F}\cup\mathcal{F}_{2,1}} \int_F \boldsymbol{c}_{\mathrm{upw}}(w_h)[\![v_h]\!]\, d\sigma.$$

The first term can directly be controlled due to $\nabla\cdot(\boldsymbol{c} w_h) = \nabla\cdot\boldsymbol{c} w_h + \boldsymbol{c}\cdot\nabla w_h$. The second and third terms eventually cancel on boundary faces. Otherwise, they can be directly controlled. On interfaces, we use that both terms sum up to the downwind flux $\int_F \boldsymbol{c}_{\mathrm{down}}(w_h)[\![v_h]\!]\, d\sigma$ (defined analogously to (7.30), (7.31)) to deduce that these terms can also be controlled by the last term in (7.105). □

All that is left to do for a convergence result in $\|\cdot\|_{V_h}$ is to show an estimate of the type

$$\|u - \Pi u\|_{V(h)} \le C h^m |u|_{m+1,\mathcal{T}_h} \quad (7.106)$$

with an appropriate comparison function $\Pi u \in V(h)$. A good choice for Π is the elementwise orthogonal L^2-projection (see Remark 3.33, 3)). To verify (7.106), we need an auxiliary proposition that relates face integrals to element integrals. In case of a shape-regular triangulation there is a constant $c > 0$ such that $h_F \ge c h_K$ for K adjacent to F (see Lemma 7.3). This, Lemma 7.4, and Young's inequality imply that there is a constant $C > 0$ independent of h such that

$$\|v\|_{V(h)}^2 \le C \sum_{K \in \mathcal{T}_h} \left[h_K^{-2} \|v\|_{0,K}^2 + |v|_{1,K}^2 + h_K^2 |v|_{2,K}^2 \right] \quad \text{for all } v \in V(h).$$

As the critical term, we handle the second one in (7.102), with $C > 0$ being a generic constant, as follows:

$$\begin{aligned} C \sum_{F \in \mathcal{F}} \frac{1}{h_F} \|[\![v]\!]\|_{0,F}^2 &\le C \sum_{K \in \mathcal{T}_h} \frac{1}{h_K} (\|\nabla v\|_{0,K} + \frac{1}{h_K} \|v\|_{0,K}) \|v\|_{0,K} \\ &\le C \sum_{K \in \mathcal{T}_h} \frac{1}{h_K^2} \|v\|_{0,K}^2 + \|\nabla v\|_{0,K}^2 , \end{aligned}$$

and the treatment of the additional term in (7.105), where $[\![v]\!]$ is replaced by $\{\!\{\boldsymbol{K}^{1/2} \nabla v\}\!\}$ and h_F^{-1} by h_F, runs similarly. With this estimate, we can formulate the main convergence result in the energy norm, which has become a corollary, since it can be obtained as in Theorem 3.32 (see Remark 3.33).

Corollary 7.33 (Convergence in the energy norm) *Assume that the conditions to the family of triangulations, formulated before Lemma 7.3, are satisfied and* $\boldsymbol{K}, \mathbf{c}$ *are smooth on each* $K \in \mathcal{T}_h$. *If* $k \ge 1$ *and* $u \in H^{m+1}(\mathcal{T}_h)$ *with* $1 \le m \le k$ *fulfils* (7.18)–(7.20), *and* $u_h \in V_h$ *is the IPG solution, then there is a constant* $C > 0$ *independent of h such that it holds*

$$\|u - u_h\|_{V_h} \le C h^m |u|_{m+1,\mathcal{T}_h} .$$

Proof According to Remark 6.9, 2) with $W \subset H^{m+1}(\mathcal{T}_h)$ and Lemmata 7.25, 7.30, 7.32 it remains to complete the estimate

$$\inf_{w_h \in V_h} \|u - w_h\|_{V(h)} \le \|u - \Pi u\|_{V(h)},$$

but the latter result is known from (7.106). □

This immediately implies the following L^2-convergence result.

Theorem 7.34 (Convergence in the $L^2(\Omega)$-norm) *Under the assumptions of Corollary 7.33, in case of Dirichlet boundary conditions or if* $\nabla \cdot c + 2r \ge r_0 > 0$ *as in Lemma 7.27, the IPG solution* $u_h \in V(h)$ *converges with order at most m to u in the sense of* $L^2(\Omega)$, *i.e.,*

$$\|u - u_h\|_0 \le C h^m |u|_{m+1,\mathcal{T}_h} .$$

Convergence analysis in weaker norms

To apply the results of Lemma 6.11, and in particular (6.26), we have to consider the adjoint problem. According to the discussion in Section 3.4.3, this corresponds to the formulation not in divergence form (3.13), (3.19)–(3.21), but with $\tilde{\boldsymbol{c}} := -\boldsymbol{c}$. The condition on V-coercivity of the form and the unique existence in V correspond each to the other. In the same way, the bilinear forms of the IPG methods are adjoint to a_h from (7.101). The considerations from Lemma 7.25 hold also true for the adjoint problem in the SIPG case, guaranteeing a consistent method with a unique solution, fulfilling the order of convergence estimate in Corollary 7.33.

Thus, the total consistency term in (6.20) reduces to

$$
\begin{aligned}
(a - a_h)(u, v_g) = &- \sum_{F \in \mathcal{F} \cup \mathcal{F}_3} \int_F \left[\theta \{\!\!\{ \boldsymbol{K} \nabla v_g \}\!\!\} \cdot [\![u]\!] - \{\!\!\{ \boldsymbol{K} \nabla u \}\!\!\} \cdot [\![v_g]\!]\right] d\sigma \\
&- \sum_{F \in \mathcal{F} \cup \mathcal{F}_3} \int_F \frac{\eta}{h_F} [\![u]\!] \cdot [\![v_g]\!] d\sigma - \sum_{F \in \mathcal{F}} \int_F \boldsymbol{c}_{\mathrm{upw}}(u) \cdot [\![v_g]\!] d\sigma
\end{aligned}
$$

for $u \in V$ being the weak solution of (7.96), v_g the solution of the adjoint problem (6.17), (6.18). Therefore we have the following result.

Theorem 7.35 *Let $u \in H^{k+1}(\mathcal{T}_h)$ and $u_h \in V(h)$ be its SIPG approximate. Assume that for any right-hand side $g \in H^m(\Omega)$, $0 \le m \le k-1$, the solution v_g of the adjoint problem is in $H^{m+2}(\mathcal{T}_h)$ such that $\|v_g\|_{m+2,\mathcal{T}_h} \le C_s \|g\|_{m,\Omega}$ for some constant $C_s > 0$. Furthermore the solutions u and v_g satisfy* (7.18)–(7.20). *Then, there is a constant $C > 0$ independent of h such that*

$$
\|u - u_h\|_{-m,\Omega} \le C h^{k+m+1} |u|_{k+1}\mathcal{T}_h.
$$

Proof Under the given regularity and boundary conditions the above error term fulfils

$$
(a - a_h)(u, v_g) = 0.
$$

□

Remark 7.36 For other methods, where $\theta \neq -1$, the discretization of the adjoint problem is no longer consistent to the adjoint problem, i.e., according to Lemma 6.11, the additional term

$$
(\theta + 1) \sum_{F \in \mathcal{F} \cup \mathcal{F}_3} \int_F \{\!\!\{ \boldsymbol{K} \nabla v_g \}\!\!\} \cdot [\![u - u_h]\!] d\sigma \tag{7.107}
$$

has to be estimated for sufficiently smooth solutions u, v_g.

7.4.2 Additional Aspects of Interior Penalty and Related Methods

This section deals with some additional properties of the IPG methods. At first, we discuss a possible approach of minimizing the computational costs. Afterwards, we deal with low regularity solutions.

Enriched interior penalty methods

The DGMs described in Section 7.4.1 have several advantages compared to standard finite elements (often denoted as *continuous Galerkin method* (*CGM*) in this context) which are analysed in Chapter 3. As opposed, the most prominent drawback of DG schemes is their larger number of degrees of freedom. Enriched Galerkin (EG) schemes try to remedy this drawback while conserving most of the advantages of the DGM. The idea of these schemes is to combine the DG bilinear form a_h and linear form ℓ_h of (7.100) with more complex test and trial spaces than V_h of (7.99). This can be expected to work out, since consistency (cf. Lemma. 7.25) and coercivity (cf. Lemma 7.30) are automatically ensured if the used discrete spaces V_h satisfy $V_h \subset \mathbb{P}_k(\mathcal{T}_h)$, which holds true for

$$V_h := V_m^k := (\mathbb{P}_k(\mathcal{T}_h) \cap C(\Omega)) + \mathbb{P}_m(\mathcal{T}_h),$$

where $-1 \leq m \leq k$ for $k \in \mathbb{N}$ and $\mathbb{P}_{-1}(\mathcal{T}_h) := \{0\}$. Thus, obviously $\mathbb{P}_m(\mathcal{T}_h) \subset V_m^k \subset \mathbb{P}_k(\mathcal{T}_h)$ and the sum is not direct if $m \neq -1$, since global constants could be incorporated in both spaces. Assume $d \leq 3$. We need the mapping

$$\widetilde{\Pi} : \ H^2(\Omega) \to V_h, \quad u \mapsto I_h^k u - \Pi_h^m I_h^k u + \Pi_h^m u,$$

utilizing the structure of V_h as direct sum of two linear spaces. Here, I_h^k is the interpolation operator for local polynomials of degree at most k (see (2.51)) and Π_h^m is the elementwise orthogonal L^2-projection to polynomials of degree at most m (see Remark 3.33, 3). Note that $\widetilde{\Pi}$ is not a projection, since $V_h \not\subset C(\Omega)$ and thus $\widetilde{\Pi}$ is not well defined on V_m^k but only on $C(\overline{\Omega}) \cap \mathbb{P}_k(\mathcal{T}_h)$. However, the choice $\widetilde{\Pi}u := I_h^k u$ will also work, but is a projection to a subspace of V_h only. $\widetilde{\Pi}$ is constructed to allow for an uniform analysis covering the continuous Galerkin, DG, and EG cases; cf. [213] (where the arguments of the previous section are transferred to EG using $\widetilde{\Pi}$). It can also be rewritten to $\widetilde{\Pi}u = I_h^k u + \Pi_h^m(u - I_h^k u)$, i.e., it can be interpreted as an interpolation operator, where the local mass is corrected to be exactly reproduced if $m = 0$ and also higher moments are reproduced if $m > 0$. Moreover, exactly as in Section 7.4.1 we set $V(h) := H^2(\mathcal{T}_h)$ which allows us to repeat all arguments and recover Corollary 7.33, Theorem 7.34, and Theorem 7.35 for the EG (with SIPG bilinear form) instead of the DG (SIPG) approximation, respectively. The enriched Galerkin method aims at obtaining similar stability and convergence properties, as DG, but using significantly fewer degrees of freedom. The reduced amount of freedom, however, has to be balanced against a significantly "less diagonal" mass matrix (the DG mass matrix is at least block-diagonal and diagonal if orthogonal test and trial functions are used). EG has proven to be useful in several practical applications. However, EG is not a finite element method in the sense that for $m \geq 0$ the global constants are included in both subspaces of V_h, i.e., the sum is not direct. This results in the fact that $\widetilde{\Pi}$ is well defined, but the representation of $\widetilde{\Pi}u$ in terms of trial functions of $\mathbb{P}_k(\mathcal{T}_h) \cap C(\overline{\Omega})$ and $\mathbb{P}_m(\mathcal{T}_h)$ is not unique and, thus, cannot be localized to single elements. This problem can be bypassed by removing the obsolete degrees of freedom from the global system of equations or using defect correction methods as local solvers (which can deal with this issue).

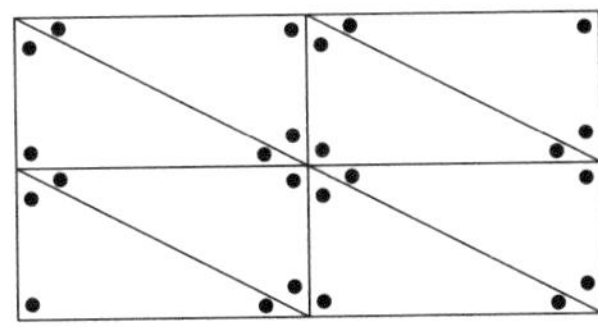
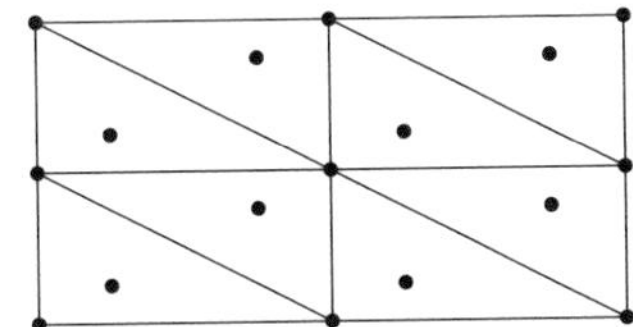

Fig. 7.10: Illustration of the degrees of freedom (bullets) for first-order DG (left) and EG with $k = 1$ and $m = 0$ (right).

Remark 7.37 1) If $m \geq 0$, V_h contains the elementwise constants which ensures local mass conservation in the same way as it holds for the DG methods of Section 7.4.1 (see discussion of Definition 7.58). The space $\mathbb{P}_0(\mathcal{T}_h) \cap C(\Omega)$ contains only constants and does not yield any reasonable approximation property.
2) The most prominent choice of m is $m = 0$, since this ensures local conservation of mass in the sense of 1), while minimizing the number of degrees of freedom. Note that in a standard proof the order of convergence is not improved by m at all, motivating to choose m as small as possible. In the case of Figure 7.9, the amount of degrees of freedom reduces from three per triangle to one per triangle and one per node ($k = 1$).
3) On the one hand, choosing $m = -1$ leads to a classical finite element method, where the enforcement of the boundary condition is executed weakly. On the other hand, $m = k$ leads to the classical DG methods.
4) For $m = 0$, the resulting scheme can be interpreted as cell-centred finite volume stabilization of the method (where the diffusion operator is discretized in a nonconsistent way (cf. Remark 7.26), which in the overall method is consistent again). Moreover, the resulting methods can be understood as a DG stabilization of continuous Galerkin methods if $m > 0$.

Minimal regularity solutions, liftings, discrete gradients, and local discontinuous Galerkin methods

The results in the above section always assume that the solution u is very regular, i.e., $u \in H^{k+1}(\mathcal{T}_h)$ for optimal convergence behaviour or at least "approximately" $u \in H^2(\Omega)$ in the sense of (7.18), (7.19), (7.20) for the method to be consistent. In the following we will consider the case that u still belongs to $H^1(\Omega)$ but is not significantly smoother, so that terms like the trace of $\boldsymbol{K}\nabla u$ on a face F are generally not defined. In such a situation, $a_h(u, v_h)$ cannot be evaluated. This is due to the discretization of the diffusion operator. Therefore, this paragraph will examine its numerical implementation in more detail. However, the analysis in this case requires an additional construct.

Definition 7.38 (Lifting operator and discrete gradient) Let $\mathcal{T}_h$ be a shape-regular family of partitions of $\Omega \subset \mathbb{R}^d$, $\boldsymbol{K} \in W_1^1(\mathcal{T}_h)^{d\times d}$, $F \in \overline{\mathcal{F}}$, and $m \in \mathbb{N}_0$.

1) The *local lifting operator* $\mathcal{L}_F^m$ is defined by

$$\mathcal{L}_F^m : L^2(F)^d \to \mathbb{P}_m(\mathcal{T}_h)^d, \quad \boldsymbol{\varphi} \mapsto \mathcal{L}_F^m(\boldsymbol{\varphi})$$

by

$$\int_\Omega \boldsymbol{K}\mathcal{L}_F^m(\boldsymbol{\varphi}) \cdot \boldsymbol{w}_h dx = \int_F \{\!\!\{\boldsymbol{K}\boldsymbol{w}_h\}\!\!\} \cdot \boldsymbol{\varphi} d\sigma \quad \text{for all } \boldsymbol{w}_h \in \mathbb{P}_m(\mathcal{T}_h)^d$$

(cf. (7.25)).

2) The *global lifting operator* $\mathcal{L}_h^m$ is defined as

$$\mathcal{L}_h^m : H^1(\mathcal{T}_h) \to \mathbb{P}_m(\mathcal{T}_h)^d, \quad v \mapsto \mathcal{L}_h^m(v) := \sum_{F \in \mathcal{F} \cup \mathcal{F}_3} \mathcal{L}_F^m([\![v]\!]).$$

3) The *discrete gradient* is defined as

$$\mathcal{G}_h^m : H^1(\mathcal{T}_h) \to L^2(\Omega)^d, \quad v \mapsto \left[\sum_{K \in \mathcal{T}_h} \nabla v|_K\right] - \mathcal{L}_h^m(v).$$

Remark 7.39 1) The unique existence of $\mathcal{L}_F^m(\boldsymbol{\varphi})$ follows from the Riesz representation theorem and $\mathbb{P}_m(\mathcal{T}_h)^d$ being finite-dimensional. We observe that the support of $\mathcal{L}_F^m(\varphi)$ consists of the partition elements of which F is part of the boundary. Its computation requires the solution of a local problem in 2 dim $\mathbb{P}_m(\widehat{K})$ variables for $F \in \mathcal{F}$. Beyond that, lifting operators and discrete gradient depend on the diffusion coefficient $\boldsymbol{K}$.

2) Lifting and discrete gradient operators can be evaluated for solutions $u \in H^1(\mathcal{T}_h)$. Moreover, we have that with the SIPG version of a_h (i.e., $\theta = -1$) the identity

$$\begin{aligned}
&a_h^{\mathrm{SIPG}}(u_h, v_h)\\
&= \int_\Omega \boldsymbol{K}\mathcal{G}_h^m(u_h) \cdot \mathcal{G}_h^m(v_h) - \boldsymbol{K}\mathcal{L}_h^m(u_h) \cdot \mathcal{L}_h^m(v_h) dx\\
&\quad + \sum_{F \in \mathcal{F} \cup \mathcal{F}_3} \int_F \frac{\eta}{h_F}[\![u_h]\!] \cdot [\![v_h]\!] d\sigma + \int_\Omega r u_h v_h dx - \sum_{K \in \mathcal{T}_h} \int_K u_h \boldsymbol{c} \cdot \nabla v_h dx\\
&\quad + \sum_{F \in \mathcal{F} \cup \mathcal{F}_{2,1}} \int_F \boldsymbol{c}_{\mathrm{upw}}(u_h) \cdot [\![v_h]\!] d\sigma + \sum_{F \in \mathcal{F}_{2,2}} \int_F \tilde{\alpha} u_h v_h d\sigma\\
&=: a_h^{\mathcal{L}_h}(u_h, v_h) \quad \text{for all } u_h, v_h \in \mathbb{P}_k(\mathcal{T}_h)
\end{aligned} \tag{7.108}$$

holds if $\nabla \mathbb{P}_k(\mathcal{T}_h) \subset \mathbb{P}_m(\mathcal{T}_h)^d$, i.e., $m \in \{k-1, k\}$ and $k \geq 1$ (if $\mathbb{P}_k(K) = \mathcal{P}_k$, where the fact that test functions for lifting operators are gradients and therefore in $\mathbb{P}_m(\mathcal{T}_h)^d$ is used). Thus, this method formulated in terms of discrete gradients and lifting operators is stable and gives the same results as the SIPG method (independent of the regularity of the solution u). Nevertheless it should not be used in numerical realizations because of the local problems to be solved. It is solely for analysis purposes. Also observe that the consistency of the solution is not guaranteed anymore since the identity (7.108) only holds for $u_h \in \mathbb{P}_k(\mathcal{T}_h)$.

3) Since Theorem 7.35 only works for SIPG methods, we can expect first order of convergence for the SIPG method applied to pure diffusion–reaction problems only, and we will not analyse $\theta \in \{0, 1\}$.

Introducing

$$\hat{\boldsymbol{p}}_h := -\boldsymbol{K}\mathcal{G}_h^m(u_h) - \boldsymbol{K} \sum_{F \in \mathcal{F}_3} \mathcal{L}_F^m(\tilde{g}_3\mathfrak{n}) \in \mathbb{P}_m(\mathcal{T}_h)^d$$

as the flux density (in a weak sense), $u_h \in \mathbb{P}_k(\mathcal{T}_h)$ motivates the *local discontinuous Galerkin (LDG)* method:

Find $\boldsymbol{p}_h \in \mathbb{P}_m(\mathcal{T}_h)^d$, $u_h \in \mathbb{P}_k(\mathcal{T}_h)$ such that

$$\int_\Omega \boldsymbol{K}^{-1}\boldsymbol{p}_h \cdot \boldsymbol{q}_h dx - \sum_{K \in \mathcal{T}_h} \int_K u_h(\nabla \cdot \boldsymbol{q}_h) dx + \sum_{F \in \overline{\mathcal{F}} \setminus \mathcal{F}_3} \int_F \{\!\{u_h\}\!\} [\![\boldsymbol{q}_h]\!] d\sigma \tag{7.109}$$

$$= - \sum_{F \in \mathcal{F}_3} \int_F \tilde{g}_3 \mathfrak{n} \cdot \boldsymbol{q}_h d\sigma \quad \text{for all } \boldsymbol{q}_h \in \mathbb{P}_m(\mathcal{T}_h)^d$$

and

$$\begin{aligned}
&- \sum_{K \in \mathcal{T}_h} \int_K \boldsymbol{p}_h \cdot \nabla v_h dx + \sum_{K \in \mathcal{T}_h} \int_K [r u_h v_h - u_h \boldsymbol{c} \cdot \nabla v_h]\, dx \\
&+ \sum_{F \in \mathcal{F} \cup \mathcal{F}_3} \int_F \left[\{\!\{\boldsymbol{p}_h\}\!\} \cdot [\![v_h]\!] + \frac{\eta}{h_F} [\![u_h]\!] \cdot [\![v_h]\!] \right] d\sigma \\
&+ \sum_{F \in \mathcal{F} \cup \mathcal{F}_{2,1}} \int_F \boldsymbol{c}_{\text{upw}}(u_h) \cdot [\![v_h]\!] d\sigma + \sum_{F \in \mathcal{F}_{2,2}} \int_F \tilde{\alpha} u_h v_h d\sigma \\
&= \int_\Omega f v_h dx + \sum_{F \in \mathcal{F}_2} \int_F \tilde{g}_1 v_h d\sigma + \sum_{F \in \mathcal{F}_1} \int_F \tilde{g}_2 v_h d\sigma \\
&- \sum_{F \in \mathcal{F}_3} \int_F \left[\boldsymbol{c} \tilde{g}_3 v_h - \frac{\eta}{h_F} \tilde{g}_3 v_h \right] d\sigma \quad \text{for all } v_h \in \mathbb{P}_k(\mathcal{T}_h).
\end{aligned} \tag{7.110}$$

This construction indicates a strong relation between the SIPG and the LDG methods: The equation (7.109) is the aforementioned weak formulation of $\hat{\boldsymbol{p}}_h$ in a formal way, and equation (7.106) corresponds to the SIPG formulation with a modified jump term $\int_F \{\!\{\boldsymbol{p}_h\}\!\} [\![v_h]\!] d\sigma$ instead of $- \int_F \{\!\{K \nabla u_h\}\!\} [\![v_h]\!] d\sigma$. This relation will be made more clear in Lemma 7.41. For the LDG methods the choice $m \in \{k-1, k\}$ is possible, where $m = k - 1$ reduces the computational costs. Moreover this relation classifies LDG as a primal method although it is widely considered to be a dual method due to its use of the auxiliary variable $\boldsymbol{p}_h$.

Lemma 7.40 *There is a unique solution of* (7.109), (7.110).

Proof The system of equations induced by (7.109), (7.110) is square, since the test and trial spaces are equal. Thus, it is sufficient to show that the system is injective, i.e., for $\tilde{g}_1 = \tilde{g}_2 = \tilde{g}_3 = f = 0$ the only possible solution is $(u_h, \boldsymbol{p}_h)$ being zero. To do so, we set $\boldsymbol{p}_h = \boldsymbol{q}_h$ and $v_h = u_h$, add the equations, integrate by parts, and use the identities

$$\begin{aligned} \{\!\{u_h\}\!\} [\![\boldsymbol{p}_h]\!] + [\![u_h]\!] \cdot \{\!\{\boldsymbol{p}_h\}\!\} &= [\![u_h \boldsymbol{p}_h]\!] \quad \text{on } F \in \mathcal{F}, \\ [\![u_h]\!] \cdot \{\!\{\boldsymbol{p}_h\}\!\} &= [\![u_h \boldsymbol{p}_h]\!] \quad \text{on } F \notin \mathcal{F}, \end{aligned} \tag{7.111}$$

(7.103), and some algebraic manipulations, to obtain

$$\int_\Omega \left[\boldsymbol{K}^{-1} \boldsymbol{p}_h \cdot \boldsymbol{p}_h + (r + \frac{1}{2} \nabla \cdot \boldsymbol{c}) u_h^2 \right] dx + \sum_{F \in \overline{\mathcal{F}}} \int_F \left[\frac{1}{2} |\boldsymbol{c} \cdot \mathfrak{n}| + \frac{\eta}{h_F} \right] [\![u_h]\!]^2 d\sigma + \sum_{F \in \mathcal{F}_{2,2}} \int_F \tilde{\alpha} u_h^2 d\sigma = 0.$$

This directly implies the result. □

Lemma 7.41 *Let* $\boldsymbol{K} = \tilde{k} \boldsymbol{I}$ *with* $\tilde{k} > 0$ *being a scalar constant. Then the solution* $u_h \in \mathbb{P}_k(\mathcal{T}_h)$ *of*

$$\begin{aligned} &a_h^{\mathcal{L}_h}(u_h, v_h) + \int_\Omega \tilde{k} \mathcal{L}_h^m(u_h) \cdot \mathcal{L}_h^m(v_h) dx \\ &= \ell_h(v_h) + \sum_{F \in \mathcal{F}_3} \int_F \tilde{g}_3 \tilde{k} \mathcal{L}_F^m(v_h) \cdot \mathfrak{n} d\sigma \quad \textit{for all } v_h \in \mathbb{P}_k(\mathcal{T}_h) \end{aligned} \tag{7.112}$$

is the solution component u_h *of the local discontinuous Galerkin method* (7.110) *for* (7.96), *and* $\boldsymbol{p}_h = -\tilde{k} \mathcal{G}_h^m(u_h) - \tilde{k} \sum_{F \in \mathcal{F}_3} \mathcal{L}_F^m(\tilde{g}_3 \mathfrak{n})$.

Proof We start with rewriting (7.112) by introducing $\tilde{k}^{-1} \boldsymbol{p}_h := -\mathcal{G}_h^m(u_h) - \sum_{F \in \mathcal{F}_3} \mathcal{L}_F^m(\tilde{g}_3 \mathfrak{n})$. Weakly enforcing this equality yields

$$\int_\Omega \tilde{k}^{-1} \boldsymbol{p}_h \cdot \boldsymbol{q}_h dx + \int_\Omega \mathcal{G}_h^m(u_h) \boldsymbol{q}_h dx = - \sum_{F \in \mathcal{F}_3} \int_F \tilde{g}_3 \boldsymbol{q}_h \cdot \mathfrak{n} d\sigma \quad \text{for all } \boldsymbol{q}_h \in \mathbb{P}_m(\mathcal{T}_h)^d,$$

and

$$\begin{aligned} &\int_\Omega \mathcal{G}_h^m(v_h) \boldsymbol{q}_h dx \\ &= - \sum_{K \in \mathcal{T}_h} \int_K v_h (\nabla \cdot \boldsymbol{q}_h) dx + \sum_{F \in \overline{\mathcal{F}}} \int_F [\![v_h \boldsymbol{q}_h]\!] d\sigma - \sum_{F \in \mathcal{F} \cup \mathcal{F}_3} \int_F [\![v_h]\!] \cdot \{\!\{\boldsymbol{q}_h\}\!\} d\sigma \end{aligned}$$

by definition of $\mathcal{G}_h^m$ and integration by parts. This together gives the identity in (7.109) if one uses the symmetry relations (7.111). We rewrite a part of (7.110) with this identity (for v_h general, $\boldsymbol{q}_h := \boldsymbol{p}_h$) as follows:

$$
\begin{aligned}
&-\sum_{K\in\mathcal{T}_h}\int_K \boldsymbol{p}_h\cdot\nabla v_h dx + \sum_{F\in\mathcal{F}\cup\mathcal{F}_3}\int_F \{\!\!\{\boldsymbol{p}_h\}\!\!\}\cdot[\![v_h]\!]d\sigma \\
&\quad= \sum_{K\in\mathcal{T}_h}\int_K(\nabla\cdot\boldsymbol{p}_h)v_h dx - \sum_{F\in\overline{\mathcal{F}}}[\![v_h\boldsymbol{p}_h]\!]d\sigma + \sum_{F\in\mathcal{F}\cup\mathcal{F}_3}\{\!\!\{\boldsymbol{p}_h\}\!\!\}\cdot[\![v_h]\!]d\sigma \\
&\quad= -\int_\Omega \mathcal{G}_h(v_h)\boldsymbol{p}_h dx \\
&\quad= \int_\Omega \tilde{k}\mathcal{G}_h(u_h)\cdot\mathcal{G}_h(v_h)dx + \sum_{F\in\mathcal{F}_3}\int_F \tilde{g}_3\tilde{k}(\nabla v_h - \mathcal{L}_F^m(v_h))\cdot\mathfrak{n}d\sigma.
\end{aligned}
$$

Substituting this relation into (7.110) shows its equivalence with (7.112). □

We will not further analyse LDG methods in more detail, since their approximation quality is very similar to that of SIPG methods, as can be suspected from Lemma 7.41. However, they impose substantially higher computational costs. However, LDG methods are a useful tool for certain specific applications, e.g., for hybridization (cf. Section 7.5). Finally we cite a result for the convergence of the SIPG method to minimal regularity solutions of the Poisson problem.

Theorem 7.42 (Convergence to minimal regularity solutions) *Let $k \geq 1$ and u_h be the solution of the discrete problem* (7.100) *for the Poisson equation (i.e., $\boldsymbol{c} = \boldsymbol{0}$, $r = 0$, $\tilde{g}_3 = 0$, $\partial\Omega = \widetilde{\Gamma}_3$ in* (7.96)*) with the SIPG method. Then*

$$
\|u - u_h\|_0 + \sum_{K\in\mathcal{T}_h}\|\nabla u - \nabla u_h\|_{0,K} + \sum_{F\in\overline{\mathcal{F}}}\frac{\eta}{h_F}\|[\![u_h]\!]\|_{L^2(F)} \to 0 \quad \text{for } h\to 0.
$$

Proof The proof can be found in [21, Theorem 5.12] and uses discrete gradients. □

Theorem 7.42 can also be extended to the full case of (7.96). Moreover, under the additional assumptions of Theorem 7.35 it can be shown that u_h converges to u superlinearly in the sense of $L^2(\Omega)$, i.e.,

$$
\frac{\|u-u_h\|_0}{h} \to 0 \qquad \text{for } h\to 0.
$$

This result is an improvement if the poor convergence behaviour of u_h is caused by a very rough right-hand side $f \in H^{-1}(\Omega)$ with $f \notin L^2(\Omega)$.

Exercises

Problem 7.17 Work out in detail the regularity of the coefficients necessary to justify Corollary 7.33.

Problem 7.18 a) Work out in detail the regularity assumptions for the LDG method to be consistent.
b) Prove stability of the LDG method.

Problem 7.19 Instead of utilizing the Lax–Milgram theorem, use inf-sup-based arguments to show stability of the IPG methods. Can you also obtain stability of the OBB method?

Problem 7.20 Investigate the choices of $\theta \in \mathbb{R}$ and $\eta \in \mathbb{R}$ for the IPG methods:

a) Are the IPG methods for arbitrary $\theta \in \mathbb{R}$ stable and convergent? Argue in favour of the exposition of this book, which only discusses $\theta \in \{-1, 0, 1\}$.
b) Since η has been assumed to be chosen large enough for the method to be stable, what are possible reasons for choosing η as small as possible?

Problem 7.21 Following the lines of Section 2.4.2, deduce an assembly strategy (including transformation formulas of the respective integrals) for the stiffness matrix of the IP methods.

Problem 7.22 a) Show that the EG method is stable and deduce its convergence properties based on the polynomial degrees k and m.
b) Compare DG, CG, and EG in one spatial dimension. Are EG and DG the same methods in one spatial dimension?
c) Based of Figures 7.9 and 7.10, calculate the sizes of the stiffness matrices of CG, DG, and EG. Also compare the number of possible nonzero entries of the respective matrices.

7.5 Hybridization

7.5.1 Hybridization in General

Mixed or nonconforming methods as developed in the previous sections are often impeded by an increase of degrees of freedom compared to the classical, i.e., continuous finite element methods of the Chapters up to 4. In addition, if coercivity is lost as in the mixed methods, the discrete problems also lose the property of a positive real system matrix, on which Krylov-type methods rely (see Section 5.4). *Hybridization* is a technique originally aiming at a reduction of degrees of freedom and thus close to *static condensation* (see after (3.78)). It turns out that additionally

- The property of the system matrix to be positive real can be restored.
- The additional unknown, to the computation of which the system is reduced, is an approximation either of the scalar or the vectorial unknown, with a comparable order of convergence, or after postprocessing, even higher (superconvergence).

This last aspect will be developed in the next subsection for the dual mixed formulation of the Darcy problem (6.54), (6.55), with the assumptions of Section 6.1 to render the problem well defined in the introduced spaces, the RT_{k-1}-discretization (7.56)/(7.65). In the following, we refer to this situation as *Darcy example*.

Using the notation of (6.56), we start with a problem of the form

$$\begin{aligned} Ap + B^T u + Cu &= \ell_1, \\ Bp - Du &= \ell_2, \end{aligned} \tag{7.113}$$

with $A: M \to M'$, $B: M \to X'$, $C: X \to M'$, $D: X \to X'$ for Hilbert spaces M and X, $\ell_1 \in M'$, $\ell_2 \in X'$, i.e., inhomogeneous boundary conditions are incorporated into the right-hand sides by a shift of the solution (compare the beginning of Subsection 7.2.1). The Galerkin discretization under consideration is formulated in the finite-dimensional spaces M_h, X_h by means of the corresponding operators A_h, B_h, C_h, D_h, and the right-hand sides ℓ_{1h}, ℓ_{2h}. Finally, after choosing the bases of M_h and X_h, the discrete variational problem is equivalent to the system of linear equations

$$\begin{pmatrix} \mathcal{A}_h & \mathcal{B}_h^T + C_h \\ \mathcal{B}_h & -\mathcal{D}_h \end{pmatrix} \begin{pmatrix} \boldsymbol{\xi} \\ \boldsymbol{\eta} \end{pmatrix} = \begin{pmatrix} \boldsymbol{q}_{\xi} \\ \boldsymbol{q}_{\eta} \end{pmatrix}. \tag{7.114}$$

A classical technique to reduce the above system is *static condensation*, i.e., the *Schur complement* applied to a specific partitioning. Under the assumption that $\mathcal{A}_h$ is invertible, which is true in the Darcy example mentioned above, (7.114) is equivalent to

$$(\mathcal{B}_h \mathcal{A}_h^{-1} (\mathcal{B}_h^T + C_h) + \mathcal{D}_h)\boldsymbol{\eta} = \mathcal{B}_h \mathcal{A}_h^{-1} \boldsymbol{q}_{\xi} - \boldsymbol{q}_{\eta} \tag{7.115}$$

and then

$$\mathcal{A}_h \boldsymbol{\xi} = \boldsymbol{q}_{\xi} - (\mathcal{B}_h^T + C_h)\boldsymbol{\eta}. \tag{7.116}$$

If $\mathcal{D}_h$ is invertible, which is fulfilled in the Darcy example if $d(x) \geq d_0 > 0$ for the reaction term, an alternative reduction is

$$((\mathcal{B}_h^T + C_h)\mathcal{D}_h^{-1}\mathcal{B}_h + \mathcal{A}_h)\boldsymbol{\xi} = \boldsymbol{q}_{\xi} + (\mathcal{B}_h^T + C_h)\mathcal{D}_h^{-1}\boldsymbol{q}_{\eta}. \tag{7.117}$$

Such an approach is only feasible if $\mathcal{A}_h$ or $\mathcal{D}_h$ is block-diagonal with a small block size, ideally related to the degrees of freedom only of one element. In such a case $\mathcal{A}_h^{-1}$ or $\mathcal{D}_h^{-1}$, respectively, can be computed with minor effort, not deteriorating the sparsity pattern too much. This requirement is guaranteed if the degrees of freedom represented by $\boldsymbol{\xi}$ or $\boldsymbol{\eta}$, respectively, are only interconnected in the local patches corresponding to the block size, i.e., ideally only at one element. This is the basis for the elimination of interior degrees of freedom for continuous finite elements (see after (3.78)). For the Darcy equation in the dual mixed form with RT- or BDM-elements the matrix $\mathcal{D}_h$ is block-diagonal, but not $\mathcal{A}_h$. In such a situation hybridization will work under the following additional assumption:

> Let $\widetilde{M}_h$ be a finite-dimensional space such that $M_h \subset \widetilde{M}_h$ (resulting, for instance, from the removal of the inter-element connections in M_h).

For the above with RT_{k-1}-elements we can choose

$$\widetilde{M}_h := RT_{k-1}(\mathcal{T}_h) := \{\boldsymbol{q} \in (L^2(\Omega))^d \mid \boldsymbol{q}|_K \in RT_{k-1}(K) \quad \text{for all } K \in \mathcal{T}_h\}. \tag{7.118}$$

We suppose that A_h can be modified to $\widetilde{A}_h : \widetilde{M}_h \to \widetilde{M}_h'$, i.e., $\widetilde{A}_h\boldsymbol{q}_h = A_h\boldsymbol{q}_h$ for all $\boldsymbol{q}_h \in M_h$ by extending the definition and restricting the image space, and correspondingly there are extensions $\widetilde{B}_h, \widetilde{C}_h^T$, and $\tilde{\ell}_{1h}$. In the specific case, i.e., for

$$a_h(\boldsymbol{p}_h, \boldsymbol{q}_h) := \int_\Omega L\boldsymbol{p}_h \cdot \boldsymbol{q}_h dx,$$

this expression is also well defined for $\boldsymbol{p}_h, \boldsymbol{q}_h \in \widetilde{M}_h$ and similar for $\widetilde{C}_h$, and for

$$b_h(\boldsymbol{q}_h, u_h) := -\int_\Omega u_h \nabla \cdot \boldsymbol{q}_h dx = -\sum_{K\in\mathcal{T}_h} \int_K u_h \nabla \cdot \boldsymbol{q}_h dx \tag{7.119}$$

the second form has to be used.

The matrix representations (for a fixed basis) are denoted with $\widetilde{\mathcal{A}}_h, \widetilde{\mathcal{B}}_h, \widetilde{C}_h$. Note that the basis used in $\widetilde{M}_h$ is usually not just an extension of the basis of M_h, as a block-diagonal structure of $\widetilde{\mathcal{A}}_h$ shall be achieved.

Furthermore we assume the following:

There exists a finite-dimensional space Y_h and a linear operator $E_h : Y_h \to \widetilde{M}_h'$ such that

$$\ker E_h^T = M_h\,. \tag{7.120}$$

Then the discrete variational equation (7.113) on $M_h \times X_h$ is equivalent to the following problem:

Find $(\tilde{\boldsymbol{p}}_h, \tilde{\boldsymbol{u}}_h, \lambda_h) \in \widetilde{M}_h \times X_h \times Y_h$ such that

$$\begin{aligned} \widetilde{A}_h\tilde{\boldsymbol{p}}_h + \widetilde{B}_h^T\tilde{u}_h + \widetilde{C}_h\tilde{u}_h + E_h\lambda_h &= \tilde{\ell}_{1h},\\ \widetilde{B}_h\tilde{\boldsymbol{p}}_h - D_h\tilde{u}_h &= \ell_{2h},\\ E_h^T\tilde{\boldsymbol{p}}_h &= 0. \end{aligned} \tag{7.121}$$

The solution component $\lambda_h \in Y_h$ is also called the *Lagrange multiplier*.

Lemma 7.43 *If* $(\boldsymbol{p}_h, u_h) \in M_h \times X_h$ *is a solution of* (7.113)*, then there exists an element* $\lambda_h \in Y_h$ *such that* $(\boldsymbol{p}_h, u_h, \lambda_h)$ *is a solution of* (7.121).

If $(\tilde{\boldsymbol{p}}_h, \tilde{\boldsymbol{u}}_h, \lambda_h) \in \widetilde{M}_h \times X_h \times Y_h$ *is a solution of* (7.121)*, then* $\tilde{\boldsymbol{p}}_h \in M_h$ *and* $(\tilde{\boldsymbol{p}}_h, \tilde{u}_h) \in M_h \times X_h$ *is a solution of* (7.113).

Proof Let $(\boldsymbol{p}_h, u_h) \in M_h \times X_h$ be a solution of (7.113), then the second equation of (7.121) is satisfied. The third equation is satisfied because of (7.120). Setting

$$\boldsymbol{r}_h := \tilde{\ell}_{1h} - \widetilde{A}_h\boldsymbol{p}_h - \widetilde{B}_h^T u_h - \widetilde{C}_h u_h,$$

we see that

$$\langle \boldsymbol{r}_h, \boldsymbol{q}_h\rangle = \langle \ell_1 - A_h\boldsymbol{p}_h - B_h^T u_h - C_h u_h, \boldsymbol{q}_h\rangle = 0 \quad \text{for all } \boldsymbol{q}_h \in M_h,$$

that is,

$$\boldsymbol{r}_h \in M_h^{\perp} = (\ker E_h^T)^{\perp} = \operatorname{im}(E_h).$$

The second part can be shown analogously. □

This kind of argument has been used already several times; see Section 6.2, and also the proof of Theorem 6.29 can be seen in this way.

In the Darcy example we can take

$$Y_h := \{\mu \in L^2(\mathcal{F}) \mid \mu|_F \in \mathcal{P}_{k-1}(F) \quad \text{for all } F \in \mathcal{F}\}, \tag{7.122}$$

and E_h is defined by

$$e_h(\mu_h, \boldsymbol{q}_h) := \sum_{F\in\mathcal{F}} \int_F \mu_h [\![\boldsymbol{q}_h]\!] d\sigma \quad \text{for all } \mu_h \in Y_h,\ \boldsymbol{q}_h \in \widetilde{M}_h. \tag{7.123}$$

Then (7.120) is a consequence of (7.56).

Lemma 7.44 *In the hybridization of the Galerkin discretization of (6.54), (6.55) with RT_{k-1}-elements by means of (7.115), the Lagrange multiplier is unique if the solution of (6.56) in $M_h \times X_h$ is unique.*

Proof We only have to show $\ker E_h = \{0\}$. Let $\mu_h \in \ker E_h$, i.e.,

$$\sum_{F\in\mathcal{F}} \int_F \mu_h [\![\boldsymbol{q}_h]\!] d\sigma = 0 \quad \text{for all } \boldsymbol{q}_h \in \widetilde{M}_h.$$

For $K \in \mathcal{T}_h$ and $F \in \mathcal{F}_K \cap \mathcal{F}$, let $\mu_h|_F = \sum_{i=1}^{a_k} \alpha_i w_i^{(k)}$, where $\{w_i^{(k)}\}$ is the basis of $\mathcal{P}_{k-1}(\mathbb{R}^{d-1})$ from (7.49). We define an element $\boldsymbol{q}_h \in \widetilde{M}_h$ as follows:

On K and for $F \in \mathcal{F}_K \cap \mathcal{F}$ we take the above coefficients α_i as values of the degrees of freedom in (7.49), for $F' \in \mathcal{F}_K \setminus \{F\}$ we take zero values. On $K' \in \mathcal{T}_h \setminus \{K\}$ we simply set $\boldsymbol{q}_h|_{K'} := \boldsymbol{0}$.

Then

$$0 = \int_F \mu_h \boldsymbol{q}_h \cdot \boldsymbol{n} d\sigma = \sum_{i=1}^{a_k} \alpha_i^2,$$

hence $\mu_h|_F = 0$ for a general face F. □

At the algebraic level, the system of equations (7.114) has been extended to

$$\begin{pmatrix} \widetilde{\mathcal{A}}_h & \widetilde{\mathcal{B}}_h^T + \widetilde{\mathcal{C}}_h & \mathcal{E}_h \\ \widetilde{\mathcal{B}}_h & -\mathcal{D}_h & \boldsymbol{0} \\ \mathcal{E}_h^T & \boldsymbol{0} & \boldsymbol{0} \end{pmatrix} \begin{pmatrix} \tilde{\xi} \\ \tilde{\eta} \\ \lambda \end{pmatrix} = \begin{pmatrix} \tilde{q}_\xi \\ q_\eta \\ \boldsymbol{0} \end{pmatrix}, \tag{7.124}$$

i.e., the size is increased even more, but for the appropriate basis and degrees of freedom the matrix $\widetilde{\mathcal{A}}_h$ has a block-diagonal structure. If—as in the Darcy example—

also X_h allows for a element-local basis, then this also applies to $\widetilde{\mathcal{B}}_h, \widetilde{C}_h, \mathcal{D}_h$ and the only global coupling resides in $\mathcal{E}_h$. In any case, we can eliminate $\tilde{\boldsymbol{\xi}}$ as described above leading to

$$\begin{pmatrix} \mathcal{A} & C \\ C^T + \mathcal{D} & \mathcal{B} \end{pmatrix} \begin{pmatrix} \tilde{\boldsymbol{\eta}} \\ \boldsymbol{\lambda} \end{pmatrix} = \begin{pmatrix} \hat{\boldsymbol{q}}_\eta \\ \hat{\boldsymbol{q}}_\lambda \end{pmatrix}, \tag{7.125}$$

where

$$\begin{aligned} \mathcal{A} &:= \widetilde{\mathcal{B}}_h \widetilde{\mathcal{A}}_h^{-1} (\widetilde{\mathcal{B}}_h^T + \widetilde{C}_h) + \mathcal{D}_h, & \mathcal{B} &:= \mathcal{E}_h^T \widetilde{\mathcal{A}}_h^{-1} \mathcal{E}_h, \\ C &:= \widetilde{\mathcal{B}}_h \widetilde{\mathcal{A}}_h^{-1} \mathcal{E}_h, & \mathcal{D} &:= \mathcal{E}_h^T \widetilde{\mathcal{A}}_h^{-1} \widetilde{C}_h, \\ \hat{\boldsymbol{q}}_\eta &:= -\boldsymbol{q}_\eta + \widetilde{\mathcal{B}}_h \widetilde{\mathcal{A}}_h^{-1} \tilde{\boldsymbol{q}}_\xi, & \hat{\boldsymbol{q}}_\lambda &:= \mathcal{E}_h^T \widetilde{\mathcal{A}}_h^{-1} \boldsymbol{q}_\xi. \end{aligned}$$

Here we have used that $\widetilde{\mathcal{A}}_h$ is symmetric and invertible. The element $\tilde{\boldsymbol{\xi}}$ can be reconstructed from $\tilde{\boldsymbol{\eta}}$ and $\boldsymbol{\lambda}$ by

$$\tilde{\boldsymbol{\xi}} = \widetilde{\mathcal{A}}_h^{-1} \boldsymbol{q}_\xi - \widetilde{\mathcal{A}}_h^{-1} (\widetilde{\mathcal{B}}_h^T + \widetilde{C}_h) \tilde{\boldsymbol{\eta}} - \widetilde{\mathcal{A}}_h^{-1} \mathcal{E}_h \boldsymbol{\lambda}\,.$$

If also $\mathcal{A}$ is block-diagonal, as in the Darcy example, and invertible, a further elimination is possible, leading to

$$\mathcal{E}\boldsymbol{\lambda} = \overline{\boldsymbol{q}}, \tag{7.126}$$

where

$$\begin{aligned} \mathcal{E} &:= -(C^T + \mathcal{D})\mathcal{A}^{-1} C + \mathcal{B}, \\ \overline{\boldsymbol{q}} &:= \hat{\boldsymbol{q}}_\lambda - (C^T + \mathcal{D})\mathcal{A}^{-1} \hat{\boldsymbol{q}}_\eta, \end{aligned}$$

with the reconstructions

$$\tilde{\boldsymbol{\eta}} := \mathcal{A}^{-1} \hat{\boldsymbol{q}}_\eta - \mathcal{A}^{-1} C \boldsymbol{\lambda}$$

and $\tilde{\boldsymbol{\xi}}$ as above.

This overall procedure first of all reduces the problem size: In the Darcy example the amount of flux degrees at the faces remains in form of the multipliers, and the volume degrees of freedom are eliminated. Even more important, the character of the system equation also changes. Let us first consider the symmetric case $C_h = 0$, i.e., there is no convection in the Darcy example.

Lemma 7.45 *Assume that $\widetilde{\mathcal{A}}_h$ is symmetric and positive definite, $C_h = 0$, and $\mathcal{D}_h$ is symmetric and positive semidefinite. If the matrix $\mathcal{D}_h$ is even positive definite or if the property*

$$\widetilde{\mathcal{B}}_h^T \boldsymbol{x} + \mathcal{E}_h \boldsymbol{y} = \boldsymbol{0} \implies \boldsymbol{x} = \boldsymbol{0},\ \boldsymbol{y} = \boldsymbol{0} \tag{7.127}$$

is satisfied, then the system matrix of (7.125) *is symmetric and positive definite.*

Proof Since the matrices $\mathcal{A}$ and $\mathcal{B}$ are symmetric, the system matrix of (7.125) is symmetric. The induced quadratic form reads as

$$
\begin{aligned}
&\begin{pmatrix} \boldsymbol{x} \\ \boldsymbol{y} \end{pmatrix}^T \begin{pmatrix} \mathcal{A} & C \\ C^T & \mathcal{B} \end{pmatrix} \begin{pmatrix} \boldsymbol{x} \\ \boldsymbol{y} \end{pmatrix} = \boldsymbol{x}^T \mathcal{A} \boldsymbol{x} + 2\boldsymbol{x}^T C \boldsymbol{y} + \boldsymbol{y}^T \mathcal{B} \boldsymbol{y} \\
&= (\widetilde{\mathcal{B}}_h^T \boldsymbol{x})^T \widetilde{\mathcal{A}}_h^{-1} (\widetilde{\mathcal{B}}_h^T \boldsymbol{x}) + \boldsymbol{x}^T \mathcal{D}_h \boldsymbol{x} + 2(\widetilde{\mathcal{B}}_h^T \boldsymbol{x})^T \widetilde{\mathcal{A}}_h^{-1} (\mathcal{E}_h \boldsymbol{y}) + (\mathcal{E}_h \boldsymbol{y})^T \widetilde{\mathcal{A}}_h^{-1} (\mathcal{E}_h \boldsymbol{y}) \\
&= (\widetilde{\mathcal{B}}_h^T \boldsymbol{x} + \mathcal{E}_h \boldsymbol{y})^T \widetilde{\mathcal{A}}_h^{-1} (\widetilde{\mathcal{B}}_h^T \boldsymbol{x} + \mathcal{E}_h \boldsymbol{y}) + \boldsymbol{x}^T \mathcal{D}_h \boldsymbol{x} \geq 0,
\end{aligned} \tag{7.128}
$$

and its definiteness follows from one of the additional assumptions. □

Remark 7.46 1) Condition (7.127) induces the full column rank of $\widetilde{\mathcal{B}}_h^T$ and $\mathcal{E}_h$ and thus the definiteness of $\mathcal{A}$ and $\mathcal{B}$, being necessary for the assertion.

2) In the Darcy example the assumption (7.127) is not satisfied. Indeed, the condition in (7.127) is equivalent to

$$b_h(\boldsymbol{q}_h, u_h) + e_h(\lambda_h, \boldsymbol{q}_h) = 0 \quad \text{for all } \boldsymbol{q}_h \in \widetilde{M}_h,$$

that is,

$$\sum_{F \in \mathcal{F}} \lambda_h [\![\boldsymbol{q}_h]\!] d\sigma - \sum_{K \in \mathcal{T}_h} \int_K \nabla \cdot \boldsymbol{q}_h \, u_h \, dx = 0 \quad \text{for all } \boldsymbol{q}_h \in \widetilde{M}_h. \tag{7.129}$$

Here $\lambda_h \in Y_h$ corresponds to the representing vector $\boldsymbol{y}$ (in the fixed basis), and correspondingly $u_h \in X_h$ to $\boldsymbol{x}$.
However, there are nontrivial elements u_h, λ_h (resp. $\boldsymbol{x}, \boldsymbol{y}$) that satisfy the relation (7.129). They can be constructed as follows:
Fix some $K \in \mathcal{T}_h$ and set $u_h|_K := 1$ and $u_h|_{K'} := 0$ on every $K' \in \mathcal{T}_h \setminus \{K\}$.
For $F \in \mathcal{F}_K \cap \mathcal{F}$, set $\lambda_h|_F := 1$ and $\lambda_h := 0$ otherwise. Then (7.129) reduces to

$$\int_K \nabla \cdot \boldsymbol{q}_h \, dx = \int_{\partial K} \boldsymbol{q}_h \cdot \mathfrak{n}_h \, dx,$$

which holds true for every $\boldsymbol{q}_h \in \widetilde{M}_h$.

Note, however, that the restricted requirement of full column rank of $\widetilde{\mathcal{B}}_h^T$ and $\mathcal{E}_h$ holds true:

$$\mathbf{0} = \mathcal{B}_h^T \boldsymbol{x} \implies 0 = \sum_{K \in \mathcal{T}_h} \int_K \nabla \cdot \boldsymbol{q}_h u dx \quad \text{for all } \boldsymbol{q}_h \in \widetilde{M}_h.$$

Because of (7.48) there exists an element $\boldsymbol{q}_h \in \widetilde{M}_h$ such that $\nabla \cdot \boldsymbol{q}_h = u_h \in \mathcal{P}_{k-1}$ and thus $u_h = 0$, i.e., $\boldsymbol{x} = \mathbf{0}$. The argument for $\mathcal{E}_h$ is similar.

In the symmetric case we have a result similar to Lemma 7.45 for the fully reduced matrix $\mathcal{E}$ from (7.126).

Lemma 7.47 *Assume that $\widetilde{\mathcal{A}}_h$ is symmetric and positive definite, $C_h = 0$, $\mathcal{D}_h$ is symmetric and positive semidefinite, and $\widetilde{\mathcal{B}}_h^T, \mathcal{E}_h$ have full column rank. If $\mathcal{D}_h$ is positive definite or $\widetilde{\mathcal{B}}_h^T(\mathcal{B}_h \widetilde{\mathcal{A}}_h^{-1} \widetilde{\mathcal{B}}_h^T + \mathcal{D}_h)^{-1} \mathcal{B}_h \widetilde{\mathcal{A}}_h^{-1}$ does not have the eigenvalue 1, then the system matrix $\mathcal{E}$ in (7.126) is symmetric and positive definite.*

Proof As in the proof of Lemma 7.45 we see that the block-matrix

$$\begin{pmatrix} \mathcal{A} & C \\ C^T & \mathcal{B} \end{pmatrix},$$

is symmetric and positive semidefinite (cf. (7.128)). Since $\widetilde{\mathcal{B}}_h^T$ has full column rank, the matrix $\mathcal{A}$ is invertible. Then the Schur complement $\mathcal{E} = \mathcal{B} - C^T \mathcal{A}^{-1} C$ exists and is symmetric, too, and it holds that

$$\boldsymbol{y}^T \mathcal{E} \boldsymbol{y} = \boldsymbol{y}^T \mathcal{B} \boldsymbol{y} + (C\boldsymbol{y})^T \boldsymbol{x} = \boldsymbol{x}^T \mathcal{A} \boldsymbol{x} + 2\boldsymbol{x}^T C \boldsymbol{y} + \boldsymbol{y}^T \mathcal{B} \boldsymbol{y} \geq 0$$

for all $\boldsymbol{y}$, where $\boldsymbol{x} := -\mathcal{A}^{-1} C \boldsymbol{y}$. In other words, the induced quadratic form of $\mathcal{E}$ coincides with the induced quadratic form of the block-matrix with the above particular choice of $\boldsymbol{x}$. In order to show the definiteness of $\mathcal{E}$ we consider the equation $\boldsymbol{y}^T \mathcal{E} \boldsymbol{y} = 0$. Then we immediately see from (7.128) that the definiteness of $\mathcal{D}_h$ implies $\boldsymbol{x} = \boldsymbol{0}$, hence $\boldsymbol{y}^T \mathcal{B} \boldsymbol{y} = 0$ and the full-rank property of $\mathcal{E}_h$ yields $\boldsymbol{y} = \boldsymbol{0}$. If $\mathcal{D}_h$ is only semidefinite, we have to investigate the case

$$\boldsymbol{0} = \widetilde{\mathcal{B}}_h^T \boldsymbol{x} + \mathcal{E}_h \boldsymbol{y} = (-\widetilde{\mathcal{B}}_h^T (\widetilde{\mathcal{B}}_h \widetilde{\mathcal{A}}_h^{-1} \widetilde{\mathcal{B}}_h^T + \mathcal{D}_h)^{-1} \widetilde{\mathcal{B}}_h \widetilde{\mathcal{A}}_h^{-1} + I) \mathcal{E}_h \boldsymbol{y}.$$

Then the assumption formulated last implies $\mathcal{E}_h \boldsymbol{y} = \boldsymbol{0}$ and thus again $\boldsymbol{y} = \boldsymbol{0}$. □

In the general case $C_h \neq 0$ we may expect the system matrices to be positive real, at least for a small convective part.

A further aspect is that the Lagrange multipliers may exhibit an asymptotic order of convergence to a solution component and even lead to a superconvergent construction. This will be investigated for the Darcy example.

7.5.2 Convergence of the Multipliers for the Hybridized Mixed *RT*-Element Discretizations of the Darcy Equation

We want to express the elimination process in terms of variational equations following [126]. The hybridized mixed approximation of the Darcy equation (6.54), (6.55) with RT_{k-1}-elements, $k \in \mathbb{N}$, (7.56), and (7.65) reads as follows:

Find $(\boldsymbol{p}_h, u_h, \lambda_h) \in \widetilde{M}_h \times X_h \times Y_h$ (see (7.118), (7.122)) such that

$$\begin{aligned} \widetilde{A}_h \boldsymbol{p}_h + \widetilde{B}_h^T u_h + \widetilde{C}_h u_h + E_h \lambda_h &= \tilde{\ell}_1, \\ \widetilde{B}_h \boldsymbol{p}_h - D_h u_h &= \ell_2, \\ E_h^T \boldsymbol{p}_h &= 0, \end{aligned} \tag{7.130}$$

where the operators are related to the forms a, b, c, d from (6.54) and e_h from (7.123). The inhomogeneous flux boundary conditions (3.19) are supposed to be eliminated by an element $\overline{p} \in H(\text{div}; \Omega)$ such that $-\overline{p} \cdot \mathfrak{n} = \tilde{g}_1$ on $\widetilde{\Gamma}_1$, and the right-hand sides ℓ_1

from (6.43), ℓ_2 from (6.52) are substituted by

$$\ell_1(\boldsymbol{q}) - \int_\Omega \boldsymbol{L}\overline{\boldsymbol{p}} \cdot \boldsymbol{q}\,dx, \quad \ell_2(v) + \int_\Omega \nabla \cdot \overline{\boldsymbol{p}}\, v\,dx.$$

Note that the discretization (7.130) is consistent in the following sense:
Let $u \in H^1(\Omega)$ be a weak solution of (7.96) fulfilling $\boldsymbol{p} := -\boldsymbol{K}\nabla u + \boldsymbol{c}u \in H(\mathrm{div};\Omega)$. Extend E_h to $L^2(\mathcal{F})$, then $(\boldsymbol{p}, u, u|_{\mathcal{F}})$ satisfies the first and third equations in (7.121). If $u \in H^2(\Omega)$, the second equation is also fulfilled. The third equation holds true by assumption, the second stems from the strong form of the differential equation in $L^2(\Omega)$, and the first equation follows on $K \in \mathcal{T}_h$ by testing $\boldsymbol{p}$ with $\boldsymbol{q} \in H(\mathrm{div};K)$ and integration by parts (see (6.51)).

The equation (7.130) can equivalently be written elementwise, i.e., for given $\lambda_h = m \in Y_h$ the first two equations on an element $K \in \mathcal{T}_h$ constitute a mixed finite element discretization on one element with boundary data either encoded in $\tilde{\ell}_1, \ell_2$ or $E_h m$ (as Dirichlet data on $\partial K \cap \Omega$). Assume that this problem is uniquely solvable in $M_h \times X_h$ (see Sections 6.2, 7.2 for corresponding conditions). Then also the first two equations in (7.130) have a unique solution for given $m \in Y_h$, which by linearity is a sum of solutions corresponding to m, i.e., the interior boundary data, and to $\tilde{\ell}_1, \ell_2$, i.e., the distributed and boundary data. We denote these solutions by

$$\boldsymbol{p}_{h,m} \in \widetilde{M}_h,\ u_{h,m} \in X_h \quad \text{and} \quad \boldsymbol{p}_{h,d} \in \widetilde{M}_h,\ u_{h,d} \in X_h,$$

respectively, i.e.,

$$\begin{aligned} \widetilde{A}_h \boldsymbol{p}_{h,m} + \widetilde{B}_h^T u_{h,m} + \widetilde{C}_h u_{h,m} &= -E_h m, \\ \widetilde{B}_h \boldsymbol{p}_{h,m} - D_h u_{h,m} &= 0, \end{aligned} \tag{7.131}$$

$$\begin{aligned} \widetilde{A}_h \boldsymbol{p}_{h,d} + \widetilde{B}_h^T u_{h,d} + \widetilde{C}_h u_{h,d} &= \tilde{\ell}_1, \\ \widetilde{B}_h \boldsymbol{p}_{h,d} - D_h u_{h,d} &= \ell_2. \end{aligned} \tag{7.132}$$

Note that also these equations can be solved elementwise and define a linear mapping from Y_h to $\widetilde{M}_h \times X_h$, i.e., a *lifting* (compare Subsection 7.4.2). The solution $(\boldsymbol{p}_{h,d}, u_{h,d})$ could be further subdivided into solutions to $\tilde{g}_1, \tilde{g}_2, \tilde{g}_3, f$, respectively, each of which being linear, i.e., (7.132) defines a linear mapping from Z to $\widetilde{M}_h \times X_h$, where Z is the data space containing $(\tilde{g}_1, \tilde{g}_2, \tilde{g}_3, f)$.

To simplify the notation, we will skip the index h, i.e., use $(\boldsymbol{p}_m, u_m)$ and $(\boldsymbol{p}_d, u_d)$, and set

$$\boldsymbol{p}_{m,d} := \boldsymbol{p}_m + \boldsymbol{p}_d, \quad u_{m,d} := u_m + u_d\,. \tag{7.133}$$

Lemma 7.48 *Let $\lambda_h, \mu_h \in Y_h$, $m = \lambda_h$ and denote by $\boldsymbol{p}_\mu, u_\mu$ the solution of (7.131) for the right-hand side μ_h, then*

1) $\langle \widetilde{A}_h \boldsymbol{p}_m, \boldsymbol{p}_\mu \rangle + \langle \widetilde{C}_h^T \boldsymbol{p}_m, u_\mu \rangle + \langle E_h^T \boldsymbol{p}_m, \mu_h \rangle + \langle D_h u_m, u_\mu \rangle = 0,$
2) $\langle \widetilde{A}_h \boldsymbol{p}_m, \boldsymbol{p}_d \rangle + \langle \widetilde{C}_h u_d, \boldsymbol{p}_m \rangle + \langle D_h u_m, u_d \rangle = \tilde{\ell}_1(\boldsymbol{p}_m),$
3) $\langle E_h m, \boldsymbol{p}_d \rangle + \langle \widetilde{C}_h u_m, \boldsymbol{p}_d \rangle - \langle \widetilde{C}_h u_d, \boldsymbol{p}_m \rangle = -\tilde{\ell}_1(\boldsymbol{p}_m) - \ell_2(u_m).$

Proof 1) Test the first equation of the problem (7.131) with right-hand side μ_h by $\boldsymbol{p}_m$ and the second equation of (7.131) by u_μ, and subtract the results.

2) Test the first equation in (7.132) by $\boldsymbol{p}_m$ and the second equation in (7.131) by u_d, and subtract the results.

3) Test the equations (7.131) by $(\boldsymbol{p}_d, u_d)$ and the equations (7.132) by $(\boldsymbol{p}_m, u_m)$, and combine the results appropriately. □

Theorem 7.49 *Let* $(\boldsymbol{p}_h, u_h, \lambda_h) \in \widetilde{M}_h \times X_h \times Y_h$ *be the solution of the hybridized* RT_{k-1}*-method (7.130). Then for* $m \in Y_h$ *the following statements are equivalent:*

(i) $E_h^T(\boldsymbol{p}_m + \boldsymbol{p}_d) = 0$,
(ii) $\boldsymbol{p}_h = \boldsymbol{p}_m + \boldsymbol{p}_d$,
(iii) $m = \lambda_h$,
(iv) $a_h(m, \mu) = b_h(\mu)$ *for all* $\mu \in Y_h$,

where

$$\begin{aligned} a_h(m, \mu) &:= \langle \widetilde{A}_h \boldsymbol{p}_m, \boldsymbol{p}_\mu \rangle + \langle \widetilde{C}_h^T \boldsymbol{p}_m, u_\mu \rangle + \langle D_h u_m, u_\mu \rangle, \\ b_h(\mu) &:= -\big(\tilde{\ell}_1(\boldsymbol{p}_\mu) + \ell_2(u_\mu) - \langle \widetilde{C}_h u_d, \boldsymbol{p}_\mu \rangle + \langle \widetilde{C}_h^T \boldsymbol{p}_d, u_\mu \rangle\big). \end{aligned} \tag{7.134}$$

Proof Problem 7.24. □

The form (7.134) indicates the coercive character of the hybridized, eliminated formulation and can (for $C_h = 0$) also be used as a basis for an error analysis. Since we already have the error analysis of Section 7.2, we can investigate the approximation power of λ_h on this basis.

Besides the L^2-norm on Y_h, we will use

$$\|\mu\|_{-1/2,h}^2 := \sum_{F \in \mathcal{F}} h_F \|\mu\|_{0,F}^2 \quad \text{for all } \mu \in Y_h. \tag{7.135}$$

In addition to the interpolation operation $I_K : H(\mathrm{div}; K) \to RT_{k-1}(K)$ and its global version I_h (see (7.57)) we will also use the L^2-orthogonal projections $\Pi_K : L^2(K) \to X_K$ and $\Pi_F : L^2(F) \to Y_F$ $(= \mathcal{P}_{k-1}(F))$ and their global versions Π and $\Pi_{\mathcal{F}}$.

Theorem 7.50 *Assume* $u \in H^1(\Omega)$, $\boldsymbol{p} \in H(\mathrm{div}; \Omega)$ *for the solution of* (7.96). *Let* $(\boldsymbol{p}_h, u_h, \lambda_h) \in \widetilde{M}_h \times X_h \times Y_h$ *be the solution of the hybridized* RT_{k-1}*-method for the Darcy problem. Then there exist constants* $C > 0$ *independent of* u_h *and* h *such that*

1) $\|\lambda_h - \Pi_F u\|_{0,F} \le C\big(h_K^{1/2}\|\boldsymbol{p} - \boldsymbol{p}_h\|_{0,K} + h_K^{-1/2}\|\Pi_K u - u_h\|_{0,K} + h_K^{1/2}\|u - u_k\|_{0,K}\big)$
for all $K \in \mathcal{T}_h$ *and* $F \in \mathcal{F}_K$.

2) If $(\mathcal{T}_h)_h$ *is shape-regular, then*

$$\|\lambda_h - \Pi_{\mathcal{F}} u\|_{-1/2,h} \le C\big(h\|\boldsymbol{p} - \boldsymbol{p}_h\|_0 + \|\Pi u - u_h\|_0 + h\|u - u_h\|_0\big).$$

Proof 1) Consider $K \in \mathcal{T}_h$ and $F \in \mathcal{F}_K$. According to the degrees of freedom (7.49), (7.50) there exists a unique element $\tilde{\boldsymbol{q}} \in RT_{k-1}(K)$ such that

$$
\begin{aligned}
\tilde{\boldsymbol{q}} \cdot \boldsymbol{n} &= \lambda_h - \Pi_F u && \text{on } F,\\
\tilde{\boldsymbol{q}} \cdot \boldsymbol{n} &= 0 && \text{on } \partial K \setminus F,\\
\tilde{\boldsymbol{q}} &\perp \mathcal{P}_{k-2}(K) && \text{in } L^2(K).
\end{aligned}
$$

From Lemma 6.14, first on the reference element $\widehat{K}$, then by scaling and an inverse estimate (see Lemma 7.5), we see that there exists a constant $C > 0$ such that

$$
h_K |\nabla \cdot \tilde{\boldsymbol{q}}|_{0,K} + \|\tilde{\boldsymbol{q}}\|_{0,K} \leq C h_K^{1/2} \|\lambda_h - \Pi_F u\|_{0,F}. \tag{7.136}
$$

We extend $\tilde{\boldsymbol{q}}$ by $\boldsymbol{0}$ on $K' \neq K$ and use it as a test function in the first equation of (7.130), leading to

$$
\int_K \boldsymbol{L}\boldsymbol{p}_h \tilde{\boldsymbol{q}} dx - \int_K u_h \nabla \cdot \tilde{\boldsymbol{q}} dx + \int_K \tilde{\boldsymbol{c}} \cdot \tilde{\boldsymbol{q}} u_h dx + \int_F \lambda_h (\lambda_h - \Pi_F u) d\sigma = \tilde{\ell}_1(\tilde{\boldsymbol{q}}).
$$

By consistency we have the same equation for the solution components $\boldsymbol{p}, u$, and

$$
\int_F u(\lambda_h - \Pi_F u) d\sigma = \int_F \Pi_F u (\lambda_h - \Pi_F u) d\sigma.
$$

Hence

$$
\|\lambda_h - \Pi_F u\|_{0,F}^2 = -\int_K \boldsymbol{L}(\boldsymbol{p}_h - \boldsymbol{p}) \tilde{\boldsymbol{q}} dx - \int_K \tilde{\boldsymbol{c}} \cdot \tilde{\boldsymbol{q}} (u_h - u) dx + \int_K (u_h - \Pi_K u) \nabla \cdot \tilde{\boldsymbol{q}} dx,
$$

i.e., the assertion follows by the Cauchy–Schwarz inequality and (7.136).
2) follows immediately from 1) together with Lemma 7.3, 1). □

Corollary 7.51 *Under the assumptions of Theorems 7.15, 7.50 and assuming $(\mathcal{T}_h)_h$ to be shape-regular, in the hybridized RT_{k-1}-method for the Darcy problem the Lagrange multiplier λ_h satisfies the estimate*

$$
\|\lambda_h - \Pi_{\mathcal{F}} u\|_{-1/2,h} \leq C h^{k+1} \left(|\boldsymbol{p}|_{k,\mathcal{T}_h} + |\nabla \cdot \boldsymbol{p}|_{k,\mathcal{T}_h} + |u|_{k,\mathcal{T}_h} \right)
$$

for some constant $C > 0$.

Proof Follows directly from Theorem 7.50, Theorem 7.14, and Theorem 7.15. □

The next step is the postprocessing of u_h, based on u_h and λ_h to render an L^2-approximation of order $k+1$. We restrict ourselves to the case $d = 2$ and odd $k \in \mathbb{N}$ and define an element $\tilde{u}_h \in \mathcal{P}_k(\mathcal{T}_h)$ by

$$
\begin{aligned}
\Pi_{\mathcal{F}}(\tilde{u}_h) &= \lambda_h,\\
\Pi^{k-3}(\tilde{u}_h - u_h) &= 0 \quad (\text{for } k \geq 3),
\end{aligned} \tag{7.137}
$$

where Π^{k-3} is the L^2-projection onto X_h for $k-3$ instead of $k-1$. By Remark 7.2, 4) $\tilde{u}_h$ exists uniquely.

Theorem 7.52 *Under the assumptions of Corollary 7.51, let $k \in \mathbb{N}$ and define $\tilde{u}_h \in \mathcal{P}_k(\mathcal{T}_h)$ as in (7.137). Then*

$$\|u - \tilde{u}_h\|_0 \leq Ch^{k+1}\left(|\boldsymbol{p}|_{k,\mathcal{T}_h} + |\nabla \cdot \boldsymbol{p}|_{k,\mathcal{T}_h} + \|u\|_{k+1,\mathcal{T}_h}\right)$$

for some constant $C > 0$.

Proof We compare $\tilde{u}_h$ with $\tilde{u}_h^* := I_h(u)$, the interpolant of u in $CR_k(\Omega)$ according to Remark 7.2, 4), then

$$\begin{aligned} \Pi_{\mathcal{F}}(\tilde{u}_h - \tilde{u}_h^*) &= \lambda_h - \Pi_{\mathcal{F}}\tilde{u}_h^* = \lambda_h - \Pi_{\mathcal{F}}u, \\ \Pi^{k-3}(\tilde{u}_h - \tilde{u}_h^*) &= \Pi^{k-3}(u_h - u) = \Pi^{k-3}(u_h - \Pi u). \end{aligned}$$

Applying (7.17) gives

$$\|\tilde{u}_h - \tilde{u}_h^*\|_{0,K} \leq C\Big(\|u_h - \Pi_K u\|_{0,K} + h_K^{1/2} \sum_{F \in \partial K} \|\lambda_h - \Pi_F u\|_{0,F}\Big).$$

Theorem 7.50 and Theorem 7.15 imply

$$\|\tilde{u}_h - \tilde{u}_h^*\|_0 \leq Ch^{k+1}\left(|\boldsymbol{p}|_k + |\nabla \cdot \boldsymbol{p}|_k + |u|_k\right),$$

and the combination with (7.16) concludes the proof. □

We continue only for the case $d = 2$ and k odd. If the element $\tilde{u}_h$ should not only belong to $\mathcal{P}_k(\mathcal{T}_h)$, but also have inter-element continuity properties as in $CR_k(\Omega)$, the missing degrees of freedom have to be compensated. This can be done by a space of bubble functions. Set

$$\begin{aligned} Z^k(\mathcal{T}_h) &:= CR_k(\Omega) \otimes B^{k+2}(\mathcal{T}_h), \\ B^{k+2}(\mathcal{T}_h) &:= \{v \in \mathcal{P}_{k+2}(\mathcal{T}_h) \mid v|_F = 0 \quad \text{for all } F \in \mathcal{F}\}, \end{aligned}$$

then, for every $u \in \mathcal{P}_{k-1}(\mathcal{T}_h)$, $\mu \in \mathcal{P}_{k-1}(\mathcal{F})$, there exists a unique element $\psi_h \in Z^k(\mathcal{T}_h)$ such that

$$\Pi\psi_h = u, \; \Pi_{\mathcal{F}}\psi_h = \mu. \tag{7.138}$$

To simplify, we assume that all coefficients $\boldsymbol{K}, \boldsymbol{c}, r$ are defined as constants on each element of $\mathcal{T}_h$ and $\boldsymbol{K} := \tilde{k}\boldsymbol{I}$ with a scalar function $\tilde{k}$. This means that they do not impede the (local) L^2-projection properties.

Let $(\boldsymbol{p}_h, u_h, \lambda_h)$ be the solution of the hybridized RT_{k-1}-method, ψ_h the postprocessed element from u_h, λ_h according to (7.138). Then $(\boldsymbol{p}_h, \psi_h)$ solves

$$\langle \widetilde{A}_h \boldsymbol{p}_h + \widetilde{C}_h \boldsymbol{\psi}_h, \boldsymbol{q}_h \rangle + \sum_{K \in \mathcal{T}_h} \int_K \nabla\psi_h \cdot \boldsymbol{q}_h dx = \tilde{\ell}_1(\boldsymbol{q}_h) \quad \text{for all } \boldsymbol{q}_h \in RT_{k-1}(\mathcal{T}_h), \tag{7.139}$$

as integration by parts does not induce boundary terms and, because of $\Pi\nabla \cdot \boldsymbol{p}_h = \nabla \cdot I_h \boldsymbol{p}_h$,

$$\ell_2(\Pi v_h) = -\sum_{K\in\mathcal{T}_h}\int_K \nabla v_h \cdot \boldsymbol{p}_h dx + \int_\Omega r\psi_h v_h dx \quad \text{for all } v_h \in Z^{k-1}(\mathcal{T}_h). \tag{7.140}$$

Let Π_{RT}^{k-1} be the L^2-orthogonal projection on $RT_{k-1}(\mathcal{T}_h)$. Then, assuming $\tilde{\ell}_1 = 0$, i.e., homogeneous boundary conditions,

$$\boldsymbol{L}\boldsymbol{p}_h - \tilde{\boldsymbol{c}}\psi_h = -\Pi_{RT}^{k-1}(\nabla\psi_h).$$

Therefore from (7.140) we get

$$\begin{aligned}\sum_{K\in\mathcal{T}_h}\int_K \tilde{k}\Pi_{RT}^{k-1}(\nabla\psi_h)\cdot\nabla v_h dx - \int_\Omega \boldsymbol{c}\psi_h\cdot\nabla v_h dx + \int_\Omega r\psi_h v_h dx \\ = \ell_2(\Pi v_h) \quad \text{for all } v_h \in Z^{k-1}(\mathcal{T}_h),\end{aligned} \tag{7.141}$$

and for homogeneous boundary conditions the right-hand side reduces to

$$\int_\Omega \Pi f v_h dx.$$

The problem can be further separated into problems in $CR_{k-1}(\Omega)$ and $B^{k+1}(\Omega)$ (Problem 7.25). Thus we could transform to a nonconforming formulation solely in a scalar variable.

7.5.3 Hybrid Discontinuous Galerkin Methods

The procedure of hybridization can also be applied to discontinuous Galerkin methods with their enormous degrees of freedom, but we cannot proceed according to Lemma 7.43 to achieve an equivalent localized version that is suitable for static condensation, since in the DG method there are no inter-element restrictions to be relaxed. Nevertheless we can mimic the procedure by introducing numerical traces and flux densities, being only approximations but allow a static condensation to a problem in those traces only, which can be viewed as a method of its own. We consider a hybrid mixed dual variant of the LDG method (denoted LDG-H method). Hybridized versions of IP methods could also be considered, but we will discover later that hybridized LDG methods have an advantage over hybridized IP methods which does not hold for their nonhybrid versions.

We generalize the definition of 7.122 (similar to the broken polynomial space $\mathbb{P}_k(\mathcal{T}_h)$ from (7.99)) for functions which live on the union of the grid faces (excluding the Dirichlet faces), denoted by the *skeleton* $\mathcal{S} := \bigcup_{F\in\overline{\mathcal{F}}\setminus\mathcal{F}_3} F$, i.e.,

$$\mathbb{P}_k(\overline{\mathcal{F}}\setminus\mathcal{F}_3) := \{\lambda_h \in L^2(\mathcal{S}) \mid \lambda_h|_F \in \mathbb{P}_k(F) \quad \text{for all } F \in \overline{\mathcal{F}}\setminus\mathcal{F}_3\}.$$

The aim is to modify the LDG method by introducing a *numerical trace* $\lambda_h = \widehat{u}_h$ on $\mathcal{S}$ in $\mathbb{P}_k(\overline{\mathcal{F}} \setminus \mathcal{F}_3)$ and a *numerical flux density* $\widehat{\boldsymbol{p}}_h$ such that

1) The original variable $u_h \in \mathcal{P}_k(\mathcal{T}_h)$, i.e., the solution of the LDG-H method, can be constructed from λ_h with minor effort.
2) There is a system of linear equations for λ_h alone with a sparsity structure similar to other FEM.
3) The method is locally mass conservative in terms of u_h and $\widehat{\boldsymbol{p}}_h$.

Concerning the first requirement, we recognize that it would be advantageous, if we constructed u_h and $\boldsymbol{p}_h$ using (7.110) on an element-by-element basis, cf. (7.133). Introducing the abbreviations

$$\widehat{u}_h := \{\!\!\{ \boldsymbol{u}_h \}\!\!\}, \qquad \widehat{\boldsymbol{p}}_h := \{\!\!\{ \boldsymbol{p}_h \}\!\!\} + \frac{\eta}{h_F} [\![u_h]\!] + \boldsymbol{c}_{\mathrm{upw}}(u_h),$$

we can rewrite (7.110) on each element $K \in \mathcal{T}_h$ as

$$\begin{aligned}
&\int_K \boldsymbol{K}^{-1} \boldsymbol{p}_h \cdot \boldsymbol{q}_h dx - \int_K u_h (\nabla \cdot \boldsymbol{q}_h) dx \\
&\qquad + \sum_{F \in \overline{\mathcal{F}}_K} \int_F \widehat{u}_h \boldsymbol{q}_h \cdot \boldsymbol{n}_K d\sigma = 0 \qquad \text{for all } \boldsymbol{q}_h \in \mathbb{P}_k^d(\mathcal{T}_h), \\
&-\int_K \boldsymbol{p}_h \cdot \nabla v_h dx + \int_K [-\boldsymbol{c} u_h \cdot \nabla v_h + r u_h v_h] \, dx \\
&\qquad + \sum_{F \in \overline{\mathcal{F}}_K} \int_F \widehat{\boldsymbol{p}}_h \cdot \boldsymbol{n}_K v_h d\sigma = \int_K f v_h dx \qquad \text{for all } v_h \in \mathbb{P}_k(\mathcal{T}_h).
\end{aligned} \tag{7.142}$$

For the modification for boundary faces, see Problem 7.26.

To use (7.142) to determine $\boldsymbol{p}_h$ and u_h element-locally from $\widehat{\boldsymbol{p}}_h$ and $\widehat{u}_h$, these latter quantities must be element-local, what is not the case in their preliminary definition. Therefore we have to consider these as new element-locally defined unknowns. This is done by choosing $\lambda_h \in \mathbb{P}_k(\overline{\mathcal{F}} \setminus \mathcal{F}_3)$ and setting, for the numerical trace,

$$\widehat{u}_h := \begin{cases} \lambda_h & \text{if } F \in \overline{\mathcal{F}} \setminus \mathcal{F}_3, \\ \tilde{g}_3 & \text{if } F \in \mathcal{F}_3. \end{cases} \tag{7.143}$$

We have to proceed in a similar way with the numerical flux density. We get an element-local definition if we write out the three contributions and substitute $\{\!\!\{ \boldsymbol{p}_h \}\!\!\}$ by $\boldsymbol{p}_h$, and $u_h|_{K'}$ for K' with $K \cap K' = F$ in the jump and in the upwind term by $\widehat{u}_h|_F$, i.e., we set

$$\widehat{\boldsymbol{p}}_h := \begin{cases} \boldsymbol{p}_h + \tau(u_h - \widehat{u}_h)\boldsymbol{n}_K + \widehat{u}_h \boldsymbol{c} & \text{if } \boldsymbol{c} \cdot \boldsymbol{n}_K < 0, \\ \boldsymbol{p}_h + \tau(u_h - \widehat{u}_h)\boldsymbol{n}_K + u_h \boldsymbol{c} & \text{if } \boldsymbol{c} \cdot \boldsymbol{n}_K \geq 0. \end{cases} \tag{7.144}$$

Note that the penalization parameter τ might depend on all kinds of grid information and is the equivalent to $\frac{\eta}{h_F}$.

As the numerical trace and the flux density always refer to a single element $K \in \mathcal{T}_h$ only, the jump $[\![\widehat{\boldsymbol{p}}_h]\!]$ and the average $\{\!\!\{\widehat{\boldsymbol{p}}_h\}\!\!\}$ of the numerical flux densities at an interface $F = K \cap K'$ have the forms

$$
\begin{aligned}
[\![\widehat{\boldsymbol{p}}_h]\!] &= [\![\boldsymbol{p}_h]\!] + 2\tau(\{\!\!\{u_h\}\!\!\} - \widehat{u}_h) + |\boldsymbol{c} \cdot \boldsymbol{n}^{\mathrm{d}}|(u_h^{\mathrm{d}} - \widehat{u}_h), \\
\{\!\!\{\widehat{\boldsymbol{p}}_h\}\!\!\} &= \{\!\!\{\boldsymbol{p}_h\}\!\!\} + \frac{\tau}{2}[\![u_h]\!] + \frac{1}{2}(u_h^{\mathrm{d}} + \widehat{u}_h)\boldsymbol{c},
\end{aligned}
\tag{7.145}
$$

where u_h^{d} refers to the value of u_h belonging to the element with $\boldsymbol{c} \cdot \boldsymbol{n}^{\mathrm{d}} \geq 0$. Analogously, we define u_h^{u} as the value of u_h belonging to the element with $\boldsymbol{c} \cdot \boldsymbol{n}^{\mathrm{u}} < 0$.

Note that $\boldsymbol{p}_h$ in the LDG method denotes an approximation of diffusive flux density, but $\widehat{\boldsymbol{p}}_h$ is an approximation of the total flux density, modified by an upwind stabilization and a jump stabilization term. Note that only $\widehat{\boldsymbol{p}}_h \cdot \boldsymbol{n}_K|_F$ appears in (7.142). Thus requiring $\widehat{\boldsymbol{p}}_h \cdot \boldsymbol{n}_K|_F \in \mathbb{P}_k(F)$, the corresponding degrees of freedom can be fixed by the fulfilment of a third requirement in terms of

$$
\sum_{F \in \overline{\mathcal{F}} \setminus \mathcal{F}_3} \int_F [\![\widehat{\boldsymbol{p}}_h]\!] s_h \, d\sigma = \sum_{F \in \mathcal{F}_2} \int_F \tilde{\alpha} \widehat{u}_h s_h \, d\sigma - \sum_{i=1}^{2} \sum_{F \in \mathcal{F}_i} \int_F \tilde{g}_i s_h \, d\sigma \qquad \text{for all } s_h \in \mathbb{P}_k(\overline{\mathcal{F}} \setminus \mathcal{F}_3),
\tag{7.146}
$$

ensuring local mass conservation (since for s_h, supported at one interface only, the right-hand side vanishes) and completing the system of equations. In the aforementioned representation (7.142) of the LDG-H method, we did not distinguish between bilinear and linear forms and will not do so in the following. However, for implementing this method, splitting up the above expressions into bilinear and linear ones is mandatory (see Problem 7.26). Moreover, the LDG-H method is clearly separated into the local (7.142) and the global (7.146) equations.

So we can use static condensation to proceed similarly to the RT-method in (7.117). Note that $\widehat{\boldsymbol{p}}_h \cdot \boldsymbol{n}_K|_F$ is originally a multivalued function on F, whether F is considered as $F \in \mathcal{F}_K$ or as $F \in \mathcal{F}_{K'}$, where $F = K \cap K'$, but (7.146) renders the component of the solution of (7.142), (7.146) to be well defined on $\overline{\mathcal{F}} \setminus \mathcal{F}_3$. Thus, using an appropriate basis, the matrices become block-diagonal, and the left-hand side of the linear system of equations corresponding to (7.142), (7.146) has the form (in continuation of the notation of (7.124))

$$
\begin{pmatrix}
\widetilde{\mathcal{A}}_h & \widetilde{\mathcal{B}}_h^T & \boldsymbol{0} & \mathcal{E}_h \\
\widetilde{\mathcal{B}}_h & -(\mathcal{D}_h + \widetilde{\mathcal{C}}_h) & \mathcal{F}_h & \boldsymbol{0} \\
\mathcal{E}_h^T & \boldsymbol{0} & \boldsymbol{0} & \boldsymbol{0}
\end{pmatrix}
\begin{pmatrix}
\tilde{\boldsymbol{\xi}} \\ \tilde{\boldsymbol{\eta}} \\ \tilde{\boldsymbol{\chi}} \\ \boldsymbol{\lambda}
\end{pmatrix},
$$

where the notation is taken from (7.124), $\tilde{\boldsymbol{\chi}}$ are the parameters of $\widehat{\boldsymbol{p}}_h \cdot \boldsymbol{n}_K|_F$ (denoted in the following by $\widehat{\boldsymbol{p}}_h$ for short), and $\mathcal{F}_h$ corresponds to the last term of the left-hand side of the second equation of (7.142). Note that (7.144) defines a local linear relationship

$$
\widehat{\boldsymbol{p}}_h = f(\boldsymbol{p}_h, u_h, \lambda_h)
$$

or for the chosen basis

$$\tilde{\chi} = \mathcal{G}_1\tilde{\xi} + \mathcal{G}_2\tilde{\eta} + \mathcal{G}_3\lambda,$$

that allows the left-hand side to be rewritten as

$$\begin{pmatrix} \widetilde{\mathcal{A}}_h & \widetilde{\mathcal{B}}_h^T & \mathcal{E}_h \\ \widetilde{\mathcal{B}}_h + \mathcal{F}_h\mathcal{G}_1 & -(\mathcal{D}_h + \widetilde{C}_h) + \mathcal{F}_h\mathcal{G}_2 & \mathcal{F}_h\mathcal{G}_3 \\ \mathcal{E}_h^T & \mathbf{0} & \mathbf{0} \end{pmatrix}\begin{pmatrix} \tilde{\xi} \\ \tilde{\eta} \\ \lambda \end{pmatrix}.$$

By means of appropriate redefinitions, we obtain the form (7.124), which, due to the invertibility of $\mathcal{A}_h$, enables us to also eliminate $\tilde{\xi}$ in order to achieve the form (7.125). Note that even in the pure diffusion case $\widetilde{C}_h = 0$ there is no symmetry. To do the next elimination step, the block-diagonal matrix

$$\mathcal{A} := (\widetilde{\mathcal{B}}_h + \mathcal{F}_h\mathcal{G}_1)\mathcal{A}_h^{-1}\widetilde{\mathcal{B}}_h^T + \mathcal{D}_h + \widetilde{C}_h - \mathcal{F}_h\mathcal{G}_2$$

has to be invertible. Here $\widetilde{\mathcal{B}}_h\mathcal{A}_h^{-1}\widetilde{\mathcal{B}}_h^T$ is symmetric positive definite (see Lemma 7.45) and similarly $\mathcal{D}_h$ for $r \geq r_0 \geq 0$.

There is a unique solution of the LDG-H method. This can be done analogously to Lemma 7.40.

Lemma 7.53 *The LDG-H method* (7.142) *admits a unique solution.*

The convergence analysis of the LDG-H method is tedious and skipped for that reason. However, in the paper [127] the following result has been shown.

Theorem 7.54 *Let* $\boldsymbol{c} = \mathbf{0}$ *and* $r = 0$ *and consider a family of triangulations. If all trial and test spaces contain polynomials of degree at most* $k \geq 1$, *and the solution* $(u, \boldsymbol{p}) \in H^{k+1}(\Omega) \times H^{k+1}(\Omega)^d$, *then there is a constant* $C(u, \boldsymbol{p}) > 0$ *that might also depend on Sobolev seminorms of* u *and* $\boldsymbol{p}$ *such that*

$$\|u - u_h\|_0 + \|\boldsymbol{p} - \boldsymbol{p}_h\|_0 \leq C(u, \boldsymbol{p})h^{k+1}.$$

Additionally, for the elementwise constant local mean $\overline{u}_h$ *of* u_h *(*$\overline{u}$ *of* u*, respectively) we have*

$$\|\overline{u} - \overline{u}_h\|_0 \leq C(u, \boldsymbol{p})h^{k+2}.$$

This result paves the way to several postprocessing approaches enhancing the numerical solution.

Postprocessing of the total flux density

The first postprocessing approach aims at constructing an element-by-element procedure to obtain an optimally convergent approximation $\boldsymbol{p}_h^*$ of $\boldsymbol{p}$ belonging to $H(\mathrm{div};\Omega)$. Thus, we project $\boldsymbol{p}_h$ to the Raviart–Thomas space $RT_k(K)$ for $k \geq 0$ (see (7.56) and note that in Sections 7.2 and 7.5.2 we have dealt with RT_{k-1}). This is done by enforcing that

$$\int_F (\boldsymbol{p}_h^* - \widehat{\boldsymbol{p}}_h) \cdot \mathfrak{n} \mu d\sigma = 0 \quad \text{for all} \quad \mu \in \mathcal{P}_k(F) \quad \text{and all} \quad F \in \mathcal{F}_K,$$

$$\int_K (\boldsymbol{p}_h^* - \boldsymbol{p}_h)\, \boldsymbol{q} dx = 0 \quad \text{for all} \quad \boldsymbol{q} \in \mathcal{P}_{k-1}(K), \quad k \geq 1,$$

locally, i.e., for all $K \in \mathcal{T}_h$ (see Theorem 7.9). Due to the unique definition of $\widehat{\boldsymbol{p}}_h \cdot \mathfrak{n}|_F$ and the considerations in Section 7.2, $\boldsymbol{p}_h^*$ is an element of $H(\text{div}; \Omega)$ and converges with the optimal order $(k+1)$.

Postprocessing of the scalar variable

Next, we execute an additional postprocessing step to obtain an approximate u_h^* that converges with order $(k+2)$ to u. To do so, we define $u_h^* \in \mathcal{P}_{k+1}(K)$ via

$$\int_K (\boldsymbol{K}\nabla u_h^* - \boldsymbol{c} u_h^*) \nabla v dx + \int_{\partial K} \widehat{\boldsymbol{p}}_h \cdot \mathfrak{n} v d\sigma = \int_K f v dx \quad \text{for all } v \in \mathcal{P}_{k+1}(K),$$

$$\int_K u_h^* dx = \int_K u_h dx,$$

which indeed can be shown to converge with order $k+2$ under the assumptions of Theorem 7.54, cf. [127].

Exercises

Problem 7.23 Prove Lemma 7.48, 3) in detail.

Problem 7.24 Prove Theorem 7.49.

Problem 7.25 Work out the separation of the problem (7.141) into problems in $CR_{k-1}(\Omega)$ and $B^{k+1}(\Omega)$ in detail.

Problem 7.26 Give a precise definition of the hybrid LDG method in (7.142), (7.146) in terms of bilinear and linear forms and spaces.

7.6 Local Mass Conservation and Flux Reconstruction

7.6.1 Approximation of Boundary Fluxes

Up till now the quality of a discretization has been measured by (orders of) convergence, but also qualitative properties such as comparison and maximum principles have been discussed for FDM (see Section 1.4) and the classical FEM from Chapters 2 and 3 (in particular, Section 3.9). They are conforming in a $H^1(\Omega)$-based formulation and thus the approximation space satisfies

$$V_h \subset C(\overline{\Omega}).$$

Therefore we will shortly speak of *continuous Galerkin methods* (*CGM*). Further important qualitative properties are related to *(mass) conservation*.

We start with the formulation of the diffusion–advection–reaction equation in divergence form (see (7.96)), i.e., the weak formulation reads

Find $u \in H^1(\Omega)$ such that $u|_{\widetilde{\Gamma}_3} = \tilde{g}_3$ and

$$-\int_\Omega \boldsymbol{p} \cdot \nabla v dx + \int_\Omega \tilde{r} uv dx - \int_{\widetilde{\Gamma}_1} \tilde{g}_1 v d\sigma + \int_{\widetilde{\Gamma}_2} \left[\tilde{\alpha} u - \tilde{g}_2\right] v d\sigma = \int_\Omega f v dx \qquad \text{for all } v \in H^1_{\widetilde{\Gamma}_3}(\Omega), \tag{7.147}$$

where $\boldsymbol{p} = \boldsymbol{p}(u) := -\boldsymbol{K}\nabla u + \boldsymbol{c}u$, or, in short,

$$a(u, v) = \ell(v) \quad \text{for all } v \in H^1_{\widetilde{\Gamma}_3}(\Omega)$$

with appropriate forms (see (3.40)). In the case of no Dirichlet boundary conditions, i.e., $\widetilde{\Gamma}_3 = \emptyset$, the choice $v = 1$ in (7.147) leads to the *global (mass) conservation law*:

$$\int_{\widetilde{\Gamma}_1} \tilde{g}_1 d\sigma + \int_{\widetilde{\Gamma}_2} \left[\tilde{g}_2 - \tilde{\alpha} u\right] d\sigma + \int_\Omega [f - \tilde{r} u] dx = 0. \tag{7.148}$$

If the test function $v_h := 1$ is also allowed in the CGM, this shows the relationship (7.148) for the discrete solution u_h—exactly or approximately if quadrature rules are applied. In the presence of Dirichlet boundary conditions, i.e., of an open system, there is a boundary flux over $\widetilde{\Gamma}_3$, denoted by

$$\hat{g}_3 := -\boldsymbol{p} \cdot \mathfrak{n} = (\boldsymbol{K}\nabla u - \boldsymbol{c}u) \cdot \mathfrak{n} \quad \text{on } \widetilde{\Gamma}_3, \tag{7.149}$$

which is not prescribed but is part of the solution. For $u \in H^1(\Omega)$, fulfilling the regularity requirements

$$(7.18), (7.19), (7.20) \tag{7.150}$$

integration by parts renders

$$a(u, v) - \ell(v) = \int_{\widetilde{\Gamma}_3} \hat{g}_3 v d\sigma \quad \text{for all } v \in H^1(\Omega). \tag{7.151}$$

Let V_h be the approximation space of the $\mathcal{P}_k$-based CGM (see Section 3.3), and $\widetilde{V}_h$ the corresponding space without (homogeneous) Dirichlet conditions. For simplicity, we assume $\tilde{g}_3 = 0$ (otherwise we have to pursue as in Section 3.5.1). By $\widetilde{V}_{h,D}$ we denote the subspace, for whose elements all degrees of freedom that are not on $\widetilde{\Gamma}_3$

vanish, thus

$$\widetilde{V}_h = V_h \oplus V_{h,D} \,.$$

Let $\overline{v}_h \in \widetilde{V}_h$ be such that $\overline{v}_h = 1$ on $\widetilde{\Gamma}_3$, then the quantity

$$F(\overline{v}_h) := a(u_h, \overline{v}_h) - \ell(\overline{v}_h) \tag{7.152}$$

can be considered as an approximation of the *total flux (in the direction of the inward normal) across* $\widetilde{\Gamma}_3$ (cf. (7.151)), which is independent of the values of $\overline{v}_h$ on $\overline{\Omega} \setminus \widetilde{\Gamma}_3$. This consideration can be refined to any subset $\widetilde{\widetilde{\Gamma}}_3 \subset \widetilde{\Gamma}_3$ that is a union of boundary elements. Note that (7.152) can be evaluated with minor effort, without requiring the solution of a system of equations. Expecting F to be a functional of a flux approximation, we can hope for superconvergence (compare Lemma 6.11), and indeed the following result was shown in the paper [94] using duality arguments.

Theorem 7.55 *Let $u \in H^{k+1}(\mathcal{T}_h)$ for some $k \in \mathbb{N}$, fulfilling* (7.150). *Assume that the solution Φ of the dual problem for Dirichlet boundary conditions equals 1 and homogeneous boundary conditions otherwise belongs to $H^{k+1}(\mathcal{T}_h)$, and consider the CGM based on $\mathcal{P}_k$ with a shape-regular family of triangulations $(\mathcal{T}_h)_h$. Then there exists a constant $C > 0$ such that the following error estimate holds:*

$$\left| F(\overline{v}_h) + \int_{\widetilde{\Gamma}_3} \boldsymbol{p}(u) \cdot \boldsymbol{n} d\sigma \right| \le C h^{2k} \|u\|_{k+1,\mathcal{T}_h} \,.$$

Proof See Problem 7.29. □

To approximate the boundary flux over $\widetilde{\Gamma}_3$ (in $L^2(\widetilde{\Gamma}_3)$), we can follow the reasoning as in Lemma 7.43. Namely, we extend V_h to $\widetilde{V}_h$ and define

$$\begin{aligned} E_h : Y_h &\to \widetilde{V}_h' \qquad \text{by} \\ e_h(\mu_h, v_h) &= \int_{\widetilde{\Gamma}_3} \mu_h v_h d\sigma \end{aligned} \tag{7.153}$$

with

$$Y_h := \left\{ \mu_h \in C(\widetilde{\Gamma}_3) \;\middle|\; \mu_h|_F \in \mathcal{P}_k(\mathbb{R}^{d-1}) \quad \text{for all } F \in \widehat{\mathcal{F}}_h \right\},$$

where $\widehat{\mathcal{F}}_h$ is the $(d-1)$-dimensional partition of $\widetilde{\Gamma}_3$, induced by $\mathcal{T}_h$. Thus the CGM solution, i.e., the element $u_h \in V_h$ such that

$$a(u_h, v_h) = \ell(v_h) \quad \text{for } v_h \in V_h$$

is equivalent to the solution of the following problem (see Lemma 7.43):

Find $(u_h, \lambda_h) \in \widetilde{V}_h \times Y_h$ such that

$$\begin{aligned} \widetilde{A}_h u_h + E_h \lambda_h &= \tilde{\ell}, \\ E_h^T u_h &= 0, \end{aligned} \tag{7.154}$$

where $\widetilde{A}_h : \widetilde{V}_h \to \widetilde{V}_h'$ is an extension of A_h induced by a on $V_h \times V_h$, and $\tilde{\ell}$ is an extension of ℓ.

The approximation (7.154) is consistent in the sense that u and $\hat{g}_3$ (see (7.149)) for $u \in H^1(\mathcal{T}_h)$ such that (7.150) holds satisfy the corresponding relation. Hence we have the following result.

Theorem 7.56 *Let $u \in H^{k+1}(\mathcal{T}_h)$ fulfill (7.150) for some $k \in \mathbb{N}$. Consider the CGM based on $\mathcal{P}_k$ with a shape-regular family of triangulations $(\mathcal{T}_h)_h$. Then there exists a constant $C > 0$ such that the following error estimate holds:*

$$\left\|\Pi_{\widetilde{\Gamma}_3}(-\boldsymbol{p} \cdot \boldsymbol{n}) - \lambda_h\right\|_{0,\widetilde{\Gamma}_3} \le Ch^{k-1/2}|u|_{k,\mathcal{T}_h},$$

where $\Pi_{\widetilde{\Gamma}_3} : L^2(\widetilde{\Gamma}_3) \to Y_h$ is the L^2-orthogonal projection.

Proof We have from (7.154) and the mentioned consistency that

$$\widetilde{A}_h(u - u_h) + E_h(-\boldsymbol{p} \cdot \boldsymbol{n} - \lambda_h) = 0.$$

As a test function $v_h \in \widetilde{V}_h$ we choose

$$v_h := \begin{cases} \Pi_{\widetilde{\Gamma}_3}(-\boldsymbol{p} \cdot \boldsymbol{n}) - \lambda_h & \text{on } \widetilde{\Gamma}_3, \\ 0 & \text{on all } K \in \mathcal{T}_h,\ K \cap \widetilde{\Gamma}_3 = \emptyset, \end{cases} \tag{7.155}$$

i.e., in detail

$$a(u - u_h, v_h) + \int_{\widetilde{\Gamma}_3} \left[\Pi_{\widetilde{\Gamma}_3}(-\boldsymbol{p} \cdot \boldsymbol{n}) - \lambda_h\right]^2 d\sigma = 0.$$

Thus

$$\|\Pi_{\widetilde{\Gamma}_3}(-\boldsymbol{p} \cdot \boldsymbol{n}) - \lambda_h\|_{0,\widetilde{\Gamma}_3}^2 \le C\|u - u_h\|_1 \|v_h\|_1 \le Ch^k |u|_{k,\mathcal{T}_h} h^{-1/2} \|\Pi_{\widetilde{\Gamma}_3}(-\boldsymbol{p} \cdot \boldsymbol{n}) - \lambda_h\|_{0,\widetilde{\Gamma}_3},$$

where we have used the estimate

$$\|v_h\|_{1,\Omega} \le Ch^{-1/2}\|v_h\|_{0,\widetilde{\Gamma}_3}$$

which has been shown in [94, Proof of Thm. 2.2]. □

Remark 7.57 1) λ_h has as many degrees of freedom as there are on $\widetilde{\Gamma}_3$ and can be computed from (7.154) choosing a corresponding basis of Y_h as ansatz and their prolongation to $\overline{\Omega}$ as in (7.155) as test functions.

2) Substituting Y_h by

$$\widetilde{Y}_h := \left\{\mu_h \in L^2(\widetilde{\Gamma}_3) \;\middle|\; \mu_h|_F \in \mathcal{P}_k(\mathbb{R}^{d-1}) \quad \text{for all } F \in \widehat{\mathcal{F}}_h\right\}$$

could give an elementwise computation of λ_h, but of unclear approximation.

7.6.2 Local Mass Conservation and Flux Reconstruction

As already discussed, not only a good approximation of the scalar quantity u, but also of the corresponding (total) flux density $\boldsymbol{p}$ is of importance. In particular for coupled problems (see Section 0.4), where this flux drives other quantities of the problem, also its consistency in terms of *local (mass) conservation* is mandatory. Due to the fact that a discrete approximation has only a finite amount of degrees of freedom to be used for approximation and qualitative purposes, it is not to be expected that all finite element methods directly give approximations both of u and $\boldsymbol{p}$, and with desired continuity properties. But it may well be that a *postprocessing* is possible to add further properties. This has been the case for RT-based mixed methods, where hybridization leads to an approximation $\hat{u}$ of better order and continuity properties (see Section 7.5.2).

To formulate our aim generally, let u_h (and possibly $\boldsymbol{p}_h$) be approximations of u and $\boldsymbol{p}$ from an FEM based on a triangulation family $(\mathcal{T}_h)_h$. Under appropriate regularity requirements the method is supposed to be of order $k \in \mathbb{N}$ in the sense that

$$\|u - u_h\|_0 \leq Ch^k \quad \text{and} \quad \|\boldsymbol{p} - \boldsymbol{p}_h\|_0 \leq Ch^k .$$

Definition 7.58 Let $\widehat{\mathcal{T}}_h$ be a polygonal partition constructed from $\mathcal{T}_h$ of the same fineness, i.e., $|\widehat{K}| = O(h^d)$ for all $\widehat{K} \in \widehat{\mathcal{T}}_h$, with $\widehat{\mathcal{F}}$ denoting the union of interior faces of $\widehat{\mathcal{T}}_h$. We call $\hat{u}_h, \hat{\boldsymbol{p}}_h$ a *postprocessing with local mass conservation with respect to* $\widehat{\mathcal{T}}_h$, if

(1) $\|u - \hat{u}_h\|_0 \leq C_1 h^k$,

(2) $\|\boldsymbol{p} - \hat{\boldsymbol{p}}_h\|_0 \leq C_2 h^k$,

(3) $[\![\hat{\boldsymbol{p}}_h]\!]_F = 0 \quad$ for all $F \in \widehat{\mathcal{F}}$,

(4) $$\left| \int_{\partial \widehat{K}} \hat{\boldsymbol{p}}_h \cdot \mathbf{n} d\sigma + \int_{\widehat{K}} \tilde{r} \hat{u} dx - \int_{\widehat{K}} f dx \right| \leq C_3 h^k \quad \text{for } \widehat{K} \in \widehat{\mathcal{T}}_h, \tag{7.156}$$

(5) The construction of $\hat{u}_h, \hat{\boldsymbol{p}}_h$ from u_h (and eventually $\boldsymbol{p}_h$) shall require minor effort.

Remark 7.59 1) The typical situation is $\widehat{\mathcal{T}}_h = \mathcal{T}_h$. The more general situation includes the node-oriented finite volume methods of Chapter 8.

2) The relaxation in (4) compared to

$$\int_{\partial \widehat{K}} \hat{\boldsymbol{p}}_h \cdot \mathbf{n} d\sigma + \int_{\widehat{K}} \tilde{r} \hat{u}_h dx = \int_{\widehat{K}} f dx \tag{7.157}$$

is to include the effect of quadrature, penalty terms, etc; the hard qualitative requirement is (3).

3) For a solution fulfilling (7.18), (7.20), the relation (7.157) is equivalent to

$$\int_{\partial \widehat{K}} \hat{\boldsymbol{p}}_h \cdot \mathbf{n} d\sigma = \int_{\partial \widehat{K}} \boldsymbol{p} \cdot \mathbf{n} d\sigma + \int_{\widehat{K}} \tilde{r}(u - \hat{u}_h) dx.$$

4) If $\hat{\boldsymbol{p}}_h|_{\widehat{K}}$ is smooth for $\widehat{K} \in \widetilde{\mathcal{T}}_h$, then (3) is equivalent to $\hat{\boldsymbol{p}}_h \in H(\text{div};\Omega)$ and $\int_{\partial \widehat{K}} \hat{\boldsymbol{p}}_h \cdot \mathbf{n} d\sigma = \int_{\widehat{K}} \nabla \cdot \hat{\boldsymbol{p}}_h dx$.
5) The informal notion of (5) means that either local elementwise systems of equations or a system of equations with condition number bounded in h have to be solved so that the total effort is at most $O(N)$, N being the number of degrees of freedom.
6) In the presence of sources $(f, r, \tilde{g}_1, \tilde{g}_2, \alpha)$ it may be better to speak of *local mass balancing*.

A method which satisfies Definition 7.58 without restrictions—and that was the motivation for its development in the dual mixed formulation—is the Darcy equation and its conforming consistent finite element discretization by RT-elements, for instance. We have for $k \in \mathbb{N}$ (see Section 7.2.2)

$$\hat{u}_h = u_h \in \mathcal{P}_{k-1}(\mathcal{T}_h), \ \hat{\boldsymbol{p}}_h = \boldsymbol{p}_h \in RT_{k-1}(\Omega) \subset H(\text{div};\Omega).$$

(1) and (2) hold true ((2) even in $\|\cdot\|_{H(\text{div})}$), (3) by definition, and (4) exactly ($C_3 = 0$), and additionally

$$K^{-1} \hat{\boldsymbol{p}}_h = -\nabla \hat{u}_h + \tilde{\boldsymbol{c}} \hat{u} \quad \text{in } H(\text{div};\Omega)'.$$

The *DG method* from (7.100), (7.101) has no intrinsic flux definition as above, similar to the CGM, but it allows local test functions χ_K (the indicator function of K) leading to (for $K \in \mathcal{T}_h$, $K \subset \Omega$)

$$\int_{\partial K} \left[\left(-\{\!\!\{ \boldsymbol{K} \nabla u_h \}\!\!\} + \boldsymbol{c}_{\text{upw}}(u_h) \right) \cdot \mathbf{n} + \frac{\eta}{h_F} [\![u_h]\!] \cdot \mathbf{n} \right] d\sigma + \int_K r u_h dx = \int_K f dx \,. \quad (7.158)$$

Thus the term in the surface integral can be defined as $\hat{\boldsymbol{p}}_h$ and (4) is satisfied exactly. If (by approximation) the integrand is in $\mathcal{P}_{k-1}(\mathcal{F})$, the prolongation to $RT_{k-1}(\Omega)$ is possible (see Theorem 7.9), fulfilling (3). The approximation statement (1) is satisfied for $k + 1$ (see Theorem 7.35), and the validity of (2) depends on the order of convergence of $\|\boldsymbol{p} - \hat{\boldsymbol{p}}_h\|_{0,\mathcal{F}}$, which is unclear due to the penalty term.

For the LDG method according to (7.110) the same reasoning holds true, now with the following definition on $\mathcal{F}$:

$$\hat{\boldsymbol{p}}_h := -\{\!\!\{ \tilde{\boldsymbol{p}}_h \}\!\!\} + \boldsymbol{c}_{\text{upw}}(u_h) + \frac{\eta}{h_F} [\![u_h]\!],$$

where $\tilde{\boldsymbol{p}}_h$ approximates the diffusive flux.

For the LDG-H method $\hat{\boldsymbol{p}}_h$ is modified according to (7.145) (substitute there $\boldsymbol{p}_h$ by $\tilde{\boldsymbol{p}}_h$).

Now let us turn to the classical CGM, i.e.,

$$V_h = \left\{ v_h \in C(\overline{\Omega}) \ \middle| \ v|_K \in \mathcal{P}_k(K), v|_{\widetilde{\Gamma}_3} = 0 \right\},$$

where for simplicity we only consider $\tilde{g}_3 = 0$.

A first definition on $\mathcal{F}$ could be

$$\overline{\boldsymbol{p}}_h := -\{\!\!\{\boldsymbol{K}\nabla u_h\}\!\!\} + \boldsymbol{c}_{\mathrm{upw}}(u_h), \tag{7.159}$$

and on Ω

$$\boldsymbol{p}_h := -\boldsymbol{K}\nabla u_h + \boldsymbol{c}u_h,$$

where the upwinding allows for coefficients $\boldsymbol{c}$ that are only smooth on each element. Then (1) to (3) can be fulfilled, with $\overline{\boldsymbol{p}}_h \in RT_l(\Omega)$ being an appropriate prolongation. The validity of (4) is unclear, as localized test functions are not allowed. We therefore apply an equivalent hybridization as in Lemma 7.43 and (7.153) by enlarging V_h to

$$\widetilde{V}_h := \mathcal{P}_k(\mathcal{T}_h).$$

Then a has to be prolongated to $\tilde{a}_h$, i.e., $\widetilde{A}_h : \widetilde{V}_h \to \widetilde{V}_h'$ on the operator level, and ℓ to $\tilde{\ell}$. Let Y_h be a finite-dimensional space and e_h a bilinear term with corresponding linear operator

$$E_h : Y_h \to (\widetilde{V}_h)'$$

such that

$$\ker E_h^T = V_h.$$

Then, according to Lemma 7.43, the variational problem is equivalent to

Find $(\tilde{u}_h, \lambda_h) \in \widetilde{V}_h \times Y_h$ such that

$$\begin{aligned} \tilde{a}_h(\tilde{u}_h, \tilde{v}_h) + e_h(\lambda_h, \tilde{v}_h) &= \tilde{\ell}_h(\tilde{v}_h) \qquad \text{for all } \tilde{v}_h \in \widetilde{V}_h, \\ e_h(\mu_h, \tilde{u}_h) &= 0 \qquad \text{for all } \tilde{\mu}_h \in Y_h, \end{aligned} \tag{7.160}$$

meaning for the solution of (7.160) $\tilde{u}_h \in V_h$ and $\tilde{u}_h = u_h$, u_h being the CGM approximation. Note that neither for $\tilde{a}_h$ nor for Y_h there is a unique choice.

A first attempt could be

$$\begin{aligned} Y_h &:= \left\{\mu \in L^2(\mathcal{F} \cup \mathcal{F}_3) \;\middle|\; \mu|_F \in \mathcal{P}_k(F) \quad \text{for all } F \in \mathcal{F} \cup \mathcal{F}_3\right\} =: \widehat{Y}_h, \\ e_h(\mu_h, \tilde{u}_h) &:= \sum_{F \in \mathcal{F} \cup \mathcal{F}_3} \int_F [\![\tilde{u}_h]\!] \mu_h \, d\sigma, \end{aligned} \tag{7.161}$$

and $\tilde{a}_h$ as given by a, with $\int_\Omega \boldsymbol{K}\nabla u \nabla v dx$ substituted by $\sum_{K \in \mathcal{T}_h} \int_K \boldsymbol{K}\nabla u \nabla v dx$.

As testing with χ_K, $K \in \mathcal{T}_h$, is allowed, (4) of Definition 7.58 is satisfied exactly:

$$-\int_{\widetilde{\Gamma}_1 \cap \partial K} \tilde{g}_1 d\sigma + \int_{\widetilde{\Gamma}_2 \cap \partial K} \left[\tilde{\alpha}\tilde{u}_h - \tilde{g}_2\right] d\sigma + \int_{\partial K \setminus (\widetilde{\Gamma}_1 \cup \widetilde{\Gamma}_2)} \lambda_h d\sigma = \int_K f dx \qquad \text{for all } K \in \mathcal{T}_h, \tag{7.162}$$

i.e., λ_h together with the boundary data on $\widetilde{\Gamma}_1 \cup \widetilde{\Gamma}_2$ seem to be an approximation of $\boldsymbol{p} \cdot \boldsymbol{n}$ on $\overline{\mathcal{F}}$. Due to its local facewise definition it is easy to compute from the first equation of (7.160), given $\tilde{u}_h$. Unfortunately its approximation properties are unclear.

Therefore we modify the approach. We choose $\tilde{a}_h$, $\tilde{\ell}_h$ according to (7.101), but without the penalty term, which is a prolongation of a to $\widetilde{V}_h \times \widetilde{V}_h$ (see (7.166) for more details).

Then (7.162) is modified by adding at the left-hand side the following terms:

$$\int_{\partial K \setminus \widetilde{\Gamma}_1 \cup \widetilde{\Gamma}_2} \big(- \{|\boldsymbol{K} \nabla \tilde{u}_h|\} + \boldsymbol{c}_{\mathrm{upw}}(\tilde{u}_h) \big) \cdot \boldsymbol{n} d\sigma$$

(note that we only consider $\tilde{g}_3 = 0$).

Thus λ_h is now a correction to (7.159) to induce local mass conservation. Still the discretization is consistent for $u \in H^1(\Omega)$ satisfying (7.150) (see Lemma 7.25), and therefore we get the error equation for the flux correction:

$$\tilde{a}_h(u - \tilde{u}_h, \tilde{v}_h) - e_h(\lambda_h, \tilde{v}_h) = 0 \quad \text{for all } \tilde{v}_h \in \widetilde{V}_h. \tag{7.163}$$

For given norms $\|\cdot\|_{\widetilde{V}_h}$ on $\widetilde{V}_h$, $\|\cdot\|^*_{\widetilde{V}_h}$ on $\widetilde{V}_h + \operatorname{span}\{u\}$, and $\|\cdot\|_{Y_h}$ on Y_h, we assume that the bilinear form $\tilde{a}_h$ is continuous, i.e.,

$$|\tilde{a}_h(u, v_h)| \le C \|u\|^*_{\widetilde{V}_h} \|v_h\|_{\widetilde{V}_h} \quad \text{for all } u \in \widetilde{V}_h + \operatorname{span}\{u\},\ v_h \in \widetilde{V}_h$$

with a constant $C > 0$, and the bilinear form e_h satisfies an inf-sup condition, i.e., there is a constant $\alpha > 0$ such that

$$\alpha := \inf_{\mu_h \in Y_h \setminus \{0\}} \sup_{\tilde{v}_h \in \widetilde{V}_h \setminus \{0\}} \frac{e_h(\mu_h, \tilde{v}_h)}{\|\mu_h\|_{Y_h} \|\tilde{v}_h\|_{\widetilde{V}_h}} > 0. \tag{7.164}$$

Then we obtain from (7.163) that

$$\begin{aligned}
\|\lambda_h\|_{Y_h} &\le \frac{1}{\alpha} \sup_{\tilde{v}_h \in \widetilde{V}_h \setminus \{0\}} \frac{e(\lambda_h, \tilde{v}_h)}{\|\tilde{v}_h\|_{Y_h}} \\
&= \frac{1}{\alpha} \sup_{\tilde{v}_h \in \widetilde{V}_h \setminus \{0\}} \frac{a_h(u - \tilde{u}_h, \tilde{v}_h)}{\|\tilde{v}_h\|_{\widetilde{V}_h}} \le C \|u - \tilde{u}_h\|^*_{\widetilde{V}_h},
\end{aligned}$$

i.e., for

$$\begin{aligned}
\hat{\boldsymbol{p}}_h &:= \big(- \{|\boldsymbol{K} \nabla \tilde{u}_h|\} + \boldsymbol{c}_{\mathrm{upw}}(\tilde{u}_h) \big) \cdot \boldsymbol{n} - \lambda_h \\
&= \overline{\boldsymbol{p}}_h - \lambda_h
\end{aligned}$$

we get the estimate

$$\|\boldsymbol{p} - \hat{\boldsymbol{p}}_h\|_{Y_h} \le C \big(\|u - u_h\|_{\widetilde{V}_h} + \|\boldsymbol{p} - \overline{\boldsymbol{p}}_h\|_{Y_h} \big).$$

To ensure (7.164), we restrict $\widehat{Y}_h$ to

$$Y_h := \Big\{\mu \in \widehat{Y}_h \ \Big| \ \sum_{F \in \mathcal{F}_a} s_{a,F}\, |F|\, \mu(a) = 0 \quad \text{for all } a \in \mathcal{N} \cup \partial\mathcal{N}_3\Big\},$$

where $\mathcal{N} \cup \partial\mathcal{N}_3$ is the set of nodes interior to Ω or on $\widetilde{\Gamma}_3$, $\mathcal{F}_a$ is the set of faces sharing the node a, and

$$s_{a,F} := \begin{cases} 1, & \boldsymbol{n}_F \text{ is in clockwise rotation around } a, \\ -1, & \text{otherwise.} \end{cases}$$

Then, for $v_h \in V_h$, we have

$$\sum_{F \in \mathcal{F}_a} s_{a,F}\, [\![v_h]\!](a) = 0$$

and therefore (see Problem 7.30)

$$\ker E_h^T = Y_h \tag{7.165}$$

remains true.

As norms we choose

$$\|v\|_{\widetilde{V}_h} := \|v\|_{V_h} \tag{7.166}$$

according to (7.102), without the penalty term

$$\sum_{F \in \mathcal{F} \cup \mathcal{F}_3} \frac{\eta}{h_F} \|[v_h]\|_{0,F}^2,$$

and

$$\|v\|_{\widetilde{V}_h}^* := \|v\|_{\widetilde{V}_h}^2 + \sum_{F \in \mathcal{F} \cup \mathcal{F}_3} h_F \|\{|\boldsymbol{K}\nabla v|\}\|_{0,F}^2,$$

and the prolongation $\tilde{a}$ and $\tilde{\ell}$, according to the DG-form from (7.101), but omitting the penalty term

$$\sum_{F \in \mathcal{F} \cup \mathcal{F}_3} \int_F \frac{\eta}{h_F} [\![u_h]\!][\![v_h]\!] d\sigma.$$

Then the continuity of $\tilde{a}$ is assured by an appropriate modification of the proof of Lemma 7.32.

The restriction of Y_h ensures the inf-sup condition (at least for $d = 2$).

Lemma 7.60 *Let $d = 2$, then* (7.164) *holds true.*

Proof See [100, Lemma 4.3]. □

Despite the additional continuity condition, the Lagrange multiplier λ_h can be computed in a local way. The idea is to write it as

$$\lambda_h = \sum_{\omega} \lambda_\omega,$$

where the sets ω form a patch of elements constituting the support of the nodal basis functions of V_h. Although the patches overlap, the elements λ_ω can be computed independent of each other. For $d = 2$, this was explained in [100, Sect. 5.3].

After having computed $\hat{\boldsymbol{p}}_h$ on $\mathcal{F} \cup \mathcal{F}_3$, it can be extended to Ω as follows. Denoting by $\Pi_F^k : L^2(F) \to \mathcal{P}_k(F)$ the $L^2(F)$-projector, we require

$$\begin{aligned} \hat{\boldsymbol{p}}_h &\in RT_k(\Omega), \\ (\hat{\boldsymbol{p}}_h \cdot \mathfrak{n})|_F &= -\Pi_F^k \tilde{g}_1 \quad \text{for all } F \in \mathcal{F}_1, \\ (\hat{\boldsymbol{p}}_h \cdot \mathfrak{n})|_F &= \Pi_F^k(-\tilde{g}_2 + \tilde{\alpha} u_h), \end{aligned} \tag{7.167}$$

$$\int_F \hat{\boldsymbol{p}}_h \cdot \mathfrak{n}\, \varphi \, d\sigma = \int_F \left[\overline{\boldsymbol{p}}_h \cdot \mathfrak{n} - \lambda_h\right] \varphi \, d\sigma \quad \text{for all } \varphi \in \mathcal{P}_k(F),\ F \in \mathcal{F} \cup \mathcal{F}_3, \tag{7.168}$$

and for $k \geq 1$ the interior degrees are fixed by

$$\int_K \hat{\boldsymbol{p}}_h \cdot \boldsymbol{w} \, dx = \int_K \big(- \boldsymbol{K} \nabla u_h + \boldsymbol{c}_{\text{upw}}(u_h) \big) \cdot \boldsymbol{w} \, dx \qquad \text{for all } \boldsymbol{w} \in \mathcal{P}_{k-1}^d(K),\ K \in \mathcal{T}_h, \tag{7.169}$$

which means elementwise computations.

The consideration are formalized by showing the convergence of the flux densities.

Theorem 7.61 *Let $u \in H^1(\Omega) \cap H^k(\mathcal{T}_h)$ fulfil (7.150), $\boldsymbol{c} \in C(\overline{\Omega})$, and $\tilde{g}_i \in H^1(\text{div}; \Omega)$, $i = 1, 2$, be prolongations. Then there exists a constant $C > 0$ such that*

$$\|\boldsymbol{p} - \hat{\boldsymbol{p}}_h\|_{0,\Omega} \leq C h^k \left(|u|_{k,\mathcal{T}_h} + \sum_{i=1}^{2} \|\tilde{g}_i\|_{H(\text{div})} \right).$$

Proof Set

$$\hat{\boldsymbol{p}}_h := -\boldsymbol{K} \nabla u_h + \boldsymbol{c} u_h + \tau_h,$$

then it suffices to prove that

$$\|\tau_h\|_{0,\Omega} \leq C h^k \left(|u|_k + \sum_{i=1}^{2} \|\tilde{g}_i\|_{H(\text{div})} \right).$$

We have by (7.167), (7.168) for $F \in \mathcal{F}_1 \cup \mathcal{F}_2$, $i = 1, 2$

$$\tau_h \cdot \mathfrak{n} = (\boldsymbol{p} - \overline{\boldsymbol{p}}_h) \cdot \mathfrak{n} - (\Pi_F^k \tilde{g}_i - \tilde{g}_i) + \begin{cases} 0, & i = 1, \\ \Pi_F^k(\alpha(u_h - u)), & i = 2. \end{cases}$$

For $F \in \mathcal{F} \cup \mathcal{F}_3$, $F \subset K \in \mathcal{T}_h$, it holds that

$$\int_F \tau_h|_K \cdot \mathfrak{n}\, \varphi \, d\sigma = - \int_F \left[\{\!\!\{ \boldsymbol{K} \nabla u_h - \boldsymbol{K} \nabla u \}\!\!\} - \lambda_h \right] \varphi \, d\sigma \quad \text{for all } \varphi \in \mathcal{P}_k(F),$$

as $[\![\boldsymbol{K} \nabla u]\!]_F = 0$ due to (7.19).

By scaling we have, for $K \in \mathcal{T}_h$,

$$
\begin{aligned}
\|\tau_h\|_{0,K} &\le C \sum_{F \in \partial K} h_F^{1/2} \|\tau_{h|K} \cdot \boldsymbol{n}\|_{0,F} \\
&\le C\Bigg(\|\lambda_h\|_{Y_h} + \sum_{F \in \overline{\mathcal{F}}} h_F^{1/2} \|\{\!\!\{\boldsymbol{K}\nabla u - \boldsymbol{K}\nabla u_h\}\!\!\}\|_{0,F}) \\
&\quad + \sum_{F \in \mathcal{F}_1 \cup \mathcal{F}_2} h_F^{1/2} \Big(\sum_{i=1}^{2} \|\tilde{g}_i - \Pi_F^k \tilde{g}_i\|_{0,F} + \|u - u_h\|_{0,F} \Big)\Bigg) \\
&\le C\Bigg(\|u - u_h\|_{V_h} + h^k \sum_{i=1}^{2} \|\tilde{g}_i\|_{H(\mathrm{div})} \Bigg).
\end{aligned}
$$

The rest follows from the considerations after Lemma 7.32. □

Remark 7.62 This section is inspired by [100], where for $d = 2$, $\boldsymbol{c} = \boldsymbol{0}$, $r = 0$, $\boldsymbol{K} = \boldsymbol{I}$, more systematically also DG and nonconforming FEM (see Section 7.1) have been treated.

Exercises

Problem 7.27 Consider the diffusion–advection–reaction equation, not in divergence form (3.36)–(3.39) with no Dirichlet boundary conditions ($\Gamma_3 = \emptyset$) and $\nabla \cdot \boldsymbol{c} = 0$.

a) Derive a formulation similar to (7.147) for global mass conservation.
b) Consider a conforming CFEM for u, with approximation u_h, and an approximation for $\boldsymbol{c}, \boldsymbol{c}_h$ as discussed in Section 7.2 or 7.3. Find requirements for the discretizations such that global mass conservation still holds.

Problem 7.28 Consider the diffusion–advection–reaction equation in divergence form (7.96) with no Dirichlet boundary conditions ($\Gamma_3 = \emptyset$) and let the flow field $\boldsymbol{c}$ (together with a pressure q) be the solution of a Darcy (of the type $\nabla \cdot \boldsymbol{c} = 0$, $\boldsymbol{c} := -\boldsymbol{K}\nabla q$, no Dirichlet boundary conditions) (6.54), (6.55) or a Stokes system (Formulation II, p. 386).

a) Formulate local and global mass conservation (of fluid and of substance).
b) Consider mixed discretizations from Sections 7.2 and 7.3. Find requirements such that global and local mass conservation holds.

Problem 7.29 Prove Theorem 7.55.

Problem 7.30 Verify the assertion (7.165).

Chapter 8
The Finite Volume Method

Over the last decades, finite volume methods (FVMs) have enjoyed great popularity in various fields of computational mathematics (for example, in computational aerodynamics, fluid mechanics, solid-state physics, semiconductor device modelling. . .). In comparison with the standard discretization tools, namely finite element methods and finite difference methods, FVMs are occupying, in some sense, an intermediate position. Actually, treatments in the context of *generalized finite difference methods* as well as the *finite element* approach can be found in the literature.

The basic idea behind the construction of finite volume schemes is to exploit the divergence form of the equation (cf. Section 0.5) by integrating it over mutually disjoint subdomains (called finite volumes, control volumes, finite boxes. . .) and to use Gauss' theorem to convert volume integrals into surface integrals, which are then discretized. In the class of second-order linear elliptic differential equations, expressions of the form

$$Lu := -\nabla \cdot (\boldsymbol{K}\,\nabla u - \boldsymbol{c}\,u) + r\,u = f \tag{8.1}$$

are typical (cf. (0.33)), where

$$\boldsymbol{K} : \Omega \to \mathbb{R}^{d,d}, \quad \boldsymbol{c} : \Omega \to \mathbb{R}^d, \quad r, f : \Omega \to \mathbb{R}.$$

The corresponding "parabolic version" is

$$\frac{\partial u}{\partial t} + Lu = f$$

and will be treated in Chapter 9.

First-order partial differential equations such as the classical conservation laws

$$\nabla \cdot \boldsymbol{p}(u) = 0,$$

where $\boldsymbol{p} : \mathbb{R} \to \mathbb{R}^d$ is a nonlinear vector field depending on u, or higher order partial differential equations (such as the biharmonic equation (3.47)), or even systems of

P. Knabner and L. Angermann, *Numerical Methods for Elliptic and Parabolic Partial Differential Equations*, Texts in Applied Mathematics 44,
https://doi.org/10.1007/978-3-030-79385-2_8

partial differential equations can be successfully discretized by the finite volume method (FVM). In this chapter, we will consider only equations of the type (8.1).

To the authors' knowledge, MacNeal [183] seems to be the first who created discrete (difference) schemes on irregular grids by means of so-called Voronoi boxes. In [139], a theory of difference schemes on irregular quadrilateral grids has been presented. Since their independent introduction by McDonald [186] and MacCormack and Paullay [182], FVM s have been widely used for the numerical simulation of transonic flows governed by conservation laws. Finite volume schemes based on central differences have become particularly popular, following the work of Jameson et al. [155]. In this formulation, usually referred to as the *cell-centred* scheme, the flow variables are associated with the centres of the computational cells, which are quadrilaterals in two dimensions. An alternative scheme has been introduced by Ni [197], where the flow variables are kept at the vertices of the computational cells. The resulting method is called the *cell-vertex* scheme, and it presents a natural generalization of the familiar box scheme to quadrilateral grids. Later, both approaches have been extended to more general grids. Differently, in a third type of finite volume schemes, the so-called *node-oriented* methods, the control volumes are associated with a given distribution of points (nodes). Here, the set of control volumes typically constitutes a partition $\mathcal{T}^*$ which is, in some sense, dual to a given partition $\mathcal{T}$. The unknowns are both associated with this given partition and, at the same time, with the control volumes.

Some further milestones in the development of FVMs

1958 Marčuk	computation of nuclear reactors
1960 Forsythe & Wasow	computation of neutrone diffusion
1971 McDonald	fluid mechanics
1972 MacCormack & Paullay	fluid mechanics
1973 Rizzi & Inouye	fluid mechanics in 3 dimensions
1977 Samarski	"integro-interpolation method", "balance method"
⋮	
1979 Jameson	"finite volume method"
1984 Heinrich	"integro-balance method", "generalized FDM"
⋮	
1977 Weiland	"finite integration technique"
1987 Bank & Rose	"box method"
⋮	

8.1 The Basic Idea of the Finite Volume Method

Now we will describe the fundamental steps in the derivation of the FVM. For simplicity, we restrict ourselves to the case $d = 2$ and $r = 0$. As in Section 6.2 we set $\boldsymbol{p} = \boldsymbol{p}(u) := -\boldsymbol{K}\,\nabla u + \boldsymbol{c}\,u$. Then equation (8.1) becomes

$$\nabla \cdot \boldsymbol{p}(u) = f\,. \tag{8.2}$$

In order to obtain a finite volume discretization, the domain Ω will be subdivided into M subdomains Ω_i such that the collection of all those subdomains forms a *partition* $\mathcal{T}_h^* := \{\Omega_i\}_{i=1}^M$ of Ω, that is

1) each Ω_i is an open, simply connected, and polygonally bounded set without slits,
2) $\Omega_i \cap \Omega_j = \emptyset \quad (i \neq j)$,
3) $\cup_{i=1}^M \overline{\Omega_i} = \overline{\Omega}$.

These subdomains Ω_i are called *control volumes* or *control domains.*

Without going into more detail, we mention that there also exist FVMs with a well-defined overlapping of the control volumes (that is, condition 2 is violated).

The next step, which is in common with all FVMs, consists in integrating equation (8.2) over each control volume Ω_i. After that, Gauss' divergence theorem is applied:

$$\int_{\partial\Omega_i} \boldsymbol{\nu} \cdot \boldsymbol{p}(u)\, d\sigma = \int_{\Omega_i} f\, dx\,, \quad i \in \{1, \ldots, M\}\,,$$

where $\boldsymbol{\nu}$ denotes the outer unit normal to $\partial\Omega_i$. By the first condition of the partition, the boundary $\partial\Omega_i$ is formed by straight-line segments Γ_{ij} $(j = 1, \ldots, n_i)$, along which the normal $\boldsymbol{\nu}|_{\Gamma_{ij}} =: \boldsymbol{\nu}_{ij}$ is constant (see Figure 8.1). So the line integral can be decomposed into a sum of line integrals from which the following equation results:

$$\sum_{j=1}^{n_i} \int_{\Gamma_{ij}} \boldsymbol{\nu}_{ij} \cdot \boldsymbol{p}(u)\, d\sigma = \int_{\Omega_i} f\, dx\,. \tag{8.3}$$

Now the integrals occurring in (8.3) have to be approximated. This can be done in very different ways, and so different final discretizations are obtained.

In general, FVMs can be distinguished by the following criteria:

- the geometric shape of the control volumes Ω_i,
- the position of the unknowns ("problem variables") with respect to the control volumes,
- the approximation of the boundary (line ($d = 2$) or surface ($d = 3$)) integrals.

Especially the second criterion divides the FVMs into two large classes: the *cell-centred* and the *cell-vertex* FVMs. In the cell-centred methods, the unknowns are associated with the control volumes (for example, any control volume corresponds

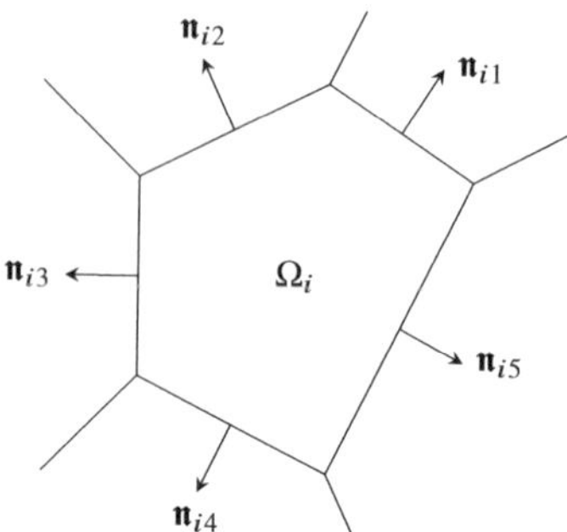

Fig. 8.1: A control volume.

to a function value at some interior point (e.g. at the barycentre)). In the cell-vertex methods, the unknowns are located at the vertices of the control volumes. Sometimes, instead of the first-mentioned class a subdivision into two classes, the so-called *cell-centred* and *node-oriented* methods, is considered. The difference is whether the problem variables are assigned to the control volumes or, given the problem variables, associated control volumes are defined.

Example 8.1 *Consider the homogeneous Dirichlet problem for the Poisson equation on the unit square (Fig. 8.2):*

$$\begin{aligned} -\Delta u &= f \quad && \text{in } \Omega = (0,1)^2\,, \\ u &= 0 && \text{on } \partial\Omega\,. \end{aligned}$$

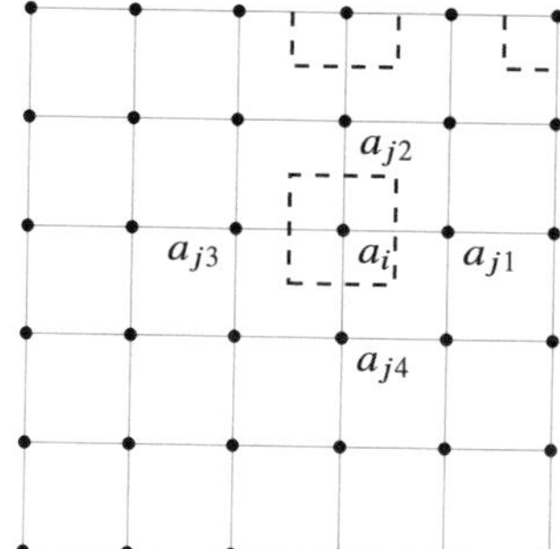

Problem variables:

Function values at nodes a_i of a square grid with grid width $h > 0$

Control volumes:

$\Omega_i := \left\{x \in \Omega \,\middle|\, |x - a_i|_\infty < \frac{h}{2}\right\}$

Fig. 8.2: Problem variables and control volumes in a cell-centred finite volume method.

For an inner control volume Ω_i (i.e. $a_i \in \Omega$), equation (8.3) takes the form

$$-\sum_{k=1}^{4} \int_{\Gamma_{ij_k}} \mathfrak{n}_{ij_k} \cdot \nabla u \, d\sigma = \int_{\Omega_i} f \, dx\,,$$

where $\Gamma_{ij_k} := \partial\Omega_i \cap \partial\Omega_{j_k}$. *A closer look at the directional derivatives shows that*

$$\begin{aligned} \mathfrak{n}_{ij_1} \cdot \nabla u &= \partial_1 u\,, \quad & \mathfrak{n}_{ij_2} \cdot \nabla u &= \partial_2 u\,, \\ \mathfrak{n}_{ij_3} \cdot \nabla u &= -\partial_1 u\,, \quad & \mathfrak{n}_{ij_4} \cdot \nabla u &= -\partial_2 u\,. \end{aligned}$$

i.e. they are just partial derivatives with respect to the first or the second variable on the corresponding parts of the boundary.

Approximating the integrals on Γ_{ij_k} *by means of the midpoint rule and replacing the derivatives by difference quotients, we have*

$$\begin{aligned} -\sum_{k=1}^{4} \int_{\Gamma_{ij_k}} \mathfrak{n}_{ij_k} \cdot \nabla u \, d\sigma &\approx -\sum_{k=1}^{4} \mathfrak{n}_{ij_k} \cdot \nabla u \left(\frac{a_i + a_{j_k}}{2} \right) h \\ &\approx -\left[\frac{u(a_{j_1}) - u(a_i)}{h} + \frac{u(a_{j_2}) - u(a_i)}{h} - \frac{u(a_i) - u(a_{j_3})}{h} - \frac{u(a_i) - u(a_{j_4})}{h} \right] h \\ &= 4\,u(a_i) - \sum_{k=1}^{4} u(a_{j_k})\,. \end{aligned}$$

Thus, we obtain exactly the expression that results from the application of a finite element method with continuous, piecewise linear ansatz and test functions on a Friedrichs–Keller triangulation (cf. Figure 2.9).

Furthermore, if we approximate the integral $\int_{\Omega_i} f\,dx$ *by* $f(a_i)h^2$, *we see that this term coincides with the trapezoidal rule applied to the right-hand side of the mentioned finite element formulation (cf. Lemma 2.13).*

Actually, it is no accident that both discretization methods lead to the same algebraic system. Later on, we will prove a more general result to confirm the above observation.

The boundary control volumes are treated as follows:

If $a_i \in \partial\Omega$, *then parts of the boundary* $\partial\Omega_i$ *lie on* $\partial\Omega$. *At these nodes, the Dirichlet boundary conditions already prescribe values of the unknown function, and so there is no need to include the boundary control volumes into the balance equations (8.3).*

A detailed description for the case of flux boundary conditions will be given later in Subsection 8.2.2.

Example 8.2 *We consider the same boundary value problem as in Example 8.1 with the problem variables and the control volumes as depicted in Figure 8.3. In the interior of* Ω, *the resulting discretization yields a 12-point stencil (in the terminology of finite difference methods).*

Remark 8.3 In the finite volume discretization of systems of partial differential equations (resulting from fluid mechanics, for example), both methods are used simultaneously for different variables; see Figure 8.4 or 8.12.

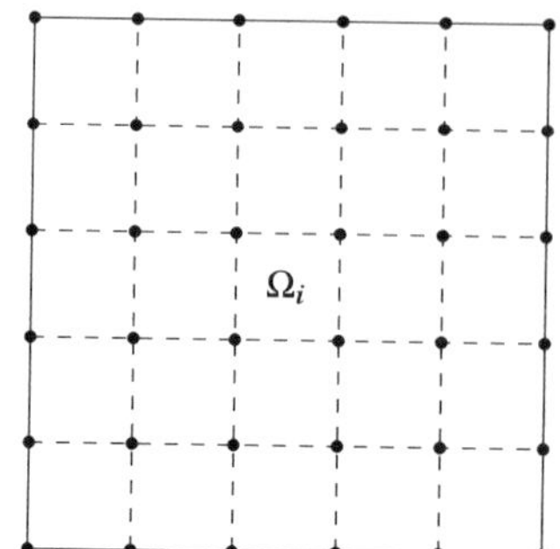

Fig. 8.3: Problem variables and control volumes in a cell-vertex finite volume method.

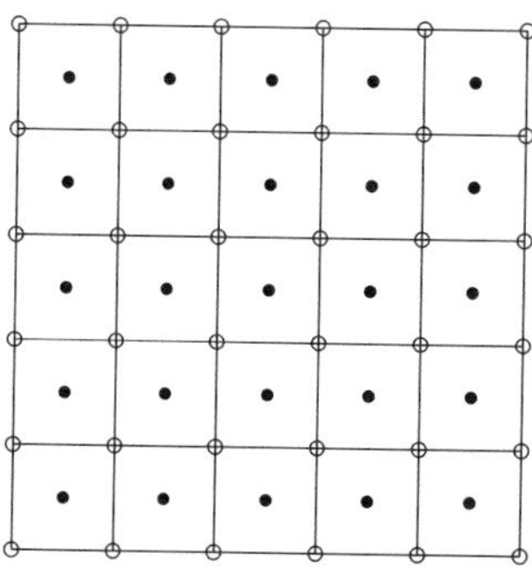

Fig. 8.4: Finite volume discretization of systems of partial differential equations.

Assets and Drawbacks of the Finite Volume Method

Assets:

- Flexibility with respect to the geometry of the domain Ω (as in finite element methods).
- Admissibility of unstructured grids (as in finite element methods, important for adaptive methods).
- Simple assembling in the case $d = 2$ and for not too irregular grids in the cases $d \geq 3$.
- Conservation of certain laws valid for the continuous problem (for example, conservation laws or maximum principles). This property is important in the numerical solution of differential equations with discontinuous coefficients or of convection-dominated diffusion–convection equations (see Section 8.2.4).
- Easy linearization of nonlinear problems (simpler than in finite element methods (Newton's method)).
- Simple discretization of boundary conditions (as in finite element methods, especially a "natural" treatment of Neumann or mixed boundary conditions).
- In principle, no restriction of the spatial dimension d of the domain Ω.

Drawbacks:

- Smaller field of applications in comparison with finite element or finite difference methods.
- Difficulties in the design of higher order methods (no so-called p-version available as in the finite element method).
- In higher spatial dimensions ($d \geq 3$), the construction of some classes or types of control volumes may be a complex task and thus may lead to a time-consuming assembling.
- Difficult mathematical analysis (stability, convergence, ...).

Exercise

Problem 8.1 Given the boundary value problem

$$-(au')' = 0 \quad \text{in } (0,1), \qquad u(0) = 1\,, \; u(1) = 0\,,$$

with piecewise constant coefficients

$$a(x) := \begin{cases} \kappa\alpha\,, & x \in (0,\xi)\,, \\ \alpha\,, & x \in (\xi,1)\,, \end{cases}$$

where α, κ are positive constants and $\xi \in (0,1) \setminus \mathbb{Q}$:

a) What is the weak solution $u \in H^1(0,1)$ of this problem?
b) For general "smooth" coefficients a, the differential equation is obviously equivalent to

$$-au'' - a'u' = 0\,.$$

Therefore, the following discretization is suggested:

$$-a_i \frac{u_{i-1} - 2u_i + u_{i+1}}{h^2} - \frac{a_{i+1} - a_{i-1}}{2h}\,\frac{u_{i+1} - u_{i-1}}{2h} = 0\,,$$

where an equidistant grid with the nodes $x_i = ih \; (i = 0, \ldots, N+1)$ and $a_i := a(x_i)$, $u_i :\approx u(x_i)$ is used.
This discretization is also formally correct in the given situation of discontinuous coefficients. Find the discrete solution $(u_i)_{i=1}^N$ in this case.
c) Under what conditions do the values u_i converge to $u(x_i)$ for $h \to 0$?

8.2 The Finite Volume Method for Linear Elliptic Differential Equations of Second Order on Triangular Grids

In this section, we will explain the development and the analysis of a FVM of *cell-centred* type for a model problem. Here, $\Omega \subset \mathbb{R}^2$ is a bounded, simply connected domain with a polygonal boundary, but without slits.

8.2.1 Admissible Control Volumes

The Voronoi Diagram

By $\{a_i\}_{i\in\overline{\Lambda}} \subset \overline{\Omega}$ we denote a consecutively numbered point set that includes all vertices of Ω, where $\overline{\Lambda}$ is the corresponding set of indices. Typically, the points a_i are placed at those positions where the values $u(a_i)$ of the exact solution u are to be approximated. The set

$$\tilde{\Omega}_i := \left\{x \in \mathbb{R}^2 \;\middle|\; |x - a_i| < \left|x - a_j\right| \quad \text{for all } j \neq i\right\}$$

is called the *Voronoi polygon* (or *Dirichlet domain, Wigner–Seitz cell, Thiessen polygon, ...*). The family $\left\{\tilde{\Omega}_i\right\}_{i\in\overline{\Lambda}}$ is called the *Voronoi diagram* of the point set $\{a_i\}_{i\in\overline{\Lambda}}$.

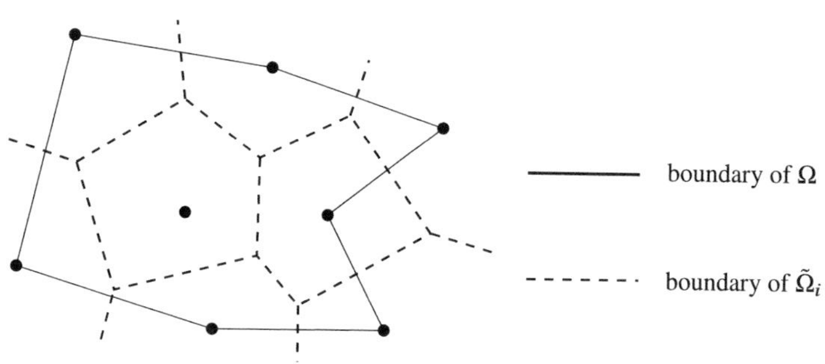

Fig. 8.5: Voronoi diagram.

The Voronoi polygons are convex (see Problem 8.2), but not necessarily bounded, sets (consider the situation near the boundary in Figure 8.5). Their boundaries are polygons. The vertices of these polygons are called *Voronoi vertices*.

Provided the set $\overline{\Lambda}$ contains at least three elements (i.e. $|\overline{\Lambda}| \geq 3$) it can be shown that at any Voronoi vertex at least three Voronoi polygons meet, unless all points a_i are collinear (see Problem 8.3). According to this property, Voronoi vertices are classified into regular and degenerate Voronoi vertices: In a *regular* Voronoi vertex, the boundaries of exactly three Voronoi polygons meet, whereas a *degenerate*

Voronoi vertex is shared by at least four Voronoi polygons. In the latter case, all the corresponding nodes a_i are located at some circle (they are "cocyclic", cf. Figure 8.6).

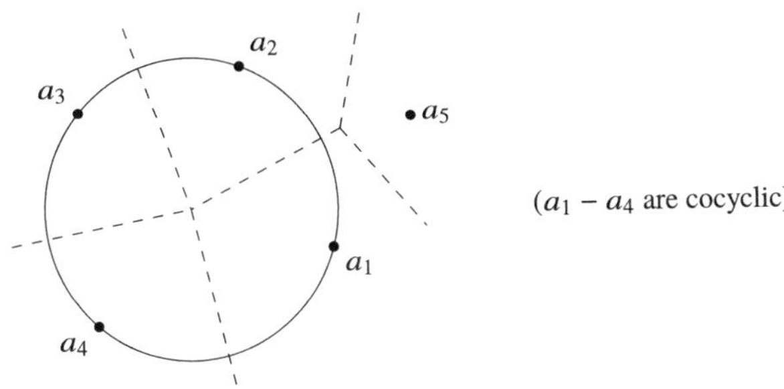

Fig. 8.6: Degenerate and regular Voronoi vertex.

Now the elements Ω_i (control volumes) of the partition of Ω required for the definition of the FVM can be introduced as follows:

$$\Omega_i := \tilde{\Omega}_i \cap \Omega, \quad i \in \overline{\Lambda}.$$

As a consequence, the domains Ω_i need not necessarily be convex if Ω is nonconvex (cf. Figure 8.5).

Furthermore, the following notation will be used:

$$\Lambda_i := \left\{ j \in \overline{\Lambda} \setminus \{i\} \;\middle|\; \partial\Omega_i \cap \partial\Omega_j \neq \emptyset \right\}, \quad i \in \overline{\Lambda},$$

for the set of indices of neighbouring nodes,

$\Gamma_{ij} := \partial\Omega_i \cap \partial\Omega_j, \quad j \in \Lambda_i$, for a joint piece of the boundaries of neighbouring control volumes,

m_{ij} for the length of Γ_{ij}, i.e. $m_{ij} := |\Gamma_{ij}|_1$.

The *dual graph* of the Voronoi diagram is defined as follows:

Any pair of points a_i, a_j such that $m_{ij} > 0$ is connected by a straight-line segment $\overline{a_i a_j}$. In this way, a further partition of Ω with an interesting property results.

Theorem 8.4 *If all Voronoi vertices are regular, then the dual graph coincides with the set of edges of a triangulation of the convex hull of the given point set.*

Proof See, for example, [137, Thm. 2.1]. □

This triangulation is called a *Delaunay triangulation.*

If among the Voronoi vertices there are degenerate ones, then a triangulation can be obtained from the dual graph by a subsequent local triangulation of the remaining m-polygons ($m \geq 4$). Due to the intersection property of Voronoi polygons mentioned above (cf. Problem 8.3), a Delaunay triangulation has the interesting property that two interior angles subtended by any given edge sum to no more than π. In this respect, Delaunay triangulations satisfy the first part of the angle condition formulated in Section 3.9 for the maximum principle in finite element methods.

Therefore, if Ω is convex, then we automatically get a triangulation together with the Voronoi diagram. In the case of a nonconvex domain Ω, certain modifications could be required to achieve a correct triangulation (Fig. 8.7).

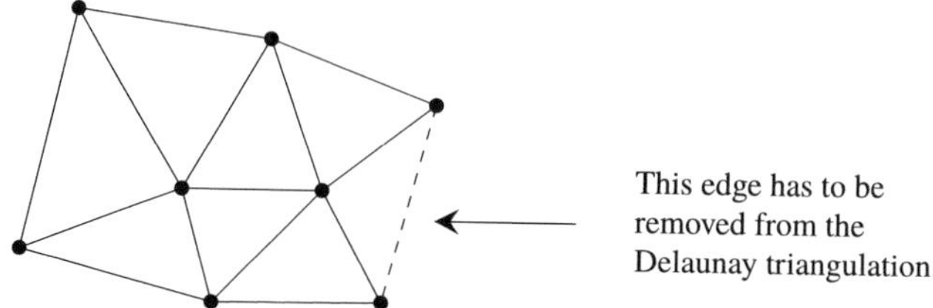

Fig. 8.7: Delaunay triangulation to the Voronoi diagram from Figure 8.5.

The implication

$$\text{Voronoi diagram} \quad \Rightarrow \quad \text{Delaunay triangulation},$$

which we have just discussed, suggests that we ask about the converse statement. We do not want to answer it completely at this point, but we give the following sufficient condition.

Theorem 8.5 *If a consistent triangulation of Ω (in the sense of Section 2.2) consists of nonobtuse triangles exclusively, then it is a Delaunay triangulation, and the corresponding Voronoi diagram can be constructed by means of the perpendicular bisectors of the triangles' edges.*

We mention that the centre of the circumcircle of a nonobtuse triangle, which becomes a Voronoi vertex, is located within the closure of that triangle.

In the analysis of the FVM, the following relation is important, using for the set of indices of nodes belonging to K the notation

$$\Lambda_K := \{i \in \overline{\Lambda} \mid a_i \in K\}$$

from now on.

Lemma 8.6 *Given a nonobtuse triangle K, then for the corresponding parts $\Omega_{i,K} := \Omega_i \cap K$ of the control volumes Ω_i, we have*

$$\frac{1}{4}|K| \leq |\Omega_{i,K}| \leq \frac{1}{2}|K|, \quad i \in \Lambda_K .$$

The Donald diagram

In contrast to the Voronoi diagram, where the construction starts from a given point set, the starting point here is a triangulation $\mathcal{T}_h$ of Ω, which is allowed to contain obtuse triangles.

Again, let K be a triangle with vertices a_{i_k}, $k \in \{1,2,3\}$. We define

$$\Omega_{i_k,K} := \left\{x \in K \,\middle|\, \lambda_j(x) < \lambda_k(x),\ j \neq k\right\},$$

where λ_k denote the barycentric coordinates with respect to a_{i_k} (cf. (3.62)).

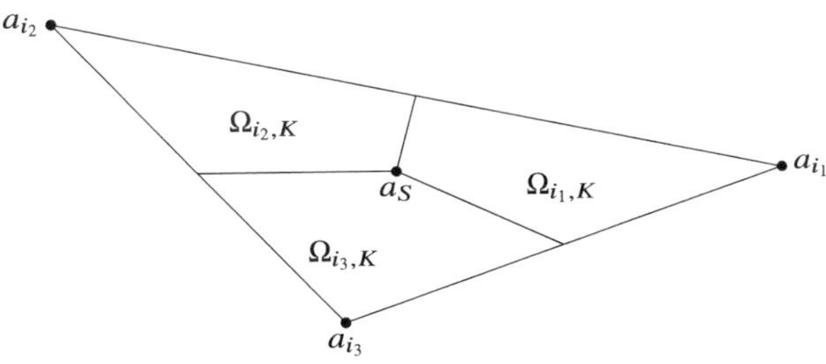

Fig. 8.8: The subdomains $\Omega_{i_k,K}$.

Obviously, the barycentre satisfies $a_S = \frac{1}{3}(a_{i_1} + a_{i_2} + a_{i_3}) = \frac{1}{3}\sum_{i\in\Lambda_K} a_i$, and (see, for comparison, Lemma 8.6)

$$3\,|\Omega_{i,K}| = |K|, \quad i \in \Lambda_K. \tag{8.4}$$

This relation is a simple consequence of the geometric interpretation of the barycentric coordinates as area coordinates given in Section 3.3. The required control volumes are defined as follows (see Figure 8.8):

$$\Omega_i := \operatorname{int}\left(\bigcup_{K:\,\Lambda_K \ni i} \overline{\Omega}_{i,K}\right), \quad i \in \overline{\Lambda}.$$

The family $\{\Omega_i\}_{i\in\overline{\Lambda}}$ is called a *Donald diagram.*

The quantities Γ_{ij}, m_{ij}, and Λ_i are defined similarly as in the case of the Voronoi diagram. We mention that the boundary pieces Γ_{ij} are not necessarily straight, but polygonal in general.

8.2.2 Finite Volume Discretization

The model under consideration is a special case of the boundary value problem (6.30)–(6.34). Instead of the matrix-valued diffusion coefficient $\boldsymbol{K}$, we will take a scalar coefficient $k : \Omega \to \mathbb{R}$, that is, $\boldsymbol{K} = k\boldsymbol{I}$. Moreover, homogeneous Dirichlet boundary conditions are to be satisfied, that is, $\widetilde{g_3} = 0$. So the boundary value problem reads as follows:

$$\begin{aligned} \nabla \cdot \boldsymbol{p} + ru &= f && \text{in } \Omega, \\ \boldsymbol{p} &= -k\nabla u + \boldsymbol{c}u && \text{in } \Omega, \\ -\boldsymbol{p} \cdot \mathfrak{n} &= \widetilde{g_1} && \text{on } \widetilde{\Gamma}_1, \\ -\boldsymbol{p} \cdot \mathfrak{n} + \alpha u &= \widetilde{g_2} && \text{on } \widetilde{\Gamma}_2, \\ u &= 0 && \text{on } \widetilde{\Gamma}_3, \end{aligned} \tag{8.5}$$

with $k, r, f : \Omega \to \mathbb{R}$, $\boldsymbol{c} : \Omega \to \mathbb{R}^2$, $\widetilde{g_1} : \widetilde{\Gamma}_1 \to \mathbb{R}$ and $\alpha, \widetilde{g_2} : \widetilde{\Gamma}_2 \to \mathbb{R}$ given.

The Case of the Voronoi Diagram

Let the domain Ω be partitioned by a Voronoi diagram and the corresponding Delaunay triangulation, where the triangulation has to satisfy the compatibility condition (T6) from Section 3.3. The structure of the boundary conditions implies a canonical decomposition of the index set $\overline{\Lambda}$ as follows:

$$\overline{\Lambda} := \Lambda \cup \left(\bigcup_{l=1}^{3} \partial\Lambda_l \right),$$

where $\Lambda := \{i \in \overline{\Lambda} \mid a_i \in \Omega\}$, $\partial\Lambda_l := \{i \in \overline{\Lambda} \mid a_i \in \widetilde{\Gamma}_l\}$, $l \in \{1, 2, 3\}$. Due to the homogeneity of the Dirichlet boundary conditions, it is sufficient to consider only those control volumes Ω_i that are associated with the nodes from $\Lambda_0 := \Lambda \cup \partial\Lambda_1 \cup \partial\Lambda_2$.

In the first step, the first differential equation in (8.5) is integrated over the single control volumes Ω_i :

$$\int_{\Omega_i} \nabla \cdot \boldsymbol{p}\, dx + \int_{\Omega_i} r\, u\, dx = \int_{\Omega_i} f\, dx, \quad i \in \Lambda_0. \tag{8.6}$$

The application of Gauss' divergence theorem to the first integral of the left-hand side of (8.6) yields

$$\int_{\Omega_i} \nabla \cdot \boldsymbol{p}\, dx = \int_{\partial\Omega_i} \mathfrak{n} \cdot \boldsymbol{p}\, d\sigma .$$

Due to $\partial\Omega_i = \bigcup_{j \in \Lambda_i} \Gamma_{ij}$ for $i \in \Lambda$ and $\partial\Omega_i = \bigcup_{j \in \Lambda_i} \Gamma_{ij} \cup (\partial\Omega_i \cap \partial\Omega)$ for $i \in \partial\Lambda_1 \cup \partial\Lambda_2$, it follows that

$$\int_{\Omega_i} \nabla \cdot \boldsymbol{p}\, dx = \sum_{j \in \Lambda_i} \int_{\Gamma_{ij}} \mathfrak{n}_{ij} \cdot \boldsymbol{p}\, d\sigma + \int_{\partial\Omega_i \cap \partial\Omega} \mathfrak{n} \cdot \boldsymbol{p}\, d\sigma , \tag{8.7}$$

where $\mathfrak{n}_{ij}$ is the (constant) outer unit normal to Γ_{ij} (with respect to Ω_i). If $i \in \Lambda$, the boundary segment Γ_{ij} belongs to two triangles, say K, K' (cf. Figure 8.9), and is subsequently divided as $\Gamma_{ij} = \Gamma_{ij}^K \cup \Gamma_{ij}^{K'}$ with $\Gamma_{ij}^K := \Gamma_{ij} \cap K$. Correspondingly, it holds that $m_{ij} = m_{ij}^K + m_{ij}^{K'}$, where $m_{ij}^K := |\Gamma_{ij}^K|_1$ is the length of the segment Γ_{ij}^K. In the case $i, j \in \partial\Lambda := \overline{\Lambda} \setminus \Lambda$, the boundary segment Γ_{ij} belongs to one triangle only, say K, and $\Gamma_{ij}^K = \Gamma_{ij}, m_{ij} = m_{ij}^K$.

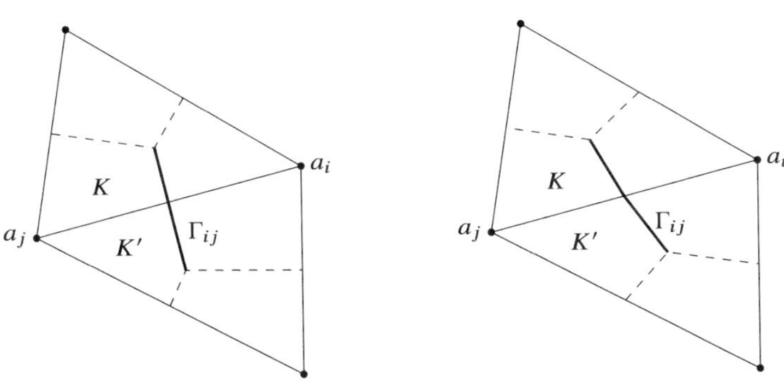

Fig. 8.9: An interior edge Γ_{ij} in the Voronoi case (left) and the Donald case (right).

In the next step, we approximate the line integrals over Γ_{ij} in (8.7). First, the coefficients k and $\mathfrak{n}_{ij} \cdot \boldsymbol{c}$ are approximated on Γ_{ij} by constants $\mu_{ij} > 0$, respectively, γ_{ij}:

$$k|_{\Gamma_{ij}} \approx \mu_{ij} = \text{const} > 0, \quad \mathfrak{n}_{ij} \cdot \boldsymbol{c}|_{\Gamma_{ij}} \approx \gamma_{ij} = \text{const}.$$

In the simplest case, the approximation can be realized by the corresponding value at the midpoint $a_{\Gamma_{ij}}$ of the straight-line segment Γ_{ij}. A better choice is

$$\gamma_{ij} := \begin{cases} \dfrac{1}{m_{ij}} \displaystyle\int_{\Gamma_{ij}} \mathfrak{n}_{ij} \cdot \boldsymbol{c}\, d\sigma, & m_{ij} > 0, \\ \mathfrak{n}_{ij} \cdot \boldsymbol{c}(a_{\Gamma_{ij}}), & m_{ij} = 0, \end{cases} \tag{8.8}$$

provided the integrals can be computed exactly or approximated by means of a convergent quadrature rule. Using these approximations in (8.7), we obtain that

$$\int_{\Omega_i} \nabla \cdot \boldsymbol{p}\, dx \approx -\sum_{j \in \Lambda_i} \int_{\Gamma_{ij}} \left[\mu_{ij} \left(\mathfrak{n}_{ij} \cdot \nabla u\right) - \gamma_{ij}\, u\right] d\sigma + \int_{\partial\Omega_i \cap \partial\Omega} \mathfrak{n} \cdot \boldsymbol{p}\, d\sigma\,. \tag{8.9}$$

The normal derivatives are approximated by difference quotients; that is,

$$\mathfrak{n}_{ij} \cdot \nabla u \approx \frac{u(a_j) - u(a_i)}{d_{ij}} \quad \text{with } d_{ij} := \left|a_i - a_j\right|.$$

This formula is exact for such functions that are linear along the straight-line segment $\overline{a_i a_j}$ between the points a_i, a_j. So it remains to approximate the integral of u over Γ_{ij}. For this, a convex combination of the values of u at the nodes a_i and a_j is taken

$$u|_{\Gamma_{ij}} \approx r_{ij}\, u(a_i) + \left(1 - r_{ij}\right)\, u(a_j),$$

where $r_{ij} \in [0, 1]$ is a parameter to be defined subsequently. In general, r_{ij} depends on μ_{ij}, γ_{ij}, and d_{ij}.

Collecting all the above approximations, we arrive at the following relation:

$$\begin{aligned}
&\int_{\Omega_i} \nabla \cdot \boldsymbol{p}\, dx \\
&\approx \sum_{j \in \Lambda_i} \left\{ \mu_{ij}\, \frac{u(a_i) - u(a_j)}{d_{ij}} + \gamma_{ij}\, \left[r_{ij}\, u(a_i) + \left(1 - r_{ij}\right)\, u(a_j) \right] \right\} m_{ij} \\
&\quad + \int_{\partial\Omega_i \cap \partial\Omega} \boldsymbol{\nu} \cdot \boldsymbol{p}\, d\sigma\,.
\end{aligned}$$

The integrals over $\partial\Omega_i \cap \partial\Omega$ are expressed by means of the boundary conditions:

$$\begin{aligned}
\int_{\partial\Omega_i \cap \partial\Omega} \boldsymbol{\nu} \cdot \boldsymbol{p}\, d\sigma &= \int_{\partial\Omega_i \cap \widetilde{\Gamma}_1} \boldsymbol{\nu} \cdot \boldsymbol{p}\, d\sigma + \int_{\partial\Omega_i \cap \widetilde{\Gamma}_2} \boldsymbol{\nu} \cdot \boldsymbol{p}\, d\sigma \\
&= - \int_{\partial\Omega_i \cap \widetilde{\Gamma}_1} \widetilde{g}_1\, d\sigma - \int_{\partial\Omega_i \cap \widetilde{\Gamma}_2} \widetilde{g}_2\, d\sigma + \int_{\partial\Omega_i \cap \widetilde{\Gamma}_2} \alpha u\, d\sigma\,,
\end{aligned}$$

and the resulting integrals are approximated as follows:

$$\begin{aligned}
&\int_{\partial\Omega_i \cap \widetilde{\Gamma}_l} \widetilde{g}_l\, d\sigma \approx \widetilde{g}_l(a_i)\, |\partial\Omega_i \cap \widetilde{\Gamma}_l|_1 =: \widetilde{g}_{li}\, \widetilde{m}_{il}, \quad \text{with } \widetilde{m}_{il} := |\partial\Omega_i \cap \widetilde{\Gamma}_l|_1,\ l \in \{1, 2\}, \\
&\int_{\partial\Omega_i \cap \widetilde{\Gamma}_2} \alpha u\, d\sigma \approx \alpha(a_i)\, u(a_i)\, \widetilde{m}_{i2} =: \alpha_i\, u(a_i)\, \widetilde{m}_{i2}\,.
\end{aligned}$$

To approximate the remaining integrals from (8.6), the following formulas are used:

$$\begin{aligned}
&\int_{\Omega_i} r\, u\, dx \approx r(a_i)\, u(a_i)\, m_i =: r_i\, u(a_i)\, m_i\,, \quad \text{with } m_i := |\Omega_i|\,, \\
&\int_{\Omega_i} f\, dx \approx f(a_i)\, m_i =: f_i\, m_i\,.
\end{aligned}$$

Instead of $\alpha_i := \alpha(a_i)$, $\widetilde{g}_{il} := \widetilde{g}_l(a_i)$, $r_i := r(a_i)$ or $f_i := f(a_i)$, the approximations

$$\begin{aligned}
\widetilde{g}_{il} &:= \frac{1}{\widetilde{m}_{il}} \int_{\partial\Omega_i \cap \widetilde{\Gamma}_l} \widetilde{g}_l\, d\sigma, & \alpha_i &:= \frac{1}{\widetilde{m}_{i2}} \int_{\partial\Omega_i \cap \widetilde{\Gamma}_2} \alpha\, d\sigma, \\
r_i &:= \frac{1}{m_i} \int_{\Omega_i} r\, dx \quad \text{or} & f_i &:= \frac{1}{m_i} \int_{\Omega_i} f\, dx
\end{aligned} \tag{8.10}$$

can also be used. As with (8.8) this requires that the integrals can be computed exactly or approximated by means of convergent quadrature rules.

Denoting the unknown approximate values for $u(a_i)$ by u_i, we obtain the following linear system of equations:

$$\begin{aligned}\sum_{j\in\Lambda_i}\left\{\mu_{ij}\frac{u_i-u_j}{d_{ij}}+\gamma_{ij}\left[r_{ij}u_i+\left(1-r_{ij}\right)u_j\right]\right\}m_{ij}+r_i\,u_i\,m_i+\alpha_i\,u_i\,\widetilde{m}_{i2}\\ =f_i m_i+\widetilde{g}_{1i}\,\widetilde{m}_{i1}+\widetilde{g}_{2i}\,\widetilde{m}_{i2}\,,\quad i\in\Lambda_0,\end{aligned}\tag{8.11}$$

and from the Dirichlet boundary condition: $u_i=0,\ i\in\partial\Lambda_3$.

The term in the braces can be interpreted as the negative normal component of a *numerical flux density* on Γ_{ij}:

$$-p_{ij}:=\mu_{ij}\frac{u_j-u_i}{d_{ij}}-\gamma_{ij}\left[r_{ij}u_i+\left(1-r_{ij}\right)u_j\right]\quad\left(\approx-\boldsymbol{n}_{ij}\cdot\boldsymbol{p}(u)|_{\Gamma_{ij}}\right).\tag{8.12}$$

The representation (8.11) clearly indicates the affinity of the FVM to the finite difference method. However, for the subsequent analysis, it is more convenient to rewrite this system of equations in terms of a discrete variational equality.

Multiplying the ith equation in (8.11) by arbitrary numbers $v_i\in\mathbb{R}$ and summing the results up over $i\in\Lambda_0$, we get

$$\begin{aligned}\sum_{i\in\Lambda_0}v_i\left\{\sum_{j\in\Lambda_i}\left\{\mu_{ij}\frac{u_i-u_j}{d_{ij}}+\gamma_{ij}\left[r_{ij}\,u_i+\left(1-r_{ij}\right)\,u_j\right]\right\}m_{ij}+r_i\,u_i\,m_i+\alpha_i\,u_i\,\widetilde{m}_{i2}\right\}\\ =\sum_{i\in\Lambda_0}\left\{f_i\,m_i+\widetilde{g}_{1i}\,\widetilde{m}_{i1}+\widetilde{g}_{2i}\,\widetilde{m}_{i2}\right\}v_i\,.\end{aligned}\tag{8.13}$$

In order to obtain a discrete variational formulation of the scheme (8.13), we have to introduce appropriate discrete bilinear forms. It is useful to do this in a somewhat larger framework, namely for the space X_h that is defined as in (2.30) but without prescribing any boundary condition

$$X_h:=\left\{u\in C(\bar{\Omega})\;\middle|\;u|_K\in\mathcal{P}_1(K)\text{ for all }K\in\mathcal{T}_h\right\}$$

(see also (3.119)). Now the following discrete bilinear forms on $X_h\times X_h$ can be defined:

$$\begin{aligned}a_h^0\left(u_h,v_h\right)&:=\sum_{i\in\Lambda_0}v_i\sum_{j\in\Lambda_i}\mu_{ij}\left(u_i-u_j\right)\frac{m_{ij}}{d_{ij}}\,,\\ b_h\left(u_h,v_h\right)&:=\sum_{i\in\Lambda_0}v_i\sum_{j\in\Lambda_i}\left[r_{ij}\,u_i+\left(1-r_{ij}\right)\,u_j\right]\gamma_{ij}\,m_{ij}\,,\\ d_h\left(u_h,v_h\right)&:=\sum_{i\in\Lambda_0}r_i\,u_i\,v_i\,m_i+\sum_{i\in\partial\Lambda_2}\alpha_i\,u_i\,v_i\,\widetilde{m}_{i2}\,,\\ a_h\left(u_h,v_h\right)&:=a_h^0\left(u_h,v_h\right)+b_h\left(u_h,v_h\right)+d_h\left(u_h,v_h\right)\,.\end{aligned}\tag{8.14}$$

Furthermore, for two continuous functions $v, w \in C(\overline{\Omega})$, we set

$$\langle w, v\rangle_{0,h} := \sum_{i\in\overline{\Lambda}} w_i\, v_i\, m_i\,, \tag{8.15}$$

where $w_i := w(a_i),\ v_i := v(a_i)$. Finally, we define a linear functional ℓ_h on X_h as

$$\ell_h(v_h) := \langle f, v_h\rangle_{0,h} + \sum_{i\in\partial\Lambda_1} \widetilde{g}_{1i}\, v_i\, \widetilde{m}_{i1} + \sum_{i\in\partial\Lambda_2} \widetilde{g}_{2i}\, v_i\, \widetilde{m}_{i2}\,.$$

Remark 8.7 $\langle\cdot,\cdot\rangle_{0,h}$ is a scalar product on X_h. In particular, the following norm can be introduced:

$$\|v_h\|_{0,h} := \sqrt{\langle v_h, v_h\rangle_{0,h}}\,, \quad v_h \in X_h\,. \tag{8.16}$$

In (3.155), a slightly different discrete L^2-norm has been defined using the same notation. This multiple use seems to be acceptable, since for shape-regular families of triangulations both norms are equivalent uniformly in h as the following remark shows.

Remark 8.8 Let the family of triangulations $(\mathcal{T}_h)_h$ be shape-regular. Then the norms defined in (3.155) and in (8.16), and also the norms $\|\cdot\|_{0,h}$ and $\|\cdot\|_0$, are equivalent on X_h uniformly with respect to h; i.e. there exist two constants $C_1, C_2 > 0$ independent of h such that

$$C_1\|v\|_0 \le \|v\|_{0,h} \le C_2\|v\|_0 \quad \text{for all } v \in X_h\,.$$

Proof Due to Theorem 3.46 (1), only the uniform equivalence of the discrete L^2-norms has to be shown. Denoting such an equivalence by $\cong$, we have for $v \in X_h$

$$\begin{aligned}
\sum_{i\in\overline{\Lambda}} |v_i|^2 m_i &= \sum_{i\in\overline{\Lambda}} |v_i|^2 \sum_{K:\, K\cap\Omega_i\neq\emptyset} |\Omega_{i,K}| \\
&\overset{\text{Lemma 8.6 or (8.4)}}{\cong} \sum_{i\in\overline{\Lambda}} \sum_{K:\, K\cap\Omega_i\neq\emptyset} |v_i|^2 |K| \\
&= \sum_{K\in\mathcal{T}_h} |K| \sum_{i\in\Lambda_K} |v_i|^2 \\
&\cong \sum_{K\in\mathcal{T}_h} h_K^2 \sum_{i\in\Lambda_K} |v_i|^2.
\end{aligned}$$

The last equivalence applies because, due to the shape regularity of $\{\mathcal{T}_h\}_h$, there is a uniform lower bound for the angles of $K \in \mathcal{T}_h$ (see (3.111)) and thus $|K| \ge Ch_K^2$. □

Setting

$$V_h := \left\{u \in C(\bar{\Omega}) \;\middle|\; u|_K \in \mathcal{P}_1(K) \text{ for all } K \in \mathcal{T}_h,\ u = 0 \text{ on } \widetilde{\Gamma}_3\right\}$$

($= X_h \cap V$ with V from (3.31)), the discrete variational formulation of the FVM is this

Find $u_h \in V_h$ such that

$$a_h(u_h, v_h) = \ell_h(v_h) \quad \text{for all } v_h \in V_h \,. \tag{8.17}$$

The FVM originating from (8.6) by subsequent approximation is based on local (mass) conservation. A comparison with Definition 7.58 for $\widehat{\mathcal{T}}_h := \{\Omega_i\}_{i\in\overline{\Lambda}}$, $\hat{u}_h := u_h$, $\hat{\boldsymbol{p}}_h \cdot \boldsymbol{n}$ at $F \in \widehat{\mathcal{F}}$ constant by (8.12), shows that even if (1), (4) are supposed to hold and (3) is satisfied by definition, there is no clear way to define $\hat{\boldsymbol{p}}_h$ on Ω. If alternatively we set $\hat{\boldsymbol{p}}_h := -\boldsymbol{K}\nabla u_h + \boldsymbol{c}u_h$, then (4) still holds, but (3) is only satisfied for $\boldsymbol{K}$ and $\boldsymbol{c}$ being continuous in each element.

Remark 8.9 Inhomogeneous Dirichlet boundary conditions are treated theoretically and practically as in the finite element method (see Section 3.2.1).

Up to now, the choice of the weighting parameters r_{ij} has remained open.

For this, two cases can be roughly distinguished

1. There exists a pair of indices $(i, j) \in \Lambda \times \overline{\Lambda}$ such that $\mu_{ij} \ll |\gamma_{ij}| d_{ij}$.
2. There is no such pair (i, j) with $\mu_{ij} \ll |\gamma_{ij}| d_{ij}$.

In the second case, an appropriate choice is $r_{ij} \equiv \frac{1}{2}$. To some extent, this can be seen as a generalization of the central difference method to nonuniform grids. The first case corresponds to a locally *convection-dominated* situation and requires a careful selection of the weighting parameters r_{ij}. This will be explained in more detail in Section 10.3.

In general, the weighting parameters are of the following structure:

$$r_{ij} = R\left(\frac{\gamma_{ij}\, d_{ij}}{\mu_{ij}}\right), \tag{8.18}$$

where $R : \mathbb{R} \to [0, 1]$ is some function to be specified. The argument $\frac{\gamma_{ij}\, d_{ij}}{\mu_{ij}}$ is called the *local Péclet number*. Typical examples for this function R are

$$\begin{aligned}
&R(z) = \frac{1}{2}\,[\operatorname{sign}(z) + 1]\,, && \textit{full upwinding,}\\
&R(z) = \begin{cases} (1-\tau)/2\,, & z < 0\,,\\ (1+\tau)/2\,, & z \ge 0\,, \end{cases} && \tau(z) := \max\left\{0, 1 - \frac{2}{|z|}\right\},\\
&R(z) = 1 - \frac{1}{z}\left(1 - \frac{z}{e^z - 1}\right), && \textit{exponential upwinding}\,.
\end{aligned} \tag{8.19}$$

All these functions possess many common properties. For example, for all $z \in \mathbb{R}$,

$$\begin{aligned} &(\text{P1}) \qquad [1 - R(z) - R(-z)]\, z = 0\,, \\ &(\text{P2}) \qquad \left[R(z) - \frac{1}{2}\right] z \geq 0\,, \\ &(\text{P3}) \qquad 1 - [1 - R(z)]\, z \geq 0\,. \end{aligned} \tag{8.20}$$

Note that the constant function $R = \frac{1}{2}$ satisfies the conditions (P1) and (P2) but not (P3).

The Case of the Donald Diagram

Let the domain Ω be triangulated as in the finite element method. Then, following the explanations given in the second part of Section 8.2.1, the corresponding Donald diagram can be created.

The discrete bilinear form in this case is defined by

$$a_h(u_h, v_h) := \langle k\, \nabla u_h, \nabla v_h \rangle_0 + b_h(u_h, v_h) + d_h(u_h, v_h)\,;$$

that is, the principal part of the differential expression is discretized as in the finite element method, where b_h, d_h, and V_h are defined as in the first part of this section. In particular, γ_{ij} is to be understood as an approximation to $\mathbf{n}_{ij} \cdot \boldsymbol{c}$ on the boundary part Γ_{ij} of the corresponding Donald control volume.

8.2.3 Comparison with the Finite Element Method

As we have already seen in Example 8.1, it may happen that a finite volume discretization coincides with a finite difference or finite element discretization. We also mention that the control volumes from that example are exactly the Voronoi polygons to the grid points (i.e. to the nodes of the triangulation).

Here, we will consider this observation in more detail. By $\{\varphi_i\}_{i\in\overline{\Lambda}}$, we denote the nodal basis of the space X_h of continuous, piecewise linear functions on a consistent triangulation of the domain Ω.

Lemma 8.10 *Let $\mathcal{T}_h$ be a consistent triangulation of Ω (in the sense of Section 2.2), all triangles of which are nonobtuse, and consider the corresponding Voronoi diagram in accordance with Theorem 8.5. Then, for an arbitrary triangle $K \in \mathcal{T}_h$, the following relation holds:*

$$\int_K \nabla\varphi_j \cdot \nabla\varphi_i \, dx = -\frac{m_{ij}^K}{d_{ij}}\,, \qquad i, j \in \Lambda_K,\ i \neq j\,.$$

Proof Here, we use some of the notation and the facts prepared at the beginning of Section 3.9. In particular, α_{ij}^K denotes the interior angle of K that is located in opposite the edge with vertices a_i, a_j. Next, the following equality is an obvious fact from elementary geometry: $2\,\sin\alpha_{ij}^K\, m_{ij}^K = \cos\alpha_{ij}^K\, d_{ij}$ (see Problem 8.9). It remains to recall the relation

$$\int_K \nabla\varphi_j \cdot \nabla\varphi_i\, dx = -\frac{1}{2}\cot\alpha_{ij}^K$$

from Lemma 3.50, and the statement immediately follows. □

Corollary 8.11 *Under the assumptions of Lemma* 8.10*, we have for* $k \equiv 1$,

$$\langle \nabla u_h, \nabla v_h\rangle_0 = a_h^0(u_h, v_h) \quad \textit{for all } u_h, v_h \in X_h\,.$$

Proof It is sufficient to verify the relation for $v_h = \varphi_i$ and arbitrary $i \in \Lambda$. First, we see that

$$\langle \nabla u_h, \nabla \varphi_i\rangle_0 = \sum_{K\subset \mathrm{supp}\varphi_i} \int_K \nabla u_h \cdot \nabla\varphi_i\, dx\,.$$

Furthermore,

$$\begin{aligned}\int_K \nabla u_h \cdot \nabla\varphi_i\, dx &= \sum_{j:\,\Lambda_K\ni j} u_j \int_K \nabla\varphi_j\cdot\nabla\varphi_i\, dx\\ &= u_i\int_K \nabla\varphi_i\cdot\nabla\varphi_i\,dx + \sum_{j\neq i:\,\Lambda_K\ni j} u_j\int_K \nabla\varphi_j\cdot\nabla\varphi_i\,dx\,.\end{aligned}$$

Since

$$1 = \sum_{j:\,\Lambda_K\ni j}\varphi_j$$

over K, it follows that

$$\nabla\varphi_i = -\sum_{j\neq i:\,\Lambda_K\ni j}\nabla\varphi_j\,; \tag{8.21}$$

that is, by means of Lemma 8.10,

$$\begin{aligned}\int_K \nabla u_h\cdot\nabla\varphi_i\,dx &= \sum_{j\neq i:\,\Lambda_K\ni j}(u_j - u_i)\int_K\nabla\varphi_j\cdot\nabla\varphi_i\,dx\\ &= \sum_{j\neq i:\,\Lambda_K\ni j}(u_i-u_j)\,\frac{m_{ij}^K}{d_{ij}}\,.\end{aligned} \tag{8.22}$$

Summing over all $K \subset \operatorname{supp}\varphi_i$, we get

$$\langle\nabla u_h,\nabla\varphi_i\rangle_0 = \sum_{j\in\Lambda_i}(u_i-u_j)\,\frac{m_{ij}}{d_{ij}} = a_h^0(u_h,\varphi_i)\,.$$

□

Remark 8.12 By a more sophisticated argumentation, it can be shown that the above corollary remains valid if the diffusion coefficient k is constant on all triangles $K \in \mathcal{T}_h$ and if the approximation μ_{ij} is chosen according to

$$\mu_{ij} := \begin{cases} \dfrac{1}{m_{ij}} \displaystyle\int_{\Gamma_{ij}} k \, d\sigma = \dfrac{k|_K \, m_{ij}^K + k|_{K'} \, m_{ij}^{K'}}{m_{ij}}, & m_{ij} > 0, \\ 0, & m_{ij} = 0, \end{cases} \tag{8.23}$$

where K, K' are both triangles sharing the vertices a_i, a_j.

Treatment of Matrix-valued Diffusion Coefficients

Corollary 8.11 and Remark 8.12 are valid only in the spatial dimension $d = 2$. However, for more general control volumes, higher spatial dimensions, or not necessarily scalar diffusion coefficients, weaker statements can be proven.

As an example, we will state the following fact. As a by-product, we also obtain an idea for how to derive discretizations in the case of matrix-valued diffusion coefficients. Namely, equation (8.22) can also be written in the form

$$\int_K \nabla u_h \cdot \nabla \varphi_i \, dx = - \int_{\partial\Omega_i \cap K} \nabla u_h \cdot \mathbf{n} \, d\sigma \,,$$

and this carries over to the general case.

Lemma 8.13 *Let $\mathcal{T}_h$ be a consistent triangulation of Ω, where in the case of the Voronoi diagram, it is additionally required that all triangles be nonobtuse. Furthermore, assume that the diffusion matrix $\boldsymbol{K} : \Omega \to \mathbb{R}^{2,2}$ is constant on the single elements of $\mathcal{T}_h$. Then for any $i \in \Lambda$ and $K \in \mathcal{T}_h$, we have*

$$\int_K (\boldsymbol{K}\nabla u_h) \cdot \nabla \varphi_i \, dx = - \int_{\partial\Omega_i \cap K} (\boldsymbol{K}\nabla u_h) \cdot \mathbf{n} \, d\sigma \quad \text{for all } u_h \in X_h,$$

where $\{\Omega_i\}_{i \in \overline{\Lambda}}$ is either a Voronoi or a Donald diagram and $\mathbf{n}$ denotes the outer unit normal with respect to Ω_i.

Without difficulties, the proof can be carried over from the proof of a related result in [39, Lemma 6.1].

Now we will show how to use this fact to formulate discretizations for the case of matrix-valued diffusion coefficients. Namely, using relation (8.21), we easily see that

$$\begin{aligned} \int_{\partial\Omega_i \cap K} (\boldsymbol{K}\nabla u_h) \cdot \mathbf{n} \, d\sigma &= \sum_{j:\, \Lambda_K \ni j} \int_{\partial\Omega_i \cap K} u_j (\boldsymbol{K}\nabla \varphi_j) \cdot \mathbf{n} \, d\sigma \\ &= \sum_{j \neq i:\, \Lambda_K \ni j} (u_j - u_i) \int_{\partial\Omega_i \cap K} (\boldsymbol{K}\nabla \varphi_j) \cdot \mathbf{n} \, d\sigma \,. \end{aligned}$$

Summing over all triangles that lie in the support of φ_i, we obtain by Lemma 8.13 the relation

$$\int_\Omega (\boldsymbol{K}\nabla u_h)\cdot\nabla\varphi_i\,dx = \sum_{j\in\Lambda_i}\left(u_i - u_j\right)\int_{\partial\Omega_i}(\boldsymbol{K}\nabla\varphi_j)\cdot\boldsymbol{\nu}\,d\sigma\,. \tag{8.24}$$

With the definition

$$\mu_{ij} := \begin{cases} \dfrac{d_{ij}}{m_{ij}}\displaystyle\int_{\partial\Omega_i}(\boldsymbol{K}\nabla\varphi_j)\cdot\boldsymbol{\nu}\,d\sigma\,, & m_{ij} > 0\,,\\ 0\,, & m_{ij} = 0\,, \end{cases} \tag{8.25}$$

it follows that

$$\int_\Omega (\boldsymbol{K}\nabla u_h)\cdot\nabla\varphi_i\,dx = \sum_{j\in\Lambda_i}\mu_{ij}\left(u_i - u_j\right)\frac{m_{ij}}{d_{ij}}\,.$$

Note that, in the case of Voronoi diagrams, (8.23) is a special case of the choice (8.25).

Consequently, in order to obtain a discretization for the case of a matrix-valued diffusion coefficient, it is sufficient to replace in the bilinear form b_h and, if the Voronoi diagram is used, also in a_h^0, the terms involving μ_{ij} according to formula (8.25) to achieve accordance in the diffusion of the FVM with a conforming consistent FEM.

Remark 8.14 The choices (8.25)/(8.23) eliminate the consistency error in the diffusion part of the node-oriented FVM viewed as a nonconsistent Galerkin approximation in the sense of Definition 6.5. In Section 8.3, for the cell-centred FVM other choices are preferred.

Implementation of the Finite Volume Method

In principle, the FVM can be implemented in different ways. If the linear system of equations is implemented in a node-oriented manner (as in finite difference methods), the entries of the system matrix A_h and the components of the right-hand side $\boldsymbol{q}_h$ can be taken directly from (8.11).

On the other hand, an element-oriented assembling is possible, too. This approach is preferable, especially in the case where an existing finite element program is to be extended by a finite volume module. The idea of how to do this is suggested by equation (8.24). Namely, for any triangle $K \in \mathcal{T}_h$, the restricted bilinear form $a_{h,K}$ with the appropriate definition of μ_{ij} according to (8.25) is defined as follows:

$$\begin{aligned} &a_{h,K}(u_h, v_h)\\ &:= \sum_{i\in\Lambda_0} v_i\left\{\sum_{j\neq i:\,\Lambda_K\ni j}\left\{\mu_{ij}\frac{u_i - u_j}{d_{ij}} + \gamma_{ij}\left[r_{ij}u_i + \left(1 - r_{ij}\right)u_j\right]\right\}m_{ij}^K + r_i u_i m_i^K\right\}\\ &\quad + \sum_{i\in\partial\Lambda_2}\alpha_i\,u_i\,v_i\,\widetilde{m}_{i2}^K, \end{aligned}$$

where $m_i^K := |\Omega_i \cap K|$, $\widetilde{m}_{i2}^K := |\partial\Omega_i \cap \widetilde{\Gamma}_2 \cap K|_1$. Then the contribution of the triangle K to the matrix entry $(A_h)_{ij}$ of the matrix A_h is equal to $a_{h,K}(\varphi_j, \varphi_i)$. In the same way, the right-hand side of (8.13) can be split elementwise.

8.2.4 Properties of the Discretization

Here, we will give a brief overview of basic properties of FVMs. For the sake of brevity, we restrict ourselves to the case of a constant scalar diffusion coefficient $k > 0$ and to pure Dirichlet boundary conditions ($\widetilde{\Gamma}_3 = \partial\Omega$). Then in the definitions (8.14), (8.15) the index sets $\Lambda_0, \overline{\Lambda}$ can be replaced by Λ, and the boundary terms are needless. In particular, the right-hand side of (8.17) takes the form $\ell_h(v_h) := \langle f, v_h \rangle_{0,h}$. Moreover, in the case of a constant diffusion coefficient, we select $\mu_{ij} := k$ for all $i \in \Lambda$, $j \in \Lambda_i$, in accordance with (8.23). The space V_h gets the form (2.30).

Lemma 8.15 *Suppose the approximations γ_{ij} of $\mathbf{n}_{ij} \cdot \mathbf{c}|_{\Gamma_{ij}}$ satisfy $\gamma_{ji} = -\gamma_{ij}$ and the r_{ij} are defined by* (8.18) *with a function R satisfying* (P1). *Then we get for all $u_h, v_h \in V_h$,*

$$
\begin{aligned}
b_h(u_h, v_h) &= \frac{1}{2} \sum_{i\in\Lambda} \sum_{j\in\Lambda_i} u_i\, v_i\, \gamma_{ij}\, m_{ij} \\
&+ \frac{1}{2} \sum_{i\in\Lambda} \sum_{j\in\Lambda_i} \left[\left(r_{ij} - \frac{1}{2} \right) (u_i - u_j)(v_i - v_j) + \frac{1}{2} (u_j v_i - u_i v_j) \right] \gamma_{ij}\, m_{ij}\,.
\end{aligned}
$$

Proof First, we observe that b_h can be rewritten as follows:

$$
\begin{aligned}
b_h(u_h, v_h) = &\sum_{i\in\Lambda} \sum_{j\in\Lambda_i} v_i \left[(1 - r_{ij})\, u_j - \left(\frac{1}{2} - r_{ij} \right) u_i \right] \gamma_{ij}\, m_{ij} \\
&+ \frac{1}{2} \sum_{i\in\Lambda} \sum_{j\in\Lambda_i} u_i\, v_i\, \gamma_{ij}\, m_{ij}\,.
\end{aligned}
\tag{8.26}
$$

In the first term, we change the order of summation and rename the indices:

$$
\begin{aligned}
b_h(u_h, v_h) = &\sum_{i\in\Lambda} \sum_{j\in\Lambda_i} v_j \left[(1 - r_{ji})\, u_i - \left(\frac{1}{2} - r_{ji} \right) u_j \right] \gamma_{ji}\, m_{ji} \\
&+ \frac{1}{2} \sum_{i\in\Lambda} \sum_{j\in\Lambda_i} u_i\, v_i\, \gamma_{ij}\, m_{ij}\,.
\end{aligned}
$$

Next, we make use of the following relations, which easily result from $d_{ji} = d_{ij}$ and the assumptions on γ_{ij} and r_{ij}:

$$
(1 - r_{ji})\, \gamma_{ji} = -r_{ij}\, \gamma_{ij}\,, \qquad \left(\frac{1}{2} - r_{ji} \right) \gamma_{ji} = \left(\frac{1}{2} - r_{ij} \right) \gamma_{ij}\,.
$$

So we get, due to $m_{ji} = m_{ij}$,

$$b_h(u_h, v_h) = \sum_{i\in\Lambda}\sum_{j\in\Lambda_i} v_j \left[-r_{ij}u_i - \left(\frac{1}{2} - r_{ij}\right)u_j\right] \gamma_{ij}\, m_{ij} + \frac{1}{2}\sum_{i\in\Lambda}\sum_{j\in\Lambda_i} u_i\, v_i\, \gamma_{ij}\, m_{ij}\,.$$

Taking the arithmetic mean of both representations of b_h, we arrive at

$$\begin{aligned} b_h(u_h, v_h) &= \frac{1}{2}\sum_{i\in\Lambda}\sum_{j\in\Lambda_i} u_i\, v_i\, \gamma_{ij}\, m_{ij} \\ &+ \frac{1}{2}\sum_{i\in\Lambda}\sum_{j\in\Lambda_i}\left[(1-r_{ij})\, u_j v_i - r_{ij}u_i v_j - \left(\frac{1}{2} - r_{ji}\right)(u_i v_i + u_j v_j)\right]\gamma_{ij} m_{ij} \\ &= \frac{1}{2}\sum_{i\in\Lambda}\sum_{j\in\Lambda_i}\left[\left(\frac{1}{2} - r_{ij}\right)(u_j v_i + u_i v_j - u_i v_i - u_j v_j)\right. \\ &\qquad \left. + \frac{1}{2}(u_j v_i - u_i v_j)\right]\gamma_{ij} m_{ij} + \frac{1}{2}\sum_{i\in\Lambda}\sum_{j\in\Lambda_i} u_i v_i \gamma_{ij} m_{ij}\,. \end{aligned}$$

□

Corollary 8.16 *Let* $c_1, c_2, \nabla\cdot\boldsymbol{c} \in C(\overline{\Omega})$. *Under the assumptions of Lemma* 8.15 *and also assuming property* (P2) *for R, the bilinear form* b_h *satisfies for all* $v_h \in V_h$ *the estimate*

$$b_h(v_h, v_h) \geq \frac{1}{2}\sum_{i\in\Lambda} v_i^2 \left[\int_{\Omega_i}\nabla\cdot\boldsymbol{c}\, dx + \sum_{j\in\Lambda_i}\int_{\Gamma_{ij}}(\gamma_{ij} - \boldsymbol{\mathfrak{n}}_{ij}\cdot\boldsymbol{c})\, d\sigma\right]. \qquad (8.27)$$

Proof Due to $\left(r_{ij} - \frac{1}{2}\right)\gamma_{ij} \geq 0$, because of property (P2) in (8.20), it immediately follows that

$$b_h(v_h, v_h) \geq \frac{1}{2}\sum_{i\in\Lambda}\sum_{j\in\Lambda_i} v_i^2\, \gamma_{ij}\, m_{ij} = \frac{1}{2}\sum_{i\in\Lambda} v_i^2 \sum_{j\in\Lambda_i}\gamma_{ij}\, m_{ij}\,.$$

For the inner sum, we can write

$$\begin{aligned} \sum_{j\in\Lambda_i}\gamma_{ij}\, m_{ij} &= \sum_{j\in\Lambda_i}\int_{\Gamma_{ij}}\gamma_{ij}\, d\sigma \\ &= \sum_{j\in\Lambda_i}\int_{\Gamma_{ij}}\boldsymbol{\mathfrak{n}}_{ij}\cdot\boldsymbol{c}\, d\sigma + \sum_{j\in\Lambda_i}\int_{\Gamma_{ij}}(\gamma_{ij} - \boldsymbol{\mathfrak{n}}_{ij}\cdot\boldsymbol{c})\, d\sigma\,. \end{aligned}$$

The first term can be rewritten as an integral over the boundary of Ω_i, i.e.

$$\sum_{j\in\Lambda_i}\int_{\Gamma_{ij}}\boldsymbol{\mathfrak{n}}_{ij}\cdot\boldsymbol{c}\, d\sigma = \int_{\partial\Omega_i}\boldsymbol{\mathfrak{n}}\cdot\boldsymbol{c}\, d\sigma\,.$$

By Gauss' divergence theorem, it follows that

$$\int_{\partial\Omega_i} \boldsymbol{n} \cdot \boldsymbol{c}\, d\sigma = \int_{\Omega_i} \nabla \cdot \boldsymbol{c}\, dx\,.$$

□

Remark 8.17 If the approximations γ_{ij} are chosen according to (8.8), then $\gamma_{ji} = -\gamma_{ij}$ holds and Corollary 8.16 can be applied. The estimate (8.27) simplifies to

$$b_h(v_h, v_h) \geq \frac{1}{2} \sum_{i \in \Lambda} v_i^2 \int_{\Omega_i} \nabla \cdot \boldsymbol{c}\, dx\,.$$

Using a similar argument as in the treatment of the term $\sum_{j\in\Lambda_i} \gamma_{ij}\, m_{ij}$ in the proof of Corollary 8.16, the value $d_h(v_h, v_h)$ can be represented as follows:

$$\begin{aligned} d_h(v_h, v_h) &= \sum_{i \in \Lambda} r_i\, v_i^2\, m_i = \sum_{i \in \Lambda} v_i^2 \int_{\Omega_i} r_i\, dx \\ &= \sum_{i \in \Lambda} v_i^2 \int_{\Omega_i} r\, dx + \sum_{i \in \Lambda} v_i^2 \int_{\Omega_i} (r_i - r)\, dx\,. \end{aligned} \tag{8.28}$$

The second term vanishes if the approximations r_i are defined as in (8.10).

Theorem 8.18 *Let the r_{ij} be defined by (8.18) with R satisfying* (P1) *and* (P2). *Suppose $k > 0$, $c_1, c_2, \nabla \cdot \boldsymbol{c}, r \in C(\overline{\Omega})$, $r + \frac{1}{2}\nabla \cdot \boldsymbol{c} \geq r_0 = const \geq 0$ on Ω and that the approximations γ_{ij}, respectively, r_i, are chosen according to* (8.8), *respectively,* (8.10). *Under the assumptions of Lemma* 8.10, *we have for all $v_h \in V_h$,*

$$a_h\,(v_h, v_h) \geq k\,\langle \nabla v_h, \nabla v_h \rangle_0 + r_0 \sum_{i \in \Lambda} v_i^2\, m_i = k\,|v_h|_1^2 + r_0 \|v_h\|_{0,h}^2\,.$$

Proof We start with the consideration of $a_h^0\,(v_h, v_h)$. Due to Corollary 8.11, the relation

$$a_h^0\,(v_h, v_h) = k\,\langle \nabla v_h, \nabla v_h \rangle_0 = k\,|v_h|_1^2$$

holds. Furthermore, by Remark 8.17 and equation (8.28), we have

$$b_h\,(v_h, v_h) + d_h\,(v_h, v_h) \geq \sum_{i \in \Lambda} v_i^2 \int_{\Omega_i} \left(\frac{1}{2}\nabla \cdot \boldsymbol{c} + r\right) dx \geq r_0 \sum_{i \in \Lambda} v_i^2\, m_i\,.$$

Since by definition,

$$a_h\,(v_h, v_h) = a_h^0\,(v_h, v_h) + b_h\,(v_h, v_h) + d_h\,(v_h, v_h),$$

both relations yield the assertion. □

Corollary 8.19 *Under the assumptions of Theorem* 8.18 *and for a shape-regular family of triangulations* $\{\mathcal{T}_h\}_h$, *there exists a constant* $\alpha > 0$ *independent of* h *such that*

$$a_h(v_h, v_h) \geq \alpha \|v_h\|_1^2 \quad \text{for all } v_h \in V_h\,.$$

Proof By Remark 8.8 and Theorem 8.18,

$$a_h(v_h, v_h) \geq k\,|v_h|_1^2 + r_0 C_1^2 \|v_h\|_0^2\,,$$

i.e. we can take $\alpha := \min\{k; r_0 C_1^2\}$. □

Theorem 8.18 (or Corollary 8.19) asserts the stability of the method. It is the fundamental result for the proof of an error estimate.

Theorem 8.20 *Let* $\{\mathcal{T}_h\}_{h\in(0,\bar h]}$ *be a shape-regular family of consistent triangulations, where in the case of the Voronoi diagram it is additionally required that all triangles be nonobtuse. Furthermore, suppose in the (simplified) problem* (8.5) *that* $k > 0$, $c_1, c_2, \nabla\cdot\boldsymbol{c}, r \in C(\overline{\Omega})$, $r + \frac{1}{2}\nabla\cdot\boldsymbol{c} \geq r_0 = \text{const} > 0$ *on* Ω, $f \in C^1(\overline{\Omega})$, *and that the approximations* γ_{ij}, *respectively*, r_i, *are chosen according to* (8.8), *respectively,* (8.10). *Let the* r_{ij} *be defined by* (8.18) *with* R *satisfying* (P1) *and* (P2). *If the exact solution* u *of* (8.5) *belongs to* $H^2(\Omega)$ *and* $u_h \in V_h$ *denotes the solution of* (8.17), *then*

$$\|u - u_h\|_1 \leq C\,h\big[\|u\|_2 + |f|_{1,\infty}\big],$$

where the constant $C > 0$ *is independent of* h.

Proof The proof is based on a similar idea as in the proof of Strang's second lemma (Theorem 6.10) in Section 6.1 with the difference that instead of $a_h(u, v_h)$ the term $a_h(I_h(u), v_h)$ is added and subtracted, where $I_h : C(\bar\Omega) \to V_h$ is the interpolation operator defined in (3.83). In contrast to the situation considered in Section 6.1, here the bilinear form a_h cannot be prolongated continuously to the space $U(h) = H_0^1(\Omega) + V_h$ but only to $\big(H_0^1(\Omega) \cap H^2(\Omega)\big) + V_h$.

Setting $v_h := u_h - I_h(u)$ and noting that $\ell_h(v_h) = \langle f, v_h\rangle_{0,h}$, we have

$$\begin{aligned} a_h(v_h, v_h) &= a_h(u_h, v_h) - a_h(I_h(u), v_h) = \ell_h(v_h) - a_h(I_h(u), v_h) \\ &= \ell_h(v_h) - \tilde\ell_h(v_h) + \tilde\ell_h(v_h) - a_h(I_h(u), v_h), \end{aligned}$$

where $\tilde\ell_h(v_h) := \sum_{i\in\Lambda} v_i \int_{\Omega_i} f\,dx$.

Now, for $f \in C^1(\overline{\Omega})$ and the choice $f_i := f(a_i)$, it is easy to see that

$$|f_i - f(x)| \leq |f|_{1,\infty} \max_{K:\,\Lambda_K \ni i} h_K \leq C\,h|f|_{1,\infty} \quad \text{for all } x \in \Omega_i\,,$$

hence

$$
\begin{aligned}
|\ell_h(v_h) - \tilde{\ell}_h(v_h)| &= \left|\sum_{i\in\Lambda} v_i \int_{\Omega_i} (f_i - f)\,dx\right| \le C\,h|f|_{1,\infty} \sum_{i\in\Lambda} |v_i| m_i \\
&\le C\,h|f|_{1,\infty}\left(\sum_{i\in\Lambda} v_i^2 m_i\right)^{1/2} \underbrace{\left(\sum_{i\in\Lambda} m_i\right)^{1/2}}_{\le\sqrt{|\Omega|}} \le C\,h|f|_{1,\infty}\|v_h\|_{0,h}.
\end{aligned}
$$

For the other choice of f_i (see (8.10)), the same estimate is trivially satisfied. The difficult part of the proof is to estimate the second consistency error

$$
|\tilde{\ell}_h(v_h) - a_h(I_h(u), v_h)| = \left|\sum_{i\in\Lambda} v_i \int_{\Omega_i} Lu\,dx - a_h(I_h(u), v_h)\right|,
$$

where $Lu = -\nabla\cdot(k\,\nabla u - \mathbf{c}\,u) + r\,u \;= f$ is considered as an equation in $L^2(\Omega)$. This is very extensive, and so we will omit the details. A complete proof of the following result is given in the paper [77]:

$$
\left|\sum_{i\in\Lambda} v_i \int_{\Omega_i} Lu\,dx - a_h(I_h(u), v_h)\right| \le C\,h\|u\|_2\left(|v_h|_1^2 + \|v_h\|_{0,h}^2\right)^{1/2}. \tag{8.29}
$$

Putting both estimates together and taking into consideration Remark 8.8, we arrive at

$$
\begin{aligned}
a_h(v_h, v_h) &\le C\,h\big[\|u\|_2 + |f|_{1,\infty}\big]\left(|v_h|_1^2 + \|v_h\|_{0,h}^2\right)^{1/2} \\
&\le C\,h\big[\|u\|_2 + |f|_{1,\infty}\big]\|v_h\|_1 .
\end{aligned}
$$

By Corollary 8.19, we conclude from this that

$$
\|v_h\|_1 \le C\,h\big[\|u\|_2 + |f|_{1,\infty}\big].
$$

It remains to apply the triangle inequality and the standard interpolation error estimate (cf. Theorem 3.32 with $k = 1$ or Theorem 3.38)

$$
\|u - u_h\|_1 \le \|u - I_h(u)\|_1 + \|v_h\|_1 \le C\,h\big[\|u\|_2 + |f|_{1,\infty}\big].
$$

□

We point out that the error measured in the H^1-seminorm is of the same order as for the finite element method with linear finite elements.

Now we will turn to the investigation of some interesting properties of the method.

Inverse Monotonicity

The so-called *inverse monotonicity* is a further important property of the boundary value problem (8.5) that is inherited by the finite volume discretization without any additional restrictive assumptions. Namely, it is well known that under appropriate assumptions on the coefficients, the solution u is nonnegative if the (continuous) right-hand side f in (8.5) is nonnegative in Ω.

We will demonstrate that this remains true for the approximative solution u_h. Only at this place is the property (P3) of the weighting function R used; the preceding results are also valid for the simple case $R(z) \equiv \frac{1}{2}$.

There is a close relation to the maximum principles investigated in Sections 1.4 and 3.9. However, the result given here is weaker, and the proof is based on a different technique.

Theorem 8.21 *Let the assumptions of Theorem* 8.18 *be satisfied, but R in (8.18) has to satisfy* (P1)–(P3). *Further, suppose that $f \in C(\overline{\Omega})$ and $f(x) \geq 0$ for all $x \in \Omega$. Moreover, in the case of the Donald diagram, only the weighting function $R(z) = \frac{1}{2}\left[\operatorname{sign}(z) + 1\right]$ is permitted.*

Then

$$u_h(x) \geq 0 \quad \textit{for all } x \in \Omega\,.$$

Proof We start with the case of the Voronoi diagram. Let u_h be the solution of (8.17) with $f(x) \geq 0$ for all $x \in \Omega$. Then we have the following additive decomposition of u_h:

$$u_h = u_h^+ - u_h^-\,, \qquad \text{where } u_h^+ := \max\left\{0, u_h\right\}\,.$$

In general, u_h^+, u_h^- do not belong to V_h. So we interpolate them in V_h and set in (8.17) $v_h := I_h(u_h^-)$, where $I_h : C\left(\bar{\Omega}\right) \to V_h$ is the interpolation operator (3.83). It follows that

$$0 \leq \langle f, v_h \rangle_{0,h} = a_h\left(u_h, v_h\right) = a_h\left(I_h(u_h^+), I_h(u_h^-)\right) - a_h\left(I_h(u_h^-), I_h(u_h^-)\right)\,.$$

By Theorem 8.18, we have

$$k\left|I_h(u_h^-)\right|_1^2 \leq a_h\left(I_h(u_h^-), I_h(u_h^-)\right) \leq a_h\left(I_h(u_h^+), I_h(u_h^-)\right)\,.$$

If we were able to show that $a_h\left(I_h(u_h^+), I_h(u_h^-)\right) \leq 0$, then the theorem would be proven, because this relation implies $\left|I_h(u_h^-)\right|_1 = 0$, and from this we immediately get $u_h^- = 0$, and so $u_h = u_h^+ \geq 0$.

Since $u_i^+ u_i^- = 0$ for all $i \in \Lambda$, it follows from (8.26) in the proof of Lemma 8.15 that

$$b_h\left(I_h(u_h^+), I_h(u_h^-)\right) = \sum_{i\in\Lambda}\sum_{j\in\Lambda_i}\left(1 - r_{ij}\right)\, u_j^+\, u_i^-\, \gamma_{ij}\, m_{ij}\,. \tag{8.30}$$

Furthermore, obviously $d_h\left(I_h(u_h^+), I_h(u_h^-)\right) = 0$ holds. Thus,

$$a_h\left(I_h(u_h^+), I_h(u_h^-)\right) = \sum_{i\in\Lambda}\sum_{j\in\Lambda_i}\left[-\frac{\mu_{ij}}{d_{ij}}u_j^+ + \gamma_{ij}\left(1-r_{ij}\right)u_j^+\right]u_i^- m_{ij}$$

$$= -\sum_{i\in\Lambda}\sum_{j\in\Lambda_i}\frac{\mu_{ij}}{d_{ij}}\left[1-\frac{\gamma_{ij}\,d_{ij}}{\mu_{ij}}\left(1-r_{ij}\right)\right]u_j^+ u_i^- m_{ij}\,.$$

Due to $1-[1-R(z)]\,z \geq 0$ for all $z \in \mathbb{R}$ (cf. property (P3) in (8.20)) and $u_j^+ u_i^- \geq 0$, it follows that

$$a_h\left(I_h(u_h^+), I_h(u_h^-)\right) \leq 0\,.$$

So it remains to investigate the case of the Donald diagram. The function $R(z) = \frac{1}{2}[\operatorname{sign}(z)+1]$ has the property

$$[1-R(z)]\,z = \frac{1}{2}[1-\operatorname{sign}(z)]\,z \leq 0 \quad \text{for all } z \in \mathbb{R},$$

that is (cf. (8.30)),

$$b_h\left(I_h(u_h^+), I_h(u_h^-)\right) \leq 0\,.$$

Taking $u_i^+ u_i^- = 0$ into consideration, we get

$$a_h\left(I_h(u_h^+), I_h(u_h^-)\right) \leq \left\langle k\,\nabla I_h(u_h^+), \nabla I_h(u_h^-)\right\rangle_0$$

$$= k\sum_{i\in\Lambda}\sum_{j\in\Lambda_i} u_j^+ u_i^- \left\langle \nabla\varphi_j, \nabla\varphi_i\right\rangle_0\,.$$

Now Lemma 3.50 implies that

$$a_h\left(I_h(u_h^+), I_h(u_h^-)\right) \leq -\frac{k}{2}\sum_{i\in\Lambda}\sum_{j\in\Lambda_i} u_j^+ u_i^- \left(\cot\alpha_{ij}^K + \cot\alpha_{ij}^{K'}\right),$$

where K and K' are a pair of triangles sharing a common edge with vertices a_i, a_j.

Since all triangles are nonobtuse, we have $\cot\alpha_{ij}^K \geq 0$, $\cot\alpha_{ij}^{K'} \geq 0$, and hence

$$a_h\left(I_h(u_h^+), I_h(u_h^-)\right) \leq 0\,.$$

□

Global Conservativity

The property of global conservativity is typically studied for the following special case of the boundary value problem (8.5), in which $r = 0$ and $\widetilde{\Gamma}_1 = \partial\Omega$:

$$\begin{aligned} \nabla\cdot\boldsymbol{p} &= f && \text{in } \Omega,\\ \boldsymbol{p} &= -k\nabla u + \boldsymbol{c}u && \text{in } \Omega,\\ -\boldsymbol{p}\cdot\boldsymbol{n} &= g && \text{on } \partial\Omega. \end{aligned} \tag{8.31}$$

Integrating the first differential equation over Ω, we conclude from Gauss' divergence theorem that

$$\int_\Omega \nabla \cdot \boldsymbol{p}\, dx = \int_{\partial\Omega} \boldsymbol{n} \cdot \boldsymbol{p}\, d\sigma = -\int_{\partial\Omega} g\, d\sigma\,,$$

and hence

$$\int_{\partial\Omega} g\, d\sigma + \int_\Omega f\, dx = 0\,. \tag{8.32}$$

This is a necessary compatibility condition for the data describing the balance between the total flow over the boundary and the distributed sources. We will demonstrate that the discretization satisfies a discretized version of this compatibility condition, which is called *discrete global conservativity.*

The finite volume discretization (8.13) of the problem (8.31) in X_h reads as follows:

$$\begin{aligned} &\sum_{i\in\overline{\Lambda}} v_i \sum_{j\in\Lambda_i} \left\{\mu_{ij} \frac{u_i - u_j}{d_{ij}} + \gamma_{ij}\left[r_{ij}\, u_i + \left(1 - r_{ij}\right)\, u_j\right]\right\} m_{ij} \\ &\qquad = \sum_{i\in\overline{\Lambda}} f_i\, v_i\, m_i + \sum_{i\in\partial\Lambda} g_i\, v_i\, \widetilde{m}_{i1} \end{aligned} \tag{8.33}$$

(note that here $\Lambda_0 = \overline{\Lambda}$ and $\partial\Lambda_1 = \partial\Lambda$).

Since the particular function $i_h :\equiv 1$ belongs to X_h, we may take $v_h = i_h$ in the discretization (8.33). Then, repeating the above symmetry argument (cf. the proof of Lemma 8.15), we get

$$\sum_{i\in\overline{\Lambda}} \sum_{j\in\Lambda_i} \mu_{ij}\, \left(u_i - u_j\right) \frac{m_{ij}}{d_{ij}} = -\sum_{i\in\overline{\Lambda}} \sum_{j\in\Lambda_i} \mu_{ij}\, \left(u_i - u_j\right) \frac{m_{ij}}{d_{ij}}\,,$$

that is,

$$\sum_{i\in\overline{\Lambda}} \sum_{j\in\Lambda_i} \mu_{ij}\, \left(u_i - u_j\right) \frac{m_{ij}}{d_{ij}} = 0\,.$$

On the other hand, using the same argument, we have

$$\begin{aligned} &\sum_{i\in\overline{\Lambda}} \sum_{j\in\Lambda_i} \left[r_{ij}\, u_i + \left(1 - r_{ij}\right) u_j\right] \gamma_{ij}\, m_{ij} \\ &= \sum_{i\in\overline{\Lambda}} \sum_{j\in\Lambda_i} \left[r_{ji}\, u_j + \left(1 - r_{ji}\right)\, u_i\right]\, \gamma_{ji}\, m_{ji} \\ &= -\sum_{i\in\overline{\Lambda}} \sum_{j\in\Lambda_i} \left[\left(1 - r_{ij}\right) u_j + r_{ij}\, u_i\right] \gamma_{ij}\, m_{ij}\,. \end{aligned} \tag{8.34}$$

Consequently, this term vanishes, too, and we arrive at

$$0 = \sum_{i\in\overline{\Lambda}} f_i\, m_i + \sum_{i\in\partial\Lambda} g_i\, \widetilde{m}_{i1}\,. \tag{8.35}$$

This is the discrete analogue of (8.32). It is precisely in this sense that the FVM is globally conservative.

We point out that if the values f_i and g_i are chosen according to (8.10), we get

$$\sum_{i\in\overline{\Lambda}} f_i\, m_i = \sum_{i\in\overline{\Lambda}} \int_{\Omega_i} f\, dx = \int_{\Omega} f\, dx, \quad \sum_{i\in\partial\Lambda} g_i\, \widetilde{m}_{i1} = \sum_{i\in\partial\Lambda} \int_{\partial\Omega_i\cap\partial\Omega} g\, d\sigma = \int_{\partial\Omega} g\, d\sigma$$

separately, that is we can recover the compatibility condition (8.32) term by term from the discrete scheme (8.33).

The relation (8.35) ensures the solvability of the discrete system (8.33).

In the case of the Donald diagram, we obviously have

$$\langle k\nabla u_h, \nabla v_h\rangle_0 = 0\,.$$

Since the proof of (8.34) does not depend on the particular type of the control volumes, the property of discrete global conservativity in the sense of (8.35) is satisfied for the Donald diagram, too.

Exercises

Problem 8.2 Show that Voronoi polygons are convex.

Problem 8.3 Suppose $|\overline{\Lambda}| \geq 3$. Show that at each Voronoi vertex at least three Voronoi polygons meet unless all elements of $\overline{\Lambda}$ are collinear

Problem 8.4 Suppose that the domain $\Omega \subset \mathbb{R}^2$ can be triangulated by means of equilateral triangles with edge length $h > 0$ in an admissible way.

a) Give the shape of the control domains in the case of the Voronoi and the Donald diagrams.
b) Using the control domains from subproblem a), discretize the Poisson equation with homogeneous Dirichlet boundary conditions by means of the FVM.

Problem 8.5 Under the simplifying assumptions of Subsection 8.2.4, formulate an existence result for the weak solution in $H_0^1(\Omega)$ of the boundary value problem (8.5) similar to Theorem 3.13. In particular, what form will condition (3.18) take?

Problem 8.6 Verify Remark 8.7; i.e. show that $\langle\cdot,\cdot\rangle_{0,h}$ possesses the properties of a scalar product on X_h.

Problem 8.7 Prove Remark 8.8 in detail.

Problem 8.8 Verify or disprove the properties (P1)–(P3) for the three weighting functions given before (8.20) and for $R \equiv \frac{1}{2}$.

Problem 8.9 Let K be a nonobtuse triangle. The length of the segments $\Gamma_{ij}^K := \Gamma_{ij} \cap K$, $i, j \in \Lambda_K$, $i \neq j$, is denoted by m_{ij}^K, and d_{ij} is the length of the edge connecting the vertex a_i with a_j. Finally, α_{ij}^K is the interior angle of K opposite that edge.

Demonstrate the following relation: $2m_{ij}^K = d_{ij} \cot \alpha_{ij}^K, \quad i, j \in \Lambda_K,\ i \neq j$.

Problem 8.10 a) Formulate problem (8.17) in terms of an algebraic system of type (1.34).

b) Show that for the resulting matrix $A_h \in \mathbb{R}^{M_1, M_1}$, where M_1 is the number of elements of the index set Λ, the following relation is valid: $A_h^T \mathbf{1} \geq \mathbf{0}$. Here, as in Section 1.4, $\mathbf{0}$ and $\mathbf{1}$ denote vectors of dimension M_1, *all* components of which are equal to 0 and 1, respectively.
(This is nothing other than the property (1.35) (3) (i) except for the transpose of A_h.)

Programming project 8.1 Add to the code from Project 3.3 (or from Problems 3.2, 3.1) a finite volume module for the case of scalar diffusion coefficients (i.e. $\boldsymbol{K} = k\boldsymbol{I}$ with $k : \Omega \to \mathbb{R}$) and Voronoi diagrams. Implement different weights according to (8.19). Test your implementation for the example from Project 3.1 (for comparison) and for the following problem: $\Omega := (0,1)^2$, $\Gamma_1 := \emptyset$, $k := 1$, $\boldsymbol{c} := (1,1)^T$, $r := 2$, and f is such that $u(x,y) = xy\big(1 - e^{(x-1)/k}\big)\big(1 - e^{(y-1)/k}\big)$ is the exact solution.

8.3 A Cell-oriented Finite Volume Method for Linear Elliptic Differential Equations of Second Order

8.3.1 The One-Dimensional Case

Without loss of generality, we consider the interval $\Omega := (0, 1)$ and points $0 =: x_{\frac{1}{2}} < x_{\frac{3}{2}} < \ldots < x_{N+\frac{1}{2}} := 1$, $N \in \mathbb{N}$, defining the control volumes

$$\Omega_i := (x_{i-\frac{1}{2}}, x_{i+\frac{1}{2}}), \quad i = 1, \ldots, N,$$

with $h_i := |\Omega_i| := x_{i+\frac{1}{2}} - x_{i-\frac{1}{2}}$, $i = 1, \ldots, N$, $h_0 := h_{N+1} := 0$, i.e.

$$\sum_{i=1}^{N} h_i = 1, \quad h := \max_{1 \leq i \leq N} h_i, \tag{8.36}$$

and "grid points" a_i selected in the control volumes,

$$a_i \in \Omega_i, \quad i = 1, \ldots, N.$$

Supplementing $a_0 := x_{\frac{1}{2}}$ and $a_{N+1} := x_{N+\frac{1}{2}}$, we have

$$
\begin{aligned}
0 = a_0 = x_{\frac{1}{2}} < a_1 < x_{\frac{3}{2}} < \ldots < x_{i-\frac{1}{2}} < a_i < x_{i+\frac{1}{2}} < \ldots \\
< a_N < x_{N+\frac{1}{2}} = a_{N+1} = 1\,.
\end{aligned}
\tag{8.37}
$$

A control volume Ω_i is subdivided in its left "half" with length $h_i^- := a_i - x_{i-\frac{1}{2}}$ and analogously $h_i^+ := x_{i+\frac{1}{2}} - a_i$, i.e. $h_i := h_i^- + h_i^+$. Denoting by

$$
h_{i+\frac{1}{2}} := a_{i+1} - a_i, \quad i = 0, \ldots, N,
$$

the distance of neighbouring grid points, we have

$$
h_{i+\frac{1}{2}} = a_{i+1} - a_i = h_{i+1}^- + h_i^+\,, \quad i = 0, \ldots, N\,,
$$

noting $h_0^+ := h_{N+1}^- := 0$. If the grid points are chosen according to

$$
a_i := \frac{1}{2}(x_{i-\frac{1}{2}} + x_{i+\frac{1}{2}}), \quad \text{i.e.} \quad h_i^- = h_i^+ = \frac{h_i}{2}\,,
\tag{8.38}
$$

the family of control volumes can be viewed as either a Voronoi- or a Donald diagram. However, in general, we avoid this assumption and rather consider the unknown approximate value u_i not only to be an approximation of $u(a_i)$ but also of

$$
\tilde{u}(a_i) := \frac{1}{h_i} \int_{\Omega_i} u(x)\, dx\,.
$$

We do not only want to consider non-equidistant grids (for which e.g. $h_{i+\frac{1}{2}} \neq h_i$ also in the case (8.38)), but also coefficients, which are only piecewise continuous on the partition $\{\Omega_i \mid i = 1, \ldots, N\}$ (reflecting microscopically heterogeneous material).

The one-dimensional version of (8.1) has the form

$$
\begin{aligned}
-(ku_x - cu)_x + ru &= f \quad \text{in } \Omega, \\
u(0) = u(1) &= 0.
\end{aligned}
\tag{8.39}
$$

The assumptions about the coefficients are (for the moment)

$$
k, c, r \in L^\infty(\Omega), \; k(x) \geq k_0 \quad \text{for almost all } x \in \Omega
\tag{8.40}
$$

(allowing, for instance, that k is piecewise constant on Ω_i).

Let $p := -ku_x + cu$ denote the total flux density and $p_{i+\frac{1}{2}}\big|_{+/-}$, $i = 1, \ldots, N$, its approximations at $x_{i+\frac{1}{2}}$ from the right and from the left, respectively. Then the integral formulation of (8.39) on each Ω_i reads as

$$
p(x_{i+\frac{1}{2}}) - p(x_{i-\frac{1}{2}}) + \int_{\Omega_i} r(x)u(x)\, dx = \int_{\Omega_i} f(x)\, dx =: h_i f_i\,, \quad i = 1, \ldots, N\,.
\tag{8.41}
$$

Since the local (mass) conservation law implies that the flux density p has to be continuous at the common boundary points $x_{i+\frac{1}{2}}$ of neighbouring control volumes, the approximations $p_{i+\frac{1}{2}}\big|_{+/-}$ should possess an analogous property so that a discrete conservation law may hold, that is,

$$p_{i+\frac{1}{2}}\big|_{+} = p_{i+\frac{1}{2}}\big|_{-} \qquad \text{for all } i = 1, \ldots, N-1 . \tag{8.42}$$

The convective part of $p_{i+\frac{1}{2}}\big|_{+/-}$ is defined, analogously to (8.8) ff., by

$$\gamma_{i+\frac{1}{2}}\big|_{+/-}\left(r^{+/-}_{i+\frac{1}{2}} u_i + \left(1 - r^{+/-}_{i+\frac{1}{2}}\right) u_{i+1}\right),$$

where $\gamma_{i+\frac{1}{2}}\big|_{+/-} \approx c(x_{i+\frac{1}{2}})\big|_{+,-}$ and $r^{+/-}_{i+\frac{1}{2}}$ allows for upwind schemes in the sense of (8.18). We will concentrate on the diffusive part and assume k to be piecewise constant on the control volumes, i.e.

$$k(x) = k_i \quad \text{for } x \in \Omega_i , \quad l = 1, \ldots, N .$$

Assigning temporarily to the point $x_{i+\frac{1}{2}}$ the auxiliary variable $u_{i+\frac{1}{2}}$, the diffusive flux approximations at $x_{i+\frac{1}{2}}$ from the left (from Ω_i) are defined by $-k_i \frac{u_{i+\frac{1}{2}} - u_i}{h_i^+}$, $i = 1, \ldots, N$, and from the right (from Ω_{i+1}) by $-k_{i+1} \frac{u_{i+1} - u_{i+\frac{1}{2}}}{h_{i+1}^-}$, $i = 0, \ldots, N-1$. Then the discrete continuity condition (8.42) leads to

$$-k_i \frac{u_{i+\frac{1}{2}} - u_i}{h_i^+} = -k_{i+1} \frac{u_{i+1} - u_{i+\frac{1}{2}}}{h_{i+1}^-} .$$

Solving this equation for $u_{i+\frac{1}{2}}$ gives

$$u_{i+\frac{1}{2}} = \frac{\frac{k_{i+1}}{h_{i+1}^-} u_{i+1} + \frac{k_i}{h_i^+} u_i}{\frac{k_i}{h_i^+} + \frac{k_{i+1}}{h_{i+1}^-}}, \qquad i = 1, \ldots, N-1, \tag{8.43}$$

therefore the diffusive part can be approximated by

$$p_{i+\frac{1}{2},\text{diff}} := -k^*_{i+\frac{1}{2}} \frac{u_{i+1} - u_i}{h_{i+\frac{1}{2}}}, \tag{8.44}$$

where

$$k^*_{i+\frac{1}{2}} := \frac{h_i^+ + h_{i+1}^-}{h_i^+ \frac{1}{k_i} + h_{i+1}^- \frac{1}{k_{i+1}}} . \tag{8.45}$$

That is, in case of a discontinuous coefficient k, it is approximated at the boundary of a control volume by a weighted harmonic mean.

Assuming that the coefficient c is continuous, let $\gamma_{i+\frac{1}{2}}\big|_{+/-}$ be chosen as continuous, too

$$\gamma_{i+\frac{1}{2}} = \gamma_{i+\frac{1}{2}}\big|_{+} = \gamma_{i+\frac{1}{2}}\big|_{-} .$$

If analogously to (8.18)

$$r_{i+\frac{1}{2}} := r^{+/-}_{i+\frac{1}{2}} = R\left(\frac{\gamma_{i+\frac{1}{2}}\, h_{i+\frac{1}{2}}}{k^*_{i+\frac{1}{2}}}\right) \tag{8.46}$$

is chosen and R satisfies (P1), then also the convective flux part $p_{i+\frac{1}{2},\mathrm{conv}}$ is continuous at $x_{i+\frac{1}{2}}$, and the continuity of the total flux is satisfied. Putting the approximations of the diffusive and convective parts together, we arrive at

$$p_{i+\frac{1}{2}} = -k^*_{i+\frac{1}{2}}\frac{u_{i+1}-u_i}{h_{i+\frac{1}{2}}} + \gamma_{i+\frac{1}{2}}\left(r_{i+\frac{1}{2}}u_i + \left(1-r_{i+\frac{1}{2}}\right)u_{i+1}\right), \quad i = 1, \dots, N-1. \tag{8.47}$$

For $i = 0$ and $i = N$, due to the homogeneous Dirichlet boundary conditions, they have the form

$$\begin{aligned} p_{\frac{1}{2}}\Big|_+ &= -k_1\tfrac{u_1-0}{h_1^-} + \gamma_{\frac{1}{2}}\left(r_{\frac{1}{2}}\cdot 0 + \left(1-r_{\frac{1}{2}}\right)u_1\right),\\ p_{N+\frac{1}{2}}\Big|_- &= -k_N\tfrac{0-u_N}{h_N^+} + \gamma_{N+\frac{1}{2}}\left(r_{N+\frac{1}{2}}u_N + \left(1-r_{N+\frac{1}{2}}\right)\cdot 0\right). \end{aligned} \tag{8.48}$$

Finally, approximating $\int_{\Omega_i} r(x)u(x)\,dx$ in (8.41) by $r_i h_i u_i$, we obtain from (8.41) the discrete equations

$$p_{i+\frac{1}{2}} - p_{i-\frac{1}{2}} + r_i h_i u_i = h_i f_i\,, \quad i = 1\dots, N. \tag{8.49}$$

Note the slight variation in the definition of γ compared with section 8.2.2, as the sign is now incorporated in the appearance of p.

Remark 8.22 1) For heterogeneous Dirichlet boundary conditions, the value 0 in (8.48) has to be substituted by the corresponding boundary value.

2) Analogously to (8.37) also $0 = a_0 < x_{\frac{1}{2}}$ and $x_{N+\frac{1}{2}} < a_{N+1} = 1$ could be considered leading to control volumes such that $\cup_{i=1}^N \Omega_i$ is a proper subset of Ω.

3) In the case of the node-oriented FVM (assuming (8.38)), we get the form (8.49) again, with the same approximation for the convective part of the flux, whereas for the diffusive part we have

$$\begin{aligned} p_{i+\frac{1}{2},\mathrm{diff}}\Big|_- &= -k_i^-\tfrac{u_{i+1}-u_i}{h_{i+\frac{1}{2}}},\\ p_{i+\frac{1}{2},\mathrm{diff}}\Big|_+ &= -k_{i+1}^+\tfrac{u_{i+1}-u_i}{h_{i+\frac{1}{2}}}, \end{aligned}$$

as the approximating function u_h is continuous and piecewise linear on the partition $\{\tilde{K}_i\}_{i=0}^N$, $\tilde{K}_i := [a_i, a_{i+1}]$. Thus, the continuity of the flux density is equivalent to $k_i^- = k_{i+1}^+ =: k^*_{i+\frac{1}{2}}$ without specifying the approximation for k at $x_{i+\frac{1}{2}}$.

In the following, we restrict ourselves to a pure diffusion problem with a constant diffusion coefficient k, i.e.

$$c = \gamma = 0\,, \quad \text{and, without loss of generality,} \quad k = 1\,,$$

but we keep the general grid. FVMs, in particular in one dimension, "look like" FDMs, but in fact they fulfill different conditions for *asymptotic consistency*. The flux density approximation, in our case

$$p_{i+\frac{1}{2}} = -\frac{u_{i+1} - u_i}{h_{i+\frac{1}{2}}},$$

is asymptotically consistent. Namely, assuming that the exact solution u of (8.39) belongs to $C^2([0, 1])$ and setting

$$\begin{aligned} p^*_{i+\frac{1}{2}} &:= -\tfrac{u(a_{i+1})-u(a_i)}{h_{i+\frac{1}{2}}}, \\ R_{i+\frac{1}{2}} &:= p^*_{i+\frac{1}{2}} - (-u_x(x_{i+\frac{1}{2}})) \end{aligned}$$

we see that the estimate

$$\left|R_{i+\frac{1}{2}}\right| \leq Ch$$

holds. This means, the FVM (here both viewed as node-oriented or cell-centred) is asymptotically *flux-consistent* (in the above sense) and *locally (mass) conservative* in the sense

$$p_{i+\frac{1}{2}} - p_{i-\frac{1}{2}} = h_i f_i .$$

However, it is not (partial differential equation) asymptotically *consistent* in the sense of Definition 1.3, i.e. as a FDM, as can be seen as follows.

The ith equation of (8.49) can be written in finite difference form as

$$\frac{1}{h_i}\left(p_{i+\frac{1}{2}} - p_{i-\frac{1}{2}}\right) = f_i, \quad i = 1, \ldots, N,$$

or, using (8.44) with $k^*_{i+\frac{1}{2}} = 1$, in more detail as

$$\frac{1}{h_i}\left(-\frac{u_{i+1} - u_i}{h_{i+\frac{1}{2}}} + \frac{u_i - u_{i-1}}{h_{i-\frac{1}{2}}}\right) = f_i . \tag{8.50}$$

Now, if for instance, $f \in C^1([0, 1])$, then the quadrature error in f_i can be estimated by

$$|f_i - f(a_i)| \leq Ch .$$

The consistency error $\tau_h(a_i)$ is obtained by substituting $u_{i\pm 1}$ in (8.50) by $u(x_{i\pm 1})$ and forming the difference.

Taylor's expansion

$$u(x_{i\pm 1}) = u(a_i) \pm h_{i\pm\frac{1}{2}} u_x(a_i) + \frac{h^2_{i\pm\frac{1}{2}}}{2} u_{xx}(a_i) \pm \frac{1}{6} h^3_{i\pm\frac{1}{2}} u_{xxx}(\xi_i^{\pm})$$

for intermediate points $\xi_i^{\pm}$ with $a_i \ldots \xi_i^{\pm} \ldots x_{i\pm 1}$ yields

$$-\tau_h(a_i) = \frac{1}{h_i}\frac{h_{i+\frac{1}{2}} + h_{i-\frac{1}{2}}}{2} u_{xx}(a_i) + f(a_i) + O(h)$$
$$= \left(-\frac{1}{h_i}\frac{h_{i+\frac{1}{2}} + h_{i-\frac{1}{2}}}{2} + 1\right) f(a_i) + O(h)\,.$$

Under the assumption (8.38) the factor in front of $f(a_i)$ also has the form

$$\alpha_i := -\frac{1}{h_i}\frac{h_{i+1} + 2h_i + h_{i-1}}{4} + 1\,.$$

For an equidistant grid this term vanishes, but for

$$h_i = h \text{ for even } i, \qquad h_i = \frac{h}{2} \text{ for odd } i$$

we get

$$\alpha_i = \begin{cases} \frac{1}{4} & \text{for even } i \\ \frac{1}{3} & \text{for odd } i\,. \end{cases}$$

Correspondingly, we get a method consistent in the sense of FDM by setting

$$\frac{4}{h_{i+1} + 2h_i + h_{i-1}}\left(-\frac{u_{i+1} - u_i}{h_{i+\frac{1}{2}}} + \frac{u_i - u_{i-1}}{h_{i-\frac{1}{2}}}\right) = f(a_i), \quad i = 2, \ldots, N-1. \quad (8.51)$$

This consideration shows that any (order of) convergence analysis for a FVM has to interpret the FVM as an approximation of a FEM, as done in Section 8.2, or use flux consistency and conservativity.

The latter procedure is demonstrated in the subsequent Theorem 8.28.

In preparation for that we introduce some related FDM-like notation, namely $\Omega_h := \{a_i \mid i = 1 \ldots N\}$ (cf. (1.5)) and $\overline{\Omega}_h := \{a_i \mid i = 0 \ldots N+1\}$. In terms of grid functions $v_h : \Omega_h \to \mathbb{R}$ with their representation vectors $\boldsymbol{v}_h = (v_i) \in \mathbb{R}^N$, $v_i := v_h(a_i)$, we use the discrete L^2-norm (see Remark 8.8)

$$\|v_h\|_{0,h} := \left(\sum_{i=1}^{N} |v_i|^2 h_i\right)^{\frac{1}{2}}. \quad (8.52)$$

The discrete analogue of the $H_0^1(\Omega)$-seminorm $|\,.\,|_1$ with a weight $\alpha : \Omega \to \mathbb{R}_+$ is then

$$|v_h|_{1,h}^{(\alpha)} := \left(\sum_{i=0}^{N+1} \alpha_{i+\frac{1}{2}} \frac{|v_{i+1} - v_i|^2}{h_{i+\frac{1}{2}}}\right)^{\frac{1}{2}} \quad \text{for } v_h : \overline{\Omega}_h \to \mathbb{R} \text{ such that } v_0 = v_{N+1} = 0, \quad (8.53)$$

and $|v_h|_{1,h} := |v_h|_{1,h}^{(1)}$. If the grid function v_h is interpolated by a continuous, piecewise linear function, the seminorm $|\cdot|_{1,h}$ coincides with $|\cdot|_1$.

An equivalent representation of the squared seminorm $|v_h|_{1,h}^{(\alpha)2}$ is

$$\sum_{i=1}^{N} \alpha_{i+\frac{1}{2}} \frac{(v_{i+1}-v_i)^2}{h_{i+\frac{1}{2}}} = \sum_{i=1}^{N} \beta_{i+\frac{1}{2}} (v_{i+1}-v_i)^2 \quad \text{with } \beta_{i+\frac{1}{2}} := \frac{\alpha_{i+\frac{1}{2}}}{h_{i+\frac{1}{2}}},$$

i.e. for

$$\begin{aligned} \alpha_{i+\frac{1}{2}} &:= k^*_{i+\frac{1}{2}}, \quad i = 1, \dots, N, \\ \alpha_{\frac{1}{2}} &:= k_1, \quad \alpha_{N+\frac{1}{2}} := k_N, \end{aligned} \tag{8.54}$$

using (8.45), we have

$$\begin{aligned} \beta_{i+\frac{1}{2}} &= \frac{k_i k_{i+1}}{h_i^+ k_{i+1} + h_{i+1}^- k_i}, \quad i = 1, \dots, N, \\ \beta_{\frac{1}{2}} &= \frac{k_1}{h_{\frac{1}{2}}}, \quad \beta_{N+\frac{1}{2}} = \frac{k_N}{h_{N+\frac{1}{2}}}. \end{aligned}$$

Analogously to Section 8.2.2, we define the discrete variational equation for the unknown grid function $u_h : \overline{\Omega}_h \to \mathbb{R}$ with $u_0 = u_{N+1} = 0$ equivalent to the set of equations (8.49), (8.47), (8.48) as follows:

$$a_h(u_h, v_h) = \langle f_h, v_h \rangle_h \quad \text{for all } v_h : \overline{\Omega}_h \to \mathbb{R} \text{ such that } v_0 = v_{N+1} = 0, \tag{8.55}$$

and $a_h := a_h^0 + b_h + d_h$ with

$$\begin{aligned} a_h^0(u_h, v_h) &:= \sum_{i=1}^{N} v_i \left(-\alpha_{i+\frac{1}{2}} \frac{u_{i+1}-u_i}{h_{i+\frac{1}{2}}} + \alpha_{i-\frac{1}{2}} \frac{u_i - u_{i-1}}{h_{i-\frac{1}{2}}} \right), \\ b_h(u_h, v_h) &:= \sum_{i=1}^{N} v_i \gamma_{i+\frac{1}{2}} \left(r_{i+\frac{1}{2}} u_i + (1 - r_{i+\frac{1}{2}}) u_{i+1} \right) - v_i \gamma_{i-\frac{1}{2}} \left(r_{i-\frac{1}{2}} u_i + (1 - r_{i-\frac{1}{2}}) u_{i-1} \right), \\ d_h(u_h, v_h) &:= \sum_{i=0}^{N+1} v_i r_i h_i u_i, \end{aligned}$$

where $\alpha_{i+\frac{1}{2}}$ is given by (8.54), $f_h : \overline{\Omega}_h \to \mathbb{R}$ with $f_i := f_h(a_i)$, and

$$\langle u_h, v_h \rangle_h := \sum_{i=0}^{N+1} h_i u_i v_i\,. \tag{8.56}$$

That is, for $v_h : \overline{\Omega}_h \to \mathbb{R}$ such that $v_0 = v_{N+1} = 0$, (8.56) induces the norm (8.52). The analysis of these forms is analogous to the one in Sections 8.2.3, 8.2.4.

In particular, the following coercivity result holds.

Lemma 8.23 *Assume that k is piecewise continuous on the partition $\{\Omega_i \mid i = 1, \dots, N\}$ and satisfies the condition from* (8.40). *Then*

$$a_h^0(v_h, v_h) = |v_h|_{1,h}^{(\alpha)} \geq k_0 |v_h|_{1,h} \quad \textit{for all } v_h : \overline{\Omega}_h \to \mathbb{R} \textit{ such that } v_0 = v_{N+1} = 0.$$

Proof

$$
\begin{aligned}
a_h^0(v_h, v_h) &= \sum_{i=0}^{N} -\alpha_{i+\frac{1}{2}} \frac{v_{i+1}-v_i}{h_{i+\frac{1}{2}}} v_i + \sum_{i=1}^{N+1} \alpha_{i-\frac{1}{2}} \frac{v_i - v_{i-1}}{h_{i-\frac{1}{2}}} v_i \\
&= \sum_{i=0}^{N} -\alpha_{i+\frac{1}{2}} \frac{v_{i+1}-v_i}{h_{i+\frac{1}{2}}} v_i + \alpha_{i+\frac{1}{2}} \frac{v_{i+1}-v_i}{h_{i+\frac{1}{2}}} v_{i+1} = \sum_{i=0}^{N} \alpha_{i+\frac{1}{2}} \frac{(v_{i+1}-v_i)^2}{h_{i+\frac{1}{2}}} .
\end{aligned}
$$

□

Next we assume c to be continuous on Ω and $\gamma_{i+\frac{1}{2}}$ be defined as

$$
\gamma_{i+\frac{1}{2}} = c(x_{i+\frac{1}{2}}) .
$$

Comparing the forms b_h and d_h with the ones for the two-dimensional node-oriented case, we observe the same structure with the following concordance:

node-oriented	cell-centred
$u_h, v_h \in X_h$	$u_h, v_h : \overline{\Omega}_h \to \mathbb{R}$
$j \in \Lambda_i$	$j = i-1, i+1$
γ_{ij}	$\gamma_{i+\frac{1}{2}}$ or $-\gamma_{i-\frac{1}{2}}$
μ_{ij}	$k^*_{1+\frac{1}{2}}$ or $k^*_{1-\frac{1}{2}}$
m_{ij}	$h_{i\mp\frac{1}{2}}$
m_i	h_i

This shows that Lemma 8.15 can be transferred to the cell-centred case, and Corollary 8.16 gets the following form.

Corollary 8.24 *Assume that $c_x \in L^1(\Omega)$ exists and R satisfies (P1), (P2), then*

$$
b_h(v_h, v_h) \geq \frac{1}{2} \sum_{i=0}^{N+1} v_i^2 \int_{\Omega_i} c_x \, dx \quad \textit{for all } v_h : \overline{\Omega}_h \to \mathbb{R} \textit{ such that } v_0 = v_{N+1} = 0.
$$

Proof Because of Lemma 8.15, we have

$$
\begin{aligned}
b_h(v_h, v_h) &\geq \frac{1}{2} \sum_{i=1}^{N} v_i^2 \left(\gamma_{i+\frac{1}{2}} - \gamma_{i-\frac{1}{2}}\right) = \frac{1}{2} \sum_{i=1}^{N} v_i^2 \left(c(x_{i+\frac{1}{2}}) - c(x_{i-\frac{1}{2}})\right) \\
&= \frac{1}{2} \sum_{i=1}^{N} v_i^2 \int_{\Omega_i} c_x \, dx .
\end{aligned}
$$

□

Similar to Theorem 8.18, we have

Theorem 8.25 *Let k satisfy* (8.40) *and $c \in C(\overline{\Omega})$, $c_x \in L^1(\Omega)$ be such that $r + \frac{1}{2}c_x \geq r_0 \geq 0$ in Ω. If the weighting parameters $r^{+/-}_{i+\frac{1}{2}}$ are chosen according to* (8.46) *with R satisfying (P1), (P2), and the approximations r_i are chosen according to* (8.10), *i.e. $r_i := \frac{1}{h_i} \int_{\Omega_i} r \, dx$, then the family of forms a_h is coercive uniformly in h with respect to $\| \, . \, \|_{1,h}$ in the sense*

$$a_h(v_h, v_h) \geq k_0 |v_h|_{1,h}^2 + r_0 \|v_h\|_{0,h}^2 \quad \textit{for all } v_h : \overline{\Omega}_h \to \mathbb{R} \textit{ such that } v_0 = v_{N+1} = 0.$$

Proof Concerning the last summand of a_h, we have

$$d_h(v_h, v_h) = \sum_{i=0}^{N+1} r_i v_i^2 h_i = \sum_{i=1}^{N} v_i^2 \int_{\Omega_i} r\, dx\,,$$

and therefore, by Lemma 8.23, Corollary 8.24,

$$\begin{aligned} a_h(v_h, v_h) &\geq k_0 |v_h|_{1,h}^2 + \sum_{i=1}^{N} v_i^2 \int_{\Omega_i} \left(\frac{1}{2} c_x + r \right) dx \\ &\geq k_0 |v_h|_{1,h}^2 + \sum_{i=1}^{N} v_i^2 r_0 h_i = k_0 |v_h|_{1,h}^2 + r_0 \|v_h\|_{0,h}^2 \,. \end{aligned}$$

□

Hence, we have the following uniform discrete H^1-stability of a discrete solution of (8.55). In particular, this implies the uniqueness of the trivial solution of the homogeneous problem ($f = 0$), thus the uniqueness in general and thus the unique existence of a discrete solution.

Theorem 8.26 *Let u_h be a solution of* (8.55), *then under the assumptions of Theorem 8.25, possibly excluding* (8.10), *we have*

$$\frac{k_0}{2} |u_h|_{1,h}^2 + r_0 \|u_h\|_{0,h}^2 \leq \frac{1}{2k_0} \|\hat{f}_h\|_{0,h}^2 \,,$$

where

$$\hat{f}_i := f_i + \hat{r}_i - r_i \,.$$

Here, the approximations r_i are the values used in the definition of (8.55), *and* $\hat{r}_i := \frac{1}{h_i} \int_{\Omega_i} r\, dx$.

Proof Setting $v_h := u_h$ in (8.55), we get

$$\hat{a}_h(u_h, u_h) = \langle f_h, u_h \rangle_h + \langle \hat{r}_h - r_h, u_h \rangle_h \,.$$

The bilinear form $\hat{a}_h$ corresponds to a_h with the approximation $\hat{r}_h$ that is chosen according to (8.10), i.e. to the form a_h used in Theorem 8.25. Then this theorem leads to the estimate

$$\begin{aligned} k_0 |u_h|_{1,h}^2 + r_0 \|u_h\|_{0,h}^2 &\leq \langle f_h + \hat{r}_h - r_h, u_h \rangle_h \leq \|\hat{f}_h\|_{0,h} \|u_h\|_{0,h} \leq \|\hat{f}_h\|_{0,h} \|u_h\|_\infty \\ &\leq \|\hat{f}_h\|_{0,h} |u_h|_{1,h} \leq \frac{k_0}{2} |u_h|_{1,h}^2 + \frac{1}{2k_0} \|\hat{f}_h\|_{0,h}^2 \end{aligned}$$

using Remark 8.27, 1) in the last but one step. □

Remark 8.27 1) Because of the continuous embedding $H_0^1(\Omega) \subset C(\overline{\Omega})$ in one dimension (see Lemma 3.4), we also expect a corresponding discrete estimate. In fact, we have that

$$\|v_h\|_\infty \leq |v_h|_{1,h} \quad \text{for all } v_h : \overline{\Omega}_h \to \mathbb{R} \text{ such that } v_0 = v_{N+1} = 0.$$

Indeed, from $v_i = \sum_{j=0}^{i-1}(v_{j+1} - v_j)$, we immediately get the estimate

$$|v_i| \leq \sum_{j=0}^{i-1} |v_{j+1} - v_j| \leq \left(\sum_{j=0}^{i-1} \frac{(v_{j+1} - v_j)^2}{h_{j+\frac{1}{2}}} \right)^{\frac{1}{2}} \left(\sum_{j=0}^{i-1} h_{j+\frac{1}{2}} \right)^{1/2} \leq |v_h|_{1,h} \cdot 1 .$$

2) In the case of a constant coefficient c, say $c \geq 0$ without loss of generality, and full upwinding, the proof of Lemma 8.15 and the subsequent result simplify considerably. In this case, b_h takes the form

$$b_h(u_h, v_h) = c \sum_{i=1}^{N+1} v_i(u_i - u_{i-1}),$$

therefore

$$b_h(u_h, u_h) = c \sum_{i=1}^{N+1} (u_i - u_{i-1})u_i .$$

Because of $\sum_{i=1}^{N+1} u_i^2 = \sum_{i=1}^{N+1} u_{i-1}^2$, we have

$$\sum_{i=1}^{N+1} (u_i - u_{i-1})^2 = \sum_{i=1}^{N+1} (u_i - u_{i-1})u_i - \sum_{i=1}^{N+1} (u_i - u_{i-1})u_{i-1} =: T_1 + T_2,$$

and

$$T_2 = - \sum_{i=1}^{N+1} u_i u_{i-1} - u_i^2 = \sum_{i=1}^{N+1} (u_i - u_{i-1})u_i = T_1 .$$

Hence

$$b_h(u_h, u_h) = c\frac{1}{2} \sum_{i=1}^{N+1} (u_i - u_{i-1})^2 \geq 0 .$$

In this sense, the following theorem is an analogue to Theorem 8.20 for cell-centred FVM in one dimension.

Theorem 8.28 *Let $\Omega := (0, 1)$ and $\mathcal{T}_h^* = (\Omega_i)_{i=1}^N$ be a partition in the sense of (8.36), let $k \in L^\infty(\Omega)$, $k(x) \geq k_0 > 0$ a.e. in Ω, $c \in C(\overline{\Omega})$, $c_x \in L^1(\Omega)$, $r \in C(\overline{\Omega})$,* *and $f \in L^1(\Omega)$. Furthermore, let the assumptions of Theorem 8.25 be fulfilled, and let k, f as well as the solution u of (8.39) be piecewise smooth in the sense $k \in C^1(\overline{\Omega_i})$, $f \in C(\overline{\Omega_i})$, $u \in C^2(\overline{\Omega_i})$, $i = 1, \ldots, N$. Then there exists a constant $C > 0$, depending only on k_0, $\|k\|_\infty$, $\|u_{xx}\|_{\infty,\Omega_i}$ such that the solution $u_h : \overline{\Omega}_h \to \mathbb{R}$ with $u_0 = u_{N+1} = 0$ of (8.49), (8.47), (8.48) uniquely exists and satisfies*

$$|u_h - U|_{1,h} + \|u_h - U\|_{0,h} \le Ch,$$

where $U : \overline{\Omega}_h \to \mathbb{R}$ is the restriction of u to $\overline{\Omega}_h$.

Remark 8.29 Consider (8.39), but with flux boundary condition at $x = 0$

$$(ku_x - cu)(0) = f_0 .$$

Then, under the conditions of Theorem 3.16, we have $u \in H_0^1(\Omega)$, i.e. $u \in C(\overline{\Omega})$ (see Lemma 3.4). Therefore also the diffusive flux density

$$-ku_x(x) = \int_0^x (f - ru)\, dx - cu(x) + f_0$$

is continuous for continuous c. If k is jumping (at the boundaries of the control volumes) the same applies to u_x, therefore the solution will not belong to $H^2(\Omega)$.

Proof (of Theorem 8.28) Step 1 (Uniqueness of the discrete solution): The uniqueness and therefore the existence follow from Theorem 8.26.

Step 2 (Asymptotic consistency of the flux approximations): In the diffusive part, we have to compare the exact flux density, which is continuous according to Remark 8.29, at $x_{i+\frac{1}{2}}$, that is

$$\overline{p}_{i+\frac{1}{2}} := -(ku_x)(x_{i+\frac{1}{2}}),$$

with

$$p^*_{i+\frac{1}{2}} := -k^*_{i+\frac{1}{2}} \frac{u(a_{i+1}) - u(a_i)}{h_{i+\frac{1}{2}}} .$$

We will show that the consistency error

$$\tau_{i+\frac{1}{2}} := p^*_{i+\frac{1}{2}} - \overline{p}_{i+\frac{1}{2}}$$

satisfies the estimate

$$|\tau_{i+\frac{1}{2}}| \le C_1 h . \tag{8.57}$$

As auxiliary values, we consider the one-sided approximations

$$p^{*,-}_{i+\frac{1}{2}} := -k_i \frac{u(x_{i+\frac{1}{2}}) - u(a_i)}{h_i^+},$$

$$p^{*,+}_{i+\frac{1}{2}} := -k_{i+1} \frac{u(a_{i+1}) - u(x_{i+\frac{1}{2}})}{h_{i+1}^-} .$$

By assumption, these approximation have the desired order of consistency

$$\begin{aligned} p^{*,-}_{i+\frac{1}{2}} &= \overline{p}_{i+\frac{1}{2}} + \tau^-_{i+\frac{1}{2}}, \quad |\tau^-_{i+\frac{1}{2}}| \le C_2 h, \\ p^{*,+}_{i+\frac{1}{2}} &= \overline{p}_{i+\frac{1}{2}} + \tau^+_{i+\frac{1}{2}}, \quad |\tau^+_{i+\frac{1}{2}}| \le C_2 h. \end{aligned} \tag{8.58}$$

These two equations for $p^{*,+/-}_{i+\frac{1}{2}}$ give us the consistency error for the approximation (8.43) of $u_{i+\frac{1}{2}}$, namely

$$u(x_{i+\frac{1}{2}}) = \frac{\frac{k_{i+1}}{h^-_{i+1}} u(a_{i+1}) + \frac{k_i}{h^+_i} u(a_i)}{\frac{k_i}{h^+_i} + \frac{k_{i+1}}{h^-_{i+1}}} + \tau^u_{i+\frac{1}{2}}, \tag{8.59}$$

where

$$\tau^u_{i+\frac{1}{2}} := \frac{\tau^+_{i+\frac{1}{2}} - \tau^-_{i+\frac{1}{2}}}{\frac{k_i}{h^+_i} + \frac{k_{i+1}}{h^-_{i+1}}}.$$

Hence,

$$|\tau^u_{i+\frac{1}{2}}| \le \frac{1}{k_0} \frac{h^+_i h^-_{i+1}}{h^+_i + h^-_{i+1}} \left| \tau^+_{i+\frac{1}{2}} - \tau^-_{i+\frac{1}{2}} \right|. \tag{8.60}$$

Now, analogously to the derivation of $k^*_{i+\frac{1}{2}}$ in (8.43), we use (8.59) for $p^{*,-}_{i+\frac{1}{2}}$ leading to

$$p^{*,-}_{i+\frac{1}{2}} = p^*_{i+\frac{1}{2}} - \frac{k_i}{h^+_i} \tau^u_{i+\frac{1}{2}}.$$

Putting the estimates (8.58), (8.60) together gives the desired estimate (8.57).

Considering the convective flux and $p_{\frac{1}{2}}, p_{N+\frac{1}{2}}$ in an analogous manner concludes this part of the proof.

Step 3 (Error estimate): The exact flux density $\overline{p}_{i+\frac{1}{2}}$ satisfies the conservation equation

$$\overline{p}_{i+\frac{1}{2}} - \overline{p}_{i-\frac{1}{2}} + r_i h_i u(a_i) + S_i = h_i f_i, \quad i = 1, \ldots, N,$$

where the consistency error in the reaction term stems from the quadrature rule applied, i.e. we know that $|S_i| \le C_3 h$. Therefore, the consistency error in the conservation equation satisfies

$$p^*_{i+\frac{1}{2}} - p^*_{i-\frac{1}{2}} + r_i h_i u(a_i) + S_i = h_i f_i + \tau_{i+\frac{1}{2}} - \tau_{i-\frac{1}{2}}.$$

If $e_i := u(a_i) - u_i = U_i - u_i$ denote the error components to be estimated, then

$$a_h(e_h, v_h) = \langle g_h, v_h \rangle_h,$$

where

$$g_i := S_i + \tau_{i+\frac{1}{2}} - \tau_{i-\frac{1}{2}}, \quad i = 1, \ldots, N,$$

and the result follows from Theorem 8.26. □

Remark 8.30 1) If instead of $f_i = \int_{\Omega_i} f(x)dx/h_i$ (from (8.41)) a quadrature rule is used, the error appears also in the right-hand side g_i, and thus the approximation order of the quadrature rule has to be 1. The same applies to another choice of r_i.

2) Let $\hat{u}_h \in V_h := \{v \in H_0^1(0,1) \,|\, v|_{[a_i,a_{i+1}]} \in \mathcal{P}_1,\ i = 0,\dots,N\}$ be the nodal interpolant of $u_h : \overline{\Omega}_h \to \mathbb{R}$ with $u_0 = u_{N+1} = 0$. Then Theorem 8.28 implies

$$|I_h(u) - \hat{u}_h|_1 \leq Ch$$

with the nodal interpolation operator I_h, and thus for $u \in H^2(0,1)$:

$$|u - \hat{u}_h|_1 \leq Ch.$$

To conclude the same estimate in the L^2- norm $\|\cdot\|_0$, the equivalence of $\|\cdot\|_{0,h}$ and $\|\cdot\|_0$ on V_h is necessary, which requires the existence of constants $C_1 > 0$, $0 < C_2 < 1$, such that $C_1 h_i \leq h_i^+ \leq C_2 h_i$, $i = 1,\dots,N$ (see Remark 8.8).
In this respect, there is a similarity between Theorem 8.28 and Theorem 6.7, the flux consistency providing the error estimate of the consistency error in the variational formulation.

8.3.2 A Cell-centred Finite Volume Method on Polygonal/Polyhedral Grids

We now repeat the exposition of Section 8.2 but exchange the "first the nodes (or first the grid), then the control volumes" point of view for the reverse "first the control volumes, then the nodes" point of view. This gives more flexibility so that we can treat two and three space dimensions simultaneously, and allows for (diffusion) coefficients only to be piecewise smooth on the partition $\mathcal{T}_h^* := \{\Omega_i\}_{i\in\overline{\Lambda}}$ generated by the control volumes, but makes it not so obvious how to treat matrix-valued diffusion coefficients. We go step by step and start with a general partition $\mathcal{T}_h^*$ with the following properties. Let $d = 2$ or $d = 3$ and Ω be an open, polygonally/polyhedrally bounded subset of $\mathbb{R}^d$.

1) The sets Ω_i, $i \in \Lambda$, are open, polygonal/polyhedral, convex subsets of Ω such that

$$\bigcup_{i\in\Lambda} \overline{\Omega}_i = \overline{\Omega}. \tag{8.61}$$

2) For each $\Omega_i \in \mathcal{T}_h^*$, there is a finite subset $\mathcal{F}_i$ of $(d-1)$-dimensional plane faces representing the boundary, i.e. $\partial\Omega_i = \bigcup_{\Gamma\in\mathcal{F}_i} \overline{\Gamma}$. The set of all such faces is denoted by $\overline{\mathcal{F}} := \bigcup_{\Omega_i\in\mathcal{T}_h^*} \mathcal{F}_i$, and it can be split as $\overline{\mathcal{F}} = \mathcal{F} \cup \partial\mathcal{F}$, where $\partial\mathcal{F}$ denotes the faces in $\partial\Omega$ and $\mathcal{F}$ the faces in Ω.
3) For any $i, j \in \Lambda$, $i \neq j$, either $|\overline{\Omega}_i \cap \overline{\Omega}_j|_{d-1} = 0$ or $\overline{\Omega}_i \cap \overline{\Omega}_j =: \Gamma_{ij} \in \mathcal{F}$, where $|\,.\,|_{d-1}$ denotes the $(d-1)$-dimensional (Lebesgue) measure.
4) Let a family of "grid points" $\{a_i \in \overline{\Omega}_i \;\big|\; i \in \Lambda\}$ be selected.

- For $\Gamma_{ij} \in \mathcal{F}$, $i \neq j$, we require $a_i \neq a_j$ and that the straight line $a_i a_j$ through a_i and a_j is orthogonal to Γ_{ij}.
- If $\Gamma \in \partial\mathcal{F}$ is a (boundary) face of Ω_i and $a_i \notin \Gamma$, then there exists a point $a_{i,\Gamma} \in \Gamma$ such that the straight line $a_i a_{i,\Gamma}$ through a_i and $a_{i,\Gamma}$ is orthogonal to Γ.

Thus, we have kept the notation of Section 8.2, but $i \in \Lambda$ now primarily refers to the control volume (for example, see Figure 8.10).

According to the above property 4), we define

$$d_{i,\Gamma} := |a_i - a_{i,\Gamma}| \quad \text{with} \quad a_{i,\Gamma} \in \begin{cases} a_i a_j \cap \partial\Omega_i & \text{if } \Gamma = \overline{\Omega}_i \cap \overline{\Omega}_j,\ i \neq j, \\ \Gamma & \text{if } \Gamma \subset \partial\Omega \cap \partial\Omega_i,\ a_i \notin \Gamma, \end{cases} \tag{8.62}$$

and

$$d_{ij} := d_{i,\Gamma} + d_{j,\Gamma}.$$

Note that $d_{i,\Gamma} > 0$.

We start with a diffusion problem with a scalar diffusion coefficient k, i.e.

$$\begin{aligned} -\nabla \cdot (k \nabla u) &= f \quad \text{in } \Omega, \\ u &= 0 \quad \text{on } \partial\Omega. \end{aligned} \tag{8.63}$$

The coefficient k is supposed to be piecewise constant on the partition $\mathcal{T}_h^*$.

Analogously to (8.41), we set as discrete conservation equation

$$\sum_{\Gamma \in \mathcal{F}_i} p_{i,\Gamma} m_\Gamma = m_i f_i, \quad i \in \Lambda, \tag{8.64}$$

where $m_\Gamma := |\Gamma|_{d-1}$, $m_i := |\Omega_i|_d$, and $f_i := \dfrac{1}{m_i} \displaystyle\int_{\Omega_i} f(x)\, dx$.

The diffusive flux density approximation is given by

$$p_{i,\Gamma} := - \begin{cases} \mu_{i,\Gamma} \dfrac{u_j - u_i}{d_{ij}} & \text{if } \Gamma = \overline{\Omega}_i \cap \overline{\Omega}_j,\ i \neq j, \\ \mu_{i,\Gamma} \dfrac{0 - u_i}{d_{i,\Gamma}} & \text{if } \Gamma \subset \partial\Omega \cap \partial\Omega_i,\ a_i \notin \Gamma, \end{cases} \tag{8.65}$$

where $\mu_{i,\Gamma}$ is chosen analogously to (8.45), i.e.

$$\mu_{i,\Gamma} := \begin{cases} k_{i,j}^* := \dfrac{d_{ij}}{\frac{1}{k_j}\frac{d_{j,\Gamma}}{d_{ij}} + \frac{1}{k_i}\frac{d_{i,\Gamma}}{d_{ij}}} & \text{if } \Gamma = \overline{\Omega}_i \cap \overline{\Omega}_j,\ i \neq j, \\ k_i & \text{if } \Gamma \subset \partial\Omega \cap \partial\Omega_i,\ a_i \notin \Gamma. \end{cases} \tag{8.66}$$

Thus the discretization resembles (8.12), but with a different choice of μ_{ij} than in (8.23). Analogously to (8.48) the Dirichlet boundary conditions are incorporated into $p_{i,\Gamma}$ if $\Gamma \subset \partial\Omega$.

For the more general equation

$$-\nabla \cdot (k\nabla u) + \boldsymbol{c} \cdot \nabla u + ru = f \tag{8.67}$$

the conservation equation (8.64) has to be substituted by

$$\sum_{\Gamma \in \mathcal{F}_i} p_{i,\Gamma} m_\Gamma + \sum_{\Gamma \in \mathcal{F}_i} \gamma_{i,\Gamma} u_{\Gamma,+} m_\Gamma + r_i m_i u_i = m_i f_i \,, \tag{8.68}$$

where $\gamma_{i,\Gamma} := \frac{1}{m_\Gamma} \int_\Gamma \mathfrak{n}_{i,\Gamma} \cdot \boldsymbol{c} \, d\sigma$, $\mathfrak{n}_{i,\Gamma}$ is the outer unit normal vector on $\Gamma \subset \partial\Omega_i$, and r_i is defined analogously to f_i. We restrict ourselves here to the full upwind method (see (8.19)) by setting

$$u_{\Gamma,+} := \begin{cases} u_i & \text{for } \gamma_{i,\Gamma} \geq 0 \,, \\ u_j & \text{for } \gamma_{i,\Gamma} < 0 \,, \ \Gamma = \overline{\Omega}_i \cap \overline{\Omega}_j \,, \\ 0 & \text{for } \gamma_{i,\Gamma} < 0 \,, \ \Gamma \subset \partial\Omega \,. \end{cases}$$

For this case an estimate of order one in a discrete H^1-norm can be shown, too.

We now extend the discretization to a matrix-valued diffusion coefficient $\boldsymbol{K}$ such that (8.45) holds. The only modification occurs in the definition of $\mu_{i,\Gamma}$ according to (8.66). First we define a piecewise constant approximation of $\boldsymbol{K}$ by

$$\tilde{\boldsymbol{K}}_i := \frac{1}{m_i} \int_{\Omega_i} \boldsymbol{K}(x) \, dx \quad \text{for } i \in \Lambda$$

and then substitute in (8.66) k_i by $k_i := |\tilde{\boldsymbol{K}}_i \cdot \mathfrak{n}_{i,\Gamma}|$.

Exercise

Problem 8.11 The methods described in Section 8.2 automatically provide the gradient of the numerical solution, including an estimate of the gradient error in the L^2-norm (see Theorem 8.20). Cell-oriented methods allow grids of more general geometry but do not inherently provide any discrete gradient. The following steps explain how a discrete gradient can be generated by postprocessing.

a) Given $v \in C(\overline{\Omega}_i) \cap C^1(\Omega_i)$, show

$$\int_{\Omega_i} \nabla v \, dx = \sum_{\Gamma \in \mathcal{F}_i} \mathfrak{n}_{i,\Gamma} \int_\Gamma v d\sigma \,.$$

b) Derive from a) the relationship

$$\nabla v(a_i) = \frac{1}{|\Omega_i|} \sum_{\Gamma \in \mathcal{F}_i} \mathfrak{n}_{i,\Gamma} v(a_\Gamma) m_\Gamma + O(h),$$

where a_Γ is the barycentre of Γ and $v \in C^2(\overline{\Omega}_i)$.

c) Prove that

$$\nabla v(a_i) = \frac{1}{2|\Omega_i|} \sum_{\Gamma = \overline{\Omega}_i \cap \overline{\Omega}_j \in \mathcal{F}_i} \mathfrak{n}_{i,\Gamma}[v(a_i) + v(a_j)] m_\Gamma + O(h),$$

provided that $a_i + a_j - 2a_\Gamma = O(h^2)$ and $v \in C^2(\overline{\Omega})$.

8.4 Multipoint Flux Approximations

The methods considered so far (and in Section 10.3) are local two-point methods, i.e. they locally approximate the normal flux through an interior control volume face Γ_{ij}, that is

$$\int_{\Gamma_{ij}} \mathfrak{n}_{ij} \cdot \boldsymbol{p} \, d\sigma \approx p_{ij} m_{ij}$$

(see (8.7)), by means of two function values of the solution, see (8.12), (8.65) and (10.27). For instance, in the case $\boldsymbol{K} = k\boldsymbol{I}$ with a scalar function k, we discussed the approximation

$$p_{ij} := t_i u(a_j) + t_j u(a_j) \tag{8.69}$$

with the coefficients

$$t_i := \frac{\mu_{ij}}{d_{ij}} + r_{ij}\gamma_{ij}, \quad t_j := -\frac{\mu_{ij}}{d_{ij}} + \left(1 - r_{ij}\right)\gamma_{ij},$$

where $\mu_{ij} \approx k|_{\Gamma_{ij}}$ and γ_{ij} was chosen according to (8.8). We mention that the coefficients satisfy the relation

$$t_i + t_j = \gamma_{ij} \,. \tag{8.70}$$

As the detailed error analysis of this approach shows (see, e.g. [77, 78]), this approximate normal component can be interpreted as a piecewise constant function with respect to a certain auxiliary partition (sometimes referred to as a diamond-type partition), which differs from both the finite volume partition $\mathcal{T}_h^*$ and, if present, the interpolation partition $\mathcal{T}_h$. In this respect it is a natural idea to improve the approximation of the flux (or the flux density) by including additional nodal values with the aim of achieving a particular continuity and thus better local conservation properties. However, the implementation of this idea differs from the gradient recovery described in Sect. 4.2, since a higher order of approximation plays the main role in the latter, while local conservation laws are of less importance.

There are basically two approaches:

- the so-called *multipoint flux approximation (MPFA) methods*, the derivation of which is more reminiscent of generalized finite difference methods and for which, due to the principle, there is no sufficiently general theory,
  and

- mixed FEM-like methods (cf. Section 8.5), partly in connection with condensation, which eliminates the flow variables for the final scheme.

Multipoint flux approximation (MPFA) methods generalize the approximation (8.69) in such a way that more than two values of the approximate solution are included. Based on the assumptions 1) – 4) from the beginning of Section 8.3.2, the following describes the general design.

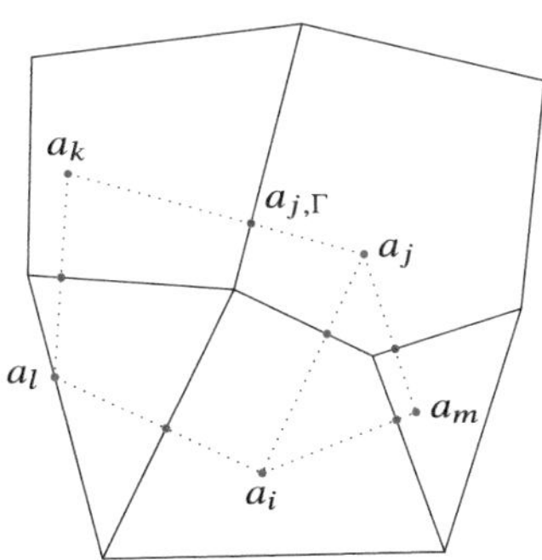

Fig. 8.10: Two control volume patches: Admisssible control volumes (solid lines) and interaction regions (dashed lines).

Denote by $\mathcal{N}^*$ the set of the interior vertices of all control volumes and by $\Lambda(a^*)$, $a^* \in \mathcal{N}^*$, the set of indices of those control volumes that have the common vertex a^*, i.e. $\Lambda(a^*) := \left\{i \in \overline{\Lambda} \;\middle|\; \partial\Omega_i \ni a^*\right\}$. These control volumes form the *finite volume patch* around the *patch center* a^*:

$$\Delta(a^*) := \bigcup_{i \in \Lambda(a^*)} \overline{\Omega}_i .$$

The nodes corresponding to the control volumes in the patch form the so-called *interaction region* of the patch, see Figure 8.10.

Let us consider at first a single control volume Ω_i, $i \in \Lambda(a^*)$. For each face $\Gamma \in \mathcal{F}_i$ we set $\mathfrak{n}_{i,\Gamma} \cdot \boldsymbol{p}|_{\Omega_i} \approx -\boldsymbol{\mu}_{i,\Gamma} \cdot \nabla u + \gamma_{i,\Gamma} u$, where $\boldsymbol{\mu}_{i,\Gamma} \in \mathbb{R}^d$, $\gamma_{i,\Gamma} \in \mathbb{R}$ are approximations to $\boldsymbol{K}|_\Gamma \mathfrak{n}_{i,\Gamma}$, $\mathfrak{n}_{i,\Gamma} \cdot \boldsymbol{c}|_\Gamma$ satisfying $\boldsymbol{\mu}_{i,\Gamma} = -\boldsymbol{\mu}_{j,\Gamma}$, $\gamma_{i,\Gamma} = -\gamma_{j,\Gamma}$ for $\Gamma = \Gamma_{ij} = \overline{\Omega}_i \cap \overline{\Omega}_j$; for instance

$$\boldsymbol{\mu}_{i,\Gamma} := \frac{1}{m_\Gamma} \int_\Gamma \boldsymbol{K} \mathfrak{n}_{i,\Gamma} \, d\sigma, \quad \gamma_{i,\Gamma} := \frac{1}{m_\Gamma} \int_\Gamma \mathfrak{n}_{i,\Gamma} \cdot \boldsymbol{c} \, d\sigma$$

(cf. (8.8)).

In the next step, u is assumed to be affine-linear on the control volume Ω_i, i.e. we have a representation of the form

$$u(x) = u(a_i) + \nabla u_i \cdot (x - a_i) \quad \text{for all } x \in \Omega_i \tag{8.71}$$

with a constant vector ∇u_i. We consider the set $\mathcal{F}(a^*) := \{\Gamma \in \mathcal{F} \mid \Gamma \ni a^*\}$ of faces that have the vertex a^* in common and choose a subset $\mathcal{F}_i(a^*) \subset \mathcal{F}_i \cap \mathcal{F}(a^*)$ of d (interior) faces of Ω_i such that the vectors $a_i - a_{i,\Gamma}$, $\Gamma \in \mathcal{F}_i(a^*)$, are linearly independent, where $a_{i,\Gamma}$ is the intersection point of the segment $a_i a_j$ and $\partial\Omega_i$ or any point on Γ if $\Gamma \subset \partial\Omega \cap \partial\Omega_i$, $a_i \notin \Gamma$ (cf. (8.62)). Then we can consider the d equations

$$\nabla u_i \cdot (a_{i,\Gamma} - a_i) = u(a_{i,\Gamma}) - u(a_i), \quad \Gamma \in \mathcal{F}_i(a^*).$$

If we knew all the d values on the right-hand side we could regard this set of equations as a system of linear equations

$$A_{i,a^*} \nabla u_i = \boldsymbol{b}_{i,a^*} \tag{8.72}$$

with

$$A_{i,a^*} := (a_{i,\Gamma} - a_i)_{\Gamma \in \mathcal{F}_i(a^*)}, \quad \boldsymbol{b}_{i,a^*} := (u(a_{i,\Gamma}) - u(a_i))_{\Gamma \in \mathcal{F}_i(a^*)},$$

which uniquely defines the d components of the (constant) vector ∇u_i. Denote by $\boldsymbol{u}_i := A_{i,a^*}^{-1} \boldsymbol{b}_{i,a^*}$ this solution. This has two consequences:

- Thanks to (8.71), the values of the affine-linear function u are known on the whole control volume Ω_i (in fact, even up to the boundary if the boundary values are understood as the limit values from the interior). In particular, the values $u(a_{i,\Gamma})$ are known for all faces $\Gamma \in \mathcal{F}_i$.
- We obtain an approximative representation $p_{i,\Gamma}$ of $\boldsymbol{\mathfrak{n}}_{i,\Gamma} \cdot \boldsymbol{p}|_{\Omega_i}$ at Γ:

$$\boldsymbol{\mathfrak{n}}_{i,\Gamma} \cdot \boldsymbol{p}|_{\Omega_i} \approx p_{i,\Gamma} := -\boldsymbol{\mu}_{i,\Gamma} \cdot \boldsymbol{u}_i + \gamma_{i,\Gamma}[r_{i,\Gamma}\, u(a_i) + (1 - r_{i,\Gamma}) u(a_j)], \tag{8.73}$$

  where the coefficients $r_{i,\Gamma}$ can be chosen as in Section 8.2.

For each of the (interior) faces $\Gamma = \Gamma_{ij} \in \mathcal{F}$ there is an equation of the form $p_{i,\Gamma} = -p_{j,\Gamma}$, equating the approximate flux densities at that face Γ which is common to the two neighbouring control volumes Ω_i, Ω_j:

$$-\boldsymbol{\mu}_{i,\Gamma} \cdot \boldsymbol{u}_i + \gamma_{i,\Gamma}[r_{i,\Gamma}\, u(a_i) + (1 - r_{i,\Gamma}) u(a_j)] = \boldsymbol{\mu}_{j,\Gamma} \cdot \boldsymbol{u}_j - \gamma_{j,\Gamma}[r_{j,\Gamma}\, u(a_j) + (1 - r_{j,\Gamma}) u(a_i)].$$

Since the right-hand side of this relation can be rewritten as

$$-\boldsymbol{\mu}_{i,\Gamma} \cdot \boldsymbol{u}_j + \gamma_{i,\Gamma}[r_{i,\Gamma}\, u(a_i) + (1 - r_{i,\Gamma}) u(a_j)]$$

(cf. (P1) in (8.20)), the equation reduces to

$$\boldsymbol{\mu}_{i,\Gamma} \cdot (\boldsymbol{u}_j - \boldsymbol{u}_i) = 0. \tag{8.74}$$

It should be noted here that the disappearance of the convective terms is a natural consequence of the definition (8.73) of $p_{i,\Gamma}$, which in turn is based on the orthogonality of $a_i a_j$ and $a_i a_{i,\Gamma}$ to $\Gamma_{ij} \in \mathcal{F}$ and $\Gamma \in \partial\mathcal{F}$, resp., see assumption 4) from the beginning of Section 8.3.2.

Now, if we select an arbitrary interior vertex $a^* \in \mathcal{N}^*$ and restrict the set of equations (8.74) to those faces that have the vertex a^* in common, i.e. we consider (8.74) for all $\Gamma \in \mathcal{F}(a^*)$, we get $|\mathcal{F}(a^*)|$ equations in which the values of u at $a_{i,\Gamma}$,

$\Gamma = \Gamma_{ij} \in \mathcal{F}(a^*)$, occur linearly. Identifying the values of u in common points $a_{i,\Gamma}$, i.e. $u(a_{i,\Gamma}) = u(a_{j,\Gamma})$ for $\Gamma = \Gamma_{ij} \in \mathcal{F}(a^*)$, one with the other (i.e. we require the continuity of u at these points), we obtain $|\bigcup_{i\in\Lambda(a^*)} \mathcal{F}_i(a^*)|$ unknown values $u(a_{i,\Gamma})$. Thus, the restricted set of equations (8.74) can be seen as a system of linear equations with respect to these unknown values. The essence of the method, but also its complications, ly in the solution of this system for these values. It is generally not clear whether this system is solvable. In particular, for $d = 3$ and unstructured grids, the case $|\bigcup_{i\in\Lambda(a^*)} \mathcal{F}_i(a^*)| < |\mathcal{F}(a^*)|$ could even occur, which makes solvability even more difficult. However, if the system can be solved (not necessarily in a unique way), then the values $u(a_{i,\Gamma})$, $\Gamma = \Gamma_{ij} \in \mathcal{F}(a^*)$, are finally available as a linear combination of the function values of u in the finite volume patch. Having obtained these expressions, we can insert them into (8.73) to get an expression of $p_{i,\Gamma}$ by means of the function values $u(a_j)$, $j \in \Lambda(a^*)$, only:

$$p_{i,\Gamma} = \sum_{j\in\Lambda(a^*)} t_{ij,a^*} u(a_j). \tag{8.75}$$

This expression can be used to approximate the flux through the corresponding portion of Γ that contains $a_{i,\Gamma}$ and a^*. The combination of the results with those of the neighbouring interaction regions allows to generate the complete contributions to the flux across Γ. In this way we arrive at a finite volume scheme.

The coefficients t_{ij,a^*} are called the *transmissibilities*. They satisfy a relation analogous to (8.70):

$$\sum_{j\in\Lambda(a^*)} t_{ij,a^*} = \gamma_{ij} .$$

Indeed, if u is constant, different from zero, on the control volume patch $\Delta(a^*)$, the right-hand side $\boldsymbol{b}_{i,a^*}$ in (8.72) is the zero vector for all control volumes from the patch, and thus $\boldsymbol{u}_i = \boldsymbol{0}$ for all $i \in \Lambda(a^*)$. So (8.73) yields $p_{i,\Gamma} = \gamma_{i,\Gamma}[r_{i,\Gamma}\, u(a_i) + (1 - r_{i,\Gamma})u(a_j)] = \gamma_{i,\Gamma}[r_{i,\Gamma} + (1 - r_{i,\Gamma})]\, u(a_i) = \gamma_{i,\Gamma} u(a_i)$, whereas (8.75) gives $p_{i,\Gamma} = \sum_{j\in\Lambda(a^*)} t_{ij,a^*} u(a_j) = u(a_i) \sum_{j\in\Lambda(a^*)} t_{ij,a^*}$.

While MPFA methods have numerically been shown to converge in many studies ([70], [175]), it has proven more difficult to provide a theoretical analysis. Even on the pure diffusion case it can happen that they loose symmetry, coercivity or inverse monotonicity. The theoretical analysis of MPFA methods is mainly based on the possibility of interpreting them as a mixed finite element method with specific quadrature rules [160], or as a so-called *mimetic finite difference method*, see, e.g. [176], [161]. In both cases, it was successful to transfer certain convergence results. However, most of the investigations are restricted to the pure diffusion case with a special focus on anisotropic, heterogeneous diffusion matrices, whereby especially in the three-dimensional case only very special (parallelepipedic) grids are considered. A proof of convergence of an MPFA method for the pure diffusion problem on more general polyhedral grids has been presented in the paper [71]. The methodology used in it is not based on similarities with the mixed finite element or mimetic finite difference methods, but rather uses typical arguments from the theory of FVMs.

Exercise

Problem 8.12 Consider the pure diffusion equation $-\nabla \cdot (\boldsymbol{K} \nabla u) = f$ in a square domain with $\boldsymbol{K} := \begin{pmatrix} 2 & 1 \\ 1 & 2 \end{pmatrix}$. Take a square control volumes grid of width $h > 0$ and calculate the transmissibilities t_{ij,a^*} for an arbitrary interior vertex $a^* \in \mathcal{N}^*$ and all faces $\Gamma = \Gamma_{ij} \in \mathcal{F}(a^*)$.

8.5 Finite Volume Methods in the Context of Mixed Finite Element Methods

It is one of the characteristic features of the FVM that directional components (typically the normal or conormal component) of the numerical flux density (for instance p_{ij} in (8.12)) occur in a fairly natural way on the boundary of the control volumes. Therefore it is not so far as to ask about separate approximations to the flux density, and this question immediately leads to the idea of creating a formulation of the finite volume method in the framework of a mixed finite element method. This idea has been pursued since around the mid-1990s, with different approaches being used.

Most of the existing papers are based on the lowest-order $H(\mathrm{div})$-conforming Raviart-Thomas elements (see Section 7.2.2, in particular (7.46)) and discretize the diffusion equation (with a matrix-valued coefficient) using cell-centred FVMs, see [92], [225], [131], to mention just a few.

At this point we want to describe another approach that is more suitable for the FVMs described above in Section 8.2.2 and is based on so-called $H(\mathrm{curl})$-*conforming* finite elements. These elements frequently arise in problems of electromagnetism. They are also interesting because they stand as a great example of performance for low-order elements that do not use the standard Lagrangian degrees of freedom. Their degrees of freedom include tangential components related to the elements of the interpolation partition. These tangential components are, however, at the same time the normal (or approximately normal) components with respect to the elements of the finite volume partition – and this makes $H(\mathrm{curl})$-conforming elements attractive.

8.5.1 The Problem and Its Mixed Formulation

The model under consideration is a special case of the boundary value problem (6.30)–(6.34), where homogeneous Dirichlet boundary conditions are to be satisfied, that is $\widetilde{g}_3 = 0$. So the boundary value problem reads as follows:

$$\begin{aligned} \nabla\cdot \boldsymbol{p} + ru &= f && \text{in } \Omega,\\ \boldsymbol{p} &= -\boldsymbol{K}\nabla u + \boldsymbol{c}u && \text{in } \Omega,\\ -\boldsymbol{p}\cdot\mathfrak{n} &= \widetilde{g_1} && \text{on } \widetilde{\Gamma}_1,\\ -\boldsymbol{p}\cdot\mathfrak{n} + \alpha u &= \widetilde{g_2} && \text{on } \widetilde{\Gamma}_2,\\ u &= 0 && \text{on } \widetilde{\Gamma}_3, \end{aligned} \tag{8.76}$$

with $r, f : \Omega \to \mathbb{R}$, $\boldsymbol{c} : \Omega \to \mathbb{R}^d$, $\boldsymbol{K} : \Omega \to \mathbb{R}^{d,d}$, $\widetilde{g_1} : \widetilde{\Gamma}_1 \to \mathbb{R}$ and $\alpha, \widetilde{g_2} : \widetilde{\Gamma}_2 \to \mathbb{R}$ given. Now we set

$$\begin{aligned} M &:= H(\mathrm{div};\Omega) := \left\{\boldsymbol{q}\in L^2(\Omega)^d : \nabla\cdot\boldsymbol{q}\in L^2(\Omega)\right\},\\ X &:= H^1_{\widetilde{\Gamma}_3}(\Omega) = \{v\in H^1(\Omega) : \gamma_{0,\widetilde{\Gamma}_3}(v) = 0\} \quad \text{(see (6.37))}, \end{aligned}$$

where M is equipped with the usual norm $\|\cdot\|_{H(\mathrm{div})}$, see (6.38).

The first equation in (8.76) is multiplied by $v \in X$ and integrated over Ω:

$$\langle \nabla\cdot\boldsymbol{p}, v\rangle + \langle ru, v\rangle = \langle f, v\rangle\,.$$

The boundary conditions are included by means of the following argument. Usually the term $\langle \nabla\cdot\boldsymbol{p}, v\rangle$ is integrated by parts leading to

$$\begin{aligned} \langle \nabla\cdot\boldsymbol{p}, v\rangle &= -\langle \boldsymbol{p}, \nabla v\rangle + \langle \mathfrak{n}\cdot\boldsymbol{p}, v\rangle_{\partial\Omega} = -\langle \boldsymbol{p}, \nabla v\rangle + \langle \mathfrak{n}\cdot\boldsymbol{p}, v\rangle_{\widetilde{\Gamma}_1} + \langle \mathfrak{n}\cdot\boldsymbol{p}, v\rangle_{\widetilde{\Gamma}_2}\\ &= -\langle \boldsymbol{p}, \nabla v\rangle - \langle \widetilde{g_1}, v\rangle_{\widetilde{\Gamma}_1} + \langle \alpha u, v\rangle_{\widetilde{\Gamma}_2} - \langle \widetilde{g_2}, v\rangle_{\widetilde{\Gamma}_2}\,. \end{aligned}$$

Applying integration by parts to the first term in the last line again, we get

$$\langle \nabla\cdot\boldsymbol{p}, v\rangle = \langle \nabla\cdot\boldsymbol{p}, v\rangle - \langle \mathfrak{n}\cdot\boldsymbol{p}, v\rangle_{\widetilde{\Gamma}_1\cup\widetilde{\Gamma}_2} - \langle \widetilde{g_1}, v\rangle_{\widetilde{\Gamma}_1} + \langle \alpha u, v\rangle_{\widetilde{\Gamma}_2} - \langle \widetilde{g_2}, v\rangle_{\widetilde{\Gamma}_2}\,.$$

Thus the variational formulation of the first equation can be written as

$$\langle \nabla\cdot\boldsymbol{p}, v\rangle - \langle \mathfrak{n}\cdot\boldsymbol{p}, v\rangle_{\widetilde{\Gamma}_1\cup\widetilde{\Gamma}_2} + \langle \alpha u, v\rangle_{\widetilde{\Gamma}_2} + \langle ru, v\rangle = \langle f, v\rangle + \langle \widetilde{g_1}, v\rangle_{\widetilde{\Gamma}_1} + \langle \widetilde{g_2}, v\rangle_{\widetilde{\Gamma}_2}\,.$$

Multiplication of the second equation in (8.76) by $\boldsymbol{q} \in M$ and integration over Ω gives

$$\langle \boldsymbol{p}, \boldsymbol{q}\rangle + \langle \boldsymbol{q}, \boldsymbol{K}\nabla u\rangle - \langle \boldsymbol{c}u, \boldsymbol{q}\rangle = 0.$$

Setting

$$\begin{aligned} &a\langle \boldsymbol{p}, \boldsymbol{q}\rangle := \langle \boldsymbol{p}, \boldsymbol{q}\rangle, \quad b_1(\boldsymbol{q}, u) := \langle \boldsymbol{q}, \boldsymbol{K}\nabla u\rangle, \quad c(u, \boldsymbol{q}) := -\langle \boldsymbol{c}u, \boldsymbol{q}\rangle,\\ &b_2(\boldsymbol{p}, v) := -\langle \nabla\cdot\boldsymbol{p}, v\rangle + \langle \mathfrak{n}\cdot\boldsymbol{p}, v\rangle_{\widetilde{\Gamma}_1\cup\widetilde{\Gamma}_2}, \quad d(u, v) := \langle \alpha u, v\rangle_{\widetilde{\Gamma}_2} + \langle ru, v\rangle,\\ &\ell_1(\boldsymbol{q}) := 0, \quad \ell_2(v) := -\langle f, v\rangle - \langle \widetilde{g_1}, v\rangle_{\widetilde{\Gamma}_1} - \langle \widetilde{g_2}, v\rangle_{\widetilde{\Gamma}_2}\,, \end{aligned} \tag{8.77}$$

the so-called *primal-dual formulation* of (8.76) is:

Find $(\boldsymbol{p}, u) \in M \times X$ such that

$$\begin{aligned} a\langle \boldsymbol{p}, \boldsymbol{q}\rangle + b_1(\boldsymbol{q}, u) + c(u, \boldsymbol{q}) &= \ell_1(\boldsymbol{q}) && \text{for all } \boldsymbol{q}\in M,\\ b_2(\boldsymbol{p}, v) - d(u, v) &= \ell_2(v) && \text{for all } v\in X. \end{aligned} \tag{8.78}$$

8.5.2 The Finite-Dimensional Aproximation

In order to approximate problem (8.78) by a finite-dimensional system of linear equations, we turn to a nonconforming Petrov–Galerkin formulation. So we assume there are given four finite-dimensional real Hilbert spaces M_h, N_h, X_h, Y_h, five bilinear forms $a_h : M_h \times N_h \to \mathbb{R}$, $b_{1h} : N_h \times X_h \to \mathbb{R}$, $b_{2h} : M_h \times Y_h \to \mathbb{R}$ $c_h : X_h \times N_h \to \mathbb{R}$ and $d_h : X_h \times Y_h \to \mathbb{R}$ as well as two linear functionals $\ell_{1h} : N_h \to \mathbb{R}$, $\ell_{2h} : Y_h \to \mathbb{R}$. Then the discrete problem reads as follows:

Find $(\boldsymbol{p}_h, u_h) \in M_h \times X_h$ such that

$$\begin{aligned} a_h(\boldsymbol{p}_h, \boldsymbol{q}_h) + b_{1h}(\boldsymbol{q}_h, u_h) + c_h(u_h, \boldsymbol{q}_h) &= \ell_{1h}(\boldsymbol{q}_h) \quad \text{for all } \boldsymbol{q}_h \in N_h, \\ b_{2h}(\boldsymbol{p}_h, v_h) - d_h(u_h, v_h) &= \ell_{2h}(v_h) \quad \text{for all } v_h \in Y_h. \end{aligned} \tag{8.79}$$

In what follows we will demonstrate how the FVM (8.11) described in Section 8.2.2 for $d = 2$ and the case of Voronoi diagrams can be reformulated as a problem of the form (8.79). While the extension of method (8.11) to the three-dimensional case is relatively straightforward and we have therefore restricted ourselves in Section 8.2.2 to the case $d = 2$, larger differences between $d = 2$ and $d = 3$ occur in the mixed formulation. Therefore we will consider both cases.

So let a Voronoi diagram $\mathcal{T}_h^* = \{\Omega_i\}_{i\in\overline{\Lambda}}$ and the corresponding Delaunay triangulation $\mathcal{T}_h$ of $\Omega \subset \mathbb{R}^d$, $d = 2, 3$, be given. For a single element $K \in \mathcal{T}_h$ we denote by Λ_K the set of indices of all nodes belonging to K. It is known that the faces Γ_{ij}, $i, j \in \Lambda_K : i < j$, intersect in some point $a_{V,K}$, which coincides with the centre of the circumcircle/circumsphere of K (see Section 8.2.1 for the case $d = 2$). This point is called the *Voronoi vertex associated with the simplex K*.

We assume that all simplices $K \in \mathcal{T}_h$ are self-centred, with a simplex being said to be *self-centred*, if the Voronoi vertex $a_{V,K}$ lies in K. Then we can introduce a third partition which will be especially important for the definition of the discrete scheme. The elements of this partition are defined as follows. Considering a single element $K \in \mathcal{T}_h$, again, we set

$$\Omega_{ij}^K := \operatorname{int}\left(\operatorname{conv}\left\{a_i, a_j, a_{V,K}\right\}\right) \quad \text{for } d = 2,$$

and

$$\Omega_{ij}^K := \operatorname{int}\left(\operatorname{conv}\left\{a_i, a_j, a_{V,F_k}, a_{V,F_j}, a_{V,K}\right\}\right) \quad \text{for } d = 3,$$

where $k, j \in \Lambda_K \setminus \{i, j\}$ and $k \neq j$, and a_{V,F_k} denotes the Voronoi vertex of the face F_k of K that lies opposite to a_k (see Figure 8.11). Hence, we have the representation

$$K = \bigcup_{i,j\in\Lambda_K:\, i<j} \overline{\Omega_{ij}^K}\,.$$

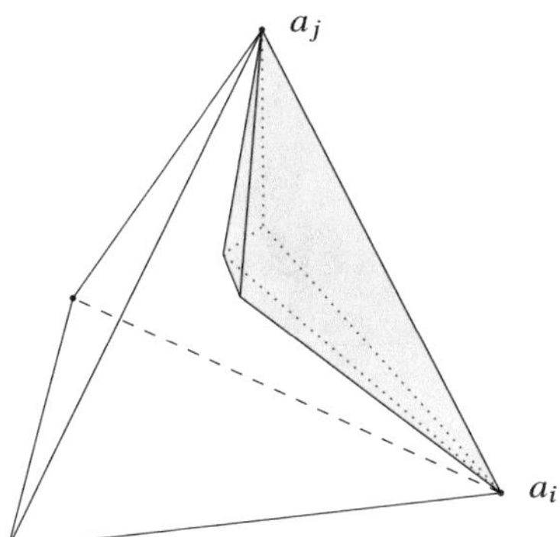

Fig. 8.11: The element Ω_{ij}^K in the case $d = 3$.

The Local Flux-Density Approximating Space R_1

For the flux-density approximation we use the affine-linear element from the so-called first family of *Nédélec elements* [196]. The idea behind is to enrich the polynomial space $\mathcal{P}_0(K)^d$ in such a way that that the rotation of the elements involved also spans $\mathcal{P}_0(K)^d$.

The rotation (curl-operator of vector calculus) is defined in the weak sense similar to Definition 2.9 and (2.18). Given a vector field $\boldsymbol{\phi} \in C_0^\infty(\Omega)^d$, its *rotation* is defined in the Cartesian form as

$$\operatorname{curl} \boldsymbol{\phi} := \begin{cases} \partial_1\phi_2 - \partial_2\phi_1, & d = 2, \\ (\partial_2\phi_3 - \partial_3\phi_2, \partial_3\phi_1 - \partial_1\phi_3, \partial_1\phi_2 - \partial_2\phi_1)^T, & d = 3. \end{cases}$$

For a vector field $\boldsymbol{q} \in L^2(\Omega)^2$, $v \in L^2(\Omega)$ is the *weak (scalar) rotation* $v = \operatorname{curl} \boldsymbol{q}$ if

$$\int_\Omega v\phi \, dx = \int_\Omega \boldsymbol{q} \cdot \operatorname{curl} \phi \, dx \quad \text{for all } \phi \in C_0^\infty(\Omega), \tag{8.80}$$

where the vector field $\operatorname{curl} \phi$ is the vector rotation of the scalar function ϕ (see (6.89)), and for a vector field $\boldsymbol{q} \in L^2(\Omega)^3$, $\boldsymbol{v} \in L^2(\Omega)^3$ is the *weak (vector) rotation* $\boldsymbol{v} = \operatorname{curl} \boldsymbol{q}$ if

$$\int_\Omega \boldsymbol{v} \cdot \boldsymbol{\phi} \, dx = \int_\Omega \boldsymbol{q} \cdot \operatorname{curl} \boldsymbol{\phi} \, dx \quad \text{for all } \boldsymbol{\phi} \in C_0^\infty(\Omega)^3. \tag{8.81}$$

The corresponding function space is defined by

$$\begin{aligned} H(\operatorname{curl}; \Omega) &:= \big\{ \boldsymbol{q} \in L^2(\Omega)^d \mid \operatorname{curl} \boldsymbol{q} \text{ exists weakly} \\ &\qquad \text{in } L^2(\Omega)\ (d = 2) \text{ or } L^2(\Omega)^d\ (d = 3) \big\}, \\ \|\boldsymbol{q}\|_{H(\operatorname{curl})} &:= \left(\|\boldsymbol{q}\|_0^2 + \|\operatorname{curl} \boldsymbol{q}\|_0^2 \right)^{1/2}. \end{aligned} \tag{8.82}$$

The lowest order edge-element $\boldsymbol{R}_1(K)$ on a triangle, respectively a tetrahedron, K is defined as follows:

- for $d = 2$:

$$\boldsymbol{R}_1(K) := \left\{ \boldsymbol{a} + b \begin{pmatrix} x_2 \\ -x_1 \end{pmatrix} : \ \boldsymbol{a} \in \mathbb{R}^2,\ b \in \mathbb{R} \right\} \quad \text{with} \quad \dim \boldsymbol{R}_1(K) = 3,$$

  for $d = 3$:

$$\boldsymbol{R}_1(K) := \{ \boldsymbol{a} + \boldsymbol{b} \times x : \ \boldsymbol{a}, \boldsymbol{b} \in \mathbb{R}^3 \} \quad \text{with} \quad \dim \boldsymbol{R}_1(K) = 6.$$

- the edge-based degrees of freedom:

$$\mathcal{M}_E : \ \boldsymbol{v} \mapsto \int_E \boldsymbol{v} \cdot \boldsymbol{\tau}_E d\sigma \quad \text{for all edges } E \text{ of } K,$$

  where $\boldsymbol{\tau}_E$ represents a fixed unit vector along the edge E.

It can be shown that these degrees of freedom define an element of $\boldsymbol{R}_1(K)$ in a unique way and that the space

$$\Big\{ \boldsymbol{q} \in L^2(\Omega)^d : \ \boldsymbol{q}|_K \in \boldsymbol{R}_1(K) \quad \text{for all } K \in \mathcal{T}_h$$
$$\text{and} \quad \mathcal{M}_E(v|_K) = \mathcal{M}_E(v|_{K'}) \quad \text{for all } E = K \cap K',\ K, K' \in \mathcal{T}_h \Big\}$$

is a subspace of $H(\text{curl}; \Omega)$ [196, Thm. 1].

Alternative degrees of freedom are:

- $$\mathcal{M}_E : \ \boldsymbol{v} \mapsto \frac{1}{|E|} \int_E \boldsymbol{v} \cdot \tilde{\boldsymbol{\tau}}_E \, d\sigma,$$

  where $\tilde{\boldsymbol{\tau}}_E$ is one of the edge vectors of edge E.

- $$\mathcal{M}_E : \ \boldsymbol{v} \mapsto (\boldsymbol{v} \cdot \boldsymbol{\tau}_E)(a_E),$$

  where a_E is the midpoint of E.

These degrees of freedom ensure that the element is invariant under affine-linear transformations. However, the second alternative set of degrees of freedom has a narrower domain of definition.

Lemma 8.31 *The elements of* $\boldsymbol{R}_1(K)$ *have constant tangential trace on the edges of* K.

Proof Let $K := \text{conv}\,\{a_1, \ldots, a_{d+1}\}$. It is sufficient to demonstrate the assertion for the local shape functions $\varphi_E := \lambda_i \nabla \lambda_j - \lambda_j \nabla \lambda_i$, where $E = \overline{a_i a_j}$ and λ_i is the barycentric coordinate associated with the node a_i (see Problem 8.14). Let $E' = \overline{a_k a_l}$ be an edge with the tangential vector $\boldsymbol{\tau}_{E'}$. From the representation

$$\nabla \lambda_i \cdot \boldsymbol{\tau}_{E'} = \frac{1}{|E'|}[\lambda_i(a_l) - \lambda_i(a_k)]$$

for the constant directional derivative of λ_i we see that

$$\varphi_E \cdot \boldsymbol{\tau}_{E'} = \frac{1}{|E'|}\left\{\lambda_i[\lambda_j(a_l) - \lambda_j(a_k)] - \lambda_j[\lambda_i(a_l) - \lambda_i(a_k)]\right\}.$$

Now we have the following cases:

1. $E' = E$, i.e. $i = k$, $j = l$. Then

$$\varphi_E \cdot \boldsymbol{\tau}_{E'} = \frac{1}{|E'|}\left\{\lambda_i + \lambda_j\right\} = \frac{1}{|E'|}.$$

2. $E' \cap E$ consists of one common node, say $E' \cap E = \{a_i\} = \{a_k\}$. Then $\lambda_j|_{E'} = 0$ and

$$\varphi_E \cdot \boldsymbol{\tau}_{E'} = \frac{1}{|E'|}\lambda_j = 0 \quad \text{on } E'.$$

   The other situations can be treated in the same way.
3. $E' \cap E = \emptyset$, then $\varphi_E \cdot \boldsymbol{\tau}_{E'} = 0$ obviously.

Summarizing all the cases, we get

$$\varphi_E \cdot \boldsymbol{\tau}_{E'} = \frac{1}{|E'|}\delta_{EE'} \quad \text{on } E' \text{ for all edges } E, E' \text{ of } K.$$

□

The Discrete Function Spaces

Using the Nédélec element introduced above, we define the following spaces:

$$\begin{aligned}
M_h &:= \Big\{\boldsymbol{q} \in H(\mathrm{curl};\Omega) : \ \boldsymbol{q}|_K \in \boldsymbol{R}_1(K) \quad \text{for all } K \in \mathcal{T}_h \\
&\qquad \text{and} \quad \boldsymbol{\tau}_E \cdot \boldsymbol{q}|_K(a_E) = 0 \quad \text{for all } E \in \partial\mathcal{E}_3\Big\}, \\
&\qquad \text{where } \partial\mathcal{E}_3 \text{ denotes the set of all edges of } \mathcal{T}_h \text{ on } \widetilde{\Gamma}_3, \\
N_h &:= \Big\{\boldsymbol{q} \in L^2(\Omega)^d : \ \boldsymbol{q}|_{\Omega_{ij}^K} \in \mathcal{P}_0(\Omega_{ij}^K)\mathfrak{n}_{ij} \quad \text{for all } K \in \mathcal{T}_h,\ i, j \in \Lambda_K \setminus \partial\Lambda_3,\ i \neq j \\
&\qquad \text{and} \quad \boldsymbol{q}|_{\Omega_{ij}^K} = 0 \quad \text{for all } K \in \mathcal{T}_h,\ i, j \in \Lambda_K \cap \partial\Lambda_3,\ i \neq j\Big\}, \\
X_h &:= \{v \in X : \ v|_K \in \mathcal{P}_1(K) \quad \text{for all } K \in \mathcal{T}_h\}, \\
Y_h &:= \left\{v \in L^2(\Omega) : \ v|_{\Omega_i} \in \mathcal{P}_0(\Omega_i) \quad \text{for all } i \in \Lambda_0\right\}.
\end{aligned}$$

Obviously, we have

$$\dim M_h = \dim N_h = d(|\overline{\mathcal{E}}| - |\partial\mathcal{E}_3|)$$

and

$$\dim X_h = \dim Y_h = |\overline{\mathcal{N}}| - |\partial\mathcal{N}_3|,$$

where $\overline{\mathcal{N}}$ denotes set of all nodes of $\mathcal{T}_h$, $\partial\mathcal{N}_3$ the set of all nodes of $\mathcal{T}_h$ on $\widetilde{\Gamma}_3$, and $\overline{\mathcal{E}}$ the set of all edges of $\mathcal{T}_h$. The spaces N_h, X_h, Y_h are subspaces of N, X, Y, respectively, and they can naturally be equipped with the norms induced from the corresponding superspaces. Since $M_h \not\subset M$, the definition of its norm is more complicated and will be done later.

The Discrete Problem

Since the bilinear form b_1 from (8.77) is well-defined on $N_h \times X_h$, we could simply set $b_{1h} := b_1$. Similarly, the linear forms ℓ_1, ℓ_2 are well-defined on N_h and Y_h, respectively, so $\ell_{1h} := \ell_1 = 0$ and $\ell_{2h} := \ell_2$.

The discrete version a_h of a is defined on $M_h \times N_h$ as follows:

$$
\begin{aligned}
a_h(\boldsymbol{p}_h, \boldsymbol{q}_h) &:= \sum_{K\in\mathcal{T}_h} \sum_{i,j\in\Lambda_K:\, i<j} \boldsymbol{p}_h|_K(a_{ij}) \cdot \boldsymbol{q}_h|_{\Omega_{ij}^K} |\Omega_{ij}^K|, \\
b_{1h}(\boldsymbol{q}_h, u_h) &:= \sum_{K\in\mathcal{T}_h} \sum_{i,j\in\Lambda_K:\, i<j} \mu_{ij} \nabla u_h|_K \cdot \boldsymbol{q}_h|_{\Omega_{ij}^K} |\Omega_{ij}^K|, \\
c_h(u_h, \boldsymbol{q}_h) &:= -\sum_{K\in\mathcal{T}_h} \sum_{i,j\in\Lambda_K:\, i<j} \left[r_{ij}u_i + \left(1 - r_{ij}\right) u_j\right] \boldsymbol{c}_{ij} \cdot \boldsymbol{q}_h|_{\Omega_{ij}^K} |\Omega_{ij}^K|, \\
b_{2h}(\boldsymbol{p}_h, v_h) &:= -\sum_{K\in\mathcal{T}_h} \sum_{i\in\Lambda_0} v_i \sum_{j\in\Lambda_K\setminus\{i\}} \mathfrak{n}_{ij} \cdot \boldsymbol{p}_h|_K(a_{ij}) m_{ij}^K, \\
d_h(u_h, v_h) &:= \sum_{K\in\mathcal{T}_h} \sum_{i\in\partial\Lambda_2} \alpha_i\, u_i\, v_i\, \widetilde{m}_{i2}^K + \sum_{K\in\mathcal{T}_h} \sum_{i\in\Lambda_0} r_i u_i v_i m_i^K .
\end{aligned}
$$

where μ_{ij} is from (8.25), $\boldsymbol{c}_{ij} := \dfrac{1}{m_{ij}} \displaystyle\int_{\Gamma_{ij}} \boldsymbol{c}\, d\sigma$ (cf. (8.8)) and $m_i^K := |\Omega_i \cap K|$, $\widetilde{m}_{i2}^K := |\partial\Omega_i \cap \widetilde{\Gamma}_2 \cap K|_1$.

Now we are also able to define the norm on M_h (even if we no longer use it):

$$\|\boldsymbol{p}_h\|_{M_h} := \|\boldsymbol{p}_h\|_0 + \sup_{z_h\in Y_h} \frac{b_{2h}(\boldsymbol{p}_h, z_h)}{\|z_h\|_Y} .$$

After all these preparations we can now recover the scheme (8.11). Indeed, recalling the definition of the elements of N_h, the first equation of the discrete system (8.79) can be written as

$$\mathfrak{n}_{ij} \cdot \boldsymbol{p}_h|_K(a_{ij}) + \mu_{ij}\, \mathfrak{n}_{ij} \cdot \nabla u_h|_K - \mathfrak{n}_{ij} \cdot \boldsymbol{c}_{ij} \left[r_{ij}u_i + \left(1 - r_{ij}\right) u_j\right] = 0,$$
$$K \in \mathcal{T}_h,\ i, j \in \Lambda_K :\ i < j,$$

hence

$$\mathfrak{n}_{ij} \cdot \boldsymbol{p}_h|_K(a_{ij}) = -\mu_{ij}\, \mathfrak{n}_{ij} \cdot \nabla u_h|_K + \gamma_{ij} \left[r_{ij} u_i + \left(1 - r_{ij}\right) u_j \right] \tag{8.83}$$

(note that $\boldsymbol{c}_{ij} \cdot \mathfrak{n}_{ij} = \gamma_{ij}$, see (8.8)). So on the one hand we obtain a representation of the tangential components of $\boldsymbol{p}_h|_K(a_{ij})$ via $u_h \in X_h$, which can be used in the second equation of the discrete system (8.79), on the other hand we are able to define a unique element $\boldsymbol{p}_h \in M_h$ which can be viewed as a discrete flux density. In particular, in the case of a constant scalar diffusion coefficient k and vanishing convection, i.e. $\boldsymbol{c} = \boldsymbol{0}$, we even recover $-k\nabla u_h|_K$, i.e. $\boldsymbol{p}_h|_K = -k\nabla u_h|_K$.

Taking into consideration the relation

$$\mathfrak{n}_{ij} \cdot \nabla u_h|_K = \frac{u_j - u_i}{d_{ij}}, \quad K \ni a_{ij},$$

we get from (8.83)

$$\begin{aligned} b_{2h}(\boldsymbol{p}_h, v_h) &= - \sum_{K \in \mathcal{T}_h} \sum_{i \in \Lambda_0} v_i \sum_{j \in \Lambda_K \setminus \{i\}} \mathfrak{n}_{ij} \cdot \boldsymbol{p}_h|_K(a_{ij}) m_{ij}^K \\ &= - \sum_{K \in \mathcal{T}_h} \sum_{i \in \Lambda_0} v_i \sum_{j \in \Lambda_K \setminus \{i\}} \left\{ \mu_{ij} \frac{u_i - u_j}{d_{ij}} + \gamma_{ij} \left[r_{ij} u_i + \left(1 - r_{ij}\right) u_j \right] \right\} m_{ij}^K, \end{aligned}$$

and from the second equation of (8.79)

$$\begin{aligned} - \sum_{K \in \mathcal{T}_h} \Bigg\{ &\sum_{i \in \Lambda_0} v_i \sum_{j \in \Lambda_K \setminus \{i\}} \left\{ \mu_{ij} \frac{u_i - u_j}{d_{ij}} + \gamma_{ij} \left[r_{ij} u_i + \left(1 - r_{ij}\right) u_j \right] \right\} m_{ij}^K \\ &+ \sum_{i \in \partial\Lambda_2} \alpha_i\, u_i\, v_i\, \widetilde{m}_{i2}^K + \sum_{i \in \Lambda_0} r_i u_i v_i m_i^K \Bigg\} = - \langle f, v \rangle - \langle \widetilde{g_1}, v \rangle_{\widetilde{\Gamma}_1} - \langle \widetilde{g_2}, v \rangle_{\widetilde{\Gamma}_2} . \end{aligned}$$

If $f, \widetilde{g_1}, \widetilde{g_2}$ are continuous on $\overline{\Omega}, \widetilde{\Gamma}_1$ and $\widetilde{\Gamma}_2$, resp., we can apply simple quadrature rules and arrive then at an approximation of the right-hand side:

$$\langle f, v \rangle + \langle \widetilde{g_1}, v \rangle_{\widetilde{\Gamma}_1} + \langle \widetilde{g_2}, v \rangle_{\widetilde{\Gamma}_2} \approx \sum_{i \in \Lambda_0} \{ f_i\, m_i + \widetilde{g}_{1i}\, \widetilde{m}_{i1} + \widetilde{g}_{2i}\, \widetilde{m}_{i2} \}\, v_i .$$

With this, we get (8.13).

The result can also be interpreted as follows: To obtain an approximate $\boldsymbol{p}_h$ of the flux density defined a.e. on the whole domain Ω, it is sufficient to interpolate the numerical flux densities p_{ij} (see (8.12)) in M_h. In this respect, from an algorithmic point of view, the bypass via the mixed formulation is not compelling. The advantage of the mixed formulation, however, is that it can also be used to obtain an error estimate for the flux density approximation. A complete analysis (solution existence and uniqueness, error estimates) of the abstract discrete problem (8.79) for the case of vanishing bilinear forms c, c_h and d, d_h, together with the application of the theoretical results to the pure diffusion equation in the case $d = 2$, can be found in the paper [79].

Exercises

Problem 8.13 Show that the space $\boldsymbol{R}_1(K)$ can be equivalently defined by

$$\boldsymbol{R}_1(K) := \left\{\boldsymbol{a} + \boldsymbol{A}x : \boldsymbol{a} \in \mathbb{R}^d,\ \boldsymbol{A} = -\boldsymbol{A}^T \in \mathbb{R}^{d,d}\right\}, \quad d = 2, 3.$$

Problem 8.14 Let $K := \operatorname{conv}\{a_1, \ldots, a_{d+1}\}$ and denote by λ_i the barycentric coordinate associated with the node a_i.

a) Show that the local shape functions $\varphi_E := \lambda_i \nabla \lambda_j - \lambda_j \nabla \lambda_i$ belong to $\boldsymbol{R}_1(K)$.
Hint: Problem 8.13.
b) Show that a basis of $\boldsymbol{R}_1(K)$ is realized by $\{\varphi_E \mid E \text{ is an edge of } K\}$.

8.6 Finite Volume Methods for the Stokes and Navier–Stokes Equations

In Section 6.3, we have analysed the stationary Stokes equations (6.79) and discussed various mixed formulations of this system. Mixed finite element methods to approximate the solution (6.79) have been explained in Section 7.3. There we could see that the straightforward application of the mixed finite element method poses the basic problem of satisfying the discrete inf-sup condition. This is indeed a restriction from the point of view of implementational effort since equal order velocity and pressure spaces do not satisfy this condition. Moreover, the minimal pair of spaces of nonequal order, namely continuous piecewise linear polynomials for the velocity components and piecewise constant polynomials for the pressure, violates this condition, too. In general, FVMs are not able to resolve this conflict either. There are therefore – as with the finite element methods – methods that attempt to compensate for the loss of the inf-sup condition by so-called stabilization, and methods in which the interpolation spaces of the solution meet a discrete inf-sup condition. For the sake of completeness, it should also be mentioned that there are a few methods which use weakly or exactly divergence free interpolation spaces for the velocity components so that the unknown pressure can be eliminated from the saddle point system, resulting in a smaller, symmetric and positive definite system.

Here we present some exemplary approaches to FVMs of both groups for the numerical solution of (6.79).

As already described at the beginning of this chapter, there is in principle a more finite-difference-oriented or a more finite-element-oriented perspective in the development and investigation of FVMs. The first group includes in particular the staggered grids methods, for which the so-called *MAC (marker and cell) method* [146] on rectangular grids and its variants on unstructured grids have been very popular among the practitioners of the FVM (Fig. 8.12).

However, for domains with complex shape and for $d \geq 2$, staggered grid methods can hardly be extended to general nonorthogonal curvilinear or unstructured grids,

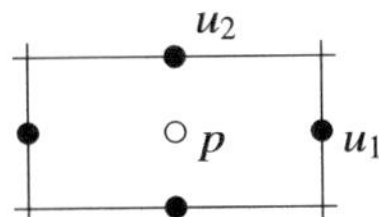

Fig. 8.12: Staggered placement of unknowns ($d = 2$) according to [146]. The velocity components are defined at the cell boundaries while the pressure values are defined at the cell centres.

which are needed to handle complex geometries. Although such attempts have been made up to the present day, they are not really convincing. Some disadvantages of generalized staggered grid approaches can be seen in the following: In d dimensions, $d + 1$ staggered grid families are required because the d velocity components have to be integrated over different volumes. In unstructured grids, besides the memory requirement to store such grid families, the use of Cartesian velocity components can lead to problems when one or more of the cell boundary surfaces become aligned with a velocity component.

In this respect, the use of so-called *collocated grids* is a more attractive alternative. In the early, simple variants, all the unknown discrete variables are defined at the same location in each control volume (typically the centroid), i.e. each scalar unknown variable has one degree of freedom per volume. The stability of these methods results from stabilization terms incorporated in the definition of the discrete bilinear forms. The newer, more sophisticated variants are formulated by amalgamating and extending concepts that are native to both FVMs and finite element methods. This type of methods is therefore also referred to as *finite volume element methods*.

An Equal-order Stabilized Finite Volume Method

As a first example, we outline an equal-order stabilized FVM [135]. The stationary Stokes equation (6.79) are considered in the mixed form (Formulation II, p. 386), where the spaces are

$$M := (H_0^1(\Omega))^d, \quad X := L_0^2(\Omega).$$

The bilinear form a is even M-coercive on the whole space.

The finite volume partitions $\mathcal{T}^* = \{\Omega_i \mid i \in \Lambda\}$ of the polygonal/polyhedral domain Ω satisfy the conditions 1)–4) from the beginning of Section 8.3.2.

The discretization of the matrix Laplacian/the bilinear form a can be adopted directly from (8.64) componentwise. The disretization of the divergence and gradient operators is motivated by the following arguments. For $(\boldsymbol{v}, q) \in M \times L^2(\Omega)$ and $q_i := \frac{1}{m_i}\int_{\Omega_i} q(x)\,dx$, $\boldsymbol{v}_i := \frac{1}{m_i}\int_{\Omega_i} \boldsymbol{v}(x)\,dx$, we write

$$\begin{aligned}
\int_\Omega q\nabla\cdot \boldsymbol{v}\,dx &\approx \sum_{i\in\Lambda}\int_{\Omega_i} q\nabla\cdot \boldsymbol{v}\,dx \approx \sum_{i\in\Lambda} q_i \int_{\Omega_i}\nabla\cdot \boldsymbol{v}\,dx = \sum_{i\in\Lambda} q_i\int_{\partial\Omega_i}\boldsymbol{\nu}\cdot\boldsymbol{v}\,d\sigma \\
&= \sum_{i\in\Lambda} q_i\int_{\partial\Omega_i\cap\Omega}\boldsymbol{\nu}\cdot\boldsymbol{v}\,d\sigma = \sum_{i\in\Lambda} q_i\sum_{\Gamma\in\mathcal{F}_i\cap\mathcal{F}}\int_\Gamma \boldsymbol{\nu}\cdot\boldsymbol{v}\,d\sigma \\
&\approx \sum_{i\in\Lambda} q_i\sum_{\Gamma\in\mathcal{F}_i\cap\mathcal{F}}\frac{d_{i,\Gamma}\boldsymbol{v}_i + d_{j,\Gamma}\boldsymbol{v}_j}{d_{ij}}\cdot\boldsymbol{\nu}_{ij}\,m_\Gamma .
\end{aligned} \tag{8.84}$$

The obtained expression is well defined for $(\boldsymbol{v}_h, q_h)\in M_h\times\tilde{X}_h$, where

$$\tilde{X}_h := \{q_h\in L^2(\Omega)\ \big|\ q_h|_{\Omega_i}\in\mathcal{P}_0(\Omega_i)\text{ for all } i\in\Lambda\} \tag{8.85}$$

is the space of functions which are piecewise constant over each control volume $\Omega_i\in\mathcal{T}^*$, and $M_h := (\tilde{X}_h)^d$. Hence it can be taken as the discrete version $-b_h(\boldsymbol{v}_h, q_h)$ of the bilinear form $-b(\boldsymbol{v}, q)$.

Next, splitting the last sum into two parts, changing the summation order and renaming the indices in the second of the obtained sum terms (cf. the proof of Lemma 8.15) leads to

$$\begin{aligned}
-b_h(\boldsymbol{v}_h, q_h) &= \sum_{i\in\Lambda} q_i\sum_{\Gamma\in\mathcal{F}_i\cap\mathcal{F}}\frac{d_{i,\Gamma}}{d_{ij}}\boldsymbol{v}_i\cdot\boldsymbol{\nu}_{ij}\,m_\Gamma + \sum_{i\in\Lambda}\sum_{\Gamma\in\mathcal{F}_i\cap\mathcal{F}}\frac{d_{i,\Gamma}}{d_{ij}}\boldsymbol{v}_i\cdot\boldsymbol{\nu}_{ji}\,q_j\,m_\Gamma \\
&= \sum_{i\in\Lambda}\sum_{\Gamma\in\mathcal{F}_i\cap\mathcal{F}}\frac{d_{i,\Gamma}}{d_{ij}}\boldsymbol{v}_i\cdot\boldsymbol{\nu}_{ij}\,(q_i - q_j)\,m_\Gamma .
\end{aligned}$$

This formula can be seen as a discrete version of the relation

$$\int_\Omega q\nabla\cdot\boldsymbol{v}\,dx = -\int_\Omega \nabla q\cdot\boldsymbol{v}\,dx \quad \text{for } (\boldsymbol{v}, q)\in M\times H^1(\Omega).$$

In this way we arrived at a discrete version of the gradient operator.

Now we are prepared to formulate a discrete version of the stationary Stokes equations (6.79) (with $\eta := \rho := 1$) w.r.t. $(\boldsymbol{v}_h, p_h)\in M_h\times\tilde{X}_h$:

$$\begin{aligned}
-\sum_{\Gamma\in\mathcal{F}_i\cap\mathcal{F}}\frac{\boldsymbol{v}_j-\boldsymbol{v}_i}{d_{ij}}m_\Gamma - \sum_{\Gamma\in\mathcal{F}_i\cap\partial\mathcal{F}}\frac{-\boldsymbol{v}_i}{d_{i,\Gamma}}m_\Gamma + \sum_{\Gamma\in\mathcal{F}_i\cap\mathcal{F}}\frac{d_{i,\Gamma}}{d_{ij}}(p_j - p_i)m_\Gamma\boldsymbol{\nu}_{ij} &= m_i f_i , \\
\sum_{\Gamma\in\mathcal{F}_i\cap\mathcal{F}}\frac{d_{i,\Gamma}\boldsymbol{v}_i + d_{j,\Gamma}\boldsymbol{v}_j}{d_{ij}}\cdot\boldsymbol{\nu}_{ij}\,m_\Gamma &= 0, \quad i\in\Lambda,
\end{aligned} \tag{8.86}$$

where $\boldsymbol{\nu}_{ij}$ denotes the normal to the face $\Gamma\in\mathcal{F}_i\cap\mathcal{F}$ oriented from Ω_i to Ω_j.

When considering this system, we make two observations. At first we easily see that this system of equations is singular, since the element $(\mathbf{0}, p_0)$ with $p_0 = \text{const.}$ belongs to the kernel of the associated discrete operator. Therefore we have to impose an additional condition, for instance

$$\sum_{i \in \Lambda} p_i m_i = 0.$$

On $\tilde{X}_h$, this relation is an exact reformulation of the constraint $\int_\Omega p dx = 0$. Consequently we restrict the above system (8.86) to $M_h \times X_h$, where $X_h := \tilde{X}_h \cap X$.

The second observation is a stability issue. If we look at the particular case $d = 2$, $\Omega = (0, 1)^2$ with a partition by square control volumes, we can identify checkerboard type solutions, completely analogous to the example ① in Section 7.3. As already mentioned, this problem can be solved by introducing a stabilization term.

So we finally arrive at the following finite volume formulation of the stationary Stokes equations (6.79) (with $\eta = \rho = 1$):

Find $(\boldsymbol{v}_h, p_h) \in M_h \times X_h$ such that

$$\begin{aligned} & -\sum_{\Gamma \in \mathcal{F}_i \cap \mathcal{F}} \frac{\boldsymbol{v}_j - \boldsymbol{v}_i}{d_{ij}} m_\Gamma - \sum_{\Gamma \in \mathcal{F}_i \cap \partial\mathcal{F}} \frac{-\boldsymbol{v}_i}{d_{i,\Gamma}} m_\Gamma + \sum_{\Gamma \in \mathcal{F}_i \cap \mathcal{F}} \frac{d_{i,\Gamma}}{d_{ij}} (p_j - p_i) m_\Gamma \boldsymbol{n}_{ij} = m_i \boldsymbol{f}_i , \\ & \sum_{\Gamma \in \mathcal{F}_i \cap \mathcal{F}} \frac{d_{i,\Gamma}\boldsymbol{v}_i + d_{j,\Gamma}\boldsymbol{v}_j}{d_{ij}} \cdot \boldsymbol{n}_{ij}\, m_\Gamma - \lambda \sum_{\Gamma \in \mathcal{F}_i \cap \mathcal{F}} (h_i^2 + h_j^2) \frac{p_j - p_i}{d_{ij}} m_\Gamma = 0 , \quad i \in \Lambda, \end{aligned} \tag{8.87}$$

where λ is a positive parameter and h_i is the diameter of Ω_i.

This method was investigated in [135] for the following particular choices of finite volume partitions:

- if $d = 2$, each control volume Ω_i is either a triangle with internal angles strictly lower than $\pi/2$ or a rectangle, and a_i is the intersection of the perpendicular bisectors of each face,
- if $d = 3$, each control volume is a rectangular parallelepiped, and a_i is the intersection of the straight lines issued from the barycentre of the face and orthogonal to the face.

In all of these cases, the points $y_i \in \Gamma$ are the barycentres of the faces Γ of Ω_i. Within this framework, stability of the scheme and error estimates for a smooth weak solution $(\boldsymbol{v}, p) \in \left(M \cap H^2(\Omega)\right) \times \left(X \cap H^1(\Omega)\right)$ of the Stokes problem were demonstrated in [135].

A Stable Finite Volume Element Method

The next example shows a method that does not require additional stabilization, since it can be understood as a nonconforming finite element method with a stable pair of spaces.

Let the domain Ω be a polygon and $\mathcal{T}_h$ be a consistent triangulation of Ω. If we define a nonconforming finite element space for the velocity by

$$M_h := \Big\{ v \in L^2(\Omega) \;\Big|\; v|_K \in \mathcal{P}_1, \int_F [\![v]\!] d\sigma = 0 \text{ for all } F \in \mathcal{F}, \\ \int_F v d\sigma = 0 \text{ for all } F \in \partial\mathcal{F} \Big\}^2 \tag{8.88}$$

(cf. (7.6)) and the pressure space by

$$X_h := \{ q_h \in L^2(\Omega) \;\big|\; q_h|_K \in \mathcal{P}_0 \text{ for all } K \in \mathcal{T}_h \}, \tag{8.89}$$

the so-called *P_1–P_0 nonconforming finite element approximation of the Stokes problem* (Formulation II, p. 386) reads as follows:

Find $(\boldsymbol{v}_h, p_h) \in M_h \times X_h$ such that

$$\begin{aligned} a_h(\boldsymbol{v}_h, \boldsymbol{w}_h) + b_h(\boldsymbol{w}_h, p_h) &= \langle \boldsymbol{f}, \boldsymbol{w}_h \rangle_0 \quad \text{for all } \boldsymbol{w}_h \in M_h, \\ b_h(\boldsymbol{v}_h, q_h) &= 0 \quad\quad\quad\quad \text{for all } q_h \in X_h, \end{aligned}$$

where

$$\begin{aligned} a_h(\boldsymbol{v}_h, \boldsymbol{w}_h) &:= \langle \nabla \boldsymbol{v}_h, \nabla \boldsymbol{w}_h \rangle_{0,h} = \sum_{K \in \mathcal{T}_h} \langle \nabla \boldsymbol{v}_h, \nabla \boldsymbol{w}_h \rangle_{0,K}, \\ b_h(\boldsymbol{v}_h, q_h) &:= -\langle \nabla \cdot \boldsymbol{v}_h, q_h \rangle_{0,h} = -\sum_{K \in \mathcal{T}_h} \langle \nabla \cdot \boldsymbol{v}_h, q_h \rangle_{0,K} \end{aligned}$$

with

$$\langle \nabla \boldsymbol{v}_h, \nabla \boldsymbol{w}_h \rangle_{0,K} := \sum_{l=1}^{d} \langle \nabla v_{lh}, \nabla w_{lh} \rangle_{0,K}$$

(analogous for $\langle \nabla \boldsymbol{v}_h, \nabla \boldsymbol{w}_h \rangle_{0,h}$). The nonconforming P_1–P_0 elements on triangles are known to satisfy the inf-sup condition (see [129]).

Following [123], the elements of the dual partition $\mathcal{T}_h^*$ are associated with the element faces $F \in \overline{\mathcal{F}}$ of the primal triangulation $\mathcal{T}_h$. If $\overline{\Lambda}$ here denotes the index set of a numbering of all faces, the set Λ contains all indices of interior faces, and $\partial\Lambda$ contains all indices of boundary faces. Then

$$\Omega_i := \begin{cases} \operatorname{int}\left(\operatorname{conv}\{a_k, a_l, a_{S,K}, a_{S,K'}\}\right), & \text{if } a_k + a_l = 2a_{S,i} \in K \cap K',\ i \in \Lambda. \\ \operatorname{int}\left(\operatorname{conv}\{a_k, a_l, a_{S,K}\}\right), & \text{if } a_k + a_l = 2a_{S,i} \in K,\ i \in \partial\Lambda, \end{cases}$$

where $a_{S,i}$ is the barycentre (midpoint) of the ith face.

To get a finite volume type discretization of the stationary Stokes equations (6.79) (with $\eta = \rho = 1$), we define a space $\widetilde{M}_h$, associated with the dual partition $\mathcal{T}_h^*$, by

$$\widetilde{M}_h := \{v \in L^2(\Omega) \mid v|_{\Omega_i} \in \mathcal{P}_0(\Omega_i) \text{ for all } i \in \overline{\Lambda} \text{ and } v|_{\Omega_i} = 0 \text{ for all } i \in \partial\Lambda\}^2,$$

multiply the momentum equation by $\boldsymbol{w}_h \in \widetilde{M}_h$ and the incompressible continuity equation by $q \in X_h$ and integrate the results over Ω:

Find $(\boldsymbol{v}_h, p_h) \in M_h \times X_h$ such that

$$\begin{aligned} \tilde{a}_h(\boldsymbol{v}_h, \boldsymbol{w}_h) + \tilde{b}_h(\boldsymbol{w}_h, p_h) &= \langle \boldsymbol{f}, \boldsymbol{w}_h \rangle_0 \qquad && \text{for all } \boldsymbol{w}_h \in \widetilde{M}_h, \\ \tilde{b}_h(\boldsymbol{v}_h, q_h) &= 0 && \text{for all } q_h \in X_h, \end{aligned}$$

where

$$\begin{aligned} \tilde{a}_h(\boldsymbol{v}_h, \boldsymbol{w}_h) &:= -\sum_{i \in \Lambda} \boldsymbol{w}_h(a_{S,i}) \cdot \int_{\partial\Omega_i} \partial_{\mathfrak{n}} \nabla \boldsymbol{v}_h \, d\sigma, \\ \tilde{b}_h(\boldsymbol{v}_h, q_h) &:= \sum_{i \in \Lambda} \boldsymbol{v}_h(a_{S,i}) \cdot \int_{\partial\Omega_i} q_h \mathfrak{n} \, d\sigma, \\ \langle \boldsymbol{f}, \boldsymbol{w}_h \rangle &:= \sum_{i \in \Lambda} \boldsymbol{w}_h(a_{S,i}) \cdot \int_{\Omega_i} \boldsymbol{f} \, dx \, . \end{aligned}$$

The relation between the FVM and the nonconforming method is given in the following lemma. It says that both methods generate the same coefficient matrix of the system of linear equations.

Lemma 8.32 *Let $\mathcal{T}_h$ be a consistent triangulation of Ω. Then it holds, for any $\boldsymbol{v}_h, \boldsymbol{w}_h \in M_h$ and $q_h \in X_h$:*

$$\begin{aligned} a_h(\boldsymbol{v}_h, \boldsymbol{w}_h) &= -\sum_{i \in \Lambda} \boldsymbol{w}_h(a_{S,i}) \cdot \int_{\partial\Omega_i} \partial_{\mathfrak{n}} \nabla \boldsymbol{v}_h \, d\sigma, \\ b_h(\boldsymbol{v}_h, q_h) &:= \sum_{i \in \Lambda} \boldsymbol{v}_h(a_{S,i}) \cdot \int_{\partial\Omega_i} q_h \mathfrak{n} \, d\sigma. \end{aligned}$$

Proof The statement is a simple consequence of Lemma 8.13 for the case of Donald diagrams and $\boldsymbol{K} = \boldsymbol{I}$. Namely, if we consider a single element $K \in \mathcal{T}_h$, say with vertices a_i, a_j and a_k, then the local nonconforming basis function ψ_i (associated with the face in opposition to the vertex a_i) can be expressed on K by means of the conforming basis functions φ_i (associated with the vertices) as follows:

$$\psi_i = \varphi_k + \varphi_j - \varphi_i \, .$$

Furthermore, for one component $v_{lh}|_K \in \mathcal{P}_1(K)$ of the velocity it holds, by Lemma 8.13:

$$\int_K \nabla v_{lh} \cdot \nabla \varphi_i \, dx = -\int_{\partial\Omega_i^D \cap K} \nabla v_{lh} \cdot \mathfrak{n} \, d\sigma,$$

where Ω_i^D is the barycentric control volume associated with a_i. Hence

$$\int_K \nabla v_{lh} \cdot \nabla \psi_i \, dx = -\int_{\partial\Omega_k^D \cap K} \nabla v_{lh} \cdot \mathfrak{n} \, d\sigma - \int_{\partial\Omega_j^D \cap K} \nabla v_{lh} \cdot \mathfrak{n} \, d\sigma + \int_{\partial\Omega_i^D \cap K} \nabla v_{lh} \cdot \mathfrak{n} \, d\sigma.$$

Now, since each of the boundary parts $\partial\Omega_k^D \cap K$ consists of two straight-line segments, that is

$$\partial\Omega_k^D \cap K = (\partial\Omega_k^D \cap \partial\Omega_j^D \cap K) \cup (\partial\Omega_k^D \cap \partial\Omega_i^D \cap K) := \Gamma_{kj}^{D,K} \cap \Gamma_{ki}^{D,K},$$

we get

$$\begin{aligned}
&\int_K \nabla v_{lh} \cdot \nabla \psi_i \, dx \\
&= -\int_{\Gamma_{kj}^{D,K}} \nabla v_{lh} \cdot \mathfrak{n}_{kj} \, d\sigma - \int_{\Gamma_{ki}^{D,K}} \nabla v_{lh} \cdot \mathfrak{n}_{ki} \, d\sigma - \int_{\Gamma_{jk}^{D,K}} \nabla v_{lh} \cdot \mathfrak{n}_{jk} \, d\sigma \\
&\quad - \int_{\Gamma_{ji}^{D,K}} \nabla v_{lh} \cdot \mathfrak{n}_{ji} \, d\sigma + \int_{\Gamma_{ik}^{D,K}} \nabla v_{lh} \cdot \mathfrak{n}_{ik} \, d\sigma + \int_{\Gamma_{ij}^{D,K}} \nabla v_{lh} \cdot \mathfrak{n}_{ij} \, d\sigma \\
&= -\int_{\Gamma_{kj}^{D,K}} \nabla v_{lh} \cdot \mathfrak{n}_{kj} \, d\sigma - \int_{\Gamma_{ki}^{D,K}} \nabla v_{lh} \cdot \mathfrak{n}_{ki} \, d\sigma + \int_{\Gamma_{kj}^{D,K}} \nabla v_{lh} \cdot \mathfrak{n}_{kj} \, d\sigma \\
&\quad - \int_{\Gamma_{ji}^{D,K}} \nabla v_{lh} \cdot \mathfrak{n}_{ji} \, d\sigma - \int_{\Gamma_{ki}^{D,K}} \nabla v_{lh} \cdot \mathfrak{n}_{ki} \, d\sigma - \int_{\Gamma_{ji}^{D,K}} \nabla v_{lh} \cdot \mathfrak{n}_{ji} \, d\sigma \\
&= -2\int_{\Gamma_{ki}^{D,K}} \nabla v_{lh} \cdot \mathfrak{n}_{ki} \, d\sigma - 2\int_{\Gamma_{ji}^{D,K}} \nabla v_{lh} \cdot \mathfrak{n}_{ji} \, d\sigma.
\end{aligned}$$

It remains to observe that the bisector $a_{S,j}a_j$ can be represented as $a_{S,j}a_S \cup a_S a_j = \Gamma_{ki}^{D,K} \cup (\partial\Omega_k \cap \partial\Omega_i)$ and $|\partial\Omega_k \cap \partial\Omega_i| = 2|\Gamma_{ki}^{D,K}|$. Since the integrands are constant along the corresponding bisectors, we arrive at

$$\begin{aligned}
\int_K \nabla v_{lh} \cdot \nabla \psi_i \, dx &= -\int_{\partial\Omega_k \cap \partial\Omega_i} \nabla v_{lh} \cdot \mathfrak{n}_{ki} \, d\sigma - \int_{\partial\Omega_j \cap \partial\Omega_i} \nabla v_{lh} \cdot \mathfrak{n}_{ji} \, d\sigma \\
&= -\int_{\partial\Omega_i \cap K} \nabla v_{lh} \cdot \mathfrak{n} \, d\sigma.
\end{aligned}$$

Concerning the bilinear form b_h, we consider at first the local terms:

$$\langle \nabla \cdot \boldsymbol{v}_h, q_h \rangle_{0,K} = q_h|_K \int_K \nabla \cdot \boldsymbol{v}_h \, dx = q_h|_K \int_{\partial K} \mathfrak{n} \cdot \boldsymbol{v}_h \, d\sigma = q_h|_K \sum_{i \in \Lambda_K} \int_{F_i} \mathfrak{n} \cdot \boldsymbol{v}_h \, d\sigma,$$

where here Λ_K is the set of indices of the faces of K. Let $\Lambda_K = \{i, j, k\}$. Since $\mathfrak{n} \cdot \boldsymbol{v}_h$ is linear on the faces, its mean value over a face is equal to the value it takes at the barycentre, hence we have

$$
\begin{aligned}
&\int_{F_i} \mathfrak{n} \cdot \boldsymbol{v}_h \, d\sigma \\
&= \left[\mathfrak{n} \cdot \boldsymbol{v}_h(a_{S,i})\, |F_i| + \mathfrak{n}_{ij} \cdot \boldsymbol{v}_h(a_{S,i})\, |\partial\Omega_i \cap \partial\Omega_j| + \mathfrak{n}_{ik} \cdot \boldsymbol{v}_h(a_{S,i})\, |\partial\Omega_i \cap \partial\Omega_k|\right] \\
&\quad - \left[\mathfrak{n}_{ij} \cdot \boldsymbol{v}_h(a_{S,i})\, |\partial\Omega_i \cap \partial\Omega_j| + \mathfrak{n}_{ik} \cdot \boldsymbol{v}_h(a_{S,i})\, |\partial\Omega_i \cap \partial\Omega_k|\right] \\
&= \int_{\partial(\Omega_i \cap K)} \mathfrak{n} \cdot \tilde{\boldsymbol{v}}_h \, d\sigma - \int_{\partial\Omega_i \cap K} \mathfrak{n} \cdot \tilde{\boldsymbol{v}}_h \, d\sigma,
\end{aligned}
$$

where $\tilde{\boldsymbol{v}}_h \in \tilde{M}_h$ is such that $\tilde{\boldsymbol{v}}_h(a_{S,i}) = \boldsymbol{v}_h(a_{S,i})$, $i \in \overline{\Lambda}$. Because of

$$
\int_{\partial(\Omega_i \cap K)} \mathfrak{n} \cdot \tilde{\boldsymbol{v}}_h \, d\sigma = \int_{\Omega_i \cap K} \nabla \cdot \tilde{\boldsymbol{v}}_h \, dx = 0
$$

we get

$$
\langle \nabla \cdot \boldsymbol{v}_h, q_h \rangle_{0,K} = -q_h|_K \sum_{i \in \Lambda_K} \int_{\partial\Omega_i \cap K} \mathfrak{n} \cdot \tilde{\boldsymbol{v}}_h \, d\sigma .
$$

It follows that

$$
\begin{aligned}
b_h(\boldsymbol{v}_h, q_h) &= -\sum_{K \in \mathcal{T}_h} \langle \nabla \cdot \boldsymbol{v}_h, q_h \rangle_{0,K} = \sum_{K \in \mathcal{T}_h} \sum_{i \in \Lambda_K} q_h|_K \int_{\partial\Omega_i \cap K} \mathfrak{n} \cdot \tilde{\boldsymbol{v}}_h \, d\sigma \\
&= \sum_{i \in \overline{\Lambda}} \sum_{K:\, K \cap \Omega_i \neq \emptyset} q_h|_K \int_{\partial\Omega_i \cap K} \mathfrak{n} \cdot \tilde{\boldsymbol{v}}_h \, d\sigma = \sum_{i \in \overline{\Lambda}} \tilde{\boldsymbol{v}}_h(a_{S,i}) \cdot \int_{\partial\Omega_i} q_h \mathfrak{n} \, d\sigma \\
&= \sum_{i \in \Lambda} \boldsymbol{v}_h(a_{S,i}) \cdot \int_{\partial\Omega_i} q_h \mathfrak{n} \, d\sigma .
\end{aligned}
$$

□

It is not difficult to check that the right-hand sides of the discrete momentum equations coincide if $\boldsymbol{f}$ is piecewise constant w.r.t. $\mathcal{T}_h$. For example, if $\boldsymbol{f} \in H^1(\Omega)^d$, then they have only an order h difference, which can be easily demonstrated using the tools provided in Section 3.4.

Finite Volume Based Methods for the Stationary Navier–Stokes Equations

In this subsection we will comment on finite-volume discretizations of the stationary form of the incompressible Navier–Stokes equations (ENS2b) (with η=const. > 0 and $\rho := 1$). These equations differ formally from the Stokes system by the nonlinear convective term. Besides the stability issue (velocity-pressure coupling) when discretizing the Stokes system, this new term brings additional difficulties. Here, the nonlinearity per se is not the main difficulty, but the (nonlinearly influenced) interaction between diffusive and convective effects, especially when the Reynolds number is not too small. Roughly speaking, all methods that work for the Stokes system are candidates for the numerical solution of the incompressible Navier–Stokes equations provided the nonlinear convective term can be handled appropriately.

The weak formulation of stationary equations (ENS2b) reads as follows:

Find $(\boldsymbol{v}, p) \in M \times X$ such that

$$\begin{aligned} a(\boldsymbol{v}, \boldsymbol{w}) + n(\boldsymbol{v}, \boldsymbol{v}, \boldsymbol{w}) + b(\boldsymbol{w}, p) &= \langle \boldsymbol{f}, \boldsymbol{w}\rangle_0 \qquad &&\text{for all } \boldsymbol{w} \in M, \\ b(\boldsymbol{v}, q) &= 0 &&\text{for all } q \in X, \end{aligned}$$

where the bilinear forms a, b are defined as in the case of the Stokes problem and the trilinear form $n : M^3 \to \mathbb{R}$ is defined as

$$n(\boldsymbol{u}, \boldsymbol{v}, \boldsymbol{w}) := \langle (\boldsymbol{u} \cdot \nabla)\boldsymbol{v}, \boldsymbol{w}\rangle_0 \,. \tag{8.90}$$

In [78] it was found that – based on the proof schemes in [53, Sect. IV.2], where a complete analysis for the P_1–P_0 nonconforming finite element approximation (8.88), (8.89) is presented – the following general result can be formulated:

If – in addition to the usual approximation properties – a family of finite element spaces $(M_h \times X_h)_h$ satisfies a uniform discrete inf-sup condition for b_h (see Remark 6.22, 2)), then the following three properties of the discrete form $n_h : M_h^3 \to \mathbb{R}$ are sufficient conditions to establish convergence of the numerical method for the incompressible Navier–Stokes equations:

Semidefiniteness: $n_h(\boldsymbol{u}_h, \boldsymbol{w}_h, \boldsymbol{w}_h) \geq 0$,

Lipschitz-continuity: $|n_h(\boldsymbol{u}_h, \boldsymbol{v}_h, \boldsymbol{w}_h) - n_h(\boldsymbol{z}_h, \boldsymbol{v}_h, \boldsymbol{w}_h)| \leq C\|\boldsymbol{u}_h - \boldsymbol{z}_h\|_h \|\boldsymbol{v}_h\|_h \|\boldsymbol{w}_h\|_h$,

Consistency: $|n(\boldsymbol{u}, \boldsymbol{v}, \boldsymbol{w}_h) - n_h(I_h\boldsymbol{u}, I_h\boldsymbol{v}, \boldsymbol{w}_h)| \leq Ch\|\boldsymbol{u}\|_2 \|\boldsymbol{v}\|_2 \|\boldsymbol{w}_h\|_h$,

$$\text{for all } \boldsymbol{u}_h, \boldsymbol{v}_h, \boldsymbol{w}_h, \boldsymbol{z}_h \in M_h, \;\; \boldsymbol{u}, \boldsymbol{v} \in H^2(\Omega)^d \cap M,$$

where $\|\cdot\|_h$ is a norm on M_h and $I_h : H^2(\Omega)^d \cap M \to M_h$ denotes some interpolation operator.

In what follows we outline the basic principles of a finite volume type discretization of the trilinear form such that these properties are satisfied. We start from (8.90) and decompose it as

$$n(\boldsymbol{u}, \boldsymbol{v}, \boldsymbol{w}) = \sum_{l=1}^{d} n_s(\boldsymbol{u}, v_l, w_l),$$

where $\boldsymbol{u}, \boldsymbol{v}, \boldsymbol{w} \in M$, $\nabla \cdot \boldsymbol{u} = 0$, and

$$n_s(\boldsymbol{u}, v, w) := \langle (\boldsymbol{u} \cdot \nabla)v, w\rangle_0 = \langle \boldsymbol{u} \cdot \nabla v, w\rangle_0, \quad v, w \in H_0^1(\Omega).$$

Furthermore, it is not difficult to show that the scalar part n_s can be rewritten in the skew-symmetric form

$$n_s(\boldsymbol{u}, v, w) = \frac{1}{2}[\langle \nabla \cdot (\boldsymbol{u}v), w\rangle_0 - \langle v, \nabla \cdot (\boldsymbol{u}w)\rangle_0],$$

which simply implies that

$$\forall \boldsymbol{u}, \boldsymbol{w} \in M_h : \quad n(\boldsymbol{u}, \boldsymbol{w}, \boldsymbol{w}) = 0. \tag{8.91}$$

However, the discretization of n_s is based on a slightly different representation, namely

$$n_s(\boldsymbol{u}, v, w) = \frac{1}{2}[c(\boldsymbol{u}, v, w) + c(-\boldsymbol{u}, w, v)],$$

where

$$c(\boldsymbol{u}, v, w) := \langle \nabla \cdot (\boldsymbol{u} v), w \rangle_0 - \frac{1}{2} \langle (\nabla \cdot \boldsymbol{u}) v, w \rangle_0 .$$

To be more concrete, we describe it in the notation from Section 8.2. Nonetheless, the discretization principle can be applied to much more general partition pairs $\mathcal{T}_h, \mathcal{T}_h^*$, see [78], including also so-called diamond-cell partitions. We set (cf. (8.8), (8.18))

$$\gamma_{ij} = \gamma_{ij}(\boldsymbol{u}_h) := \begin{cases} \dfrac{1}{m_{ij}} \displaystyle\int_{\Gamma_{ij}} \mathfrak{n} \cdot \boldsymbol{u}_h \, d\sigma , & m_{ij} > 0 , \\ \mathfrak{n} \cdot \boldsymbol{u}_h(a_{\Gamma_{ij}}), & m_{ij} = 0 , \end{cases}$$

$$r_{ij} = r_{ij}(\boldsymbol{u}_h) := R\left(\frac{\gamma_{ij} \, d_{ij}}{\eta} \right),$$

where $R : \mathbb{R} \to [0, 1]$ is a function satisfying the properties (P1)–(P3) in (8.20).

Then, by analogous arguments to the derivation of finite volume discretizations in Section 8.2, we get

$$\begin{aligned} c(\boldsymbol{u}, v, w) &\approx \sum_{i \in \Lambda} w_i \sum_{j \in \Lambda} [r_{ij} v_i + (1 - r_{ij}) v_j] \gamma_{ij} m_{ij} - \frac{1}{2} \sum_{i \in \Lambda} w_i v_i \sum_{j \in \Lambda} \gamma_{ij} m_{ij} \\ &= \sum_{i \in \Lambda} w_i \sum_{j \in \Lambda} \left[(1 - r_{ij}) v_j - \left(\frac{1}{2} - r_{ij} \right) v_i \right] \gamma_{ij} m_{ij} \end{aligned}$$

(cf. Lemma 8.15) and

$$c(-\boldsymbol{u}, w, v) \approx \sum_{i \in \Lambda} v_i \sum_{j \in \Lambda} \left[(1 - r_{ij}(-\boldsymbol{u}_h)) w_j - \left(\frac{1}{2} - r_{ij}(-\boldsymbol{u}_h) \right) w_i \right] \gamma_{ij}(-\boldsymbol{u}_h) m_{ij} .$$

The property (P1) in (8.20) enables to express the parameters $r_{ij}(-\boldsymbol{u}_h)$, $\gamma_{ij}(-\boldsymbol{u}_h)$ through $r_{ij}(\boldsymbol{u}_h)$, $\gamma_{ij}(\boldsymbol{u}_h)$, hence we obtain

$$\begin{aligned} n_s(\boldsymbol{u}, v, w) \approx \frac{1}{2} \sum_{i \in \Lambda} \sum_{j \in \Lambda} \Bigg[& (1 - r_{ij}) v_j w_i - \left(\frac{1}{2} - r_{ij} \right) v_i w_i \\ & - r_{ij} w_j v_i - \left(\frac{1}{2} - r_{ij} \right) v_i w_i \Bigg] \gamma_{ij} m_{ij} . \end{aligned}$$

Finally, the properties $\gamma_{ji} = -\gamma_{ij}$ and (P1) in (8.20) together with a symmetry argument lead to the final definition of n_{sh}:

$$\begin{aligned} n_s(\boldsymbol{u}, v, w) \approx n_{sh}(\boldsymbol{u}_h, v_h, w_h) := \frac{1}{2} \sum_{i \in \Lambda} \sum_{j \in \Lambda} \Bigg[& \left(r_{ij} - \frac{1}{2} \right) (v_i - v_j)(w_i - w_j) \\ & + \frac{1}{2} (v_j w_i - v_i w_j) \Bigg] \gamma_{ij} m_{ij} . \end{aligned}$$

Thus the discretization of the complete trilinear form n is clear:

$$n(\boldsymbol{u}, \boldsymbol{v}, \boldsymbol{w}) \approx n_h(\boldsymbol{u}_h, \boldsymbol{v}_h, \boldsymbol{w}_h) := \sum_{l=1}^{d} n_{sh}(\boldsymbol{u}_h, v_{lh}, w_{lh}) .$$

This definition makes it obvious that the discrete trilinear form n_h fulfills the first of the above three requirements to n_h: If the control function R satisfies (P1), (P2) in (8.20), then it holds

$$\forall \boldsymbol{u}_h, \boldsymbol{w}_h \in M_h : \quad n_h(\boldsymbol{u}_h, \boldsymbol{w}_h, \boldsymbol{w}_h) \geq 0 .$$

Hence n_h does not possess the property (8.91) but it violates it with the right sign. The verification of the other two requirements to n_h is more involved and depends much more heavily on the particular constellations. A complete analysis within a general finite volume approach in the dimensions $d = 2, 3$ for the discretization of the trilinear form that is guaranteed to have all the three properties can be found in [78].

With regard to the initially described equal-order finite volume discretization of the Stokes equations (8.87), it should be mentioned that, following the same ideas as in [135], in the work [136] a discretization for the transient incompressible Navier–Stokes equations on more general grids has been studied.

The discrete bilinear form introduced there does not have our sharp semi-definiteness property, even in the case of upwind stabilization, but only a lower bound of the type

$$\forall \boldsymbol{u}_h, \boldsymbol{w}_h \in M_h : \quad n_h(\boldsymbol{u}_h, \boldsymbol{w}_h, \boldsymbol{w}_h) \geq -\frac{1}{2} \int |\boldsymbol{w}_h|^2 \operatorname{div}_h(u_h) dx,$$

where $M_h := (\tilde{X}_h)^d$ (see (8.85)) and div_h is a discrete divergence operator (similar to (8.84)). Nevertheless this property is sufficient to get the required stability and convergence estimates.

Finally we mention that a careful treatment of the trilinear form in the context of discontinuous Galerkin discretizations allows to recover the skew-symmetry (8.91) of n to its discrete counterpart, i.e. it holds (with $M_h := (\mathcal{P}_k(\mathcal{T}_h))^d$, cf. (7.99))

$$\forall \boldsymbol{u}_h, \boldsymbol{w}_h \in M_h : \quad n_h(\boldsymbol{u}_h, \boldsymbol{w}_h, \boldsymbol{w}_h) = 0$$

(see [21, Lemma 6.39]).

Exercise

Problem 8.15 Examine the instability of the system (8.86) in detail analogously to the example ① in Section 7.3.

Chapter 9
Discretization Methods for Parabolic Initial Boundary Value Problems

9.1 Problem Setting and Solution Concept

In this section, initial boundary value problems for the linear case of the differential equation (0.33) are considered. We choose the form (3.13) together with the boundary conditions (3.19)–(3.21), which have already been discussed in Section 0.5. In Section 3.2, conditions have been developed to ensure a unique weak solution of the stationary boundary value problem. In contrast to Chapter 3, the heterogeneities are now allowed also to depend on time t, but for the sake of simplicity, we do not do so for the coefficients in the differential equations and the boundary conditions, which covers most of the applications, in which linear models are sufficient, for example, from Chapter 0. From time to time, we will restrict attention to homogeneous Dirichlet boundary conditions for further ease of exposition. Thus, the problem reads as follows:

The domain Ω is assumed to be a bounded Lipschitz domain and we suppose that $\Gamma_1, \Gamma_2, \Gamma_3$ form a disjoint decomposition of the boundary $\partial\Omega$ (cf. (0.39)):

$$\partial\Omega = \Gamma_1 \cup \Gamma_2 \cup \Gamma_3 \,,$$

where Γ_3 is a relatively closed subset of the boundary.

In the *space-time cylinder* $Q_T := \Omega \times (0,T)$, $T > 0$, and its boundary $S_T := \partial\Omega \times (0,T)$, there are given functions $f : Q_T \to \mathbb{R}$, $g : S_T \to \mathbb{R}$, $g(x,t) = g_i(x,t)$ for $x \in \Gamma_i$, $i = 1,2,3$, and $u_0 : \Omega \to \mathbb{R}$. The problem is to find a function $u : Q_T \to \mathbb{R}$ such that

$$\begin{aligned} \frac{\partial}{\partial t}(bu) + Lu &= f \quad \text{in } Q_T\,, \\ Ru &= g \quad \text{on } S_T\,, \\ u &= u_0 \quad \text{on } \Omega \times \{0\}\,, \end{aligned} \tag{9.1}$$

where Lv denotes the differential expression for some function $v : \Omega \to \mathbb{R}$,

P. Knabner and L. Angermann, *Numerical Methods for Elliptic and Parabolic Partial Differential Equations*, Texts in Applied Mathematics 44,
https://doi.org/10.1007/978-3-030-79385-2_9

$$(Lv)(x) := -\nabla \cdot (\boldsymbol{K}(x)\nabla v(x)) + \boldsymbol{c}(x) \cdot \nabla v(x) + r(x)v(x) \tag{9.2}$$

with sufficiently smooth, time-independent coefficients

$$\boldsymbol{K} : \Omega \to \mathbb{R}^{d,d}, \quad \boldsymbol{c} : \Omega \to \mathbb{R}^d, \quad r, b : \Omega \to \mathbb{R}.$$

The boundary condition is expressed by the shorthand notation $Ru = g$, which means, for a function $\alpha : \Gamma_2 \to \mathbb{R}$ on $\partial\Omega$,

- Neumann boundary condition (cf. (0.41) or (0.36))

$$\boldsymbol{K}\nabla u \cdot \mathfrak{n} = \partial_{\mathfrak{n}_{\boldsymbol{K}}} u = g_1 \quad \text{on } \Gamma_1 \times (0,T), \tag{9.3}$$

- mixed boundary condition (cf. (0.37))

$$\boldsymbol{K}\nabla u \cdot \mathfrak{n} + \alpha u = \partial_{\mathfrak{n}_{\boldsymbol{K}}} u + \alpha u = g_2 \quad \text{on } \Gamma_2 \times (0,T), \tag{9.4}$$

- Dirichlet boundary condition (cf. (0.38))

$$u = g_3 \quad \text{on } \Gamma_3 \times (0,T). \tag{9.5}$$

Thus, the stationary boundary problem considered so far reads

$$\begin{aligned} Lu(x) &= f(x) \quad \text{for } x \in \Omega, \\ Ru(x) &= g(x) \quad \text{for } x \in \partial\Omega. \end{aligned} \tag{9.6}$$

It is to be expected that both for the analysis and the discretization there are strong links between (9.6) and (9.1). The formulation (9.1) in particular includes the heat equation (cf. (0.20))

$$\frac{\partial u}{\partial t} - \nabla \cdot (\boldsymbol{K}\nabla u) = f \quad \text{in } Q_T, \tag{9.7}$$

or, for constant scalar coefficients, the normal form of the heat equation (cf. (0.19))

$$\frac{\partial u}{\partial t} - \Delta u = f \quad \text{in } Q_T \tag{9.8}$$

with appropriate initial and boundary conditions.

Again as in Chapter 1, one of the simplest cases will be, in two space dimensions ($d = 2$), the case of a rectangle $\Omega := (0,a) \times (0,b)$, $a, b > 0$, or even the one-dimensional case ($d = 1$) with $\Omega := (0,a)$. For the latter, equation (9.8) further reduces to

$$\frac{\partial u}{\partial t} - \frac{\partial^2}{\partial x^2} u = 0 \quad \text{in } Q_T = (0,a) \times (0,T). \tag{9.9}$$

For problem (9.1), the following typical analytical questions arise:

- existence of (classical) solutions,
- properties of the (classical) solutions,
- weaker concepts of the solution.

As in the case of elliptic boundary value problems, the theory of classical solutions requires comparatively strong assumptions on the data of the initial boundary value problem. In particular, along the edge $\partial\Omega \times \{0\}$ of the space-time cylinder initial and boundary conditions meet so that additional compatibility conditions have to be taken into account.

There are also applications where the coefficient functions $b, \boldsymbol{K}, \boldsymbol{c}, r$ are also time dependent. This is most obvious for the underlying flow field described by $\boldsymbol{c}$, which need not to be stationary. If one considers nonlinear models, where $\boldsymbol{K}$ or $\boldsymbol{c}$ or r depend on the solution (see 0.3 for example), then within a linearizing iteration (see Chapter 11) the discretized version of a model of the type (9.1) with

$$\boldsymbol{K} : Q_T \to \mathbb{R}^{d,d}, \quad \boldsymbol{c} : Q_T \to \mathbb{R}^d, \quad r, b : Q_T \to \mathbb{R} \tag{9.10}$$

has to be solved.

We will not consider this problem in theory, but take it into account in the formulation of the algorithms.

In Sections 6.2 and 7.2, we have investigated a mixed formulation for basically the same elliptic boundary value problem (with an operator according to (6.29) and boundary conditions to (6.32)–(6.34)). Then the time-dependent version reads as

$$\begin{aligned} \boldsymbol{p} &= -\boldsymbol{K}\nabla u + \boldsymbol{c}u && \text{in } Q_T, \\ \partial_t(bu) + \nabla \cdot \boldsymbol{p} + ru &= f && \text{in } Q_T, \\ u &= u_0 && \text{on } \Omega \times \{0\}, \end{aligned} \tag{9.11}$$

with boundary conditions similar to (9.3)–(9.5), but based on (6.32)–(6.34). On the other hand, the instationary Stokes equation from Section 6.3 (equations (ENS1a), (ENS2c); note the change in notation $u \leftrightarrow p$, $\boldsymbol{p} \leftrightarrow \boldsymbol{v}$)

$$\begin{aligned} \partial_t \boldsymbol{v} - \eta\Delta\boldsymbol{v} + \tfrac{1}{\varrho}\nabla p &= f && \text{in } Q_T, \\ \nabla \cdot \boldsymbol{v} &= 0 && \text{in } Q_T, \\ \boldsymbol{v} &= \boldsymbol{0} && \text{on } S_T, \\ \boldsymbol{v} &= \boldsymbol{v}_0 && \text{on } \Omega \times \{0\} \end{aligned} \tag{9.12}$$

also has a time derivative added, but of the vectorial quantity, and as the time derivative always appears in the conservation equation considered, it is the "second" in (9.11), but the "first" in (9.12).

We first concentrate on the primal formation (9.1)–(9.5).

Representation of Solutions in a Special Case

To enhance the familiarity with the problem and for further comparison, we briefly sketch a method, named *separation of variables*, by which closed-form solutions in the form of infinite series can be obtained for special cases. Also in these cases, the representations are not meant to be a numerical method (by its evaluation), but only serve as a theoretical tool. We assume from now on

$$b \in L^\infty(\Omega), \quad b(x) \geq b_0 > 0 \quad \text{for } x \in \Omega. \tag{9.13}$$

We start with the case of homogeneous data, i.e. $f = 0$, $g_i = 0$ $(i = 1, 2, 3)$ so that the process is determined only by the initial data u_0.

We allow also complex-valued solutions and assume a solution of (9.1) to have the factored form $u(x,t) = v(t)w(x)$ with $v \neq 0$, $w \neq 0$. This leads to

$$\frac{v'(t)}{v(t)} = \frac{-Lw(x)}{(bw)(x)} \quad \text{for all } (x,t) \in Q_T \text{ where } v(t) \neq 0,\ w(x) \neq 0. \tag{9.14}$$

Therefore, both sides of the equation (9.14) must be constant, for example, equal to $-\lambda$ for some $\lambda \in \mathbb{C}$. Therefore, equation (9.14) splits into two problems, namely

$$v'(t) = -\lambda v(t), \quad t \in (0,T), \tag{9.15}$$

which for the initial condition $v(0) = 1$ has the solution

$$v(t) = e^{-\lambda t},$$

and

$$\begin{aligned} Lw(x) &= \lambda(bw)(x),\ x \in \Omega, \\ Rw(x) &= 0, \qquad\qquad x \in \partial\Omega. \end{aligned} \tag{9.16}$$

A nontrivial solution $w : \overline{\Omega} \to \mathbb{C}$ of the boundary value problem (9.16) is called an *eigenfunction* to the (generalized) *eigenvalue* $\lambda \in \mathbb{C}$ of the problem (9.6). If the pairs (w_i, λ_i), $i = 1, \ldots, N$, are eigenfunctions/values for (9.6), then the superposition principle states that the function

$$u(x,t) := \sum_{i=1}^{N} c_i e^{-\lambda_i t} w_i(x) \tag{9.17}$$

is a (complex) solution of the homogeneous initial boundary value problem (9.1) for the initial value

$$u_0(x) := \sum_{i=1}^{N} c_i w_i(x), \tag{9.18}$$

where the coefficients $c_i \in \mathbb{C}$ are arbitrary. If there is a countably infinite number of eigenfunctions, we not only have to require the pointwise convergence of the sums (9.17) and (9.18) for $N \to \infty$ but also ensure the uniform convergence of (9.17) after term by term differentiation, for each type of derivatives which occurs in (9.6). Then also

$$u(x,t) := \sum_{i=1}^{\infty} c_i e^{-\lambda_i t} w_i(x) \tag{9.19}$$

is a solution to the initial date

$$u_0(x) := \sum_{i=1}^{\infty} c_i w_i(x). \tag{9.20}$$

As a (formal) shorthand notation for (9.19)–(9.20) ,we may introduce

$$u(x,t) = e^{tL} u_0(x). \tag{9.21}$$

For an inhomogeneous right-hand side of the form

$$f(x,t) = b(x) \sum_{i=1}^{N} f_i(t) w_i(x), \tag{9.22}$$

the solution representation can be extended to *(variation-of-constants formula)*

$$u(x,t) := \sum_{i=1}^{N} c_i e^{-\lambda_i t} w_i(x) + \sum_{i=1}^{N} \int_0^t f_i(s) e^{-\lambda_i (t-s)} ds \, w_i(x), \tag{9.23}$$

and at least formally the sum can be replaced by the infinite series. To verify (9.23) it suffices to consider the case $u_0 = 0$, for which we have

$$\begin{aligned}
(\partial_t (bu))(x,t) &= \sum_{i=1}^{N} f_i(t)(bw_i)(x) - \sum_{i=1}^{N} \int_0^t f_i(s) e^{-\lambda_i (t-s)} ds \, \lambda_i (bw_i)(x) \\
&= f(x,t) - L\left(\sum_{i=1}^{N} \int_0^t f_i(s) e^{-\lambda_i (t-s)} ds \, w_i \right)(x) \\
&= f(x,t) - Lu(x,t).
\end{aligned}$$

From these solution representations, we can conclude that initial data (and thus also perturbances contained in it) and also the influence of the right-hand side act only exponentially damped if the real parts of all eigenvalues are positive.

For $d := 1$, $\Omega := (0,a)$, $b := 1$, $Lv := -v''$, and (homogeneous) Dirichlet boundary conditions we get the real eigenvalues

$$\lambda^\nu := \left(\frac{\nu \pi}{a} \right)^2, \quad \nu \in \mathbb{N}, \tag{9.24}$$

and the corresponding eigenfunctions

$$w^\nu(x) := \sin\left(\nu \frac{\pi}{a} x \right). \tag{9.25}$$

If the initial date u_0 has the representation

$$u_0(x) = \sum_{\nu=1}^{\infty} c_\nu \sin\left(\nu \frac{\pi}{a} x\right), \tag{9.26}$$

then, for example, the (formal) solution of the homogeneous ($f = 0$) differential equation reads

$$u(x,t) = \sum_{\nu=1}^{\infty} c_\nu e^{-\lambda_\nu t} \sin\left(\nu \frac{\pi}{a} x\right). \tag{9.27}$$

The eigenfunctions w^ν are orthogonal with respect to the scalar product $\langle \cdot, \cdot \rangle_0$ in $L^2(\Omega)$, since they satisfy

$$\left\langle \sin\left(\nu \frac{\pi}{a} \cdot\right), \sin\left(\mu \frac{\pi}{a} \cdot\right)\right\rangle_0 = \begin{cases} 0 & \text{for } \nu \neq \mu, \\ \dfrac{a}{2} & \text{for } \nu = \mu, \end{cases} \tag{9.28}$$

which can by checked by means of well-known identities for the trigonometric functions.

Therefore (see below (9.68)),

$$c_\nu = \frac{\langle u_0, w^\nu \rangle_0}{\langle w^\nu, w^\nu \rangle_0}, \tag{9.29}$$

which is called the *Fourier coefficient* in the *Fourier expansion* (9.26) of u_0.

Of course, the pairs (w^ν, λ^ν) depend on the type of boundary conditions. For Neumann boundary conditions in $x = 0$ and $x = a$, we have

$$\begin{aligned} \lambda^\nu &= \left(\nu \frac{\pi}{a}\right)^2, \\ w^\nu(x) &= \cos\left(\nu \frac{\pi}{a} x\right), \quad \nu \in \mathbb{N}_0. \end{aligned} \tag{9.30}$$

The occurrence of $w^0 = 1$, $\lambda^0 = 0$ reflects the nontrivial solvability of the pure Neumann problem (which therefore is excluded by the conditions of Theorem 3.16).

For $d := 2$, $\Omega := (0,a)\times(0,b)$, the constant coefficient function $b := 1$, and $Lu := -\Delta u$ the eigenvalues and eigenfunctions can be derived from the one-dimensional case because of the factorization

$$\begin{aligned} -\Delta(v^\nu(x)\, \tilde{v}^\mu(y)) &= -\left(v^\nu\right)''(x)\, \tilde{v}^\mu(y) - v^\nu(x) \left(\tilde{v}^\mu\right)''(y) \\ &= (\lambda^\nu + \tilde{\lambda}^\mu) v^\nu(x)\, \tilde{v}^\mu(y). \end{aligned}$$

Therefore, one has to choose the eigenfunctions/values (v^ν, λ^ν) (w.r.t. x on $(0,a)$) for the required boundary conditions at $x = 0$ and $x = a$, and $(\tilde{v}^\mu, \tilde{\lambda}^\mu)$ (w.r.t. y on $(0,b)$) for the required boundary conditions at $y = 0$, $y = b$.

For (homogeneous) Dirichlet boundary conditions everywhere on $\partial\Omega$, this leads to the eigenvalues

$$\lambda^{\nu\mu} = \left(\nu \frac{\pi}{a}\right)^2 + \left(\mu \frac{\pi}{b}\right)^2 \tag{9.31}$$

and the eigenfunctions

$$w^{\nu\mu}(x,y) = \sin\left(\nu\frac{\pi}{a}x\right)\sin\left(\mu\frac{\pi}{b}y\right), \quad \nu,\mu \in \mathbb{N}. \tag{9.32}$$

Note that the smallest eigenvalue is $\left(\frac{\pi}{a}\right)^2 + \left(\frac{\pi}{b}\right)^2$, and $\lambda^{\nu\mu} \to \infty$ for $\nu \to \infty$ or $\mu \to \infty$.

As a further concluding example, we note the case of boundary conditions of mixed type:

$$\begin{aligned} u(0,y) = u(a,y) &= 0 && \text{for } y \in [0,b],\\ \underbrace{\nabla u\cdot\mathfrak{n}(x,0)}_{=-\partial_2 u(x,0)} = \underbrace{\nabla u\cdot\mathfrak{n}(x,b)}_{=\partial_2 u(x,b)} &= 0 && \text{for } x \in (0,a). \end{aligned}$$

Here, we have the eigenvalues

$$\lambda^{\nu\mu} = \left(\nu\frac{\pi}{a}\right)^2 + \left(\mu\frac{\pi}{b}\right)^2$$

and the eigenfunctions

$$w^{\nu\mu}(x,y) = \sin\left(\nu\frac{\pi}{a}x\right)\cos\left(\mu\frac{\pi}{b}y\right), \quad \nu \in \mathbb{N},\ \mu \in \mathbb{N}_0.$$

A Sketch of the Theory of Weak Solutions

As in the study of the elliptic boundary value problems (3.13), (3.19)–(3.21), for equation (9.1), a weak formulation can be given that reduces the requirements with respect to the differentiability properties of the solution.

The idea is to treat time and space variables in a different way:

1) • For fixed $t \in (0,T)$, the function $x \mapsto u(x,t)$ is interpreted as a parameter-dependent element $u(t)$ of some space V whose elements are functions of $x \in \Omega$. An obvious choice for the FEM of Chapter 3 is (see Subsection 3.2.1, (III)) the space

$$V := H^1_{\Gamma_3}(\Omega) = \{v \in H^1(\Omega) \mid \gamma_{0,\Gamma_3}(v) = 0\},$$

see (6.37).

   • In a next step, that is, for varying t, a function $t \mapsto u(t)$ results with values in the (function) space V.

2) In addition to V, a further space $H = L^2(\Omega)$ occurs, from which the initial value u_0 is taken and which contains V as a dense subspace. A subspace V is called *dense* in H if the closure of V with respect to the norm on H coincides with H. Then we have the dense embeddings

$$V \hookrightarrow H \hookrightarrow V',$$

i.e. a scalar product $\langle\cdot,\cdot\rangle$ on H can be uniquely extended to a *duality pairing* $\langle\cdot,\cdot\rangle : V' \times V \to \mathbb{R}$. One also speaks of the *Gelfand-triple* (V, H, V').
3) The time derivative is understood in a generalized sense; see (9.33).
4) The generalized solution $t \mapsto u(t)$ is sought as an element of a function space, the elements of which are "function-valued" (cf. (1)).

Definition 9.1 Let X denote one of the spaces H or V (in particular, this means that the elements of X are functions on $\Omega \subset \mathbb{R}^d$).

1) The space $C^l([0,T], X)$, $l \in \mathbb{N}_0$, consists of all continuous functions $v : [0,T] \to X$ that have continuous derivatives up to the order l on $[0,T]$ with the norm

$$\sum_{i=0}^{l} \sup_{t\in(0,T)} \|v^{(i)}(t)\|_X .$$

For the sake of simplicity, the notation $C([0,T], X) := C^0([0,T], X)$ is used.
2) The space $L^p((0,T), X)$ with $1 \le p \le \infty$ consists of all functions on $(0,T) \times \Omega$ with the following properties:

$$v(t,\cdot) \in X \text{ for almost any } t \in (0,T),\;\; F \in L^p(0,T) \text{ with } F(t) := \|v(t,\cdot)\|_X .$$

Furthermore,

$$\|v\|_{L^p((0,T),X)} := \|F\|_{L^p(0,T)}.$$

Remark 9.2 $f \in L^2(Q_T) \Rightarrow f \in L^2\left((0,T), H\right)$.

Proof The proof is essentially a consequence of Fubini's theorem (see [1, Thm. 1.54]). □

Concerning the interpretation of the time derivative and of the weak formulation, a comprehensive treatment is possible only within the framework of the theory of distributions; thus, a detailed explanation is beyond the scope of this book. A short but mathematically rigorous introduction can be found in the book [69, Chapter 23].

The basic idea consists in the following definition:

A function $u \in L^2\left((0,T), V\right)$ is said to have a *weak derivative* w if the following holds:

$$\int_0^T u(t)\,\Psi'(t)\,dt = -\int_0^T w(t)\,\Psi(t)\,dt \quad \text{for all } \Psi \in C_0^\infty(0,T). \tag{9.33}$$

Usually, this derivative w is denoted by $\dfrac{du}{dt}$ or u'.

Remark 9.3 The integrals occurring above are to be understood as so-called *Bochner* integrals and are extensions of the Lebesgue integral to function-valued mappings. Therefore, equation (9.33) is an equality of functions.

An important property to be used frequently in the following is the estimate

$$\left\| \int_0^T u(t)dt \right\|_X \leq \int_0^T \|u(t)\|_X dt \qquad \text{for } u \in L^1((0,T),X)$$

Before, we give a weak formulation of (9.1), the following notion is worth recalling:

$$a(u,v) := \int_\Omega [\boldsymbol{K}\nabla u \cdot \nabla v + (\boldsymbol{c} \cdot \nabla u + ru)\, v]\, dx + \int_{\Gamma_2} \alpha u v \, d\sigma, \quad u, v \in V. \qquad (9.34)$$

Furthermore, we set

$$\ell(t; v) := \langle f(t), v\rangle_0 + \int_{\Gamma_1} g_1(\cdot, t) v \, d\sigma + \int_{\Gamma_2} g_2(\cdot, t) v \, d\sigma, \quad v \in V \qquad (9.35)$$

and

$$\langle u, v\rangle_0^{(b)} := \langle bu, v\rangle_0, \quad \|u\|_0^{(b)} := \left(\langle u, u\rangle_0^{(b)}\right)^{\frac{1}{2}}, \quad u, v \in H. \qquad (9.36)$$

Obviously, (9.13) guarantees that

$$b_0^{\frac{1}{2}} \|v\|_0 \leq \|v\|_0^{(b)} \leq \|b\|_\infty^{\frac{1}{2}} \|v\|_0 \quad \text{for all } v \in H,$$

i.e. the norms are equivalent, therefore in all subsequent estimates, the norms can be interchanged with a corresponding change of constants.

Let $u_0 \in H$, $f \in L^2((0,T),H)$, and in case of Dirichlet conditions, we restrict ourselves to the homogeneous case.

An element $u \in L^2((0,T),V)$ is called a *weak solution* of (9.1) if it has a weak derivative $\frac{du}{dt} = u' \in L^2((0,T),H)$ and the following holds

$$\begin{aligned} \left\langle \frac{du}{dt}(t), v\right\rangle_0^{(b)} + a(u(t), v) = \ell(t; v) \quad &\text{for all } v \in V \text{ and } t \in (0,T), \\ u(0) = u_0\,. & \end{aligned} \qquad (9.37)$$

Also the weaker requirement $u' \in L^2((0,T),V')$ is possible, then the term $\left\langle \frac{du}{dt}(t), v\right\rangle_0^{(b)}$ has to be interpreted as the application of the duality pairing, with $\langle \cdot, \cdot \rangle^{(b)}$ from (9.36). Due to $u \in L^2((0,T),V)$ and $u' \in L^2((0,T),V')$, we also have $u \in C([0,T],H)$ (see [25, Sect. 5.9, Thm. 3]) so that the initial condition is meaningful in the classical sense.

In what follows, the bilinear form a is assumed to be continuous on $V \times V$ (see (3.2)) and V-coercive (see (3.3)). The latter means that there exists a constant $\alpha > 0$ such that

$$a(v,v) \geq \alpha \|v\|_V^2 \quad \text{for all } v \in V. \qquad (9.38)$$

We consider only homogeneous boundary conditions, i.e. $g_1 = g_2 = 0$, too, and the right-hand side functional ℓ reduces to $\ell(t; v) = \langle f(t), v\rangle_0$.

Lemma 9.4 *Let a be a V-coercive, continuous bilinear form, $u_0 \in H$, and $f \in C([0,T], H)$, and suppose the considered boundary conditions are homogeneous. Then, for the solution u of (9.37) the following estimate holds:*

$$\|u(t)\|_0^{(b)} \le \|u_0\|_0^{(b)} e^{-\tilde{\alpha}t} + \int_0^t \|f(s)\|_0^{(b^{-1})} e^{-\tilde{\alpha}(t-s)}\, ds \quad \textit{for all } t \in (0,T),$$

where $\tilde{\alpha} := \alpha\|b\|_\infty^{-1}$.

Proof Without loss of generality we may suppose that $\|u(t)\|_0^{(b)} > 0$, otherwise, the estimate is trivially satisfied. The following equations are valid almost everywhere in $(0, T)$. Setting $v = u(t)$, (9.37) reads as

$$\langle u'(t), u(t)\rangle_0^{(b)} + a(u(t), u(t)) = \langle f(t), u(t)\rangle_0 .$$

Using the relation

$$\begin{aligned}\langle u'(t), u(t)\rangle_0^{(b)} &= \frac{1}{2}\frac{d}{dt}\langle u(t), u(t)\rangle_0^{(b)} = \frac{1}{2}\frac{d}{dt}\left(\|u(t)\|_0^{(b)}\right)^2 \\ &= \|u(t)\|_0^{(b)} \frac{d}{dt}\|u(t)\|_0^{(b)}\end{aligned}$$

and the V-coercivity, it follows that

$$\|u(t)\|_0^{(b)} \frac{d}{dt}\|u(t)\|_0^{(b)} + \alpha \|u(t)\|_V^2 \le \langle f(t), u(t)\rangle_0 .$$

Now the simple inequality

$$\|u(t)\|_0^{(b)} \le \|b\|_\infty^{1/2}\|u(t)\|_V$$

and the Cauchy–Schwarz inequality

$$\langle f(t), u(t)\rangle_0 \le \|f(t)\|_0^{(b^{-1})}\|u(t)\|_0^{(b)}$$

yield, after division by $\|u(t)\|_0^{(b)}$, the estimate

$$\frac{d}{dt}\|u(t)\|_0^{(b)} + \alpha\|b\|_\infty^{-1}\|u(t)\|_0^{(b)} \le \|f(t)\|_0^{(b^{-1})}.$$

Setting $\tilde{\alpha} := \alpha\|b\|_\infty^{-1}$ and multiplying this relation by $e^{\tilde{\alpha}t}$, the relation

$$\frac{d}{dt}(e^{\tilde{\alpha}t}\|u(t)\|_0^{(b)}) = e^{\tilde{\alpha}t}\frac{d}{dt}\|u(t)\|_0^{(b)} + \tilde{\alpha}e^{\tilde{\alpha}t}\|u(t)\|_0^{(b)}$$

leads to

$$\frac{d}{dt}(e^{\tilde{\alpha}t}\|u(t)\|_0^{(b)}) \le e^{\tilde{\alpha}t}\|f(t)\|_0^{(b^{-1})}.$$

The integration over $(0, t)$ results in

$$e^{\tilde{\alpha} t}\|u(t)\|_0^{(b)} - \|u(0)\|_0^{(b)} \le \int_0^t e^{\tilde{\alpha} s}\|f(s)\|_0^{(b^{-1})}\,ds$$

for all $t \in (0, T)$. Multiplying this by $e^{-\tilde{\alpha} t}$ and taking into consideration the initial condition, we get the asserted relation

$$\|u(t)\|_0^{(b)} \le \|u_0\|_0^{(b)}\,e^{-\tilde{\alpha} t} + \int_0^t \|b^{-1/2} f(s)\|_0\,e^{-\tilde{\alpha}(t-s)}\,ds\,.$$

□

Remark 9.5 1) The result of Lemma 9.4 can be extended to inhomogeneous boundary conditions leading to

$$\|u(t)\|_0^{(b)} \le \left(\left(\|u_0\|_0^{(b)}\right)^2 e^{-\hat{\alpha} t} + \frac{1}{\|b\|_\infty (2\tilde{\alpha} - \hat{\alpha})} \int_0^t \|\ell(s)\|_{V'}^2\,e^{-\hat{\alpha}(t-s)}\,ds \right)^{1/2}$$

for every $0 < \hat{\alpha} < 2\tilde{\alpha}$.

2) In fact, here and in the following, we can also treat more general underlying elliptic problems, the bilinear form a of which only satisfies a so-called *Gårding inequality* instead of the above V-coercivity condition (9.38):
There are constants $\alpha > 0$ and $\kappa \ge 0$ such that

$$a(v, v) \ge \alpha \|v\|_V^2 - \kappa \left(\|v\|_0^{(b)}\right)^2 \quad \text{for all } v \in V. \tag{9.39}$$

Note that u is weak solution of (9.34) if and only if

$$\tilde{u}(t) := e^{-\kappa t} u(t)$$

is a weak solution of (9.34) with the bilinear form a substituted by

$$\tilde{a}(u, v) := a(u, v) + \kappa \langle u, v \rangle_0^{(b)}$$

and f, g_1, g_2 by $f e^{-\kappa t}$, $g_1 e^{-\kappa t}$, $g_2 e^{-\kappa t}$, respectively. If a satisfies the inequality (9.39), then $\tilde{a}$ is V-coercive. This means that Lemma 9.4 holds true for $\tilde{u}$, and thus for u with $\tilde{\alpha}$ substituted by $\tilde{\alpha} - \kappa$. Therefore, possible damping properties may got lost or be substituted by exponential increase. Apart of that, the subsequent stability and error estimates hold true also for this more general case. In this sense, condition (9.39) allows us to remove the sign requirements developed for V-coercivity (see Theorem 3.16), which are not necessary anymore (see Problem 9.6).

3) For more regular solutions also, a stability estimate in $L^\infty(0, T; V)$ is possible. Assume $u_0 \in V$ and $\ell(t; v) = \langle f(t), v \rangle_0$, i.e. $f(t) \in H$, and for solution $u' \in L^2((0, T), V)$, then also testing with u' is possible too, leading to

$$a(u(t),u(t)) \leq a(u_0,u_0) + \frac{1}{4}\int_0^t \left(\|f(s)\|_0^{(b)}\right)^2 ds.$$

More precise versions are formulated in Problem 9.5.

There is a close relation between Lemma 9.4 and solution representations such as (9.23) (with the sum being infinite). The eigenvalue problem (9.16) is generalized to the following variational form (see also the end of Section 2.2):

Definition 9.6 A number $\lambda \in \mathbb{C}$ is called an *eigenvalue* to the *eigenvector* $w \in V\setminus\{0\}$ if

$$a(w,v) = \lambda\langle w,v\rangle_0^{(b)} \quad \text{for all } v \in V.$$

Assume in addition to our assumptions that the bilinear form is symmetric and the embedding of V into H is compact (see [51]), which is the case here. Then all the eigenvalues are real and there are enough eigenvectors in the sense that there exists a sequence of pairs (w_i,λ_i) such that $0 < \lambda_1 \leq \lambda_2 \leq \ldots$, the eigenvectors w_i are orthonormal with respect to the L^2-scalar product $\langle\cdot,\cdot\rangle_0^{(b)}$, and every $v \in V$ has a unique representation (in H) as

$$v = \sum_{i=1}^{\infty} c_i w_i\,. \tag{9.40}$$

As in (9.29), the Fourier coefficients c_i are given by

$$c_i = \langle v, w_i\rangle_0^{(b)}\,. \tag{9.41}$$

In fact, (9.40) gives a rigorous framework to the specific considerations in (9.20) and subsequent formulas. From (9.40) and (9.41), we conclude *Parseval's identity*

$$\|v\|_0^{(b)2} = \sum_{i=1}^{\infty} |\langle v, w_i\rangle_0^{(b)}|^2. \tag{9.42}$$

Furthermore, the sequence $v_i := \lambda_i^{-1/2} w_i$ is orthonormal with respect to $a(\cdot,\cdot)$, and a representation corresponding to (9.40), (9.41) holds such that

$$a(v,v) = \sum_{i=1}^{\infty} |a(v,v_i)|^2 = \sum_{i=1}^{\infty} \lambda_i^{-1}|a(v,w_i)|^2 = \sum_{i=1}^{\infty} \lambda_i |\langle v,w_i\rangle_0^{(b)}|^2. \tag{9.43}$$

From (9.42) and (9.43), we see that the coercivity constant is related to the smallest eigenvalue λ_1.

For example, if

$$\|v\|_V^2 = a(v,v) + \langle v,v\rangle_0$$

as for the Poisson equation, then if $\alpha \in (0,1)$ is maximal in the relation

$$a(v,v) \geq \alpha\|v\|_V^2 \quad \text{for all } v \in V,$$

then $\alpha' := \alpha/(1-\alpha)$ is maximal in the relation

$$a(v, v) \geq \alpha' \|v\|_0^2 \quad \text{for all } v \in V,$$

i.e.

$$\lambda_{\min} = \lambda_1 = \frac{\alpha}{1-\alpha} \,.$$

Furthermore, the solution representation (9.23) (with the sum being infinite in H) also holds true under the assumptions mentioned and also leads to the estimate of Lemma 9.4. Note, however, that the proof there does not require the symmetry of the bilinear form.

Exercises

Problem 9.1 Consider the initial boundary value problem

$$\begin{aligned} u_t - u_{xx} &= 0 && \text{in } (0, \infty) \times (0, \infty), \\ u(0, t) &= h(t), && t \in (0, \infty), \\ u(x, 0) &= 0, && x \in (0, \infty), \end{aligned}$$

where $h : (0, \infty) \to \mathbb{R}$ is a differentiable function, the derivative of which has at most exponential growth.

a) Show that the function

$$u(x, t) = \sqrt{\frac{2}{\pi}} \int_{x/\sqrt{2t}}^{\infty} e^{-s^2/2} h\left(t - \frac{x^2}{2s^2}\right) ds$$

is a solution.

b) Is u_t bounded in the domain of definition? If not, give conditions on h that guarantee the boundedness of u_t.

Problem 9.2 Consider the initial boundary value problem in one space dimension

$$\begin{aligned} u_t - u_{xx} &= 0 && \text{in } (0, \pi) \times (0, \infty), \\ u(0, t) = u(\pi, t) &= 0, && t \in (0, \infty), \\ u(x, 0) &= u_0(x), && x \in (0, \pi). \end{aligned}$$

a) Solve it by means of the method of separation.
b) Give a representation for $\|u_t(t)\|_0$.
c) Consider the particular initial condition $u_0(x) = \pi - x$ and investigate, using the result from subproblem (b), the asymptotic behaviour of $\|u_t(t)\|_0$ near $t = 0$.

Problem 9.3 a) Show that the weak formulation (9.37) with $u' \in L^2((0,T),V')$ is equivalent to the following variational equation:

Find $u \in X$ such that

$$c(u, y) = \langle \tilde{\ell}, y\rangle_Y \quad \text{for all } y \in Y,$$

where

$$c(v, y) := \int_0^T \left[\left\langle b\frac{dv}{dt}, y\right\rangle_V + a(v, y)\right] dt,$$

$$\langle \tilde{\ell}, y\rangle_Y := \int_0^T \langle \ell, y\rangle_V \, dt,$$

$$X := \left\{ y \in L^2((0,T),V) \;\middle|\; \frac{dy}{dt} \in L^2((0,T),V'),\ v(0) = 0 \right\},$$

$$Y := L^2((0,T),V).$$

b) Assuming (9.13), (9.38) show unique existence of the formulation of 9.3) by using Theorem 6.2.

c) Can the above considerations extended to t-dependent a (see (9.53))? Discuss also t-dependent b.

Problem 9.4 For H, V being Hilbert spaces, V dense in H, show that $u, v \in L^2((0,T),V)$ such that $\frac{du}{dt}, \frac{dv}{dt} \in L^2((0,T),V')$ satisfy the relationship

$$\int_0^T \langle b\frac{d}{dt}u, v\rangle_V \, dt = \langle bu(t), v(t)\rangle_H \big|_0^T - \int_0^T \langle b\frac{d}{dt}v, u\rangle_V \, dt.$$

Problem 9.5 Let the domain $\Omega \subset \mathbb{R}^d$ be bounded by a sufficiently smooth boundary and set $V := H_0^1(\Omega)$, $H := L^2(\Omega)$. Furthermore, $a : V \times V \to \mathbb{R}$ is a continuous, V-coercive, symmetric bilinear form and $u_0 \in H$. Prove by using the so-called *energy method* (cf. the proof of Lemma 9.4), the following a priori estimates for the solution u of the initial boundary value problem

$$\left\langle \frac{du}{dt}(t), v\right\rangle_0 + a(u(t), v) = 0 \qquad \text{for all } v \in V,\ t \in (0,T),$$

$$u(0) = u_0 .$$

a) $\alpha t \|u(t)\|_1^2 + 2\int_0^t s \left\|\frac{du}{dt}(s)\right\|_0^2 ds \le M \int_0^t \|u(s)\|_1^2 \, ds .$

b) $\left\|\frac{du}{dt}(t)\right\|_0 \le \sqrt{\frac{M}{2\alpha}\frac{1}{t}}\|u_0\|_0 .$

Here, M and α denote the corresponding constants in the continuity and coercivity conditions, respectively.

Problem 9.6 Relax the condition of Theorem 3.16 such that (9.39) is still satisfied.

9.2 Semidiscretization by the Vertical Method of Lines

For solving parabolic equations numerically, a wide variety of methods exists. The most important classes of these methods are the following:

- *Full discretizations:*
  - Application of finite difference methods to the classical initial boundary value problem (as of the form (9.1)).
  - Application of so-called space-time finite element methods to a variational formulation that includes the time variable, too.
- *Semidiscretizations:*
  - *The vertical method of lines:* Here, the discretization starts with respect to the spatial variable(s) (e.g. by means of the finite difference method, the (mixed) finite element method, or one of the finite volume methods).
  - *The horizontal method of lines* (Rothe's method): Here, the discretization starts with respect to the time variable.

As the name indicates, a semidiscretization has to be followed by a further discretization step to obtain a full discretization, which may be one of the above-mentioned or not. The idea behind semidiscretization methods is to have intermediate problems that are of a well-known structure. In the case of the vertical method of lines, a system of ordinary differential equations arises for the solution of which appropriate solvers are often available. Rothe's method generates a sequence of elliptic boundary value problems for which efficient solution methods have been developed in the former chapters.

The attributes "vertical" and "horizontal" of the semidiscretizations are motivated by the graphical representation of the domain of definition of the unknown function $u = u(x, t)$ in one space dimension (i.e. $d = 1$), namely, assigning the abscissa (horizontal axis) of the coordinate system to the variable x and the ordinate (vertical axis) to the variable t so that the spatial discretization yields problems that are setted along vertical lines.

In this section, and similarly in the subsequent sections, the vertical method of lines will be considered in more detail. We will develop the analogous (semi)discretization approaches for the finite difference method, the (mixed) finite element method, and the node-oriented finite volume method. This will allow us to analyse these methods in a uniform way, as far as only the emerging (matrix) structure of the discrete problems will play a role. On the other hand, different techniques of analysis as in Chapters 1, 3 and 8 will further elucidate advantages and disadvantages of the methods. Readers who are interested only in a specific approach may skip some of the following subsections.

The Vertical Method of Lines for the Finite Difference Method

As a first example, we start with the heat equation (9.8) with Dirichlet boundary conditions on a rectangle $\Omega = (0, a) \times (0, b)$. As in Section 1.2, we apply the five-point stencil discretizations at the grid points $x \in \Omega_h$ (according to (1.5)) for every fixed $t \in [0, T]$. This leads to the approximation

$$\begin{aligned} \partial_t u_{ij}(t) &+ \frac{1}{h^2}\Big(- u_{i,j-1}(t) - u_{i-1,j}(t) + 4u_{ij}(t) - u_{i+1,j}(t) - u_{i,j+1}(t)\Big) \\ &= f_{ij}(t), \qquad i = 1, \dots, l-1,\ j = 1, \dots, m-1,\ t \in (0, T), \end{aligned} \tag{9.44}$$

$$u_{ij}(t) = g_{ij}(t), \quad i \in \{0, l\},\ j = 0, \dots, m, \quad j \in \{0, m\},\ i = 0, \dots, l. \tag{9.45}$$

Here, we use the notation

$$\begin{aligned} f_{ij}(t) &:= f(ih, jh, t), \\ g_{ij}(t) &:= g(ih, jh, t), \end{aligned} \tag{9.46}$$

and the subscript 3 in the boundary condition is omitted. Additionally, the initial condition (at the grid points) has to be prescribed, that is,

$$u_{ij}(0) = u_0(ih, jh), \quad (ih, jh) \in \overline{\Omega}_h\,. \tag{9.47}$$

The system (9.44), (9.45), (9.47) is a system of (linear) ordinary differential equations (in the "index" (i, j)). If, as in Section 1.2, we fix an ordering of the grid points, the system takes the form

$$\begin{aligned} \frac{d}{dt}\boldsymbol{u}_h(t) + A_h \boldsymbol{u}_h(t) &= \boldsymbol{q}_h(t), \quad t \in (0, T), \\ \boldsymbol{u}_h(0) &= \boldsymbol{u}_0, \end{aligned} \tag{9.48}$$

with $A_h, \boldsymbol{u}_h, \boldsymbol{q}_h$ as in (1.10), (1.11) (but now $\boldsymbol{u}_h = \boldsymbol{u}_h(t)$ and $\boldsymbol{q}_h = \boldsymbol{q}_h(t)$ because of the t-dependence of f and g).

The unknown is the function

$$\boldsymbol{u}_h : [0, T] \to \mathbb{R}^{M_1}, \tag{9.49}$$

which means that the Dirichlet boundary conditions are eliminated as in Section 1.2.

For a simplification of the notation, we use in the following the symbol M instead of M_1, which also includes the eliminated degrees of freedom. Only in Sections 9.7 and 9.8, we will return to the original notation.

More generally, if we consider a finite difference approximation, the application of which to the stationary problem (9.6) will lead to the system of equations

$$A_h \boldsymbol{u}_h = \boldsymbol{q}_h,$$

where $\boldsymbol{u}_h \in \mathbb{R}^M$, then the same method applied to (9.1) for every fixed $t \in (0, T)$ leads to (9.48) with the diagonal coefficient matrix $B_h = \operatorname{diag}\big(b(x_i)_i\big)$ in the time derivative. In particular, the system (9.48) has a unique solution due to the theorem of Picard–Lindelöf (cf. [51]).

The Vertical Method of Lines for the Primal Finite Element Method

We proceed as for the finite difference method by now applying the finite element method to (9.1) in its weak formulation (9.37) for every fixed $t \in (0,T)$, with a, ℓ defined by (9.34), (9.35). So let $V_h \subset V$ denote a finite-dimensional subspace with $\dim V_h := M(h) := M$ and let $u_{0h} \in V_h$ be some approximation to u_0. Then the *semidiscrete problem* reads as follows:

Find $u_h \in L^2((0,T),V_h)$ with $u_h' \in L^2((0,T),H)$, $u_h(0) = u_{0h}$ such that

$$\left\langle \frac{d}{dt} u_h(t), v_h \right\rangle_0^{(b)} + a(u_h(t), v_h) = \ell(t; v_h) \quad \text{for all } v_h \in V_h\,,\ t \in (0,T). \tag{9.50}$$

To gain a more specific form of (9.50), again we represent the unknown $u_h(t)$ by its degrees of freedom. So let $\{\varphi_i\}_{i=1}^M$ be a basis of V_h, $u_h(t) = \sum_{i=1}^M \xi_i(t)\,\varphi_i$ and $u_{0h} = \sum_{i=1}^M \xi_{0i}\,\varphi_i$. In particular, because of

$$\frac{d}{dt} u_h(t) = \sum_{i=1}^M \frac{d}{dt} \xi_i(t)\varphi_i\,,$$

we have $u_h' \in L^2((0,T),V_h)$. Then for any $t \in (0,T)$, the discrete variational equation (9.50) is equivalent to

$$\sum_{j=1}^M \langle \varphi_j, \varphi_i \rangle_0^{(b)} \frac{d\xi_j(t)}{dt} + \sum_{j=1}^M a(\varphi_j, \varphi_i)\,\xi_j(t) = \ell(t;\varphi_i) \quad \text{for all } i \in \{1,\dots,M\}\,.$$

Denoting by $\hat{A}_h := \left(a(\varphi_j,\varphi_i)\right)_{ij}$ the *stiffness matrix*, by $B_h := \left(\langle \varphi_j, \varphi_i \rangle_0^{(b)}\right)_{ij}$ the *weighted mass matrix*, and by

$$\boldsymbol{\beta}_h(t) := (\ell(t;\varphi_i))_i\,,$$

respectively, $\boldsymbol{\xi}_{0h} := (\xi_{0i})_i$, the vectors of the right-hand side and of the initial value, we obtain for $\boldsymbol{\xi}_h(t) := (\xi_i(t))_i$ the following system of linear ordinary differential equations with constant coefficients:

$$\begin{aligned} B_h \frac{d}{dt} \boldsymbol{\xi}_h(t) + \hat{A}_h\, \boldsymbol{\xi}_h(t) &= \boldsymbol{\beta}_h(t), \quad t \in (0,T), \\ \boldsymbol{\xi}_h(0) &= \boldsymbol{\xi}_{0h}\,. \end{aligned} \tag{9.51}$$

Note that because of (9.51) $\boldsymbol{\xi}_h \in C^1([0,T],\mathbb{R}^M)$ and thus $u_h' \in C([0,T],V_h)$.

Since the matrix B_h is symmetric and positive definite, it can be factored (e.g. by means of Cholesky's decomposition) as $B_h = E_h^T E_h$. Introducing the new variable

$\boldsymbol{u}_h := E_h \boldsymbol{\xi}_h$ (to preserve the possible definiteness of A_h), the above system (9.51) can be written as follows:

$$\begin{aligned} \frac{d}{dt}\boldsymbol{u}_h(t) + A_h\, \boldsymbol{u}_h(t) &= \boldsymbol{q}_h(t), \quad t \in (0,T), \\ \boldsymbol{u}_h(0) &= \boldsymbol{u}_{h0}, \end{aligned} \tag{9.52}$$

where $A_h := E_h^{-T} \hat{A}_h E_h^{-1}$ is an $\mathbb{R}^M$-coercive matrix and $\boldsymbol{q}_h := E_h^{-T} \boldsymbol{\beta}_h$, $\boldsymbol{u}_{h0} := E_h \boldsymbol{\xi}_{0h}$.

Thus, again the discretization leads us to a system (9.48).

In the case of time-dependent coefficients $b, \boldsymbol{K}, \alpha, \boldsymbol{c}$, or r instead of (9.51), we get

$$\left(\left\langle \sum_{j=1}^{M} \frac{d}{dt}\big(b(x,t)\xi_j(t)\big)\varphi_j, \varphi_i \right\rangle_0 \right)_i + \hat{A}_h(t)\boldsymbol{\xi}_h(t) = \boldsymbol{\beta}_h(t), \tag{9.53}$$

where $\hat{A}_h(t) := (a(t;\varphi_j,\varphi_i))_{ij}$ and $a(t;u,v)$ corresponds to (9.34) with t-dependent coefficients. We do not rewrite the first term by using the product rule $\frac{d}{dt}\big(b(x,t)\xi_j(t)\big) = \frac{d}{dt}b(x,t)\xi_j(t) + b(x,t)\frac{d}{dt}\xi_j(t)$, even if the regularity would allow it, and thus (9.53) by

$$B_h(t)\frac{d}{dt}\boldsymbol{\xi}_h(t) + (\hat{A}_h(t) + \widetilde{B}_h(t))\boldsymbol{\xi}_h(t) = \boldsymbol{\beta}_h(t),$$

where $\widetilde{B}_h(t)$ is the mass matrix weighted with $\frac{d}{dt}b$, as this formulation would destroy the conservation form.

Remark 9.7 By means of the same arguments as in the proof of Lemma 9.4, an estimate of $\|u_h(t)\|_0$ can be derived, and also Remark 9.5 holds true.

The Vertical Method of Lines for the Mixed Finite Element Method

According to the formulation and notations of (6.55), we may add a time derivative at two places

$$\begin{aligned} \alpha_1 \left\langle \frac{d}{dt}\boldsymbol{p}, \boldsymbol{q} \right\rangle_0 + a(\boldsymbol{p},\boldsymbol{q}) + b(\boldsymbol{q},u) + c(u,\boldsymbol{q}) &= \ell_1(t;\boldsymbol{q}) \quad \text{for all } \boldsymbol{q} \in M, \\ -\alpha_2 \left\langle \frac{d}{dt}u, v \right\rangle_0 + b(\boldsymbol{p},v) - d(u,v) &= \ell_2(t;v) \\ &\quad \text{for all } v \in X,\ t \in (0,T). \end{aligned}$$

The L^2-scalar products may be also weighted.

According to (9.11), (9.12), $\alpha_1 = 0, \alpha_2 = 1$ corresponds to the Darcy formulation, $\alpha_1 = 1, \alpha_2 = 0$ to the instationary Stokes equation. Restricting to the first case, the *semidiscrete dual mixed finite element method* reads as follows for finite-dimensional subspaces

$$X_h \subset X, \quad \dim X_h := M(h) := M, M_h \subset M, \quad \dim M_h := N(h) := N,$$

and some approximation $u_{0h} \in X_h$ to u_0:

Find $u_h \in L^2(0,T;X_h)$, $u_h' \in L^2((0,T),X_h)$, $\boldsymbol{p}_h \in L^2(0,T;M_h)$ such that $u_h(0) = u_{0h}$ and

$$\begin{aligned} a(\boldsymbol{p}_h(t), \boldsymbol{q}_h) + b(\boldsymbol{q}_h, u_h(t)) + c(u_h(t), \boldsymbol{q}_h) &= \ell_1(t; \boldsymbol{q}_h) \\ &\qquad \text{for all } \boldsymbol{q}_h \in M_h, \\ -\left\langle \frac{d}{dt} u_h(t), v_h \right\rangle_0^{(b)} + b(\boldsymbol{p}_h(t), v_h) - d(u_h(t), v) &= \ell_2(t; v) \\ &\qquad \text{for all } v_h \in X_h,\ t \in (0,T) \end{aligned} \tag{9.54}$$

(note the double usage of the letter b, which however is clear from the context) with the forms as in (6.54), (6.51), (6.43) and the spaces as in (6.53). For simplicity, the flux boundary condition is assumed to be homogeneous, i.e. $\tilde{g}_1 = 0$. The stability of this formulation will be investigated later (Theorem 9.21).

Choosing a basis of M_h and of X_h, then similar to Remark 6.22, 4), using the notation from there, we get the following *linear differential algebraic system*:

$$\begin{aligned} \mathcal{A}_h \boldsymbol{\xi}(t) + (\mathcal{B}_h^T + C_h)\boldsymbol{\eta}_h(t) &= \boldsymbol{q}_h^{(1)}(t) \\ \mathcal{E}_h \frac{d}{dt}\boldsymbol{\eta}_h(t) - \mathcal{B}_h \boldsymbol{\xi}(t) + \mathcal{D}_h \boldsymbol{\eta}_h(t) &= -\boldsymbol{q}_h^{(2)}(t), \quad t \in (0,T), \\ \boldsymbol{\eta}_h(0) &= \boldsymbol{\eta}_{0h}, \end{aligned} \tag{9.55}$$

where now $\mathcal{E}_h$ denotes the weighted mass matrix (cf. (9.51)).

The Vertical Method of Lines for the (Node-Oriented) Finite Volume Method

Based on the finite volume methods introduced in Section 8.2, in this subsection, a finite volume semidiscretization is given for the problem (9.1) in its weak formulation (9.37) for every fixed $t \in (0,T)$ in the special case $\Gamma_3 = \partial\Omega$ and of homogeneous Dirichlet boundary conditions. As in Chapter 8, the only essential difference to problem (9.1) is that here the differential expression L is in divergence form, i.e.

$$Lu := -\nabla \cdot (\boldsymbol{K}\, \nabla u - \boldsymbol{c}\, u) + r\, u = f\,,$$

where the data $\boldsymbol{K}, \boldsymbol{c}, r$, and f are as in (9.2).

Correspondingly, the bilinear form a in the weak formulation (9.37) is to be replaced by

$$a(u,v) = \int_\Omega [(\boldsymbol{K}\, \nabla u - \boldsymbol{c}\, u) \cdot \nabla v + ruv]\, dx\,. \tag{9.56}$$

In order to obtain a finite volume semidiscretization of the problem (9.1) in divergence form, and of (9.37) with the modification (9.56), we recall the way that it was done in the elliptic situation. Namely, comparing the weak formulation of the elliptic problem (see Definition 2.2) with the finite volume method in the discrete variational formulation (8.17), we see that the bilinear form a and the linear form $\ell(\cdot) := \langle f, \cdot \rangle_0$ have been replaced by certain discrete forms a_h and $\ell_h(\cdot) := \langle f, \cdot \rangle_{0,h}$, respectively. This formal procedure can be applied to the weak formulation (9.37) of the parabolic problem, too.

So let $V_h \subset V$ denote a finite-dimensional subspace as introduced in Section 8.2 with $\dim V_h := M(h) := M$ and let $u_{0h} \in V_h$ be some approximation to u_0. Then, the *semidiscrete finite volume method* reads as follows:

Find $u_h \in L^2\left((0,T), V_h\right)$ with $u_h' \in L^2\left((0,T), H\right)$, $u_h(0) = u_{0h}$ such that

$$\left\langle \frac{d}{dt} u_h(t), v_h \right\rangle_{0,h}^{(b)} + a_h(u_h(t), v_h) = \langle f(t), v_h \rangle_{0,h} \qquad \text{for all } v_h \in V_h\,,\ t \in (0,T), \tag{9.57}$$

where both the bilinear form a_h and the form $\langle \cdot, \cdot \rangle_{0,h}$ have been formally defined in Section 8.2, and here we use the weighted version

$$\langle w, v \rangle_{0,h}^{(b)} := \sum_{i \in \Lambda} b(a_i) w_i v_i m_i\,, \tag{9.58}$$

assuming that b is continuous. However, to facilitate the comparison of the finite volume discretization with the previously described methods, here we set $\Lambda := \{1, \ldots, M\}$.

In analogy to the case of the finite element method (cf. Remark 9.7), a stability estimate for the finite volume method can be obtained. Namely, under the assumptions of Theorem 8.18, we have that

$$a_h(v_h, v_h) \ge \alpha \|v_h\|_{0,h}^2 \quad \text{for all } v_h \in V_h$$

with some constant $\alpha > 0$ independent of h. Then, taking $v_h = u_h(t)$ in (9.57), we get

$$\left\langle \frac{d}{dt} u_h(t), u_h(t) \right\rangle_{0,h}^{(b)} + a_h(u_h(t), u_h(t)) = \langle f(t), u_h(t) \rangle_{0,h}\,,$$

and, after some calculations,

$$\frac{d}{dt} \|u_h(t)\|_{0,h}^{(b)} + \tilde{\alpha} \|u_h(t)\|_{0,h}^{(b)} \le \|f(t)\|_{0,h}^{(b^{-1})}$$

with $\tilde{\alpha} := \alpha \|b\|_\infty^{-1}$.

The subsequent arguments are as in the proof of Lemma 9.4, i.e. we obtain

$$\|u_h(t)\|_{0,h}^{(b)} \le \|u_0\|_{0,h}^{(b)}\, e^{-\tilde{\alpha} t} + \int_0^t \|f(s)\|_{0,h}^{(b^{-1})}\, e^{-\tilde{\alpha}(t-s)}\, ds\,. \tag{9.59}$$

If the right-hand side of (9.57) is a general bounded linear form, i.e. instead of $\langle f(t), v_h \rangle_{0,h}$, we have the term $\ell_h(t; v_h)$, where $\ell_h : \; (0, T) \times V_h \to \mathbb{R}$ is such that

$$|\ell_h(t; v)| \le \|\ell_h(t)\|_{V_h'} \|v\|_{0,h}^{(b)} \quad \text{for all } v \in V_h \,,\ t \in (0, T),$$

with $\|\ell_h(t)\|_{V_h'} < \infty$ for all $t \in (0, T)$, then an analogous estimate holds:

$$\|u_h(t)\|_{0,h}^{(b)} \le \|u_0\|_{0,h}^{(b)} e^{-\tilde{\alpha} t} + \int_0^t \|\ell_h(s)\|_{V_h'} e^{-\tilde{\alpha}(t-s)} \, ds \,. \tag{9.60}$$

As in the previous subsection, we now want to give a more specific form of (9.57).

Given a basis $\{\varphi_i\}_{i=1}^M$ of the space V_h, such that $\varphi_i(a_j) = \delta_{ij}$ for the underlying nodes, we have the unique expansions

$$u_h(t) = \sum_{i=1}^M \xi_i(t) \, \varphi_i \quad \text{and} \quad u_{0h} = \sum_{i=1}^M \xi_{0i} \, \varphi_i \,.$$

Then for any $t \in (0, T)$, the discrete variational equation (9.57) is equivalent to

$$b(a_i) m_i \frac{d\xi_i(t)}{dt} + \sum_{j=1}^M a_h(\varphi_j, \varphi_i) \, \xi_j(t) = \langle f(t), \varphi_i \rangle_{0,h}$$
$$\text{for all } i \in \{1, \ldots, M\},$$

where $m_i = |\Omega_i|$. Using the notation $\hat{A}_h := \big(a_h(\varphi_j, \varphi_i)\big)_{ij}$ for the finite volume *stiffness matrix*, $B_h := \operatorname{diag}\,(b(a_i) m_i)$ for the finite volume *weighted mass matrix* $\boldsymbol{\beta}_h(t) := \big(\langle f(t), \varphi_i \rangle_{0,h}\big)_i$ for the vector of the right-hand side, and $\boldsymbol{\xi}_{0h} := (\xi_{0i})_i$ for the vector of the initial value, we obtain for the unknown vector function $\boldsymbol{\xi}_h(t) := (\xi_i(t))_i$ the following system of linear ordinary differential equations with constant coefficients:

$$\begin{aligned} B_h \frac{d}{dt} \boldsymbol{\xi}_h(t) + \hat{A}_h \, \boldsymbol{\xi}_h(t) &= \boldsymbol{\beta}_h(t) \,, \quad t \in (0, T), \\ \boldsymbol{\xi}_h(0) &= \boldsymbol{\xi}_{0h} \,. \end{aligned} \tag{9.61}$$

In contrast to the system (9.51) arising in the finite element semidiscretization, here the matrix B_h is diagonal. Therefore, it is very easy to introduce the new variable $\boldsymbol{u}_h := E_h \boldsymbol{\xi}$ with $E_h := \operatorname{diag}\big(\sqrt{b(a_i) m_i}\,\big)$, and the above system (9.61) can be written as follows:

$$\begin{aligned} \frac{d}{dt} \boldsymbol{u}_h(t) + A_h \, \boldsymbol{u}_h(t) &= \boldsymbol{q}_h(t) \,, \quad t \in (0, T), \\ \boldsymbol{u}_h(0) &= \boldsymbol{u}_{h0} \,, \end{aligned} \tag{9.62}$$

where $A_h := E_h^{-1} \hat{A}_h E_h^{-1}$ is an $\mathbb{R}^M$-coercive matrix and $\boldsymbol{q}_h := E_h^{-1} \boldsymbol{\beta}_h$, $\boldsymbol{u}_{h0} := E_h \boldsymbol{\xi}_{0h}$.

Thus, again we have arrived at a system of the form (9.48).

Representation of Solutions in a Special Case

The solution of the system (9.48) can be represented explicitly if there is a basis of $\mathbb{R}^M$ consisting of eigenvectors of the matrix A_h. For simplicity, we assume that all eigenvalues λ_i of A_h are real. This will be explained in the following, but is not meant for numerical use, since eigenvectors and values can only be specified explicitly in special cases. Rather, it will serve as a tool for comparison with the continuous and the fully discrete cases.

Let $(\boldsymbol{w}_i, \lambda_i)$, $i = 1, \ldots, M$, be the eigenvectors and real eigenvalues of A_h. Then the following representation exists uniquely:

$$\boldsymbol{u}_0 = \sum_{i=1}^{M} c_i \boldsymbol{w}_i \qquad \text{and} \qquad \boldsymbol{q}_h(t) = \sum_{i=1}^{M} q_h^i(t) \boldsymbol{w}_i \,. \tag{9.63}$$

Again by a separation of variables approach (cf. (9.23)), we see that

$$\boldsymbol{u}_h(t) = \sum_{i=1}^{M} \left[c_i e^{-\lambda_i t} + \int_0^t q_h^i(s) e^{-\lambda_i (t-s)} ds \right] \boldsymbol{w}_i \,. \tag{9.64}$$

A more compact notation is given by

$$\boldsymbol{u}_h(t) = e^{-tA_h} \boldsymbol{u}_0 + \int_0^t e^{-(t-s)A_h} \boldsymbol{q}_h(s) ds \tag{9.65}$$

if we define, for a matrix $B \in \mathbb{R}^{M,M}$, the *matrix exponential*

$$e^B := \sum_{\nu=0}^{\infty} \frac{B^\nu}{\nu!} \,.$$

This can be seen as follows. Let

$$\begin{aligned} T &:= (\boldsymbol{w}_1, \ldots, \boldsymbol{w}_M) \in \mathbb{R}^{M,M}, \\ \Lambda &:= \operatorname{diag}(\lambda_i) \in \mathbb{R}^{M,M}. \end{aligned}$$

Then

$$A_h T = T\Lambda, \quad \text{or equivalently,} \quad T^{-1} A_h T = \Lambda,$$

and therefore

$$T^{-1} e^{-tA_h} T = \sum_{\nu=0}^{\infty} \frac{t^\nu}{\nu!} T^{-1} (-A_h)^\nu T = \sum_{\nu=0}^{\infty} \frac{t^\nu}{\nu!} (-\Lambda)^\nu,$$

since $T^{-1}(-A_h)^\nu T = T^{-1}(-A_h) T T^{-1} (-A_h) T T^{-1} \ldots T$. Hence,

$$T^{-1}e^{-tA_h}T = \operatorname{diag}\left(\sum_{\nu=0}^{\infty} \frac{(-\lambda_i t)^\nu}{\nu!}\right) = \operatorname{diag}\left(e^{-\lambda_i t}\right). \tag{9.66}$$

Since we have the relation $T\boldsymbol{\gamma} = \boldsymbol{u}_0$ for $\boldsymbol{\gamma} := (c_1, \dots, c_M)^T \in \mathbb{R}^M$, we conclude from (9.64) in the case $\boldsymbol{q}_h = 0$ that

$$\boldsymbol{u}_h(t) = T\operatorname{diag}(e^{-\lambda_i t})\boldsymbol{\gamma} = TT^{-1}e^{-tA_h}T\boldsymbol{\gamma} = e^{-tA_h}\boldsymbol{u}_0,$$

and similarly in the general inhomogeneous case.

A basis of eigenvalues exists if A_h is *self-adjoint* with respect to a scalar product $\langle\cdot,\cdot\rangle_h$ in $\mathbb{R}^M$, meaning that

$$\langle \boldsymbol{v}, A_h\boldsymbol{u}\rangle_h = \langle A_h\boldsymbol{v}, \boldsymbol{u}\rangle_h \quad \text{for all } \boldsymbol{u}, \boldsymbol{v} \in \mathbb{R}^M.$$

Then the eigenvectors are even *orthogonal*, that is,

$$\langle \boldsymbol{w}_i, \boldsymbol{w}_j\rangle_h = 0 \quad \text{for } i \neq j \tag{9.67}$$

because of

$$\lambda_i\langle \boldsymbol{w}_i, \boldsymbol{w}_j\rangle_h = \langle A_h\boldsymbol{w}_i, \boldsymbol{w}_j\rangle_h = \langle \boldsymbol{w}_i, A_h\boldsymbol{w}_j\rangle_h = \lambda_j\langle \boldsymbol{w}_i, \boldsymbol{w}_j\rangle_h,$$

and thus (9.67) holds if $\lambda_i \neq \lambda_j$. However, eigenvectors corresponding to one (multiple) eigenvalue can always be orthonormalized. For orthogonal eigenvectors $\boldsymbol{w}_i$, the coefficient c_i from (9.63) has the form

$$c_i = \frac{\langle \boldsymbol{u}_0, \boldsymbol{w}_i\rangle_h}{\langle \boldsymbol{w}_i, \boldsymbol{w}_i\rangle_h}, \tag{9.68}$$

and analogously for q_h^i.

Order of Convergence Estimates for the Finite Difference Method in a Special Case

As an illustrative example, we consider a case where the eigenvalues and eigenvectors of A_h are known explicitly—the five-point stencil discretization of the Poisson equation with Dirichlet conditions in $\Omega := (0, a) \times (0, b)$, see Subsection 1.2. Instead of considering a fixed ordering of the grid points, we prefer to use the "natural" two-dimensional indexing, i.e. we regard the eigenvectors as grid functions. As seen in Section 1.2, A_h is symmetric and thus self-adjoint with respect to the Euclidean scalar product weighted by the factor h^d in the general case $\Omega \subset \mathbb{R}^d$:

$$\langle \boldsymbol{u}, \boldsymbol{v}\rangle_h := h^d \sum_{i=1}^{M} u_i v_i. \tag{9.69}$$

The norm induced by this scalar product is exactly the discrete L^2-norm defined in (1.18) (for $d = 2$ and for the vectors representing the grid functions):

$$|\boldsymbol{u}|_{0,h} := \langle \boldsymbol{u}, \boldsymbol{u}\rangle_h^{1/2} = h^{d/2}\left(\sum_{i=1}^{M} |u_i|^2\right)^{1/2}. \tag{9.70}$$

If we operate with the grid function $U : \Omega_h \to \mathbb{R}$, we denote its norm by $\|U\|_{0,h}$.

The eigenvectors, which have already been noted for a special case after Theorem 5.4, are written as grid functions

$$\begin{aligned} u^{\nu\mu}(x, y) := \sin\left(\nu\frac{\pi}{a}x\right)\sin\left(\mu\frac{\pi}{b}y\right) \quad &\text{for } (x, y) \in \Omega_h \\ &\text{and } \nu = 1, \ldots, l-1,\ \mu = 1, \ldots, m-1, \end{aligned} \tag{9.71}$$

and correspond to the eigenvalues

$$\lambda_h^{\nu\mu} := \frac{2}{h^2}\left(2 - \cos\left(\nu\frac{\pi}{a}h\right) - \cos\left(\mu\frac{\pi}{b}h\right)\right).$$

Note that the eigenvectors (9.71) are the eigenfunctions (9.32) of the continuous problem evaluated at the grid points, but the grid points allow to distinguish only the maximal frequencies $\frac{l-1}{2}$ and $\frac{m-1}{2}$ so that for other indices ν, μ the given grid functions would be repeated.

Due to $2\sin^2\left(\frac{\xi}{2}\right) = 1 - \cos(\xi)$, an alternative representation of the eigenvalues is

$$\lambda_h^{\nu\mu} = \frac{4}{h^2}\left(\sin^2\left(\nu\frac{\pi}{a}\frac{h}{2}\right) + \sin^2\left(\mu\frac{\pi}{b}\frac{h}{2}\right)\right),$$

so that, for $h \to 0$,

$$\lambda_h^{\nu\mu} = \left(\nu\frac{\pi}{a}\right)^2\left(\frac{\sin\left(\nu\frac{\pi}{a}\frac{h}{2}\right)}{\left(\nu\frac{\pi}{a}\frac{h}{2}\right)}\right)^2 + \left(\mu\frac{\pi}{b}\right)^2\left(\frac{\sin\left(\mu\frac{\pi}{b}\frac{h}{2}\right)}{\left(\mu\frac{\pi}{b}\frac{h}{2}\right)}\right)^2 \to \left(\nu\frac{\pi}{a}\right)^2 + \left(\mu\frac{\pi}{b}\right)^2 \tag{9.72}$$

holds, i.e. the eigenvalues converge to the eigenvalues (9.31) of the boundary value problem with an order of convergence estimate of $O(h^2)$.

The eigenvectors are orthogonal with respect to $\langle\cdot,\cdot\rangle_h$, since they correspond to different eigenvalues (see (9.67)). To specify the Fourier coefficients according to (9.68), we need the relation

$$\langle \boldsymbol{u}^{\nu\mu}, \boldsymbol{u}^{\nu\mu}\rangle_h = \frac{ab}{4} \tag{9.73}$$

(see Problem 9.8).

To investigate the accuracy of the semidiscrete approximation, the solution representations can be compared. To simplify the exposition, we consider only the case $f = 0$ so that, because of (9.23), (9.32) we have

$$u(x, y, t) = \sum_{\nu,\mu=1}^{\infty} c_{\nu\mu} e^{-\lambda^{\nu\mu} t} \sin\left(\nu \frac{\pi}{a} x\right) \sin\left(\mu \frac{\pi}{b} y\right),$$

and

$$c_{\nu\mu} = \frac{4}{ab} \int_0^b \int_0^a u_0(x, y) \sin\left(\nu \frac{\pi}{a} x\right) \sin\left(\mu \frac{\pi}{b} y\right) dx dy$$

because of (9.29) and (9.28) (applied in every space direction), and finally

$$\lambda^{\nu\mu} = \left(\nu \frac{\pi}{a}\right)^2 + \left(\mu \frac{\pi}{b}\right)^2$$

for the continuous solution. For the semidiscrete approximation at a grid point $(x, y) \in \Omega_h$, we have, due to (9.64),

$$u_h(x, y, t) = \sum_{\nu=1}^{l-1} \sum_{\mu=1}^{m-1} c_{\nu\mu}^h e^{-\lambda_h^{\nu\mu} t} \sin\left(\nu \frac{\pi}{a} x\right) \sin\left(\mu \frac{\pi}{b} y\right)$$

and

$$\begin{aligned} c_{\nu\mu}^h &= \frac{4}{ab} h^2 \sum_{i=1}^{l-1} \sum_{j=1}^{m-1} u_0(ih, jh) \sin\left(\nu \frac{\pi}{a} ih\right) \sin\left(\mu \frac{\pi}{b} jh\right), \\ \lambda_h^{\nu\mu} &= \frac{4}{h^2} \left(\sin^2\left(\nu \frac{\pi}{a} \frac{h}{2}\right) + \sin^2\left(\mu \frac{\pi}{b} \frac{h}{2}\right) \right). \end{aligned}$$

Comparing the representations of u and u_h at the grid points, we see that the solution u has additionally the terms for $\nu = l, \ldots$ and $\mu = m, \ldots$ in its infinite series.

They can be estimated by

$$\begin{aligned} &\left| \left(\sum_{\nu=l}^{\infty} \sum_{\mu=1}^{\infty} + \sum_{\nu=1}^{\infty} \sum_{\mu=m}^{\infty} \right) c_{\nu\mu} e^{-\lambda^{\nu\mu} t} \sin\left(\nu \frac{\pi}{a} x\right) \sin\left(\mu \frac{\pi}{b} y\right) \right| \\ &\quad \le C_1 \left(\sum_{\nu=l}^{\infty} \sum_{\mu=1}^{\infty} + \sum_{\nu=1}^{\infty} \sum_{\mu=m}^{\infty} \right) e^{-\lambda^{\nu\mu} t} \end{aligned}$$

with $C_1 := \max\left\{ |c_{\nu\mu}| \;\middle|\; \nu, \mu \in \mathbb{N}, \nu \notin \{1, \ldots, l-1\} \text{ or } \mu \notin \{1, \ldots, m-1\} \right\}$. Furthermore,

$$\left(\sum_{\nu=l}^{\infty} \sum_{\mu=1}^{\infty} + \sum_{\nu=1}^{\infty} \sum_{\mu=m}^{\infty} \right) e^{-\lambda^{\nu\mu} t} \le C_2 \sum_{\nu=l}^{\infty} e^{-\left(\nu \frac{\pi}{a}\right)^2 t} + C_3 \sum_{\mu=m}^{\infty} e^{-\left(\mu \frac{\pi}{b}\right)^2 t}$$

with $C_2 := \sum_{\mu=1}^{\infty} e^{-\left(\mu\frac{\pi}{b}\right)^2 t} \leq \frac{q_2}{1-q_2}$ for $t \geq \bar{t} > 0$, where $q_2 := e^{-\left(\frac{\pi}{b}\right)^2 \bar{t}}$ because of $\sum_{\mu=1}^{\infty} q^\mu = \frac{q}{1-q}$ for $|q| < 1$, and C_3 is defined analogously ($\mu \leftrightarrow \nu$, $a \leftrightarrow b$) with an upper bound $\frac{q_1}{1-q_1}$, $q_1 := e^{-\left(\frac{\pi}{a}\right)^2 \bar{t}}$.

Finally, we conclude the estimate because of $\sum_{\mu=l}^{\infty} q^\mu = \frac{q^l}{1-q}$ by

$$\leq C_1 \left(C_2 \frac{q_1^l}{1-q_1} + C_3 \frac{q_2^m}{1-q_2} \right) .$$

Therefore, this error contribution for $t \geq \bar{t}$ (for a fixed $\bar{t} > 0$) approaches 0 for $l \longrightarrow \infty$ and $m \longrightarrow \infty$. The larger $\bar{t}$ is, the more this error term will decrease. Because of, for example, $l = a/h$ and thus $q_1^l = \exp\left(-\frac{\pi^2}{a}\bar{t}\frac{1}{h}\right)$, the decay in h is exponential and thus much stronger than a term like $O(h^2)$. Therefore, we have to compare the terms in the sum only for $\nu = 1, \ldots, l-1, \quad \mu = 1, \ldots, m-1$, i.e. the error in the Fourier coefficient and in the eigenvalue:

$$c_{\nu\mu} e^{-\lambda^{\nu\mu} t} - c_{\nu\mu}^h e^{-\lambda_h^{\nu\mu} t} = \left(c_{\nu\mu} - c_{\nu\mu}^h \right) e^{-\lambda^{\nu\mu} t} + c_{\nu\mu}^h \left(e^{-\lambda^{\nu\mu} t} - e^{-\lambda_h^{\nu\mu} t} \right) .$$

Note that $c_{\nu\mu}^h$ can be perceived as an approximation of $c_{\nu\mu}$ by the trapezoidal sum with step size h in each spatial direction (see, e.g. [56, p. 149]), since the integrand in the definition of $c_{\nu\mu}$ vanishes for $x = 0$ or $x = a$ and $y \in [0, b]$, and $y = 0$ or $y = b$ and $x \in [0, a]$. Thus we have for $u_0 \in C^2(\overline{\Omega})$,

$$|c_{\nu\mu} - c_{\nu\mu}^h| = O(h^2).$$

Because of

$$e^{-\lambda^{\nu\mu} t} - e^{-\lambda_h^{\nu\mu} t} = e^{-\lambda^{\nu\mu} t} \left(1 - e^{-(\lambda_h^{\nu\mu} - \lambda^{\nu\mu}) t} \right),$$

and $|\lambda_h^{\nu\mu} - \lambda^{\nu\mu}| = O(h^2)$ (see (9.72)), also this term is of order $O(h^2)$ and will be damped exponentially (depending on t and the size of the smallest eigenvalue $\lambda^{\nu\mu}$).

In summary, we expect that the dominating error in the discrete maximum norm $\|\cdot\|_\infty$ at the grid points (cf. Definition 1.17) is of size

$$O(h^2),$$

which will also be damped exponentially for increasing t. Note that we have given only heuristic arguments and that the considerations cannot be transferred to the Neumann case, where the eigenvalue $\lambda = 0$ occurs.

Now we turn to a rigorous treatment of the finite element method..

Order of Convergence Estimates for the (Primal) Finite Element Method

We will investigate the finite element method on a more abstract level as in the previous subsection, but we will achieve a result (in different norms) of similar character. As explained at the end of Section 9.1, there is a strong relation between the V-coercivity of the bilinear form a with the parameter α and a positive lower bound of the eigenvalues. Here, we rely on the results already achieved in Section 2.3 and Section 3.4 for the stationary case.

For that, we introduce the so-called *elliptic projection* of the solution $u(t)$ of (9.37) as a very important tool in the proof.

Definition 9.8 For a V-coercive, continuous bilinear form $a : V \times V \to \mathbb{R}$, the *elliptic*, or *Ritz, projection* $R_h : V \to V_h$ is defined by

$$v \mapsto R_h v \quad \Longleftrightarrow \quad a(R_h v - v, v_h) = 0 \quad \text{for all } v_h \in V_h\,.$$

Theorem 9.9 *Under the assumptions of Definition 9.8:*

1) $R_h : V \to V_h$ is linear and continuous.
2) R_h yields quasi-optimal *approximations; that is,*

$$\|v - R_h v\|_V \le \frac{M}{\alpha} \inf_{v_h \in V_h} \|v - v_h\|_V\,,$$

where M and α are the Lipschitz and coercivity constants, respectively, according to (2.46) *and* (2.47).

Proof The linearity of the operator R_h is obvious. The remaining statements immediately follow from Lemma 2.16 and Theorem 2.17, see Problem 9.9. □

Making use of the elliptic projection, we are able to prove the following result.

Theorem 9.10 *Suppose a is a V-coercive, continuous bilinear form, $f \in C([0,T], H)$, $u_0 \in V$, and $u_{0h} \in V_h$. If $u \in C^1([0,T], V)$, then:*

$$\begin{aligned}\|u_h(t) - u(t)\|_0^{(b)} \le \|u_{0h} - R_h u_0\|_0^{(b)}\, e^{-\tilde{\alpha} t} + \|(I - R_h)u(t)\|_0^{(b)} \\ + \int_0^t \|(I - R_h)u'(s)\|_0^{(b^{-1})}\, e^{-\tilde{\alpha}(t-s)}\, ds\,.\end{aligned}$$

Proof First, the error is decomposed as follows:

$$u_h(t) - u(t) = u_h(t) - R_h u(t) + R_h u(t) - u(t) =: \theta(t) + \varrho(t)\,.$$

We take $v = v_h \in V_h$ in (9.37) and obtain by the definition of R_h,

$$\langle u'(t), v_h \rangle_0^{(b)} + a(u(t), v_h) = \langle u'(t), v_h \rangle_0^{(b)} + a(R_h u(t), v_h) = \ell(t; v_h),$$

where $\ell(t;\cdot)$ has been defined in (9.35). For ϱ an estimate is available by Theorem 9.9, the estimate for $\theta(t) \in V_h$ will be achieved by deriving a variational equation for θ and applying a stability result.

Subtracting this equation from (9.50), we get

$$\langle u_h'(t), v_h\rangle_0^{(b)} - \langle u'(t), v_h\rangle_0^{(b)} + a(\theta(t), v_h) = 0,$$

and thus

$$\langle \theta'(t), v_h\rangle_0^{(b)} + a(\theta(t), v_h) = \langle u'(t), v_h\rangle_0^{(b)} - \Big\langle \frac{d}{dt} R_h u(t), v_h\Big\rangle_0^{(b)} = -\langle \varrho'(t), v_h\rangle_0^{(b)} . \tag{9.74}$$

The application of Lemma 9.4 yields

$$\|\theta(t)\|_0^{(b)} \le \|\theta(0)\|_0^{(b)} e^{-\widetilde{\alpha} t} + \int_0^t \|\varrho'(s)\|_0^{(b^{-1})} e^{-\widetilde{\alpha}(t-s)}\, ds .$$

Since the elliptic projection is continuous (Theorem 9.9, 1)) and $u(t)$ is sufficiently smooth, the operator R_h and the time derivative $\frac{d}{dt}$ commute (see Problem 9.10), that is, $\varrho'(t) = (R_h - I)u'(t)$. It remains to apply the triangle inequality to get the stated result.

□

Remark 9.11 In the proof, we have used the the following, easy-to-prove assertions:

1) If $u \in C([0,T], V)$, then $R_h u \in C([0,T], V)$.
2) If $u \in C^1([0,T], V)$, then $R_h u \in C^1([0,T], V)$ and $(R_h u)' = R_h(u')$ (see Problem 9.10).

Theorem 9.10 has the following interpretation:
The error norm $\|u_h(t) - u(t)\|_0^{(b)}$ is estimated by

- the initial error (exponentially decaying in t), which occurs only if u_{0h} does not coincide with the elliptic projection of u_0,
- the projection error of the exact solution $u(t)$ measured in the weighted norm of H,
- the projection error of $u'(t)$ measured in the weighted norm of H and integrally weighted by the factor $e^{-\alpha(t-s)}$ on $(0, t)$.

Besides error estimates in $\|\cdot\|_0^{(b)}$, also estimates in $H^1(\Omega)$ are possible. As an example, we mention the following result.

Theorem 9.12 *In addition to the assumptions of Theorem 9.10, let the bilinear form a be symmetric. Then there exists a constant $C = C(\alpha, M) > 0$ independent of u such that*

$$\begin{aligned}&\|u_h(t) - u(t)\|_V\\ &\le C\left(\|u_{0h} - R_h u_0\|_V + \left(\int_0^t \left(\|(I - R_h)u'(s)\|_0^{(b)}\right)^2 ds\right)^{1/2}\right) + \|(I - R_h)u(t)\|_V .\end{aligned}$$

Proof We proceed as in the proof of Theorem 9.10. Because of Remark 9.11, we may take $v_h = \theta'(t)$ in (9.74), which leads to

$$\left(\|\theta'(t)\|_0^{(b)}\right)^2 + a(\theta(t), \theta'(t)) = -\langle \varrho'(t), \theta'(t)\rangle_0^{(b)}$$

and thus to

$$\left(\|\theta'(t)\|_0^{(b)}\right)^2 + \frac{1}{2}\frac{d}{dt}a(\theta(t), \theta(t)) \le \frac{1}{2}\left(\|\varrho'(t)\|_0^{(b)}\right)^2 + \frac{1}{2}\left(\|\theta'(t)\|_0^{(b)}\right)^2.$$

In particular, we get

$$\frac{d}{dt}a(\theta(t), \theta(t)) \le \left(\|\varrho'(t)\|_0^{(b)}\right)^2$$

and therefore, after integration,

$$\begin{aligned}\alpha\|\theta(t)\|_V^2 \le a(\theta(t), \theta(t)) &\le a(\theta(0), \theta(0)) + \int_0^t \left(\|\varrho'(t)\|_0^{(b)}\right)^2 ds \\ &\le M\|\theta(0)\|_V^2 + \int_0^t \left(\|\varrho'(t)\|_0^{(b)}\right)^2 ds\,.\end{aligned}$$

Proceeding as in the proof of Theorem 9.10 concludes the assertion. □

We see that in order to obtain semidiscrete error estimates, we need estimates of the projection error measured in the weighted norm of $H = L^2(\Omega)$ in the case of Theorem 9.10 and, additionally, in $V = H^1(\Omega)$ in the case of Theorem 9.12. Due to $\|\cdot\|_0 \le \|\cdot\|_V$, the quasi-optimality of R_h (Theorem 9.9, 2)) in conjunction with the corresponding approximation error estimates (Theorem 3.32) already yield some error estimate. Unfortunately, this result is not optimal. However, if the adjoint boundary value problem is regular in the sense of Definition 3.39, the duality argument (Theorem 3.40) can be successfully used to derive an optimal result.

Theorem 9.13 *Suppose that the bilinear form a is V-coercive. Furthermore, let the space $V_h \subset V$ be such that for any function $w \in V \cap H^2(\Omega)$,*

$$\inf_{v_h \in V_h} \|w - v_h\|_V \le C\,h\,|w|_2\,,$$

where the constant $C > 0$ does not depend on h and w. If $u_0 \in V \cap H^2(\Omega)$, then for a sufficiently smooth solution u of (9.50), *we have the following estimates:*

1) If the solution of the adjoint elliptic boundary value problem is regular, then

$$\begin{aligned}\|u_h(t) - u(t)\|_0^{(b)} &\le \|u_{0h} - u_0\|_0^{(b)}\, e^{-\tilde{\alpha}t} \\ &\quad + C\,h^2\left(|u_0|_2\, e^{-\tilde{\alpha}t} + |u(t)|_2 + \int_0^t |u'(s)|_2\, e^{-\tilde{\alpha}(t-s)}\, ds\right).\end{aligned}$$

2) $\|u_h(t) - u(t)\|_1 \le C\left(\|u_{0h} - u_0\|_1 + h\left(|u_0|_2 + |u(t)|_2 + \left(\int_0^t |u'(s)|_2^2 ds\right)^{1/2}\right)\right).$

Proof The first term in the error bound from Theorem 9.10 is estimated by means of the triangle inequality:

$$\|u_{0h} - R_h u_0\|_0^{(b)} \le \|u_{0h} - u_0\|_0^{(b)} + \|(I - R_h)u_0\|_0^{(b)}.$$

Then the projection error estimate (Theorem 3.40, 1)) yields the given bounds of the resulting second term as well as of the remaining two terms in the error bound from Theorem 9.10. The second assertion is proven analogously. □

We have now applied twice basically the same reasoning to deduce order of convergence estimates for the semidiscrete time-dependent problem from the corresponding ones for the stationary situation based on a stability estimate. For further usage, we cast this in a more general form.

An Abstract Order of Convergence Estimate for Semidiscrete Galerkin Approximations

Let U, V, H be Hilbert spaces (U, V can even only be Banach spaces) such that $U \hookrightarrow H \hookrightarrow V'$, $a : U \times V \to \mathbb{R}$ a continuous bilinear form, $\ell \in V'$, and consider the problem (6.1') again

$$\text{Find} \quad u \in U \quad \text{such that} \quad a(u, v) = \ell(v) \quad \text{for all } v \in V. \tag{9.75}$$

We assume
① Unique solvability of (9.75) for every $\ell \in V'$ in the sense that the linear solution operator

$$S : V' \to U, \ \ell \mapsto u$$

is well defined, an isomorphism and continuous, i.e. there exists a constant $C > 0$ such that

$$\|S\ell\|_U \le C\|\ell\|_{V'}. \tag{9.76}$$

Next we consider a general Galerkin-type discretization, i.e. finite-dimensional spaces $U_h \subset H$, V_h depending on a discretization parameter $h > 0$, a continuous bilinear form $a_h : U_h \times V_h \to \mathbb{R}$, and a continuous linear form $\ell_h \in V_h'$.

For the discrete problems

$$\text{Find} \quad u_h \in U_h \quad \text{such that} \quad a_h(u_h, v_h) = \ell_h(v_h) \quad \text{for all } v_h \in V_h$$

we assume
② Unique solvability for each $\ell_h \in V_h'$ in the sense that the linear solution operator

$$S_h : V_h' \to U_h, \ \ell_h \mapsto u_h$$

is well defined, an isomorphism and continuous uniformly in h, i.e. there exists a constant $C_s > 0$ independent of h such that

$$\|S_h \ell_h\|_{U(h)} \leq C_s \|\ell_h\|_{V_h'}, \tag{9.77}$$

where $U(h) := U + U_h$ is equipped with a norm $\|\cdot\|_{U(h)}$ to be specified later.

The next postulate formulates the validity of a suitable error estimate, which can be written either in terms of the space approximation error and the consistency errors or together in terms of some power of h. Recall that $A : U \to V_h$ is the continuous linear operator defined by the bilinear form a (see (6.3')).

③ There are a Hilbert space $W \subset U$ and a seminorm $\widehat{\|\cdot\|}$ on $W + U_h$ with the property

$$\widehat{\|w_h\|} \leq C_h^{(1)} \|w_h\|_{U(h)} \quad \text{for all } w_h \in U_h, \tag{9.78}$$

where $C_h^{(1)} > 0$ is a constant. Furthermore, there exist a subspace $\widehat{V} \subset AW$ ($\subset V'$), a linear continuous mapping

$$\widehat{Q}_h : \widehat{V} \to V_h' \tag{9.79}$$

and parameters $\alpha > 0$, $C_a = C_a(u) = C_a(\|u\|_W) > 0$ independent of h such that it holds, for $u = S\ell \in W$ and $u_h = S_h \widehat{Q}_h \ell$:

$$\widehat{\|u - u_h\|} = \widehat{\|S\ell - S_h \widehat{Q}_h \ell\|} = \widehat{\|(S - S_h \widehat{Q}_h) A u\|} \leq C_a(u) h^\alpha.$$

Finally, to introduce a generalization of the Ritz projection from Definition 9.8, we assume that there is a continuous linear operator $Q_h : V' \to V_h'$ representing a discretization of the elements from V'.

Definition 9.14 Under the above conditions ② – ③ define

$$R_h : U \to U_h, \; u \mapsto R_h u := S_h Q_h A u.$$

It is not difficult to show that the assumed stationary convergence result carries over to $u - R_h u$.

Lemma 9.15 *Under the above conditions, we have for $u \in W$:*

$$\widehat{\|u - R_h u\|} \leq C_a(u) h^\alpha + C_h^{(1)} C_s \|(\widehat{Q}_h - Q_h) A u\|_{V_h'}.$$

Proof By the triangle inequality, we immediately see that

$$\widehat{\|u - R_h u\|} \leq \widehat{\|u - S_h \widehat{Q}_h A u\|} + \widehat{\|S_h \widehat{Q}_h A u - R_h u\|}.$$

The first term in the bound can be estimated using condition ③. To estimate the second term, we apply (9.77), (9.78):

$$\begin{aligned} &\widehat{\|S_h \widehat{Q}_h A u - R_h u\|} = \widehat{\|S_h (\widehat{Q}_h - Q_h) A u\|} \\ &\leq C_h^{(1)} \|S_h (\widehat{Q}_h - Q_h) A u\|_{U(h)} \leq C_h^{(1)} C_s \|(\widehat{Q}_h - Q_h) A u\|_{V_h'}. \end{aligned}$$

□

The distinction between $\widehat{Q}_h$ and Q_h makes it possible, for example, to consider a stationary approximation without quadrature in the right-hand side satisfying ③ and having the quadrature (or another perturbation) only in the time-dependent situation.

We now turn to the time-dependent case. Let $e: H \times V \to \mathbb{R}$ be a bilinear form, $\ell \in C((0,T], V')$, and $u_0 \in H$.

The time-dependent formulation reads as follows:

$$\begin{aligned} &\text{Find} \quad u \in C^1((0,T],H) \cap L^2((0,T),V) \quad \text{such that} \\ &e\left(\frac{d}{dt}u(t), v\right) + a\,(u(t), v) = \ell(t; v) =: \ell(t)v \quad \text{for all } v \in V,\ t \in (0,T), \\ &u(0) = u_0\,. \end{aligned} \tag{9.80}$$

The existence of a unique solution u is also assumed here.

To formulate a semidiscrete version, we consider a further Banach space $\widetilde{W} \supset U_h$, a bilinear form $e_h: \widetilde{W} \times V_h \to \mathbb{R}$, a linear form $\ell_h \in C((0,T], V_h')$ and an initial value $u_{0h} \in U_h$. Then we want to solve the problem:

$$\begin{aligned} &\text{Find} \quad u_h \in C^1([0,T],U_h) \quad \text{such that} \\ &e_h\left(\frac{d}{dt}u_h(t), v_h\right) + a_h\,(u_h(t), v_h) = \ell_h(t; v_h) =: \ell_h(t)v_h \\ &\qquad\qquad \text{for all } v_h \in V_h,\ t \in (0,T], \\ &u_h(0) = u_{0h}\,. \end{aligned} \tag{9.81}$$

In this case too, the existence of a unique solution u_h is assumed.

Having in mind quadrature rules for e_h (see e.g. (9.57)), $\widetilde{W}$ should be a space of piecewise continuous functions; for $e = e_h$, an L^2-space can be chosen. Thus, the semidiscrete problems are equivalent to finding functions of the stated regularity and satisfying the initial condition such that it holds, for each $t \in (0,T)$:

$$u(t) = S\left(\ell(t) - E\frac{d}{dt}u(t)\right) \quad \text{and} \quad u_h(t) = S_h\left(\ell_h(t) - E_h\frac{d}{dt}u_h(t)\right), \tag{9.82}$$

respectively, where $E: H \to V'$ and $E_h: \widetilde{W} \to V_h'$ are the operators corresponding to the forms e and e_h.

The decisive final requirement is the following uniform stability estimate for the semidiscrete problem:

④ There are parameters $C_s^{(1)}, C_s^{(2)}$ depending on t or (t,s), $t,s \in (0,T]$, such that the solution of (9.81) satisfies

$$\|u_h(t)\widehat{\|} \le C_s^{(1)}(t)\|u_{0h}\widehat{\|} + \int_0^t C_s^{(2)}(t,s)\|\ell_h(s)\|_{V_h'}\,ds \quad \text{for all } t \in (0,T].$$

We point out that this specific form is chosen for definiteness but the following conclusion also holds true for other stability estimates.

Theorem 9.16 *Under the assumptions above, in particular ① – ④, assume* $\ell \in \widehat{V}$ *and the following regularity properties of the solution of* (9.80):

$$u \in C^1((0,T],U), \quad u(t) \in W, \quad u'(t) \in W \cap \widetilde{W} \quad \text{for } t \in (0,T).$$

There is a constant $C_h^{(2)} > 0$ *such that*

$$\|E_h\|_{\widetilde{W},V_h'} \leq C_h^{(2)}$$

with $\widetilde{W}$ *endowed with the norm* $\|\cdot\widehat{\|}$, *then*

$$\begin{aligned}
&\|u(t) - u_h(t)\widehat{\|} \\
&\leq C_a\,(u(t))\,h^\alpha + C_h^{(1)} C_s \|(\widehat{Q}_h - Q_h)Au(t)\|_{V_h'} + C_s^{(1)}(t)\|u_{0h} - R_h u_0\widehat{\|} \\
&\quad + \int_0^t C_s^{(2)}(t,s)\Big[\|Q_h\ell(s) - \ell_h(s)\|_{V_h'} + \|(Q_h E - E_h)u'(s)\|_{V_h'} \\
&\qquad + C_h^{(2)} C_a\,(u'(s))\,h^\alpha + C_h^{(1)} C_h^{(2)} C_s \|(\widehat{Q}_h - Q_h)Au'(s)\|_{V_h'}\Big]ds.
\end{aligned}$$

Proof As in the proof of Theorem 9.10, we start with the decomposition

$$u_h(t) - u(t) = u_h(t) - R_h u(t) + R_h u(t) - u(t) =: \theta(t) + \varrho(t), \quad t \in (0,T],$$

where R_h is now the operator introduced in Definition 9.14. The Ritz projection error can be estimated immediately using Lemma 9.15:

$$\|\varrho(t)\widehat{\|} \leq C_a\,(u(t))\,h^\alpha + C_h^{(1)} C_s \|(\widehat{Q}_h - Q_h)Au(t)\|_{V_h'}\,.$$

For θ it holds, by (9.82)

$$\begin{aligned}
-\theta(t) &= R_h u(t) - u_h(t) = S_h\left(Q_h Au(t) - \ell_h(t) + E_h u_h'(t)\right) \\
&= S_h\left(Q_h\left(\ell(t) - Eu'(t)\right) - \ell_h(t) + E_h u_h'(t)\right) \\
&= S_h\left(Q_h\ell(t) - \ell_h(t) - Q_h Eu'(t) + E_h R_h u'(t) - E_h\left(R_h u'(t) - u_h'(t)\right)\right).
\end{aligned}$$

Due to Remark 9.11, the last difference can be rewritten as

$$E_h\left(\frac{d}{dt}(R_h u(t) - u_h(t))\right) = E_h\left(-\frac{d}{dt}\theta(t)\right),$$

meaning by (9.82) that $-\theta(t)$ solves (9.81) for the right-hand side

$$\tilde{\ell}_h(t) := Q_h\ell(t) - \ell_h(t) - \left(Q_h Eu'(t) - E_h R_h u'(t)\right).$$

Thus, we can apply the assumed stability estimate ④:

$$\|\theta(t)\widehat{\|} \leq C_s^{(1)}(t)\|\theta(0)\widehat{\|} + \int_0^t C_s^{(2)}(t,s)\|\tilde{\ell}_h(s)\|_{V_h'}\, ds.$$

To estimate the integrand, we write

$$\tilde{\ell}_h(s) = Q_h\ell(s) - \ell_h(s) - (Q_hE - E_h)u'(s) - E_h\big(u'(s) - R_hu'(s)\big)$$

and observe that, by assumption and Lemma 9.15,

$$\begin{aligned}\|E_h\big(u'(s) - R_hu'(s)\big)\|_{V_h'} &\leq C_h^{(2)}\|u'(s) - R_hu'(s)\widehat{\|}\\ \leq C_h^{(2)}C_a\,(u'(s))\,h^\alpha &+ C_h^{(2)}C_h^{(1)}C_s\|(\widehat{Q}_h - Q_h)Au'(s)\|_{V_h'}.\end{aligned}$$

Summarizing all the estimates yields the statement. □

Remark 9.17 1) If $u_0 \in W$, then

$$\begin{aligned}\|u_{0h} - R_hu_0\widehat{\|} &\leq \|u_{0h} - u_0\widehat{\|}\\ &\quad + C_a\,(u_0)\,h^\alpha + C_h^{(1)}C_s\|(\widehat{Q}_h - Q_h)Au_0\|_{V_h'}.\end{aligned}$$

2) Theorem 9.10 can be recovered from Theorem 9.16 by the following settings: $H := L^2(\Omega)$, $U := V \subset H^1(\Omega)$, $U_h := V_h \subset V$ (then $U(h) = U$), $\|\cdot\|_{U(h)} := \|\cdot\|_U := \|\cdot\|_1$,
$\widehat{Q}_h := Q_h : V' \to V_h'$, $\ell \mapsto \ell|_{V_h}$,
$W := H^2(\Omega)$, $\alpha = 2$, $\widetilde{W} := L^2(\Omega)$, $\|\cdot\widehat{\|} := \|\cdot\|_0^{(b)}$ (then $\|\cdot\widehat{\|}$ is even a norm and $C_h^{(1)} = \|b\|_\infty^{1/2}$),
e and $\|\cdot\|_{\widetilde{W}}$ are the weighted scalar product and norm, resp. (then $C_h^{(2)} = \|b\|_\infty^{1/2}$). For $\ell_h := Q_h\ell$, $E_h := Q_hE$, the corresponding consistency terms vanish.
3) As an example for a more general situation, quadrature in the mass matrix can be considered, e.g. to render it diagonal (*mass lumping*). Theorem 9.16 shows that a quadrature error of the convergence order α is sufficient to preserve the order of convergence estimate.

Order of Convergence Estimates for a Mixed Finite Element Method

We consider a general mixed variational formulation in the form (6.51), with the Darcy equation (6.49) and the Stokes equation (6.86) as major examples. For both, conditions for unique stable existence have been specified and likewise for a Galerkin approximation. To simplify the notation, in the following, we only consider a conforming and consistent approximation by assuming

$$\begin{gathered}M_h \subset M,\ X_h \subset X,\\ a_h := a,\ b := b_h,\ c := c_h,\ d := d_h,\ \ell_{ih} := \ell_i,\ i = 1, 2,\end{gathered}$$

meaning

Find $(\boldsymbol{p}_h, u_h) \in M_h \times X_h$ such that

$$\begin{aligned} a(\boldsymbol{p}_h, \boldsymbol{q}_h) + b(\boldsymbol{q}_h, u_h) + c(u_h, \boldsymbol{q}_h) &= \ell_1(\boldsymbol{q}_h) \quad \text{for all } \boldsymbol{q}_h \in M_h\,, \\ b(\boldsymbol{p}_h, v_h) - d(u_h, v_h) &= \ell_2(v_h) \quad \text{for all } v_h \in X_h\,. \end{aligned}$$

For this variational equation on $M_h \times X_h$, we note that sufficient conditions for the unique stable solvability have been provided in Sections 7.2 and 7.3, and also order of convergence estimates for the finite element spaces discussed there. Therefore, we can assume that the requirements ① – ③ are met with norms and spaces to be reconsidered later. The time-dependent versions read (note the redefinition of ℓ_2):

$$\begin{aligned} a(\boldsymbol{p}(t), \boldsymbol{q}) + b(\boldsymbol{q}, u(t)) + c(u(t), \boldsymbol{q}) &= \ell_1(t; \boldsymbol{q}) \quad \text{for all } \boldsymbol{q} \in M, \\ e\left(\frac{d}{dt}u(t), v\right) - b(\boldsymbol{p}(t), v) + d(u(t), v) &= \ell_2(t; v) \quad \text{for all } v \in X, \quad t \in (0,T), \\ u(0) &= u_0 \end{aligned} \tag{9.83}$$

with $\boldsymbol{p} \in C([0,T], M)$, $u \in C([0,T], X)$, $u' \in C((0,T], X)$, assuming $\ell_1 \in C([0,T], M')$, $\ell_2 \in C([0,T], X')$ using the abbreviation $\ell_i(t) := \ell_i(t; \cdot)$, $i = 1, 2$.

The semidiscrete version is

$$\begin{aligned} a(\boldsymbol{p}_h(t), \boldsymbol{q}_h) + b(\boldsymbol{q}_h, u_h(t)) + c(u_h(t), \boldsymbol{q}) &= \ell_1(t; \boldsymbol{q}_h) =: \ell_{1h}(t)\boldsymbol{q}_h \\ &\qquad \text{for all } \boldsymbol{q}_h \in M_h, \\ e\left(\frac{d}{dt}u_h(t), v_h\right) - b(\boldsymbol{p}_h(t), v_h) + d(u_h(t), v_h) &= \ell_2(t; v_h) =: \ell_{2h}(t)v_h \\ &\qquad \text{for all } v_h \in X,\ t \in (0,T), \\ u_h(0) &= u_{0h} \end{aligned} \tag{9.84}$$

with similar regularity requirements as above.

The missing time derivative in the first equation could be added formally by a zero form, but nevertheless the differential–algebraic equation (9.84) needs to be investigated concerning unique existence and stability, i.e. ④. To this end, we rewrite (9.84) as an ordinary differential equation solely in u_h. The components of the corresponding solution operator S from ① are written as $S_{\boldsymbol{p}}$, S_u,

$$S = \begin{pmatrix} S_{\boldsymbol{p}} \\ S_u \end{pmatrix} : \; M' \times X' \to M \times X,$$

and analogously for the discrete operator S_h. To incorporate the Darcy equation, we make the following weak assumptions.

⑤ The bilinear forms a and e are symmetric, a is positive definite and e is X_h-coercive:

$$\begin{aligned} &a(\boldsymbol{q}_h, \boldsymbol{q}_h) > 0 && \text{for all } \boldsymbol{q}_h \in M_h \setminus \{\boldsymbol{0}\}, \\ &e(v_h, v_h) \geq \alpha \|v_h\|_X^2 && \text{for all } v_h \in X_h, \text{ with } \alpha > 0 \text{ independent of } h. \end{aligned}$$

These properties ensure that A_h and E_h are invertible, and

$$\|\boldsymbol{q}_h\|_a := a(\boldsymbol{q}_h, \boldsymbol{q}_h)^{1/2}, \quad \|v_h\|_e := e(v_h, v_h)^{1/2}$$

are norms on M_h and X_h, respectively.

Furthermore, there exists a constant $C > 0$, independent of h, such that

$$\begin{aligned} |c(u_h, \boldsymbol{p}_h)| &\le C\|\boldsymbol{p}_h\|_a\|u_h\|_X \quad \text{for all } (\boldsymbol{p}_h, u_h) \in M_h \times X_h, \\ |d(u_h, v_h)| &\le C\|u_h\|_X\|v_h\|_X \quad \text{for all } u_h, v_h \in X_h, \\ |e(u_h, v_h)| &\le C\|u_h\|_X\|v_h\|_X \quad \text{for all } u_h, v_h \in X_h. \end{aligned}$$

Define an operator

$$T_h : X_h \to X_h, \; v_h \mapsto S_{h,u}\begin{pmatrix} \mathbf{0} \\ E_h v_h \end{pmatrix}, \tag{9.85}$$

where $E_h : X_h \to X_h'$ is the operator corresponding to $e : X_h \times X_h \to \mathbb{R}$, $S_{h,u}$ is the second component of the corresponding solution operator S_h from ②, and analogously A_h, B_h, C_h, D_h, and $S_{h,\boldsymbol{p}}$ are to be understood.

Lemma 9.18 *The operator T_h from* (9.85) *is an isomorphism.*

Proof As X_h is finite dimensional, only the injectivity has to be proven.

Let $u_h := T_h v_h = 0$, $\boldsymbol{p}_h := S_{h,\boldsymbol{p}}\begin{pmatrix} \mathbf{0} \\ E_h v_h \end{pmatrix}$, i.e. $A_h\boldsymbol{p}_h + (B_h^T + C_h)u_h = 0$ and thus $\boldsymbol{p}_h = \mathbf{0}$.

Therefore,

$$E_h v_h = B_h\boldsymbol{p}_h - D_h u_h = 0, \text{ i.e. } v_h = 0.$$

□

Lemma 9.19 $(\boldsymbol{p}_h, u_h) : [0,T] \to M_h \times X_h$ *is a solution of* (9.84) *in* $[0,T]$ *if and only if u_h is a solution in* $[0,T]$ *of*

$$\begin{aligned} y'(t) &= -T_h^{-1}y(t) + F(t), \quad t \in (0,T), \\ y(0) &= u_{0h}, \end{aligned} \tag{9.86}$$

where $F(t) := T_h^{-1}S_{h,u}\begin{pmatrix} \ell_{1h} \\ \ell_{2h} \end{pmatrix}$, *and $\boldsymbol{p}_h$ is recovered from* $A_h\boldsymbol{p}_h(t) + (B_h^T + C_h)u_h(t) = \ell_{1h}(t)$.

If $\ell_h \in C^k([0,T], (M\times X)')$ *for* $k \in \mathbb{N}_0$, *then a solution fulfills* $\boldsymbol{p}_h \in C^k([0,T], M_h)$, $u_h \in C^{k+1}([0,T], X_h)$.

If $\ell_{1h} \in C^{k+1}([0,T], M_h')$, *then* $\boldsymbol{p} \in C^{k+1}([0,T], M_h)$.

Proof The following statements are equivalent for functions of the noted regularity:

(i) $(\boldsymbol{p}_h, u_h)$ is a solution of (9.84),

(ii) $\begin{pmatrix} \boldsymbol{p}_h(t) \\ u_h(t) \end{pmatrix} = S_h \begin{pmatrix} \ell_{1h} \\ \ell_{2h} - E_h u_h'(t) \end{pmatrix}$,

(iii) The first equation and

$$T_h^{-1} u_h(t) = T_h^{-1} S_{h,u} \begin{pmatrix} \ell_{1h} \\ \ell_{2h} \end{pmatrix} - T_h^{-1} S_{h,u} \begin{pmatrix} \boldsymbol{0} \\ E_h u_h'(t) \end{pmatrix} = F(t) - T_h^{-1} T_h u_h'(t).$$

The regularity follows directly from the theory of ordinary differential equations and the algebraic equation. □

Lemma 9.20 *The problem* (9.84) *has a unique solution on* $[0, T]$.

Proof Since the problem for y in (9.86) has a unique solution in $[0, T]$, we can define

$$\begin{pmatrix} \boldsymbol{p}_h(t) \\ u_h(t) \end{pmatrix} := S_h \begin{pmatrix} \ell_{1h} \\ \ell_{2h} - E_h y'(t) \end{pmatrix}. \tag{9.87}$$

The reasoning for Lemma 9.19 shows $u_h = y$ and the solution property. If $\ell_{1h} = \ell_{2h} = 0$, $u_{0h} = 0$, then $u_h = y = 0$ and thus by (9.87) also $\boldsymbol{p}_h = \boldsymbol{0}$. □

Theorem 9.21 *Let* $\widehat{\|(\boldsymbol{q}_h, v_h)\|} := \left(a(\boldsymbol{q}_h, \boldsymbol{q}_h) + e(v_h, v_h)\right)^{1/2}$, $\ell_1 \in C^1([0, T], \widetilde{M}_h')$, *where* $\widetilde{M}_h'$ *is the dual space of* $(M_h, \|\cdot\|_a)$.

1) *If* $\ell_1 = 0$, *then there are constants* $C_s^{(1)}, C_s^{(2)} > 0$ *such that a stability estimate of type* ④ *holds for problem* (9.84).
2) *In the general case, the term* $\|\ell_{1h}(s)\|_{\widetilde{M}_h'} + \|\ell_{1h}'(s)\|_{\widetilde{M}_h'}$ *additionally occurs in the integral of the stability estimate.*

Proof Due to Lemma 9.19, we have $\boldsymbol{p}_h \in C^1([0, T], M_h)$, $u_h \in C^1([0, T], X_h)$. Differentiating the first equation of (9.84) w.r.t. time and testing the result with $\boldsymbol{q} := \boldsymbol{p}_h(t)$ yields

$$\frac{1}{2}\frac{d}{dt} a\left(\boldsymbol{p}_h(t), \boldsymbol{p}_h(t)\right) + b\left(\boldsymbol{p}_h(t), u_h'(t)\right) + c(u_h'(t), \boldsymbol{p}_h(t)) = \ell_{1h}'(t)\left(\boldsymbol{p}_h(t)\right).$$

Testing the second equation with $v := u_h'(t)$ gives

$$e\left(u_h'(t), u_h'(t)\right) - b\left(\boldsymbol{p}_h(t), u_h'(t)\right) + d\left(u_h(t), u_h'(t)\right) = \ell_{2h}(t)\left(u_h'(t)\right).$$

Adding both equations, using the estimates from ⑤, separating the products in such a way that the terms $\|u_h'\|_X^2$ from the right-hand side can be absorbed, we arrive at an estimate of the form

$$\begin{aligned} &\frac{d}{dt} a\left(\boldsymbol{p}_h(t), \boldsymbol{p}_h(t)\right) + \tilde{\alpha}\|u_h'(t)\|_X^2 \\ &\leq C\left(\|\boldsymbol{p}_h(t)\|_a^2 + \|u_h(t)\|_X^2 + \|\ell_{1h}'(t)\|_{\widetilde{M}_h'}^2 + \|\ell_{2h}(t)\|_{X_h'}^2\right), \end{aligned} \tag{9.88}$$

where $0 < \tilde{\alpha} \leq \alpha$.

Next, testing the first equation with $\boldsymbol{q} := \boldsymbol{p}_h(t)$, the second with $v := u_h(t)$ and adding the results gives

$$a\left(\boldsymbol{p}_h(t), \boldsymbol{p}_h(t)\right) + c\left(u_h(t), \boldsymbol{p}_h(t)\right) + \frac{1}{2}\frac{d}{dt} e\left(u_h(t), u_h(t)\right) + d\left(u_h(t), u_h(t)\right)$$
$$= \ell_{1h}(t)\left(p_h(t)\right) + \ell_{2h}(t)\left(u_h(t)\right).$$

As above, we get

$$\frac{d}{dt} e\left(u_h(t), u_h(t)\right) + \|\boldsymbol{p}_h\|_a^2 \leq C\left(\|u_h(t)\|_X^2 + \|\ell_{1h}(t)\|_{\widetilde{M}_h'}^2 + \|\ell_{2h}(t)\|_{X_h'}^2\right).$$

From ⑤, we conclude that the norms $\|\cdot\|_X$ and $\|\cdot\|_e$ are uniformly equivalent on X_h. Therefore, the terms $\|u_h(t)\|_X^2$ on the right-hand side here and in (9.88) can be replaced by $\|u_h(t)\|_e^2$.
Summing up both inequalities gives

$$\frac{d}{dt} z(t) \leq C\left(z(t) + \|\ell_{1h}(t)\|_{\widetilde{M}_h'}^2 + \|\ell_{1h}'(t)\|_{\widetilde{M}_h'}^2 + \|\ell_{2h}(t)\|_{X_h'}^2\right)$$

for $z(t) := \widehat{\|(\boldsymbol{p}_h(t), u_h(t))\|}^2$.

From this, both assertions can be concluded by Gronwall's Lemma (see, e.g. [48, Thm. 1.2.1]). □

Remark 9.22 The Darcy equation (6.54) is covered by the above theorem, assuming that the existence conditions ①, ② and the convergence condition ③ are satisfied. This is guaranteed in the case $\boldsymbol{c} = \boldsymbol{0}$, $r = 0$ by the inf–sup conditions for $\tilde{b}$ and $\tilde{b}_h$ and appropriate regularity conditions (see Theorems 6.17, 6.21, 6.25 in the specifications in Section 7.2).

The assumptions of ④ are satisfied because the matrix $\boldsymbol{L}$ and thus the bilinear form

$$a(\boldsymbol{q}, \boldsymbol{q}) = \int_\Omega \boldsymbol{L}\boldsymbol{q} \cdot \boldsymbol{q} dx$$

are uniformly positive definite. For the bilinear form

$$c(u, \boldsymbol{q}) = -\int_\Omega \boldsymbol{L}\boldsymbol{c}u \cdot \boldsymbol{q} dx,$$

it easily follows that

$$|c(u, \boldsymbol{q})| \leq \|\boldsymbol{L}^{1/2}\boldsymbol{c}\|_\infty \|u\|_X \|\boldsymbol{q}\|_a,$$

and for

$$d(u, v) = \int_\Omega ruv dx,$$

we have the estimate

$$|d(u, v)| \leq \|r\|_\infty \|u\|_X \|v\|_X \,.$$

The requirements $\ell_{1h}(t), \ell'_{1h} \in \widetilde{M}'_h$ would render boundary conditions to be homogeneous. This can be avoided by an refinement of the estimate. Because of

$$b(\boldsymbol{p}, u) = \int_\Omega \nabla \boldsymbol{p} u dx$$

and $\nabla \cdot \boldsymbol{p}_h \in X_h$ for $\boldsymbol{p}_h \in M_h$, we can test the second equation with $\nabla \cdot \boldsymbol{p}_h$ leading to the further estimate

$$\|\nabla \cdot \boldsymbol{p}_h\|_0^2 \leq C(\|u_h\|_0^2 + \|\ell_2\|_{X'_h}^2).$$

In this way, the full norm $\|\boldsymbol{p}_h\|_{H(\text{div})}^2$ is available on the left-hand side of the resulting total estimate. Hence, only $\ell_{1h}(t), \ell'_{1h}(t) \in M'_h$ is necessary.

Order of Convergence Estimates for the (Node-Oriented) Finite Volume Method

For simplicity, we restrict attention to pure homogeneous Dirichlet conditions ($\Gamma_3 = \partial\Omega$). To cast the problem into the general framework (9.76) ff., set $U := V := H_0^1(\Omega)$ (all equipped with the H^1-norm), S being the solution operator of the variational equation defined by (9.56), $V_h := U_h$ is the finite-dimensional space according to Section 8.2, and S_h the discrete solution operator (see (8.13) ff.). For $f \in C(\overline{\Omega})$, we set $\ell \in V'$ by $\ell(v) := \langle f, v\rangle_0$ for all $v \in V$. The operator $\widehat{Q}_h : \hat{V} \to V'_h$ is defined by

$$\left\langle \widehat{Q}_h \ell, v_h \right\rangle = \langle f, v_h\rangle_{0,h} \quad \text{for all } \ell = \langle f, \cdot\rangle_0 \in \hat{V},\ v_h \in V_h$$

with $\langle\cdot,\cdot\rangle_{0,h}$ defined in (8.15), and $\hat{V} := C(\overline{\Omega}) \cap V'$.

Under the assumptions of Corollary 8.19, the uniform stability ① of S_h according to (9.77) is valid with

$$C_s := 1/\alpha$$

(with the constant α from Corollary 8.19). The order of convergence ③ is valid for $W := H^2(\Omega)$ under the assumption of Theorem 8.20 with order $\alpha = 1$ and

$$C_a(u) = C(\|u\|_2 + |f|_{1,\infty}).$$

We define the operator $R_h : V \to V_h$ by

$$a_h(R_h v, v_h) = a(v, v_h) \quad \text{for all } v \in V,\ v_h \in V_h, \tag{9.89}$$

i.e. in Definition 9.14, we choose

$$Q_h : V' \to V_h', \ \ell \mapsto \ell|_{V_h}.$$

Thus, Lemma 9.15 assumes the quasi-optimality of R_h, with consistency term $\|(\hat{Q}_h - Q_h)\ell\|_{V_h'}$. For the semidiscrete version, we have

$$e_h(u_h, v_h) := \langle u_h, v_h \rangle_{0,h}^{(b)}, \quad \widetilde{W} := C(\overline{\Omega}).$$

The stability estimate ④ is given by (9.60) with

$$\|\cdot\widehat{\|} := \|\cdot\|_{0,h}^{(b)}, \quad C_s^{(1)}(t) := e^{-\tilde{\alpha}t}, \quad C_s^{(2)}(t,s) := e^{-\tilde{\alpha}(t-s)},$$

hence $C_h^{(1)} = C_h^{(2)} = C_2\|b\|_\infty^{1/2}$, where C_2 is the constant from Remark 8.8. To apply the error estimate of Theorem 9.16, estimates of the consistency errors are still required:

$$\|(\widehat{Q}_h - Q_h)Au(t)\|_{V_h'}, \quad \|(\widehat{Q}_h - Q_h)Au'(s)\|_{V_h'}, \tag{9.90}$$

$$\|Q_h\ell(s) - \ell_h(s)\|_{V_h'}, \quad \|(Q_hE - E_h)u'(s)\|_{V_h'}. \tag{9.91}$$

The first three estimates are delivered, in the case of the Donald diagram, by the following lemma. If the weight b belongs to $C^1(\overline{\Omega})$, then the fourth estimate can also be obtained from this result by replacing w by bw.

Lemma 9.23 *Assume $w \in C^1(\overline{\Omega})$ and $v \in V_h$. Then, if the finite volume partition of Ω is a Donald diagram, there is a constant $C > 0$ independent of h such that*

$$|\langle w, v\rangle_{0,h} - \langle w, v\rangle_0| \le Ch|w|_{1,\infty}\|v\|_{0,h}.$$

Proof We start with a simple rearrangement of the order of summation:

$$\langle w, v\rangle_{0,h} = \sum_{j=1}^{M} w_j v_j m_j = \sum_{K\in\mathcal{T}_h} \sum_{j:\, \partial K \ni a_j} w_j v_j |\Omega_{j,K}|,$$

where $\Omega_{j,K} = \Omega_j \cap \operatorname{int} K$. First, we will consider the inner sum. For any triangle $K \in \mathcal{T}_h$ with barycentre $a_{S,K}$, we can write

$$\begin{aligned}
\sum_{j:\, \partial K \ni a_j} w_j v_j |\Omega_{j,K}| &= \sum_{j:\, \partial K \ni a_j} [w_j - w(a_{S,K})] v_j |\Omega_{j,K}| \\
&\quad + \sum_{j:\, \partial K \ni a_j} w(a_{S,K}) \left[v_j |\Omega_{j,K}| - \int_{\Omega_{j,K}} v\, dx \right] \\
&\quad + \sum_{j:\, \partial K \ni a_j} \int_{\Omega_{j,K}} [w(a_{S,K}) - w] v\, dx + \sum_{j:\, \partial K \ni a_j} \int_{\Omega_{j,K}} wv\, dx \\
&=: I_{1,K} + I_{2,K} + I_{3,K} + \int_K wv\, dx\,.
\end{aligned}$$

To estimate $I_{1,K}$, we apply the Cauchy–Schwarz inequality and get

$$|I_{1,K}| \leq \left(\sum_{j:\,\partial K \ni a_j} |w_j - w(a_{S,K})|^2 |\Omega_{j,K}| \right)^{1/2} \|v\|_{0,h,K},$$

where

$$\|v\|_{0,h,K} := \left(\sum_{j:\,\partial K \ni a_j} v_j^2 |\Omega_{j,K}| \right)^{1/2}.$$

Since $|w_j - w(a_{S,K})| \leq h_K |w|_{1,\infty}$, it follows that

$$|I_{1,K}| \leq h_K |w|_{1,\infty} \sqrt{|K|} \|v\|_{0,h,K}.$$

Similarly, for $I_{3,K}$, we easily get

$$\begin{aligned} |I_{3,K}| &= \left| \int_{\Omega_K} [w(a_{S,K}) - w] v \, dx \right| \\ &\leq \|w(a_{S,K}) - w\|_{0,K} \|v\|_{0,K} \leq h_K |w|_{1,\infty} \sqrt{|K|} \|v\|_{0,K}. \end{aligned}$$

So, it remains to consider $I_{2,K}$. Obviously,

$$I_{2,K} = w(a_{S,K}) \sum_{j:\,\partial K \ni a_j} \int_{\Omega_{j,K}} [v_j - v] \, dx.$$

We will show that if Ω_j belongs to a Donald diagram, then the sum vanishes. To do so, let us suppose that the triangle under consideration has the vertices a_i, a_j, and a_k. The set $\Omega_{j,K}$ can be decomposed into two subtriangles by drawing a straight line between $a_{S,K}$ and a_j. We will denote the interior of these triangles by $\Omega_{j,K,i}$ and $\Omega_{j,K,k}$; i.e.

$$\Omega_{j,K,i} := \text{int}\big(\text{conv}\{a_j, a_{S,K}, a_{ij}\}\big), \quad \Omega_{j,K,k} := \text{int}\big(\text{conv}\{a_j, a_{S,K}, a_{kj}\}\big).$$

On each subtriangle, the integral of v can be calculated exactly by means of the trapezoidal rule. Since $|\Omega_{j,K,i}| = |\Omega_{j,K,k}| = |K|/6$ in the case of the Donald diagram (cf. also (8.4)), we have

$$\begin{aligned} \int_{\Omega_{j,K,i}} v \, dx &= \frac{|K|}{18} \left[v_j + \frac{v_j + v_i}{2} + \frac{v_j + v_i + v_k}{3} \right] \\ &= \frac{|K|}{18} \left[\frac{11}{6} v_j + \frac{5}{6} v_i + \frac{1}{3} v_k \right], \\ \int_{\Omega_{j,K,k}} v \, dx &= \frac{|K|}{18} \left[\frac{11}{6} v_j + \frac{5}{6} v_k + \frac{1}{3} v_i \right]. \end{aligned}$$

Consequently,

$$\int_{\Omega_{j,K}} v\,dx = \frac{|K|}{18}\left[\frac{11}{3}v_j + \frac{7}{6}v_i + \frac{7}{6}v_k\right],$$

and thus

$$\sum_{j:\,\partial K \ni a_j} \int_{\Omega_{j,K}} v\,dx = \frac{|K|}{3} \sum_{j:\,\partial K \ni a_j} v_j\,.$$

On the other hand, since $3|\Omega_{j,K}| = |K|$ (cf. (8.4)), we have

$$\sum_{j:\,\partial K \ni a_j} \int_{\Omega_{j,K}} v_j\,dx = \frac{|K|}{3} \sum_{j:\,\partial K \ni a_j} v_j\,,$$

and so $I_{2,K} = 0$. In summary, we have obtained the following estimate:

$$|I_{1,K} + I_{2,K} + I_{3,K}| \le h_K |w|_{1,\infty}\sqrt{|K|}\left[\|v\|_{0,h,K} + \|v\|_{0,K}\right].$$

So, it follows that

$$\begin{aligned}
|\,\langle w, v\rangle_{0,h} - \langle w, v\rangle_0\,| &\le \sum_{K\in\mathcal{T}_h} |I_{1,K} + I_{2,K} + I_{3,K}| \\
&\le h|w|_{1,\infty} \sum_{K\in\mathcal{T}_h} \sqrt{|K|}\left[\|v\|_{0,h,K} + \|v\|_{0,K}\right].
\end{aligned}$$

By the Cauchy–Schwarz inequality,

$$\sum_{K\in\mathcal{T}_h} \sqrt{|K|}\|v\|_{0,h,K} \le \left(\sum_{K\in\mathcal{T}_h} |K|\right)^{1/2}\left(\sum_{K\in\mathcal{T}_h} \|v\|_{0,h,K}^2\right)^{1/2} = \sqrt{|\Omega|}\|v\|_{0,h}$$

and, similarly,

$$\sum_{K\in\mathcal{T}_h} \sqrt{|K|}\|v\|_{0,K} \le \sqrt{|\Omega|}\|v\|_0\,.$$

So, we finally arrive at

$$|\,\langle w, v\rangle_{0,h} - \langle w, v\rangle_0\,| \le Ch|w|_{1,\infty}\left[\|v\|_{0,h} + \|v\|_0\right].$$

Since the norms $\|\cdot\|_{0,h}$ and $\|\cdot\|_0$ are equivalent on V_h (see Remark 8.8), we get

$$|\,\langle w, v\rangle_{0,h} - \langle w, v\rangle_0\,| \le Ch|w|_{1,\infty}\|v\|_{0,h}.$$

□

Theorem 9.24 *In addition to the assumptions of Theorem 8.18, let $b \in C^1(\overline{\Omega})$, $f \in C([0,T], C^1(\overline{\Omega}))$, $u_0 \in V \cap H^2(\Omega)$, and $u_{0h} \in V_h$. Then if $u(t)$ is sufficiently smooth, the solution $u_h(t)$ of the semidiscrete finite volume method* (9.57)) *on Donald diagrams satisfies the following estimate:*

$$\begin{aligned}
&\|u_h(t) - u(t)\|_{0,h}^{(b)} \le \|u_{h0} - u_0\|_{0,h}^{(b)} e^{-\tilde{\alpha}t} \\
&+ Ch\Big[\|u_0\|_2 e^{-\alpha t} + \|u(t)\|_2 + |Au(t)|_{1,\infty} \\
&+ \int_0^t \left[|f(s)|_{1,\infty} + |u'(s)|_{1,\infty} + \|u'(s)\|_2 + |Au'(s)|_{1,\infty}\right] e^{-\tilde{\alpha}(t-s)} ds\Big]
\end{aligned}$$

with $\tilde{\alpha} := \alpha\|b\|_\infty^{-1}$.

Remark 9.25 1) In comparison with the finite element method, the result is not optimal in h. The reason is that, in general, the finite volume method does not yield optimal L^2-error estimates even in the elliptic case, but this type of result is necessary to obtain optimal estimates. The loss of order can be understood by looking at Lemma 6.11. The critical component is the fourth term, the order of which does not exceed $\gamma = 1$, in general (see (6.25)).

2) The terms $|Au(t)|_{1,\infty}$, $|Au'(s)|_{1,\infty}$ in the upper bound can be avoided by showing directly an estimate of the elliptic projection error:

$$\|(R_h - I)v\|_{0,h}^{(b)} \le Ch\|v\|_2 \quad \text{for all } v \in V \cap H^2(\Omega).$$

Indeed, for an arbitrary element $v \in V \cap H^2(\Omega)$ and with $v_h := R_h v - I_h(v)$, we have by (9.89) that

$$a_h(v_h, v_h) = a_h(R_h v, v_h) - a_h(I_h(v), v_h) = a(v, v_h) - a_h(I_h(v), v_h).$$

By integration by parts in the first term of the right-hand side, it follows that

$$a_h(v_h, v_h) = \langle Lv, v_h\rangle_0 - a_h(I_h(v), v_h).$$

From [77], an estimate of the right-hand side is known (cf. also (8.29)); thus,

$$a_h(v_h, v_h) \le Ch\|v\|_2 \left(|v_h|_1^2 + \|v_h\|_{0,h}^2\right)^{1/2}.$$

So Theorem 8.18 yields

$$\left(|v_h|_1^2 + \|v_h\|_{0,h}^2\right)^{1/2} \le Ch\|v\|_2 .$$

By the triangle inequality,

$$\|(R_h - I)v\|_{0,h} \le \|R_h v - I_h(v)\|_{0,h} + \|I_h(v) - v\|_{0,h}.$$

Since the second term vanishes by the definitions of $\|\cdot\|_{0,h}$ and I_h, we get the assertion.

An outlook on the handling of nonlinear equations and systems using the example of semilinear systems of diffusion–convection reaction equations will be given in Section 11.4.

Exercises

Problem 9.7 Let $A \in \mathbb{R}^{M,M}$ be an $\mathbb{R}^M$-coercive matrix and let the symmetric positive definite matrix $B \in \mathbb{R}^{M,M}$ have the Cholesky decomposition $B = E^T E$. Show that the matrix $\hat{A} := E^{-T} A E^{-1}$ is $\mathbb{R}^M$-coercive.

Problem 9.8 Prove identity (9.73) by first proving the corresponding identity for one space dimension:

$$h \sum_{i=1}^{l-1} \sin^2\left(\nu \frac{\pi}{a} ih\right) = \frac{a}{2}.$$

Problem 9.9 Let V be a Banach space and $a : V \times V \to \mathbb{R}$ a V-coercive, continuous bilinear form. Show that the Ritz projection $R_h : V \to V_h$ into a subspace $V_h \subset V$ (cf. Definition 9.8) has the following properties:

a) $R_h : V \to V_h$ is continuous because of $\|R_h u\|_V \leq \frac{M}{\alpha}\|u\|_V$,
b) R_h yields quasi-optimal approximations, that is,

$$\|u - R_h u\|_V \leq \frac{M}{\alpha} \inf_{v_h \in V_h} \|u - v_h\|_V .$$

Here, M and α denote the constants in the continuity and coercivity conditions, respectively.

Problem 9.10 Let $u \in C^1([0, T], V)$. Show that $R_h u \in C^1([0, T], V)$ and $\frac{d}{dt} R_h u(t) = R_h \frac{d}{dt} u(t)$.

Problem 9.11 Consider the (primal) finite element method, but with quadrature, i.e. (9.50) and $\langle \cdot, \cdot \rangle_0^{(b)}$ substituted by $\langle \cdot, \cdot \rangle_{0,h}^{(b)}$, and a by a_h, ℓ by ℓ_h, as in Section 3.6 to include quadrature rules for the approximate evaluations of the integrals. Show an order of convergence estimate on the basis of the general Theorem 9.16.

Problem 9.12 Consider the semidiscrete approximation of (9.37) by the DG method (7.100) applied in space. Show an stability estimate for the semidiscrete problem in $L^\infty((0, T), V_h)$, similar to Lemma 9.4, with an appropriate norm on the ansatz space.

Problem 9.13 Consider the semidiscrete approximation of (9.37) by the DG method (7.100) applied in space. Investigate the validity of an order of convergence result on the basis of Theorem 9.16 and of Problem 9.12.

Problem 9.14 Consider the time-dependent Stokes equation (ENS1a), (ENS2a) on $\Omega \times (0, T]$ with (6.82) and initial data $\boldsymbol{u}(0) = \boldsymbol{u}_0$.

a) Work out the constrained weak formulation (Formulation I) for

$$V := \{\boldsymbol{v} \in H_0^1(\Omega)^d \mid \nabla\cdot\boldsymbol{v} = 0\} \text{ and } H := \{\boldsymbol{v} \in H(\mathrm{div};\Omega) \mid \nabla\cdot\boldsymbol{v} = 0, \boldsymbol{v}\cdot\boldsymbol{n}_{|\partial\Omega} = 0\}.$$

Note that V is densely embedded in H.

b) For $\boldsymbol{u}_0 \in H, \boldsymbol{f} \in L^2((0,T], H^{-1}(\Omega)^d)$, show unique existence on $[0,T]$.
Hint: Use Problem 9.3.

Problem 9.15 Consider a semidiscrete conforming discretization of the initial boundary value problem of Problem 9.14, a).

a) Investigate the validity of a stability estimate of the type ④.
b) Investigate the validity of error estimate based on Theorem 9.16.

Problem 9.16 Consider the time-dependent Stokes equation as in Problem 9.14.

a) Work out the mixed weak formulation (Formulation II) for $M := H_0^1(\Omega)^d$, $X := L_0^2(\Omega)$.
b) For $\boldsymbol{u}_0 \in V$, $\boldsymbol{f} \in L^2(Q_T)^d$, show the unique existence of a solution $\boldsymbol{u}, p$.

Hint: Start with Problem 9.14, b) to find $\boldsymbol{u}$, then construct the pressure p similar to Section 6.3, using Problem 6.18.

Problem 9.17 Consider a semidiscrete conforming discretization of the initial boundary value problem from Problem 9.16, a).

a) Investigate the validity of a stability estimate of the type ④.
b) Investigate the validity of error estimate based on Theorem 9.16.

Problem 9.18 Transfer the derivation of the finite volume method given in Section 8.2.2 for the case of an coercive boundary value problem to the parabolic initial boundary value problem (9.1) in divergence form; i.e. convince yourself that the formalism of obtaining (9.57) indeed can be interpreted as a finite volume semidiscretization of (9.1).

9.3 Fully Discrete Schemes

As we have seen, the application of the vertical method of lines results in the following situation:

- There is a linear system of ordinary differential equations of high order (dimension) to be solved.
- There is an error estimate for the solution u of the initial boundary value problem (9.1) by means of the exact solution u_h of the system (9.48).

A difficulty in the choice and in the analysis of an appropriate discretization method for systems of ordinary differential equations is that many standard estimates involve the Lipschitz constant of the corresponding right-hand side, here $\boldsymbol{q}_h - A_h\boldsymbol{u}_h$ (cf. (9.48)). But this constant is typically large for small spatial parameters h, and so we would obtain nonrealistic error estimates (cf. Theorem 3.48).

There are two alternatives. For comparatively simple time discretizations, certain estimates can be derived in a direct way (i.e. without using standard estimates for systems of ordinary differential equations). The second way is to apply specific time discretizations in conjunction with refined methods of proof.

Here, we will explain the first way for the so-called *one-step-theta method.*

One-Step Discretizations in Time, in Particular for the Finite Difference Method

We start from the problem (9.48), which resulted from spatial discretization techniques. Provided that $T < \infty$, the time interval $(0, T)$ is subdivided into $N \in \mathbb{N}$ subintervals of equal length $\tau := T/N$. Furthermore, we set $t_n := n\tau$ for $n \in \{0, \ldots, N\}$ and $\boldsymbol{u}_h^n \in \mathbb{R}^M$ for an approximation of $\boldsymbol{u}_h(t_n)$. If the time interval is unbounded, the *time step* $\tau > 0$ is given, and the number $n \in \mathbb{N}$ is allowed to increase without limitation; that is, we set formally $N = \infty$.

The values $t = t_n$, where an approximation is to be determined, are called *time levels*. The restriction to equidistant time steps is only for the sake of simplicity. We approximate $\frac{d}{dt}\boldsymbol{u}_h$ by the difference quotient

$$\frac{d}{dt}\boldsymbol{u}_h(t) \sim \frac{1}{\tau}(\boldsymbol{u}_h(t_{n+1}) - \boldsymbol{u}_h(t_n)).$$

If we interpret this approximation to be at $t = t_n$, we take the forward difference quotient; at $t = t_{n+1}$, we take the backward difference quotient; at $t = t_n + \frac{1}{2}\tau$, we take the symmetric difference quotient. Again, we obtain a generalization and unification by introducing a parameter $\Theta \in [0, 1]$ and interpreting the approximation to be taken at $t = t_n + \Theta\tau$. As for $\Theta \neq 0$ or 1, we are not at a time level, and so we need the further approximation

$$A_h\boldsymbol{u}_h\left((n+\Theta)\,\tau\right) \sim \overline{\Theta}A_h\boldsymbol{u}_h(t_n) + \Theta A_h\boldsymbol{u}_h(t_{n+1}).$$

Here, we use the abbreviation $\overline{\Theta} := 1 - \Theta$. The *(one-step-)theta method* applied to (9.48) defines a sequence of vectors $\boldsymbol{u}_h^0, \ldots, \boldsymbol{u}_h^N$ by, for $n = 0, 1, \ldots, N-1$,

$$\begin{aligned} \frac{1}{\tau}\left(\boldsymbol{u}_h^{n+1} - \boldsymbol{u}_h^n\right) + \overline{\Theta}A_h\boldsymbol{u}_h^n + \Theta A_h\boldsymbol{u}_h^{n+1} &= \boldsymbol{q}_h((n+\Theta)\tau), \\ \boldsymbol{u}_h^0 &= \boldsymbol{u}_0\,. \end{aligned} \tag{9.92}$$

If we apply this discretization to the more general form (9.51), we get correspondingly

$$\frac{1}{\tau}\left(B_h \boldsymbol{u}_h^{n+1} - B_h \boldsymbol{u}_h^n\right) + \overline{\Theta}\hat{A}_h \boldsymbol{u}_h^n + \Theta \hat{A}_h \boldsymbol{u}_h^{n+1} = \boldsymbol{q}_h((n+\Theta)\tau)\,. \tag{9.93}$$

Analogously to (9.51), the more general form can be transformed to (9.92), assuming that B_h is regular: either by multiplying (9.93) by B_h^{-1} or in the case of a decomposition $B_h = E_h^T E_h$ (for a symmetric positive definite B_h) by multiplying by E_h^{-T} and a change of variables to $E_h \boldsymbol{u}_h^n$. We will apply two techniques in the following:

One is based on the eigenvector decomposition of A_h; thus for (9.93), this means to consider the generalized eigenvalue problem

$$\hat{A}_h \boldsymbol{v} = \lambda B_h \boldsymbol{v}\,. \tag{9.94}$$

Note that the Galerkin approach for the eigenvalue problems according to Definition 9.6 leads to such a generalized eigenvalue problem with the stiffness matrix $\hat{A}_h$ and the mass matrix B_h.

The other approach is based on the matrix properties (1.35)* or (1.35). For the most important case,

$$B_h = \operatorname{diag}(b_i), \quad b_i > 0 \quad \text{for} \quad i = 1, \ldots, M\,, \tag{9.95}$$

which corresponds to the mass lumping procedure, the above-mentioned transformation reduces to a diagonal scaling, which does not influence any of the properties (1.35)* or (1.35).

Having this in mind, in the following we will consider explicitly only the formulation (9.92).

In the case $\Theta = 0$, the *explicit Euler method*, u_h^n can be determined explicitly by

$$\boldsymbol{u}_h^{n+1} = \tau(\boldsymbol{q}_h(t_n) - A_h \boldsymbol{u}_h^n) + \boldsymbol{u}_h^n = (I - \tau A_h)\boldsymbol{u}_h^n + \tau \boldsymbol{q}(t_n)\,.$$

Thus, the effort for one time step consists of an AXPY operation, a vector addition, and a matrix–vector operation. For dimension M, the first of these is of complexity $O(M)$, and also the last one if the matrix is sparse in the sense defined at the beginning of Chapter 5. On the other hand, for $\Theta \neq 0$, the method is *implicit*, since for each time step a system of linear equations has to be solved with the system matrix $I + \Theta\tau A_h$. Here, the cases $\Theta = 1$, *the implicit Euler method* (IE), and $\Theta = \frac{1}{2}$, the *Crank–Nicolson method* (CN), will be of interest. Due to our restriction to time-independent coefficients, the matrix is the same for every time step (for constant τ). If direct methods (see Section 2.5) are used, then the LU factorization has to be computed only once, and only forward and backward substitutions with changing right-hand sides are necessary, where computation for $\Theta \neq 1$ also requires a matrix–vector operation. For band matrices, for example, operations of the complexity bandwidth × dimension are necessary, which means for the basic example of the heat equation on a rectangle $O(M^{3/2})$ operations instead of $O(M)$ for the explicit method. Iterative methods for the resolution of (9.92) cannot make use of the constant matrix, but with $\boldsymbol{u}_h^n$ there is a good initial iterate if τ is not too large.

Although the explicit Euler method $\Theta = 0$ seems to be attractive, we will see later that with respect to accuracy or stability one may prefer $\Theta = \frac{1}{2}$ or $\Theta = 1$.

To investigate further the theta method, we resolve recursively the relations (9.92) to gain the representation

$$\begin{aligned} \boldsymbol{u}_h^n &= \left((I + \Theta\tau A_h)^{-1}\left(I - \overline{\Theta}\tau A_h\right)\right)^n \boldsymbol{u}_0 \\ &\quad + \tau \sum_{k=1}^{n} \left((I + \Theta\tau A_h)^{-1}\left(I - \overline{\Theta}\tau A_h\right)\right)^{n-k} (I + \Theta\tau A_h)^{-1} \boldsymbol{q}_h\left(t_k - \overline{\Theta}\tau\right). \end{aligned} \tag{9.96}$$

Here, we use the abbreviation $A^{-n} = (A^{-1})^n$ for a matrix A. Comparing this with the solution (9.65) of the semidiscrete problem, we see the approximations

$$e^{-t_n A_h} \sim E_{h,\tau}^n,$$

where

$$E_{h,\tau} := (I + \Theta\tau A_h)^{-1}\left(I - \overline{\Theta}\tau A_h\right)$$

and

$$\begin{aligned} \int_0^{t_n} e^{-(t_n - s)A_h} \boldsymbol{q}_h(s)ds &= \int_0^{t_n} \left(e^{-\tau A_h}\right)^{(t_n - s)/\tau} \boldsymbol{q}_h(s)ds \\ &\sim \tau \sum_{\substack{k=1 \\ s=k\tau}}^{n} E_{h,\tau}^{(t_n - s)/\tau} (I + \Theta\tau A_h)^{-1} \boldsymbol{q}_h\left(s - \overline{\Theta}\tau\right). \end{aligned} \tag{9.97}$$

The matrix $E_{h,\tau}$ thus is the solution operator of (9.92) for one time step and homogeneous boundary conditions and right-hand side. It is to be expected that it has to capture the qualitative behaviour of $e^{-\tau A_h}$ that it is approximating. This will be investigated in the next section.

One-Step Discretizations for the Finite Element Method

The fully discrete scheme can be achieved in two ways: Besides applying (9.92) to (9.48) in the transformed variable or in the form (9.93), the discretization approach can be applied directly to (9.50):

With $\partial U^{n+1} := (U^{n+1} - U^n)/\tau$, $f^{n+s} := sf(t_{n+1}) + (1-s)f(t_n)$, $\ell^{n+s}(v) := s\ell(t_{n+1}, v) + (1-s)\ell(t_n, v)$, b according to (9.35), $s \in [0, 1]$, and with a fixed number $\Theta \in [0, 1]$, the fully discrete method for (9.50) then reads as follows:

Find a sequence $U^0, \ldots, U^N \in V_h$ such that for $n \in \{0, \ldots, N-1\}$,

$$\begin{aligned}\left\langle\partial U^{n+1}, v_h\right\rangle_0^{(b)} + a(\Theta U^{n+1} + \overline{\Theta} U^n, v_h) &= \ell^{n+\Theta}(v_h) \\ &\quad \text{for all } v_h \in V_h\,, \\ U^0 &= u_{0h}\,. \end{aligned} \tag{9.98}$$

An alternative choice for the right-hand side, closer to the finite difference method, is the direct evaluation at $t_n + \Theta\tau$, e.g. $f(t_n + \Theta\tau)$. The version here is chosen to simplify the order of convergence estimate in Section 9.8.

By representing the U^n by means of a basis of V_h as after (9.50), again we get the form (9.93) (or (9.92) in the transformed variable). Note that also for $\Theta = 0$ the problem here is implicit if B_h is not diagonal. Therefore, *mass lumping* is often applied, and the scalar product $\langle\cdot,\cdot\rangle_0$ in (9.98) is replaced by an approximation due to numerical quadrature, i.e.

$$\begin{aligned}\langle\partial U^{n+1}, v_h\rangle_{0,h}^{(b)} + a(\Theta U^{n+1} + \overline{\Theta} U^n, v_h) &= \ell^{n+\Theta}(v_h) \\ &\quad \text{for all } v_h \in V_h\,, \\ U^0 &= u_{0h}\,. \end{aligned} \tag{9.99}$$

As explained in Section 3.5.2, $\langle u_h, v_h\rangle_{0,h}$ is the sum over all contributions from elements $K \in \mathcal{T}_h$, which takes the form (3.130) for the reference element. In the case of Lagrange elements and a nodal quadrature formula, we have for the nodal basis functions φ_i:

$$\langle\varphi_j, \varphi_i\rangle_{0,h}^{(b)} = \langle\varphi_i, \varphi_i\rangle_{0,h}^{(b)}\,\delta_{ij} =: b_i\delta_{ij} \quad \text{for} \quad i,j = 1,\dots,M, \tag{9.100}$$

since for $i \neq j$ the integrand $\varphi_i\varphi_j$ vanishes at all quadrature points. In this case, we arrive at the form (9.93) with a matrix B_h satisfying (9.95).

Here, we use the same notation as in (9.58) for the FVM. Note that for shape-regular triangulations, both induced norms are equivalent according to Remark 8.8.

One-Step Discretizations for the Node-oriented Finite Volume Method

As in the previous subsection on the finite element approach, the semidiscrete formulation (9.57) can be discretized in time directly:

Find a sequence $U^0, \dots, U^N \in V_h$ such that for $n \in \{0, \dots, N-1\}$,

$$\begin{aligned}\left\langle\partial U^{n+1}, v_h\right\rangle_{0,h}^{(b)} + a_h(\Theta U^{n+1} + \overline{\Theta} U^n, v_h) &= \left\langle f^{n+\Theta}, v_h\right\rangle_{0,h} \\ &\quad \text{for all } v_h \in V_h, \\ U^0 &= u_{0h}\,, \end{aligned} \tag{9.101}$$

where ∂U^{n+1}, Θ, $f^{n+\Theta}$ are defined as before in (9.98), $\langle\cdot,\cdot\rangle_{0,h}^{(b)}$ as in (9.58).

Remember that here we consider only homogeneous boundary conditions.

If the elements U^n, U^{n+1} are represented by means of a basis of V_h, we recover the form (9.93).

Since the mass matrix B_h is diagonal, the problem can be regarded as being explicit for $\Theta = 0$.

In the case of time-dependent coefficients, the formulation (9.93) (and correspondingly (9.101)) has to be modified to

$$\frac{1}{\tau}\left(B_h^{n+1}\boldsymbol{u}_h^{n+1} - B_h^n\boldsymbol{u}_h^n\right) + \overline{\Theta}\hat{A}_h^n\boldsymbol{u}_h^n + \Theta\hat{A}_h^{n+1}\boldsymbol{u}_h^{n+1} = \boldsymbol{q}_h((n+\Theta)\tau), \tag{9.102}$$

where B_h^k and $\hat{A}_h^k$ are the analogs to B_h and $\hat{A}_h$ at the time level t_k. Note that here, also for equidistant time stepping in general, the matrix of the linear system of equations to be solved will change from time level to time level such that there are no advantages in using direct solvers anymore.

Exercises

Problem 9.19 Consider linear simplicial elements defined on a general conforming triangulation of a polygonally bounded domain $\Omega \subset \mathbb{R}^2$.

a) Determine the entries of the mass matrix B_h.
b) Using the trapezoidal rule, determine the entries of the lumped mass matrix $\operatorname{diag}(b_i)$.

Problem 9.20 Consider (9.37) with a bilinear form $a := a_d + a_c$ with corresponding operators A_d, A_c, where a_d is V-coercive and, without loss of generality, $\alpha = 1$ as coercivity constant, $A_d \in L[V, V']$, but A_c only from $L[V, H]$. Set $c := \max_{u \in V\setminus\{0\}} \|A_c u\|_H / \|u\|_V$. Consider the following *semi-explicit scheme* (compare (9.98)):

$$\langle \partial U^{n+1}, v_h\rangle_0^{(b)} + a_d(U^{n+1}, v_h) + a_c(U^n, v_h) = \langle \ell^{n+1}, v_h\rangle_0 \quad \text{for all } v_h \in V_h,$$

where $\ell^n := \int_{t_n-1}^{t_n} \ell(s)ds$. Show for $\tau \leq \frac{\alpha}{2c^2}$ with some $\alpha > 0$ that

$$\|U^{n+1}\|_H^2 + \alpha\tau\|U^{n+1}\|_V^2 \leq \|U^n\|_H^2 + \alpha\frac{\tau}{2}\|U^n\|_V^2 + \frac{\tau}{\alpha}\|\ell^{n+1}\|_{V'}.$$

Conclude a discrete stability estimate.

Programming project 9.1 Consider the spatially one-dimensional initial boundary value problem

$$\begin{aligned} u_t - ku_{xx} &= f \quad \text{in } (a,b)\times(0,T), \\ u(\cdot,0) &= u_0 \quad \text{in } (a,b), \\ u(a) = u(b) &= 0 \quad \text{in } (0,T). \end{aligned}$$

Use the conforming linear finite element method (or the finite difference method) and the one-step-theta method to solve the problem numerically.

Take the particular problem $a := 0$, $b := 1$, $T := 1$, $k := 1$, where f is chosen in such a way that the exact solution is $u(x, t) := \sin(2\pi x)\cos(2\pi t)$, and test your implementation for an equidistant spatial grid of size $h := 1/500$ and time step sizes $\tau := 10^{-i}$, $i = 1, 2, 3, 4$.

Compute and visualize (logarithmic plot) the l^2-errors as τ tends to zero for the cases $\Theta := 0, 1/2, 1$ (explicit Euler/Crank–Nicolson/implicit Euler methods).

You may now further analyse the behaviour of the implicit Euler and Crank–Nicolson method in the case of rough initial values $u_0 \in L^2(0, 1)$ (use $f \equiv 0$) or incompatible data.

Programming project 9.2 Modify your finite element solver from Project 3.1 (or Project 3.2) to solve the following initial boundary value problem for the time-dependent advection–diffusion equation

$$\begin{aligned} \frac{\partial u}{\partial t} + \nabla \cdot (\boldsymbol{c} u) - k\Delta u &= 0 && \text{in } \Omega \times (0, T), \\ \nabla u \cdot \boldsymbol{n} &= 0 && \text{on } \Gamma_1 \times (0, T), \\ u &= g && \text{on } \Gamma_3 \times (0, T), \\ u(\cdot, 0) &= u_0 && \text{on } \Omega \times \{0\}, \end{aligned}$$

where $\Gamma_3 := \partial\Omega \setminus \Gamma_1$.

For the time discretization, use the one-step-theta method.

Test the implementation with $\Theta := 0, 1/2, 1$ for the following particular problem: $\Omega := (0, 1)^2$ with $\Gamma_3 = 1 \times [0, 1]$ and $T := 1$. The advection and diffusion coefficients are given by

$$\boldsymbol{c} := (-0.5, 0)^T \text{ and } k := 0.01\,.$$

The initial and boundary conditions are

$$u_0 := 0, \quad g(t) := \begin{cases} 5\sin(3\pi t), & 0 \le t \le 0.2, \\ 0, & \text{otherwise}\,. \end{cases}$$

Choose different grid sizes and visualize the results. In some cases, a smaller element size might require a finer time step. Change the values of $\boldsymbol{c}$ and k and observe the changes in the solution.

9.4 Stability

In Section 9.3, we have seen that at least if a basis of eigenvectors of the discretization matrix A_h allows for the solution representation (9.65) for the semidiscrete method, the qualitative behaviour of $e^{-\tau A_h}\boldsymbol{u}_0$ should be preserved by $E_{h,\tau}\boldsymbol{u}_0$, being one time step τ for homogeneous boundary conditions and right-hand side ($\boldsymbol{q}_h = 0$) in the semi- and fully discrete cases. It is sufficient to consider the eigenvector $\boldsymbol{w}_i$ instead of a general $\boldsymbol{u}_0$. Thus, we have to compare

$$\left(e^{-\tau A_h}\right) \boldsymbol{w}_i = (e^{-\lambda_i \tau}) \boldsymbol{w}_i \tag{9.103}$$

with

$$\left((I + \Theta\tau A_h)^{-1} \left(I - \overline{\Theta}\tau A_h\right)\right) \boldsymbol{w}_i = \left(\frac{1 - \overline{\Theta}\tau\lambda_i}{1 + \Theta\tau\lambda_i}\right) \boldsymbol{w}_i \,. \tag{9.104}$$

We see that the exponential function is approximated by

$$R(z) = \frac{1 + (1 - \Theta)z}{1 - \Theta z}, \tag{9.105}$$

the *stability function*, at the points $z = -\lambda_i \tau \in \mathbb{C}$, given by the eigenvalues λ_i, and the time step size τ.

For n time steps and $\boldsymbol{q}_h = 0$, we have

$$\left(e^{-\tau A_h}\right)^n \boldsymbol{w}_i = e^{-\lambda_i t_n} \boldsymbol{w}_i \sim R(-\lambda_i \tau)^n \boldsymbol{w}_i \,. \tag{9.106}$$

Thus, the restriction to eigenvectors $\boldsymbol{w}_i$ with eigenvalues λ_i has diagonalized the system of ordinary differential equations (9.48) for $\boldsymbol{q}_h = 0$ to the scalar problems

$$\begin{aligned} \xi' + \lambda_i \xi &= 0, \quad t \in (0, T), \\ \xi(0) &= \xi_0 \end{aligned} \tag{9.107}$$

(for $\xi_0 = 1$) with its solution $\xi(t) = e^{-\lambda_i t}\xi_0$, for which the one-step-theta method gives the approximation

$$\xi_{n+1} = R(-\lambda_i \tau)\xi_n = (R(-\lambda_i \tau))^{n+1} \xi_0 \tag{9.108}$$

at $t = t_{n+1}$. A basic requirement for a discretization method is the following:

Definition 9.26 A one-step method is called *nonexpansive* if for two numerical approximations $\boldsymbol{u}_h^n$ and $\tilde{\boldsymbol{u}}_h^n$, generated under the same conditions except for two discrete initial values $\boldsymbol{u}_0$ and $\tilde{\boldsymbol{u}}_0$, respectively, the following estimate is valid:

$$|\boldsymbol{u}_h^{n+1} - \tilde{\boldsymbol{u}}_h^{n+1}| \leq |\boldsymbol{u}_h^n - \tilde{\boldsymbol{u}}_h^n|, \quad n \in \{0, \ldots, N-1\}\,.$$

A recursive application of this estimate immediately results in

$$|\boldsymbol{u}_h^n - \tilde{\boldsymbol{u}}_h^n| \leq |\boldsymbol{u}_0 - \tilde{\boldsymbol{u}}_0|, \quad n \in \{1, \ldots, N\}\,.$$

Here, a general one-step method has the form

$$\boldsymbol{u}_h^{n+1} = \boldsymbol{u}_h^n + \tau \boldsymbol{\Phi}(\tau, t_n, \boldsymbol{u}_h^n), \quad n \in \{0, \ldots, N-1\},$$

with $\boldsymbol{u}_h^0 = \boldsymbol{u}_0$ and a so-called *generating function* $\boldsymbol{\Phi} : \mathbb{R}_+ \times [0, T) \times \mathbb{R}^M \to \mathbb{R}^M$ that characterizes the particular method. The generating function of the one-step-theta method applied to the system (9.48) is

$$\boldsymbol{\Phi}(\tau, t, \boldsymbol{\xi}) = -(I + \tau\Theta A_h)^{-1} \left[A_h \boldsymbol{\xi} - \boldsymbol{q}_h(t + \Theta\tau)\right] .$$

Thus, nonexpansiveness models the fact that perturbances, in particular errors, are not amplified in time by the numerical method. This is considerably weaker than the exponential decay in the continuous solution (see (9.23)), which would be too strong a request.

Having in mind (9.106)–(9.108), and expecting the (real parts of the) eigenvalues to be positive, the following restriction is sufficient:

Definition 9.27 A one-step method is called *A-stable* if its application to the scalar model problem (9.107)

$$\begin{aligned} \xi' + \lambda\xi &= 0, \quad t \in (0, T), \\ \xi(0) &= \xi_0, \end{aligned}$$

yields a nonexpansive method for all complex parameters λ with $\operatorname{Re}\lambda > 0$ and arbitrary step sizes $\tau > 0$.

Because of (9.108), we have

$$\xi_{n+1} - \tilde{\xi}_{n+1} = R(-\lambda\tau)[\xi_n - \tilde{\xi}_n]$$

for two approximations of the one-step-theta method applied to (9.107). This shows that the condition

$$|R(z)| \le 1 \quad \text{for all } z \text{ with } \operatorname{Re} z < 0$$

is sufficient for the A-stability of the method. More generally, any one-step method that can be written for (9.107) in the form

$$\xi_{n+1} = R(-\lambda_i\tau)\xi_n \tag{9.109}$$

is nonexpansive iff

$$|R(-\lambda_i\tau)| \le 1 . \tag{9.110}$$

The one-step-theta method is nonexpansive for (9.48) in the case of an eigenvector basis if (9.110) holds for all eigenvalues λ_i and step size τ. A convenient formulation can be achieved by the notion of the domain of stability.

Definition 9.28 Given a stability function $R : \mathbb{C} \to \mathbb{C}$, the set

$$S_R := \{z \in \mathbb{C} : |R(z)| < 1\}$$

is called a *domain of (absolute) stability* of the one-step method $\xi_{n+1} = R(-\lambda\tau)\xi_n$.

Example 9.29 *For the one-step-theta method, we have*

1) For $\Theta = 0$, S_R is the (open) unit disk with centre $z = -1$.

2) For $\Theta = \frac{1}{2}$, S_R coincides with the left complex half-plane (except for the imaginary axis).
3) For $\Theta = 1$, S_R is the whole complex plane except for the closed unit disk with centre $z = 1$.

The notion of A-stability reflects the fact that the property $|e^{-\lambda\tau}| \le 1$ for $\operatorname{Re}\lambda > 0$ is satisfied by the function $R(-\lambda\tau)$, too

Corollary 9.30 *For a continuous stability function R, the one-step method $\xi^{n+1} = R(-\lambda\tau)\xi^n$ is A-stable if the closure $\overline{S}_R$ of its domain of stability contains the left complex half-plane.*

Thus, the Crank–Nicolson and the implicit Euler methods are A-stable, but not the explicit Euler method. To have nonexpansiveness, we need the requirement

$$|1 - \lambda_i\tau| = |R(-\lambda_i\tau)| \le 1\,, \tag{9.111}$$

which is a *step size restriction:* For positive λ_i, it reads

$$\tau \le 2/\max\{\lambda_i \mid i = 1, \dots, M\}\,. \tag{9.112}$$

For the example of the five-point stencil discretization of the heat equation on a rectangle with Dirichlet boundary conditions according to (9.44)–(9.46), equation (9.111) reads

$$\left|1 - \frac{\tau}{h^2}2\left(2 - \cos\left(\nu\frac{\pi}{a}h\right) - \cos\left(\mu\frac{\pi}{b}h\right)\right)\right| \le 1 \tag{9.113}$$

for all $\nu = 1, \dots, l-1$, $\mu = 1, \dots, m-1$.

The following condition is sufficient (and for $l, m \to \infty$ also necessary):

$$\frac{\tau}{h^2} \le \frac{1}{4}\,. \tag{9.114}$$

For the finite element method, a similar estimate holds in a more general context. Under the assumptions of Theorem 3.48, we conclude from its proof (see (3.160)) that the following holds:

$$\max\{\lambda_i \mid i = 1, \dots, M\} \le C\left(\min_{K\in\mathcal{T}_h} h_K\right)^{-2}$$

for the eigenvalues of $B_h^{-1}\hat{A}_h$, where $B_h = E_h^T E_h$ is the mass matrix and $\hat{A}_h$ the stiffness matrix, and thus also for $A_h = E_h B_h^{-1}\hat{A}_h E_h^{-1}$. Here, C is a constant independent of h.

Therefore, we have

$$\tau\Big/\left(\min_{K\in\mathcal{T}_h} h_K\right)^2 \le 2/C \tag{9.115}$$

as a sufficient condition for the nonexpansiveness of the method with a specific constant depending on the stability constant of the bilinear form and the constant from Theorem 3.46, 2).

These step size restrictions impede the attractivity of the explicit Euler method, and so implicit versions are often used. But also in the A-stable case there are distinctions in the behaviour (of the stability functions). Comparing them, we see that

$$\begin{aligned} &\text{for } \Theta = \tfrac{1}{2} : R(-x) \to -1 \text{ for } x \to \infty ; \\ &\text{for } \Theta = 1 : R(-x) \to 0 \quad \text{for } x \to \infty . \end{aligned} \tag{9.116}$$

This means that for the implicit Euler method the influence of large eigenvalues will be more greatly damped, the larger they are, corresponding to the exponential function to be approximated, but the Crank–Nicolson method preserves these components nearly undamped in an oscillatory manner. This may lead to a problem for "rough" initial data or discontinuities between initial data and Dirichlet boundary conditions. On the other hand, the IE method also may damp solution components too strongly, making the solution "too" smooth.

These (and other) observations motivate the following concepts.

Definition 9.31 1) One-step methods whose stability function satisfies

$$\lim_{\operatorname{Re} z \to -\infty} R(z) = 0$$

are called *L-stable.*

2) If the stability function satisfies

- $|R(z)| < 1 \quad$ for all z with $\operatorname{Re} z < 0$,
- $\lim_{\operatorname{Re} z \to -\infty} |R(z)| < 1$,

the methods are called *strongly A-stable*.

Thus, strong A-stability takes an intermediate position.

Example 9.32 *1) Among the one-step-theta methods, only the implicit Euler method ($\Theta = 1$) is L-stable.*

2) The Crank–Nicolson method ($\Theta = \frac{1}{2}$) is not strongly A-stable, because of (9.116).

The nonexpansiveness of a one-step method can also be characterized by a norm condition for the solution operator $E_{h,\tau}$.

Theorem 9.33 *Let the spatial discretization matrix A_h have a basis of eigenvectors $\boldsymbol{w}_i$ orthogonal with respect to the scalar product $\langle \cdot, \cdot \rangle_h$, with eigenvalues λ_i, $i = 1, \ldots, M$. Consider the problem* (9.48) *and its discretization in time by a one-step method with a linear solution representation*

$$\boldsymbol{u}_h^n = E_{h,\tau}^n \boldsymbol{u}_0 \tag{9.117}$$

for $\boldsymbol{q}_h = 0$, where $E_{h,\tau} \in \mathbb{R}^{M,M}$, and a stability function R such that (9.109) *and*

$$E_{h,\tau} \boldsymbol{w}_i = R(-\lambda_i \tau) \boldsymbol{w}_i \tag{9.118}$$

for $i = 1, \ldots, M$. Then the following statements are equivalent:

(i) The one-step method is nonexpansive for the model problem (9.107) *and all eigenvalues λ_i of A_h.*

(ii) The one-step method is nonexpansive for the problem (9.48)*, with respect to the norm $\|\cdot\|_h$ induced by $\langle\cdot,\cdot\rangle_h$.*

(iii) $\|E_{h,\tau}\|_h \leq 1$ in the matrix norm $\|\cdot\|_h$ induced by the vector norm $\|\cdot\|_h$.

Proof We prove (i) $\Rightarrow$ (iii) $\Rightarrow$ (ii) $\Rightarrow$ (i):

(i) $\Rightarrow$ (iii): According to (9.110), the property (i) is characterized by

$$|R(-\lambda_i\tau)| \leq 1\,, \tag{9.119}$$

for the eigenvalues λ_i.

For the eigenvector $\boldsymbol{w}_i$ with eigenvalue λ_i, we have (9.118) and thus, for an arbitrary $\boldsymbol{u}_0 = \sum\limits_{i=1}^{M} c_i \boldsymbol{w}_i$,

$$\begin{aligned}\|E_{h,\tau}\boldsymbol{u}_0\|_h^2 &= \left\|\sum_{i=1}^{M} c_i E_{h,\tau}\boldsymbol{w}_i\right\|_h^2 \\ &= \left\|\sum_{i=1}^{M} c_i R(-\lambda_i\tau)\boldsymbol{w}_i\right\|_h^2 = \sum_{i=1}^{M} c_i^2 |R(-\lambda_i\tau)|^2\,\|\boldsymbol{w}_i\|_h^2\,,\end{aligned}$$

because of the orthogonality of the $\boldsymbol{w}_i$, and analogously,

$$\|\boldsymbol{u}_0\|_h^2 = \sum_{i=1}^{M} c_i^2 \|\boldsymbol{w}_i\|_h^2\,,$$

and finally, because of (9.119),

$$\|E_{h,\tau}\boldsymbol{u}_0\|_h^2 \leq \sum_{i=1}^{M} c_i^2 \|\boldsymbol{w}_i\|_h^2 = \|\boldsymbol{u}_0\|_h^2\,,$$

which is assertion (iii).

(iii) $\Rightarrow$ (ii): is obvious.

(ii) $\Rightarrow$ (i):

$$|R(-\lambda_i\tau)|\,\|\boldsymbol{w}_i\|_h = \|R(-\lambda_i\tau)\boldsymbol{w}_i\|_h = \|E_{h,\tau}\boldsymbol{w}_i\|_h \leq \|\boldsymbol{w}_i\|_h\,.$$

□

Thus, nonexpansiveness is often identical to what is (vaguely) called *stability*:

Definition 9.34 A one-step method with a solution representation $E_{h,\tau}$ for $q_h = 0$ is called *stable* with respect to the vector norm $\|\cdot\|_h$ if

$$\|E_{h,\tau}\|_h \leq 1$$

in the induced matrix norm $\|\cdot\|_h$.

Till now, we have considered only homogeneous boundary data and right-hand sides. At least for the one-step-theta method this is not a restriction:

Theorem 9.35 *Consider the one-step-theta method under the assumption of Theorem 9.33, with* $\operatorname{Re}\lambda_i \geq 0,\ i = 1, \ldots, M$, *and with* τ *such that the method is stable. Then the solution is stable in initial condition* $\boldsymbol{u}_0$ *and right-hand side* $\boldsymbol{q}_h$ *in the following sense:*

$$\|\boldsymbol{u}_h^n\|_h \leq \|\boldsymbol{u}_0\|_h + \tau \sum_{k=1}^{n} \left\| \boldsymbol{q}_h \left(t_k - \overline{\Theta}\tau \right) \right\|_h . \tag{9.120}$$

Proof From the solution representation (9.96), we conclude that

$$\|\boldsymbol{u}_h^n\|_h \leq \|E_{h,\tau}\|_h^n \|\boldsymbol{u}_0\|_h + \tau \sum_{k=1}^{n} \|E_{h,\tau}\|_h^{n-k} \|(I + \tau\Theta A_h)^{-1}\|_h \|\boldsymbol{q}_h(t_k - \overline{\Theta}\tau)\|_h \tag{9.121}$$

using the submultiplicativity of the matrix norm.

We have the estimate

$$\|(I + \Theta\tau A_h)^{-1} \boldsymbol{w}_i\|_h = \left| \frac{1}{1 + \Theta\, \tau\lambda_i} \right| \|\boldsymbol{w}_i\|_h \ \leq\ \|\boldsymbol{w}_i\|_h ,$$

and thus as in the proof of Theorem 9.33, (i) $\Rightarrow$ (iii),

$$\|(I + \Theta\, \tau A_h)^{-1}\|_h \ \leq\ 1,$$

which concludes the proof. □

The stability condition requires step size restrictions for $\Theta < \frac{1}{2}$, which have been discussed above for $\Theta = 0$.

The requirement of stability can be weakened to

$$\|E_{h,\tau}\|_h \leq 1 + K\tau \tag{9.122}$$

for some constant $K > 0$, which in the situation of Theorem 9.33 is equivalent to

$$|R(-\lambda\tau)| \leq 1 + K\tau,$$

for all eigenvalues λ of A_h. Because of

$$(1 + K\tau)^n \leq e^{Kn\tau},$$

in (9.120), the additional factor e^{KT} appears and correspondingly $e^{K(n-k)\tau}$ in the sum. If the process is to be considered only in a small time interval, this becomes part of the constant, but for large time horizons, the estimate becomes inconclusive.

On the other hand, for the one-step-theta method for $\frac{1}{2} < \Theta \leq 1$ the estimate $\|E_{h,\tau}\|_h \leq 1$ and thus the constants in (9.120) can be sharpened to $\|E_{h,\tau}\|_h \leq R(-\lambda_{\min}\tau)$, where $\lambda_{\min}$ is the smallest eigenvalue of A_h, reflecting the exponential decay. For example, for $\Theta = 1$, the (error in the) initial data is damped with the factor

$$\|E_{h,\tau}\|_h^n = R(-\lambda_{\min}\tau)^n = \frac{1}{(1+\lambda_{\min}\tau)^n},$$

which for $\tau \leq \tau_0$ for some fixed $\tau_0 > 0$ can be estimated by

$$e^{-\lambda n\tau} \quad \text{for some } \lambda > 0.$$

We expect that the degree of approximation of e^z by $R(z)$ is reflected in the order of approximation in time. We will show that this is true, if the scheme is sufficiently stable or the initial data are sufficiently smooth or time step restrictions hold.

To do so we assume that, for some $q \in \mathbb{N}$, we have

$$R(z) - e^z = O(z^{q+1}) \quad \text{for } z \in \mathbb{C},\ |z| \to 0, \tag{9.123}$$

in particular $q = 1$ for $\Theta \neq \frac{1}{2}$ and $q = 2$ for $\Theta = \frac{1}{2}$ for the one-step-theta method. In addition to the assumption of Theorem 9.33 concerning the basis of eigenvectors $\boldsymbol{w}_i$ corresponding to the eigenvalues λ_i, let $\|\boldsymbol{w}_i\|_h = 1$ and $\operatorname{Re}\lambda_i > 0$ for $i = 1, \ldots, M$ in the following.

We want to estimate

$$\boldsymbol{u}_h(t) - \boldsymbol{u}_h^n = \sum_{i=1}^{M} c_i \left(e^{-\lambda_i n\tau} - R(-\lambda_i\tau)^n\right) \boldsymbol{w}_i, \tag{9.124}$$

where $\boldsymbol{u}_{0h} = \sum_{i=1}^{M} c_i \boldsymbol{w}_i$, but independent of M (and also the time horizon T). For this, either stability or a step size constraint is necessary.

The smoothness of the initial data can be measured in the parameter-dependent weighted norm

$$\|\boldsymbol{v}\|_{h,s} := \left\langle A_h^s \boldsymbol{v}, \boldsymbol{v}\right\rangle_h^{\frac{1}{2}} = \|A_h^{\frac{s}{2}}\boldsymbol{v}\|_h = \left(\sum_{i=1}^{M} \lambda_i^s \langle \boldsymbol{v}, \boldsymbol{w}_i\rangle_0\right)^{\frac{1}{2}}, \quad s \geq 0, \tag{9.125}$$

i.e. $\|\cdot\|_{h,0} = \|\cdot\|_h$.

This might be considered as the approximation of a $H^s(\Omega)$-type norm as the eigenvector/eigenvalues $\boldsymbol{w}_i$, λ_i can be interpreted as approximations of the eigenfunctions/eigenvalues of the elliptic boundary value problem (as investigated in a special case in Section 5.1.2, etc.). The following result shows that the consistency order q is preserved by stability and sufficient smoothness of the initial value.

Theorem 9.36 *Assume that the spectrum of the spatial discretization matrix A_h has the property that there exists numbers $\delta > 0$, $\alpha \in \left[0, \frac{\pi}{2}\right)$ such that it holds:*

$$|\lambda_i| \le \delta \Longrightarrow \lambda_i \in \left\{z = -re^{i\varphi} \in \mathbb{C} \;\middle|\; r > 0,\ |\varphi| \le \alpha\right\}. \tag{9.126}$$

Consider a one-step method satisfying (9.117), (9.118), *which is accurate of order q (according to* (9.123)) *and A-stable. Then there is a constant $C = C(q) > 0$ independent of M, T, τ and h such that*

$$\|\boldsymbol{u}_h^n - \boldsymbol{u}_h(t)_n\|_h \le C\tau^q \|\boldsymbol{u}_{0h}\|_{h,2q} \quad \textit{for all } n = 1, \ldots, N.$$

Proof According to (9.124), the assertion is equivalent to

$$\|F_n(-\tau A_h)\boldsymbol{u}_{0h}\|_h \le C\tau^q \|\boldsymbol{u}_{0h}\|_{h,2q},$$

where

$$F_n(B) := e^{nB} - R(B)^n.$$

This again is equivalent to

$$\|F_n(-\tau A_h)\|_{2q,0} \le C\tau^q,$$

where $\|B\|_{2q,0}$ indicates the operator norm of $B : (\mathbb{R}^M, \|\cdot\|_{h,2q}) \to (\mathbb{R}^M, \|\cdot\|_h)$ on $\mathbb{R}^{M,M}$.

It is easy to see that for $B \in \mathbb{R}^{M,M}$ commuting with A_h, as A_h is an isomorphism, we have

$$\|B\|_{2q,0} = \|A_h^{-q} B\|_{0,0}. \tag{9.127}$$

Therefore, as $F_n(-\tau A_h)$ is commuting with A_h, it is sufficient to prove

$$\|(\tau A_h)^{-q} F_n(-\tau A_h)\|_{0,0} \le C,$$

which is equivalent to (see Theorem 9.33)

$$|(\tau\lambda_i)^{-q} F_n(-\tau\lambda_i)| \le C \quad \text{or} \quad |F_n(-\tau\lambda_i)| \le C\left(\tau|\lambda_i|\right)^q.$$

In the following, $C > 0$ denotes a generic constant independent of M, T, τ and h. Because of (9.123), there are numbers $b_0 > 0$, $C > 0$ such that

$$|R(z) - e^z| \le C\,|z|^{q+1} \quad \text{for } |z| \le b_0,$$

and thus, for a possibly smaller $b_0 > 0$ and some $0 < \alpha < 1$:

$$|R(z)| \le e^{\alpha \operatorname{Re} z} \quad \text{for } |z| \le b_0,\ \operatorname{Re} z \le 0.$$

Therefore, for $z_i := \lambda_i \tau$ and $i \in \{1, \ldots, M\}$ such that $|z_i| \le b_0$, $\operatorname{Re} z_i \ge 0$:

$$
\begin{aligned}
|F_n(-z_i)| &= |R(-z_i)^n - (e^{-z_i})^n| = |R(-z_i) - e^{-z_i}| \left| \sum_{j=0}^{n-1} R(-z_i)^{n-1-j} e^{-jz_i} \right| \\
&\leq C\,|z_i|^{q+1} \sum_{j=0}^{n-1} e^{-\alpha(n-1)\operatorname{Re} z_i} e^{-j(1-\alpha)\operatorname{Re} z_i} \leq C\,|z_i|^{q+1} n e^{-\alpha(n-1)\operatorname{Re} z_i}.
\end{aligned}
\tag{9.128}
$$

Since the function $x \mapsto xe^{-ax+b}$ with $a, b \geq 0$ is bounded for $x \geq 0$, we obtain, taking into consideration (9.126),

$$
|F_n(-z_i)| \leq C\,|z_i|^q.
$$

Note that no stability property has been used so far. This property is now essential in that range of z_i which has not yet been considered. Namely, for z_i satisfying $|z_i| > b_0$, $\operatorname{Re} z_i \geq 0$, the stability implies the estimate

$$
\begin{aligned}
|F_n(-z_i)| &\leq |R(-z_i)^n| + |e^{-nz_i}| \\
&\leq 1 + e^{-n\operatorname{Re} z_i} \leq 2 \leq Cb_0^q \leq C\,|z_i|^q.
\end{aligned}
$$

□

Theorem 9.37 *Consider a one-step method satisfying* (9.117), (9.118), *which is accurate of order q (according to* (9.123)*) and strongly A-stable. Then there is a constant $C = C(q) > 0$ independent of M, T, τ and h such that*

$$
\|\boldsymbol{u}_h^n - \boldsymbol{u}_h(t_n)\|_h \leq C\left(\frac{\tau}{t_n}\right)^q \|\boldsymbol{u}_{0h}\|_h \quad \textit{for all } n = 1, \ldots, N.
$$

Proof In the notation of Theorem 9.36, we have to show

$$
\|F_n(-\lambda_i A_h)\|_{0,0} \leq C\left(\frac{\tau}{t_n}\right)^q = Cn^{-q},
$$

being equivalent to

$$
|F_n(-\lambda_i \tau)| \leq Cn^{-q} \quad \text{for all } i = 1, \ldots, M.
$$

For $|z_i| \leq b_0$, $\operatorname{Re} z_i \geq 0$, we modify the estimate (9.128) to

$$
|F_n(-z_i)| \leq Cn^{-q} \left(n|z_i|\right)^{q+1} e^{-\alpha(n-1)\operatorname{Re} z_i} \leq Cn^{-q}.
$$

For $\operatorname{Re} z_i \geq b_0$, we have $\sup_{\operatorname{Re} z \geq b_0} |R(z)| =: k < 1$, due to the strong A-stability.

Setting $c := -\ln k > 0$, we therefore have for $\operatorname{Re} z \geq b_0$:

$$
\begin{aligned}
|R(z)|^n &\leq k^n = e^{-cn} \leq Cn^{-q}, \\
|e^{-z}|^n &= e^{-\operatorname{Re} zn} \leq e^{-b_0 n} \leq Cn^{-q},
\end{aligned}
$$

and we can conclude the proof as above. □

Remark 9.38 Note that if the step size restriction

$$|\tau\lambda_i| \le b_0, \ \operatorname{Re}\lambda_i \ge 0$$

(see (9.128)) holds, no stability assumption is necessary.

Thus, for the Crank–Nicolson scheme, either a step size restriction is required or no error damping in time is to be expected.

In fact, there is a simple remedy to make also the Crank–Nicolson method error smoothing:

Theorem 9.39 *Consider the Crank–Nicolson method, started with* two *steps of the implicit Euler method, then*

$$\|\boldsymbol{u}_h - \boldsymbol{u}_h(t_n)\|_h \le C\left(\frac{\tau}{t_n}\right)^2 \|\boldsymbol{u}_{0h}\|_h \quad \textit{for } n = 1, \ldots, N.$$

Proof See [59, Thm. 7.4]. □

Remark 9.40 We also could have reversed the reasoning by applying Rothe's method: Then the above considerations have to be proven for the discretization in time compared to $u(t)$, and then the additional spatial error has to be considered (see [59, Ch. 7]).

We conclude this section with an example.

Example 9.41 (Prothero-Robinson model) *Let* $g \in C^1[0, T]$ *be given. We consider the initial value problem*

$$\begin{aligned} \xi' + \lambda(\xi - g) &= g', \quad t \in (0, T), \\ \xi(0) &= \xi_0 . \end{aligned}$$

Obviously, g *is a particular solution of the differential equation, so the general solution is*

$$\xi(t) = e^{-\lambda t}[\xi_0 - g(0)] + g(t) .$$

In the special case $g(t) := \arctan t$, $\lambda := 500$, *and for the indicated values of* ξ_0, *Figure 9.1 shows the qualitative behaviour of the solution.*

It is worth mentioning that the figure is extremely scaled: The continuous line (to $\xi_0 = 0$*) seems to be straight, but it is the graph of* g.

The explicit Euler method for this model is

$$\xi^{n+1} = (1 - \lambda\tau)\xi^n + \tau\,[g'(t_n) + \lambda g(t_n)] \ .$$

According to the above considerations, it is nonexpansive only if $\lambda\tau \le 1$ *holds. For large numbers* λ, *this is a very restrictive step size condition; see also the discussion of* (9.112) *to* (9.114).

Due to their better stability properties, implicit methods such as the Crank–Nicolson and the implicit Euler methods do not have such step size restrictions. Nevertheless, the application of implicit methods is not free from surprises. For example, in the case of large numbers λ, *an order reduction can occur.*

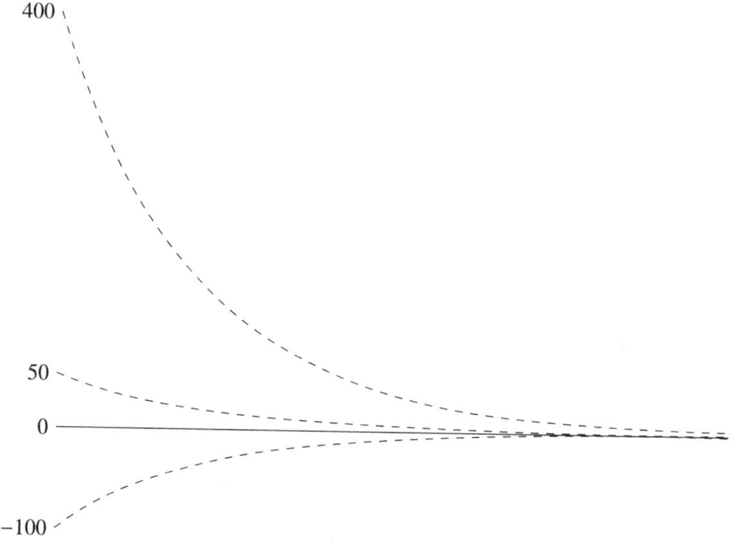

Fig. 9.1: Solutions of the Prothero–Robinson model for different initial values.

Exercises

Problem 9.21 Determine the corresponding domain of stability S_R of the one-step-theta method for the following values of the parameter Θ : $0, \frac{1}{2}, 1$.

Problem 9.22 Show the L-stability of the implicit Euler method.

Problem 9.23 a) Show that the discretization

$$\xi^n = \xi^{n-2} + 2\tau f(t_{n-1}, \xi^{n-1}), \quad n = 2, \dots N \quad (\textit{midpoint rule}),$$

applied to the model equation $\xi' = f(t, \xi)$ with $f(t, \xi) = -\lambda\xi$ and $\lambda > 0$ leads, for a sufficiently small step size $\tau > 0$, to a general solution that can be additively decomposed into a decaying and an increasing (by absolute value) oscillating component.

b) Show that the oscillating component can be damped if additionally the quantity ξ_*^N is computed:

$$\xi_*^N = \frac{1}{2}\left[\xi^N + \xi^{N-1} + \tau f(t_N, \xi^N)\right] \quad (\textit{modified midpoint rule}).$$

Problem 9.24 Let $m \in \mathbb{N}$ be given. Find a polynomial $R_m(z) = 1 + z + \sum_{j=2}^{m} \gamma_j z^j$ ($\gamma_j \in \mathbb{R}$) such that the corresponding domain of absolute stability for $R(z) := R_m(z)$ contains an interval of the negative real axis that is as large as possible.

9.5 High-Order One-Step and Multistep Methods

9.5.1 One-Step Methods

The implicit Euler and the Crank–Nicolson methods do not only have different stability behaviour but also different order of approximation (see also Section 9.7), as can be seen from their stability functions

$$R_{IE}(z) = \frac{1}{1-z}, \quad R_{CN}(z) = \frac{1+\frac{1}{2}z}{1-\frac{1}{2}z}.$$

Both are approximations to e^z in the sense of (9.123) with $q = 1$ for R_{IE} and $q = 2$ for R_{CN}. To reach a higher order scheme, one approach can consist of choosing a function R satisfying (9.123) with $q \geq 2$ such that the corresponding scheme (for the homogeneous case)

$$\boldsymbol{u}_h^{n+1} = E_{h,\tau}\boldsymbol{u}_h^n, \quad E_{h,\tau} := R(-\tau A_h) \tag{9.129}$$

can be evaluated with an effort that is in reasonable relation to the gain of accuracy. Furthermore, the scheme has to be stable, which in general means A-stable.

If R is a rational function, i.e.

$$R_{jk}(z) = \frac{P_{jk}(z)}{Q_{jk}(z)}, \quad P_{jk} \in \mathcal{P}_k(\mathbb{C}),\ Q_{jk} \in \mathcal{P}_j(\mathbb{C}),\ j,k \in \mathbb{N}_0, \tag{9.130}$$

then the one-step method reads

$$Q_{jk}(-\tau A_h)\boldsymbol{u}_h^{n+1} = P_{jk}(-\tau A_h)\boldsymbol{u}_h^n, \tag{9.131}$$

i.e. a matrix polynomial has to be multiplied with a vector, a corresponding matrix has to be set up and a system of linear equations has to be solved.

This seems to be critical, as besides the effort also sparsity patterns at least get widened by an increase of band width and fill-in (Problem 9.35). If a decomposition of Q_{jk} is given in the form

$$Q_{jk}(-\tau A_h) = \prod_{i=1}^{j}(\lambda_i I + \tau A_h)$$

(note that all matrix polynomials with the same argument commute), then (9.131) is equivalent to a sequence of systems of linear equations, i.e. a *fractional-step method* emerges. One possible variant is

$$\begin{aligned}(\lambda_1 I + \tau A_h)\boldsymbol{u}_h^{n+1/j} &= P_{jk}(-\tau A_h)\boldsymbol{u}_h^n,\\ (\lambda_i I + \tau A_h)\boldsymbol{u}_h^{n+i/j} &= \boldsymbol{u}_h^{n+(i-1)/j} \qquad \text{for } i = 2,\dots,j.\end{aligned}$$

If additionally $P_{jk}(-\tau A_h) = \prod_{i=1}^{k}(\mu_i I + \tau A)$, then also the number of multiplications at the right-hand side can be reduced, i.e. for $j = k$:

$$(\lambda_i I + \tau A_h)\boldsymbol{u}_h^{n+i/k} = (\mu_i I + \tau A_h)\boldsymbol{u}_h^{n+(i-1)/k} \quad \text{for } i = 1, \ldots, k. \tag{9.132}$$

Now k systems of linear equations of dimension M of the known "implicit Euler" type have to be solved, but the setup of the matrices in (9.132) becomes negligible

Rational stability functions which provide optimal approximations satisfy the condition

$$q = j + k$$

in (9.123) and these *Padé-approximations* are uniquely existing and explicitly known. The first ones are depicted in Fig. 9.2.

$j \setminus k$	0	1	2
0	1	$1 + z$	$1 + z + \frac{1}{2}z^2$
1	$\frac{1}{1 - z}$	$\frac{1 + \frac{1}{2}z}{1 - \frac{1}{2}z}$	$\frac{1 + \frac{2}{3}z + \frac{1}{6}z^2}{1 - \frac{1}{3}z}$
2	$\frac{1}{1 - z + \frac{1}{2}z^2}$	$\frac{1 + \frac{1}{3}z}{1 + \frac{2}{3}z + \frac{1}{6}z^2}$	$\frac{1 + \frac{1}{2}z + \frac{1}{12}z^2}{1 - \frac{1}{2}z + \frac{1}{12}z^2}$
3	$\frac{1}{1 - z + \frac{1}{2}z^2 - \frac{1}{6}z^3}$	$\frac{1 + \frac{1}{4}z}{1 - \frac{3}{4}z + \frac{1}{4}z^2 - \frac{1}{24}z^3}$	

Fig. 9.2: The Padé tableau for e^z.

We rediscover the implicit Euler method as the pair (1,0), the Crank–Nicolson method as (1,1) and the explicit Euler scheme as (0,1). To achieve A-stability, we have to choose an implicit scheme ($j \geq k$). In particular, the following result holds.

Theorem 9.42 *The Padé approximations from Figure 9.2 for e^z have the properties:*

1) For $j = k$, $j = k + 1$, $j = k + 2$:

$$|R_{jk}(z)| < 1 \quad \textit{for } \operatorname{Re} z < 0.$$

2) For $j = k$,

$$|R_{kk}(z)| \to 1 \quad \textit{for } |z| \to \infty,\ \operatorname{Re} z < 0.$$

3) For $j = k + 1,\ j = k + 2$,

$$|R_{jk}(z)| \to 0 \quad \textit{for } |z| \to \infty,\ \operatorname{Re} z < 0.$$

Proof See [36, Sect. IV.4]. □

So we get A-stable methods from the diagonal and the two subdiagonals of the Padé tableau, but the diagonal variants show the same lack of damping as the Crank–Nicolson method, and the subdiagonal ones are L-stable and have the same overdamping as the implicit Euler scheme. Additionally some of the $Q_{jk}(-\tau A_h)$ (for $j \geq 2$) have not only real zeros, such that the form (9.131) would require complex arithmetics.

Alternatively one could choose a rational ansatz for R composed of real linear factors with the aim to give up parts of the possible accuracy for improved stability, like being strongly A-stable. To this aim, we make the ansatz

$$R_\Theta(z) := \frac{(1 + \alpha\Theta' z)(1 + \overline{\alpha}\Theta z)^2}{(1 - \alpha\Theta z)^2(1 - \overline{\alpha}\Theta' z)}, \tag{9.133}$$

where $\Theta := 1 - \frac{1}{2}\sqrt{2} \approx 0.29$ is fixed, $\Theta' := 1 - 2\Theta \approx 0.42$ and $\alpha \in \left(\frac{1}{2}, 2\right]$ to be chosen with $\overline{\alpha} := 1 - \alpha$. Then the general representation

$$\begin{aligned} R_\Theta(z) = 1 + z + \frac{1}{2}z^2\big(1 - (2\alpha - 1)(2\Theta^2 - 4\Theta + 1)\big) \\ + \frac{1}{6}r(\Theta, \alpha)z^3 + 0(|z|^4), \end{aligned}$$

shows that this choice results in the order $q = 2$ ("only" instead of the optimal order $q = 6$). However, because of

$$|R_\Theta(z)| < 1 \quad \text{for } \operatorname{Re} z < 0, \qquad \lim_{\substack{\operatorname{Re} z > 0 \\ |z| \to \infty}} |R_\Theta(z)| = \frac{\overline{\alpha}}{\alpha} < 1 \tag{9.134}$$

the method is strongly A-stable, but not L-stable, if $\alpha > \frac{1}{2}$. Furthermore, an expanded Taylor expansion shows

$$|r(\Theta, \alpha)| \leq \frac{1}{2},$$

and this is to be compared to the corresponding factor of the Crank-Nicolson method

$$r_{CN} = \frac{1}{12},$$

which is the best possible for all second-order schemes.

According to (9.132), the scheme can be written as a fractional-step scheme, the *fractional-step-theta method* (not to be mixed up with the one-step-theta method of (9.48)). We can combine the factors in such a way that each "substep" is a one-step-theta method, with the parameters α for step size $\Theta\tau$ and $\overline{\alpha}$ for step size $\Theta'\tau$. This allows a natural extension to inhomogeneous problems:

$$
\begin{aligned}
(I + \alpha\Theta\tau A_h)\boldsymbol{u}_h^{n+\frac{1}{3}} &= (I - \overline{\alpha}\Theta\tau A_h)\boldsymbol{u}_h^n + \Theta\tau\boldsymbol{\ell}_h^{n+\alpha\Theta}, \\
(I + \overline{\alpha}\Theta'\tau A_h)\boldsymbol{u}_h^{n+\frac{2}{3}} &= (I - \alpha\Theta'\tau)\boldsymbol{u}_h^{n+\frac{1}{3}} + \Theta'\tau\boldsymbol{\ell}_h^{n+\alpha\Theta+\overline{\alpha}\Theta'}, \\
(I + \alpha\Theta\tau A_h)\boldsymbol{u}_h^{n+1} &= (I - \overline{\alpha}\Theta\tau A_h)\boldsymbol{u}_h^{n+\frac{2}{3}} + \Theta\tau\boldsymbol{\ell}_h^{n+2\alpha\Theta+\overline{\alpha}\Theta'} .
\end{aligned}
\tag{9.135}
$$

Note that in [50, (5.1.26)–(5.1.28)] another evaluation of the right hand side is proposed, namely at $n, n+1-\Theta, n+1-\Theta$.

A general way to get a high-order discretization is the (classical) *Runge–Kutta* approach. As a reminder, for a system of ordinary differential equations

$$
\frac{d}{dt}\boldsymbol{u}(t) = \boldsymbol{f}(t, \boldsymbol{u}(t)), \quad \boldsymbol{u}(0) = \boldsymbol{u}_0 \tag{9.136}
$$

w.r.t. a function $\boldsymbol{u} : [0, T] \to \mathbb{R}^M$, an *(implicit) Runge–Kutta scheme* with l stages is defined by a *Butcher tableau* (see Fig. 9.3) via the solution of the system of equations in lM variables $\boldsymbol{k} = (\boldsymbol{k}_1^T, \ldots, \boldsymbol{k}_l^T)^T$:

$$
\begin{array}{c|ccc}
\alpha_1 & \beta_{11} & \cdots & \beta_{1l} \\
\vdots & \vdots & & \vdots \\
\alpha_l & \beta_{l1} & \cdots & \beta_{ll} \\
\hline
 & \gamma_1 & \cdots & \gamma_l
\end{array}
\qquad \text{or in short} \qquad
\begin{array}{c|c}
\alpha & B \\
\hline
 & \gamma
\end{array}
$$

Fig. 9.3: A Butcher tableau.

$$
\boldsymbol{k}_i = \boldsymbol{f}\left(t_n + \alpha_i\tau, \boldsymbol{u}^n + \tau\sum_{\nu=1}^{l}\beta_{i\nu}\,\boldsymbol{k}_\nu\right), \quad i = 1, \ldots, l,
$$

and then one step is

$$
\boldsymbol{u}^{n+1} := \boldsymbol{u}^n + \tau\sum_{i=1}^{l}\gamma_i\boldsymbol{k}_i .
$$

In particular, for a linear, homogeneous time-independent system, where $f(t, \boldsymbol{u}) = -A_h\boldsymbol{u}$, we have the representation

$$
\boldsymbol{k} = -I \otimes A_h\boldsymbol{u}^n - B \otimes A_h\boldsymbol{k}, \tag{9.137}
$$

where

$$A \otimes B = \begin{pmatrix} a_{11}B & \dots & a_{1n}B \\ a_{n1}B & \dots & a_{mn}B \end{pmatrix} \in \mathbb{R}^{mp,nr}$$

denotes the *Kronecker product* for matrices $A \in \mathbb{R}^{m,n}$, $B \in \mathbb{R}^{p,r}$ (see Appendix A.3).

Here, a (consistency) order of $2l$ is possible, and the stability function is rational with degrees $j, k \leq l$ (see (9.130)). The diagonal entries of the Padé tableau can be rediscovered as *Gauss–Legendre methods* derived by a collocation ansatz with the (transformed) zeros of the Legendre polynomials as collocation points.

The original formulation above is not a fractional-step method, as the stage vectors $\boldsymbol{k}_i$ are rather intermediate derivatives. The change of variables

$$z_i := \boldsymbol{u}^n + \tau \sum_{j=1}^{l} \beta_{ij} \boldsymbol{k}_j$$

leads to the system to be solved

$$z_i = \boldsymbol{u}^n + \tau \sum_{j=1}^{l} \beta_{ij} \boldsymbol{f}\left(t_n + \alpha_j \tau, z_j\right).$$

In particular, in the linear case the system

$$z = \mathbf{1} \otimes \boldsymbol{u}^n + \tau B \otimes A_h z \tag{9.138}$$

with $\mathbf{1} := (1, \dots, 1)^T$ has to be solved. In this way, we obtain the scheme

$$\boldsymbol{u}^{n+1} = \boldsymbol{u}^n + \tau \sum_{i=1}^{l} \gamma_i \boldsymbol{f}(t_n + \alpha_i \tau, z_i)$$

in the general case, and, for invertible B in the linear case,

$$\boldsymbol{u}^{n+1} = \boldsymbol{u}^n + (\gamma^T B^{-1} \otimes I)(z - \mathbf{1} \otimes \boldsymbol{u}^n),$$

avoiding further evaluations of $\boldsymbol{f}$, i.e. in the linear case multiplication with A_h. So to perform a "fractional-step" method, a system of linear equations of dimension Ml has to be solved, but with the special structure (9.137)/(9.138).

An alternative approach for the numerical solution of ordinary differential equations are linear multistep methods.

9.5.2 Linear Multistep Methods

A *linear multistep method* with l steps (in short: *linear l-step method*) for (9.136) is given by

$$\frac{1}{\tau}\sum_{k=0}^{l} a_k \boldsymbol{u}^{n+k} = \sum_{k=0}^{l} b_k \boldsymbol{f}(t_{n+k}, \boldsymbol{u}^{n+k}),$$

i.e. for the linear, time-independent homogeneous case

$$\frac{1}{\tau}\sum_{k=0}^{l} a_k \boldsymbol{u}^{n+1} + \sum_{k=0}^{l} b_k A_h \boldsymbol{u}^{n+k} = 0,$$

which only requires the solution of systems of linear equation of implicit Euler type (with $a_l I + b_l \tau A_h$ as system matrix) to compute $\boldsymbol{u}^{n+l}$ from the known $\boldsymbol{u}^{n+l-1}, \ldots, \boldsymbol{u}^n$, such that an additional method is necessary to provide $\boldsymbol{u}^1, \ldots, \boldsymbol{u}^{l-1}$. Linear multistep methods can be either achieved by numerical quadrature leading to *Adams–Bashforth–* or *Nyström-* methods (explicit methods) or to *Adams–Moulton–* or *Milne–Simpson–*methods (implicit methods). Alternatively, numerical differentiation can be used leading to *backward-differentiation formulas* (BDF). To investigate the A-stability for these methods all the zeros $y_i(z)$ of the *stability polynomial*

$$\chi(y; z) := \sum_{k=0}^{l}(a_k - z b_k) y^k, \quad y, z \in \mathbb{C},$$

have to be considered, leading to the requirement

$$|y_i(z)| \leq 1, \quad i = 1, \ldots, l,$$

which is called the *root condition*. This condition together with the consistency property ensures convergence, but contrary to one-step methods we need the sharpened condition

$$|y_i(z)| < 1, \quad i = 1, \ldots, l,$$

to guarantee the proper qualitative behaviour. For $l = 1$, we rediscover the condition $|R(z)| < 1$. Therefore, the Definitions 9.27, 9.30, etc., have to be modified by substituting $R(z)$ by the zeros $y_i(z)$.

Although BDF formulas are very attractive in terms of effort, the following *Dahlquist barrier* is a big disappointment.

Theorem 9.43 *A linear A-stable multistep method is implicit and has at most order of convergence two. From these, the Crank–Nicolson method has the smallest error constant.*

Proof [36, Ch. V, Thm. 1.3] □

However, the concept of A-stability can be weakened to obtain a wider range of methods.

Definition 9.44 1) A method is called *A(α)–stable*, $\alpha \in \left[0, \frac{\pi}{2}\right]$, if the (unbounded) sector

$$\left\{z = -re^{i\varphi} \in \mathbb{C} \;\middle|\; r > 0,\ |\varphi| \leq \alpha\right\}$$

belongs to the domain of stability.

2) A method is called *stiffly stable*, if there exist parameters $a, b, d > 0$ such that the sets

$$R_1 := \{z \in \mathbb{C} \mid \operatorname{Re} z \le -d\},$$
$$R_2 := \{z \in \mathbb{C} \mid -d \le \operatorname{Re} z \le -a, \ |\operatorname{Im} z| \le b\}$$

both belong to the domain of stability.

In this sense, the BDF formulas

BDF2: $$\boldsymbol{u}^{n+2} - \frac{4}{3}\boldsymbol{u}^{n+1} + \frac{1}{3}\boldsymbol{u}_n + \frac{2\tau}{3}A_h\boldsymbol{u}^{n+2} = 0,$$

BDF3: $$\boldsymbol{u}^{n+3} - \frac{18}{11}\boldsymbol{u}^{n+2} + \frac{9}{11}\boldsymbol{u}^{n+1} - \frac{2}{11}\boldsymbol{u}^n + \frac{6\tau}{11}A_h\boldsymbol{u}^{n+3} = 0,$$

etc.

up to the order $l \le 6$ are a good choice, as they have full convergence order l and satisfy the conditions of Definition 9.44, i.e. they both are A(α)- and stiffly stable, as can be seen from Fig. 9.4 and Fig. 9.5.

l	1	2	3	4	5	6
α (in degree)	90	90	≈ 86	≈ 73	≈ 51	≈ 17

Fig. 9.4: A(α)-stability of BDF methods [36, Ch. V, (2.7)].

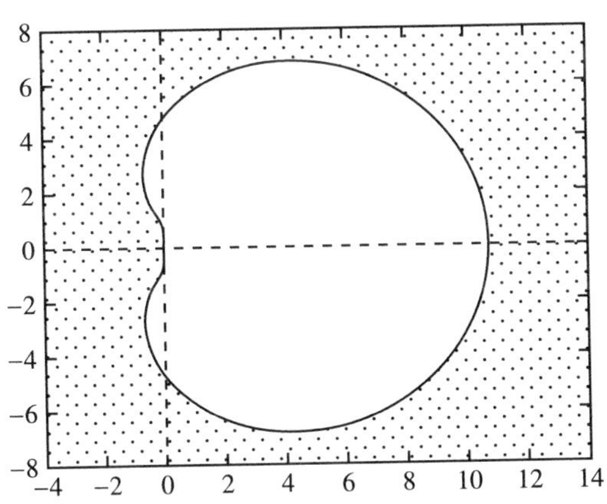

Fig. 9.5: Stability region for the BDF4 method, $d \approx 0.667$.

9.5.3 Discontinuous Galerkin Method (DGM) in Time

Instead of using a finite difference scheme in time, we can also achieve a time discretization by means of a Galerkin approach. We want to end up with a *time marching scheme*, i.e. preserve the essential property that the solution at $t = \bar{t} \le T$ only depends on the solution for $t \le \bar{t}$. In the definition (9.37), we could weaken the notion of solution to $u' \in L^2((0,T), V')$, where

$$\left\langle \frac{d}{dt}u(t), v \right\rangle_{V',V}$$

is to be understood as the application of a functional, but even then at most $u \in C([0,T], H)$ holds true, i.e. a conforming Galerkin approach is generally not possible. Therefore, we will develop a nonconforming approach by replacing (9.33) by a modified integrated form. So for a given time stepping $t_n = n\tau$, $n = 0, \ldots, N$, we set $I_n := (t_n, t_{n+1}]$, $n = 0, \ldots, N-1$, and consider the local variational formulations

$$\int_{I_n} \left[\langle u'(t), v(t) \rangle + a(u(t), v(t)) \right] dt + \left\langle [\![u]\!]_n, v_n^+ \right\rangle_0 = \int_{I_n} \ell(t; v(t))\, dt \qquad (9.139)$$

for all $n = 0, \ldots, N-1$, where $v_n^- := \lim_{t \to t_n^-} v(t)$, $v_n^+ := \lim_{t \to t_n^+} v(t)$ are the one-sided limits and

$$[\![v_n]\!] := v_n^+ - v_n^-$$

the jump of v at $t = t_n$, with $u_0^- := u_0$ for the solution. Here, $u, v \in L^2((0,T), V)$, $u', v' \in L^2(I_n, H)$ for $n = 0, \ldots, N-1$ such that also discontinuous elements u, v are allowed.

This defines the underlying solution space, which is also taken as the test space. The jump term in (9.139) may be considered as s stability term (compare after (7.97)). Let

$$P_{I_n} := \mathcal{P}_k(I_n, V) := \{v : I_n \to V \mid v(t) = \sum_{j=0}^{k} w_j t^j \quad \text{with } w_j \in V\},$$
$$V_\tau := \{v \in L^2((0,T), V) \mid v|_{I_n} \in \mathcal{P}_k(I_n, V),\ n = 0, \ldots, N-1\},$$

where the degrees of freedom Σ_{I_n} can be chosen quite arbitrarily.

The *(time-)discontinuous Galerkin method, dG(k)* for short, consists in the (conforming) Galerkin approach in V_τ for the fomulation (9.139). For $k = 0$, a variant (concerning the treatment of the right-hand side) of the implicit Euler method is recovered and for $k = 1$ the (2,1) Padé approximation of e^z (Problem 9.26).

For the following, we follow closely [97]. The choice of basis and degrees of freedom is quite arbitrary. One can take advantage of this by selecting any basis $\{\varphi_{nj}\}_{j=1}^{k+1}$ of $\mathcal{P}_k(I_n, \mathbb{R})$. Then the solution $u \in V_\tau$ on I_n is given by

$$u(t) = \sum_{j=1}^{k+1} u_n^j \, \varphi_{nj}(t), \; t \in I_n ,$$

with $u_n^j \in V$ to be determined, and elements $v \in V_\tau$ with

$$v(t) = w \, \varphi_{ni}(t), \; w \in V, \; i = 1, \ldots, k+1,$$

are sufficient as general test functions. Furthermore, the appearing integrals in time can be evaluated exactly by choosing a quadrature rule

$$J_n(f) = \sum_{q=1}^{N} \omega_{nq} \, f(t_{nq}) \tag{9.140}$$

on I_n which is exact for $f \in \mathcal{P}_{2k}(I_n, \mathbb{R})$. In this way, we get the following (stationary) variational equations

$$\begin{aligned} &\sum_{j=1}^{k+1} \left[\gamma_{ij} \left\langle u_n^j, w \right\rangle_0 + \alpha_{ij} a(u_n^j, w) \right] \\ &\quad = \left\langle u_{n-1}^-, w \right\rangle_0 \varphi_{ni}(t_{n-1}) + J_n(\ell(\cdot; w)\, \varphi_{ni}(t)) =: b_i(w) \\ &\qquad\qquad\qquad\qquad \text{for all } w \in V, \; i = 1, \ldots, k+1, \end{aligned} \tag{9.141}$$

where

$$\begin{aligned} \gamma_{ij} &:= J_n(\varphi_{nj}' \varphi_{ni}) + \varphi_{nj}(t_{n-1}) \varphi_{ni}(t_{n-1}), \\ \alpha_{ij} &:= J_n(\varphi_{nj} \varphi_{ni}). \end{aligned}$$

By transforming to the reference element $\hat{I} := (0, 1]$, we get

$$\gamma_{ij} = \hat{\gamma}_{ij}, \; \alpha_{ij} = \tau \hat{\alpha}_{ij} ,$$

i.e. these coefficients only have to be built once (compare Section 2.2).

A quadrature of the required accuracy is delivered by *Gauss-formulas* (see [56, Sect. 3.6]). If we use Gauss quadrature in $k+1$ nodes, there is one degree of freedom which can be fixed to achieve the required accuracy for elements from $\mathcal{P}_{2k}(\mathbb{R})$. Hence, selecting the corresponding nodal basis function to these quadrature points for the basis, the degrees of freedom are obtained. Selecting a $(k+1)$-point right-sided *Gauss–Radau-formula* (with $t_{n,k+1} = t_{n+1}$) finally gives $u(t_{n+1}) = u_n^{k+1}$. Now the "elliptic" problems (9.141) can be approximated by a conforming Galerkin method in space given by $V_h \subset V$, leading to the ansatz space

$$V_{\tau h} := \{ v \in L^2((0,T), V) \; \big| \; V|_{I_n} \in \mathcal{P}_k(I_n, V_h), \; n = 0, \ldots, N-1 \}$$

and to a system of linear equation for the restriction of the approximate solution $u_{\tau h}$ to I_n in terms of

$$\boldsymbol{u}_{nh} = \left(u_h^1, \ldots, u_h^{k+1}\right)^T \in \mathbb{R}^{(k+1)M} : \qquad (G_h + B_{\tau h})\boldsymbol{u}_{nh} = \boldsymbol{\beta}_h, \tag{9.142}$$

where

$$\begin{aligned} M &:= \dim V_h \quad \text{and} \quad G_h := \hat{\gamma} \otimes I_h, \quad B_{\tau h} := \tau\hat{\alpha} \otimes A_h, \\ &\quad \text{with } \hat{\gamma} := (\hat{\gamma}_{ij}),\ \hat{\alpha} := \hat{\alpha}_{ij} \in \mathbb{R}^{k+1,k+1}, \\ I_h &:= (\langle \varphi_j, \varphi_i \rangle)_{ij} \in \mathbb{R}^{M,M}, \quad A_h := (a(\varphi_j, \varphi_i))_{ij} \in \mathbb{R}^{M,M}, \quad \boldsymbol{\beta}_h = (b_i(\varphi_j)), \end{aligned}$$

and again $\boldsymbol{\beta}_h$ written as a vector in $\mathbb{R}^{(k+1)M}$ for a basis $\varphi_1, \ldots, \varphi_M$ of V_h. Thus, we have again arrived at a system of linear equations of the same structure as (9.137)/(9.138).

Exercises

Problem 9.25 Determine the sector of stability

$$\left\{z = -re^{i\varphi} \in \mathbb{C} \;\middle|\; r > 0,\ |z| \leq \delta,\ |\varphi| \leq \alpha\right\} \subset S_R$$

of the one-step-theta method for the parameter $\Theta \in \left[0, \frac{1}{2}\right]$ and for the methods discussed in the above section.

Problem 9.26 Consider the dG(k) method for (9.139) for $k = 0$ and $k = 1$ in detail and establish a connection to the Padé approximations of e^z.

Problem 9.27 To construct time discretizations for ODEs, classical DG methods (see Section 7.4) can be used. To this end, consider the pure advection equation $u' + ru = f$ in one space dimension in the interval $(0, T)$ and with Dirichlet boundary conditions at $t = 0$.

a) Show that the application of the dG(0) method with full upwinding to the above ODE is equivalent to the implicit Euler scheme. Which quadrature rule has to be used?
b) Show that the dG(k) method is L-stable.
Hint: Rewrite the linear test problem in matrix form.
c) Let u be the weak solution of the ODE $u' + ru = f$ on $[0, T]$ with $u(0) = u_0$. Let u_h be the result of applying the the dG(k) method to the ODE. Show the following error estimates:

$$\|u - u_h\|_0 \leq C\tau^{k+1}|u|_{k+1}, \qquad |u(T) - u_h(T)| \leq C\tau^{k+1}|u|_{k+1},$$

where $\tau > 0$ is the time step size.
Hint: Use the approaches of Section 7.4.

Problem 9.28 Consider the ordinary differential equation

$$\frac{d}{dt}\boldsymbol{u}(t) = (A_1 + A_2)\boldsymbol{u}(t)$$
$$\boldsymbol{u}(0) = \boldsymbol{u}_0,$$

which may be semidiscrete version of e.g. (9.37), with A_1 being the diffusion part and A_2 the convection part. Assume for simplicity $A_1 = \alpha A$, $A_2 = \overline{\alpha} A$, A symmetric, positive definite, and consider the (splitting) variant of the fractional-step-theta method (9.135)

$$(I + \Theta\tau A_1)\boldsymbol{u}_h^{n+\frac{1}{3}} = (I - \Theta\tau A_2)\boldsymbol{u}^n$$
$$(I + \Theta'\tau A_2)\boldsymbol{u}_h^{n+\frac{2}{3}} = (I - \Theta'\tau A_1)\boldsymbol{u}^{n+\frac{1}{3}}$$
$$(I + \Theta\tau A_1)\boldsymbol{u}_h^{n} = (I - \Theta\tau A_2)\boldsymbol{u}^{n+\frac{2}{3}}$$

Derive the stability function analogous to (9.133).

Programming project 9.3 Extend the code from Project 9.2 to solve the initial boundary value problem by means of the fractional-step-theta method (9.135) and linear multistep methods in time: BDF1, BDF2, and BDF3. (Remind that BDF1 is the implicit Euler method, which is already implemented in the code from Project 9.2). Test the implemantation for the example from Project 9.2.

9.6 Exponential Integrators

In this section, we return to the system of ordinary differential equations (see (9.136))

$$\frac{d}{dt}\boldsymbol{u} = \boldsymbol{f}(t, \boldsymbol{u}), \quad \boldsymbol{u}(0) = \boldsymbol{u}_0 \tag{9.143}$$

w.r.t. a function $\boldsymbol{u} : [0, T] \to \mathbb{R}^M$, and describe the concept of *exponential integrators*. Roughly speaking, the (unavoidable) approximation step in the development of practically applicable exponential integrators starts at a different (later) point in comparison to the conventional methods. This leads to remarkable property of exponential integrators that they are exact for linear constant coefficient problems, provided the occurring exponential terms are computed exactly. In consequence, since the above-discussed stability properties are based on such model problems, (exact) exponential integrators possess those stability properties. Of course, the unavoidable approximations destroy this ideal situation, but the only limits result from the computation of the exponential terms. Practical applications nevertheless show a number of advantages in terms of accuracy, stability, and efficiency. For instance, performance results of exponential integrators for advancing stiff, semidiscrete formulations of the unsaturated Richards equation in time can be found in [114], and for the simulation of geothermal processes in heterogeneous porous media in [222].

In general, there are different ways to create exponential integrators. One approach consists in the linearization of the right-hand side in the neighbourhood of $(t_n, \boldsymbol{u}^n)$:

$$\boldsymbol{f}(t, \boldsymbol{u}(t)) = \boldsymbol{f}(t_n, \boldsymbol{u}^n) + B^n(\boldsymbol{u}(t) - \boldsymbol{u}^n) + \boldsymbol{b}^n(t - t_n) + \boldsymbol{r}^n(t, \boldsymbol{u}(t)),$$

where $\boldsymbol{b}^n \in \mathbb{R}^M$ is a vector (typically an approximation to the first column of the Jacobi matrix of $\boldsymbol{f}$ at (t_n, U^n)), $B^n \in \mathbb{R}^{M,M}$ is a matrix (typically an approximation to the last M column of the Jacobi matrix of $\boldsymbol{f}$ at (t_n, U^n)), and $\boldsymbol{r}^n$ is the remainder. Such an approach can be successful in situations where $\boldsymbol{f}$ does not have a special structure, in particular where the spectral properties of B^n are not known in sufficient detail or no appropriate decomposition of B^n is available (as in the case below).

In other cases, for example, in the discretization of semilinear parabolic equations, the right-hand side already has a special structure, for instance,

$$f(t, \boldsymbol{u}) = -A_h(t)\boldsymbol{u} + \boldsymbol{q}_h(t, \boldsymbol{u}), \tag{9.144}$$

where $A_h : (0,T] \to \mathbb{R}^{M,M}$ and $\boldsymbol{q}_h : (0,T] \times \mathbb{R}^M \to \mathbb{R}^M$, or, in the autonomous case,

$$f(t, \boldsymbol{u}) = -A_h\boldsymbol{u} + \boldsymbol{q}_h(\boldsymbol{u}) \tag{9.145}$$

with $A_h \in \mathbb{R}^{M,M}$ and $\boldsymbol{q}_h : \mathbb{R}^M \to \mathbb{R}^M$.

The design of Runge–Kutta methods and quadrature-type multistep methods is based on the well-known representation of the exact solution $\boldsymbol{u}$ of (9.143) at $t = t_{n+l}$ with $l = 1$ for Runge–Kutta methods and $l \geq 2$ for multistep methods:

$$\boldsymbol{u}(t_{n+l}) = \boldsymbol{u}(t_n) + \int_{t_n}^{t_{n+l}} f(s, \boldsymbol{u}(s))ds\,.$$

In what follows we will focus on the case $l = 1$ and (9.145), i.e. on Runge–Kutta methods for the differential equation

$$\frac{d}{dt}\boldsymbol{u} + A_h\boldsymbol{u} = \boldsymbol{q}_h(t, \boldsymbol{u}),$$

and refer to the literature for more general situations.

Multiplying the equation by the integrating factor e^{tA_h} and integrating the result from t_n to t_{n+1}, we obtain the *variation-of-constants formula*:

$$\boldsymbol{u}(t_{n+1}) = e^{-\tau A_h}\boldsymbol{u}(t_n) + \int_{t_n}^{t_{n+1}} e^{-(t_{n+1}-s)A_h}\boldsymbol{q}_h(s, \boldsymbol{u}(s))ds\,. \tag{9.146}$$

If $\boldsymbol{q}_h$ is not present, a numerical method can be obtained by replacing the matrix exponential $e^{-\tau A_h}$ by a suitable (rational) approximation as described above (see e.g. (9.106)).

The evaluation of $e^{-\tau A_h}\boldsymbol{v}$ for some $\boldsymbol{v} \in \mathbb{R}^M$ is nothing else than the solving the initial-value problem

$$\frac{d}{dt}\boldsymbol{u} + A_h\boldsymbol{u} = 0 \quad \boldsymbol{u}(0) = \boldsymbol{v}.$$

Therefore, if this evaluation is done exactly, the method is exact for linear systems of ODEs with constant coefficients and thus trivially A-stable. However, if this were done using a standard numerical method, no real benefit would be obtained. The point is that there is only the need to evaluate the action of $e^{-\tau A_h}$ on a vector $\boldsymbol{v}$, and this can be done efficiently (but with a certain error, see below).

If $\boldsymbol{q}_h$ is present, the integral has to be evaluated or approximated. Now it is well known from the theory of quadrature rules (for instance, in the integration of rapidly oscillating functions [56, Sect. 3.7]) that it can be advantageous to exploit a particular multiplicative structure of the integrand in such a way that only one factor is approximated, say by a polynomial, and then the resulting (complete) integrand is such that it can be evaluated exactly, for instance, by integration by parts.

$$\frac{1}{\tau}\int_{t_n}^{t_{n+1}} e^{-(t_{n+1}-s)A_h}\boldsymbol{q}_h(s,\boldsymbol{u}(s))ds \approx \sum_{i=1}^{l}\gamma_i(-\tau A_h)\boldsymbol{q}_h(t_n+\alpha_i\tau,\boldsymbol{u}^{ni}),$$

where $\boldsymbol{u}^{ni} :\approx \boldsymbol{u}(t_n + \alpha_i\tau))$.

To determine the internal stages $\boldsymbol{u}^{ni}$, the formula (9.146) is approximated on the interval $[t_n, t_n + \alpha_i\tau]$, i.e. t_{n+1} is replaced by $t_n + \alpha_i\tau$, $i = 1, \ldots, l$, and the resulting integrals are approximated by quadrature formulas.

In summary, the following *one-step exponential Runge–Kutta method* is obtained:

$$\begin{aligned}
\boldsymbol{u}^{n+1} &:= e^{-\tau A_h}\boldsymbol{u}^n + \tau\sum_{i=1}^{l}\gamma_i(-\tau A_h)\boldsymbol{q}_h(t_n+\alpha_i\tau,\boldsymbol{u}^{ni}),\\
\boldsymbol{u}^{ni} &= e^{-\alpha_i\tau A_h}\boldsymbol{u}^n + \tau\sum_{\nu=1}^{l}\beta_{i\nu}(-\tau A_h)\boldsymbol{q}_h(t_n+\alpha_\nu\tau,\boldsymbol{u}^{n\nu}).
\end{aligned} \tag{9.147}$$

Similar to conventional Runge–Kutta methods, here a nonlinear system w.r.t. the internal stages $\boldsymbol{u}^{ni}$ (i.e. in lM Variables) is to be solved. In principle, this could be done a by fixed-point iteration.

However, due to their good stability properties, the restriction to explicit methods is sufficient in many cases. Their Butcher tableau is shown in Fig. 9.6.

The limit $A_h \to 0$ (in $\mathbb{R}^{M,M}$) in Fig. 9.6 generates the standard explicit Runge–Kutta scheme for the equation $\frac{d}{dt}\boldsymbol{u} = \boldsymbol{q}_h(t,\boldsymbol{u})$ (cf. Problem 9.30).

Example 9.45 *We consider the simplest case $l = 1$, where a constant polynomial approximation at $t = t_n$ (i.e. $\alpha_1 = 0$) is used:*

$$
\begin{array}{c|ccccc}
\alpha_1 & 0 & 0 & \cdots & & 0 \\
\alpha_2 & \beta_{21}(-\tau A_h) & 0 & \cdots & & 0 \\
\vdots & \vdots & \ddots & & & \vdots \\
\alpha_l & \beta_{l1}(-\tau A_h) & \cdots & \beta_{l,l-1}(-\tau A_h) & & 0 \\
\hline
 & \gamma_1(-\tau A_h) & \cdots & \gamma_{l-1}(-\tau A_h) & \gamma_l(-\tau A_h) &
\end{array}
$$

Fig. 9.6: The Butcher tableau of the explicit variants of (9.147).

$$
\begin{aligned}
\frac{1}{\tau}\int_{t_n}^{t_{n+1}} e^{-(t_{n+1}-s)A_h}\boldsymbol{q}_h(s,\boldsymbol{u}(s))ds &\approx \frac{1}{\tau}\int_{t_n}^{t_{n+1}} e^{-(t_{n+1}-s)A_h}\boldsymbol{q}_h(t_n,\boldsymbol{u}(t_n))ds \\
&= (e^{-\tau A_h}-I)(-\tau A_h)^{-1}\boldsymbol{q}_h(t_n,\boldsymbol{u}(t_n)).
\end{aligned}
$$

Then, with

$$
\varphi_1(z) := \frac{e^z-1}{z} \qquad (=: \gamma_1(z)),
$$

it follows:

$$
\begin{aligned}
\boldsymbol{u}^{n+1} &:= e^{-\tau A_h}\boldsymbol{u}^n + \tau\varphi_1(-\tau A_h)\boldsymbol{q}_h(t_n,\boldsymbol{u}^n) \\
&= \boldsymbol{u}^n + \tau\varphi_1(-\tau A_h)(\boldsymbol{q}_h(t_n,\boldsymbol{u}^n) - A_h\boldsymbol{u}^n) \\
&= \boldsymbol{u}^n + \tau\varphi_1(-\tau A_h)\boldsymbol{f}(t_n,\boldsymbol{u}^n).
\end{aligned}
$$

The last formula shows that it is sufficient for the feasibility of the method to be able to compute or at least to approximate terms of the type $\varphi_1(-\tau A_h)\boldsymbol{v}$.

A general way to derive higher order methods is to use the Lagrange interpolation of $\boldsymbol{q}_h$. To underline the main ideas, we first consider the linear case, i.e. $\boldsymbol{q}_h = \boldsymbol{q}_h(t)$. Then

$$
\boldsymbol{q}_h(s) \approx \sum_{i=1}^{l} \boldsymbol{q}_h(t_n+\alpha_i\tau)p_i(s),
$$

where

$$
p_i(s) := \prod_{\substack{j=1\\ j\neq i}}^{l} \frac{s-t_n-\alpha_j\tau}{(\alpha_i-\alpha_j)\tau} \underset{s=t_n+\alpha\tau}{=} \prod_{\substack{j=1\\ j\neq i}}^{l} \frac{\alpha-\alpha_j}{\alpha_i-\alpha_j}
$$

are the standard univariate Lagrange basis polynomials (cf. (3.71)).

Hence,

$$
\gamma_i(-\tau A_h) := \frac{1}{\tau}\int_{t_n}^{t_{n+1}} e^{-(t_{n+1}-s)A_h}p_i(s)ds. \tag{9.148}
$$

In order to analyse the scheme (9.147) with this or other choices of γ_i, we expand the right-hand side of (9.146) into a Taylor polynomial plus integral remainder:

$$\begin{aligned}\boldsymbol{q}_h(s) &= \sum_{j=0}^{l-1} \frac{1}{j!}\, \boldsymbol{q}_h^{(j)}(t_n)(s-t_n)^j + \frac{1}{(l-1)!}\int_{t_n}^{s} \boldsymbol{q}_h^{(l)}(r)(s-r)^{l-1}\,dr \\ &= \sum_{j=1}^{l} \frac{1}{(j-1)!}\, \boldsymbol{q}_h^{(j-1)}(t_n)(s-t_n)^{j-1} + \frac{1}{(l-1)!}\int_{t_n}^{s} \boldsymbol{q}_h^{(l)}(r)(s-r)^{l-1}\,dr.\end{aligned} \tag{9.149}$$

Hence,

$$\int_{t_n}^{t_{n+1}} e^{-(t_{n+1}-s)A_h}\boldsymbol{q}_h(s)ds = I_1 + I_2 \tag{9.150}$$

with

$$\begin{aligned}I_1 &:= \sum_{j=1}^{l} \frac{1}{(j-1)!}\, \boldsymbol{q}_h^{(j-1)}(t_n)\int_{t_n}^{t_{n+1}} e^{-(t_{n+1}-s)A_h}(s-t_n)^{j-1}ds \\ &= \tau\sum_{j=1}^{l} \boldsymbol{q}_h^{(j-1)}(t_n)\tau^{j-1}\varphi_j(-\tau A_h),\end{aligned}$$

where, using the substitution $s =: t_n + \alpha\tau$,

$$\varphi_j(z) := \frac{1}{\tau(j-1)!}\int_{t_n}^{t_{n+1}} e^{(t_{n+1}-s)z/\tau}\left(\frac{s-t_n}{\tau}\right)^{j-1} ds = \frac{1}{(j-1)!}\int_0^1 e^{(1-\alpha)z}\alpha^{j-1}d\alpha. \tag{9.151}$$

An analogous substitution of r allows to rewrite the second integral as follows:

$$\begin{aligned}I_2 &:= \frac{1}{(l-1)!}\int_{t_n}^{t_{n+1}} e^{-(t_{n+1}-s)A_h}\int_{t_n}^{s} \boldsymbol{q}_h^{(l)}(r)(s-r)^{l-1}\,dr\,ds \\ &= \frac{\tau^{l+1}}{(l-1)!}\int_0^1 e^{-(1-\alpha)\tau A_h}\int_0^{\alpha} \boldsymbol{q}_h^{(l)}(t_n+\xi\tau)(\alpha-\xi)^{l-1}\,d\xi\, d\alpha.\end{aligned}$$

Then it follows from (9.146), (9.147) that

$$\begin{aligned}\boldsymbol{u}^{n+1} - \boldsymbol{u}(t_{n+1}) &= e^{-\tau A_h}\boldsymbol{u}^n - e^{-\tau A_h}\boldsymbol{u}(t_n) \\ &\quad + \tau\sum_{i=1}^{l}\gamma_i(-\tau A_h)\boldsymbol{q}_h(t_n+\alpha_i\tau) - \int_{t_n}^{t_{n+1}} e^{-(t_{n+1}-s)A_h}\boldsymbol{q}_h(s)ds.\end{aligned}$$

In this way, we get an equation for the discrete time evolution of the error $\boldsymbol{e}^n := \boldsymbol{u}^n - \boldsymbol{u}(t_n)$:

$$\boldsymbol{e}^{n+1} = e^{-\tau A_h}\boldsymbol{e}^n + \boldsymbol{\varepsilon}_l^n. \tag{9.152}$$

In order to express the first term in $\boldsymbol{\varepsilon}_l^n$, we use the representation (9.149) again:

$$\begin{aligned}
&\boldsymbol{q}_h(t_n + \alpha_i \tau) \\
&= \sum_{j=1}^{l} \frac{1}{(j-1)!} \boldsymbol{q}_h^{(j-1)}(t_n)(\alpha_i \tau)^{j-1} + \frac{1}{(l-1)!} \int_{t_n}^{t_n+\alpha_i\tau} \boldsymbol{q}_h^{(l)}(r)(t_n + \alpha_i \tau - r)^{l-1} dr \\
&= \sum_{j=1}^{l} \frac{1}{(j-1)!} \boldsymbol{q}_h^{(j-1)}(t_n)(\alpha_i \tau)^{j-1} + \frac{\tau^l}{(l-1)!} \int_0^{\alpha_i} \boldsymbol{q}_h^{(l)}(t_n + \alpha\tau)(\alpha_i - \alpha)^{l-1} d\alpha.
\end{aligned}$$

Then, by (9.150),

$$\begin{aligned}
\boldsymbol{\varepsilon}_l^n &= \tau \sum_{i=1}^{l} \gamma_i(-\tau A_h) \sum_{j=1}^{l} \frac{1}{(j-1)!} \boldsymbol{q}_h^{(j-1)}(t_n)(\alpha_i\tau)^{j-1} - \tau \sum_{j=1}^{l} \boldsymbol{q}_h^{(j-1)}(t_n)\tau^{j-1}\varphi_j(-\tau A_h) \\
&\quad + \tau \sum_{i=1}^{l} \gamma_i(-\tau A_h) \frac{\tau^l}{(l-1)!} \int_0^{\alpha_i} \boldsymbol{q}_h^{(l)}(t_n + \alpha\tau)(\alpha_i - \alpha)^{l-1} d\alpha \\
&\quad - \frac{\tau^{l+1}}{(l-1)!} \int_0^1 e^{-(1-\alpha)\tau A_h} \int_0^{\alpha} \boldsymbol{q}_h^{(l)}(t_n + \xi\tau)(\alpha - \xi)^{l-1} d\xi d\alpha.
\end{aligned} \tag{9.153}$$

The first difference allows the representation

$$\sum_{j=1}^{l} \left[\sum_{i=1}^{l} \gamma_i(-\tau A_h) \frac{\alpha_i^{j-1}}{(j-1)!} - \varphi_j(-\tau A_h) \right] \tau^j \boldsymbol{q}_h^{(j-1)}(t_n).$$

If the terms in the square brackets are successively set to zero, we get corresponding order conditions. Note that this statement is valid for the general form (9.147) but not only for the quadrature base on Lagrange interpolation.

The next theorem is essentially based on an uniform estimate of the spectral norm of the matrix exponential of the form

$$\|e^{-tA_h}\| \le C_{A_h} e^{\omega t} \quad \text{for all } t \ge 0, \tag{9.154}$$

where $C_{A_h} \ge 1$ and $\omega \in \mathbb{R}$ are constants independent of h. In this way, we get error estimates which do not depend on h (or, equally, on M).

For an example, we can go back to the special situation considered in Section 9.2, where the matrix T is formed by orthonormal eigenvectors of a symmetric, uniformly (w.r.t. h) positive definite matrix A_h. Then, since T is an orthogonal matrix, the relation (9.66) shows that $\|e^{-tA_h}\| \le e^{\omega t}$ with $-\lambda_i \le \omega < 0$ for all eigenvalues λ_i of A_h. Since the matrix is uniformly (w.r.t. h) positive definite, ω does not depend on h.

Theorem 9.46 *Assume that the matrix A_h satisfies condition (9.154). Let $\boldsymbol{q}_h \in C^p([0,T], R^M)$ for some $p \in \mathbb{N}$, $p \le l$. In addition, let the parameters of method (9.147) have the following properties:*

1) The weights $\gamma_i(-\tau A_h)$ are uniformly bounded for $h \geq 0$, that is $\|\gamma_i(-\tau A_h)\| \leq C_w$ for some constant $C_w > 0$ independent of τ and h.

2) $\displaystyle\sum_{i=1}^{l} \gamma_i(-\tau A_h)\frac{\alpha_i^{j-1}}{(j-1)!} - \varphi_j(-\tau A_h) = 0$ for $j = 1, \ldots p$.

Then there exists a constant $C = C(T) > 0$ independent of τ such that

$$\|\boldsymbol{u}^n - \boldsymbol{u}(t_n)\| \leq C\tau^p \int_0^{t_n} \|\boldsymbol{q}_h^{(l)}(s)\| ds$$

holds uniformly for all $t_n \in [0, T]$.

Proof The recursive error equation (9.152) shows that

$$\begin{aligned}
\boldsymbol{e}^n &= e^{-n\tau A_h}\boldsymbol{e}^0 + \sum_{j=0}^{n-1} e^{-(n-(j+1))\tau A_h}\boldsymbol{\varepsilon}_l^j \\
&= e^{-t_n A_h}\underbrace{\boldsymbol{e}^0}_{=0} + \sum_{j=0}^{n-1} e^{-t_{n-(j+1)}A_h}\boldsymbol{\varepsilon}_l^j.
\end{aligned}$$

Hence,

$$\|\boldsymbol{e}^n\| \leq \sum_{j=0}^{n-1} \|e^{-t_{n-(j+1)}A_h}\|\|\boldsymbol{\varepsilon}_l^j\| \leq C_{A_h}\sum_{j=0}^{n-1} e^{\omega t_{n-(j+1)}}\|\boldsymbol{\varepsilon}_l^j\|.$$

If ω is nonpositive, we can estimate the terms $e^{\omega t_{n-(j+1)}}$ by one, otherwise by $e^{\omega T}$. In both cases, we get

$$\|\boldsymbol{e}^n\| \leq \tilde{C}\sum_{j=0}^{n-1} \|\boldsymbol{\varepsilon}_l^j\|$$

with a constant $\tilde{C} > 0$ possibly dependent on T.

Now it can be seen from (9.153) with $l = p$, $n = j$ that

$$\begin{aligned}
\boldsymbol{\varepsilon}_p^j = \tau\sum_{i=1}^{p}\gamma_i(-\tau A_h)\frac{\tau^p}{(p-1)!}\int_0^{\alpha_i} \boldsymbol{q}_h^{(p)}(t_j + \alpha\tau)(\alpha_i - \alpha)^{p-1}d\alpha \\
- \frac{\tau^{p+1}}{(p-1)!}\int_0^1 e^{-(1-\alpha)\tau A_h}\int_0^{\alpha} \boldsymbol{q}_h^{(p)}(t_j + \xi\tau)(\alpha - \xi)^{p-1}d\xi d\alpha.
\end{aligned}$$

Since

$$\begin{aligned}
\left\|\int_0^{\alpha} \boldsymbol{q}_h^{(p)}(t_j + \xi\tau)(\alpha - \xi)^{p-1}d\xi\right\| &\leq \int_0^{\alpha} \|\boldsymbol{q}_h^{(p)}(t_j + \xi\tau)\|(\alpha - \xi)^{p-1}d\xi \\
&\leq \int_0^1 \|\boldsymbol{q}_h^{(p)}(t_j + \xi\tau)\|d\xi \\
&\leq \frac{1}{\tau}\int_{t_j}^{t_{j+1}} \|\boldsymbol{q}_h^{(p)}(s)\|ds,
\end{aligned}$$

the stated bound follows from (9.154) and the assumptions 1),2) on the weights. □

Corollary 9.47 *Let the condition* (9.154) *be satisfied. Then the exponential quadrature rule with the weights* (9.148) *satisfies the order conditions up to order l, thus it is convergent of order l.*

Proof Since the quadrature rule is exact for polynomials of maximal degree $l-1$, the weights γ_i of the exponential quadrature rule (9.147) satisfy the order conditions 2) of the above theorem for $p = l$. The boundedness of the weights follows from the fact that the set of order conditions 2) can be interpreted as a system of linear equations w.r.t. $\gamma_i(z)$ with the right-hand sides $(j-1)!\varphi_j(z)$. The matrix of this system is a transposed Vandermonde matrix, and it is well known that it is a regular matrix. Hence every weight $\gamma_i(-\tau A_h)$ is a linear combination of the l terms $(j-1)!\varphi_j(-\tau A_h)$, where all the coefficients can be bounded by a constant which depends only on l and the relative step sizes α_i.

By the definition (9.151) of φ_j and condition (9.154), we see that

$$\begin{aligned}\|\varphi_j(-\tau A_h)\| &\leq \frac{1}{(j-1)!}\int_0^1 \|e^{-(1-\alpha)\tau A_h}\|\alpha^{j-1}d\alpha \\ &\leq \frac{C_{A_h}}{(j-1)!}\int_0^1 e^{\omega(1-\alpha)\tau}\alpha^{j-1}d\alpha \\ &\leq \frac{C_{A_h}}{(j-1)!}\int_0^1 e^{\omega(1-\alpha)\tau}d\alpha = \frac{C_{A_h}}{(j-1)!}\frac{e^{\omega\tau}-1}{\omega\tau}.\end{aligned}$$

Since we can assume that τ is sufficiently small in the sense that $\omega\tau \leq C$, for some constant $C > 0$, we finally obtain

$$\|\varphi_j(-\tau A_h)\| \leq \frac{C_{A_h}}{(j-1)!}\frac{e^C-1}{C}.$$

□

The analysis of the general scheme (9.147) is much more involved than in the linear case; its presentation would exceed the purpose of this book. In this respect, we refer, for example, to the overview papers [189], [149], on which our presentation of this topic is strongly based. Therefore, we move on to the still open question of the computation of the terms $e^{-\tau A_h}\boldsymbol{v}$ or $\varphi_j(-\tau A_h)\boldsymbol{v}$.

In the case where the matrix A_h is large and sparse, these matrix–vector products can be approximated efficiently by Krylov subspace methods. So let us consider a function $\Phi : \mathbb{C} \to \mathbb{C}$ with the particular choices $\Phi(z) = e^z$ or $\Phi(z) = \varphi_j(z)$ in mind.

In Section 5.4, we got to know the Arnoldi algorithm which generates recursively an orthonormal basis $\boldsymbol{v}_1, \ldots, \boldsymbol{v}_k$ of the Krylov subspace $K_k(A; \boldsymbol{g}^{(0)})$ for a matrix $A \in \mathbb{R}^{M,M}$ and a vector $\boldsymbol{g}^{(0)} \in \mathbb{R}^M$.

It is not difficult to transfer many of those results from the real case $\mathbb{R}^M$ to the complex case $\mathbb{C}^M$. This extension is required by the fact that, based on a function-theoretic approach, the matrix function $\Phi(B)$ of a matrix $B \in \mathbb{C}^{M,M}$ can be defined via the so-called *Cauchy integral formula*

$$\Phi(B) := \frac{1}{2\pi} \int_\Gamma \Phi(\lambda)(\lambda I - B)^{-1} d\lambda, \tag{9.155}$$

where Γ is a closed contour in the complex plane which surrounds the *numerical range* (or *field of values*) $\{\boldsymbol{w}^* B \boldsymbol{w} \mid \boldsymbol{w} \in \mathbb{C}^M,\ |\boldsymbol{w}\| = 1\}$ of B. To use this formula to calculate $\Phi(B)\boldsymbol{v}$, systems of the form

$$(\lambda I - B)\boldsymbol{y}_\lambda = \boldsymbol{v}, \quad \lambda \in \Gamma \subset \mathbb{C}, \tag{9.156}$$

have to be solved.

In the complex-valued setting of Arnoldi's method to generate an orthonormal basis of the Krylov space $K_k(B; \boldsymbol{v})$, the relations $BV_k = V_k H_k + \boldsymbol{w}_k \boldsymbol{e}_k^T$ (see (5.96)) and $V_k^* V_k = I$ remain true, where $V_k = (\boldsymbol{v}_1, \ldots, \boldsymbol{v}_k) \in \mathbb{C}^{M,k}$ and $H_k \in \mathbb{C}^{k,k}$. Furthermore, it can be shown that, under certain assumptions, an appropriate approximation to $\Phi(B)v$ is obtained by $V_k \Phi(H_k) \boldsymbol{e}_1$. The matrix function $\Phi(H_k)$ is well defined by the formula (9.155) for all k because the property (5.97) implies that the numerical range of H_k is a subset of the numerical range of B.

When comparing the systems (9.156) of dimension M with the corresponding systems (involving H_k) of dimension k, it is not surprising that, for $k \ll M$, the computation of $\Phi(H_k)\boldsymbol{e}_1$ is often much easier in practice than the computation of $\Phi(B)\boldsymbol{v}$, e.g. by diagonalization or Padé approximation of H_k. Indeed, comparatively early results from the literature showed that exponential methods have an advantage over the conventional implicit approach, due to faster Krylov convergence of exponential functions, compared with the rational functions arising from the implicit approach [148].

Exercises

Problem 9.29 With $\varphi_0(z) := e^z$, show that the functions φ_j, $j \in \mathbb{N}_0$, satisfy $\varphi_j(0) = 1/j!$ and the recurrence relation

$$\varphi_{j+1}(z) = \frac{1}{z}\left[\varphi_j(z) - \varphi_j(0)\right], \quad j = 0, 1, 2, \ldots$$

Problem 9.30 Show that the method from Example 9.45 (the so-called *exponential Euler method*) recovers the standard explicit Euler method for the equation $\frac{d}{dt}\boldsymbol{u} = \boldsymbol{q}_h(t, \boldsymbol{u})$ as $A_h \to 0$ (in $\mathbb{R}^{M,M}$).

9.7 The Maximum Principle

We start with the semidiscretization as described in Section 9.2 and discuss the validity of comparison and maximum principles. Taking (9.51)/(9.52) as a starting point, where inhomogeneous Dirichlet conditions were eliminated and included in $\boldsymbol{q}_h$, for the above purpose we want to have the dependence explicitly as in Section 1.4, i.e. in addition to the system matrix $A_h \in \mathbb{R}^{M_1,M_1}$, we have a rectangular matrix $\widehat{A}_h \in \mathbb{R}^{M_1,M-M_1}$ to incorporate the Dirichlet data:

$$\widetilde{A}_h := (A_h\ \widehat{A}_h) \in \mathbb{R}^{M_1,M}, \tag{9.157}$$

where M_1 is the number of degrees of freedom to be computed, and $M > M_1$ includes the Dirichlet degrees of freedom.
– Note the change in notation using $\widehat{A}_h$ compared to (9.51). –
To include also the FEM of (9.50) in this formulation, there V_h has to be substituted by X_h from (3.119) which is in accordance with Section 3.5.1. Thus, the unknowns $\boldsymbol{u}_h(t) \in \mathbb{R}^{M_1}$ are to be amended by the Dirichlet degrees of freedom $\hat{\boldsymbol{u}}_h(t) \in \mathbb{R}^{M-M_1}$ giving the extended vector

$$\tilde{\boldsymbol{u}}_h(t) = \begin{pmatrix} \boldsymbol{u}_h(t) \\ \hat{\boldsymbol{u}}_h(t) \end{pmatrix} \in \mathbb{R}^M. \tag{9.158}$$

In the FEM case, the ansatz is now, for $u_h(t) \in X_h = \operatorname{span}\{\varphi_1, \cdots, \varphi_M\}$,

$$u_h(t) = \sum_{i=1}^{M_1} \boldsymbol{u}_{hi}(t)\varphi_i + \sum_{i=M_1+1}^{M} \hat{\boldsymbol{u}}_{hi}(t)\varphi_i =: \sum_{i=1}^{M} \tilde{\boldsymbol{u}}_{hi}(t)\varphi_i.$$

If the mass matrix B_h is not diagonal, also the contributions of the Dirichlet data have to be considered in the time derivative, i.e. with

$$\widetilde{B}_h := (B_h\ \widehat{B}_h) \in \mathbb{R}^{M_1,M}$$

the ODE system reads

$$\begin{aligned} \widetilde{B}_h \frac{d}{dt}\tilde{\boldsymbol{u}}_h(t) + \widetilde{A}_h \tilde{\boldsymbol{u}}_h(t) &= \boldsymbol{\beta}_h(t), \\ \boldsymbol{u}_h(0) &= \boldsymbol{u}_{h0}. \end{aligned} \tag{9.159}$$

To show a comparison principle, we restrict ourselves to a diagonal matrix B_h (i.e. $\widehat{B}_h = 0$):

$$B_h = \operatorname{diag}(b_i) \quad \text{with } b_i > 0,\ i = 1, \cdots, M_1. \tag{9.160}$$

Theorem 9.48 *Consider two solutions $\boldsymbol{u}_{h1}, \boldsymbol{u}_{h2}$ of* (9.159) *for the data*

$$\boldsymbol{\beta}_1(t) \le \boldsymbol{\beta}_2(t),\ \hat{\boldsymbol{u}}_{h1}(t) \le \hat{\boldsymbol{u}}_{h2}(t) \quad \textit{for all } t \in (0,T],$$
$$\boldsymbol{u}_{01} \le \boldsymbol{u}_{02},$$

and assume that $\widetilde{A}_h$ satisfies

$$(\widetilde{A}_h)_{ij} \le 0 \quad \textit{for all } i \ne j. \tag{9.161}$$

Then

$$\boldsymbol{u}_{h1}(t) \le \boldsymbol{u}_{h2}(t) \quad \textit{for all } t \in (0,T].$$

In addition, if $\boldsymbol{u}_{01} < \boldsymbol{u}_{02}$, then

$$\boldsymbol{u}_{h1}(t) < \boldsymbol{u}_{h2}(t) \quad \textit{for all } t \in (0,T].$$

Proof By linearity, it is equivalent to show the corresponding nonnegativity statement, i.e. we set $\boldsymbol{u}_h := \boldsymbol{u}_{h2}$ and $\boldsymbol{\beta}_1(t) := \boldsymbol{0}$, $\hat{\boldsymbol{u}}_{h1} := \boldsymbol{0}$, $\boldsymbol{u}_{01} := \boldsymbol{0}$, then $\boldsymbol{u}_{h1} = \boldsymbol{0}$. Furthermore, it is sufficient to show only the second statement, as the solution of the linear ODE system depends pointwise continuously on the initial data (see e.g. [64, Thm. 14.VI]).

Therefore, we assume that $\boldsymbol{\beta}(t) := \boldsymbol{\beta}_2(t) \ge 0$, $\hat{\boldsymbol{u}}_h(t) := \hat{\boldsymbol{u}}_{h2}(t) \ge 0$ for all $t \in (0,T]$ and $\boldsymbol{u}_{h0} := \boldsymbol{u}_{02} > \boldsymbol{0}$, where in general $\boldsymbol{x} = (\boldsymbol{x}_i) > \boldsymbol{0}$ means that $x_i > 0$ for all indices i. We want to show that $\boldsymbol{u}_h(t) > \boldsymbol{0}$ for all $t \in (0,T]$.

Assume on the contrary that there is a point $t_1 \in (0,T]$ and an index $i \in \{1,\cdots,M_1\}$ such that

$$u_h^i(t_i) \le 0,$$

where $\boldsymbol{u}_h(t) = (\boldsymbol{u}_h^i(t))_i$.

Then the set $\left\{t \in [0,t_1] \;\middle|\; u_h^j(t) = 0 \text{ for some } j \in \{1,\cdots,M_1\}\right\}$ is nonempty. Since it is compact, it has a minimal element $t_2 \in (0,t_1]$, i.e.

$$u_h^j(t_2) = 0,\ u_h^j(t) > 0 \quad \text{for } t \in [0,t_2),$$

$$u_h^i(t) \ge 0 \quad \text{for all } i \in \{i,\cdots,M\} \setminus \{j\},\ t \in [0,t_2].$$

The jth equation in (9.159) reads as

$$y' = -\gamma y + \alpha(t),$$

where $y := u_h^j$, $\boldsymbol{\beta} := (\beta_i)$, $\widetilde{A}_h := (a_{ij})$, $\gamma := a_{jj}/b_j$, $\alpha(t) := -\dfrac{1}{b_j}\Big(\sum_{\substack{k=1\\k\ne j}}^{M} a_{jk}\tilde{u}_h^k(t) - \beta_j(t)\Big)$, i.e. $\alpha(t) \ge 0$ for $t \in [0,t_2]$. Furthermore $y(0) > 0$.

It follows that

$$y' \ge -\gamma y,$$

i.e.

$$\ln\left(\frac{y(t)}{y(0)}\right) = \int_0^t \frac{y'(s)}{y(s)}ds \ge -\gamma t \quad \text{for all } t < t_2,$$

and thus by continuity

$$y(t_2) \geq \exp(-\gamma t_2)y(0) > 0,$$

a contradiction. □

Remark 9.49 The following version of a local maximum principle holds:
Assume for some $t \in (0, T]$ and some $i \in \{1, \cdots, M_1\}$ that

$$\beta_i(t) \leq 0,\ u_h^i(t) \geq \tilde{u}_h^j(t) \quad \text{for all } j \in \{1, \cdots, M\}.$$

If additionally to (9.161)

$$\sum_{j=1}^{M} (\widetilde{A}_h)_{ij} = 0$$

holds, then

$$\frac{d}{dt} u_h^i(t) \leq 0. \tag{9.162}$$

Indeed, considering the ith equation in (9.159), we see that

$$\begin{aligned}
\frac{d}{dt} u_h^i(t) &= -\frac{1}{b_i}\left[\sum_{\substack{j=1\\ j\neq i}}^{M} a_{ij}\tilde{u}_h^j(t) + a_{ii}u_h^i(t) - \beta_i(t)\right] \\
&= -\frac{1}{b_i}\sum_{\substack{j=1\\ j\neq i}}^{M} a_{ij}\left[\tilde{u}_h^j(t) - u_h^i(t) - \beta_i(t)\right] \leq 0.
\end{aligned}$$

There is an analogous property for the minimum, too. The assertion (9.162) is also called *local extremum diminishing (LED) property*.

The conditions which have been used are in fact (1.39), (2), (5) and then (6)*. We now consider the one-step-theta method for (9.159), further assuming (9.160), first with $b_i := 1$, as in the FDM.

In Section 1.4, we have seen that, for a discrete problem of the form (1.34), there is a hierarchy of properties ranging from a comparison principle to a strong maximum principle, which is in turn implied by a hierarchy of conditions, partly summarized as (1.35) or (1.35)*. To remind the reader, we regroup these conditions accordingly:
The collection of conditions (1.35) (1), (2), (3)(i), (4)* is called *(IM)*.
(IM) implies the inverse monotonicity of A_h (Theorem 1.13, (1.42)).
The collection of conditions *(IM)*, (5) is called *(CP)*.
(CP) implies a comparison principle in the sense of Corollary 1.14.
The collection of conditions *(CP)*, (6)* is called $(MP)^*$.
$(MP)^*$ implies a maximum principle in the form of Theorem 1.11 (1.41).
Alternatively, the collection of conditions *(CP)* (6)# (see Problem 1.13) is called *(MP)*.
(MP) implies a maximum principle in the form of Theorem 1.10 (1.37).

Finally, the collection of conditions *(CP)*, (6), (4) (instead of (4)*), (7) is called *(SMP)*.

(SMP) implies a strong maximum principle in the sense of Theorem 1.10.

An L^∞-stability estimate in the sense of Theorem 1.15 is closely related. This will be taken up in the next section.

In the following, we will discuss the above-mentioned properties for the one-step-theta method, cast into the form (1.34), on the basis of corresponding properties of the underlying elliptic problem and its discretization. It will turn out that under a reasonable condition (see (9.167)), condition (4)* (and thus (3)(ii)) will not be necessary for the elliptic problem. This reflects the fact that contrary to the elliptic problem, for the parabolic problem also the case of a pure Neumann boundary condition (where no degrees of freedom are given and thus eliminated) is allowed, since the initial condition acts as a Dirichlet boundary condition.

In assuming that the discretization of the underlying elliptic problem is of the form (1.34), we return to the notation $M = M_1 + M_2$, where M_2 is the number of degrees of freedom eliminated, and thus $A_h, B_h \in \mathbb{R}^{M_1,M_1}$.

We write the discrete problem according to (9.92) as one large system of equations for the unknown

$$\boldsymbol{u}_h = \begin{pmatrix} \boldsymbol{u}_h^1 \\ \boldsymbol{u}_h^2 \\ \vdots \\ \boldsymbol{u}_h^N \end{pmatrix}, \tag{9.163}$$

in which the vector of grid values $\boldsymbol{u}_h^i \in \mathbb{R}^{M_1}$ are collected to one large vector of dimension $\overline{M}_1 := N \cdot M_1$. This is necessary to deal with a global extremum, i.e. a strong maximum principle, otherwise one level, i.e. $N = 1$ is sufficient. Thus, the grid points in $\Omega \times (0, T)$ are the points (x_j, t_n), $n = 1, \ldots, N$, $x_j \in \Omega_h$, e.g. for the finite difference method. The defining system of equations has the form

$$C_h \boldsymbol{u}_h = \boldsymbol{p}_h, \tag{9.164}$$

where

$$C_h = \begin{pmatrix} I + \tau\Theta A_h & & & 0 \\ -I + \tau\overline{\Theta} A_h & \ddots & & \\ & \ddots & \ddots & \\ 0 & & \ddots & \ddots \\ & & -I + \tau\overline{\Theta} A_h & I + \tau\Theta A_h \end{pmatrix},$$

again with $\overline{\Theta} := 1 - \Theta$,

$$\boldsymbol{p}_h = \begin{pmatrix} \tau \boldsymbol{q}_h(\Theta\tau) + (I - \tau\overline{\Theta}A_h)\boldsymbol{u}_0 \\ \tau \boldsymbol{q}_h((1+\Theta)\tau) \\ \vdots \\ \vdots \\ \tau \boldsymbol{q}_h(N-1+\Theta)\tau) \end{pmatrix}.$$

Since the spatial discretization is performed as in the stationary case, and in the nth step the discretization relates to $t = t_{n-1} + \Theta\tau$ and also the approximation

$$A_h u(t_{n-1} + \Theta\tau) \sim \overline{\Theta} A_h u(t_{n-1}) + \Theta A_h u(t_n)$$

enters the formulation (9.92), we can assume to have the following structure of the right-hand side of (9.92):

$$\boldsymbol{q}_h((n-1+\Theta)\tau) = -\hat{A}_h(\overline{\Theta}\hat{\boldsymbol{u}}_h^{n-1} + \Theta\hat{\boldsymbol{u}}_h^n) + \boldsymbol{f}((n-1+\Theta)\tau) \text{ for } n = 1, \dots, N. \quad (9.165)$$

Here the $\hat{\boldsymbol{u}}_h^n \in \mathbb{R}^{M_2}$ are the known spatial boundary values on time level t_n, which have been eliminated from the equation as explained, e.g. in Chapter 1 for the finite difference method. But as noted, we allow also for the case where such values do not appear (i.e. $M_2 = 0$) then (9.165) reduces to

$$\boldsymbol{q}_h((n-1+\Theta)\tau) = \boldsymbol{f}((n-1+\Theta)\tau) \quad \text{for} \quad n = 1, \dots, N.$$

For the continuous problem, data are prescribed at the *parabolic boundary* $\Omega \times \{0\} \cup \partial\Omega \times [0, T]$; correspondingly, the known values $\hat{\boldsymbol{u}}_h^i$ are collected with the initial data $\boldsymbol{u}_0 \in \mathbb{R}^{M_1}$ to a large vector

$$\hat{\boldsymbol{u}}_h = \begin{pmatrix} \boldsymbol{u}_0 \\ \hat{\boldsymbol{u}}_h^0 \\ \hat{\boldsymbol{u}}_h^1 \\ \vdots \\ \hat{\boldsymbol{u}}_h^N \end{pmatrix},$$

i.e. a vector of dimension $\overline{M}_2 := M_1 + (N+1)M_2$, which may reduce to $\hat{\boldsymbol{u}}_h = \boldsymbol{u}_0 \in \mathbb{R}^{M_1}$.

With this notation, we have

$$\boldsymbol{p}_h = -\hat{C}_h \hat{\boldsymbol{u}}_h + \boldsymbol{e} \quad (9.166)$$

if we define

$$\hat{C}_h = \begin{pmatrix} -I + \tau\overline{\Theta}A_h & \tau\overline{\Theta}\hat{A}_h & \tau\Theta\hat{A}_h & & 0 \\ 0 & \ddots & \ddots & & \vdots \\ \vdots & & \ddots & \ddots & \vdots \\ 0 & & & \tau\overline{\Theta}\hat{A}_h & \tau\Theta\hat{A}_h \end{pmatrix},$$

$$e = \begin{pmatrix} \tau f(\Theta\tau) \\ \tau f((1+\Theta)\tau) \\ \vdots \\ \vdots \\ \tau f((N-1+\Theta)\tau) \end{pmatrix}.$$

In the following, the validity of (1.35)* or (1.35) for

$$\widetilde{C}_h = (C_h, \hat{C}_h)$$

will be investigated on the basis of corresponding properties of

$$\widetilde{A}_h = (A_h, \hat{A}_h)\,.$$

Note that even if A_h is irreducible, the matrix C_h is always reducible, since $\boldsymbol{u}_h^n$ depends only on $\boldsymbol{u}_h^1, \ldots, \boldsymbol{u}_h^{n-1}$, but not on the future time levels. (Therefore, (9.164) serves only for the theoretical analysis, but not for the actual computation.)

In the following, we assume that

$$\tau\overline{\Theta}(A_h)_{jj} < 1 \quad \text{for } j = 1, \ldots, M_1\,, \tag{9.167}$$

which is always satisfied for the implicit Euler method ($\Theta = 1$). Then:

(1) $(C_h)_{rr} > 0$ for $r = 1, \ldots, \overline{M}_1$
holds if (1) is valid for A_h. Actually, also $(A_h)_{jj} > -1/(\tau\Theta)$ would be sufficient.

(2) $(C_h)_{rs} \le 0$ for $r, s = 1, \ldots, \overline{M}_1,\ r \ne s$:
If (2) is valid for A_h, then (2) holds for $N = 1$, and for $N > 1$ only the nonpositivity of the diagonal elements of the off-diagonal block of C_h, i.e. $-I + \tau\overline{\Theta}A_h$, is in question. This is ensured by (9.167) (weakened to "$\le$").

(3) (i) $C_r := \sum_{s=1}^{\overline{M}_1} (C_h)_{rs} \ge 0$ for $r = 1, \ldots, \overline{M}_1$,
(ii) $C_r > 0$ for at least one $r \in \{1, \ldots, \overline{M}_1\}$:
We set

$$A_j := \sum_{k=1}^{M_1} (A_h)_{jk}$$

so that condition (3)(i) for A_h means that $A_j \ge 0$ for $j = 1, \ldots, M_1$. Therefore, we have

$$C_r = 1 + \tau\Theta A_j > 0 \tag{9.168}$$

for the indices r of the first time level, where the "global" index r corresponds to the "local" spatial index j. If this is the only case, i.e. for $N = 1$ the condition (3)(i) for A_h can be weakened to

$$(3)^*(i) \quad 1 + \tau\Theta \sum_{k=1}^{M_1} (A_h)_{jk} \ge 0. \tag{9.169}$$

For the following time levels, the relation

$$C_r = 1 - 1 + \tau(\Theta + \overline{\Theta})A_j = \tau A_j \geq 0 \tag{9.170}$$

holds, i.e. (3)(i) and (ii).

(4)* *For every* $r_1 \in \{1, \ldots, \overline{M}_1\}$ *satisfying*

$$\sum_{r=1}^{\overline{M}_1} (C_h)_{rs} = 0 \tag{9.171}$$

there exist indices $r_2, \ldots, r_{l+1}$ *such that*

$$(C_h)_{r_i r_{i+1}} \neq 0 \quad \text{for} \quad i = 1, \ldots, l$$

and

$$\sum_{s=1}^{\overline{M}_1} (C_h)_{r_{l+1} s} > 0\,. \tag{9.172}$$

To avoid too many technicalities, we adopt the background of a finite difference method. Actually, only matrix properties enter the reasoning. We call (space-time) grid points satisfying (9.171) *far from the boundary*, and those satisfying (9.172) *close to the boundary*. Due to (9.168), all points of the first time level are close to the boundary (consistent with the fact that the grid points for $t_0 = 0$ belong to the parabolic boundary). This concludes the case $N = 1$. For the subsequent time level n, due to (9.170), a point (x_i, t_n) is close to the boundary if x_i is close to the boundary with respect to A_h. Therefore, the requirement of (4)*, that a point far from the boundary can be connected via a chain of neighbours to a point close to the boundary, can be realized in two ways: Firstly, within the time level n, i.e. the diagonal block of C_h if A_h satisfies condition (4)*. Secondly, without this assumption a chain of neighbours exist by (x, t_n), (x, t_{n-1}) up to (x, t_1), i.e. a point close to the boundary, since the diagonal element of $-I + \tau\Theta A_h$ does not vanish due to (9.167). This reasoning additionally has established the following:

(4)# *If* A_h *is irreducible, then a grid point* (x, t_n), $x \in \Omega_h$, *can be connected via a chain of neighbours to every grid point* (y, t_k), $y \in \Omega_h$ *and* $0 \leq k \leq n$.

(5) $(\hat{C}_h)_{rs} \leq 0$ *for* $r = 1, \ldots, \overline{M}_1$, $s = \overline{M}_1 + 1, \ldots, \overline{M}_2$:
Analogously to (2), this follows from (5) for $\hat{A}_h$ and (9.167).

(6)* $\widetilde{C}_r := \sum_{s=1}^{M} (\widetilde{C}_h)_{rs} = 0$ *for* $r = 1, \ldots, M$:
Analogously to (9.170), we have

$$\widetilde{C}_r = \tau \widetilde{A}_j := \tau \sum_{k=1}^{M} (\widetilde{A}_h)_{jk}$$

so that the property is equivalent to the corresponding one of $\widetilde{A}_h$.

(6) $\widetilde{C}_r \geq 0$ *for* $r = 1, \ldots, \overline{M}$
is equivalent to (6) for $\widetilde{A}_h$ by the above argument.
(7) *For every* $s \in \overline{M}_1 + 1, \ldots, \overline{M}$ *there exists an* $r \in \{1, \ldots, \overline{M}_1\}$ *such that* $(\hat{C}_h)_{rs} \neq 0$:
Every listed boundary value should influence the solution: For the values from $\hat{\boldsymbol{u}}_h^0, \ldots, \hat{\boldsymbol{u}}_h^N$ this is the case iff $\hat{A}_h$ satisfies this property. Furthermore, the "local" indices of the equation, where the boundary values appear, are the same for each time level. For the values from $\boldsymbol{u}_0 \in \mathbb{R}^{M_1}$ the assertion follows from (9.167).

The above considerations lead to the following result.

Theorem 9.50 *Consider the one-step-theta method in the form* (9.92). *Let* (9.167) *hold.*

1) *If the spatial discretization* $\widetilde{A}_h$ *satisfies* (1.35) (1), (2), (3)*(i) *(see* (9.169)*),* *and* (5), *then a* comparison principle *holds: If for two sets of data* $\boldsymbol{f}_i$, $\boldsymbol{u}_{0i}$ *and* $\hat{\boldsymbol{u}}_{hi}^n$, $n = 0, \ldots, N$ *and* $i = 1, 2$, *we have*

$$\boldsymbol{f}_1((n-1+\Theta)\tau) \leq \boldsymbol{f}_2((n-1+\Theta)\tau) \quad \textit{for } n = 1, \ldots, N,$$

and

$$\boldsymbol{u}_{01} \leq \boldsymbol{u}_{02}, \quad \hat{\boldsymbol{u}}_{h1}^n \leq \hat{\boldsymbol{u}}_{h2}^n \quad \textit{for } n = 0, \ldots, N,$$

then

$$\boldsymbol{u}_{h1}^n \leq \boldsymbol{u}_{h2}^n \quad \textit{for } n = 1, \ldots, N$$

for the corresponding solutions.
If $\hat{\boldsymbol{u}}_{h1}^n = \hat{\boldsymbol{u}}_{h2}^n$ *for* $n = 1, \ldots, N$, *then condition* (1.35) (5) *can be omitted.*

2) *Assume* $\boldsymbol{f} \leq \boldsymbol{0}$. *If* $\widetilde{A}_h$ *additionally satisfies* (1.35) (6)*, *then the following* weak maximum principle *holds:*

$$\max_{\substack{r \in \{1,\ldots,M\} \\ n=0,\ldots,N}} (\tilde{\boldsymbol{u}}_h^n)_r \leq \max \left\{ \max_{r \in \{1,\ldots,M_1\}} (\boldsymbol{u}_0)_r, \max_{\substack{r \in \{M_1+1,\ldots,M\} \\ n=0,\ldots,N}} (\hat{\boldsymbol{u}}_h^n)_r \right\},$$

where

$$\tilde{\boldsymbol{u}}_h^n := \begin{pmatrix} \boldsymbol{u}_h^n \\ \hat{\boldsymbol{u}}_h \end{pmatrix}.$$

3) *Assume* $\boldsymbol{f} \leq \boldsymbol{0}$. *If* $\widetilde{A}_h$ *satisfies* (1.35) (1), (2), (3)(i), (4), (5), (6), (7), *then a* strong maximum principle *in the following sense holds:*
If the components of $\tilde{\boldsymbol{u}}_h^n$, $n = 0, \ldots, N$, *attain a nonnegative maximum for some spatial index* $r \in \{1, \ldots, M_1\}$ *and at some time level* $k \in \{1, \ldots, N\}$, *then all components for the time levels* $n = 0, \ldots, k$ *are equal.*

Proof For (1), it is sufficient to consider $N = 1$, i.e. to prove

$$\boldsymbol{f}((n+\Theta)\tau) \geq \boldsymbol{0}, \quad \tilde{\boldsymbol{u}}_h^n \geq \boldsymbol{0}, \quad \hat{u}_h^{n+1} \geq \boldsymbol{0} \quad \text{for } n = 1, \cdots, N$$

imply

$$\tilde{\boldsymbol{u}}_h^{n+1} \geq \boldsymbol{0},$$

then the result follows by induction over n. Similarly, also for (2), the case $N = 1$ is sufficient, i.e. it has to be proved:

If in addition $\boldsymbol{f}((n-1+\Theta)\tau) \leq \boldsymbol{0}$ and $\widetilde{A}_h$ satisfies (1.35)(6)*, then

$$\max_{r\in\{1,\ldots,M_1\}} (\tilde{\boldsymbol{u}}_h^{n+1})_r \leq \max\left\{ \max_{r\in\{1,\ldots,M\}} (\tilde{\boldsymbol{u}}_h^{n})_r, \max_{r\in\{M_1+1,\ldots,M\}} (\hat{\boldsymbol{u}}_h^{N+1})_r \right\}$$

for $n = 0, \cdots, N$.

For (3), observe that Theorem 1.10 cannot be applied directly to (9.164), since C_h is reducible. Therefore, the proof of Theorem 1.10 has to be repeated: We conclude that the solution is constant at all points that are connected via a chain of neighbours to the point where the maximum is attained. According to (4)#, these include all grid points (x, t_l) with $x \in \Omega_h$ and $l \in \{0, \ldots, k\}$. From (9.167) and the discussion of (7) we see that the connection can also be continued to boundary values up to level k. □

The additional condition (9.167), which may be weakened to nonstrict inequality, as seen above, actually is a time step restriction: Consider again the example of the five-point stencil discretization of the heat equation on a rectangle, for which we have $(A_h)_{jj} = 4/h^2$. Then the condition takes the form

$$\frac{\tau}{h^2} < \frac{1}{4(1-\Theta)} \tag{9.173}$$

for $\Theta < 1$. This is very similar to the condition (9.114), (9.115) for the explicit Euler method, but the background is different.

As already noted, the results above also apply to the more general form (9.93) under the assumption (9.95). The condition (9.167) then takes the form

$$\tau\overline{\Theta}(A_h)_{jj} \leq b_j \quad \text{for} \quad j = 1, \ldots, M_1 . \tag{9.174}$$

Remark 9.51 1) The comparison principle 1) of Theorem 9.50 is equivalent to the following *nonnegativity preservation statement*:

$$\boldsymbol{f}\big((n-1+\Theta)\tau\big) \geq \boldsymbol{0} \quad \text{for all } n = 1, \cdots, N,$$

$$\boldsymbol{u}_0 \geq \boldsymbol{0},\ \hat{\boldsymbol{u}}_h^n \geq \boldsymbol{0} \quad \text{for all } n = 0, \cdots, N$$

imply

$$\boldsymbol{u}_n \geq \boldsymbol{0} \quad \text{for all } n = 1, \cdots, N.$$

2) For a nondiagonal matrix B_h the difference quotient has to be substituted by

$$\frac{1}{\tau}\big(B_h(\boldsymbol{u}_h^n - \boldsymbol{u}_h^{n-1}) + \hat{B}_h(\hat{\boldsymbol{u}}_h^n - \hat{\boldsymbol{u}}_h^{n-1})\big).$$

Then, in order to satisfy (5), the additional requirements are:

$$\tau\bar{\Theta}\widehat{A}_h \le \widehat{B}_h, \quad \tau\Theta\widehat{A}_h + \widehat{B}_h \le 0.$$

Assuming $\widehat{B}_h \ge 0$ and having $(\widehat{A}_h)_{ij} \le 0$ for $i \ne j$ in mind, the first condition is fulfilled, if (9.174) holds, but the second seems to be non-valid in general.

3) For $\boldsymbol{f} \ge \mathbf{0}$, the assertions 2) and 3) of Theorem 9.50 are valid after exchanging "max" by "min" and "nonnegative" by "nonpositive".

4) An equivalent formulation of Theorem 9.50, 2) is

$$(\tilde{\boldsymbol{u}}_h^n)_r \;\le\; \max\left\{ \max_{r\in\{1,\ldots,M_1\}} (\boldsymbol{u}_0)_r, \max_{\substack{r\in\{M_1+1,\ldots,M\}\\ n=0,\ldots,N}} (\hat{\boldsymbol{u}}_h^n)_r \right\} \mathbf{1},$$

where $\mathbf{1} = (1, \cdots, 1)^T$.

The conditions to the row sums ((1.35) (3), (4)*, (1.35) (6), (6)*) play an important role. In the case of (primal) FEM, the condition (4)* reads as

$$\sum_{j=1}^{M} (\widetilde{A}_h)_{ij} = 0. \tag{9.175}$$

If the constant function $u_h := 1$ belongs to X_h, the left-hand side of (9.175) is related to the ith equation of the discrete problem, i.e. for the formulation (3.13), (3.19)–(3.21):

$$\sum_{j=1}^{M} (\widetilde{A}_h)_{ij} = a(1, \varphi_i) = \int_\Omega r\varphi_i dx + \int_{\Gamma_2} \alpha\varphi_i d\sigma.$$

Note that if we start from a formulation like (0.29), where $r = \nabla \cdot \boldsymbol{c}$, then (9.175) requires the absence of sources, of mixed boundary conditions, and $\nabla \cdot \boldsymbol{c} = 0$.

For the formulation (3.36)–(3.39), i.e. for the divergence form, we have

$$\hat{a}(1, \varphi) = -\int_\Omega \left[\boldsymbol{c} \cdot \nabla\varphi_i - r\varphi_i\right] dx + \int_{\widetilde{\Gamma}_2} \tilde{\alpha}\varphi_i d\sigma$$

(see (3.40)) and

$$-\int_\Omega \boldsymbol{c} \cdot \nabla\varphi_i dx = \int_\Omega \nabla \cdot \boldsymbol{c}\varphi_i dx - \int_{\partial\Omega} \boldsymbol{c} \cdot \boldsymbol{n}\varphi_i d\sigma.$$

If the boundary condition (3.38) is given by $\tilde{\alpha} = \boldsymbol{c} \cdot \boldsymbol{n}$, then

$$\hat{a}(1, \varphi) = \int_\Omega \left[\nabla \cdot \boldsymbol{c} + r\right] \varphi_i dx - \int_{\widetilde{\Gamma}_1 \cup \widetilde{\Gamma}_3} \boldsymbol{c} \cdot \boldsymbol{n}\varphi_i d\sigma.$$

Exercises

Problem 9.31 Formulate the results of this section, in particular condition (9.167), for the problem in the form (9.93) with B_h according to (9.95) (i.e. appropriate for finite element discretizations with mass lumping, see (9.100)).

Problem 9.32 Show the validity of $(6)^{\#}$ from Problem 1.13 for C_h if it holds here for A_h and conclude as in Problem 1.13 a weak maximum principle for the one-step-theta method.

Problem 9.33 Consider the initial boundary value problem in one space dimension

$$\begin{cases} u_t - \varepsilon u_{xx} + c u_x = f \quad \text{in } (0,1)\times(0,T), \\ u(0,t) = g_-(t),\ u(1,t) = g_+(t),\ t \in (0,T), \\ u(x,0) = u_0(x),\ x \in (0,1), \end{cases}$$

where $T > 0$ and $\varepsilon > 0$ are constants, and $c, f : (0,1)\times(0,T) \to \mathbb{R}$, $u_0 : (0,1) \to \mathbb{R}$, and $g_-, g_+ : (0,T) \to \mathbb{R}$ are sufficiently smooth functions such that the problem has a classical solution.

Define $h := 1/m$ and $\tau = T/N$ for some numbers $m, N \in \mathbb{N}$. Then the so-called *full upwind finite difference method* for this problem reads as follows: Find a sequence of vectors $\boldsymbol{u}_h^0, \dots, \boldsymbol{u}_h^N$ by

$$\begin{aligned} \frac{u_i^{n+1} - u_i^n}{\tau} - \varepsilon \frac{u_{i+1}^{n+1} - 2u_i^{n+1} + u_{i-1}^{n+1}}{h^2} - c^- \frac{u_{i+1}^{n+1} - u_i^{n+1}}{h} + c^+ \frac{u_i^{n+1} - u_{i-1}^{n+1}}{h} \\ = f_i^{n+1}, \qquad i = 1, \dots, m-1,\ n = 0, \dots, N-1, \end{aligned}$$

where $c = c^+ - c^-$ with $c^+ = \max\{c, 0\}$, $f_i^n = f(ih, n\tau)$, $u_i^0 = u_0(ih)$, $u_0^n = g_-(n\tau)$ and $u_m^n = g_+(n\tau)$.

Prove that a weak maximum principle holds for this method.

9.8 Order of Convergence Estimates in Space and Time

Based on the stability results already derived, we will investigate the (order of) convergence properties of the one-step-theta method for different discretization approaches. Although the results will be comparable, they will be in different norms, appropriate for the specific discretization method, as already seen in Chapters 1, 3, and 8.

Order of Convergence Estimates for the Finite Difference Method

From Section 1.4 we know that the investigation of the (order of) convergence of a finite difference method consists of two ingredients:

- (order of) convergence of the consistency error,
- stability estimates.

The last tool has already been provided by Theorem 9.35 and by Theorem 1.15, which together with the considerations of Section 9.7 allow us to concentrate on the consistency error. Certain smoothness properties will be required for the classical solution u of the initial boundary value problem (9.1), which in particular makes its evaluation possible at the grid points $x_i \in \overline{\Omega}_h$ at each instance of time $t \in [0, T]$ and also of various derivatives. The vector representing the corresponding grid function (for a fixed ordering of the grid points) will be denoted by $\boldsymbol{U}(t)$, or for short by $\boldsymbol{U}^n := \boldsymbol{U}(t_n)$ for $t = t_n$. The corresponding grid points depend on the boundary condition. For a pure Dirichlet problem, the grid points will be from Ω_h, but if Neumann or mixed boundary conditions appear, they belong to the enlarged set

$$\widetilde{\Omega}_h := \overline{\Omega}_h \cap (\Omega \cup \Gamma_1 \cup \Gamma_2) . \tag{9.176}$$

Then the error at the grid points and each time level is given by

$$\boldsymbol{e}_h^n := \boldsymbol{U}^n - \boldsymbol{u}_h^n \quad \text{for} \quad n = 0, \dots, N , \tag{9.177}$$

where $\boldsymbol{u}_h^n$ is the solution of the one-step-theta method according to (9.92). The consistency error $\hat{q}_h$ as a grid function on $\Omega_h \times \{t_1, \dots, t_N\}$ or correspondingly a sequence of vectors $\hat{\boldsymbol{q}}_h^n$ in $\mathbb{R}^{M_1}$ for $n = 1, \dots, N$ is then defined by

$$\begin{aligned} \hat{\boldsymbol{q}}_h^{n+1} := & \frac{1}{\tau}\left(\boldsymbol{U}^{n+1} - \boldsymbol{U}^n\right) + \Theta A_h \boldsymbol{U}^{n+1} \\ & + \overline{\Theta} A_h \boldsymbol{U}^n - \boldsymbol{q}_h((n+\Theta)\tau) \end{aligned} \tag{9.178}$$

for $n = 0, \dots, N-1$. Then the error grid function obviously satisfies

$$\begin{aligned} \frac{1}{\tau}\left(\boldsymbol{e}_h^{n+1} - \boldsymbol{e}_h^n\right) + \Theta A_h \boldsymbol{e}_h^{n+1} + \overline{\Theta} A_h \boldsymbol{e}_h^n &= \hat{\boldsymbol{q}}_h^{n+1} \quad \text{for} \ \ n = 0, \dots, N-1 , \\ \boldsymbol{e}_h^0 &= \boldsymbol{0} \end{aligned} \tag{9.179}$$

(or nonvanishing initial data if the initial condition is not evaluated exactly at the grid points). In the following we estimate the grid function $\hat{q}_h$ in the discrete maximum norm

$$\begin{aligned} \|\hat{q}_h\|_\infty &:= \max\{|(\hat{\boldsymbol{q}}_h^n)_r| \mid r \in \{1, \dots, M_1\}, n \in \{1, \dots, N\}\} \\ &= \max\{|\hat{\boldsymbol{q}}_h^n|_\infty \mid n \in \{1, \dots, N\}\}, \end{aligned} \tag{9.180}$$

i.e. pointwise in space and time. An alternative norm would be the discrete L^2-norm, i.e.

$$\|\hat{q}_h\|_{0,h} := \left(\tau \sum_{n=1}^{N} h^d \sum_{r=1}^{M_1} \left|(\hat{\boldsymbol{q}}_h^n)_r\right|^2\right)^{1/2} = \left(\tau \sum_{n=1}^{N} |\hat{\boldsymbol{q}}_h^n|_{0,h}^2\right)^{1/2} , \tag{9.181}$$

using the spatial discrete L^2-norm from (9.70), where the same notation is employed. If for the sequence of underlying grid points considered there is a constant $C > 0$ independent of the discretization parameter h such that

$$M_1 = M_1(h) \leq Ch^{-d}, \tag{9.182}$$

then obviously,

$$\|\hat{q}_h\|_{0,h} \leq (CT)^{1/2}\|\hat{q}_h\|_\infty$$

so that the L^2-norm is weaker than the maximum norm. Condition (9.182) is satisfied for such uniform grids, as considered in Section 1.2. A norm in between is defined by

$$\|\hat{q}_h\|_{\infty,0,h} := \max\left\{|\hat{\boldsymbol{q}}_h^n|_{0,h} \mid n = 1,\ldots,N\right\}, \tag{9.183}$$

which is stronger than (9.181) and in the case of (9.182) weaker than the maximum norm.

Analogously to Section 1.4, we denote $\boldsymbol{U}^n$ amended by the eliminated boundary values $\hat{\boldsymbol{U}}_h^n \in \mathbb{R}^{M_2}$ by the vector $\widetilde{\boldsymbol{U}}^n \in \mathbb{R}^M$.

For simplicity we restrict attention, at the beginning, to the case of pure Dirichlet data. Taking into account (9.165) and assuming that $\boldsymbol{f}((n-1+\Theta)\tau)$ is derived from the continuous right-hand side by evaluation at the grid points, we get

$$\begin{aligned}\hat{\boldsymbol{q}}_h^{n+1} &= \frac{1}{\tau}(\boldsymbol{U}^{n+1} - \boldsymbol{U}^n) - \left(\frac{d}{dt}\boldsymbol{U}\right)(t_n + \Theta\tau)\\ &\quad + \Theta\widetilde{A}_h\widetilde{\boldsymbol{U}}^{n+1} + \overline{\Theta}\widetilde{A}_h\widetilde{\boldsymbol{U}}^n - (L\boldsymbol{U})(t_n + \Theta\tau)\\ &=: \boldsymbol{S}_1 + \boldsymbol{S}_2,\end{aligned} \tag{9.184}$$

so that $\boldsymbol{S}_1$, consisting of the first two terms, is the consistency error for the time discretization.

Here $\frac{d}{dt}\boldsymbol{U}$ and $L\boldsymbol{U}$ are the vectors representing the grid functions corresponding to $\frac{d}{dt}u$ and Lu, which requires the continuity of these functions as in the notion of a classical solution.

We make the following assumption: The spatial discretization has the order of consistency α measured in $\|\cdot\|_\infty$ (according to (1.17)) if the solution of the stationary problem (9.6) is in $C^p(\overline{\Omega})$ for some $\alpha > 0$ and $p \in \mathbb{N}$.

For example, for the Dirichlet problem of the Poisson equation and the five-point stencil discretization on a rectangle, we have seen in Chapter 1 that $\alpha = 2$ is valid for $p = 4$. If we assume for $u(\cdot, t)$, u being the solution of (9.1), that

$$\text{the spatial derivatives up to order } p \text{ exist continuously and are bounded uniformly in } t \in [0, T], \tag{9.185}$$

then there exists a constant $C > 0$ such that

$$|(\widetilde{A}_h\widetilde{\boldsymbol{U}}(t))_i - (Lu(\cdot,t))(x_i)| \leq Ch^\alpha \tag{9.186}$$

for every grid point $x_i \in \Omega_h$ and $t \in [0, T]$.

In the case of Neumann or mixed boundary conditions, some of the equations will correspond to discretizations of these boundary conditions.

This discretization may be directly a discretization of (9.3) or (9.4) (typically, if one-sided difference quotients are used) or a linear combination of the discretizations of the differential operator at $x_i \in \widetilde{\Omega}_h$ and of the boundary differential operator of (9.3) or (9.4) (to eliminate "artificial" grid points) (see Section 1.3).

Thus we have to take $x_i \in \widetilde{\Omega}_h$ and interpret Lu in (9.186) as this modified differential operator for $x_i \in \Gamma_1 \cup \Gamma_2$ just described to extend all the above reasoning to the general case.

The estimation of the contribution $\boldsymbol{S}_2$ on the basis of (9.186) is directly possible for $\Theta = 0$ or $\Theta = 1$, but requires further smoothness for $\Theta \in (0, 1)$.

We have

$$\boldsymbol{S}_2 = \boldsymbol{S}_3 + \boldsymbol{S}_4 ,$$

where

$$\begin{aligned} \boldsymbol{S}_3 &:= \Theta(\widetilde{A}_h \widetilde{\boldsymbol{U}}^{n+1} - (L\boldsymbol{U})(t_{n+1})) + \overline{\Theta}(\widetilde{A}_h \widetilde{\boldsymbol{U}}^{n} - (L\boldsymbol{U})(t_n)), \\ \boldsymbol{S}_4 &:= \Theta(L\boldsymbol{U})(t_{n+1}) + \overline{\Theta}(L\boldsymbol{U})(t_n) - (L\boldsymbol{U})(t_n + \Theta\tau) . \end{aligned}$$

By Taylor expansion we conclude for a function $v \in C^2[0, T]$ that

$$\Theta v(t_{n+1}) + \overline{\Theta} v(t_n) = v(t_n + \Theta\tau) + \tau^2 \left(\overline{\Theta} \frac{\Theta^2}{2} v''(t_n^1) + \Theta \frac{\overline{\Theta}^2}{2} v''(t_n^2) \right)$$

for some $t_n^1 \in (t_n, t_n + \Theta\tau)$, $t_n^2 \in (t_n + \Theta\tau,\ t_{n+1})$ so that

$$|\boldsymbol{S}_4|_\infty \le C\tau^2 \tag{9.187}$$

for some constant $C > 0$ independent of τ and h if for $\Theta \in (0, 1)$ the solution u of (9.1) satisfies

$$\frac{\partial}{\partial t} Lu, \quad \frac{\partial^2}{\partial t^2} Lu \in C(\overline{Q}_T) . \tag{9.188}$$

This is a quite severe regularity assumption, which often does not hold.

For $\boldsymbol{S}_3$ we conclude directly from (9.186) that

$$|\boldsymbol{S}_3|_\infty \le Ch^\alpha . \tag{9.189}$$

To estimate $\boldsymbol{S}_1$ we have to distinguish between $\Theta = \frac{1}{2}$ and $\Theta \neq \frac{1}{2}$: If

$$\frac{\partial}{\partial t} u, \ \frac{\partial^2}{\partial t^2} u \in C(\overline{Q}_T) \text{ and for } \Theta = \frac{1}{2} \text{ also } \frac{\partial^3}{\partial t^3} u \in C(\overline{Q}_T), \tag{9.190}$$

then Lemma 1.2 implies (for $\Theta = 0, 1, \frac{1}{2}$, for $\Theta \in (0, 1)$ again with a Taylor expansion)

$$|\boldsymbol{S}_1|_\infty \le C\tau^\beta \tag{9.191}$$

for some constant C, independent of τ and h, with $\beta = 1$ for $\Theta \neq \frac{1}{2}$ and $\beta = 2$ for $\Theta = \frac{1}{2}$.

Thus, under the additional regularity assumptions (9.185), (9.188), (9.190), and if the spatial discretization has order of consistency α in the maximum norm, i.e. (9.186), then the one-step-theta method has the following order of consistency:

$$\|\hat{q}_h\|_\infty \leq C(h^\alpha + \tau^\beta) \tag{9.192}$$

for some constant C, independent of τ and h, with β as in (9.191).

By using a weaker norm one might hope to achieve a higher order of convergence. If this is, for example, the case for the spatial discretization, e.g. by considering the discrete L^2-norm $\|\cdot\|_{0,h}$ instead of $\|\cdot\|_\infty$, then instead of (9.186) we have

$$\|\widetilde{A}_h \widetilde{\boldsymbol{U}}(t) - Lu(\cdot, t)\|_{0,h} \leq \ Ch^\alpha , \tag{9.193}$$

where the terms in the norm denote the corresponding grid functions.

Then again under (weaker forms of) the additional regularity assumptions (9.185), (9.188), (9.190) and assuming (9.182), we have

$$\|\hat{q}_h\|_{0,h} \leq C(h^\alpha + \tau^\beta) . \tag{9.194}$$

By means of Theorem 9.35 we can conclude the first order of convergence result.

Theorem 9.52 *Consider the one-step-theta method and assume that the spatial discretization matrix A_h has a basis of eigenvectors $\boldsymbol{w}_i$ with eigenvalues $\lambda_i \geq 0$, $i = 1, \ldots, M_1$, orthogonal with respect to the scalar product $\langle \cdot, \cdot \rangle_h$, defined in* (9.69)*. The spatial discretization has order of consistency α in $\|\cdot\|_{0,h}$ for solutions in $C^p(\overline{\Omega})$. If τ is such that the method is stable according to* (9.122)*, then for a sufficiently smooth solution u of* (9.1) *(e.g.* (9.185), (9.188), (9.190)*), and for a sequence of grid points satisfying* (9.182)*, the method converges in the norm $\|\cdot\|_{\infty,0,h}$ with the order*

$$O(h^\alpha + \tau^\beta),$$

where $\beta = 2$ for $\Theta = \frac{1}{2}$ and $\beta = 1$ otherwise.

Proof Due to Theorem 9.35 and (9.179) we have to estimate the consistency error in a norm defined by $\tau \sum_{n=1}^{N} |\hat{\boldsymbol{q}}_h^n|_{0,h}$ (i.e. a discrete L^1-L^2-norm), which is weaker than $\|\hat{q}_h\|_{0,h}$, in which the estimate has been verified in (9.194). □

Again we see here a smoothing effect in time: The consistency error has to be controlled only in a discrete L^1-sense to gain a convergence result in a discrete L^∞-sense.

If a consistency estimate is provided in $\|\cdot\|_\infty$ as in (9.192), a convergence estimate still needs the corresponding stability. Instead of constructing a vector as in Theorem 1.15 for the formulation (9.164), we will argue directly with the help of the comparison principle (Theorem 9.50, 1)), which would have been possible also in Section 1.4 (see Problem 1.14).

Theorem 9.53 *Consider the one-step-theta method and assume that the spatial discretization matrix A_h satisfies (1.35) (1), (2), (3) (i) and assume its L^∞-stability by the existence of vectors $\boldsymbol{w}_h \in \mathbb{R}^{M_1}$ and a constant $C > 0$ independent of h such that*

$$A_h \boldsymbol{w}_h \geq \mathbf{1} \quad \textit{and} \quad |\boldsymbol{w}_h|_\infty \leq C\,. \tag{9.195}$$

The spatial discretization has order of consistency α in $\|\cdot\|_\infty$ for solutions in $C^p(\overline{\Omega})$. If (9.167) is satisfied, then for a sufficiently smooth solution u of (9.1) (e.g. (9.185), (9.188), (9.190)) the method converges in the norm $\|\cdot\|_\infty$ with the order

$$O(h^\alpha + \tau^\beta),$$

where $\beta = 2$ for $\Theta = \frac{1}{2}$ and $\beta = 1$ otherwise.

Proof From (9.192) we conclude that

$$-\hat{C}(h^\alpha + \tau^\beta)\mathbf{1} \leq \hat{\boldsymbol{q}}_h^n \leq \hat{C}(h^\alpha + \tau^\beta)\mathbf{1} \quad \text{for} \quad n = 1, \dots, N$$

for some constant $\hat{C}$ independent of h and τ.

Thus (9.179) implies

$$\begin{aligned} \frac{1}{\tau}\left(\boldsymbol{e}_h^{n+1} - \boldsymbol{e}_h^n\right) + \Theta A_h \boldsymbol{e}_h^{n+1} + \overline{\Theta} A_h \boldsymbol{e}_h^n &\leq \hat{C}(h^\alpha + \tau^\beta)\mathbf{1}, \\ \boldsymbol{e}_h^0 &= 0\,. \end{aligned}$$

Setting $\boldsymbol{w}_h^n := \hat{C}(h^\alpha + \tau^\beta)\boldsymbol{w}_h$ with $\boldsymbol{w}_h$ from (9.195), this constant sequence of vectors satisfies

$$\frac{1}{\tau}\left(\boldsymbol{w}_h^{n+1} - \boldsymbol{w}_h^n\right) + \Theta A_h \boldsymbol{w}_h^{n+1} + \overline{\Theta} A_h \boldsymbol{w}_h^n \geq \hat{C}(h^\alpha + \tau^\beta)\mathbf{1}\,.$$

Therefore, the comparison principle (Theorem 9.50, (1)) implies

$$\boldsymbol{e}_h^n \leq \boldsymbol{w}_h^n = \hat{C}(h^\alpha + \tau^\beta)\boldsymbol{w}_h$$

for $n = 0, \dots, N$, and analogously, we see that

$$-\hat{C}(h^\alpha + \tau^\beta)\boldsymbol{w}_h \leq \boldsymbol{e}_h^n$$

so that

$$\left|\left(\boldsymbol{e}_h^n\right)_j\right| \leq \hat{C}(h^\alpha + \tau^\beta)(\boldsymbol{w}_h)_j \tag{9.196}$$

for all $n = 0, \dots, N$ and $j = 1, \dots, M_1$, and finally,

$$|\boldsymbol{e}_h^n|_\infty \leq \hat{C}(h^\alpha + \tau^\beta)|\boldsymbol{w}_h|_\infty \leq \hat{C}(h^\alpha + \tau^\beta)C$$

with the constant C from (9.195). □

Note that the pointwise estimate (9.196) is more precise, since it also takes into account the shape of $\boldsymbol{w}_h$. In the example of the five-point stencil with Dirichlet conditions on the rectangle (see the discussion around (1.46)) the error bound is smaller in the vicinity of the boundary (which is to be expected due to the exactly fulfilled boundary conditions).

Order of Convergence Estimates for the (Primal) Finite Element Method

We now return to the one-step-theta method for the finite element method as introduced in (9.98). In particular, instead of considering grid functions as for the finite difference method, the finite element method allows us to consider directly a function U^n from the finite-dimensional approximation space V_h and thus from the underlying function space V. For the sake of simplicity we suppress the appearance of the coefficient b in the time derivative.

In the following, an error analysis for the case $\Theta \in [\frac{1}{2}, 1]$ under the assumption $u \in C^2([0,T], H)$ will be given. In analogy with the decomposition of the error in the semidiscrete situation, we write

$$u(t_n) - U^n = u(t_n) - R_h u(t_n) + R_h u(t_n) - U^n =: \varrho(t_n) + \theta^n .$$

The first term of the right-hand side is the error of the elliptic projection at the time t_n, and for this term an estimate is already known. The following identity is used to estimate the second member of the right-hand side, which immediately results from the definition of the elliptic projection:

$$\begin{aligned}
&\left\langle \frac{1}{\tau}(\theta^{n+1} - \theta^n), v_h \right\rangle_0 + a(\Theta\theta^{n+1} + \overline{\Theta}\theta^n, v_h) \\
&= \left\langle \frac{1}{\tau}((R_h u(t_{n+1}) - R_h u(t_n)), v_h \right\rangle_0 + a(\Theta R_h u(t_{n+1}) + \overline{\Theta} R_h u(t_n), v_h) \\
&\quad - \left\langle \frac{1}{\tau}(U^{n+1} - U^n), v_h \right\rangle_0 - a(\Theta U^{n+1} + \overline{\Theta} U^n, v_h) \\
&= \left\langle \frac{1}{\tau}(R_h u(t_{n+1}) - R_h u(t_n)), v_h \right\rangle_0 + a(\Theta u(t_{n+1}) + \overline{\Theta} u(t_n), v_h) \\
&\quad - \ell^{n+\Theta}(v_h) \\
&= \left\langle \frac{1}{\tau}(R_h u(t_{n+1}) - R_h u(t_n)), v_h \right\rangle_0 - \langle \Theta u'(t_{n+1}) + \overline{\Theta} u'(t_n), v_h \rangle_0 \\
&= \langle w^n, v_h \rangle_0 ,
\end{aligned}$$

where

$$w^n := \frac{1}{\tau}(R_h u(t_{n+1}) - R_h u(t_n)) - \Theta u'(t_{n+1}) - \overline{\Theta} u'(t_n) .$$

Taking into consideration the inequality $a(v_h, v_h) \geq 0$, the particular choice of the test function as $v_h = \Theta\theta^{n+1} + \overline{\Theta}\theta^n$ yields

$$\Theta \|\theta^{n+1}\|_0^2 + (1 - 2\Theta)\langle \theta^n, \theta^{n+1} \rangle_0 - \overline{\Theta} \|\theta^n\|_0^2 \leq \tau \langle w^n, \Theta\theta^{n+1} + \overline{\Theta}\theta^n \rangle_0 .$$

For $\Theta \in [\frac{1}{2}, 1]$ we have $(1 - 2\Theta) \leq 0$, and hence

$$\begin{aligned}\left(\|\theta^{n+1}\|_0 - \|\theta^n\|_0\right) & \left[\Theta\|\theta^{n+1}\|_0 + \overline{\Theta}\|\theta^n\|_0\right] \\ &= \Theta\|\theta^{n+1}\|_0^2 + (1-2\Theta)\|\theta^n\|_0\|\theta^{n+1}\|_0 - \overline{\Theta}\|\theta^n\|_0^2 \\ &\leq \Theta\|\theta^{n+1}\|_0^2 + (1-2\Theta)\langle\theta^n, \theta^{n+1}\rangle_0 - \overline{\Theta}\|\theta^n\|_0^2 \\ &\leq \tau\|w^n\|_0 \left[\Theta\|\theta^{n+1}\|_0 + \overline{\Theta}\|\theta^n\|_0\right] .\end{aligned}$$

Dividing each side by the expression in the square brackets, we get

$$\|\theta^{n+1}\|_0 \leq \|\theta^n\|_0 + \tau\|w^n\|_0 .$$

The recursive application of this inequality leads to

$$\|\theta^{n+1}\|_0 \leq \|\theta^0\|_0 + \tau \sum_{j=0}^{n} \|w^j\|_0 . \tag{9.197}$$

That is, it remains to estimate the terms $\|w^j\|_0$. A simple algebraic manipulation yields

$$\begin{aligned} w^n := & \left[\frac{1}{\tau}((R_h - I)u(t_{n+1}) - (R_h - I)u(t_n))\right] + \left[\frac{1}{\tau}(u(t_{n+1}) - u(t_n))\right. \\ & \left. -\Theta u'(t_{n+1}) - \overline{\Theta}u'(t_n)\right] =: g_1 + g_2 . \end{aligned} \tag{9.198}$$

Taylor expansion with integral remainder implies

$$u(t_{n+1}) = u(t_n) + u'(t_n)\tau + \int_{t_n}^{t_{n+1}} (t_{n+1} - s)u''(s)\,ds$$

and

$$u(t_n) = u(t_{n+1}) - u'(t_{n+1})\tau + \int_{t_{n+1}}^{t_n} (t_n - s)u''(s)\,ds .$$

Using the above relations we get the following useful representations of the difference quotient of u in t_n :

$$\begin{aligned}\frac{1}{\tau}(u(t_{n+1}) - u(t_n)) &= u'(t_n) + \frac{1}{\tau}\int_{t_n}^{t_{n+1}} (t_{n+1} - s)u''(s)\,ds , \\ \frac{1}{\tau}(u(t_{n+1}) - u(t_n)) &= u'(t_{n+1}) + \frac{1}{\tau}\int_{t_n}^{t_{n+1}} (t_n - s)u''(s)\,ds .\end{aligned}$$

Multiplying the first equation by $\overline{\Theta}$ and the second one by Θ, the summation of the results yields

$$\begin{aligned}\frac{1}{\tau}(u(t_{n+1}) - u(t_n)) &= \Theta u'(t_{n+1}) + \overline{\Theta}u'(t_n) \\ &\quad + \frac{1}{\tau}\int_{t_n}^{t_{n+1}} [\Theta t_n + \overline{\Theta}t_{n+1} - s]u''(s)\,ds .\end{aligned}$$

Since $|\Theta t_n + \overline{\Theta} t_{n+1} - s| \le \tau$, the term g_2 in the decomposition (9.198) of w^n can be estimated as

$$\left\| \frac{1}{\tau}(u(t_{n+1}) - u(t_n)) - \Theta u'(t_{n+1}) - \overline{\Theta} u'(t_n) \right\|_0 \le \int_{t_n}^{t_{n+1}} \|u''(s)\|_0 \, ds \, .$$

To estimate the term g_1 in (9.198), Taylor expansion with integral remainder is applied to the function $v(t) := (R_h - I)u(t)$. Then we have

$$\frac{1}{\tau}((R_h - I)u(t_{n+1}) - (R_h - I)u(t_n)) = \frac{1}{\tau} \int_{t_n}^{t_{n+1}} [(R_h - I)u(s)]' \, ds \, .$$

With the assumption on u using the fact that the derivative and the elliptic projection commute, we get

$$\left\| \frac{1}{\tau}((R_h - I)u(t_{n+1}) - (R_h - I)u(t_n)) \right\|_0 \le \frac{1}{\tau} \int_{t_n}^{t_{n+1}} \|(R_h - I)u'(s)\|_0 \, ds \, .$$

With (9.197) and summing the estimates for $\|w^n\|_0$ we obtain the following result.

Theorem 9.54 *Let a be a V-coercive, continuous bilinear form, $u_{0h} \in V_h$, $u_0 \in V$, $\Theta \in [\frac{1}{2}, 1]$. If $u \in C^2([0, T], H)$, then*

$$\begin{aligned} \|u(t_n) - U^n\|_0 \le{}& \|u_{0h} - R_h u_0\|_0 + \|(I - R_h)u(t_n)\|_0 \\ &+ \int_0^{t_n} \|(I - R_h)u'(s)\|_0 \, ds + \tau \int_0^{t_n} \|u''(s)\|_0 \, ds \, . \end{aligned}$$

Remark 9.55 1) Under stronger smoothness assumptions on u and by detailed considerations it can also be shown that the Crank–Nicolson method ($\Theta = \frac{1}{2}$) is of order 2 in τ.

In fact, the term g_2 in the above proof by considering the Taylor expansion with one term more can be estimated by

$$\|g_2\|_0 \le \frac{3}{4}\tau \int_{t_n}^{t_{n+1}} \|u'''(s)\|_0 \, ds \, ,$$

provided $u \in C^3([0, T], H)$ and therefore the last term in the estimate of Theorem 9.54 can be substituted for $\Theta = \frac{1}{2}$ by

$$\tau^2 \frac{3}{4} \int_0^{t_n} \|u'''(s)\|_0 \, ds \, .$$

2) Contrary to the semidiscrete situation (Theorem 9.13), the fully discrete estimate does not reflect any exponential decay in time.

Utilizing the error estimate for the elliptic projection as in Section 9.2 (cf. Theorem 9.13) and assuming $u_0 \in V \cap H^2(\Omega)$, we have

$$\|u(t_n) - U^n\|_0 \le \|u_{0h} - u_0\|_0 + Ch^2 \left[\|u_0\|_2 + \|u(t_n)\|_2 + \int_0^{t_n} \|u'(s)\|_2 \, ds \right] + \tau^\alpha \int_0^{t_n} \|u''(s)\|_0 \, ds \, .$$

If, in addition, $\|u_{0h} - u_0\|_0 \le Ch^2 \|u_0\|_2$, we obtain

$$\|u(t_n) - U^n\|_0 \le C(u)(h^2 + \tau^\alpha),$$

with $C(u) > 0$ depending on the solution u (and thus on u_0) but not depending on h and τ.

Hereby $\alpha = 1$ or $\alpha = 2$, if $\Theta = \frac{1}{2}$ and $u \in C^3([0, T], H)$.

To conclude this section we give without proof a summary of error estimates for all possible values of Θ:

$$\|u(t_n) - U^n\|_0 \le \begin{cases} C(u)(h^2 + \tau), & \text{if } \Theta \in [\frac{1}{2}, 1], \\ C(u)(h^2 + \tau^2), & \text{if } \Theta = \frac{1}{2}, \\ C(u)h^2, & \text{if } \Theta \in [0, 1] \text{ and } \tau \le \vartheta h^2, \end{cases} \tag{9.199}$$

where $\vartheta > 0$ is a constant upper bound of the step size relation τ/h^2.

The occurrence of such a restriction is not surprising, since similar requirements have already appeared for the finite difference method.

We also mention that the above restriction to a constant step size τ is only for simplicity of the notation. If a variable step size τ_{n+1} is used (which is typically determined by a step size control strategy), then the number τ in Theorem 9.54 is to be replaced by $\max_{n=0,\ldots,N-1} \tau_n$.

Order of Convergence Estimates for the (Node-Oriented) Finite Volume Method

We now consider the one-step-theta method for the finite volume method as introduced in (9.101).

The error analysis will run in a similar way as for the finite element method.

We write

$$u(t_n) - U^n = u(t_n) - R_h u(t_n) + R_h u(t_n) - U^n =: \varrho(t_n) + \theta^n \, ,$$

where R_h is the auxiliary operator defined in (9.89). So for the first term of the right-hand side, an estimate is already known.

From the definition (9.89) and (9.37), we immediately derive the following identity:

$$\begin{aligned}
&\left\langle \frac{1}{\tau}(\theta^{n+1} - \theta^n), v_h \right\rangle_{0,h} + a_h(\Theta\theta^{n+1} + \overline{\Theta}\theta^n, v_h) \\
&= \left\langle \frac{1}{\tau}(R_h u(t_{n+1}) - R_h u(t_n)), v_h \right\rangle_{0,h} + a_h(\Theta R_h u(t_{n+1}) + \overline{\Theta} R_h u(t_n), v_h) \\
&\quad - \left\langle \frac{1}{\tau}(U^{n+1} - U^n), v_h \right\rangle_{0,h} - a_h(\Theta U^{n+1} + \overline{\Theta} U^n, v_h) \\
&= \left\langle \frac{1}{\tau}(R_h u(t_{n+1}) - R_h u(t_n)), v_h \right\rangle_{0,h} + a(\Theta u(t_{n+1}) + \overline{\Theta} u(t_n), v_h) \\
&\quad - \langle f^{n+\Theta}, v_h \rangle_{0,h} \\
&= \left\langle \frac{1}{\tau}(R_h u(t_{n+1}) - R_h u(t_n)), v_h \right\rangle_{0,h} - \langle \Theta u'(t_{n+1}) + \overline{\Theta} u'(t_n), v_h \rangle_0 \\
&\quad + \langle f^{n+\Theta}, v_h \rangle_0 - \langle f^{n+\Theta}, v_h \rangle_{0,h} \\
&= \left\langle \frac{1}{\tau}(R_h u(t_{n+1}) - R_h u(t_n)), v_h \right\rangle_{0,h} - \langle \Theta u'(t_{n+1}) + \overline{\Theta} u'(t_n), v_h \rangle_{0,h} \\
&\quad + \langle \Theta u'(t_{n+1}) + \overline{\Theta} u'(t_n), v_h \rangle_{0,h} - \langle \Theta u'(t_{n+1}) + \overline{\Theta} u'(t_n), v_h \rangle_0 \\
&\quad + \langle f^{n+\Theta}, v_h \rangle_0 - \langle f^{n+\Theta}, v_h \rangle_{0,h} \\
&= \langle w^n, v_h \rangle_{0,h} + r^n(v_h),
\end{aligned}$$

where

$$w^n := \frac{1}{\tau}(R_h u(t_{n+1}) - R_h u(t_n)) - \Theta u'(t_{n+1}) - \overline{\Theta} u'(t_n)$$

and

$$\begin{aligned}
r^n(v_h) := {} & \langle \Theta u'(t_{n+1}) + \overline{\Theta} u'(t_n), v_h \rangle_{0,h} - \langle \Theta u'(t_{n+1}) + \overline{\Theta} u'(t_n), v_h \rangle_0 \\
& + \langle f^{n+\Theta}, v_h \rangle_0 - \langle f^{n+\Theta}, v_h \rangle_{0,h} .
\end{aligned}$$

Under the assumptions of Theorem 8.18, we know that $a_h(v_h, v_h) \geq 0$ for all $v_h \in V_h$. The particular choice of the test function as $v_h = v_h^\Theta := \Theta\theta^{n+1} + \overline{\Theta}\theta^n$ yields, similarly to the finite element case, for $\Theta \in [\frac{1}{2}, 1]$ the estimate

$$\begin{aligned}
&\left[\|\theta^{n+1}\|_{0,h} - \|\theta^n\|_{0,h}\right] \left[\Theta\|\theta^{n+1}\|_{0,h} + \overline{\Theta}\|\theta^n\|_{0,h}\right] \\
&\leq \tau \left(\langle w^n, v_h^\Theta \rangle_{0,h} + r^n(v_h^\Theta) \right) \\
&\leq \tau \left(\|w^n\|_{0,h} + \sup_{v_h \in V_h} \frac{r^n(v_h)}{\|v_h\|_{0,h}} \right) \|v_h^\Theta\|_{0,h} \\
&\leq \tau \left(\|w^n\|_{0,h} + \sup_{v_h \in V_h} \frac{r^n(v_h)}{\|v_h\|_{0,h}} \right) \left[\Theta\|\theta^{n+1}\|_{0,h} + \overline{\Theta}\|\theta^n\|_{0,h}\right] .
\end{aligned}$$

Dividing each side by the expression in the square brackets, we get

$$\|\theta^{n+1}\|_{0,h} \leq \|\theta^n\|_{0,h} + \tau \left(\|w^n\|_{0,h} + \sup_{v_h \in V_h} \frac{r^n(v_h)}{\|v_h\|_{0,h}} \right) .$$

The recursive application of this inequality leads to

$$\|\theta^{n+1}\|_{0,h} \le \|\theta^0\|_{0,h} + \tau \sum_{j=0}^{n} \|w^j\|_{0,h} + \tau \sum_{j=0}^{n} \sup_{v_h \in V_h} \frac{r^j(v_h)}{\|v_h\|_{0,h}} . \tag{9.200}$$

The representation of w^j obtained in the subsection on the finite element method yields the following estimate:

$$\|w^j\|_{0,h} \le \frac{1}{\tau} \int_{t_j}^{t_{j+1}} \|(R_h - I)u'(s)\|_{0,h}\, ds + \int_{t_j}^{t_{j+1}} \|u''(s)\|_{0,h}\, ds .$$

or of higher order as in Remark (9.55),(1) for more smooth u in the case $\Theta = \frac{1}{2}$. Furthermore, by Lemma 9.23, we have

$$|r^j(v_h)| \le Ch \left[\Theta |u'(t_{j+1})|_{1,\infty} + \overline{\Theta} |u'(t_j)|_{1,\infty} + |f^{j+\Theta}|_{1,\infty}\right] \|v_h\|_{0,h} .$$

Using both estimates in (9.200), we obtain

$$\begin{aligned}
&\|\theta^{n+1}\|_{0,h} \\
&\le \|\theta^0\|_{0,h} + C\left[\int_0^{t_{n+1}} \|(R_h - I)u'(s)\|_{0,h}\, ds + \tau \int_0^{t_{n+1}} \|u''(s)\|_{0,h}\, ds\right] \\
&\quad + Ch\tau\left[\overline{\Theta}|u'(0)|_{1,\infty} + \sum_{j=1}^{n} |u'(t_j)|_{1,\infty} + \Theta|u'(t_{n+1})|_{1,\infty} \right.\\
&\qquad\qquad \left. + \sum_{j=0}^{n} |f^{j+\Theta}|_{1,\infty}\right] \\
&\le \|\theta^0\|_{0,h} + C\left[\int_0^{t_{n+1}} \|(R_h - I)u'(s)\|_{0,h}\, ds + \tau \int_0^{t_{n+1}} \|u''(s)\|_{0,h}\, ds\right] \\
&\quad + Ch\left[\sup_{s\in(0,t_{n+1})} |u'(s)|_{1,\infty} + \sup_{s\in(0,t_{n+1})} |f(s)|_{1,\infty}\right].
\end{aligned}$$

This is the basic estimate. The final estimate is easily obtained by the same approach as in the finite element method. In summary, we have the following result.

Theorem 9.56 *In addition to the assumptions of Theorem* 8.18, *consider the finite volume method on Donald diagrams. Furthermore, let* $u_{0h} \in V_h$, $u_0 \in V \cap H^2(\Omega)$, $f \in C([0,T], C^1(\overline{\Omega}))$, $\Theta \in [\frac{1}{2}, 1]$. *Then if* $u \in C^2([0,T], C^2(\overline{\Omega}))$, *the following estimate is valid:*

$$\begin{aligned}
\|u(t_n) - U^n\|_{0,h} \le \|u_{0h} - u_0\|_{0,h} + Ch\Big[&\|u_0\|_2 + \|u(t_n)\|_2 \\
&+ \int_0^{t_n} \|u'(s)\|_2\, ds + \sup_{s\in(0,t_n)} |u'(s)|_{1,\infty} \\
&+ \sup_{s\in(0,t_n)} |f(s)|_{1,\infty}\Big] + C\tau \int_0^{t_n} \|u''(s)\|_{0,h}\, ds .
\end{aligned}$$

For $\Theta = \frac{1}{2}$ and $u \in C^3([0,T], C(\overline{\Omega}))$ the last term can be replaced by $c\tau^2 \int_0^{t_n} \|u'''(\tau)\|_{0,h} ds$.

Exercises

Problem 9.34 Verify Remark 9.55.

Problem 9.35 Consider a matrix $A \in \mathbb{R}^{M,M}$ with given profile and bandwidth. For $n \in \mathbb{N}$ investigate profile and bandwidth of A^n.

Problem 9.36 Consider the DG method (7.100) combined with the one-step-theta method for $\Theta = 0, 1/2, 1$. Follow the proof of Theorem 9.54 (see also Remark 9.55) to show an order of convergence estimate in space and time, assuming appropriate regularity.

Problem 9.37 Consider the time-dependent Darcy equation (9.83) with a conforming discretization in space (e.g. by RT-elements) and the one-step-theta method for $\Theta = 0, 1/2, 1$. Follow the exposition from (9.84) on to show an order of convergence estimate in space and time, assuming appropriate regularity.

Problem 9.38 Consider the mixed formulation in space of the time-dependent Stokes equation and a conforming discretization in space as in Problem 9.17, use the one-step-theta method for $\Theta = 0, 1/2, 1$ in time. Show an order of convergence estimate in space and time, assuming appropriate regularity.

Chapter 10
Discretization Methods for Convection-Dominated Problems

10.1 Standard Methods and Convection-Dominated Problems

As we have seen in the introductory Chapter 0, the modelling of transport and reaction processes in porous media results in differential equations of the form

$$\partial_t u - \nabla \cdot (\boldsymbol{K} \nabla u - \boldsymbol{c} u) = f \,,$$

which is a special case of the form (0.33), and just the time-dependent version of the formulation in divergence form (3.36) with $b = 1$. Similar equations occur in the modelling of the heat transport in flowing water, the carrier transport in semiconductors, and the propagation of epidemics. These application-specific equations often share the property that their so-called *global Péclet number*

$$\mathrm{Pe} := \frac{\|\boldsymbol{c}\|_\infty \operatorname{diam}(\Omega)}{\|\boldsymbol{K}\|_\infty} \tag{10.1}$$

is significantly larger than one. For example, representative values range from 25 (transport of a dissolved substance in ground water) up to about 10^7 (modelling of semiconductors). In such cases, the equations are called *convection-dominated.*

Therefore, in what follows, the Dirichlet boundary value problem introduced by (3.13), (3.21) ($\Gamma_3 = \partial\Omega$) will be looked at from the point of view of large global Péclet numbers, whereas in Section 10.4, the initial boundary value problem from Chapter 9 will be considered from this aspect.

Let $\Omega \subset \mathbb{R}^d$ denote a bounded domain with a Lipschitz continuous boundary. Given a function $f : \Omega \to \mathbb{R}$, a function $u : \Omega \to \mathbb{R}$ is to be determined such that

$$\begin{aligned} Lu &= f \quad \text{in } \Omega\,, \\ u &= 0 \quad \text{on } \Gamma\,, \end{aligned} \tag{10.2}$$

where again

$$Lu := -\nabla \cdot (\boldsymbol{K} \nabla u) + \boldsymbol{c} \cdot \nabla u + ru \,,$$

P. Knabner and L. Angermann, *Numerical Methods for Elliptic and Parabolic Partial Differential Equations*, Texts in Applied Mathematics 44,
https://doi.org/10.1007/978-3-030-79385-2_10

with sufficiently smooth coefficients

$$\boldsymbol{K} : \Omega \to \mathbb{R}^{d,d}, \quad \boldsymbol{c} : \Omega \to \mathbb{R}^d, \quad r : \Omega \to \mathbb{R}.$$

Unfortunately, standard discretization methods (finite difference, finite element, and finite volume methods) fail when applied to convection-dominated equations. At first glance, this seems to be a contradiction to the theory of these methods presented in the preceding chapters, because there we did not have any restriction on the global Péclet number. This apparent contradiction may be explained as follows: On the one hand, the theoretical results are still true for the convection-dominated case, but on the other hand, some assumptions of the statements therein (such as "for sufficiently small h") lack sharpness. This, in turn, may lead to practically unrealistic conditions (cf. the later discussion of the estimate (10.14)). For example, it may happen that the theoretically required step sizes are so small that the resulting discrete problems are too expensive or even untreatable.

So one can ask whether the theory is insufficient or not. The following example will show that this is not necessarily the case.

Example 10.1 *Given a constant diffusion coefficient $k > 0$, consider the boundary value problem*

$$\begin{aligned} (-ku' + u)' &= 0 \quad \textit{in } \Omega := (0, 1), \\ u(0) = u(1) - 1 &= 0. \end{aligned} \tag{10.3}$$

Its solution is

$$u(x) = \frac{1 - \exp(x/k)}{1 - \exp(1/k)}.$$

A rough sketch of the graph (Figure 10.1) shows that this function has a significant boundary layer at the right boundary of the interval even for the comparatively small global Péclet number Pe $= 100$. *In the larger subinterval (about* $(0, 0.95)$*) it is very smooth (nearly constant), whereas in the remaining small subinterval (about* $(0.95, 1)$*) the absolute value of its first derivative is large.*

Given an equidistant grid of width $h = 1/(M+1)$, $M \in \mathbb{N}$, a discretization by means of symmetric difference quotients yields the difference equations

$$\begin{aligned} -k\frac{u_{i-1} - 2u_i + u_{i+1}}{h^2} + \frac{u_{i+1} - u_{i-1}}{2h} &= 0, \quad i \in \{1, \ldots, M\} =: \Lambda, \\ u_0 = u_{M+1} - 1 &= 0. \end{aligned}$$

Collecting the coefficients and multiplying the result by $2h$, we arrive at

$$\left(-\frac{2k}{h} - 1\right) u_{i-1} + \frac{4k}{h} u_i + \left(-\frac{2k}{h} + 1\right) u_{i+1} = 0, \quad i \in \Lambda.$$

If we make the ansatz $u_i = \lambda^i$, the difference equations can be solved exactly:

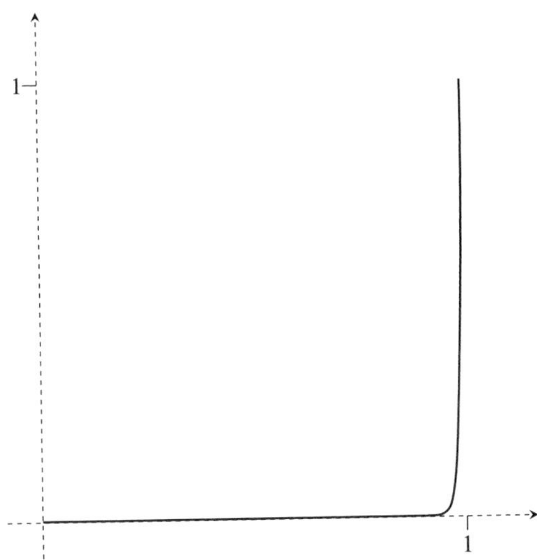

Fig. 10.1: Solution of (10.3) for $k = 0.01$.

$$u_i = \frac{1 - \left(\frac{2k+h}{2k-h}\right)^i}{1 - \left(\frac{2k+h}{2k-h}\right)^{M+1}} .$$

In the case $2k < h$, *which is by no means unrealistic (e.g., for the typical value* $k = 10^{-7}$*), the numerical solution considerably oscillates, in contrast to the behaviour of the exact solution* u. *These oscillations do not disappear until* $h < 2k$ *is reached, but this condition is very restrictive for small values of* k.

But even if the condition $h < 2k$ *is satisfied, undesirable effects can be observed. For example, in the special case* $h = k$ *we have at the node* $a_M = Mh$ *that*

$$\begin{aligned} u(a_M) &= \frac{1 - \exp(Mh/k)}{1 - \exp(1/k)} = \frac{1 - \exp(M)}{1 - \exp(M+1)} = \frac{\exp(-M) - 1}{\exp(-M) - \exp(1)} \\ &\to \exp(-1) \quad \textit{for } h \to 0, \end{aligned}$$

whereas the numerical solution at this point asymptotically behaves like (note that $\lambda = (2k+h)/(2k-h) = 3$*)*

$$u_M = \frac{1 - \lambda^M}{1 - \lambda^{M+1}} = \frac{\lambda^{-M} - 1}{\lambda^{-M} - \lambda} \to \frac{1}{\lambda} = \frac{1}{3} \quad \textit{for } h \to 0 .$$

So the numerical solution does not converge to the exact solution at the node a_M.

Again this is no contradiction to possible convergence results for the finite difference method in the discrete maximum norm, since now the diffusion coefficient is *not fixed, but rather the discretization is to be viewed as belonging to the limit case $k = 0$, with an artificial diffusion part in the discretization (see (10.9) below).*

Finite Difference Methods with Symmetric and One-Sided Difference Quotients

The oscillations in Example 10.1 show that in this case no comparison principle as in Corollary 1.14 is valid. Such a comparison principle, or more strongly a maximum principle, will lead to nonnegative solutions in the case of nonnegative right-hand side and Dirichlet data. This avoids for a homogeneous right-hand side an *undershooting*, as observed in Example 10.1, i.e., negative solution values in this case, and also an *overshooting*, i.e., solution values larger than the maximum of the Dirichlet data, provided that condition (1.35) (6)* holds.

In the following we will examine how the convective part influences the matrix properties (1.35) and thus the validity of a maximum or comparison principle and also conclude a first simple remedy.

We consider the model problem (10.2), for simplicity on a rectangle $\Omega = (0, a) \times (0, b)$, with constant, scalar $\boldsymbol{K} = k\boldsymbol{I}$ and equipped with an equidistant grid Ω_h. To maintain the order of consistency 2 of a spatial discretization of $-\nabla \cdot (\boldsymbol{K}\nabla u) = -k\Delta u$ by the five-point stencil, the use of the symmetric difference quotient for the discretization of

$$(\boldsymbol{c} \cdot \nabla u)(x) = c_1(x)\partial_1 u(x) + c_2(x)\partial_2 u(x)$$

suggests itself, i.e., for a grid point $x \in \Omega_h$,

$$c_1(x)\partial_1 u(x) \sim c_1(x)\frac{1}{2h}(u_{i+1,j} - u_{i-1,j}), \tag{10.4}$$

and similarly for $c_2(x)\partial_2 u(x)$ (cf. (1.7) for the notation). This leads to the following entries of the system matrix $\tilde{A}_h$, for example, in a rowwise numbering (compare (1.13)):

left secondary diagonal:	$-\dfrac{c_1(x)}{2h} - \dfrac{k}{h^2}\,;$
right secondary diagonal:	$+\dfrac{c_1(x)}{2h} - \dfrac{k}{h^2}\,;$
$l+1$ positions to the left:	$-\dfrac{c_2(x)}{2h} - \dfrac{k}{h^2}\,;$
$l+1$ positions to the right:	$+\dfrac{c_2(x)}{2h} - \dfrac{k}{h^2}\,;$
diagonal:	$\dfrac{4k}{h^2}\,.$

Conditions (1.35) (1) and (1.35) (6)* obviously hold.

We check the conditions sufficient for a comparison principle (Corollary 1.14). To satisfy condition (1.35) (2) we require

$$-\frac{k}{h^2} + \frac{|c_1(x)|}{2h} < 0\,,$$
$$-\frac{k}{h^2} + \frac{|c_2(x)|}{2h} < 0\,.$$

Denoting the *grid Péclet number* by

$$\mathrm{Pe}_h := \frac{\|\boldsymbol{c}\|_\infty h}{2k}\,, \tag{10.5}$$

the above conditions are satisfied if

$$\mathrm{Pe}_h < 1 \tag{10.6}$$

is satisfied. Under this assumption also the conditions (1.35) (5) and (7) are satisfied, and thus also (3), i.e., (10.6), is sufficient for the validity of a comparison principle. In Example 10.1 this is just the condition $h < 2k$.

The grid Péclet number is obviously related to the global Péclet number from (10.1) by

$$\mathrm{Pe}_h = \mathrm{Pe}\,\frac{h}{2\,\mathrm{diam}(\Omega)}\,.$$

The requirement (10.6) can always be met by choosing h sufficiently small, but for large Pe this may be a severe requirement, necessary for the sake of stability of the method, whereas for the accuracy desired a larger step size may be sufficient. A simple remedy to ensure condition (1.35) (2) is to use a one-sided (*upwind*) discretization of $c_1\partial_1 u$ and $c_2\partial_2 u$, which is selected against the stream direction defined by c_1 and c_2, respectively:

$$\begin{aligned} &\text{For } c_1(x) \geq 0: \quad && c_1(x)\partial_1 u(x) \sim c_1(x)\frac{1}{h}(u_{i,j} - u_{i-1,j}),\\ &\text{for } c_1(x) < 0: \quad && c_1(x)\partial_1 u(x) \sim c_1(x)\frac{1}{h}(u_{i+1,j} - u_{i,j}), \end{aligned} \tag{10.7}$$

and analogously for $c_2\partial_2 u$.

Due to this choice there are only additional nonnegative addends to the diagonal position and nonpositive ones to the off-diagonal positions compared to the five-point stencil or another discretization of a diffusive part. Thus all properties (1.35) (1)–(7), (4)*, and (6)* remain unaffected; i.e., the upwind discretization satisfies all qualitative properties of Section 1.4 from the inverse monotonicity to the strong maximum principle, without any restrictions to the local Péclet number.

The drawback lies in the reduced accuracy, since the one-sided difference quotient has only order of consistency 1. In Section 10.3 we will develop more refined upwind discretizations.

Due to

$$c_1 \frac{u(x,y)-u(x-h,y)}{h} = \tag{10.8}$$
$$\frac{c_1 h}{2} \frac{-u(x-h,y)+2u(x,y)-u(x+h,y)}{h^2} + c_1 \frac{u(x+h,y)-u(x-h,y)}{2h},$$

and analogously for the forward difference quotient, the upwind discretization can be perceived as a discretization with symmetric difference quotients if a step-size-dependent diffusive part, also discretized with $\partial^- \partial^+$, is added with the diffusion coefficient

$$\boldsymbol{K}_h(x) := \frac{h}{2} \begin{pmatrix} |c_1(x)| & 0 \\ 0 & |c_2(x)| \end{pmatrix}. \tag{10.9}$$

Therefore, one also speaks of adding *artificial diffusion* (or *viscosity*). The disadvantage of this *full* upwind method is that it recognizes the flow direction only if the flow is aligned to one of the coordinate axes. This will be improved in Section 10.2.

Error Estimates for the Standard Finite Element Method

In order to demonstrate the theoretical deficiencies, we will again reproduce the way for obtaining standard error estimates for a model problem. So let $\boldsymbol{K}(x) \equiv \varepsilon \boldsymbol{I}$ with a constant coefficient $\varepsilon > 0$, $\boldsymbol{c} \in C^1(\overline{\Omega}, \mathbb{R}^d)$, $r \in C(\overline{\Omega})$, $f \in L^2(\Omega)$. Furthermore, assume that the following inequality is valid in Ω, where $r_0 > 0$ is a constant: $r - \frac{1}{2}\nabla \cdot \boldsymbol{c} \geq r_0$.

Then the bilinear form $a : V \times V \to \mathbb{R}$, $V := H_0^1(\Omega)$, corresponding to the boundary value problem (10.2), reads as (cf. (3.24))

$$a(u,v) := \int_\Omega [\varepsilon \nabla u \cdot \nabla v + \boldsymbol{c} \cdot \nabla u\, v + r\, uv\,]\, dx\,, \quad u, v \in V\,. \tag{10.10}$$

To get a coercivity estimate of a, we set $u = v \in V$ in (10.10) and take the relation $2v(\boldsymbol{c} \cdot \nabla v) = \boldsymbol{c} \cdot \nabla v^2$ into account. Integrating by parts the middle term, we obtain

$$\begin{aligned} a(v,v) &= \varepsilon |v|_1^2 + \langle \boldsymbol{c} \cdot \nabla v, v \rangle_0 + \langle rv, v \rangle_0 \\ &= \varepsilon |v|_1^2 - \left\langle \frac{1}{2} \nabla \cdot \boldsymbol{c}, v^2 \right\rangle_0 + \langle rv, v \rangle_0 = \varepsilon |v|_1^2 + \left\langle r - \frac{1}{2} \nabla \cdot \boldsymbol{c}, v^2 \right\rangle_0. \end{aligned}$$

Introducing the so-called *ε-weighted H^1-norm* by

$$\|v\|_\varepsilon := \left(\varepsilon |v|_1^2 + \|v\|_0^2 \right)^{1/2}, \tag{10.11}$$

we immediately arrive at the estimate

$$a(v,v) \geq \varepsilon |v|_1^2 + r_0 \|v\|_0^2 \geq \tilde{\alpha} \|v\|_\varepsilon^2, \tag{10.12}$$

where $\tilde{\alpha} := \min\{1, r_0\}$ does not depend on ε.

Due to $\boldsymbol{c} \cdot \nabla u = \nabla \cdot (\boldsymbol{c}u) - (\nabla \cdot \boldsymbol{c})u$, integration by parts yields for arbitrary $u, v \in V$ the identity

$$\langle \boldsymbol{c} \cdot \nabla u, v \rangle_0 = -\langle u, \boldsymbol{c} \cdot \nabla v \rangle_0 - \langle (\nabla \cdot \boldsymbol{c})u, v \rangle_0 \, .$$

So we get the continuity estimate

$$\begin{aligned} |a(u,v)| &\leq \varepsilon |u|_1 |v|_1 + \|\boldsymbol{c}\|_{0,\infty} \|u\|_0 |v|_1 + \left(|\boldsymbol{c}|_{1,\infty} + \|r\|_{0,\infty} \right) \|u\|_0 \|v\|_0 \\ &\leq \left(\sqrt{\varepsilon}\, |u|_1 + \|u\|_0 \right) \left\{ \left(\sqrt{\varepsilon} + \|\boldsymbol{c}\|_{0,\infty} \right) |v|_1 + \left(|\boldsymbol{c}|_{1,\infty} + \|r\|_{0,\infty} \right) \|v\|_0 \right\} \qquad (10.13) \\ &\leq \tilde{M} \|u\|_\varepsilon \|v\|_1 \, , \end{aligned}$$

where $\tilde{M} := 2 \max\{\sqrt{\varepsilon} + \|\boldsymbol{c}\|_{0,\infty}, |\boldsymbol{c}|_{1,\infty} + \|r\|_{0,\infty}\}$.

Since we are interested in the case of small diffusion $\varepsilon > 0$ and present convection (i.e., $\|\boldsymbol{c}\|_{0,\infty} > 0$), the continuity constant $\tilde{M}$ can be bounded independent of ε. It is not very surprising that the obtained continuity estimate is nonsymmetric, since also the differential expression L behaves like that. Passing over to a symmetric estimate results in the following relation:

$$|a(u,v)| \leq \frac{\tilde{M}}{\sqrt{\varepsilon}} \|u\|_\varepsilon \|v\|_\varepsilon \, .$$

Now, if $V_h \subset V$ denotes a finite element space, we can argue as in the proof of Céa's lemma (Theorem 2.17) and get an error estimate for the corresponding finite element solution $u_h \in V_h$. To do this, the nonsymmetric continuity estimate (10.13) is sufficient. Indeed, for arbitrary $v_h \in V_h$, we have

$$\tilde{\alpha} \|u - u_h\|_\varepsilon^2 \leq a(u - u_h, u - u_h) = a(u - u_h, u - v_h) \leq \tilde{M} \|u - u_h\|_\varepsilon \|u - v_h\|_1 \, .$$

Thus

$$\|u - u_h\|_\varepsilon \leq \frac{\tilde{M}}{\tilde{\alpha}} \inf_{v_h \in V_h} \|u - v_h\|_1 \, .$$

Here the constant $\tilde{M}/\tilde{\alpha}$ does not depend on ε, h, and u. This estimate is weaker than the standard estimate, because the ε-weighted H^1-norm is weaker than the H^1-norm. Moreover, the error of the best approximation is *not* independent of ε, in general. For example, if we apply continuous, piecewise linear elements, then, under the additional assumption $u \in H^2(\Omega)$, Theorem 3.32 yields the estimate

$$\inf_{v_h \in V_h} \|u - v_h\|_1 \leq \|u - I_h(u)\|_1 \leq Ch|u|_2 \, ,$$

where the constant $C > 0$ does not depend on ε, h, and u. So, we finally arrive at the relation

$$\|u - u_h\|_\varepsilon \leq Ch|u|_2 \, . \qquad (10.14)$$

However, the H^2-seminorm of the solution u depends on ε in a disadvantageous manner; for example, it may be (cf. also [53, Lemma III.1.18]) that

$$|u|_2 = O(\varepsilon^{-3/2}) \qquad (\varepsilon \to 0) \, .$$

This result is sharp, since for examples of boundary value problems for ordinary linear differential equations the error of the best approximation already exhibits this asymptotic behaviour.

So the practical as well as the theoretical problems mentioned above indicate the necessity to use special numerical methods for solving convection-dominated equations. In the next sections, a small collection of these methods will be depicted.

Exercises

Problem 10.1 a) Given a constant diffusion coefficient $\varepsilon > 0$, rewrite the ordinary boundary value problem

$$\begin{aligned} (-\varepsilon u' + u)' &= 0 \quad \text{in } \Omega := (0,1), \\ u(0) &= 0, \quad u(1) = 1 \end{aligned}$$

into an equivalent form but with nonnegative right-hand side and homogeneous Dirichlet boundary conditions.

b) Compute the $H^2(0,1)$-seminorm of the solution of the transformed problem and investigate its dependence on ε.

Programming project 10.1 Write a function that solves the problem

$$\begin{aligned} -u'' + cu' &= 0 \quad \text{in } \Omega := (0,1), \\ u(0) = 0, &\quad u(1) = 1 \end{aligned}$$

numerically using finite differences on an equidistant grid with grid size $h > 0$ (you may extend the code from Project 1.1). To discretize the first-order term, implement the forward, backward, and central difference quotient. Pass a parameter to your function that defines which one to use.

Compare the results for $h := 2^{-i}$, $i = 3, \ldots, 8$, and $c := 10^j$, $j = -1, \ldots, 2$. The exact solution is $u(x) = (e^{cx} - 1)/(e^c - 1)$.

Programming project 10.2 Consider the two-dimensional situation of Project 10.1 with general Dirichlet boundary conditions:

$$\begin{aligned} -\varepsilon \Delta u + \boldsymbol{c} \cdot \nabla u &= f \quad \text{in } \Omega := (0,a) \times (0,b), \\ u &= g \quad \text{on } \partial\Omega. \end{aligned}$$

Write a function that solves this problem numerically by means of a finite difference scheme on a quadratic grid with grid size $h := 1/m$, $m \in \mathbb{N}$ (you may start from the code of Project 1.2 or Project 1.3).

The approximation of the derivatives in x/y-direction in the first-order term should be made depending on the value of $\boldsymbol{c}(x, y)$ as follows:

- Backward difference quotient if the respective component is positive.
- Forward difference quotient if the respective component is negative.
- Central difference quotient if the respective component is zero.

Test your implementation for $\Omega := (0,1)^2$ with $\varepsilon := 10^{-i}$, $i = 1, 3, 5, 7$, $f := 0$, and

- $\boldsymbol{c} := \begin{pmatrix} 2 \\ 1 \end{pmatrix}, \quad g(x,y) := \begin{cases} 0, & y < \frac{3}{4}x + \frac{1}{4}, \\ \frac{1}{2}, & x = \left(0, \frac{1}{4}\right), \\ 1, & \text{otherwise}, \end{cases}$
- $\boldsymbol{c} := \begin{pmatrix} 9.975x - 19.22(y - \frac{1}{2}) \\ -9.975(y - \frac{1}{2}) + 6.41x \end{pmatrix}, \quad g(x,y) := \begin{cases} 1 - x, & \frac{1}{4} \le y \le \frac{9}{20}, \\ 0, & \text{otherwise}. \end{cases}$

10.2 The Streamline-Diffusion Method

The streamline-diffusion method is the prevalent method in the numerical treatment of stationary convection-dominated problems. The basic idea is due to Brooks and Hughes [113], who called the method the *streamline upwind Petrov–Galerkin method (SUPG method)*.

We describe the idea of the method for a special case of boundary value problem (10.2) under consideration. Let the domain $\Omega \subset \mathbb{R}^d$ be a bounded polyhedron. We consider the same model as in the preceding section, that is, $\boldsymbol{K}(x) \equiv \varepsilon \boldsymbol{I}$ with a constant coefficient $\varepsilon > 0$, $\boldsymbol{c} \in C^1(\overline{\Omega}, \mathbb{R}^d)$, $r \in C(\overline{\Omega})$, $f \in L^2(\Omega)$. We also assume that the inequality $r - \frac{1}{2}\nabla \cdot \boldsymbol{c} \ge r_0$ is valid in Ω, where $r_0 > 0$ is a constant. Then the variational formulation of (10.2) reads as follows:

Find $u \in V$ such that

$$a(u, v) = \langle f, v \rangle_0 \quad \text{for all } v \in V, \tag{10.15}$$

where a is the bilinear form (10.10).

Given a shape-regular family of triangulations $\{\mathcal{T}_h\}$, let $V_h \subset V$ denote the set of continuous functions that are piecewise polynomial of degree $k \in \mathbb{N}$ and satisfy the boundary conditions, i.e.,

$$V_h := \left\{ v_h \in V \;\middle|\; v_h|_K \in \mathcal{P}_k(K) \text{ for all } K \in \mathcal{T}_h \right\} . \tag{10.16}$$

If in addition the solution $u \in V$ of (10.15) belongs to the space $H^{k+1}(\Omega)$, we have, by (3.100), the following error estimate for the interpolant $I_h(u)$:

$$\|u - I_h(u)\|_{l,K} \le c_{\text{int}} h_K^{k+1-l} |u|_{k+1,K} \tag{10.17}$$

for $0 \le l \le k+1$ and all $K \in \mathcal{T}_h$. Since the spaces V_h are of finite dimension, a so-called *inverse inequality* can be proven (cf. Theorem 3.46, 2) and Problem 10.2):

$$\|\Delta v_h\|_{0,K} \le \frac{c_{\text{inv}}}{h_K} |v_h|_{1,K} \tag{10.18}$$

for all $v_h \in V_h$ and all $K \in \mathcal{T}_h$. Here it is important that the constants $c_{\mathrm{int}}, c_{\mathrm{inv}} > 0$ from (10.17) and (10.18), respectively, do not depend on u or v_h and on the particular elements $K \in \mathcal{T}_h$.

The basic idea of the streamline-diffusion method consists in the addition of suitably weighted residuals to the variational formulation (10.15). Because of the assumption $u \in H^{k+1}(\Omega)$, $k \in \mathbb{N}$, the differential equation can be interpreted as an equation in $L^2(\Omega)$. In particular, it is valid on any element $K \in \mathcal{T}_h$ in the sense of $L^2(K)$, i.e.,

$$-\varepsilon\Delta u + \boldsymbol{c}\cdot\nabla u + ru = f \quad \text{almost everywhere in } K \text{ and for all } K \in \mathcal{T}_h\,.$$

Next we take an elementwise defined mapping $\tau : V_h \to L^2(\Omega)$ and multiply the local differential equation in $L^2(K)$ by the restriction of $\tau(v_h)$ to K. Scaling by a parameter $\delta_K \in \mathbb{R}$ and summing the results over all elements $K \in \mathcal{T}_h$, we obtain

$$\sum_{K\in\mathcal{T}_h} \delta_K \,\langle -\varepsilon\Delta u + \boldsymbol{c}\cdot\nabla u + ru, \tau(v_h)\rangle_{0,K} = \sum_{K\in\mathcal{T}_h} \delta_K \,\langle f, \tau(v_h)\rangle_{0,K}\,.$$

If we add this relation to equation (10.15) restricted to V_h, we see that the weak solution $u \in V \cap H^{k+1}(\Omega)$ satisfies the following variational equation:

$$a_h(u, v_h) = \langle f, v_h\rangle_h \quad \text{for all } v_h \in V_h\,,$$

where

$$a_h(u, v_h) := a(u, v_h) + \sum_{K\in\mathcal{T}_h} \delta_K \,\langle -\varepsilon\Delta u + \boldsymbol{c}\cdot\nabla u + ru, \tau(v_h)\rangle_{0,K}\,,$$

$$\langle f, v_h\rangle_h := \langle f, v_h\rangle_0 + \sum_{K\in\mathcal{T}_h} \delta_K \,\langle f, \tau(v_h)\rangle_{0,K}\,.$$

Then the corresponding discretization reads as follows:

Find $u_h \in V_h$ such that

$$a_h(u_h, v_h) = \langle f, v_h\rangle_h \quad \text{for all } v_h \in V_h\,. \tag{10.19}$$

In this way we have constructed a consistent and conforming Galerkin discretization (see Definition (6.5)), i.e., we have Galerkin orthogonality (6.16):

Corollary 10.2 *Suppose the problems* (10.15) *and* (10.19) *have a solution* $u \in V \cap H^{k+1}(\Omega)$ *and* $u_h \in V_h$, *respectively. Then the following error equation is valid:*

$$a_h(u - u_h, v_h) = 0 \quad \textit{for all } v_h \in V_h\,. \tag{10.20}$$

In the *streamline-diffusion method (sdFEM)*, the mapping τ used in (10.19) is chosen as $\tau(v_h) := \boldsymbol{c}\cdot\nabla v_h$.

Without going into details, we mention that a further option is to set $\tau(v_h) := -\varepsilon\Delta v_h + \boldsymbol{c}\cdot\nabla v_h + rv_h$. This results in the so-called *Galerkin/least squares–FEM (GLSFEM)* [150].

Especially with regard to the extension of the method to other finite element spaces, the discussion of how to choose τ and δ_K is not yet complete.

Interpretation of the Additional Term in the Case of Linear Elements

If the finite element spaces V_h are formed by piecewise linear functions (i.e., in the above definition (10.16) of V_h we have $k = 1$), we get $\Delta v_h|_K = 0$ for all $K \in \mathcal{T}_h$. If in addition there is no reactive term (i.e., $r = 0$), the discrete bilinear form is

$$a_h(u_h, v_h) = \int_\Omega \varepsilon \nabla u_h \cdot \nabla v_h \, dx + \langle \boldsymbol{c} \cdot \nabla u_h, v_h \rangle_0 + \sum_{K \in \mathcal{T}_h} \delta_K \, \langle \boldsymbol{c} \cdot \nabla u_h, \boldsymbol{c} \cdot \nabla v_h \rangle_{0,K} \, .$$

Since the scalar product appearing in the sum can be rewritten as $\langle \boldsymbol{c} \cdot \nabla u_h, \boldsymbol{c} \cdot \nabla v_h \rangle_{0,K} = \int_K (\boldsymbol{c}\boldsymbol{c}^T \nabla u_h) \cdot \nabla v_h \, dx$, we obtain the following equivalent representation:

$$a_h(u_h, v_h) = \sum_{K \in \mathcal{T}_h} \int_K \left((\varepsilon I + \delta_K \boldsymbol{c}\boldsymbol{c}^T) \nabla u_h \right) \cdot \nabla v_h \, dx + \langle \boldsymbol{c} \cdot \nabla u_h, v_h \rangle_0 \, .$$

This shows that the additional term introduces an element-dependent extra diffusion in the direction of the convective field $\boldsymbol{c}$ (cf. also Problem 0.3), which motivates the name of the method. In this respect, the streamline-diffusion method can be understood as an improved version of the full upwind method, as seen, for example, in (10.7).

Analysis of the Streamline-Diffusion Method

To start the analysis of stability and convergence properties of the streamline-diffusion method, we consider the term $a_h(v_h, v_h)$ for arbitrary $v_h \in V_h$.

As in Section 3.2.1, the structure of the discrete bilinear form a_h allows us to derive the estimate

$$a_h(v_h, v_h) \geq \varepsilon |v_h|_1^2 + r_0 \|v_h\|_0^2 + \sum_{K \in \mathcal{T}_h} \delta_K \, \langle -\varepsilon \Delta v_h + \boldsymbol{c} \cdot \nabla v_h + r v_h, \boldsymbol{c} \cdot \nabla v_h \rangle_{0,K} \, .$$

Furthermore, neglecting for a moment the second term in the sum on the right-hand side and using the elementary inequality $ab \leq a^2 + b^2/4$ for arbitrary $a, b \in \mathbb{R}$, we get

$$\begin{aligned}
&\left|\sum_{K\in\mathcal{T}_h}\delta_K\left\langle-\varepsilon\Delta v_h+rv_h,\boldsymbol{c}\cdot\nabla v_h\right\rangle_{0,K}\right| \\
&\qquad\le\sum_{K\in\mathcal{T}_h}\left\{\left|\left\langle-\varepsilon\sqrt{|\delta_K|}\,\Delta v_h,\sqrt{|\delta_K|}\,\boldsymbol{c}\cdot\nabla v_h\right\rangle_{0,K}\right|\right. \\
&\qquad\qquad\left.+\left|\left\langle\sqrt{|\delta_K|}\,rv_h,\sqrt{|\delta_K|}\,\boldsymbol{c}\cdot\nabla v_h\right\rangle_{0,K}\right|\right\} \\
&\qquad\le\sum_{K\in\mathcal{T}_h}\left\{\varepsilon^2|\delta_K|\,\|\Delta v_h\|_{0,K}^2+|\delta_K|\,\|r\|_{0,\infty,K}^2\|v_h\|_{0,K}^2\right. \\
&\qquad\qquad\left.+\frac{|\delta_K|}{2}\,\|\boldsymbol{c}\cdot\nabla v_h\|_{0,K}^2\right\}.
\end{aligned}$$

By means of the inverse inequality (10.18) it follows that

$$\begin{aligned}
\left|\sum_{K\in\mathcal{T}_h}\delta_K\left\langle-\varepsilon\Delta v_h+rv_h,\boldsymbol{c}\cdot\nabla v_h\right\rangle_{0,K}\right| &\le\sum_{K\in\mathcal{T}_h}\left\{\varepsilon^2|\delta_K|\,\frac{c_{\mathrm{inv}}^2}{h_K^2}|v_h|_{1,K}^2\right. \\
&\left.+|\delta_K|\,\|r\|_{0,\infty,K}^2\|v_h\|_{0,K}^2+\frac{|\delta_K|}{2}\,\|\boldsymbol{c}\cdot\nabla v_h\|_{0,K}^2\right\}.
\end{aligned}$$

Putting things together, we obtain

$$\begin{aligned}
a_h(v_h,v_h)\ge\sum_{K\in\mathcal{T}_h}&\left\{\left(\varepsilon-\varepsilon^2|\delta_K|\,\frac{c_{\mathrm{inv}}^2}{h_K^2}\right)|v_h|_{1,K}^2\right. \\
&\left.+\left(r_0-|\delta_K|\,\|r\|_{0,\infty,K}^2\right)\|v_h\|_{0,K}^2+\left(\delta_K-\frac{|\delta_K|}{2}\right)\|\boldsymbol{c}\cdot\nabla v_h\|_{0,K}^2\right\}.
\end{aligned}$$

The choice

$$0<\delta_K\le\frac{1}{2}\min\left\{\frac{h_K^2}{\varepsilon c_{\mathrm{inv}}^2},\frac{r_0}{\|r\|_{0,\infty,K}^2}\right\} \tag{10.21}$$

leads to

$$a_h(v_h,v_h)\ge\frac{\varepsilon}{2}|v_h|_1^2+\frac{r_0}{2}\|v_h\|_0^2+\frac{1}{2}\sum_{K\in\mathcal{T}_h}\delta_K\|\boldsymbol{c}\cdot\nabla v_h\|_{0,K}^2\,.$$

Therefore, if the so-called *streamline-diffusion norm* is defined by

$$\|v\|_{\mathrm{sd}}:=\left(\varepsilon|v|_1^2+r_0\|v\|_0^2+\sum_{K\in\mathcal{T}_h}\delta_K\|\boldsymbol{c}\cdot\nabla v\|_{0,K}^2\right)^{1/2},\qquad v\in V,$$

then the choice (10.21) implies the estimate

$$\frac{1}{2}\|v_h\|_{\mathrm{sd}}^2\le a_h(v_h,v_h)\quad\text{for all } v_h\in V_h\,. \tag{10.22}$$

Obviously, the streamline-diffusion norm $\|\cdot\|_{\mathrm{sd}}$ is stronger than the ε-weighted H^1-norm (10.11); i.e.,

$$\min\{1, \sqrt{r_0}\}\|v\|_\varepsilon \le \|v\|_{\mathrm{sd}} \quad \text{for all } v \in V\,.$$

Now an error estimate can be proven. Since estimate (10.22) holds only on the finite element spaces V_h, we consider first the norm of $I_h(u) - u_h \in V_h$ and make use of the error equation (10.20):

$$\frac{1}{2}\|I_h(u) - u_h\|_{\mathrm{sd}}^2 \le a_h(I_h(u) - u_h, I_h(u) - u_h) = a_h(I_h(u) - u, I_h(u) - u_h)\,.$$

In particular, under the assumption $u \in V \cap H^{k+1}(\Omega)$ the following three estimates are valid:

$$\begin{aligned}\varepsilon \int_\Omega \nabla(I_h(u) - u)\cdot\nabla(I_h(u) - u_h)\,dx &\le \sqrt{\varepsilon}\,|I_h(u) - u|_1\|I_h(u) - u_h\|_{\mathrm{sd}}\\ &\le c_{\mathrm{int}}\sqrt{\varepsilon}\,h^k|u|_{k+1}\|I_h(u) - u_h\|_{\mathrm{sd}}\,,\end{aligned}$$

$$\begin{aligned}&\int_\Omega [\boldsymbol{c}\cdot\nabla(I_h(u) - u) + r(I_h(u) - u)](I_h(u) - u_h)\,dx\\ &= \int_\Omega (r - \nabla\cdot\boldsymbol{c})(I_h(u) - u)(I_h(u) - u_h)\,dx - \int_\Omega (I_h(u) - u)\,\boldsymbol{c}\cdot\nabla(I_h(u) - u_h)\,dx\\ &\le \|r - \nabla\cdot\boldsymbol{c}\|_{0,\infty}\|I_h(u) - u\|_0\|I_h(u) - u_h\|_0 + \|I_h(u) - u\|_0\|\boldsymbol{c}\cdot\nabla(I_h(u) - u_h)\|_0\\ &\le C\left\{\left(\sum_{K\in\mathcal{T}_h}\|I_h(u) - u\|_{0,K}^2\right)^{1/2} + \left(\sum_{K\in\mathcal{T}_h}\delta_K^{-1}\|I_h(u) - u\|_{0,K}^2\right)^{1/2}\right\}\|I_h(u) - u_h\|_{\mathrm{sd}}\\ &\le Ch^k\left(\sum_{K\in\mathcal{T}_h}\left(1 + \delta_K^{-1}\right)h_K^2|u|_{k+1,K}^2\right)^{1/2}\|I_h(u) - u_h\|_{\mathrm{sd}}\,,\end{aligned}$$

and

$$\begin{aligned}&\left|\sum_{K\in\mathcal{T}_h}\delta_K\,\langle -\varepsilon\Delta(I_h(u) - u) + \boldsymbol{c}\cdot\nabla(I_h(u) - u) + r(I_h(u) - u), \boldsymbol{c}\cdot\nabla(I_h(u) - u_h)\rangle_{0,K}\right|\\ &\le \sum_{K\in\mathcal{T}_h} c_{\mathrm{int}}\sqrt{\delta_K}\,\left[\varepsilon h_K^{k-1} + \|\boldsymbol{c}\|_{0,\infty,K}h_K^k + \|r\|_{0,\infty,K}h_K^{k+1}\right]\\ &\qquad\qquad \times |u|_{k+1,K}\sqrt{\delta_K}\,\|\boldsymbol{c}\cdot\nabla(I_h(u) - u_h)\|_{0,K}\\ &\le C\left(\sum_{K\in\mathcal{T}_h}\delta_K\left[\varepsilon h_K^{k-1} + h_K^k + h_K^{k+1}\right]^2|u|_{k+1,K}^2\right)^{1/2}\|I_h(u) - u_h\|_{\mathrm{sd}}\,.\end{aligned}$$

Condition (10.21), which was already required for estimate (10.22), implies that

$$\varepsilon\delta_K \le \frac{h_K^2}{c_{\mathrm{inv}}^2},$$

and so the application to the first term of the last bound leads to

$$\begin{aligned}&\Big|\sum_{K\in\mathcal{T}_h}\delta_K\big\langle -\varepsilon\Delta(I_h(u)-u)+\boldsymbol{c}\cdot\nabla(I_h(u)-u)\\&\qquad\qquad +r(I_h(u)-u),\boldsymbol{c}\cdot\nabla(I_h(u)-u_h)\big\rangle_{0,K}\Big|\\&\qquad\qquad\le Ch^k\Big(\sum_{K\in\mathcal{T}_h}[\varepsilon+\delta_K]\,|u|_{k+1,K}^2\Big)^{1/2}\|I_h(u)-u_h\|_{\mathrm{sd}}\,.\end{aligned}$$

Collecting the estimates and dividing by $\|I_h(u)-u_h\|_{\mathrm{sd}}$, we obtain the relation

$$\|I_h(u)-u_h\|_{\mathrm{sd}} \le Ch^k\Big(\sum_{K\in\mathcal{T}_h}\Big[\varepsilon+\frac{h_K^2}{\delta_K}+h_K^2+\delta_K\Big]|u|_{k+1,K}^2\Big)^{1/2}.$$

Finally, the terms in the square brackets will be equilibrated with the help of condition (10.21). We rewrite the ε-dependent term in this condition as

$$\frac{h_K^2}{\varepsilon c_{\mathrm{inv}}^2} = \frac{2}{c_{\mathrm{inv}}^2\|\boldsymbol{c}\|_{\infty,K}}\mathrm{Pe}_K h_K$$

with

$$\mathrm{Pe}_K := \frac{\|\boldsymbol{c}\|_{\infty,K}h_K}{2\varepsilon}. \tag{10.23}$$

This *local Péclet number* is a refinement of the definition (10.5).

The following distinctions concerning Pe_K are convenient:

$$\mathrm{Pe}_K \le 1 \quad \text{and} \quad \mathrm{Pe}_K > 1\,.$$

In the first case, we choose

$$\delta_K = \delta_0\mathrm{Pe}_K h_K = \delta_1\frac{h_K^2}{\varepsilon}, \qquad \delta_0 = \frac{2}{\|\boldsymbol{c}\|_{\infty,K}}\delta_1\,,$$

with appropriate constants $\delta_0 > 0$ and $\delta_1 > 0$, respectively, which are independent of K and ε. Then we have

$$\varepsilon+\frac{h_K^2}{\delta_K}+h_K^2+\delta_K = \Big(1+\frac{1}{\delta_1}\Big)\varepsilon+h_K^2+\delta_1\frac{2\mathrm{Pe}_K}{\|\boldsymbol{c}\|_{0,\infty,K}}h_K \le C(\varepsilon+h_K),$$

where $C > 0$ is independent of K and ε. In the second case, it is sufficient to choose $\delta_K = \delta_2 h_K$ with an appropriate constant $\delta_2 > 0$ that is independent of K and ε. Then

$$\delta_K = \frac{\delta_2}{\text{Pe}_K}\text{Pe}_K h_K = \frac{\delta_2 \|\boldsymbol{c}\|_{0,\infty,K}}{2\text{Pe}_K}\frac{h_K^2}{\varepsilon}$$

and

$$\varepsilon + \frac{h_K^2}{\delta_K} + h_K^2 + \delta_K = \varepsilon + \left(\frac{1}{\delta_2} + \delta_2\right) h_K + h_K^2 \leq C(\varepsilon + h_K),$$

with $C > 0$ independent of K and ε. Note that in both cases the constants can be chosen sufficiently small, independent of Pe_K, that the condition (10.21) is satisfied. Now we are prepared to prove the following error estimate.

Theorem 10.3 *Let the parameters δ_K be given by*

$$\delta_K = \begin{cases} \delta_1 \dfrac{h_K^2}{\varepsilon}, & \text{Pe}_K \leq 1, \\ \delta_2 h_K, & \text{Pe}_K > 1, \end{cases}$$

where $\delta_1, \delta_2 > 0$ do not depend on K and ε and are chosen such that condition (10.21) is satisfied. If the weak solution u of (10.15) belongs to $H^{k+1}(\Omega)$, then

$$\|u - u_h\|_{\text{sd}} \leq C\left(\sqrt{\varepsilon} + \sqrt{h}\right) h^k |u|_{k+1},$$

where the constant $C > 0$ is independent of ε, h, and u.

Proof By the triangle inequality, we get

$$\|u - u_h\|_{\text{sd}} \leq \|u - I_h(u)\|_{\text{sd}} + \|I_h(u) - u_h\|_{\text{sd}}.$$

An estimate of the second addend is already known. To deal with the first term, the estimates of the interpolation error (10.17) are used directly:

$$\begin{aligned}
&\|u - I_h(u)\|_{\text{sd}}^2 \\
&= \varepsilon|u - I_h(u)|_1^2 + r_0\|u - I_h(u)\|_0^2 + \sum_{K\in\mathcal{T}_h} \delta_K \|\boldsymbol{c}\cdot\nabla(u - I_h(u))\|_{0,K}^2 \\
&\leq c_{\text{int}}^2 \sum_{K\in\mathcal{T}_h} \left[\varepsilon h_K^{2k} + r_0 h_K^{2(k+1)} + \delta_K \|\boldsymbol{c}\|_{0,\infty,K}^2 h_K^{2k}\right] |u|_{k+1,K}^2 \\
&\leq C h^{2k} \sum_{K\in\mathcal{T}_h} \left[\varepsilon + h_K^2 + \delta_K\right] |u|_{k+1,K}^2 \leq C(\varepsilon + h) h^{2k} |u|_{k+1}^2.
\end{aligned}$$

□

Remark 10.4 1) In the case of large local Péclet numbers, we have $\varepsilon \leq \frac{1}{2}\|\boldsymbol{c}\|_{\infty,K} h_K$ and thus

$$\|u - u_h\|_0 + \left(\delta_2 \sum_{K\in\mathcal{T}_h} h_K \|\boldsymbol{c}\cdot\nabla(u - u_h)\|_{0,K}^2\right)^{1/2} \leq C h^{k+1/2}|u|_{k+1}.$$

So the L^2-error of the solution is not optimal in comparison with the estimate of the interpolation error

$$\|u - I_h(u)\|_0 \le Ch^{k+1}|u|_{k+1},$$

whereas the L^2-error of the directional derivative of u in the direction of $\boldsymbol{c}$ is optimal.

2) In general, the seminorm $|u|_{k+1}$ depends on negative powers of ε. Therefore, if $h \to 0$, the convergence in Theorem 10.3 is not uniform with respect to ε.
3) The proof can be viewed as a refined version of Theorem 6.6: We have $\|\cdot\|_V = \|\cdot\|_{sd}$, which by the choice of δ_K becomes h-dependent, $\|\cdot\|_U = \|\cdot\|_1$, but the stability condition (NB2h) is fulfilled thanks to the coercivity estimate (10.22).

Comparing the estimate from Theorem 10.3 for the special case of continuous linear elements with the estimate (10.14) for the corresponding standard method given at the end of the introduction, i.e.,

$$\|u - u_h\|_\varepsilon \le Ch|u|_2,$$

we see that the error of the streamline-diffusion method is measured in a stronger norm than the $\|\cdot\|_\varepsilon$-norm and additionally, that the error bound is asymptotically better in the interesting case $\varepsilon < h$. A further advantage of the streamline-diffusion method is to be seen in the fact that its implementation is not much more difficult than that of the standard finite element method.

However, there are also some disadvantages: Since the error bound involves the H^{k+1}-seminorm of the solution u, it may depend on negative powers of ε. Furthermore, there is no general rule to determine the parameters δ_1, δ_2. Usually, they are chosen more or less empirically. This may be a problem when the streamline-diffusion method is embedded into more complex programs (for example, for solving nonlinear problems); see Problem 10.5. Finally, in contrast to the finite volume methods described in the next section, the property of inverse monotonicity (cf. Theorem 8.21) cannot be proven in general, and in fact computational experiments show that spurious oscillations of the numerical solution can occur.

Exercise

Problem 10.2 Given an arbitrary, but fixed, triangle K with diameter h_K, prove the inequality

$$\|\Delta p\|_{0,K} \le \frac{c_{\text{inv}}}{h_K}|p|_{1,K}$$

for arbitrary polynomials $p \in \mathcal{P}_k(K)$, $k \in \mathbb{N}$, where the constant $c_{\text{inv}} > 0$ is independent of K and p.

Problem 10.3 Verify that the streamline-diffusion norm $\|\cdot\|_{\text{sd}}$ is indeed a norm.

Problem 10.4 Investigate global and local mass conservation for the streamline-diffusion method.

10.3 Finite Volume Methods

In the convection-dominated situation, the finite volume method introduced in Chapter 8 proves to be a very stable, but not so accurate, method. One reason for this stability lies in an appropriate asymptotic behaviour of the weighting function R for large absolute values of its argument.

Namely, if we consider the examples of nonconstant weighting functions given in Section 8.2.2, we see that

$$\text{(P4)} \qquad \lim_{z\to-\infty} R(z) = 0\,, \quad \lim_{z\to\infty} R(z) = 1\,.$$

In the general case of the model problem (8.5) with $k = \varepsilon > 0$, (P4) implies that for $\frac{\gamma_{ij} d_{ij}}{\varepsilon} \ll -1$ the term $r_{ij}u_i + (1 - r_{ij})u_j$ in the bilinear form b_h effectively equals u_j, whereas in the case $\frac{\gamma_{ij} d_{ij}}{\varepsilon} \gg 1$ the quantity u_i remains.

In other words, in the case of dominating convection, the approximation b_h evaluates the "information" (u_j or u_i) *upwind*, i.e., just at that node (a_j or a_i) from which "the flow is coming".

This essentially contributes to the stabilization of the method and makes it possible to prove properties such as global conservativity or inverse monotonicity (cf. Section 8.2.4) *without* any restrictions on the size of the local Péclet number $\frac{\gamma_{ij} d_{ij}}{\varepsilon}$ and thus without any restrictions on the ratio of h and ε. This local Péclet number (note the missing factor 2 in comparison to (10.23)) also takes the direction of the flow compared to the edge $a_i a_j$ into account.

The Choice of Weighting Parameters

In order to motivate the choice of the weighting parameters in the case of the Voronoi diagram, we recall the essential step in the derivation of the finite volume method, namely the approximation of the integral

$$I_{ij} := \int_{\Gamma_{ij}} \left[\mu_{ij}\left(\boldsymbol{\nu}_{ij} \cdot \nabla u\right) - \gamma_{ij}\, u\right] d\sigma = -\int_{\Gamma_{ij}} \boldsymbol{\nu}_{ij} \cdot \boldsymbol{p}\, d\sigma\,.$$

It first suggests itself to apply a simple quadrature rule, for example,

$$I_{ij} \approx -p_{ij} m_{ij}\,,$$

where p_{ij} is a constant approximation to $\mathfrak{n}_{ij} \cdot \boldsymbol{p}(u)|_{\Gamma_{ij}}$. Next, if the edge $\overline{a_i a_j}$ (with midpoint a_{ij}) is parameterized according to

$$x = x(\tau) = a_{ij} + \tau d_{ij} \mathfrak{n}_{ij}, \quad \tau \in \left[-\frac{1}{2}, \frac{1}{2}\right],$$

and if we introduce the composite function $w(\tau) := -u(x(\tau))$, then we can write

$$\left[\mu_{ij} \left(\mathfrak{n}_{ij} \cdot \nabla u\right) - \gamma_{ij}\, u\right]\Big|_{a_{ij}} = -p(0) \quad \text{with} \quad p(\tau) := \frac{\mu_{ij}}{d_{ij}} \frac{dw}{d\tau}(\tau) - \gamma_{ij} w(\tau).$$

The relation defining the function p can be interpreted as a linear ordinary differential equation for the unknown function $w : \left[-\frac{1}{2}, \frac{1}{2}\right] \to \mathbb{R}$. Provided that p is continuous on the interval $\left[-\frac{1}{2}, \frac{1}{2}\right]$, the differential equation can be solved exactly:

$$\begin{aligned} w(\tau) = &\left\{ \frac{d_{ij}}{\mu_{ij}} \int_{-1/2}^{\tau} p(s) \exp\left(-\frac{\gamma_{ij} d_{ij}}{\mu_{ij}} \left(s + \frac{1}{2}\right)\right) ds + w\left(-\frac{1}{2}\right)\right\} \\ &\times \exp\left(\frac{\gamma_{ij} d_{ij}}{\mu_{ij}} \left(\tau + \frac{1}{2}\right)\right). \end{aligned}$$

Replacing p by the constant $p(0)$, we get, in the case $\gamma_{ij} \neq 0$,

$$\begin{aligned} w(\tau) \approx &\left\{ \frac{p(0)}{\gamma_{ij}} \left[1 - \exp\left(-\frac{\gamma_{ij} d_{ij}}{\mu_{ij}} \left(\tau + \frac{1}{2}\right)\right)\right] + w\left(-\frac{1}{2}\right)\right\} \\ &\times \exp\left(\frac{\gamma_{ij} d_{ij}}{\mu_{ij}} \left(\tau + \frac{1}{2}\right)\right). \end{aligned}$$

In particular,

$$w\left(\frac{1}{2}\right) \approx \left\{ \frac{p(0)}{\gamma_{ij}} \left[1 - \exp\left(-\frac{\gamma_{ij} d_{ij}}{\mu_{ij}}\right)\right] + w\left(-\frac{1}{2}\right)\right\} \exp\left(\frac{\gamma_{ij} d_{ij}}{\mu_{ij}}\right); \tag{10.24}$$

that is, the value $p(0)$ can be approximately expressed by means of the values $w(\pm\frac{1}{2})$:

$$p(0) \approx \gamma_{ij} \frac{w(\frac{1}{2}) - w(-\frac{1}{2}) \exp\left(\frac{\gamma_{ij} d_{ij}}{\mu_{ij}}\right)}{\exp\left(\frac{\gamma_{ij} d_{ij}}{\mu_{ij}}\right) - 1}. \tag{10.25}$$

In the case $\gamma_{ij} = 0$, it immediately follows from the exact solution and the approximation $p \approx p(0)$ that

$$p(0) \approx \mu_{ij} \frac{w(\frac{1}{2}) - w(-\frac{1}{2})}{d_{ij}}.$$

Since this is equal to the limit of (10.25) for $\gamma_{ij} \to 0$, we can exclusively work with the representation (10.25).

If we define the weighting function $R : \mathbb{R} \to [0,1]$ by

$$R(z) := 1 - \frac{1}{z}\left(1 - \frac{z}{e^z - 1}\right), \tag{10.26}$$

then, with the choice $r_{ij} := R\left(\frac{\gamma_{ij} d_{ij}}{\mu_{ij}}\right)$, the relationship (10.25) after a simple algebraic manipulation can be written as

$$p(0) \approx p_{ij} := \mu_{ij}\frac{u_i - u_j}{d_{ij}} + \left[r_{ij}u_i + \left(1 - r_{ij}\right)u_j\right]\gamma_{ij}\,. \tag{10.27}$$

This shows that this is exactly the approximation scheme given in Section 8.2.2; cf. (8.12), (8.19).

The use of the weighting function (10.26) yields a discretization method that can be interpreted as a generalization of the so-called Il'in–Allen–Southwell scheme. However, in order to avoid the comparatively expensive computation of the function values r_{ij} of (10.26), often simpler functions $R : \mathbb{R} \to [0,1]$ are used (see (8.19)), which are to some extent approximations of (10.26) keeping the properties (P1) to (P4).

At the end of this paragraph we will illustrate the importance of the properties (P1) to (P3), especially for convection-dominated problems. Property (P2) has been used in the proof of the basic stability estimate (8.27). On the other hand, we have seen at several places (e.g., in Section 1.4 or in Chapter 5) that the matrix A_h of the corresponding system of linear algebraic equations should have positive diagonal entries. For example, if in the differential equation from (10.2) the reaction term disappears, then properties (P1) and (P3) guarantee that the diagonal entries are at least nonnegative. This can be seen as follows:

From (8.11) we conclude the following formula:

$$(A_h)_{ii} = \left[\frac{\mu_{ij}}{d_{ij}} + \gamma_{ij}r_{ij}\right]m_{ij} = \frac{\mu_{ij}}{d_{ij}}\left[1 + \frac{\gamma_{ij}d_{ij}}{\mu_{ij}}r_{ij}\right]m_{ij}, \quad i \in \Lambda\,.$$

If we replace in property (P3) the number z by $-z$, then we get, by property (P1),

$$0 \le 1 + [1 - R(-z)]\,z = 1 + zR(z)\,.$$

Therefore, if the weighting function R satisfies (P1) and (P3), then we have that $(A_h)_{ii} \ge 0$ for all $i \in \Lambda$.

The simple choice $r_{ij} \equiv \frac{1}{2}$ does not satisfy property (P3). In this case, the condition $(A_h)_{ii} \ge 0$ leads to the requirement

$$-\frac{\gamma_{ij}d_{ij}}{2\mu_{ij}} \le 1\,,$$

which in the case $\gamma_{ij} \le 0$, i.e., for a local flow from a_j to a_i, is a restriction to the ratio of h and ε, and this is analogous to the condition (10.6) on the grid Péclet number, where only the sizes of K, $\|\boldsymbol{c}\|_{0,\infty,K}$, and h enter.

Similarly, it can be shown that property (P3) implies the nonpositivity of the off-diagonal entries of A_h.

An Error Estimate

At the end of this section an error estimate will be cited, which can be derived similarly to the corresponding estimate of the standard method. The only special aspect is that the dependence of the occurring quantities on ε is carefully tracked (see [77]).

Theorem 10.5 *Let $(\mathcal{T}_h)_h$ be a shape-regular family of conforming triangulations, all triangles of which are nonobtuse. Furthermore, in addition to the assumptions on the coefficients of the bilinear form (10.10), let $f \in C^1(\overline{\Omega})$.*

If the exact solution u of the model problem belongs to $H^2(\Omega)$ and if $u_h \in V_h$ denotes the approximative solution of the finite volume method (8.17), where the approximations γ_{ij}, respectively r_i, are chosen according to (8.8), respectively (8.10), then for sufficiently small $\bar{h} > 0$ the estimate

$$\|u - u_h\|_\varepsilon \le C \frac{h}{\sqrt{\varepsilon}} \left[\|u\|_2 + |f|_{1,\infty} \right], \quad h \in (0, \bar{h}],$$

holds, where both the constant $C > 0$ and $\bar{h} > 0$ do not depend on ε.

In special, but practically not so relevant, cases (for example, if the triangulations are of Friedrichs–Keller type), it is possible to remove the factor $\frac{1}{\sqrt{\varepsilon}}$ in the bound above.

Comparing the finite volume method with the streamline-diffusion method, we see that the finite volume method is less accurate. However, it is globally conservative and inverse monotone.

Exercise

Problem 10.5 The parameters δ_K in Theorem 10.3 depend on the constants δ_1, δ_2 to be chosen by the user. In some cases this is undesirable. A parameter-free choice is motivated by the comparison with the finite volume discretization of the one-dimensional problem (10.2) for the case of constant coefficients $k := \varepsilon > 0$, $c \neq 0$, and $r := 0$.

Given an equidistant grid and assuming that δ_K does not depend on K, show that the choice

$$\delta_K := \frac{h}{2c}\left(\coth \frac{ch}{2\varepsilon} - \frac{2\varepsilon}{ch} \right)$$

yields the exponential upwind scheme.

Programming project 10.3 Test the finite volume solver from Project 8.1 for the second example in the case of small diffusion, i.e., for $k := 10^{-i}$, $i = 1, 3, 5, 7$.

10.4 The Lagrange–Galerkin Method

In the previous sections, discretization methods for stationary diffusion–convection equations were presented. In conjunction with the method of lines, these methods can also be applied to parabolic problems. However, since the method of lines decouples spatial and temporal variables, it cannot be expected that the peculiarities of nonstationary diffusion–convection equations are reflected adequately.

The so-called *Lagrange–Galerkin method* attempts to bypass this problem by means of an intermediate change from the Eulerian coordinates (considered up to now) to the so-called *Lagrangian coordinates*. The latter are chosen in such a way that the origin of the coordinate system (i.e., the position of the observer) is moved with the convective field, and in the new coordinates no convection occurs.

To illustrate the basic idea, the following initial boundary value problem will be considered, where $\Omega \subset \mathbb{R}^d$ is a bounded domain With Lipschitz continuous boundary and $T > 0$:

For given functions $f : Q_T \to \mathbb{R}$ and $u_0 : \Omega \to \mathbb{R}$, find a function $u : Q_T \to \mathbb{R}$ such that

$$\begin{aligned} \frac{\partial u}{\partial t} + Lu &= f \quad \text{in } Q_T\,, \\ u &= 0 \quad \text{on } S_T\,, \\ u &= u_0 \quad \text{on } \Omega \times \{0\}\,, \end{aligned} \tag{10.28}$$

where

$$(Lu)(x,t) := -\nabla \cdot (\boldsymbol{K}(x)\,\nabla u(x,t)) + \boldsymbol{c}(x,t)\cdot \nabla u(x,t) + r(x,t)u(x,t), \tag{10.29}$$

with sufficiently smooth coefficients

$$\boldsymbol{K} : \Omega \to \mathbb{R}^{d,d}, \quad \boldsymbol{c} : Q_T \to \mathbb{R}^d, \quad r : Q_T \to \mathbb{R}.$$

As usual, the differential operators ∇ and $\nabla\cdot$ act only with respect to the spatial variables.

The new coordinate system is obtained by solving the following parameter-dependent auxiliary problem:
Given $(x,s) \in \overline{Q}_T$, find a vector field $\boldsymbol{X} : \overline{\Omega} \times [0,T]^2 \to \mathbb{R}^d$ such that

$$\begin{aligned} \frac{d}{dt}\boldsymbol{X}(x,s,t) &= \boldsymbol{c}(\boldsymbol{X}(x,s,t),t), \quad t \in (0,T), \\ \boldsymbol{X}(x,s,s) &= x\,. \end{aligned} \tag{10.30}$$

The trajectories $\boldsymbol{X}(x,s,\cdot)$ are called *characteristics* (through (x,s)). If $\boldsymbol{c}$ is continuous on $\overline{Q}_T$ and, for fixed $t \in [0,T]$, Lipschitz continuous with respect to the first argument on $\overline{\Omega}$, then there exists a unique solution $\boldsymbol{X} = \boldsymbol{X}(x,s,t)$. Denoting by u the sufficiently smooth solution of (10.28) and setting

$$\hat{u}(x,t) := u(\boldsymbol{X}(x,s,t),t) \quad \text{for fixed } s \in [0,T],$$

then the chain rule implies that

$$\frac{\partial \hat{u}}{\partial t}(x,t) = \left(\frac{\partial u}{\partial t} + \boldsymbol{c} \cdot \nabla u\right)(\boldsymbol{X}(x,s,t),t)\,.$$

The particular value

$$\frac{\partial \hat{u}}{\partial t}(x,s) = \frac{\partial u}{\partial t}(x,s) + \boldsymbol{c}(x,s) \cdot \nabla u(x,s)$$

is called the *material derivative* of u at (x,s). Thus the differential equation reads as

$$\frac{\partial \hat{u}}{\partial t} - \nabla \cdot (\boldsymbol{K}\nabla u) + ru = f\,;$$

i.e., it is formally free of any convective terms.

Now the equation will be semidiscretized by means of the horizontal method of lines. A typical way is to approximate the time derivative by backward difference quotients. So let an equidistant partition of the time interval $(0,T)$ with step size $\tau := T/N$, $N \in \mathbb{N}$ (provided that $T < \infty$), be given.

Tracking the characteristics backwards in time, in the strip $\Omega \times [t_n, t_{n+1})$, $n \in \{0, 1, \dots, N-1\}$, with $x = \boldsymbol{X}(x, t_{n+1}, t_{n+1})$ the following approximation results:

$$\frac{\partial \hat{u}}{\partial t} \approx \frac{1}{\tau}[\hat{u}(x,t_{n+1}) - \hat{u}(x,t_n)] = \frac{1}{\tau}[u(x,t_{n+1}) - u(\boldsymbol{X}(x,t_{n+1},t_n),t_n)]\,.$$

Further, if V_h denotes a finite-dimensional subspace of V in which we want to find the approximations to $u(\cdot, t_n)$, the method reads as follows:

Given $u_{0h} \in V_h$, find an element $U^{n+1} \in V_h$, $n \in \{0, \dots, N-1\}$, such that

$$\begin{aligned} \frac{1}{\tau}\left\langle U^{n+1} - U^n(\boldsymbol{X}(\cdot,t_{n+1},t_n)), v_h\right\rangle_0 + \left\langle \boldsymbol{K}\nabla U^{n+1} \cdot \nabla v_h, 1\right\rangle_0 + \left\langle r(\cdot,t_{n+1})U^{n+1}, v_h\right\rangle_0 \\ = \langle f(\cdot,t_{n+1}), v_h\rangle_0 \quad \text{for all } v_h \in V_h, \\ U^0 = u_{0h}\,. \end{aligned} \tag{10.31}$$

A possible extension of the method is to use time-dependent subspaces; that is, given a sequence of subspaces $V_h^n \subset V$, $n \in \{0, \dots, N\}$, the approximations U^n to $u(\cdot, t_n)$ are chosen from V_h^n.

So the basic idea of the Lagrange–Galerkin method, namely the elimination of convective terms by means of an appropriate transformation of coordinates, allows the application of standard discretization methods and makes the method attractive for situations where convection is dominating.

In fact, there exists a whole variety of papers dealing with error estimates for the method in the convection-dominated case, but often under the condition that the system (10.30) is integrated exactly.

In practice, the exact integration is impossible, and the system (10.30) has to be solved numerically (cf. [195]). This may lead to stability problems, so there is still a considerable need in the theoretical foundation of Lagrange–Galerkin methods.

For a model situation it has been possible to prove order of convergence estimates uniformly in the Péclet number for (10.31) (see [98]). The key is the consequent use of Lagrangian coordinates, revealing that (10.31) is just the application of the implicit Euler method to an equation arising from a transformation by characteristics defined piecewise backward in time. This equation is a pure diffusion problem, but with a coefficient reflecting the transformation. In conjunction with the backward Euler method this is not visible in the elliptic part to be discretized. Thus tracking the characteristics backward in time turns out to be important.

Exercise

Problem 10.6 Let $d := 3$, $\boldsymbol{c} \in C^1(\overline{Q}_T)^3$, and denote by $J(x,s,t)$ the Jacobian determinant of the mapping $x \mapsto \boldsymbol{X}(x,s,t)$.

a) Show that

$$\frac{\partial J}{\partial t}(x,s,t) = \nabla \cdot \boldsymbol{c}\big(\boldsymbol{X}(x,s,t),t\big)J(x,s,t), \quad (x,s,t) \in \Omega \times (0,T)^2.$$

Hints: Make use of the fact that the determinant of a matrix is multilinear in the columns (this means that if we fix all but one column of the matrix, the determinant function is linear in the remaining column) and Schwarz's theorem of the symmetry of the second partial derivatives.

b) Show that $\nabla \cdot \boldsymbol{c} = 0$ in Q_T implies $J = 1$.

c) Let $w \in C(Q_T)$ and $\nabla \cdot \boldsymbol{c} = 0$ in Q_T. Verify the following conservation law:

$$\int_G w(x,t)dx = \int_{\boldsymbol{X}(G,s,t)} \hat{w}(y,t)dy$$

for any nonempty open subset $G \subset \Omega$.

10.5 Algebraic Flux Correction and Limiting Methods

The stabilization methods introduced so far may prevent spurious oscillations (e.g., in the vicinity of large gradients) of the solutions. However, we are interested in a physical solution to a physically motivated problem. In particular, this means that the numerical solution should preserve characteristic properties of the analytical solution of the underlying physical problem such as, for example,

1) nonnegativity/comparison principle,
2) minimum/maximum principle.

Using the SUPG or the finite volume method, the properties 1) and 2) do not necessarily hold or the method has only a low order. Therefore we are interested in schemes that ensure these properties and at the same time are sufficiently accurate. In addition, the first property is a weaker version of the second (and is therefore implied by it), at least in the linear case. We therefore expect the first property to be valid in a more general setting than the second.

However, it is well known from the literature that the absence of spurious (non-physical) oscillations and a high accuracy are generally two contradicting requirements for numerical methods, especially in connection with the properties 1) and 2). One example is Godunov's theorem [141, § 3], which states that a linear, monotonicity-preserving difference method for the linear advection equation $u_t = u_x$ is at most first-order accurate. Since such equations serve as model equations for the so-called reduced problem of (9.1)–(9.5) (i.e., $\boldsymbol{K} = \boldsymbol{0}$), accurate numerical methods for (9.1)–(9.5) cannot be considered in isolation from those methods that provide an accurate numerical approximation to the solution of the reduced problem. A further example is Kershaw's result [159], according to which no finite difference approximation of second-order partial derivatives on general unstructured grids that is both linear and second-order or higher order accurate can have an algebraic representation in the form of an M-matrix. Therefore, in order to achieve a high-order monotone method, such a numerical method has to include nonlinear terms. Hence it is not surprising that the methods described in the later Sections 10.5.3–10.5.5 and 10.6 are nonlinear, even if the boundary value problem (9.1)–(9.5) itself is linear.

The starting point of the *algebraic flux correction (AFC) method* applied to (9.1)–(9.5) is the ODE system that we obtain after applying a spatial discretization procedure, e.g., a finite element method, see (9.51). This system might violate the properties 1) and 2), but we will see that it can be modified in such a way that the properties are nevertheless satisfied. This results in a second ODE system of the same size, called a *low-order scheme*. The solution of the low-order scheme is supposed to be smooth and diffusive.

The idea of the AFC method is to switch between those two systems and to construct a solution that combines the desired properties of the two systems. On the one hand, we need the property that the solution of the original system approximates the solution of the PDE. On the other hand, we want the properties 1) and 2) of the low-order scheme to hold. It has been applied successfully only for lowest order finite elements (cf. [93, p. 2428]), i.e., linear finite elements. The only exception is [181], where Bernstein finite elements are used for discretization. A derivation of the AFC scheme for quadratic finite elements analogous to the case of linear finite elements is not so easily possible, since it causes problems when using row sum mass lumping (cf. Section 10.5.1).

To avoid too many technical details, we will only consider a simplified version of the initial boundary value problem (9.1)–(9.5), namely the case $\Gamma_2 = \emptyset$, $r = f = 0$, and $g_1 = 0$, that is,

$$\begin{aligned} \partial_t u - \nabla \cdot (\boldsymbol{K}\nabla u - \boldsymbol{c}u) &= 0 \quad \text{in } Q_T, \\ \boldsymbol{K}\nabla u \cdot \mathfrak{n} &= 0 \quad \text{on } \widetilde{\Gamma}_2 \times (0,T), \\ u &= \tilde{g} \quad \text{on } \widetilde{\Gamma}_3 \times (0,T), \\ u &= u_0 \quad \text{on } \Omega \times \{0\}. \end{aligned} \tag{10.32}$$

Note that the diffusive flux condition corresponds to a mixed condition with the full flux density and $\tilde{\alpha} := \boldsymbol{c} \cdot \mathfrak{n}$. According to (3.40) the (bi)linear forms resulting from the weak formulation of (10.32) are

$$a(u,v) = \int_\Omega \left[\boldsymbol{K}\nabla u \cdot \nabla v + \boldsymbol{c} \cdot \nabla u\, v\right] dx + \int_{\widetilde{\Gamma}_2} \boldsymbol{c} \cdot \mathfrak{n}\, uv d\sigma, \quad \ell(v) = 0.$$

Restricting for simplicity for the moment to pure Dirichlet conditions we remind: The initial boundary problem (10.32) satisfies a weak and strong maximum principle of the type (1.41) or Theorem 9.50, 2), if no solution-dependent sources or sinks are present. For (10.32) this means that $\nabla \cdot \boldsymbol{c} = 0$ has to be required. For a weak maximum principle of the type (1.37) no such requirement is necessary. From such a weak maximum principle a comparison principle or equivalently *nonnegativity preservation* can be concluded (compare Theorem 9.48 and 9.50,1)).

Performing the semidiscretization of (10.32) by means of primal FEM leads to (9.159), for which a comparison principle (Theorem 9.48) and a version of a weak maximum/minimum principle (Remark 9.49) hold, if the extended stiffness matrix $\widetilde{A}_h$ (see (9.157)) has specific properties, in particular (1.35), (2), (5) and also (1.35), (6), or (6)*.

10.5.1 Construction of a Low-order Semidiscrete Scheme

Our goal is to find a system of ODEs that does not differ significantly from (9.159), but preserves nonnegativity and fulfils the minimum/maximum principle. To this end, the matrices $\widetilde{A}_h$ and $\widetilde{B}_h$ will be modified in such a way that Theorem 9.48 and Remark 9.49 can be applied.

Modification of the matrices

In order to fulfil the requirements of Theorem 9.48, the matrices $\widetilde{B}_h$ and $\widetilde{A}_h$ may have to be modified. We split up the matrix $\widetilde{A}_h$ into the matrices $\widetilde{L}$ and $\widetilde{K}$:

$$\widetilde{A}_h = \widetilde{L} + \widetilde{K}.$$

The matrix $\widetilde{L}$ discretizes the diffusion operator and $\widetilde{K}$ the advection operator. The submatrix L is symmetric, but K is not. We need this splitting, since we modify the matrices $\widetilde{L}$ and $\widetilde{K}$ separately as follows:

- Row sum mass lumping. The entries of the row sum lumped matrix $B^\ell \in \mathbb{R}^{M_1,M}$ are defined as

$$B^\ell_{ij} := 0 \quad \text{for all } i, j,\ i \neq j, \qquad B^\ell_{ii} := \sum_j \widetilde{B}_{ij}.$$

- Adding diffusion operators. The diffusion operators $\widetilde{D}^L, \widetilde{D}^K \in \mathbb{R}^{M_1,M}$ are elementwise defined as

$$\begin{aligned}
\widetilde{D}^K_{ij} &:= \begin{cases} -\max\{\widetilde{K}_{ij}, \widetilde{K}_{ji}\}^+ & \text{for } i, j = 1, \ldots, M_1,\ i \neq j, \\ -(\widetilde{K}_{ij})^+ & \text{for } i = 1, \ldots, M_1,\ j = M_1 + 1, \ldots, M, \end{cases} \\
\widetilde{D}^L_{ij} &:= -(L_{ij})^+ \quad \text{for all } i, j, i \neq j, \\
\widetilde{D}^K_{ii} &:= -\sum_{j \neq i} D^K_{ij}, \qquad \widetilde{D}^L_{ii} := -\sum_{j \neq i} D^L_{ij},
\end{aligned}$$

where $a^+ := \max\{a, 0\}$ for $a \in \mathbb{R}$. We obtain the modified matrices $\widetilde{L}^\ell$ and $\widetilde{K}^\ell$ by

$$\widetilde{L}^\ell := \widetilde{L} + \widetilde{D}^L \quad \text{and} \quad \widetilde{K}^\ell := \widetilde{K} + \widetilde{D}^K.$$

The matrices $\widetilde{D}^L$ and $\widetilde{D}^K$ are called *diffusion operators*, as adding them to the system of ODEs (9.159) produces a smooth and diffusive solution. It might be interesting to notice that we do not have to modify $\widetilde{L}$, if the triangulation satisfies some angle and connectivity conditions (cf. Subsection 3.9). Let $\widetilde{A}^\ell$ be the modification of the matrix $\widetilde{A}$ that is defined as the sum of the modified matrices $\widetilde{L}^\ell$ and $\widetilde{K}^\ell$, i.e.,

$$\widetilde{A}^\ell := \widetilde{L}^\ell + \widetilde{K}^\ell.$$

The construction of $\widetilde{A}^l$ is such that the sign conditions of (1.35) (2) and (5) are always fulfilled without disturbing the row sum condition (1.35) (3) and (6):

We have for $r = 1, \ldots, M_1$, $s = 1, \ldots, M$, $r \neq s$:

$$\begin{aligned}
(\widetilde{L} + \widetilde{D}^L)_{rs} &= \widetilde{L}_{rs} - (\widetilde{L}_{rs})^+ \leq 0, \\
(\widetilde{K} + \widetilde{D}^K)_{rs} &= \widetilde{K}_{rs} - \max\{\widetilde{K}_{rs}, \widetilde{K}_{sr}\}^+ \leq 0,
\end{aligned}$$

i.e., (1.35) (2) and (5) are satisfied individually for $\widetilde{L}^\ell$ and $\widetilde{K}^\ell$ and thus for $\widetilde{A}^\ell$. Thus we have the following condition for (1.35) for $\widetilde{A}^\ell$:

(1.35) (1): $(\widetilde{A}^\ell)_{ii} \geq 0$ follows from (2) and (3)(i), for $(\widetilde{A}^\ell)_{ii} > 0$.

$$(A_h)_{ii} \geq 0 \tag{10.33}$$

is sufficient, as

$$0 \leq (\widetilde{A}^\ell)_{ii} = (A_h)_{ii} + \sum_{\substack{j=1 \\ j \neq i}}^{M} (L_{ij})^+ + \max\{\widetilde{K}_{ij}, \widetilde{K}_{ji}\}^+.$$

(1.35) (2) is always satisfied.
(1.35) (3)(i): Because of

$$\sum_{j=1}^{M} D_{ij}^{K} = 0, \quad \sum_{j=1}^{M} D_{ij}^{L} = 0$$

and

$$D_{ij}^{K} \leq 0, \quad D_{ij}^{L} \leq 0 \quad \text{for } i \neq j,$$

$$\sum_{j=1}^{M_1} (A_h)_{ij} \geq 0 \quad \text{for } i = 1, \ldots, M_1 \tag{10.34}$$

is sufficient

For (3)(ii) the same condition for A_h is sufficient.

(1.35) (4) and (4)*: There seems to be no simple sufficient condition as by the modification $(A^\ell)_{ij} = 0$ for $(A_h)_{ij} \neq 0$ is possible.

(1.35) (5) is always satisfied.

(1.35) (6) is equivalent to

$$\sum_{i=1}^{M} (\widetilde{A}_h)_{ij} \geq 0 \quad \text{for } i = 1, \ldots, M_1. \tag{10.35}$$

(1.35) (6)* is equivalent to

$$\sum_{i=1}^{M} (\widetilde{A}_h)_{ij} = 0 \quad \text{for } i = 1, \ldots, M_1. \tag{10.36}$$

For linear finite elements, the row sum mass lumping is equivalent to an inexact evaluation of B using a node-oriented Newton–Côtes quadrature rule (cf. Subsection 3.5.2). Let $\overline{B}^{\ell}$ denote the approximated matrix. We obtain

$$B \approx \overline{B}_{ij}^{\ell} := \sum_{k=1}^{M_1} \omega_k \varphi_j(a_k)\varphi_i(a_k) = \omega_i \delta_{ij},$$

where ω_k are the quadrature weights. For $k = 1, \ldots, M_1$ the quadrature weights are given by $\omega_k = \int_\Omega \varphi_k \, dx$ (cf. (3.134)). We can compute the diagonal of the row sum lumped mass matrix by

$$B_{ii}^{\ell} = \sum_{j=1}^{M} \widetilde{B}_{ij} = \sum_{j=1}^{M} \int_\Omega \varphi_j \varphi_i \, dx = \int_\Omega \varphi_i \Big(\sum_{j=1}^{M} \varphi_j \Big) \, dx = \int_\Omega \varphi_i \, dx = \omega_i.$$

This shows the above-stated equivalence and furthermore that $\widetilde{B}_{ii}^{\ell} > 0$ in the case of linear finite elements. In the case of quadratic finite elements, we cannot guarantee that $\int_\Omega \varphi_k \, dx > 0$. This makes it impossible to use the row sum mass lumping strategy as defined above for quadratic finite elements. A strategy to circumvent this problem is presented in [181].

The low-order scheme

The modified ODE system from (9.159) reads as follows:

$$\widetilde{B}^{\ell}\partial_t\tilde{\boldsymbol{u}}_h + \widetilde{A}^{\ell}\tilde{\boldsymbol{u}}_h = \boldsymbol{0}, \quad t \in (0,T]. \tag{10.37}$$

By construction, it fulfils the property to be nonnegativity preserving. For the minimum/maximum principle, the additional property of vanishing row sums of $\tilde{A}$ must be fulfilled. If the row sums are nonzero, it does not make sense to require that the discretization is LED.

10.5.2 The Fully Discrete System

In this section, we carry out the time discretization of equation (10.37) as in Section 9.3. We analyse under which conditions the desired properties of the semidiscrete system (i.e., nonnegativity preservation, minimum/maximum principle) still remain valid after time discretization. Applying the one-step-theta method to (10.37), we obtain for $n = 0, \dots, N-1$

$$\left(\widetilde{B}^{\ell} + \tau\Theta\widetilde{A}^{\ell}\right)\tilde{\boldsymbol{U}}^{n+1} = \left(\widetilde{B}^{\ell} - \tau(1-\Theta)\widetilde{A}^{\ell}\right)\tilde{\boldsymbol{U}}^{n}. \tag{10.38}$$

Conditions to the fully discrete scheme

From Theorem 9.50 and Remark 9.51 we conclude the following sufficient conditions for the one-step-theta discretization (10.38) of the modified semidiscrete ODE (10.37):

We always assume (9.174), i.e., the step size restriction.

For nonnegativity/comparison principle: (10.33), and

$$b_i + \tau\Theta\sum_{j=1}^{M_1}(A_h)_{ij} \geq 0 \quad \text{for } i = 1, \dots, M_1. \tag{10.39}$$

As $b_i > 0$, this condition is satisfied for sufficiently small τ.

For a weak maximum/minimum principle: additionally (10.36).

10.5.3 Algebraic Flux Correction

The low-order scheme (10.37) constructed in Section 10.5.1 may preserve nonnegativity or satisfy the minimum/maximum principle, but its solution is not a sufficiently accurate approximation to the solution of the original problem in most cases. As described in the introduction to the AFC method, our goal is to restore as much of the original problem as possible while keeping the nonnegativity preservation and the minimum/maximum principle. This idea leads to the following nonlinear system of ODEs:

$$\begin{aligned} \widetilde{B}^{\ell}\partial_t\tilde{\boldsymbol{U}} + \widetilde{A}^{\ell}\tilde{\boldsymbol{U}} &= \overline{\boldsymbol{f}}(\tilde{\boldsymbol{U}}, \partial_t\tilde{\boldsymbol{U}}), \quad t \in (0, T), \\ \boldsymbol{U}(0) &= \boldsymbol{U}_0, \end{aligned} \tag{10.40}$$

where $\overline{\boldsymbol{f}}$ denotes a modified ("limited") right-hand side $\boldsymbol{f}$ defined by

$$\boldsymbol{f}(\tilde{\boldsymbol{U}}, \partial_t\tilde{\boldsymbol{U}}) := (\tilde{B}^{\ell} - \widetilde{B})\partial_t\tilde{\boldsymbol{U}} + \widetilde{D}^{L}\tilde{\boldsymbol{U}} + \widetilde{D}^{K}\tilde{\boldsymbol{U}}.$$

$\boldsymbol{f}(\tilde{\boldsymbol{U}}, \partial_t\tilde{\boldsymbol{U}})$ is the difference between the low-order scheme (10.37) and the original problem (9.159). The nonlinearity of (10.40) is hidden in the limiting procedure used to obtain $\overline{\boldsymbol{f}}$ from $\boldsymbol{f}$.

Decomposition into local flux densities

In order to carry out the limiting procedure, we first decompose $\boldsymbol{f}$ into discrete anti-diffusive flux densities f_{ij}. Using the zero row sum property (1.35) (6)* of the matrices $\widetilde{B}^{\ell} - \widetilde{B}$, $\widetilde{D}^{L}$, and $\widetilde{D}^{K}$, we can write

$$\begin{aligned} (\widetilde{B}^{\ell}\tilde{\boldsymbol{U}} - \widetilde{B}\tilde{\boldsymbol{U}})_i &= \sum_{j\neq i} \widetilde{B}_{ij}(\widetilde{U}_i - \widetilde{U}_j), \\ (\widetilde{D}^{L}\tilde{\boldsymbol{U}})_i &= \sum_{j\neq i} \widetilde{D}^{L}_{ij}(\widetilde{U}_j - \widetilde{U}_i), \\ (\widetilde{D}^{K}\tilde{\boldsymbol{U}})_i &= \sum_{j\neq i} \widetilde{D}^{K}_{ij}(\widetilde{U}_j - \widetilde{U}_i). \end{aligned}$$

This yields the representation

$$f_i = \sum_{j\neq i} f_{ij}, \quad \text{where } f_{ij} := (B_{ij}\partial_t - D^{L}_{ij} - D^{K}_{ij})(U_i - U_j) \quad \text{for } j \neq i,$$

and f_i denotes the ith component of the vector $\boldsymbol{f} = (f_i)_{i=1,\ldots,M_1}$. The flux density f_{ij} describes a mass exchange from node a_j into node a_i. Since $f_{ij} = -f_{ji}$ for $i, j = 1, \ldots, M_1$, $i \neq j$, the mass exchange between non-Dirichlet nodes is conservative. The value f_i is the entire anti-diffusive flux density into node a_i.

Flux limiting

We carry out flux limiting by finding solution-dependent limiting factors $\alpha_{ij} \in [0, 1]$ that limit the anti-diffusive flux densities f_{ij}. The limited flux densities $\overline{f}_{ij}$ are defined as $\overline{f}_{ij} := \alpha_{ij} f_{ij}$. Then the components of the right-hand side of (10.40) are given by

$$\overline{f}_i = \sum_{j \neq i} \overline{f}_{ij} .$$

The limited mass exchange, described by the limited anti-diffusive flux densities $\overline{f}_{ij}$, is still conservative in the case $\alpha_{ij} = \alpha_{ji}$. The choice $\alpha_{ij} = 1$ for all indices under consideration recovers the original unlimited system, whereas $\alpha_{ij} = 0$ for all indices yields the low-order scheme. A reasonable choice of the limiting factors should have the following properties:

- In smooth regions we can accept the maximum amount of anti-diffusion. Hence, we choose $\alpha_{ij} = 1$ for flux densities that correspond to nodes in those regions, as we do not need any diffusion smoothing of the solution to prevent undershoots, etc.
- In regions with sharp fronts and steep gradients, the original scheme can produce a physically incorrect solution, for example, undershoots. There we may add an artificial diffusion to the scheme. This leads to the choice of $0 \leq \alpha_{ij} < 1$ for limiting factors that correspond to nodes in such regions.

We use a limiting procedure that is designed in such a way that the following estimate holds:

$$c_i(U_i^{\min} - U_i) \leq \sum_{j \neq i} \overline{f}_{ij} \leq c_i(U_i^{\max} - U_i), \tag{10.41}$$

where c_i are nonnegative coefficients to be specified later,

$$U_i^{\min} := \min_{j \in \Lambda_i \cup \{i\}} U_j, \quad U_i^{\max} := \max_{j \in \Lambda_i \cup \{i\}} U_j,$$

and $\Lambda_i := \{j \neq i \,|\, B_{ij} \neq 0 \text{ or } C_{ij} \neq 0\}$. Condition (10.41) leads to a cancellation of all positive fluxes into the node a_i, if a local maximum is attained at this node. Analogously, all negative fluxes into a node, where a local minimum is reached, are cancelled. This ensures the LED property of (9.160), i.e., a local maximum cannot increase and a local minimum cannot decrease. As a consequence, undershoots and overshoots are successfully prevented. Provided that all nodes are connected via neighbourhood relations, condition (10.41) ensures that the nonnegativity property is preserved after flux limiting and that the minimum/maximum principle still holds.

10.5.4 The Nonlinear AFC Scheme

Next we investigate how to solve the nonlinear flux-limited system of ODEs numerically. The application of the one-step-theta method for time discretization of (10.40) leads to the nonlinear system

$$\frac{1}{\tau}\widetilde{B}^{\ell}(\tilde{\boldsymbol{U}}^{n+1}-\tilde{\boldsymbol{U}}^{n})+\Theta\widetilde{A}^{\ell}\tilde{\boldsymbol{U}}^{n+1}+(1-\Theta)\widetilde{A}^{\ell}\tilde{\boldsymbol{U}}^{n}=\overline{\boldsymbol{f}}(\Theta\tilde{\boldsymbol{U}}^{n+1}+(1-\Theta)\tilde{\boldsymbol{U}}^{n},\frac{1}{\tau}(\tilde{\boldsymbol{U}}^{n+1}-\tilde{\boldsymbol{U}}^{n})).$$

Using $\widetilde{C}^{\ell}=\widetilde{B}^{\ell}+\tau\Theta\widetilde{A}^{\ell}$ and $\widetilde{D}^{\ell}=\widetilde{B}^{\ell}-\tau(1-\Theta)\widetilde{A}^{\ell}$, we can equivalently write

$$\widetilde{C}^{\ell}\tilde{\boldsymbol{U}}^{n+1}=\widetilde{D}^{\ell}\tilde{\boldsymbol{U}}^{n}+\tau\overline{\boldsymbol{f}}(\Theta\tilde{\boldsymbol{U}}^{n+1}+(1-\Theta)\tilde{\boldsymbol{U}}^{n},(\tilde{\boldsymbol{U}}^{n+1}-\tilde{\boldsymbol{U}}^{n})/\tau).$$

In order to solve this nonlinear equation, we apply a fixed-point iteration. Let $(\tilde{\boldsymbol{U}}^{(k)})_{k\in\mathbb{N}_0}$ be a sequence that approximates $\tilde{\boldsymbol{U}}^{n+1}$. We obtain the next iterate $\tilde{\boldsymbol{U}}^{(k+1)}$ from the following linear equation:

$$\widetilde{C}^{\ell}\tilde{\boldsymbol{U}}^{(k+1)}=\widetilde{D}^{\ell}\tilde{\boldsymbol{U}}^{n}+\tau\overline{\boldsymbol{f}}(\Theta\tilde{\boldsymbol{U}}^{(k)}+(1-\Theta)\tilde{\boldsymbol{U}}^{n},(\tilde{\boldsymbol{U}}^{(k)}-\tilde{\boldsymbol{U}}^{n})/\tau). \tag{10.42}$$

As initial value of the fixed-point iteration we may set $\tilde{\boldsymbol{U}}^{(0)}:=\tilde{\boldsymbol{U}}^{n}$.

The iteration terminates if the Euclidean norm of the difference between two successive iterates, i.e., $|\tilde{\boldsymbol{U}}^{(k+1)}-\tilde{\boldsymbol{U}}^{(k)}|$, is less than a given tolerance. The iteration applied corresponds to (11.18).

Theorem 10.6 *Assume* (9.174), (10.33), (10.36), (10.39), *and a limiting procedure satisfying* (10.41) *with* $c_i:=\widetilde{B}^{\ell}_{ii}/\tau$, *then each iterate defined by* (10.42) *fulfils nonnegativity preservation and a weak maximum/minimum principle.*

Proof To show that the update (10.42) is nonnegativity preserving and keeps the minimum/maximum principle, we split the proof into three steps and use Theorem 9.50 for each step. We can show that every step preserves these properties, using the properties of the low-order scheme and the estimate (10.41).

1) Compute an explicit low-order approximation $\boldsymbol{U}^{+}$:

$$\widetilde{B}^{\ell}\tilde{\boldsymbol{U}}^{+}=\widetilde{D}^{\ell}\tilde{\boldsymbol{U}}^{n}.$$

2) Add the limited anti-diffusive flux densities to the intermediate solution $\boldsymbol{U}^{+}$:

$$\widetilde{B}^{\ell}\tilde{\boldsymbol{U}}^{++}=\widetilde{B}^{\ell}\boldsymbol{U}^{+}+\tau\overline{\boldsymbol{f}}(\theta\tilde{\boldsymbol{U}}^{(k)}+(1-\theta)\tilde{\boldsymbol{U}}^{n},(\tilde{\boldsymbol{U}}^{(k)}-\tilde{\boldsymbol{U}}^{n})/\tau).$$

3) Solve the linear system of equations for the next iterate $\boldsymbol{U}^{(k+1)}$:

$$\widetilde{C}^{\ell}\tilde{\boldsymbol{U}}^{(k+1)}=\widetilde{B}^{\ell}\tilde{\boldsymbol{U}}^{++}.$$

Next, we show that the steps 1)–3) preserve the desired, above-mentioned properties. Due to (10.35) we have

$$C_{ii}^{\ell} = B_{ii} + \tau\Theta A_{ii}^{\ell} > -\tau\Theta \sum_{\substack{j=1 \\ j\neq i}}^{M_1} A_{ij}^{\ell} \geq 0 \quad \text{for all } i. \tag{10.43}$$

Therefore C^{ℓ} satisfies

$$\sum_{j=1}^{M_1} C_{ij}^{\ell} > 0 \quad \text{for } i = 1, \dots, M,$$

i.e., the conditions (1.35)(1), (2), (3)(i), (4)* are fulfilled and thus C^{ℓ} is inverse monotone. Furthermore as $\widetilde{B}^{\ell} = \text{diag}\,(b_i)$, $b_i > 0$, and (1.35) (6)* holds, step 3) satisfies both properties. Due to (9.174) we have

$$D_{ii}^{\ell} = B_{ii}^{\ell} - \tau(1-\Theta)A_{ii}^{\ell} \geq 0 \quad \text{for all } i \quad \Longleftrightarrow \quad (1-\Theta)\tau \leq \frac{B_{ii}^{\ell}}{A_{ii}^{\ell}} \quad \text{for all } i, \tag{10.44}$$

and thus the same is true for step 1). Additionally step 2) preserves nonnegativity and the minimum/maximum principle if we choose $c_i := \widetilde{B}_{ii}^{\ell}/\tau \geq 0$ in (10.41). This choice leads to

$$U_i^{+\min} \leq U_i^{+} + \frac{\tau}{\widetilde{B}_{ii}} \sum_{j\neq i} \overline{f}_{ij} \leq U_i^{+\max},$$

which bounds the ith component of the update in step 2) from above by the local maximum and from below by the local minimum of the vector to be updated. These bounds and the connectivity condition yield nonnegativity preservation and the minimum/maximum principle. □

10.5.5 A Limiting Strategy

The following algorithm determines the limiting factors α_{ij}. It is taken from [169, Sect. 6.4].

1) Compute the sums of the positive/negative anti-diffusive flux densities into node a_i:

$$P_i^{+} := \sum_{j\neq i} \max\{0, f_{ij}\}, \quad P_i^{-} := \sum_{j\neq i} \min\{0, f_{ij}\}.$$

2) Determine the distance to a local maximum/minimum and the bounds:

$$Q_i^{+} := \frac{\widetilde{B}_{ii}^{\ell}}{\tau}(U_i^{+\max} - U_i^{+}), \quad Q_i^{-} := \frac{\widetilde{B}_{ii}^{\ell}}{\tau}(U_i^{+\min} - U_i^{+}).$$

3) Evaluate the nodal correction factors for the net increment to node a_i:

$$R_i^+ := \min\left\{1, \frac{Q_i^+}{P_i^+}\right\}, \quad R_i^- := \min\left\{1, \frac{Q_i^-}{P_i^-}\right\}.$$

If the node a_i is a Dirichlet node, we set $R_i^\pm := 1$.

4) Check the sign of the raw anti-diffusive flux density f_{ij} and multiply by

$$\alpha_{ij} := \begin{cases} \min\{R_i^+, R_j^-\} & \text{if } f_{ij} > 0, \\ \min\{R_i^-, R_j^+\} & \text{if } f_{ij} < 0. \end{cases}$$

It can be proven that this algorithm produces limited flux densities that satisfy (10.41) with $c_i := \widetilde{B}_{ii}^\ell/\tau$ (cf. [169, Sect. 6.4]).

Exercise

Problem 10.7 a) Compare the SUPG method and AFC: What are the respective (dis)advantages? Which method should be chosen for which kind of application?

b) If a global bound needs to be satisfied, cutting off the finite element approximation might serve as stabilization technique. What are the (dis)advantages of this procedure compared to SUPG and AFC?

Problem 10.8 Extend the AFC method to the finite difference, finite volume, and discontinuous Galerkin methods. Can the AFC method stabilize the three methods? How does the physical interpretation of the respective corrected fluxes change?

10.6 Slope Limitation Techniques

For convection-dominated problems, finite elements and especially discontinuous Galerkin methods might generate spurious oscillations in the higher order ($p \geq 1$ apart from the mean value) parts that can also lead to local mean values violating a bound preserving property although the exact solution satisfies such principles. This can be corrected utilizing techniques borrowed from Section 10.5, but more discontinuous Galerkin method-specific limiting techniques (exploiting the fact that the element local basis of trial space might be chosen as orthonormal) can be applied. Among these techniques one can distinguish, in general, the following:

- Algebraic flux correcting techniques as discussed in Section 10.5.
- Slope limiting techniques that correct the higher order parts without changing the local means (cf. [207]).
- Monolithic limiting techniques that are included in the variational formulation—for example, a constraint (e.g., $u_h \geq 0$) might be added leading to minimization problems of the type

$$\frac{1}{2}a_h(u_h, u_h) - \ell_h(u_h) \to \min \quad \text{such that } u_h \geq 0.$$

Monolithic limiting techniques often are computationally very challenging and will be skipped in this section. However, we point out that the class of monolithic convex limiting approaches discussed in [172] is related to the algebraic flux correction approach of Section 10.5.3.

We will concentrate in the following on DG methods, i.e., methods being locally mass conservative, as discussed in Section 7.6. Note that all aforementioned approaches have to be engineered in such a way that the local mass conservation property still holds. Beyond this, it is desirable to preserve the maximum order of convergence while ensuring that all nonphysical effects are eliminated.

In the following, we will consider the slope limiting techniques, since they are constructed solely for DG methods, while the other approaches can be applied in a more general setting, as developed by Aizinger [72] and Kuzmin [170].

A *slope limiter* is a local postprocessing filter that constrains linear and higher order parts of local solutions. The local mean values are left unchanged by this procedure and thus no mass is lost or gained in the case of a hierarchical and orthogonal basis (where the constant part solely represents the local mean, since due to orthogonality $\int \varphi_1 \varphi_j \, dx = 0$ for $j > 1$). On the other hand, slope limiting introduces artificial diffusion. Thus, slope limiting is used to enforce numerical admissibility conditions (local bounds) and flux limiters/correctors to enforce physical admissibility conditions (global bounds). The use of local bounds for slope limiting purposes is appropriate since it results in more reliable detection of nonphysical artifacts and does not produce additional consistency errors in the DG approximation to the mean values.

By Taylor expansion around the centroid $a_K := \frac{1}{|K|}\int_K x \, dx$ of K, a function $u_h|_K \in \mathcal{P}_k(K)$ has the local representation [207]

$$u_h|_K(x) = \overline{u}_{h,K} + \nabla u_h(a_K) \cdot (x - a_K) + \sum_{2 \leq |\boldsymbol{j}| \leq k} \partial^{\boldsymbol{j}} u_h(a_K) \frac{(x - a_K)^{\boldsymbol{j}} - \overline{(x - a_K)^{\boldsymbol{j}}}}{\boldsymbol{j}!}$$

with multi-index $\boldsymbol{j} \in \mathbb{N}^d$ and $\overline{(x - a_K)^{\boldsymbol{j}}}$ denoting the mean of $(x - a_K)^{\boldsymbol{j}}$. Notice that the local mean $\overline{v}(x) = \int_K v(x) dx$ of all higher order parts is zero [207].

For "troubled elements" K (the definition and detection is discussed below), a linear slope limiter replaces the local solution $u_h|_K \in \mathcal{P}_k(K)$ by the linear reconstruction $v_h \in \mathcal{P}_1(K)$ with

$$v_h(x) = \overline{u}_{h,K} + \beta_K \nabla u_h(a_K) \cdot (x - a_K) \tag{10.45}$$

and determines the maximum admissible slope by choosing values $\beta_K \in [0, 1]$ such that v_h is bounded in all vertices $a_{K,i} \in K$, i. e.,

$$u_*^{K,i} \leq v_h(a_{K,i}) \leq u_{K,i}^* \tag{10.46}$$

with some local upper and lower bounds $u_*^{K,i} < u_{K,i}^*$ depending on the design of the limiter. Since v_h is linear, $\overline{v}_h = v_h(a_K)$. Using the linear reconstruction, $\overline{u}_{h,K}$ remains unchanged since $\overline{u}_{h,K} = \overline{v}_h$, i. e., mass is conserved locally. However, the choice $\beta_K < 1$ introduces a consistency error.

The linear vertex-based limiter of Kuzmin [170] is an improved version of the Barth–Jespersen limiter [95] by using a larger, vertex-oriented stencil (see [140] for a theoretical analysis that explains the poor performance of stencils based on edge/face neighbourhoods). It also is extendable to higher orders; cf. [170, 171] for this extension.

For the linear vertex-based limiter, the bounds $u_*^{K,i}$, $u_{K,i}^*$ in (10.46) are defined as the minimum and maximum integral means of the triangular neighbourhood of $a_{K,i} \in K$ (see (4.19)), denoted by $\Delta(a_{K,i})$:

$$u_*^{K,i} := \min_{K' \subset \Delta(a_{K,i})} \overline{u}_{h,K'}, \qquad u_{K,i}^* := \max_{K' \subset \Delta(a_{K,i})} \overline{u}_{h,K'}.$$

Due to $v_h \in \mathcal{P}_1(K)$, the extrema of v_h are attained at $a_{K,i}$. To enforce (10.46), the correction factor β_K is defined as

$$\beta_K := \min_i \begin{cases} \dfrac{u_{K,i}^* - \overline{u}_{h,K}}{u_{K,i} - \overline{u}_{h,K}} & \text{if} \qquad u_{K,i} > u_{K,i}^*, \\ 1 & \text{if } u_*^{K,i} \leq u_{K,i} \leq u_{K,i}^*, \\ \dfrac{u_*^{K,i} - \overline{u}_{h,K}}{u_{K,i} - \overline{u}_{h,K}} & \text{if} \qquad u_{K,i} < u_*^{K,i}, \end{cases}$$

where $u_{K,i} := u_h(a_{K,i})$ guaranteeing that $v_h(a_{K,i}) \in \left[u_*^{K,i}, u_{K,i}^*\right]$. Therefore, we call the vertices $a_{K,i}$ *control points*. The generally used indicator for *troubled elements* is the condition that the local reconstruction (10.45) requires a correction factor $\beta_K < 1$ to satisfy (10.46). If this condition is not met, the local solution $u_h|_K \in \mathcal{P}_k(K)$ is left unchanged.

In the case $p > 1$, the vertex-based slope limiting can be carried out in a hierarchical manner by applying the linear slope limiter to the linear part of the Taylor polynomial derivatives [170]. The correction factors of higher order derivatives serve as lower bounds for the correction factors of lower order derivatives, which is necessary for the preservation of the high order of the method. In contrast to the case $p = 1$, the limited high-order Taylor polynomials will not be provably bounded by the means and may even violate global bounds in some pathological cases.

We emphasize that the purpose of slope limiting in general is not to produce pointwise bounded DG approximations but to prelimit high-order Taylor polynomials in such a way that the resulting numerical solution does not include spurious oscillations.

Exercise

Problem 10.9 Can the slope limiting approach be generalized to (continuous) finite elements by ensuring that nodal values must be bound by local means? What important properties of slope limiters will be lost if they are applied to continuous finite elements?

Chapter 11
An Outlook to Nonlinear Partial Differential Equations

11.1 Nonlinear Problems and Iterative Methods

Despite the concentration of most of the book on linear boundary value problems and their discretizations, many real-world models are rather nonlinear elliptic or parabolic boundary value problems with various types of nonlinearities. To be specific, we restrict ourselves to the formulation in divergence form (3.36) and its time-dependent counterpart and the classical FEM based on Chapter 2 and Chapter 3. Let a, ℓ be the forms from (3.36) to (3.39), a first nonlinear (here: semilinear) boundary value problem reads:

Find $u \in V \subset H^1(\Omega)$ such that

$$a(u, v) + \int_\Omega \psi(u(x))v(x)dx = \ell(v) \quad \text{for all } v \in V, \tag{11.1}$$

corresponding to the nonlinear equation

$$f(u) := Au + G(u) = \ell$$

in V', where $H_0^1(\Omega) \subset V \subset H^1(\Omega)$ according to the type of boundary conditions, $A : V \to V'$ is defined as in (6.3) and $G : U \to V'$ is defined on some open subset $U \subset V$ by

$$\big(G(u)\big)(v) := \int_\Omega \psi(u)v\,dx \quad \text{for all } u \in U,\ v \in V. \tag{11.2}$$

We assume that the problem (11.1) is well defined, in particular that the operator G is continuous (e.g., ψ is sublinear and $U = V$ or L^∞-bounds are incorporated into U) and that a unique solution $u \in U$ exists. To link to linear boundary value problems and (FEM) discretizations, two approaches are possible. In the *first discretize – then iterate (discit) approach*, the FEM discretization is applied to (11.1) leading to

P. Knabner and L. Angermann, *Numerical Methods for Elliptic and Parabolic Partial Differential Equations*, Texts in Applied Mathematics 44,
https://doi.org/10.1007/978-3-030-79385-2_11

Find $u_h \in V_h$ such that

$$a(u_h, v_h) + \int_\Omega \psi(u_h(x))v_h(x)dx = \ell(v_h) \quad \text{for all } v_h \in V_h,$$

giving a nonlinear system of equations

$$f_h(u_h) := A_h u_h + G_h(u_h) = \ell_h, \tag{11.3}$$

cf. (6.15).

The quality of approximation can be investigated as in the previous chapters and we expect—at least for smooth nonlinearities—comparable results. So we have the discretization error

$$\|u - u_h\|_V,$$

(in addition to a modelling error), and a *solution error*

$$\|u_h - \hat{u}_h\|_V$$

stemming from the approximate solution of (11.3) (which already holds true for the linear case). One may argue that these (or all) errors should be balanced. The approximate solution of (11.3) is usually done by an iterative method, e.g., of fixed-point form:

For given $U_h \subset V_h$, $\widehat{\Phi}_h : U_h \to U_h$, $\hat{u}_h^0 \in U_h$, a sequence $\big(\hat{u}_h^{(k)}\big)_k$ is defined by

$$\hat{u}_h^{(k+1)} = \widehat{\Phi}_h\big(\hat{u}_h^{(k)}\big), \quad k = 0, 1, \ldots. \tag{11.4}$$

We expect convergence and will stop the iteration for a solution error ($\hat{u}_h := \hat{u}_h^{(k)}$ for the final k) in the desired quantity. As discussed in Chapter 5, the workload for the solution is a consequence of the number of iterations, i.e., the speed of convergence, and the amount of operations for one iteration. While the dependence of the latter on h can be estimated (if, e.g., it requires a corresponding linear system of equations to be solved), the dependence of the former is unclear. It is highly desirable that the amount of iterations is bounded in h.

On the other hand, in the *first iterate – then discretize (itdisc) approach*, the iteration is set up in the function space V:

$$u^{(k+1)} := \Phi\big(u^{(k)}\big) \tag{11.5}$$

with $\Phi : U \to U$, $U \subset V$, $u^{(0)} \in U$ and made operational by discretizations of the fixed-point function by $\Phi_h : U_h \to U_h$ for given $U_h \subset V_h$ and $u_h^{(0)}$ $(:= I_h(u^{(0)}))$ and performing the iteration

$$u_h^{(k+1)} = \Phi_h\big(u^{(k)}\big). \tag{11.6}$$

There is *asymptotic mesh independence* of the iteration, if the sequence $\big(u_h^{(k)}\big)_h$ and $u^{(k)}$ are close together exhibiting the same convergence behaviour for small $h > 0$

(see [20, Sect. 8.1]). In the consistent, conforming Galerkin setting only considered here, we expect exact or approximate commutativity of the approaches in the sense

$$u_h^{(k)} \approx \hat{u}_h^{(k)}$$

(see Figure 11.1).

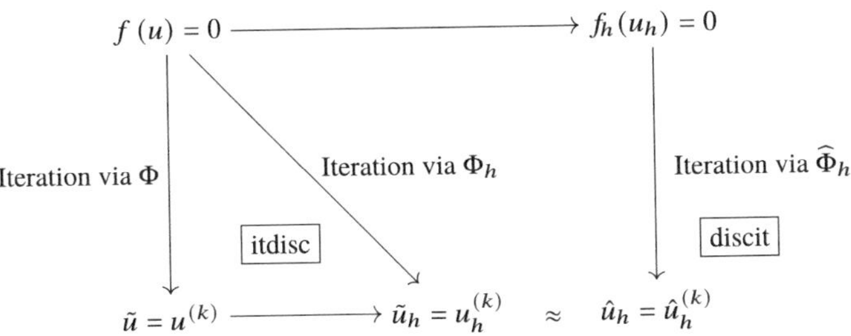

Fig. 11.1: The discit and the itdisc approach.

In general, the problem may be formulated in different equivalent settings, namely:

$$\text{Find} \quad x \in U \quad \text{with} \quad f(x) = 0\,. \tag{11.7}$$

Then x is called a *root* of (11.7) and a *zero* of f.

$$\text{Find} \quad x \in U \quad \text{with} \quad f(x) = x\,. \tag{11.8}$$

Then x is called a *fixed point*.

Here $U \subset V$, $f: U \to V'$ is a mapping, where V is the underlying function space, either V according to (11.1) in the first iterate – then discretize approach or $V := \mathbb{R}^m$ otherwise. For the fixed-point formulation, the property $f: U \to V$ is necessary. This is possible for (11.1), by rewriting it as

$$u = \bar{f}(u) := A^{-1}(\ell - G(u)). \tag{11.9}$$

The transition from one formulation to another follows by redefining f in evident ways.

In most cases, a root or a fixed point cannot be computed (within exact arithmetic) in a finite number of operations, but only by an *iterative method*, i.e., by a mapping

$$\Phi: U \to U,$$

so that (as in (5.7)) for the sequence

$$x^{(k+1)} := \Phi\bigl(x^{(k)}\bigr) \tag{11.10}$$

with given initial value $x^{(0)}$ we get

$$x^{(k)} \to x \quad \text{for} \quad k \to \infty, \tag{11.11}$$

where x is the solution of (11.7) or (11.8).

As we already stated in Section 5.1, in the case of a continuous mapping Φ it follows from (11.10), (11.11) that the limit x satisfies

$$x = \Phi(x). \tag{11.12}$$

This means that (11.12) should imply that x is a solution of (11.7) or (11.8). The extension of the definition of consistency in Section 5.1 requires the inverse implication.

The example (11.3) in fixed-point form shows that in the discit-approach an evaluation of Φ may mean in addition to function evaluation the solution of systems of linear equations (of same dimension as the discretization) or of nonlinear equations (of reduced dimension). In addition to the criteria of comparison for linear stationary methods we now have to take into account the following: Methods may, if they do at all, converge only locally, which leads to the following definition:

Definition 11.1 If in the above situation (11.11) holds for all $x^{(0)} \in U$ (i.e., for arbitrary starting values from U), then the sequence $\left(x^{(k)}\right)_k$ is called *globally convergent*. If an open subset $\tilde{U} \subset U$ exists such that (11.11) holds for all $x^{(0)} \in \tilde{U}$, then $\left(x^{(k)}\right)_k$ is called *locally convergent*. In the latter case, the set $\tilde{U}$ is called the *range of convergence* of the iteration.

On the other hand, we may observe a faster convergence than the linear convergence introduced in (5.3):

Definition 11.2 Let $\left(x^{(k)}\right)_k$ be a sequence in V, $x \in V$, and $\|\cdot\|$ a norm on V. The sequence $\left(x^{(k)}\right)_k$ *converges linearly* to x with respect to $\|\cdot\|$ if there exists a constant C with $0 < C < 1$ such that

$$\left\|x^{(k+1)} - x\right\| \le C\left\|x^{(k)} - x\right\| \quad \text{for all } k \in \mathbb{N}.$$

The sequence $\left(x^{(k)}\right)_k$ *converges with order of convergence* $p > 1$ to x if $x^{(k)} \to x$ for $k \to \infty$ and if there exists a constant $C > 0$ such that

$$\left\|x^{(k+1)} - x\right\| \le C\left\|x^{(k)} - x\right\|^p \quad \text{for all } k \in \mathbb{N}.$$

The sequence $\left(x^{(k)}\right)_k$ *converges superlinearly* to x if

$$\lim_{k\to\infty} \frac{\|x^{(k+1)} - x\|}{\|x^{(k)} - x\|} = 0.$$

The case $p = 2$ is also called *quadratic convergence*. Thus, while a linearly converging method guarantees a reduction of the error by a constant factor $C \in (0, 1)$, this reduction is improved step by step in the case of superlinear or higher order convergence. When we encounter quadratic convergence, for example, the number of significant digits is doubled in every step (minus a fixed number), so that usually only a small number of iterations will be necessary. For this reason variants of the quadratically converging *Newton's method* (Section 11.3) are attractive. But the restriction of local convergence may require modifications to enlarge the range of convergence.

To evaluate the complexity of a numerical method, the number of elementary operations for an iteration has to be considered. By an elementary operation we want also to understand the evaluation of functions like the sine, although this is much more costly than an ordinary floating-point operation. A typical subproblem during an iteration cycle is the solution of a system of linear equations, analogously to the simpler systems in the form (5.10) occurring in linear stationary problems. Besides the effort to *assemble* this system of equations, we have to account for the work to solve it, which can be done using one of the methods described in Section 2.5 and Chapter 5, i.e., in particular, again by means of an iterative method. We call this a *secondary* or *inner iteration*, which is attractive because of the sparse structure of the matrices originating from the discretization, as already discussed in Chapter 5. Here an *inexact* variant may be useful, in which the inner iteration is performed only up to a precision that conserves the convergence properties of the *outer iteration*. The numerical cost for the assembling may, in fact, be more expensive than the cost for the inner iteration. Hence methods with low cost for the assembling (but worse convergence) should also be considered. Keeping this in mind, we devote an introductory chapter to the *fixed-point iterations*, which are, roughly speaking, methods in which the iteration mapping Φ coincides with the mapping f.

As a guideline for the following sections, direct fixed-point methods do not require differentiability of the nonlinearity opposite to Newton's methods, which in general have better convergence properties. A further approach replaces differentiability by convexity (and symmetry) to write the problem as a (non-smooth) optimization problem. We do not pursue this line, but give some introduction by a sequence of exercises (Problems 11.1–11.4, 11.14–11.18). Our excuse may be seen in the fact that with a semismooth Newton's method we can integrate also such problems into our approach.

Exercises

Problem 11.1 The Poisson equation (2.1), (2.2) may be interpreted as the displacement of a clamped membrane. Assume the further constraint $u \geq \varphi$ (where $\varphi \in C(\overline{\Omega})$ with $\varphi \leq 0$ on $\partial\Omega$ is a given function) such that the set of admissible functions is reduced to

$$K := \{v \in H_0^1(\Omega) \mid v \geq \varphi \quad \text{a.e. in } \Omega\}.$$

Argue that a weak formulation should be the minimization of F from (2.11) over K and show that this problem is equivalent to the following problem:

Find $u \in K$ such that

$$a(u, v-u) \geq \langle f, v-u \rangle_0 \quad \text{for all } v \in K. \tag{11.13}$$

Show that for solutions $u \in H^2(\Omega)$ the *variational inequality* (11.13) is equivalent to

$$\begin{aligned} -\Delta u \geq f,\ u \geq \varphi \quad & \text{in } \Omega, \\ (u-\varphi)(-\Delta u - f) = 0 \quad & \text{in } \Omega, \end{aligned}$$

i.e., a *pointwise complementarity problem*, and also to:

There is a disjoint subdivision $\Omega = \Omega_0 \cup \Omega_+$ with the *coincidence set*

$$\Omega_0 := \{x \in \Omega \mid u(x) = \varphi(x)\}$$

and

$$-\Delta u = f \text{ on } \Omega_+ .$$

Problem 11.2 Consider a variational inequality of the form (11.13) in a Hilbert space V with a bilinear form a fulfilling (2.46), (2.47), and $f \in V'$. Let $K \subset V$ be a convex cone, i.e., a nonempty convex set such that $\lambda K \subset K$ for all $\lambda \geq 0$. Show:

a) (11.13) is equivalent to

$$\begin{aligned} u \in K,\ a(u, v) \geq \langle f, v \rangle \quad & \text{for all } v \in K, \\ a(u, u) = \langle f, u \rangle. & \end{aligned}$$

b) Two solutions u_i corresponding to the data $f_i, i = 1, 2$, satisfy the stability estimate

$$\|u_1 - u_2\|_V \leq \frac{M}{\alpha}\|f_1 - f_2\|_{V'}.$$

Problem 11.3 Under the assumptions of Problem 11.2, let $j : V \to \mathbb{R} \cup \{-\infty, \infty\}$ be a convex, lower semicontinuous functional, $j \not\equiv -\infty$ (recall that j is called *lower semicontinuous* at $v \in V$ if for all $s \in (-\infty, j(v))$ there is a number $\delta > 0$ such that $j(w) > s$ whenever $\|w - v\|_V < \delta$). Consider the following variational inequality problem:

Find $u \in V$ such that

$$a(u, v-u) + j(v) - j(u) \geq \langle f, v-u \rangle \quad \text{for all } v \in V. \tag{11.14}$$

a) Show that (11.14) reduces to (11.13) for

$$j(v) := \begin{cases} 0, & v \in K, \\ +\infty & \text{elsewhere.} \end{cases}$$

b) Show: If a is symmetric, then (11.14) is equivalent to

Minimize J on V, where

$$J(v) := \frac{1}{2}a(v,v) + j(v) - \langle f, v\rangle.$$

Argue with basic results from convex optimization that u exists uniquely.

Problem 11.4 Formulate the contact problem with a contact function φ for (11.1) (for symmetric a). Which conditions the nonlinearity ψ must fulfil to apply Problem 11.3?

11.2 Fixed-Point Iterations

For the fixed-point formulation (11.8) the choice $\Phi := f$ is evident according to (11.12); in other words, the *fixed-point iteration* reads as

$$x^{(k+1)} := f\big(x^{(k)}\big), \quad k = 0, 1, \ldots . \tag{11.15}$$

To diminish the distance between two consecutive members of the sequence, i.e.,

$$\big\|\Phi\big(x^{(k+1)}\big) - \Phi\big(x^{(k)}\big)\big\| = \big\|x^{(k+2)} - x^{(k+1)}\big\| < \big\|x^{(k+1)} - x^{(k)}\big\|,$$

it is sufficient that the iteration mapping (here $\Phi = f$) is contractive (see Appendix A.4).

Sufficient conditions for a contraction are given by the following lemma.

Lemma 11.3 *Let $U \subset \mathbb{R}^m$ be open and convex, and $g: U \to \mathbb{R}^m$ continuously differentiable. If*

$$\sup_{x\in U} \|Dg(x)\| =: L < 1$$

holds, where the matrix norm $\|\cdot\|$ in $\mathbb{R}^{m,m}$ is compatible with the vector $\|\cdot\|$ in $\mathbb{R}^m$, then g is contractive in U.

Proof Problem 11.5. □

Therefore, if $U \subset \mathbb{R}^m$ is open, $f: U \subset \mathbb{R}^m \to \mathbb{R}^m$ is continuously differentiable, and if there exists some $\tilde{x} \in U$ with $\|Df(\tilde{x})\| < 1$, then there exists a closed convex neighbourhood $\tilde{U}$ of $\tilde{x}$ such that

$$\|Df(x)\| \le L < 1 \quad \text{for all } x \in \tilde{U}$$

and, for example, $L := \frac{1}{2}(\|Df(\tilde{x})\| + 1)$, guaranteeing the contractivity of f in $\tilde{U}$.

The unique existence of a fixed point and the convergence of (11.15) is guaranteed if the set U, on which f is a contraction, is mapped into itself.

Theorem 11.4 (Banach's fixed-point theorem) *Let $(V, \|\cdot\|)$ be a Banach space, $U \subset V$, $U \neq \emptyset$, and U be closed. If $f: U \to V$ is contractive with Lipschitz constant $L < 1$ and $f(U) \subset U$, then we have*

1) There exists one and only one fixed point $x \in U$ of f.
2) For arbitrary $x^{(0)} \in U$ the fixed-point iteration (11.15) converges to x, and we have

$$
\begin{aligned}
\left\|x^{(k)} - x\right\| &\le \frac{L}{1-L}\left\|x^{(k)} - x^{(k-1)}\right\| \qquad \text{(a posteriori error estimate)} \\
&\le \frac{L^k}{1-L}\left\|x^{(1)} - x^{(0)}\right\| \qquad \text{(a priori error estimate).}
\end{aligned}
$$

Proof The iterates $x^{(k+1)} := f(x^{(k)})$ are well defined because of $f(U) \subset U$.
We prove that $\left(x^{(k)}\right)_k$ is a Cauchy sequence (see Appendix A.4):

$$
\begin{aligned}
\left\|x^{(k+1)} - x^{(k)}\right\| &= \left\|f(x^{(k)}) - f(x^{(k-1)})\right\| \le L\left\|x^{(k)} - x^{(k-1)}\right\| \\
&\le L^2\left\|x^{(k-1)} - x^{(k-2)}\right\| \le \ldots \le L^k\left\|x^{(1)} - x^{(0)}\right\|,
\end{aligned}
\tag{11.16}
$$

so that for any $k, l \in \mathbb{N}$

$$
\begin{aligned}
\left\|x^{(k+l)} - x^{(k)}\right\| &\le \left\|x^{(k+l)} - x^{(k+l-1)}\right\| + \left\|x^{(k+l-1)} - x^{(k+l-2)}\right\| + \ldots + \left\|x^{(k+1)} - x^{(k)}\right\| \\
&\le (L^{k+l-1} + L^{k+l-2} + \ldots + L^k)\left\|x^{(1)} - x^{(0)}\right\| \\
&= L^k(1 + L + \ldots + L^{l-1})\left\|x^{(1)} - x^{(0)}\right\| \\
&\le L^k \sum_{l=0}^{\infty} L^l \|x^{(1)} - x^{(0)}\| = L^k \frac{1}{1-L}\left\|x^{(1)} - x^{(0)}\right\|.
\end{aligned}
$$

Thus we have $\left\|x^{(k+l)} - x^{(k)}\right\| \to 0$ for $k \to \infty$; i.e., $\left(x^{(k)}\right)_k$ is a Cauchy sequence and thus converges to some $x \in V$ because of the completeness of V. Due to the closedness of U we conclude that $x \in U$. Since we have

$$
x^{(k+1)} \to x, \quad f\left(x^{(k)}\right) \to f(x) \quad \text{for } k \to \infty,
$$

x is also a fixed point of f.

The fixed point is unique, because for two fixed points $x, \overline{x}$ it holds that

$$
\|x - \overline{x}\| = \|f(x) - f(\overline{x})\| \le L\|x - \overline{x}\|,
$$

which immediately implies $x = \overline{x}$ because of $L < 1$. Moreover, we have

$$
\begin{aligned}
\left\|x^{(k)} - x\right\| &= \left\|f(x^{(k-1)}) - f(x)\right\| \le L\|x^{(k-1)} - x\| \\
&\le L\left(\left\|x^{(k-1)} - x^{(k)}\right\| + \left\|x^{(k)} - x\right\|\right),
\end{aligned}
$$

and thus from (11.16),

$$
\left\|x^{(k)} - x\right\| \le \frac{L}{1-L}\left\|x^{(k)} - x^{(k-1)}\right\| \le \frac{L}{1-L}L^{k-1}\left\|x^{(1)} - x^{(0)}\right\|.
$$

□

Remark 11.5 1) If an additional iteration step is carried out, a lower a posteriori error estimate is possible:

$$\frac{1}{1+L}\left\|x^{(k+1)}-x^{(k)}\right\| \leq \left\|x^{(k)}-x\right\| .$$

2) If f consists additively of a linear and nonlinear part

$$f(x) := Ax + g(x) \tag{11.17}$$

and

$$A := A_2 - A_1$$

with an "easy" to invert A_1, then

$$\Phi(x) = A_1^{-1}(A_2 x + g(x)) \tag{11.18}$$

is possible—a preconditioning with A_1.

Proof (of Remark 11.5, *1))* Since $\left\|x^{(k+1)}-x\right\| \leq L\left\|x^{(k)}-x\right\|$, the triangle inequality yields

$$\left\|x^{(k+1)}-x^{(k)}\right\| \leq \left\|x^{(k+1)}-x\right\| + \left\|x^{(k)}-x\right\| \leq (1+L)\left\|x^{(k)}-x\right\| .$$

□

According to Lemma 11.3, we can often construct a closed set U such that f is contractive on U. Then it remains to verify that $f(U) \subset U$. For this, the following result is helpful.

Lemma 11.6 *Let $U \subset V$, $f: U \to V$. If there exists an element $y \in U$ and a number $r > 0$ with*

$$\overline{B}_r(y) \subset U,$$

with f contractive on $\overline{B}_r(y)$ with Lipschitz constant $L < 1$, so that

$$\|y - f(y)\| \leq r(1-L),$$

then f has one and only one fixed point in $\overline{B}_r(y)$, and the iteration (11.15) *converges.*

Proof Problem 11.6. □

In the setting of Theorem 11.4 the fixed-point iteration is thus globally convergent in U. In the setting of Lemma 11.6 it is locally convergent in U (globally in $\overline{B}_r(y)$). We see that in the situation of Theorem 11.4 the sequence $\left(x^{(k)}\right)$ has a linear order of convergence (and in general not a better one) because of

$$\|x^{(k+1)} - x\| = \|f(x^{(k)}) - f(x)\| \leq L\|x^{(k)} - x\|.$$

A sufficient condition for local convergence of the corresponding order is given by the following theorem.

Theorem 11.7 *Let $U \subset V$ be open and $\Phi : U \to U$ be continuous. If there exist some element $\overline{x} \in U$, an open set $\widetilde{U} \subset U$ with $\overline{x} \in \widetilde{U}$, and constants C, p with $p \geq 1$, $C \geq 0$, and $C < 1$ for $p = 1$, such that*

$$\|\Phi(x) - \overline{x}\| \leq C\|x - \overline{x}\|^p \quad \textit{for all } x \in \widetilde{U}$$

holds, then the iteration defined by $x^{(k+1)} := \Phi\big(x^{(k)}\big)$ for a given $x^{(0)} \in U$ converges locally to $\overline{x}$ of order at least p, and $\overline{x}$ is a fixed point of Φ.

Proof Choose $W := B_r(\overline{x}) \subset \widetilde{U}$, with $r > 0$ sufficiently small, such that $W \subset V$ and

$$Cr^{p-1} =: L < 1\,.$$

If $x^{(k)} \in W$, then

$$\left\|x^{(k+1)} - \overline{x}\right\| = \left\|\Phi\big(x^{(k)}\big) - \overline{x}\right\| \leq C\left\|x^{(k)} - \overline{x}\right\|^p < Cr^p < r,$$

hence $x^{(k+1)} \in W$, too. This shows that $x^{(0)} \in W$ implies $x^{(k)} \in W$ for all $k \in \mathbb{N}$. Furthermore, we have

$$\left\|x^{(k+1)} - \overline{x}\right\| \leq C\left\|x^{(k)} - \overline{x}\right\|^p < C\,r^{p-1}\left\|x^{(k)} - \overline{x}\right\| = L\left\|x^{(k)} - \overline{x}\right\|,$$

therefore

$$x^{(k)} \to \overline{x} \quad \text{for } k \to \infty,$$

and consequently,

$$\overline{x} = \lim_{k\to\infty} x^{(k+1)} = \lim_{k\to\infty} \Phi\big(x^{(k)}\big) = \Phi(\overline{x})\,.$$

□

The special case of a scalar equation shows that we can expect at most linear convergence for $\Phi = f$:

Corollary 11.8 *Let $U \subset \mathbb{R}$ be an open subset, $\Phi : U \to U$ p-times continuously differentiable on U, and $\overline{x} \in U$ a fixed point of Φ.*

If $\Phi'(\overline{x}) \neq 0$, $|\Phi'(\overline{x})| < 1$ for $p = 1$ and $\Phi'(\overline{x}) = \ldots = \Phi^{(p-1)}(\overline{x}) = 0$, $\Phi^{(p)}(\overline{x}) \neq 0$ for $p > 1$, then the iteration defined by Φ is locally convergent to $\overline{x}$ with the order of convergence p, but not better.

Proof Taylor's expansion of Φ at $\overline{x}$ gives, for $x \in U$,

$$\Phi(x) = \Phi(\overline{x}) + \frac{\Phi^{(p)}(\xi)}{p!}(x - \overline{x})^p \quad \text{with } \xi \in (x, \overline{x}),$$

and in the case $p = 1$ we have $|\Phi'(\xi)| < 1$ for sufficiently small $|x - \overline{x}|$. Thus, there exists a neighbourhood V of $\overline{x}$ such that $|\Phi(x) - \overline{x}| \leq C|x - \overline{x}|^p$ for all $x \in V$ and $C < 1$ for $p = 1$. Theorem 11.7 implies the order of convergence p. The example $\Phi(x) := Lx^p$ with $L < 1$ for $p = 1$ with the fixed point $x = 0$ shows that no improvement is possible.

□

Exercises

Problem 11.5 Prove Lemma 11.3 by the help of the mean-value theorem.

Problem 11.6 Prove Lemma 11.6.

11.3 Newton's Method and Its Variants

11.3.1 The Standard Form of Newton's Method

In the following we want to study the formulation stated in (11.7), i.e., the problem of finding the solutions of

$$f(x) = 0 .$$

The simplest method of Chapter 5, the Richardson iteration (cf. (5.28)), suggests the direct application of the fixed-point iteration for, e.g., $\Phi(x) := -f(x) + x$. This approach is successful only if, in the case of a differentiable f, the Jacobi matrix $I - Df(x)$ is small near the solution in the sense of Lemma 11.3, where $Df(x) := \left(\partial_j f_i(x)\right)_{ij}$. A relaxation method similar to (5.30) leads to *damped* variants, which will be treated later.

The method in its *corrector formulation*, analogously to (5.10) with

$$\delta^{(k)} := x^{(k+1)} - x^{(k)},$$

is

$$\delta^{(k)} = -f\left(x^{(k)}\right), \tag{11.19}$$

or in its relaxation formulation, with a relaxation parameter $\omega > 0$,

$$\delta^{(k)} = -\omega f(x^{(k)})$$

with the fixed-point mapping

$$\Phi_\omega(x) := x - \omega f(x),$$

which for a Lipschitz continuous function f may be contractive for a (small) range $\omega \subset (0, 1)$. Now we want to introduce another approach to define Φ that is not based on relaxation and also provides a better rate of convergence:

Let $x^{(0)}$ be an approximation to a zero. An improved approximation could likely be achieved by the following:

- Replace f by a simpler function g that approximates f near $x^{(0)}$ and whose zero is to be determined.
- Find $x^{(1)}$ as the solution of $g(x) = 0$.

Newton's method needs the (Fréchet) differentiability of f, and g is chosen to be the approximating affine-linear function given by $Df(x^{(0)})$, i.e.,

$$g(x) := f(x^{(0)}) + Df(x^{(0)})(x - x^{(0)}) .$$

Under the assumption that $Df(x^{(0)})$ is invertible, the new iterate $x^{(1)}$ is determined by solving the *linear* equation

$$Df(x^{(0)})(x^{(1)} - x^{(0)}) = -f(x^{(0)}), \tag{11.20}$$

or formally by

$$x^{(1)} := x^{(0)} - Df(x^{(0)})^{-1} f(x^{(0)}) .$$

This suggests the following definition:

$$\Phi(f)(x) := x - Df(x)^{-1} f(x) . \tag{11.21}$$

Here Φ is well defined only if $Df(x)$ is invertible in a neighbourhood of the zero, and $x \in V$ is a zero of f if and only if x is a fixed point of Φ. When executing the iteration (in $\mathbb{R}^m$), we do not compute $Df(x^{(k)})^{-1}$ but solve only the linear system of equations similar to (11.20).

Thus, the kth iteration of *Newton's method* reads as follows: Solve

$$Df(x^{(k)})\,\delta^{(k)} = -f(x^{(k)}) \tag{11.22}$$

and set

$$x^{(k+1)} := x^{(k)} + \delta^{(k)} . \tag{11.23}$$

Equation (11.23) has the same form as (5.10) with $W := Df(x^{(k)})$, with the *residual* at $x^{(k)}$

$$d^{(k)} := f(x^{(k)}) .$$

Thus (in the discit approach) the subproblem of the kth iteration is easier in the sense that it consists of a system of linear equations (with the same structure of dependence as f; see Problem 11.10). In the same sense the system of equations (5.10) in the case of linear stationary methods is "easier" to solve than the original problem of the same type. Furthermore, W depends on k, in general.

An application of (11.22), (11.23) to the linear problem $Ax = b$, for which $Df(x) = A$ for all $x \in \mathbb{R}^m$, results in (5.10) with $W := A$, a method converging in one step, which just reformulates the original problem:

$$A(x - x^{(0)}) = -(Ax^{(0)} - b).$$

The convergence range of the iteration may be very small, as can be shown already by one-dimensional examples. But in this neighbourhood of the solution we have, e.g., for $m = 1$, the following fact.

Corollary 11.9 *Let $f \in C^3(\mathbb{R})$ and let $\overline{x}$ be a simple zero of f (i.e., $f'(\overline{x}) \neq 0$). Then Newton's method converges locally to $\overline{x}$, of order at least 2.*

Proof There exists an open neighbourhood V of $\overline{x}$ such that $f'(x) \neq 0$ for all $x \in V$; i.e., Φ is well defined by (11.21), continuous on V, and $\overline{x}$ is a fixed point of Φ. According to Corollary 11.8 it suffices to show that $\Phi'(\overline{x}) = 0$:

$$\Phi'(x) = 1 - \frac{f'(x)^2 - f(x)f''(x)}{f'(x)^2} = f(x)\frac{f''(x)}{f'(x)^2} = 0 \quad \text{for } x = \overline{x},$$

and Φ'' exists continuously, because $f \in C^3(\mathbb{R})$. □

In the following we want to develop a general local theorem of convergence for Newton's method (according to L.V. Kantorovich). It necessitates only the Lipschitz continuity of Df and ensures the existence of a zero, too. Here we always consider a fixed vector norm on $\mathbb{R}^m$ and consider a compatible matrix norm on $\mathbb{R}^{m,m}$. As a prerequisite we need the following result.

Lemma 11.10 *Let $C_0 \subset \mathbb{R}^m$ be convex and open, $f : C_0 \to \mathbb{R}^m$ differentiable, and suppose there exists a number $\gamma > 0$ such that*

$$\|Df(x) - Df(y)\| \leq \gamma \|x - y\| \quad \textit{for all } x, y \in C_0 . \tag{11.24}$$

Then

$$\|f(x) - f(y) - Df(y)(x - y)\| \leq \frac{\gamma}{2} \|x - y\|^2 \quad \textit{for all } x, y \in C_0 .$$

Proof Let $\varphi : [0, 1] \to \mathbb{R}^m$ be defined by $\varphi(t) := f(y + t(x - y))$ for arbitrary but fixed $x, y \in C_0$. Then φ is differentiable on $[0, 1]$ and

$$\varphi'(t) = Df(y + t(x - y))(x - y) .$$

Thus for $t \in [0, 1]$ we have

$$\begin{aligned}
\|\varphi'(t) - \varphi'(0)\| &= \|(Df(y + t(x - y)) - Df(y))(x - y)\| \\
&\leq \|Df(y + t(x - y)) - Df(y)\|\|x - y\| \leq \gamma t \|x - y\|^2 .
\end{aligned}$$

For

$$\Delta := f(x) - f(y) - Df(y)(x - y) = \varphi(1) - \varphi(0) - \varphi'(0) = \int_0^1 (\varphi'(t) - \varphi'(0))\, dt$$

we also get

$$\|\Delta\| \leq \int_0^1 \|\varphi'(t) - \varphi'(0)\|\, dt \leq \gamma\|x - y\|^2 \int_0^1 t\, dt = \frac{1}{2}\gamma \|x - y\|^2 .$$

□

Now we are able to conclude local, quadratic convergence.

Theorem 11.11 *Let $C \subset \mathbb{R}^m$ be convex, open and $f : C \to \mathbb{R}^m$ differentiable. Assume that, for $x^{(0)} \in C$, there exist numbers $\alpha, \beta, \gamma > 0$ such that*

$$h := \alpha\,\beta\,\gamma/2 < 1\,,$$

$$r := \alpha/(1-h)\,,$$

$$\overline{B}_r\left(x^{(0)}\right) \subset C\,.$$

Furthermore, we require

1) *Df is Lipschitz continuous on $C_0 := B_{r+\varepsilon}\left(x^{(0)}\right)$ for some $\varepsilon > 0$ with a constant γ in the sense of* (11.24).
2) *For all $x \in B_r\left(x^{(0)}\right)$ there exists $Df(x)^{-1}$, and $\left\|Df(x)^{-1}\right\| \le \beta$.*
3) *$\left\|Df\left(x^{(0)}\right)^{-1} f\left(x^{(0)}\right)\right\| \le \alpha$.*

Then:

a) *The Newton iteration*

$$x^{(k+1)} := x^{(k)} - Df\left(x^{(k)}\right)^{-1} f\left(x^{(k)}\right)$$

is well defined and

$$x^{(k)} \in B_r\left(x^{(0)}\right) \quad \textit{for all } k \in \mathbb{N}\,.$$

b) *$x^{(k)} \to \overline{x}$ for $k \to \infty$ and $f(\overline{x}) = 0$.*

c) *$\|x^{(k+1)} - \overline{x}\| \le \dfrac{\beta\gamma}{2}\|x^{(k)} - \overline{x}\|^2$ and $\|x^{(k)} - \overline{x}\| \le \alpha \dfrac{h^{2^k - 1}}{1 - h^{2k}}$ for $k \in \mathbb{N}$.*

Proof a): To show that $x^{(k+1)}$ is well defined it is sufficient to verify

$$x^{(k)} \in B_r\left(x^{(0)}\right)\ (\subset C) \quad \text{for all } k \in \mathbb{N}\,.$$

By induction we prove the extended proposition

$$x^{(k)} \in B_r\left(x^{(0)}\right) \text{ and } \left\|x^{(k)} - x^{(k-1)}\right\| \le \alpha\, h^{2^{k-1}-1} \text{ for all } k \in \mathbb{N}\,. \tag{11.25}$$

The proposition (11.25) holds for $k = 1$, because according to 3)

$$\left\|x^{(1)} - x^{(0)}\right\| = \left\|Df\left(x^{(0)}\right)^{-1} f\left(x^{(0)}\right)\right\| \le \alpha < r\,.$$

Let (11.25) be valid for $l = 1, \ldots, k$. Then $x^{(k+1)}$ is well defined, and by the application of the Newton iteration for $k-1$ we get

$$\begin{aligned}
\left\|x^{(k+1)} - x^{(k)}\right\| &= \left\|Df\left(x^{(k)}\right)^{-1} f\left(x^{(k)}\right)\right\| \le \beta \left\|f\left(x^{(k)}\right)\right\| \\
&= \beta \left\|f\left(x^{(k)}\right) - f\left(x^{(k-1)}\right) - Df\left(x^{(k-1)}\right)\left(x^{(k)} - x^{(k-1)}\right)\right\| \\
&\le \frac{\beta\gamma}{2}\left\|x^{(k)} - x^{(k-1)}\right\|^2
\end{aligned}$$

according to Lemma 11.10 with $C_0 := B_r\left(x^{(0)}\right)$, and

$$\left\|x^{(k+1)} - x^{(k)}\right\| \leq \frac{\beta\gamma}{2}\left\|x^{(k)} - x^{(k-1)}\right\|^2 \leq \frac{\beta\gamma}{2}\alpha^2 h^{2^k-2} = \alpha h^{2^k-1}.$$

Thus the second part of (11.25) holds for $k+1$, and also

$$\begin{aligned}\left\|x^{(k+1)} - x^{(0)}\right\| &\leq \left\|x^{(k+1)} - x^{(k)}\right\| + \left\|x^{(k)} - x^{(k-1)}\right\| + \ldots + \left\|x^{(1)} - x^{(0)}\right\| \\ &\leq \alpha\left(h^{2^k-1} + h^{2^{k-1}-1} + \ldots + h^7 + h^3 + h + 1\right) \\ &< \alpha/(1-h) = r.\end{aligned}$$

Hence (11.25) holds for $k+1$.

b): Using (11.25) we are able to verify that $\left(x^{(k)}\right)_k$ is a Cauchy sequence, because for $l \geq k$ we have

$$\begin{aligned}\left\|x^{(l+1)} - x^{(k)}\right\| &\leq \left\|x^{(l+1)} - x^{(l)}\right\| + \left\|x^{(l)} - x^{(l-1)}\right\| + \ldots + \left\|x^{(k+1)} - x^{(k)}\right\| \\ &\leq \alpha\, h^{2^k-1}\left(1 + h^{2^k} + \left(h^{2^k}\right)^3 + \ldots\right) \\ &< \frac{\alpha\, h^{2^k-1}}{1-h^{2^k}} \to 0 \quad \text{for } k \to \infty,\end{aligned} \tag{11.26}$$

since $h < 1$. Hence there exists $\overline{x} := \lim_{k\to\infty} x^{(k)}$ and $\overline{x} \in \overline{B}_r\left(x^{(0)}\right)$.

Furthermore, $f(\overline{x}) = 0$, because we can conclude from $x^{(k)} \in B_r\left(x^{(0)}\right)$ that

$$\left\|Df\left(x^{(k)}\right) - Df\left(x^{(0)}\right)\right\| \leq \gamma\left\|x^{(k)} - x^{(0)}\right\| < \gamma r.$$

Thus

$$\left\|Df\left(x^{(k)}\right)\right\| \leq \gamma r + \left\|Df\left(x^{(0)}\right)\right\| =: K,$$

and from $f\left(x^{(k)}\right) = -Df\left(x^{(k)}\right)\left(x^{(k+1)} - x^{(k)}\right)$ we obtain

$$\left\|f\left(x^{(k)}\right)\right\| \leq K\left\|x^{(k+1)} - x^{(k)}\right\| \to 0 \quad \text{for } k \to \infty.$$

Thus we also have

$$f(\overline{x}) = \lim_{k\to\infty} f(x^{(k)}) = 0.$$

c): Letting $l \to \infty$ in (11.26) we can prove the second part in c); the first part follows from

$$\begin{aligned}x^{(k+1)} - \overline{x} &= x^{(k)} - Df\left(x^{(k)}\right)^{-1} f\left(x^{(k)}\right) - \overline{x} \\ &= x^{(k)} - \overline{x} - Df\left(x^{(k)}\right)^{-1}\left(f\left(x^{(k)}\right) - f(\overline{x})\right) \\ &= Df\left(x^{(k)}\right)^{-1}\left(f(\overline{x}) - f\left(x^{(k)}\right) - Df\left(x^{(k)}\right)\left(\overline{x} - x^{(k)}\right)\right),\end{aligned}$$

which implies, according to Lemma 11.10 with $C_0 := B_{r+\varepsilon}\left(x^{(0)}\right) \subset C$,

$$\left\|x^{(k+1)} - \overline{x}\right\| \le \beta \frac{\gamma}{2} \left\|x^{(k)} - \overline{x}\right\|^2 .$$

□

The termination criterion (5.15), which is oriented at the residual, may also be used for the nonlinear problem (and not just for the Newton iteration). This can be deduced in analogy to (5.16):

Theorem 11.12 *Let the following be valid:*

There exists a zero $\overline{x}$ of f such that $Df(\overline{x})$ is nonsingular and Df is Lipschitz continuous in an open neighbourhood C of $\overline{x}$. (11.27)

Then for every $\varrho > 0$ there exists a number $\delta > 0$ such that

$$\|f(y)\|\,\|x - \overline{x}\| \le (1 + \varrho)\,\kappa(Df(\overline{x}))\,\|f(x)\|\,\|y - \overline{x}\| \quad \textit{for all } x, y \in B_\delta(\overline{x}).$$

Proof See [41, p. 69, p. 72] and Problem 11.8. □

Here κ is the condition number in a matrix norm that is compatible with the chosen vector norm. For $x := x^{(k)}$ and $y := x^{(0)}$ we get (locally) the generalization of (5.16).

11.3.2 Modifications of Newton's Method

Modifications of Newton's method aim in two directions:

- Reduction of the cost of the assembling and the solution of the system of equations (11.22) (without a significant deterioration of the properties of convergence).
- Enlargement of the range of convergence.

We can account for the first aspect by simplifying the matrix in (11.22) (*modified* or *simplified Newton's method*). The extreme case is the replacement of $Df\left(x^{(k)}\right)$ by the identity matrix; this leads us to the fixed-point iteration (11.19). If the mapping f consists of a nonlinear and a linear part (see (11.17)), then the system of equations (11.22) of the Newton iteration reads as

$$\left(A + Dg\left(x^{(k)}\right)\right) \delta^{(k)} = -f\left(x^{(k)}\right).$$

A straightforward simplification in this case is the fixed-point iteration

$$A\,\delta^{(k)} = -f\left(x^{(k)}\right). \tag{11.28}$$

It may be interpreted as the fixed-point iteration (11.18) of the system that is preconditioned with $A_l := A$. In (11.28) the matrix is identical in every iteration step; therefore it has to be assembled only once, and if we use a direct method (cf. Section 2.5), the LU factorization has to be carried out only once. Thus with forward and backward substitution we have only to perform methods with lower computational cost. For iterative methods we cannot rely on this advantage, but we can expect that $x^{(k+1)}$ is close to $x^{(k)}$, and consequently $\delta^{(k,0)} := 0$ constitutes a good initial guess. Accordingly, the assembling of the matrix gets more important with respect to the overall computational cost, and savings during the assembling become relevant.

We get a system of equations similar to (11.28) by applying the *chord method* (see Problem 11.7), where the linear approximation of the initial iterate is maintained, i.e.,

$$Df\big(x^{(0)}\big)\,\delta^{(k)} = -f\big(x^{(k)}\big)\,. \tag{11.29}$$

If the matrix $B\big(x^{(k)}\big)$, which approximates $Df\big(x^{(k)}\big)$, is changing in each iteration step, i.e.,

$$B\big(x^{(k)}\big)\,\delta^{(k)} = -f\big(x^{(k)}\big)\,, \tag{11.30}$$

then the only advantage can be a possibly easier assembling or solvability of the system of equations. If the partial derivatives $\partial_j f_i(x)$ are more difficult to evaluate than the function $f_i(y)$ itself (or possibly not evaluable at all), then the approximation of $Df(x^{(k)})$ by difference quotients can be taken into consideration. This corresponds to

$$B\big(x^{(k)}\big)\,e_j = \frac{1}{h}\big(f(x+he_j) - f(x)\big) \tag{11.31}$$

for the jth column of $B\big(x^{(k)}\big)$ with a fixed $h > 0$. The number of computations for the assembling of the matrix remains the same: m^2 for the full matrix and analogously for the sparse matrix (see Problem 11.10). Observe that numerical differentiation is an ill-posed problem, which means that we should ideally choose $h \sim \delta^{1/2}$, where $\delta > 0$ is the error level in the evaluation of f. Even then we can merely expect

$$\big\|Df\big(x^{(k)}\big) - B\big(x^{(k)}\big)\big\| \le C\delta^{1/2}$$

(see [41, pp. 80]). Thus in the best case we can expect only half of the significant digits of the machine precision. The second aspect of facilitated solvability of (11.30) occurs if there appear "small" entries in the Jacobi matrix, due to a problem-dependent weak coupling of the components, and these entries may be skipped. Take, for example, a block-structured Jacobi matrix $Df\big(x^{(k)}\big)$ as in (5.39):

$$Df\big(x^{(k)}\big) = \big(A_{ij}\big)_{ij}, \quad A_{ij} \in \mathbb{R}^{m_i,m_j}\,,$$

such that the blocks A_{ij} may be neglected for $j > i$. Then there results a nested system of equations of the dimensions $m_1, m_2, \ldots, m_p$.

The possible advantages of such simplified Newton's methods have to be weighted against the disadvantage of a deterioration in the order of convergence: Instead of an estimate like that in Theorem 11.11, c), we have to expect an additional term

$$\big\|B\big(x^{(k)}\big) - Df\big(x^{(k)}\big)\big\|\,\big\|x^{(k)} - x\big\|.$$

This means only linear or—by successive improvement of the approximation — superlinear convergence (see [41, pp. 75]). If we have a good initial iterate, it may often be advantageous to perform a small number of steps of Newton's method.

So in the following we will treat again Newton's method, although the subsequent considerations can also be transferred to its modifications.

If the linear problems (11.22) are solved by means of an iterative scheme, we have the possibility to adjust the accuracy of the algorithm in order to reduce the number of inner iterations, without a (severe) deterioration of the convergence of the outer iteration of the Newton iteration. So dealing with such *inexact Newton's methods*, we determine instead of $\delta^{(k)}$ from (11.22) only some correction $\tilde{\delta}^{(k)}$, which satisfies (11.22) only up to an *inner residual* $r^{(k)}$, i.e.,

$$Df\left(x^{(k)}\right)\tilde{\delta}^{(k)} = -f\left(x^{(k)}\right) + r^{(k)} .$$

The new iterate is given by

$$x^{(k+1)} := x^{(k)} + \tilde{\delta}^{(k)} .$$

The accuracy of $\tilde{\delta}^{(k)}$ is estimated by the requirement

$$\left\|r^{(k)}\right\| \le \eta_k \left\|f\left(x^{(k)}\right)\right\| \tag{11.32}$$

with certain properties for the sequence $\left(\eta_k\right)_k$ that still have to be determined. Since the natural choice of the initial iterate for solving (11.22) is $\delta^{(k,0)} := 0$, (11.32) corresponds to the termination criterion (5.15). Conditions for η_k can be deduced from the following result.

Theorem 11.13 *Let* (11.27) *hold and consider compatible matrix and vector norms. Then there exists for every* $\varrho > 0$ *a number* $\delta > 0$ *such that for* $x^{(k)} \in B_\delta(\overline{x})$,

$$\begin{aligned}\left\|x^{(k+1)} - \overline{x}\right\| \le{}& \left\|x^{(k)} - Df\left(x^{(k)}\right)^{-1} f\left(x^{(k)}\right) - \overline{x}\right\| \\ &+ (1+\varrho)\,\kappa\left(Df(\overline{x})\right)\eta_k \left\|x^{(k)} - \overline{x}\right\| .\end{aligned} \tag{11.33}$$

Proof By the choice of δ we can ensure the nonsingularity of $Df(x^{(k)})$. From

$$\tilde{\delta}^{(k)} = -Df\left(x^{(k)}\right)^{-1} f\left(x^{(k)}\right) + Df\left(x^{(k)}\right)^{-1} r^{(k)}$$

it follows that

$$\begin{aligned}\left\|x^{(k+1)} - \overline{x}\right\| &= \left\|x^{(k)} - \overline{x} + \tilde{\delta}^{(k)}\right\| \\ &\le \left\|x^{(k)} - \overline{x} - Df\left(x^{(k)}\right)^{-1} f\left(x^{(k)}\right)\right\| + \left\|Df\left(x^{(k)}\right)^{-1} r^{(k)}\right\| .\end{aligned}$$

The assertion can be deduced from the estimate

$$\begin{aligned}\left\|Df\left(x^{(k)}\right)^{-1} r^{(k)}\right\| &\le (1+\varrho)^{1/2}\left\|Df(\overline{x})^{-1}\right\|\left\|r^{(k)}\right\| \\ &\le (1+\varrho)^{1/2}\left\|Df(\overline{x})^{-1}\right\|\eta_k\,(1+\varrho)^{1/2}\left\|Df(\overline{x})\right\|\left\|x^{(k)} - \overline{x}\right\| .\end{aligned}$$

Here we used Problem 11.8 b), c) and (11.32). □

The first part of the approximation corresponds to the error of the exact Newton step, which can be estimated using the same argument as in Theorem 11.11, c) (with Problem 11.8, b)) by

$$\left\|x^{(k)} - Df\left(x^{(k)}\right)^{-1} f\left(x^{(k)}\right) - \overline{x}\right\| \leq (1+\varrho)^{1/2} \left\|Df(\overline{x})^{-1}\right\| \frac{\gamma}{2} \left\|x^{(k)} - \overline{x}\right\|^2 .$$

This implies the following result.

Corollary 11.14 *Let the assumptions of Theorem 11.13 be satisfied. Then there exist numbers $\delta > 0$ and $\overline{\eta} > 0$ such that, for $x^{(0)} \in B_\delta(\overline{x})$ and $\eta_k \leq \overline{\eta}$ for all $k \in \mathbb{N}$, the inexact Newton's method has the following properties:*

1) The sequence $\left(x^{(k)}\right)_k$ converges linearly to $\overline{x}$.
2) If $\eta_k \to 0$ for $k \to \infty$, then $\left(x^{(k)}\right)_k$ converges superlinearly.
3) If $\eta_k \leq K\left\|f\left(x^{(k)}\right)\right\|$ for a constant $K > 0$, then $\left(x^{(k)}\right)_k$ converges quadratically.

Proof Problem 11.9. □

The estimate (11.33) suggests a careful choice of a very fine accuracy level $\overline{\eta}$ of the inner iteration to guarantee the above statements of convergence. This is particularly true for ill-conditioned Jacobi matrices $Df(\overline{x})$ (which is common for discretization matrices, see (5.34)). In fact, the analysis in the weighted norm $\|\cdot\| := \|Df(\overline{x})\cdot\|$ shows that only $\eta_k \leq \overline{\eta} < 1$ has to be ensured (cf. [41, pp. 97]). With this and on the basis of

$$\tilde{\eta}_k = \alpha \left\|f\left(x^{(k)}\right)\right\|^2 / \left\|f\left(x^{(k-1)}\right)\right\|^2$$

for some $\alpha \leq 1$ we can construct η_k in an adaptive way (see [41, p. 105]).

Most of the iterative methods introduced in Chapter 5 do not require the explicit knowledge of the matrix $Df\left(x^{(k)}\right)$. It suffices that the matrix-vector multiplication $Df\left(x^{(k)}\right)y$ is feasible for vectors y, in general for fewer than m of them; i.e., the directional derivative of f in $x^{(k)}$ in the direction y is needed. Thus in case a difference scheme for the derivatives of f should be necessary or reasonable, it is more convenient to choose directly a difference scheme for the directional derivative.

Since we cannot expect convergence of Newton's method in general, we require indicators for the convergence behaviour of the iteration. The solution $\overline{x}$ is in particular also the solution of

$$\text{Minimize} \quad \|f(x)\|^2 \quad \text{for } x \in \mathbb{R}^m .$$

Thus we could expect a descent of the sequence of iterates $\left(x^{(k)}\right)$ in this functional, i.e.,

$$\left\|f(x^{(k+1)})\right\| \leq \overline{\Theta}\left\|f(x^{(k)})\right\| \quad \text{for a } \overline{\Theta} < 1 .$$

If this *monotonicity test* is not fulfilled, the iteration is terminated. Such an example of an inexact Newton's method is given in Table 11.1.

In order to avoid the termination of the method due to divergence, the *continuation* or *homotopy methods* have been developed. They attribute the problem $f(x) = 0$ to a family of problems to provide successively good initial iterates. The approach

Let $x^{(0)}, \tau > 0, \eta_0, \overline{\Theta} \in (0,1), k = 0, i = 0$ be given.

(1) $$\tilde{\delta}^{(k,0)} := 0\,,\ i := 1\,.$$

(2) Determine the ith iterate $\tilde{\delta}^{(k,i)}$ for $Df(x^{(k)})\tilde{\delta}^{(k)} = -f(x^{(k)})$ and compute

$$r^{(i)} := Df(x^{(k)})\tilde{\delta}^{(k,i)} + f(x^{(k)})\,.$$

(3) If $\|r^{(i)}\| \le \eta_k \|f(x^{(k)})\|$, then go to (4), else set $i := i + 1$ and go to (2).

(4) $$\tilde{\delta}^{(k)} := \tilde{\delta}^{(k,i)}\,.$$

(5) $$x^{(k+1)} := x^{(k)} + \tilde{\delta}^{(k)}\,.$$

(6) If $\|f(x^{(k+1)})\| > \overline{\Theta}\|f(x^{(k)})\|$, interrupt.

(7) If $\|f(x^{(k+1)})\| \le \tau \|f(x^{(0)})\|$, end. Else compute η_{k+1}, set $k := k + 1$, and go to (1).

Table 11.1: Inexact Newton's method with monotonicity test.

presented at the end of Section 11.4 for time-dependent problems is similar to the continuation methods. Another approach (which can be combined with the latter) modifies the (inexact) Newton's method, so that the range of convergence is enlarged: Applying the *damped (inexact) Newton's method* means reducing the step length of $x^{(k)}$ to $x^{(k+1)}$ as long as we observe a decrease conformable to the monotonicity test. One strategy of damping, termed *Armijo's rule*, is described in Table 11.2 and replaces the steps (1), (5), and (6) in Table 11.1.

Let additionally $\alpha, \beta \in (0,1)$ be given.

(1) $\tilde{\delta}^{(k,0)} := 0,\ i := 1,\ \lambda_k := 1.$

(5) If $\|f(x^{(k)} + \lambda_k \tilde{\delta}^{(k)})\| \ge (1 - \alpha\lambda_k)\|f(x^{(k)})\|$, set $\lambda_k := \beta\lambda_k$ and go to (5).

(6) $x^{(k+1)} := x^{(k)} + \lambda_k \tilde{\delta}^{(k)}\,.$

Table 11.2: Damped inexact Newton step according to Armijo's rule.

Thus damping Newton's method means also a relaxation similar to (5.30), where $\omega := \lambda_k$ is being adjusted to the iteration step as in (5.41).

In the formulation of Table 11.2 the iteration may eventually not terminate if in step (5) λ_k is successively reduced. This must be avoided in a practical implementation of the method. But except for situations where divergence is obvious, this situation will not appear, because we have the following theorem:

Theorem 11.15 *Let constants $\alpha, \beta, \gamma > 0$ exist such that the conditions 1), 2) of Theorem 11.11 on $\bigcup_{k\in\mathbb{N}} B_r(x^{(k)})$ hold for the sequence $\left(x^{(k)}\right)_k$ defined according to Table 11.2. Let $\eta_k \leq \overline{\eta}$ for an $\overline{\eta} < 1 - \alpha$.*

Then, if $f\left(x^{(0)}\right) \neq 0$, there exists a number $\overline{\lambda} > 0$ such that $\lambda_k \geq \overline{\lambda}$ for all $k \in \mathbb{N}$. If furthermore the sequence $\left(x^{(k)}\right)_k$ is bounded, then there exists a zero $\overline{x}$ satisfying (11.27), and

$$x^{(k)} \to \overline{x} \quad \text{for } k \to \infty .$$

There exists a number $k_0 \in \mathbb{N}$ such that the relation

$$\lambda_k = 1 \quad \text{for all } k \geq k_0$$

holds.

Proof See [41, pp. 139]. □

We see that in the final stage of the iteration we again deal with the (inexact) Newton's method with the previously described behaviour of convergence.

Finally, the following should be mentioned: The problem $f(x) = 0$ and Newton's method are *affine-invariant* in the sense that a transition to the problem $Af(x) = 0$ with a nonsingular $A \in \mathbb{R}^{m,m}$ changes neither the problem nor the iterative method, since

$$D(Af)(x)^{-1}Af(x) = Df(x)^{-1}f(x) .$$

Among the assumptions of Theorem 11.11, the inequality (11.24) is not affine-invariant. A possible alternative would be

$$\|Df(y)^{-1}(Df(x) - Df(y))\| \leq \gamma\|x - y\| ,$$

which fulfils the requirement. With the proof of Lemma 11.10 it follows that

$$\|Df(y)^{-1}(f(x) - f(y) - Df(y)(x - y))\| \leq \frac{\gamma}{2}\|x - y\|^2 .$$

With this argument a similar variant of Theorem 11.11 can be proven.

The test of monotonicity is not affine-invariant, so probably the *natural test of monotonicity*

$$\left\|Df\left(x^{(k)}\right)^{-1} f\left(x^{(k+1)}\right)\right\| \leq \overline{\Theta}\left\|Df\left(x^{(k)}\right)^{-1} f\left(x^{(k)}\right)\right\|$$

has to be preferred. The vector on the right-hand side has already been computed, being, except for the sign, the Newton correction $\delta^{(k)}$. But for the vector in the left-hand side, $-\overline{\delta}^{(k+1)}$, the system of equations

$$Df\left(x^{(k)}\right)\overline{\delta}^{(k+1)} = -f\left(x^{(k+1)}\right)$$

additionally has to be solved.

Exercises

Problem 11.7 Consider the chord method as described in (11.29). Prove the convergence of this method to the solution $\overline{x}$ under the following assumptions:

a) Let (11.27) be satisfied with $\overline{B}_r(\overline{x}) \subset C$,
b) $\left\| \left[Df(x^{(0)}) \right]^{-1} \right\| \le \beta$,
c) $2\beta\gamma r < 1$,
d) $x^{(0)} \in \overline{B}_r(\overline{x})$.

Problem 11.8 Let the assumption (11.27) hold. Prove for compatible matrix and vector norms that for every $\varrho > 0$ there exists a number $\delta > 0$ such that, for every $x \in B_\delta(\overline{x})$,

a) $\|Df(x)\| \le (1+\varrho)^{1/2} \|Df(\overline{x})\|$,
b) $\|Df(x)^{-1}\| \le (1+\varrho)^{1/2} \|Df(\overline{x})^{-1}\|$

(*Hint:* Employ $\|(I-M)^{-1}\| \le 1/(1-\|M\|)$ for $\|M\| < 1$),
c) $(1+\varrho)^{-1/2} \|Df(\overline{x})^{-1}\|^{-1} \|x-\overline{x}\| \le \|f(x)\| \le (1+\varrho)^{1/2} \|Df(\overline{x})\| \|x-\overline{x}\|$,
d) Theorem 11.12.

Problem 11.9 Prove Corollary 11.14.

Problem 11.10 Let $U \subset \mathbb{R}^m$ be open and convex. Consider the problem (11.7) with a continuously differentiable mapping $f : U \to \mathbb{R}^m$. For $i = 1, \ldots, m$, let $J_i \subset \{1, \ldots, m\}$ be defined by

$$\partial_j f_i(x) = 0 \quad \text{for } j \notin J_i \text{ and every } x \in U .$$

The mapping f is called *sparse*, if $l_i := |J_i| < m$, or *sparse in the strict sense*, if $l_i \le l$ for all $i = 1, \ldots, m$, and $l < m$ is independent of m for a sequence of problems (11.7) of dimension m.

Then the evaluation of $Df(x)$ and its approximation according to (11.31) both need $\sum_{k=1}^m l_k$ evaluations of $\partial_j f_i$ or of f_i, respectively. What is the computational effort for a difference approximation

$$\frac{f(x + h\delta/\|\delta\|) - f(x)}{h} \|\delta\|$$

of the directional derivative $Df(x)\delta$?

11.4 Semilinear Boundary Value Problems for Elliptic and Parabolic Equations

In this section we treat *semilinear* problems as the simplest nonlinear case, where nonlinearities do not occur in parts containing derivatives. Hence we want to examine differential equations of the form (0.33) that satisfy (0.42) and (0.43).

Stationary Diffusion–Convection–Reaction Problems

As a stationary problem we consider the differential equation (see (11.1))

$$Lu(x) + \psi(x, u(x)) = 0 \quad \text{for } x \in \Omega \tag{11.34}$$

with the linear elliptic differential operator L in divergence form according to (3.36) and linear boundary conditions on $\partial\Omega$ according to (3.37)–(3.39). Here $\psi : \Omega\times\mathbb{R} \to \mathbb{R}$ denotes a mapping that is supposed to be continuously differentiable in its second argument. The differentiability of ψ carries over to the (Fréchet) differentiability of $G : U \to V'$ (see (11.2)):

$$\big(G(u)\big)(v) := \langle \psi(\cdot, u), v\rangle_0 \quad \text{for all } u \in U,\ v \in V. \tag{11.35}$$

Lemma 11.16 *Assume $\overline{u} \in U$ and ψ is differentiable in its second argument on a set $\Omega \times W$, $W \subset \mathbb{R}$, where*

1) $\overline{u} \in L^\infty(\Omega)$ and $W \supset [\operatorname{ess\,inf} \overline{u}, \operatorname{ess\,sup} \overline{u}]$ is bounded, or
2) $W = \mathbb{R}$ and there exist constants $C_1, C_2 \geq 0$ such that

$$\left|\frac{\partial}{\partial u}\psi(x, u)\right| \leq C_1 + C_2|u|^{p-2} \quad \textit{for } p > 1,\ d \leq 2 \textit{ or } p = 2d/(d-2). \tag{11.36}$$

Then the operator G is (Fréchet) differentiable at $\overline{u}$ and

$$\langle DG(\overline{u})w, v\rangle = \int_\Omega \frac{\partial}{\partial u}\psi(\cdot, \overline{u})wv\,dx \quad \textit{for all } w \in U,\ v \in V.$$

Proof We only consider the case 2), for 1) the proof can be repeated in a simplified way.

To show that the operator $DG(\overline{u}) : U \to V'$ is well defined and then also continuous (see [3, Thm. 3.12]), it is sufficient to take $p \geq 1$ such that

$$V \subset L^p(\Omega) \tag{11.37}$$

and thus

$$L^{p'}(\Omega) \subset V' \quad \text{with} \quad \frac{1}{p} + \frac{1}{p'} = 1,$$

and show the differentiability of

$$G : (U, \|\cdot\|_{L^p(\Omega)}) \to L^{p'}(\Omega)$$

at $\overline{u}$.

According to Sobolev's embedding theorem [1, Thm. 5.4], the embedding (11.37) holds for every $p < \infty$, if $d \le 2$, and for $p \le 2d/(d-2)$, if $d \ge 3$. To show that

$$DG(\overline{u}) : L^p(\Omega) \to L^{p'}(\Omega)$$

is well defined, it is sufficient to show that the superposition operator induced by $\frac{\partial}{\partial u}\psi$ is well defined from $L^p(\Omega)$ to $L^s(\Omega)$ with $1/p + 1/s = 1/p'$, meaning $s = p/(p-2)$. This follows from (11.36).

To show that $DG(\overline{u})$ is the derivative, because of Lemma 11.10 it is sufficient to estimate

$$I(v) := \int_\Omega z(x)^2 |v(x)| dx$$

for $\overline{u} + z \in U$, $v \in V$:

$$\begin{aligned} I(v) &\le \|z\|_{L^p} \|z\|_{L^s} \|v\|_{L^p} && \text{where } 2/p + 1/s = 1 \text{ and } p \text{ from 2)} \\ &\le C\|z\|_1 \|z\|_{L^s} \|v\|_1 && \text{by Sobolev's embedding theorem [1, Thm. 5.4],} \end{aligned}$$

and thus

$$\frac{\|I(v)\|_{V'}}{\|z\|_1} \to 0 \quad \text{for } \|z\|_1 \to 0.$$

□

According to the type of boundary conditions, a Galerkin discretization in $V_h \subset V$ and $V_h = \text{span}\,\{\varphi_1, \ldots, \varphi_M\}$ with the approximative solution $u_h \in V_h$ in the representation $u_h = \sum_{i=1}^{M} \xi_i \varphi_i$ gives

$$S\boldsymbol{\xi} + G_h(\boldsymbol{\xi}) = \boldsymbol{b} \tag{11.38}$$

with the stiffness matrix $S := \big(a\,(\varphi_j, \varphi_i)\,\big)_{i,j}$ and a vector $\boldsymbol{b}$ that contains the contributions of the inhomogeneous boundary conditions. Here the nonlinear mapping $G_h : \mathbb{R}^M \to \mathbb{R}^M$ is defined by

$$G_h(\boldsymbol{\xi}) := \big(G_{h,j}(\boldsymbol{\xi})\big)_j \quad \text{with} \quad G_{h,j}(\boldsymbol{\xi}) := \int_\Omega \psi\Big(x, \sum_{i=1}^{M} \xi_i \varphi_i(x)\Big) \varphi_j(x)\, dx\,.$$

Note that this notation differs from that in Section 2.2 and the subsequent chapters: There we denoted S by A_h and $\boldsymbol{b} - G_h(\boldsymbol{\xi})$ by $\boldsymbol{q}_h$. For reasons of brevity we partially omit the index h.

For the moment we want to suppose that the mapping G_h can be evaluated exactly. The system of equations (11.38) with

$$A := S \quad \text{and} \quad g(\boldsymbol{\xi}) := G_h(\boldsymbol{\xi}) - \boldsymbol{b}$$

is of the type introduced in (11.17) in the variable $\boldsymbol{\xi}$. Thus we may apply, besides the Newton iteration, the fixed-point iteration, introduced in (11.28), and the variants of Newton's method, namely the modified and inexact versions with their already discussed advantages and drawbacks. We have to examine the question of how the properties of the matrix will change by the transition from A to $A + DG_h(\overline{\boldsymbol{\xi}})$, where $\overline{\boldsymbol{\xi}}$ stands for the current iterate. As the discretization is conforming, i.e.,

$$G_h = i_{V_h'} \circ G \circ i_{V_h}$$

with the embeddings i_{V_h} and $i_{V_h'} = i_{V_h}^T$, we have from Lemma 11.16

$$\left(DG_h(\overline{\boldsymbol{\xi}})\right)_{ij} = \int_\Omega \frac{\partial}{\partial u}\psi(\cdot,\overline{u})\varphi_i\varphi_j\,dx\,, \tag{11.39}$$

where $\overline{\boldsymbol{u}} := P\overline{\boldsymbol{\xi}} = \sum_{i=1}^{M} \overline{\xi}_i\varphi_i \in V_h$ denotes the function belonging to the representing vector $\overline{\boldsymbol{\xi}}$. This shows that a conforming Galerkin discretization leads to the commutativity in the diagram of Figure 11.1.

The identity (11.39) shows that $DG_h(\overline{\boldsymbol{\xi}})$ is symmetric and positive semidefinite, respectively definite, if the following condition for $\alpha = 0$, respectively $\alpha > 0$, holds:

$$\text{There exists some number } \alpha \geq 0 \text{ such that } \frac{\partial}{\partial u}\psi(\cdot,u) \geq \alpha \text{ for all } u \in \mathbb{R}\,. \tag{11.40}$$

More precisely, we have for $\boldsymbol{\eta} \in \mathbb{R}^M$, if (11.40) is valid,

$$\boldsymbol{\eta}^T DG_h(\overline{\boldsymbol{\xi}})\boldsymbol{\eta} = \int_\Omega \frac{\partial}{\partial u}\psi(\cdot,\overline{u})\,|P\boldsymbol{\eta}|^2\,dx \geq \alpha\,\|P\boldsymbol{\eta}\|_0^2\,.$$

For such a monotone nonlinearity the properties of definiteness of the stiffness matrix S may be "enforced". If, on the other hand, we want to make use of the properties of an M-matrix that can be ensured by the conditions (1.35) or (1.35)*, then it is not clear whether these properties are conserved after addition of $DG_h(\overline{\boldsymbol{\xi}})$. This is due to the fact that $DG_h(\overline{\boldsymbol{\xi}})$ is a sparse matrix of the same structure as S, but it also entails a spatial coupling that is not contained in the continuous formulation (11.34).

Numerical Quadrature

Owing to the above reason, the use of a node-oriented quadrature rule for the approximation of $G_h(\boldsymbol{\xi})$ is suggested, i.e., a quadrature formula of the type

$$Q(f) := \sum_{i=1}^{M} \omega_i f(a_i) \quad \text{for } f \in C(\overline{\Omega}) \tag{11.41}$$

with weights $\omega_i \in \mathbb{R}$. Such a quadrature formula results from

$$Q(f) := \int_\Omega I(f)\,dx \quad \text{for } f \in C(\overline{\Omega}), \tag{11.42}$$

where

$$I: C(\overline{\Omega}) \to V_h, \quad I(f) := \sum_{i=1}^{M} f(a_i)\varphi_i,$$

is the interpolation operator of the degrees of freedom. For this consideration we thus assume that only Lagrangian elements enter the definition of V_h. In the case of (11.42) the weights in (11.41) are hence given by

$$\omega_i = \int_\Omega \varphi_i\,dx .$$

This corresponds to the local description (3.134). More specifically, we get, for example, for the linear approach on simplices as a generalization of the *composite trapezoidal rule*,

$$\omega_i = \frac{1}{d+1} \sum_{K \in \mathcal{T}_h:\, K \ni a_i} |K| . \tag{11.43}$$

Approximation of the mapping G_h by a quadrature rule of the type (11.41) gives

$$\tilde{G}_h(\boldsymbol{\xi}) = \left(\tilde{G}_{h,j}(\boldsymbol{\xi})\right)_j \quad \text{with} \quad \tilde{G}_{h,j}(\boldsymbol{\xi}) := \omega_j \psi(a_j, \xi_j),$$

because of $\varphi_j(a_i) = \delta_{ij}$. We see that the approximation $\tilde{G}_h$ has the property that $\tilde{G}_{h,j}$ depends only on ξ_j. We call such a $\tilde{G}_h$ a *diagonal field*. Qualitatively, this corresponds better to the continuous formulation (11.34) and leads to the fact that $D\tilde{G}_h(\overline{\boldsymbol{\xi}})$ is diagonal:

$$D\tilde{G}_h(\overline{\boldsymbol{\xi}})_{ij} = \omega_j \frac{\partial}{\partial u}\psi(a_j, \overline{\xi}_j)\delta_{ij} . \tag{11.44}$$

If we impose that all quadrature weights ω_i are positive, which is the case in (11.43) and also in other examples in Section 3.5.2, all of the above considerations about the properties of $D\tilde{G}_h(\overline{\boldsymbol{\xi}})$ and $S + D\tilde{G}_h(\overline{\boldsymbol{\xi}})$ remain valid; additionally, if S is an M-matrix, because the conditions (1.35) or (1.35)* are fulfilled, then $S + D\tilde{G}_h(\overline{\boldsymbol{\xi}})$ remains an M-matrix, too. This is justified by the following fact (compare [61] and [6], cf. (1.36) for the notation):

If A is an M-matrix and $B \geq A$ with $b_{ij} \leq 0$ for $i \neq j$, then B is an M-matrix as well. (11.45)

Conditions of Convergence for Iterative Schemes

Comparing the requirements for the fixed-point iteration and Newton's method stated in the Theorems 11.4 and 11.11, we observe that the conditions in Theorem 11.4 can be fulfilled only in special cases, where $S^{-1}D\tilde{G}_h(\overline{\boldsymbol{\xi}})$ is small according to a suitable matrix norm (see Lemma 11.3). But it is also difficult to draw general conclusions about requirement 3) in Theorem 11.11, which together with $h < 1$ quantifies the closeness of the initial iterate to the solution. The condition 1), on the other hand, is met for (11.39) and (11.44) if $\frac{\partial}{\partial u}\psi$ is Lipschitz continuous in its second argument (see Problem 11.11). Concerning the condition 2) we have the following: Let ψ be monotone nondecreasing in its second argument (i.e., (11.40) holds with $\alpha \geq 0$) and let S be symmetric and positive definite, which is true for a problem without convection terms (compare (3.28)). Then we have in the spectral norm

$$\left\|S^{-1}\right\|_2 = 1/\lambda_{\min}(S),$$

where $\lambda_{\min}(S) > 0$ denotes the smallest eigenvalue of S. Hence

$$\left\|(S + DG_h(\boldsymbol{\xi}))^{-1}\right\|_2 = 1/\lambda_{\min}\left(S + DG_h(\overline{\boldsymbol{\xi}})\right) \leq 1/\lambda_{\min}(S) = \left\|S^{-1}\right\|_2,$$

and consequently, requirement 2) is valid with $\beta := \left\|S^{-1}\right\|_2$.

Concerning the choice of the initial iterate, there is no generally successful strategy. We may choose the solution of the linear subproblem, i.e.,

$$S\boldsymbol{\xi}^{(0)} = \boldsymbol{b}\,. \tag{11.46}$$

In case it fails to converge even with damping, then we may apply, as a generalization of (11.46), the continuation method to the family of problems

$$f(\boldsymbol{\xi};\lambda) := S + \lambda G_h(\boldsymbol{\xi}) - \boldsymbol{b} = 0$$

with a continuation parameter $\lambda \in [0, 1]$. If all these problems have solutions $\boldsymbol{\xi} = \boldsymbol{\xi}_\lambda$ so that $Df(\boldsymbol{\xi};\lambda)$ exists and is nonsingular in a neighbourhood of $\boldsymbol{\xi}_\lambda$, and if there exists a continuous solution trajectory without bifurcation, then the interval $[0, 1]$ can be discretized by $0 = \lambda_0 < \lambda_1 < \ldots < \lambda_N = 1$, and solutions $\boldsymbol{\xi}_{\lambda_i}$ of $f(\boldsymbol{\xi};\lambda_i) = 0$ can be obtained by performing a Newton iteration with the (approximative) solution for $\lambda = \lambda_{i-1}$ as starting iterate. Since the solutions $\boldsymbol{\xi}_{\lambda_i}$ for $i < N$ are just auxiliary means, they should be obtained rather coarsely, i.e., with one or two Newton steps. The stated conditions are fulfilled under the supposition (11.40). If this condition of monotonicity does not hold, we may encounter a *bifurcation* of the continuous solution (see, for example, [55, pp. 28]).

Instationary Diffusion–Convection–Reaction Problems

The elliptic boundary value problem (11.34) corresponds to the parabolic initial value problem

$$\partial_t \big(bu(x,t)\big) + Lu(x,t) + \psi(x, u(x,t)) = 0 \quad \text{for } (x,t) \in Q_T \tag{11.47}$$

with linear boundary conditions according to (3.19)–(3.21) and b as in (9.13) and the initial condition

$$u(x,0) = u_0(x) \quad \text{for } x \in \Omega\,. \tag{11.48}$$

We have already met an example for (11.47), (11.48) in (0.32). Analogously to (11.38) and (9.51), the Galerkin discretization in V_h (i.e., the semidiscretization) leads to the nonlinear system of ordinary differential equations

$$B\frac{d}{dt}\boldsymbol{\xi}(t) + S\boldsymbol{\xi}(t) + G_h(\boldsymbol{\xi}(t)) = \boldsymbol{\beta}(t) \quad \text{for } t \in (0,T]\,, \quad \boldsymbol{\xi}(0) = \boldsymbol{\xi}_0$$

for the representing vector $\boldsymbol{\xi}(t)$ of the approximation $u_h(\cdot,t) = \sum_{i=1}^{M} \xi_i(t)\varphi_i$, where $u_{0h} = \sum_{i=1}^{M} \xi_{0i}\varphi_i$ is an approximation of the initial value u_0 (see Section 9.2). The matrix B is the mass matrix

$$B = \Big(\big\langle b\varphi_j, \varphi_i\big\rangle_0\Big)_{ij}\,,$$

and $\boldsymbol{\beta}(t)$ contains the contributions of the inhomogeneous boundary conditions analogously to $\boldsymbol{b}$ in (11.38).

To obtain the fully discrete scheme we use the one-step-theta method as in Section 9.3. Here we allow the time step size τ_n to vary in each step, in particular determined by a time step control before the execution of the nth time step. So, if the approximation U^n is known for $t = t_n$, then the approximation U^{n+1} for $t = t_{n+1} := t_n + \tau_n$ is given in generalization of (9.98) as the solution of

$$\begin{aligned}\left\langle \frac{1}{\tau_n}\Big(U^{n+1} - U^n\Big), v_h\right\rangle_0^{(b)} &+ a\Big(\Theta U^{n+1} + (1-\Theta)U^n, v_h\Big) \\ &+ \big\langle \psi^{n+\Theta}, v_h\big\rangle = \Theta\boldsymbol{\beta}(t_{n+1}) + (1-\Theta)\boldsymbol{\beta}(t_n).\end{aligned} \tag{11.49}$$

Here $\Theta \in [0,1]$ is the fixed parameter of implicity. For the choice of $\psi^{n+\Theta}$ we have two possibilities:

$$\psi^{n+\Theta} = \Theta\psi(\cdot, U^{n+1}) + (1-\Theta)\psi(\cdot, U^n) \tag{11.50}$$

or

$$\psi^{n+\Theta} = \psi\Big(\cdot, \Theta U^{n+1} + (1-\Theta)U^n\Big)\,. \tag{11.51}$$

In the explicit case, i.e., $\Theta = 0$, (11.49) represents a linear system of equations for U^{n+1} (with the system matrix B) and does not have to be treated further here. In the implicit case $\Theta \in (0, 1]$ we obtain again a nonlinear system of the type (11.17), i.e.,

$$A\boldsymbol{\xi} + g(\boldsymbol{\xi}) = 0\,,$$

in the variable $\boldsymbol{\xi} = \boldsymbol{\xi}^{n+1}$, where $\boldsymbol{\xi}^{n+1}$ is the representing vector of U^{n+1}, i.e., $U^{n+1} = \sum_{i=1}^{M} \xi_i^{n+1}\varphi_i$. Now we have for the variant (11.50),

$$\begin{aligned} A &:= B + \Theta\tau_n S, \\ g(\boldsymbol{\xi}) &:= \Theta\tau_n G_h(\boldsymbol{\xi}) - \boldsymbol{b}, \end{aligned} \tag{11.52}$$

with

$$\begin{aligned} \boldsymbol{b} := {} & (B - (1-\Theta)\tau_n S)\,\boldsymbol{\xi}^n - (1-\Theta)\tau_n G_h(\boldsymbol{\xi}^n) \\ & + \Theta\boldsymbol{\beta}(t_{n+1}) + (1-\Theta)\boldsymbol{\beta}(t_n)\,. \end{aligned} \tag{11.53}$$

For the variant (11.51) g changes to

$$g(\boldsymbol{\xi}) := \tau_n G_h\left(\Theta\boldsymbol{\xi} + (1-\Theta)\boldsymbol{\xi}^n\right) - \boldsymbol{b}\,,$$

and in the definition of $\boldsymbol{b}$ the second summation term drops out. The vector $\boldsymbol{\xi}^n$ is the representing vector of the already known approximation U^n.

Numerical Quadrature

As in the stationary case we can approximate g by a quadrature rule of the form (11.41), which leads to

$$\tilde{g}(\boldsymbol{\xi}) = \Theta\tau_n\tilde{G}_h(\boldsymbol{\xi}) - \boldsymbol{b}$$

in (11.50) and to

$$\tilde{g}(\boldsymbol{\xi}) = \tau_n\tilde{G}_h\left(\Theta\boldsymbol{\xi} + (1-\Theta)\boldsymbol{\xi}^n\right) - \boldsymbol{b}$$

in (11.51). The functional matrices of g and $\tilde{g}$ are thus equal for (11.50) and (11.51), except to the point where $\frac{\partial}{\partial u}\psi$ is being evaluated. Consequently, it suffices in the following to refer to (11.50). Based on the same motivation, a quadrature rule of the form (11.41) can be applied to the mass matrix B. Such a *mass lumping* results in a diagonal approximation of the mass matrix

$$\tilde{B} := \operatorname{diag}(b(a_i)\omega_i)\,.$$

In contrast to the stationary case we get the factor $\Theta\tau_n$ in front of the nonlinearity, where the time step size τ_n may be chosen arbitrarily small. Of course, we have to take into account that the number of time steps necessary to achieve a fixed time T is, respectively, raised. All of the above considerations about the matrix properties of $A + Dg(\overline{\boldsymbol{\xi}})$ are conserved, where A is no longer the stiffness matrix, but represents the linear combination (11.52) with the mass matrix. This reduces the requirements concerning the V-coercivity of a (see (3.28)) and thus the positive definiteness of A.

Admittedly, A is not necessarily an M-matrix if S is one, because the conditions (1.35) or (1.35)* are not valid. Here the approximation $\tilde{B}$ is advantageous, because using nonnegative weights will conserve this property due to (11.45).

Conditions of Convergence for Iterative Schemes

Clear differences arise in answering the question of how to ensure the convergence of the iterative schemes. Even for the fixed-point iteration it is true that the method converges globally if only the time step size τ_n is chosen small enough. We want to demonstrate this in the following by an example of a quadrature rule with nonnegative weights in the mass matrix and the nonlinearity. Therefore, the Lipschitz constant of $A^{-1}g$ is estimated according to Lemma 11.3. Let the norm be a matrix norm induced by a p-norm $|\cdot|_p$ (see (A3.2)) and let A be nonsingular. We get

$$\begin{aligned}\|A^{-1}\| \sup_{\xi\in\mathbb{R}^M} \|D\tilde{g}(\xi)\| &\leq \left\|\left(I+\Theta\tau_n\tilde{B}^{-1}S\right)^{-1}\tilde{B}^{-1}\right\| \Theta\tau_n \sup_{s\in\mathbb{R}}\left|\frac{\partial}{\partial u}\psi(\cdot,s)\right| \|\tilde{B}\| \\ &\leq \Theta\tau_n \sup_{s\in\mathbb{R}}\left|\frac{\partial}{\partial u}\psi(\cdot,s)\right| \kappa(\tilde{B}) \left\|\left(I+\Theta\tau_n\tilde{B}^{-1}S\right)^{-1}\right\| \\ &=: C\tau_n \left\|\left(I+\Theta\tau_n\tilde{B}^{-1}S\right)^{-1}\right\|.\end{aligned}$$

Thus we assume the boundedness of $\frac{\partial}{\partial u}\psi$ on $\Omega\times\mathbb{R}$ (which may be weakened, cf. Lemma 11.16). For a given $\vartheta\in(0,1)$ choose τ_n sufficiently small such that

$$\Theta\tau_n\|\tilde{B}^{-1}S\| \leq \vartheta$$

holds. By Lemma A3.11, it follows that

$$\left\|\left(I+\Theta\tau_n\tilde{B}S\right)^{-1}\right\| \leq \frac{1}{1-\vartheta},$$

and thus we obtain

$$\gamma := \frac{C\tau_n}{1-\vartheta}$$

as a Lipschitz constant for $A^{-1}g$. We see that by choosing τ_n sufficiently small, the contraction property of $A^{-1}g$ can be guaranteed. From this fact a (heuristic) step size control can be deduced that reduces the step size when a lack of convergence is detected and repeats the step, and in case of satisfactory convergence increases the time step size.

Nevertheless, in general, Newton's method is to be preferred: Here we can expect that the quality of the initial iterate $\xi^{(0)} := \xi^n$ for time step $(n+1)$ improves the smaller we choose τ_n. The step size control mentioned above may thus be chosen here, too (in conjunction with the enlargement of the range of convergence via damping). Nonetheless, a problem only to be solved in numerical practice consists in coordinating the control parameters of the time step control, the damping strategy, and eventually the termination of the inner iteration in such a way that overall, an efficient algorithm is obtained.

Before we pass to more complicated problems, we also refer to Section 9.6 in which we introduced—to a certain extent in advance—exponential integrators for semilinear parabolic problems.

Diffusion–Convection–Reaction Systems

We consider a system of diffusion–convection–reaction equations, i.e., instead of solving the scalar equation (11.34) we search for an unknown vector field $\boldsymbol{u} : (0,T) \to V^N$ with components $\boldsymbol{u} = (u_1,\ldots,u_N)^T$, where V is the basic function space. The equations to be fulfilled for N reacting *species (or components)* are

$$\frac{\partial}{\partial t}(b_k u_k) + L_k u_k + \psi_k(\cdot, u_1,\ldots,u_N) = f_k \quad \text{in } Q_T,\ k = 1,\ldots,N, \tag{11.54}$$

where the spatial operators L_k are as in (3.13) but—without loss of generality—$r = 0$. The coefficients $b_k : \Omega \to \mathbb{R}$ satisfy the conditions (9.13). The boundary conditions are considered in the generality of (3.19)–(3.21), i.e.,

$$R_k u_k = g_k\,, \quad k = 1,\ldots,N.$$

Although in general the subdivisions of $\partial\Omega$ may differ between the components, we only want to consider situations here in which the weak formulation can be given in V^N.

The nonlinearities $\psi_k : \Omega \times \mathbb{R}^N \to \mathbb{R}$, $k = 1\ldots,N$, are supposed to be continuously differentiable in all arguments except for the first one. Analogously to the beginning of this section (see (11.35)) we define the superposition operators $G_k : U^N \to V'$ by

$$\big(G_k(\boldsymbol{u})\big)(v) := \langle \psi_k(\cdot, u_1,\ldots,u_N), v\rangle_0 \quad \text{for all } \boldsymbol{u} \in U^N,\ v \in V, \tag{11.55}$$

where $U \subset V$ is some open subset.

We assume that the problem (11.54) is well defined, in particular that the operators G_k are continuously differentiable (cf. Lemma 11.16) and that a unique solution $\boldsymbol{u} \in U^N$ exists.

To prove a stability estimate for (11.54) we have to investigate differences of solutions $\boldsymbol{u} := \boldsymbol{u}^{(1)} - \boldsymbol{u}^{(2)}$. The mean-value theorem leads to the term

$$\langle DG_k(\hat{\boldsymbol{u}})\boldsymbol{u}, v\rangle =: \tilde{G}_k(v),$$

where $\hat{\boldsymbol{u}}$ is a convex combination of $\boldsymbol{u}^{(1)}$ and $\boldsymbol{u}^{(2)}$, and DG_k is the kth row of the Jacobi matrix DG. Repeating the proof of Lemma 9.4, i.e., testing with $v := u_k$ and adding up the equations (with $b_k := b := 1$ for simplicity and $\tilde{\alpha} := \min\limits_{k=1,\ldots,N} \alpha_k$) leads to

$$\frac{d}{dt}\left(\sum_{k=1}^{N}\|u_k(t)\|_0\right)+\tilde{\alpha}\sum_{k=1}^{N}\|u_k(t)\|_0+\sum_{k=1}^{N}\frac{1}{\|u_k(t)\|_0}\tilde{G}_k(u_k(t))\leq\sum_{k=1}^{N}\|f_k(t)\|_\infty$$

and

$$\sum_{k=1}^{N}\frac{1}{\|u_k(t)\|_0}\tilde{G}_k(u_k(t))\geq-\sum_{k=1}^{N}\|DG_k(\hat{\boldsymbol{u}}(t))\boldsymbol{u}(t)\|_0\geq-\beta\sum_{k=1}^{N}\|u_k(t)\|_0,$$

assuming an appropriately bounded $\beta > 0$. Then we can conclude the following result (compare Remark 9.5).

Theorem 11.17 *Consider the diffusion–convection–reaction system* (11.54) *with* $b_k := 1$ *and solutions* $\boldsymbol{u}^{(1)}, \boldsymbol{u}^{(2)}$ *and assume that the convex hull* C *of all attained values of* $\boldsymbol{u}^{(1)}, \boldsymbol{u}^{(2)}$ *is contained in* U^N. *Set*

$$\gamma := \sup\left\{\left|\left(\|DG_{kl}(\hat{\boldsymbol{u}})\|_{L^\infty}\right)\right|_F \;\middle|\; \hat{\boldsymbol{u}}\in C\right\},$$

where $|\cdot|_F$ *is the Frobenius norm. If* $\gamma < \infty$, *then, for* $\hat{\alpha} := \alpha - \gamma$,

$$\sum_{k=1}^{N}\|u_k(t)\|_0\leq\left(\sum_{k=1}^{N}\|u_{k,0}\|_0\right)e^{-\hat{\alpha}t}+\int_0^t\sum_{k=1}^{N}\|f_k(s)\|_0\,e^{-\hat{\alpha}(t-s)}\,ds.$$

Proof A feasible constant β is given by

$$\tilde{\beta} := \sup\left\{\max\left\{\|DG_k(\hat{\boldsymbol{u}})\|_0 \mid k=1,\ldots,N\right\} \;\middle|\; \hat{\boldsymbol{u}}\in C\right\}.$$

We have to show that $\tilde{\beta} \leq \gamma$. This follows from estimates of the type

$$\begin{aligned}\int_\Omega\left(\sum_{k=1}^{N}g_k u_k\right)^2 dx &\leq \int_\Omega\left(\sum_{k=1}^{N}g_k^2\right)\left(\sum_{k=1}^{N}u_k^2\right)dx\\ &\leq\left(\sum_{k=1}^{N}\|g_k\|_\infty^2\right)\sum_{k=1}^{N}\int_\Omega u_k^2\,dx\leq\left(\sum_{k=1}^{N}\|g_k\|_\infty^2\right)\left(\sum_{k=1}^{N}\|u_k\|_0\right)^2.\end{aligned}$$

□

Now we want to consider some aspects of semidiscretizations of (11.54). Let $V_h := \operatorname{span}\{\varphi_1,\ldots,\varphi_M\}$, then the conforming and consistent semidiscrete finite element discretization of (11.54) reads as follows:

Find $\boldsymbol{u}_h = (u_{h,1},\ldots,u_{h,N}) \in V_h^N$ such that

$$\frac{\partial}{\partial t}\left\langle u_{h,k}, v_h\right\rangle_0^{(b)} + a_k(u_{h,k},v_h) + \langle G_k(\boldsymbol{u}_h), v_h\rangle = \ell_k(v_k) \quad \text{for all } v_h\in V_h,\quad k=1,\ldots,N. \tag{11.56}$$

Here a_k are the bilinear forms corresponding to L_k and R_k, and ℓ_k the linear forms stemming from the right-hand sides.

Let $u_{h,k} = \sum_{i=1}^{M} \xi_{i,k}\varphi_i \in V_h$, $k = 1, \ldots, M$ be the representation in the nodal basis, assuming Lagrangian elements. The local nonlinear part has the form

$$\left\langle G_k(\boldsymbol{u}_h), \varphi_j \right\rangle := \int_\Omega \psi_k \left(x, \left(\sum_{i=1}^{M} \xi_{i,l}\,\varphi_i \right)_l \right) \varphi_j dx\,. \tag{11.57}$$

As discussed in the lines following (11.41), it is advisable to use a node-oriented quadrature rule leading to the approximation

$$\left\langle G_k(\boldsymbol{u}_h), \varphi_j \right\rangle \sim \omega_j \psi_k \left(a_j, \left(\xi_{j,l}\right)_l\right), \tag{11.58}$$

where the coefficients ω_i are the quadrature weights and thus

$$\langle G_k(\boldsymbol{u}_h), v_h \rangle \sim \sum_{i=1}^{M} \omega_i \psi_k \left(a_i, \left(\xi_{i,l}\right)_l\right) \eta_i \quad \text{for} \quad v_h = \sum_{i=1}^{M} \eta_i \varphi_i \in V_h$$

(see the similarity to the finite volume situation, cf. (8.10)).

A typical reaction rate representation is

$$\psi(x, \boldsymbol{u}) = S\boldsymbol{R}(x, \boldsymbol{u}),$$

where $S \in \mathbb{R}^{N,N}$ is the *stoichiometric matrix*, and $\boldsymbol{R}$ the rate vector, depending on $\boldsymbol{u}$ and other variables like temperature, here indicated by the dependence on x. To understand better the connectivity and sparsity pattern of the emerging system of equations, we restrict to the linear(ized) situation

$$\psi(x, \boldsymbol{u}) = S\hat{R}(x)\boldsymbol{u} \tag{11.59}$$

with $S\hat{R}(x) \in \mathbb{R}^{N,N}$. A constant part may be incorporated into $\boldsymbol{f} := (f_i)_i$. Here we have

$$\langle G_k(\boldsymbol{u}_h), v_h \rangle \sim \sum_{i=1}^{M} \omega_i \left(S\hat{R}(a_i)(\xi_{i,\cdot}) \right)_k \eta_i\,.$$

Thus here the degrees of freedom are ordered as

$$\underbrace{\xi_{1,1}, \ldots, \xi_{M,1}}_{\text{component } 1}, \ldots, \underbrace{\xi_{1,N}, \ldots, \xi_{M,N}}_{\text{component } N}.$$

The overall discretization matrix consists of the matrix $A \in \mathbb{R}^{NM,NM}$, which results from the discretization of the transport process and is of the following form:

$$A := \operatorname{diag}(A_k), \quad k = 1, \ldots, N,$$

$A_k \in \mathbb{R}^{M,M}$ being the discretization matrix of the scalar transport part, and the coupling across the components comes from the reaction part

$$\begin{aligned} B &:= (B_{kl}), & k,l &= 1,\dots,N,\\ B_{kl} \in \mathbb{R}^{MM},\ B_{kl} &:= \operatorname{diag}(\beta_{kl,i}), & i &= 1,\dots,M, \end{aligned} \tag{11.60}$$

and

$$\beta_{kl,i} := \omega_i\ (S\hat{R}(a_i))_{kl}\,.$$

If $\hat{R}$ is independent of a_i, i.e.,

$$\beta_{kl,i} := (S\hat{R})_{kl}\ \omega_i\,,$$

B has the form

$$B = S\hat{R} \otimes D$$

with $D := \operatorname{diag}(\omega_i) \in \mathbb{R}^{M,M}$, $\otimes$ being the Kronecker product (see (A3.12)).

In this case the eigenvalues of B are exactly the products of the eigenvalues of $S\hat{R}$ and of D, i.e.,

$$\kappa(B) = \kappa(S\hat{R}) \cdot \kappa(D) \tag{11.61}$$

for the spectral condition number κ. $\kappa(D)$ is independent of h under approximate conditions (see Theorem 3.47) similar to the condition number of the mass matrix M.

If we change to the alternative ordering

$$\underbrace{\xi_{1,1},\dots,\xi_{1,N}}_{\text{node } 1},\dots,\underbrace{\xi_{M,1},\dots,\xi_{M,N}}_{\text{node } M},,$$

then the matrix B gets the block structure

$$\tilde{B} = \operatorname{diag}(\tilde{B}_i), \quad i = 1,\dots,M,$$

and $\tilde{B}_i \in \mathbb{R}^{N,N}$ is the dense matrix

$$\tilde{B}_i = \omega_i\ S\hat{R}(a_i).$$

As the permutation of the order corresponds to a simultaneous permutation of rows and columns of B, i.e., a similarity transformation, the eigenvalues remain unchanged and then the eigenvalues of B are exactly those of $\tilde{B}_i$, leading to the following generalization of (11.61):

$$\kappa(B) = \frac{\lambda_{\max}(\omega_i\ S\hat{R}(a_i))}{\lambda_{\min}(\omega_i\ S\hat{R}(a_i))},$$

where $\lambda_{\max}$ denotes the maximum of all eigenvalues over all $i = 1,\dots,M$, $\lambda_{\min}$ analogously the minimum, and thus

$$\kappa(B) \leq \max_{i=1,\ldots,M} \kappa(S\hat{R}(a_i))\,\kappa(D). \tag{11.62}$$

The condition number of $S\hat{R}(a_i)$ may be large in the stiff case, i.e., if the aspect of reaction rates is large. Concerning the solution of such a system of equations one should pay attention to the different conditioning of A and B. Direct methods cannot take advantage of the special structuring explained above. Krylov-subspace methods, which only need the computation of a residuum

$$\boldsymbol{r} = (A + B)\boldsymbol{u} - \boldsymbol{b},$$

can do: For $A\boldsymbol{u}$ the block-diagonality of A can be used, and for $B\boldsymbol{u}$ the same applies after reordering. The handling of the single blocks can be done in parallel. This is discussed in Section 11.6.

The time-discrete version can be discussed as above for the scalar equation.

Singular Limit: Immobile Species

In particular in macroscopic models in porous media also immobile species u_p appear meaning $L_p u_p = 0$.

Setting $N := N_{\mathrm{m}} + N_{\mathrm{im}}$ with $N_{\mathrm{m}}, N_{\mathrm{im}} \in \mathbb{N}$, the operators L_k and R_k are as above for $k = 1, \ldots, N_{\mathrm{m}}$, whereas $L_k := 0$ and $R_k := 0$ for $k = N_{\mathrm{m}} + 1, \ldots, N_{\mathrm{im}}$. The system of differential equations and boundary conditions has the form

$$\begin{aligned}
\frac{\partial}{\partial t}(b_k u_k) + L_k u_k + \psi_k(\cdot, u_1, \ldots, u_N) &= f_k\,, \\
R_k u_k &= g_k\,, \qquad k = 1, \ldots, N_{\mathrm{m}}\,, \\
\frac{\partial}{\partial t}(b_k u_k) + \psi_k(\cdot, u_1, \ldots, u_N) &= f_k\,, \qquad k = N_{\mathrm{m}} + 1, \ldots, N.
\end{aligned} \tag{11.63}$$

So the last N_{im} PDEs have degenerated to ODEs, meaning that they are spatially local and the coupling is only via the species. The same applies to the time-discrete version.

Singular Limit: Reactions in Equilibrium

Often the relaxation time of the single reactions are quite different, which suggests that some of the reactions are to be considered to be in equilibrium. After an appropriate transformation (see [166]) some of the PDEs or ODEs can be substituted by algebraic equations, i.e., with

$$N := N_{\mathrm{m}} + N_{\mathrm{im}} + N_{\mathrm{eq}}$$

the system reads as (keeping the notation)

$$\frac{\partial}{\partial t}(b_k u_k) + L_k u_k + \psi_k(\cdot, u_1, \ldots, u_N) = f_k\,, \tag{11.64}$$

$$R_k u_k = g_k\,, \quad k = 1, \ldots, N_{\mathrm{m}}\,,$$

$$\frac{\partial}{\partial t}(b_k u_k) + \psi_k(\cdot, u_1, \ldots, u_N) = f_k\,, \quad k = N_{\mathrm{m}} + 1, \ldots, N_{\mathrm{m}} + N_{\mathrm{im}}\,, \tag{11.65}$$

$$\psi_k(\cdot, u_1, \ldots, u_N) = 0, \quad k = N_{\mathrm{m}} + N_{\mathrm{im}} + 1, \ldots, N. \tag{11.66}$$

The last set of equations is a fortiori local in space such that the last and also the middle subsystem after time discretization, that is

$$\overline{\psi}_k(\cdot, u_1, \ldots, u_N) := b_k u_k + \tau\psi_k(\cdot, u_1, \ldots, u_N) - \overline{f}_k = 0, \\ k = N_{\mathrm{m}} + 1, \ldots, N_{\mathrm{m}} + N_{\mathrm{im}}\,, \tag{11.67}$$

can be used to distinguish between the *primary variables* $u_1, \ldots, u_{N_{\mathrm{m}}}$ and the *secondary variables* $u_{N_{\mathrm{m}}+1}, \ldots, u_N$:
It requires that, given $u_1, \ldots, u_{N_{\mathrm{m}}} \in \mathbb{R}$ (≥ 0, as solutions, being concentrations, in general have to satisfy $u_k \geq 0$), the equations (11.67), (11.66) uniquely determine

$$u_{N_{\mathrm{m}}+1}, \ldots, u_N \in \mathbb{R}\ (\geq 0),$$

or, equivalently, that the existence of a *resolution function*

$$(u_{N_{\mathrm{m}}+1}, \ldots, u_N) = R(u_1, \ldots, u_{N_{\mathrm{m}}})$$

can be assured. Then, by substitution the problem reduces to the first N_{m} PDEs (*substitution method*). But note that in general R is not given explicitly, so that the evaluation of R requires the (approximate) solution of the nonlinear system of $N_{\mathrm{im}} + N_{\mathrm{eq}}$ equations at every spatial (discretization) point. This can be done in an inner iteration of an overall iterative scheme for the discrete version of (11.63) after substitution, e.g., by Newton's method. Note that usually $M \gg N > N_{\mathrm{im}} + N_{\mathrm{eq}}$, i.e., the nonlinear systems to be solved are "small" compared to the overall complexity. If the outer iteration is of Newton type, also the Jacobi matrix of R has to be evaluated in every outer iteration step. Using the implicit function theorem, this can be done only evaluating the partial derivatives of ψ_k and $\overline{\psi}_k$ (see Section 11.6).

Exercises

Problem 11.11 Study the Lipschitz property of DG defined by (11.39) and of $D\tilde{G}$ defined by (11.44), provided $\frac{\partial}{\partial u}\psi$ is Lipschitz continuous in its second argument.

Problem 11.12 Decide whether $A^{-1}g$ is contractive in case of (11.52)–(11.53).

Problem 11.13 The boundary value problem

$$-u'' + e^u = 0 \quad \text{in } (0,1), \qquad u(0) = u(1) = 0,$$

is to be discretized by a finite element method using continuous, piecewise linear functions on equidistant grids. Quadrature is to be done with the trapezoidal rule.

a) Compute the matrix $A_h \in \mathbb{R}^{m,m}$ and the nonlinear vector field $F_h : \mathbb{R}^m \to \mathbb{R}^m$ in a matrix-vector notation

$$A_h U_h + F_h(U_h) = 0$$

of the discretization. Here $U_h \in \mathbb{R}^m$ denotes the vector of unknown nodal values of the approximate solution and, for uniqueness of the representation, the elements of A_h are independent of the discretization parameter h.

b) Study the convergence of the iterative procedure

$$\begin{aligned} &1) \quad (2+h^2)U_h^{(k+1)} = \left((2+h^2)I - A_h\right) U_h^{(k)} - F_h\left(U_h^{(k)}\right), \\ &2) \quad 2U_h^{(k+1)} + F_h\left(U_h^{(k+1)}\right) = (2I - A_h)\, U_h^{(k)}. \end{aligned}$$

Programming project 11.1 Write a function that solves

$$\begin{aligned} -\Delta u &= -u^2 && \text{in } \Omega := (0,1)^2, \\ \partial_{\mathbf{n}} u &= 0 && \text{on } \Gamma_1 := (0,1) \times \{0\}, \\ u &= 1 && \text{on } \Gamma_3 := \partial\Omega \setminus \Gamma_1 \end{aligned}$$

using conforming linear finite elements and Newton's method. Use $u_h =: (0, \ldots, 0)^T$ as initial value for all non-Dirichlet nodes. Use quadrature rules of order 1. (You may start from the code of Project 3.1 or Project 2.5.)

11.5 Quasilinear Equations

In this section we have a look at nonlinearities with terms involving derivatives, i.e., the time-dependent generalization of (3.36) reads

$$\frac{\partial}{\partial t}\big(b(\cdot,u)\big) - \nabla \cdot \big(\boldsymbol{K}(\cdot,u)\nabla u - \boldsymbol{c}(\cdot,u)u\big) + \psi(\cdot,u) = f, \qquad (11.68)$$

i.e., here we consider the divergence form with a more direct appearance of a driving flux $\boldsymbol{c}$. Note that the flux and mixed boundary conditions are correspondingly nonlinear by having as flux density

$$\boldsymbol{p} = \boldsymbol{p}(u) := -\boldsymbol{K}(\cdot,u)\nabla u + \boldsymbol{c}(\cdot,u)u\,.$$

Examples for the appearance of such nonlinearities are (0.20) or (0.25).

Nonlinearity in Time Derivative

This is similar to the reaction term from Section 11.4 and takes its form after time discretization, but note that no convergence enhancing factor τ is induced. This also shows the difference between a kinetically described reaction leading to (11.34) or (11.64), (11.65) and an equilibrium reaction as in (11.66).

Nonlinearity in the Conductivity

We compute the derivative in the weak formulation, i.e., of

$$E(u)v := \int_\Omega \big(\boldsymbol{K}(x,u(x))\nabla u(x)\big)\cdot\nabla v(x)dx \quad \text{for all } u,v\in V$$

only formally reaching at

$$\langle DE(\overline{u})w, v\rangle = \int_\Omega \left[\boldsymbol{K}(\overline{u})\nabla w + \frac{\partial}{\partial u}\boldsymbol{K}(\cdot,\overline{u})\nabla\overline{u}\, w\right]\cdot\nabla v\, dx\,. \tag{11.69}$$

Different from the reaction term, here the derivative not only is constituted by a linear diffusive term, but in addition also by a convective term with the given flow field

$$\hat{\boldsymbol{c}} := -\frac{\partial}{\partial u}\boldsymbol{K}(\cdot,\overline{u})\nabla\overline{u}\,.$$

This term may be large compared to $\boldsymbol{K}(\overline{u})$, i.e., the linearized equation may become convection-dominated. In such a situation it may be advisable not to use the classical conforming discretization, but to apply stabilization techniques as discussed in Chapter 10.

A typical case is

$$\boldsymbol{K}(x,u) := k(u)\mathbb{K}(x) \tag{11.70}$$

with $k:\ \mathbb{R}\to\mathbb{R}$ and $\mathbb{K}:\ \Omega\to\mathbb{R}^{d,d}$, separating the anisotropy, originating from the domain, and the nonlinearity, originating from the process. Here the variable transformation

$$U(x,t) := \mathcal{K}(u(x,t)),$$

where

$$\mathcal{K}:\ \mathbb{R}\to\mathbb{R},\ \xi\mapsto\mathcal{K}(\xi) := \int_0^\xi k(s)ds \tag{11.71}$$

is the *Kirchhoff transform*, leads to

$$\nabla U(x,t) = k(x,u(x,t))\nabla u(x,t)$$

and thus to

$$\boldsymbol{K}(x,u(x,t))\nabla u(x,t) = \mathbb{K}(x)\nabla U(x,t).$$

If $k(u) > 0$ for $u \in \mathbb{R}$, the mapping $\mathcal{K}$ is invertible, and the equation (11.68) can be written in the variable U:

$$\frac{\partial}{\partial t}(b(\cdot, \mathcal{K}^{-1}(U))) - \nabla \cdot (\mathbb{K}\nabla U) - \nabla \cdot (\boldsymbol{c}(\cdot, \mathcal{K}^{-1}(U))\mathcal{K}^{-1}(U)) + \psi(\cdot, \mathcal{K}^{-1}(u)) = f. \tag{11.72}$$

In this way the diffusive part becomes linear to the expense of (further) nonlinearities in the time derivative, the convective, and the reaction part.

Nonlinearity in the Convective Term

We compute the derivative in the weak formulation, i.e., of

$$F(u)v = -\int_\Omega u(x)\,\boldsymbol{c}(x, u(x)) \cdot \nabla v(x) dx \quad \text{for all } u, v \in V$$

only formally, reaching at

$$\langle DF(\overline{u})w, v\rangle = -\int_\Omega \left[\boldsymbol{c}(\cdot, \overline{u})w + \frac{\partial}{\partial u}\boldsymbol{c}(\cdot, \overline{u})\overline{u}\, w\right] \cdot \nabla v\, dx,$$

i.e., this is a linear convective term with the underlying flow field

$$\hat{\boldsymbol{c}} := \boldsymbol{c}(\cdot, \overline{u}) + \frac{\partial}{\partial u}\boldsymbol{c}(\cdot, \overline{u})\overline{u}\,.$$

In the case $\|\hat{\boldsymbol{c}}\| > \|\boldsymbol{c}\|$, similar to above, the application of stabilization techniques from Chapter 10 may be advisable.

The iterative methods may range between variants of Newton's method and the fixed-point method. For simplicity we omit the reaction term. The full Newton's method corresponds to a discretized version of

$$\begin{aligned}
&\frac{\partial}{\partial t}\left(\frac{\partial}{\partial u}b(\cdot, u^{(k)})\delta^{(k)}\right) - \nabla \cdot \left(\boldsymbol{K}(\cdot, u^{(k)})\nabla\delta^{(k)} - \boldsymbol{c}(\cdot, u^{(k)})\delta^{(k)}\right) \\
&\qquad + \nabla \cdot \left(\left[-\frac{\partial}{\partial u}\boldsymbol{K}(\cdot, u^{(k)})\nabla u^{(k)} + \frac{\partial}{\partial u}\boldsymbol{c}(\cdot, u^{(k)})u^{(k)}\right]\delta^{(k)}\right) \\
&= f - \frac{\partial}{\partial t}b(\cdot, u^{(k)}) + \nabla \cdot (\boldsymbol{K}(\cdot, u^{(k)})\nabla u^{(k)}) - \nabla \cdot (\boldsymbol{c}(\cdot, u^{(k)})\nabla u^{(k)})
\end{aligned} \tag{11.73}$$

with linear boundary conditions having in their flux and mixed part a flux defined by $u^{(k)}$ as the term above under the divergence operator, and then

$$u^{(k+1)} := u^{(k)} + \delta^{(k)}.$$

Assuming the form

$$b(\cdot, u) = \tilde{b}(\cdot, u)u \tag{11.74}$$

we have the same structure as in the other terms.

Neglecting the terms

$$\frac{\partial}{\partial u}\tilde{b}(\cdot, u^{(k)})\delta^{(k)},$$

$$\nabla \cdot \left(\left[-\frac{\partial}{\partial u}\boldsymbol{K}(\cdot, u^{(k)})\nabla u^{(k)} + \frac{\partial}{\partial u}\boldsymbol{c}(\cdot, u^{(k)})u^{(k)}\right]\delta^{(k)}\right),$$

(11.73) reduces to the fixed-point iteration

$$\frac{\partial}{\partial t}\left(b(\cdot, u^{(k)})u^{(k+1)}\right) - \nabla \cdot \left(\boldsymbol{K}(\cdot, u^{(k)})\nabla u^{(k+1)} - \boldsymbol{c}(\cdot, u^{(k)})u^{(k+1)}\right) = f \qquad (11.75)$$

and boundary condition based on the linear flux density

$$-\boldsymbol{K}(\cdot, u^{(k)})\nabla u - \boldsymbol{c}(\cdot, u^{(k)})u,$$

in which the coefficients are "frozen" at the old iterate. Thus this method is an approximate Newton's method.

Exercises

Problem 11.14 As in Sections 5.5, 5.6, consider a hierarchy of nonlinear problems

$$f_k(x_k) = 0, \quad k = 0, \ldots, \ell.$$

Under the conditions of Section 5.6 investigate a nested Newton iteration similar to Table 5.10. Under which conditions for global (linear) or local (quadratic) convergence a result similar to Theorem 5.28 holds true?

Problem 11.15 Consider for $f : \mathbb{R}^k \to \mathbb{R}$ the complementarity problem in two variables $x \in \mathbb{R}^k$, $y \in \mathbb{R}$:

$$f(x)y = 0, \; f(x) \geq 0, \; y \geq 0. \qquad (11.76)$$

A function $\zeta : \mathbb{R}^2 \to \mathbb{R}$ is called a *complementarity function*, if

$$\zeta(a, b) = 0 \Leftrightarrow ab = 0, \; a \geq 0, \; b \geq 0.$$

a) Show that for any complementarity function (11.76) is equivalent to

$$\zeta(f(x), y) = 0.$$

b) Show that the following functions are complementarity functions:

1) $\zeta(a, b) := a + b - (a^2 + b^2)^{1/2}$,
2) $\zeta(a, b) := \min(a, b)$.

Problem 11.16 Consider the situation of Problem 11.15. Show that the complementarity functions from b) are Lipschitz continuous, but not differentiable everywhere. Is a differentiable complementarity function possible?

11.6 Iterative Methods for Semilinear Differential Systems

In Section 11.4 different systems have been introduced reflecting the interplay of different processes and time scales (*multiphysics problems*) such as diffusion and convection for the transport processes, and kinetic or equilibrium description for local reaction processes. We know from Chapter 10 that for convection-dominated transport special discretizations are necessary, and processes are combined with a wide range of relaxation times up to zero in the extreme of equilibrium reactions. If it comes to the efficient solution of the emerging overall discrete system, there are two philosophies. One aims at an (iterative) decoupling of the mentioned process, to take advantage of specific connectivity structures (see Section 11.4), existing algorithms or even software. This usually has the form of a fixed-point iteration (or *Picard iteration*), sometimes also called *operator splitting*, but see Section 11.7 for a more precise usage of this notion. On the other hand, one may argue that a fixed-point iteration has unclear convergence properties and is at most linearly convergent. This line of thought suggests to consider the system as a whole and apply (a variant) of Newton's method (*all-in-one* or *monolithic* or *fully implicit* approach).

We will compare these approaches for a model problem indicating that we do not have a dichotomy, but rather a continuously varying family of methods connecting two extreme ends.

As a model problem we consider the system of operator equations

$$\begin{aligned} A_1 u_1 + G_1(u_1, u_2) &= 0, \\ A_2 u_2 + G_2(u_1, u_2) &= 0 \end{aligned} \tag{11.77}$$

for $u_i \in V$, $A_i : V \to V'$, $G : V \times V \to V' \times V'$.

Here A_i stands either for the (time-discretized) elliptic transport operator $A_i := L$ or—in case of equilibrium reactions—$A_i := 0$. If (11.77) is a time-discretized parabolic problem, then either

$$A_i := B + \tau L,$$

where $\langle Bw, v\rangle := \int_\Omega bwv dx$, or

$$A := B$$

for the ODE part (11.65), and in these cases the operator G_i contains the factor τ in its u_i−dependent part ensuring contractivity for small τ. The nonlinearities G_i may be viewed as superposition operators stemming from chemical reactions

$$\langle G_i(u_1, u_2), v_i\rangle := \int_\Omega \psi_i(u_1, u_2) v_i dx\,.$$

We write the iterations on a continuous level, but discuss their complexity having a conforming Galerkin discretization in mind. The full Newton's method reads for the increment $\delta_i^{(k)}$, i.e.,

$$u_i^{(k+1)} = u_i^{(k)} + \delta_i^{(k)}$$

as follows:

$$\begin{aligned} A_1\delta_1^{(k)} + DG_{1,1}(\boldsymbol{u}^{(k)})\delta_1^{(k)} + DG_{1,2}(\boldsymbol{u}^{(k)})\delta_2^{(k)} &= -A_1u_1^{(k)} - G_1(\boldsymbol{u}^{(k)}),\\ A_2\delta_2^{(k)} + DG_{2,1}(\boldsymbol{u}^{(k)})\delta_1^{(k)} + DG_{2,2}(\boldsymbol{u}^{(k)})\delta_2^{(k)} &= -A_2u_2^{(k)} - G_2(\boldsymbol{u}^{(k)}),\end{aligned} \tag{11.78}$$

where $DG_{i,j}(\boldsymbol{u}) := \frac{\partial}{\partial u_j} G_i(\boldsymbol{u})$, $\boldsymbol{u} = (u_1, u_2)^T$.

Thus the block-diagonal structure of the transport part

$$\begin{pmatrix} A_1 & 0 \\ 0 & A_2 \end{pmatrix}$$

is overlaid by the linearized reaction description rendering a linear system of dimension $2M$ (resp. NM for N species) to be solved in each iteration. If $A_2 = 0$ or $A_2 = B$, the second subsystem decouples to M systems of the (small) size N such that their resolution (in each iteration) seems to be affordable. This static condensation reduces the problem to a system of size M with the system matrix

$$A_1 + DG_{1,1}(\boldsymbol{u}^{(k)}) - DG_{1,2}(\boldsymbol{u}^{(k)})\big(\alpha B + DG_{2,2}(\boldsymbol{u}^{(k)})\big)^{-1} DG_{2,1}(\boldsymbol{u}^{(k)}), \tag{11.79}$$

$\alpha = 0$ or $\alpha = 1$ as discussed above.

On the other hand, to decouple the two transport blocks, the fixed-point iteration with $A_i^{-1}G_i$ is the most obvious method. It reads as (see (11.18)):

$$A_i\delta_i^{(k)} = -A_iu_i^{(k)} - G_i(u_1^{(k)}, u_2^{(k)}), \quad i = 1, 2. \tag{11.80}$$

Thus it may be viewed as an approximate Newton's method, where all contributions $DG_{i,j}$ are cancelled from the system matrix. A method in between is achieved by the *block-Jacobi method* (see (5.40)):

$$\begin{aligned} A_1\delta_1^{(k)} + G_1(u_1^{(k)} + \delta_1^{(k)}, u_2^{(k)}) &= -A_1u_1^{(k)},\\ A_2\delta_2^{(k)} + G_2(u_1^{(k)}, u_2^{(k)} + \delta_2^{(k)}) &= -A_2u_2^{(k)}.\end{aligned} \tag{11.81}$$

The decoupling reduces the size, but leaves nonlinear systems to be solved. One can again think of Newton's method with a fixed number of iterations, at the extreme with one iteration, with the start iterate $\delta_i^{(k,0)} := 0$.

This leads to

$$A_i\delta_i^{(k)} + DG_{i,i}(\boldsymbol{u}^{(k)})\delta_i^{(k)} = -A_iu_i^{(k)} - G_i(\boldsymbol{u}^{(k)}), \quad i = 1, 2, \tag{11.82}$$

which is again an approximate Newton's method. Another one with less cancellation in the Jacobi matrix, but still decoupling into two subsystems may be

$$\begin{aligned}&\text{for } i = 1: \quad \text{equation (11.82)},\\ &\text{for } i = 2: \quad A_2\delta_2^{(k)} + DG_{2,2}(\boldsymbol{u}^{(k)})\delta_2^{(k)} + DG_{2,1}(\boldsymbol{u}^{(k)})\delta_1^{(k)} = -A_2 u_2^{(k)} - G_2(\boldsymbol{u}^{(k)}).\end{aligned} \tag{11.83}$$

The block-Jacobi method of (11.81) can be modified to a block-Gauss–Seidel method by substituting the second equation by

$$A_2\delta_2^{(k)} + G_2(u_1^{(k)} + \delta_1^{(k)}, u_2^{(k)} + \delta_2^{(k)}) = -A_2 u_2^{(k)},$$

and analogously also in (11.82).

The consequences are identical to the above, with a modified residuum in the second equation, and thus no interpretation as approximate Newton's method.

We return to the case where the second equation is spatially purely local. We may do the resolution of this second block for the "secondary" variables u_2 not within the linearization, but "exactly", if possible, i.e., if the *resolution function* H exists such that

$$\alpha B u_2 + G_2(u_1, u_2) = 0 \iff u_2 = H(u_1).$$

Then the system reduces to

$$A_1 u_1 + \widetilde{G}_1(u_1) = 0$$

with $\widetilde{G}_1(u_1) := G_1(u_1, H(u_1)) =: G_1(\tilde{\boldsymbol{u}})$ with the Newton's method

$$A_1\delta_1^{(k)} + \left(DG_{1,1}(\tilde{\boldsymbol{u}}^{(k)}) + DG_{1,2}(\tilde{\boldsymbol{u}}^{(k)})\frac{d}{du_1}H(u_1^{(k)})\right)\delta_1^{(k)} = -A_1 u_1^{(k)} - G_1(\tilde{\boldsymbol{u}}^{(k)}), \tag{11.84}$$

and if $\alpha B + DG_{2,2}(\tilde{\boldsymbol{u}})$ is invertible:

$$\frac{d}{du_1}H(u_1) = -\left(\alpha B + DG_{2,2}(\tilde{\boldsymbol{u}})\right)^{-1} DG_{2,1}(\tilde{\boldsymbol{u}}). \tag{11.85}$$

Note that it is not required that H is available in an explicit manner: $u_2 = H(u_1)$ means to solve the local equations of the second block, which can be done (approximately) by a Newton's method as inner iteration, i.e., for given u_1:

$$\left(\alpha B + DG_{2,2}(u_1, u_2^{(k)})\right)\delta_2^{(k)} = -G_2(u_1, u_2^{(k)}),$$

which requires the solution of M "small" systems of size N. Then not only $H(u_1)$ is computed approximately, but also the derivative by (11.85)—a task of identical kind.

Note that the structure of the system matrix from (11.84), (11.85) is the same as in (11.82), but with different points of evaluation and different residuals. Thus we have here a more flexible approach, where we can control the accuracy of the evaluation of the resolution function, also in the sense of an inexact Newton's method.

Exercises

Problem 11.17 Let $g : \mathbb{R}^n \to \mathbb{R}^m$ be locally Lipschitz continuous such that Dg exists for a dense subset M. Define the *B-subdifferential* as

$$\partial_B g(x) := \Big\{ J \in \mathbb{R}^{m,n} \;\Big|\; J = \lim_{\substack{x_k \in M \\ x_k \to x}} Dg(x_k) \Big\},$$

and the *generalized Jacobian* as

$$\partial g(x) := \operatorname{conv}\big(\partial_B g(x)\big).$$

g is called *strongly semismooth*, if all directional derivatives $\frac{\partial g}{\partial d}$ in x for directions $d \in \mathbb{R}^n$ exist and it holds, for some $c \geq 0$:

$$\|Jd - \frac{\partial g}{\partial d}\| \leq c\|d\|^2.$$

Show that the complementarity functions from Problem 11.15 are strongly semismooth and compute $\partial g(x)$.

Problem 11.18 Let $f : \mathbb{R}^n \to \mathbb{R}^n$ be strongly semismooth. To find $\overline{x} \in \mathbb{R}^n$ such that $f(\overline{x}) = 0$ we consider the *semismooth Newton's method*:

$$J(x^{(k)})\delta^{(k)} = -f(x^{(k)}), \quad x^{(k+1)} := x^{(k)} + \delta^{(k)}$$

for an arbitrary $J(x^{(k)}) \in \partial g(x^{(k)})$.

Assume that $\overline{x}$ exists and J is invertible for every $J \in \partial g(x^{(k)})$. Investigate the convergence behaviour.

11.7 Splitting Methods

11.7.1 Noniterative Operator Splitting

Usually a PDE model not only describes one single process, but rather the interplay of several ones like diffusion, convection, and reaction as in the standard formulations (3.36) or (9.1) or more general as discussed in Section 11.4. The following considerations in particular apply to nonlinear problems and therefore are placed in this chapter. For sake of simplicity we concentrate on the linear(ized) situation (e.g., after applying a Newton's method, see Section 11.3). In the most simple linear situation, after space discretization this leads to a system of ODEs of the form (see Section 9.2)

$$\frac{d}{dt}\boldsymbol{u}(t) + A\boldsymbol{u}(t) + B\boldsymbol{u}(t) = 0, \quad t > 0, \qquad (11.86)$$
$$\boldsymbol{u}(0) = \boldsymbol{u}_0,$$

where for simplicity we omit distributed sources and set the coefficient in the time derivative equal to 1. Here, for instance, A stands for the discretization matrix of diffusion (and convection) and B for the discretization matrix of the local reaction (compare Section 11.4). In general we have $\boldsymbol{u}(t) \in \mathbb{R}^M$, $A, B \in \mathbb{R}^{M,M}$. To consider general linear(ized) situations, actually the time dependence of A, B should be allowed. Also an extension to 3 or more processes is possible.

It may be computationally simpler to consider the processes separately as the separated problems decouple further (see (11.54) and after), i.e., instead of (11.86) we solve

$$\frac{d}{dt}\boldsymbol{u}_1(t) + A\boldsymbol{u}_1(t) = 0, \quad t \in (0, \tau], \qquad (11.87)$$
$$\boldsymbol{u}_1(0) = \boldsymbol{u}_0,$$

$$\frac{d}{dt}\boldsymbol{u}_2(t) + B\boldsymbol{u}_2(t) = 0, \quad t > 0, \qquad (11.88)$$
$$\boldsymbol{u}_2(0) = \boldsymbol{u}_1(\tau),$$

and set

$$\boldsymbol{w}(\tau) := \boldsymbol{u}_2(\tau).$$

This approach is also called *Lie-splitting*. Comparing the solutions of (11.86) and (11.87)–(11.88) (see (9.65)), that is

$$\boldsymbol{u}(\tau) = e^{-(A+B)\tau}\boldsymbol{u}_0,$$
$$\boldsymbol{w}(\tau) := \boldsymbol{u}_2(\tau) = e^{-B\tau}\boldsymbol{u}_1(\tau) = e^{-B\tau}e^{-A\tau}\boldsymbol{u}_0,$$

we see that $\boldsymbol{u}(\tau) = \boldsymbol{u}_2(\tau)$, if the *commutator* of A and B, defined by

$$[A, B] := AB - BA,$$

vanishes. This commutativity is rarely the case, i.e., in general a *splitting error* occurs. Note that this approach, also called *fractional stepping*, is not a discretization method, as the *split steps* are still infinite-dimensional problems, but more a *model reduction approach*, which has to be amended by a time discretization. Concerning the splitting errors, Taylor expansion of the exponentials leads to

$$\tau E_s \boldsymbol{u}_0 := \left(e^{-(A+B)\tau} - e^{-B\tau}e^{A\tau}\right)\boldsymbol{u} = \frac{1}{2}[A, B]\tau^2\boldsymbol{u}_0 + O(\tau^3).$$

Analogously to the local consistency error for initial value problems for ODEs, we speak of a *local splitting error* of order 1. A splitting error estimate seemingly does not depend on the ordering of A and B, i.e., considering a B-A-splitting instead of A-B-splitting of (11.87), even in the common situation of one operator A (e.g., stemming from a diffusion process) with $\|A\|$ unbounded for the discretization parameter $h \to 0$ (see Theorem 3.48) and one operator B (e.g., stemming from a reaction process) with $\|B\|$ bounded in h. Combining the splitting approach with time stepping discretizations of (11.87), (11.88) we have time steps $\tau_A, \tau_B \le \tau$, i.e., the splitting error cannot be controlled independent of the time discretization. On the other hand, the time stepping methods can be different with different τ_A, τ_B, e.g., an implicit (one-step-) method with $\tau_A = \tau$ and an explicit multistep method with $\tau_B \ll \tau$. As we can view (11.87), (11.88), applied n-times such that $n\tau = T$ is the desired time horizon, as a solution of

$$\frac{d}{dt}\boldsymbol{v}(t) + C(t)\boldsymbol{v}(t) = 0, \quad t \in (0, 2T],$$
$$\boldsymbol{v}(0) = \boldsymbol{u}_0,$$

$$\text{with} \quad C(T) := \begin{cases} A, \; t \in \big(2k\tau, (2k+1)\tau\big] \\ B, \; t \in \big((2k+1)\tau, (2k+2)\tau\big] \end{cases}, \; k = 0, \ldots, n-1,$$

the additional discretization error and its temporal behaviour is at most the worse case of the two methods applied. Therefore, concentrating on the splitting error, it is desirable to have a splitting of higher order. A first possibility is an *additive splitting* in the following form:

- Perform an A-B-splitting step according to (11.87), (11.88) leading to $\boldsymbol{u}_1$.
- Perform an B-A-splitting step leading to $\boldsymbol{v}_1$ and set

$$\boldsymbol{w}(\tau) := \frac{1}{2}\left(\boldsymbol{u}_1(\tau) + \boldsymbol{v}_1(\tau)\right), \quad \text{i.e.}$$

$$\boldsymbol{w}(\tau) = \frac{1}{2}\left(e^{-B\tau}e^{-A\tau} + e^{-A\tau}e^{-B\tau}\right)\boldsymbol{u}_0. \tag{11.89}$$

Here the effort is doubled, but the two splitting procedures can be performed in parallel. As the leading error term, we have a method of order 2. This also holds true for the *Strang-splitting* with the splitting sequence $\frac{1}{2}A - B - \frac{1}{2}A$, i.e.,

$$\boldsymbol{w}(\tau) := e^{-A\frac{\tau}{2}}e^{-B\tau}e^{-A\frac{\tau}{2}}\boldsymbol{u}_0 .$$

The local splitting error turns out to be

$$\tau E_S \boldsymbol{u}_0 = \frac{1}{24}\left(\big[A, [A, B]\big] + 2\big[B, [A, B]\big]\right)\tau^3 \boldsymbol{w}\left(\frac{\tau}{2}\right) + O\left(\tau^5\right).$$

More precisely, the following result can be shown.

Theorem 11.18 *Let there exist two parameters α, β with $0 \le \alpha \le 1 \le \beta$ and two constants $c_1, c_2 \ge 0$ such that, in some compatible matrix and vector norms, the following holds:*

1) $\|[A, B]\boldsymbol{v}\| \le c_1\|(-A)^\alpha \boldsymbol{v}\|$ *for all* $\boldsymbol{v} \in \mathbb{R}^M$,
2) $\big\|\big[A, [A, B]\big]\boldsymbol{v}\big\| \le c_2\|(-A)^\beta \boldsymbol{v}\|$ *for all* $\boldsymbol{v} \in \mathbb{R}^M$.

Then

$$\|e^{A\frac{\tau}{2}} e^{B\tau} e^{A\frac{\tau}{2}} \boldsymbol{v} - e^{(A+B)\tau}\| \le C\tau^3 \|(-A)^\beta \boldsymbol{v}\|$$

with a constant $C \ge 0$ only depending on c_1, c_2, $\|B\|$.

Proof See [154, Thm. 2.1]. □

The complexity of the Strang-splitting is comparable to the Lie-splitting, as for an interval $[0, T]$ with $T = n\tau$ it only requires $n+1$ substeps (with step size $\frac{\tau}{2}, \tau \ldots, \tau, \frac{\tau}{2}$) instead of $2n$.

By recursive application similar splitting methods for models with N instead of 2 components are available.

Concerning the stability, we get immediately (compare Theorem 9.33):

Theorem 11.19 *Let*

1) $\|e^{-A\tau}\| \le \exp(\alpha\tau)$,
2) $\|e^{-B\tau}\| \le \exp(\beta\tau)$,
3) $\|e^{-(A+B)\tau}\| \le \exp(\gamma\tau)$

with $\alpha, \beta, \gamma \in \mathbb{R}$, $T := n\tau$. Then the following error estimate holds for the Lie-splitting, where $\delta := \max\{\alpha + \beta, \gamma\}$:

$$\|\boldsymbol{u}(k\tau) - \boldsymbol{w}(k\tau)\| \le \exp\big(\delta(k-1)\tau\big) k\tau E_S \boldsymbol{u}_0, \quad 1 \le k \le n.$$

Proof Set $R := e^{-(A+B)\tau}$, $S := e^{-B\tau}e^{-A\tau}$, then it is sufficient to estimate

$$\|R^k - S^k\| = \|\sum_{j=0}^{k-1} R^j (R-S) S^{k-j-1}\| \le \sum_{j=0}^{k-1} \|R\|^j \|R-S\| \|S\|^{k-j-1}$$

$$\le \sum_{j=0}^{k-1} \exp(\gamma j\tau) \|R-S\| \exp\big((\alpha+\beta)(k-j-1)\tau\big) \le k \exp\big(\delta(k-1)\tau\big) \|R-S\|.$$

□

Similar results hold true for the symmetric versions (Problem 11.21).

This gives a stable method if all methods are stable ($\alpha, \beta, \gamma \le 0$) with linear growth in T, otherwise the growth behaviour of the original problem is expected. The approaches can be generalized to non-autonomous and nonlinear situations:

$$\frac{d}{dt}\boldsymbol{u}(t) = \boldsymbol{F}(t, \boldsymbol{u}(t)),\ \boldsymbol{u}(0) = \boldsymbol{u}_0,$$

where

$$\boldsymbol{F}(t, \boldsymbol{u}) := \boldsymbol{F}_1(t, \boldsymbol{u}) + \boldsymbol{F}_2(t, \boldsymbol{u})$$

may be approximated on $(t_n, t_{n+1}]$, $t_{n+1} := t_n + \tau$, with the Lie-splitting

$$\begin{aligned}
&\frac{d}{dt}\boldsymbol{u}_1(t) = F_1(t, \boldsymbol{u}_1(t)), && \boldsymbol{u}_1(t_n) = \boldsymbol{w}(t_n),\\
&\frac{d}{dt}\boldsymbol{u}_2(t) = F_2(t, \boldsymbol{u}_2(t)), && \boldsymbol{u}_2(t_n) = \boldsymbol{u}_1(t_{n+1}),\\
&\boldsymbol{w}(t_{n+1}) := \boldsymbol{u}_2(t_{n+1}),
\end{aligned}$$

or with the Strang-splitting

$$\begin{aligned}
&\frac{d}{dt}\boldsymbol{u}_1(t) = F_1(t, \boldsymbol{u}_1(t)), && t_n < t \le t_n + \frac{\tau}{2}, && \boldsymbol{u}_1(t_n) = \boldsymbol{w}(t_n),\\
&\frac{d}{dt}\boldsymbol{u}_2(t) = F_2(t, \boldsymbol{u}_2(t)), && t_n < t \le t_{n+1}, && \boldsymbol{u}_2(t_n) = \boldsymbol{u}_1\left(t_n + \frac{\tau}{2}\right),\\
&\frac{d}{dt}\boldsymbol{u}_3(t) = F_1(t, \boldsymbol{u}_3(t)), && t_n + \frac{\tau}{2} < t \le t_{n+1}, && \boldsymbol{u}_3\left(t_n + \frac{\tau}{2}\right) = \boldsymbol{u}_2(t_{n+1}),\\
&\boldsymbol{w}(t_{n+1}) := \boldsymbol{u}_3(t_{n+1})
\end{aligned}$$

with corresponding (formal) local splitting errors of the order as seen, e.g., for the Lie-splitting

$$\tau E_S(\boldsymbol{u}(t_n)) = \frac{1}{2}\left[\frac{\partial F_1}{\partial u}F_2 - \frac{\partial F_2}{\partial u}F_1\right](t_n, u(t_n))\tau^2.$$

Going back to the linear case, one may ask whether higher order splittings are possible by more general multiplicative versions of the form

$$\prod_{j=1}^{r} e^{-B\beta_j\tau}e^{-A\gamma_j\tau},$$

where we recover the Lie-splitting by $r := 1$, $\beta_1 := \gamma_1 := 1$, and the Strang-splitting by $r := 2$, $\beta_1 := 0$, $\beta_2 := 1$, $\gamma_1 := \gamma_2 := \frac{1}{2}$, or by additive versions built of such terms, meaning

$$\sum_{i=1}^{\gamma} \alpha_i \prod_{j=1}^{\gamma} e^{-B\beta_{ij}\tau}e^{-A\gamma_{ij}\tau}, \quad \sum_{i=1}^{\gamma} \alpha_i = 1. \tag{11.90}$$

To conclude a stability behaviour as in Theorem 11.19 from single terms, $\alpha_i \ge 0$ is reasonable, also (at least for diffusion problems) no backward steps should be allowed, i.e., $\beta_{ij}, \gamma_{ij} \ge 0$. It turns out that under such conditions no splitting order greater than two is possible [215].

A further aspect, which may increase the splitting error, are inhomogeneous boundary conditions: Consider the ideal case of commuting operators A, B, i.e., no splitting error so far, and inhomogeneous boundary conditions $\boldsymbol{g}$ solely for A, then we add in (11.86) and (11.87) the term $\boldsymbol{g}$ and $\tilde{\boldsymbol{g}}$, respectively, on the right-hand side, i.e., for the Lie-splitting in the A-B form we get

$$\boldsymbol{u}(\tau) = e^{-(A+B)\tau}\boldsymbol{u}_0 + \int_0^{\tau} e^{-(A+B)(\tau-s)}\boldsymbol{g}(s)\,ds,$$

and

$$w(\tau) = e^{-B\tau} e^{-A\tau} u_0 + e^{-B\tau} \int_0^\tau e^{-A(\tau-s)} \tilde{g}(s)\, ds\,.$$

Hence there is no equality for $\tilde{g} = g$, but only for

$$\tilde{g}(t) := e^{Bt} g(t),$$

meaning to solve an auxiliary problem backward in time. For the B-A-splitting we get

$$w(\tau) = e^{-A\tau} e^{-B\tau} u_0 + \int_0^\tau e^{-A(\tau-s)} \tilde{g}(s)\, ds,$$

i.e., the equality with $u(\tau)$ requires

$$\tilde{g}(t) := e^{-B(\tau-t)} g(t).$$

For a reaction–diffusion problem, B being the reaction term, we expect a nontrivial kernel reflecting the reaction invariants, and in general positive and negative eigenvalues.

Reactions approaching equilibrium quickly will lead to large positive eigenvalues, i.e., we get a *stiff system*, meaning

$$B = \frac{1}{\varepsilon}(B_0 + \varepsilon B_1),$$

where one does not want to resolve the ε-term scale, i.e., $\tau \gg \varepsilon$ such that deteriorations of the results mentioned above are to be expected. By a refined analysis, which looks for an $O(\varepsilon)$-accurate reduced problem, in which the ODEs for the fast components are substituted by the algebraic requirement belonging to the corresponding solution manifold (the *reduced manifold*, compare with (11.66)), it can be shown that the local splitting error estimates are reduced by one order for the Lie-splittings A-B or B-A, and for the Strang-splittings $\frac{1}{2}A - B - \frac{1}{2}A$ or $\frac{1}{2}B - A - \frac{1}{2}B$, but the leading error terms are now different. For the B-A-splitting only a compatibility consideration for u is necessary such that the leading term vanishes, but the Strang-splitting $\frac{1}{2}B - A - \frac{1}{2}B$ only has order one. The reason lies in the amount the splitting solution approaches the reduced manifold as indicated in Fig. 11.2, which leads to a preference for the B-A or $\frac{1}{2}B - A - \frac{1}{2}B$ versions. For more details see [217].

As mentioned, splitting methods can be used to separate processes of different behaviour, like diffusion and convection, or, as below, diffusion–convection and reaction. A more traditional area is the *dimensional splitting*, where a single process, e.g., diffusion, is subdivided into processes of the same type but of lower dimensions, typically of dimension one. This requires a Cartesian structure of the spatial domain Ω and a corresponding separation of the process, which is quite restrictive, and opposite to the general approach of this book, therefore we do not consider details but refer, e.g., to the book [38] and the literature cited there. Note, however, that the following methods can also be used for other splitting approaches. There are *fractional-step* or *locally one-dimensional (LOD) methods*, which combine directly

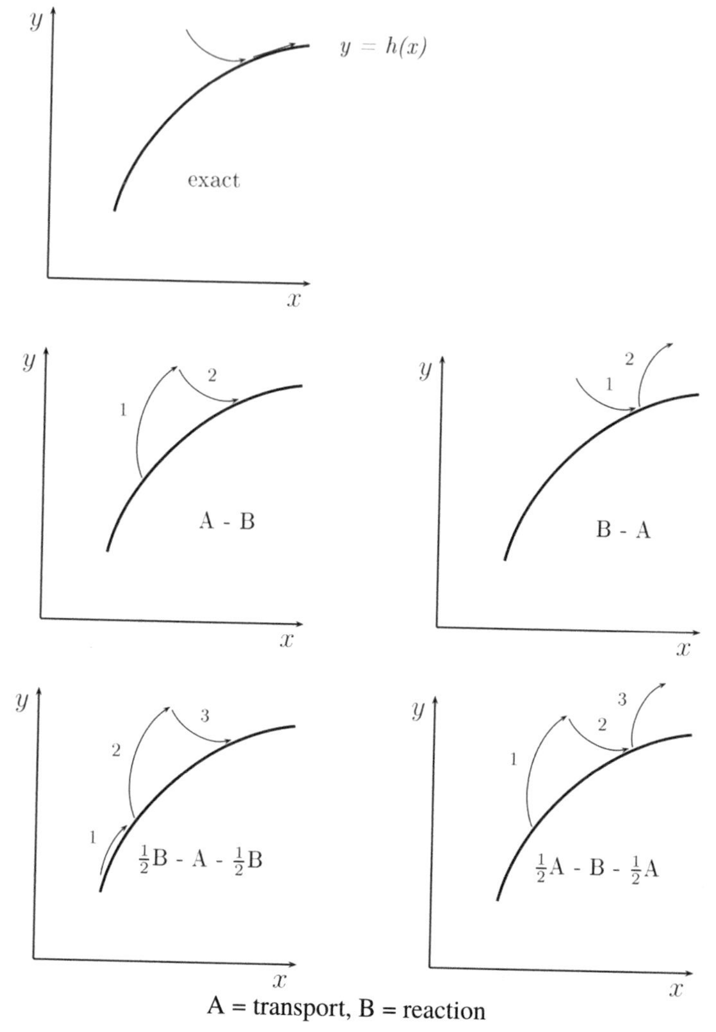

Fig. 11.2: Dynamical behaviour of splitting schemes (based on [217]).

a Lie-splitting with a one-step method with step size τ. Using the implicit Euler method, this reads in the above notation:

$$\begin{aligned} \boldsymbol{u}_1 &= \boldsymbol{u}_0 + \tau A \boldsymbol{u}_1\,, \\ \boldsymbol{u}_2 &= \boldsymbol{u}_1 + \tau B \boldsymbol{u}_2\,. \end{aligned} \tag{11.91}$$

This method is not consistent with the equilibrium in the sense that an element $\boldsymbol{u}_0$ such that $(A + B)\boldsymbol{u}_0 = 0$ does not imply this property for $\boldsymbol{u}_1$ or $\boldsymbol{u}_2$. The *LOD-Crank-Nicolson method* reads as

$$\begin{aligned} u_1 &= u_0 + \frac{1}{2}\tau(Au_0 + Au_1), \\ u_2 &= u_1 + \frac{1}{2}\tau(Bu_1 + Bu_2). \end{aligned} \tag{11.92}$$

The *alternating direction implicit (ADI)* methods try to restore the consistency with the equilibrium. The *Peaceman-Rachford method* reads as

$$\begin{aligned} u_1 &= u_0 + \frac{1}{2}\tau(Au_0 + Bu_1), \\ u_2 &= u_1 + \frac{1}{2}\tau(Bu_1 + Au_2). \end{aligned} \tag{11.93}$$

This method can be viewed as a Strang-splitting discretized by two implicit (step size $\tau/2$ and τ) and an explicit (step size $\tau/2$) Euler steps.

11.7.2 Iterative Operator Splitting

A principal disadvantage of the methods of the previous section is that they couple the splitting error and the discretization error and thus impede the use of an implicit time stepping method with large time steps. We have noticed that with the form (11.90) no higher consistency order than two is possible under general stability conditions. Therefore we try an additive form ($r = 2$), but with interdependence of the summands.

Given an approximation $\tilde{u}$ of u on $[0, \tau]$, we define the following splitting:

$$\begin{aligned} \frac{d}{dt}u_1(t) + Au_1(t) + B\tilde{u}(t) &= 0, \qquad u(0) = u_0, \\ \frac{d}{dt}u_2(t) + Au_1(t) + Bu_2(t) &= 0, \qquad u(0) = u_0. \end{aligned} \tag{11.94}$$

Let $\|u\|_\infty := \max\{\|u(t)\| \mid t \in [0, \tau]\}$ for bounded functions $u: [0, \tau] \to \mathbb{R}^M$.

Theorem 11.20 *Under the assumptions 1), 2) of Theorem 11.19, the local splitting error of* (11.94) *satisfies*

$$\|u_2 - u\|_\infty \le C_\alpha C_\beta \|A\|\|B\|\|\tilde{u} - u\|_\infty \tau^2,$$

where $C_\alpha := 1$ *for* $\alpha \le 0$, $C_\alpha := \exp(\alpha\tau)$ *for* $\alpha \ge 0$.

Proof Consider the error functions

$$\tilde{e} := \tilde{u} - u, \quad e_i := u_i - u \quad \text{on } [0, \tau],$$

thus

$$\frac{d}{dt}\boldsymbol{e}_1 + A\boldsymbol{e}_1 = -B\tilde{\boldsymbol{e}}, \qquad \boldsymbol{e}_1(0) = \boldsymbol{0},$$
$$\frac{d}{dt}\boldsymbol{e}_2 + B\boldsymbol{e}_2 = -A\boldsymbol{e}_1, \qquad \boldsymbol{e}_1(0) = \boldsymbol{0},$$

and by the variation-of-constants formula

$$\boldsymbol{e}_1(t) = -\int_0^t e^{-A(t-s)} B\tilde{\boldsymbol{e}}(s)\, ds \quad \text{for } 0 \le t \le \tau,$$

i.e.,

$$\|\boldsymbol{e}_1(t)\| \le \int_0^t \exp(\alpha(t-s))\|B\|\|\tilde{\boldsymbol{e}}\|_\infty\, ds = \frac{1}{\alpha}\big(\exp(\alpha t) - 1\big)\|B\|\|\tilde{\boldsymbol{e}}\|_\infty .$$

Since

$$\frac{1}{\alpha}\big(\exp(\alpha t) - 1\big) = \frac{\exp(\alpha t) - 1}{\alpha t}\, t = \exp(\alpha\tilde{t})\, t$$

with $0 \le \tilde{t} \le t$ by the mean-value theorem we get

$$\|\boldsymbol{e}_1(t)\| \le C_\alpha \|B\|\|\tilde{\boldsymbol{e}}\|_\infty\, t \quad \text{for } t \in [0, \tau],$$

and thus

$$\|\boldsymbol{e}_1\|_\infty \le C_\alpha \|B\|\|\tilde{\boldsymbol{e}}\|_\infty\, \tau.$$

Analogously:

$$\begin{aligned} \|\boldsymbol{e}_2(t)\| &\le \int_0^t \exp(\beta(t-s))\|A\|\|\boldsymbol{e}_1(s)\|\, ds \\ &\le C_\alpha\|A\|\|B\|\|\tilde{\boldsymbol{e}}\|_\infty \int_0^t \exp(\beta(t-s)) s\, ds \le C_\alpha C_\beta \|A\|\|B\|\|\tilde{\boldsymbol{e}}\|_\infty\, \tau^2. \end{aligned} \tag{11.95}$$

□

Now, an iteration over (11.94) is possible, i.e., we can set $\tilde{\boldsymbol{u}} := \boldsymbol{u}^{(k)}$ and $\boldsymbol{u}^{(k+1)} := \boldsymbol{u}_2$, and we have from Theorem 11.20 that

$$\|(\boldsymbol{u}^{(k)} - \boldsymbol{u})(\tau)\| \le \|\boldsymbol{u}^{(k)} - \boldsymbol{u}\|_\infty \le (C_\alpha C_\beta \|A\|\|B\|)^k \|\boldsymbol{u} - \boldsymbol{u}^{(0)}\|_\infty \tau^{2k}.$$

If $C_\alpha C_\beta \|A\|\|B\|\tau^2$ is small and k is taken large enough, this might be a possibility to get the splitting error (significantly) less than the time discretization error. In the case that A represents a diffusion–convection process and B a reaction process, the values of $\alpha \le 0$ and C_α do not seem to be critical. However, from Theorem 3.48 we know that $\|A\| = O(h^{-2})$, so there may be a problem if h is small, and again a small τ may be required to ensure error reduction. Note that the requirement

$$C_\alpha C_\beta \|A\|\|B\|\tau^2 < 1$$

is close to a step size requirement for explicit time stepping methods similar to (9.112), but only on τ/h.

Now we consider the iteration together with an implicit Euler time discretization similar to (11.91), i.e., we want to solve the problem

$$\mathcal{A}\boldsymbol{u} := (I + \tau(A + B))\boldsymbol{u} = \boldsymbol{u}_0 \tag{11.96}$$

and employ the iteration

$$\begin{aligned} (I + \tau A)\boldsymbol{u}' &= -\tau B\boldsymbol{u}^{(k)} + \boldsymbol{u}_0, \\ (I + \tau B)\boldsymbol{u}'' &= -\tau A\boldsymbol{u}' + \boldsymbol{u}_0, \\ u^{(k+1)} &:= \boldsymbol{u}''. \end{aligned} \tag{11.97}$$

Because of

$$\begin{aligned} -\tau A\boldsymbol{u}' + \boldsymbol{u}_0 &= -(I + \tau A)\boldsymbol{u}' + \boldsymbol{u}' + \boldsymbol{u}_0 = \tau B\boldsymbol{u}^{(k)} + \boldsymbol{u}' \\ &= \tau B\boldsymbol{u}^{(k)} + (I + \tau A)^{-1}(-\tau B\boldsymbol{u}^{(k)} + \boldsymbol{u}_0) \\ &= (I - (I + \tau A)^{-1})\tau B\boldsymbol{u}^{(k)} + (I + \tau A)^{-1}\boldsymbol{u}_0 \end{aligned}$$

we have a linear stationary method

$$\boldsymbol{u}^{(k+1)} = M\boldsymbol{u}^{(k)} + N\boldsymbol{u}_0$$

with

$$M := (I + \tau B)^{-1}(I - (I + \tau A)^{-1})\tau B = (I + \tau B)^{-1}(I + \tau A)^{-1}AB\tau^2$$

and

$$N := (I + \tau B)^{-1}(I + \tau A)^{-1}.$$

Because of

$$N\mathcal{A} = (I + \tau B)^{-1}(I + \tau A)^{-1}(1 + \tau A + \tau B) = (I + \tau B)^{-1}(I + (I + \tau A)^{-1}\tau B)$$

we have

$$M + N\mathcal{A} = (I + \tau B)^{-1}(I + \tau B) = I,$$

i.e., the consistency of the method. To assure convergence, we need (see Theorem 5.1)

$$\varrho(M) < 1.$$

If τ is sufficiently small so that

$$\tau\|A\| < 1,\ \tau\|B\| < 1, \tag{11.98}$$

we have (see (A3.11))

$$\|M\| \le \tau^2 \frac{\|A\|}{1-\tau\|A\|} \frac{\|B\|}{1-\tau\|B\|}, \tag{11.99}$$

i.e., there is convergence for sufficiently small τ with a rate proportional to τ^2, however, for instance, $\|A\|$ can be large.

Note that (11.98) for a diffusion problem has the same form as the step size restriction for the stability of an explicit time stepping method, which raises the question whether iterative operator splitting and implicit time discretization are compatible. An alternative to the multiplicative method (11.97) is the additive variant

$$\begin{aligned} (I+\tau A)\boldsymbol{u}' &= -\tau B\boldsymbol{u}^{(k)} + \boldsymbol{u}_0, \\ (I+\tau B)\boldsymbol{u}'' &= -\tau A\boldsymbol{u}^{(k)} + \boldsymbol{u}_0, \\ u^{(k+1)} &:= \frac{1}{2}(\boldsymbol{u}' + \boldsymbol{u}''), \end{aligned} \tag{11.100}$$

i.e., here are

$$\begin{aligned} M &:= -\frac{1}{2}((I+\tau A)^{-1}\tau B + (I+\tau B)^{-1}\tau A), \\ N &:= \frac{1}{2}((I+\tau A)^{-1} + (I+\tau B)^{-1}), \end{aligned}$$

and

$$N\mathcal{A} = \frac{1}{2}(2I + (I+\tau A)^{-1}\tau B + (I+\tau B)^{-1}\tau A).$$

Therefore $N\mathcal{A} + M = I$, i.e., there is consistency. Assuming (11.98), we have

$$\|M\| \le \frac{1}{2}\left(\frac{\|B\|}{1-\tau\|A\|} + \frac{\|A\|}{1-\tau\|B\|}\right)\tau.$$

An iteration over the Peaceman–Rachford approach (11.93) does not seem to give a consistent method. For further simplification the additive version (11.100) can be reduced to half of it ("delaying" the other process), e.g.,

$$(I+\tau A)\boldsymbol{u}^{(k+1)} = -\tau B\boldsymbol{u}^{(k)} + \boldsymbol{u}_0, \tag{11.101}$$

which with

$$M = -(I+\tau A)^{-1}\tau B,\ N = (I+\tau A)^{-1}$$

gives a consistent method.

Taking into account the many objections concerning (iterative) operator splitting as basic iterative method, it can rather be used as a preconditioner (see Theorem 5.22). In case of one splitting step as a preconditioner we have to estimate

$$\kappa(N\mathcal{A}).$$

For (11.97) we get

$$\begin{aligned} N\mathcal{A} = I - M &= (I + \tau B)^{-1}(I + (I + \tau A)^{-1}\tau B) \\ &= (I + \tau B)^{-1}(I + \tau A)^{-1}(I + \tau A + \tau B). \end{aligned}$$

In the additive version (11.100) we have

$$N\mathcal{A} = I - M = I + \frac{1}{2}\left((I + \tau A)^{-1}\tau B + (I + \tau B)^{-1}\tau A\right).$$

Here we have an implicit Euler diffusion and reaction step, either separately one after the other or in parallel. The reaction problems often are so small that direct solvers are sufficient. For diffusion-type problems the iterative schemes and preconditioners of Sections 5.3 to 5.7 can be applied. In the overall treatment of a problem with nonlinear reactions by means of a version of Newton's method (Section 11.3 and 11.5), one (implicit Euler) step could unfold in the hierarchy of iterations:

- Primary: Newton's method,
- Secondary, for linearized problems: Krylov-subspace method with splitting preconditioning,
- Tertiary, for linear diffusion problems: Krylov-subspace method with a preconditioning from Sections 5.3 to 5.7.

Exercises

Problem 11.19 Consider the formulation (9.1), (9.2) with $g := 0$, $r := 0$. Compute the error, splitting diffusion and convection by means of the Lie-splitting and show that the splitting error vanishes if $\boldsymbol{c}$ and $\boldsymbol{K}$ are independent of x.

Problem 11.20 Consider the formulation (11.54) (general transport-reaction system), consider the error, splitting diffusion/convection and reaction by means of the Lie-splitting and show that the splitting error vanishes in each of the following cases:

a) Linear reaction (11.59), and $\hat{R}$ is independent of x,
b) ψ is independent of x, $\boldsymbol{K} := \boldsymbol{0}$, $\nabla \cdot \boldsymbol{c} := 0$.

Problem 11.21 Show a stability result similar to Theorem 11.19 for the Strang-splitting.

A
Appendices

A.1 Notation

This overview is a compilation of the most important and frequently used symbols. We also refer to Appendices A.2–A.5. The references at the end of each entry indicate the first occurrence in the text.

$\mathbb{C}$	set of complex numbers – p. 32
$\mathbb{N}$	set of natural numbers – p. 18
$\mathbb{N}_0$	$:= \mathbb{N} \cup \{0\}$ – p. 58
$\mathbb{Q}$	set of rational numbers – p. 493
$\mathbb{R}$	set of real numbers – p. 1
$\mathbb{R}_+$	set of positive real numbers – p. 299
$\mathbb{Z}$	set of integers – p. 21
$\operatorname{Re} z$	real part of the complex number z – p. 32
$\operatorname{Im} z$	imaginary part of the complex number z – p. 32
x^T	transpose of the vector $x \in \mathbb{R}^d$, $d \in \mathbb{N}$ – p. 1
$\lvert x\rvert_p$	$:= \left(\sum_{j=1}^d \lvert x_j\rvert^p\right)^{1/p}$, $x = (x_1, \ldots, x_d)^T \in \mathbb{R}^d$, $d \in \mathbb{N}$, $p \in [1, \infty)$ – p. 726
$\lvert x\rvert_\infty$	$:= \max_{j=1,\ldots,d} \lvert x_j\rvert$ maximum norm of the vector $x \in \mathbb{R}^d$, $d \in \mathbb{N}$ – p. 28
$\lvert x\rvert$	$:= \lvert x\rvert_2$ Euclidean norm of the vector $x \in \mathbb{R}^d$, $d \in \mathbb{N}$ – p. 121
$x \cdot y$	$:= x^T y = \sum_{j=1}^d x_j y_j$ scalar product of the vectors $x, y \in \mathbb{R}^d$ – p. 2
$\langle x, y\rangle_A$	$:= y^T Ax = y \cdot Ax$ energy product of the vectors $x, y \in \mathbb{R}^d$ w.r.t. a symmetric, positive definite matrix A – (5.46)
$\lVert x\rVert_A$	energy norm of $x \in \mathbb{R}^d$ w.r.t. a symmetric, positive definite matrix A – p. 241
$\lvert\alpha\rvert$	$:= \lvert\alpha\rvert_1$ order (or length) of the multi-index $\alpha \in \mathbb{N}_0^d$, $d \in \mathbb{N}$ – p. 58
I	identity matrix or identity operator – p. 3
e_j	jth unit vector in $\mathbb{R}^m$, $j = 1, \ldots, m$ – p. 115
$\operatorname{diag}(\lambda_i)$	$= \operatorname{diag}(\lambda_1, \ldots, \lambda_m)$ diagonal matrix in $\mathbb{R}^{m,m}$ with diagonal entries $\lambda_1, \ldots, \lambda_m \in \mathbb{C}$ – p. 94

P. Knabner and L. Angermann, *Numerical Methods for Elliptic and Parabolic Partial Differential Equations*, Texts in Applied Mathematics 44,
https://doi.org/10.1007/978-3-030-79385-2_A

A^T	transpose of the matrix A – p. 27
A^{-T}	transpose of the inverse matrix A^{-1} – p. 86
$\det A$	determinant of the square matrix A – p. 86
$\lambda_{\min}(A)$	minimum eigenvalue of a matrix A with real eigenvalues – p. 243
$\lambda_{\max}(A)$	maximum eigenvalue of a matrix A with real eigenvalues – p. 243
$\kappa(A)$	condition number of the square matrix A – p. 239
$\sigma(A)$	set of eigenvalues (spectrum) of the square matrix A – p. 253
$\varrho(A)$	spectral radius of the square matrix A – p. 238
$\|A\|_2$	spectral norm of the square matrix A – p. 121
$m(A)$	bandwidth of the symmetric matrix A – p. 95
$\operatorname{Env}(A)$	hull of the square matrix A – p. 95
$p(A)$	profile of the square matrix A – p. 95
$B_\varrho(x_0)$	$:= \{x \mid \|x - x_0\| < \varrho\}$ open ball in a normed space – p. 114
$\overline{B}_\varrho(x_0)$	$:= \{x \mid \|x - x_0\| \le \varrho\}$ closed ball in a normed space – p. 705
$\operatorname{diam}(G)$	diameter of the set $G \subset \mathbb{R}^d$ – p. 62
$\|G\|_n$	n-dimensional (Lebesgue) measure of the $G \subset \mathbb{R}^n$, $n \in \{1, \ldots, d\}$ – p. 126
$\|G\|$	$:= \|G\|_d$ d-dimensional (Lebesgue) measure of the set $G \subset \mathbb{R}^d$ – p. 53
$\operatorname{vol}(G)$	length ($d = 1$), area ($d = 2$), volume ($d = 3$) of "geometric bodies" $G \subset \mathbb{R}^d$ – (2.42)
$\operatorname{int} G$	interior of the set G – (2.28)
∂G	boundary of the set G – p. 2
$\overline{G}$	closure of the set G – p. 19
$\operatorname{span} G$	linear hull of the set G – p. 73
$\operatorname{conv} G$	convex hull of the set G – p. 88
$\|G\|$	cardinal number of the discrete set G – p. 141
$\mathfrak{n}, \mathfrak{n}_G$	outer unit normal w.r.t. the set $G \subset \mathbb{R}^d$ – p. 2
Ω	nonempty domain of $\mathbb{R}^d$, typically polygonally or polyhedrally bounded – p. 1
Γ	$:= \partial\Omega$ boundary of the domain $\Omega \subset \mathbb{R}^d$ – (0.39)
Γ_i	$\subset \Gamma$ boundary component ($i = 1, 2, 3$, for different types of boundary conditions) for formulations not in divergence form – (0.39)
$\widetilde{\Gamma}_i$	$\subset \Gamma$ boundary component ($i = 1, 2, 3$, for different types of boundary conditions) for formulations in divergence form – p. 127
Q_T	$:= \Omega \times (0, T)$ space-time cylinder ($T > 0$) – p. 14
S_T	$:= \partial\Omega \times (0, T)$ lateral surface of Q_T ($T > 0$) – p. 15
$\operatorname{supp} \varphi$	support of the function φ – p. 69
$\{\!\{\cdot\}\!\}$	average of a scalar function or a vector field (typically across an interface) – (7.24), (7.25)
$[\![\cdot]\!]$	jump of a scalar function or a vector field (typically across an interface) – (7.24), (7.25)
f^{-1}	inverse of the mapping f – p. 10
$f[G]$	image of the set G under the mapping f – p. 65
$\operatorname{im} f$	$:= f[G]$ if G is the domain of definition of f – p. 343

$\ker f$	kernel (nullspace) of the linear mapping f – p. 343
$f\|_K$	restriction of $f : G \to \mathbb{R}$ to a subset $K \subset G$ – p. 333
$\\|v\\|_X$	norm of the element v of the normed space X – p. 76
$\dim X$	dimension of the finite-dimensional linear space X – p. 65
$L[X,Y]$	set of linear, continuous operators acting from the normed space X in the normed space Y – p. 118
$\\| \cdot \\|_{X,Y}$	operator norm on $L[X,Y]$ – p. 342
$\langle \cdot, \cdot \rangle_X$	duality pairing on the Banach space X – p. 342
X'	$:= L[X,\mathbb{R}]$ dual space of the real normed space X – p. 341
$O(\cdot)$, $o(\cdot)$	Landau symbols of asymptotic analysis – p. 99
δ_{ij}	$(i,j \in \mathbb{N}_0)$ Kronecker symbol, i.e., $\delta_{ii} = 1$ and $\delta_{ij} = 0$ if $i \neq j$ – p. 64
χ_G	indicator function of a set $G \subset \mathbb{R}^d$ – p. 424
S_R	domain of (absolute) stability – p. 609

Differential expressions

∂_l	$(l \in \mathbb{N})$ symbol for the partial derivative w.r.t. the lth variable – p. 20
∂_t	$(t \in \mathbb{R})$ symbol for the partial derivative w.r.t. the variable t – (0.1)
∂^α	$(\alpha \in \mathbb{N}_0^d$ multi-index) αth partial derivative – (2.17)
$\nabla = \text{grad}$	$:= (\partial_1, \ldots, \partial_d)^T$ Nabla operator (symbolic vector) – p. 20
$\nabla \cdot = \text{div}$	$:= \sum_{i=1}^d \partial_i$ divergence operator – p. 20
curl	scalar or vector rotation – (6.89), p. 539
$\partial_{\boldsymbol{\mu}}$	$:= \boldsymbol{\mu} \cdot \nabla$ directional derivative w.r.t. the vector $\boldsymbol{\mu}$ – (0.41)
$D\Phi$	$:= \frac{\partial \Phi}{\partial x} := (\partial_j \Phi_i)_{i,j=1}^m$ Jacobi matrix or functional matrix of a differentiable mapping $\Phi : \mathbb{R}^m \to \mathbb{R}^m$ – p. 18
Δ	Laplace operator – p. 20

Coefficients in differential expressions

$\boldsymbol{K}$	diffusion coefficient (a square matrix function) – p. 3
$\boldsymbol{c}$	convection coefficient (a vector function) – (0.6)
r	reaction coefficient – p. 16
α, g_i	$(i = 1,2,3)$ boundary data for formulations not in divergence form – p. 14
$\widetilde{\alpha}, \widetilde{g}_i$	$(i = 1,2,3)$ boundary data for formulations in divergence form – p. 127

Discretization methods

$\mathcal{T}, \mathcal{T}_h$	partition (not necessarily simplicial) – p. 61
$\mathbb{T}$	set of consistent refinements of an initial partition – p. 230
K	element of $\mathcal{T}_h$ (geometric element, cell) – p. 61
$\hat{K}$	reference element – p. 64
$\overline{\mathcal{F}}, \overline{\mathcal{F}}_h$	set of all faces of a partition – p. 395
$\mathcal{F}, \mathcal{F}_h$	set of all interior faces – p. 220
$\partial\mathcal{F}, \partial\mathcal{F}_h$	set of all boundary faces – p. 395
$\mathcal{F}_i$	set of faces belonging to the boundary component Γ_i or $\widetilde{\Gamma}_i$ (except of Section 8.3) – p. 395
$\mathcal{F}_K$	set of faces belonging to an element K – p. 369

F	face of a geometric element – p. 136
$\overline{\mathcal{E}}$	set of all edges – p. 542
$\overline{\mathcal{N}}$	set of all nodes – p. 214
$\mathcal{N}$	set of all interior nodes – p. 429
a_i	numbered node of a partition – p. 62
$a_S, a_{S,K}$	barycentre of a simplex K – (3.63)
$a_{S,i}$	barycentre of a numbered face F_i – p. 394
V_h	ansatz space – p. 61
X_h	extended ansatz space without any homogeneous Dirichlet boundary conditions – p. 63
a_h	approximated bilinear form a – p. 173
a_h^{loc}	local approximated bilinear form a – p. 444
b_h, c_h, d_h, e_h	approximated bilinear forms b, c, d, e in mixed formulations – p. 375
ℓ_h	approximated linear form ℓ – p. 173
ℓ_h^{loc}	local approximated linear form ℓ – p. 444
ℓ_{1h}, ℓ_{2h}	approximated linear forms ℓ_1, ℓ_2 in mixed formulations – p. 375
$\boldsymbol{c}_{\text{upw}}(v_h)$	upwind evaluation of $\boldsymbol{c} v_h$ – (7.30)
P_K	local ansatz space – (3.51)
Σ, Σ_K	set of local degrees of freedom – p. 136
$a_i a_j$	straight line through the nodes a_i, a_j – p. 530
$\overline{a_i a_j}$	straight-line segment through the nodes a_i, a_j – p. 100

Special notation for finite volume methods

$\mathcal{T}_h^*$	control volume partition – p. 489
Ω_i	control volume associated with the node a_i – p. 489
m_i	$:= \vert\Omega_i\vert_d$ (volume of Ω_i) – p. 500
$\mathcal{F}_i$	set of faces of the control volume Ω_i – p. 529
$\overline{\mathcal{F}}$	set of faces of all control volumes – p. 529
$\mathcal{F}$	set of interior (w.r.t. Ω) faces of the control volume Ω_i – p. 529
$\partial\mathcal{F}$	set of control volume faces in $\partial\Omega$ – p. 529
$\mathcal{N}^*$	set of the interior vertices of all control volumes – p. 533
$\overline{\Lambda}$	index set of all nodes to which a control volume is assigned – p. 494
Λ	index set of all nodes to which an interior control volume is assigned – p. 498
$\partial\Lambda$	index set of all nodes to which a boundary control volume is assigned – p. 499
$\partial\Lambda_i$	index set of all nodes to which a boundary control volume related to the boundary component Γ_i or $\widetilde{\Gamma}_i$ is assigned – p. 498
Λ_i	set of indices of neighbouring nodes to the control volume Ω_i – p. 495
λ_K	set of indices of nodes belonging to K – p. 496
Γ_{ij}	common boundary of two neighbouring control volumes Ω_i, Ω_j – p. 495
γ_{ij}	approximate normal component of the convective vector field $\boldsymbol{c}$ on Γ_{ij} – p. 499
μ_{ij}	approximate diffusion coefficient on Γ_{ij} – p. 499

d_{ij} $:= |a_i - a_j|$ (Euclidean distance between the nodes a_i, a_j) – p. 499
m_{ij} $:= |\Gamma_{ij}|_{d-1}$ (length or area of Γ_{ij}) – p. 495
p_{ij} normal component of the numerical flux density at Γ_{ij} – (8.12)
$\mathfrak{n}_{ij}$ outer unit normal w.r.t. Ω_i on Γ_{ij} – p. 499
a_h, a_h^0, b_h, d_h discrete bilinear forms (8.14)
$\gamma_{i,\Gamma}$ approximate normal component of the convective vector field $\boldsymbol{c}$ on $\Gamma \in \mathcal{F}_i$ – p. 531
$\mu_{i,\Gamma}$ approximate diffusion coefficient on $\Gamma \in \mathcal{F}_i$ – (8.66)
$a_{i,\Gamma}$ foot of the perpendicular through a_i at the boundary element $\Gamma \subset \partial\mathcal{F} \cap \mathcal{F}_i$ – 530
$d_{i,\Gamma}$ $:= |a_i - a_{i,\Gamma}|$ – (8.62)
m_Γ $|\Gamma|_{d-1}$ (length or area of the boundary element $\Gamma \in \mathcal{F}_i$) – p. 530
$p_{i,\Gamma}$ normal component of the numerical flux density at $\Gamma \in \mathcal{F}_i$ – (8.65)
$\mathfrak{n}_{i,\Gamma}$ outer unit normal w.r.t. Ω_i on $\Gamma \in \mathcal{F}_i$ – p. 531

Function spaces (see also Appendix A.5)

$\mathcal{P}_k(G)$ set of polynomials of maximum degree k on $G \subset \mathbb{R}^d$ – p. 63
$\mathcal{P}_k(\mathcal{T}_h)$ set of piecewise polynomials of maximum degree k on a partition $\mathcal{T}_h$ – (3.107)
$\mathcal{Q}_k(G)$ set of tensor polynomials of maximum degree k on $G \subset \mathbb{R}^d$ – p. 144
$\mathbb{P}_k(G)$ $\in \{\mathcal{P}_k(G), \mathcal{Q}_k(G)\}$ – p. 442
$C(G) = C^0(G)$ set of continuous functions on G – p. 20
$C^l(G)$ ($l \in \mathbb{N}$) set of l-times continuously differentiable functions on G – p. 20
$C^\infty(G)$ set of infinitely often continuously differentiable functions on G – p. 115
$C(\overline{G}) = C^0(\overline{G})$ set of bounded and uniformly continuous functions on G – p. 20
$C^l(\overline{G})$ ($l \in \mathbb{N}$) set of functions with bounded and uniformly continuous derivatives up to the order l on G – p. 20
$C_0^\infty(G)$ set of infinitely often continuously differentiable functions on G with compact support – p. 58
$L^p(G)$ ($p \in [1, \infty)$) set of Lebesgue-measurable functions whose pth power of their absolute value is Lebesgue-integrable on G – p. 27
$L^\infty(G)$ set of measurable, essentially bounded functions – p. 27
$\langle \cdot, \cdot \rangle_{0,G}$ scalar product in $L^2(G)$ † – p. 52
$\| \cdot \|_{0,G}$ norm on $L^2(G)$ † – p. 27
$\| \cdot \|_{0,p,G}$ ($p \in [1, \infty]$) norm on $L^p(G)$ † – p. 772
$\| \cdot \|_{\infty,G}$ norm on $L^\infty(G)$ † – p. 27
$W_p^l(G)$ ($l \in \mathbb{N}$, $p \in [1, \infty]$) set of l-times weakly differentiable functions from $L^p(G)$, with derivatives in $L^p(G)$ – p. 113
$\| \cdot \|_{l,p,G}$ ($l \in \mathbb{N}$, $p \in [1, \infty]$) norm on $W_p^l(G)$ † – p. 773
$| \cdot |_{l,p,G}$ ($l \in \mathbb{N}$, $p \in [1, \infty]$) seminorm on $W_p^l(G)$ † – p. 157
$H^l(G)$ $:= W_2^l(G)$ ($l \in \mathbb{N}$) – p. 112
$H_0^1(G)$ set of functions from $H^1(G)$ with vanishing trace on ∂G – p. 116
$H_\Gamma^1(\Omega)$ $:= \{v \in H^1(\Omega) \mid \gamma_{0,\Gamma}(v) = 0\}, \Gamma \subset \partial\Omega$ – (6.37)

$\langle\cdot,\cdot\rangle_{l,G}$ — ($l \in \mathbb{N}$) scalar product in $H^l(G)$ † – (3.6)
$\|\cdot\|_{l,G}$ — ($l \in \mathbb{N}$) norm on $H^l(G)$ † – (3.6)
$|\cdot|_{l,G}$ — ($l \in \mathbb{N}$) seminorm on $H^l(G)$ † – p. 113
$\langle\cdot,\cdot\rangle_{0,h}$ — discrete $L^2(\Omega)$-scalar product – (8.15)
$\|\cdot\|_{0,h}$ — discrete $L^2(\Omega)$-norm – (1.18)
$L^2(\partial G)$ — set of square Lebesgue-integrable functions on the boundary ∂G – p. 114
$L^2(\Gamma)$ — set of square Lebesgue-integrable functions on the boundary part $\Gamma \subset \partial G$ – (3.23)
$H^{1/2}(\partial\Omega) := \operatorname{im}(\gamma_0) = \gamma_0\left(H^1(\Omega)\right)$ — trace space – p. 358
$H^{-1/2}(\partial\Omega) := \left(H^{1/2}(\partial\Omega)\right)'$ — dual of the trace space $H^{1/2}(\partial\Omega)$ – p. 358
$H^{1/2}_{00}(\Gamma)$ — subset of $L^2(\Gamma)$ such that the extension to $\partial\Omega$ by zero of the elements belongs to $H^{1/2}(\partial\Omega)$ – p. 359
$H^{1/2}(\Gamma)$ — localized trace space – p. 359
$H^{-1/2}(\Gamma) := \left(H^{1/2}_{00}(\Gamma)\right)'$ — dual space – p. 359
$H^l(\mathcal{T}_h)$ — broken Sobolev space on a partition $\mathcal{T}_h$ – p. 396
$\|\cdot\|_{l,\mathcal{T}_h}$ — norm of $H^l(\mathcal{T}_h)$ – p. 396
$\|\cdot\|_{l,p,\mathcal{T}_h}$ — broken Sobolev norm – p. 160
$|\cdot|_{l,p,\mathcal{T}_h}$ — broken Sobolev seminorm – p. 160
$H(\mathrm{curl};G)$ — subspace of $L^2(G)^d$ the elements of which have a weak rotation in $L^2(G)$ ($d=2$) or $L^2(G)^3$ ($d=3$), resp. – (8.82)
$\|\cdot\|_{H(\mathrm{curl})}$ — norm of $H(\mathrm{curl};G)$ – (8.82)
$H(\mathrm{div};G)$ — subspace of $L^2(G)^d$ the elements of which have a weak divergence in $L^2(G)$ – (6.38)
$\|\cdot\|_{H(\mathrm{div})}$ — norm of $H(\mathrm{div};G)$ – (6.38)
$\mathcal{H}(\mathrm{div};G)$ — subset of $H(\mathrm{div};G)$ the elements of which have a normal component trace in $L^2(\Gamma)$, where $\Gamma \subset \partial G$ – (6.41)
$\|\cdot\|_{\mathcal{H}(\mathrm{div})}$ — norm of $\mathcal{H}(\mathrm{div};G)$ – 361
$C([0,T],X) = C^0([0,T],X)$ — set of continuous functions on $[0,T]$ with values in the normed space X – p. 564
$C^l([0,T],X)$ — ($l \in \mathbb{N}$) set of l-times continuously differentiable functions on $[0,T]$ with values in the normed space X – p. 564
$L^p((0,T),X)$ — ($p \in [1,\infty]$) Lebesgue-space of functions on $[0,T]$ with values in the normed space X – p. 564

In general:

$X(\mathcal{T}_h)$ — $:= \{v : \overline{\Omega} \to \mathbb{R} \mid v|_K \in X(K) \text{ for all } K \in \mathcal{T}_h\}$, where X stands for one of the function spaces on G defined above
$X^d, X(G)^d$ — $:= \{\boldsymbol{v} = (v_i) : G \to \mathbb{R}^d \mid v_i \in X \text{ for all } i \in \{1,\ldots,d\}\}$, where X stands for one of the function spaces on G defined above

Operators

γ_0 — $: H^1(\Omega) \to L^2(\partial\Omega)$ — trace operator – p. 115
$\gamma_{0,\Gamma}$ — localized trace operator on $\Gamma \subset \partial\Omega$ – (6.36)

$\gamma_{n,\Gamma}$	normal component trace operator on $\Gamma \subset \partial\Omega$ – (6.39)
I_K	local interpolation operator on a geometric element K – p. 155
I_h	global interpolation operator – (3.83)
Π_K	local approximation operator on a geometric element K, typically an $L^2(K)$-projector – p. 161
Π, Π_h	global approximation operator, typically an $L^2(\Omega)$-projector – p. 161

† **Convention:** In the case $G = \Omega$, this specification is omitted.

A.2 Basic Concepts of Analysis

A subset $G \subset \mathbb{R}^d$ is called a *set of measure zero* if, for any number $\varepsilon > 0$, a countable family of balls B_j with d-dimensional volume $\varepsilon_j > 0$ exists such that

$$\sum_{j=1}^{\infty} \varepsilon_j < \varepsilon \qquad \text{and} \qquad G \subset \bigcup_{j=1}^{\infty} B_j\,.$$

Two functions $f, g : G \to \mathbb{R}$ are called *equal almost everywhere* (in short: *equal a.e.*, notation: $f \equiv g$) if the set $\{x \in G \mid f(x) \neq g(x)\}$ is of measure zero.

In particular, a function $f : G \to \mathbb{R}$ is called *vanishing almost everywhere* if it is equal to the constant function zero almost everywhere.

A function $f : G \to \mathbb{R}$ is called *measurable* if there exists a sequence $(f_i)_i$ of step functions $f_i : G \to \mathbb{R}$ such that $f_i \to f$ for $i \to \infty$ almost everywhere.

In what follows, G denotes a subset of $\mathbb{R}^d$, $d \in \mathbb{N}$.

(i) A point $x = (x_1, x_2, \dots, x_d)^T \in \mathbb{R}^d$ is called a *boundary point* of G if every open neighbourhood (perhaps an open ball) of x contains a point of G as well as a point of the complementary set $\mathbb{R} \setminus G$.
(ii) The collection of all boundary points of G is called the *boundary* of G and is denoted by ∂G.
(iii) The set $\overline{G} := G \cup \partial G$ is called the *closure* of G.
(iv) The set G is called *closed* if $\overline{G} = G$.
(v) The set G is called *open* if $G \cap \partial G = \emptyset$.
(vi) The set $G \setminus \partial G$ is called the *interior* of G and is denoted by $\operatorname{int} G$.

A subset $G \subset \mathbb{R}^d$ is called *connected* if for arbitrary distinct points $x_1, x_2 \in G$ there exists a continuous curve in G connecting them.

The set G is called *convex* if any two points from G can be connected by a straight-line segment in G.

A nonempty, open, and connected set $G \subset \mathbb{R}^d$ is called a *domain* in $\mathbb{R}^d$.

By $\alpha = (\alpha_1, \dots, \alpha_d)^T \in \mathbb{N}_0^d$ a so-called *multi-index* is denoted. Multi-indices are a popular tool to abbreviate some elaborate notation. For example,

$$\partial^\alpha := \prod_{i=1}^{d} \partial_i^{\alpha_i}, \quad \alpha! := \prod_{i=1}^{d} \alpha_i!, \quad |\alpha| := \sum_{i=1}^{d} \alpha_i .$$

The number $|\alpha|$ is called the *order* (or *length*) of the multi-index α.

For a continuous function $\varphi : G \to \mathbb{R}$, the set $\operatorname{supp}\varphi := \overline{\{x \in G \mid \varphi(x) \neq 0\}}$ denotes the *support* of φ.

A.3 Basic Concepts of Linear Algebra

A square matrix $A \in \mathbb{R}^{n,n}$ with entries a_{ij} is called *symmetric* if $a_{ij} = a_{ji}$ holds for all $i, j \in \{1, \ldots, n\}$, i.e., $A = A^T$.

A matrix $A \in \mathbb{R}^{n,n}$ is called *positive definite* if $x \cdot Ax > 0$ for all $x \in \mathbb{R}^n \setminus \{0\}$.

Given a polynomial $p \in \mathcal{P}_k$, $k \in \mathbb{N}_0$, of the form

$$p(z) = \sum_{j=0}^{k} a_j z^j \quad \text{with} \quad a_j \in \mathbb{C},\ j \in \{0, \ldots, k\}$$

and a matrix $A \in \mathbb{C}^{n,n}$, then the following *matrix polynomial* of A can be established:

$$p(A) := \sum_{j=0}^{k} a_j A^j .$$

Eigenvalues and Eigenvectors

Let $A \in \mathbb{C}^{n,n}$. A number $\lambda \in \mathbb{C}$ is called an *eigenvalue* of A if

$$\det(A - \lambda I) = 0 .$$

If λ is an eigenvalue of A, then any vector $x \in \mathbb{C}^n \setminus \{0\}$ such that

$$Ax = \lambda x \qquad (\Leftrightarrow (A - \lambda I)x = 0)$$

is called an *eigenvector* of A associated with the eigenvalue λ.

The polynomial $p_A(\lambda) := \det(A - \lambda I)$ is called the *characteristic polynomial* of A.

The set of all eigenvalues of a matrix A is called the *spectrum* of A, denoted by $\sigma(A)$.

If all eigenvalues of a matrix A are real, then the numbers $\lambda_{\max}(A)$ and $\lambda_{\min}(A)$ denote the largest, respectively, smallest, of these eigenvalues.

The number $\varrho(A) = \max_{\lambda\in\sigma(A)} |\lambda|$ is called the *spectral radius* of A.

Norms of Vectors and Matrices

The *norm of a vector* $x \in \mathbb{R}^n$, $n \in \mathbb{N}$, is a real-valued function $x \mapsto |x|$ satisfying the following three properties:

(i) $|x| \geq 0$ for all $x \in \mathbb{R}^n$, $|x| = 0 \Leftrightarrow x = 0$,
(ii) $|\alpha x| = |\alpha|\,|x|$ for all $\alpha \in \mathbb{R}$, $x \in \mathbb{R}^n$,
(iii) $|x + y| \leq |x| + |y|$ for all $x, y \in \mathbb{R}^n$.

For example, the most frequently used vector norms are

(a) the *maximum norm*:

$$|x|_\infty := \max_{j=1\ldots n} |x_j|\,. \tag{A3.1}$$

(b) the ℓ_p-*norm*, $p \in [1, \infty)$:

$$|x|_p := \left(\sum_{j=1}^{n} |x_j|^p\right)^{1/p} . \tag{A3.2}$$

The important case $p = 2$ yields the so-called *Euclidean norm*:

$$|x| := |x|_2 := \left(\sum_{j=1}^{n} x_j^2\right)^{1/2} . \tag{A3.3}$$

The three most important norms (that is, $p = 1, 2, \infty$) in $\mathbb{R}^n$ are *equivalent* in the following sense: The inequalities

$$\begin{aligned}
\frac{1}{\sqrt{n}}\,|x|_2 &\leq |x|_\infty \leq |x|_2 \leq \sqrt{n}\,|x|_\infty\,, \\
\frac{1}{n}\,|x|_1 &\leq |x|_\infty \leq |x|_1 \leq n\,|x|_\infty\,, \\
\frac{1}{\sqrt{n}}\,|x|_1 &\leq |x|_2 \leq |x|_1 \leq \sqrt{n}\,|x|_2
\end{aligned}$$

are valid for all $x \in \mathbb{R}^n$.

The *norm of the matrix* $A \in \mathbb{R}^{n,n}$ is a real-valued function $A \mapsto \|A\|$ satisfying the following four properties:

(i) $\|A\| \geq 0$ for all $A \in \mathbb{R}^{n,n}$, $\|A\| = 0 \Leftrightarrow A = 0$,
(ii) $\|\alpha A\| = |\alpha|\,\|A\|$ for all $\alpha \in \mathbb{R}$, $A \in \mathbb{R}^{n,n}$,
(iii) $\|A + B\| \leq \|A\| + \|B\|$ for all $A, B \in \mathbb{R}^{n,n}$,
(iv) $\|AB\| \leq \|A\|\,\|B\|$ for all $A, B \in \mathbb{R}^{n,n}$.

In comparison with the definition of a vector norm, we include here an additional property (iv), which is called the *submultiplicative property*. It restricts the general set of matrix norms to the practically important class of *submultiplicative norms*.

The most common matrix norms are

(a) the *total norm*:

$$\|A\|_G := n \max_{1\le i,k\le n} |a_{ik}|\,, \tag{A3.4}$$

(b) the *Frobenius norm*:

$$\|A\|_F := \left(\sum_{i,k=1}^{n} a_{ik}^2 \right)^{1/2}, \tag{A3.5}$$

(c) the *maximum row sum*:

$$\|A\|_\infty := \max_{1\le i\le n} \sum_{k=1}^{n} |a_{ik}|\,, \tag{A3.6}$$

(d) the *maximum column sum*:

$$\|A\|_1 := \max_{1\le k\le n} \sum_{i=1}^{n} |a_{ik}|\,. \tag{A3.7}$$

All these matrix norms are equivalent. For example, we have

$$\frac{1}{n}\|A\|_G \le \|A\|_p \le \|A\|_G \le n\|A\|_p\,, \quad p \in \{1, \infty\}\,,$$

or

$$\frac{1}{n}\|A\|_G \le \|A\|_F \le \|A\|_G \le n\|A\|_F\,.$$

Note that the spectral radius $\varrho(A)$ is not a matrix norm, as the following simple example shows:

For $A = \begin{pmatrix} 0 & 1 \\ 0 & 0 \end{pmatrix}$, we have that $A \ne 0$ but $\varrho(A) = 0$.

However, for any matrix norm $\|\cdot\|$ the following relation is valid:

$$\varrho(A) \le \|A\|\,. \tag{A3.8}$$

Very often, matrices and vectors simultaneously appear as a product Ax. In order to be able to handle such situations, there should be a certain correlation between matrix and vector norms.

A matrix norm $\|\cdot\|$ is called *mutually consistent* or *compatible* with the vector norm $|\cdot|$ if the inequality

$$|Ax| \le \|A\|\,|x| \tag{A3.9}$$

is valid for all $x \in \mathbb{R}^n$ and all $A \in \mathbb{R}^{n,n}$.

Examples of mutually consistent norms are

$$\|A\|_G \quad \text{or} \quad \|A\|_\infty \quad \text{with} \quad |x|_\infty\,,$$

$$\|A\|_G \quad \text{or} \quad \|A\|_1 \quad \text{with} \quad |x|_1\,,$$

$$\|A\|_G \quad \text{or} \quad \|A\|_F \quad \text{with} \quad |x|_2\,.$$

In many cases, the bound for $|Ax|$ given by (A3.9) is not sharp enough, i.e., for $x \neq 0$ we just have that

$$|Ax| < \|A\| \, |x| \, .$$

Therefore, the question arises of how to find, for a given vector norm, a compatible matrix norm such that in (A3.9) the equality holds for at least one element $x \neq 0$.

Given a vector norm $|x|$, the number

$$\|A\| := \sup_{x \in \mathbb{R}^n \setminus \{0\}} \frac{|Ax|}{|x|} = \sup_{x \in \mathbb{R}^n :\, |x|=1} |Ax|$$

is called the *induced* or *subordinate* matrix norm.

The induced norm is a compatible matrix norm with the given vector norm. It is the smallest norm among all matrix norms that are compatible with the given vector norm $|x|$.

To illustrate the definition of the induced matrix norm, the matrix norm induced by the Euclidean vector norm is derived:

$$\|A\|_2 := \max_{|x|_2=1} |Ax|_2 = \max_{|x|_2=1} \sqrt{x^T(A^TA)x} = \sqrt{\lambda_{\max}(A^TA)} = \sqrt{\varrho(A^TA)} \, . \quad \text{(A3.10)}$$

The matrix norm $\|A\|_2$ induced by the Euclidean vector norm is also called the *spectral norm.* This term becomes understandable in the special case of a symmetric matrix A. If $\lambda_1, \ldots, \lambda_n$ denote the real eigenvalues of A, then the matrix $A^TA = A^2$ has the eigenvalues λ_i^2 satisfying

$$\|A\|_2 = |\lambda_{\max}(A)| \, .$$

For symmetric matrices, the spectral norm coincides with the spectral radius. Because of (A3.8), it is the smallest possible matrix norm in that case.

As a further example, the maximum row sum $\|A\|_\infty$ is the matrix norm induced by the maximum norm $|x|_\infty$.

The number

$$\kappa(A) := \|A\| \, \|A^{-1}\|$$

is called the *condition number* of the matrix A with respect to the matrix norm under consideration.

The following relation holds:

$$1 \leq \|I\| = \|AA^{-1}\| \leq \|A\| \, \|A^{-1}\| \, .$$

For $|\cdot| = |\cdot|_p$, the condition number is also denoted by $\kappa_p(A)$. If all eigenvalues of A are real, the number

$$\kappa(A) := \lambda_{\max}(A)/\lambda_{\min}(A)$$

is called the *spectral condition number.* Hence, for a symmetric matrix A the equality $\kappa(A) = \kappa_2(A)$ is valid.

Occasionally, it is necessary to estimate small perturbations of nonsingular matrices. For this purpose, the following result is useful (*perturbation lemma* or *Neumann's lemma*). Let $A \in \mathbb{R}^{n,n}$ satisfy $\|A\| < 1$ with respect to an arbitrary, but fixed, matrix norm. Then the inverse of $I - A$ exists and can be represented as a convergent power series of the form

$$(I - A)^{-1} = \sum_{j=0}^{\infty} A^j,$$

with

$$\|(I - A)^{-1}\| \le \frac{1}{1 - \|A\|}. \tag{A3.11}$$

The Kronecker Product

Let $A \in \mathbb{R}^{m,n}$ and $B \in \mathbb{R}^{p,r}$, then the *Kronecker product* $A \otimes B \in \mathbb{R}^{mp,nr}$ is defined as

$$A \otimes B := \begin{pmatrix} a_{11}B & \dots & a_{1n}B \\ a_{n1}B & \dots & a_{mn}B \end{pmatrix}. \tag{A3.12}$$

A special case is $A := \boldsymbol{x}$ with $\boldsymbol{x} \in \mathbb{R}^m$, $B := \boldsymbol{y}^T$ with $\boldsymbol{y} \in \mathbb{R}^p$, for which

$$A \otimes B = \boldsymbol{x} \otimes \boldsymbol{y} = \boldsymbol{x}\boldsymbol{y}^T \in \mathbb{R}^{m,p}$$

becomes the *tensor product* of $\boldsymbol{x}, \boldsymbol{y}$.

In the case of square matrices $A \in \mathbb{R}^{m,m}$ and $B \in \mathbb{R}^{p,p}$, the Kronecker product has important special properties:
1) If A, B are invertible, then $A \otimes B$ is invertible, too, and

$$(A \otimes B)^{-1} = A^{-1} \otimes B^{-1}.$$

2) If $\lambda_i \in \mathbb{C}$, $i = 1, \dots, m$, and $\mu_j \in \mathbb{C}$, $j = 1, \dots, p$, are the eigenvalues of A, B, respectively, then

$$\lambda_i \mu_j, \quad i = 1, \dots, m, \ j = 1, \dots, p$$

are the eigenvalues of $A \otimes B$. Therefore

$$\|A \otimes B\|_2 = \|A\|_2 \|B\|_2 \quad \text{and} \quad \kappa_2(A \otimes B) = \kappa_2(A)\kappa_2(B).$$

Special Matrices

The matrix $A \in \mathbb{R}^{n,n}$ is called an *upper*, respectively *lower*, *triangular matrix* if its entries satisfy $a_{ij} = 0$ for $i > j$, respectively, $a_{ij} = 0$ for $i < j$.

A matrix $H \in \mathbb{R}^{n,n}$ is called an *(upper) Hessenberg matrix* if it has the following structure:

$$H := \begin{pmatrix} h_{11} & & & & \\ h_{21} & \ddots & & * & \\ & \ddots & \ddots & & \\ & & \ddots & \ddots & \\ & 0 & & h_{nn-1} & h_{nn} \end{pmatrix}$$

(that is, $h_{ij} = 0$ for $i > j + 1$).

The matrix $A \in \mathbb{R}^{n,n}$ satisfies the *strict row sum criterion* (or is *strictly row diagonally dominant*) if it satisfies

$$\sum_{\substack{j=1 \\ j \neq i}}^{n} |a_{ij}| < |a_{ii}| \quad \text{for all } i = 1, \ldots, n\,. \tag{A3.13}$$

It satisfies the *strict column sum criterion* if the following relation holds:

$$\sum_{\substack{i=1 \\ i \neq j}}^{n} |a_{ij}| < |a_{jj}| \quad \text{for all } j = 1, \ldots, n\,.$$

The matrix $A \in \mathbb{R}^{n,n}$ satisfies the *weak row sum criterion* (or is *weakly row diagonally dominant*) if

$$\sum_{\substack{j=1 \\ j \neq i}}^{n} |a_{ij}| \leq |a_{ii}| \quad \text{holds for all } i = 1, \ldots, n$$

and the strict inequality "$<$" is valid for at least one number $i \in \{1, \ldots, n\}$.

The weak column sum criterion is defined similarly.

The matrix $A \in \mathbb{R}^{n,n}$ is called *reducible* if there exist subsets $N_1, N_2 \subset \{1, \ldots, n\}$ with $N_1 \cap N_2 = \emptyset$, $N_1 \neq \emptyset \neq N_2$, and $N_1 \cup N_2 = \{1, \ldots, n\}$ such that the following property is satisfied:

$$\text{For all } i \in N_1,\ j \in N_2:\ a_{ij} = 0\,.$$

A matrix that is not reducible is called *irreducible*.

A matrix $A \in \mathbb{R}^{n,n}$ is called *monotone* (or *of monotone type*) if the relation $Ax \leq Ay$ for two (otherwise arbitrary) elements $x, y \in \mathbb{R}^n$ implies $x \leq y$. Here the relation sign is to be understood componentwise.

A matrix of monotone type is invertible.

A matrix $A \in \mathbb{R}^{n,n}$ is a matrix of monotone type if it is invertible and all entries of the inverse are nonnegative.

An important subclass of matrices of monotone type is formed by the so-called M-matrices.

A monotone matrix A with $a_{ij} \leq 0$ for $i \neq j$ is called an *M-matrix*.

Let $A \in \mathbb{R}^{n,n}$ be a matrix with $a_{ij} \leq 0$ for $i \neq j$ and $a_{ii} \geq 0$ $(i, j \in \{1, \ldots, n\})$. In addition, let A satisfy one of the following conditions:

1) A satisfies the strict row sum criterion.
2) A satisfies the weak row sum criterion and is irreducible.

Then A is an M-matrix.

A.4 Some Definitions and Arguments of Linear Functional Analysis

Working with vector spaces whose elements are (classical or generalized) functions, it is desirable to have a measure for the "length" or "magnitude" of a function, and, as a consequence, for the distance of two functions.

Let V be a real vector space (in short, an $\mathbb{R}$ vector space) and let $\|\cdot\|$ be a real-valued mapping $\|\cdot\| : V \to \mathbb{R}$.

The pair $(V, \|\cdot\|)$ is called a *normed space* ("V is endowed with the *norm* $\|\cdot\|$") if the following properties hold:

$$\|u\| \geq 0 \quad \text{for all } u \in V, \quad \|u\| = 0 \Leftrightarrow u = 0, \tag{A4.1}$$

$$\|\alpha u\| = |\alpha| \, \|u\| \quad \text{for all } \alpha \in \mathbb{R}, \ u \in V, \tag{A4.2}$$

$$\|u + v\| \leq \|u\| + \|v\| \quad \text{for all } u, v \in V. \tag{A4.3}$$

The property (A4.1) is called *definiteness*; (A4.3) is called the *triangle inequality*. If a mapping $\|\cdot\| : V \to \mathbb{R}$ satisfies only (A4.2) and (A4.3), it is called a *seminorm*. Due to (A4.2), we still have $\|0\| = 0$, but there may exist elements $u \neq 0$ with $\|u\| = 0$.

A particularly interesting example of a norm can be obtained if the space V is equipped with a so-called *scalar product*. This is a mapping $\langle \cdot, \cdot \rangle : V \times V \to \mathbb{R}$ with the following properties:

1) $\langle \cdot, \cdot \rangle$ is a *bilinear form*, that is,

$$\begin{aligned} \langle u, v_1 + v_2 \rangle &= \langle u, v_1 \rangle + \langle u, v_2 \rangle \quad &&\text{for all} \quad u, v_1, v_2 \in V, \\ \langle u, \alpha v \rangle &= \alpha \langle u, v \rangle &&\text{for all} \quad u, v \in V, \ \alpha \in \mathbb{R}, \end{aligned} \tag{A4.4}$$

and an analogous relation is valid for the first argument.

2) $\langle \cdot, \cdot \rangle$ is *symmetric*, that is,

$$\langle u, v \rangle = \langle v, u \rangle \quad \text{for all} \quad u, v \in V. \tag{A4.5}$$

3) $\langle \cdot, \cdot \rangle$ is *positive*, that is,

$$\langle u, u \rangle \geq 0 \quad \text{for all} \quad u \in V. \tag{A4.6}$$

4) $\langle\cdot,\cdot\rangle$ is *definite*, that is,

$$\langle u, u\rangle = 0 \Leftrightarrow u = 0\,. \tag{A4.7}$$

A positive and definite bilinear form is called *positive definite*.

A scalar product $\langle\cdot,\cdot\rangle$ defines a norm on V in a natural way if we set

$$\|v\| := \langle v, v\rangle^{1/2}\,. \tag{A4.8}$$

In absence of the definiteness (A4.7), only a seminorm is induced.

A norm (or a seminorm) induced by a scalar product (respectively, by a symmetric and positive bilinear form) has some interesting properties. For example, it satisfies the *Cauchy–Schwarz inequality*, that is,

$$|\langle u, v\rangle| \le \|u\|\,\|v\| \quad \text{for all } u, v \in V\,, \tag{A4.9}$$

and the *parallelogram identity*

$$\|u+v\|^2 + \|u-v\|^2 = 2(\|u\|^2 + \|v\|^2) \quad \text{for all } u, v \in V\,. \tag{A4.10}$$

Typical examples of normed spaces are the spaces $\mathbb{R}^n$ equipped with one of the ℓ^p-norms (for some fixed $p \in [1,\infty]$). In particular, the Euclidean norm (A3.3) is induced by the *Euclidean scalar product*

$$(x, y) \mapsto x \cdot y \quad \text{for all } x, y \in \mathbb{R}^n\,. \tag{A4.11}$$

On the other hand, infinite-dimensional function spaces play an important role (see Appendix A.5).

If a vector space V is equipped with a scalar product $\langle\cdot,\cdot\rangle$, then, in analogy to $\mathbb{R}^n$, an element $u \in V$ is said to be *orthogonal* to $v \in V$ if

$$\langle u, v\rangle = 0\,. \tag{A4.12}$$

For $G \subset V$, the set

$$G^\perp := \{u \in V \mid \langle u, v\rangle = 0 \text{ for all } v \in G\} \tag{A4.13}$$

is called the *orthogonal complement* of G.

Given a normed space $(V, \|\cdot\|)$, it is easy to define the concept of *convergence* of a sequence $(u_i)_i$ in V to $u \in V$:

$$u_i \to u \quad \text{for } i \to \infty \qquad \Longleftrightarrow \qquad \|u_i - u\| \to 0 \quad \text{for } i \to \infty\,. \tag{A4.14}$$

Often, it is necessary to consider function spaces endowed with different norms. In such situations, different kinds of convergence may occur. However, if the corresponding norms are equivalent, then there is no change in the type of convergence. Two norms $\|\cdot\|_1$ and $\|\cdot\|_2$ in V are called *equivalent* if there exist constants $C_1, C_2 > 0$ such that

$$C_1\|u\|_1 \le \|u\|_2 \le C_2\|u\|_1 \quad \text{for all } u \in V\,. \tag{A4.15}$$

If there is only a one-sided inequality of the form

$$\|u\|_2 \le C\|u\|_1 \quad \text{for all } u \in V \tag{A4.16}$$

with a constant $C > 0$, then the norm $\|\cdot\|_1$ is called *stronger* than the norm $\|\cdot\|_2$.

In a finite-dimensional vector space, all norms are equivalent. Examples can be found in Appendix A.3. In particular, it is important to observe that the constants may depend on the dimension n of the finite-dimensional vector space. This observation also indicates that in the case of infinite-dimensional vector spaces, the equivalence of two different norms cannot be expected, in general.

As a consequence of (A4.15), two equivalent norms $\|\cdot\|_1, \|\cdot\|_2$ in V yield the same type of convergence:

$$\begin{aligned} u_i \to u \text{ w.r.t. } \|\cdot\|_1 \ &\Leftrightarrow \|u_i - u\|_1 \to 0 \\ &\Leftrightarrow \|u_i - u\|_2 \to 0 \ \Leftrightarrow u_i \to u \text{ w.r.t. } \|\cdot\|_2 \,. \end{aligned} \tag{A4.17}$$

In this book, the finite-dimensional vector space $\mathbb{R}^n$ is used in two aspects: For $n = d$, it is the basic space of independent variables, and for $n = M$ or $n = m$ it represents the finite-dimensional trial space. In the first case, the equivalence of all norms can be used in all estimates without any side effects, whereas in the second case the aim is to obtain uniform estimates with respect to all M and m, and so the dependence of the equivalence constants on M and m has to be followed thoroughly.

Now we consider two normed spaces $(V, \|\cdot\|_V)$ and $(W, \|\cdot\|_W)$. A mapping $f : V \to W$ is called *continuous* in $v \in V$ if for all sequences $(v_i)_i$ in V with $v_i \to v$ for $i \to \infty$ we get

$$f(v_i) \to f(v) \quad \text{for} \quad i \to \infty \,.$$

Note that the first convergence is measured in $\|\cdot\|_V$ and the second one in $\|\cdot\|_W$. Hence a change of the norm may have an influence on the continuity. As in classical analysis, we can say that

$$\begin{aligned} &f \text{ is continuous in all } v \in V \iff \\ &f^{-1}[G] \text{ is closed for each closed } G \subset W \,. \end{aligned} \tag{A4.18}$$

Here, a subset $G \subset W$ of a normed space W is called *closed* if for any sequence $(u_i)_i$ from G such that $u_i \to u$ for $i \to \infty$ the inclusion $u \in G$ follows. Because of (A4.18), the closedness of a set can be verified by showing that it is a continuous preimage of a closed set. Since open sets are the complements of closed sets, the statement (A4.18) can be expressed equivalently as

$$\begin{aligned} &f \text{ is continuous in all } v \in V \iff \\ &f^{-1}[G] \text{ is open for each open } G \subset W \,. \end{aligned} \tag{A4.19}$$

The concept of continuity is a qualitative relation between the preimage and the image. A quantitative relation is given by the stronger notion of Lipschitz continuity:

A mapping $f: V \to W$ is called *Lipschitz continuous* if there exists a constant $L > 0$, the *Lipschitz constant*, such that

$$\|f(u) - f(v)\|_W \leq L\|u - v\|_V \quad \text{for all } u, v \in V\,. \tag{A4.20}$$

A Lipschitz continuous mapping with $L < 1$ is called *contractive* or a *contraction*; cf. Figure A.1.

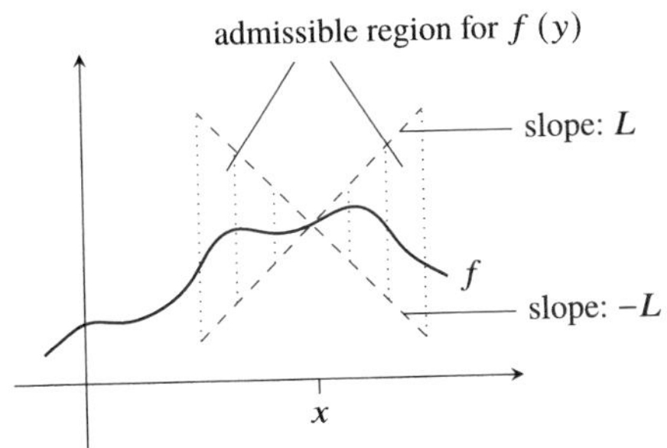

Fig. A.1: Lipschitz continuity (for $V = W = \mathbb{R}$).

Most of the mappings used are *linear*, that is, they satisfy

$$\left.\begin{aligned} f(u+v) &= f(u) + f(v)\,, \\ f(\lambda u) &= \lambda f(u)\,, \end{aligned}\right\} \quad \text{for all } u, v \in V \text{ and } \lambda \in \mathbb{R}\,. \tag{A4.21}$$

For a linear mapping, the Lipschitz continuity is equivalent to the *boundedness*, that is, there exists a constant $C > 0$ such that

$$\|f(u)\|_W \leq C\|u\|_V \quad \text{for all } u \in V\,. \tag{A4.22}$$

In fact, for a linear mapping f, the continuity at one point is equivalent to (A4.22). Linear, continuous mappings acting from V to W are also called (linear, continuous) *operators* and are denoted by capital letters, for example, $S, T, \ldots$.

In the case $V = W = \mathbb{R}^n$, the linear, continuous operators in $\mathbb{R}^n$ are the mappings $x \mapsto Ax$ defined by matrices $A \in \mathbb{R}^{n,n}$. Their boundedness, for example, with respect to $\|\cdot\|_V = \|\cdot\|_W = \|\cdot\|_\infty$ is an immediate consequence of the compatibility property (A3.9) of the $\|\cdot\|_\infty$-norm. Moreover, since all norms in $\mathbb{R}^n$ are equivalent, these mappings are bounded with respect to any norms in $\mathbb{R}^n$.

Similar to (A4.22), a bilinear form $f: V \times V \to \mathbb{R}$ is continuous if it is *bounded*, that is, if there exists a constant $C > 0$ such that

$$|f(u, v)| \leq C\,\|u\|_V\|v\|_V \quad \text{for all } u, v \in V\,. \tag{A4.23}$$

In particular, due to (A4.9) any scalar product is continuous with respect to the induced norm of V, that is,

$$u_i \to u\,,\; v_i \to v \quad \Rightarrow \quad \langle u_i, v_i \rangle \to \langle u, v \rangle\,. \tag{A4.24}$$

Now let $(V, \|\cdot\|_V)$ be a normed space and W a subspace that is (additionally to $\|\cdot\|_V$) endowed with the norm $\|\cdot\|_W$. The *embedding* from $(W, \|\cdot\|_W)$ to $(V, \|\cdot\|_V)$, i.e., the linear mapping that assigns any element of W to itself but considered as an element of V, is continuous iff the norm $\|\cdot\|_W$ is stronger than the norm $\|\cdot\|_V$ (cf. (A4.16)).

If $(V, \|\cdot\|_V)$ is a normed space and W a linear subspace, the *equivalence class* of an element $u \in V$ is the set $[u] := \{v \in V \mid u - v \in W\}$. The *quotient space* is the set of equivalence classes of elements of V, i.e., $V/W := \{[u] \mid u \in V\}$. The quotient space can be equipped with the norm $\|[u]\|_{V/W} := \inf\{\|v\|_V \mid v \in [u]\}$. The use of quotient spaces makes it possible to identify different elements of V that belong to the same equivalence class.

The collection of linear, continuous operators from $(V, \|\cdot\|_V)$ to $(W, \|\cdot\|_W)$ forms an $\mathbb{R}$ vector space with the following (argumentwise) operations:

$$\begin{aligned}
(T + S)(u) &:= T(u) + S(u) \text{ for all } u \in V\,,\\
(\lambda T)(u) &:= \lambda T(u) \qquad\quad \text{for all } u \in V\,,
\end{aligned}$$

for all operators T, S, and $\lambda \in \mathbb{R}$. This space is denoted by

$$L[V, W]\,. \tag{A4.25}$$

In the special case $W = \mathbb{R}$, the corresponding operators are called linear, continuous *functionals*, and the notation

$$V' := L[V, \mathbb{R}] \tag{A4.26}$$

is used. The $\mathbb{R}$ vector space $L[V, W]$ can be equipped with a norm, the so-called *operator norm*, by

$$\|T\| := \sup\left\{\|T(u)\|_W \;\middle|\; u \in V,\; \|u\|_V \le 1\right\} \quad \text{for} \quad T \in L[V, W]\,. \tag{A4.27}$$

Here $\|T\|$ is the smallest constant such that (A4.22) holds. Specifically, for a functional $f \in V'$, we have that

$$\|f\| = \sup\left\{|f(u)| \;\middle|\; \|u\|_V \le 1\right\}\,.$$

For example, in the case $V = W = \mathbb{R}^n$ and $\|u\|_V = \|u\|_W$, the norm of a linear, bounded operator that is represented by a matrix $A \in \mathbb{R}^{n,n}$ coincides with the corresponding induced matrix norm (cf. Appendix A.3).

Let $(V, \|\cdot\|_V)$ be a normed space. A sequence $(u_i)_i$ in V is called a *Cauchy sequence* if for any $\varepsilon > 0$ there exists a number $n_0 \in \mathbb{N}$ such that

$$\|u_i - u_j\|_V \le \varepsilon \quad \text{for all } i, j \in \mathbb{N} \text{ with } i, j \ge n_0\,.$$

The space V is called *complete* or a *Banach space* if for any Cauchy sequence $(u_i)_i$ in V there exists an element $u \in V$ such that $u_i \to u$ for $i \to \infty$. If the norm $\|\cdot\|_V$ of a Banach space V is induced by a scalar product, then V is called a *Hilbert space*.

A subspace W of a Banach space is complete iff it is closed. A basic problem in the variational treatment of boundary value problems consists in the fact that the space of continuous functions (cf. the preliminary definition (2.7)), which is required to be taken as a basis, is not complete with respect to the norm ($\|\cdot\|_l$, $l = 0$ or $l = 1$). However, if in addition to the normed space $(W, \|\cdot\|)$, a larger space V is given that is complete with respect to the norm $\|\cdot\|$, then that space or the closure

$$\widetilde{W} := \overline{W} \tag{A4.28}$$

(as the smallest Banach space containing W) can be used. Such a *completion* can be introduced for any normed space in an abstract way. The problem is that the "nature" of the limit elements remains vague.

If relation (A4.28) is valid for some normed space W, then W is called *dense* in $\widetilde{W}$. In fact, given W, all "essential" elements of $\widetilde{W}$ are already captured. For example, if T is a linear, continuous operator T from $(\widetilde{W}, \|\cdot\|)$ to another normed space, then the identity

$$T(u) = 0 \quad \text{for all} \quad u \in W \tag{A4.29}$$

is sufficient for

$$T(u) = 0 \quad \text{for all} \quad u \in \widetilde{W}\,. \tag{A4.30}$$

The space of linear, bounded operators is complete if the image space is complete. In particular, the space V' of linear, bounded functionals on the normed space V is always complete.

To formulate the next result, the following terms are required, which extend definition (A4.13) to the case of normed spaces V.

The *annihilator* of a subset $G \subset V$ in V' is defined by

$$G^\perp := \{f \in V' \mid f(v) = 0 \text{ for all } v \in G\}, \tag{A4.31}$$

and the *pre-annihilator* of $Q \subset V'$ in V is given by

$$Q_\perp := \{v \in V \mid f(v) = 0 \text{ for all } f \in Q\}.$$

Given two Banach spaces V, W and $A \in L[V, W]$, the *Closed Range Theorem* tells that the following statements are equivalent:

(i) $\operatorname{im}(A)$ is closed.
(ii) $\operatorname{im}(A) = \left(\ker(A^T)\right)_\perp$.
(iii) $\operatorname{im}(A^T)$ is closed.
(iv) $\operatorname{im}(A^T) = \left(\ker(A)\right)^\perp$,

where the *dual operator* $A^T : W' \to V'$ is defined by $A^T\psi := \psi \circ A$ for all $\psi \in W'$.

For a proof we refer to [67, Sect. VII.5].

If in addition to the assumptions of the Closed Range Theorem the operator $T \in L[V, W]$ is surjective, the *Open Mapping Theorem* states that T maps every open set $G \subset V$ onto an open set of W.

A proof can be found in [67, Sect. II.5].

A.5 Function Spaces

In this section $G \subset \mathbb{R}^d$ denotes a bounded domain.

The function space $C(G)$ contains all (real-valued) functions defined on G that are continuous in G. By $C^l(G)$, $l \in \mathbb{N}$, the set of l-times continuously differentiable functions on G is denoted. Usually, for the sake of consistency, the conventions $C^0(G) := C(G)$ and $C^\infty(G) := \bigcap_{l=0}^\infty C^l(G)$ are used.

Functions from $C^l(G)$, $l \in \mathbb{N}_0$, and $C^\infty(G)$ need not be bounded, as for $d = 1$ the example $f(x) := x^{-1}$, $x \in (0, 1)$ shows.

To overcome this difficulty, further spaces of continuous functions are introduced. The space $C(\overline{G})$ contains all bounded and uniformly continuous functions on G, whereas $C^l(\overline{G})$, $l \in \mathbb{N}$, consists of functions with bounded and uniformly continuous derivatives up to order l on G. Here the conventions $C^0(\overline{G}) := C(\overline{G})$ and $C^\infty(\overline{G}) := \bigcap_{l=0}^\infty C^l(\overline{G})$ are used, too.

The space $C_0(G)$, respectively $C_0^l(G)$, $l \in \mathbb{N}$, denotes the set of all those continuous, respectively, l-times continuously differentiable, functions, the supports of which are contained in G. Often this set is called the set of functions with compact support in G. Since G is bounded, this means that the supports do not intersect boundary points of G. We also set $C_0^0(G) := C_0(G)$ and $C_0^\infty(G) := C_0(G) \cap C^\infty(G)$.

The linear space $L^p(G)$, $p \in [1, \infty)$, contains all Lebesgue-measurable functions defined on G whose pth power of their absolute value is Lebesgue integrable on G. The norm in $L^p(G)$ is defined as follows:

$$\|u\|_{0,p,G} := \left\{ \int_G |u|^p \, dx \right\}^{1/p}, \quad p \in [1, \infty) .$$

In the case $p = 2$, the specification of p is frequently omitted, that is, $\|u\|_{0,G} = \|u\|_{0,2,G}$. The $L^2(G)$-scalar product

$$\langle u, v \rangle_{0,G} := \int_G uv \, dx , \quad u, v \in L^2(G) ,$$

induces the $L^2(G)$-norm by setting $\|u\|_{0,G} := \sqrt{\langle u, u \rangle_{0,G}}$.

The space $L^\infty(G)$ contains all measurable, essentially bounded functions on G, where a function $u : G \to \mathbb{R}$ is called *essentially bounded* if the quantity

$$\|u\|_{\infty,G} := \operatorname*{ess\,sup}_G |u| := \inf_{G_0 \subset G: \, |G_0|_d = 0} \; \sup_{x \in G \setminus G_0} |u(x)|$$

is finite. For continuous functions, this norm coincides with the usual maximum norm:

$$\|u\|_{\infty,G} = \max_{x\in\overline{G}} |u(x)| , \quad u \in C(\overline{G}) .$$

For $1 \leq q \leq p \leq \infty$, we have $L^p(G) \subset L^q(G)$, and the embedding is continuous.

The space $W_p^l(G)$, $l \in \mathbb{N}$, $p \in [1, \infty]$, consists of all l-times weakly differentiable functions from $L_p(G)$ with derivatives in $L^p(G)$. In the special case $p = 2$, we also write $H^l(G) := W_2^l(G)$. In analogy to the case of continuous functions, the convention $H^0(G) := L^2(G)$ is used. The norm in $W_p^l(G)$ is defined as follows:

$$\|u\|_{l,p,G} := \left\{ \sum_{|\alpha|\leq l} \int_G |\partial^\alpha u|^p \, dx \right\}^{1/p} , \quad p \in [1, \infty) ,$$

$$\|u\|_{l,\infty,G} := \max_{|\alpha|\leq l} |\partial^\alpha u|_{\infty,G} .$$

In $H^l(G)$ a scalar product can be defined by

$$\langle u, v \rangle_{l,G} := \sum_{|\alpha|\leq l} \int_G \partial^\alpha u \partial^\alpha v \, dx , \quad u, v \in H^l(G) .$$

The norm induced by this scalar product is denoted by $\| \cdot \|_{l,G}$, $l \in \mathbb{N}$:

$$\|u\|_{l,G} := \sqrt{\langle u, u \rangle_{l,G}} .$$

For $l \in \mathbb{N}$, the symbol $| \cdot |_{l,G}$ stands for the corresponding $H^l(G)$-*seminorm*:

$$|u|_{l,G} := \sqrt{\sum_{|\alpha|=l} \int_G |\partial^\alpha u|^2 \, dx} .$$

The space $H_0^1(G)$ is defined as the closure (or completion) of $C_0^\infty(G)$ in the norm $\| \cdot \|_1$ of $H^1(G)$.

Convention: Usually, in the case $G = \Omega$ the specification of the domain in the above norms and scalar products is omitted.

In the study of partial differential equations, it is often desirable to speak of boundary values of functions defined on the domain G. In this respect, the Lebesgue spaces of functions that are square integrable at the boundary of G are important. To introduce these spaces, some preparations are necessary.

In what follows, a point $x \in \mathbb{R}^d$ is written in the form $x = \binom{x'}{x_d}$ with $x' = (x_1, \ldots, x_{d-1})^T \in \mathbb{R}^{d-1}$.

A domain $G \subset \mathbb{R}^d$ is said to be *located at one side of* ∂G if for any $x \in \partial G$ there exist an open neighbourhood $U_x \subset \mathbb{R}^d$ and an orthogonal mapping Q_x in $\mathbb{R}^d$ such that the point x is mapped to a point $\hat{x} = (\hat{x}_1, \ldots, \hat{x}_d)^T$, and so U_x is mapped onto a neighbourhood $U_{\hat{x}} \subset \mathbb{R}^d$ of $\hat{x}$, where in the neighbourhood $U_{\hat{x}}$ the following properties hold:

1) The image of $U_x \cap \partial G$ is the graph of some function $\Psi_x : Y_x \subset \mathbb{R}^{d-1} \to \mathbb{R}$, that is, $\hat{x}_d = \Psi_x(\hat{x}_1, \ldots, \hat{x}_{d-1}) = \Psi_x(\hat{x}')$ for $\hat{x}' \in Y_x$.
2) The image of $U_x \cap G$ is "above this graph" (i.e., the points in $U_x \cap G$ correspond to $\hat{x}_d > 0$).
3) The image of $U_x \cap (\mathbb{R}^d \setminus \overline{G})$ is "below this graph" (i.e., the points in $U_x \cap (\mathbb{R}^d \setminus \overline{G})$ correspond to $\hat{x}_d < 0$).

A domain G that is located at one side of ∂G is called a C^l *domain*, $l \in \mathbb{N}$, respectively, a *Lipschitz(ian) domain*, if all Ψ_x are l-times continuously differentiable, respectively, Lipschitz continuous, in Y_x.

Bounded Lipschitz domains are also called *strongly Lipschitz*.

For bounded domains located at one side of ∂G, it is well known (cf., e.g., [66]) that from the whole set of neighbourhoods $\{U_x\}_{x \in \partial G}$ there can be selected a family $\{U_i\}_{i=1}^n$ of finitely many neighbourhoods covering ∂G, i.e., $n \in \mathbb{N}$ and $\partial G \subset \bigcup_{i=1}^n U_i$. Furthermore, for any such family there exists a system of functions $\{\varphi_i\}_{i=1}^n$ with the properties $\varphi_i \in C_0^\infty(U_i)$, $\varphi_i(x) \in [0,1]$ for all $x \in U_i$ and $\sum_{i=1}^n \varphi_i(x) = 1$ for all $x \in \partial G$. Such a system is called a *partition of unity*.

If the domain G is at least Lipschitzian, then Lebesgue's integral over the boundary of G is defined by means of those partitions of unity. In correspondence to the definition of a Lipschitz domain, Q_i, Ψ_i, and Y_i denote the orthogonal mapping on U_i, the function describing the corresponding local boundary, and the preimage of $Q_i(U_i \cap \partial G)$ with respect to Ψ_i.

A function $v : \partial G \to \mathbb{R}$ is called *Lebesgue integrable over* ∂G if the composite functions $\hat{x}' \mapsto v\left(Q_i^T \begin{pmatrix} \hat{x}' \\ \Psi_i(\hat{x}') \end{pmatrix}\right)$ belong to $L^1(Y_i)$. The integral is defined as follows:

$$\begin{aligned}
\int_{\partial G} v(s)\,ds &:= \sum_{i=1}^n \int_{\partial G} v(s)\varphi_i(s)\,ds \\
&:= \sum_{i=1}^n \int_{Y_i} v\left(Q_i^T \begin{pmatrix} \hat{x}' \\ \Psi_i(\hat{x}') \end{pmatrix}\right) \varphi_i\left(Q_i^T \begin{pmatrix} \hat{x}' \\ \Psi_i(\hat{x}') \end{pmatrix}\right) \\
&\qquad \times \sqrt{|\det(\partial_j \Psi_i(\hat{x}')\partial_k \Psi_i(\hat{x}'))_{j,k=1}^{d-1}|}\,d\hat{x}' .
\end{aligned}$$

A function $v : \partial G \to \mathbb{R}$ belongs to $L^2(\partial G)$ iff both v and v^2 are Lebesgue integrable over ∂G.

In the investigation of time-dependent partial differential equations, linear spaces whose elements are functions of the time variable $t \in [0,T]$, $T > 0$, with values in a normed space X are of interest.

A function $v : [0,T] \to X$ is called *continuous on* $[0,T]$ if for all $t \in [0,T]$ the convergence $\|v(t+k) - v(t)\|_X \to 0$ as $k \to 0$ holds.

The space $C([0,T],X) = C^0([0,T],X)$ consists of all continuous functions $v : [0,T] \to X$ such that

$$\sup_{t \in (0,T)} \|v(t)\|_X < \infty .$$

The space $C^l([0, T], X)$, $l \in \mathbb{N}$, consists of all continuous functions $v : [0, T] \to X$ that have continuous derivatives up to order l on $[0, T]$ with the norm

$$\sum_{i=0}^{l} \sup_{t \in (0,T)} \|v^{(i)}(t)\|_X .$$

The space $L^p((0, T), X)$ with $1 \le p \le \infty$ consists of all functions on $(0, T) \times \Omega$ for which

$$v(t, \cdot) \in X \text{ for any } t \in (0, T), \quad F \in L^p(0, T) \quad \text{with } F(t) := \|v(t, \cdot)\|_X .$$

Furthermore,

$$\|v\|_{L^p((0,T),X)} := \|F\|_{L^p(0,T)} .$$

References: Textbooks and Monographs

1. R.A. Adams, J.J.F. Fournier, *Sobolev Spaces*, Pure and Applied Mathematics, vol. 140, 2nd edn. (Elsevier/Academic Press, Amsterdam, 2003), Publisher's website
2. M. Ainsworth, J.T. Oden, *A Posteriori Error Estimation in Finite Element Analysis* (Wiley, New York, 2000). https://doi.org/10.1002/9781118032824
3. J. Appel, P.P. Zabrejko, *Nonlinear Superposition Operators*. Cambridge Tracts in Mathematics, 95 (Cambridge University Press, Cambridge-New York, 1990). https://doi.org/10.1017/CBO9780511897450
4. O. Axelsson, V.A. Barker, *Finite Element Solution of Boundary Value Problems: Theory and Computation* (Academic Press, Orlando, 1984). https://doi.org/10.1137/1.9780898719253
5. R.E. Bank, PLTMG: A Software Package for Solving Elliptic Partial Differential Equations: Users' Guide 13.0. Department of Mathematics, University of California at San Diego, La Jolla, CA (2018), http://netlib.org/pltmg/guide.pdf
6. A. Berman, R.J. Plemmons, *Nonnegative Matrices in the Mathematical Sciences* (SIAM, Philadelphia, 1994) (corrected republication of the work first published in 1979 by Academic Press, San Diego). https://doi.org/10.1137/1.9781611971262
7. D. Boffi, F. Brezzi, M. Fortin, *Mixed Finite Element Methods and Applications* (Springer, New York, 2013). https://doi.org/10.1007/978-3-642-36519-5
8. H. Borouchaki, P.L. George, *Meshing, Geometric Modeling and Numerical Simulation*, vol. 1 (ISTE/Wiley, London/Hoboken, 2017). https://doi.org/10.1002/9781119384335. P.L. George, H. Borouchaki, F. Alauzet, P. Laug, A. Loseille, L. Maréchal, *Meshing, Geometric Modeling and Numerical Simulation*, vol. 2 (ISTE/Wiley, London/Hoboken, 2019). https://doi.org/10.1002/9781119384380
9. D. Braess, *Finite Elements: Theory, Fast Solvers, and Applications in Solid Mechanics*, 3rd edn. (Cambridge University Press, Cambridge, 2007). https://doi.org/10.1017/CBO9780511618635
10. S.C. Brenner, L.R. Scott, *The Mathematical Theory of Finite Element Methods*, Texts in Applied Mathematics, vol. 15, 3rd edn. (Springer, New York, 2008). https://doi.org/10.1007/978-0-387-75934-0
11. V.I. Burenkov, *Sobolev Spaces on Domains* (Teubner, Stuttgart, 1998). https://doi.org/10.1007/978-3-663-11374-4
12. G.F. Carey, *Computational Grids: Generation, Adaption and Solution Strategies* (Taylor& Francis, Washington-London, 1997), Publisher's website
13. Z. Chen, *Finite Element Methods and Their Applications* (Springer, Heidelberg, 2005). https://doi.org/10.1007/3-540-28078-2

P. Knabner and L. Angermann, *Numerical Methods for Elliptic and Parabolic Partial Differential Equations*, Texts in Applied Mathematics 44,
https://doi.org/10.1007/978-3-030-79385-2_A

14. S..-W. Cheng, T..K. Dey, J.R. Shewchuk, *Delaunay Mesh Generation* (CRC Press, Boca Raton, 2013). https://doi.org/10.1201/b12987
15. A.J. Chorin, J.E. Marsden, *A Mathematical Introduction to Fluid Mechanics* (Springer, New York, 1993). https://doi.org/10.1007/978-1-4612-0883-9
16. P.G. Ciarlet, Basic error estimates for elliptic problems, in *Handbook of Numerical Analysis, Volume II: Finite Element Methods (Part 1)*, ed. by P.G. Ciarlet, J.L. Lions (North-Holland, Amsterdam, 1991). https://doi.org/10.1016/S1570-8659(05)80039-0
17. P.G. Ciarlet, *Linear and Nonlinear Functional Analysis with Applications* (SIAM, Philadelphia, 2015), Publisher's website
18. R. Dautray, J.-L. Lions, *Mathematical Analysis and Numerical Methods For Science and Technology. Volume 3: Spectral Theory and Applications* (Springer, New York, 2000). https://doi.org/10.1007/978-3-642-61529-0
19. R. Dautray, J.-L. Lions, *Mathematical Analysis and Numerical Methods for Science and Technology. Volume 4: Integral Equations and Numerical Methods* (Springer, New York, 2000). https://doi.org/10.1007/978-3-642-61531-3
20. P. Deuflhard, *Newton Methods for Nonlinear Problems*, Springer Series in Computational Mathematics, vol. 35 (Springer, Berlin, 2011). https://doi.org/10.1007/978-3-642-23899-4
21. D.A. Di Pietro, A. Ern, *Mathematical Aspects of Discontinuous Galerkin Methods* (Springer, Berlin, 2012). https://doi.org/10.1007/978-3-642-22980-0
22. V. Dolejši, M. Feistauer, *Discontinuous Galerkin Method* (Springer International Publishing, 2015). https://doi.org/10.1007/978-3-319-19267-3
23. C. Eck, H. Garcke, P. Knabner, *Mathematical Modeling* (Springer International Publishing, 2017). https://doi.org/10.1007/978-3-319-55161-6
24. A. Ern, J.-L. Guermond, *Theory and Practice of Finite Elements* (Springer, New York, 2004). https://doi.org/10.1007/978-1-4757-4355-5
25. L.C. Evans, *Partial Differential Equations*, 2nd edn. (American Mathematical Society, Providence, 2010). https://doi.org/10.1090/gsm/019
26. P.J. Frey, P.L. George, *Mesh Generation*, 2nd edn. (ISTE, Wiley, London/Hoboken, 2008). https://doi.org/10.1002/9780470611166
27. P.L. George, H. Borouchaki, *Delaunay Triangulation and Meshing* (Hermes, Paris, 1998). Translated from the 1997 French original by the authors, J.P. Frey, S.A. Canann
28. D. Gilbarg, N.S. Trudinger, *Elliptic Partial Differential Equations of Second Order* (Springer, New York, 2001) (reprint of the 1998 edn.). https://doi.org/10.1007/978-3-642-61798-0
29. V. Girault, P.-A. Raviart, *Finite Element Methods for Navier-Stokes Equations* (Springer, New York, 1986). https://doi.org/10.1007/978-3-642-61623-5
30. A. Greenbaum, *Iterative Methods for Solving Linear Systems*, Society for Industrial and Applied Mathematics, Philadelphia, Pa., 1997. https://doi.org/10.1137/1.9781611970937
31. P. Grisvard, *Elliptic Problems in Nonsmooth Domains* (SIAM, Philadelphia, 2002). (reprint of the 1985 edn.). https://doi.org/10.1137/1.9781611972030
32. W. Hackbusch, *Elliptic Differential Equations. Theory and Numerical Treatment*, 2nd edn. (Springer, New York, 2017). https://doi.org/10.1007/978-3-662-54961-2
33. W. Hackbusch, *Iterative Solution of Large Sparse Systems of Equations*, 2nd edn. (Springer, Cham, 2016). https://doi.org/10.1007/978-3-319-28483-5
34. W. Hackbusch, *Multi-Grid Methods and Applications* (Springer, New York, 2003) (2nd printing of the 1985 ed.). https://doi.org/10.1007/978-3-662-02427-0
35. L.A. Hageman, D.M. Young, *Applied Iterative Methods* (Academic Press, New York, 1981). https://doi.org/10.1016/C2009-0-21990-8
36. E. Hairer, G. Wanner, *Solving Ordinary Differential Equations II. Stiff and Differential-Algebraic Problems*, 2nd edn. (Springer, Berlin, 1996). https://doi.org/10.1007/978-3-642-05221-7

37. U. Hornung (ed.), *Homogenization and Porous Media* (Springer, New York, 1997). https://doi.org/10.1007/978-1-4612-1920-0
38. W. Hundsdorfer, J.G. Verwer, *Numerical Solution of Time-Dependent Advection-Diffusion-Reaction Equations*, Springer Series in Computational Mathematics, vol. 33 (Springer, New York, 2003). https://doi.org/10.1007/978-3-662-09017-6
39. T. Ikeda, *Maximum Principle in Finite Element Models for Convection–Diffusion Phenomena* (North-Holland, Amsterdam, 1983). https://doi.org/10.1016/s0304-0208(08)x7173-7
40. C. Johnson, *Numerical Solution of Partial Differential Equations by the Finite Element Method* (Cambridge University Press, Cambridge, 1987). https://doi.org/10.1007/BF00046566
41. C.T. Kelley, *Iterative Methods for Linear and Nonlinear Equations* (SIAM, Philadelphia, 1995). https://doi.org/10.1137/1.9781611970944
42. D. Kuzmin, *A Guide to Numerical Methods for Transport Equations*. Free CFD Book, Friedrich-Alexander-Universität, Erlangen-Nürnberg, 2010
43. D.S.H. Lo, *Finite Element Mesh Generation* (CRC Press, Boca Raton, 2015). https://doi.org/10.1201/b17713
44. J.D. Logan, *Transport Modeling in Hydrogeochemical Systems* (Springer, New York, 2001). https://doi.org/10.1007/978-1-4757-3518-5
45. W. McLean, *Strongly Elliptic Systems and Boundary Integral Equations* (Cambridge University Press, Cambridge, 2000). https://doi.org/10.2307/3621632
46. J. Nečas, *Les Méthodes Directes En Théorie Des Équations Elliptiques* (Masson/Academia, Paris/Prague, 1967)
47. M.A. Olshanskii, E.E. Tyrtyshnikov, *Iterative Methods for Linear Systems* (SIAM, Philadelphia, 2014). https://doi.org/10.1137/1.9781611973464
48. B.G. Pachpatte, *Inequalities for Differential and Integral Equations* (Academic Press, San Diego, 1998), Publisher's website
49. A. Quarteroni, A. Valli, *Domain Decomposition Methods for Partial Differential Equations* (Oxford University Press, Oxford, 1999), Publisher's website
50. R. Rannacher, *Numerik partieller Differentialgleichungen* (Heidelberg University Publishing, Heidelberg, 2017). https://doi.org/10.17885/heiup.281.370
51. M. Renardy, R.C. Rogers, *An Introduction to Partial Differential Equations*, 2nd edn. (Springer, New York, 2004). https://doi.org/10.1007/b97427
52. B. Riviere, *Discontinuous Galerkin Methods for Solving Elliptic and Parabolic Equations: Theory and Implementation* (Society for Industrial and Applied Mathematics, Philadelphia, 2008). https://doi.org/10.1137/1.9780898717440
53. H.-G. Roos, M. Stynes, L. Tobiska, *Robust Numerical Methods for Singularly Perturbed Differential Equations*, Springer Series in Computational Mathematics, vol. 24, 2nd edn. (Springer, New York, 2008). https://doi.org/10.1007/978-3-540-34467-4
54. Y. Saad, *Iterative Methods for Sparse Linear Systems*, 2nd edn. (SIAM, Philadelphia, 2003). https://doi.org/10.1137/1.9780898718003
55. D.H. Sattinger, *Topics in Stability and Bifurcation Theory* (Springer, New York, 1973). https://doi.org/10.1007/BFb0060079
56. J. Stoer, *Introduction to Numerical Analysis*, 3rd edn. (Springer, New York, 2002). https://doi.org/10.1007/978-0-387-21738-3
57. G. Strang, G.J. Fix, *An Analysis of the Finite Element Method. New Edition*, 2nd edn. (Wellesley-Cambridge Press, Wellesley, 2008), Publisher's website
58. J.C. Strikwerda, *Finite Difference Schemes and Partial Differential Equations*, 2nd edn. (SIAM, Philadelphia, 2004). https://doi.org/10.1137/1.9780898717938
59. V. Thomeé, *Galerkin Finite Element Methods for Parabolic Problems*, 2nd edn. (Springer, Berlin, 2006). https://doi.org/10.1007/3-540-33122-0

60. J.F. Thompson, Z.U.A. Warsi, C.W. Mastin, *Numerical Grid Generation: Foundations and Applications* (North-Holland, Amsterdam, 1985)
61. R.S. Varga, *Matrix Iterative Analysis* (Springer, New York, 2000). https://doi.org/10.1007/978-3-642-05156-2
62. J.L. Vázquez, *The Porous Medium Equation. Mathematical Theory* (Clarendon Press, Oxford, 2007). https://doi.org/10.1093/acprof:oso/9780198569039.001.0001
63. R. Verfürth, *A Posteriori Error Estimation Techniques for Finite Element Methods* (University Press, Oxford, 2013). https://doi.org/10.1093/acprof:oso/9780199679423.001.0001
64. W. Walter, *Ordinary Differential Equations*, Graduate Texts in Mathematics, vol. 182 (Springer, New York, 1998). https://doi.org/10.1007/978-1-4612-0601-9
65. S. Whitaker, *The Method of Volume Averaging* (Kluwer Academic Publishers, Dordrecht, 1998). https://doi.org/10.1007/978-94-017-3389-2
66. J. Wloka, *Partial Differential Equations* (Cambridge University Press, New York, 1987). https://doi.org/10.1017/CBO9781139171755
67. K. Yosida, *Functional Analysis* (Springer, Berlin, 1994) (reprint of the 6th edn.). https://doi.org/10.1007/978-3-642-61859-8
68. D.M. Young, *Iterative Solution of Large Linear Systems* (Academic Press, New York, 1971). https://doi.org/10.1016/C2013-0-11733-3
69. E. Zeidler, *Nonlinear Functional Analysis and Its Applications. II/A: Linear Monotone Operators* (Springer, New York, 1990). https://doi.org/10.1007/978-1-4612-0985-0

References: Journal Papers and Other Resources

70. I. Aavatsmark, G.T. Eigestad, R.A. Klausen, M.F. Wheeler, I. Yotov, Convergence of a symmetric MPFA method on quadrilateral grids. Comput. Geosci. **11**(4), 333–345 (2007). https://doi.org/10.1007/s10596-007-9056-8
71. L. Agélas, C. Guichard, R. Masson, Convergence of finite volume MPFA O type schemes for heterogeneous anisotropic diffusion problems on general meshes. Int. J. Finite Vol. **7**(2), 1–34 (2010). https://doi.org/10.1016/j.crma.2008.07.015
72. V. Aizinger, A geometry independent slope limiter for the discontinuous Galerkin method, in *Computational Science and High Performance Computing IV*, ed. by E. Krause, Y. Shokin, M. Resch, D. Kröner, N. Shokina (Springer, Berlin, 2011), pp. 207–217. https://doi.org/10.1007/978-3-642-17770-5_16
73. M. Allkämper, F. Gaspoz, R. Klöfkorn, A weak compatibility condition for Newest Vertex Bisection in any dimension. SIAM J. Sci. Comput. **40**(6), A3853–A3872 (2018). https://doi.org/10.1137/17M1156137
74. M.S. Alnæs, J. Blechta, J. Hake, et al., The FEniCS project version 1.5. Arch. Numer. Softw. **3**(100), 9–23 (2015). https://doi.org/10.11588/ans.2015.100.20553, https://fenicsproject.org/
75. R. Anderson, J. Andrej, A. Barker et al., MFEM: A modular finite element methods library. Comput. Math. Appl. **81**, 42–74 (2021). https://doi.org/10.1016/j.camwa.2020.06.009, https://mfem.org/
76. L. Angermann, An a-posteriori estimation for the solution of elliptic boundary value problems by means of upwind FEM. IMA J. Numer. Anal. **12**, 201–215 (1992). https://doi.org/10.1093/imanum/12.2.201
77. L. Angermann, Error estimates for the finite-element solution of an elliptic singularly perturbed problem. IMA J. Numer. Anal. **15**, 161–196 (1995). https://doi.org/10.1093/imanum/15.2.161
78. L. Angermann, Error analysis of upwind-discretizations for the steady-state incompressible Navier-Stokes equations. Adv. Comput. Math. **13**, 167–198 (2000). https://doi.org/10.1023/A:1018902210295
79. L. Angermann, Node-centered finite volume schemes and primal-dual mixed formulations. Commun. Appl. Anal. **7**(4), 529–566 (2003)
80. L. Angermann, A posteriori estimates for errors of functionals on finite volume approximations to solutions of elliptic boundary-value problems, in *Inverse Problems and Large-Scale Computations*, ed. by L. Beilina, Y.V. Shestopalov (Springer, New York, 2013), pp. 57–68. Springer Proceedings in Mathematics & Statistics, vol. 52. https://doi.org/10.1007/978-3-319-00660-4_5

P. Knabner and L. Angermann, *Numerical Methods for Elliptic and Parabolic Partial Differential Equations*, Texts in Applied Mathematics 44,
https://doi.org/10.1007/978-3-030-79385-2

81. T. Apel, M. Dobrowolski, Anisotropic interpolation with applications to the finite element method. Computing **47**, 277–293 (1992). https://doi.org/10.1007/BF02320197
82. D. Arndt, W. Bangerth, W. Blais, et al., The `deal.II` library, version 9.2. J. Numer. Math. **28**(3), 131–146 (2020). https://doi.org/10.1515/jnma-2020-0043, https://dealii.org/
83. D.C. Arney, J.E. Flaherty, An adaptive mesh-moving and local refinement method for time-dependent partial differential equations. ACM Trans. Math. Softw. **16**(1), 48–71 (1990). https://doi.org/10.1145/77626.77631
84. D.N. Arnold, A. Mukherjee, L. Pouly, Locally adapted tetrahedral meshes using bisection. SIAM J. Sci. Comput. **22**(2), 431–448 (2000). https://doi.org/10.1137/S1064827597323373
85. D.G. Aronson, The porous medium equation, in *Nonlinear Diffusion Problems*, ed. by A. Fasano, M. Primicerio. Lecture Notes in Mathematics, vol. 1224 (1986), pp. 1–46. https://doi.org/10.1007/BFb0072687
86. M.A. Awad, A.A. Rushdi, M.A. Abbas, S.A. Mitchell, A.H. Mahmoud, C.L. Bajaj, M.S. Ebeida, All-hex meshing of multiple-region domains without cleanup. Procedia Eng. **163**, 251–261 (2016). https://doi.org/10.1016/j.proeng.2016.11.055
87. I. Babuška, M. Vogelius, Feedback and adaptive finite element solution of one-dimensional boundary value problems. Numer. Math. **44**(1), 75–102 (1984). https://doi.org/10.1007/BF01389757
88. I. Babuška, A. Miller, A feedback finite element method with a posteriori error estimation: part I. The finite element method and some basic properties of the a posteriori error estimator. Comput. Methods Appl. Mech. Eng. **61**, 1–40 (1987). https://doi.org/10.1016/0045-7825(87)90114-9
89. E. Bänsch, An adaptive finite-element strategy for the three-dimensional time-dependent Navier-Stokes equations. J. Comput. Appl. Math. **36**, 3–28 (1991). https://doi.org/10.1016/0377-0427(91)90224-8
90. E. Bänsch, Local mesh refinement in 2 and 3 dimensions. IMPACT Comput. Sci. Eng. **3**, 181–191 (1991). https://doi.org/10.1016/0899-8248(91)90006-G
91. R.E. Bank, A.H. Sherman, An adaptive, multi-level method for elliptic boundary value problems. Computing **26**, 91–105 (1981). https://doi.org/10.1007/BF02241777
92. J. Baranger, J.-F. Maitre, F. Oudin, Connection between finite volume and mixed finite element methods. RAIRO Modél. Math. Anal. Numér. **30**(4), 445–465 (1996). https://doi.org/10.1051/m2an/1996300404451
93. G.R. Barrenechea, V. John, P. Knobloch, Analysis of algebraic flux correction schemes. SIAM J. Numer. Anal. **54**(4), 2427–2451 (2016). https://doi.org/10.1137/15M1018216
94. J.W. Barrett, C.M. Elliott, Total flux estimates for a finite-element approximation of elliptic equations. IMA J. Numer. Anal. **7**(2), 129–148 (1987). https://doi.org/10.1093/imanum/7.2.129
95. T.J. Barth, D.C. Jesperson, The design and application of upwind schemes on unstructured meshes. Paper 89-0366. AIAA (1989). https://doi.org/10.2514/6.1989-366
96. P. Bastian, M. Blatt, A. Dedner et al., The DUNE framework: basic concepts and recent developments. Comput. Math. Appl. **81**, 75–112 (2021). https://doi.org/10.1016/j.camwa.2020.06.007, https://dune-project.org/
97. S. Basting, E. Bänsch, *Preconditioners for the discontinuous Galerkin time-stepping method of arbitrary order*. ESAIM: Math. Model. Numer. Anal. **51**(4), 1173-1-195 (2016). https://doi.org/10.1051/m2an/2016055
98. M. Bause, P. Knabner, Uniform error analysis for Lagrange-Galerkin approximations of convection-dominated problems. SIAM J. Numer. Anal. **39**(6), 1954–1984 (2002). https://doi.org/10.1137/S0036142900367478
99. R. Becker, R. Rannacher, A feed-back approach to error control in finite element methods: basic analysis and examples. East-West J. Numer. Math. **4**(4), 237–264 (1996)

100. R. Becker, D. Capatina, R. Luce, Local flux reconstructions for standard finite element methods on triangular meshes. SIAM J. Numer. Anal. **54**(4), 2684–2706 (2016). https://doi.org/10.1137/16M1064817
101. C. Bernardi, Optimal finite-element interpolation on curved domains. SIAM J. Numer. Anal. **26**(5), 1212–1240 (1989). https://doi.org/10.1137/0726068
102. C. Bernardi, Y. Maday, A.T. Patera, A new nonconforming approach to domain decomposition: the mortar element method, in *Nonlinear Partial Differential Equations and Their Applications*, ed. by H. Brezis, J.-L. Lions (Longman, 1994)
103. C. Bernardi, V. Girault, A local regularization operator for triangular and quadrilateral finite elements. SIAM J. Numer. Anal. **35**(5), 1893–1916 (1998). https://doi.org/10.1137/S0036142995293766
104. J. Bey, Tetrahedral grid refinement. Computing **55**, 355–378 (1995). https://doi.org/10.1007/BF02238487
105. J. Bey, Simplicial grid refinement: on Freudenthal's algorithm and the optimal number of congruence classes. Numer. Math. **85**(1), 1–29 (2000). https://doi.org/10.1007/s002110050475
106. P. Binev, W. Dahmen, R. DeVore, Adaptive finite element methods with convergence rates. Numer. Math. **97**(2), 219–268 (2004). https://doi.org/10.1007/s00211-003-0492-7
107. T.D. Blacker, R.J. Meyers, Seams and wedges in plastering: a 3-D hexahedral mesh generation algorithm. Eng. Comput. **9**, 83–93 (1993). https://doi.org/10.1007/BF01199047
108. T.D. Blacker, M.B. Stephenson, Paving: a new approach to automated quadrilateral mesh generation. Int. J. Numer. Methods Eng. **32**, 811–847 (1991). https://doi.org/10.1002/nme.1620320410
109. J.M. Boland, R.A. Nicolaides, Stable and semistable low order finite elements for viscous flows. SIAM J. Numer. Anal. **22**(3), 474–492 (1985). https://doi.org/10.1137/0722028
110. F.A. Bornemann, B. Erdmann, R. Kornhuber, Adaptive multilevel methods in three space dimensions. Int. J. Numer. Methods Eng. **36**(18), 3187–3203 (1993). https://doi.org/10.1002/nme.1620361808
111. A. Bowyer, Computing Dirichlet tessellations. Comput. J. **24**(2), 162–166 (1981). https://doi.org/10.1093/comjnl/24.2.162
112. J.H. Bramble, A.H. Schatz, Higher order local accuracy by averaging in the finite element method. Math. Comput. **31**, 94–111 (1977). https://doi.org/10.2307/2005782
113. A.N. Brooks, T.J.R. Hughes, Streamline-upwind/Petrov-Galerkin formulations for convection dominated flows with particular emphasis on the incompressible Navier-Stokes equations. Comput. Methods Appl. Mech. Eng. **32**, 199–259 (1982). https://doi.org/10.1016/0045-7825(82)90071-8
114. E.J. Carr, T.J. Moroney, I.W. Turner, Efficient simulation of unsaturated flow using exponential time integration. Appl. Math. Comput. **217**, 6587–6596 (2011). https://doi.org/10.1016/j.amc.2011.01.041
115. C. Carstensen, S.A. Funken, Constants in Clément-interpolation error and residual based a posteriori estimates in finite element methods. East-West J. Numer. Math. **8**(3), 153–175 (2000)
116. C. Carstensen, Reliable and efficient averaging techniques as universal tool for a posteriori finite element error control on unstructured grids. Int. J. Numer. Anal. Model. **3**(3), 333–347 (2006)
117. C. Carstensen, L. Demkowicz, J. Gopalakrishnan, A posteriori error control for DPG methods. SIAM J. Numer. Anal. **52**(3), 1335–1353 (2014). https://doi.org/10.1137/130924913
118. C. Carstensen, M. Feischl, M. Page, D. Praetorius, Axioms of adaptivity. Comput. Math. Appl. **7**(6), 1195–1253 (2014). https://doi.org/10.1016/j.camwa.2013.12.003
119. J.M. Cascon, C. Kreuzer, R.H. Nochetto, K.G. Siebert, Quasi-optimal convergence rate for an adaptive finite element method. SIAM J. Numer. Anal. **46**(5), 2524–2550 (2008). https://doi.org/10.1137/07069047X

120. J.C. Cavendish, Automatic triangulation of arbitrary planar domains for the finite element method. Int. J. Numer. Methods Eng. **8**(4), 679–696 (1974). https://doi.org/10.1002/nme.1620080402
121. W.M. Chan, Overset grid technology development at NASA Ames Research Center. Comput. Fluids **38**, 496–503 (2009). https://doi.org/10.1016/j.compfluid.2008.06.009
122. L. Chen, C. Zhang, A coarsening algorithm on adaptive grids by newest vertex bisection and its applications. J. Comput. Math. **28**(6), 767–789 (2010). https://doi.org/10.4208/jcm.1004-m3172
123. S.-H. Chou, Analysis and convergence of a covolume method for the generalized Stokes problem. Math. Comput. **66**(217), 85–104 (1997). https://doi.org/10.1090/S0025-5718-97-00792-8
124. P. Ciarlet Jr., J. Huang, J. Zou, Some observations on generalized saddle-point problems. SIAM J. Matrix Anal. Appl. **25**(1), 224–236 (2003). https://doi.org/10.1137/S0895479802410827
125. P. Clément, Approximation by finite element functions using local regularization. RAIRO Anal. Numér. **9**(R-2), 77–84 (1975). https://doi.org/10.1051/m2an/197509R200771
126. B. Cockburn, J. Gopalakrishnan, A characterization of hybridized mixed methods for second order elliptic problems. SIAM J. Numer. Anal. **42**(1), 283–301 (2004). https://doi.org/10.1137/S0036142902417893
127. B. Cockburn, B. Dong, J. Guzmán, M. Restelli, R. Sacco, A hybridizable discontinuous Galerkin method for steady-state convection-diffusion-reaction problems. SIAM J. Sci. Comput. **31**(5), 3827–3846 (2009). https://doi.org/10.1137/080728810
128. COMSOL Multiphysics. https://www.comsol.com/
129. M. Crouzeix, P.-A. Raviart, Conforming and nonconforming finite element methods for solving the stationary Stokes equations, I. RAIRO Anal. Numér. **R-7**(3), 33–76 (1973). https://doi.org/10.1051/m2an/197307R300331
130. W. Dörfler, A convergent adaptive algorithm for Poisson's equation. SIAM J. Numer. Anal. **33**(3), 1106–1124 (1996). https://doi.org/10.1137/0733054
131. M.G. Edwards, M. Pal, Positive-definite q-families of continuous subcell Darcy-flux CVD(MPFA) finite-volume schemes and the mixed finite element method. Int. J. Numer. Methods Fluids **57**(4), 355–387 (2008). https://doi.org/10.1002/fld.1586
132. T. Elbinger, Accurate and Efficient Approximation of Micro-Macro Models by Means of Mixed Finite Elements with Applications to Cell Biology. Dissertation, Naturwissenschaftliche Fakultät, Universität Erlangen-Nürnberg, 2021
133. C. Erath, D. Praetorius, Adaptive vertex-centered finite volume methods with convergence rates. SIAM J. Numer. Anal. **54**(4), 2228–2255 (2016). https://doi.org/10.1137/15M1036701
134. C. Erath, D. Praetorius, Adaptive vertex-centered finite volume methods for general second-order linear elliptic PDEs. IMA J. Numer. Anal. **39**(2), 983–1008 (2019). https://doi.org/10.1093/imanum/dry006
135. R. Eymard, R. Herbin, J.C. Latché, On a stabilized collocated finite volume scheme for the Stokes problem. ESAIM: Math. Model. Numer. Anal. **40**(3), 501–527 (2006). https://doi.org/10.1051/m2an:2006024
136. R. Eymard, R. Herbin, J.C. Latché, Convergence analysis of a collocated finite volume scheme for the incompressible Navier-Stokes equations on general 2D or 3D meshes. SIAM J. Numer. Anal. **45**(1), 1–36 (2007). https://doi.org/10.1137/040613081
137. S. Fortune, Voronoi diagrams and Delaunay triangulations, in *Computing in Euclidean Geometry*, 2nd edn., ed. by D.-Z. Du, F.K. Hwang (World Scientific, Singapore, 1995), pp. 225–265. https://doi.org/10.1142/9789812831699_0007
138. E. Gagliardo, Caratterizzazioni delle Tracce sulla Frontiera Relative ad Alcune Classi di Funzioni in n Variabili. Rend. Sem. Mat. Padova **27**, 284–305 (1957)

139. V. Girault, Theory of a finite difference method on irregular networks. SIAM J. Numer. Anal. **11**(2), 260–282 (1974). https://doi.org/10.1137/0711026
140. A. Giuliani, L. Krivodonova, Analysis of slope limiters on unstructured triangular meshes. J. Comput. Phys. **374**, 1–26 (2018). https://doi.org/10.1016/j.jcp.2018.07.031
141. S.K. Godunov, A difference method for numerical calculation of discontinuous solutions of the equations of hydrodynamics. Mat. Sb. (N.S.) **47**(89), 271–306 (1959)
142. S. Gross, A. Reusken, Parallel multilevel tetrahedral grid refinement. SIAM J. Sci. Comput. **26**(4), 1261–1288 (2005). https://doi.org/10.1137/S1064827503425237
143. J. Grande, Red-green refinement of simplicial meshes in d dimensions. Math. Comput. **88**(316), 751–782 (2018). https://doi.org/10.1090/mcom/3383
144. T. Gudi, A new error analysis for discontinuous finite element methods for linear elliptic problems. Math. Comput. **79**(272), 2169–2189 (2010). https://doi.org/10.1090/S0025-5718-10-02360-4
145. P.C. Hammer, A.H. Stroud, Numerical integration over simplexes and cones. Math. Tables Aids Comput. **10**, 130–137 (1956). https://doi.org/10.2307/2002483
146. F.H. Harlow, J.E. Welch, Numerical calculation of time-dependent viscous incompressible flow of fluid with free surface. Phys. Fluids **8**(12), 2182–2189 (1965). https://doi.org/10.1063/1.1761178
147. C.R. Harris, K.J.K. Millman, S.J. van der Walt et al., Array programming with NumPy. Nature **585**(7825), 357–362 (2020). https://doi.org/10.1038/s41586-020-2649-2, https://numpy.org/
148. M. Hochbruck, C. Lubich, On Krylov subspace approximations to the matrix exponential operator. SIAM J. Numer. Anal. **34**(5), 1911–1925 (1997). https://doi.org/10.1137/S0036142995280572
149. M. Hochbruck, A. Ostermann, Exponential integrators. Acta Numer. **19**, 209–286 (2010). https://doi.org/10.1017/S0962492910000048
150. T.J.R. Hughes, L.P. Franca, G.M. Hulbert, A new finite element formulation for computational fluid dynamics: VIII. The Galerkin/least-squares method for advective-diffusive equations. Comput. Methods Appl. Mech. Eng. **73**(2), 173–189 (1989). https://doi.org/10.1016/0045-7825(89)90111-4
151. HYPRE. https://computing.llnl.gov/projects/hypre-scalable-linear-solvers-multigrid-methods/software
152. S. Hyun, L.-E. Lindgren, Smoothing and adaptive remeshing schemes for graded element. Commun. Numer. Methods Eng. **17**(1), 1–17 (2001)
153. Y. Ito, A.M. Shih, B.K. Soni, Octree-based reasonable-quality hexahedral mesh generation using a new set of refinement templates. Int. J. Numer. Methods Eng. **77**(13), 1809–1833 (2009). https://doi.org/10.1002/nme.2470
154. T. Jahnke, C. Lubich, Error bounds for exponential operator splittings. BIT **40**(4), 735–744 (2000). https://doi.org/10.1023/A:1022396519656
155. A. Jameson, Schmidt W., E. Turkel, Numerical solutions of the Euler equations by finite volume methods using Runge-Kutta time-marching schemes. Paper 81-1259, AIAA, New York, 1981
156. P. Jamet, Estimation of the interpolation error for quadrilateral finite elements which can degenerate into triangles. SIAM J. Numer. Anal. **14**, 925–930 (1977). https://doi.org/10.1137/0714062
157. H. Jin, R. Tanner, Generation of unstructured tetrahedral meshes by advancing front technique. Int. J. Numer. Methods Eng. **36**, 1805–1823 (1993). https://doi.org/10.1002/nme.1620361103
158. B. Keith, A. Astaneh, L. Demkowicz, Goal-oriented adaptive mesh refinement for discontinuous Petrov-Galerkin methods. SIAM J. Numer. Anal. **57**(4), 1649–1676 (2019). https://doi.org/10.1137/18M1181754

159. D.S. Kershaw, Differencing of the diffusion equation in Lagrangian hydrodynamic codes. J. Comput. Phys. **39**, 375–395 (1981). https://doi.org/10.1016/0021-9991(81)90158-3
160. R. Klausen, R. Winther, Robust convergence of multi point flux approximation on rough grids. Numer. Math. **104**, 317–337 (2006). https://doi.org/10.1007/s00211-006-0023-4
161. R. Klausen, A. Stephansen, Convergence of multi-point flux approximations on general grids and media. Int. J. Numer. Anal. Model. **9**(3), 584–606 (2012)
162. P. Knabner, G. Summ, The invertibility of the isoparametric mapping for pyramidal and prismatic finite elements. Numer. Math. **88**(4), 661–681 (2001). https://doi.org/10.1007/PL00005454
163. S. Korotov, M. Křížek, A. Kropáč, Strong regularity of a family of face-to-face partitions generated by the longest-edge bisection algorithm. Comput. Math. Math. Phys. **48**(9), 1687–1698 (2008). https://doi.org/10.1134/S0965542508090170
164. S. Korotov, Á. Plaza, J.P. Suárez, Longest-edge n-section algorithms: properties and open problems. J. Comput. Appl. Math. **293**, 139–146 (2016). https://doi.org/10.1016/j.cam.2015.03.046
165. I. Kossaczky, A recursive approach to local mesh refinement in two and three dimensions. J. Comput. Appl. Math. **55**, 275–288 (1994). https://doi.org/10.1016/0377-0427(94)90034-5
166. S. Kräutle, P. Knabner, A new numerical reduction scheme for fully coupled multicomponent transport-reaction problems in porous media. Water Resour. Res. **41**(9), 1–17 (2005). https://doi.org/10.1029/2004WR003624
167. M. Křížek, On the maximum angle condition for linear tetrahedral elements. SIAM J. Numer. Anal. **29**, 513–520 (1992). https://doi.org/10.1137/0729031
168. M. Kížek, T. Strouboulis, How to generate local refinements of unstructured tetrahedral meshes satisfying a regularity ball condition. Numer. Methods PDE **13**, 201–214 (1997). https://doi.org/10.1002/(SICI)1098-2426(199703)13:2<201::AID-NUM5>3.0.CO;2-T
169. D. Kuzmin, M. Möller, Algebraic flux correction I. Scalar conservation laws, in *Flux-Corrected Transport: Principles, Algorithms, and Applications*, ed. by D. Kuzmin, R. Löhner, S. Turek (Springer, Berlin, 2005), pp. 155–206. https://doi.org/10.1007/3-540-27206-2_6
170. D. Kuzmin, A vertex-based hierarchical slope limiter for p-adaptive discontinuous Galerkin methods. J. Comput. Appl. Math. **233**(12), 3077–3085 (2010). https://doi.org/10.1016/j.cam.2009.05.028
171. D. Kuzmin, Slope limiting for discontinuous Galerkin approximations with a possibly non-orthogonal Taylor basis. Int. J. Numer. Methods Fluids **71**(9), 1178–1190 (2013). https://doi.org/10.1002/fld.3707
172. D. Kuzmin, Monolithic convex limiting for continuous finite element discretizations of hyperbolic conservation laws. Comput. Methods Appl. Mech. Eng. **361**, 112804 (2020). https://doi.org/10.1016/j.cma.2019.112804
173. C.L. Lawson, Software for C^1 surface interpolation, in *Mathematical Software III*, ed. by J.R. Rice (Academic Press, New York, 1977), pp. 161–194. https://doi.org/10.1016/B978-0-12-587260-7.50011-X
174. X. Liang, Y.S. Zhang, An octree-based dual contouring method for triangular and tetrahedral mesh generation with guaranteed angle range. Eng. Comput. **30**(2), 211–222 (2013). https://doi.org/10.1007/s00366-013-0328-8
175. K. Lie, S. Krogstad, I.S. Ligaarden, J.R. Natvig, H.M. Nilsen, B. Skaflestad, Open-source MATLAB implementation of consistent discretisations on complex grids. Comput. Geosci. **16**(2), 297–322 (2012). https://doi.org/10.1007/s10596-011-9244-4
176. K. Lipnikov, M. Shashkov, I. Yotov, Local flux mimetic finite difference methods. Numer. Math. **112**, 115–152 (2009). https://doi.org/10.1007/s00211-008-0203-5
177. J.W.H. Liu, The multifrontal method for sparse matrix solution: theory and practice. SIAM Rev. **34**(1), 82–109 (1992). https://doi.org/10.1137/1034004

178. A. Liu, B. Joe, On the shape of tetrahedra from bisection. Math. Comput. **63**(207), 141–154 (1994). https://doi.org/10.1090/S0025-5718-1994-1240660-4
179. A. Liu, B. Joe, Quality local refinement of tetrahedral meshes based on bisection. SIAM J. Sci. Comput. **16**(6), 1269–1291 (1995). https://doi.org/10.1137/0916074
180. A. Logg, K.-A. Mardal, G.N. Wells et al., *Automated Solution of Differential Equations by the Finite Element Method* (Springer, 2012). https://doi.org/10.1007/978-3-642-23099-8
181. C. Lohmann, D. Kuzmin, J.N. Shadid, S. Mabuza, Flux-corrected transport algorithms for continuous Galerkin methods based on high order Bernstein finite elements. J. Comput. Phys. **344**, 151–186 (2017). https://doi.org/10.1016/j.jcp.2017.04.059
182. R.W. MacCormack, A.J. Paullay, Computational efficiency achieved by time splitting of finite difference operators. Paper 72-154, AIAA, New York, 1972. https://doi.org/10.2514/6.1972-154
183. R.H. MacNeal, An asymmetric finite difference network. Q. Appl. Math. **11**(3), 295–310 (1953)
184. J.M. Maubach, Iterative methods for non-linear partial differential equations. Ph.D. thesis, Department of Mathematics, University of Nijmegen, 1991
185. J.M. Maubach, Local bisection refinement for n-simplicial grids generated by reflection. SIAM J. Sci. Comput. **16**(1), 210–227 (1995). https://doi.org/10.1137/0916014
186. P.W. McDonald, The computation of transonic flow through two-dimensional gas turbine cascades. Paper 71-GT-89, ASME, New York, 1971. https://doi.org/10.1115/71-GT-89
187. K. Mekchay, R.H. Nochetto, Convergence of adaptive finite element methods for general second order linear elliptic PDEs. SIAM J. Numer. Anal. **43**(5), 1803–1827 (2005). https://doi.org/10.1137/04060929X
188. S. Meshkat, D. Talmor, Generating a mixed mesh of hexahedra, pentahedra and tetrahedra from an underlying tetrahedral mesh. Int. J. Numer. Methods Eng. **49**(1–2), 17–30 (2000). https://doi.org/10.1002/1097-0207(20000910/20)49:1/2<17::AID-NME920>3.0.CO;2-U
189. B.V. Minchev, W.M. Wright, A review of exponential integrators for first order semi-linear problems. Preprint Numerics No. 2/2005, Norwegian University of Science and Technology, Trondheim, 2005
190. W.F. Mitchell, Unified multilevel adaptive finite element methods for elliptic problems. Ph.D. thesis, University of Illinois at Urbana-Champaign, 1988
191. W.F. Mitchell, A comparison of adaptive refinement techniques for elliptic problems. ACM Trans. Math. Softw. **15**(4), 326–347 (1989). https://doi.org/10.1145/76909.76912
192. W.F. Mitchell, Adaptive refinement for arbitrary finite element spaces with hierarchical bases. J. Comput. Appl. Math. **36**, 65–78 (1991). https://doi.org/10.1016/0377-0427(91)90226-A
193. W.F. Mitchell, 30 years of newest vertex bisection. J. Numer. Anal. Ind. Appl. Math. **11**(1–2), 11–22 (2017). https://doi.org/10.1063/1.4951755
194. P. Möller, P. Hansbo, On advancing front mesh generation in three dimensions. Int. J. Numer. Methods Eng. **38**, 3551–3569 (1995). https://doi.org/10.1002/nme.1620382102
195. K.W. Morton, A. Priestley, E. Süli, Stability of the Lagrange-Galerkin method with non-exact integration. RAIRO Modél. Math. Anal. Numér. **22**(4), 625–653 (1988). https://doi.org/10.1051/m2an/1988220406251
196. J.C. Nédélec, Mixed finite elements in $\mathbb{R}^3$. Numer. Math. **35**(3), 315–341 (1980). https://doi.org/10.1007/BF01396415
197. R.-H. Ni, A multiple grid scheme for solving the Euler equations. AIAA J. **20**, 1565–1571 (1981). https://doi.org/10.2514/6.1981-1025
198. J.T. Oden, I. Babuška, C.E. Baumann, A discontinuous hp finite element method for diffusion problems. J. Comput. Phys. **146**, 491–519 (1998). https://doi.org/10.1016/S0898-1221(99)00117-0
199. OpenFOAM. https://openfoam.org/

200. S.J. Owen, S.A. Canann,S. Saigal, Pyramid elements for maintaining tetrahedra and hexahedra conformability, in *Trends in Unstructured Mesh Generation: Presented at the 1997 Joint ASME/ASCE/SES Summer Meeting, June 29-July 2, 1997, Evanston, Illinois*, vol. 220, ed. by S.A. Canann, S. Saigal (American Society of Mechanical Engineers, AMD, New York, 1997), pp. 123–129
201. S. Park, K. Lee, A new approach to automated multiblock decomposition for grid generation: a hypercube++ approach, in *Handbook of Grid Generation*, ed. by J.F. Thompson, B.K. Soni, N.P. Weatherill (CRC Press, Boca Raton, 1999), pp. 10.1–10.22
202. J. Pellerin, A. Johnen, K. Verhetsel, J.-F. Remacle, Identifying combinations of tetrahedra into hexahedra: a vertex based strategy. Comput.-Aided Des. **105**, 1–10 (2018). https://doi.org/10.1016/j.proeng.2017.09.779
203. J. Peraire, J. Peiró, K. Morgan, Adaptive remeshing for three-dimensional compressible flow computations. J. Comput. Phys. **103**, 269–285 (1992). https://doi.org/10.1016/0021-9991(92)90401-J
204. V.T. Rajan, Optimality of the Delaunay triangulation in $\mathbb{R}^d$, in *Proceedings of the 7th ACM Symposium on Computational Geometry*, pp. 357–363, 1991. https://doi.org/10.1145/109648.109688
205. V.T. Rajan, Optimality of the Delaunay triangulation in $\mathbb{R}^d$. Discret. Comput. Geom. **12**(1), 189–202 (1994). https://doi.org/10.1007/BF02574375
206. S.I. Repin, A posteriori error estimation for approximate solutions of variational problems by duality theory, in *Proceedings of ENUMATH 97*, ed. by H.G. Bock et al. (World Scientific Publishing, Singapore, 1998), pp. 524–531
207. B. Reuter, V. Aizinger, M. Wieland, F. Frank, P. Knabner, FESTUNG: a MATLAB/GNU Octave toolbox for the discontinuous Galerkin method, Part II: advection operator and slope limiting. Comput. Math. Appl. **72**(7), 1896–1925 (2016). https://doi.org/10.1016/j.camwa.2016.08.006
208. M.-C. Rivara, Mesh refinement processes based on the generalized bisection of simplices. SIAM J. Numer. Anal. **21**(3), 604–613 (1984). https://doi.org/10.1137/0721042
209. M.-C. Rivara, Algorithms for refining triangular grids suitable for adaptive and multigrid techniques. Int. J. Numer. Methods Eng. **20**, 745–756 (1984). https://doi.org/10.1002/nme.1620200412
210. M.-C. Rivara, Local modification of meshes for adaptive and/or multigrid finite-element methods. J. Comput. Appl. Math. **36**, 79–89 (1991). https://doi.org/10.1016/0377-0427(91)90227-B
211. R. Rodríguez, Some remarks on Zienkiewicz-Zhu estimator. Numer. Methods PDE **10**(5), 625–635 (1994). https://doi.org/10.1002/num.1690100509
212. W. Ruge, K. Stueben, Algebraic multigrid, in *Multigrid Methods*, ed. by S.F. McCormick (SIAM, Philadelphia, 1987), pp. 73–130. https://doi.org/10.1137/1.9781611971057.ch4
213. A. Rupp, S. Lee, Continuous Galerkin and enriched Galerkin methods with arbitrary order discontinuous trial functions for the elliptic and parabolic problems with jump conditions. J. Sci. Comput. **84**(9), 25 (2020). https://doi.org/10.1007/s10915-020-01255-4
214. L.R. Scott, S. Zhang, Finite element interpolation of nonsmooth functions satisfying boundary conditions. Math. Comput. **54**(190), 483–493 (1990). https://doi.org/10.2307/2008497
215. Q. Sheng, Solving linear partial differential equations by exponential splitting. IMA J. Numer. Anal. **9**(2), 199–212 (1989). https://doi.org/10.1093/imanum/9.2.199
216. M.S. Shephard, M.K. Georges, Automatic three-dimensional mesh generation by the finite octree technique. Int. J. Numer. Methods Eng. **32**, 709–749 (1991). https://doi.org/10.1002/nme.1620320406
217. B. Sportisse, An analysis of operator splitting techniques in the stiff case. J. Comput. Phys. **161**, 140–168 (2000). https://doi.org/10.1006/jcph.2000.6495

218. R. Stevenson, The completion of locally refined simplicial partitions created by bisection. Math. Comput. **77**(261), 227–241 (2008). https://doi.org/10.1090/S0025-5718-07-01959-X
219. G. Summ, *Quantitative Interpolationsfehlerabschätzungen Für Triangulierungen Mit Allgemeinen Tetraeder- Und Hexaederelementen*. Diplomarbeit, Friedrich-Alexander-Universität Erlangen-Nürnberg, 1996
220. L. Sun, G. Zhao, X. Ma, Adaptive generation and local refinement methods of three-dimensional hexahedral element mesh. Finite Elem. Anal. Des. **50**, 184–200 (2012). https://doi.org/10.1016/j.finel.2011.09.009
221. Ch. Tapp, Anisotrope Gitter – Generierung und Verfeinerung. Dissertation, Friedrich–Alexander–Universität Erlangen–Nürnberg, 1999
222. A. Tambue, I. Berre, J.M. Nordbotten, Efficient simulation of geothermal processes in heterogeneous porous media based on the exponential Rosenbrock-Euler and Rosenbrock-type methods. Adv. Water Resour. **53**, 250–262 (2013). https://doi.org/10.1016/j.advwatres.2012.12.004
223. C.T. Traxler, An algorithm for adaptive mesh refinement in n dimensions. Computing **59**(2), 115–137 (1997). https://doi.org/10.1007/BF02684475
224. P. Virtanen, R. Gommers, T.E. Oliphant, et al., SciPy 1.0: fundamental algorithms for scientific computing in Python. Nat. Methods **17**, 261–272 (2020). https://doi.org/10.1038/s41592-019-0686-2, https://scipy.org/
225. M. Vohralík, Equivalence between lowest-order mixed finite element and multi-point finite volume methods on simplicial meshes. ESAIM: Math. Model. Numer. Anal. **40**(2), 367–391 (2006). https://doi.org/10.1051/m2an:2006013
226. D.F. Watson, Computing the n-dimensional Delaunay tessellation with application to Voronoi polytopes. Comput. J. **24**(2), 167–172 (1981). https://doi.org/10.1093/comjnl/24.2.167
227. Wikipedia. Comparison of linear algebra libraries, https://en.wikipedia.org/wiki/Comparison_of_linear_algebra_libraries
228. M.A. Yerry, M.S. Shephard, Automatic three-dimensional mesh generation by the modified-octree technique. Int. J. Numer. Methods Eng. **20**, 1965–1990 (1984). https://doi.org/10.1002/nme.1620201103
229. S. Zhang, Successive subdivisions of tetrahedra and multigrid methods on tetrahedral meshes. Houst. J. Math. **21**(3), 541–556 (1995)
230. X. Zhao, S. Mao, Z. Shi, Adaptive finite element methods on quadrilateral meshes without hanging nodes. SIAM J. Sci. Comput. **32**(4), 2099–2120 (2010). https://doi.org/10.1137/090772022
231. J.Z. Zhu, O.C. Zienkiewicz, E. Hinton, J. Wu, A new approach to the development of automatic quadrilateral mesh generation. Int. J. Numer. Methods Eng. **32**, 849–866 (1991). https://doi.org/10.1002/nme.1620320411
232. O.C. Zienkiewicz, J.Z. Zhu, The superconvergent patch recovery and a posteriori error estimates. Parts I, II. Int. J. Numer. Methods Eng. **33**(7), 1331–1364, 1365–1382 (1992). https://doi.org/10.1002/nme.1620330702
233. J. Xu, Iterative methods by space decomposition and subspace correction. SIAM Rev. **34**(4), 581–613 (1992). https://doi.org/10.1137/1034116

Index

P. Knabner and L. Angermann, *Numerical Methods for Elliptic and Parabolic Partial Differential Equations*, Texts in Applied Mathematics 44,
https://doi.org/10.1007/978-3-030-79385-2

E

F

Q

R

S

T

U

V

W

Z

Printed in the United States
by Baker & Taylor Publisher Services